12th European Workshop on Advanced Control and Diagnosis (ACD 2015)

Journal of Physics: Conference Series Volume 659

Pilsen, Czech Republic
19 - 20 November 2015

ISBN: 978-1-5108-1733-3
ISSN: 1742-6588

Printed from e-media with permission by:

Curran Associates, Inc.
57 Morehouse Lane
Red Hook, NY 12571

Some format issues inherent in the e-media version may also appear in this print version.

Printed by Curran Associates, Inc. (2016)

For permission requests, please contact the Institute of Physics
at the address below.

Institute of Physics
Dirac House, Temple Back
Bristol BS1 6BE UK

Phone: 44 1 17 929 7481
Fax: 44 1 17 920 0979

techtracking@iop.org

Additional copies of this publication are available from:

Curran Associates, Inc.
57 Morehouse Lane
Red Hook, NY 12571 USA
Phone: 845-758-0400
Fax: 845-758-2634
Email: curran@proceedings.com
Web: www.proceedings.com

Table of contents

Volume 659

12th European Workshop on Advanced Control and Diagnosis (ACD 2015)
19–20 November 2015, Pilsen, Czech Republic

Accepted papers received: 3 November 2015
Published online: 19 November 2015

Preface

Papers

Plenary session

Control theory

Identification and signal processing

Control applications I

Estimation and filtering

Control applications II

Fault detection

Poster session

Journal of Physics: Conference Series **659** (2015) 011001 doi:10.1088/1742-6596/659/1/011001

12th European Workshop on Advanced Control and Diagnosis

The 12th European Workshop on Advanced Control and Diagnosis (ACD 2015) took place at the Research Centre NTIS - New Technologies for the Information Society, Faculty of Applied Sciences, University of West Bohemia, Pilsen, Czech Republic, on November 19 – 20, 2015.

The annual European Workshop on Advanced Control and Diagnosis has been organized since 2003 by Control Engineering departments of several European universities in Germany, France, the UK, Poland, Italy, Hungary, and Denmark to bring together senior and junior academics and engineers from diverse fields of automatic control, fault detection, and signal processing. The workshop provides an opportunity for researchers and developers to present their recent theoretical developments, practical applications, or even open problems. It also offers a great opportunity for industrial partners to express their needs and priorities and to review the current activities in the fields.

A total of 74 papers have been submitted for ACD 2015. Based on the peer reviews 48 papers were accepted for the oral presentation and 10 papers for the poster presentation. The accepted papers covered areas of control theory and applications, identification, estimation, signal processing, and fault detection. In addition, four excellent plenary lectures were delivered by Prof. Fredrik Gustafsson (Automotive Sensor Mining for Tire Pressure Monitoring), Prof. Vladimír Havlena (Advanced Process Control for Energy Efficiency), Prof. Silvio Simani (Advanced Issues on Wind Turbine Modelling and Control), and Prof. Robert Babuška (Learning Control in Robotics). The ACD 2015 was for the first time in the workshop history co-sponsored by the International Federation of Automatic Control (IFAC).

On behalf of the ACD 2015 organising committee, we would like to thank all those who prepared and submitted papers, participated in the peer review process, supported, and attended the workshop.

Ondřej Straka	Ivo Punčochář	Jindřich Duník
General Chair	IPC Co-Chair	IPC Co-Chair

Journal of Physics: Conference Series **659** (2015) 011001 doi:10.1088/1742-6596/659/1/011001

Workshop Organisation

General Chair
Ondřej Straka (Czech Republic)

Honorary General Chair (In Memoriam)
Miroslav Šimandl (Czech Republic)

International Programme Committee Co-Chairs
Jindřich Duník (Czech Republic)
Ivo Punčochář (Czech Republic)

Steering Committee
Józef Korbicz (Poland)
Didier Theilliol (France)
Silvio Simani (Italy)
Hans Henrik Niemann (Denmark)
Keith Burnham (United Kingdom)
Horst Schulte (Germany)

International Programme Committee
Harald Aschemann (Germany)
Christophe Aubrun (France)
Andrzej Bartoszewicz (Poland)
Alessandro Casavola (Italy)
Philippe Goupil (France)
Paolo Castaldi (Italy)
Vincent Cocquempot (France)
Giuseppe Franzè (Italy)
Kevin Guelton (France)
Robert Haber (Germany)
David Henry (France)
Sirkka-Liisa Jämsä-Jounela (Finland)
Józef Korbicz (Poland)
Jan M. Kościelny (Poland)
Balázs Kulcsár (Sweden)
Zsófia Lendek (Romania)
Didier Maquin (France)
Andrés Marcos (Spain)
Miroslav Fikar (Slovak Republic)
Andreas Rauh (Germany)
Hans Henrik Niemann (Denmark)
Krzysztof Patan (Poland)
Ronald J. Patton (United Kingdom)
Vicenç Puig (Spain)
José Ragot (France)
Dominique Sauter (France)
Silvio Simani (Italy)
Ralf Stetter (Germany)
Didier Theilliol (France)
Marcin Witczak (Poland)
Elena Zattoni (Italy)
Ali Zolghadri (France)
Keith Burnham (United Kingdom)
Ole Ravn (Denmark)
Niels Kjølstad Poulsen (Denmark)
Vladimír Havlena (Czech Republic)
Ondřej Straka (Czech Republic)
Michał Bartyś (Poland)
Thomas Parisini (United Kingdom)

Local Organising Committee
Zdeněk Franče (Czech Republic)
Jindřich Havlík (Czech Republic)
Marek Fehér (Czech Republic)
Oliver Kost (Czech Republic)
Ladislav Král (Czech Republic)
Jan Škach (Czech Republic)
Miroslav Flídr (Czech Republic)
Jakub Prüher (Czech Republic)

Journal of Physics: Conference Series **659** (2015) 011001 doi:10.1088/1742-6596/659/1/011001

Hosting Institution

Research Centre NTIS - New Technologies for the Information Society, Faculty of Applied Sciences, University of West Bohemia, Pilsen, Czech Republic.

Workshop Web Page

http://www.acd2015.cz/

Sponsors

Co-Sponsor

Plenary Lectures

Fredrik Gustafsson (Department of Electrical Engineering, Linköping University, Sweden)
Title: Automotive Sensor Mining for Tire Pressure Monitoring

Abstract: The wheel speed sensor provides a rich source of information for monitoring the state of the car, in particular the driveline, as well as the road condition. The presentation gives examples of how wheel speed and other on-board sensors can be used to detect and grade different faults in the car, as well as changes in the road surface. Particular attention is given to a tire pressure monitoring system, today deployed in more than 10 million cars, that is based on extensive sensor fusion and diagnosis algorithms.

Biography: Fredrik Gustafsson is professor in Sensor Informatics at Department of Electrical Engineering, Linkþing University. His research interests are in stochastic signal processing, adaptive filtering and change detection, with applications to communication, vehicular, airborne, and audio systems. He is a co-founder of the companies NIRA Dynamics (automotive safety systems), Softube (audio effects) and SenionLab (indoor positioning systems). He was an associate editor for IEEE Transactions of Signal Processing 2000-2006, IEEE Transactions on Aerospace and Electronic Systems 2010-2012, and EURASIP Journal on Applied Signal Processing 2007-2012. He was awarded the Arnberg prize by the Royal Swedish Academy of Science (KVA) 2004, elected member of the Royal Academy of Engineering Sciences (IVA) 2007, and elevated to IEEE Fellow 2011. He is the author of the Wiley textbook 'Adaptive Filtering and Change Detection'. He was awarded the Harry Rowe Mimno Award 2011 for the tutorial "Particle Filter Theory and Practice with Positioning Applications", which was published in the AESS Magazine in July 2010, and was co-author of "Smoothed state estimates under abrupt changes using sum-of-norms regularization" that received the Automatica paper prize in 2014.

Journal of Physics: Conference Series **659** (2015) 011001 doi:10.1088/1742-6596/659/1/011001

Vladimír Havlena (Honeywell ACS Global Laboratory Prague, Czech Republic)
Title: Advanced Process Control for Energy Efficiency

Abstract: Improved energy efficiency is a global trend and a motivation for many advanced process control and optimization projects. In energy intensive process industries, energy creates up to 70% of production costs. Application of standard advanced control packages may not cover all aspects of the problem. More holistic approach spanning better responsiveness of the plant to the varying demands, adaptation to process ageing and varying economic environment like real-time electricity/gas pricing, and environmental constraints, requires further development of advanced process control and real time optimization technology. The talk will cover the state-of-the-art of advanced process control technology as well as several application examples from processing industries.

Biography: Prof. Vladimir Havlena graduated from the Faculty of electrical engineering of the Czech Technical University of Prague, in 1982. He spent over 30 years in academic and research positions in control theory and process control applications. Currently, he is a professor at the Faculty of electrical engineering, CTU, lecturing Estimation, filtering and detection MSc course and supervising a number of PhD students, and also a Senior Fellow and Technology manager of Process Control and Optimization group in Honeywell ACS Global laboratory in Prague. His areas of interest cover model predictive control, real time optimization, system identification, Kalman filtering and inferential sensing.

Silvio Simani (Department of Engineering, University of Ferrara, Italy)
Title: Advanced Issues on Wind Turbine Modelling and Control

Abstract: The motivation for this talk comes from a real need to have an overview about the challenges of modelling and control for very demanding systems, such as wind turbine systems, which require reliability, availability, maintainability, and safety over power conversion efficiency. These issues have begun to stimulate research and development in the wide control community particularly for these installations that need a high degree of "sustainability". Note that this topic represents a key point for offshore wind turbines, since they are characterised by expensive and/or safety critical maintenance work. In this case, a clear conflict exists between ensuring a high degree of availability and reducing maintenance times, which affect the final energy cost. On the other hand, wind turbines have highly nonlinear dynamics, with a stochastic and uncontrollable driving force as input in the form of wind speed, thus representing an interesting challenge also from the modelling point of view. Suitable control methods can provide a sustainable optimisation of the energy conversion efficiency over wider than normally expected working conditions. Moreover, a proper mathematical description of the wind turbine system should be able to capture the complete behaviour of the process under monitoring, thus providing an important impact on the control design itself. In this way, the control scheme could guarantee prescribed performance, whilst also giving a degree of "tolerance" to possible deviation of characteristic properties or system parameters from standard conditions, if properly included in the wind turbine model itself. The most important developments in advanced controllers for wind turbines are addressed, and open problems in the areas of modelling of wind turbines are also outlined.

Biography: Silvio Simani received the Laurea degree (cum laude) in Electronic Engineering in 1996, and the Ph.D. in Information Science (Automatic Control) in 2000. Since 2006 Dr. Simani has been an IEEE Senior Member, and from 2000 a member of the SAFEPROCESS Technical Committee. Since 2002 Dr. Simani has been an Assistant Professor at the Department of Engineering of the University of Ferrara. Dr. Simani's research interests include fault diagnosis, fault tolerant control, and system identification, on which he published about 160 refereed journal and conference papers, as well as three books and chapters. In 2010 Dr. Simani was the general chair of the 1st EACD – European Workshop on Advanced Control and Diagnosis held in Ferrara. Since 2009, Dr. Simani has been member of the SysTol and EACD Steering Committees, as well as Associate Editor of the International Journal of Applied Mathematics and Computer Sciences.

Robert Babuška (DCSC, Delft University of Technology, Netherlands)
Title: Learning Control in Robotics

Abstract: Developments in robotics are coming at an accelerated rate and robots are quickly being adopted for tasks in unstructured environments. In such settings, there is a need for effective learning control methods that can deal with the uncertainty in the robot's environment, handle new situations and perform robustly under changing conditions. In addition, learning is essential for effective interaction with humans, so that robots can be trained by means of demonstration, become more responsive and cooperative with humans. There is a variety of learning control approaches, ranging from the acquisition of initial skills via teaching by a human, through transfer learning methods that enable the reuse of previously learned skills and behaviors, to learning from experience, which aims at continual improvement of performance. The talk will focus on the last category of methods, addressing mainly techniques based on reinforcement learning, which is a form of data-driven optimal control. We will cover sample-efficient algorithms suitable for real-time learning in continuous state and action spaces and show examples of applications in robot motion control. We will also address recently introduced learning methods based on deep neural networks and discuss their potential for robot control.

Biography: Robert Babuška received the M.Sc. degree (cum laude) in control engineering from the Czech Technical University in Prague, in 1990, and the Ph.D. degree (cum laude) from the Delft University of Technology, the Netherlands, in 1997. He has had faculty appointments at Czech Technical University Prague and at the Delft University of Technology. Currently, he is a Professor of Intelligent Control and Robotics and the scientific director of the TU Delft Robotics Institute. His research interests include reinforcement learning, neural and fuzzy systems, identification and state-estimation, nonlinear model-based and adaptive control and dynamic multi-agent systems. He has been working on applications of these techniques in the fields of robotics, mechatronics, and aerospace.

Peer review statement

All papers published in this volume of *Journal of Physics: Conference Series* have been peer reviewed through processes administered by the proceedings Editors. Reviews were conducted by expert referees to the professional and scientific standards expected of a proceedings journal published by IOP Publishing.

Journal of Physics: Conference Series **659** (2015) 012001 doi:10.1088/1742-6596/659/1/012001

Advanced Issues of Wind Turbine Modelling and Control[1]

Silvio Simani[2]

Department of Engineering, University of Ferrara. Via Saragat 1E – 44123 Ferrara (FE), Italy

E-mail: silvio.simani@unife.it

Abstract. The motivation for this paper comes from a real need to have an overview about the challenges of modelling and control for very demanding systems, such as wind turbine systems, which require reliability, availability, maintainability, and safety over power conversion efficiency. These issues have begun to stimulate research and development in the wide control community particularly for these installations that need a high degree of "sustainability". Note that this topic represents a key point mainly for offshore wind turbines with very large rotors, since they are characterised by challenging modelling and control problems, as well as expensive and safety critical maintenance works. In this case, a clear conflict exists between ensuring a high degree of availability and reducing maintenance times, which affect the final energy cost. On the other hand, wind turbines have highly nonlinear dynamics, with a stochastic and uncontrollable driving force as input in the form of wind speed, thus representing an interesting challenge also from the modelling point of view. Suitable control methods can provide a sustainable optimisation of the energy conversion efficiency over wider than normally expected working conditions. Moreover, a proper mathematical description of the wind turbine system should be able to capture the complete behaviour of the process under monitoring, thus providing an important impact on the control design itself. In this way, the control scheme could guarantee prescribed performance, whilst also giving a degree of "tolerance" to possible deviation of characteristic properties or system parameters from standard conditions, if properly included in the wind turbine model itself. The most important developments in advanced controllers for wind turbines are addressed, and open problems in the areas of modelling of wind turbines are also outlined.

1. Introduction

Wind energy represents a fast–developing interdisciplinary field comprising many different branches of engineering and science. According to the National Renewable Energy Laboratory (NREL), from 2002 to 2007 the installed capacity of wind grew at a rate of about 30% from 2002 to 2007 [1]. This situation is depicted in Fig. 1.

It can be seen how global wind power installations increased by 35, 467 in 2013, bringing total installed capacity up to 318,137 MW. During 2010–2011 more than half of all new wind power was added outside of the traditional markets of Europe and North America, mainly driven by the continuing boom in China which accounted for nearly half of all of the installations at 18,000 MW in 2011. China now has 91,424 MW of wind power installed. Several countries have

[1] Invited plenary paper.

[2] Invited plenary speaker.

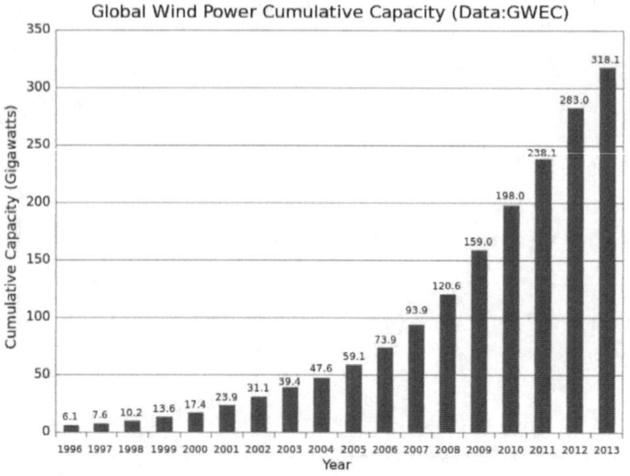

Figure 1. The installed wind energy capacity [2].

achieved relatively high levels of wind power penetration, such as 21% of stationary electricity production in Denmark, 18% in Portugal, 16% in Spain, 14% in Ireland, and 9% in Germany in 2010 [3, 4]. As of 2011, 83 countries around the world are using wind power on a commercial basis. It is clear why wind power is recognised as an effective and green solution for energy harvesting. However, even if the U.S. receives less than 2% of its electrical energy from wind, NREL report lays the framework for achieving 20% of the U.S. electrical energy generation from wind in U.S. by the year 2030 [3, 4]. Despite the expected growth in the installed capacity of wind turbines in recent years, engineering and science challenges still exist. Since wind turbine installations must guarantee both power capture and economical advantages, also the size of wind turbines has grown dramatically from 1980 [3, 4].

Modern wind turbines have large, flexible structures operating in uncertain environments, thus representing interesting cases for advanced control solutions [5]. Advanced controllers can help to achieve the desirable goal of decreasing the wind energy cost by increasing the efficiency, and thus the energy capture, or by reducing structural loading and increasing the lifetimes of the components and turbine structures [5]. This review paper aims also at sketching the main challenges that exist in the wind industry and to stimulate new research topics in this area. Although wind turbines come in both vertical–axis and horizontal–axis configurations, this work will focus only on horizontal–axis wind turbines, since they represent the most commonly produced large–scale installations today.

Horizontal–axis wind turbines have the advantage that the rotor is placed atop a tall tower, where it can take advantage of larger wind speeds higher above the ground. Moreover, horizontal–axis wind turbines used for utility–scale installations include pitchable blades, improved power capture and structural performance, as well as no need for tensioned cables used to add structural stability [5]. Vertical–axis solutions are more common for smaller turbines, where these disadvantages become less important and the benefits of reduced noise and omni–directionality become more pronounced. Note finally that the generating capacity of modern and commercial turbines ranges from less than 1kW to several MW. The proper wind turbine system modelling oriented to the design of a suitable control strategy is more cost–effective for large wind turbines, and therefore this work will focus on wind turbines with capacities of several MW.

Another important issue derives from the steadily increasing sizes and a growing complexity of wind turbines, thus giving rise to more severe requirements regarding the system safety, reliability and availability [6]. The safety demand can be commonly achieved by introducing redundancy in the system architecture, like additional sensors, which become vital for a safer operation of wind turbines. The classic example regards the pitch system for adjusting the angles of a rotor blade. For each of the three blades, one totally independent pitch system is used, such that in the worst case of a malfunction in one or two pitch systems, the remaining one or two would still be able to bring the turbine to a standstill. This solution improves the system safety, but it generates additional costs and possibly additional turbine downtimes due to faults in the redundant system parts. The enhanced safety may lead to reduce the system availability.

Even when reducing hardware redundancies, large wind turbines are prone to unexpected malfunctions or alterations of the nominal working conditions. Many of these anomalies, even if not critical, often lead to turbine shutdowns, again for safety reasons. Especially in offshore wind turbines, this may result in a substantially reduced availability, because rough weather conditions may prevent the prompt replacement of the damaged system parts. The need for reliability and availability that guarantees the continuous energy production requires the so–called sustainable control solutions. These schemes enable to keep the turbine in operation in the presence of anomalous situations, perhaps with reduced performance, while managing the maintenance operations. Apart from increasing availability and reducing turbine downtimes, sustainable control schemes might also obviate the need for more hardware redundancy, if virtual sensors could replace redundant hardware sensors [7, 8]. These schemes currently employed in wind turbines are typically on the level of the supervisory control, where commonly used strategies include sensor comparison, model comparison and thresholding tests [7, 8]. These strategies enable safe turbine operations, which involve shutdowns in case of critical situations, but they are not able to actively counteract anomalous working conditions. Therefore, the goal of this work is also to investigate these so–called sustainable control strategies, which allow to obtain a system behaviour that is close to the nominal situation in presence of unpermitted deviations of any characteristic properties or system parameters from standard conditions (*i.e.* a fault) [9]. Moreover, these schemes should provide the reconstruction of the equivalent unknown input that represents the effect of a fault, thus achieving the so–called fault diagnosis task [7, 8].

The rest of this paper is organised as follows. Section 2 describes the configurations and basic operation of wind turbines. Section 3 explains the layout of the wind turbine main control loops, including wind inflow characteristics and available sensors and actuators for use in control. Section 4 describes the current state of wind turbine control, which is then followed by a discussion of advanced control opportunities in Section 5. On the other hand, Section 5.1 outlines the main sustainable control strategies recently proposed for wind turbines. Concluding remarks are finally summarised in Section 6.

2. Wind Turbine Modelling Issues

Prior to apply any new control strategies on a real wind turbine, the efficacy of the control scheme has to be tested in detailed aero–elastic simulation model. Several simulation packages exist that are commonly used in academia and industry for wind turbine load simulation. This paper recalls one of the most used simulation package, that is the Fatigue, Aerodynamics, Structures, and Turbulence (FAST) code [10] provided by NREL, since it represents a reference simulation environment for the development of high–fidelity wind turbine prototypes that are taken as a reference test–cases for many practical studies.

FAST provides a high–fidelity wind turbine model with 24 degrees of freedom, which is appropriate for testing the developed control algorithms but not for control design. For the latter purpose, a reduced–order dynamic wind turbine model, which captures only dynamic effects directly influenced by the control, is recalled in this section and it can be used for model–

based control design. It almost corresponds to the model presented in [5].

The main components of a horizontal-axis wind turbine that are visible from the ground are its tower, nacelle, and rotor, as can be seen in Fig. 2. The nacelle houses the generator, which is driven by the high–speed shaft. The high-speed shaft is in turn usually driven by a gear box, which steps up the rotational speed from the low–speed shaft. The low-speed shaft is connected to the rotor, which includes the airfoil–shaped blades. These blades capture the kinetic energy in the wind and transform it into the rotational kinetic energy of the wind turbine.

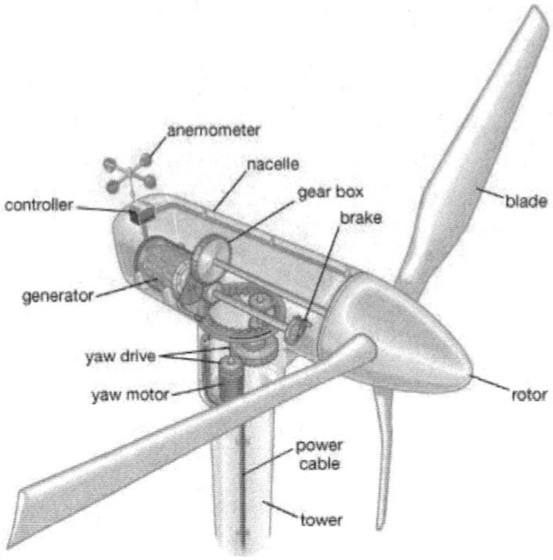

Figure 2. Main wind turbine components.

The complete wind turbine model consists of several submodels for the mechanical structure, the aerodynamics, as well as the dynamics of the pitch system and the generator/ converter system, as sketched in Fig. 3. The generator/converter dynamics are usually described as a first order delay system. However, when the delay time constant is very small, an ideal converter can be assumed, such that the reference generator torque signal is equal to the actual generator torque. In this situation, the generator torque can be considered as a system input.

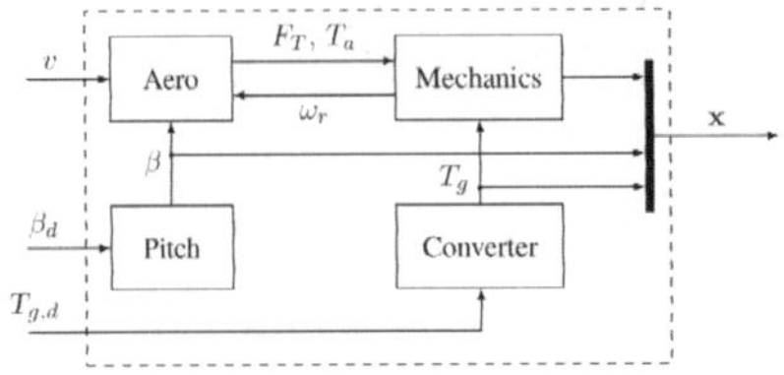

Figure 3. Block diagram of the complete wind turbine model.

Fig. 3 reports also the wind turbine inputs and outputs. In particular, v is wind speed, F_T and T_a correspond to the rotor thrust force and rotor torque, respectively; ω_r is the rotor angular velocity, x the state vector, T_g the generator torque, and $T_{g,d}$ the demanded generator torque. β is the pitch angle, whilst β_d its demanded value.

2.1. Wind Turbine Tower and Blade Models

As an example, a mechanical wind turbine model with four degrees of freedom is considered, since these degrees of freedom are the most strongly affected by the wind turbine control. In particular, the represent the fore–aft tower bending, the flap–wise blade bending, the rotor rotation, and the generator rotation [11].

Both the tower and blade bending are not modelled by means of bending beam models, but only the the translational displacement of the tower top and the blade tip are considered, where the bending stiffness parameters are transformed into equivalent translational stiffness parameters, as depicted in Fig. 4.

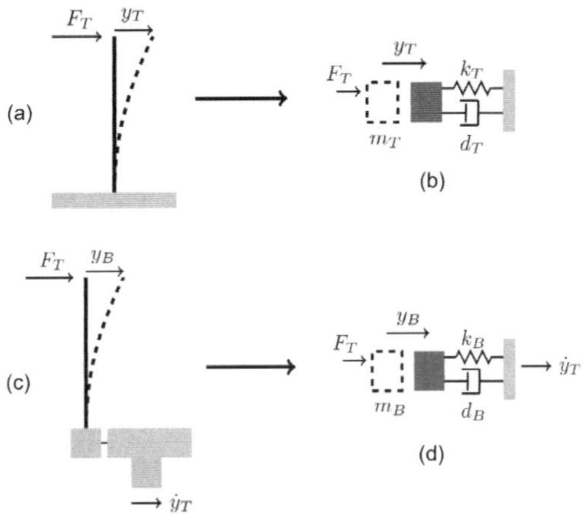

Figure 4. (a) tower bending, (b) mechanical model with spring and damper. (c) blade bending and (d) mechanical model [11].

For the tower, the equivalent translational stiffness parameter is derived by means of a direct stiffness method common in structural mechanics calculations [11].

Since the blades move with the tower, the blade tip displacement is considered in the moving tower coordinate system and the tower motion must be taken into account for the derivation of the kinetic energy of the blade. The force F_T acts both on the tower and on N blades. Only one collective blade degree of freedom is considered. Note that the N blade degrees of freedom would have to be considered individually if control strategies for load reduction involving individual blade pitch control were designed. The assumption that the same external force F_T acts on both the tower and the blade degrees of freedom (with N blades) is a simplification. It is reasonable, however, because the rotor thrust force, which is caused by the aerodynamic lift forces acting on the blade elements, acts on the tower top, thus causing a distributed load on each blade. This distributed load generates a bending of the blade, which could be modelled as a bending beam. A beam subjected not to a distributed load but to a concentrated load at the upper point must have a higher bending beam stiffness, in order that the same displacement results at the upper point. However, a reduced–order wind turbine model considers only the blade tip displacement, which requires the assumption of a translational stiffness. To obtain an adequate translational stiffness constant, the bending stiffness of the bending beam must thus be larger than the case of a distributed load.

On the other hand, the drivetrain consisting of rotor, shaft and generator is modelled as a two–mass inertia system, including shaft torsion, where the two inertias are connected with a torsional spring with spring constant k_S and a torsional damper with damping constant d_S, as illustrated in Fig. 5.

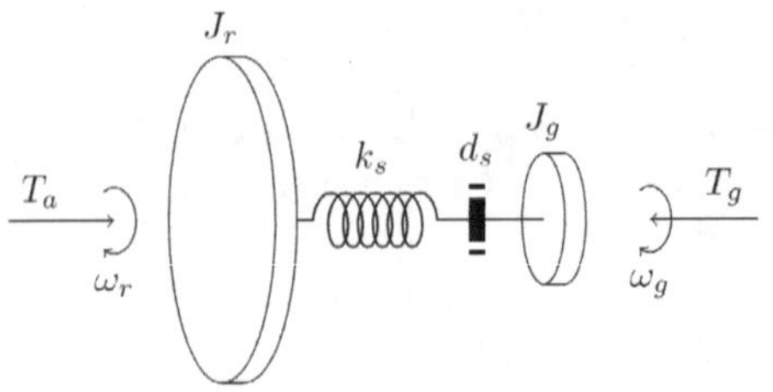

Figure 5. Model of the drivetrain [11].

With reference to Fig. 5, the angular velocities ω_r and ω_g are the time derivatives of the rotation angles θ_r and θ_g. In this case, the rotor torque T_a is generated by the lift forces on the individual blade elements, whilst T_g represents the generator torque. The ideal gearbox effect can be simply included in the generator model by multiplying the generator inertia J_g by the square of the gearbox ratio n_g.

The motion equations are derived by means of Lagrangian dynamics, which first requires to define the generalised coordinates and generalised external forces. In this way, the energy terms of the system are derived, as well as the motion equations. The vector of generalised coordinates is given by: $\mathbf{q} = [y_T,\, y_B,\, \theta_r,\, \theta_g]^T$, whilst the vector of external forces is defined as $\mathbf{f} = [F_T,\, F_T,\, T_a,\, -T_g]$.

The generalised force F_T represents the rotor thrust force, which can be computed from the wind speed at the blade and from the aerodynamic map of the thrust coefficient. On the other hand, the generalised force T_a is given by the aerodynamic rotor torque, which can be calculated from the wind speed and from the aerodynamic map of the torque coefficient described in Section 2.4. By considering the tower dynamics, the complete blade tip displacement is given by $y_T + y_B$, and the kinetic energy has the following form:

$$E_K = \frac{1}{2} m_T\, \dot{y}_T^2 + \frac{1}{2} N\, m_B\, (\dot{y}_T + \dot{y}_B)^2 + \frac{1}{2}\, J_r \dot{\theta}_r^2 + \frac{1}{2}\, J_g\, \dot{\theta}_g^2 \tag{1}$$

In the same way, the potential energy has the form:

$$E_P = \frac{1}{2}\, k_T\, y_T^2 + \frac{1}{2}\, N\, k_B\, y_B^2 + \frac{1}{2}\, k_S \left(\theta_r - \frac{1}{n_g}\, \theta_g\right)^2 \tag{2}$$

with n_g the gearbox ratio. The dampings in the system produce generalised friction forces, which can be written as derivatives of a quadratic form, *e.g.* the dissipation function. In this case, it assumes the form:

$$P_D = \frac{1}{2}\, d_T\, \dot{y}_T^2 + \frac{1}{2}\, N\, d_B\, \dot{y}_B^2 + \frac{1}{2}\, d_S \left(\dot{\theta}_r - \frac{1}{n_g}\, \dot{\theta}_g\right)^2 \tag{3}$$

The Lagrangian equations of second order including the dissipation term are given by:

$$\frac{d}{dt}\left(\frac{\partial L}{\partial \dot{q}_i}\right) - \frac{\partial L}{\partial q_i} = f_i - \frac{\partial P_D}{\partial \dot{q}_i} \tag{4}$$

where the Lagrangian function L denotes the difference between kinetic and potential energy. As the kinetic energy in (1) does not depend on the generalised coordinates and the potential energy in (2) does not depend on the generalised velocities, the motion equations in the following form are obtained:

$$\left\{ \begin{array}{rcl} (m_T + N\,m_B)\,\ddot{y}_T + N\,m_B\,\ddot{y}_B + d_T\,\dot{y}_T + k_T\,y_T & = & F_T \\ N\,m_B\,\ddot{y}_T + N\,m_B\,\ddot{y}_B + N\,dB\,\dot{y}_B + N\,k_B\,y_B & = & F_T \\ J_r\,\ddot{\theta}_r + d_S\left(\omega_r - \frac{1}{n_g}\,\omega_g\right) + k_S\left(\theta_r - \frac{1}{n_g}\,\theta_g\right) & = & T_a \\ J_r\,\frac{1}{n_g}\,\ddot{\theta}_g - d_S\left(\omega_r - \frac{1}{n_g}\,\omega_g\right) - k_S\left(\theta_r - \frac{1}{n_g}\,\theta_g\right) & = & -T_g \end{array} \right. \tag{5}$$

The system (5) can be rewritten in matrix form as:

$$\mathbf{M}\,\ddot{q} + \mathbf{D}\,\dot{q} + \mathbf{K}\,q = \mathbf{f} \tag{6}$$

where the mass matrix $\mathbf{M}$, the damping matrix $\mathbf{M}$ and the stiffness matrix $\mathbf{K}$ have the form:

$$\mathbf{M} = \begin{bmatrix} m_T + N\,m_B & N\,m_B & 0 & 0 \\ N\,m_B & N\,m_B & 0 & 0 \\ 0 & 0 & J_r & 0 \\ 0 & 0 & 0 & J_g \end{bmatrix}, \quad \mathbf{D} = \begin{bmatrix} d_T & 0 & 0 & 0 \\ 0 & N\,d_B & 0 & 0 \\ 0 & 0 & d_S & -\frac{d_S}{n_g} \\ 0 & 0 & -\frac{d_S}{n_g} & \frac{d_S}{n_g^2} \end{bmatrix}$$
$$\mathbf{K} = \begin{bmatrix} k_T & 0 & 0 & 0 \\ 0 & N\,k_B & 0 & 0 \\ 0 & 0 & k_S & -\frac{k_S}{n_g} \\ 0 & 0 & -\frac{k_S}{n_g} & \frac{k_S}{n_g^2} \end{bmatrix} \tag{7}$$

The second order system of differential equations (6) can be transformed into a first order state–space model by introducing the state vector $x = [q,\,\dot{q}]^T$. To this aim, the expression (6) is solved with respect to the second time derivative of the coordinate vector $\mathbf{q}$. The equivalent state–space model is thus obtained in the form:

$$\left\{ \begin{array}{rcl} \dot{\mathbf{x}} & = & \mathbf{A}_m\,\mathbf{x} + \mathbf{B}_m\,\mathbf{u}_m \\ \mathbf{y} & = & \mathbf{C}_m\,\mathbf{x} \end{array} \right. \tag{8}$$

where the state vector is given by $\mathbf{x} = \left[y_T,\,y_B,\,\theta_r,\,\theta_g,\,\dot{y}_T,\,\dot{y}_B,\,\dot{\theta}_r,\,\dot{\theta}_g\right]^T$, the input vector is $\mathbf{u}_m = [F_T,\,T_a,\,T_g]^T$, whilst the system matrices have the form:

$$\mathbf{A}_m = \begin{bmatrix} 0_{4\times4} & I_{4\times4} \\ -\mathbf{M}^{-1}\mathbf{K} & -\mathbf{M}^{-1}\mathbf{D} \end{bmatrix}, \quad \mathbf{B}_m = \begin{bmatrix} 0_{4\times3} \\ \mathbf{M}^{-1}\mathbf{Q} \end{bmatrix}, \quad \mathbf{C}_m = \mathbf{I}_{8\times8}, \quad \text{with } \mathbf{Q} = \begin{bmatrix} 1 & 0 & 0 \\ 1 & 0 & 0 \\ 0 & 1 & 0 \\ 0 & 0 & -1 \end{bmatrix} \tag{9}$$

2.2. Pitch Model

In pitch–regulated wind turbines, the pitch angle of the blades is controlled only in the full load region to reduce the aerodynamic rotor torque, thus maintaining the turbine at the desired rotor speed. Moreover, the pitching of the blades to feather position (*i.e.* 90°) is used as main braking system to bring the turbine to standstill in critical situations. Two different types of pitch technologies are usually exploited in wind turbines, *i.e.* hydraulic and electromechanical pitch systems. For hydraulic pitch systems, the dynamics can be modelled by means of a second–order delay model [6], which is able to display oscillatory behaviour. For electromechanical pitch systems, which are more commonly used, a first–order delay model is sufficient. In this work, the first–order delay model is recalled:

$$\dot{\beta} = -\frac{1}{\tau}\,\beta + \frac{1}{\tau}\,\beta_d \tag{10}$$

where β and β_d are the physical and the demanded pitch angle, respectively. The parameter τ denotes the delay time constant.

2.3. Generator/Converter Dynamic Model

An explicit model for the generator/converter dynamics can be included into the complete wind turbine system model. Note that for mere simulation purposes, this is not necessary, since the generator/converter dynamics are relatively fast. However, when advanced control designs are considered, an explicit generator/converter model might be required in order to take into account generator torque fast dynamics. In this case, a simple first order delay model can be sufficient, as described *e.g.* in [6]:

$$\dot{T}_g = -\frac{1}{\tau_g}\,T_g + \frac{1}{\tau_g}\,T_{g,d} \tag{11}$$

where $T_{g,d}$ represents the demanded generator torque, whilst τ_g the delay time constant.

2.4. Aerodynamic Model

The aerodynamic submodel consists of the expressions for the thrust force F_T acting on the rotor and the aerodynamic rotor torque T_a. They are determined by the reference force F_{st} and by the aerodynamic rotor thrust and torque coefficients C_T and C_Q:

$$\left\{\begin{array}{lcl} F_T &=& F_{st}\,C_T\,(\lambda,\,\beta) \\ T_a &=& F_{st}\,R\,C_Q\,(\lambda,\,\beta) \end{array}\right. \tag{12}$$

The reference force F_st is defined from the impact pressure $\frac{1}{2}\,\rho\,v^2$ and the rotor swept area $\pi\,R^2$ (with rotor radius R), where ρ denotes the air density:

$$F_{st} = \frac{1}{2}\,\rho\,\pi\,R^2\,v^2 \tag{13}$$

It is worth noting that for simulation purpose, the static wind speed v is used. However, a more accurate model should exploit the effective wind speed $v_e = v - (\dot{y}_T + \dot{y}_B)$, *i.e.* the static wind speed corrected with by the tower and blade motion effects. However, the aerodynamic maps used for the calculation of the rotor thrust and torque are usually represented as static 2–dimensional tables, which already take into account the dynamic contributions of both the tower and the blade motions.

As highlighted in the expressions (12), the rotor thrust and torque coefficients (C_T, C_Q) depend on the tip speed ratio $\lambda = \frac{\omega_r\,R}{v}$ and the pitch angle β. Therefore, the rotor thrust F_T and torque T_a assume the following expressions:

$$\left\{\begin{array}{lcl} F_T &=& \frac{1}{2}\,\rho\,\pi\,R^2\,C_T\,(\lambda,\,\beta)\;v^2 \\ T_a &=& \frac{1}{2}\,\rho\,\pi\,R^3\,C_Q\,(\lambda,\,\beta)\;v^2 \end{array}\right. \tag{14}$$

The expressions (14) highlight that the rotor thrust F_T and torque T_a are nonlinear functions dependent on the wind speed v, the rotor speed ω_r, and the pitch angle β. These functions are usually expressed as two–dimensional maps, which must be known for the whole range of variation of both the pitch angles and tip speed ratios. These maps are usually a static approximation of more detailed aerodynamic computations that can be obtained using for example the Blade Element Momentum (BEM) method. In this case, the aerodynamic lift and drag forces at each blade section are calculated and integrated in order to obtain the rotor thrust and torque.

It is worth noting that for simulation purposes, the tabulated versions of the aerodynamic maps C_Q and C_T are sufficient. On the other hand, for control design, the derivatives of the rotor torque (and thrust) are needed, thus requiring a description of the aerodynamic maps as analytical functions. Therefore, these maps can be approximated using combinations of polynomial and exponential functions, whose powers and coefficients are estimated via *e.g.* identification [12] approaches.

2.5. Wind Turbine Overall Model

By replacing the expressions (14) for the rotor thrust and torque into the mechanical model (8) and adding the models (10) and (11) for the pitch and the generator/converter dynamics, a nonlinear state–space model is obtained:

$$\begin{cases} \dot{\mathbf{x}} &= \mathbf{A}\,\mathbf{x} + \mathbf{B}\,\mathbf{u} + g\,(\mathbf{x},\, v) \\ \mathbf{y} &= \mathbf{C}\,\mathbf{x} \end{cases} \tag{15}$$

with a state vector that now includes the pitch angle and the generator torque: $\mathbf{x} = \left[y_T,\, y_B,\, \theta_r - \theta_g,\, \dot{y}_T,\, \dot{y}_B,\, \dot{\theta}_r,\, \dot{\theta}_g,\, \beta\right]^T$. Since the rotor thrust force and the rotor torque have been used as inputs for the vector $\mathbf{u}_m$ in the mechanical submodel (8), a new input vector is defined for the complete state–space model (15), *i.e.* $\mathbf{u} = [\beta_d,\, T_g]^T$, whose components are the demanded pitch angle and the generator torque, respectively. The wind speed is normally considered as a disturbance input. The linear part of the state–space model (15) is defined by the matrices:

$$\mathbf{A} = \begin{bmatrix} 0_{3\times3} & \tilde{\mathbf{L}} & 0_{3\times1} & 0_{3\times1} \\ \mathbf{M}^{-1}\tilde{\mathbf{K}} & \mathbf{M}^{-1}\mathbf{D} & 0_{4\times1} & \begin{bmatrix} 0_{3\times1} \\ -\frac{1}{J_g} \end{bmatrix} \\ 0_{1\times3} & 0_{1\times4} & -\frac{1}{\tau} & 0 \\ 0_{1\times3} & 0_{1\times4} & 0 & -\frac{1}{\tau_g} \end{bmatrix}, \quad \mathbf{B} = \begin{bmatrix} 0_{7\times1} & 0_{7\times1} \\ \frac{1}{\tau} & 0 \\ 0 & -\frac{1}{\tau_g} \end{bmatrix} \tag{16}$$

with:

$$\tilde{\mathbf{L}} = \begin{bmatrix} 1 & 0 & 0 & 0 \\ 0 & 1 & 0 & 0 \\ 0 & 0 & 1 & -1 \end{bmatrix} \text{ and } \tilde{\mathbf{K}} = \begin{bmatrix} k_T & 0 & 0 \\ 0 & N\,k_B & 0 \\ 0 & 0 & k_S \\ 0 & 0 & -k_S \end{bmatrix} \tag{17}$$

Moreover, the system vector in (15) nonlinearly depends on the state and input vector:

$$g\,(\mathbf{x},\, v) = \begin{bmatrix} 0_{4\times1} \\ \frac{1}{m_B}\,F_T\,(\mathbf{x},\, v) \\ \frac{1}{J_r}\,T_a\,(\mathbf{x},\, v) \\ 0_{3\times1} \end{bmatrix} \tag{18}$$

Here, the rotor thrust and torque expressions are given in (14), whilst the mass and damping matrices are defined in (7).

It is worth noting that in a real wind turbine, the centrifugal forces acting on the rotating rotor blades lead to a stiffening of the blades. As a consequence, the bending behaviour of the

Journal of Physics: Conference Series **659** (2015) 012001 doi:10.1088/1742-6596/659/1/012001

rotor blades depends on the rotor speed itself. By considering again the translational spring–mass system of the blade–tip displacement, this second–order effect can be included in the model (15) by introducing a translational blade stiffness parameter k_B dependent on the rotor speed, *i.e.* $k_B(\omega_r) = \alpha\, m_B\, r_B\, \omega_r^2$. r_B denotes the distance from the blade root to the blade centre of mass and α tuning parameter. In this way, by including the centrifugal stiffening correction, the nonlinear system vector $g(\mathbf{x}, v)$ in (18) has the form:

$$g(\mathbf{x}, v) = \begin{bmatrix} 0_{3\times 1} \\ \frac{N}{m_T}\, k_B(\omega_r)\, y_B \\ \frac{1}{N\, m_B}\, F_T(\mathbf{x}, v) + \frac{m_T + N\, m_B}{m_T\, m_B}\, k_B(\omega_r)\, y_B \\ \frac{1}{J_r}\, T_a(\mathbf{x}, v) \\ 0_{3\times 1} \end{bmatrix} \qquad (19)$$

The inclusion of the centrifugal term is inspired from the FAST code, in order to obtain a high–fidelity wind turbine simulation model. For example, the translational blade bending model could be required when overspeed scenarios shall be taken into account. However, for the usual operating regimes of a wind turbine, the corrections induced by the centrifugal blade stiffening have only minor effects on the final results. Therefore, the centrifugal correction has been recalled here for the sake of completeness, but it has limited interest in real cases.

2.6. Measurement Errors

Wind turbine high–fidelity simulators, which were described for example in [6, 13], consider white noise added to all measurements. This relies on the assumption that noisy sensor signals should represent more realistic scenarios. However, this is not the case, as a realistic simulation would require an accurate knowledge of each sensor and its measurement reliability. To the best of the authors' knowledge and from their experience with wind turbine systems, all main measurements acquired from the wind turbine process (rotor and generator speed, pitch angle, generator torque), are virtually noise–free or affected by very weak noise.

3. Wind Turbine Control Strategies

Wind turbine control goals and strategies are affected by turbine configuration. Horizontal–axis wind turbine may be "upwind", with the rotor on the upwind side of the tower, or "downwind". The choice of upwind versus downwind configuration affects the choice of yaw controller and the turbine dynamics, and thus the structural design. Wind turbines may also be variable pitch or fixed pitch, meaning that the blades may or may not be able to rotate along their longitudinal axes. Although fixed–pitch machines are less expensive initially, the reduced ability to control loads and change the aerodynamic torque means that they are becoming less common within the realm of large wind turbines. Variable-pitch turbines may allow all or part of their blades to rotate along the pitch axis.

Moreover, wind turbines can be variable speed or fixed speed. Variable–speed turbines tend to operate closer to their maximum aerodynamic efficiency for a higher percentage of the time, but require electrical power processing so that the generated electricity can be fed into the electrical grid at the proper frequency. As generator and power electronics technologies improve and costs decrease, variable–speed turbines are becoming more popular than constant–speed turbines at the utility scale.

Fig. 6 shows an example power curve for a variable–speed wind turbine. When the wind speed is low (usually below 6 m/s), the power available in the wind is low compared to losses in the turbine system so the turbine is not running. This operational region is sometimes known as Region 1. When the wind speed is high, Region 3 (above 11.7 m/s), power is limited to avoid exceeding safe electrical and mechanical load limits.

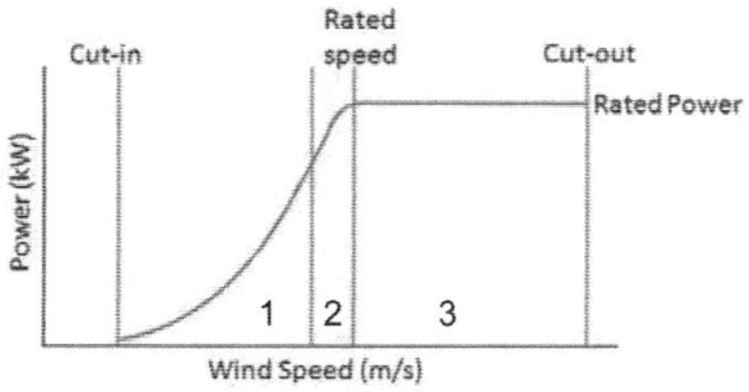

Figure 6. Example of wind turbine power curve.

The main difference between fixed–speed and variable speed wind turbines appears for mid–range wind speeds, the Region 2 in Fig. 6 , which normally encompasses wind speeds between 6 and 11.7 m/s. Except for one design operating point (10 m/s), a variable–speed turbine captures more power than a fixed–speed turbine. The reason for the discrepancy is that variable–speed turbines can operate at maximum aerodynamic efficiency over a wider range of wind speeds than fixed–speed turbines. The maximum difference between the generated power of the two wind turbines in Region 2 can be about 150 kW.

For typical wind speed situations, the variable–speed turbine can capture 2.3% more energy per year than the constant–speed turbine, which is considered to be a significant difference in the wind industry. Not shown in Fig. 6 is the "high wind cut–out", a wind speed above which the turbine is powered down and stopped to avoid excessive operating loads. High wind cut–out typically occurs at wind speeds above 20 – 30 m/s for large turbines, with many factors determining the exact value. Even a perfect wind turbine cannot fully capture the power available in the wind. In fact, actuator disc theory shows that the theoretical maximum aerodynamic efficiency, which is called the Betz Limit, is approximately 60% of the wind power. The reason that an efficiency of 100% cannot be achieved is that the wind must have some kinetic energy remaining after passing through the rotor disc. If it did not, the wind would by definition be stopped and no more wind would be able to pass through the rotor to provide energy to the turbine.

In designing controllers for wind turbines, it is often assumed (as in (14)) that the wind speed is uniform across the rotor plane. However, as indicated by the "instantaneous wind field" in Fig. 7, the wind input can vary substantially in space and time as it approaches the rotor plane. The deviations of the wind speed from the expected nominal wind speed across the rotor plane are considered disturbances for control design. It is virtually impossible to obtain a good measurement of the wind speed encountering the blades because of the spatial and temporal variability and also because the rotor interacts with and changes the wind input. Not only does turbulent wind cause the wind to be different for the different blades, but the wind speed input is different at different positions along each blade. Utility–scale wind turbines have several levels of control, which can be called "supervisory control", "operational control", and "subsystem control".

The top–level supervisory control determines when the turbine starts and stops in response to changes in the wind speed, and also monitors the health of the turbine. The operational control determines how the turbine achieves its control objectives in Regions 2 and 3. The subsystem controllers cause the generator, power electronics, yaw drive, pitch drive, and other actuators to perform as desired. In this section, the operational control loops and the controllers shown in Fig. 7, which exploit the submodels described in Section 2. In particular, the main control objectives, which are recalled in Section 3.1, will be exploited for illustrating the pitch and torque controllers in Section 4.

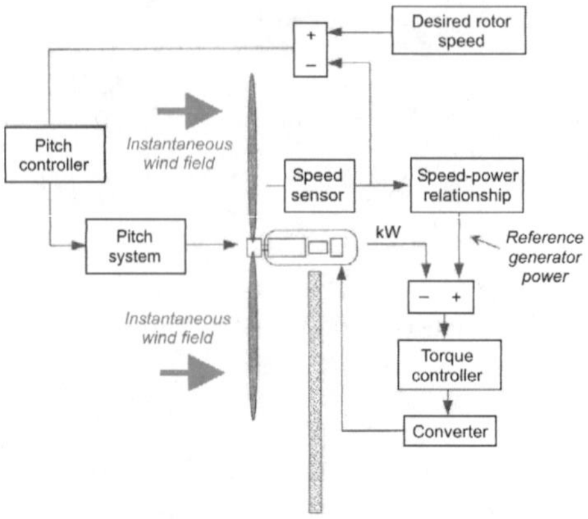

Figure 7. Block diagram of the wind turbine control loops.

3.1. Control Loops and Objectives

The primary Region 2 control objective for a variable–speed wind turbine is to maximise the power coefficient, and in particular the C_Q map in (12). As already shown in Section 2.4, this power coefficient is a function of the turbine s tip–speed ratio λ, which is defined in Section 2.4. Thus, the tip–speed ratio is the ratio of the linear (tangential) speed of the blade tip to the wind speed, v is always time-varying, and ω_r is time–varying for a variable–speed turbine. The relationship between C_Q and the tip–speed ratio λ is a turbine–specific nonlinear function. As already discussed, C_Q also depends on the blade pitch angle in a nonlinear fashion, and these relationships have the same basic shape for most modern wind turbines. An example of C_Q surface is shown in Fig. 8 for a generic wind turbine.

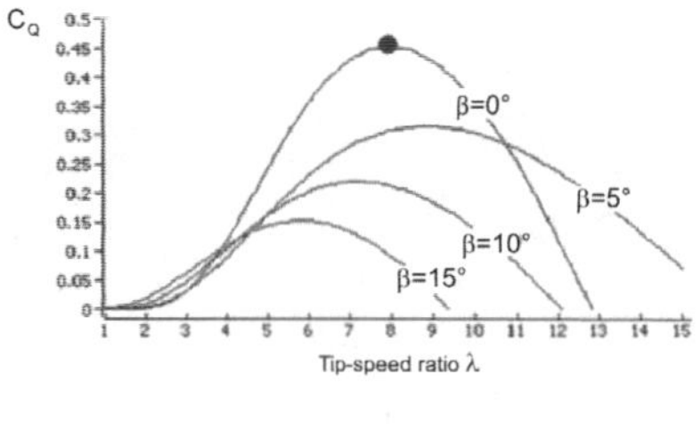

Figure 8. Example of power coefficient curve.

As shown in Fig. 8, the turbine will operate at its highest aerodynamic efficiency point, $C_{\max}$, at a certain pitch angle and tip–speed ratio. The pitch angle is easy to control, and can be reliably maintained at the optimal efficiency point. However, the tip–speed ratio depends on the incoming wind speed v and therefore is constantly changing. Thus, the Region 2 control is primarily concerned with varying the turbine speed to track the wind speed. Section 4 will explain how this control objective can be achieved.

On utility–scale wind turbines, Region 3 control is typically performed via a separate pitch control loop, as shown in Fig. 3 of Section 2. In the Region 3, the primary objective is to limit the turbine power so that safe electrical and mechanical loads are not exceeded. Power limitation can be achieved by pitching the blades or by yawing the turbine out of the wind, both of which can reduce the aerodynamic torque below what is theoretically available from an increase in wind speed. Note that the power P is related to rotor speed ω_r and aerodynamic torque T_a by the relation:

$$P = T_a \, \omega_r \tag{20}$$

If the power and rotor speed are held constant, the aerodynamic torque must also be constant even as wind speed varies. It is desirable to produce as much power as the turbine can safely produce, the limit of which is known as the turbine s rated power. In the Region 3, the pitch control loop regulates the rotor speed ω_r (at the turbine s "rated speed") so that the turbine operates at its rated power.

It is worth noting that the wind turbine blades may be controlled to all turn collectively or to each turn independently or individually. As outlined in Section 2.2, suitable pitch systems can be used to change the aerodynamic torque from the wind input, and are often fast enough to be of interest to the control community. Typical maximum pitch rates range from 18 deg/s for 600 kW research turbines down to 8 deg/s for 5 MW turbines. Variable–pitch turbines can limit power either by pitching to "stall" or to "feather", and fixed–pitch turbines typically limit power by entering the aerodynamic stall regime above rated wind speed. A blade in full feather is one in which the leading edge of the blade points directly into the wind. A discussion of the benefits of pitching to feather versus pitching to stall is outside the scope of this review paper, but more information can be found *e.g.* in [5].

4. Feedback Controls for Wind Turbines

This section provides further information regarding what control strategies are typically used for the torque control and the pitch control blocks in Fig. 3 of Section 2. As depicted in Fig. 3, both control loops typically only use rotor speed feedback. The other sensors and measurements acquired from the wind turbine can be used for advanced control purposes, as outlined in Section 5.1.

As shown in Fig. 6, the nominal operating trajectory of the wind turbine is created to satisfy different demands below and above a certain wind speed. Since the classical control approach deals only with SISO transfer functions, and because several references exist, the control task is split into the design of multiple separate compensators. The design of the complete wind turbine controller is thus divided into four main control boxes and design steps, as listed below:

(i) Controller operating in partial load condition: it refers to the design of the generator torque controller. This controller operates in the partial load (Region 2), and should maximise the energy production while minimising mechanical stress and actuator usage;

(ii) Controller operating in full load condition: it concerns the speed controller and power controller. These controllers operate in the full load (Region 3), and should track the rated generator speed and limiting the output power;

(iii) Bumpless transfer: it describe the design of the mechanism that eliminates bumps on the control signals, when switching between the controllers in the partial load and full load regions;

(iv) Structural stress damper: it regards the design of structure and drivetrain stress damper. The purpose of the module is to dampen drivetrain oscillations and reduce structural stress that could affect the wind turbine tower.

The first two items are the main control loops, whilst the two other tasks concern advanced control issues, which can enhance both the control and system performances. Note also that the transfer functions outlined throughout this section need to be discretised to allow the implementation of the controllers and filters in real–time conditions. In this way, the overall controller design consists of using two different controllers for the partial load region and the full load region. When the wind speed is below the rated value, the control system should maintain the pitch angle at its optimal value and control the generator torque in order to achieve the optimal tip–speed ratio (switch to Region 2).

Above the rated wind speed the output power is kept constant by pitching the rotor blades, while using a power controller that manipulates the generator torque around a constant value to remove steady–state errors on the output power. In both regions a drivetrain stress damper is exploited to dampen drivetrain oscillations actively. Together, the two sets of controllers are able to solve the control task of tracking the ideal power curve in Fig. 8. In order to switch smoothly between the two sets of controllers a bumpless transfer mechanism is implemented.

4.1. Partial Load Operation Controller

At low wind speeds, *i.e.* in partial load operation, variable–speed control is implemented to track the optimum point on the C_Q–surface for maximising the power output. The speed of the generator is controlled by regulating the torque on the generator through the generator torque controller. The purpose of this section is to go through the design of the generator torque controller. In partial load operation it is chosen to operate the wind turbine at $\beta = 0^o$ since the maximum power coefficient is obtained at this pitch angle. This means that the highest efficiency is achieved for:

$$\lambda_{\text{opt}} = \frac{\omega_{r,\text{opt}}\, R}{v} \tag{21}$$

where λ_{opt} is the tip–speed ratio maximising the C_Q–value for $\beta = 0^o$, and $\omega_{r,\text{opt}}$ is the optimum rotor speed. In order to obtain the optimal tip–speed ratio a method is used, which suggests to apply a certain generator torque as a function of the generator speed [14]. The advantage of this approach is that only the measurement of the rotor speed or generator speed is required. When utilising this approach, the controller structure for partial load operation is illustrated in Fig. 9.

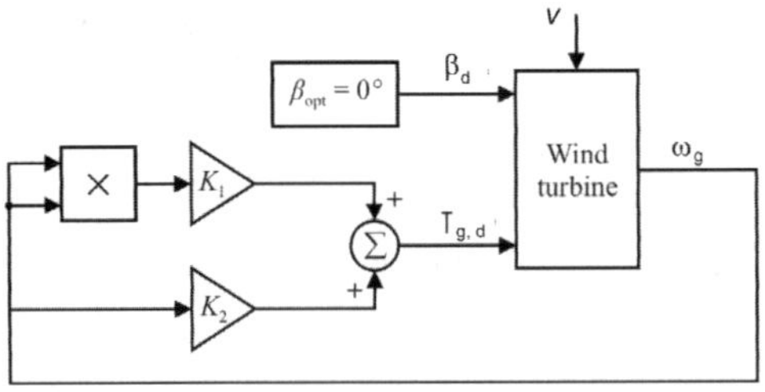

Figure 9. Generator torque controller for operation in partial load region (Region 2).

The principle of the standard control law is to calculate the wind speed in the definition of the tip–speed ratio, and replace it into the expression for the aerodynamic torque in (14). Hence, the relation can be obtained expressing the required generator torque based on the maximum power coefficient and the optimal tip–speed ratio:

$$v(t) = \frac{\omega_r(t)\, R}{\lambda(t)} \tag{22}$$

This expression is inserted into (14) describing the aerodynamic torque:

$$T_a(t) = \frac{1}{2}\rho\,\pi\,R^2\,\frac{R^3}{\lambda^3(t)}C_Q\left(\lambda(t),\,\beta(t)\right)\omega_r^2(t) \tag{23}$$

Since the wind turbine includes a transmission system, the gear ratio and friction components of the drivetrain have to be considered when determining the generator torque corresponding to a certain aerodynamic torque. In order to describe the generator torque only as function of the generator speed, the system has to be assumed in steady–state, where $\dot{\omega}_r(t) = \dot{\omega}_g(t) = 0$ and $\omega_g(t) = n_g\,\omega_r(t)$. In this way, by considering the drivetrain equations in (5), the following expression is obtained:

$$T_{g,d} = \frac{1}{2}\,\rho\,\pi\,R^2\,\frac{R^3}{n_g^3\,\lambda_{\text{opt}}^3}\,C_{\max}\,\omega_g^2(t) - d_S\left(\frac{1}{n_g^2}+1\right)\omega_g \tag{24}$$

with:

$$K_1 = \tfrac{1}{2}\,\rho\,\pi\,R^2\,\tfrac{R^3}{n_g^3\,\lambda_{\text{opt}}^3}, \quad K_2 = -d_S\left(\tfrac{1}{n_g^2}+1\right) \tag{25}$$

4.2. Full Load Operation Controller
For the high wind speeds, *i.e.* in full load operation, the desired operation of the wind turbine is to keep the rotor speed and the generated power at constant values. The main idea is to use the pitch system to control the efficiency of the aerodynamics while applying the rated generator torque. However, in order to improve tracking of the power reference and cancel steady–state errors on the output power, a power controller is also introduced. Therefore, in this section the design of the speed controller and the power controller is sketched. The structure of the controllers operating above the rated wind speed is shown in Fig. 10.

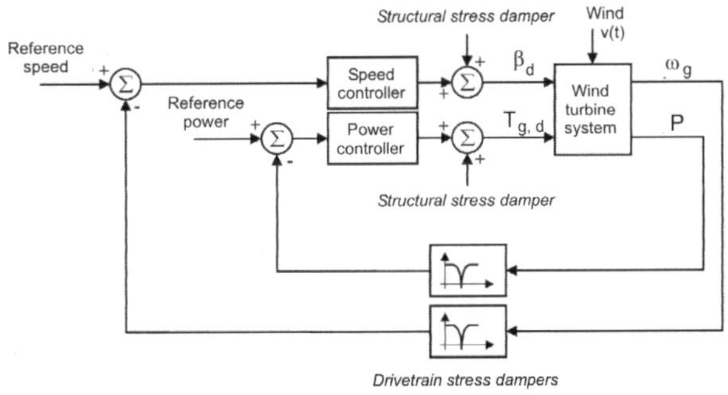

Figure 10. Speed controller and power controller for operation in the full load region (Region 3).

The wind speed is considered the disturbance input to the system. However, higher frequency components such as the resonant frequency of the drive train are also apparent on the measured generator speed. Therefore, the measured generator speed is band–stop filtered before it is fed to the controller, to remove the drivetrain eigenfrequency from the measurement. This solution is also found in other wind turbine control schemes to mitigate the effects of structural oscillations, by injecting suitable signals in the control loops. In the following, the design of the speed controller and the power controller is presented.

With reference to the speed controller of Fig. 10, it is implemented as a PI controller that is able to track the speed reference and cancel possible steady–state errors on the generator speed. The speed controller transfer function $D_s(s)$ has the form:

$$D_s(s) = K_{ps}\left(1 + \frac{1}{T_{is}}\frac{1}{s}\right) \tag{26}$$

where K_{ps} is the PI proportional gain and T_{is} is the reset rate of the integrator. It can be shown that pitching the blades has a larger influence on the aerodynamic torque at higher wind speeds. For this reason, the gain K_{ps} of the speed controller should be large near the rated wind speed but smaller at higher wind speeds. The optimal gain of the speed controller associated with a certain wind speed can make the system become unstable at higher wind speeds due to the increasing gain of the system. Therefore, the speed controller is configured with one set of parameters in the region corresponding to stationary wind speeds in the interval $12 - -15$ m/s, while a smaller gain is utilised for the region covering wind speeds of $15 - -25$ m/s. Although the system has different gains in these two working regions, it is possible to design the controllers so that similar transient responses of the controlled system are obtained.

On the other hand, with reference to the power controller of Fig. 10, it is implemented again in order to cancel possible steady–state errors on the output power. This suggests using slow integral control for the power controller, as this will eventually cancel steady–state errors on the output power without interfering with the speed controller. However, it may be beneficial to make the power controller faster to improve accuracy in the tracking of the rated power. The power controller is realized as a PI controller, whose transfer function $D_p(s)$ has the form:

$$D_p(s) = K_{pp}\left(1 + \frac{1}{T_{ip}}\frac{1}{s}\right) \tag{27}$$

where K_{pp} is the proportional gain of the PI regulator, whilst T_{ip} is the reset rate of the integrator. By exploiting the measured output power directly can be a problem, since the measurement is very noisy. This means that the measurement noise has to be take into account in the design and yields that the proportional gain has to be sufficiently small. The proportional gain is usually chosen using a trial and error approach while the reset rate is selected large enough to avoid overshoot on the step response.

4.3. Structural and Drivetrain Stress Damper

Active stress damping solutions are deployed in large horizontal–axis wind turbines to mitigate fatigue damage due to drivetrain and structural oscillations and vibrations. The idea is to add proper components to the wind turbine control signals to compensate for the oscillations in the drivetrain and the tower vibrations. These signals should have frequencies equal to the eigenfrequencies of the drivetrain and the wind turbine structure, which can be found by filtering the measurement of the generator speed and the generated power. When the outputs from these filters are added to the generator torque and the pitch command, the phase of the filters must be zero at the resonant frequency to achieve the desired damping effects. These oscillation and vibration dampers are thus implemented to add compensating signals, as shown in Fig. 10.

Second order filter structures for the stress and the structural damping have been proposed and can be applied to dampen the eigenfrequency of both the drivetrain and the tower structure [5]. In general, the filter time constant introduces a zero in the filter that can be used to compensate for time lags in the system. To determine the gain of the filter, the root loci are plotted for the transfer functions from the wind turbine inputs to its outputs. More details on the design of these filters, which are beyond the scope of this paper, can be found in [5].

Note that, due to the higher loads at higher wind speeds, it is favourable if the filter gains depend on the point of operation. A simple way of fulfilling this property is to apply different gains in the partial and full load configurations of the wind turbine controller. Therefore, a bumpless transfer mechanism must ensure that no bumps exist on the control signals in the switch between two different controllers.

5. Advanced Control of Wind Turbines

There are many aspects of wind turbine performance that can be improved with more advanced control development. Researchers have developed methods for using adaptive control to compensate for unknown or time–varying parameters [15]. Other researchers have also begun to investigate the addition of feed-forward control to improve the disturbance rejection performance when the incoming wind profile deviates from that expected [16]. Most of these feed-forward controllers use estimates of the disturbance (or wind deviation). New sensing technologies will enable various avenues of advanced control research. For instance, there has been recent interest in evaluating the potential of LIDAR (which stands for "LIght Detection And Ranging") sensors for wind turbine control [16]. LIDAR is a remote optical sensing technology that has been used since the 1970s for meteorology for measuring wind speed profiles for monitoring hurricanes and wind conditions around airports. New lidar systems based on solid–state sources and off–the–shelf telecommunications equipment allow for inexpensive deployment, modularity, and improved reliability. Depending on the particular type of technology used, lidar sensors can provide quantities representing the wind speed and direction and various wind turbulence and shear parameters. An accurate measurement of the wind profile over the entire rotor plane in Fig. 7 can enable feed-forward pitch control and feed-forward torque control to improve performance dramatically. Advanced wind turbine controllers are further discussed and compared in [4, 16]. As turbines get larger and blades get longer, it is possible that turbine manufacturers will build turbines that allow for different pitch angles at different radial positions along the blades relative to the standard blade twist angle. In this case, separate actuators and controllers may be necessary, opening up even more control opportunities [16].

Note finally that, the need of advanced control solutions for these very demanding systems, motivated also the requirement of reliability, availability, maintainability, and safety over power conversion efficiency. These issues have begun to stimulate research and development of the so–called sustainable control, which is outlined in Section 5.1.

5.1. Sustainable Control Issues

In general, wind turbines in the megawatt size are expensive, and hence their availability and reliability must be high in order to maximise the energy production. This issue could be particularly important for offshore installations, where Operation and Maintenance (O & M) services have to be minimised, since they represent one of the main factors of the energy cost. The capital cost, as well as the wind turbine foundation and installation determine the basic term in the cost of the produced energy, which constitute the energy 'fixed cost'. The O & M represent a 'variable cost' that can increase the energy cost up to about the 30%. At the same time, industrial systems have become more complex and expensive, with less tolerance for performance degradation, productivity decrease and safety hazards. This leads also to an ever increasing requirement on reliability and safety of control systems subjected to process abnormalities and component faults. As a result, it is extremely important the Fault Detection and Diagnosis (FDD) or the Fault Detection and Isolation (FDI) tasks, as well as the achievement of fault–tolerant features for minimising possible performance degradation and avoiding dangerous situations. With the advent of computerised control, communication networks and information techniques, it makes possible to develop novel real–time monitoring and fault–tolerant design techniques for industrial processes, but brings challenges.

In the last years, many works have been proposed on wind turbine FDI/FDD, and the most relevant are *e.g.* in [6]. On the other hand, regarding the FTC problem for wind turbines, it was recently analysed with reference to an offshore wind turbine benchmark *e.g.* in [6]. In general, FTC methods are classified into two types, *i.e.* Passive Fault Tolerant Control (PFTC) scheme and Active Fault Tolerant Control (AFTC) scheme [17]. In PFTC, controllers are fixed and are designed to be robust against a class of presumed faults. In contrast to PFTC, AFTC reacts to the system component failures actively by reconfiguring control actions so that the stability and acceptable performance of the entire system can be maintained. In particular for wind turbines, FTC designs were considered and compared in [6]. These processes are nonlinear dynamic systems, whose aerodynamics are nonlinear and unsteady, whilst their rotors are subject to complicated turbulent wind inflow fields driving fatigue loading. Therefore, the so–called wind turbine 'sustainable' control represents a complex and challenging task [18].

Therefore, the purpose of this section is outline the basic solutions to sustainable control design, which are able of handling faults affecting the controlled wind turbine. For example, changing dynamics of the pitch system due a fault cannot be accommodated by signal correction. Therefore, it should be considered in the controller design, to guarantee stability and a satisfactory performance. Among the possible causes for changed dynamics of the pitch system, they can due to a change in the air content of the hydraulic system oil. This fault is considered since it is the most likely to occur, and since the reference controller becomes unstable when the hydraulic oil has a high air content. Another issue raises when the generator speed measurement is unavailable, and the controller should rely on the measurement of the rotor speed, which is contaminated with much more noise than the generator speed measurement. This makes it necessary to reconfigure the controller to obtain a reasonable performance of the control system.

Section 5.2 outlines the main differences between active and passive fault–tolerant control systems and suggests how they are applied to the considered system.

5.2. Active and Passive Fault–Tolerant Control Systems

In order to outline and compare the controllers developed using active and passive fault–tolerant design approaches, they should be derived using the same procedures in the fault–free case. In this way, any differences in their performance or design complexity would be caused only by the fault tolerance approach, rather than the underlying controller solutions. Furthermore, the controllers should manage the parameter–varying nature of the wind turbine along its nominal operating trajectory caused by the aerodynamic nonlinearities. Usually, in order to comply with these requirements, the controllers are usually designed for example using Linear Parameter–Varying (LPV) modelling or fuzzy descriptions [5].

The two fault–tolerant control solutions have different structures as shown in Fig. 11. Note that only the active fault–tolerant controller (AFTC) relies on a fault diagnosis algorithm (FDD). This represents the main difference between the two control schemes.

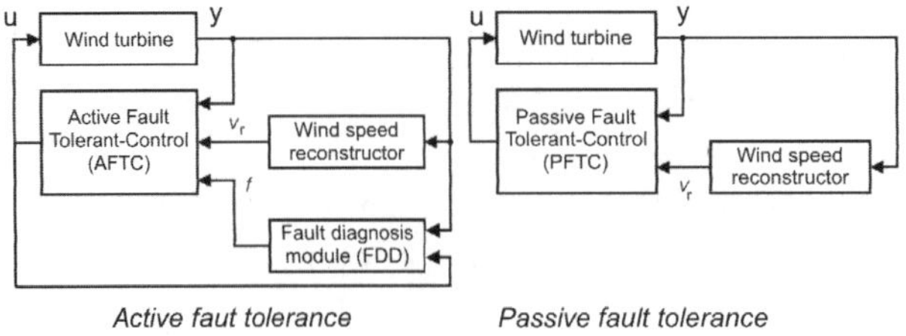

Figure 11. Structures of the active and passive fault–tolerant control systems.

The main point between AFTC and PFTC schemes is that an active fault–tolerant controller relies on a fault diagnosis system, which provides information about the faults f to the controller. In the considered case the fault diagnosis system FDD contains the estimation of the unknown input (fault) affecting the system under control. The knowledge of the fault f allows the AFTC to reconfigure the current state of the system. On the other hand, the FDD is able to improve the controller performance in fault–free conditions, since it can compensate *e.g.* the modelling errors, uncertainty and disturbances. On the other hand, the PFTC scheme does not rely on a fault diagnosis algorithm, but is designed to be robust towards any possible faults. This is accomplished by designing a controller that is optimised for the fault–free situation, while satisfying some graceful degradation requirements in the faulty cases. However, with respect to the robust control design, the PFTC strategy provides reliable controllers that guarantee the same performance with no risk of false FDI or reconfigurations.

In general, the methods used in the fault–tolerant controller designs should rely on output feedback, since only part of the state vector is measured. Additionally, they should take the measurement noise into account. Moreover, the design methods should be suited for nonlinear systems or linear systems with varying parameters. The latest proposed solutions for the derivation of both active and the passive fault–tolerant controllers rely on LPV and fuzzy descriptions, to which the fault–tolerance properties are added, since these frameworks methods are able to provide stability and guaranteed performance with respect to parameter variations, uncertainty and disturbance. Additionally, LPV and fuzzy controller design methods are well–established in multiple applications including wind turbines [5]. To add fault–tolerance to the common LPV and fuzzy controller formulation, different approaches can be exploited. For example, the AFTC scheme can use the parameters of both the LPV and fuzzy structures estimated by the FDD module for scheduling the controllers [19, 15]. On the other hand, different approaches can be used to obtain fault–tolerance in the PFTC methods. For this purpose, the design methods described in [5, 20] can be modified to cope with parametric uncertainties, as addressed *e.g.* in [21]. Alternatively, other methods could have been used such as [22], which preserves the nominal performance. Generally, these approaches rely on solving some optimisation problems where a controller is calculated subject to maximising the disturbance attenuation.

6. Conclusion

This paper revised the most important modelling and control issues on wind turbines from a systems and control engineering point of view. A walk around the wind turbine control loops discussed the goals of the most common solutions and overviews the typical actuation and sensing available on commercial turbines. The work intended to provide also an updated and broader perspective by covering not only the modelling and control of individual wind turbines, but also outlining a number of areas for further research, and anticipating new issues that can open up new paradigms for advanced control approaches. In summary, wind energy is a fast growing industry, and this growth has led to a large demand for better modelling and control of wind turbines. Uncertainty, disturbance and other deviations from normal working conditions of the wind turbines make the control challenging, thus motivating the need for advanced modelling and a number of so–called sustainable control approaches that should be explored to reduce the cost of wind energy. By enabling this clean renewable energy source to provide and reliably meet the world's electricity needs, the tremendous challenge of solving the world's energy requirements in the future will be enhanced. The wind resource available worldwide is large, and much of the world's future electrical energy needs can be provided by wind energy alone if the technological barriers are overcome. The application of sustainable controls for wind energy systems is still in its infancy, and there are many fundamental and applied issues that can be addressed by the systems and control community to significantly improve the efficiency, operation, and lifetimes of wind turbines.

References

[1] Johnson K E and Fleming P A 2011 *Mechatronics* **21** 728–736 DOI: 10.1016/j.mechatronics.2010.11.010
[2] Global Wind Energy Council 2014 Wind Energy Statistics 2013 Report
[3] Pao L Y and Johnson K E 2009 *Proceedings of the American Control Conference, 2009 – ACC'09*

(St. Louis, MO, USA: IEEE) pp 2076–2089 ISSN: 0743–1619. ISBN: 978–1–4244–4523–3. DOI: 10.1109/ACC.2009.5160195

[4] Laks J H, Pao L Y and Wright A D 2009 *Proceedings of the American Control Conference, 2009 – ACC'09* (St. Louis, MO, USA: IEEE) pp 2096–2103 ISSN: 0743–1619. ISBN: 978–1–4244–4523–3. DOI: 10.1109/ACC.2009.5160590

[5] Bianchi F D, Battista H D and Mantz R J 2007 *Wind Turbine Control Systems: Principles, Modelling and Gain Scheduling Design* 1st ed Advances in Industrial Control (Springer) ISBN: 1–84628–492–9

[6] Odgaard P F, Stoustrup J and Kinnaert M 2013 *IEEE Transactions on Control Systems Technology* **21** 1168–1182 ISSN: 1063–6536. DOI: 10.1109/TCST.2013.2259235

[7] Chen J and Patton R J 1999 *Robust Model-Based Fault Diagnosis for Dynamic Systems* (Boston, MA, USA: Kluwer Academic Publishers)

[8] Ding S X 2008 *Model-based Fault Diagnosis Techniques: Design Schemes, Algorithms, and Tools* 1st ed (Berlin Heidelberg: Springer) ISBN: 978–3540763031

[9] Isermann R 2005 *Fault-Diagnosis Systems: An Introduction from Fault Detection to Fault Tolerance* 1st ed (Weinheim, Germany: Springer–Verlag) ISBN: 3540241124

[10] Jonkman J M and Buhl Jr M L 2005 FAST User's Guide Tech. Rep. NREL/EL–500–38230 National Renewable Energy Laboratory Golden, CO, USA

[11] Georg S, Schulte H and Aschemann H 2012 *Proceedings of the IEEE International Conference on Fuzzy Systems* (Brisbane, Australia: IEEE) pp 1737–1744 ISSN: 1098–7584. ISBN: 978–1–4673–1507–4. DOI: 10.1109/FUZZ–IEEE.2012.6251302

[12] Simani S and Castaldi P 2014 *International Journal of Robust and Nonlinear Control* **24** 1283–1303 John Wiley. DOI: 10.1002/rnc.2993

[13] Odgaard P F and Johnson K 2013 *Proc. of the 2013 American Control Conference – ACC* (Washington DC, USA: IEEE Control Systems Society & American Automatic Control Council) pp 4447–4452 ISSN: 0743–1619. ISBN: 978–1–4799–0177–7

[14] Johnson K E, Pao L Y, Balas M J and Fingersh L J 2006 *IEEE Control Systems Magazine* **26** 70–81 DOI: 10.1109/MCS.2006.1636311

[15] Simani S and Castaldi P 2013 *Control Engineering Practice* **21** 1678–1693 special Issue Invited Paper. ISSN: 0967–0661. PII: S0967–0661(13)00155–X. DOI: http://dx.doi.org/10.1016/j.conengprac.2013.08.009

[16] Pao L Y and Johnson K E 2011 *IEEE Control Systems Magazine* **31** 44–62

[17] Mahmoud M, Jiang J and Zhang Y 2003 *Active Fault Tolerant Control Systems: Stochastic Analysis and Synthesis* Lecture Notes in Control and Information Sciences (Berlin, Germany: Springer–Verlag) ISBN: 3540003185

[18] Odgaard P F and Stoustrup J 2013 *Proceedings of the IEEE Multiconference on Systems and Control – MSC2013* (Hyderabad, India) pp 1–6

[19] Sami M and Patton R J 2012 *Proceedings of the 8th IFAC Symposium on Fault Detection, Supervision and Safety of Technical Processes – SAFEPROCESS 2012* vol 8 ed Verde C, Astorga Zaragoza C M and Molina A (National Autonomous University of Mexico, Mexico City, Mexico) pp 349–354 DOI: 10.3182/20120829–3–MX–2028.00131

[20] Galdi V, Piccolo A and Siano P 2008 *IEEE Transactions on Energy Conversion* **23**

[21] Puig V 2010 *International Journal of Applied Mathematics and Computer Science* **20** 619–635 DOI: 10.2478/v10006–010–0046–y

[22] Niemann H and Stoustrup J 2005 *International Journal of Control* **78** 1091–1110

12th European Workshop on Advanced Control and Diagnosis (ACD 2015) IOP Publishing
Journal of Physics: Conference Series **659** (2015) 012002 doi:10.1088/1742-6596/659/1/012002

Control design for discrete-time state-multiplicative noise stochastic systems

Dušan Krokavec and Anna Filasová

Department of Cybernetics and Artificial Intelligence, Faculty of Electrical Engineering and Informatics, Technical University of Košice, Letná 9, 042 00 Košice, Slovakia

E-mail: dusan.krokavec@tuke.sk; anna.filasova@tuke.sk

Abstract. Design conditions for existence of the H_∞ linear state feedback control for discrete-time stochastic systems with state-multiplicative noise and polytopic uncertainties are presented in the paper. Using an enhanced form of the bounded real lemma for discrete-time stochastic systems with state-multiplicative noise, the LMI-based procedure is provided for computation of the gains of linear, as well as nonlinear, state control law. The approach is illustrated on an example demonstrating the validity of the proposed method.

1. Introduction

The control design for systems with stochastic uncertainties is an area of study for several decades (see [7] and the references therein) and so for this practically and theoretically attractive field were derived numerous solutions. Modeling some parameter uncertainties as white noise processes in a linear system, this approach was applied for discrete-time [10], as well as for continuous-time systems [3], [14], [17]. Regrettably, these ways of doing it on linear quadratic optimal control principle led mainly to nonstandard Riccati equations. Representing the system uncertainty by unknown disturbance signal, H_∞ principle stayed the most prominent method of dealing with the corresponding control problem of disturbance relaxation [21]. Combining the both mentioned representations of uncertainty, adapting H_∞ approach towards the disturbance attenuation for stochastic systems and including those that ensure a performance bound in the H_∞ sense, these efforts have been contributed in design conditions. By that way, new accessions have been made in diverse practical problems for finite and infinite time horizon for continuous-time [4], [9], as well as for discrete-time stochastic systems [2], [5], denoting later such system as linear stochastic systems with multiplicative noise. It should be noted, the stochastic control for linear and nonlinear systems with multiplicative noises has been finding its applications in many fields in control theory, mainly in constrain control [11] and in gain scheduling when the scheduling parameters are corrupted with measurement noise [19].

Considering new results in control of systems with polytopic uncertainties [1], [8], [18], as well as in bounded real lemma forms for discrete-time stochastic systems [6], [12], [15], new design conditions for H_∞ linear control for linear discrete-time stochastic systems with state-multiplicative noise and polytopic uncertainties are derived in the paper. To present this, the paper is divided in these sections. Following the introduction in Sec. 1, the control design task for discrete-time stochastic systems with state-multiplicative noise is presented in Sec. 2. The preliminary results focused on two bounded real lemma forms for such stochastic systems are presented in Sec. 3 and, subsequently, in Sec. 4 there are derived the design conditions for H_∞ linear control of state-multiplicative noise stochastic systems with polytopic uncertainties. Sec. 5 illustrates the constrained control design task by a numerical solution and Sec. 6 draws some conclusions.

Throughout the paper, the notations are narrowly standard in such a way that $\boldsymbol{x}^T$, $\boldsymbol{X}^T$ denotes the transpose of the vector $\boldsymbol{x}$ and matrix $\boldsymbol{X}$, respectively, $\boldsymbol{X} = \boldsymbol{X}^T > 0$, (≥ 0), means that $\boldsymbol{X}$ is a symmetric positive definite (semi-definite) matrix, diag$[\cdot]$ designates a block diagonal matrix, the symbol $\boldsymbol{I}_n$ indicates the n-th order identity matrix, $E_o\{\cdot\}$ remits expectation with respect to the variable o, $\mathcal{Z}_+$ is the set of all positive integers, $I\!R$ denotes the set of real numbers, $I\!R^{n\times r}$ refers to the set of all $n \times r$ real matrices and $L_2\langle 0, +\infty)$ entails the space of square summable discrete vector random sequence over $\langle 0, +\infty)$.

2. Problem formulation

Throughout the paper, the task is concerned with the feedback design to control the stochastic discrete-time linear dynamic system given by the set of equations

$$\boldsymbol{q}(i+1) = (\boldsymbol{F} + \boldsymbol{V}o(i))\boldsymbol{q}(i) + \boldsymbol{G}\boldsymbol{u}(i) + \boldsymbol{W}\boldsymbol{d}(i)\,, \tag{1}$$

$$\boldsymbol{y}(i) = \boldsymbol{C}\boldsymbol{q}(i)\,, \tag{2}$$

where $\boldsymbol{q}(i) \in I\!R^n$, $\boldsymbol{u}(i) \in I\!R^r$, $\boldsymbol{y}(i) \in I\!R^m$ is the state, input, and output vector, respectively, $\boldsymbol{d}(i) \in I\!R^{r_w}$ is an exogenous disturbance vector, and $\boldsymbol{F} \in I\!R^{n\times n}$, $\boldsymbol{G} \in I\!R^{n\times r}$, $\boldsymbol{C} \in I\!R^{m\times n}$, $\boldsymbol{W} \in I\!R^{n\times r_w}$, $\boldsymbol{V} \in I\!R^{n\times n}$ are real matrices.

It is assumed that the multiplicative noise $o(i)$, $0 \leq i \leq j$ satisfies the conditions

$$E_o\{o(i)\} = 0, \quad E_o\{o(i)o(j)\} = \delta_{ij}, \quad \delta_{ij} = \begin{cases} 1, & i = j, \\ 0, & i \neq j, \end{cases} \tag{3}$$

where $E_o\{\cdot\}$ denotes expectation with respect to $o(i)$ and δ_{ij} is the Kronecker delta function. Disturbance is a non-anticipative process, where $\{\boldsymbol{d}(i)\} \in L_2(\langle 0, \infty); I\!R^{r_v})$.

The problem of the interest is to design in the mean square sense stable the closed-loop system by the state feedback control.

3. Basic preliminaries

In order to create more convenient space for new design task conditions, two new modifications of the stability solution for stochastic systems (compare, e.g., [3], [9], [4], [5]), are presented first. Since these modifications constitute the first stage in the solution of the considered problem, to the best of authors' belief, inclusion of the complete proofs was an easier way than to interpret in detail how to adapt the original bounded real lemma (BRL) formulation for the discrete-time stochastic systems with multiplicative noise.

Lemma 1 *[12] If $\boldsymbol{X}$ is a symmetric positive definite matrix, $\boldsymbol{L}_l$ are matrices of appropriate dimension, $a_l \geq 0$ is a real scalar, $l = 1, 2, \ldots s$ and s is a positive integer, then*

$$\left(\sum_{h=1}^{s} a_h \boldsymbol{L}_h^T\right) \boldsymbol{X} \left(\sum_{k=1}^{s} a_k \boldsymbol{L}_k\right) \leq \sum_{h=1}^{s} a_h \boldsymbol{L}_h^T \boldsymbol{X} \boldsymbol{L}_h \left(\sum_{k=1}^{s} a_k\right) \tag{4}$$

and if $\sum_{k=1}^{s} a_k = 1$,

$$\left(\sum_{h=1}^{s} a_h \boldsymbol{L}_h^T\right) \boldsymbol{X} \left(\sum_{k=1}^{s} a_k \boldsymbol{L}_k\right) \leq \sum_{h=1}^{s} a_h \boldsymbol{L}_h^T \boldsymbol{X} \boldsymbol{L}_h\,. \tag{5}$$

Journal of Physics: Conference Series **659** (2015) 012002 doi:10.1088/1742-6596/659/1/012002

Proposition 1 *(BRL for discrete-time MIMO state-multiplicative noise stochastic systems) The autonomous system (1), (2) is stable in the mean square sense and with the quadratic performance γ if there exists a positive definite matrix $\boldsymbol{P} \in I\!R^{n \times n}$ and a positive scalar $\gamma \in I\!R$ such that*

$$\boldsymbol{P} = \boldsymbol{P}^T > 0\,, \qquad \gamma > 0\,, \tag{6}$$

$$\begin{bmatrix} -\boldsymbol{P} & * & * & * & * \\ \boldsymbol{0} & -\gamma \boldsymbol{I}_{r_w} & * & * & * \\ \boldsymbol{PF} & \boldsymbol{PW} & -\boldsymbol{P} & * & * \\ \boldsymbol{PV} & \boldsymbol{0} & \boldsymbol{0} & -\boldsymbol{P} & * \\ \boldsymbol{C} & \boldsymbol{0} & \boldsymbol{0} & \boldsymbol{0} & -\boldsymbol{I}_m \end{bmatrix} < 0\,, \tag{7}$$

Hereafter, $*$ denotes the symmetric item in a symmetric matrix.

Proof: (compare [13]) Let the Lyapunov function candidate takes the form

$$v(\boldsymbol{q}(i)) = \boldsymbol{q}^T(i)\boldsymbol{P}\boldsymbol{q}(i) + \sum_{j=0}^{i-1}(\boldsymbol{y}^T(j)\boldsymbol{y}(j) - \gamma \boldsymbol{d}^T(j)\boldsymbol{d}(j))\,, \tag{8}$$

where $\gamma \in I\!R$ is square of the H_∞ norm of the transfer function matrix defined for the disturbance input $\boldsymbol{d}$ and the system output $\boldsymbol{y}$. Taking expectation with respect to $o(i)$ it yields

$$\begin{aligned} E_o\{v(\boldsymbol{q}(i+1))\} - v(\boldsymbol{q}(i)) \;&=\; E_o\{\boldsymbol{q}^T(i+1)\boldsymbol{P}\boldsymbol{q}(i+1)\} - \boldsymbol{q}^T(i)\boldsymbol{P}\boldsymbol{q}(i) \\ &+\; \boldsymbol{y}^T(i)\boldsymbol{y}(i) - \gamma \boldsymbol{d}^T(i)\boldsymbol{d}(i) < 0\,. \end{aligned} \tag{9}$$

Solving for the deterministically unforced system, then

$$\begin{aligned} E_o\{\boldsymbol{q}^T(i+1)\boldsymbol{P}\boldsymbol{q}(i+1)\} \;&=\; \boldsymbol{d}^T(i)\boldsymbol{W}^T\boldsymbol{P}\boldsymbol{W}\boldsymbol{d}(i) \\ &+\; \boldsymbol{q}^T(i)(\boldsymbol{F}^T\boldsymbol{P}\boldsymbol{F} + E_o\{o^2(i)\}\boldsymbol{V}^T\boldsymbol{P}\boldsymbol{V})\boldsymbol{q}(i) \\ &+\; \boldsymbol{q}^T(i)\boldsymbol{F}^T\boldsymbol{P}\boldsymbol{W}\boldsymbol{d}(i) + \boldsymbol{d}^T(i)\boldsymbol{W}^T\boldsymbol{P}\boldsymbol{F}\boldsymbol{q}(i) \end{aligned} \tag{10}$$

and, since $E_o\{o^2(i)\} = 1$, substituting (2) and (10) into (9), then also it yields

$$\begin{aligned} E_o\{v(\boldsymbol{q}(i+1))\} - v(\boldsymbol{q}(i)) \;&=\; \boldsymbol{q}^T(i)(-\boldsymbol{P} + \boldsymbol{F}^T\boldsymbol{P}\boldsymbol{F} + \boldsymbol{V}^T\boldsymbol{P}\boldsymbol{V} + \boldsymbol{C}^T\boldsymbol{C})\boldsymbol{q}(i) \\ &+\; \boldsymbol{q}^T(i)\boldsymbol{F}^T\boldsymbol{P}\boldsymbol{W}\boldsymbol{d}(i) + \boldsymbol{d}^T(i)\boldsymbol{W}^T\boldsymbol{P}\boldsymbol{F}\boldsymbol{q}(i) \\ &+\; \boldsymbol{d}^T(i)(\boldsymbol{W}^T\boldsymbol{P}\boldsymbol{W} - \gamma \boldsymbol{I}_{r_w})\boldsymbol{d}(i) + \,. \end{aligned} \tag{11}$$

Defining the composed vector

$$\boldsymbol{q}^{\circ T}(i) = \begin{bmatrix} \boldsymbol{q}^T(i) & \boldsymbol{d}^T(i) \end{bmatrix}, \tag{12}$$

the inequality

$$E_o\{v(\boldsymbol{q}(i+1))\} - v(\boldsymbol{q}(i)) = \boldsymbol{q}^{\circ T}(i)\boldsymbol{P}^\circ \boldsymbol{q}^\circ(i) < 0 \tag{13}$$

is satisfied only if

$$\boldsymbol{P}^\circ = \begin{bmatrix} -\boldsymbol{P} + \boldsymbol{F}^T\boldsymbol{P}\boldsymbol{F} + \boldsymbol{V}^T\boldsymbol{P}\boldsymbol{V} + \boldsymbol{C}^T\boldsymbol{C} & \boldsymbol{F}^T\boldsymbol{P}\boldsymbol{W} \\ \boldsymbol{W}^T\boldsymbol{P}\boldsymbol{F} & -\gamma \boldsymbol{I}_{r_w} + \boldsymbol{W}^T\boldsymbol{P}\boldsymbol{W} \end{bmatrix} < 0\,. \tag{14}$$

Writing (14) as follows

$$\begin{bmatrix} -\boldsymbol{P} + \boldsymbol{V}^T\boldsymbol{P}\boldsymbol{V} + \boldsymbol{C}^T\boldsymbol{C} & \boldsymbol{0} \\ \boldsymbol{0} & -\gamma \boldsymbol{I}_{r_w} \end{bmatrix} + \begin{bmatrix} \boldsymbol{F}^T\boldsymbol{P} \\ \boldsymbol{W}^T\boldsymbol{P} \end{bmatrix} \boldsymbol{P}^{-1} \begin{bmatrix} \boldsymbol{PF} & \boldsymbol{PW} \end{bmatrix} < 0 \tag{15}$$

and using the Schur complement property, (15) can be rewritten in the following form

$$\begin{bmatrix} -\boldsymbol{P}+\boldsymbol{V}^T\boldsymbol{P}\boldsymbol{V}+\boldsymbol{C}^T\boldsymbol{C} & \boldsymbol{0} & \boldsymbol{F}^T\boldsymbol{P} \\ \boldsymbol{0} & -\gamma\boldsymbol{I}_{r_w} & \boldsymbol{W}^T\boldsymbol{P} \\ \boldsymbol{P}\boldsymbol{F} & \boldsymbol{P}\boldsymbol{W} & -\boldsymbol{P} \end{bmatrix} < 0. \tag{16}$$

Applying twice the Schur complement property, (16) implies (7). This concludes the proof. ■

Proposition 2 *(enhanced BRL for discrete-time state-multiplicative noise stochastic systems) The autonomous system (1), (2) is stable in the mean square sense and with the quadratic performance γ if there exist positive definite matrices $\boldsymbol{P},\boldsymbol{Q} \in I\!R^{n\times n}$ and a positive scalar $\gamma \in I\!R$ such that*

$$\boldsymbol{P}=\boldsymbol{P}^T>0\,,\quad \boldsymbol{Q}=\boldsymbol{Q}^T>0\,,\quad \gamma>0\,, \tag{17}$$

$$\begin{bmatrix} -\boldsymbol{P} & * & * & * & * \\ \boldsymbol{0} & -\gamma\boldsymbol{I}_{r_v} & * & * & * \\ \boldsymbol{Q}\boldsymbol{F} & \boldsymbol{Q}\boldsymbol{W} & \boldsymbol{P}-2\boldsymbol{Q} & * & * \\ \boldsymbol{Q}\boldsymbol{V} & \boldsymbol{0} & \boldsymbol{0} & -\boldsymbol{Q} & * \\ \boldsymbol{C} & \boldsymbol{0} & \boldsymbol{0} & \boldsymbol{0} & -\boldsymbol{I}_m \end{bmatrix} < 0. \tag{18}$$

Proof: (compare [13]) Since the system (1) in the autonomous mode implies

$$\boldsymbol{F}\boldsymbol{q}(i)+\boldsymbol{V}o(i)\boldsymbol{q}(i)+\boldsymbol{W}\boldsymbol{d}(i)-\boldsymbol{q}(i+1)=\boldsymbol{0}\,, \tag{19}$$

then, with an arbitrary symmetric positive definite matrix $\boldsymbol{Q} \in I\!R^{n\times n}$, it yields

$$\left(\boldsymbol{q}^T(i+1)+0.5\boldsymbol{q}^T(i)\boldsymbol{V}^To(i)\right)\boldsymbol{Q}\left(\boldsymbol{F}\boldsymbol{q}(i)+\boldsymbol{V}o(i)\boldsymbol{q}(i)+\boldsymbol{W}\boldsymbol{d}(i)-\boldsymbol{q}(i+1)\right)=\boldsymbol{0}\,, \tag{20}$$

$$E_o\{\boldsymbol{q}^T(i+1)\}\boldsymbol{Q}\left(\boldsymbol{F}\boldsymbol{q}(i)+\boldsymbol{W}\boldsymbol{d}(i)-E_o\{\boldsymbol{q}(i+1)\}\right)+0.5E_o\{o^2(i)\}\boldsymbol{q}^T(i)\boldsymbol{V}^T\boldsymbol{Q}\boldsymbol{V}\boldsymbol{q}(i)=\boldsymbol{0}\,, \tag{21}$$

respectively. Adding (21) and its transposition to (9) gives

$$\begin{aligned} E_o\{v(\boldsymbol{q}(i+1))\}-v(\boldsymbol{q}(i)) \;=\; & E_o\{\boldsymbol{q}^T(i+1)\}\boldsymbol{P}E_o\{\boldsymbol{q}(i+1)\}+\boldsymbol{q}^T(i)(\boldsymbol{C}^T\boldsymbol{C}-\boldsymbol{P})\boldsymbol{q}(i) \\ & + E_o\{\boldsymbol{q}^T(i+1)\}\boldsymbol{Q}\left(\boldsymbol{F}\boldsymbol{q}(i)+\boldsymbol{W}\boldsymbol{d}(i)-E_o\{\boldsymbol{q}(i+1)\}\right) \\ & + \left(\boldsymbol{F}\boldsymbol{q}(i)+\boldsymbol{W}\boldsymbol{d}(i)-E_o\{\boldsymbol{q}(i+1)\}\right)^T\boldsymbol{Q}E_o\{\boldsymbol{q}(i+1)\} \\ & + \boldsymbol{q}^T(i)\boldsymbol{V}^T\boldsymbol{Q}\boldsymbol{V}\boldsymbol{q}(i)-\gamma\boldsymbol{d}^T(i)\boldsymbol{d}(i)<0\,. \end{aligned} \tag{22}$$

Defining the composed vector

$$\boldsymbol{q}^{\bullet T}(i)=\begin{bmatrix} \boldsymbol{q}^T(i) & \boldsymbol{d}^T(i) & E_o\{\boldsymbol{q}^T(i+1)\} \end{bmatrix}, \tag{23}$$

then (22) can be written as

$$E_o\{v(\boldsymbol{q}(i+1))\}-v(\boldsymbol{q}(i))=\boldsymbol{q}^{\bullet T}(i)\boldsymbol{P}^{\bullet}\boldsymbol{q}^{\bullet}(i)<0\,, \tag{24}$$

where

$$\boldsymbol{P}^{\bullet}=\begin{bmatrix} -\boldsymbol{P}+\boldsymbol{C}^T\boldsymbol{C}+\boldsymbol{V}^T\boldsymbol{Q}\boldsymbol{V} & \boldsymbol{0} & \boldsymbol{F}^T\boldsymbol{Q} \\ \boldsymbol{0} & -\gamma\boldsymbol{I}_{r_w} & \boldsymbol{W}^T\boldsymbol{Q} \\ \boldsymbol{Q}\boldsymbol{F} & \boldsymbol{Q}\boldsymbol{W} & \boldsymbol{P}-2\boldsymbol{Q} \end{bmatrix} < 0. \tag{25}$$

To find the LMI form, (25) can be rewritten using Schur complement property as (18). This concludes the proof. ■

The inequality (18) is an enhanced representation of BRL for the given class of stochastic systems. It is linear with respect to the system variables but does not involve any product of the Lyapunov matrix $\boldsymbol{P}$ and the system matrices $\boldsymbol{F}$, $\boldsymbol{V}$, $\boldsymbol{W}$. This offers mainly its possibility to be applied in a singular task analysis.

4. Uncertain state-multiplicative noise stochastic systems

The importance of Proposition 2 is that it separates $\boldsymbol{P}$ from $\boldsymbol{F}$, $\boldsymbol{V}$, $\boldsymbol{W}$ and, consequently, also from $\boldsymbol{G}$. This enables to derive design conditions for a system with polytopic uncertainties.

Assuming that the matrices $\boldsymbol{F}$, $\boldsymbol{G}$, $\boldsymbol{V}$, $\boldsymbol{W}$ of (1) are not precisely known but belong to a polytopic uncertainty domain

$$\mathcal{O} := \left\{ \begin{array}{c} \boldsymbol{a} \in \mathcal{Q} \\ (\boldsymbol{F},\boldsymbol{G},\boldsymbol{V},\boldsymbol{W})(\boldsymbol{a}) : (\boldsymbol{F},\boldsymbol{G},\boldsymbol{V},\boldsymbol{W})(\boldsymbol{a}) = \sum_{l=1}^{s} a_l (\boldsymbol{F}_l,\boldsymbol{G}_l,\boldsymbol{V}_l,\boldsymbol{W}_l) \end{array} \right\}, \tag{26}$$

$$\mathcal{Q} = \{(a_1,\cdots,a_s) : \sum_{l=1}^{s} a_l = 1;\ a_l > 0,\ i = 1,2,\ldots,s\}, \tag{27}$$

where $\mathcal{Q}$ is the unit simplex, $\boldsymbol{F}_l$, $\boldsymbol{G}_l$, $\boldsymbol{V}_l$ and $\boldsymbol{W}_l$ are constant matrices with appropriate dimensions, and a_l, $l = 1, 2, \ldots, s$ are time-invariant uncertainties.

Since $\boldsymbol{a}$ is constrained to the unit simplex as (27), the matrices $(\boldsymbol{F},\boldsymbol{G},\boldsymbol{V},\boldsymbol{W})(\boldsymbol{a})$ are affine functions of the uncertain parameter vector $\boldsymbol{a} \in I\!R^n$ described by the convex combination of the vertex matrices $(\boldsymbol{F}_l,\boldsymbol{G}_l,\boldsymbol{V}_l,\boldsymbol{W}_l)$, $l = 1,2,\ldots,s$.

Theorem 1 *The uncertain autonomous system (1), (2), (26) is stable in the mean square sense and with the quadratic performance γ if there exist symmetric positive definite matrices $\boldsymbol{P},\boldsymbol{Q} \in I\!R^{n\times n}$ and a positive scalar $\gamma \in I\!R$ such that*

$$\boldsymbol{P} = \boldsymbol{P}^T > 0, \quad \boldsymbol{Q} = \boldsymbol{Q}^T > 0, \quad \gamma > 0, \tag{28}$$

$$\begin{bmatrix} -\boldsymbol{P} & * & * & * & * \\ \boldsymbol{0} & -\gamma \boldsymbol{I}_{r_v} & * & * & * \\ \boldsymbol{Q}\boldsymbol{F}_l & \boldsymbol{Q}\boldsymbol{W}_l & \boldsymbol{P}-2\boldsymbol{Q} & * & * \\ \boldsymbol{Q}\boldsymbol{V}_l & \boldsymbol{0} & \boldsymbol{0} & -\boldsymbol{Q} & * \\ \boldsymbol{C} & \boldsymbol{0} & \boldsymbol{0} & \boldsymbol{0} & -\boldsymbol{I}_m \end{bmatrix} < 0. \tag{29}$$

for all $l = 1,2,\ldots,s$.

Proof: Now, since $\sum_{l=1}^{s} a_l = 1$, the first state difference equation of the uncertain system (1), (26) in the autonomous mode implies

$$\sum_{l=1}^{s} a_l \left(\boldsymbol{F}_l\boldsymbol{q}(i) + \boldsymbol{V}_l o(i)\boldsymbol{q}(i) + \boldsymbol{W}_l\boldsymbol{d}(i) - \boldsymbol{q}(i+1)\right) = \boldsymbol{0}, \tag{30}$$

Then, with an arbitrary symmetric positive definite matrix $\boldsymbol{Q} \in I\!R^{n\times n}$, it yields

$$\sum_{k=1}^{s} a_k \left(\boldsymbol{q}^T(i+1) + 0.5\boldsymbol{q}^T(i)\boldsymbol{V}_k^T o(i)\right) \boldsymbol{Q} \sum_{l=1}^{s} a_l \left(\boldsymbol{F}_l\boldsymbol{q}(i) + \boldsymbol{V}_l o(i)\boldsymbol{q}(i) + \boldsymbol{W}_l\boldsymbol{d}(i) - \boldsymbol{q}(i+1)\right) = \boldsymbol{0}, \tag{31}$$

which implies

$$\begin{aligned} \boldsymbol{0} &= E_o\{\boldsymbol{q}^T(i+1)\}\boldsymbol{Q} \sum_{l=1}^{s} a_l \left(\boldsymbol{F}_l\boldsymbol{q}(i) + \boldsymbol{W}_l\boldsymbol{d}(i) - E_o\{\boldsymbol{q}(i+1)\}\right) \\ &\quad + 0.5E_o\{o^2(i)\} \sum_{h=1}^{s}\sum_{k=1}^{s} a_k a_l \boldsymbol{q}^T(i)\boldsymbol{V}_k^T\boldsymbol{Q}\boldsymbol{V}_l\boldsymbol{q}(i). \end{aligned} \tag{32}$$

Since the Lyapunov function candidate can takes the form (8) then also yields (9). Thus, by adding (32) and its transposition to (9), it can obtain

$$\begin{aligned} E_o\{v(\boldsymbol{q}(i+1))\} - v(\boldsymbol{q}(i)) &= E_o\{\boldsymbol{q}^T(i+1)\}\boldsymbol{P}E_o\{\boldsymbol{q}(i+1)\} + \boldsymbol{q}^T(i)(\boldsymbol{C}^T\boldsymbol{C} - \boldsymbol{P})\boldsymbol{q}(i) \\ &+ \sum_{l=1}^{s} a_l E_o\{\boldsymbol{q}^T(i+1)\}\boldsymbol{Q}\left(\boldsymbol{F}_l\boldsymbol{q}(i) + \boldsymbol{W}_l\boldsymbol{d}(i) - E_o\{\boldsymbol{q}(i+1)\}\right) \\ &+ \sum_{l=1}^{s} a_l \left(\boldsymbol{F}_l\boldsymbol{q}(i) + \boldsymbol{W}_l\boldsymbol{d}(i) - E_o\{\boldsymbol{q}(i+1)\}\right)^T \boldsymbol{Q}E_o\{\boldsymbol{q}(i+1)\} \\ &+ \sum_{k=1}^{s}\sum_{l=1}^{s} a_k a_l \boldsymbol{q}^T(i)\boldsymbol{V}_k^T\boldsymbol{Q}\boldsymbol{V}_l\boldsymbol{q}(i) - \gamma\boldsymbol{d}^T(i)\boldsymbol{d}(i) < 0\,. \end{aligned} \tag{33}$$

Using the inequality (5) double summation elements can be eliminated from (33 and so

$$\begin{aligned} E_o\{v(\boldsymbol{q}(i+1))\} - v(\boldsymbol{q}(i)) &\le E_o\{\boldsymbol{q}^T(i+1)\}\boldsymbol{P}E_o\{\boldsymbol{q}(i+1)\} + \boldsymbol{q}^T(i)(\boldsymbol{C}^T\boldsymbol{C} - \boldsymbol{P})\boldsymbol{q}(i) \\ &+ \sum_{l=1}^{s} a_l E_o\{\boldsymbol{q}^T(i+1)\}\boldsymbol{Q}\left(\boldsymbol{F}_l\boldsymbol{q}(i) + \boldsymbol{W}_l\boldsymbol{d}(i) - E_o\{\boldsymbol{q}(i+1)\}\right) \\ &+ \sum_{l=1}^{s} a_l \left(\boldsymbol{F}_l\boldsymbol{q}(i) + \boldsymbol{W}_l\boldsymbol{d}(i) - E_o\{\boldsymbol{q}(i+1)\}\right)^T \boldsymbol{Q}E_o\{\boldsymbol{q}(i+1)\} \\ &+ \sum_{l=1}^{s} a_l \boldsymbol{q}^T(i)\boldsymbol{V}_l^T\boldsymbol{Q}\boldsymbol{V}_l\boldsymbol{q}(i) - \gamma\boldsymbol{d}^T(i)\boldsymbol{d}(i) < 0\,. \end{aligned} \tag{34}$$

Exploiting the composed vector (23) then (34) can be written as

$$E_o\{v(\boldsymbol{q}(i+1))\} - v(\boldsymbol{q}(i)) \le \sum_{l=1}^{s} a_l \boldsymbol{q}^{\bullet T}(i)\boldsymbol{P}_l^{\bullet}\boldsymbol{q}^{\bullet}(i) < 0\,, \tag{35}$$

where

$$\boldsymbol{P}_l^{\bullet} = \begin{bmatrix} -\boldsymbol{P} + \boldsymbol{C}^T\boldsymbol{C} + \boldsymbol{V}_l^T\boldsymbol{Q}\boldsymbol{V}_l & \boldsymbol{0} & \boldsymbol{F}_l^T\boldsymbol{Q} \\ \boldsymbol{0} & -\gamma\boldsymbol{I}_{r_w} & \boldsymbol{W}_l^T\boldsymbol{Q} \\ \boldsymbol{Q}\boldsymbol{F}_l & \boldsymbol{Q}\boldsymbol{W}_l & \boldsymbol{P} - 2\boldsymbol{Q} \end{bmatrix} < 0\,. \tag{36}$$

Thus, using Schur complement property, the inequality (36) can be rewritten as (29). This concludes the proof. ■

The following parts address the problem of finding the state-feedback control law that stabilize the system (1), (2), (26) and achieve the optimal level of unknown input disturbance attenuation.

Considering the linear state-feedback control law

$$\boldsymbol{u}(i) = -\boldsymbol{K}\boldsymbol{q}(i) \tag{37}$$

where $\boldsymbol{K} \in I\!R^{r\times n}$ is the feedback controller gain matrix, the design conditions are given by the following theorem.

Theorem 2 *The control law (37) to the system (1), (2), (26) exists if there exist symmetric positive definite matrices* $\boldsymbol{X}, \boldsymbol{Z} \in I\!R^{n\times n}$*, a matrix* $\boldsymbol{Y} \in I\!R^{r\times n}$ *and a positive scalar* $\gamma \in I\!R$ *such that for all* $l = 1, 2, \ldots, s$

$$\boldsymbol{X} = \boldsymbol{X}^T > 0\,, \quad \boldsymbol{Z} = \boldsymbol{Z}^T > 0\,, \quad \gamma > 0\,, \tag{38}$$

$$\begin{bmatrix} -\boldsymbol{Z} & * & * & * & * \\ \boldsymbol{0} & -\gamma\boldsymbol{I}_{r_w} & * & * & * \\ \boldsymbol{F}_l\boldsymbol{X} - \boldsymbol{G}_l\boldsymbol{Y} & \boldsymbol{W}_l & \boldsymbol{Z} - 2\boldsymbol{X} & * & * \\ \boldsymbol{V}_l\boldsymbol{X} & \boldsymbol{0} & \boldsymbol{0} & -\boldsymbol{X} & * \\ \boldsymbol{C}\boldsymbol{X} & \boldsymbol{0} & \boldsymbol{0} & \boldsymbol{0} & -\boldsymbol{I}_m \end{bmatrix} < 0\,. \tag{39}$$

When the above conditions hold,

$$\boldsymbol{K} = \boldsymbol{Y}\boldsymbol{X}^{-1}\,. \tag{40}$$

Proof: Since it was supposed that $\boldsymbol{Q}$ is a positive definite matrix, defining the transform matrix

$$\boldsymbol{T} = \text{diag}\left[\ \boldsymbol{X} \quad \boldsymbol{I}_{r_v} \quad \boldsymbol{X} \quad \boldsymbol{X} \quad \boldsymbol{I}_m\ \right], \quad \boldsymbol{X} = \boldsymbol{Q}^{-1}, \tag{41}$$

then, pre-multiplying the left-hand side and post-multiplying the right-hand side of (29) by (41), it yields

$$\begin{bmatrix} -\boldsymbol{XPX} & * & * & * & * \\ \boldsymbol{0} & -\gamma\boldsymbol{I}_{r_v} & * & * & * \\ \boldsymbol{F}_l\boldsymbol{X} & \boldsymbol{W}_l & \boldsymbol{XPX}-2\boldsymbol{X} & * & * \\ \boldsymbol{V}_l\boldsymbol{X} & \boldsymbol{0} & \boldsymbol{0} & -\boldsymbol{X} & * \\ \boldsymbol{CX} & \boldsymbol{0} & \boldsymbol{0} & \boldsymbol{0} & -\boldsymbol{I}_m \end{bmatrix} < 0\,. \tag{42}$$

Replacing $\boldsymbol{F}_l$ in (29) by the closed-loop system matrix

$$\boldsymbol{F}_{cl} = \boldsymbol{F}_l - \boldsymbol{G}_l\boldsymbol{K}\,, \tag{43}$$

then (42) takes the form

$$\begin{bmatrix} -\boldsymbol{XPX} & * & * & * & * \\ \boldsymbol{0} & -\gamma\boldsymbol{I}_{r_v} & * & * & * \\ \boldsymbol{F}_l\boldsymbol{X}-\boldsymbol{G}_l\boldsymbol{KX} & \boldsymbol{W}_l & \boldsymbol{XPX}-2\boldsymbol{X} & * & * \\ \boldsymbol{V}_l\boldsymbol{X} & \boldsymbol{0} & \boldsymbol{0} & -\boldsymbol{X} & * \\ \boldsymbol{CX} & \boldsymbol{0} & \boldsymbol{0} & \boldsymbol{0} & -\boldsymbol{I}_m \end{bmatrix} < 0 \tag{44}$$

and with the notations

$$\boldsymbol{Z} = \boldsymbol{XPX}\,, \quad \boldsymbol{Y} = \boldsymbol{KX}\,, \tag{45}$$

(44) implies (39). This concludes the proof. ■

If the uncertain stochastic state-multiplicative systems with polytopic uncertainties (1), (2), (26) under control (37) is not stable for given set of α_l, the weighted sum of linear state-feedback control laws (the parallel distributed compensation) of the form

$$\boldsymbol{u}(i) = -\sum_{h=1}^{s} a_h\boldsymbol{K}_h\boldsymbol{q}(i) \tag{46}$$

is proposed, where $\boldsymbol{K}_h \in I\!R^{r\times n}$, $h = 1, 2, \ldots, s$ is the set of feedback controller gain matrices to be designed.

Theorem 3 *The control law (46) to the system (1), (2), (26) exists if there exist symmetric positive definite matrices $\boldsymbol{X}_h, \boldsymbol{Z}_h \in I\!R^{n\times n}$, matrices $\boldsymbol{Y}_h \in I\!R^{r\times n}$ and a positive scalar $\gamma \in I\!R$ such that for all $l, h = 1, 2, \ldots, s$*

$$\boldsymbol{X}_h = \boldsymbol{X}_h^T > 0\,, \quad \boldsymbol{Z}_h = \boldsymbol{Z}_h^T > 0\,, \quad \gamma > 0\,, \tag{47}$$

$$\begin{bmatrix} -\boldsymbol{Z}_h & * & * & * & * \\ \boldsymbol{0} & -\gamma\boldsymbol{I}_{r_w} & * & * & * \\ \boldsymbol{F}_l\boldsymbol{X}_h-\boldsymbol{G}_l\boldsymbol{Y}_h & \boldsymbol{W}_l & \boldsymbol{Z}_h-2\boldsymbol{X}_h & * & * \\ \boldsymbol{V}_l\boldsymbol{X}_h & \boldsymbol{0} & \boldsymbol{0} & -\boldsymbol{X}_h & * \\ \boldsymbol{CX}_h & \boldsymbol{0} & \boldsymbol{0} & \boldsymbol{0} & -\boldsymbol{I}_m \end{bmatrix} < 0\,. \tag{48}$$

When the above conditions hold,

$$\boldsymbol{K}_h = \boldsymbol{Y}_h\boldsymbol{X}_h^{-1}\,. \tag{49}$$

Proof: With respect to (46), the closed-loop system matrix $\boldsymbol{F}_l$ is considered in the form

$$\boldsymbol{F}_{clh} = \boldsymbol{F}_l - \boldsymbol{G}_l\boldsymbol{K}_h\,, \tag{50}$$

and so (34) can be modified to describe the closed loop stability as

$$\begin{aligned} E_o\{v(\boldsymbol{q}(i+1))\} - v(\boldsymbol{q}(i)) \le\; & E_o\{\boldsymbol{q}^T(i+1)\}\boldsymbol{P}E_o\{\boldsymbol{q}(i+1)\} + \boldsymbol{q}^T(i)(\boldsymbol{C}^T\boldsymbol{C} - \boldsymbol{P})\boldsymbol{q}(i) \\ & + \sum_{l=1}^{s}\sum_{h=1}^{s} a_l a_h E_o\{\boldsymbol{q}^T(i+1)\}\boldsymbol{Q}_h\left(\boldsymbol{F}_{clh}\boldsymbol{q}(i) + \boldsymbol{W}_l\boldsymbol{d}(i) - E_o\{\boldsymbol{q}(i+1)\}\right) \\ & + \sum_{l=1}^{s}\sum_{h=1}^{s} a_l a_h\left(\boldsymbol{F}_{clh}\boldsymbol{q}(i) + \boldsymbol{W}_l\boldsymbol{d}(i) - E_o\{\boldsymbol{q}(i+1)\}\right)^T\boldsymbol{Q}_h E_o\{\boldsymbol{q}(i+1)\} \\ & + \sum_{l=1}^{s}\sum_{h=1}^{s} a_l a_h \boldsymbol{q}^T(i)\boldsymbol{V}_l^T\boldsymbol{Q}_h\boldsymbol{V}_l\boldsymbol{q}(i) - \gamma\boldsymbol{d}^T(i)\boldsymbol{d}(i) < 0\,. \end{aligned} \tag{51}$$

Exploiting the composed vector (23) then the condition (51) can be reformulated as

$$E_o\{v(\boldsymbol{q}(i+1))\} - v(\boldsymbol{q}(i)) \le \sum_{l=1}^{s}\sum_{h=l}^{s} a_l a_h \boldsymbol{q}^{\bullet T}(i)\boldsymbol{P}^{\bullet}_{clh}\boldsymbol{q}^{\bullet}(i) < 0\,, \tag{52}$$

where

$$\boldsymbol{P}^{\bullet}_{lh} = \begin{bmatrix} -\boldsymbol{P} + \boldsymbol{C}^T\boldsymbol{C} + \boldsymbol{V}_l^T\boldsymbol{Q}_h\boldsymbol{V}_l & \boldsymbol{0} & \boldsymbol{F}_{clh}^T\boldsymbol{Q}_h \\ \boldsymbol{0} & -\gamma\boldsymbol{I}_{r_w} & \boldsymbol{W}_l^T\boldsymbol{Q}_h \\ \boldsymbol{Q}_h\boldsymbol{F}_{clh} & \boldsymbol{Q}_h\boldsymbol{W}_l & \boldsymbol{P} - 2\boldsymbol{Q}_h \end{bmatrix} < 0 \tag{53}$$

and, using the Schur complement property, (53) can be rewritten as

$$\begin{bmatrix} -\boldsymbol{P} & \boldsymbol{0} & \boldsymbol{F}_{clh}^T\boldsymbol{Q}_h & \boldsymbol{V}_l^T\boldsymbol{Q}_h & \boldsymbol{C}^T \\ \boldsymbol{0} & -\gamma\boldsymbol{I}_{r_v} & \boldsymbol{W}_l^T\boldsymbol{Q}_h & \boldsymbol{0} & \boldsymbol{0} \\ \boldsymbol{Q}_h\boldsymbol{F}_{clh} & \boldsymbol{Q}_h\boldsymbol{W}_l & \boldsymbol{P} - 2\boldsymbol{Q}_h & \boldsymbol{0} & \boldsymbol{0} \\ \boldsymbol{Q}_h\boldsymbol{V}_l & \boldsymbol{0} & \boldsymbol{0} & -\boldsymbol{Q}_h & \boldsymbol{0} \\ \boldsymbol{C} & \boldsymbol{0} & \boldsymbol{0} & \boldsymbol{0} & -\boldsymbol{I}_m \end{bmatrix} < 0\,. \tag{54}$$

Since it was supposed that $\boldsymbol{Q}_h$ is a positive definite matrix, defining the transform matrix

$$\boldsymbol{T}_h = \mathrm{diag}\left[\; \boldsymbol{X}_h \quad \boldsymbol{I}_{r_v} \quad \boldsymbol{X}_h \quad \boldsymbol{X}_h \quad \boldsymbol{I}_m \;\right], \quad \boldsymbol{X}_h = \boldsymbol{Q}_h^{-1}, \tag{55}$$

then, pre-multiplying the left-hand side and post-multiplying the right-hand side of (54) by (55), and substituting (50) it yields

$$\begin{bmatrix} -\boldsymbol{X}_h\boldsymbol{P}\boldsymbol{X}_h & \boldsymbol{0} & \boldsymbol{X}_h\boldsymbol{F}_l^T - \boldsymbol{X}_h\boldsymbol{K}_h^T\boldsymbol{G}_l^T & \boldsymbol{X}_h\boldsymbol{V}_l^T & \boldsymbol{X}_h\boldsymbol{C}^T \\ \boldsymbol{0} & -\gamma\boldsymbol{I}_{r_v} & \boldsymbol{W}_l^T & \boldsymbol{0} & \boldsymbol{0} \\ \boldsymbol{F}_l\boldsymbol{X}_h - \boldsymbol{G}_l\boldsymbol{K}_h\boldsymbol{X}_h & \boldsymbol{W}_l & \boldsymbol{X}_h\boldsymbol{P}\boldsymbol{X}_h - 2\boldsymbol{X}_h & \boldsymbol{0} & \boldsymbol{0} \\ \boldsymbol{V}_l\boldsymbol{X}_h & \boldsymbol{0} & \boldsymbol{0} & -\boldsymbol{X}_h & \boldsymbol{0} \\ \boldsymbol{C}\boldsymbol{X}_h & \boldsymbol{0} & \boldsymbol{0} & \boldsymbol{0} & -\boldsymbol{I}_m \end{bmatrix} < 0\,. \tag{56}$$

Thus, with the notations

$$\boldsymbol{Z}_h = \boldsymbol{X}_h\boldsymbol{P}\boldsymbol{X}_h\,, \quad \boldsymbol{Y}_h = \boldsymbol{K}_h\boldsymbol{X}_h\,, \tag{57}$$

then (56) implies (48). This concludes the proof. ■

Generally speaking, there is no method which enables a Lyapunov function to be chosen from all possible candidate functions. Thus, Lyapunovs theory leads to sufficient stability conditions, the conservatism of which is dependent on the particular form $v_p(\boldsymbol{q}(i)) = \boldsymbol{q}^T(i)\boldsymbol{P}\boldsymbol{q}(i)$ and on the structure of the system. The disadvantage of application of this type of particular form function for multiple models is that very conservative stability conditions are obtained.

Note, the above systems belong to a class of systems called bilinear stochastic systems, since the system is linear in the states and in the stochastic multiplicative terms. Such systems, analyzed in the sense of negative definiteness of the difference $E_o\{v(\boldsymbol{q}(i+1))\} - v(\boldsymbol{q}(i))$ can be only asymptotically mean square stable, and so only a solution of the stationary BRL can be achieved. It is evident that while the particular form of Lyapunov function accounts mean square properties, the second part of Lypunov function reflects the worst case of the disturbance effect.

Consider the case $r = m$ (square plants), where with each output signal is associated a reference signal. Such regime is called the forced regime and is defined as follows:

Definition 1 *The forced regime for (1), (2) with the controller (46) is given by the control policy*

$$\boldsymbol{u}(i) = -\sum_{h=1}^{s} a_h \boldsymbol{K}_h \boldsymbol{q}(i) + \sum_{l=1}^{s}\sum_{h=1}^{s} a_l a_h \boldsymbol{G}_{lh} \boldsymbol{w}(i)\,, \tag{58}$$

where $\boldsymbol{w}(i) \in \mathbb{R}^m$ is desired output signal vector, and $\boldsymbol{G}_{lh} \in \mathbb{R}^{m\times m}$ is the set of signal gain matrices.

Theorem 4 *If a square system (1), (2) is stabilizable by the control policy (58) and for all l are satisfied the conditions [20]*

$$\texttt{rank}\begin{bmatrix} \boldsymbol{A}_l & \boldsymbol{B}_l \\ \boldsymbol{C} & \boldsymbol{0} \end{bmatrix} = n + m\,, \tag{59}$$

then the matrices $\boldsymbol{G}_{lh}$ take the forms

$$\boldsymbol{G}_{hl} = \Big(\boldsymbol{C}(\boldsymbol{I}_n - (\boldsymbol{A}_l - \boldsymbol{B}_l\boldsymbol{K}_h))^{-1}\boldsymbol{B}_l\Big)^{-1}. \tag{60}$$

Proof: In a steady-state, the disturbance-free equations (1), (2) and the control law (58) imply

$$E_o\{v(\boldsymbol{q}(i+1))\} = \boldsymbol{q}_s = \sum_{l=1}^{s}\sum_{h=1}^{s} a_l a_h (\boldsymbol{A}_l - \boldsymbol{B}_l\boldsymbol{K}_h)\boldsymbol{q}_s + \sum_{l=1}^{s}\sum_{h=1}^{s} a_l a_h \boldsymbol{B}_l \boldsymbol{G}_{lh} \boldsymbol{w}_s\,, \tag{61}$$

$$\boldsymbol{y}_s = \boldsymbol{C}\boldsymbol{q}_s\,, \tag{62}$$

where $\boldsymbol{q}_s$, $\boldsymbol{y}_s$, $\boldsymbol{w}_s$ are steady-state values of the vectors $\boldsymbol{q}(t)$, $\boldsymbol{y}(t)$, $\boldsymbol{w}(t)$, respectively.

Since in a steady-state (61) implies

$$\boldsymbol{0} = -\sum_{l=1}^{s}\sum_{h=1}^{s} a_l a_h (\boldsymbol{I}_n - (\boldsymbol{A}_l - \boldsymbol{B}_l\boldsymbol{K}_h))\boldsymbol{q}_s + \sum_{l=1}^{s}\sum_{h=1}^{s} a_l a_h \boldsymbol{B}_l \boldsymbol{G}_{lh} \boldsymbol{w}_o\,, \tag{63}$$

then (63) gives the equation

$$\boldsymbol{q}_s = \sum_{l=1}^{s}\sum_{h=1}^{s} a_l a_h (\boldsymbol{I}_n - (\boldsymbol{A}_l - \boldsymbol{B}_l\boldsymbol{K}_h))^{-1}\boldsymbol{B}_l \boldsymbol{G}_{lh} \boldsymbol{w}_s \tag{64}$$

and, according to (62),

$$\boldsymbol{y}_s = \sum_{l=1}^{s}\sum_{h=1}^{s} a_l a_h \boldsymbol{C}(\boldsymbol{I}_n - (\boldsymbol{A}_l - \boldsymbol{B}_l\boldsymbol{K}_h))^{-1}\boldsymbol{B}_l \boldsymbol{G}_{lh} \boldsymbol{w}_s\,. \tag{65}$$

Thus, considering $\boldsymbol{y}_s = \boldsymbol{w}_s$, then (65) implies (60). This concludes the proof. ■

Note, the static gains realized by the $\boldsymbol{G}_{lh}$ matrices are ideal in control if the plant parameters, on which the values of $\boldsymbol{W}_{lh}$ depend, are known and do not vary with time. The forced regime is basically designed for constant references and is very closely related to shift of origin.

5. Illustrative Example

The considered system is represented by the model (1), (2), (26) with the system model parameters and the sampling period T_s

$$\boldsymbol{F}_a = \begin{bmatrix} 1.0393 & 0.0073 & 0.0106 \\ 0.0224 & 1.0402 & 0.0364 \\ 0.0365 & 0.0114 & 1.0303 \end{bmatrix}, \ \boldsymbol{F}_b = \begin{bmatrix} 1.0457 & 0.0160 & 0.0227 \\ 0.0244 & 1.0537 & 0.0564 \\ 0.0742 & 0.0395 & 1.0345 \end{bmatrix},$$

$$\boldsymbol{G}_a = \begin{bmatrix} 0.0403 \\ 0.0245 \\ 0.0180 \end{bmatrix}, \ \boldsymbol{G}_b = \begin{bmatrix} 0.0622 \\ 0.0393 \\ 0.0377 \end{bmatrix}, \ \boldsymbol{W}_a = \begin{bmatrix} 0.0095 \\ 0.0243 \\ 0.0312 \end{bmatrix}, \ \boldsymbol{W}_b = \begin{bmatrix} 0.0004 \\ 0.0010 \\ 0.0013 \end{bmatrix},$$

$$\boldsymbol{C} = \begin{bmatrix} 1 & 1 & 0 \\ 0 & 0 & 1 \end{bmatrix}, \ \boldsymbol{V} = \begin{bmatrix} 0 & 0 & 0.020 \\ 0 & 0 & 0 \\ 0.015 & 0 & 0 \end{bmatrix}, \ T_s = 0.05 \ s,$$

so that considering for s = 4, the parameter matrices were included in the form

$$\begin{aligned} (\boldsymbol{F}_1, \boldsymbol{G}_1, \boldsymbol{W}_1) &= (\boldsymbol{F}_a, \boldsymbol{G}_a, \boldsymbol{W}_a), \\ (\boldsymbol{F}_2, \boldsymbol{G}_2, \boldsymbol{W}_2) &= (\boldsymbol{F}_b, \boldsymbol{G}_a, \boldsymbol{W}_a), \\ (\boldsymbol{F}_3, \boldsymbol{G}_3, \boldsymbol{W}_3) &= (\boldsymbol{F}_a, \boldsymbol{G}_b, \boldsymbol{W}_b), \\ (\boldsymbol{F}_4, \boldsymbol{G}_4, \boldsymbol{W}_4) &= (\boldsymbol{F}_b, \boldsymbol{G}_b, \boldsymbol{W}_b). \end{aligned}$$

where

$$a_1 = 0.1, \quad a_2 = 0.4, \quad a_3 = 0.2, \quad a_4 = 0.3.$$

Note, such limitation was chosen to define less than $N = 25$ inequalities with respect to (47), (48) (really, $N = 4 + 4 + 4^2 = 24$).

Solving the variables γ, $\boldsymbol{X}_i$, $\boldsymbol{Z}_i$, $\boldsymbol{Y}_i$, $i = 1, 2, 3, 4$, satisfying (47), (48) via the LMI technique using toolbox SeDuMi [16], the following results were obtained for $\gamma = 6.8564$:

$$\boldsymbol{K}_1 = \boldsymbol{K}_2 = \boldsymbol{K}_3 = \boldsymbol{K}_4 = \boldsymbol{K} = \begin{bmatrix} -29.9607 & 80.8889 & 0.6409 \end{bmatrix},$$

guaranteing the stable eigenvalues spectra of all local system matrices in such a way that

$$\begin{aligned} \rho(\boldsymbol{A}_{c1}) &= \rho(\boldsymbol{A}_1 - \boldsymbol{G}_1\boldsymbol{K}) = \{0.4363, \ 0.9096, \ 0.9779\}, \\ \rho(\boldsymbol{A}_{c2}) &= \rho(\boldsymbol{A}_2 - \boldsymbol{G}_2\boldsymbol{K}) = \{0.4899, \ 0.9291 \pm 0.1039\,\mathrm{i}\}, \\ \rho(\boldsymbol{A}_{c3}) &= \rho(\boldsymbol{A}_3 - \boldsymbol{G}_3\boldsymbol{K}) = \{-0.1275, \ 0.9068, \ 0.9910\}, \\ \rho(\boldsymbol{A}_{c4}) &= \rho(\boldsymbol{A}_4 - \boldsymbol{G}_4\boldsymbol{K}) = \{-0.0620, \ 0.9282 \pm 0.0715\,\mathrm{i}\}. \end{aligned}$$

It is evident that the uncertain state-multiplicative noise stochastic systems with the above given matrix parameters can be stabilized by the linear state controller.

Applying the designed results to the system model with the initial conditions

$$\boldsymbol{q}^T(0) = \begin{bmatrix} 0.1 & -0.1 & 0.0 \end{bmatrix}, \quad \sigma_w^2 = 0.05, \quad \sigma_o^2 = 1,$$

the simulation result of the system at the second vertex of vertices is stated in Fig. 1.

To illustrate the system behavior as the control parameters were computed using Theorem 3, the system state response of the closed-loop system in the equilibrium state control with impact of the disturbance and multiplicative noise is shown. Since analogous results were obtained for all system vertices structure, as well as for system with prescribed $\boldsymbol{a}$, in view of this, it can be concluded that, besides the rise-time, generally all the design specifications were reasonably well satisfied. It is simply verifiable that adequate results can be obtained for the forced mode.

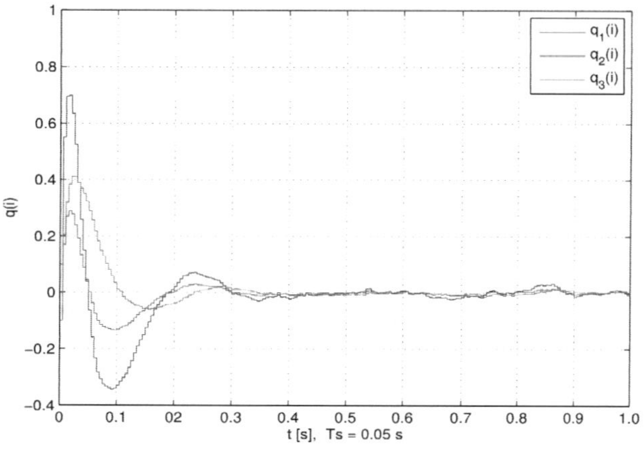

Figure 1. State response of the system at the second vertex

6. Concluding remarks

The paper presents new control design principle for discrete-time stochastic multi-variable dynamic systems with state-multiplicative noise. The stability of the control scheme in the mean square sense and with the quadratic performance is established using an enhanced representation of bounded real lemma, where, in the resulting LMIs, are decoupled potential pairs of the Lyapunov matrix and the system parameter matrices. This provides a suitable way for determination of linear state controllers by solving this naturally affine LMI tasks. Since the required state feedback gains has been obtained by solving an LMI feasibility problem, comparing with the previous results, the number of assumptions is reduced since the solutions obtained according to the conditions of both presented theorems tends to receive a linear state controller.

Presented applications can be considered as a task concerned the class of H_∞ stabilization control problems where the design conditions were newly formulated. This formulation poses the problem as a stabilization problem whose control law gain matrices take no special structures and allows to find a solution (if exists) to the control law without restrictive assumptions and additional specifications on the design parameters. The procedures used here for discrete time stochastic systems with state-multiplicative noise could be similarly extended for the continuous-time case. Moreover, a reasonable tracking performance could now be achieved.

Acknowledgments

The work presented in this paper was supported by VEGA, the Grant Agency of the Ministry of Education and the Academy of Science of Slovak Republic under Grant No. 1/0348/14. This support is very gratefully acknowledged.

References

[1] Cai G Hu C Yin B He H and Han X 2014 Gain-scheduled H_2 controller synthesis for continuous-time polytopic LPV systems *Mathematical Problems in Engineering* **2014** Article ID 972624 14p

[2] El Bouhtouri A Hinrichsen D and Pritchard A J 1999 H_∞-type control for discrete-time stochasic systems *Int. J. Robust and Nonlinear Control* **9**(13) 923-48

[3] El Ghaoui L 1995 State-feedback control of systems with multiplicative noise via linear matrix inequalities *Systems & Control Letters* **24**(3) 223-8

[4] Gao H and Wang C 2004 Improved bounded real lemma for continuous-time stochastic systems with polytopic uncertainties *Proc. 2004 American Control Conference* Boston MA USA 2654-8

[5] Gershon E and Shaked U 2008 H_∞ output-feedback control of discrete-time systems with state-multiplicative noise *Automatica* **44**(2) 574-9
[6] Gershon E and Shaked U 2013 *Advanced Topics in Control and Estimation of State-Multiplicative Noisy Systems* (London:Springer)
[7] Gershon E Shaked U and Yaesh I 2005 *H_∞ control and Estimation of State-multiplicative Linear Systems* (London:Springer)
[8] He Y Wu M and She J H 2005 Improved bounded-real-lemma representation and H_∞ control of systems with polytopic uncertainties *IEEE Tran. Circuits and Systems* **52**(7) 380-3
[9] Hinrichsen D and Pritchard A J 1998 Stochastic H_∞ *SIAM J. Control and Optimization* **36**(5) 1504-38
[10] Joshi S M 1976 On optimal control of linear systems in the presence of multiplicative noise *IEEE Tran. Aerospace and Electronic Systems* **12**(1) 80-5
[11] Krokavec D and Filasová A 2008 Constrained control of discrete-time stochastic systems *Proc. 17th World Congress IFAC* Seoul Korea 15315-20
[12] Krokavec D and Filasová A 2014 H_∞ enhanced control design of discrete-time Takagi-Sugeno state-multiplicative noisy systems *Mathematical Problems in Engineering* **2014** Article ID 151095 12p
[13] Krokavec D and Filasová A 2014 H_∞ control of discrete-time stochastic state-multiplicative systems constrained in state by equality constraints *Proc. 19th World Congress IFAC* Cape Town South Africa 8699-704
[14] McLane P 1971 Optimal stochastic control of linear systems with state- and control-dependent disturbances *IEEE Tran. Automatic Control* **16**(6) 793-8
[15] Niamsup P and Rajchakit G 2013 New results on robust stability and stabilization of linear discrete-time stochastic systems with convex polytopic uncertainties *J. Applied Mathematics* **2013** Article ID 368259 10p
[16] Peaucelle D Henrion D Labit Y and Taitz K 2002 *User's Guide for SeDuMi Interface 1.04* (Toulouse:LAAS-CNRS)
[17] Phillis Y A 1985 Controller design of systems with multiplicative noise *IEEE Tran. Automatic Control* **30**(10) 1017-9
[18] de Souza W A Teixeira M C M Santim M P A Cardim R and Assuncao E 2013 On switched control design of linear time-invariant systems with polytopic uncertainties *Mathematical Problems in Engineering* **2013** Article ID 595029 10p
[19] Todorov E 2005 Stochastic optimal control and estimation methods adapted to the noise characteristics of the sensorimotor system *Neural Computation* **17**(5) 1084–1108
[20] Wang Q G 2003 *Decoupling Control* (Berlin:Springer)
[21] Zhou K Doyle J C and Glover K 1996 *Robust and Optimal Control* (Englewood Cliffs:Prentice Hall)

Journal of Physics: Conference Series **659** (2015) 012003 doi:10.1088/1742-6596/659/1/012003

Computation of equivalent poles placement for class of 2nd order discrete bilinear systems

Lukasz Gadek [1], Leszek Koszalka [2], Keith Burnham [3]

Abstract. This paper introduces an adaptation of the classical linear control theory representation of zeros, poles and gain into a bilinear approach. The placement of poles at the complex plane is a complete description of plants dynamics; hence it is a convenient form from which calculation of various properties, e.g. rise time, settling time, is plausible. Such technique can be adjusted into the bilinear structure if poles of a quasi-linear representation (linear with respect to input) are concerned. The research outcomes with conclusion on the equivalent poles displacement and generalized rules for a 2nd order bilinear system equivalent poles input dependent loci. The proposed approach seems to be promising, as simplification of design and identification of a bilinear system increases transparency during modelling and control in practical applications and hence it may be followed by applicability of such structure in common industrial problems.

1. Introduction

Bilinear structure allows approximation of a non-linear (NL) plant into a decomposable form of linear model with NL term [1]. Initial industrial application utilizes mostly the property of bent steady state gain which improves modelling of water flow systems and industrial furnaces [2] where response is saturated gradually for high operating point (OP). By extending with the bilinear term, properties of the response become time variant with respect to current state (exceeding simplification to gain slide) as described in Section 2 and therefore a robust stability and behaviour prediction is required - pole-placement method, e.g. [3].

To achieve satisfactory performance in designing a bilinear plant controller ([4] and [3]), an efficient identification of the plant must be performed. The method on bilinear plant varying properties prediction is an extension allowing of more comprehensive understanding of the bilinear design. A similarity of equivalent poles movement with respect to OP and root locus of gain feedback system is observed in Section 3. Correlation of the classical root locus [5], pole-placement with output feedback [6] and equivalent (bilinear) poles locus will be a following stage of the research as highlighted in Section 4.1.

2. Bilinear structure

This chapter is aimed to introduce the bilinear structure in terms of mathematical representations (Section 2.1 and 2.2) and capabilities overview (Section 2.3). By extending

[1] Department of Systems and Computer Networks , Wroclaw University of Technology, Poland, e-mail: lukasz.gadek@pwr.edu.pl

[2] Department of Systems and Computer Networks , Wroclaw University of Technology, Poland

[3] Control Theory and Application Centre, Coventry University, United Kingdom

a linear model with a bi-linear term, the bilinear model is obtained. It can be illustrated with an example of auxiliary function $f(x, u)$ where x and u are time-dependant entities.

$$y = f(x, u) \qquad x \in \Re^n \qquad u \in \Re^m$$
$$y_{lin} = \sum_{i=1}^{n} x_i + \sum_{j=1}^{m} u_j \tag{1}$$
$$y_{bil} = y_{lin} + \sum_{i=1}^{n}\sum_{j=1}^{m} x_i u_j$$

In (1) y_{lin} is an output of linear model which fulfils superposition rule. The following equation contains bilinear term $\sum_{i=1}^{n}\sum_{j=1}^{m} x_i u_j$ being attached to y_{lin}. A similar approach can be applied to the State Space form and other representations.

2.1. State space

State Space is the most popular in a context of bilinear modelling and can be found in recent publications, e.g. [7]. The formulation of a bilinear MIMO term from [8] is as presented in (2).

$$Fu \otimes x \tag{2}$$
$$F = [F_1 \; F_2 \; F_3 \; ... \; F_m]$$
$$F_i \in \Re^{nxn} \qquad u \in \Re^m \qquad x \in \Re^n$$

In this paper only SISO systems are considered, hence (2) can be simplified into form of (3).

$$Fu \otimes x \rightarrow Nux \tag{3}$$
$$N \in \Re^{nxn} \qquad u \in \Re \qquad x \in \Re^n$$

With (3) SISO State Space is established in (4).

$$x_{k+1} = Ax_k + Bu_k + Nu_k x_k \tag{4}$$
$$y_k = Cx_k$$
$$y_k, u_k \in \Re$$
$$x_k, B, C \in \Re^n$$
$$A, N \in \Re^{nxn}$$

The root locus method introduced in this paper is based on the Transfer Function (TF) equivalent (presented in Section 2.2). The transmittance from state space is explicit if a canonical form [9] is utilized. Assuming observability and diagonal bilinear matrix, coefficient matrices from (5) are used.

$$A = \begin{bmatrix} -a_1 & 1 & 0 & \cdots & 0 \\ -a_2 & 0 & 1 & \cdots & 0 \\ \vdots & \vdots & \vdots & \ddots & \vdots \\ -a_n & 0 & 0 & 0 & \end{bmatrix} \qquad N = \begin{bmatrix} n_1 & 0 & 0 & \cdots & 0 \\ n_2 & 0 & 0 & \cdots & 0 \\ \vdots & \vdots & \vdots & \ddots & \vdots \\ n_n & 0 & 0 & 0 & \end{bmatrix} \tag{5}$$
$$B = [\; b_1 \;\; b_2 \;\; \cdots \;\; b_m \;\; 0 \;\; \cdots \;]^T \qquad C = [\; 1 \;\; 0 \;\; \cdots \;\; 0 \;]$$

Canonical form coefficients (5) are interchangeably used in difference equation (DE). Eq. 6 presents according Bilinear DE.

$$y_k = -\sum_{i=1}^{n} a_i y_{k-i} + \sum_{i=1}^{m} b_i u_{k-i} + \sum_{i=1}^{n} n_i u_{k-i} y_{k-i} \tag{6}$$

Under strong assumption of constant input u the bilinear form may be approximated with a linear equivalent where $\tilde{a}_i = a_i - n_i u_{k-i}$. However such approximation can be used in the open loop control only, therefore it is impractical in the majority of industrial application.

2.2. Equivalent transfer function

Transfer Function (TF) can be achieved from DE by introducing z^i - time shift operator where i is quantity of samples shifted forward. In the case of bilinear model two types of the equivalent TF are computable:

- Assuming sluggish change of input (eg. slow PID or manual control) followed by u in denominator as in (7),
- For fast controllers or systems with high inertia where y assumed to be sufficiently constant be approximate as coefficient in the numerator.

Both forms are equivalents of TF as in each denominator or numerator contains time-varying variable while in classical approach these are static.

$$\frac{Y}{U} = \frac{b_1 z^{-1} + \cdots + b_m z^{-m}}{1 + a_1 z^{-1} \cdots + a_n z^{-n} - n_1 z^{-1} u_{k-1} - \cdots - n_n z^{-n} u_{k-n}} \tag{7}$$

Natural step to recapture polynomial of static coefficients, is to assume u or y as approximately constant, e.g. if input change is negligible within last m samples then $U \approx u_{k-1} \approx \cdots \approx u_{k-m}$ (8). Such reasoning is a root of establishing the two types mentioned above.

$$\frac{Y}{U} = \frac{b_1 z^{-1} + \cdots + b_m z^{-m}}{1 + (a_1 - n_1 u_{k-1}) z^{-1} \cdots + (a_n - n_n u_{k-n}) z^{-n}} \tag{8}$$

In this paper u in the denominator approach is used, although following the same procedure for both types leads to the identical conclusion. However, the latter approach requires more significant mathematical effort.

2.3. Properties

From equivalent TF (8) two main properties of bilinear system can be highlighted. Time shift operator z is a discrete substitute of Laplace's e^{st_s} where t_s is a sampling time of discrete system.

Therefore, knowing that for $t \to \inf$ Laplace operator is converging to zero, z assembles to 1. Input/output gain of a discrete system can be calculated as $\frac{\sum b}{\sum a}$. In case of bilinear system denominator coefficient is impacted by input. If input does not change for a certain number of samples after excitation (e.g. step change) then bilinear system gain can be represented as in Fig. 1 where ratio of steady state output to input (Y_{ss}/U_{ss}) is shaped depending on N.

Eigenvalues of the bilinear system are also impacted by bilinearity. Characteristic Equation (CE) from equivalent TF (8) has a form of:

$$z^n + (a_1 - n_1 u_{k-1}) z^{n-1} \cdots + (a_n - n_n u_{k-n}) = 0 \tag{9}$$

Where z satisfying (9) are poles of the system. Correlation between u and dislocation of poles is indisputable, its prediction and estimation is presented in Section 3.

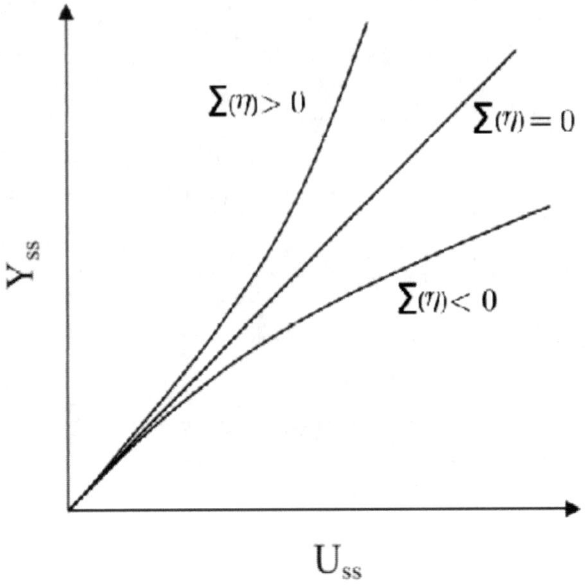

Figure 1. Steady State gain of bilinear continuous system accordingly to $\sum(n)$ - where n is first column of N from (5)

3. Equivalent poles loci

The aim of research is establishing a simple set of rules to predict perturbation of dynamical behaviour, i.e. movement of the equivalent poles, in a bilinear system. Derivation of LMI description of eigenvalues is performed in Section 3.1 while exemplary pole regions are plotted and commented in Section 3.2.

The reasoning is performed on diagonal bilinear and SISO system as described in Section 2.1. Due to simplification, initial derivation is utilizing approximated constant $U = u_{k-1\cdots n}$ which is valid if $u_{k-i} \approx u_{k-j} \forall i, j \in [0, m]$.

3.1. Analytical approach

In this section, solution of (9) with respect to U is derived. For simplicity 2nd order model with $n = 2$ is utilized. Pole computation (z) of a linear system is located in (10). Bilinear equivalent pole $\tilde{z}$ location is extended with n coefficient and U as presented in (11).

$$z = \frac{-a_1 \pm \sqrt{a_1^2 + 4a_2}}{2} \tag{10}$$

$$\tilde{z} = \frac{-a_1 + n_1 U \pm \sqrt{(a_1 - n_1 U)^2 - 4(a_2 - n_2 U)}}{2} \tag{11}$$

Initial step is calculation of eigenvalues with respect to imaginary plane, i.e. under condition $\Im(\tilde{z}) = 0$ - expanded in (12).

$$\begin{aligned} \Im(\tilde{z}) = 0 &\iff 0 \leq (a_1 - n_1 U)^2 - 4(a_2 - n_2 U) \\ \Im(\tilde{z}) = 0 &\iff 0 \leq n_1^2 U^2 - (2n_1 a_1 - 4n_2)U - 4a_2 + a_1^2 \end{aligned} \tag{12}$$

$$\Im(\tilde{z}) = 0 \iff U \in (-\infty, U^{out}] \vee U \in [U^{in}, \infty)$$

Basing on (12) break-out U^{out} and break-in U^{in} inputs can be calculated when (12) is equated to zero in. The result is presented in (13).

$$U^{out}, U^{in} = \frac{2n_1a_1 - 4n_2 \pm \sqrt{(-2n_1a_1 + 4n_2)^2 - 4(-4a_2 + a_1^2)n_1^2}}{2n_1^2} \tag{13}$$

Replacing U in (11) with (13) results with calculation of the break-out and -in points in the complex plane in (14).

$$\begin{aligned} z^{out}, z^{in} &= \frac{-a_1 + n_1 \left[\frac{2n_1a_1 - 4n_2 \pm \sqrt{(-2n_1a_1+4n_2)^2 - 4(-4a_2+a_1^2)n_1^2}}{2n_1^2}\right]}{2} \\ &= -\frac{n_2 \pm \sqrt{n_2^2 - n_1n_2a_1 + n_1^2a_2}}{n_1} \end{aligned} \tag{14}$$

From (12) it can be seen that the square root of (11) is a quadratic function. Therefore, $\Re(\tilde{z}) = \frac{|z^{out} - z^{in}|}{2}$ denotes a point on real axis at which extrema of pole imaginary part is achieved. Hence $U = -\frac{n_2}{n_1}$ is inserted into the square root in (11) resulting with (15).

$$\max|\Im(\tilde{z})| = \frac{\pm\sqrt{n_2^2 + n_1n_2a_1 - n_1^2a_2}}{n_1} i \tag{15}$$

Basing on extrema points of equivalent poles loci form (14) and (15) it can be deduced that the locus trajectory is a circle with centre in ($x = -\frac{n_2}{n_1}$, $y = 0$) (where x is real and y is imaginary axis) and range $r = \Im(\frac{\sqrt{n_2^2 - n_1n_2a_1 + n_1^2a_2}}{n_1})$ if $U \in [U^{out}, U^{in}]$. The circle equation in (16) for 2nd order bilinear system is formed.

$$\begin{aligned} (\Re(\tilde{z}) - x)^2 + (\Im(\tilde{z}) - y)^2 = r^2 \\ (\frac{-a_1 + n_1U}{2} + \frac{n_2}{n_1})^2 + \frac{(a_1 - n_1U)^2 - 4(a_2 - n_2U)}{4} = \frac{n_2^2 - n_1n_2a_1 + n_1^2a_2}{n_1^2} \end{aligned} \tag{16}$$

The equality is satisfied for all $U \in [U^{out}, U^{in}]$. Exemplary results obtained with numerical simulation are presented in Section 3.2.

Second order loci is always based on circle in range of $U \in [U^{out}, U^{in}]$ for the assumption of constant U. If U is changing within known range then the equivalent pole position must be amened accordingly to rate of change of input (17). However, according to LMI approach [10] it does not include internal instabilities which may result in discrepancies between prediction and actual state.

$$u_1 = Uu_2 = U + \Delta \tag{17}$$

Replacing U in (11) with (17) leads to amendment in the following calculations. Resulting locus formula is a circle based accordingly to (16) amended into (18).

$$\begin{aligned} (\Re_\Delta(\tilde{z}) - x_\Delta)^2 + (\Im_\Delta(\tilde{z}) - y_\Delta)^2 = r_\Delta^2 \\ (\frac{-a_1 + n_1U}{2} + \frac{n_2}{n_1})^2 + \frac{(a_1 - n_1U)^2 - 4a_2 + 4n_2(U + \Delta)}{4} = \\ = \frac{n_2^2 - n_1n_2a_1 + n_1^2a_2 - n_1^2n_2\Delta}{n_1^2} \end{aligned} \tag{18}$$

Range of the circle is impacted with Δ while centre position is not. Moreover, it can be observed that the occurrence of Δ has impact exclusively on the vertical position of the pole; regardless to the rate of change, the position is not changed respect to the real axis. This allows to represent equivalent poles as a region in which (19) holds if maximum Δ is definable.

$$r - |n_2\Delta| \leq \left[\Re(\tilde{z}) + \frac{n_2}{n_1}\right]^2 + \Im_\Delta(\tilde{z})^2 \leq r + |n_2\Delta| \tag{19}$$

Where in (19) properties r, $\Re(\tilde{z})$ and centre of gravity are identical to (16).

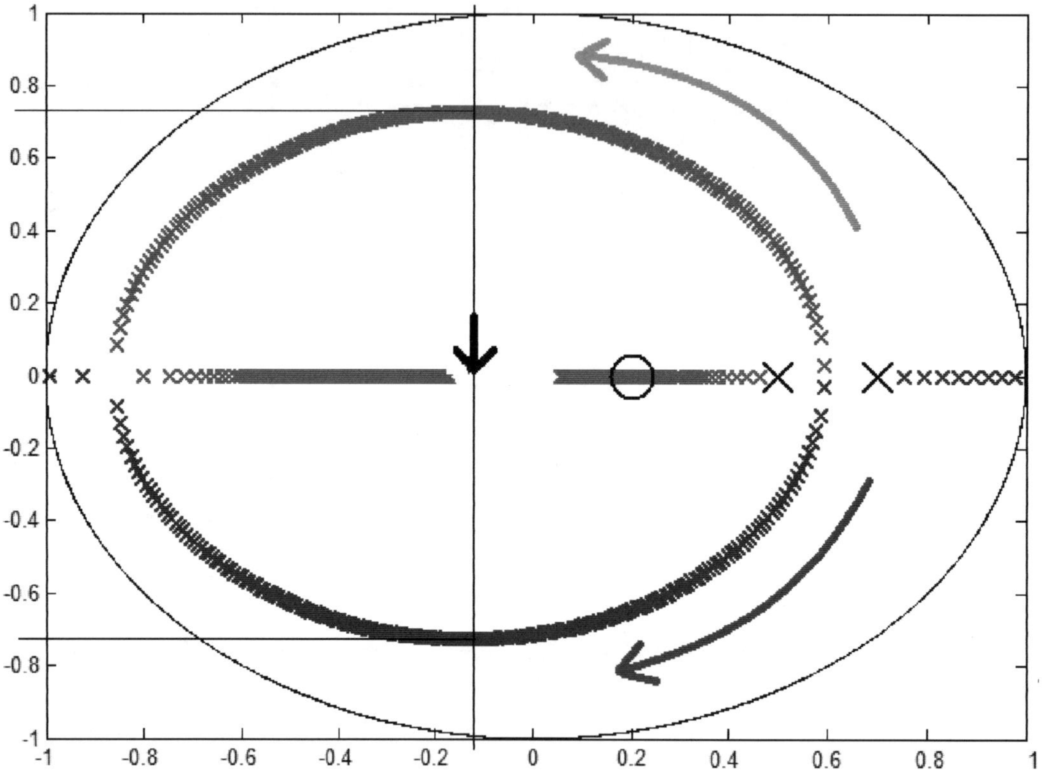

Figure 2. Equivalent poles loci with respect to increasing step input (arrow direction shows location for incrementing u)

3.2. Simulation

Method described in Section 3.1 has been validated numerically using MATLAB environment. Exemplary result of the locus for second order system with SS coefficients $A = \begin{bmatrix} 1.2 & 1 \\ -0.35 & 0 \end{bmatrix}$ $B = \begin{bmatrix} 1 & -0.2 \end{bmatrix}^T$ and $N = \begin{bmatrix} 0.015 & 0 \\ 0.002 & 0 \end{bmatrix}$ is presented in Fig. 2. According to the (16) the trace of the poles is within circle with range of $r = \Im(\frac{\sqrt{0.002^2+0.002\cdot 0.015\cdot 1.2+0.015^2\cdot 0.35}}{0.015}) = 0.72$ and centre in $x = -\frac{0.002}{0.015} = -0.13, y = 0$.

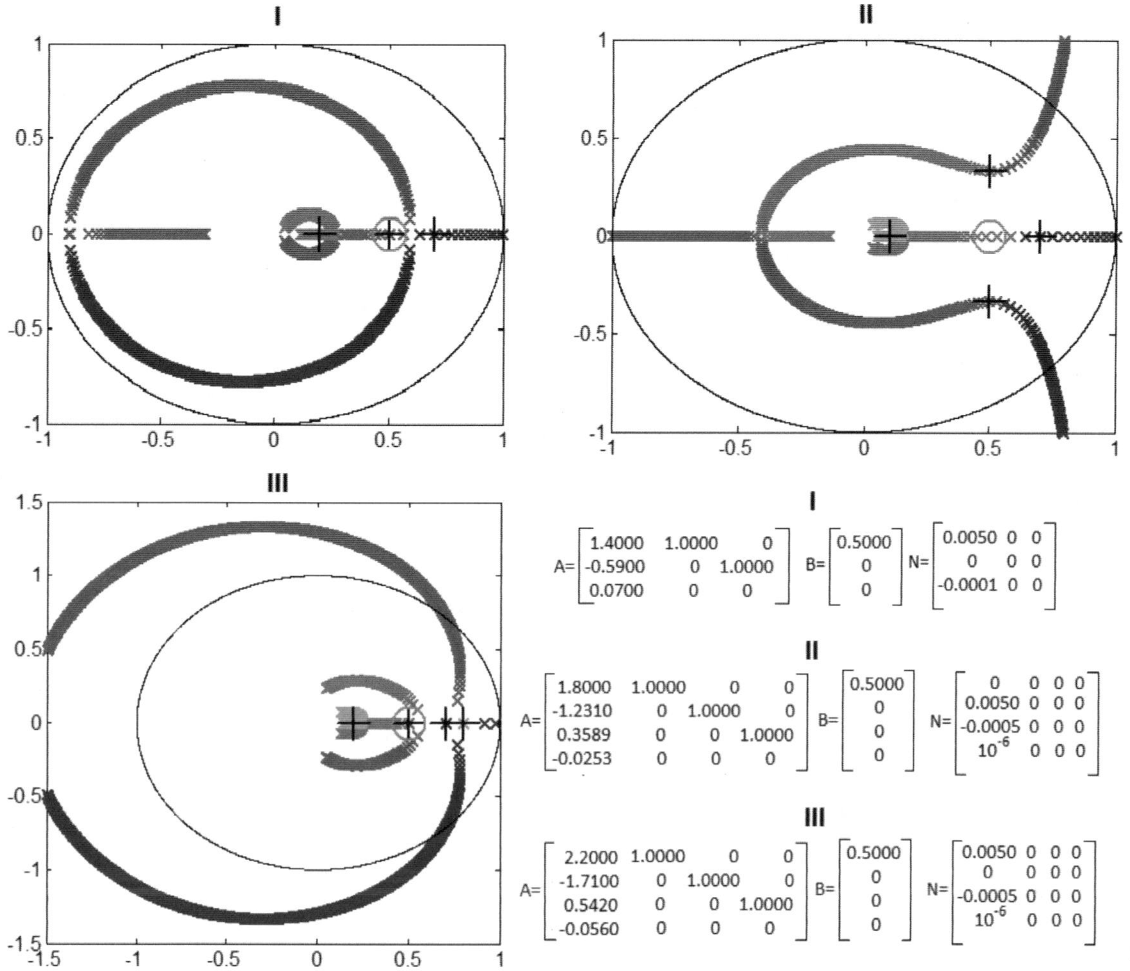

Figure 3. Exemplary Loci of higher order systems

For systems with higher orders the circular movement applies on corresponding basis. However, depending on number of poles and bilinear coefficient both: centres and trajectories might be marginally perturbed as presented in Figure 3 where:

- Plant I has two loci trajectories with $r_1 = 0.7842, r_2 = 0.08407$ and respective centres in $(-0.159, 0)$ and $(0.159, 0)$,
- Plant II poles starting in complex plane are moving perpendicular to the real axis for descending u (a loci defined as $r = \infty$?),
- In plant III around break away point trajectory is perturbed.

Further investigation on the higher order systems is proposed in Section 4.1

Fig. 4 represents a region of the equivalent poles against locus of U approximated as a constant. The system is described by following: $A = \begin{bmatrix} 1.2 & 1 \\ -0.35 & 0 \end{bmatrix}$, $B = \begin{bmatrix} 1 & -0.2 \end{bmatrix}^T$, $N = \begin{bmatrix} 0.010 & 0 \\ 0.002 & 0 \end{bmatrix}$ and $|\Delta| < 100$. Hence, basing on (18), locus range perturbation is

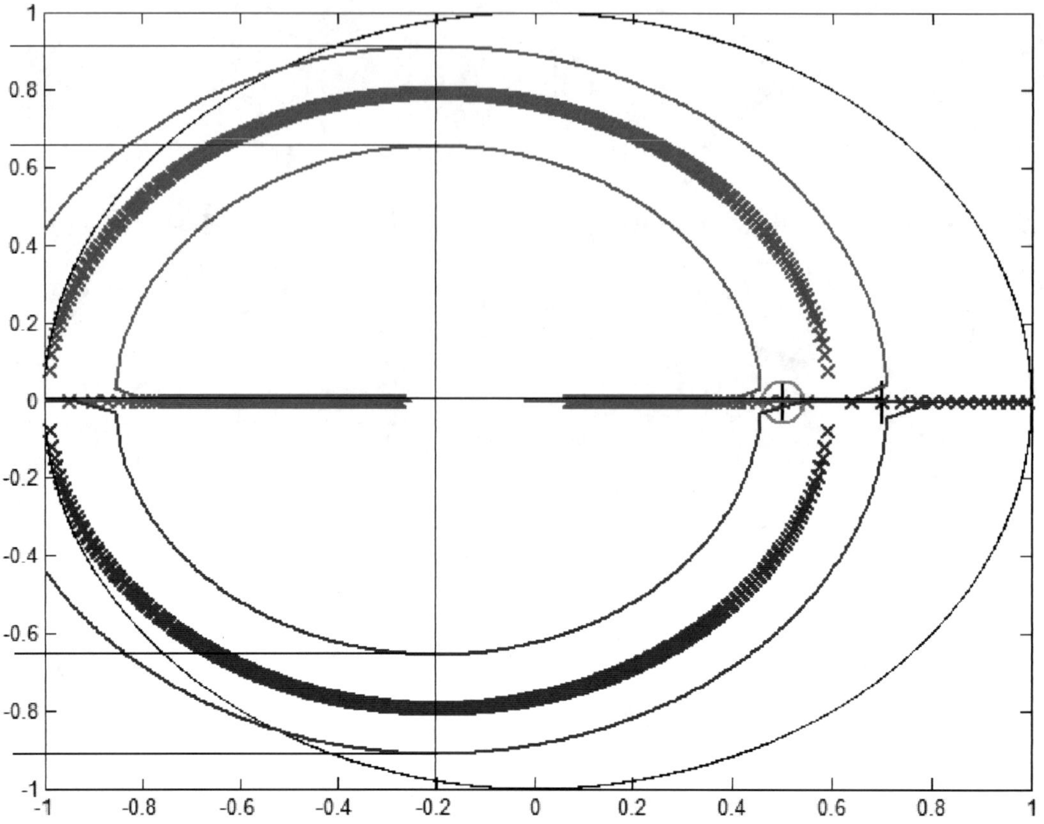

Figure 4. Equivalent poles loci as inequality region where $\Delta|u_{k-1..m}| < 100$

$r \in [0.65, 0.91]$ around the gravity centre in $(x = -0.2, y = 0)$.

Method derived in Section 3.1 has been validated for second order systems in numerical tests presented in this section (Fig. 2 and 4). The resemblance of equivalent poles and the gain feedback root loci can be observed due to the similar rotational behaviour and poles convergence at inf. It applies when close loop gain is mapped as U. The question to be asked is if this similarity is fully convertible.

4. Conclusion

Method derived for a second order bilinear system in Section 3.1 has been validated and presented graphically in Section 3.2. Moreover, similar vein pattern for third and higher order systems can be observed. The method is visually similar to the root locus of closed loop gain control [5]. Hence, parts of the method may be applied interchangeably between bilinear and linear structures.

Due to the more accurate dynamics prediction, a calculation of stable bilinear plant range of operation can be performed. It may be done utilising a correspondent method in the gain feedback loop locus. The critical (boundary) operating region of the bilinear model will be defined as a curve with respect to U and Δ.

Improved predictability is followed by enhanced identification capabilities - when observing trace of both: gain and dynamical properties (e.g. rise time and overshoot) engineer would have additional resource for assessing plant's structure type and the parameter estimation.

4.1. Future works

The intuitive continuation of the research is expanding of the reasoning for higher order systems. Although obtaining analytical solution will be less feasible due to mathematical complexity, results of simulation may be used to back calculate mathematical principle.

Another solution would be to compare equivalent poles with the classical P-gain feedback loop root locus as a correspondence between these methods has been observed (eg. in Figure 3). The potential of integration of the output feedback stabilization method [6] and bilinear model stability problem should be investigated.

Prediction of the pole location or the region of displacement may be useful for developing model based controllers. This approach contrary to the existing methods:

- frequent update of the linearised model around current OP,
- linearisation of the actual plant with a compensator [4]

has potential of detecting the internal instabilities and therefore ensuring robustness.

References

[1] Tan N and Atherton D P 2002 Stability margin computation for nonlinear systems: a parametric approach *15th IFAC Triennial world congr.* vol 15 ed L Basanez and J A de la Puente (Barcelona: IFAC) p 257

[2] Martineau S, Burnham K J, Haas O, Andrews G and Heeley A 2004 Four-term bi-linear PID controller applied to an industrial furnace *Cont. Eng. Pract.* **12** 457-64

[3] Taylor C J, Chotai A and Burnham K J 2011 Controllable forms for stabilising pole assignment design of generalised bilinear systems *Electr. Lett.* **47** 437-9

[4] Gadek L, Koszalka L and Burnham K J 2015 Compensation of n-th order nilinearity applied with model based controller *Progress in systems Engineering, Proc. of 23rd Int. Conf. on Systems Engineering* vol 330, ed H Selvaraj, D Zydek and G Chmaj (Las Vegas: Springer) pp 49-54

[5] Eydgahi A M 2001 Complementary root locus revisited *IEEE Trans. on Edu.* **44** 137-43

[6] Konigorski U 2012 Pole placement by parametric output feedback *Sys. & Cont. Lett.* **61** 292-7

[7] Lee C H and Juang J N 2012 Nonlinear system identificationa continuous-time bilinear state space approach *J. of the Astr. Sc.* **59** 398-420

[8] Verdult V and Verhaegen M 2000 Bilinear state space systems for nonlinear dynamical modelling *Th. in Biosc* **119** 1-9

[9] Reuter H 1994 State space identification of bilinear canonical forms Int. Conf. on Control vol 1 Coventry, IET pp 833-6

[10] Chilali M, Gahinet P and Apkarin P 1999 Robust Pole Placement in LMI Regions *IEE Trans. on Aut. Cont.* **44** 2257-70

Application of a repetitive process setting to design of monotonically convergent iterative learning control

Marcin Boski[1], **Wojciech Paszke**[2]

[1] Faculty of Computer Science, Electrical Engineering and Control Engineering, University of Zielona Góra,ul. Szafrana 2, 65-516 Zielona Góra, Poland.

[2] Institute of Control and Computation Engineering, University of Zielona Góra, ul. Szafrana 2, 65-516 Zielona Góra, Poland.

E-mail: marcinboski90@gmail.com, w.paszke@issi.uz.zgora.pl

Abstract. This paper deals with the problem of designing an iterative learning control algorithm for discrete linear systems using repetitive process stability theory. The resulting design produces a stabilizing output feedback controller in the time domain and a feedforward controller that guarantees monotonic convergence in the trial-to-trial domain. The results are also extended to limited frequency range design specification. New design procedure is introduced in terms of linear matrix inequality (LMI) representations, which guarantee the prescribed performances of ILC scheme. A simulation example is given to illustrate the theoretical developments.

1. Introduction

Iterative learning control (ILC) is a method of iteratively updating the control signal to a given system that repeats the same task. Each execution is known as a trial, or pass, and the sequence of operations is that a trial is completed, the system resets to the starting position and then the next trial begins, either immediately after the resetting is complete or a period of time has elapsed since completion of the resetting. The novel feature of this control law design method is the use information from the previous trial to update the control input applied on the next trial and thereby improve performance from trial-to-trial. In particular, the control objective is to find a control input so that the corresponding output precisely tracks a reference signal that is specified over a finite time interval.

Since the original work of Arimoto [2], ILC has remained as a significant area of control systems research with many algorithms experimentally verified in the research laboratory and applied in industrial applications. An overview of developments can be found in, e.g., the survey papers [1, 3, 12], where the last of these has a special focus on run-to-run control as found in the chemical process industries. Applications areas include industrial robotics, see, e.g., [7], where the pick and place operation common in many mass manufacturing processes is an immediate fit to ILC, and wafer stage motion systems, see, for example, [5].

One common approach, see, for example, [1, 3] as starting points for the literature, to ILC design is to first apply a feedback control law to stabilize and/or produce acceptable along the trial dynamics and then apply ILC to force monotonic trial-to-trial error convergence of the resulting system. This is a two step design approach with separate design of the feedback and learning filters where, for example, the ILC learning update is calculated as the inverse of the dynamics resulting from the feedback controller design.

An alternative approach to ILC design is to use the two-dimensional/repetitive systems setting, i.e., systems that propagate information in two independent directions where for ILC these directions are from trial-to-trial and along each trial respectively. Repetitive processes [10] are a distinct class of two-dimensional systems where information in the temporal domain is limited to a finite duration and hence a more natural match to ILC. These processes make a series of sweeps, termed passes or trials in the ILC setting, through a set of dynamics defined over a finite duration and once each pass is complete the process resets to the starting location. On each pass and output, termed the pass profile, is produced that acts as a forcing function on, and hence contributes to, the dynamics of the next pass profile. The result can be oscillations that increase in amplitude from pass-to-pass.

Repetitive processes cannot be controlled by direct application of standard systems theory and algorithms and this has led to the development of a stability theory for them and substantial progress on control system specification and design [10, 14, 15]. The result of repetitive process setting is a one step design for trial-to-trial error convergence and transient response along the trials and hence simultaneous treatment of the trial-to-trial error and transient response along the trials is possible.

This paper deals with simultaneous synthesis of both feedback and learning controllers in an ILC scheme for error convergence and performance, starting with a new result for monotonic trial-to-trial error convergence. This result is achieved by converting the problem to one of stability along the trial for a discrete linear repetitive process, leading to design based on Linear Matrix Inequality (LMI) based computations. Another contribution of the paper is the development of a method deal with the case when the system to be controlled has relative degree greater than unity through the use of an anticipative feedforward control law.

Throughout this paper, the null and identity matrices with the required dimensions are denoted by 0 and I, respectively, and the notation $X \succ Y$ means that the matrix $X - Y$ is positive definite. Also sym{M} is used to denote the symmetric matrix $M + M^\top$ and $\rho(\cdot)$ denotes the spectral radius of its matrix argument, i.e., if $\lambda_i, 1 \leq i \leq q$, denotes the eigenvalues of a $q \times q$ matrix, say H, $\rho(H) = \max_{1\leq i\leq q} |\lambda_i|$. The superscript $*$ denotes the complex conjugate transpose of a matrix and $\otimes$ the matrix Kronecker product.

Use will also be made of the following results, where the first is the generalized KYP lemma and the second the Projection Lemma.

Lemma 1. *[6] Consider matrices* $\mathbb{A}$, $\mathbb{B}_0$, Θ *and*

$$\Phi = \begin{bmatrix} -1 & 0 \\ 0 & 1 \end{bmatrix}, \Psi = \begin{bmatrix} 0 & \mathrm{e}^{\mathrm{j}\omega_c} \\ \mathrm{e}^{-\mathrm{j}\omega_c} & -2\cos(\omega_d) \end{bmatrix}, \tag{1}$$

with $\omega_c = (\omega_l + \omega_u)/2$, $\omega_d = (\omega_u - \omega_l)/2$ *and* ω_l, ω_u *satisfying* $-\pi \leq \omega_l \leq \omega_u \leq \pi$. *Suppose also that* $\det(\mathrm{e}^{\mathrm{j}\omega} I - \mathbb{A}) \neq 0$ *for all* $\omega \in [\omega_l, \omega_u]$. *Then the following statements are equivalent.*

i)

$$\begin{bmatrix} (\mathrm{e}^{\mathrm{j}\omega} I - \mathbb{A})^{-1}\mathbb{B}_0 \\ I \end{bmatrix}^* \Theta \begin{bmatrix} (\mathrm{e}^{\mathrm{j}\omega} I - \mathbb{A})^{-1}\mathbb{B}_0 \\ I \end{bmatrix} \prec 0, \quad \forall \omega \in [\omega_l, \omega_u]. \tag{2}$$

ii) There exist $\mathcal{Q} \succ 0$ *and a symmetric* $\mathcal{P}$ *such that*

$$\begin{bmatrix} \mathbb{A} & \mathbb{B}_0 \\ I & 0 \end{bmatrix}^\top (\Phi \otimes \mathcal{P} + \Psi \otimes \mathcal{Q}) \begin{bmatrix} \mathbb{A} & \mathbb{B}_0 \\ I & 0 \end{bmatrix} + \Theta \prec 0. \tag{3}$$

Lemma 2. *[4] Given a symmetric matrix $\Gamma \in \mathbb{R}^{p\times p}$ and two matrices Λ, Σ of column dimension p, there exists a matrix $\mathcal{W}$ such that the following inequality holds*

$$\Gamma + \mathrm{sym}\{\Lambda^\top \mathcal{W}\Sigma\} \prec 0, \tag{4}$$

if, and only if the following two projection inequalities with respect to $\mathcal{W}$ are satisfied

$$\Lambda^{\perp^\top}\Gamma\Lambda^{\perp} \prec 0,\ \Sigma^{\perp^\top}\Gamma\Sigma^{\perp} \prec 0, \tag{5}$$

where $\Lambda^{\perp}$ and $\Sigma^{\perp}$ are arbitrary matrices whose columns form a basis of nullspaces of Λ and Σ, respectively.

2. Background and problem formulation

The notation for variables in this paper is of the form $w_k(p)$, $0 \le p \le \alpha < \infty$, $k \ge 0$, where w is the scalar or vector-valued variable under consideration, α is the finite trial length and k is the trial number. The plant dynamics is assumed to be discrete linear time-invariant and described in the ILC setting as

$$\begin{aligned} x_k(p+1) =& Ax_k(p) + Bu_k(p), \\ y_k(p) =& Cx_k(p), \end{aligned} \tag{6}$$

where $x_k(p) \in \mathbb{R}^n$ is the state vector, $y_k(p) \in \mathbb{R}^m$ is the output vector and $u_k(p) \in \mathbb{R}^l$ is the control input vector. Also we define z as the standard forward shift operator along discrete-time axis, i.e.,

$$zx_k(p) = x_k(p+1),$$

see, e.g., the cited papers [3] for the details of how the z-transform can be applied over the finite trial length without errors arising from the basic definition of this transform over an infinite interval.

This paper considers system models where the input has no direct affect on the output and hence they have relative degree $r \ge 1$, i.e.,

$$CA^{r-1}B \ne 0, \text{ and } CA^iB = 0, \forall i = 1, 2, \ldots, r-2. \tag{7}$$

The form of ILC considered in this paper is shown in the block diagram of Figure 1 and consists of a unity negative feedback control loop with controller C applied on the current trial k and the ILC law as shown within the shaded part of this figure. In this paper no loss of generality arising from assuming that the initial state vector on each trial is zero and the memory block in Figure 1 represents the use of previous trial information in the computation of the current trial control input and y_d denotes the supplied reference. In the literature L is often termed the learning filter and Q the robustness filter. All computations within the shaded part of Figure 1 are completed in the time elapsed between the end of one trial and the beginning of the next, i.e., off-line.

The first task is to find guidelines for the choice of the C, L and Q filters, where to improve learning and robustness a basic rule is to choose Q as a low-pass filter whose magnitude is unity for the frequency range where reference tracking is required and zero at all other frequencies. Based on the block diagram of Figure 1, the ILC law is

$$F_{k+1}(z) = Q(z)\left(F_k(z) + L(z)E_k(z)\right),$$

and

$$E_k(z) = (I + G(z)C(z))^{-1}Y_d(z) - (I + G(z)C(z))^{-1}G(z)F_k(z).$$

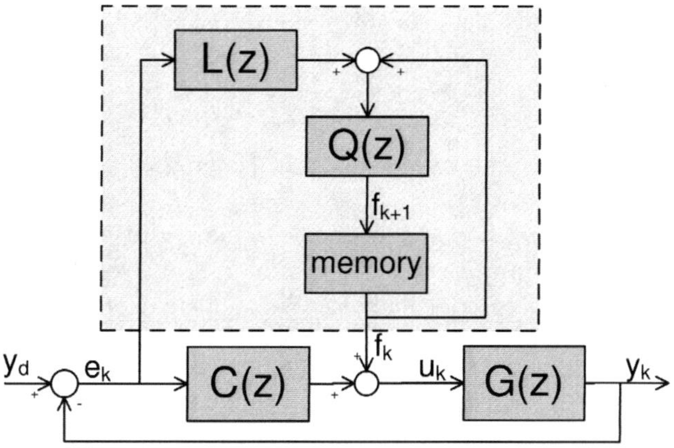

Figure 1. ILC Block Diagram Representation.

Hence, the previous trial error feedforward contribution (assuming $Y_d(z) = 0$) to the current trial error is

$$E_k(z) = -\left[(I + G(z)C(z))^{-1}G(z)\right] F(z) = -S_P(z)F_k(z),$$

where $S_P(z) = (I + G(z)C(z))^{-1}G(z)$ denotes the sensitivity function and the propagation of the error from trial-to-trial is

$$E_{k+1}(z) = M(z)E_k(z), \tag{8}$$

where

$$M(z) = Q(z)\left(I - S_P(z)L(z)\right). \tag{9}$$

One concludes from (8) is that the tracking error converges as $k \to \infty$, i.e. the error reduces between successive trials, if and only if all eigenvalues of $M(z)$ are less than one in magnitude, i.e.

$$\rho\left(M(\mathrm{e}^{\mathrm{j}\omega})\right) < 1,\ \forall \omega \in [-\pi, \pi]. \tag{10}$$

However, practical experience shows that some ILC laws have poor transients during the convergence process even if the above condition is satisfied (e.g. the tracking error may grow over some number of trials). To avoid these problems, a stronger convergence criteria is required for engineering practice. One can ensure that the Euclidean norm of the tracking error decreases monotonically for every trial if $M(\mathrm{e}^{\mathrm{j}\omega})$ satisfies the sufficient stability condition

$$\overline{\sigma}(M(\mathrm{e}^{\mathrm{j}\omega})) < 1,\ \forall \omega \in [-\pi, \pi], \tag{11}$$

where $\overline{\sigma}(\cdot)$ stands for the maximum singular value of its matrix argument. Although (10) is the true stability condition, (11) is more practical and commonly used in practise. From the same reason the condition (11) is used for this paper developments.

2.1. ILC as a repetitive process
This section formulates the ILC design problem considered in the repetitive process setting, starting with the dynamics of these processes. The state-space model of a discrete linear repetitive process [10] over the pass length α is

$$\begin{aligned} x_{k+1}(p+1) &= \mathbb{A}x_{k+1}(p) + \mathbb{B}u_{k+1}(p) + \mathbb{B}_0 y_k(p), \\ y_{k+1}(p) &= \mathbb{C}x_{k+1}(p) + \mathbb{D}u_{k+1}(p) + \mathbb{D}_0 y_k(p), \end{aligned} \tag{12}$$

where on pass k where $x_k(p) \in \mathbb{R}^n$ is the state vector, $y_k(p) \in \mathbb{R}^m$ is the output vector and $u_k(p) \in \mathbb{R}^l$ is the control input vector. Furthermore, it is necessary to specify the boundary conditions, i.e., the pass state initial vector sequence $x_k(0)$, $k \geq 1$, and the initial pass profile $y_0(p)$. In this paper, the initial state vector on each pass is taken as the zero vector and the initial pass profile vector entries are assumed to be specified over the pass length as known functions of p.

The terms $\mathbb{B}_0 y_k(p)$ and $\mathbb{D}_0 y_k(p)$ in (12) represent the contributions of the previous pass profile vector to the current pass state and pass profile vectors respectively. This inter-pass interaction is the source of the unique control problem for these processes where the sequence of pass profiles $\{y_k\}_{k\geq 1}$ can contain oscillations that increase in amplitude from pass-to-pass, i.e., with increasing k. The stability theory [10] for these processes requires that a bounded initial pass profile produces a bounded sequence of pass profiles, where in the strongest form this property is required for all possible values of the pass length. In the remainder of this paper trial is replaced by pass to conform with the repetitive process terminology.

To formulate the ILC design problem in the repetitive process terms assume $Q = I$ (at this stage frequency attenuation using the Q-filer is not considered) and rewrite (9) as

$$M(z) = I - G(z)\begin{bmatrix} C(z) & L(z) \end{bmatrix}\left(\begin{bmatrix} (I + G(z)C(z)) & G(z)L(z) \\ 0 & I \end{bmatrix}\right)^{-1}\begin{bmatrix} 0 \\ I \end{bmatrix}.$$

It is seen from the above formula that $M(z)$ can be represented as a particular case of a general control configuration [11], where the generalized plant P, i.e., the interconnection system of the controlled system, is

$$P(z) = \left[\begin{array}{c|c} P_{11}(z) & P_{12}(z) \\ \hline P_{21}(z) & P_{22}(z) \end{array}\right] = \left[\begin{array}{c|c} I & -G(z) \\ \hline 0 & -G(z) \\ I & 0 \end{array}\right], \tag{13}$$

which is the transfer-function matrix from $[e_k^\top \ u_{k+1}^\top]^\top$ to $[e_{k+1}^\top \ y_{k+1}^\top]^\top$, where $y_{k+1} = [(\overline{y}_{k+1})^\top (\widehat{y}_{k+1})^\top]^\top$. The generalized ILC controller to be found is given by $K(z) = \begin{bmatrix} C(z) & L(z) \end{bmatrix}$ and hence $M(z)$ can be rewritten as

$$M(z) = P_{11}(z) + P_{12}(z)K(z)\left(I - P_{22}(z)K(z)\right)^{-1} P_{21}(z).$$

Furthermore, the state-space representation of (13) is

$$\begin{aligned} x_{k+1}(p+1) &= Ax_{k+1}(p) + Bu_{k+1}(p), \\ \overline{y}_{k+1}(p) &= -Cx_{k+1}(p), \\ \widehat{y}_{k+1}(p) &= e_k(p), \\ e_{k+1}(p) &= -Cx_{k+1}(p) + e_k(p). \end{aligned} \tag{14}$$

Also, suppose that the ILC controller K is static, i.e.

$$u_{k+1}(p) = \begin{bmatrix} K_1 & K_2 \end{bmatrix}\begin{bmatrix} \overline{y}_{k+1}(p) \\ \widehat{y}_{k+1}(p) \end{bmatrix}. \tag{15}$$

The resulting ILC dynamics can now be written as

$$\begin{bmatrix} x_{k+1}(p+1) \\ e_{k+1}(p) \end{bmatrix} = \begin{bmatrix} A - BK_1C & BK_2 \\ -C & I \end{bmatrix}\begin{bmatrix} x_{k+1}(p) \\ e_k(p) \end{bmatrix}, \tag{16}$$

which is a *discrete linear repetitive process* state-space model of the form (12). Treated as a repetitive process, the stability along the pass property will force the trial-to-trial error to converge monotonically to zero - see the next section.

If an example described by (6) has relative degree $r > 0$, an r-step delay between its input and output results. To overcome this problem, the previous trial error is shifted by r samples to form an the anticipative feedforward control law [13]. Obviously, this is possible in practice since the error signal from trial k is available once this trial is complete. As a consequence this anticipative control law, the transfer-function matrix of the learning controller is taken as $z^r L$ instead of L and the generalized plant P resulting from the this modification is

$$P(z) = \left[\begin{array}{c|c} P_{11}(z) & P_{12}(z) \\ \hline P_{21}(z) & P_{22}(z) \end{array}\right] = \left[\begin{array}{c|c} I & -G(z) \\ \hline 0 & -z^r G(z) \\ I & 0 \end{array}\right], \tag{17}$$

where the block P_{22} includes anticipatory operator and the forward time shift is applied to the error signal transmitted through $\widehat{y}_{k+1}$ — see the 3rd equation of (14). Then, on introducing the following notation for (16)

$$\overline{A} = A - BK_1 C,\ \overline{B}_0 = BK_2,\ \overline{C} = -C,\ \overline{D}_0 = I, \tag{18}$$

and the transfer-function matrix $M(z)$ from e_k to e_{k+1} is

$$M(z) = \overline{C}(zI - \overline{A})^{-1}\overline{B}_0 z^r + \overline{D}_0 = \overline{C} z^r (zI - \overline{A})^{-1}\overline{B}_0 + \overline{D}_0. \tag{19}$$

Consider the term $z^r(zI - \overline{A})^{-1}$ and assuming $(zI - \overline{A})^{-1}$ is nonsingular $(zI - \overline{A})(zI - \overline{A})^{-1} = I$, and hence $z(zI - \overline{A})^{-1} = I + \overline{A}(zI - \overline{A})^{-1}$. Also

$$z^2(zI - \overline{A})^{-1} = z(I + \overline{A}(zI - \overline{A})^{-1}) = zI + \overline{A} + \overline{A}^2(zI - \overline{A})^{-1},$$

and hence

$$z^r(zI - \overline{A})^{-1} = \sum_{j=0}^{r-1} z^{r-1-j}\overline{A}^j + \overline{A}^r(zI - \overline{A})^{-1}. \tag{20}$$

In view of (7) and (18)

$$\overline{CA}^{r-1} = \overline{C}A^{r-1} = -CA^{r-1},\ \overline{CA}^r = -CA^{r-1}\overline{A} = -CA^{r-1}(A - BK_1 C),$$

and then $M(z)$ in (19) can be rewritten as

$$\begin{aligned} M(z) &= \overline{C}\left(\sum_{j=0}^{r-1} z^{r-1-j}\overline{A}^j + \overline{A}^r(zI - \overline{A})^{-1}\right)\overline{B}_0 + \overline{D}_0 \\ &= -CA^{r-1}\overline{B}_0 - CA^{r-1}(A - BK_1 C)(zI - (A - BK_1 C)^{-1})\overline{B}_0 + \overline{D}_0. \end{aligned}$$

or

$$M(z) = \mathbb{C}(zI - \mathbb{A})^{-1}\mathbb{B}_0 + \mathbb{D}_0, \tag{21}$$

where the matrices in the state-space quadruple $\{\mathbb{A}, \mathbb{B}_0, \mathbb{C}, \mathbb{D}_0\}$ are

$$\mathbb{A} = A - BK_1 C,\ \mathbb{B}_0 = BK_2,\ \mathbb{D}_0 = I - CA^{r-1}BK_2,\ \mathbb{C} = -CA^r + CA^{r-1}BK_1 C. \tag{22}$$

These matrices define a *discrete linear repetitive* process state-space model of the form (12) and will be used to formulate the ILC design problem in terms of repetitive process stability theory.

3. ILC design

In this section, the stability theory for discrete linear repetitive processes is used to develop a new ILC design algorithm. It has to be emphasized that several sets and concepts for stability of discrete linear repetitive processes are known [10]. From the practical point of view, the most important are those which guarantee asymptotic stability or stability along the pass (trial). Asymptotic stability guarantees a bounded sequence of pass profiles (i.e. output signals) for a bounded initial pass profile over the finite and fixed pass length α, whereas stability along the pass is stronger since it requires this property uniformly, that is, for all possible values of the pass length and hence, it is not surprising that asymptotic stability is a necessary condition for stability along the pass.

The following result characterizes stability along the pass of examples described by (12).

Lemma 3. *[10] A discrete linear repetitive process described by (12) and (22) is stable along the pass if and only if*

i) $\rho(\mathbb{D}_0) < 1$,
ii) $\rho(\mathbb{A}) < 1$,
iii) *all eigenvalues of* $M(\mathrm{e}^{\mathrm{j}\omega}) = \mathbb{C}(\mathrm{e}^{\mathrm{j}\omega} I - \mathbb{A})^{-1}\mathbb{B}_0 + \mathbb{D}_0$, $\forall \omega \in [-\pi, \pi]$ *have modulus strictly less than unity.*

Obviously, in terms of checking the conditions of the above result, only the third condition is a problem because it requires to check if all eigenvalues of $M(\mathrm{e}^{\mathrm{j}\omega})$, that is the characteristic loci, lie inside of the unit circle. Simply, the condition *iii)* of Lemma 3 requires to check if (10) holds. However, as discussed previously, we focus our attention on sufficient condition for (10), that is given by (11). Again, it is important to stress that (11) must hold for entire frequency range. In such the case only, i.e. when $\omega \in [-\pi, \pi]$, one can take $\Psi = 0$ in (1) and then making the similar steps to these shown in [9, 8] it is possible to obtain the following result.

Theorem 1. *An ILC scheme described as a discrete linear repetitive process of the form (12) and (22) is stable along the pass and hence monotonic trial-to-trial error convergence occurs if there exist matrices $S \succ 0$, $\mathcal{W}$ and a symmetric matrix $\mathcal{P}$ such that the following matrix inequalities are feasible*

$$\begin{bmatrix} S - \mathcal{W} - \mathcal{W}^\top & \mathcal{W}^\top \mathbb{A} \\ \mathbb{A}^\top \mathcal{W} & -S \end{bmatrix} \prec 0, \tag{23}$$

$$\begin{bmatrix} -\mathcal{P} & -\mathcal{W} & 0 & 0 \\ -\mathcal{W}^\top & \mathcal{P} + \mathbb{A}^\top \mathcal{W} + \mathcal{W}^\top \mathbb{A} & \mathcal{W}^\top \mathbb{B}_0 & \mathbb{C}^\top \\ 0 & \mathbb{B}_0^\top \mathcal{W} & -I & \mathbb{D}_0^\top \\ 0 & \mathbb{C} & \mathbb{D}_0 & -I \end{bmatrix} \prec 0. \tag{24}$$

Proof. Assume that (23) and (24) hold. Then $\mathcal{W} + \mathcal{W}^\top \succ S \succ 0$ which implies that $\mathcal{W}$ is non-singular and hence invertible. Next, rewrite (23) as

$$\begin{bmatrix} S & 0 \\ 0 & -S \end{bmatrix} + \mathrm{sym}\left\{ \begin{bmatrix} -I \\ \mathbb{A}^\top \end{bmatrix} \mathcal{W} \begin{bmatrix} I & 0 \end{bmatrix} \right\} \prec 0,$$

and by virtue of Lemma 2 yields $\mathbb{A}^\top S \mathbb{A} - S \prec 0$. Hence, from the Lyapunov stability theory for standard discrete linear systems, condition *ii)* of Lemma 3 holds. Next, application of the Schur's complement formula enables (24) to be rewritten as

$$\begin{bmatrix} -\mathcal{P} & 0 & 0 \\ 0 & \mathcal{P} + \mathbb{C}^\top \mathbb{C} & \mathbb{C}^\top \mathbb{D}_0 \\ 0 & \mathbb{D}_0^\top \mathbb{C} & \mathbb{D}_0^\top \mathbb{D}_0 - I \end{bmatrix} + \mathrm{sym}\left\{ \begin{bmatrix} -I \\ \mathbb{A}^\top \\ \mathbb{B}_0^\top \end{bmatrix} \mathcal{W} \begin{bmatrix} 0 & I & 0 \end{bmatrix} \right\} \prec 0, \tag{25}$$

where $\mathcal{W}$ is a slack matrix variable. Hence by the result of Lemma 2 one can find that (25) holds if and only if

$$\begin{bmatrix} \mathbb{A}^\top & I & 0 \\ \mathbb{B}_0^\top & 0 & I \end{bmatrix} \begin{bmatrix} -\mathcal{P} & 0 & 0 \\ 0 & \mathcal{P}+\mathbb{C}^\top\mathbb{C} & \mathbb{C}^\top\mathbb{D}_0 \\ 0 & \mathbb{D}_0^\top\mathbb{C} & \mathbb{D}_0^\top\mathbb{D}_0-I \end{bmatrix} \begin{bmatrix} \mathbb{A} & \mathbb{B}_0 \\ I & 0 \\ 0 & I \end{bmatrix} \prec 0. \tag{26}$$

For (26) to hold requires $\mathbb{D}_0^\top\mathbb{D}_0 - I \prec 0$ and hence that condition *i)* of Lemma 3 holds. Moreover, (26) can be written in the form (3) where

$$\Theta = \begin{bmatrix} \mathbb{C} & \mathbb{D}_0 \\ 0 & I \end{bmatrix}^\top \begin{bmatrix} I & 0 \\ 0 & -I \end{bmatrix} \begin{bmatrix} \mathbb{C} & \mathbb{D}_0 \\ 0 & I \end{bmatrix}, \tag{27}$$

and hence (2) for this case with Θ defined above is equivalent to

$$M(e^{j\omega})^* M(e^{j\omega}) \prec I, \quad \forall \omega \in [-\pi, \pi]. \tag{28}$$

Hence condition *iii)* of Lemma 3 holds and the proof is complete. □

Remark 1. *The inequality (28) can be rewritten as*

$$M(e^{j\omega})^* M(e^{j\omega}) \prec \mu^2 I, \quad \forall \omega \in [-\pi, \pi], \tag{29}$$

where $0 < \mu \leq 1$. *Then* $\overline{\sigma}(M(e^{j\omega})) < \mu$ *and by minimizing* μ *a higher speed of monotonic trial-to-trial error convergence is obtained.*

The inequalities in Theorem 1 are not LMIs as they are bilinear in $\mathcal{W}$ and the matrices defining the controller K. This means no computationally effective method exists for checking (23) and (24). However, by using a particular set of transformations, the inequality (24) can be converted into an LMI. To overcome problems with conversions of (24) into an LMI it is sufficient to find a matrix X such that

$$C\mathcal{W} = XC. \tag{30}$$

It is known that the above equality is equivalent to imposing a block diagonal structural constraint on the matrix variable $\mathcal{W}$. To proceed, observe that for a full rank matrix C, the singular value decomposition (SVD) of C can be represented as $C = U[R\ 0]V^T$ where U and V are unitary matrices and R is a diagonal matrix with positive diagonal elements in decreasing order. Application of SVD allows us to find the equivalent representation of the equality $C\mathcal{W} = XC$. Simply, take

$$\mathcal{W} = V \begin{bmatrix} \mathcal{W}_{11} & 0 \\ \mathcal{W}_{21} & \mathcal{W}_{22} \end{bmatrix} V^T, \tag{31}$$

and suppose X is chosen as $X = UR\mathcal{W}_{11}R^{-1}U^T$ and then

$$XC = \overbrace{UR\mathcal{W}_{11}R^{-1}U^T}^{X}\overbrace{U[R\ 0]V^T}^{C} = UR\mathcal{W}_{11}[I\ 0]V^T = UR[I\ 0]V^TV \begin{bmatrix} \mathcal{W}_{11} & 0 \\ \mathcal{W}_{21} & \mathcal{W}_{22} \end{bmatrix} V^T = C\mathcal{W}.$$

Hence, the matrix structure as in (31) allows us to reformulate the result of Theorem 1 to an LMI-based characterization for the existence of an ILC controller K of (15).

Theorem 2. *An ILC scheme described as a discrete linear repetitive process of the form (12) and (22) is stable along the pass and hence monotonic trial-to-trial error convergence occurs if there exist matrices* $\widehat{\mathcal{W}}$ *of (31),* K_2*,* N_1*,* $\widehat{\mathcal{S}} \succ 0$ *and* $\widehat{\mathcal{P}} \succ 0$ *such that the following LMIs are feasible*

$$\begin{bmatrix} \widehat{\mathcal{S}} - \widehat{\mathcal{W}} - \widehat{\mathcal{W}}^\top & A\widehat{\mathcal{W}} - BN_1C \\ (A\widehat{\mathcal{W}} - BN_1C)^\top & -\widehat{\mathcal{S}} \end{bmatrix} \prec 0, \tag{32}$$

$$\begin{bmatrix} -\widehat{\mathcal{P}} & -\widehat{\mathcal{W}}^\top & 0 & 0 \\ -\widehat{\mathcal{W}} & \widehat{\mathcal{P}} + \mathrm{sym}\left\{A\widehat{\mathcal{W}} - BN_1C\right\} & BK_2 & (-CA^r + CA^{r-1}BN_1C)^\top \\ 0 & K_2^\top B^\top & -I & (I - CA^{r-1}BK_2)^\top \\ 0 & -CA^r + CA^{r-1}BN_1C & I - CA^{r-1}BK_2 & -I \end{bmatrix} \prec 0. \tag{33}$$

Moreover, if the above LMIs are feasible, K_1 *if obtained from*

$$K_1 = N_1 X^{-1}. \tag{34}$$

Proof. Suppose that LMIs (32) and (33) are feasible. Then $\widehat{\mathcal{W}} + \widehat{\mathcal{W}}^\top \succ \widehat{\mathcal{S}} \succ 0$ which implies that $\widehat{\mathcal{W}}$ with the structure defined by (31) is non-singular and hence invertible. Next post- and pre-multiply (32) by $\mathrm{diag}\left\{\widehat{\mathcal{W}}^{-1}, \widehat{\mathcal{W}}^{-1}\right\}$ to obtain (23) by setting $\mathcal{W} = \widehat{\mathcal{W}}^{-1}$. Moreover, the LMI (33) results from left and right multiplying (24) by $\mathrm{diag}\{\mathcal{W}, \mathcal{W}, I, I\}$ and its transpose, respectively. Setting $\widehat{\mathcal{P}} = \mathcal{W}^\top \mathcal{P} \mathcal{W}$, $\widehat{\mathcal{S}} = \mathcal{W}^\top \mathcal{S} \mathcal{W}$, $N_1 = K_1 X$ and applying (30) to the resulting inequality completes the proof. □

3.1. ILC with design specifications over finite frequency ranges

The objective now considered is to develop an approach for shaping the maximum singular value of $M(\mathrm{e}^{\mathrm{j}\omega})$ to meet the control specifications in different frequency ranges without introducing weighting filters, where instead of considering the entire frequency range $[-\pi, \pi]$ attention can be limited to $\omega \in [0, \pi]$ only. Moreover, the results that follow or based on the fact that if this frequency range into H intervals (not necessary with the same length) such that

$$[0, \pi] = \bigcup_{h=1}^{H} [\omega_{h-1}, \omega_h], \tag{35}$$

where $\omega_0 = 0$ and $\omega_H = \pi$ then Lemma 1 can be applied at any frequency in any of these intervals. The possibility to specify different performance specifications has considerable practical significance since common performance issues occur over different frequency ranges. For example, trial-to-trial error convergence rate is in the '*low*' frequency range whereas low sensitivity to disturbances and sensor noise are in '*high*' frequency range. These objectives are included in the requirement that

$$\overline{\sigma}(M(\mathrm{e}^{\mathrm{j}\omega})) < \mu_h, \ \forall \omega \in [\omega_{h-1}, \omega_h], h = 1, \ldots, H, \tag{36}$$

where entire frequency range is partitioned according to (35) and

$$0 < \mu_h \le 1, h = 1, \ldots, H. \tag{37}$$

The choice of the frequency intervals and the μ_h are determined by knowledge of the application under consideration. Making use of Lemma 1 gives the following result for design to meet the specifications (36).

Journal of Physics: Conference Series **659** (2015) 012004 doi:10.1088/1742-6596/659/1/012004

Theorem 3. *Consider an ILC scheme described as a discrete linear repetitive process of the form (12) and (22). Furthermore, suppose that the entire frequency range is arbitrarily divided into H possible different frequency intervals as in (35). Then i) the resulting repetitive process is stable along the pass, ii) monotonic trial-to-trial error convergence occurs and iii) the finite frequency performance specifications (36) are satisfied if there exist matrices $\widehat{\mathcal{W}}$ of the structure defined in (31), K_2, N_1, $\widehat{S} \succ 0$, $\widehat{\mathcal{Q}}_h \succ 0$, $\widehat{\mathcal{P}}_h \succ 0$ and arbitrary chosen scalars μ_h satisfying (37) such that the following LMIs are feasible*

$$\begin{bmatrix} \widehat{S} - \widehat{\mathcal{W}} - \widehat{\mathcal{W}}^\top & A\widehat{\mathcal{W}} - BN_1C \\ (A\widehat{\mathcal{W}} - BN_1C)^\top & -\widehat{S} \end{bmatrix} \prec 0, \tag{38}$$

$$\begin{bmatrix} -\widehat{\mathcal{P}}_h & e^{j\omega_{ch}}\widehat{\mathcal{Q}}_h - \widehat{\mathcal{W}}^\top & 0 & 0 \\ e^{-j\omega_{ch}}\widehat{\mathcal{Q}}_h - \widehat{\mathcal{W}} & \Lambda_{22} & BK_2 & (-CA^r + CA^{r-1}BN_1C)^\top \\ 0 & K_2^\top B^\top & -\mu_h^2 I & (I - CA^{r-1}BK_2)^\top \\ 0 & -CA^r + CA^{r-1}BN_1C & I - CA^{r-1}BK_2 & -I \end{bmatrix} \prec 0 \tag{39}$$

for all $h = 1, \ldots, H$, where

$$\Lambda_{22} = \widehat{\mathcal{P}}_h - 2\cos(\omega_{dh})\widehat{\mathcal{Q}}_h + \mathrm{sym}\left\{A\widehat{\mathcal{W}} - BN_1C\right\}, \ \omega_{ch} = \frac{\omega_{h-1} + \omega_h}{2}, \ \omega_{dh} = \frac{\omega_h - \omega_{h-1}}{2}. \tag{40}$$

Moreover, if these LMIs are feasible, the controller gain K_1 can be obtained from (34).

Proof. This follows that for Theorem 2 where Ψ is chosen as in (1) for each of H frequency ranges satisfying (35) with frequency specifications given in (36). Since $\overline{\sigma}(M(e^{j\omega})) < 1$ in each frequency interval, it follows from (35) that $\overline{\sigma}(M(e^{j\omega})) < 1$ holds over entire frequency range the proof is complete. □

3.2. Q-filter design procedure

Theorem 3 allows us to determine the frequency range where $\overline{\sigma}(M(e^{j\omega})) < 1$ and perform learning over this frequency range only. The remaining frequencies should be cut-off by the Q-filter because the inverse of the plant G is not sufficiently matched at higher frequencies. Therefore, the Q-filter can be chosen as a low-pass filter (which can be implemented as the zero-phase filter, e.g., by using the `filtfilt` routine in MATLAB) with cut-off frequency equal to ω_H, i.e., the highest frequency for which the result of Theorem 3 is valid.

4. Numerical example

In this section a numerical example is considered that illustrate the effectiveness and application and of the new result in this paper. The example given in this section uses a model of laboratory servomechanism system. The system consists of DC motor and the inertial mass which are connected through the rigid shaft - see Figure 2 where a configuration scheme of the modular servo used in this example is given. Rotational motion of the mass is exited by the DC motor. As

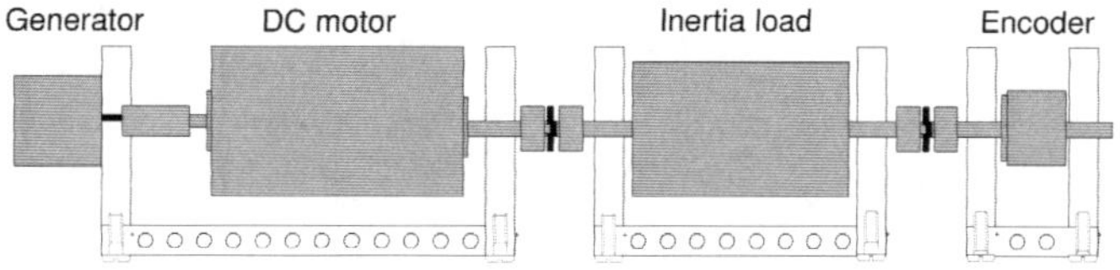

Figure 2. Diagram configuration of the modular servo.

it is in common, the rotational speed of the mass is considered as the output and the armature voltage is considered as the input and hence the following transfer-function represents model of the controlled plant

$$G(s) = \frac{\dot{\Theta}(s)}{V(s)} = \frac{K}{(Js+b)(Ls+R)+K^2}, \tag{41}$$

where K represents both the motor torque constant (K_t) and the back emf constant (K_e), J is the total moment of inertia of the rotor and the mass, b is the motor viscous friction constant, L is the electric inductance and R stands for the electric resistance. By choosing the following values $J = 0.001118$, $b = 3.5077 \cdot 10^{-6}$, $K = 0.056$, $R = 2$ and $L = 0.001$ the model of (41) results in

$$G(s) = \frac{0.056}{1.118 \cdot 10^{-6}s^2 + 0.002236s + 0.003143}. \tag{42}$$

The above transfer-function has been discretized with a sampling time of $T_s = 0.01$ seconds to give a discrete linear state-space model of the form (6) with

$$A = \begin{bmatrix} 0.986 & -3.3769 \cdot 10^{-5} \\ 0.0001 & 0 \end{bmatrix}, \; B = \begin{bmatrix} 16 \\ 0 \end{bmatrix}, C = \begin{bmatrix} 0.0148 & 12.6705 \end{bmatrix}. \tag{43}$$

Executing the design procedure of Theorem 3 for frequency range from 0 to 2[Hz] gives

$$K_2 = 0.0013, \; N_1 = 3.2071 \cdot 10^{-4}, \; W_{11} = 25.8652, \; X = 25.8652,$$

and hence by (34) gives

$$K_1 = N_1 X^{-1} = 1.2399 \cdot 10^{-5}.$$

The resulting singular value plot of $M(e^{j\omega})$ is given in Figure 3 and confirms that the design specifications are met, i.e $M(e^{j\omega}) < 1$ in chosen frequency range.

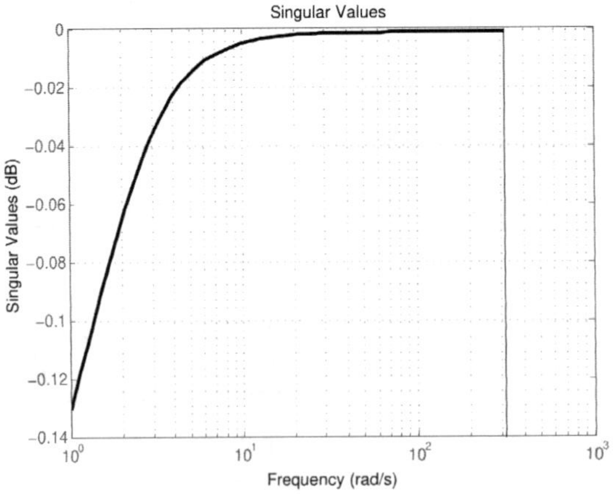

Figure 3. Singular value plot of $M(e^{j\omega})$

5. Conclusions

In this paper a systematic procedure for design of ILC schemes using the repetitive process setting has been developed. Using this design it is possible to avoid adding weighting filters to obtain the required frequency response matrix to meet differing performance specifications.

The results have been established by using the generalized KYP lemma to transform frequency domain inequalities over finite and/or semi-finite frequency ranges for a transfer-function to LMIs. A simulation study using a model for a servomechanism system illustrates the effectiveness of the design.

Acknowledgments
This work is partially supported by National Science Centre in Poland, grant No. 2014/15/B/ST7/03208.

References

[1] Ahn, H.S., Chen, Y.Q., and Moore, K.L. (2007), "Iterative learning control: brief survey and categorization," *IEEE Transactions on Systems, Man and Cybernetics, Part C*, vol. 37, no 6, pp. 1109–1121.

[2] Arimoto, S., Kawamura, S., and Miyazaki, F. (1984), "Bettering operation of robots by learning," *Journal of Robotic Systems*, vol. 1, pp. 123–140.

[3] Bristow, D.A., Tharayil, M., and Alleyne, A. (2006), "A survey of Iterative Learning Control," *IEEE Control Systems Magazine*, vol. 26, no 3, pp. 96–114.

[4] Gahinet, P., and Apkarian, P. (1994), "A linear matrix inequality approach to $\mathcal{H}_\infty$ control," *International Journal of Robust and Nonlinear Control*, vol. 4, pp. 421–448.

[5] Heertjes, M., and Tso, T. (2007), "Nonlinear iterative learning control with applications to lithographic machinery," *Control Engineering Practice*, vol. 15, pp. 1545–1555.

[6] Iwasaki, T., and Hara, S. (2005), "Generalized KYP lemma: unified frequency domain inequalities with design applications," *IEEE Transactions on Automatic Control*, vol. 50, no 1, pp. 41–59.

[7] Norrlöf, M. (2002), "An adaptive iterative learning control algorithm with experiments on an industrial robot," *IEEE Transactions on Robotics and Automation*, vol. 18, no 2, pp. 245–251.

[8] Paszke, W., Rogers, E., Gałkowski, K. and Cai, Z. (2013), "Robust finite frequency range iterative learning control design and experimental verification," *Control Engineering Practice*, vol. 21, no 10, pp. 1310–1320.

[9] Paszke, W., and Bachelier, O. (2013), "Robust control with finite frequency specification for uncertain discrete linear repetitive processes," *Multidimensional Systems and Signal Processing*, vol. 24, no 4, pp. 727–745.

[10] Rogers, E., Gałkowski, K., and Owens, D.H., *Control Systems Theory and Applications for Linear Repetitive Processes*, Vol. 349 of *Lecture Notes in Control and Information Sciences*, Berlin, Germany: Springer-Verlag (2007).

[11] Skogestad, S., and Postlethwaite, I., *Multivariable Feedback Control: Analysis and Design*, 2nd ed., John Wiley & Sons, Inc (2005).

[12] Wang, Y., Gao, F., and Doyle, F.J. (2009), "Survey on iterative learning control, repetitive control, and run-to-run control," *Journal of Process Control*, vol. 19, no. 10, pp. 1589–1600.

[13] Wang, D. (2000), "On D-type and P-type ILC designs and anticipatory approach," *International Journal of Control*, vol. 73, no 10, pp. 890–901.

[14] Wu, L., Lam, J., Paszke, W., Gałkowski, K., Rogers, E., Kummert, A. (2009), "Control and filtering for discrete linear repetitive processes with H_∞ and l_2–l_∞ performance", vol. 20, no 3, pp. 235-264.

[15] Zhou, L., She, J., Zhou, S. (2014), "A 2D system approach to the design of a robust modified repetitive-control system with a dynamic output-feedback controller," *International Journal of Applied Mathematics and Computer Science*, vol. 24, no 2, pp. 325-334.

Fault tolerant controller for a class of additive faults: a quasi-continuous high-order sliding mode approach

J. Dávila[1] J. Cieslak[2], D. Henry[2], A. Zolghadri[2]

[1] National Polytechnic Institute, México, ESIME - UPT, Section of Graduate Studies and Research. Av. Ticoman 600, Col. San Jose Ticoman, Gustavo A. Madero, México D.F.
[2] University of Bordeaux, IMS-Lab, Automatic Control Group, 351 cours de la liberation, 33405 Talence, France.

E-mail: jadavila@ipn.mx, jerome.cieslak@ims-bordeaux.fr, david.henry@ims-bordeaux.fr, ali.zolghadri@ims-bordeaux.fr

Abstract. In this paper a fault tolerant control strategy that combines the backstepping procedure and the quasi-continuous high-order sliding mode controller is proposed. The fault tolerance principle is based on a hierarchical application of the backstepping methodology ensuring the finite time convergence of the desired system states, in spite of the considered fault situations. The additive effect of the faults and disturbances is canceled out by the hierarchical application of the quasi-continuous controller ensuring fault-tolerance. The effect of Lebesgue measurable noise over the precision of the proposed controller is studied. Simulation results based on a nonlinear model of the F16 jet fighter show the efficiency of the proposed techniques.

1. Introduction

Automatic control reconfiguration and Fault Tolerant Control (FTC) can help avoid a potentially hazardous, out-of-tolerance or dangerous behavior of the controlled system, for example, see [1, 2, 3] and the references therein.

Designing a FTC scheme associated with each fault is usually carried out through passive or active FTC schemes [4]. In the passive approach, a unique control algorithm based on robust control techniques is designed to achieve the given objectives in healthy as well as in faulty situations. This is a simple and computationally attractive way to achieve fault tolerance without a fault diagnosis scheme. However, increasing robustness to some faults deteriorates the performance level in the fault-free mode. In contrast to passive designs, active FTC scheme exchanges some of the simplicity for an increased performance level, since it designs control laws that are dedicated to each fault of interest. Under the assumption that the fault is recoverable [5], stability and performance are therefore guaranteed in the presence of each fault by means of a reconfiguration mechanism that is activated when a Fault Detection an Isolation (FDI) unit detects and diagnoses any relevant signal deviation or failure. The available design methodologies in active FTC area include those based on H_∞ control [6], linear quadratic control [7], switching control [8] and fault hiding paradigm [9] to name a few. They have the major advantage that the performance level for the nominal (no fault) operating mode can be preserved. However, the majority of the available methods are built on the unrealistic assumption that the FDI and FTC outputs are instantaneously available to provide decisions and/or actions to other subsystems. The problem of guaranteeing stability and performance of the overall active FTC scheme taking into account both the FDI performances (detection delay, ...) and re-configuration mechanism, has not been considered due to its complexity. In order to address this shortcoming, an attractive technique based on the dwell-time concept has recently been introduced in [10] to show how FDI and FTC performance interact and can be managed to have a global optimal solution.

To avoid such type of complex management [10], considerable effort has been made in the area of adaptive control methods [11]. Due to its prominent adapting ability to handle the structural and parametric uncertainties and variations, adaptive control design schemes can be applied with neither control restructuring nor FDI processing. Based on backstepping technique the results of [12], [13] deal with actuator fault compensation for linear systems. These studies have been next extended to nonlinear systems in previous works [14, 15, 16]. A new backstepping scheme for parametric strict feedback nonlinear systems has been introduced in [17]. A smooth backstepping control based on exact identification of the matched and unmatched uncertainties and disturbances has been proposed for the nonlinear systems [18]. This technique exploits a High-Order Sliding Mode (HOSM) differentiator to identify the perturbations and additional HOSM differentiators are involved to facilitate the control law computation. In all works, the boundedness of all closed-loop signals can still be ensured. However, to the best of our knowledge, there is still no rigorous analysis of measurement noise effects in the area of fault tolerant backstepping technique.

In such a setting, the main contribution of this work consists of development of a technologically viable solution in a bounded noise environment. The contribution of this work is twofold: i) Firstly, the quasi-continuous High-Order Sliding Mode (HOSM) control theory [19] is applied to FTC context. It is shown that finite-time convergence can be achieved in spite of the existence of a class of faults. ii) Secondly, the effect of bounded measurement noises is analyzed.

The paper is organized as follows. Section 2 gives the problem statement. Section 3 is devoted to the proposed FTC strategy where all design and analysis steps are presented. Finally, section 4 shows the efficiency of the proposed scheme on a nonlinear model of the F16 jet aircraft.

2. Fundamentals

The proposed fault tolerant algorithm is designed using the backstepping procedure, and combining two main discontinuous algorithms, the quasi-continuous high-order sliding mode controller and the high-order sliding-modes arbitrary order differentiator. These algorithms are described in this Section.

2.1. Quasi-continuous controller [20]

Let us consider the nonlinear system described by the equation:

$$\dot{x} = a(t,x) + b(t,x)u$$

where $x \in \mathbb{R}^n$, $u \in \mathbb{R}$ and $a(t,x)$, $b(t,x) : \mathbb{R}_+ \times \mathbb{R}^n \to \mathbb{R}$ are smooth nonlinear functions.

The "quasi-continuous" arbitrary-order sliding mode controller was developed to provide finite time establishment of sliding surfaces with relative degree greater than one [20, 21].

Let us introduce the variable $\sigma(t,x) : \mathbb{R}_+ \times \mathbb{R}^n \to \mathbb{R}$ that possesses relative degree n with respect to the control input u, i.e. it satisfies the following equality:

$$\sigma^{(n)} = h(t,x) + g(t,x)u$$

where the superscript (n) stands for the n-th derivative of σ, $h(t,x) = \sigma^{(n)}|_{u=0}$ and $g(t,x) = \frac{\partial}{\partial u}\sigma^{(n)}$ and satisfy the following inequalities $|h(t,x)| \leq C, \quad 0 < K_m \leq g(t,x) \leq K_M$. The quasi-continuous arbitrary-order sliding mode controller is applied to bring to zero, in finite time, the generic scalar sliding variable σ. The recursive representation of the quasi-continuous controller for a generic variable σ is given as follows:

$$\begin{aligned}
\varphi_{0,n} &= \sigma, \; N_{0,n} = |\sigma| && (1)\\
\psi_{0,n} &= \varphi_{0,n}/N_{0,n} = \text{sign}\,\sigma, && (2)\\
\varphi_{i,n} &= \sigma^{(i)} + \beta_i N_{i-1,n}^{(n-i)/(n-i+1)} \psi_{i-1,n}, && (3)\\
N_{i,n} &= |\sigma^{(i)}| + \beta_i N_{i-1,n}^{(n-i)/(n-i+1)}, && (4)\\
\psi_{i,n} &= \varphi_{i,n}/N_{i,n} && (5)
\end{aligned}$$

where $i = 1, ..., n-1$, and $\beta_1, ..., \beta_{n-1}$ are positive numbers. Notice that, the quasi-continuous n-sliding controller is obtained as

$$u = -\alpha \psi_{n-1,n}(\sigma, \dot{\sigma}, ..., \sigma^{(n-1)}). \qquad (6)$$

It was shown in [20] that provided $\alpha > C$, and the tuning parameters $\beta_1, ..., \beta_{n-1}$ are chosen sufficiently large *in the given order* then the control law defined by (1)-(6) enforce the following equality $\sigma = \dot{\sigma} = ... = \sigma^{(n)} = 0$ in finite time.

2.2. *Arbitrary-order exact robust differentiator [22]*

The above n-th order quasi-continuous controller requires the availability of the successive derivatives, up to the order $n-1$, of the sliding variables σ. In order to reconstruct such derivatives exactly and in finite time, the well known Arbitrary-Order Exact Robust Differentiator [22] can be used.

The n-th order differentiator can be expressed in the following non-recursive form

$$\begin{aligned}
\dot{z}_0 &= z_1 - \kappa_n M^{1/n} |z_0 - \sigma|^{\frac{n}{n+1}}\, sign(z_0 - \sigma),\\
\dot{z}_1 &= z_2 - \kappa_{n-1} M^{1/(n-1)} |z_1 - \dot{z}_0|^{\frac{n-1}{n}}\, sign(z_1 - \dot{z}_0),\\
&\vdots\\
\dot{z}_i &= z_i - \kappa_{n-i} M^{1/(n-i)} |z_i - \dot{z}_{i-1}|^{\frac{n-i}{n-i+1}}\, sign(z_i - \dot{z}_{i-1}),\\
&\vdots\\
\dot{z}_n &= -\kappa_0 M\, sign(z_n - \dot{z}_{n-1})
\end{aligned} \qquad (7)$$

for suitable positive constant coefficients κ_i and M to be chosen recursively large in the given order [22]. Under the assumption that the n-th derivative of the sliding variable σ is bounded, the gains M must satisfy the following constraint: $M \geq |\sigma^{(n)}|$. According to [22] $\kappa_0 = 1.1$, $\kappa_1 = 1.5$, $\kappa_2 = 2$, $\kappa_3 = 3$, $\kappa_4 = 5$, $\kappa_5 = 8$ is one of the possible selections for the gains κ_i. Hence, the differentiator (7) can be tuned by a suitable choice of parameters M. Notice that in the absence of measurement noise, the following equalities are true after a finite time transient process:

$$|z_i - \sigma^{(i)}| = 0 \qquad i = 0, ..., n \qquad (8)$$

The presence of non-idealities like measurement noise and finite frequency commutation causes a bounded error in the estimated derivatives [22] and, as a result, a bounded loss of accuracy for the controller that uses the "noisy" derivative estimates [20, 21, 22].

The separation and robustness results relevant to the combined use of the differentiator and any n-sliding homogenous controller has been previously discussed [22], in these works the separation property is proven as a consequence of the finite-time convergence of the high-order sliding-mode differentiator.

3. Problem statement

Let us consider a nonlinear system in the following lower triangular form:

$$\begin{aligned}\dot{x}_1 &= g_1(x_1,t)+B_1(x_1,t)x_2+f_{c_1}(t)+w_1(t)\\ &\vdots\\ \dot{x}_i &= g_i(x_1,...,x_i,t)+B_i(x_1,...,x_i,t)x_{i+1}+f_{c_i}(t)+w_i(t)\\ &\vdots\\ \dot{x}_n &= g_n(x_1,...,x_n,t)+B_n(x_1,...,x_n,t)u+f_{c_n}(t)+f_a(t)+w_n(t)\end{aligned} \tag{9}$$

where $x_i \in \mathbb{R}^p$, $i = 2,...,n-1$; $x^T = [\ x_1^T \ \ x_2^T \ \ ... \ \ x_n^T\ ]$, $x \in \mathcal{X} \subseteq \mathbb{R}^{n \cdot p}$, $u \in \mathbb{R}^p$ is the control signal, $g_i(\cdot)$ are known nonlinear smooth vector fields, $B_i(x_1,...,x_i,t) \in \mathbb{R}^{p \times p}$ are known nonsingular function matrices, $w_i(t)$ are disturbances, $f_{c_i}(t)$ and $f_a(t)$ are faults. In particular, $f_{c_1}(t),...,f_{c_i}(t)$ denotes the component fault (in an additive representation) and $f_a(t)$ can model a class of actuator fault.

The considered faults satisfy the following restrictions

$$\left\| \frac{d^{(n-i)}}{dt^{(n-i)}}\left(f_{c_i}(x_1,...,x_i,t)+w_i(t)\right)\right\|_\infty \leq \Gamma_i^+ \tag{10}$$

$$f_{c_i}(x_1,...,x_i,t)+w_i(t) \in \mathcal{C}^{n-i}, \qquad i=1,...,n-1 \tag{11}$$

$$||f_{c_i}(t)+f_a(t)+w_n(t)|| \leq \Gamma_n^+ \tag{12}$$

where Γ_i^+ is a known positive scalar constant and, $\mathcal{C}^k$ denotes the set of functions with at least k bounded derivatives.

Differential equations are understood in the Filippov sense [23] in order to provide for the possibility to use discontinuous signals in controls. Filippov solutions coincide with the usual solutions, when the right-hand sides are continuous. It is assumed also that all considered inputs allow the existence of solutions and their extension to the whole semi-axis $t \geq 0$.

Given the control output $y = x_1$, the aim of this work is to provide a design methodology able to guarantee, the exact finite-time tracking of a smooth desired reference y_c, in spite of the occurrence of the faults $f_{c_i}(t)$, $f_a(t)$ and the disturbances $w_i(t)$. The whole state vector x and the control signal u are assumed available.

4. Fault tolerant control design

In this section, the backstepping strategy based on the hierarchical application of quasi-continuous controllers [19] is applied to the fault tolerant control design. The design is applied to each one of the components of the control output.

4.1. Control Design

The control objective is the tracking of a smooth command signal y_c using the control signal u. The control signal u is designed through a series of virtual controls ϕ_i, defined recursively, in such a way that at each step the virtual control signal provides the tracking of the preceding coordinate of the system.

Step 1. Define as a virtual control $x_2 = \phi_1$, with

$$\phi_1 = B_1^{-1}(x_1,t)\left[-g_1(x_1,t)+v_1\right] \tag{13}$$

$$v_1^{(n-1)} = \begin{bmatrix} -\alpha_{1_1}\psi_{n-1,n}(\sigma_{1_1},...,\sigma_{1_1}^{(n-1)}) \\ \vdots \\ -\alpha_{1_p}\psi_{n-1,n}(\sigma_{1_p},...,\sigma_{1_p}^{(n-1)}) \end{bmatrix} \tag{14}$$

where $\sigma_1 = y - y_c$, the superscript $(n-1)$ stands for the $(n-1)$th derivative of the signal v_1, and the nonlinear term $\psi_{n-1,n}(\cdot)$ will be designed further using the quasi-continuous high-order sliding mode controller in order to compensate for the fault and disturbance $f_{c_1}(t) + w_1$.

Step i. Define recursively the virtual controls $x_i = \phi_i$ as

$$\phi_i = B_i^{-1}(x_1, ..., x_i, t)\left[-g_i(x_1, ..., x_i, t) + v_i\right] \tag{15}$$

$$v_i^{(i-1)} = \begin{bmatrix} -\alpha_{i_1}\psi_{n-i,n-i+1}(\sigma_{i_1}, ..., \sigma_{i_1}^{(n-i)}) \\ \vdots \\ -\alpha_{i_p}\psi_{n-i,n-i+1}(\sigma_{i_p}, ..., \sigma_{i_p}^{(n-i)}) \end{bmatrix} \tag{16}$$

where $\sigma_i = x_i - \phi_{i-1}$ and the nonlinear control $\psi_{n-i,n-i+1}(\cdot)$ will be designed further in order to compensate for the fault and disturbance $f_{c_i}(t) + w_i$.

Step n. The control is designed as

$$u = B_n^{-1}(x_1, ..., x_n, t)\left[-g_n(x_1, ..., x_n, t) + v_n\right] \tag{17}$$

$$v_n = \begin{bmatrix} -\alpha_{n_1}\psi_{0,1}(\sigma_{n_1}) \\ \vdots \\ -\alpha_{n_p}\psi_{0,1}(\sigma_{n_p}) \end{bmatrix} \tag{18}$$

where $\sigma_n = x_n - \phi_{n-1}$.

Theorem 1 *Let us consider that inequalities (10) to (12) are satisfied for system (9). The backstepping based design (13)-(18) gurantees, after a finite-time transient, the exact tracking of the command signal $y = y_c$, even in the presence of faults and disturbances.*

Proof.

Before proceeding with the proof of convergence, it is interesting to notice that the algorithm is composed by several instances of the quasi-continuous sliding mode controller. All those instances use the variable σ and its derivatives, these derivatives can be obtained by means of several instances of the High-Order Sliding Mode differentiator.

First, we prove the convergence of the estimations of the high-order derivatives for a generic variable σ, after that, the proof of convergence of the output y to the command signal y_c is given assuming that the high-order derivatives of the surfaces have been already obtained by means of the differentiator.

Let us define the estimation error as $e_0 = z_0 - \sigma$ and $e_i = z_i - \sigma^{(i)}$. The estimation error dynamics is given by the following expression:

$$\begin{aligned} \dot{e}_0 &= e_1 - \kappa_n M^{1/n}|e_0|^{\frac{n}{n+1}}\, sign(e_0), \\ \dot{e}_1 &= e_2 - \kappa_{n-1} M^{1/(n-1)}|e_1 - \dot{e}_0|^{\frac{n-1}{n}}\, sign(e_1 - \dot{e}_0), \\ &\vdots \\ \dot{e}_i &= e_{(i+1)} - \kappa_{n-i} M^{1/(n-i)}|e_i - \dot{e}_{(i-1)}|^{\frac{n-i}{n-i+1}}\, sign(e_i - \dot{e}_{(i-1)}), \\ &\vdots \\ \dot{e}_{(n-1)} &= -\kappa_0 M sign(e_{(n-1)} - \dot{e}_{(n-2)}) + \sigma^{(n)} \end{aligned} \tag{19}$$

It is important to remark that, given the appearance of the unknown term $\sigma^{(n)}$ in the last equation of (19), this equation is uncertain, hence the estimation error can be written in a differential inclusion form as:

$$
\begin{aligned}
\dot{e}_0 &= e_1 - \kappa_n M^{1/n} |e_0|^{\frac{n}{n+1}}\, sign(e_0), \\
\dot{e}_1 &= e_2 - \kappa_{n-1} M^{1/(n-1)} |e_1 - \dot{e}_0|^{\frac{n-1}{n}}\, sign(e_1 - \dot{e}_0), \\
&\vdots \\
\dot{e}_i &= e_{(i+1)} - \kappa_{n-i} M^{1/(n-i)} |e_i - \dot{e}_{(i-1)}|^{\frac{n-i}{n-i+1}}\, sign(e_i - \dot{e}_{(i-1)}), \\
&\vdots \\
\dot{e}_n &\in -\kappa_0 M\, sign(e_n - \dot{e}_{(n-1)}) \pm [-|\sigma^{(n)}|, \quad |\sigma^{(n)}|]
\end{aligned}
\tag{20}
$$

Above given differential inclusion concides with the one presented in [22], therefore the equalities (8) are stablished in finite time.

Step 1. Define the tracking error as $\sigma_1 = y - y_c = x_1 - x_{1_c}$. *By differentiating this term is obtained:*

$$\dot{\sigma}_1 = g_1(x_1, t) + B_1(x_1, t)x_2 + f_{c_1}(t) + w_1 - \dot{x}_{c_1}$$

Using the virtual control $x_2 = \phi_1$ *and by differentiating the above given equation (n-1) times the following equation is obtained:*

$$\sigma_1^{(n)} = \left[\frac{d^{n-1}}{dt^{n-1}}(f_{c_1}(t) + w_1)\right] + v_1^{(n-1)}$$

Given that the inequality (10) is satisfied for all $i = 0, ..., n-1$, *the above given equality could be written in a differential inclusion form as:*

$$\sigma_1^{(n)} \in \left[-\Gamma_1^+ \quad \Gamma_1^+\right] + v_1^{(n-1)}$$

Then, by design, each α_{1_i}, $i = 1, ..., p$ *satisfies* $\alpha_{1_i} > \Gamma_1^+$. *The rest of the proof is a direct consequence of [20].*

Step i. Let us introduce the ith component of the virtual tracking error as $\sigma_i = x_i - \phi_{i-1}$. *The first derivative of* σ_i *is given by:*

$$\dot{\sigma}_i = g_i(x_1, ..., x_i, t) + B_i(x_1, ..., x_i, t)x_{i+1} + f_{c_i}(t) + w_i - \dot{\phi}_{i-1}$$

Replacing x_{i+1} *by the virtual control (15) and differentiating the equation until the control appears (i.e. differentiating n-i-1 times), the following equality is obtained:*

$$\sigma_i^{(n-i+1)} = \left[\frac{d^{n-i}}{dt^{n-i}}(f_{c_i}(t) + w_i)\right] + v_i^{(n-i)}$$

Using again inequality (10), the above given equality can be written in a differential inclusion form as:

$$\sigma_i^{(n-i+1)} \in \left[-\Gamma_i^+ \quad \Gamma_i^+\right] + v_i^{(n-i)}$$

Notice again that α_i *is designed to compensate for the bounded fault and disturbance, the rest of the proof is also a direct consequence of [20].*

Step n. Finally, the nth component of the virtual tracking error takes on the form:

$$\sigma_n = x_n - \phi_{n-1}$$

Its dynamics is given by the following expression:

$$\dot{\sigma}_n = g_n(x_1, ..., x_n, t) + B_n(x_1, ..., x_n, t)u + f_{c_n}(t) + f_a(t) + w_n - \dot{\phi}_{n-1}$$

By substitution of the control (17) the following equality is obtained:

$$\dot{\sigma}_n = f_{c_n}(t) + f_a(t) + w_n - \dot{\phi}_{n-1} + v_n$$

As in the above described $n-1$ steps, the fault is bounded according to (12), therefore α_n should be chosen such that $\alpha_n > \Gamma_n^+$, then the equality could be written as a differential inclusion:

$$\dot{\sigma}_n \in \left[\Gamma_n^+ \;\; \Gamma_n^+\right] + v_n$$

Notice that in Step n, the discontinuous control v_n compensates for the faults $f_{c_n}(t) + f_a(t)$ and the disturbance w_n that are only need to be bounded. The rest of the proof is a consequence of [20].

■

4.2. Noise effect analysis

The following result is taken from [22] and [20]. Let g_0 be measured with a bounded noise ϵ, $|\epsilon| < \epsilon^+$ for a known positive scalar ϵ^+. Then, following inequalities are satisfied by differentiator (7):

$$|z_i - g_0^{(i)}| \leq \mu_i \left(\varepsilon^+\right)^{(n-i+1)/(n+1)}, \quad i = 0, ..., n$$

where μ_i depends on the differentiator parameters. Therefore, the accuracy of the derivatives estimation depends on the selection of M in (7).

The application of the algorithm (1)-(17) requires the estimation of $\dot{\sigma}_{j_k}, \ddot{\sigma}_{j_k}, ..., \sigma_{j_k}^{(n-1)}$, this problem could be solved by means of (7), however the existence of noise inhibits an exact estimation of the derivatives. Therefore, the noise effect is transferred through the differentiator to the controller, in this sense, it follows from [20] that the accuracy of the controller in the presence of a bounded deterministic noise is given as follows:

$$|\sigma_{j_k}| \leq \mu_{j_k}(\epsilon^+)^{n-i/n}, \quad i = 1, ..., n-1$$

where μ_{j_k} are positive scalars that depends on the parameters of the differentiator and the controller.

5. Simulation example

5.1. System and fault modeling

In order to illustrate the application of the proposed methodology, the algorithm is applied to the model presented by [24] and [25], that corresponds to the nominal nonlinear model of an F-16 jet fighter at Mach=0.7, h=10000 ft, $\alpha_{trim} = \theta_{trim} = 0.106803$ rad, $\delta_{e_{trim}} = -0.0295$ rad, $\beta_{trim} = p_{trim} = q_{trim} = r_{trim} = \varphi_{trim} = \delta_{a_{trim}} = \delta_{r_{trim}} = 0$. The dynamics is given by

$$\dot{y} = g_1(y, t) + B_1(y, t)z + f_{c_1} + w_1 \quad (21)$$
$$\dot{z} = g_2(y, z, t) + B_2(y, z, t)\delta + f_{c_2} \quad (22)$$
$$\dot{\delta} = -A_\delta(\delta - u) \quad (23)$$

where the vector $y \in \mathbb{R}^3$ is composed of the command angles $y = [\bar{\varphi} \;\; \bar{\alpha} \;\; \bar{\beta}]^T$ in [rad] (roll angle, angle of attack, sideslip angle, respectively); $z \in \mathbb{R}^3$ is a vector of angular rates in [rad/s] ($z = [p \;\; q \;\; r]^T$, roll, pitch and yaw rates, respectively); $\delta \in \mathbb{R}^3$ is a vector of control deflections in

[rad] ($\delta = [\delta_u, \delta_e, \delta_r]^T$, aileron, elevator and rudder deflections, respectively); $u \in \mathbb{R}^3$ is a control vector applied to actuator inputs. f_{c_1}, $f_{c_2} \in \mathbb{R}^3$ are the component faults caused by damage, in this example. $w_1 \in \mathbb{R}^3$ is the vector of disturbances and $f_a \in \mathbb{R}^3$ is the vector of actuator faults. The nonlinear vector-functions $g_1(y,t) \in \mathbb{R}^3$ and $g_2(y,z,t) \in \mathbb{R}^3$ are known and differentiable and, $B_1(y,t) \in \mathbb{R}^{3\times3}$, $B_2(y,z,t) \in \mathbb{R}^{3\times3}$ are known and nonsingular matrices. It is considered that all the variables y, z, δ are measurable.

The nonlinear functions are given as follows:

$$
\begin{aligned}
g_1(y,t) &= \begin{bmatrix} 0 \\ 0.083589 - 1.15y_2 \\ -0.297y_3 \end{bmatrix}, \; B_1(y,t) = \begin{bmatrix} 1 & 0 & 0 \\ -y_3 & 0.9937 & 0 \\ y_2 + 0.00085 & 0 & -0.9973 \end{bmatrix} \\
g_2(y,z,t) &= \begin{bmatrix} g_{2_1} \\ g_{2_2} \\ g_{2_3} \end{bmatrix}, \; B_2(y,z,t) = \begin{bmatrix} -50.933 & 0 & 10.177 \\ 0 & -19.5 & 0 \\ 4.125 & 0 & -6.155 \end{bmatrix} \\
g_{2_1} &= -0.1345z_1z_2 - 0.8225z_2z_3 - 53.48y_3 - 4.324z_1 - 0.224z_3; \\
g_{2_2} &= 0.9586z_1z_3 - 0.0833(z_3^2 - z_1^2) - 1.94166 + 3.724y_2 - 1.26z_2; \\
g_{2_3} &= -0.7256z_1z_2 + 0.1345z_2z_3 + 17.67y_3 + 0.234z_1 - .649z_3; \\
A_\delta &= diag\{20,\ 20,\ 20\}
\end{aligned} \tag{24}
$$

According to technical reports [26], physical limitations of actuators are $\delta_u \in [-0.52, 0.34]$, $\delta_e \in [-0.52, 0.52]$ and $\delta_r \in [-0.52, 0.52]$ radians. The operating range of the vector y is $\bar{\varphi} \in [-1.15, 1.15]$, $\bar{\alpha} \in [-0.34, 0.34]$ and $\bar{\beta} \in [-0.34, 0.34]$ radians.

Let $y_c(t)$ be a real-time command reference profile. Desired flying qualities are given in terms of second order linear systems as follows:

$$\frac{y_{c,1}}{y_1} = \frac{2}{s^2+2.2s+2}, \quad \frac{y_{c,2}}{y_2} = \frac{4}{s^2+3s+4}, \quad \frac{y_{c,3}}{y_3} = \frac{16.98}{s^2+6.18s+16.98}, \tag{25}$$

where y_1, y_2 and y_3 correspond to $\bar{\varphi}, \bar{\alpha}$ and $\bar{\beta}$ respectively. The control goal is to ensure the convergence of the output tracking errors to zero, i.e. $\lim_{t\to\infty} |y_{c_i} - y_i| = 0, \quad \forall i = 1,2,3$, using the control signal u designed as in (17) in spite of external disturbances and fault occurrence.

To assess the proposed scheme, the following simulation commands are considered:

(i) The angle-of-attack command (with respect to a trim value $\bar{\alpha}_{trim} = 0.106803rad$) is two 3 seconds of duration pulses of 0.1 rad amplitude and polarities + and − during time intervals [0, 3] and [6, 9] s, respectively.

(ii) The roll-angle command is one pulse of $1.0rad$ amplitude during the time interval of [1, 9] s.

(iii) The sideslip angle command is zero for all time.

The flying quality filters in equation (25) are used to filter the commands.

In this work, an aircraft battle damage is considered to illustrate the workability of the proposed fault tolerant controller. According to the previous work [25], the effect of such kind of damage can be seen like a component fault where f_{c_1} and f_{c_2} of (26)-(27) can be computed by using the following modelling equations:

$$f_{c_1} = (\Delta g_{1_i}(y,z,\delta,t) + \Delta B_{1_i}(y,z,\delta,t)z) \tag{26}$$

$$f_{c_2} = (\Delta g_{2_i}(y,z,\delta,t) + \Delta B_{2_i}(y,z,\delta,t)\delta) \tag{27}$$

It is considered that the matrices $(B_1(y,t) + \Delta B_1(y,z,\delta,t))$ and $(B_2(y,z,t) + \Delta B_2(y,z,\delta,t))$ are not singular.

$$
\begin{aligned}
\Delta g_1(\cdot) &= 1\cdot(t-1.5)\Delta g_{1_1}(\cdot) + 1\cdot(t-5)\Delta g_{1_2}(\cdot) \\
\Delta B_1(\cdot) &= 1\cdot(t-1.5)\Delta B_{1_1}(\cdot) + 1\cdot(t-5)\Delta B_{1_2}(\cdot) \\
\Delta g_2(\cdot) &= 1\cdot(t-1.5)\Delta g_{2_1}(\cdot) + 1\cdot(t-5)\Delta g_{2_2}(\cdot) \\
\Delta B_2(\cdot) &= 1\cdot(t-1.5)\Delta B_{2_1}(\cdot) + 1\cdot(t-5)\Delta B_{2_2}(\cdot)
\end{aligned}
$$

where $1\cdot(\tau)$ is a unit step function and

$$\Delta g_{1_1}(\cdot) = \begin{bmatrix} 0 \\ 0 \\ 0.0534\beta + 0.002466\delta_u + 0.0186\delta_r \end{bmatrix}, \quad \Delta g_{1_2}(\cdot) = \begin{bmatrix} 0 \\ 0.04\alpha - 0.0885\delta_e \\ 0 \end{bmatrix}$$

$$\Delta B_{1_1}(\cdot) = 0_{3\times 3}, \quad \Delta B_{1_2}(\cdot) = \begin{bmatrix} 0 & 0 & 0 \\ 0 & 0.0031 & 0 \\ -0.00005 & 0 & 0 \end{bmatrix}$$

$$\Delta g_{2_1}(\cdot) = \begin{bmatrix} 8.024\beta + 0.071p + 0.055r \\ 0 \\ -5.82\beta + 0.01p + 0.133r \end{bmatrix}, \quad \Delta g_{2_2}(\cdot) = \begin{bmatrix} 0 \\ 1.856\alpha + 0.42q \\ 0 \end{bmatrix}$$

$$\Delta B_{2_1}(\cdot) = \begin{bmatrix} 0 & 0 & -5.089 \\ 0 & 0 & 0 \\ 0 & 0 & 3.077 \end{bmatrix}, \quad \Delta B_{2_2}(\cdot) = \begin{bmatrix} 0 & 0 & 0 \\ 0 & 9.75 & 0 \\ 0 & 0 & 0 \end{bmatrix}$$

Remark 1 *The cases of total failure of ailerons, elevator and rudder are not considered. For the total failure case, the control surfaces are completely disabled and can not generate rotational torques. This faulty situation cannot be recoverable [5].*

The command signals and the corresponding response of the aircraft for the three cases are shown in Fig. 1. Notice that, even in the presence of disturbances, actuator faults and component faults, the proposed controller provides an appropriate tracking of the command signals. Notice that the tracking of the command signals is not affected by the appearance of faults (see t=1.5 and t=5), the proposed controller is insensitive with respect to the actuator faults. It is important to remark that for all the cases the set of parameters is the same, i.e. $\alpha_\delta = 15$, $\alpha_z = 60$, $\alpha_y = 15$, $M_z = 200$, $M_y = 200$, that correspond to all the gains α, for the different instances of the quasi-continuous controller, and M, for the different instances of the High-Order Sliding Modes differentitator, for each set of coordinates δ, z and y. Here, the sub-indexes have been included to avoid any confusion. The tracking errors are presented in Fig. 2, notice that all the tracking errors are of the order 10^{-3}. In the Fig. 3 the ailerons, elevators and rudder deflections are shown, all the deflections are in the valid rank of the aircraft.

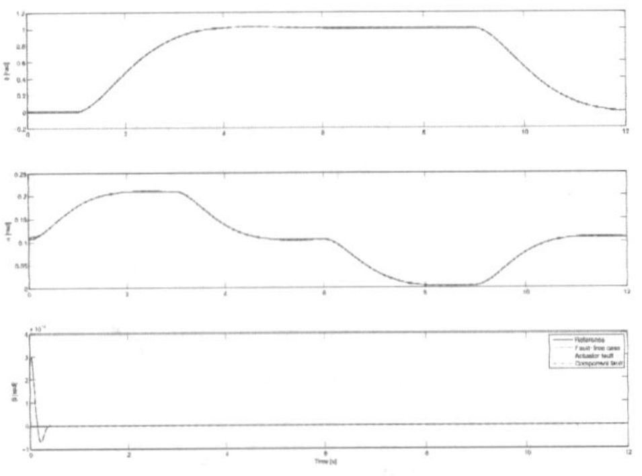

Figure 1. Tracking of the angles ($\bar{\varphi}$, $\bar{\alpha}$, $\bar{\beta}$).

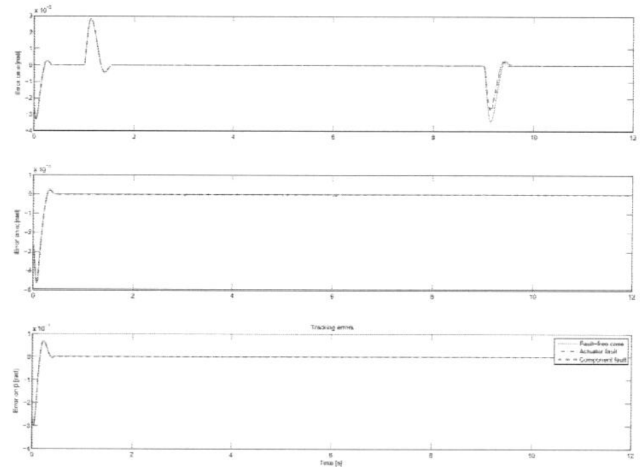

Figure 2. Tracking errors for the angles ($\bar{\varphi}$, $\bar{\alpha}$, $\bar{\beta}$).

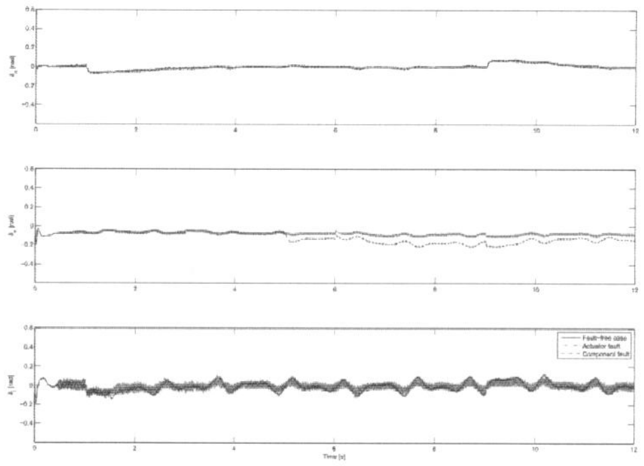

Figure 3. Vector of deflections (δ_a ailerons, δ_e elevators, δ_r rudder).

5.2. Noise effect

Let us consider now that the variables are affected by a bounded noise. For simulation purposes the measured signals, including the noise effect, takes the following form:

$$\begin{aligned}\bar{y} &= y + 0.003\sin(2.5 + 2\sin(1.8t)) + 0.003\cos(5 + 4\sin(2.5t)) \\ \bar{z} &= z + 0.002\sin(4 + 3.9\sin(2.3t)) + 0.004\cos(2 + \sin(1.5t)) \\ \bar{\delta} &= \delta + 0.006\sin(6 + 4\sin(0.5t))\end{aligned}$$

For the case with noisy measurements the corresponding response of the fault tolerant controller is shown in the Figure 4. Notice that almost perfect tracking is obtained in spite of the bounded noise. The tracking error is shown in Figure 5, it is possible to verify that the maximal tracking error is proportional to ϵ^+, as it was explained in Subsection 4.2. The commanded deflections are presented in the Figure 6. It is clear that the noise effect is not amplified by the controller.

6. Acknowledgments

The authors acknowledge the financial support from CONACyT grant 151855, CONACyT-CNRS (Fondo Institucional del CONACYT under grant 209247; CNRS, France: grant 36364) and to SIP-IPN under grant 20150771.

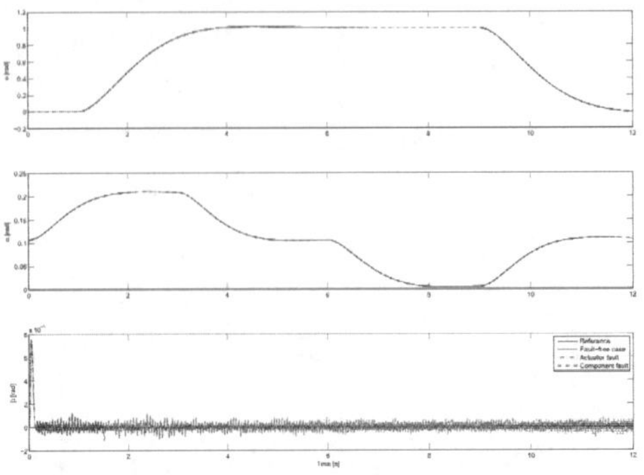

Figure 4. Tracking of the variables $\bar{\varphi}$, $\bar{\alpha}$ and $\bar{\beta}$ in the presence of bounded noise.

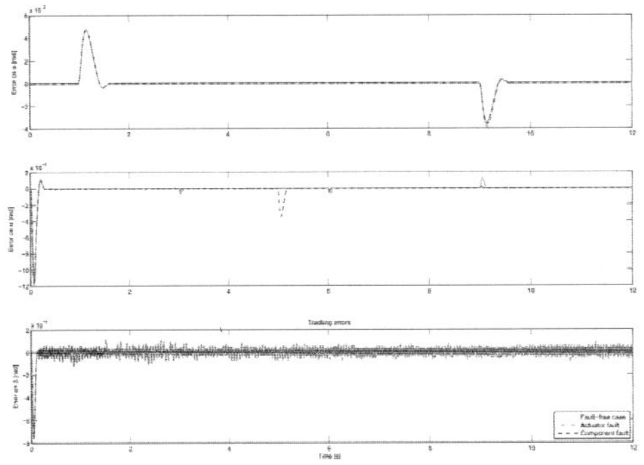

Figure 5. Tracking errors for the angles $\bar{\varphi}$, $\bar{\alpha}$ and $\bar{\beta}$ in the presence of bounded noise.

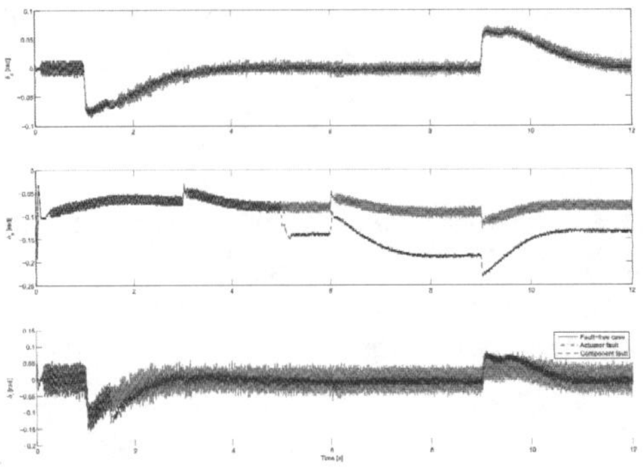

Figure 6. Vector of deflections (δ_a ailerons, δ_e elevators, δ_r rudder) in the presence of bounded noise.

References

[1] M. Blanke, M. Kinnaert, J. Lunze, M. Staroswiecki, and J. Schroder, *Diagnosis and fault tolerant control.* New York, USA: Springer, 2006.

[2] H. Noura, D. Theilliol, J. Ponsart, and A. Chamseddine, *Fault-tolerant control systems: Design and practical applications.* London, UK: Springer-Verlag, 2009.

[3] A. Zolghadri, D. Henry, J. Cieslak, D. Efimov, and P. Goupil, *Fault Diagnosis and Fault-Tolerant Control and Guidance for Aerospace Vehicles: From Theory to Application*, 1st ed., ser. Advances in Industrial Control. London: Springer-Verlag, 2014.

[4] Y. Zhang and J. Jiang, "Bibliographical review on reconfigurable fault-tolerant control systems," *Annual Reviews in Control*, vol. 32, no. 2, pp. 229–252, 2008.

[5] M. Staroswiecki and D. Berdjag, "A general fault tolerant linear quadratic control strategy under actuator outages," *International Journal of Systems Science*, vol. 41, no. 8, pp. 971–985, 2010.

[6] K. Zhou and Z. Ren, "A new controller architecture for high performance, robust, and fault-tolerant control," *IEEE Trans. Automat. Contr.*, vol. 46, no. 10, pp. 1613–1618, 2001.

[7] M. Staroswiecki, H. Yang, and B. Jiang, "Progressive accommodation of parametric faults in linear quadratic control," *Automatica*, vol. 43, no. 12, pp. 2070–2076, 2007.

[8] J. Cieslak, D. Efimov, and D. Henry, "Transient management of a supervisory fault-tolerant control scheme based on dwell-time conditions," 2014, published online DOI: 10.1002/acs.2465.

[9] J. Cieslak, D. Henry, and A. Zolghadri, "Fault tolerant flight control: From theory to piloted flight simulator experiments," *IET Control Theory & Applications*, vol. 4, no. 8, pp. 1451–1464, 2010.

[10] D. Efimov, J. Cieslak, and D. Henry, "Supervisory fault-tolerant control with mutual performance optimization," *International Journal of Adaptive Control and Signal Processing*, vol. 27, no. 4, pp. 251–279, 2013.

[11] M. Bodson and J. Groszkiewicz, "Multivariable adaptive algorithms for reconfigurable flight control," *IEEE Trans. Contr. Syst. Technol.*, vol. 5, no. 2, pp. 217–229, 1997.

[12] G. Tao, S. Joshi, and X. Ma, "Adaptive state feedback and tracking control of systems with actuator failures," *IEEE Trans. Automat. Contr.*, vol. 46, no. 1, pp. 78–95, 2001.

[13] G. Tao, S. Chen, and S. Joshi, "An adaptive actuator failure compensation controller using output feedback," *IEEE Trans. Automat. Contr.*, vol. 47, no. 3, pp. 506–511, 2002.

[14] X. Tang and G. Tao, "An adaptive nonlinear output feedback controller using dynamic bounding with an aircraft control application," *International Journal of Adaptive Control and Signal Processing*, vol. 23, no. 7, pp. 609–639, 2009.

[15] Z. Zhang and W. Chen, "Adaptive output feedback control of nonlinear systems with actuator failures," *Information Sciences*, vol. 179, no. 24, pp. 4249–4260, 2009.

[16] Z. Zhang, S. Xu, Y. Guo, and Y. Chu, "Robust adaptive output-feedback control for a class of nonlinear systems with time-varying actuator faults," *International Journal of Adaptive Control and Signal Processing*, vol. 24, no. 9, pp. 743–759, 2010.

[17] W. Wang and C. Wen, "Adaptive actuator failure compensation control of uncertain nonlinear systems with guaranteed transient performance," *Automatica*, vol. 46, no. 12, pp. 2082–2091, 2010.

[18] J. Davila, "Exact tracking using backstepping control design and high-order sliding modes," *IEEE Trans. Automat. Contr.*, vol. 58, no. 8, pp. 2077–2081, 2013.

[19] A. Estrada and L. Fridman, "Quasi-continuous HOSM control for systems with unmatched perturbations," *Automatica*, vol. 46, no. 11, pp. 1916–1919, November 2010.

[20] A. Levant, "Quasi-continuous high-order sliding-mode controllers," *IEEE Trans. Automat. Contr.*, vol. 50, no. 11, pp. 1812–1816, 2005.

[21] ——, "Homogeneous quasi-continuous sliding-mode control," *Lecture Notes in Control and Information Sciences*, vol. 334, pp. 143–168, 2006.

[22] ——, "High-order sliding modes: differentiation and output-feedback control," *International Journal of Control*, vol. 76, no. 9-10, pp. 924–941, 2003.

[23] A. Filippov, *Differential Equations with Discontinuos Right-hand Sides.* Dordrecht, The Netherlands: Kluwer Academic Publishers, 1988.

[24] M. Pachter, P. Chandler, and M. Mears, "Reconfigurable tracking control with saturation," *AIAA Journal of Guidance, Control, and Dynamics*, vol. 18, no. 5, pp. 1016–1022, 1995.

[25] Y. Shtessel, J. Buffington, and S. Banda, "Multiple timescale flight control using reconfigurable sliding modes," *AIAA Journal of Guidance, Control, and Dynamics*, vol. 22, no. 6, pp. 873–883, 1999.

[26] S. Stachowiak and J. Bosworth, "Flight test results for the f-16xl with a digital flight control system," NASA Dryden Flight Research Center, Edwards, California, Tech. Rep. NASA/TP-2004-212046, March 2004.

Journal of Physics: Conference Series **659** (2015) 012006 doi:10.1088/1742-6596/659/1/012006

Functional Dual Adaptive Control with Recursive Gaussian Process Model

Jakub Prüher and Ladislav Král
University of West Bohemia, Pilsen, Czech Republic
E-mail: jacobnzw@ntis.zcu.cz, ladkral@ntis.zcu.cz

Abstract. The paper deals with dual adaptive control problem, where the functional uncertainties in the system description are modelled by a non-parametric Gaussian process regression model. Current approaches to adaptive control based on Gaussian process models are severely limited in their practical applicability, because the model is re-adjusted using all the currently available data, which keeps growing with every time step. We propose the use of recursive Gaussian process regression algorithm for significant reduction in computational requirements, thus bringing the Gaussian process-based adaptive controllers closer to their practical applicability. In this work, we design a bi-criterial dual controller based on recursive Gaussian process model for discrete-time stochastic dynamic systems given in an affine-in-control form. Using Monte Carlo simulations, we show that the proposed controller achieves comparable performance with the full Gaussian process-based controller in terms of control quality while keeping the computational demands bounded.

1. Introduction

The problem of adaptive control of non-linear stochastic systems with unknown functions offers an interesting challenge [1–4]. It can be understood as a natural extension of the adaptive control from a class of linear systems and non-linear systems with unknown parameters to the complex systems with functional uncertainty. Sometimes this attractive direction of adaptive control is also called a functional adaptive control [1].

Since an optimal solution to this problem is practically impossible to find, a considerable attention has been focused on various sub-optimal solutions, which preserve a key aspects of the stochastic control principles originated from the work of Feldbaum [5]. It means, that a controller should respect the following conflicting goals:

- achieve tracking performance - a controller should be cautious and respects the uncertainty in model knowledge,
- reduce the uncertainty in the future - a controller should probe the system by an appropriate exciting signal.

and is referred to as a dual control [5, 6].

In recent years, several sub-optimal dual controller methods have been applied successfully in the functional approach for their positive qualities (superior control quality and admissible computational demands) and subsequently extended in several directions [7–11]. Although a partial progress was achieved in a functional adaptive control, one of the open and challenging issues remains; namely, finding of a suitable model of the controlled non-linear system. A common feature of most of the above mentioned solutions to the functional approach is that they are based on parametric models represented by various types of neural networks. These models bring difficulties such as the need to optimise the model structure or the necessity to solve a non-linear estimation problem.

Recently, in the system identification community, non-parametric Gaussian process (GP) regression models have attracted a marked interest as an alternative tool to the parametric models. A GP can be effectively thought of as a prior distribution over functions themselves, where functions can be informally regarded as infinite dimensional vectors of function values. Gaussian process (GP) models do not possess a fixed structure and thus are very flexible. They allow to combine tractable Bayesian inference with non-parametric nature of the model and are highly effective in modelling strongly non-linear phenomena. GP models have been utilised in many applications including non-linear filtering [12–15], model predictive control [16], non-linear system identification [17], time series forecasting [18], reinforcement learning [19, 20] or numerical quadrature [21]. Great advantage of parametric models is that they lend themselves nicely to the possibility of devising a recursive estimation algorithm. As GPs are non-parametric models, the development of a recursive algorithm is much more complicated. Typically, it is necessary to consider some kind of GP approximation to counteract the problem of increasing data size. Recently increased effort has been given to the design of original algorithms, which eliminate this drawback [22–26].

So far, GP modelling was utilised in functional adaptive approach only in a few pioneering works [3, 27]. It should be noted, that the GP models were implemented exclusively in a non-recursive form, causing a continuous increase of the computational demands with time, which significantly reduces their practical applicability. In another words, an effective dual controller based on GP model has not been realised yet.

Based on the motivation point, the main goal of the paper is twofold: 1) Design a functional adaptive control using the recursive GP model suggested in [28] and 2) Compare the proposed solution with the non-recursive case presented in [27] through extensive Monte-Carlo analysis in a numerical example.

The rest of the paper will present a description of the problem formulation in section 2, followed by a recursive GP model based non-linear system identification in section 3. In section 4, dual adaptive control design is completed. A numerical illustration is presented in section 5 and finally, section 6 concludes the paper.

2. Problem Formulation

The dynamical system to be controlled is a non-linear stochastic discrete time-invariant system given in an input-output representation

$$\mathcal{S}: \qquad y_{k+1} = f(\boldsymbol{x}_k^a) + g(\boldsymbol{x}_k^a)u_k + e_{k+1}, \tag{1}$$

where $f, g : \mathbb{R}^{n_y+n_u+1} \rightarrow \mathbb{R}$ are unknown non-linear functions, $\boldsymbol{x}_k^a = [y_k, \dots, y_{k-n_y}, u_{k-1}, \dots, u_{k-n_u}]^\top \in \mathbb{R}^{n_y+n_u+1}$ is the state vector, u_k and y_k are input and output signals at discrete time instants $k \in 0, 1, \dots,$ N-1 and $\{e_k\}$ is an additive noise and the following assumptions A1-A4 are considered:

A1: The non-linear functions are smooth (infinitely differentiable), i.e. $f(\boldsymbol{x}_k^a)$, $g(\boldsymbol{x}_k^a) \in \mathcal{C}^\infty$.

A2: The structural parameters n_y and n_u of the system are known.

A3: The system has a globally uniformly asymptotically stable zero dynamics and the non-linear function $g(\boldsymbol{x}_k^a)$ is bounded away from zero for all $\boldsymbol{x}_k^a$ [29].

A4: $\{e_k\} \in \mathbb{R}$ is a known Gaussian sequence with zero mean and variance σ_e^2.

The proposed control law is of an explicit-type, suboptimal dual cost function based on the bi-criterial dual (BD) controller developed by Filatov et al. [30] for linear systems. This cost function explicitly includes two separate criteria, where each criterion introduces one of the mutually opposing goals between estimation and control; *probing* and *caution*. The final control law will be obtained by a subsequent minimisation of the criteria (2) and (4).

The first criterion in the bi-criterial approach is suggested in the following form

$$J_k^c = \mathrm{E}\left\{ (y_{k+1} - r_{k+1})^2 + q\, u_k^2 \middle| \boldsymbol{I}^k \right\}, \tag{2}$$

where $\boldsymbol{I}^k$ is the information state containing all measurable inputs and outputs available up to time instant k and r_{k+1} is a bounded reference signal and $q \geq 0$ is a design parameter penalising the control action.

The criterion (2) evaluates quality of the control and involves minimisation of the expected value of the tracking error. The resulting control

$$u_k^c = \underset{\mathrm{u}_k}{\operatorname{argmin}}\, J_k^c \tag{3}$$

respects uncertainties in the knowledge of the unknown system functions (1), and is equal to *cautious control.*

The second criterion in the bi-criterial approach is chosen as

$$J_k^a = -\mathrm{E}\left\{ (y_{k+1} - \hat{y}_{k+1})^2 \middle| \boldsymbol{I}^k \right\}, \tag{4}$$

where $\hat{y}_{k+1}$ is a one step prediction of the output, which will be obtained from the GP model. This criterion evaluates the quality. It should accelerate the estimation process for future control improvement by increasing the predictive error value. The controller provides an optimal excitation added to the cautious control and determines magnitude of the *probing signal.*

Finally, the BD control u_k is then obtained by

$$u_k = \underset{\mathrm{u}_k \in \Omega}{\operatorname{argmin}}\, J_k^a, \tag{5}$$

where Ω defines a Pareto set as a region of permissible values of u_k representing an efficient trade-off between criteria J_k^a and J_k^c. The region Ω is a design parameter. The J_k^c contribution should be proportional to the uncertainty of the model to provide a probing component of the control signal u_k rich enough.

From relationships describing adaptive controller (2) – (5) it is clear that a suitable model description is necessary to complete the control design. The system model will be based on a non-parametric GP regression and control design will be performed as follows. First, a GP model of the system (1) will be specified. Then a recursive algorithm for the GP model will be proposed. Subsequently, the recursive GP model will be used to complete derivation of the control law. Finally, the whole procedure will be algorithmically summarised.

3. Gaussian Process Regression Model

This section covers the identification of a non-linear stochastic system (1) by means of the GP regression algorithm. The full Gaussian process (FGP) model is described first, in order to show its main difficulties, when applied in situations where data arrive sequentially (which is exactly the case in adaptive control). Introducing the non-recursive FGP model also serves as a motivation for the recursive GP regression algorithm proposed in the subsequent part.

3.1. Full Gaussian process model

The GP can be seen as an extension of the multivariate normal distribution to the normal distribution over infinite-dimensional random vectors (random functions). This means, that the GP, akin to the Gaussian distribution, is fully determined by its first two moments; mean and variance. Let $\boldsymbol{x}_k = [\,(\boldsymbol{x}_k^a)^\top,\, u_k\,]^\top$ be a vector of dimension $D = ny + nu + 2$ and let $\boldsymbol{y}_k = [y_1, \ldots, y_k]^\top$, $\boldsymbol{u}_k = [\,u_0,\, \ldots,\, u_{k-1}\,]$ and $\boldsymbol{X}_k = [\,\boldsymbol{x}_0,\, \ldots,\, \boldsymbol{x}_{k-1}]$ be a $D \times k$ matrix. Then, a non-parametric GP model of the unknown non-linear stochastic system (1) can be defined in the following general form

$$\mathcal{M}: \qquad \hat{y}_{k+1} = \mathcal{GP}(m^+,\, k^+), \tag{6}$$

where posterior mean m^+ and variance k^+ fully describe the GP posterior $\mathcal{GP}\,(m^+,\, k^+)$ and are given by

$$m^+ = m(\boldsymbol{x}_k\,;\,\boldsymbol{\theta}) \;+\; k(\boldsymbol{x}_k,\, \boldsymbol{X}_k\,;\,\boldsymbol{\theta})\, k(\boldsymbol{X}_k,\, \boldsymbol{X}_k\,;\,\boldsymbol{\theta})^{-1}\boldsymbol{y}_k, \tag{7}$$

$$k^+ = k(\boldsymbol{x}_k,\, \boldsymbol{x}_k\,;\,\boldsymbol{\theta}) \;-\; k(\boldsymbol{x}_k,\, \boldsymbol{X}_k\,;\,\boldsymbol{\theta})\, k(\boldsymbol{X}_k,\, \boldsymbol{X}_k\,;\,\boldsymbol{\theta})^{-1} k(\boldsymbol{X}_k,\, \boldsymbol{x}_k\,;\,\boldsymbol{\theta}), \tag{8}$$

where $m(\boldsymbol{x}_k \mid \boldsymbol{\theta}) : \mathbb{R}^D \to \mathbb{R}$ is a *mean function* and $k(\boldsymbol{x}_k,\, \boldsymbol{x}_k \mid \boldsymbol{\theta}) : \mathbb{R}^D \times \mathbb{R}^D \to \mathbb{R}$ is a *covariance function*. The mean and covariance function are parametrised by a vector of free *hyper-parameters* $\boldsymbol{\theta}$. The setting of the optimal values of $\boldsymbol{\theta}$ is discussed in Section 3.2.2 and the dependence of $m(\cdot)$ and $k(\cdot,\cdot)$ on $\boldsymbol{\theta}$ will be omitted for a brevity of notation. Moreover, it holds that $k(\boldsymbol{x}_k,\, \boldsymbol{X}_k) = [\,k(\boldsymbol{x}_k,\, \boldsymbol{x}_0),\, \ldots,\, k(\boldsymbol{x}_k,\, \boldsymbol{x}_{k-1})\,]$, $k(\boldsymbol{X}_k,\, \boldsymbol{x}_k) = k(\boldsymbol{x}_k,\, \boldsymbol{X}_k)^\top$ and $k(\boldsymbol{X}_k,\, \boldsymbol{X}_k)$ is an $k \times k$ prior covariance matrix of elements $k(\boldsymbol{x}_i,\, \boldsymbol{x}_j)$ for $i, j \in 0, \ldots, N-1$.

The mean and covariance function are chosen by the designer and imposes certain assumptions on the model. Compared to the parametric regression models, where the assumptions on the modelled function are typically expressed in a fixed model structure, the assumptions of non-parametric GP models, expressed by the choice of covariance function, are much weaker. There are many covariance functions to choose from, each expressing different assumptions about the modelled phenomena. For comprehensive account of covariance functions and their relationship to other models see [31].

Without loss of generality, zero mean function $m(\cdot)$ will be further assumed. The covariance function $k(\cdot,\cdot)$ is chosen to ease the design of the adaptive BD controller in later sections and is inspired by [32]. The mean and covariance functions are thus given by

$$m(\boldsymbol{x}_k) \equiv 0, \tag{9}$$

$$k(\boldsymbol{x}_i,\, \boldsymbol{x}_j) = k_f(\boldsymbol{x}_i^a,\, \boldsymbol{x}_j^a) \;+\; k_g(\boldsymbol{x}_i^a,\, \boldsymbol{x}_j^a) u_i u_j \;+\; k_n(\boldsymbol{x}_i^a,\, \boldsymbol{x}_j^a), \tag{10}$$

where

$$k_f(\boldsymbol{x}_i^a,\, \boldsymbol{x}_j^a) = \ell_f^2 \exp\left(-\frac{1}{2}(\boldsymbol{x}_i^a - \boldsymbol{x}_j^a)^\top \boldsymbol{\Lambda}_f^{-1}(\boldsymbol{x}_i^a - \boldsymbol{x}_j^a)\right), \tag{11}$$

$$k_g(\boldsymbol{x}_i^a,\, \boldsymbol{x}_j^a) = \ell_g^2 \exp\left(-\frac{1}{2}(\boldsymbol{x}_i^a - \boldsymbol{x}_j^a)^\top \boldsymbol{\Lambda}_g^{-1}(\boldsymbol{x}_i^a - \boldsymbol{x}_j^a)\right), \tag{12}$$

$$k_n(\boldsymbol{x}_i,\, \boldsymbol{x}_j) = \sigma_e^2\, \delta_{ij}, \tag{13}$$

and where ℓ_f, ℓ_g are vertical lengthscales, $\boldsymbol{\Lambda}_f = \mathrm{diag}(\lambda_{f1}^2,\, \ldots,\, \lambda_{fD}^2)$ (analogously for $\boldsymbol{\Lambda}_g$) are horizontal lengthscales and σ_e^2 is a noise variance. The symbol δ_{ij} is Kronecker delta. The above mentioned quantities are hyper-parameters forming a vector $\boldsymbol{\theta} = [\,\ell_f^2,\, \lambda_{f1},\, \ldots,\, \lambda_{fD},\, \ell_g^2,\, \lambda_{g1},\, \ldots,\, \lambda_{gD}\,]^\top$. The term k_n is a covariance function of a white noise with variance σ_e^2 and the covariance functions k_f and k_g are selected as *squared exponential* covariance function with *automatic relevance determination*, which is a common choice in a number of applications [14, 17].

The equations (6)–(13) describe model of the system (1) and provide a procedure how to obtain an one-step prediction $\hat{y}_{k+1}$ and variance $\mathrm{var}(y_{k+1})$, which may be useful component in the control law derivation [27]. Note that, the equations (7)–(8) show that the prediction is calculated using *all* the currently available data. This is the distinguishing feature of non-parametric models in general. In offline settings, where all the data are readily available, calculating model prediction calls for the inversion of large covariance matrix $k(\boldsymbol{X}_k^a, \boldsymbol{X}_k^a)$. In cases where the data arrives sequentially, the situation is even worse, because the size of the covariance matrix $k(\boldsymbol{X}_k, \boldsymbol{X}_k)$ grows with every new measurement, leading to ever increasing computational demands for matrix inversion. This is why controllers based on this non-recursive model form cannot be seriously considered for practical applications. To alleviate this problem, the recursive Gaussian process regression algorithm is proposed in the next section.

3.2. Recursive Gaussian process model

The proposed recursive Gaussian process algorithm is described in two steps. The recursive relations for mean and variance functions calculation is briefly outlined first. This is then followed by a procedure for hyper-parameter learning. To keep the presentation succinct, only the basic results obtained in [25] and [28] are described. For detailed derivation of the algorithm, interested reader is referred to the previously cited references.

3.2.1. RGP algorithm Huber [25] proposed the recursive Gaussian process (RGP) regression algorithm as a solution to the problem of GP regression for sequential data processing. The main idea employed in RGP is to use a predefined set of *basis vectors*, which are sequentially updated and effectively summarise the information obtained from currently available measurements about the approximated function on a user-defined domain. Instead of using all the data for prediction, as in the case of FGP, predictions of RGP model are computed from the basis vectors. The number of basis vectors $s \ll N$ (where N is the number of maximum data points) is fixed throughout the operation of the algorithm, which enables to keep the computational demands in check. On the other hand, using only fixed amount of basis vectors acts as an approximation to the full GP approach characterised by the equations (7) and (8), because the prediction is no longer a function of all the previously seen data.

Let $\tilde{\boldsymbol{X}} = [\tilde{\boldsymbol{x}}_1, \tilde{\boldsymbol{x}}_2, \ldots, \tilde{\boldsymbol{x}}_s]$ denote a $D \times s$ matrix of the basis vectors. The RGP algorithm then calculates moments m^+ and k^+ at every time step $k > 0$ by incorporating information from the new observation y_k. In summary, the RGP algorithm operates by means of the following two sets of equations

$$\text{update}\begin{cases} \boldsymbol{G}_k = \boldsymbol{C}_{k-1}^h \boldsymbol{J}_{k-1}^\top \left(C_k^p + \sigma_e^2\right)^{-1}, & \text{(14a)} \\ \boldsymbol{\mu}_k^h = \boldsymbol{\mu}_{k-1}^h + \boldsymbol{G}_k \left(y_k - \mu_k^p\right), & \text{(14b)} \\ \boldsymbol{C}_k^h = \boldsymbol{C}_{k-1}^h - \boldsymbol{G}_k \boldsymbol{J}_{k-1} \boldsymbol{C}_{k-1}^h, & \text{(14c)} \end{cases}$$

$$\text{predict}\begin{cases} \boldsymbol{J}_k = \boldsymbol{k}_k^s \boldsymbol{K}^{-1}, & \text{(15a)} \\ \mu_{k+1}^p = \boldsymbol{J}_k \boldsymbol{\mu}_k^h, & \text{(15b)} \\ C_{k+1}^p = k_k^{ss} + \boldsymbol{J}_k \left(\boldsymbol{C}_k^h - \boldsymbol{K}\right) \boldsymbol{J}_k^\top, & \text{(15c)} \end{cases}$$

where $\boldsymbol{K} = k(\tilde{\boldsymbol{X}}, \tilde{\boldsymbol{X}})$ is an $s \times s$ matrix of elements $k(\tilde{\boldsymbol{x}}_p, \tilde{\boldsymbol{x}}_q)$ for $p, q \in 1, \ldots, s$. The quantities $\boldsymbol{k}_k^s$ and k_k^{ss} are computed as

$$\boldsymbol{k}_k^s = k(\boldsymbol{x}_k, \tilde{\boldsymbol{X}}) = \boldsymbol{k}_k^{sf} + \boldsymbol{k}_k^{sg} u_k, \tag{16}$$

$$k_k^{ss} = k(\boldsymbol{x}_k, \boldsymbol{x}_k) = k_k^{ssf} + k_k^{ssg} u_k^2 + k_k^{ssn} \tag{17}$$

with

$$\boldsymbol{k}_k^{sf} = k_f(\boldsymbol{x}_k^a,\, \tilde{\boldsymbol{X}}^a), \quad (18) \qquad k_k^{ssf} = k_f(\boldsymbol{x}_k^a,\, \boldsymbol{x}_k^a), \quad (20)$$

$$\boldsymbol{k}_k^{sg} = k_g(\boldsymbol{x}_k^a,\, \tilde{\boldsymbol{X}}^a) \circ \boldsymbol{u}, \quad (19) \qquad k_k^{ssg} = k_g(\boldsymbol{x}_k^a,\, \boldsymbol{x}_k^a), \quad (21)$$

$$k_k^{ssn} = k_n(\boldsymbol{x}_k^a,\, \boldsymbol{x}_k^a), \quad (22)$$

where the symbol $\circ$ denotes element-wise product of two vectors, $\tilde{\boldsymbol{X}}^a$ is the matrix of $D-1$ rows of $\tilde{\boldsymbol{X}}$ and $\boldsymbol{u}$ denotes the last row of $\tilde{\boldsymbol{X}}$. The equations (14a) – (14c) update the function value estimates $\boldsymbol{\mu}_k^h$ and their covariance $\boldsymbol{C}_k^h$ using the last output measurement y_k. The equations (15a) – (15c) are used to calculate the RGP model output prediction $\hat{y}_{k+1} = \mu_{k+1}^p$ and predictive variance $\mathrm{var}(y_{k+1}) = C_{k+1}^p$.

3.2.2. Hyper-parameter learning procedure The RGP algorithm is able to reduce the computational demands for GP regression in sequential data processing. Note however, that in order to compute the RGP model prediction given by equations (15a) – (15c), the hyper-parameters $\boldsymbol{\theta}$ have to be determined.

A preferred method for setting of the hyper-parameters is based on maximisation of the marginal likelihood. In practice it is convenient to work with the logarithm of marginal likelihood, which for a Gaussian likelihood, has an analytic form

$$\log p(\boldsymbol{y}_k \mid \boldsymbol{X}_k, \boldsymbol{\theta}) = \underbrace{-\frac{1}{2}\boldsymbol{y}_k^\top \left(\boldsymbol{K}(\boldsymbol{\theta}) + \sigma_e^2\boldsymbol{I}\right)^{-1}\boldsymbol{y}_k}_{\text{data fit}} \underbrace{-\frac{1}{2}\log\left|\boldsymbol{K}(\boldsymbol{\theta}) + \sigma_e^2\boldsymbol{I}\right|}_{\text{complexity penalty}} - \frac{N}{2}\log 2\pi, \tag{23}$$

where $\boldsymbol{K}(\boldsymbol{\theta}) = k(\boldsymbol{X}_k,\, \boldsymbol{X}_k;\, \boldsymbol{\theta})$. The marginal likelihood consists of a *data fit* term, which encourages complex models that fit data well, and a *complexity penalty* term, which penalises models that are too complex. This ensures that maximisation of (23) will not result in model over-fitting or under-fitting [33]. The disadvantage is that the optimisation problem is non-linear, which makes finding of the global maxima very difficult. It is helpful to point out, that the objective (23) is a function of $\boldsymbol{\theta}$, where the data remain fixed. Note, that using this technique the optimal hyper-parameters are determined directly from the data and therefore, their optimisation takes place before the system output prediction is generated.

Using (23) for setting of hyper-parameters in RGP algorithm is certainly possible, but note, that this would again introduce the problem of increasing computational demands as the covariance matrix $\boldsymbol{K}(\boldsymbol{\theta})$ in the marginal likelihood is constructed from all available data. Evaluating (23) would thus become harder with every new measurement processed.

We propose an approximation of (23) by replacing the noisy observations $\boldsymbol{y}_k$ with estimates of the function values $\boldsymbol{\mu}_k^h$ at the basis vectors $\tilde{\boldsymbol{X}}$. The approximate marginal likelihood is then given by

$$\log p(\boldsymbol{y}_k \mid \boldsymbol{X}_k, \boldsymbol{\theta}) \approx \log p\left(\boldsymbol{\mu}_k^h \mid \tilde{\boldsymbol{X}}, \boldsymbol{\theta}\right) = -\frac{1}{2}\left[(\boldsymbol{\mu}_k^h)^\top \tilde{\boldsymbol{K}}(\boldsymbol{\theta})^{-1}\,\boldsymbol{\mu}_k^h + \log\left|\tilde{\boldsymbol{K}}(\boldsymbol{\theta})\right| + s\log(2\pi)\right], \tag{24}$$

where $\tilde{\boldsymbol{K}}(\boldsymbol{\theta}) = k(\tilde{\boldsymbol{X}},\, \tilde{\boldsymbol{X}};\, \boldsymbol{\theta})$. Since the number of basis vectors s is pre-defined by the user and fixed throughout the operation of the algorithm, the computational demands for evaluating the approximated marginal likelihood (24) do not grow with time. In practical implementation, the objective (24) is maximised in every time step using an iterative optimisation solver, where the initial guess are the hyper-parameters from the previous time step. To keep the computational requirements in check, we also limit the maximum number of solver iterations.

4. Bi-criterial Dual Control Design

This section is focused on the BD control derivation based on the idea of a multi-objective optimization introduced in Section 2.

4.1. Control law derivation

Let $\hat{f}_{k+1} = \boldsymbol{k}_k^{sf} \boldsymbol{K}^{-1} \boldsymbol{\mu}_k^h$ and $\hat{g}_{k+1} = \boldsymbol{k}_k^{sg} \boldsymbol{K}^{-1} \boldsymbol{\mu}_k^h$. Then the GP model mean (15b) can be rewritten to the following form

$$\hat{y}_{k+1} = \hat{f}_{k+1} + \hat{g}_{k+1} u_k. \tag{25}$$

The GP model variance (15c) can be rearranged as

$$\mathrm{var}(y_{k+1}) = \mu_k u_k^2 + \nu_k u_k + c_k, \tag{26}$$

with $\boldsymbol{\Gamma} = \boldsymbol{K}^{-1}(\boldsymbol{C}_k^h - \boldsymbol{K})(\boldsymbol{K}^{-1})^\top$, $\mu_k = k_k^{ssg} + \boldsymbol{k}_k^{sg}\boldsymbol{\Gamma}(\boldsymbol{k}_k^{sg})^\top$, $\nu_k = \boldsymbol{k}_k^{sg}\boldsymbol{\Gamma}(\boldsymbol{k}_k^{sf})^\top + \boldsymbol{k}^{sf}\boldsymbol{\Gamma}(\boldsymbol{k}_k^{sg})^\top$ and $c_k = k_k^{ssn} + k_k^{ssf} + \boldsymbol{k}_k^{sf}\boldsymbol{\Gamma}(\boldsymbol{k}_k^{sf})^\top$. The first of the BD criteria (2) can be minimized using (25), (26) and the cautious control component u_k^c is given as

$$u_k^c = \frac{\left(r_{k+1} - \hat{f}_{k+1}\right)\hat{g}_{k+1} - \frac{1}{2}\nu_k}{\hat{g}_{k+1}^2 + \mu_k + q}. \tag{27}$$

The second BD criterion (4) can be rewritten by reusing (26) as

$$J_k^a = -\,\mathrm{var}(y_{k+1}) = -\mu_k u_k^2 - \nu_k u_k - c_k. \tag{28}$$

In order to complete the derivation we need to specify the region Ω. A suitable choice is a symmetrically distributed region around the caution control u_k^c defined as

$$\Omega_k = [u_k^c - \delta,\, u_k^c + \delta], \tag{29}$$

where the optional parameter δ fixes the domain Ω as a constant.

It is obvious, that the criterion J_k^a is a concave function of variable u_k. Hence, the extreme will inevitably be found within the boundary of the domain Ω. Therefore, it is possible to state the following relation

$$u_k = u_k^c + \eta \cdot \mathrm{sign}\left(J_k^a(u_k^c - \delta) - J_k^a(u_k^c + \delta)\right), \tag{30}$$

where $\eta > 0$ is a designer parameter, which represents the gain of the probing component. The expression in brackets can be expressed as

$$J_k^a(u_k^c - \delta) - J_k^a(u_k^c + \delta) = \delta(2\mu_k u_k^c + \nu_k). \tag{31}$$

The control law derivation is finalized using (31) in (30) as

$$u_k = u_k^c + \eta \cdot \mathrm{sign}\left(2\mu_k u_k^c + \nu_k\right). \tag{32}$$

The equations (27) and (32) represent the final BD adaptive control law, where the first term is the cautious control component and second term denotes probing control component. The probing part is a function of variable μ_k and ν_k which represent the uncertainty in the system functions $f(\boldsymbol{x}_k^a)$ and $g(\boldsymbol{x}_k^a)$. As a consequence, a higher value of the probing control component is generated in the case of high model uncertainty and vice versa.

4.2. BD-RGP control algorithm

The following few basic steps summarize the operation of the proposed BD-RGP adaptive controller

RGP-based Bicriterial Dual Control Algorithm	
Step 0:	Initialize GP model hyper-parameters $\boldsymbol{\theta} = \boldsymbol{\theta}_0$, basis vector locations $\tilde{\boldsymbol{X}}$ and set $\boldsymbol{\mu}_0^h \sim \text{Uniform}(0,1)$, $\boldsymbol{C}_0^h = k(\tilde{\boldsymbol{X}}, \tilde{\boldsymbol{X}}\,;\, \boldsymbol{\theta}_0)$; set system initial condition and controller parameters, set $k = 1$
Step 1:	Measure the system output y_k.
Step 2:	Update the RGP model $\boldsymbol{\mu}_k^h$, $\boldsymbol{C}_k^h$ using (14a) – (14c).
Step 3:	Optimise hyper-parameters as $\boldsymbol{\theta} = \arg\max_{\boldsymbol{\theta}} \log\left(p\left(\boldsymbol{\mu}_k^h \,\vert\, \tilde{\boldsymbol{X}}, \boldsymbol{\theta}\right)\right)$.
Step 4:	Generate control action u_k from relations (27) and (32).
Step 5:	Predict system output μ_{k+1}^p and predictive variance C_{k+1}^p by (15a)–(15c).
Step 6:	$k = k + 1$, go to Step 1.

5. Numerical Example

We tested the proposed bi-criterial dual controller on the following example of non-linear stochastic discrete-time system

$$y_{k+1} = \frac{1.5 y_k}{1 + y_k^2} + (2 + \cos(y_k))u_k + e_{k+1}, \tag{33}$$

where e_{k+1} is a white Gaussian noise with variance $\sigma_e^2 = 0.0025$. Note, that the vector of regressors $\boldsymbol{x}_k = [y_k, u_k]$ is now two-dimensional. The bi-criterial dual controller with the full Gaussian process model (BD-FGP) as well as the recursive GP model (BD-RGP) used the same set of initial hyperparameter values $\boldsymbol{\theta} = [0, 0, 0, 0]$. The set of basis vectors for the BD-RGP was initialized as a rectangular grid of 7×5 points in the interval $[-2, 2] \times [-1, 1]$.

The main motivation for designing the BD-RGP algorithm was to alleviate the prohibitive computational demands of the BD-FGP. The Figure 1 is a result of averaging execution time measured in each control loop iteration of both algorithms over 100 Monte Carlo runs. Clearly, the demands of the BD-RGP algorithm remain constant, thus making it superior alternative to the BD-FGP in this regard.

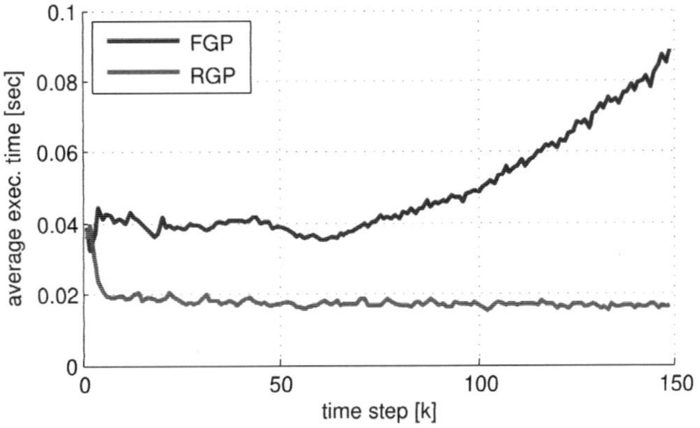

Figure 1: Average execution time per time step of the BD-FGP and BD-RGP control algorithms. The computational requirements of the BD-RGP controller remain constant, while the demands BD-FGP controller continue to grow indefinitely with every time step.

Table 1: Control quality of the bicriterial dual controller based on full (BD-FGP) and recursive Gaussian process (BD-RGP) model. Results are averaged over 100 Monte Carlo simulations. Variance estimates of the means of the criteria were obtained by bootstrap method.

Controller	$\hat{J}_1$	$\mathrm{var}(\hat{J}_1)$	$\hat{J}_{15}$	$\mathrm{var}(\hat{J}_{15})$
BD-FGP	1.08e-01	2.02e-04	1.12e-02	3.79e-05
BD-RGP	1.07e+01	9.66e+00	1.42e-02	1.01e-04

The other experiment was focused on the assessment of control quality, which was measured by

$$J_m = \frac{1}{N} \sum_{k=m}^{N} (y_{k+1} - r_{k+1})^2, \tag{34}$$

where $m \geq 1$ is the offset. We compared the control quality of the BD-FGP controller, which uses the exact non-recursive GP model, with the BD-RGP controller, which uses the approximate recursive GP model. Ideally, we would like the quality of the approximate BD-RGP to remain close enough to the exact BD-FGP. Each controller was simulated for $N = 60$ time steps and the reference signal r_k was chosen such that, a one half is a square wave and the other half is a sine wave.

Table 1 compares the control quality of the BD-FGP and the BD-RGP controllers for offsets $m = 1$, which evaluates the criterion using the whole trajectory, and $m = 15$, which ignores the first few steps to assess the behaviour after the GP model adaptation. The quality of the BD-RGP is two orders of magnitude worse than that of the BD-FGP, when the whole trajectory is taken into account. This is caused by the longer adaptation of the RGP model. However, the values of the offset criteria indicate that this adaptation quickly disappears (after about 15 time steps) and the BD-RGP control quality is then close to the BD-FGP. The reason for this behaviour is the fact, that the RGP model is an approximation of the exact FGP model, and as such takes longer to adapt (slower convergence). This intuition is corroborated by our earlier results published in [34].

Figure 2 compares the two dual controllers in terms of the system output response to the reference signal. The output response for the BD-RGP is more erratic at the start, which explains the results in the Table 1, but then stays close to the BD-FGP output response. Figure 3 shows the evolution of the FGP and the RGP model predictive uncertainty. The RGP model provides more conservative predictive variances across the whole trajectory, which is a desirable behaviour considering the RGP is an approximate model to the FGP.

6. Conclusions

In this work, we proposed the use of the RGP regression model for reducing computational demands of the BD controller based on full GP model, when applied for control of non-linear discrete time-invariant stochastic systems with functional uncertainties. The GP regression models are applied for the approximation of unknown functions in the system description. We utilize an approximation to the marginal likelihood, which makes it possible to keep the computational demands fixed when learning the GP model hyper-parameters. We derived the bi-criterial dual controller with RGP model, whose computational demands do not increase with time. After the model adaptation, the control quality of the proposed BD-RGP controller stays close to the BD-FGP controller.

Acknowledgments

This work was supported by the Czech Science Foundation, project no. GA 15-12068S.

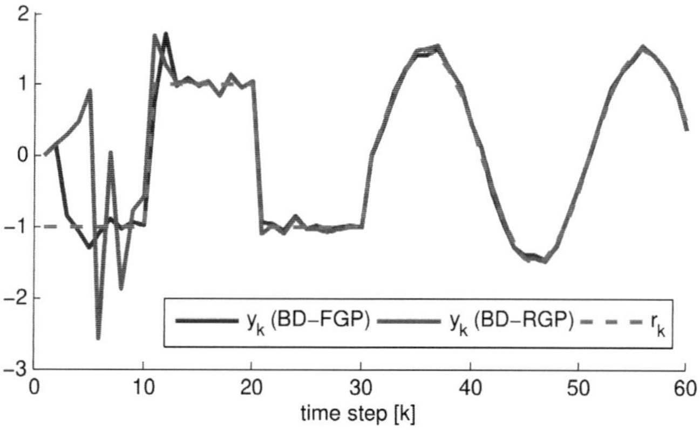

Figure 2: Comparison of the typical system output response y_{k+1} to a reference signal r_k for the BD-FGP and BD-RGP controllers. The controller using the approximate recursive GP model (BD-RGP) exhibits slower adaptation than the BD-FGP.

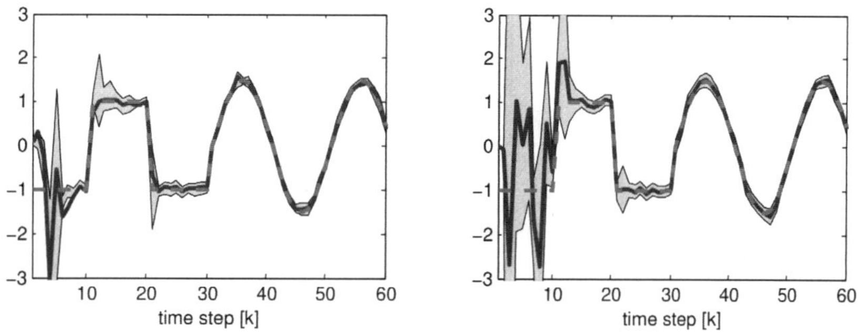

Figure 3: Comparison of model variances (gray band); the RGP model predictive variance (right) is more conservative than that of the FGP model (left). Since RGP model is an approximation of the FGP, this is a desirable behaviour.

References

[1] Fabri S G and Kadirkamanathan V 2001 *Functional Adaptive Control: An Intelligent Systems Approach* (Springer-Verlag)

[2] Herzallah R and Lowe D 2002 A novel approach to modelling and exploiting uncertainty in stochastic control systems *Artificial Neural Networks - ICANN 2002* (Springer Berlin Heidelberg) pp 801–806

[3] Sbarbaro D, Murray-Smith R and Valdes A 2004 *Proceedings of International Symposium on Neural Networks - ISNN 2004* 52–58

[4] Sarangapani J 2006 *Neural Network Control of Nonlinear Disrete-Time Systems* (Taylor & Francis)

[5] Fel'dbaum A A 1965 *Optimal Control Systems* (Academic Press, New York)

[6] Bar-Shalom Y and Tse E T 1976 *Control and dynamic systems* **12** 99–172

[7] Fabri S G and Kadirkamanathan V 1998 *Automatica* **34** 245–253

[8] Šimandl M, Král L and Hering P 2005 Neural network based bicriterial dual control of nonlinear systems *Proceedings of the* 16th *IFAC World Congress* (Prague: IFAC)

[9] Bugeja M K, Fabri S G and Camilleri L 2009 *IEEE Transactions on Systems, Man and Cybernetics - Part B* **39** 129–141

[10] Král L and Šimandl M 2011 *Int. J. Adapt. Control Signal Process.* **25** 949–964

[11] Fabri S G and Bugeja M K 2013 Unscented transform-based dual adaptive control of nonlinear MIMO systems *Proceedings of* 13th *European Control Conference* pp 392–397 Zürich

[12] Ko J and Fox D 2008 GP-BayesFilters: Bayesian Filtering Using Gaussian Process Prediction and Observation Models *Intelligent Robots and Systems, 2008. IROS 2008. IEEE/RSJ International Conference on* vol 27

(Springer) pp 3471–3476 ISBN 9781424420582

[13] Ko J, Klein D J, Fox D and Haehnel D 2007 GP-UKF: Unscented Kalman Filters with Gaussian Process Prediction and Observation Models *Intelligent Robots and Systems, 2007. IROS 2007. IEEE/RSJ International Conference on* vol 54 (IEEE) pp 1901–1907 ISBN 1424409128

[14] Deisenroth M P, Turner R D, Huber M F, Hanebeck U D and Rasmussen C E 2012 *IEEE Trans. on Automatic Control* **57** 1865–1871

[15] Särkkä S, Hartikainen J, Svensson L and Sandblom F 2014 Gaussian Process Quadratures in Nonlinear Sigma-Point Filtering and Smoothing *Information Fusion (FUSION), 2014 17th International Conference on* pp 1–8

[16] Kocijan J, Murray-Smith R, Rasmussen C and Girard A 2004 Gaussian process model based predictive control *Proc. of the American Control Conf.* vol 3 pp 2214–2219

[17] Kocijan J, Girard A, Banko B and Murray-Smith R 2005 *Mathematical and Computer Modelling of Dynamic Systems* **11** 411–424

[18] Girard A, Rasmussen C E, Quiñonero Candela J and Murray-Smith R 2003 Gaussian Process Priors With Uncertain Inputs Application to Multiple-Step Ahead Time Series Forecasting *Advances in Neural Information Processing Systems 15* ed Becker S, Thrun S and Obermayer K (MIT Press) pp 545–552

[19] Deisenroth M and Rasmussen C E 2011 Pilco: A model-based and data-efficient approach to policy search *Proc. of the 28th Int. Conf. on Machine Learning (ICML-11)* pp 465–472

[20] Park S, Mustafa S K and Shimada K 2013 Learning Based Robot Control with Sequential Gaussian Process *Robotic Intelligence In Informationally Structured Space (RiiSS), 2013 IEEE Workshop on* (IEEE) pp 120–127 ISBN 978-1-4673-5877-4

[21] Rasmussen C E and Ghahramani Z 2003 Bayesian Monte Carlo *Advances in Neural Information Processing Systems 15* 1 ed Becker S, Thrun S and Obermayer K (MIT Press) pp 505—-512

[22] Csató L and Opper M 2002 *Neural Computation* **14** 641–668

[23] Ranganathan A, Yang M H and Ho J 2011 *IEEE Transactions on Image Processing* **20** 391–404

[24] Särkkä S, Solin A and Hartikainen J 2013 *IEEE Signal Processing Magazine* **30** 51–61 ISSN 1053-5888

[25] Huber M 2013 Recursive gaussian process regression *Acoustics, Speech and Signal Processing (ICASSP), 2013 IEEE Int. Conf. on* pp 3362–3366 ISSN 1520-6149

[26] Chowdhary G, Kingravi H A, How J P and Vela P A 2014 *Neural Networks and Learning Systems, IEEE Transactions on* **PP** 1–1 ISSN 2162-237X

[27] Král L, Prüher J and Šimandl M 2014 Gaussian process based dual adaptive control of nonlinear stochastic systems *Proceedings of the 22nd Mediterranean Conference on Control and Automation* (Palermo)

[28] Prüher J and Šimandl M 2014 Gaussian process based recursive system identification *Proceedings of the* 11th *European Workshop on Advanced Control and Diagnosis* (Berlin)

[29] Chen F C and Khalil H K 1995 *IEEE Transaction on Automatic Control* **40** 791–801

[30] Filatov N M and Unbehauen H 2004 *Adaptive Dual Control* (Springer-Verlag)

[31] Rasmussen C E and Williams C K 2006 *Gaussian Processes for Machine Learning* (The MIT Press) ISBN 978-262-18253-9

[32] Sbarbaro D and Murray-Smith R 2003 Self-tuning Control of Non-linear Systems Using Gaussian Process Prior Models *Switching and Learning in Feedback Systems* ed Murray-Smith R and Shorten R (Berlin Heidelberg: Springer) pp 140–157

[33] MacKay D J C 2003 *Information Theory, Inference, and Learning Algorithms* (Cambridge University Press) ISBN 978-0521642989

[34] Prüher J and Šimandl M 2014 *Journal of Physics: Conference Series* **570** 1–9

Journal of Physics: Conference Series **659** (2015) 012007 doi:10.1088/1742-6596/659/1/012007

Design of FTC structures with PI virtual actuators

Anna Filasová, Dušan Krokavec and Vladimír Serbák

Department of Cybernetics and Artificial Intelligence, Faculty of Electrical Engineering and Informatics, Technical University of Košice, Letná 9, 042 00 Košice, Slovakia

E-mail: dusan.krokavec@tuke.sk; anna.filasova@tuke.sk; vlado.serbak@gmail.com

Abstract. The paper presents new conditions, adequate in design of proportional-integral virtual actuators and utilizable in fault-tolerant control structures which are stabilizable by dynamic output controllers. Taking into account disturbance conditions and changes of variables after the virtual actuator activation, the design conditions are outlined in terms of the linear matrix inequalities within the bounded real lemma forms. Using tuning parameters in design, and with suitable choice of the order of dynamic output controller, the approach provides a way to obtain acceptable dynamics of the closed loop system after activation of the virtual actuator.

1. Introduction

To increase the reliability of systems, fault-tolerant control structures (FTC) usually fix a system with faults so that it can continue its mission with certain limitations of functionality and quality. Considering this, the different approaches were studied in FTC design (see, e.g., [1], [15] and the references therein). The standard way of control reconfiguration discards the nominal controller from the control loop and replace it with a new one so that its parameters are re-tuned in occurred fault conditions and, in dependency on the remaining set of sensors and actuators, to recover in a certain extent the performance given on the fault-free control system [16], [25], [27].

By contrast, instead of adapting the controller to the faulty plant, the virtual approach keeps the nominal controller in the reconfigured closed-loop system and virtually adapt the faulty plant to the nominal controller in such a way that the activated virtual reconfiguration block, together with the faulty plant, imitates the fault-free plant. Since in healthy conditions the virtual blocks are not active, and control is realized by the nominal controller, the design of the virtual reconfiguration blocks may seem to be independent of the controller. Designated to sensor faults the reconfiguration block is termed virtual sensor (VS), while in the case of actuator faults is named virtual actuator (VA). Especially, an FTC strategy based on virtual actuator approach for linear piecewise affine systems with actuator faults is presented in [21], for non-linear systems that can be approximated by linear parameter-varying (LPV) models in discrete-time or continuous-time description, this policy is proposed in [17], [22], [24], respectively, and applying to continuous-time Lipschitz nonlinear systems, this practise is introduced in [6].

Until the first ideas of control reconfiguration by using of VA [2], [13] were summarized in the books [19], [23], several aspects have been used in VA design. Introducing the generalized VA it was shown [14] that reconfiguration after an actuator fault can be related to disturbance decoupling. Then, subsequently, H_∞-based VA synthesis was presented and the dual principle was conveyed in [20]. In general, these conditions for VA design are formulated in terms of a finite set of linear matrix inequalities (LMI) for proportional VAs.

The vast ma ority of applications is realized in con unction with static output controllers, but this structure causes that the response of fault isolation has to be fast and the peaks of the system variables, immediately after the activation of a VA, must be restrained [12]. Considering characteristic changes of variables in FTC after VA activation, the synthesis of dynamic virtual actuators (D C) was stated in [8], [9], [10], [11]. The approach in the paper respects specifically the generalized disturbance transfer matrix in proportional-integral (PI) VA design.

The paper is organized as follows. In Sec. 2, the H_∞ approach is presented with results on bounded real lemmas (BRL) for D C design. Formulating the separation principle in Sec. 3, the BRL based design methods are outlined for PI VA. Finally, the illustrating example is given in Sec. 4 and some concluding remarks are stated in Sec. 5.

Throughout the paper, the following notations are used: $\boldsymbol{x}^T$, $\boldsymbol{X}^T$ denotes the transpose of the vector $\boldsymbol{x}$ and the matrix $\boldsymbol{X}$, respectively, $rank(\cdot)$ remits the rank of a matrix, for a square matrix $\boldsymbol{X} < 0$ means that $\boldsymbol{X}$ is a symmetric negative definite matrix, the symbol $\boldsymbol{I}_n$ indicates the n-th order unit matrix, $I\!R$ denotes the set of real numbers, $I\!R^n$, $I\!R^{n\times r}$ refer to the set of all n-dimensional real vectors and $n \times r$ real matrices, respectively.

2. Dynamic Output Controllers

In the paper, there are taken into account systems described in fault-free conditions as

$$\boldsymbol{q}(t) = \boldsymbol{A}\boldsymbol{q}(t) + \boldsymbol{B}\boldsymbol{u}_c(t) + \boldsymbol{V}\boldsymbol{v}(t)\,, \tag{1}$$

$$\boldsymbol{y}(t) = \boldsymbol{C}\boldsymbol{q}(t)\,, \tag{2}$$

where $\boldsymbol{q}(t) \in I\!R^n$ stands up for the system state, $\boldsymbol{u}_c(t) \in I\!R^r$ denotes the control input, $\boldsymbol{y}(t) \in I\!R^m$ is the measurable output, $\boldsymbol{v}(t) \in I\!R^{r_v}$ is the vector of unknown disturbance, the matrices $\boldsymbol{A} \in I\!R^{n\times n}$, $\boldsymbol{B} \in I\!R^{n\times r}$, $\boldsymbol{C} \in I\!R^{m\times n}$ and $\boldsymbol{V} \in I\!R^{n\times r_v}$ are finite valued.

It is assumed that the system is controlled by bi-proper D C of the form

$$\boldsymbol{p}(t) = \boldsymbol{J}\boldsymbol{p}(t) + \boldsymbol{L}\boldsymbol{y}(t)\,, \tag{3}$$

$$\boldsymbol{u}_c(t) = \boldsymbol{M}\boldsymbol{p}(t) + \boldsymbol{N}\boldsymbol{y}(t) \tag{4}$$

and of an order p, where it can be accepted $1 \le p < n$ (reduced order), $p = n$ (full order) and $n < p \le p_m$ (upgraded order), while $\boldsymbol{p}(t) \in I\!R^p$ is the vector of the controller state variables and $\boldsymbol{J} \in I\!R^{p\times p}$, $\boldsymbol{L} \in I\!R^{p\times m}$, $\boldsymbol{M} \in I\!R^{r\times p}$, $\boldsymbol{N} \in I\!R^{r\times m}$.

To analyze the stability of the closed-loop system structure it can be formulated

$$\begin{bmatrix} \boldsymbol{q}(t) \\ \boldsymbol{p}(t) \end{bmatrix} = \begin{bmatrix} \boldsymbol{A}+\boldsymbol{BNC} & \boldsymbol{BM} \\ \boldsymbol{LC} & \boldsymbol{J} \end{bmatrix} \begin{bmatrix} \boldsymbol{q}(t) \\ \boldsymbol{p}(t) \end{bmatrix} + \begin{bmatrix} \boldsymbol{V} \\ \boldsymbol{0} \end{bmatrix} \boldsymbol{v}(t)\,, \tag{5}$$

$$\boldsymbol{y}(t) = \begin{bmatrix} \boldsymbol{0} & \boldsymbol{I}_m \end{bmatrix} \begin{bmatrix} \boldsymbol{0} & \boldsymbol{I}_p \\ \boldsymbol{C} & \boldsymbol{0} \end{bmatrix} \begin{bmatrix} \boldsymbol{q}(t) \\ \boldsymbol{p}(t) \end{bmatrix}. \tag{6}$$

Introducing the notations

$$\boldsymbol{q}^{\bullet T}(t) = \begin{bmatrix} \boldsymbol{q}^T(t) & \boldsymbol{p}^T(t) \end{bmatrix}, \ \boldsymbol{V}^{\bullet T} = \begin{bmatrix} \boldsymbol{V}^T & \boldsymbol{0} \end{bmatrix}, \ \boldsymbol{I}^{\bullet} = \begin{bmatrix} \boldsymbol{0} & \boldsymbol{I}_m \end{bmatrix}, \tag{7}$$

$$\boldsymbol{A}^{\bullet} = \begin{bmatrix} \boldsymbol{A} & \boldsymbol{0} \\ \boldsymbol{0} & \boldsymbol{0} \end{bmatrix}, \ \boldsymbol{B}^{\bullet} = \begin{bmatrix} \boldsymbol{0} & \boldsymbol{B} \\ \boldsymbol{I}_p & \boldsymbol{0} \end{bmatrix}, \ \boldsymbol{C}^{\bullet} = \begin{bmatrix} \boldsymbol{0} & \boldsymbol{I}_p \\ \boldsymbol{C} & \boldsymbol{0} \end{bmatrix}, \boldsymbol{K}^{\bullet} = \begin{bmatrix} \boldsymbol{J} & \boldsymbol{L} \\ \boldsymbol{M} & \boldsymbol{N} \end{bmatrix}, \tag{8}$$

where $\boldsymbol{A}^{\bullet} \in I\!R^{(n+p)\times(n+p)}$, $\boldsymbol{B}^{\bullet} \in I\!R^{(n+p)\times(p+r)}$, $\boldsymbol{C}^{\bullet} \in I\!R^{(p+m)\times(n+p)}$, $\boldsymbol{V}^{\bullet} \in I\!R^{(n+p)\times r_v}$, $\boldsymbol{I}^{\bullet} \in I\!R^{m\times(p+m)}$, it yields for stabilizable $(\boldsymbol{A}^{\bullet}, \boldsymbol{B}^{\bullet})$ and detectable $(\boldsymbol{A}^{\bullet}, \boldsymbol{C}^{\bullet})$ [3], [11],

$$\boldsymbol{q}^{\bullet}(t) = \boldsymbol{A}^{\bullet}_c\boldsymbol{q}^{\bullet}(t) + \boldsymbol{V}^{\bullet}\boldsymbol{v}(t)\,, \quad \boldsymbol{A}^{\bullet}_c = \boldsymbol{A}^{\bullet} + \boldsymbol{B}^{\bullet}\boldsymbol{K}^{\bullet}\boldsymbol{C}^{\bullet}. \tag{9}$$

$$\boldsymbol{y}^{\bullet}(t) = \boldsymbol{I}^{\bullet}\boldsymbol{C}^{\bullet}\boldsymbol{q}^{\bullet}(t) \tag{10}$$

Proposition 1 *[8] (BRL) The closed-loop system, consisting of the plant (1), (2) and DOC (3), (4), is stable with the quadratic performance $\gamma^\bullet$ if there exist a symmetric positive definite matrix $\boldsymbol{Q}^\bullet \in I\!R^{(n+p)\times(n+p)}$, a regular matrix $\boldsymbol{H}^\bullet \in I\!R^{(p+m)\times(p+m)}$, a matrix $\boldsymbol{Y}^\bullet \in I\!R^{(p+r)\times(p+m)}$ and a positive scalar $\gamma^\bullet \in I\!R$ such that*

$$\boldsymbol{Q}^\bullet = \boldsymbol{Q}^{\bullet T} > 0, \quad \gamma^\bullet > 0, \tag{11}$$

$$\begin{bmatrix} \boldsymbol{A}^\bullet\boldsymbol{Q}^\bullet + \boldsymbol{Q}^\bullet\boldsymbol{A}^{\bullet T} + \boldsymbol{B}^\bullet\boldsymbol{Y}^\bullet\boldsymbol{C}^\bullet + \boldsymbol{C}^{\bullet T}\boldsymbol{Y}^{\bullet T}\boldsymbol{B}^{\bullet T} & * & * \\ \boldsymbol{V}^{\bullet T} & -\gamma^\bullet\boldsymbol{I}_{r_v} & * \\ \boldsymbol{I}^\bullet\boldsymbol{C}^\bullet\boldsymbol{Q}^\bullet & \boldsymbol{0} & -\boldsymbol{I}_m \end{bmatrix} < 0, \tag{12}$$

$$\boldsymbol{C}^\bullet\boldsymbol{Q}^\bullet = \boldsymbol{H}^\bullet\boldsymbol{C}^\bullet, \tag{13}$$

where the generalized system matrices are defined in (7), (8).

When the above conditions hold

$$\boldsymbol{K}^\bullet = \boldsymbol{Y}^\bullet(\boldsymbol{H}^\bullet)^{-1}. \tag{14}$$

Here and hereafter $*$ denotes the symmetric item in a symmetric matrix.

In order to ad ust fault detection and isolation time to the dynamics of the closed-loop system, the selection of the order p of the D C be provided with a free tuning parameter in control design. ne serviceable method is based on incorporation a slack matrix into LMI design conditions. This augmentation is proposed in the following theorem.

Proposition 2 *[11] (enhanced BRL) The closed-loop system, consisting of the plant (1), (2) and the DOC (3), (4), is stable with the quadratic performance $\gamma^\bullet$ if for the given positive scalar $\delta^\bullet \in I\!R$ there exist symmetric positive definite matrices $\boldsymbol{R}^\bullet$, $\boldsymbol{U}^\bullet \in I\!R^{(n+p)\times(n+p)}$, a regular matrix $\boldsymbol{H}^\bullet \in I\!R^{(p+m)\times(p+m)}$, a matrix $\boldsymbol{Y}^\bullet \in I\!R^{(p+r)\times(p+m)}$ and a positive scalar $\gamma^\bullet \in I\!R$ such that*

$$\boldsymbol{R}^\bullet = \boldsymbol{R}^{\bullet T} > 0, \quad \boldsymbol{U}^\bullet = \boldsymbol{U}^{\bullet T} > 0, \quad \gamma^\bullet > 0, \tag{15}$$

$$\begin{bmatrix} \boldsymbol{A}^\bullet\boldsymbol{R}^\bullet + \boldsymbol{R}^\bullet\boldsymbol{A}^{\bullet T} + \boldsymbol{B}^\bullet\boldsymbol{Y}^\bullet\boldsymbol{C}^\bullet + \boldsymbol{C}^{\bullet T}\boldsymbol{Y}^{\bullet T}\boldsymbol{B}^{\bullet T} & * & * & * \\ \boldsymbol{U}^\bullet - \boldsymbol{R}^\bullet + \delta\boldsymbol{A}^\bullet\boldsymbol{R}^\bullet + \delta\boldsymbol{B}^\bullet\boldsymbol{Y}^\bullet\boldsymbol{C}^\bullet & -2\delta^\bullet\boldsymbol{R}^\bullet & * & * \\ \boldsymbol{V}^{\bullet T} & \delta^\bullet\boldsymbol{V}^{\bullet T} & -\gamma^\bullet\boldsymbol{I}_{r_v} & * \\ \boldsymbol{I}^\bullet\boldsymbol{C}^\bullet\boldsymbol{R}^\bullet & \boldsymbol{0} & \boldsymbol{0} & -\boldsymbol{I}_m \end{bmatrix} < 0, \tag{16}$$

$$\boldsymbol{C}^\bullet\boldsymbol{R}^\bullet = \boldsymbol{H}^\bullet\boldsymbol{C}^\bullet, \tag{17}$$

where the generalized system matrices are defined in (7), (8) and the positive $\delta^\bullet \in I\!R$ is the tuning parameter.

When the above conditions hold then 14 yields.

Consider the case $r = m$ (square plants), where with each output signal is associated a reference signal. Such regime is called the forced regime and for D C is defined as follows:

Definition 1 *The forced regime for (1), (2) with DOC (3), (4) is given by the control policy*

$$\boldsymbol{p}(t) = \boldsymbol{J}\boldsymbol{p}(t) + \boldsymbol{L}\boldsymbol{y}(t), \tag{18}$$

$$\boldsymbol{u}(t) = \boldsymbol{M}\boldsymbol{p}(t) + \boldsymbol{N}\boldsymbol{y}(t) + \boldsymbol{W}\boldsymbol{w}(t), \tag{19}$$

where $\boldsymbol{w}(t) \in I\!R^m$ is desired output signal vector, and $\boldsymbol{W} \in I\!R^{m\times m}$ is the signal gain matrix.

Theorem 1 *[10] If a square system (1), (2) is stabilizable by the control policy (18), (19), and*

$$\texttt{rank}\begin{bmatrix} \boldsymbol{A} & \boldsymbol{B} \\ \boldsymbol{C} & \boldsymbol{0} \end{bmatrix} = n + m, \tag{20}$$

then the matrix $\boldsymbol{W}$ takes the form

$$\boldsymbol{W} = -\left(\boldsymbol{C}(\boldsymbol{A} - \boldsymbol{B}\boldsymbol{M}\boldsymbol{J}^{-1}\boldsymbol{L}\boldsymbol{C} + \boldsymbol{B}\boldsymbol{N}\boldsymbol{C})^{-1}\boldsymbol{B}\right)^{-1}. \tag{21}$$

3. PI Virtual Actuators

The state-space description of the system with a single actuator fault is considered as follows

$$\boldsymbol{q}_{fa}(t) = \boldsymbol{A}\boldsymbol{q}_{fa}(t) + \boldsymbol{B}_f\boldsymbol{u}_{fa}(t) + \boldsymbol{V}\boldsymbol{v}(t)\,, \tag{22}$$

$$\boldsymbol{y}_{fa}(t) = \boldsymbol{C}\boldsymbol{q}_{fa}(t)\,, \tag{23}$$

where $\boldsymbol{q}_{fa}(t) \in I\!R^n$, $\boldsymbol{u}_{fa}(t) \in I\!R^r$, $\boldsymbol{y}_{fa}(t) \in I\!R^m$ denote the faulty system state variables vector, the vector of the acting control input variables and the vector of faulty output variables, respectively, and the matrix $\boldsymbol{B}_f \in I\!R^{n\times r}$ is finite valued, while $rank(\boldsymbol{B}_f) < rank(\boldsymbol{B})$. Moreover, it is supposed that $(\boldsymbol{A}, \boldsymbol{B}_f)$ is controllable and $\boldsymbol{u}_{fa}(t)$ is available for reconfiguration.

Analogously, using the same system variable notations, the state-space description of D C, acting on the system with a single actuator fault, but without DVA, is of the form

$$\boldsymbol{p}_{fa}(t) = \boldsymbol{J}\boldsymbol{p}_{fa}(t) + \boldsymbol{L}\boldsymbol{y}_{fa}(t)\,, \tag{24}$$

$$\boldsymbol{u}_c(t) = \boldsymbol{M}\boldsymbol{p}_{fa}(t) + \boldsymbol{N}\boldsymbol{y}_{fa}(t)\,, \tag{25}$$

where $\boldsymbol{p}_{fa}(t) \in I\!R^p$ denotes the controller state variables vector in the faulty system control.

To obtain the DVA state-space description, the following theorem is proven at first.

Theorem 2 *(separation principle) The dynamic virtual actuator for the system with a single actuator fault (22), (23) takes the form*

$$\boldsymbol{e}_{fa}(t) = (\boldsymbol{A} - \boldsymbol{B}_f\boldsymbol{G}_P)\boldsymbol{e}_{fa}(t) - \boldsymbol{B}_f\boldsymbol{G}_I\boldsymbol{e}_{pfa}(t) - \boldsymbol{B}\boldsymbol{u}_c(t)\,, \tag{26}$$

$$\boldsymbol{z}_{fa}(t) = \boldsymbol{C}\boldsymbol{e}_{fa}(t)\,. \tag{27}$$

$$\boldsymbol{u}_{fa}(t) = -\boldsymbol{G}_P\boldsymbol{e}_{fa}(t) - \boldsymbol{G}_I\boldsymbol{z}_{fa}(t)\,, \tag{28}$$

where

$$\boldsymbol{e}_{fa}(t) = \boldsymbol{q}_{fa}(t) - \boldsymbol{q}(t)\,, \quad \boldsymbol{e}_{pfa}(t) = \boldsymbol{p}_{fa}(t) - \boldsymbol{p}(t)\,, \tag{29}$$

$\boldsymbol{e}_{pfa}(t) \in I\!R^p$, $\boldsymbol{e}_{fa}(t) \in I\!R^n$ and $\boldsymbol{G}_P \in I\!R^{r\times n}$, $\boldsymbol{G}_I \in I\!R^{r\times m}$ are real matrices.

Proof: Using (1), (2), (3) and (22), (23), (24) and proposing integral part of VA as follows

$$\boldsymbol{z}_{fa}(t) = \int_0^t \boldsymbol{C}(\boldsymbol{q}_{fa}(r) - \boldsymbol{q}(r))\mathrm{d}r, \quad \boldsymbol{z}_{fa}(t) = \boldsymbol{C}(\boldsymbol{q}_{fa}(t) - \boldsymbol{q}(t))\,, \tag{30}$$

then the expression of the common system variable model is

$$\begin{bmatrix} \boldsymbol{q}_{fa}(t) \\ \boldsymbol{q}(t) \\ \boldsymbol{p}_{fa}(t) \\ \boldsymbol{p}(t) \\ \boldsymbol{z}_{fa}(t) \end{bmatrix} = \begin{bmatrix} \boldsymbol{A} & \boldsymbol{0} & \boldsymbol{0} & \boldsymbol{0} & \boldsymbol{0} \\ \boldsymbol{0} & \boldsymbol{A} & \boldsymbol{0} & \boldsymbol{0} & \boldsymbol{0} \\ \boldsymbol{L}\boldsymbol{C} & \boldsymbol{0} & \boldsymbol{J} & \boldsymbol{0} & \boldsymbol{0} \\ \boldsymbol{0} & \boldsymbol{L}\boldsymbol{C} & \boldsymbol{0} & \boldsymbol{J} & \boldsymbol{0} \\ \boldsymbol{C} & -\boldsymbol{C} & \boldsymbol{0} & \boldsymbol{0} & \boldsymbol{0} \end{bmatrix} \begin{bmatrix} \boldsymbol{q}_{fa}(t) \\ \boldsymbol{q}(t) \\ \boldsymbol{p}_{fa}(t) \\ \boldsymbol{p}(t) \\ \boldsymbol{z}_{fa}(t) \end{bmatrix} + \begin{bmatrix} \boldsymbol{B}_f & \boldsymbol{0} \\ \boldsymbol{0} & \boldsymbol{B} \\ \boldsymbol{0} & \boldsymbol{0} \\ \boldsymbol{0} & \boldsymbol{0} \\ \boldsymbol{0} & \boldsymbol{0} \end{bmatrix} \begin{bmatrix} \boldsymbol{u}_{fa}(t) \\ \boldsymbol{u}_c(t) \end{bmatrix} + \begin{bmatrix} \boldsymbol{V} \\ \boldsymbol{V} \\ \boldsymbol{0} \\ \boldsymbol{0} \\ \boldsymbol{0} \end{bmatrix} \boldsymbol{v}(t)\,. \tag{31}$$

Using the transform matrix $\boldsymbol{T}$ of the form

$$\boldsymbol{T} = \boldsymbol{T}^{-1} = \begin{bmatrix} \boldsymbol{I}_n & \boldsymbol{0} & \boldsymbol{0} & \boldsymbol{0} & \boldsymbol{0} \\ \boldsymbol{I}_n & -\boldsymbol{I}_n & \boldsymbol{0} & \boldsymbol{0} & \boldsymbol{0} \\ \boldsymbol{0} & \boldsymbol{0} & \boldsymbol{I}_p & \boldsymbol{0} & \boldsymbol{0} \\ \boldsymbol{0} & \boldsymbol{0} & \boldsymbol{I}_p & -\boldsymbol{I}_p & \boldsymbol{0} \\ \boldsymbol{0} & \boldsymbol{0} & \boldsymbol{0} & \boldsymbol{0} & \boldsymbol{I}_m \end{bmatrix}, \quad \boldsymbol{T} \begin{bmatrix} \boldsymbol{q}_{fa}(t) \\ \boldsymbol{q}(t) \\ \boldsymbol{p}_{fa}(t) \\ \boldsymbol{p}(t) \\ \boldsymbol{z}_{fa}(t) \end{bmatrix} = \begin{bmatrix} \boldsymbol{q}_{fa}(t) \\ \boldsymbol{e}_{fa}(t) \\ \boldsymbol{p}_{fa}(t) \\ \boldsymbol{e}_{pfa}(t) \\ \boldsymbol{z}_{fa}(t) \end{bmatrix}, \tag{32}$$

where $\boldsymbol{e}_{fa}(t)$, $\boldsymbol{e}_{fa}(t)$ are defined in (29), then (31) can be rewritten as

$$\begin{bmatrix} \boldsymbol{q}_{fa}(t) \\ \boldsymbol{e}_{fa}(t) \\ \boldsymbol{p}_{fa}(t) \\ \boldsymbol{e}_{pfa}(t) \\ \boldsymbol{z}_{fa}(t) \end{bmatrix} = \begin{bmatrix} \boldsymbol{A} & \boldsymbol{0} & \boldsymbol{0} & \boldsymbol{0} & \boldsymbol{0} \\ \boldsymbol{0} & \boldsymbol{A} & \boldsymbol{0} & \boldsymbol{0} & \boldsymbol{0} \\ \boldsymbol{LC} & \boldsymbol{0} & \boldsymbol{J} & \boldsymbol{0} & \boldsymbol{0} \\ \boldsymbol{0} & \boldsymbol{LC} & \boldsymbol{0} & \boldsymbol{J} & \boldsymbol{0} \\ \boldsymbol{0} & \boldsymbol{C} & \boldsymbol{0} & \boldsymbol{0} & \boldsymbol{0} \end{bmatrix} \begin{bmatrix} \boldsymbol{q}_{fa}(t) \\ \boldsymbol{e}_{fa}(t) \\ \boldsymbol{p}_{fa}(t) \\ \boldsymbol{e}_{pfa}(t) \\ \boldsymbol{z}_{fa}(t) \end{bmatrix} + \begin{bmatrix} \boldsymbol{B}_f & \boldsymbol{0} \\ \boldsymbol{B}_f & -\boldsymbol{B} \\ \boldsymbol{0} & \boldsymbol{0} \\ \boldsymbol{0} & \boldsymbol{0} \\ \boldsymbol{0} & \boldsymbol{0} \end{bmatrix} \begin{bmatrix} \boldsymbol{u}_{fa}(t) \\ \boldsymbol{u}_c(t) \end{bmatrix} + \begin{bmatrix} \boldsymbol{V} \\ \boldsymbol{0} \\ \boldsymbol{0} \\ \boldsymbol{0} \\ \boldsymbol{0} \end{bmatrix} \boldsymbol{v}(t)\,. \tag{33}$$

Defining the covering of the faulty control input as (28), the substitution of (28) in (33) leads to

$$\begin{aligned} \begin{bmatrix} \boldsymbol{q}_{fa}(t) \\ \boldsymbol{e}_{fa}(t) \\ \boldsymbol{p}_{fa}(t) \\ \boldsymbol{e}_{pfa}(t) \\ \boldsymbol{z}_{fa}(t) \end{bmatrix} &= \begin{bmatrix} \boldsymbol{A} & -\boldsymbol{B}_f\boldsymbol{G}_P & \boldsymbol{0} & -\boldsymbol{B}_f\boldsymbol{G}_I & \boldsymbol{0} \\ \boldsymbol{0} & \boldsymbol{A}-\boldsymbol{B}_f\boldsymbol{G}_P & \boldsymbol{0} & -\boldsymbol{B}_f\boldsymbol{G}_I & \boldsymbol{0} \\ \boldsymbol{LC} & \boldsymbol{0} & \boldsymbol{J} & \boldsymbol{0} & \boldsymbol{0} \\ \boldsymbol{0} & \boldsymbol{LC} & \boldsymbol{0} & \boldsymbol{J} & \boldsymbol{0} \\ \boldsymbol{0} & \boldsymbol{C} & \boldsymbol{0} & \boldsymbol{0} & \boldsymbol{O} \end{bmatrix} \begin{bmatrix} \boldsymbol{q}_{fa}(t) \\ \boldsymbol{e}_{fa}(t) \\ \boldsymbol{p}_{fa}(t) \\ \boldsymbol{e}_{pfa}(t) \\ \boldsymbol{z}_{fa}(t) \end{bmatrix} \\ &- \begin{bmatrix} \boldsymbol{0} \\ \boldsymbol{B} \\ \boldsymbol{0} \\ \boldsymbol{0} \\ \boldsymbol{0} \end{bmatrix} \boldsymbol{u}_c(t) + \begin{bmatrix} \boldsymbol{V} \\ \boldsymbol{0} \\ \boldsymbol{0} \\ \boldsymbol{0} \\ \boldsymbol{0} \end{bmatrix} \boldsymbol{v}(t)\,. \end{aligned} \tag{34}$$

Thus, the second and the fifth row of the equation (34) imply (26), (27). bviously, in view of the block structure of (34), the separation principle yields. This concludes the proof. ■

Corollary 1 *The equations of the closed-loop faulty system with activated PI VA are as follows*

$$\boldsymbol{q}^{\circ}_{fa}(t) = \boldsymbol{A}^{\circ}_{c}\boldsymbol{q}^{\circ}_{fa}(t) + \boldsymbol{V}^{\circ}_{fa}\boldsymbol{d}^{\circ}_{fa}(t)\,, \quad \boldsymbol{A}^{\circ}_{c} = \boldsymbol{A}^{\circ} - \boldsymbol{B}^{\circ}_{f}\boldsymbol{G}^{\circ}\,, \tag{35}$$

$$\boldsymbol{y}_{fa}(t) = \boldsymbol{C}^{\circ}\boldsymbol{q}^{\circ}_{fa}(t)\,, \tag{36}$$

$$\boldsymbol{q}^{\circ}_{fa}(t) = \begin{bmatrix} \boldsymbol{q}_{fa}(t) \\ \boldsymbol{z}_{fa}(t) \end{bmatrix}, \ \boldsymbol{d}^{\circ}_{fa}(t) = \begin{bmatrix} \boldsymbol{G}_P\boldsymbol{q}(t) \\ \boldsymbol{v}(t) \\ \boldsymbol{q}(t) \end{bmatrix}, \ \boldsymbol{V}^{\circ}_{fa} = \begin{bmatrix} \boldsymbol{B}_f & \boldsymbol{V} & \boldsymbol{0} \\ \boldsymbol{0} & \boldsymbol{0} & -\boldsymbol{C} \end{bmatrix}, \tag{37}$$

$$\boldsymbol{A}^{\circ} = \begin{bmatrix} \boldsymbol{A} & \boldsymbol{0} \\ \boldsymbol{0} & \boldsymbol{0} \end{bmatrix}, \ \boldsymbol{B}^{\circ}_{f} = \begin{bmatrix} \boldsymbol{B}_f & \boldsymbol{B}_f \\ \boldsymbol{0} & \boldsymbol{0} \end{bmatrix}, \ \boldsymbol{G}^{\circ} = \begin{bmatrix} \boldsymbol{G}_P \\ \boldsymbol{G}_I \end{bmatrix}, \ \boldsymbol{C}^{\circ} = \begin{bmatrix} \boldsymbol{C} & \boldsymbol{0} \end{bmatrix}, \tag{38}$$

and $\boldsymbol{q}^{\circ}_{fa}(t) \in I\!R^{(n+m)}$, $\boldsymbol{d}^{\circ}_{fa}(t) \in I\!R^{(r+r_v+n)}$, $\boldsymbol{A}^{\circ} \in I\!R^{(n+m)\times(n+m)}$, $\boldsymbol{B}^{\circ} \in I\!R^{(n+m)\times 2r}$, $\boldsymbol{C}^{\circ} \in I\!R^{m\times(n+m)}$, $\boldsymbol{G}^{\circ} \in I\!R^{2r\times(n+m)}$, $\boldsymbol{V}^{\circ}_{fa} \in I\!R^{(n+m)\times(r+r_v+n)}$. *Obviously, the stability of the closed-loop system in the reconfiguration regime is determined by the system matrix* $\boldsymbol{A}^{\circ}_{c}$.

Proof: Using (29), the first and the fifth row of the equation (34) as well as (23) give

$$\begin{bmatrix} \boldsymbol{q}_{fa}(t) \\ \boldsymbol{z}_{fa}(t) \end{bmatrix} = \begin{bmatrix} \boldsymbol{A}-\boldsymbol{B}_f\boldsymbol{G}_P & -\boldsymbol{B}_f\boldsymbol{G}_I \\ \boldsymbol{C} & \boldsymbol{0} \end{bmatrix} \begin{bmatrix} \boldsymbol{q}_{fa}(t) \\ \boldsymbol{z}_{fa}(t) \end{bmatrix} + \begin{bmatrix} \boldsymbol{B}_f & \boldsymbol{V} & \boldsymbol{0} \\ \boldsymbol{0} & \boldsymbol{0} & -\boldsymbol{C} \end{bmatrix} \begin{bmatrix} \boldsymbol{G}_P\boldsymbol{q}(t) \\ \boldsymbol{v}(t) \\ \boldsymbol{q}(t) \end{bmatrix}, \tag{39}$$

$$\boldsymbol{y}_{fa}(t) = \boldsymbol{C}\boldsymbol{q}_{fa}(t) = \begin{bmatrix} \boldsymbol{C} & \boldsymbol{0} \end{bmatrix} \begin{bmatrix} \boldsymbol{q}_{fa}(t) \\ \boldsymbol{z}_{fa}(t) \end{bmatrix} \tag{40}$$

and using the relation

$$\begin{bmatrix} \boldsymbol{A}-\boldsymbol{B}_f\boldsymbol{G}_P & -\boldsymbol{B}_f\boldsymbol{G}_I \\ \boldsymbol{C} & \boldsymbol{O} \end{bmatrix} = \begin{bmatrix} \boldsymbol{A} & \boldsymbol{0} \\ \boldsymbol{C} & \boldsymbol{0} \end{bmatrix} - \begin{bmatrix} \boldsymbol{B}_f & \boldsymbol{B}_f \\ \boldsymbol{0} & \boldsymbol{0} \end{bmatrix} \begin{bmatrix} \boldsymbol{G}_P \\ \boldsymbol{G}_I \end{bmatrix}, \tag{41}$$

with the notations (37), (38) then (39), (40) imply (35), (36), respectively. This concludes the proof. ■

Corollary 2 *The state-space description of the PI VA (26), (27) with the covering of faulty control input (28) is as follows*

$$\boldsymbol{e}^{\circ}_{fa}(t) = (\boldsymbol{A}^{\circ} - \boldsymbol{B}^{\circ}_{f}\boldsymbol{G}^{\circ})\boldsymbol{e}^{\circ}_{fa}(t) - \boldsymbol{B}^{\circ}\boldsymbol{u}_{c}(t)\,, \tag{42}$$

$$\boldsymbol{u}_{fa}(t) = -\boldsymbol{G}^{\circ}\boldsymbol{e}^{\circ}_{fa}(t)\,, \tag{43}$$

where

$$\boldsymbol{e}^{\circ}_{fa}(t) = \begin{bmatrix} \boldsymbol{e}_{fa}(t) \\ \boldsymbol{z}_{fa}(t) \end{bmatrix}, \ \boldsymbol{B}^{\circ} = \begin{bmatrix} \boldsymbol{B} \\ \boldsymbol{0} \end{bmatrix}. \tag{44}$$

In the autonomous regime the stability of the PI VA is determined by the same system matrix $\boldsymbol{A}^{\circ}_{c}$ *as stability of the closed-loop system in the reconfiguration regime.*

Proof: Writing the equations (26)-(28) in the following form

$$\begin{bmatrix} \boldsymbol{e}_{fa}(t) \\ \boldsymbol{z}_{fa}(t) \end{bmatrix} = \begin{bmatrix} \boldsymbol{A} - \boldsymbol{B}_{f}\boldsymbol{G}_{P} & -\boldsymbol{B}_{f}\boldsymbol{G}_{I} \\ \boldsymbol{C} & \boldsymbol{O} \end{bmatrix} \begin{bmatrix} \boldsymbol{e}_{fa}(t) \\ \boldsymbol{z}_{fa}(t) \end{bmatrix} - \begin{bmatrix} \boldsymbol{B} \\ \boldsymbol{0} \end{bmatrix} \boldsymbol{u}_{c}(t)\,, \tag{45}$$

$$\boldsymbol{u}_{fa}(t) = -\begin{bmatrix} \boldsymbol{G}_{P} & \boldsymbol{G}_{I} \end{bmatrix} \begin{bmatrix} \boldsymbol{e}_{fa}(t) \\ \boldsymbol{z}_{fa}(t) \end{bmatrix}, \tag{46}$$

respectively, and using the notations (38), (44) as well as the relation (41), then (45), (46) imply (42), (43). This concludes the proof. ■

Corollary 3 *The state-space description of DOC masked in inputs by PI VA and acting on the system with a single actuator fault is of the form*

$$\begin{bmatrix} \boldsymbol{p}_{fa}(t) \\ \boldsymbol{u}_{c}(t) \end{bmatrix} = \begin{bmatrix} \boldsymbol{J} & \boldsymbol{L} \\ \boldsymbol{M} & \boldsymbol{N} \end{bmatrix} \begin{bmatrix} \boldsymbol{p}_{fa}(t) \\ \boldsymbol{y}_{fa}(t) \end{bmatrix} - \begin{bmatrix} \boldsymbol{0} & \boldsymbol{0} \\ \boldsymbol{M} & \boldsymbol{N} \end{bmatrix} \begin{bmatrix} \boldsymbol{e}_{pfa}(t) \\ \boldsymbol{C}\boldsymbol{e}_{fa}(t) \end{bmatrix}, \tag{47}$$

where $\boldsymbol{y}_{fa}(t)$ *is the measurable output of the closed-loop faulty system.*

Proof: Using (23) and (29), then (2) can be rewritten as

$$\boldsymbol{y}(t) = \boldsymbol{C}\boldsymbol{q}(t) = \boldsymbol{C}(\boldsymbol{q}_{fa}(t) - (\boldsymbol{q}_{fa}(t) - \boldsymbol{q}(t))) = \boldsymbol{y}_{fa}(t) - \boldsymbol{C}\boldsymbol{e}_{fa}(t)\,. \tag{48}$$

Considering (48) as the input to the nominal D C, by using (29) and (48) then (3), (4) imply

$$\begin{bmatrix} \boldsymbol{p}(t) \\ \boldsymbol{u}_{c}(t) \end{bmatrix} = \begin{bmatrix} \boldsymbol{J} & \boldsymbol{L} \\ \boldsymbol{M} & \boldsymbol{N} \end{bmatrix} \begin{bmatrix} \boldsymbol{p}(t) \\ \boldsymbol{y}(t) \end{bmatrix} = \begin{bmatrix} \boldsymbol{J} & \boldsymbol{L} \\ \boldsymbol{M} & \boldsymbol{N} \end{bmatrix} \begin{bmatrix} \boldsymbol{p}_{fa}(t) - \boldsymbol{e}_{pfa}(t) \\ \boldsymbol{y}_{fa}(t) - \boldsymbol{C}\boldsymbol{e}_{fa}(t) \end{bmatrix}, \tag{49}$$

$$\begin{bmatrix} \boldsymbol{p}(t) \\ \boldsymbol{u}_{c}(t) \end{bmatrix} = \begin{bmatrix} \boldsymbol{J} & \boldsymbol{L} \\ \boldsymbol{M} & \boldsymbol{N} \end{bmatrix} \begin{bmatrix} \boldsymbol{p}_{fa}(t) \\ \boldsymbol{y}_{fa}(t) \end{bmatrix} - \begin{bmatrix} \boldsymbol{J} & \boldsymbol{L} \\ \boldsymbol{M} & \boldsymbol{N} \end{bmatrix} \begin{bmatrix} \boldsymbol{e}_{pfa}(t) \\ \boldsymbol{C}\boldsymbol{e}_{fa}(t) \end{bmatrix}, \tag{50}$$

respectively. Separating the equation given by the forth row of (34) as follows

$$\boldsymbol{e}_{pfa}(t) = \boldsymbol{J}\boldsymbol{e}_{pfa}(t) + \boldsymbol{L}\boldsymbol{C}\boldsymbol{e}_{fa}(t)\,, \tag{51}$$

(50) can be rewritten as

$$\begin{bmatrix} \boldsymbol{p}(t) \\ \boldsymbol{u}_{c}(t) \end{bmatrix} = \begin{bmatrix} \boldsymbol{J} & \boldsymbol{L} \\ \boldsymbol{M} & \boldsymbol{N} \end{bmatrix} \begin{bmatrix} \boldsymbol{p}_{fa}(t) \\ \boldsymbol{y}_{fa}(t) \end{bmatrix} - \begin{bmatrix} \boldsymbol{I} \\ \boldsymbol{0} \end{bmatrix} \boldsymbol{e}_{pfa}(t) - \begin{bmatrix} \boldsymbol{0} & \boldsymbol{0} \\ \boldsymbol{M} & \boldsymbol{N} \end{bmatrix} \begin{bmatrix} \boldsymbol{e}_{pfa}(t) \\ \boldsymbol{C}\boldsymbol{e}_{fa}(t) \end{bmatrix}. \tag{52}$$

Since the time derivative of (29) takes the form

$$\boldsymbol{e}_{pfa}(t) = \boldsymbol{p}_{fa}(t) - \boldsymbol{p}(t)\,, \tag{53}$$

using (53) then (52) implies the equations of D C covered by PI VA (47). This concludes the proof. ■

4. Design of PI VA

To offer a tuning parameter in PI VA design the following theorem is proven.

Theorem 3 *(enhanced BRL) The closed-loop system, consisting of the plant with a single actuator fault (22), (23), DOC (3), (4) and DVA (26), (27), is stable with the quadratic performance γ° if for given positive $\delta^\circ \in I\!R$ there exist symmetric positive definite matrices $\boldsymbol{R}^\circ, \boldsymbol{U}^\circ \in I\!R^{(n+m)\times(n+m)}$, a matrix $\boldsymbol{Y}^\circ \in I\!R^{2r\times(n+m)}$ and a positive scalar $\gamma^\circ \in I\!R$ such that*

$$\boldsymbol{R}^\circ = \boldsymbol{R}^{\circ T} > 0, \quad \boldsymbol{U}^\circ = \boldsymbol{U}^{\circ T} > 0, \quad \gamma^\circ > 0, \tag{54}$$

$$\begin{bmatrix} \boldsymbol{A}^\circ\boldsymbol{R}^\circ + \boldsymbol{R}^\circ\boldsymbol{A}^{\circ T} + \boldsymbol{B}_f^\circ\boldsymbol{Y}^\circ + \boldsymbol{Y}^{\circ T}\boldsymbol{B}_f^{\circ T} & * & * & * \\ \boldsymbol{U}^\circ - \boldsymbol{R}^\circ + \delta^\circ\boldsymbol{A}^\circ\boldsymbol{R}^\circ + \delta^\circ\boldsymbol{B}_f^\circ\boldsymbol{Y}^\circ & -2\delta^\circ\boldsymbol{R}^\circ & * & * \\ \boldsymbol{V}_{fa}^{\circ T} & \delta^\circ\boldsymbol{V}_{fa}^{\circ T} & -\gamma^\circ\boldsymbol{I}_{r+r_v+n} & * \\ \boldsymbol{C}^\circ\boldsymbol{R}^\circ & \boldsymbol{0} & \boldsymbol{0} & -\boldsymbol{I}_m \end{bmatrix} < 0, \tag{55}$$

where the system matrices are defined in (37), (38) and positive δ° is the tuning parameter. When the above conditions hold

$$\boldsymbol{G}^\circ = \boldsymbol{Y}^\circ(\boldsymbol{R}^\circ)^{-1}. \tag{56}$$

Proof: Since (35) implies

$$\boldsymbol{A}_c^\circ\boldsymbol{q}_{fa}^\circ(t) + \boldsymbol{V}_{fa}^\circ\boldsymbol{d}_{fa}^\circ(t) - \dot{\boldsymbol{q}}_{fa}^\circ(t) = \boldsymbol{0}, \tag{57}$$

$$(\boldsymbol{q}_{fa}^{\circ T}(t)\boldsymbol{S}^\circ + \delta^\circ\dot{\boldsymbol{q}}_{fa}^{\circ T}(t)\boldsymbol{S}^\circ)(\boldsymbol{A}_c^\circ\boldsymbol{q}_{fa}^\circ(t) + \boldsymbol{V}_{fa}^\circ\boldsymbol{d}_{fa}^\circ(t) - \dot{\boldsymbol{q}}_{fa}^\circ(t)) = 0, \tag{58}$$

where $\boldsymbol{S}^\circ \in I\!R^{(n+m)\times(n+m)}$ is symmetric positive definite and $\delta^\circ \in I\!R$ is a positive scalar and by considering the Lyapunov function candidate as follows

$$v(\boldsymbol{q}_{fa}^\circ(t)) = \boldsymbol{q}_{fa}^{\circ T}(t)\boldsymbol{P}^\circ\boldsymbol{q}_{fa}^\circ(t) + \int_0^t(\boldsymbol{y}_{fa}^T(\tau)\boldsymbol{y}_{fa}(\tau) - \gamma^\circ\boldsymbol{d}_{fa}^{\circ T}(\tau)\boldsymbol{d}_{fa}^\circ(\tau))\mathrm{d}\tau > 0, \tag{59}$$

where $\boldsymbol{P}^\circ \in I\!R^{(n+m)\times(n+m)}$ is symmetric positive definite matrix and $\gamma^\circ \in I\!R$ is square of the H_∞ norm of the transfer function matrix of the disturbance $\boldsymbol{d}_{fa}^\circ$, then

$$\dot{v}(\boldsymbol{q}_{fa}^\circ(t)) = \dot{\boldsymbol{q}}_{fa}^{\circ T}(t)\boldsymbol{P}^\circ\boldsymbol{q}_{fa}^\circ(t) + \boldsymbol{q}_{fa}^{\circ T}(t)\boldsymbol{P}^\circ\dot{\boldsymbol{q}}_{fa}^\circ(t) + \boldsymbol{y}_{fa}^T(t)\boldsymbol{y}_{fa}(t) - \gamma^\circ\boldsymbol{d}_{fa}^{\circ T}(t)\boldsymbol{d}_{fa}^\circ(t) < 0. \tag{60}$$

and by adding (58) as well as its transpose to (60) and then inserting (36), it can see that

$$\begin{aligned} \dot{v}(\boldsymbol{q}_{fa}^\circ(t)) &= \dot{\boldsymbol{q}}_{fa}^{\circ T}(t)\boldsymbol{P}^\circ\boldsymbol{q}_{fa}^\circ(t) + \boldsymbol{q}_{fa}^{\circ T}(t)\boldsymbol{P}^\circ\dot{\boldsymbol{q}}_{fa}^\circ(t) \\ &+ (\boldsymbol{q}_{fa}^{\circ T}(t)\boldsymbol{S}^\circ + \delta^\circ\dot{\boldsymbol{q}}_{fa}^{\circ T}(t)\boldsymbol{S}^\circ)(\boldsymbol{A}_c^\circ\boldsymbol{q}_{fa}^\circ(t) + \boldsymbol{V}_{fa}^\circ\boldsymbol{d}_{fa}^\circ(t) - \dot{\boldsymbol{q}}_{fa}^\circ(t)) \\ &+ (\boldsymbol{A}_c^\circ\boldsymbol{q}_{fa}^\circ(t) + \boldsymbol{V}_{fa}^\circ\boldsymbol{d}_{fa}^\circ(t) - \dot{\boldsymbol{q}}_{fa}^\circ(t))^T(\boldsymbol{S}^\circ\boldsymbol{q}_{fa}^\circ(t) + \delta^\circ\boldsymbol{S}^\circ\dot{\boldsymbol{q}}_{fa}^\circ(t)) \\ &+ \boldsymbol{q}_{fa}^{\circ T}(t)\boldsymbol{C}^{\circ T}\boldsymbol{C}^\circ\boldsymbol{q}_{fa}^\circ(t) - \gamma^\circ\boldsymbol{d}_{fa}^{\circ T}(t)\boldsymbol{d}_{fa}^\circ < 0. \end{aligned} \tag{61}$$

The inequality (59) can be written as

$$\dot{v}(\boldsymbol{q}_c^\circ(t)) = \boldsymbol{q}_{ce}^{\circ T}(t)\boldsymbol{P}_{ce}^\circ\boldsymbol{q}_{ce}^\circ(t) < 0, \tag{62}$$

where

$$\boldsymbol{P}_{ce}^\circ = \begin{bmatrix} \boldsymbol{S}^\circ\boldsymbol{A}_c^\circ + \boldsymbol{A}_c^{\circ T}\boldsymbol{S}^\circ + \boldsymbol{C}^{\circ T}\boldsymbol{C}^\circ & * & * \\ \boldsymbol{P}^\circ - \boldsymbol{S}^\circ + \delta^\circ\boldsymbol{S}^\circ\boldsymbol{A}_c^\circ & -2\delta^\circ\boldsymbol{S}^\circ & * \\ \boldsymbol{V}_{fa}^{\circ T}\boldsymbol{S}^\circ & \delta^\circ\boldsymbol{V}_{fa}^{\circ T}\boldsymbol{S}^\circ & -\gamma^\circ\boldsymbol{I}_{r+r_v+n} \end{bmatrix}, \quad \boldsymbol{q}_{ce}^\circ(t) = \begin{bmatrix} \boldsymbol{q}_{fa}^\circ(t) \\ \dot{\boldsymbol{q}}_{fa}^\circ(t) \\ \boldsymbol{d}_{fa}^\circ \end{bmatrix}. \tag{63}$$

Since, using the Schur complement property, (63) can be rewritten as

$$\begin{bmatrix} \boldsymbol{S}^{\circ}\boldsymbol{A}_c^{\circ} + \boldsymbol{A}^{\circ T}\boldsymbol{S}^{\circ} & * & * & * \\ \boldsymbol{P}^{\circ} - \boldsymbol{S}^{\circ} + \delta^{\circ}\boldsymbol{S}^{\circ}\boldsymbol{A}_c^{\circ} & -2\delta^{\circ}\boldsymbol{S}^{\circ} & * & * \\ \boldsymbol{V}_{fa}^{\circ T}\boldsymbol{S}^{\circ} & \delta^{\circ}\boldsymbol{V}_{fa}^{\circ T}\boldsymbol{S}^{\circ} & -\gamma^{\circ}\boldsymbol{I}_{r+r_v+n} & * \\ \boldsymbol{C}^{\circ} & \boldsymbol{0} & \boldsymbol{0} & -\boldsymbol{I}_m \end{bmatrix} < 0\,, \tag{64}$$

inserting (35) in (64) and then pre-multiplying and post-multiplying the result by the matrix

$$\boldsymbol{T}_{fe}^{\circ} = \mathrm{diag}\left[\, \boldsymbol{R}^{\circ} \quad \boldsymbol{R}^{\circ} \quad \boldsymbol{I}_{r+r_v+n} \quad \boldsymbol{I}_m \,\right], \quad \boldsymbol{R}^{\circ} = (\boldsymbol{S}^{\circ})^{-1}, \tag{65}$$

it can be obtained

$$\begin{bmatrix} (\boldsymbol{A}^{\circ} + \boldsymbol{B}_f^{\circ}\boldsymbol{G}^{\circ})\boldsymbol{R}^{\circ} + \boldsymbol{R}^{\circ}(\boldsymbol{A}^{\circ} + \boldsymbol{B}_f^{\circ}\boldsymbol{G}^{\circ})^T & * & * & * \\ \boldsymbol{R}^{\circ}\boldsymbol{P}^{\circ}\boldsymbol{R}^{\circ} - \boldsymbol{R}^{\circ} + \delta^{\circ}(\boldsymbol{A}^{\circ} + \boldsymbol{B}_f^{\circ}\boldsymbol{G}^{\circ})\boldsymbol{R}^{\circ} & -2\delta^{\circ}\boldsymbol{R}^{\circ} & * & * \\ \boldsymbol{V}_{fa}^{\circ T} & \delta^{\circ}\boldsymbol{V}_{fa}^{\circ T} & -\gamma^{\circ}\boldsymbol{I}_{r+r_v+n} & * \\ \boldsymbol{C}^{\circ}\boldsymbol{R}^{\circ} & \boldsymbol{0} & \boldsymbol{0} & -\boldsymbol{I}_m \end{bmatrix} < 0\,. \tag{66}$$

Introducing the notations

$$\boldsymbol{U}^{\circ} = \boldsymbol{R}^{\circ}\boldsymbol{P}^{\circ}\boldsymbol{R}^{\circ}, \quad \boldsymbol{Y}^{\circ} = \boldsymbol{G}^{\circ}\boldsymbol{R}^{\circ} \tag{67}$$

then (66), (67) implies (55), (56), respectively. This concludes the proof. ■

Theorem 4 *(BRL) The closed-loop system, consisting of the plant with a single actuator fault (22), (23), DOC (3), (4) and PI VA (26), (27), is stable with the quadratic performance γ° if there exist a symmetric positive definite matrix $\boldsymbol{X}^{\circ} \in \mathbb{R}^{(n+m)\times(n+m)}$, a matrix $\boldsymbol{Y}^{\circ} \in \mathbb{R}^{2r\times(n+m)}$ and a positive scalar $\gamma^{\circ} \in \mathbb{R}$ such that*

$$\boldsymbol{X}^{\circ} = \boldsymbol{X}^{\circ T} > 0, \quad \gamma^{\circ} > 0\,, \tag{68}$$

$$\begin{bmatrix} \boldsymbol{A}^{\circ}\boldsymbol{X}^{\circ} + \boldsymbol{X}^{\circ}\boldsymbol{A}^{\circ T} + \boldsymbol{B}_f^{\circ}\boldsymbol{Y}^{\circ} + \boldsymbol{Y}^{\circ T}\boldsymbol{B}_f^{\circ T} & * & * \\ \boldsymbol{V}_{fa}^{\circ T} & -\gamma^{\circ}\boldsymbol{I}_{r+r_v+n} & * \\ \boldsymbol{C}^{\circ}\boldsymbol{X}^{\circ} & \boldsymbol{0} & -\boldsymbol{I}_m \end{bmatrix} < 0\,, \tag{69}$$

where the generalized system matrices are defined in (37), (38).

When the above conditions hold

$$\boldsymbol{G}^{\circ} = \boldsymbol{Y}^{\circ}(\boldsymbol{X}^{\circ})^{-1}. \tag{70}$$

Proof: (compare [12]) Setting in (55) that $\boldsymbol{U}^{\circ} = \boldsymbol{R}^{\circ} = \boldsymbol{X}^{\circ}$, $\delta^{\circ} = 0$ and by eliminating the resulting zero row and zero column, then (55) implies (69). This concludes the proof. ■

5. Illustrative Example

The considered system is represented by the model (1), (2) with the matrix parameters [26]

$$\boldsymbol{A} = \begin{bmatrix} 0.5432 & 0.0137 & 0 & 0.9778 & 0 \\ 0 & -0.1178 & 0.2215 & 0 & -0.9661 \\ 0 & -10.5130 & -0.9967 & 0 & 0.6176 \\ 2.6221 & -0.0030 & 0 & -0.5057 & 0 \\ 0 & 0.7075 & -0.0939 & 0 & -0.2120 \end{bmatrix}, \quad \boldsymbol{C} = \begin{bmatrix} 1 & 0 & 0 & 0 & 0 \\ 0 & 1 & 0 & 0 & 0 \\ 0 & 0 & 0 & 1 & 0 \\ 0 & 0 & 0 & 0 & 1 \end{bmatrix},$$

$$\boldsymbol{B} = \begin{bmatrix} -0.0318 & -0.0548 & -0.0548 & -0.0318 & 0.0004 \\ 0.0024 & 0.0095 & -0.0095 & 0.0024 & 0.0287 \\ -2.2849 & -1.9574 & 1.9574 & 2.2849 & 1.4871 \\ -0.4628 & -0.8107 & 0.8107 & -0.4628 & 0.0024 \\ 0.0944 & -0.1861 & -0.1861 & 0.0944 & -0.8823 \end{bmatrix}, \quad \boldsymbol{V} = \begin{bmatrix} 0.7593 \\ 0.4116 \\ 0.8793 \\ 0.0272 \\ 0.0389 \end{bmatrix}, \quad \sigma_v^2 = 7.1\times 10^{-3}.$$

The system is controlled by D C (5), (6), whose parameters were determined using (15)-(17) for order $p = 1$ and the tuning parameter $\delta^\bullet = 10$. Using the SeDuMi package [18] then with $\gamma^\bullet = 13.4697$ the D C matrix parameters take the values

$$\boldsymbol{J} = -1.3767\,, \qquad \boldsymbol{L} = 10^{-8}\begin{bmatrix} -0.0673 & -0.0049 & -0.1764 & -0.0886 \end{bmatrix},$$

$$\boldsymbol{M} = 10^{-8}\begin{bmatrix} -0.2501 \\ 0.3877 \\ 0.0318 \\ 0.0714 \\ -0.0043 \end{bmatrix}, \quad \boldsymbol{N} = \begin{bmatrix} 9.9037 & -18.8161 & 5.2467 & 16.0287 \\ 8.8524 & 11.2393 & 4.6291 & -10.3231 \\ 14.8863 & 0.4394 & 7.7691 & -0.7511 \\ 6.5785 & -0.7130 & 3.4584 & 1.3946 \\ -2.8555 & -6.3480 & -1.4051 & 9.3270 \end{bmatrix}$$

and the signal gain matrix $\boldsymbol{W}$ is calculated by (21) as

$$\boldsymbol{W} = \begin{bmatrix} -5.1176 & -0.1008 & 0.0521 & -0.2320 \\ -7.0937 & 0.5788 & -0.4934 & -1.4523 \\ -11.5871 & -0.2253 & 1.4234 & 0.6632 \\ -3.0013 & -0.7243 & -1.0022 & 3.3029 \\ 2.6838 & 2.3520 & -0.6184 & -5.3107 \end{bmatrix}.$$

The closed-loop system is stable with the closed-loop system matrix eigenvalue spectrum

$$\rho(\boldsymbol{A}^\bullet + \boldsymbol{B}^\bullet\boldsymbol{G}^\bullet\boldsymbol{C}^\bullet) = \{\, -4.3447 \quad -1.9657 \quad -0.5713 \quad -1.2897 \quad -1.0802 \quad -1.3767 \,\}.$$

The control reconstruction by PI VA is illustrated for single fault of the second actuator, where the extended system matrices are constructed as follows:

$$\boldsymbol{A}^\circ = \begin{bmatrix} 0.5432 & 0.0137 & 0 & 0.9778 & 0 & 0 & 0 & 0 & 0 \\ 0 & -0.1178 & 0.2215 & 0 & -0.9661 & 0 & 0 & 0 & 0 \\ 0 & -10.5130 & -0.9967 & 0 & 0.6176 & 0 & 0 & 0 & 0 \\ 2.6221 & -0.0030 & 0 & -0.5057 & 0 & 0 & 0 & 0 & 0 \\ 0 & 0.7075 & -0.0939 & 0 & -0.2120 & 0 & 0 & 0 & 0 \\ 1 & 0 & 0 & 0 & 0 & 0 & 0 & 0 & 0 \\ 0 & 1 & 0 & 0 & 0 & 0 & 0 & 0 & 0 \\ 0 & 0 & 0 & 1 & 0 & 0 & 0 & 0 & 0 \\ 0 & 0 & 0 & 0 & 1 & 0 & 0 & 0 & 0 \end{bmatrix},$$

$$\boldsymbol{B}_f^\circ = \begin{bmatrix} -0.0318 & 0 & -0.0548 & -0.0318 & 0.0004 \\ 0.0024 & 0 & -0.0095 & 0.0024 & 0.0287 \\ -2.2849 & 0 & 1.9574 & 2.2849 & 1.4871 \\ -0.4628 & 0 & 0.8107 & -0.4628 & 0.0024 \\ 0.0944 & 0 & -0.1861 & 0.0944 & -0.8823 \\ 0 & 0 & 0 & 0 & 0 \\ 0 & 0 & 0 & 0 & 0 \\ 0 & 0 & 0 & 0 & 0 \\ 0 & 0 & 0 & 0 & 0 \end{bmatrix}, \quad \boldsymbol{C}^\circ = \begin{bmatrix} 1 & 0 & 0 & 0 & 0 & 0 & 0 & 0 & 0 \\ 0 & 1 & 0 & 0 & 0 & 0 & 0 & 0 & 0 \\ 0 & 0 & 0 & 1 & 0 & 0 & 0 & 0 & 0 \\ 0 & 0 & 0 & 0 & 1 & 0 & 0 & 0 & 0 \end{bmatrix},$$

$$\boldsymbol{V}_{fa}^\circ = \begin{bmatrix} -0.0318 & 0 & -0.0548 & -0.0318 & 0.0004 & 0.7593 & 0 & 0 & 0 & 0 & 0 \\ 0.0024 & 0 & -0.0095 & 0.0024 & 0.0287 & 0.4116 & 0 & 0 & 0 & 0 & 0 \\ -2.2849 & 0 & 1.9574 & 2.2849 & 1.4871 & 0.8793 & 0 & 0 & 0 & 0 & 0 \\ -0.4628 & 0 & 0.8107 & -0.4628 & 0.0024 & 0.0272 & 0 & 0 & 0 & 0 & 0 \\ 0.0944 & 0 & -0.1861 & 0.0944 & -0.8823 & 0.0389 & 0 & 0 & 0 & 0 & 0 \\ 0 & 0 & 0 & 0 & 0 & 0 & -1 & 0 & 0 & 0 & 0 \\ 0 & 0 & 0 & 0 & 0 & 0 & 0 & -1 & 0 & 0 & 0 \\ 0 & 0 & 0 & 0 & 0 & 0 & 0 & 0 & 0 & -1 & 0 \\ 0 & 0 & 0 & 0 & 0 & 0 & 0 & 0 & 0 & 0 & -1 \end{bmatrix}.$$

For this fault scenario the closed-loop system with above designed parameters is instable and PI VA has to be activated to stabilize the faulty system.

The parameters of PI VA were determined using (54), (55), conditioned by the tuning parameter setting $\delta^\circ = 0.217$. Exploiting the SeDuMi package, with $\gamma_\circ = 4.3700$ the PI VA gain matrices are

$$\boldsymbol{G}_P = \begin{bmatrix} -18.3055 & 0.3354 & -0.1140 & -4.0384 & -0.5380 \\ 0 & 0 & 0 & 0 & 0 \\ -10.1172 & -1.6384 & -0.0834 & -1.1921 & 0.2553 \\ -9.0562 & -3.1597 & -0.1456 & -2.6616 & 1.0467 \\ -0.7671 & 2.5586 & 0.3847 & -0.4279 & -3.0123 \end{bmatrix},$$

$$\boldsymbol{G}_I = \begin{bmatrix} -2.3650 & -0.2736 & 1.1260 & -0.3141 \\ 0 & 0 & 0 & 0 \\ -1.6237 & -0.2461 & 1.2897 & -0.0650 \\ -0.9579 & -0.1727 & 0.1167 & 0.1702 \\ -0.0191 & 0.5438 & -0.1140 & -0.3134 \end{bmatrix},$$

The eigenvalue spectrum of the matrix $\boldsymbol{A}_c^\circ = \boldsymbol{A}^\circ - \boldsymbol{B}_f^\circ \boldsymbol{G}^\circ$ is

$$\rho(\boldsymbol{A}_c^\circ) = \left\{ \begin{matrix} -1.1166 & -0.1636 & -0.1433 \\ -1.4956 \pm 1.0372i & -1.5616 \pm 0.7682i & -0.1928 \pm 0.0352i \end{matrix} \right\}$$

and this spectrum determines the dynamics of the closed-loop system after activation of PI VA.

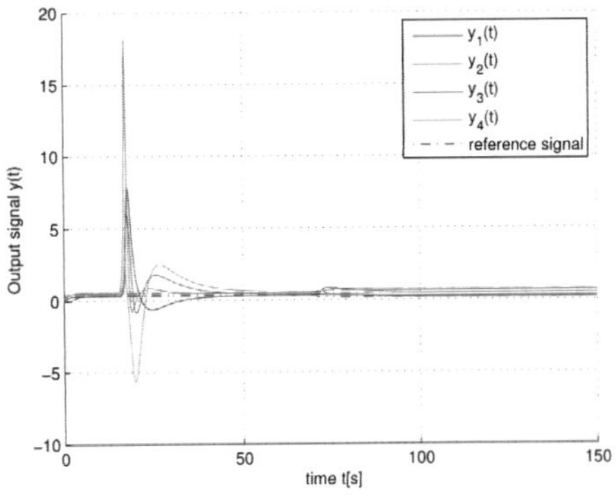

Figure 1. System output response

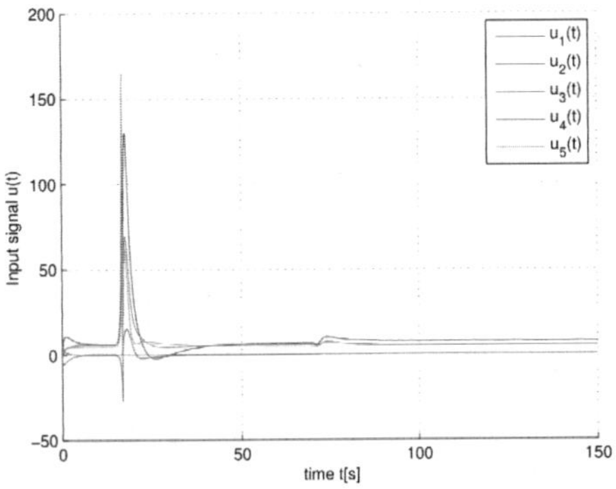

Figure 2. Control variables

In Fig. 1 and Fig. 2 are shown the time responses of the system output and control variables for the control realized by D C of order $p = 1$ in the controller forced mode and PI VA acting on the faulty system. The single second actuator fault occurred at time instant $t = 15\,s$ and PI VA was activated at time instant $t = 17\,s$. In simulation, the initial condition was set $\boldsymbol{q}_0 = 0$ and the desired output values $\boldsymbol{w}(t) = [0.3\ 0.4\ 0.5\ 0.6]$ were changed step-wise at the time instant $t = 70\,s$ to $\boldsymbol{w}(t) = [0.7\ 0.2\ 0.3\ 0.5]$. Although the values of output variables were achieved, the output and control variables peaks after the activation of PI VA are excessively high.

Within the same simulation conditions, Fig. 3 and Fig. 4 show the time responses of the system output and control variables for control realized by D C of order $p = 4$ and Fig. 5 and Fig. 6 show the time response of the same variables for control with D C of order $p = 7$ and PI VA. It is obvious that an appropriate con unction of orders of D C gives the possibility to significantly reduce the variables peaks after activation of VAs. Note, there are no rules for the optimal order selection of D C [10].

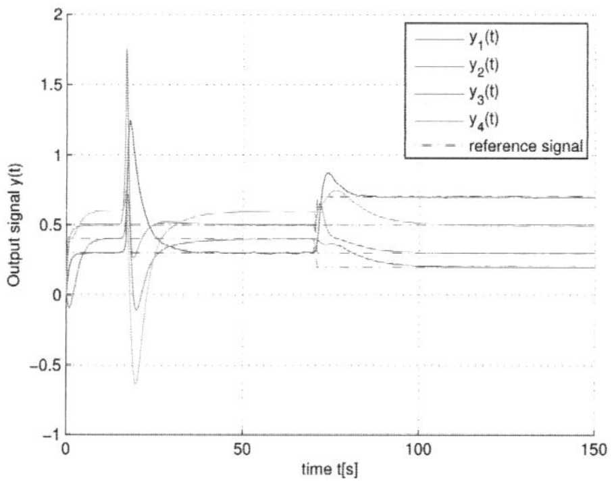

Figure 3. System output response

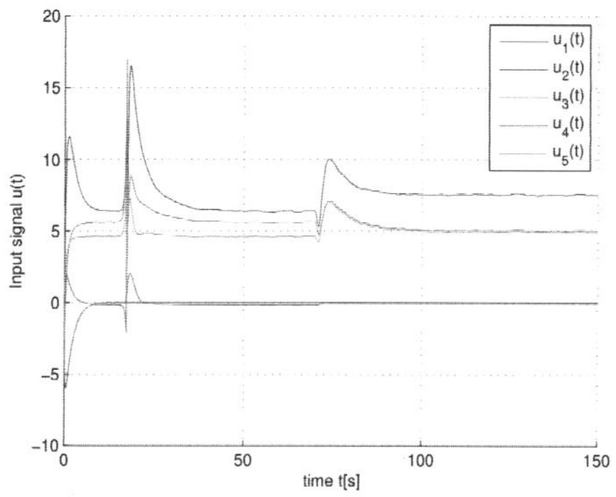

Figure 4. Control variables

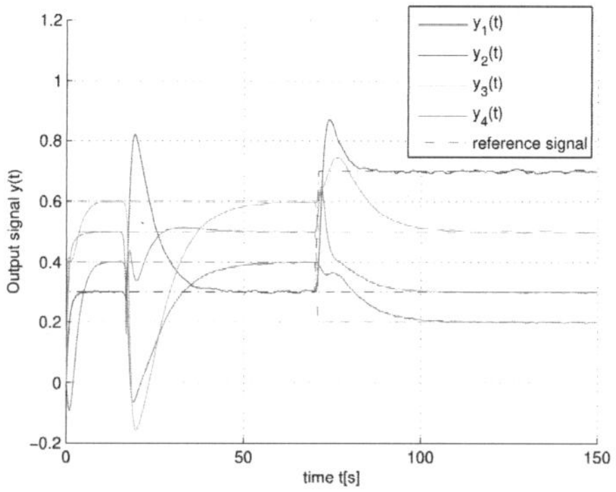

Figure 5. System output response

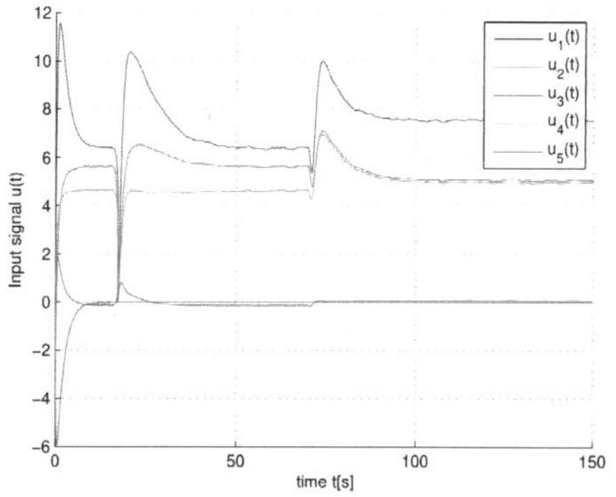

Figure 6. Control variables

6. Concluding Remarks

A key contribution of the proposed approach is the blending of the virtual actuator technique and the output control principle in a unique dynamic scheme, able to provide fault tolerance against actuator faults with such acceptable responses of the system variables, primarily after activation of DVA, which cannot by reached by applying a static output controller on the exactly same plant.

The proposed model of the dynamic effect of a virtual actuator in the FTC structure relies on newly introduced generalized disturbance patterns, reflecting fading of the nominal system state variables after DVA activation. This allows to include in the DVA design conditions the disturbance input/system output model property by H_∞ norm approach.

The application of the proposed approach requires that a fault detection and isolation subsystem is available. However, it becomes clear that the desired performances depend on the fault isolation time, but a suitable order con unction of the both dynamic components allows significantly extend the time limit of fault detection.

Acknowledgments

The work presented in this paper was supported by VEGA, the Grant Agency of the Ministry of Education and the Academy of Science of the Slovak Republic under Grant No. 1/0348/14. This support is very gratefully acknowledged.

References

[1] Alwi H Edwards C and Tan C P 2011 *Fault Detection and Fault-Tolerant Control Using Sliding Modes* (London:Springer)

[2] Blanke M Kinnaert M Lunze J and Staroswiecki M 2003 *Diagnosis and Fault-Tolerant Control* (Berlin: Springer-Verlag)

[3] Doyle J C Glover K Khargonekar P P and Francis B A 1989 State-space solutions to standard H_2 and H_∞ control problems *IEEE Tran. Automatic Control* **34**(8) 831-47

[4] Ebihara Y Peaucelle D and Arzelier D 2012 *S-Variable Approach to LMI-Based Robust Control* (London: Springer)

[5] Kailath T 1980 *Linear Systems* (Englewood Cliffs:Prentice-Hall)

[6] Khosrowjerdi M J and Barzegary S 2014 Fault tolerant control using virtual actuator for continuous-time Lipschitz nonlinear systems *Int. J. Robust and Nonlinear Control* **24**(6) 2597-607

[7] Krokavec D and Filasová A 2012 Optimal fuzzy control for a class of nonlinear systems *Mathematical Problems in Engineering* **2012** Article ID 481942 29p

[8] Krokavec D and Filasová A 2015 LMI constraints on system eigenvalues placement in dynamic output control design *Proc. 2015 IEEE Multi-conference on Systems and Control MSC 2015* Sydney Australia 1749-1754

[9] Krokavec D Filasová A and Serbák V 2015 On virtual observers in fault tolerant control *Proc. 20th Int. Conf. Process Control PC'2015* Štrbské Pleso Slovakia 266-71

[10] Krokavec D Filasová A and Serbák V 2015 FTC Structures with virtual actuators and dynamic output controllers *Proc. 9th IFAC Symp. Fault Detection, Supervision and Safety for Technical Processes SAFEPROCESS 2015* Paris France 511-516

[11] Krokavec D Filasová A and Serbák V 2016 FTC with Dynamic Virtual Actuators. Characterization via Dynamic Output Controllers and H_∞ Approachs *Mathematical Problems in Engineering* **2016** Article ID 259056 16p

[12] Krokavec D Filasová A Serbák V and Liščinský P 2016 H_∞ approach to virtual actuators design *Advanced and Intelligent Computations in Diagnosis and Control* Kowalczuk Z (Ed.) (Cham:Springer-Verlag) 209-212

[13] Lunze J and Steffen T 2002 Hybrid reconfigurable control *Modelling, Analysis, and Design of Hybrid Systems* Engell S Frehse G and Schnieder E (eds) (Berlin:Springer-Verlag) 267-284

[14] Lunze J and Steffen T 2006 Control reconfiguration after actuator failures using disturbance decoupling methods *IEEE Tran. Automatic Control* **51**(10) 1590-1601

[15] Noura H Theilliol D Ponsart J C and Chamseddine A 2009 *Fault-Tolerant Control Systems. Design and Practical Applications* (London:Springer)

[16] Mahmoud M S and Xia Y 2014 *Analysis and Synthesis of Fault-Tolerant Control Systems* (Chichester:John Wiley & Sons)

[17] Montes de Oca S Puig V Witczak M and Dziekan L 2012 Fault-tolerant control strategy for actuator faults using LPV techniques. Application to a two degree of freedom helicopter *Int. J. Applied Mathematics and Computer Science* **22**(1) 161-71

[18] Peaucelle D Henrion D Labit Y and Taitz K 2002 *User's Guide for SeDuMi Interface 1.04* (Toulouse:LAAS-CNRS)

[19] Richter J H 2011 *Reconfigurable Control of Nonlinear Dynamical Systems. A Fault-Hiding Approach* (Berlin:Springer-Verlag)

[20] Richter J H Lunze J and Steffen T 2010 The relation between the virtual actuator and the dual observer *European J. Control* **16**(5) 525-31

[21] Richter J H Heemels W P M H van de Wouw N and Lunze J 2011 Reconfigurable control of piecewise a ne systems with actuator and sensor faults. Stability and tracking *Automatica* **47**(4) 678-91

[22] Rotondo D Nejjaria F and Puig V 2014 A virtual actuator and sensor approach for fault tolerant control of LPV systems *J. Process Control* **24**(3) 203-22

[23] Steffen T 2005 *Control Reconfiguration of Dynamical Systems. Linear Approaches and Structural Tests* (Berlin:Springer-Verlag)

[24] Tabatabaeipour S M Stoustrup J and Bak T 2015 Fault-tolerant control of discrete-time LPV systems using virtual actuators and sensors *Int. J. Robust and Nonlinear Control* **25**(5) 707-34

[25] Yang I Lee D and Han D S 2015 Reconfiguration criterion for fault-tolerant control *Mathematical Problems in Engineering* **2015** Article ID 218384 11p

[26] Yu B Zhang Y Yan J Qu Y and Liu Z 2013 Fault tolerant control using linear quadratic technique against actuator faults in a UAV *Proc. 32nd Chinese Control Conference* Xi'an P.R. China 6294-99

[27] Zolghadri A Henry D Cieslak J Efimov D and Goupil P 2014 *Fault Diagnosis and Fault-Tolerant Control and Guidance for Aerospace Vehicles. From Theory to Application* (London:Springer)

SSNN toolbox for non-linear system identification

Marcel Luzar, Andrzej Czajkowski

Institute of Control and Computation Engineering, University of Zielona Góra, ul. Podgórna 50, 65–246 Zielona Góra, Poland

E-mail: {M.Luzar,A.Czajkowski}@issi.uz.zgora.pl

Abstract.
The aim of this paper is to develop and design a State Space Neural Network toolbox for a non-linear system identification with an artificial state-space neural networks, which can be used in a model-based robust fault diagnosis and control. Such toolbox is implemented in the MATLAB environment and it uses some of its predefined functions. It is designed in the way that any non-linear multi-input multi-output system is identified and represented in the classical state-space form. The novelty of the proposed approach is that the final result of the identification process is the state, input and output matrices, not only the neural network parameters. Moreover, the toolbox is equipped with the graphical user interface, which makes it useful for the users not familiar with the neural networks theory.

1. Introduction

The identification of dynamic models out of experimental data has very often been motivated and supported by the presumed ability to use the resulting models as a basis for model-based Fault Diagnosis (FD) and control design [7, 15, 19, 21]. As such, FD and control design is considered an important intended-application area for identified models. On the other hand, model-based FD and control design is built upon the assumption that a reliable model of the plant under consideration is available. Without model, there is no model-based FD and control design. Thus, there is a need for good quality system identification methods for the control design purposes.

In this paper, the system identification method which use the Artificial Neural Networks (ANNs) is proposed. Such technique is known from many years [2, 6, 9, 13, 14]. It already proven its usefulness in the tasks on linear systems identification [3] or prediction purposes [10]. However, for the non-linear ones, especially high-order and dynamic systems, it not always gives the satisfactory results [8, 20]. Moreover, there is a need for a tool, which helps the researchers to identify the system model based on the input/output measured data, without the knowledge about the neural networks. There is some predefined toolboxes i.e. in MATLAB, however any of them uses the state-space ANNs, which nowadays becomes more and more popular form of the non-linear system representation [5, 4]. A good example that proves that the such toolbox is needed is the *GMDH Toolbox for neural network-based modelling*, which was downloaded from the *Researchgate* portal around 1300 times [12] and was implemented by the author of this paper. Thus, in this paper, the implementation of the State-Space Neural Network (SSNN) toolbox is presented, which overcomes above mentioned problems. It requires from the users only basic experience with the ANNs and Matlab. It is equipped with some automatic methods and functions, which allows to clearly identify any Mulit-Input Multi-Output (MIMO) non-linear dynamic system. Some of them based on the algorithms implemented in the *NNSYSID toolbox*, designed by Nørgaard [16]. Obviously, they were suitable modified to identify and linearize the system direct in the state-space form. Note that implementation of this toolbox is the first stage of realizing the grant *Artificial neural networks for robust fault diagnosis and control of non-linear systems* conducted by the author of this paper.

The paper is organised as follows. After short introduction, Section 2 presents the design assumptions of the toolbox. In Section 3, the theoretical background is given. Section 3 describes the implementation of the toolbox with explanation of its functionalities. The system identification results, which verify the proposed approach are given in the Section 5. Finnaly, Section 6 concludes the paper.

2. Design assumptions

Prior to implementation, the following assumptions are discussed with the interested researches and taken into account:

- possibility to identify any non-linear, dynamic MIMO system with SSNN,
- transformation of the SSNN parameters into state-space matrices,
- possibility to choose options for the neural network design in the professional mode such as:
  (i) neuron number in the output and hidden network layer,
  (ii) neural network training stopping criteria,
  (iii) the weight decay,
  (iv) backpropagation parameters,
- possiblity to import training and validation data direct from the file or Matlab's workspace,
- error handling,
- Graphical User Interface (GUI).

Moreover, the SSNNtool has to be equipped with the help files which allows to understand used techniques in details.

3. Theoretical background

3.1. Process model

As was mentioned in the introduction of this paper, in the toolbox for model identification, the dynamic ANN is implemented. Among many different dynamic ANNs there is a special class of network, called the State Space Neural Network (SSNN), which scheme is presented in Fig. 1. The hidden layer outputs are propagated back to the input layer through a bank of unit delays.

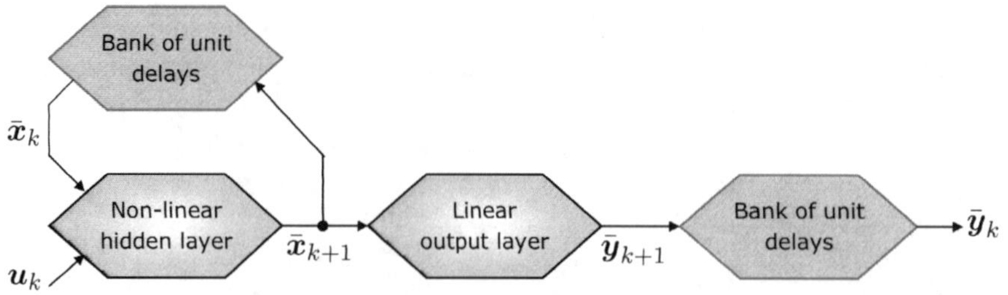

Figure 1. The state-space neural network with one hidden layer

The number of such delays defines the order of the system. In general, a user may decide how many neurons are used to produce feedback. Let $\boldsymbol{u}(k) \in \mathbb{R}^n$ be the input vector, $\bar{\boldsymbol{x}}(k) \in \mathbb{R}^q$ - the output of the hidden layer at time k, and $\bar{\boldsymbol{y}}(k) \in \mathbb{R}^m$ - the output vector. Then the state-space form of the neural model given in Fig. 1 is defined by the following equations

$$\begin{aligned} \bar{\boldsymbol{x}}(k+1) &= \bar{g}(\bar{\boldsymbol{x}}(k), \boldsymbol{u}(k)) \\ \bar{\boldsymbol{y}}(k) &= \boldsymbol{C}\bar{\boldsymbol{x}}(k) \end{aligned}, \tag{1}$$

where $\bar{g}(\cdot)$ is a non-linear function characterizing the hidden layer, and $\boldsymbol{C}$ represents synaptic weights between hidden and output neurons.

Introducing the weight matrix between input and hidden layers $\boldsymbol{W}^u$ and the matrix of recurrent links $\boldsymbol{W}^x$, the previous formulae (1) can be rewritten as follows:

$$\begin{aligned} \bar{\boldsymbol{x}}(k+1) &= h(\boldsymbol{W}^x\bar{\boldsymbol{x}}(k) + \boldsymbol{W}^u\boldsymbol{u}(k)) \\ \bar{\boldsymbol{y}}(k) &= \boldsymbol{C}\bar{\boldsymbol{x}}(k) \end{aligned}, \tag{2}$$

where $h(\cdot)$ denotes the activation function of the hidden neurons. In the most cases, when the hyperbolic tangent activation function is chosen, the modelling results are satisfactory. For the state space model the outputs of hidden neurons which constitute feedbacks are, in general, unknown during training. Thus, state space neural models can be trained only by minimizing the simulation error. If state measurements are available, the training can be carried out much easier using the series-parallel identification scheme, similarly as for the external dynamic approach (the feedforward network with tapped delay lines). In spite of this drawback, the state-space neural networks popularity was build due to its number of advantages, contrary to fully and partially recurrent networks ([17], and they are as follows:

- The determination of the states number (model order) may be done independently from the hidden neuron number. The obvious result of this feature is that the responsibility for defining the state of the network is on the neurons that propagate their outputs back to the input layer through delays. As a consequence, the output neurons are eliminated from the state definition [17]. In the recurrent networks, e.g. Williams-Zipser, Elman, locally recurrent networks, the model order is directly influenced by the number of the neurons, which significantly impedes the phase of modelling.
- Model states feed the network input (which makes them easily accessible). In case when state measurements are available at some time instants, this property can be very useful.
- State space neural models are useful in the fault diagnosis and tolerant control frameworks. State representation allows to approximate size of a fault effect and to handle different kind of faults including multiplicative and additive ones.

Mentioned above advantages of SSNN make models of this kind a very interesting and promising tool used to solve different engineering issues i.e. the fault diagnosis problem. Also the class of non-linear state space models is strongly evaluated in different scientific approaches as a nominal model.

3.2. Linearisation of the state space model

In the toolbox, to deal with the model non-linearity, the instantaneous linearisation is implemented. The primary idea is very simple. At each discrete time sample a linear model is extracted from the state space neural model. Model is linearised using Taylor series expansion around the current operating point $(\boldsymbol{x}, \boldsymbol{u}) = (\boldsymbol{x}(\tau), \boldsymbol{u}(\tau))$ and rejection of the non-linear components. The first order Taylor series expansion of the SSNN model is presented in the following equations:

$$\begin{aligned} h(\boldsymbol{x}(k),\boldsymbol{u}(k)) &= h(\boldsymbol{x}(\tau),\boldsymbol{u}(\tau)) + \tfrac{\partial h}{\partial \boldsymbol{x}}\big|_{(x,u)}\Delta x + \tfrac{\partial h}{\partial \boldsymbol{u}}\big|_{(x,u)}\Delta u \\ &= h(\boldsymbol{x}(\tau),\boldsymbol{u}(\tau)) + h'\boldsymbol{W}^x(\boldsymbol{x}(k)-\boldsymbol{x}(\tau)) \\ &\quad + h'\boldsymbol{W}^u(\boldsymbol{u}(k)-\boldsymbol{u}(\tau)). \end{aligned} \tag{3}$$

The following linear state space model is obtained:

$$\begin{aligned} \bar{\boldsymbol{x}}(k+1) &= \boldsymbol{A}\bar{\boldsymbol{x}}(k) + \boldsymbol{B}\boldsymbol{u}(k) + \boldsymbol{D} \\ \bar{\boldsymbol{y}} &= \boldsymbol{C}\bar{\boldsymbol{x}}(k) \end{aligned}, \tag{4}$$

is obtained, where $\boldsymbol{A} = h'\boldsymbol{W}^x$, $\boldsymbol{B} = h'\boldsymbol{W}^u$, $\boldsymbol{D} = \boldsymbol{x}(\tau) - \boldsymbol{A}\bar{\boldsymbol{x}}(\tau-1) - \boldsymbol{B}\boldsymbol{u}(\tau-1)$. The h' is the activation function first derivative. If the activation function is implemented in the form of hyperbolic tangent, this derivative can be simply calculated as follows:

$$h' = 1 - \tanh^2. \tag{5}$$

This property allows the linearisation to be performed very quickly. This is crucial in the terms of computation burden especially when calculations need to be carried out in real-time.

4. SSNN Toolbox design

The entire implementation of the toolbox was done using the Matlab environment. The graphical view and functionality is inspired with the Neural Network toolbox, which is predefined in Matlab. Thus, the SSNNtool consist of two windows: the main window called SSNN Toolbox and the import data window. In the import data window presented in Fig. 2, there are three panels: the data source selection panel, the variable selection panel as well as the training input data, the validation input data, the training output data and, finally, the validation output data panel. The data can be imported either from an existing file with extension *.mat or from the Matlab workspace, respectively. After assigning the variables from the list as the relevant data (by the selection button *Set*), the import data window is closed. The main window presented

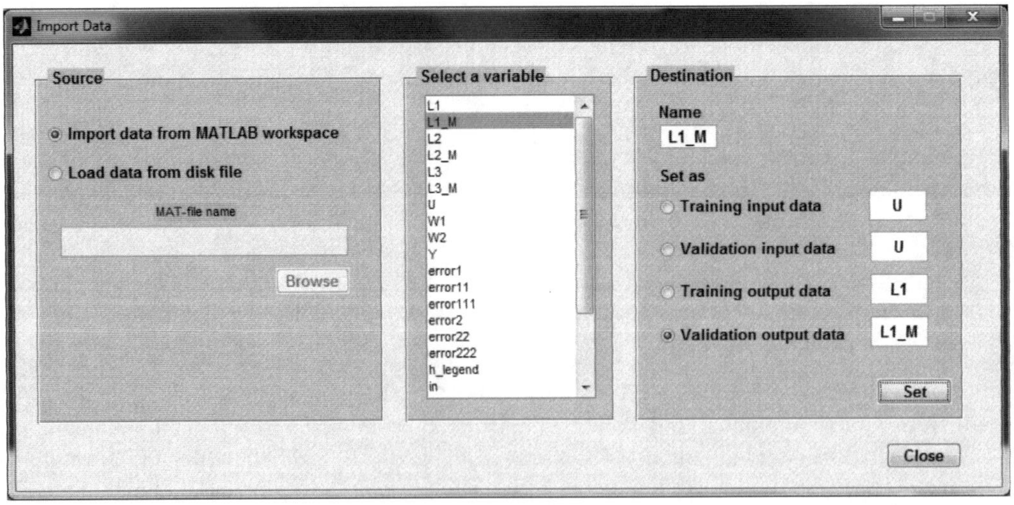

Figure 2. SSNN Toolbox data import window

in Fig. 3 is equipped with the tools for choosing the neural network training parameters. First, user can select its experience in the NN field. When the *Beginer* option is selected, user can define only neuron number in the hidden and output layer. Other necessary parameters are set as default and their values are given in the help file. In the *Advanced* mode, user can define above mentioned parameters and moreover, weight decay, stopping criteria and backpropagation parameters.

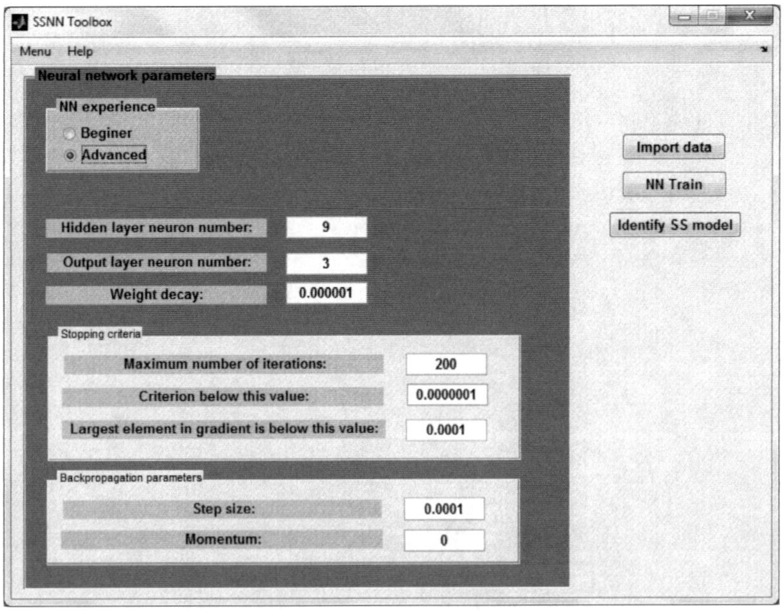

Figure 3. SSNN Toolbox main window

If the training data are imported, the SSNN training can be initiated by pressing the *NN Train* button. In the case, when data are not imported, or if are not correctly assigned as training/validation input/output data, the warning window is shown in Fig. 4.

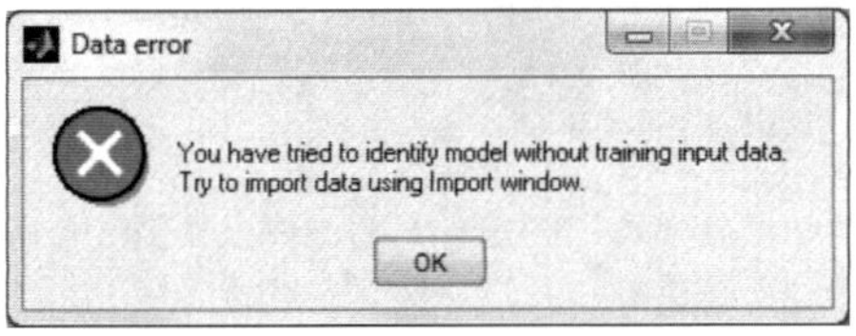

Figure 4. Warning message

When the all NN parameters are set and the data is imported correctly, the NN training is started. During this process, number of iterations with associated modelling errors are displayed in the command window. When the NN training is finished, the number of figures are displayed automatically. In this figures, the results of the NN modelling are presented. If the SSNN is trained in the satisfaction way and it reflects the real system, the neural model can be linearized and presented in the classic state-space form described in Sec. 3.2 by pressing the *Identify SS model* button. The state-space matrices **A**, **B**, **C** and **D** obtained during linearization process are automatically saved in the current folder.

It is worth to note that the SSNNtool is equipped with the help files, which explains all used methods in details. Moreover, the SSNN theoretical background and linearization approach used in the toolbox is presented. To open the help file user have to press the *Help* button in the *Main* window.

5. Illustrative example

To clearly exhibit the effectiveness of the proposed approach, two different scenarios of the system identification is performed. First, the neural network modelling of the multi-tank system is presented. Such laboratory equipment may be perceived as a Single Input Multiple Output (SIMO) non-linear system. It is chosen because its analytical model is known and given by the producer, thus it is easy to compare it with model obtained with the SSNN approach proposed in this paper. Second experiments regards the modelling of the wind turbine, which is highly non-linear MIMO system.

5.1. Multi-tank system identification

The multi-tank system presented in Fig. 5 is chosen for show the effectiveness of the developed toolbox in the system identification task. Experimental results obtained with the toolbox are then compared with the analytical model given by the manufacturer.

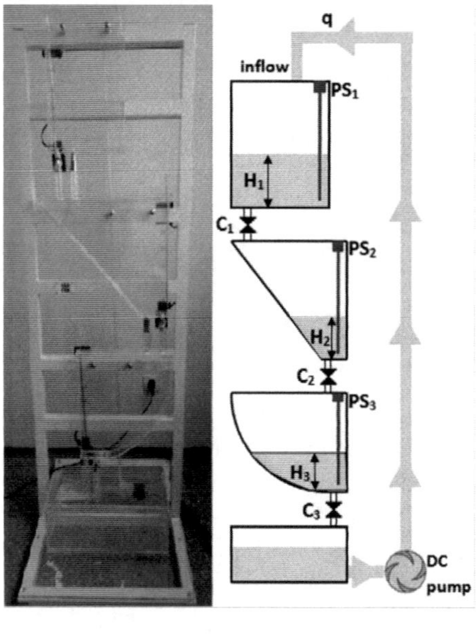

Figure 5. Multi-tank system

It consists of three separate tanks placed each above other and equipped with drain valves and level sensors based on a hydraulic pressure measurement. Each of them has a different cross-section in order to reflect system non-linearities. The lower bottom tank is a water reservoir for the system. A variable speed water pump is used to fill the upper tank. The water outflows the tanks due to gravity. The considered multi-tank system has been designed to operate with an external, PC-based digital controller. The control computer communicates with the level sensors, valves and a pump by a dedicated I/O board and the power interface. The I/O board is controlled by the real-time software, which operates in a Matlab/Simulink environment. The description of the parameters presented in Fig. 5 are given in Tab. 1.

For laminar outflow, where turbulence and acceleration of the liquid are commuted and inflow rate is constant $q = const$, the flow equation is as follows:

$$Q = \mu S \sqrt{2gH}, \tag{6}$$

Signal	Description	Units	Operating range
q	Flow rate of the fluid	$\frac{m^3}{s}$	$0 \leq q \leq 1.5e^{-4}$
H_1, H_2, H_3	Level of the tanks	m	$0.01 \leq H_1, H_2, H_3 \leq 0.35$
C_1	Open level of C_1 valve	raw (%)	$0 \leq C_1 \leq 1.0057\text{e}^{-4}$
C_2	Open level of C_2 valve	raw (%)	$0 \leq C_2 \leq 1.1963\text{e}^{-4}$
C_3	Open level of C_3 valve	raw (%)	$0 \leq C_3 \leq 9.8008\text{e}^{-5}$

Table 1. Specification of system parameters

where Q is outflow rate from the tank, μ stands for orifice outflow coefficient, S denotes the area of the orifice and H is the level of the fluid measured from level 0.

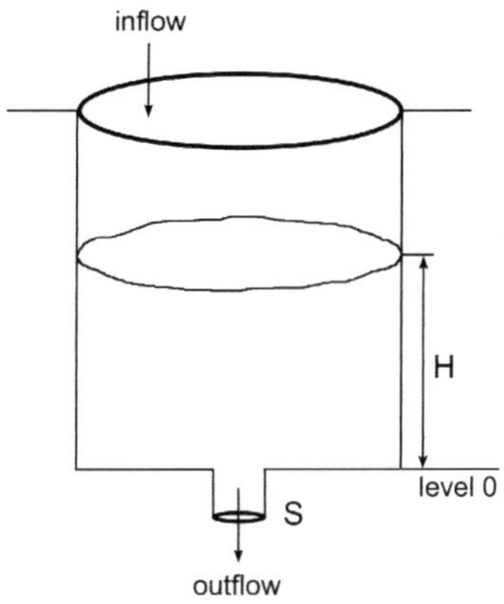

Figure 6. Laminar outflow from the tank

The outflow rate of the system is described by:

$$\begin{aligned}\frac{dV_1}{dt} &= q - C_1 H_1^{\alpha_1},\\ \frac{dV_2}{dt} &= C_1 H_1^{\alpha_1} - C_2 H_2^{\alpha_2},\\ \frac{dV_3}{dt} &= C_2 H_2^{\alpha_2} - C_3 H_3^{\alpha_3},\end{aligned} \tag{7}$$

where α_i denotes the i-th tank dynamic flow rate coefficient.
The equation (7) can be transformed into:

$$\begin{aligned}\frac{dH_1}{dt} &= \frac{1}{\beta_1(H_1)} q - \frac{1}{\beta_1(H_1)} C_1 H_1^{\alpha_1},\\ \frac{dH_2}{dt} &= \frac{1}{\beta_2(H_2)} C_1 H_1^{\alpha_1} - \frac{1}{\beta_2(H_2)} C_2 H_2^{\alpha_2},\\ \frac{dH_3}{dt} &= \frac{1}{\beta_3(H_3)} C_2 H_2^{\alpha_2} - \frac{1}{\beta_3(H_3)} C_3 H_3^{\alpha_3},\end{aligned} \tag{8}$$

where H_i is i-th tank fluid level and $\beta_i(H_i)$ stands for cross-sectional area of i-th tank at level H_i.
Cross-sectional areas for each of the tank are as follows:

$\beta_1(H_1) = aw$ - constant cross-sectional area of the upper tank,

$\beta_2(H_2) = cw + \frac{H_2}{H_{2max}}bw$ - variable cross-sectional area for the middle tank (conical),

$\beta_3(H_3) = w\sqrt{R^2 - (R - H_3)^2}$ - variable cross-sectional area of the lower tank (spherical),

where C_i denotes the resistance of output orifice of i-th tank.
Variable cross-sectional areas of the tanks introduce main nonlinearities and require particular approach.
The set of equations (8) is equivalent to:

$$\begin{aligned} F_1(u, x_1) &= \frac{1}{\beta_1(x_1)}u - \frac{1}{\beta_1(x_1)}C_1 x_1^{\alpha_1}, \\ F_2(x_1, x_2) &= \frac{1}{\beta_2(x_2)}C_1 x_1^{\alpha_1} - \frac{1}{\beta_2(x_2)}C_2 x_2^{\alpha_2}, \\ F_2(x_2, x_3) &= \frac{1}{\beta_3(x_3)}C_2 x_2^{\alpha_2} - \frac{1}{\beta_3(x_3)}C_3 x_3^{\alpha_3}, \end{aligned} \tag{9}$$

where x_i denotes the i-th tank fluid level and u is the control input.

Calculated mathematical model is a set of ordinary differential equations and it is ready to be compared with the non-linear SSNN model.

To obtain the SSNN model with the toolbox, 7500 samples of the training and validation data from the real system were collected. The control signal were chosen randomly from the following interval

$$0 \leq \boldsymbol{u}_k \leq 0.0001 \tag{10}$$

and suitably rescaled.

The result of liquid level modelling in the first tank is presented in Fig. 7 with associated modelling errors. It is clear, that both models reflects real system behaviour with good quality. Despite this, based on modelling error presented in the second graph in Fig. 7 it is easy to see that model obtained with the SSNN toolbox reflects the real system better than analytical one. The same conclusion can be drawn by analysing Fig. 8 and Fig. 9, in which the result of liquid level modelling in second and third tank is depicted, respectively. Still, the SSNN model is better than analytical one which proves the usefulness of the proposed approach in i.e. fault detection and control.

5.2. Wind turbine model identification

The presented second example is based on the wind turbine description provided in [1] with suitable modifications. Note that an analytical model is not defined. The network feeding data is as follows:

$$\boldsymbol{u}_k = \begin{bmatrix} \beta_k \\ \rho_k \\ v_{w_k} \end{bmatrix}, \boldsymbol{y}_k = \begin{bmatrix} \tau_k \\ \omega_{r_k} \\ \omega_{g_k} \\ \beta_k \end{bmatrix}, \tag{11}$$

where β is the pitch angle, ρ stands for zero-torque speed, v_w is the wind speed, τ denotes torsion angle, ω_r and ω_g is the rotor and generator velocity, respectively. The identification process was done with the *Beginer* mode, thus the neural network training parameters was set as default ones. The number of neurons was set to 13 and 6 in the hidden and output layer, respectively.

Graphs a), b), c) and d) in Fig. 10 presents the results of modelling for the rotor velocity, generator velocity and pitch angle, respectively. It is clear, that the neural model reflects real system satisfactorily.

Moreover, the histogram depicted in Fig. 11 presents the modelling errors.

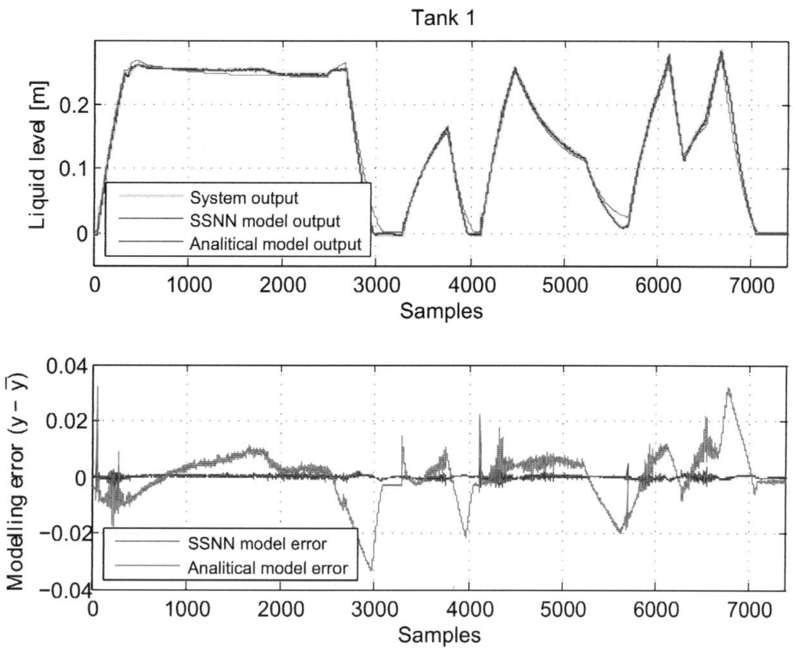

Figure 7. Comparison of the real system output with analytical and SSNN model outputs in first tank

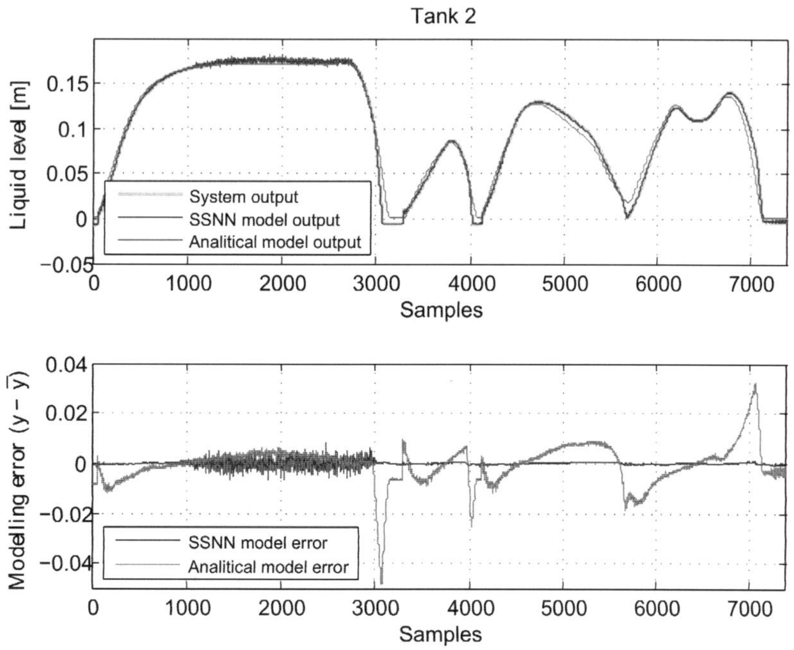

Figure 8. Comparison of the real system output with analytical and SSNN model outputs in second tank

6. Conclusions

The main objective of this paper was to design and implement the SSNNtool, which allows to clearly identify any non-linear dynamic system with the state-space neural networks. The

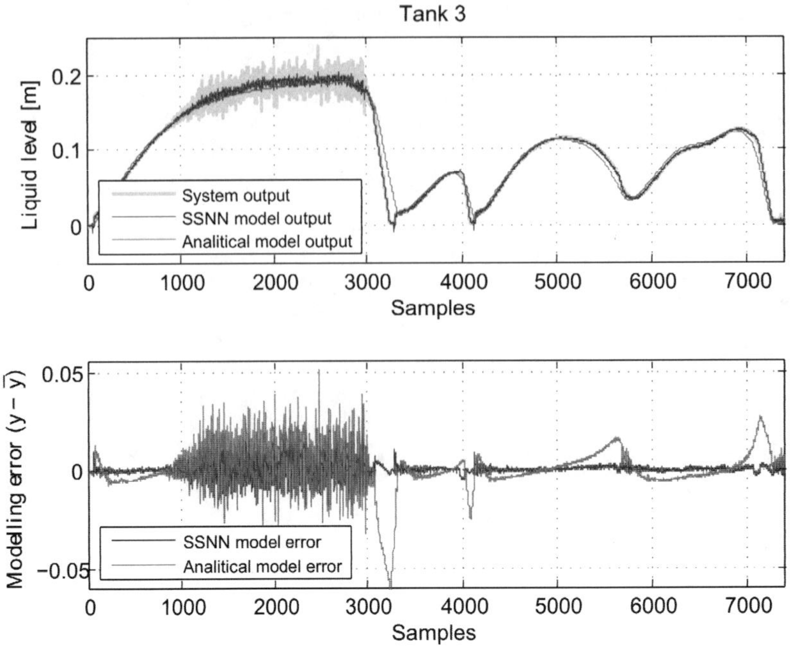

Figure 9. Comparison of the real system output with analytical and SSNN model outputs in third tank

motivation for design such a tool was the success of previously implemented by author of this paper *GMDHToolbox*. It proves that there is a need for automatic tool which can be used without any expertise in the NN field. The effectiveness of designed toolbox was verified with two experiments scenarios. In first, the obtained SSNN model was compared with the analytical model. In second one, the non-linear SS model was obtained based on the data gathered from the high-order wind turbine model. In both cases, the SSNN models reflects the real system in a very good way. Moreover, the models may be linearized and described in the classic state-space form, with the state-space matrices, not with the weights parameters which can be found in the other approaches. Such a form can be used in the FD and control design.

Designed in this paper toolbox is the first stage of realizing the grant *Artificial neural networks for robust fault diagnosis and control of non-linear systems*. The next task of this research is to extend the functionality of the toolbox by adding the possibility to identify non-linear systems in the quasi-LPV form without linearization. Such form is widely used in modern fault-tolerant controls schemes [11, 18].

6.1. *Acknowledgments*

The work was supported by the National Science Centre of Poland under grants: UMO-2014/15/N/ST7/00749 and UMO-2012/07/N/ST7/03316.

References

[1] F.D. Bianchi, H. De Battista, and R.J. Mantz. *Wind turbine control systems: principles, modelling and gain scheduling design.* Advances in industrial control. Springer, 2007.

[2] S. Chen and P. Billings, S.and Grant. Non-linear system identification using neural networks. *International journal of control*, 51(6):1191–1214, 1990.

[3] S.R. Chu, R. Shoureshi, and M. Tenorio. Neural networks for system identification. *Control Systems Magazine, IEEE*, 10(3):31–35, 1990.

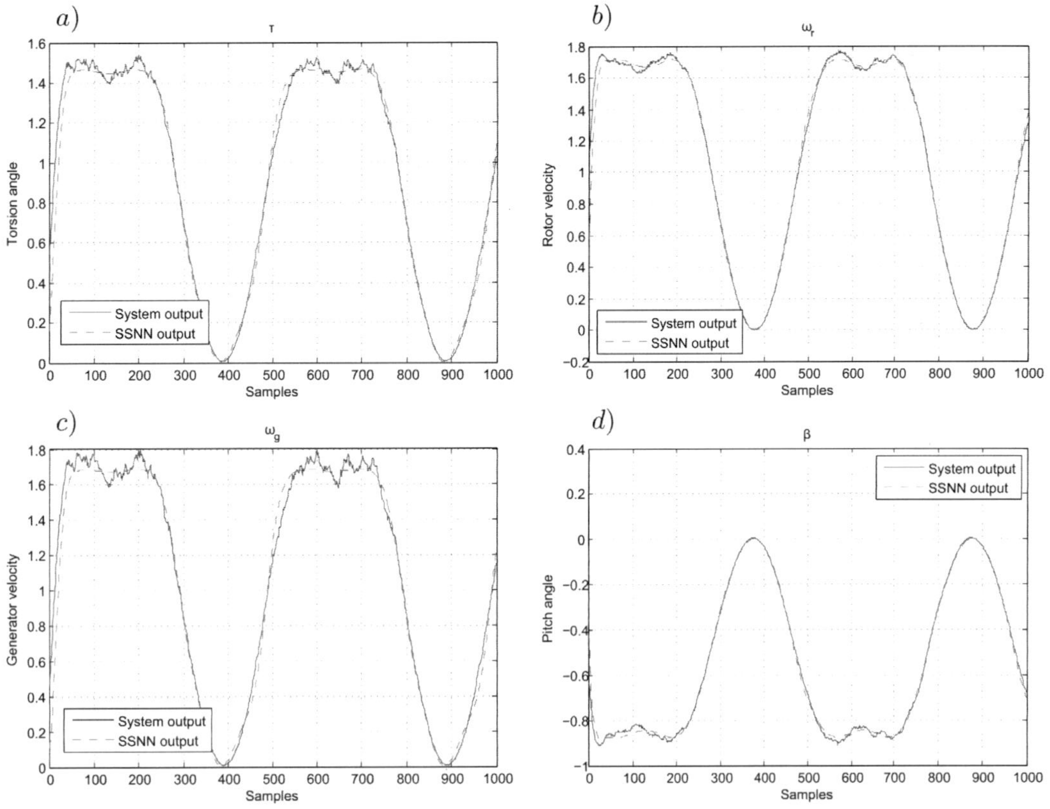

Figure 10. Results of the neural network modelling for: a) torsion angle τ, b) rotor velocity ω_r c) generator velocity ω_g, d) pitch angle β

[4] A. Czajkowski and K. Patan. Model predictive control of the two rotor aero-dynamical system using state space neural networks with delays. In *Intelligent Systems in Technical and Medical Diagnostics*, volume 230, pages 113–124. Springer Berlin Heidelberg, 2014.

[5] A. Czajkowski, K. Patan, and M. Szymański. Application of the state space neural network to the fault tolerant control system of the plc-controlled laboratory stand. *Engineering Applications of Artificial Intelligence*, 30:168–178, 2014.

[6] E. B. Kosmatopoulos, M. M. Polycarpou, M. Christodoulou, P. Ioannou, et al. High-order neural network structures for identification of dynamical systems. *Neural Networks, IEEE Transactions on*, 6(2):422–431, 1995.

[7] Dziekan Ł., Witczak M., and J. Korbicz. Active fault-tolerant control design for takagi-sugeno fuzzy systems. *Bulletin of the Polish Academy of Sciences-Technical Sciences*, 59(1):93–102, 2011.

[8] G. P. Liu. *Nonlinear identification and control: a neural network approach.* Springer Science & Business Media, 2012.

[9] M. Luzar, A. Czajkowski, M. Witczak, and M. Mrugalski. Actuators ans sensors fault diagnosis with dynamic, state-space neural networks. In *Methods and Models in Automation and Robotics - MMAR 2012 : Proceedings of the 17th IEEE International Conference*, pages 196–201, 2012.

[10] M. Luzar, Ł. Sobolewski, W. Miczulski, and J. Korbicz. Prediction of corrections for the polish time scale UTC(PL) using artificial neural networks. *Bulletin of the Polish Academy of Sciences: Technical Sciences*, 61(3):589–594, 2013.

[11] M. Luzar and Witczak. Robust MPC for a non-linear system - a neural network approach. *Journal of Physics: Conference Series*, 570(Article no: 032002), 2014.

[12] M. Luzar and M. Witczak. A GMDH toolbox for neural network-based modelling. `https://www.researchgate.net/publication/235901341_A_GMDH_TOOLBOX_FOR_NEURAL_NETWORK-BASED_MODELLING`, 2010. [Online; accessed October-2010].

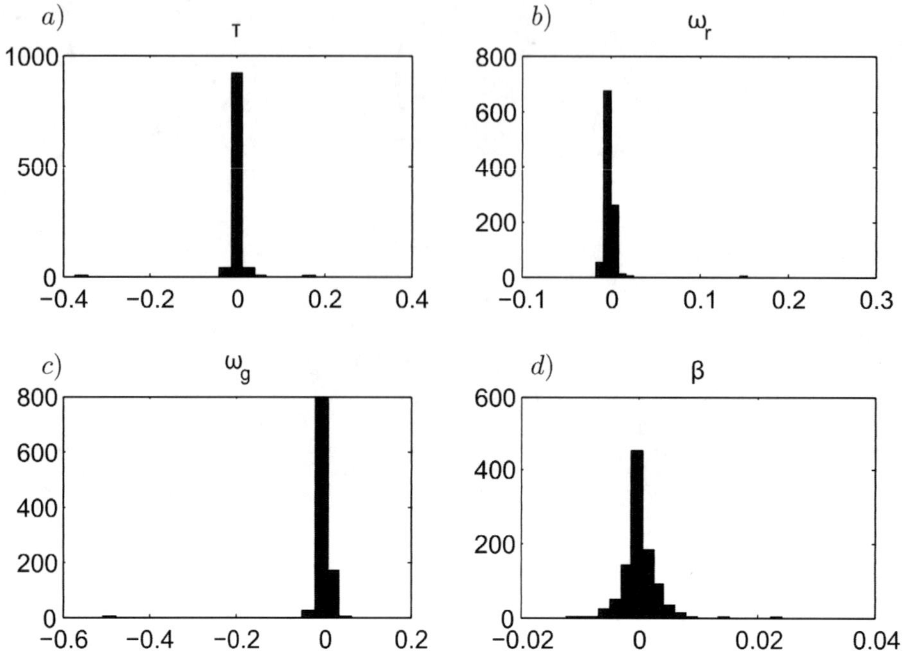

Figure 11. Neural network modelling errors for: a) torsion angle τ, b) rotor velocity ω_r c) generator velocity ω_g, d) pitch angle β

[13] M. Luzar, M. Witczak, and C. Aubrun. Robust $\mathcal{H}_\infty$ sensor fault diagnosis with neural networks. In *Intelligent Systems in Technical and Medical Diagnosis. (Korbicz J., Kowal M., (Eds.))*, pages 125–136. Springer, Berlin, 2013.

[14] M. Luzar, M. Witczak, and P. Witczak. Neural-network based robust FTC: Application to wind turbines. In *Artificial Intelligence and Soft Computing (Rutkowski L. et. all, (Eds.))*, pages 97–108. Springer, Berlin, 2014.

[15] M. Luzar, M. Witczak, P. Witczak, and C. Auburn. Neural-network based robust predictive fault-tolerant control for multi-tank system. In *Proc. of the 13th European Control Conference (ECC 2014)*, pages 276–281, Strasbourg, 2014.

[16] M. Nørgaard, O. Ravn, N.K. Poulsen, and L.K. Hansen. *Neural Networks for Modelling and Control of Dynamic Systems.* Springer-Verlag, London, 2014.

[17] K. Patan. *Artificial neural networks for the modelling and fault diagnosis of technical processes.* Springer-Verlag, Berlin, 2008.

[18] T. Peni, B. Vanek, Z. Szabo, and J. Bokor. Supervisory fault tolerant control of the GTM UAV using LPV methods. *International Journal of Applied Mathematics and Computer Science*, 25(1):117–131, 2015.

[19] D. Rotondo, F. Nejjari, and V. Puig. Robust quasi-lpv model reference ftc of a quadrotor uav subject to actuator faults. *International Journal of Applied Mathematics and Computer Science*, 25(1):7–22, 2015.

[20] J. Suykens, J. Vandewalle, and B. de Moor. *Artificial neural networks for modelling and control of non-linear systems.* Springer Science & Business Media, 2012.

[21] M. Witczak, V. Puig, and S. Montes De Oca. A fault-tolerant control strategy for non-linear discrete-time systems: application to the twin-rotor system. *International Journal of Control*, 86(10):1788–1799, 2013.

Journal of Physics: Conference Series **659** (2015) 012009 doi:10.1088/1742-6596/659/1/012009

Closed loop identification using a modified Hansen scheme

A Sekunda[1], H Niemann[1], N Kjølstad Poulsen[3] and I Santos[4]

[1] Department of Electrical Engineering, Technical Unversity of Denmark, Kongen Lyngby, Denmark

[3] Department of Applied Mathematics and Computer Science, Technical Unversity of Denmark, Kongen Lyngby, Denmark

[4] Department of Mechanical Engineering, Technical Unversity of Denmark, Kongen Lyngby, Denmark

E-mail: `aksek@elektro.dtu.dk`, `hhn@elektro.dtu.dk`, `nkpo@dtu.dk`, `ifs@mek.dtu.dk`

Abstract. It is often not feasible or even impossible to identify a plant in open loop. This might be because the plant contains unstable poles, or it is simply too expensive to remove the plant from its intended operation, among other possibilities. There are several methods for identifying a plant in closed loop [4], and one such method is the Hansen scheme [1]. Standard identification using Hansen scheme demands generating the identification signals indirectly. In this paper it is instead proposed to use the relationship between the Youla factorization of a plant and its stabilizing controller to directly measure the signals used for identification. A simulation example and identification of a gas bearing is given to show the method in action. Rotors supported by controllable gas bearings are open loop stable systems. However as the rotational speed is increased feedback control is necessary in order to keep the system stable. Furthermore because the dynamics of such a system depends on the rotational speed it is needed to conduct an identification while the system is part of a closed loop scheme. The authors believe the paper able to contribute towards a simpler and more direct way of identifying closed loop plants using Hansen scheme.

1. Introduction

Identification of closed loop systems is of great interest for many practical application. It is often not possible to conduct open loop identification on a plant. This can be due to several reasons, such as the system being open loop unstable, or disconnecting the controller being too costly. The Hansen scheme is a method for identification of closed loop systems, proposed by Fred Hansen [1]. The method takes advantage of the parametrization of all plants stabilized by a specific controller.

The experimental design procedure proposed by Fred Hansen demands the use of signals, which are not directly measurable from the plant G. It is therefore necessary to recreate the signals from known signals sent through predefined filters. The advantage of the method proposed in this paper is to get rid of the step of recreating the signals and instead measure directly the equivalent signals from the controller. Such a change will eliminate any need of a priori knowledge of the system states and reduce the amount of numerical inaccuracies related to the identification procedure.

In this paper the control architecture is based on a Luenberger observer design. The identification method proposed will be shown theoretically as well as experimentally, using data from a rotor-bearing system linked to a controllable gas bearing. However the method can easily be used with other controller architectures.

The paper is structured as follows: Section 2 introduces Hansen scheme and how to conduct identification of closed loop systems; Section 3 introduces the coprime factorization of plant and controller together with a Youla parametrization; Section 4 presents results from a simulation example for the reader; Section 5 presents an identification example of a gas bearing; Section 6 contains a conclusion on the results presented in the paper.

2. Closed loop identification using the Hansen Scheme

The Hansen scheme is a method originally introduced by Fred Hansen [1] for identification of plants connected as part of a closed loop scheme. In this section a brief description of the Hansen scheme is given together with a motivation for modification. The method takes advantage of the coprime factorisation of plant and controller and is based on the theory outlined in [7]. A standard setup for a plant as part of a closed loop scheme can be seen in Fig. 1. Here the signals r_1 and r_2 can be both known or unknown disturbance input. Let G be the nominal plant and K

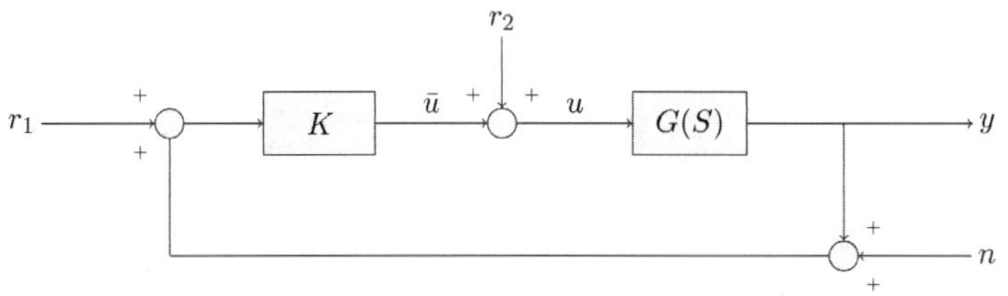

Figure 1. Closed-loop system representation.

be the nominal controller, a coprime factorization of the nominal plant is then given in Eq. (1) and a coprime factorization of the nominal controller is given in Eq. (2), given the 8 matrices satisfies the double Bezout identity shown in Eq. (3).

$$G = NM^{-1} = \tilde{M}^{-1}\tilde{N} \tag{1}$$

$$K = UV^{-1} = \tilde{V}^{-1}\tilde{U} \tag{2}$$

$$\begin{bmatrix} I & 0 \\ 0 & I \end{bmatrix} = \begin{bmatrix} \tilde{V} & -\tilde{U} \\ -\tilde{N} & \tilde{M} \end{bmatrix} \begin{bmatrix} M & U \\ N & V \end{bmatrix} = \begin{bmatrix} M & U \\ N & V \end{bmatrix} \begin{bmatrix} \tilde{V} & -\tilde{U} \\ -\tilde{N} & \tilde{M} \end{bmatrix} \in \mathbf{RH}_\infty \tag{3}$$

With a coprime factorization of the nominal plant G and of the controller K stabilizing the nominal plant $G(0)$, Eq. (4) gives a parametrization of all plants stabilized by the nominal controller K [3].

$$G(S) = (N + VS)(M + US)^{-1} = (\tilde{M} + S\tilde{U})^{-1}(\tilde{N} + S\tilde{V}) \tag{4}$$

The goal of the identification method is to find a signal uncorrelated with the output disturbance, n, to use for identification of the open loop error system, denoted S in Eq. (4).

The real plant can be described using the nominal controller and nominal plant together with the open loop error. Such a description is shown in Fig. 2 [1]. Here the output noise have been

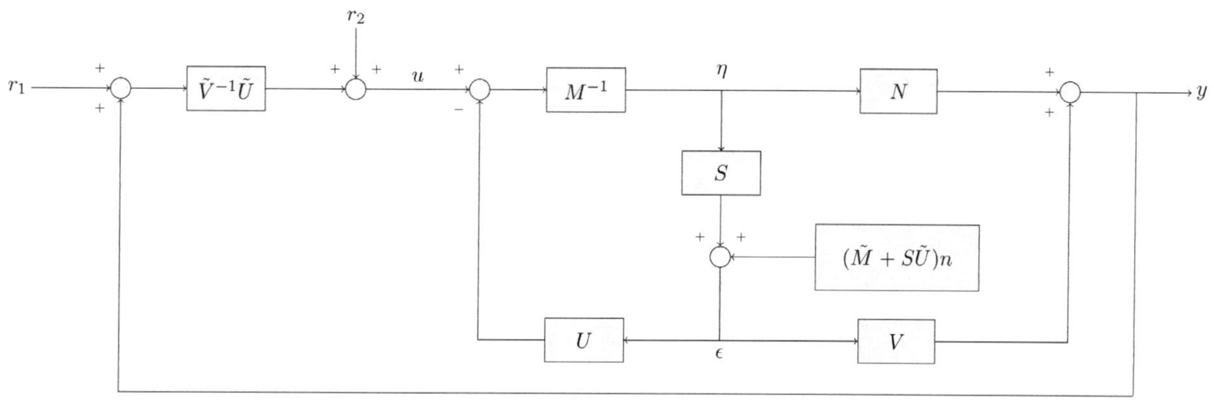

Figure 2. Block representation of the Hansen Scheme.

moved to inside the plant. This representation is believed to make it easier to determine the impact of the noise on the identification process.

For the representation shown in Fig. 2 it is possible to recreate the signal η from the 2 input, r_1 and r_2 which are independent of the input and output noise. The calculation of η is given by Eq. (5) to Eq. (10).

$$u(t) = (M + US)\eta(t) + U(\tilde{M} + S\tilde{U})n(t) \quad (5)$$
$$y(t) = (N + VS)\eta(t) + V(\tilde{M} + S\tilde{U})n(t) \quad (6)$$
$$\tilde{V}u(t) - \tilde{V}(M + US)\eta(t) = \tilde{U}y(t) - \tilde{U}(N + VS)\eta(t) \quad (7)$$
$$\tilde{V}(\tilde{V}^{-1}U(r_1(t) + y(t)) + r_2(t)) - \tilde{V}(M + US)\eta(t) = \tilde{U}y(t) - \tilde{U}(N + VS)\eta(t) \quad (8)$$
$$\tilde{U}r_1(t) + \tilde{V}r_2(t) = \tilde{V}(M + US)\eta(t) - \tilde{U}(N + VS)\eta(t) \quad (9)$$
$$\eta(t) = \tilde{U}r_1(t) + \tilde{V}r_2(t) \quad (10)$$

In a similar fashion it is possible to calculate ϵ using only the input measurements u and the output measurements y. How to calculate ϵ is derived in Eq. (11) to Eq. (15).

$$M\eta(t) = u(t) - U\epsilon(t) \quad (11)$$
$$N\eta(t) = y(t) - V\epsilon(t) \quad (12)$$
$$\tilde{N}u(t) - \tilde{N}U = \tilde{M}y(t) - \tilde{M}V\epsilon(t) \quad (13)$$
$$\tilde{M}y(t) - \tilde{N}u(t) = (\tilde{M}V - \tilde{N}U)\epsilon(t) \quad (14)$$
$$\epsilon(t) = \tilde{M}y(t) - \tilde{N}u(t) \quad (15)$$

With the signals η and ϵ it is possible to identify the open loop error system S as shown in Eq. (18), which is an open loop identification problem.

$$\eta(t) = \tilde{U}r_1(t) + \tilde{V}r_2(t) \quad (16)$$
$$\epsilon(t) = \tilde{M}y(t) + \tilde{N}u(t) \quad (17)$$
$$\epsilon(t) = S\eta(t) + d(t) \quad (18)$$

Because the signals η and ϵ are internal signals in the plant it is impossible to measure them directly. It is therefore necessary to create them from the already known signals, as shown in Eq. (16) and Eq. (17). This approach might give lead to numerical problems because of how the signals are calculated, while the input signals are also led through a filter shaping the excitation signal. Furthermore initial conditions might increase the uncertainty of the signal estimation. It is therefore believed by the authors that it is a huge advantage to measure the signals directly instead.

The rest of the paper will outline a method for how to directly apply a signal equivalent to η and how to measure a signal equivalent to ϵ using a full order observer based controller. A simulation example will be given and results from applying it to a gas bearing is finally presented.

3. Controller plant relationship

In this section the relationship between the parametrization of all controllers stabilised by a specific plant and all plants stabilized by a specific controller investigated. The results stated in the following section can also be found in [3] [8].

As shown in [3] [7] all controllers stabilising a specific plant is given in Eq. (19).

$$K(Q) = (U + MQ)(V + NQ)^{-1} = (\tilde{V} + Q\tilde{N})^{-1}(\tilde{U} + Q\tilde{M}) \mid Q \in \mathbf{RH}_\infty \tag{19}$$

$$K(Q) = \mathcal{F}_l(J_k, Q) \tag{20}$$

$$J_k = \begin{bmatrix} K & \tilde{V}^{-1} \\ V^{-1} & -V^{-1}N \end{bmatrix} \tag{21}$$

It is easy to see that the parametrization can be represented as a lower linear fractional transformation [5]. Such a controller can be implemented as seen in Fig. 3. Here Q is chosen to be zero, a set up which will be used throughout the rest of the paper. It is easy to see in Fig. 3 that by closing the loop from β to α using Q, the controller is given as in Eq. (19).

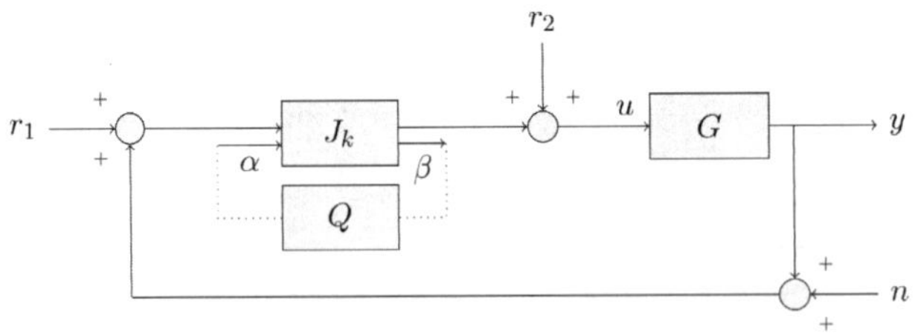

Figure 3. Closed loop description with a controller given as in Eq. (19), where the system Q is disconnected.

Equivalently it is possible to state all plants stabilised by a specific controller as shown in Eq. (22).

$$G(S) = (N + VS)(M + US)^{-1} = (\tilde{M} + S\tilde{U})^{-1}(\tilde{N} + S\tilde{V}) \mid S \in \mathbf{RH}_\infty \tag{22}$$

$$G(S) = \mathcal{F}_u(J_G, S) \tag{23}$$

$$J_G = \begin{bmatrix} -M^{-1}U & M^{-1} \\ \tilde{M}^{-1} & G \end{bmatrix} \tag{24}$$

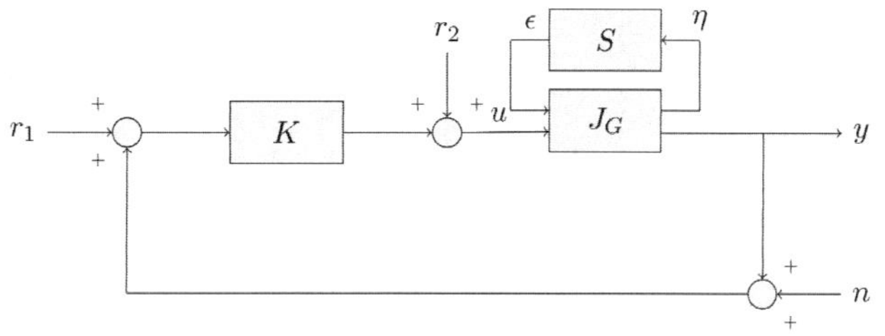

Figure 4. General lower LFT description of $G(S)$.

Again it is possible to represent $G(S)$ as an upper LFT as shown in Fig. 4. It is important to notice that the signals η and ϵ shown in Fig. 4 are the same η and ϵ signals as shown in Fig. 2.

Combining J_k and J_G as shown in Fig. 5 it is possible to calculate the transfer function from each input to the two outputs.

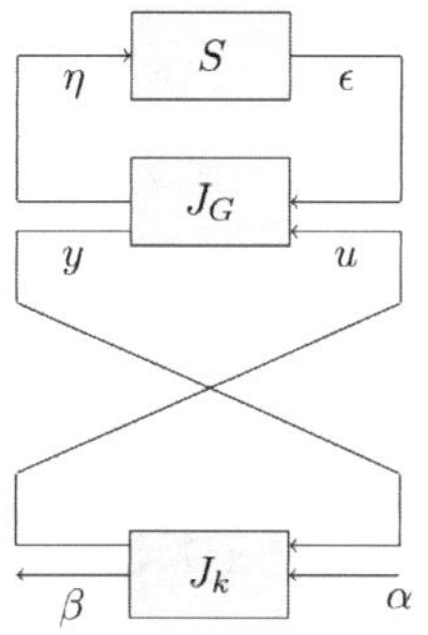

Figure 5. Block representation of the relationship between $G(S)$ and $K(Q)$.

It is easy to see that the cross coupling of J_k and J_G in Fig. 5 can be calculated as a redheffer star product. The gains of the block is shown in Eq. (25).

$$J_G \star J_k = \begin{bmatrix} \mathcal{F}_l(J_G, K) & M^{-1}(\mathbf{I} - KG)^{-1}\tilde{V}^{-1} \\ V^{-1}(\mathbf{I} - GK)^{-1}\tilde{M}^{-1} & \mathcal{F}_u(J_k, G) \end{bmatrix} \tag{25}$$

In order to determine the relationship between the signals on each side of the system Q and the system S the gains of the redheffer starproduct shown in Eq. (25) are calculated. First the transfer function from the output of S to the input of S is found:

$$\begin{aligned}
\mathcal{F}_l(J_G, K) &= -M^{-1}U + M^{-1}K(I - GK)^{-1}\tilde{M}^{-1} && (26)\\
&= -M^{-1}U + M^{-1}UV^{-1}(I - GK)^{-1}\tilde{M}^{-1} && (27)\\
&= -M^{-1}U + M^{-1}UV^{-1}(I - \tilde{M}^{-1}\tilde{N}UV^{-1})^{-1}\tilde{M}^{-1} && (28)\\
&= -M^{-1}U + M^{-1}UV^{-1}V(\tilde{M}V - \tilde{N}U)^{-1}\tilde{M}\tilde{M}^{-1} && (29)\\
&= -M^{-1}U + M^{-1}U(\tilde{M}V - \tilde{N}U)^{-1} && (30)\\
&= -M^{-1}U + M^{-1}U(\mathbf{I})^{-1} && (31)\\
&= 0 && (32)
\end{aligned}$$

Journal of Physics: Conference Series **659** (2015) 012009 doi:10.1088/1742-6596/659/1/012009

The transfer function from the output of Q to the input of Q can in a similar fashion be proved to be 0. It is therefore only left to show the transfer function respectively from the output of Q to the input of S, which can be seen in Eq. (34) and from the output of S to the input of Q, which can be seen in Eq. (33).

$$\begin{aligned} M^{-1}(\mathbf{I}-KG)^{-1}\tilde{V}^{-1} &= M^{-1}(\mathbf{I}-\tilde{V}^{-1}\tilde{U}NM^{-1})^{-1}\tilde{V}^{-1} \\ &= M^{-1}M(\tilde{V}M-\tilde{U}N)^{-1}\tilde{V}\tilde{V}^{-1} \\ &= (\tilde{V}M-\tilde{U}N)^{-1} = \mathbf{I} \qquad (33) \\ V^{-1}(\mathbf{I}-GK)^{-1}\tilde{M}^{-1} &= V^{-1}(\mathbf{I}-\tilde{M}^{-1}\tilde{N}UV^{-1})^{-1}\tilde{M}^{-1} \\ &= V^{-1}V(\tilde{M}V-\tilde{N}U)^{-1}\tilde{M}\tilde{M}^{-1} \\ &= (\tilde{M}V-\tilde{N}U)^{-1} = \mathbf{I} \qquad (34) \end{aligned}$$

Combining the results from Eq. (32),(33) and (34) the relationship between α and β can be described as in Eq. (35). The result can be found in [3] where the relationship between Q and S, to the best of the authors knowledge, was examined for the first time.

$$\beta = S\alpha \qquad (35)$$

It is thus proved that the input signal of S is equal to the output signal of Q and the output signal of S is equal to the input signal of Q. It is therefore possible to directly measure the signals equivalent to η and ϵ from the parametrized controller as the signals α and β respectively. To the best knowledge of the authors this approach is not limited to a specific controller architecture, but can be applied to any controller implemented as showed in Fig. 6.

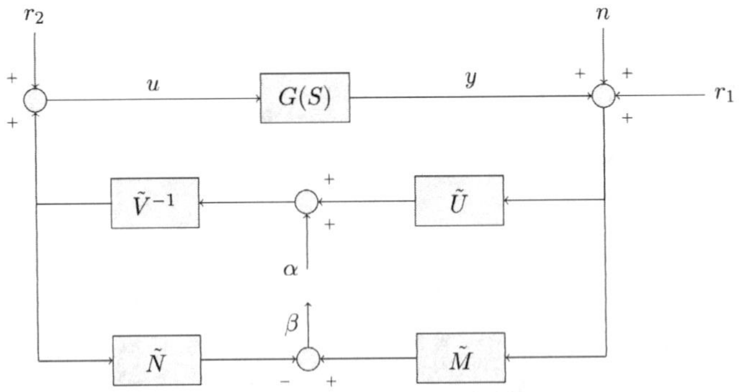

Figure 6. General controller scheme using a coprime plant and controller description.

The approach demands that it is possible to find the signals α and β in the controller. This is in this paper done using an observer based controller. It is important to notice the order of the recast controller is increased by twice the order of the nominal plant model. It can therefore be convenient to use an observer based control scheme, where the order of the controller is 2 times the order of the nominal plant, which is the minimal order possible for such a controller set up.

For a controller given as in Fig. 6 as part of a closed loop as shown in Fig. 1 the closed loop transfer functions are given in Eq. (36), where α is a free signal to choose for identification purposes.

$$\begin{bmatrix} y \\ u \\ \beta \end{bmatrix} = \begin{bmatrix} (N+VS)\tilde{U} & (N+VS)\tilde{U} & (N+VS)\tilde{V} & N+VS \\ (M+US)\tilde{U} & (M+US)\tilde{U} & (M+US)\tilde{V} & M+US \\ \tilde{M}+S\tilde{U} & \tilde{M}+S\tilde{U} & \tilde{N}+S\tilde{V} & S \end{bmatrix} \begin{bmatrix} n \\ r_1 \\ r_2 \\ \alpha \end{bmatrix} \quad (36)$$

By applying a known disturbance signal on both the observer input and the plant input, hence the α signal, it is possible to measure β as the difference between the predicted output by the observer and the output from the plant $G(S)$. This method is seen as superior in regards to the freedom of the identification signal compared to the method proposed in [1], where the excitation signal is imposed through r_1 and r_2. It is clear from Eq. (36) that the 3 excitation signals and the disturbance signal are uncorrelated why open loop identification can be conducted.

4. Simulation example

A simulation example is presented to give the reader some insight into the identification procedure. A random open loop stable plant with 2 inputs, 2 outputs and 4 states is generated. The nominal plant is chosen such that the system matrix (A_{model}) is 70% of the actual system matrix (A_{real}) as shown in Eq. (37).

$$A_{model} = 0.7 \cdot A_{real} \quad (37)$$

The nominal plant have thus both an dynamic and steady state error that needs to be identified. The test set up is shown in Fig. 7 where both the unknown disturbance signal, n, and the excitation signal (α) are implemented as white noise signals. The amplitude of the noise was chosen to be 10% of the β signal. Furthermore only output noise, r_1, was used for this simulation.

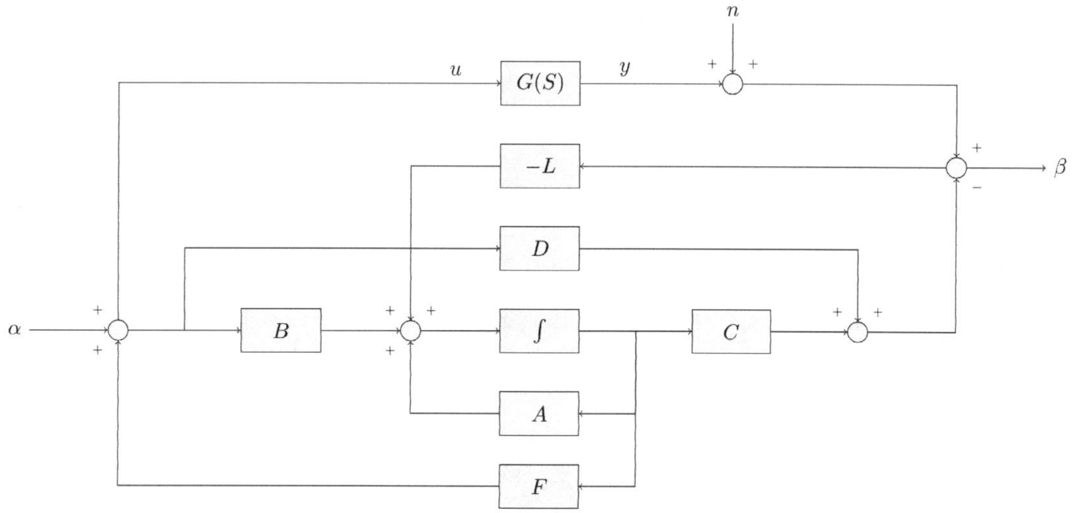

Figure 7. Block diagram of simulation example. The observer is designed for a plant where the elements of the system matrix is only 70 % of the actual values of the elements of the system matrix. The state feedback is designed using LQR design and the observer poles are placed to have insignificant dynamics compared to the state feedback poles.

The order of the state space model is chosen by calculating the real open loop error, which was found to be a 12'th order system. It is therefore possible to identify the real open loop error, S, as a 12'th order system, why the open loop error is identified as a 12'th order state space model.
In order to be able to determine the goodness of a model an error signal is calculated using Eq. (38).

$$\phi(t) = G_{real}u(t) - G_{model}u(t) \tag{38}$$

An error signal sequence of respectively the nominal model and the identified plant, using Eq (38), can be seen in Fig. 8. It is clearly seen that the identified plant reduce the output error why it can be concluded possible to improve the nominal model, which was on purpose designed to be imperfect. The dynamics are identified although there is still a small deviation between the identified and the real plant, which is due to numerical issues and presence of output noise on the identification signal. It is important to note that there is not applied any noise to the validation data shown in Fig. 8 why any deviation from zero corresponds to a deviation between the real plant and respectively the nominal or identified plant.

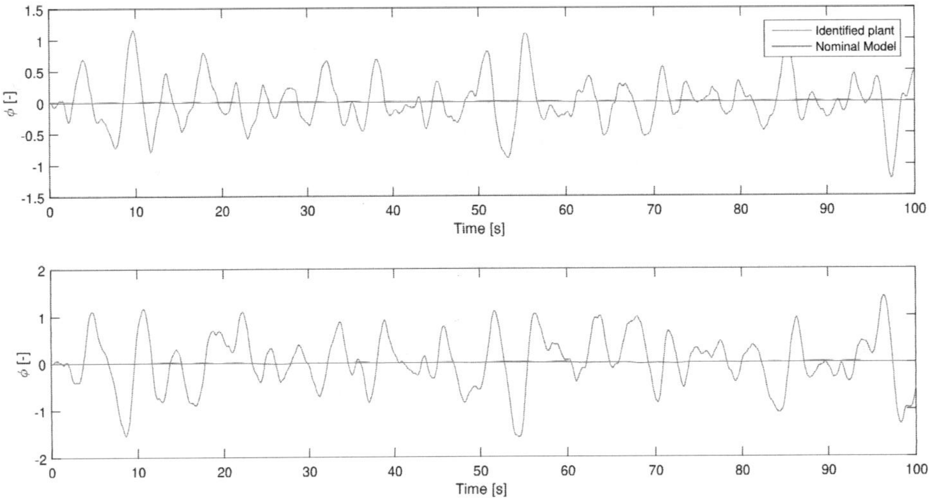

Figure 8. Goodness of fit of Nominal and identified plant. Each signal is calculated using Eq. (38). The top plot shows respectively nominal and identified model for the first output, while the lower plot shows the signal for output 2. The identified plant have a better goodness fit which can be seen from the variance in the signal being lower than for the nominal model while the mean is at 0.

A bode plot of respectively the real plant, the identified plant and the plant model is shown in Fig. 9. It is easily seen from the bode plot that the model does not agree with the real plant. However the identified plant have successfully minimized the error over the whole frequency span.
For validation of the identification procedure it is also possible to compare the real S with the one identified through the identification procedure. A bode plot of the identified and real S are shown in Fig. 10.

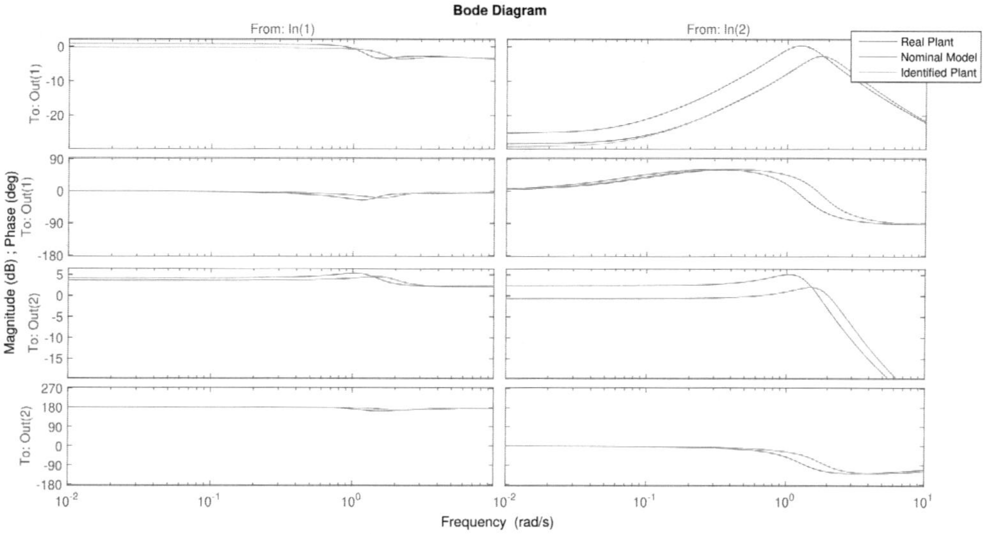

Figure 9. Bode plot of the real plant, the identified plant and the plant model. The plant model and real plant are clearly not the same, while the identified plant have almost same frequency response as the real plant.

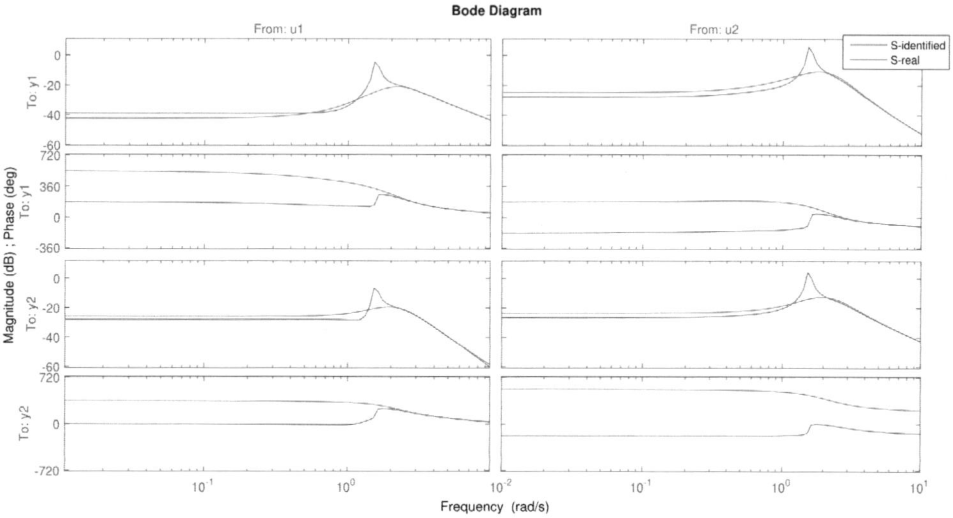

Figure 10. Bode plot of the real S calculated as described in [3] and the identified S using the procedure described in section 3.

5. Lab experiment

Identification of a gas bearing test rig is given to show the identification procedure with data obtained from a real system. A photo of the gas bearing test rig is shown in Fig. 11.

A block diagram of the test rig is shown in Fig. 12. Here $\mathbf{K_s}$ describes the direct stiffness and cross stiffness of the gas bearing, $\mathbf{D_d}$ the direct damping and cross damping of the gas bearing and $\mathbf{B}$ is the direct input gain and cross input gain of the gas bearing.

Figure 11. Photo of the gas bearing test rig. Here (1) is the turbine, (2) is the flexible shaft, (3) is a ball bearing, (4) is the gas bearing, (5) is a disc used for preload the journal and (6) is the displacement sensors.

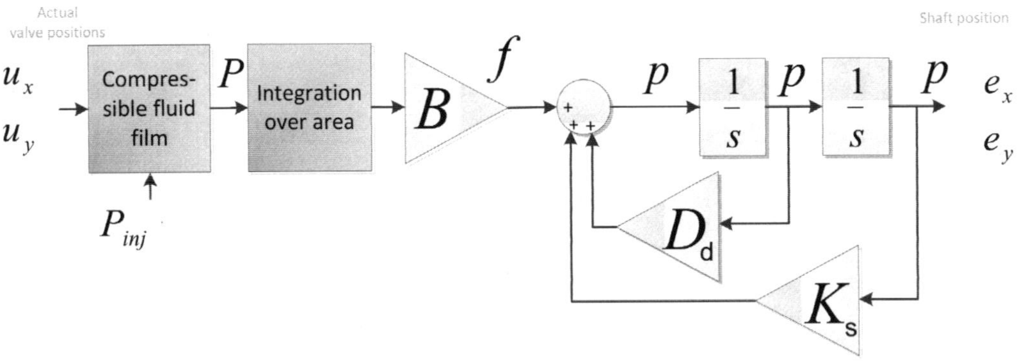

Figure 12. Block diagram representation of the gas bearing test rig.

The rotor-bearing system thus consists of 8 states, 2 inputs and 2 outputs. A model for this set up have been derived and identified in [9] where the gas bearing is not a part of a closed loop Scheme. The identification shown in this section is based on non-rotating shaft levitated by the externally presented gas bearing shown in Fig. 11, as part of a closed loop scheme. The controller implemented is designed as a standard Luenberger observer where the observer gain and state feedback gain is chosen using lqr design. It is important to note that the gas bearing is open loop stable at zero rotational speed. It is therefore possible to identify the plant using open loop techniques. However the point is to show that the method is applicable why it will be possible to use the same method for higher rotational speeds. At higher rotational speed the rotor-bearing is open loop unstable why only closed loop identification will be possible.

For identification an observer with state feedback is designed and as for the simulation example showed in section 4 a white noise excitation signal is applied as the η signal in the horizontal and vertical direction. The open loop error, S, is estimated as a 8'th order system, same order as the nominal model, which is found to produce a good identification result.

An error plot as described in Eq. (38) is presented in Fig. 13 using both the nominal model and the identified model. For validation the excitation signal is only applied in the horizontal direction. It is worth noticing that the nominal model is a good representation of the test rig, why the error is relative small compared to the output noise which is measured to have an amplitude of 1 μm.

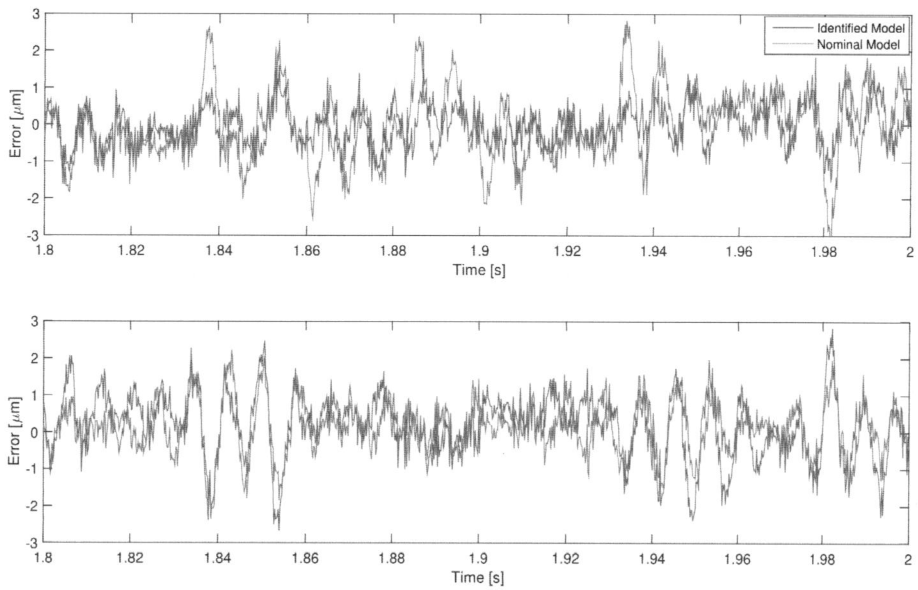

Figure 13. Plot of the error between the predicted output and measured output for horizontal and vertical direction. The top plot shows the error in the horizontal direction, while the bottom plot shows the error in the vertical direction.

From Fig. 13 it can be seen that the identification have produced a more accurate model for prediction of the horizontal direction (the top plot), while there is no significant improvement in the vertical direction. The error in the horizontal direction is reduced to close to 1 μm which is the level of the output noise induced by the displacement sensors. The result suggests that it has been possible to identify the direct gains from horizontal input to horizontal output while the cross coupling between the two inputs have not been improved through the identification. It is believed possible to identify the direct gain in the vertical direction with a pure vertical excitation signal in a likewise manner.

6. Conclussion

Identification using Hansen scheme has so far been conducted using indirect excitation signals for identification. Such an approach makes it difficult to determine the frequency response of the excitation signal. In this paper a method is given, on how to use the equivalent signals in the controller, for identification of the open loop error system S, thus letting the excitation be imposed without pre filtering which is the case with the original method proposed in [1]. It is therefore believed that the Hansen scheme is a more limiting special case of the identification approach outlined in this paper. The paper outline an approach for how to excite a plant and measure the needed signals for identification, given an observer based controller. The paper furthermore shows how it is possible to recast any controller such that it is possible to excite and measure the signals used here for identification. Using a simulation example and a Gas Bearing system for identification the method is proven to work.

References

[1] HANSEN, Fred; FRANKLIN, Gene; KOSUT, Robert. Closed-loop identification via the fractional representation: Experiment design. In: 1989 American Control Conference. 1989. p. 1422-1427.

[2] ANDERSON, Brian DO. From YoulaKucera to identification, adaptive and nonlinear control. Automatica, 1998, 34.12: 1485-1506.

[3] TAY, Teng-Tiow; MAREELS, Iven; MOORE, John B. High performance control. Springer Science & Business Media, 2012.

[4] SÖDERSTRÖM, Torsten; STOICA, Petre. System identification. Prentice-Hall, Inc., 1988.

[5] ZHOU, Kemin, et al. Robust and optimal control. New Jersey: Prentice hall, 1996.

[6] NIEMANN, Henrik; POULSEN, Niels Kjolstad. Estimation of parametric fault in closed-loop systems. In: American Control Conference (ACC), 2015. IEEE, 2015. p. 201-206.

[7] VIDYASAGAR, M. Control system synthesis: a factorization approach, 1985. Cambridge, Massachusetts.

[8] NIEMANN H., "Architecture for fault diagnosis and fault-tolerant control", Department of Electrical Engineering DTU, 2015.

[9] THEISEN, Lukas Roy Svane, et al. Experimental Grey box model identification and control of an active gas bearing. In: Vibration Engineering and Technology of Machinery. Springer International Publishing, 2015. p. 963-976.

On-line Model Structure Selection for Estimation of Plasma Boundary in a Tokamak

Vít Škvára[1], Václav Šmídl[1], Jakub Urban[2]
[1]Institute of Information Theory and Automation, Prague, Czech Republic,
[2]Institute of Plasma Physics, Prague, Czech Republic,

Abstract. Control of the plasma field in the tokamak requires reliable estimation of the plasma boundary. The plasma boundary is given by a complex mathematical model and the only available measurements are responses of induction coils around the plasma. For the purpose of boundary estimation the model can be reduced to simple linear regression with potentially infinitely many elements. The number of elements must be selected manually and this choice significantly influences the resulting shape. In this paper, we investigate the use of formal model structure estimation techniques for the problem. Specifically, we formulate a sparse least squares estimator using the automatic relevance principle. The resulting algorithm is a repetitive evaluation of the least squares problem which could be computed in real time. Performance of the resulting algorithm is illustrated on simulated data and evaluated with respect to a more detailed and computationally costly model FREEBIE.

1. Introduction

Tokamak is one the most promising concepts for future thermonuclear fusion reactors. In a tokamak, hot plasma is confined by a combination of strong toroidal and poloidal magnetic fields. The almost negligible weight of the plasma dictates that magnetohydrodynamic (MHD) equilibrium, i.e. a force-free state, describes its macroscopic evolution. MHD decsription includes also the plasma boundary (the so-called last closed flux surface), which must be controlled during a tokamak discharge because of performance and safety reasons. Real-time feedback control is necessary since the plasma behaviour cannot be reliably predicted, particularly because the equilibrium is unstable. Such a feedback control requires the plasma boundary as an input in real-time. See e.g. [1] or [2] for more details.

One of the possibilities for real-time plasma shape reconstruction is offered by magnetic sensors near the plasma boundary. However, full reconstruction of the equilibrium magnetic field (which includes the plasma shape) is a complex task due to non-linear behavior of the plasma. In [3], it was shown that the plasma boundary can be modeled by a linear combination of its base functions. Since the base functions are known, the estimation problem can be formulated as a simple linear regression with unknown number of regressors. Our task is to select the relevant subset of base functions of the model.

Selection of the structure of the linear model is a well studied problem in statistics and strong point of the Bayesian hypothesis testing. The conventional least squares problem is equivalent to maximum likelihood estimation of the linear Gaussian model. Selection of structure of this model is a conventional problem requiring to formalize two steps: i) to choose prior distribution of the

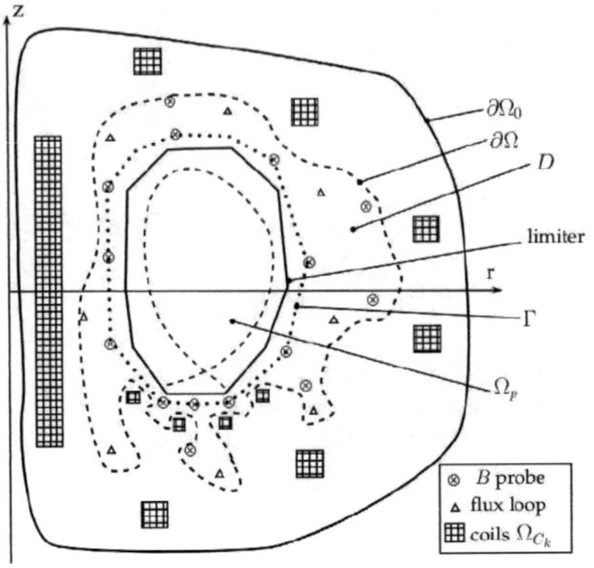

Figure 1. Tokamak section schema with measuring and shaping coils. *B probe* measures local magnetic field size, *flux loop* is a coil measuring magnetic flux ψ and Ω_{C_k} is a plasma shaping coil. Different contours important for plasma boundary reconstruction are denoted. *Limiter* is the inner surface of tokamak chamber, Ω_p is the plasma itself, delimited by dashed line. In the bottom of plasma we can observe an x-point, which is crucial for ensuring plasma stability. From [3].

parameters, and ii) to evaluate likelihoods corresponding to structures consisting of all significant combinations of regressors whose supply grows exponentially with the number of candidates.

A range of techniques arise by combination of different options for each step. This step is very important since the choice of a non-informative prior is inappropriate as it yields posterior Bayes factors dependent on an arbitrary multiplicative constant [4]. Many possibilities, including ridge regression, mixture priors, or g priors were tried, see review in [5] . Once the prior is chosen, many algorithms searching the space of solution are available using e.g. stochastic search [6], dynamic programming [7], MCMC approach [8] or variational approaches [9, 10].

In this paper, we are interested in estimation of the structure in real time for the purpose of tokamak control. Therefore, we seek an approach with simple implementation. One such possibility is the automatic relevance determination principle [11] which uses a hierarchical prior to suppress the regressors that are redundant. This technique uses the Variational Bayes approximation [12] which can be easily extended to dynamical systems [13].

The paper is organized as follows. In the second Section we introduce model of the plasma boundary is introduced in the form of linear regression. A model of Bayesian regularization of the linear regression model is introduced in the third Section and its Variational Bayes solution is presented. Experimental validation of the approach on simulated data from a real tokamak are presented in the fourth Section.

2. Model of plasma Boundary

Feedback control of plasma in a tokamak requires on-line reconstruction of the plasma boundary. The quality of the reconstruction is highly significant for tokamak operation safety and performance. Two different kinds of measurements, which are typically available on a tokamak, are considered here to determine the shape: magnetic coils measuring local magnetic field B and flux loops measuring the poloidal magnetic flux ψ.

Following [3], we will formulate the model in the standard cylindric coordinates (r, ϕ, z) that are used in tokamak geometry description. Using the assumption of symmetry of plasma in toroidal direction of plasma chamber, we omit ϕ. Then, we carry out a transformation from (r, z) to toroidal coordinates $(\zeta, \eta) \in \mathbb{R}^+ \times \langle 0, 2\pi)$ around a pole $F_0 = (r_0, z_0)$, given by [14]

$$r = \frac{r_0 \sinh \zeta}{\cosh \zeta - \cos \eta}, \tag{1}$$

$$z = z_0 + \frac{r_0 \sin \eta}{\cosh \zeta - \cos \eta}. \tag{2}$$

Now let us assume that we have three sets of measurements available:

(i) N_f measurements of magnetic flux in coordinates $x_i^f = (r_i^f, z_i^f)$, $\psi_i^{meas} \approx \psi(x_i^f)$,

(ii) N_s measurements of magnetic flux gradient between two points x_i^1, x_i^2, $\delta_i \psi^{meas} = \psi(x_i^1) - \psi(x_i^2)$,

(iii) N_B measurements of magnetic field in points x_i^B and directions d_i, $B_i^{meas} \approx B(x_i^B) d_i$.

Next, we utilize the following decomposition of the flux estimate $\hat{\psi}(\zeta, \eta)$ into a system of toroidal harmonics

$$\hat{\psi} = \hat{\psi}_{ext} + \hat{\psi}_{int}, \tag{3}$$

$$\begin{aligned} \hat{\psi}_{ext}(\zeta, \eta) =& \frac{r_0 \sinh \zeta}{\sqrt{\cosh \zeta - \cos \eta}} \times \\ & \left\{ \sum_{n=0}^{M_{ea}} a_n^e Q_{n-1/2}^1(\cosh \zeta) \cos(n\eta) + \sum_{n=1}^{M_{eb}} b_n^e Q_{n-1/2}^1(\cosh \zeta) \sin(n\eta) \right\}, \end{aligned} \tag{4}$$

$$\begin{aligned} \hat{\psi}_{int}(\zeta, \eta) =& \frac{r_0 \sinh \zeta}{\sqrt{\cosh \zeta - \cos \eta}} \times \\ & \left\{ \sum_{n=0}^{M_{ia}} a_n^i P_{n-1/2}^1(\cosh \zeta) \cos(n\eta) + \sum_{n=1}^{M_{ib}} b_n^i P_{n-1/2}^1(\cosh \zeta) \sin(n\eta) \right\}. \end{aligned} \tag{5}$$

Functions $P_{n-1/2}^1$ and $Q_{n-1/2}^1$are the associated Legendre functions of the first and second kind, of degree one and half integer order [14], that are called toroidal harmonics when evaluated at $\cosh \zeta$. To carry out the decomposition, one must choose the basis functions degrees appropriately by defining integers M_{ia}, M_{ib},M_{ea}, M_{eb}. Numerical stability of the final solution is highly dependent on this setting. Furthermore, complete decomposition is solved by determining a vector u of unknown decomposition coefficients in (4), (5), given by

$$\theta = (a_0^e, \ldots, a_{M_{ea}}^e, b_1^e, \ldots, b_{M_{eb}}^e, a_0^i, \ldots, a_{M_{ia}}^i, b_1^i, \ldots, b_{M_{ib}}^i). \tag{6}$$

This is done by minimizing cost function

$$J(\theta) = \sum_{i=1}^{N_f} \frac{(\hat{\psi}_i(\theta) - \tilde{\psi}_i^{meas})^2}{\sigma_f^2} + \sum_{i=1}^{N_s} \frac{(\delta_i \hat{\psi}(\theta) - \delta_i \tilde{\psi}^{meas})^2}{\sigma_s^2} + \sum_{i=1}^{N_B} \frac{(\hat{B}_i(\theta) - \tilde{B}_i^{meas})^2}{\sigma_B^2}. \tag{7}$$

Here, $\hat{B}$ is determined from$\hat{\psi}$ using

$$B = \frac{1}{r} \left(-\partial_z \psi, \partial_r \psi \right), \tag{8}$$

where σ_f^2, σ_s^2 and σ_B^2 are known variances of the measurement errors on the corresponding coils. Quantities $\tilde{\psi}_i^{meas}, \delta_i\tilde{\psi}^{meas}$ and $\tilde{B}_i^{meas}$ are computed from the measurements described above by subtracting the contribution of individual poloidal field coils. Vector θ can be utilized to reconstruct ψ in every point of coordinate system (r, z). Plasma boundary is identified as largest closed isoline of ψ (see Fig. 1).

Minimalization of (7) translates into equation

$$\mathrm{diag}(w)\tilde{y}(\tilde{\psi}^{meas}, \tilde{B}^{meas})^T = \mathrm{diag}(w)\tilde{A}(\hat{\psi}, \hat{B})\theta + e, \tag{9}$$

where $\tilde{y}(\tilde{\psi}^{meas}, \tilde{B}^{meas})^T \in \mathbb{R}^N, (N = N_f + N_s + N_B)$ are measurements extracted from individual coils,

$$w = (\underbrace{\sigma_f^{-1}, \cdots, \sigma_f^{-1}}_{N_f}, \underbrace{\sigma_s^{-1}, \cdots, \sigma_s^{-1}}_{N_s}, \underbrace{\sigma_B^{-1}, \cdots, \sigma_B^{-1}}_{N_B}) \tag{10}$$

is a vector of precisions of coils and diag(.) is a diagonal matrix with its argument on the main diagonal. Design matrix $\tilde{A}(\hat{\psi}, \hat{B}) \in \mathbb{R}^{N\times M}, (M = M_{ea}+M_{eb}+M_{ia}+M_{ib})$ comprises of evaluation of toroidal harmonics (4) and (5) in coordinates given by position of the measurement coils in chosen coordinate system, $e \in \mathbb{R}^N$ is the white noise with precision equal to β^{-1}. For further use, we rewrite equation (9) into a concise form

$$y = A\theta + e. \tag{11}$$

Because it was shown that precision of fit measured by MSE criterion does not translate directly into quality of final plasma boundary estimate, we shall use a different metric to assess it. We choose the *Hausdorff distance* with L_2 norm [15], which is for $O = \{o_1, ..., o_k\}$, $P = \{p_1, ..., p_l\}$, $o_i, p_i \in \mathbb{R}^m, \forall i$ defined as

$$D_H(O, P) = \max\left(\max_{o\,\in O}\min_{p\in P}\|o-p\|_2, \max_{p\in P}\min_{o\,\in O}\|o-p\|_2\right), \tag{12}$$

where $\|.\|_2$ is L_2 norm defined on $\mathbb{R}^m$. Further in the text, this metric is used to evaluate and compare output of the classical and regularized solution of the boundary reconstruction problem and a simulation.

It has been shown that the order of decomposition into toroidal harmonics significantly influences the quality of plasma boundary reconstruction. Furthermore, tested data showed that the best approach may not always be to estimate as many coefficients of the toroidal harmonics decomposition as possible. Also, the results of estimation are sensitive to the number of external and internal toroidal harmonics in a different way. We seek an optimal selection of the relevant elements of the model (11).

3. Bayesian Model Structure Identification

In this chapter, we will introduce theoretical principles of regularization, used to solve least square problem outlined in previous chapter. The conventional least squares formulation 11 corresponds to maximum likelihood estimation of likelihood function

$$p(y|A, \theta, \beta) = \mathcal{N}(y|A\theta, \beta^{-1}I). \tag{13}$$

where β is a precision parameter of the model error. The likelihood model alone cannot decide which structure of the model is relevant. This can be achieved by Bayesian treatment with carefully chosen prior distribution on unknown parameters [16].

3.1. Hierarchical prior - unknown prior variance

Hierarchical priors are a commonly used tool to achieve model structure selection. In this text, we use the automatic relevance determination mechanism [11] which is based on hierarchical prior of zero mean Gaussian with unknown variance. Specifically, we assume that the elements of vector θ are apriori independent with each vector having its own prior variance. Formally, we define

$$p(\theta|\alpha) = \mathcal{N}(0, \alpha^{-1}), \tag{14}$$
$$\alpha = \mathrm{diag}(\alpha_1, \alpha_2, \cdots, \alpha_M), \tag{15}$$
$$p(\alpha_i) = \mathcal{G}(a_0, b_0), \tag{16}$$
$$p(\beta) = \mathcal{G}(c_0, d_0), \tag{17}$$

where diag(.) is a diagonal matrix with elements of the argument vector on its diagonal.

Joint distribution of the likelihood (13) and the prior (14)–(16) is

$$p(y, \theta, \alpha, \beta|A) = p(y|A, \theta, \beta)p(\theta|\alpha)\prod_{i=1}^{M} p(\alpha_i)p(\beta). \tag{18}$$

The aim is to obtain posterior distribution of all unknown parameters, i.e. θ, α, β using the Bayes rule

$$p(\theta, \alpha, \beta|y, A) = \frac{p(y, \theta, \alpha, \beta|A)}{\int p(y, \theta, \alpha, \beta|A) d\theta d\alpha d\beta}. \tag{19}$$

Evaluation of the exact formula (19) is analytically intractable, therefore, we seek only aproximate marginals of the posterior distribution using Varitional Bayes. In further text, we drop A from all the conditional probabilities, as it is known and not a subject of estimation.

3.2. Variational Bayes Posterior

The Variational Bayes approximation is a specific form of the general divergence minimization framework [12]. It is based on the following theorem that provides conditions for analytical solution of a specific form of approximation.

Theorem 1. *Let $f(\theta|y)$ be the posterior distribution of multivariate parameter, $\theta = [\theta_1^T, \theta_2^T]^T$, and $\breve{p}(\theta|y)$ be an approximate distribution restricted to the set of conditionally independent distributions:*

$$\breve{p}(\theta|y) = \breve{p}(\theta_1, \theta_2|y) = \breve{p}(\theta_1|y)\breve{p}(\theta_2|y). \tag{20}$$

Any minimum of the Kullback-Leibler divergence from $\breve{p}(\cdot)$ to exact solution $p(\cdot)$

$$KL(\breve{p}(\theta|y)||p(\theta|y)) = \int \breve{p}(\theta|y) \ln \frac{\breve{p}(\theta|y)}{f(\theta|y)} d\theta, \tag{21}$$

is achieved when $\breve{p}(\cdot) = \tilde{p}(\cdot)$ where

$$\tilde{p}(\theta_i|y) \propto \exp\left(\mathsf{E}_{\tilde{p}(\theta_{/i}|y)}\{\ln(p(\theta, y))\}\right), \; i = 1, 2. \tag{22}$$

Here $\theta_{/i}$ denotes the complement of θ_i in θ. We will refer to $\tilde{p}(\theta_i|y)$ (22) as the VB-marginals. *Here, $\mathsf{E}_{f(\theta)}\{g(\theta)\}$ denotes expected value of function $g(\theta)$ with respect to distribution $p(\theta)$.*

Theorem 1 provides a powerful tool for approximation of joint pdfs in separable form [12]. A solution satisfying conditions (22) is typically found using an iterative algorithm.

The Variational Bayes method will be applied for the following conditional independence condition:

$$p(\theta, \alpha, \beta|y) = p(\theta|y)p(\alpha|y)p(\beta|y),$$

yileding conditions of optimality (22) in the form

$$p(\theta|y) \propto \exp\left(\mathsf{E}_{\tilde{p}(\alpha|y)\tilde{p}(\beta|y)}\left\{\ln\left(p\left(\theta, \alpha, \beta, y\right)\right)\right\}\right), \tag{23}$$

and equivalently for $\tilde{p}(\alpha|y)$ and $\tilde{p}(\beta|y)$. The functional forms of (23) are recognized to be of well known types

$$\begin{aligned}\tilde{p}(\alpha_i|y) &= \mathcal{G}(a_{\alpha i}, b_{\alpha i}),\\ \tilde{p}(\beta|y) &= \mathcal{G}(c, d),\\ \tilde{p}(\theta|y) &= \mathcal{N}(\hat{\theta}, S),\end{aligned}$$

with shaping parameters

$$a_{\alpha i} = a_{0i} + \frac{1}{2}, \qquad b_{\alpha i} = b_{0i} + \frac{1}{2}\widehat{\theta_i^2}. \tag{24}$$

$$c = c_0 + \frac{N}{2}, \qquad d = d_0 + \frac{1}{2}y^T y - y^T A\widehat{\theta} + \frac{1}{2}\widehat{\theta^T A^T A\theta} \tag{25}$$

$$\hat{\theta} = \widehat{\beta} S A^T y, \qquad S = (\widehat{\alpha} + \beta A^T A)^{-1} \tag{26}$$

The required moments of these distributions are:

$$\hat{\alpha}_i = \frac{a_{\alpha i}}{b_{\alpha i}}, \qquad \hat{\beta} = \frac{c}{d}$$

$$\widehat{\theta_i^2} = S_{ii} + \widehat{\theta}_i^{\,2}, \qquad \widehat{\theta\theta^T} = \hat{\theta}\hat{\theta}^T + S$$

The final estimation algorithm is described in Algorithm 1.

Algorithm 1 Iterative algorithm for structure selection of least squares model.

Off-line:

- select prior parameters a_{0i}, b_{0i}, c_0, d_0.
- select the number of iterations i_{max}, set iteration counter $i = 0$.

On-line:

- Initiate $p(\theta|y, x)$ by ordinary least squares solution: $\hat{\theta}^{(0)} = (A^T A)^{-1} A^T y$, $S^{(0)} = \frac{(y - A\hat{\theta}^{(0)})^T (y - A\hat{\theta}^{(0)})^T}{M-(N+1)} (A^T A)^{-1}$
- For i from 1 to i_{max}
  - update shaping parameters of α, using (24),
  - update shaping parameters of β, using (25),
  - update shaping parameters of θ,using (26),
- Report final estimate $\hat{\theta}^{(i_{max})}$ and compute plasma reconstruction

4. Experimental Validation

4.1. Tokamak COMPASS and data from a validated numerical model

Tokamak COMPASS is in operation in IPP CAS CR in Prague. On COMPASS, the process of plasma reconstruction through decomposition into a series of toroidal harmonics is implemented in VacTH code [17]. VacTH uses data measured on different types of magnetic coils to extrapolate plasma characteristics into the domain of the tokamak chamber [3].

In this paper, the performance of VacTH is evaluated on simulated data, for which the real shape of plasma boundary is known. These data are generated by the FREEBIE code [18] in inverse mode. To obtain such data, the geometry of the measuring coils and shape of the plasma boundary is chosen by the operator. Following that, appropriate feedback of every coil is computed and the corresponding set of magnetic measurements is generated. It is then used to recompute the plasma boundary via VacTH. The approach implemented in the VacTH to solve 11 is the ordinary least squares (OLS) without regularization. Among other characteristics, plasma boundary shape is obtained and can be directly compared with the original FREEBIE input (e.g. using Hausdorff distance 12).

For the purpose of this paper, all data were simulated with configuration of 16 magnetic coils and 4 saddle loops with parameters $\sigma_B = 10^{-3}, \sigma_f = \frac{10^{-1}}{2\pi}$, which agrees with the real configuration of the tokamak.

4.2. Results

Based on analysis of a training data set from the COMPASS tokamak, prior parameters of the VB approximation were determined as in table 1. This was done by substituting the OLS by LSARD (least squares automatic relevance determination) algorithm in VacTH, which produced different value of $\hat{\theta}$. With this new value, VacTH reconstructed a different shape of the plasma boundary. To asses performance of LSARD , Hausdorff distance between FREEBIE boundary and this reconstruction was computed. All boundaries were numerically interpolated into a large ($\sim 10^3$) number of points to assure more precise evaluation of similarity.

parameter	a_0	b_0	c_0	d_0	α_0	β_0
value	10^{-10}	10^{-10}	10^0	10^0	10^0	10^4

Table 1. Prior parameter settings for VB regularization.

Using prior parameters setting from table 1, performance of the LSARD algorithm was assessed on a test data set. On every tested equilibrium, plasma boundary was reconstructed using OLS, LSARD and RLS (regularized least squares) approximations. This was done for model of size$M = M_{ea} + M_{eb} + M_{ia} + M_{ib}$ (M is order of the toroidal harmonics decomposition), where $M \in \{5, ..., 22\}$ as the individual sums in (4) a (5) were extended with varying values of $M_{ea} \in \{0, ..., 6\}, M_{eb} \in \{1, ..., 6\}, M_{ia} \in \{0, ..., 4\}, M_{ib} \in \{1, ..., 4\}$. A comparison of methods is in Figure 4.2 for a number of different equilibria.

Because there are 20 measurements, we observe a clear deterioration of reconstruction with OLS as the problem of solving (11) becomes ill-conditioned for higher order M. On the other hand, reconstruction with LSARD or RLS is not affected by this fact. Although the performance of LSARD is worse in terms of the MSE statistic, plasma boundary reconstruction is more precise. This hints us that some sort of regularization is important for plasma boundary reconstruction problem, as some terms in the decomposition become obsolete with higher order of the problem. LSARD regularization clearly addresses this problem by iteratively minimizing the size of redundant terms of $\hat{\theta}$. Another important fact is that the quality of reconstruction using LSARD does not vary rapidly with changing value of M. This is important for actual implementation in VacTH code, as for computation in real time the exact choice of estimated

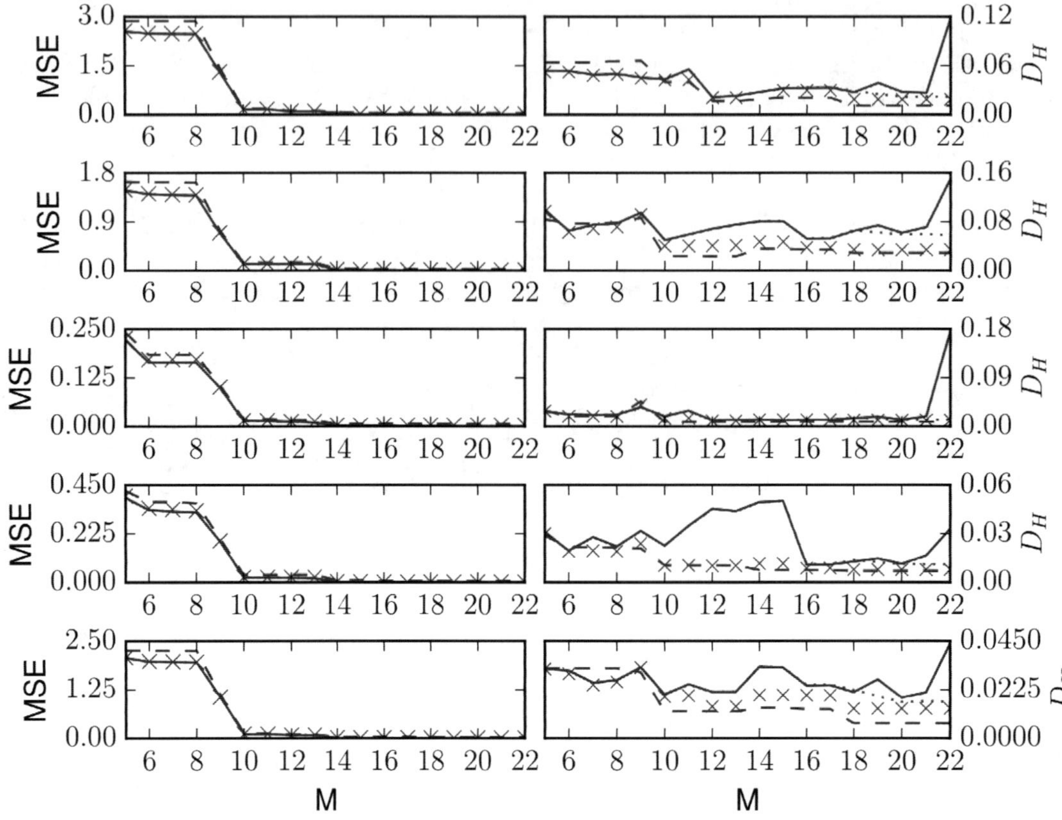

Figure 2. Results of plasma boundary reconstruction for varying order of toroidal harmonics decomposistion M for different plasma equilibria. Straight line is for OLS solution, dashed line for LSARD algorithm. Dotted line marks solution obtained with regularized least squares (RLS) with parameter $\lambda = 10$. Crossed line marks solution obtained with LASSO algorithm with parameter $\alpha = 1$. Besides MSE of fit, Hausdorff distance D_H between ground truth and solution obtained with respective solver is shown. Although the fit of LSARD method is worse than that of OLS, RLS or LASSO in terms of MSE, precision of boundary reconstruction is consistently better.

terms must be made in advance. Thus for LSARD we can simply choose M to be as large as possible to estimate with largest amount of information.

When analyzing more spherical shapes of plasma, prior parameter settings from 1 did not lead to significant improvement. It was shown that setting $a_0 = b_0 = 10^2$, which does not penalize large values of $\hat{\theta}$ as strictly, leads to improved results. These were still only slightly better than results obtained with OLS.

Finally, a comparison of plasma boundaries reconstructed using OLS, LSARD, LASSO regularization[19] and the original FREEBIE boundary is shown in Figure 3.

4.3. Offline model selection
Computational complexity of the current implementation of LSARD is too high and does not allow to replace OLS as the internal solver for real-time application in plasma reconstruction. Therefore, an attempt to determine an optimal model for which OLS is computed was made.

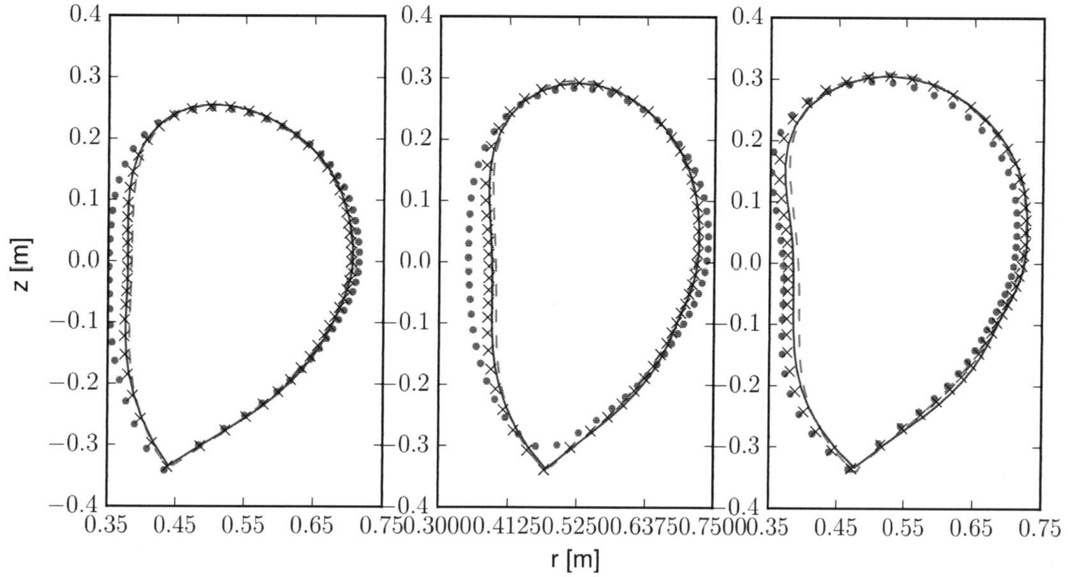

Figure 3. Comparison of plasma boundary reconstruction for three different equilibria for OLS implemented in VacTH (dotted line), LSARD (full line) and LASSO regularization with parameter $\alpha = 1$ (crossed line) using the same order of decomposition M. Ground truth, generated by FREEBIE algorithm, is denoted by dashed line.

Using sparsity property of ARD on a training data set, an optimal model was sequentially estimated by leaving out those terms of the model whose absolute value was smaller than a given constant $\delta \approx 10^{-4}$, until there were no more terms to omit in the estimation. Through this effort, a set of indices for sums (4) and (5) was obtained. This set is in Table 2. It holds that $n \in e_a$for the first sum in (4) etc. The ARD selected model is indeed very sparse when compared with the full model.

The three models were validated against each other on a distinct set of equilibria. OLS implemented in VacTH were used to compute $\hat{\theta}$ for each respective model structure. Then, plasma boundary was reconstructed. Graphical evaluation is in Figure 4. No distinct improvement from the full model to ARD selected model can be observed, although the results of ARD selected model are much more consistent.

	full model	model selected using ARD	control model
e_a	$(0, 1, 2, 3, 4, 5)$	-	(0, 1, 2)
e_b	$(1, 2, 3, 4, 5)$	-	(1, 2)
i_a	$(0, 1, 2, 3)$	(0, 1, 2)	(0, 1, 2, 3)
i_b	$(1, 2, 3)$	(1)	(1, 2, 3)

Table 2. Sets of indices, determining structure of estimated model.

5. Conclusion

The presented method of model selection has been shown to provide consistently better results on the problem of plasma boundary estimation compared to the conventional approach. It has been shown that a model with fewer carefully selected parameters avoids the overfitting problem

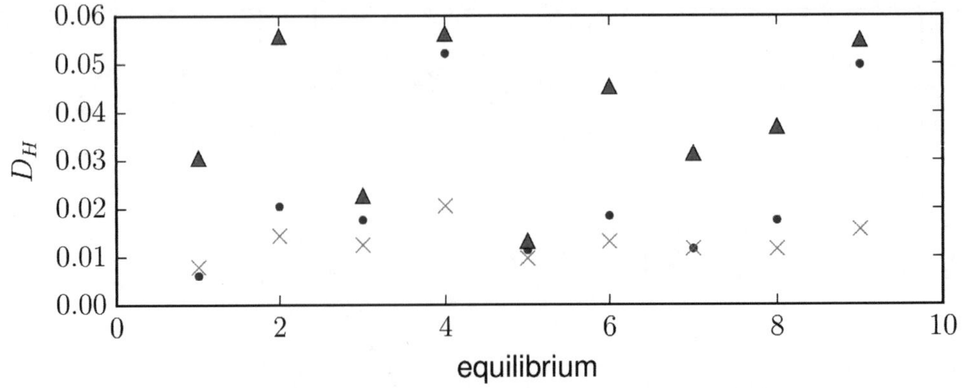

Figure 4. Comparison of full model (denoted by dots), control model (triangles) and model selected via the ARD principle (crosses) performance on different equilibria. D_H is the Hausdorff distance between the original FREEBIE boundary and the reconstruction. Notably more consistent performance of the sparse model structure selected by the ARD principle can be observed in comparison to full and control model as defined in 2.

and thus achieves better reconstruction of the overall shape of the plasma boundary. This has been achieved on simulated data from a detailed model. The complexity of the computation is intentionally kept low so that the algorithm can be run in real time on existing hardware.

References

[1] Ambrosino G and Albanese R 2005 *IEEE Control Systems Magazine* **25** 76–92
[2] Ariola M and Pironti A 2008 *Magnetic control of tokamak plasmas* (Springer Science & Business Media)
[3] Faugeras B, Blum J, Boulbe C, Moreau P and Nardon E 2014 *Plasma Physics and Controlled Fusion* **56** 114010
[4] Scott J G, Berger J O *et al.* 2010 *The Annals of Statistics* **38** 2587–2619
[5] O'Hara R B, Sillanpää M J *et al.* 2009 *Bayesian analysis* **4** 85–117
[6] George E I and McCulloch R E 1993 *Journal of the American Statistical Association* **88** 881–889
[7] Ruggieri E and Lawrence C E 2012 *Computational Statistics & Data Analysis* **56** 1319 – 1332 ISSN 0167-9473
[8] Ghosh J and Tan A 2015 *Computational Statistics & Data Analysis* **81** 76 – 88 ISSN 0167-9473
[9] Carbonetto P, Stephens M *et al.* 2012 *Bayesian Analysis* **7** 73–108
[10] Zhao K and Lian H 2014 *Computational Statistics & Data Analysis* **80** 223 – 239 ISSN 0167-9473
[11] Tipping M 2001 *The journal of machine learning research* **1** 211–244
[12] Šmídl V and Quinn A 2005 *The Variational Bayes Method in Signal Processing* (Springer)
[13] Šmídl V and Quinn A 2008 *IEEE Transactions on Signal Processing* **56** 5020–5030
[14] Lebedev N N and Silverman R A 1972 *Special functions and their applications* (Courier Corporation)
[15] Huttenlocher D P, Klanderman G, Rucklidge W J *et al.* 1993 *Pattern Analysis and Machine Intelligence, IEEE Transactions on* **15** 850–863
[16] Bishop C 2006 *Pattern recognition and machine learning* vol 1 (springer New York)
[17] Urban J, Appel L C, Artaud J F, Faugeras B, Havlicek J, Komm M, Lupelli I and Peterka M 2015 *Fusion Engineering and Design* **96-67** 998–1001
[18] Artaud J and Kim S 2012 *EPS Conference on Plasma Physics*
[19] Tibshirani R 1996 *Journal of the Royal Statistical Society. Series B (Methodological)* 267–288

Journal of Physics: Conference Series **659** (2015) 012011 doi:10.1088/1742-6596/659/1/012011

Robot Kinematics Identification: KUKA LWR4+ Redundant Manipulator Example

Sergey Kolyubin, Leonid Paramonov, and Anton Shiriaev

Dept. of Engineering Cybernetics, Norwegian University of Science and Technology, NO-7491, Trondheim, NORWAY

E-mail: {sergey.kolyubin, leonid.paramonov, anton.shiriaev}@ntnu.no

Abstract. This work is aimed at a comprehensive discussion of algorithms for the kinematic parameters identification of robotic manipulators. We deal with an open-loop geometric calibration task, when a full 6D robot's end-effector pose is measured. Effective solutions of such a task is of high interest in many practical applications, because it can dramatically improve key robot characteristics. On the first step, we select optimal calibration configurations. A comparative analysis of three different algorithms and two observability indexes used for numerical optimization is provided. Afterwards, using the acquired and pre-processed experimental data we identify modified Denavit-Hartenberg parameters of the manipulator. Estimates are obtained resolving original nonlinear forward kinematics relations. Finally, we compare nominal and calibrated geometric parameters and show how much deviations in these parameters affect robot positioning accuracy. To the best of our knowledge, such integrated efforts are new for the KUKA LWR4+ robot and Nikon K610 optical coordinate measuring machine (CMM), which were used in the study. Discussion of practical issues on how to organise the experiment is an additional contribution of this work. The proposed procedure is highly automated and can be implemented to improve manipulator's performance on a periodic basis.

1. Introduction

Absolute accuracy of a robot's tool positioning is one of its key performance characteristics that has a dedicated test standard [1]. It matters for almost all applications of industrial manipulators, but precise positioning is even more important for such tasks as laser welding, riveting, or automated fibre placement in composite manufacturing as well as for contact operations like grinding or drilling.

This is an example, where an effective parameters identification technique finds a good application, and competence areas of control and robotics communities perfectly complement each other.

While the industrial manipulators mechanics and manufacturing techniques are constantly improving, there is still a crucial trade-off between price and quality, therefore manufacturing and assembly defects still exist and cause significant deviation of the real robot parameters from its nominal values specified on the design phase. Such offsets in parameters used by a motion planner or a motion controller can significantly degrade the overall system performance and even cause the equipment damage. A way to handle the problem and improve the robot accuracy just on a software level without changing the mechanical structure of a robot is the proper calibration.

Some of the industrial robots' manufactures now offer an option, when a manipulator comes with pre-calibrated parameters. However, there is a need for an identification procedure that can be implemented on-site on a periodic basis in case if a robot was relocated, and simply because parameters can drift with time.

The calibration problem was extensively developed in fundamental works of Hollerbach, Khalil, Born and Menq, Daney, and others, and still remains an active research area [10, 11].

There are sources of robot positioning errors of different nature such as joint's and link's flexibility or gear backlash and friction, but according to [12], poor estimation of basic parameters defining robot geometry could be responsible for up to 90% of these errors. Here we discuss an *open-loop geometric calibration* meaning that we estimate robot's geometric parameters from experiments when the robot end-effector is not physically fixed in space, but its position is defined by an external measurement device.

The calibration procedure should be properly organized. Indeed, from the very beginning an engineer is required to choose a number of robot's configurations (poses) that will be most informative and insensitive to noise in further processing of measurement records. Such poses are typically found numerically via optimizing one or several aggregated criteria describing sensitivity of parameters used in solution of a forwards kinematics for a robot. Moreover, this preparatory step if often used to define sets of identifiable and non- or poorly identifiable parameters. These sets depend on the arm kinematics and setup in particular, and can be analysed in advance by detecting linear dependent columns of the parametric Jacobian for a random set of configurations [9]. In this paper we provide the comparative analysis of three different numerical optimization algorithms and two observability indexes for optimal calibration poses selection.

After we conducted a series of experiments and acquired and pre-processed necessary data, the identification itself can be organised. In most of the related works it is done by linearising forward kinematics relations and then iteratively applying the least-squares algorithm (or its modifications) [2, 6, 8]. While in general this approach should provide better convergence compared to nonlinear optimization, it requires system parametric Jacobian to satisfy certain conditions. In this study we organize search for the original nonlinear problem formulation. The reason for that is described later in the text.

This work is aimed at a comprehensive discussion of experiments and numerical procedures for the open-loop geometric calibration of the KUKA LWR4+ redundant robotic arm, when a full 6D end-effector's pose is measured using Nikon K610 optical CMM. While there are a number of studies reporting successful results on its dynamic identification, we have not been able to identify any work focused on calibration of the robot geometrical parameters. As every external measuring system either a laser tracker or a photogrammetric equipment has it special features and concomitant requirements, an experiment should be adjusted and design of an additional tooling is usually needed. In our case, we designed a special LED fixture and imposed additional constraints for calibration poses selection.

The rest of the paper is organized as follows. In Section 2 we describe the setup and practical issues on organising experiments with the optical CMM. In Section 3 we discuss the problem of selecting optimal calibration configurations and provide comparative analysis of different optimization algorithms and observability indexes followed by its results. In Section 4 we solve the problem of robot's geometric parameters identification. Finally, we conclude the paper with analysis of the obtained results and discussion on future steps.

2. Optimization of the Calibration Configurations

2.1. Problem Statement

An outcome of the geometric calibration is estimations of parameters describing relative location of neighbouring coordinate systems associated with robot links. Basically, it means identifying parameters of the forward kinematics equation

$$Y^l = F(\Phi, q)|_{q=q^l}, \tag{1}$$

where $Y^l = [x^l y^l z^l \phi^l \theta^l \psi^l]^T$ and $q^l = [q_1^l q_2^l \ldots q_n^l]^T$ are the end-effector's Cartesian pose and relative to the measurement device and the vector of joint angles for robot's configuration l respectively, n is the number of robot joints, and Φ is the vector of unknown geometric parameters.

It is clear that the calibration is feasible only if a total number of constraint equations (1) from all measurements $l = 1, 2 \ldots M$ is bigger than a number of estimated parameters. However, a good selection of the configurations is more important than its number to obtain good identification results.

As a quantitative measure of the configurations optimality, several so-called observability indices $O(Q)$, $Q = \{q^1, q^2 \ldots q^M\}$, were introduced (refer to [4, 6] for more details). However, all of these indexes rely on properties of the Jacobian evaluated in $l = 1, 2 \ldots M$ configurations

$$J(Q) = \left[J^1; J^2; \ldots; J^M\right], \tag{2}$$

where J^l in the Jacobian defined in the configuration q^l

$$J^l = \left. \frac{\partial F(\Phi, q)}{\partial \Phi} \right|_{q=q^l}.$$

Thus, the configurations optimization task can be formulated as a nonlinear constraint optimization problem

$$\max_Q O(Q), \quad \texttt{subject to} \tag{3}$$

$$\begin{cases} q_i^l \in (q_{i,min}; q_{i,max}), \\ Y^l \in (Y_{min}; Y_{max}), \end{cases} \quad \forall i = 1 \ldots n,\ l = 1 \ldots M \tag{4}$$

The most common way of composing the forward kinematics equation (1) is to use a set of modified Denavit-Hartenberg (mDH) parameters

$$\Phi = [\alpha_0, \ldots \alpha_{n+1}, d_0, \ldots d_{n+1}, \theta_0, \ldots \theta_{n+1}, r_0, \ldots r_{n+1}]$$

and a homogeneous transformation

$$^{j-1}T_j = Rot_{x,\alpha_j} Trans_{x,d_j} Rot_{z,\theta_j} Trans_{z,r_j}.$$

For more detailed explanation, please, refer to [9].

2.2. Observability Indices

In this work we validated two different observation indices [6] . The first on is the inverse condition number of the Jacobian (2)

$$O_1 = \frac{\sigma_{min}}{\sigma_{max}},$$

where σ_{min} and σ_{max} are Jacobian minimal and maximum singular values respectively (see [5]).

The second index was introduced in [2]

$$O_2 = \frac{\sqrt[R]{\sigma_R \cdot \ldots \cdot \sigma_i \ldots \sigma_1}}{\sqrt{M}},$$

where R is the Jacobian rank, M is the number of configurations, and σ_i, $i = 1 \ldots R$ are Jacobian singular values.

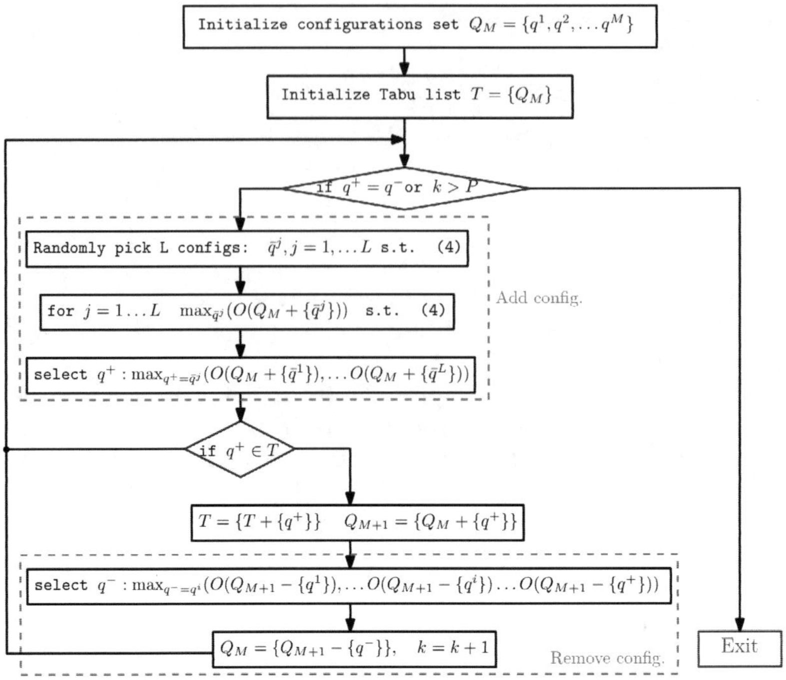

Figure 1: Diagram for the iterative meta-heuristic algorithm algorithm with the Tabu rule

2.3. Eliminating Non-Identifiable Parameters

While a set of mDH parameters is known to be the minimal for robot kinematics representation, depending on a particular calibration setup, some of the parameters in Φ can be non- or poorly identifiable. These parameters are associated with linearly dependent columns of the Jacobian J and consequently with zero of very small Jacobian singular values σ_i.

In order to recursively calculate Jacobian (2) and simultaneously eliminate non-identifiable parameters, we used an approach described in details in [9].

2.4. Pose Optimization Algorithms

We compared performance of three different algorithms for calibration poses optimization.

The first one is conjugate-type deterministic algorithm similar to one described in [9]. It tries to optimize the entire set of configuration on every iteration. However, algorithms of this type have significant limitations: strong dependence on initial conditions and sticking in a local minimum [4].

To avoid these issues we implemented a modification of the previous algorithm utilizing multistart optimization, i.e. when the same algorithm starts N times (we selected $N = 50$) consequently from randomly uniformly distributed initial conditions.

As an alternative to conjugate-type algorithms, we tested an iterative meta-heuristic algorithm with the Tabu rule [4]. Instead of optimizing the entire configurations set, it implements a 2-stage iterative procedure (see Fig. 1). Tabu list T here is a set of already verified configurations, which is introduced to prevent a premature convergence of the algorithm to local minima. This scheme is quite similar to genetic algorithms, while a stochastic behaviour is introduced here by randomly selecting candidate configurations on each *'Add config.'* step.

We selected $L = 10$ and $P = 2M$. Compared to the conjugate-type algorithm, the optimization task takes only n variables instead of $n \times M$, but should be solved $L \times P$ times.

We used a SQP (sequential quadratic programming) solver for all aforementioned algorithms.

3. Identification of the Geometric Parameters

Let us introduce a vector of discrepancies between the predicted and measured end-effector's Cartesian position

$$\Delta Y = Y_m - Y_c, \tag{5}$$

where $Y_m = \left[Y_m^1, Y_m^2 \dots Y_m^M\right]$ is the vector of measured poses and $Y_c = \left[Y_c^1, Y_c^2 \dots Y_c^M\right]$ is the vector of poses calculated from (1).

In the most of classical works forward kinematics relations (1) are linearised using Taylor series expansion such that

$$\Delta Y = J_B \Delta \Phi_B, \tag{6}$$

where the base parametric Jacobian J_B is defined from (2) eliminating linearly dependent columns and $\Delta\Phi_B$ is the vector of base parameters' correction.

In this case parameters estimates $\hat{\Phi}_B$ can be found iteratively applying ordinary least-squares solution or its damped or weighted modifications (for details refer to [6]) that minimizes the performance index $\Delta Y^T \Delta Y$

$$\begin{aligned} \Delta\Phi_{B,k} &= J_B^+ \Delta Y_k, \\ \hat{\Phi}_{B,k} &= \hat{\Phi}_{B,k-1} + \Delta\Phi_{B,k}, \end{aligned} \tag{7}$$

where J_B^+ is the pseudo-inverse of the matrix J_B.

However, convergence of the algorithm (7) is guaranteed only for well-conditioned problems when $O_1 \geq 0.01$ [6]. As we can see from Table 3 we were not able to meet this requirement for the selected setup, even applying different non-identifiable parameters elimination and parameter scaling techniques [2, 6, 8, 9]. Therefore, instead of applying (5)-(7), we formulated the identification task as a nonlinear optimization problem of finding a vector of parameters estimates $\hat{\Phi}$ such that it minimizes the norm of output residual (5)

$$\left|\left|Y_m - F(\hat{\Phi}, Q)\right|\right| \rightarrow \min.$$

4. Experimental Results

4.1. Setup Description

KUKA LWR4+ is a redundant robotic arm with 7 rotational joints. It offers unique features via its low-latency Fast Research Interface (FRI) developed with focus on research purposes [7]. At the same time, to the best of our knowledge, the only source of geometrical parameters' data for this robot is a CAD model, which provides quite rough estimations.

There are different kinds of metrological equipment suitable for calibration experiments, including laser trackers [10] and photogrammetric systems [11], which are considerably less expensive. We use the latter type Nikon K610 system. This optical CMM allows us localizing end-effector in 3D space with sampling rate up to 1kHz and volumetric accuracy of 60 μm for measurement volume of 17 m^3. This way, having complete 6D pose measurements in calibration experiments we are able to identify the maximum number of robot's geometric parameters [6, 8]. Tracking movable objects is enabled through dynamic frames concept [3]. The system defines positions of infrared LEDs by means of three linear CCD cameras scanning in different planes. Diodes on/off phases are synchronized via strobers such that the camera 'distinguish' between different LEDs.

One more element, which is needed for calibration, is a fixture with LEDs that should be attached to a robot tool plate. Based on Nikon specifications, an LED is guaranteed to be visible only if an angle between a normal to its surface and a line of sight with the camera's central point does not exceed 30°. It gives us 'a visibility cone' of each LED. Thereby, a pyramid geometry was optimized based on criteria:

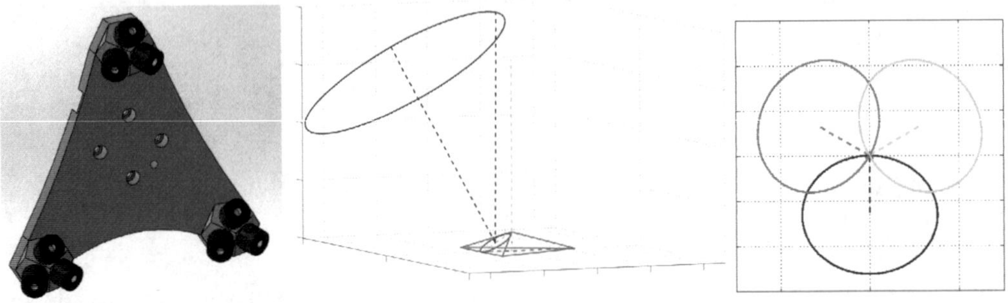

(a) Fixture CAD model (b) Pyramid geometry (c) LEDs' visibility cones: top view

Figure 2: LEDs fixture geometry

Table 1: Nominal mDH parameters

Frame	α_i, [rad]	d_i, [m]	θ_i, [rad]	r_i, [m]
0	0	0	1.549	−4.407
1	1.528	0.067	$q_1 - 0.815$	−0.137
2	$\frac{\pi}{2}$	0	q_2	0
3	$-\frac{\pi}{2}$	0	q_3	0.4
4	$-\frac{\pi}{2}$	0	q_4	0
5	$\frac{\pi}{2}$	0	q_5	0.39
6	$\frac{\pi}{2}$	0	q_6	0
7	$-\frac{\pi}{2}$	0	$q_7 - 0.268$	$2.2 \cdot 10^{-3}$
8	0	0.014	1.0523	0.079

- all three 'visibility cones' have at least one common point on a distance of $1.5m$[1] from the pyramid top, i.e. there is no 'blind spot' in between
- pyramids' sides allows to inscribe a circle about a size of a LED

Geometry behind the calculation is illustrated on Fig. 2b–2c. We have chosen a design with nine LEDs arranged in triples on sides of three similar pyramids, which are distributed on the plate surface (see Fig. 2a).

Fig. 3 illustrates relations between different coordinate frames for the calibration setup.

Nominal mDH parameters values for the considered system are in the Table 1 [2].

Except for physical limitations on joint angles, we should impose additional constraints on the LEDs fixture Cartesian coordinates. Despite the fact that we can place the camera to cover the entire robot workspace, the end-effector pose can be recovered only if at least 3 LEDs remain

[1] The distance of 1.5 m was selected based on the relative location of the camera and the robot and parameters of the camera's field of view.

[2] While the robot parameters ($i = 1 \ldots 7$) were taken from the CAD model, parameters of the transformation between the camera and the robot base frames (indexed with z) were calculated backwards from camera-based measurements, and parameters for the transformation between the robot last joint and LEDs fixture frames (indexed with e) were directly measured.In general, the Nikon camera allows direct measurements of the end-effector's pose relative to the robot base, therefore transformation between the camera and the robot base coordinate frames can be excluded from consideration.

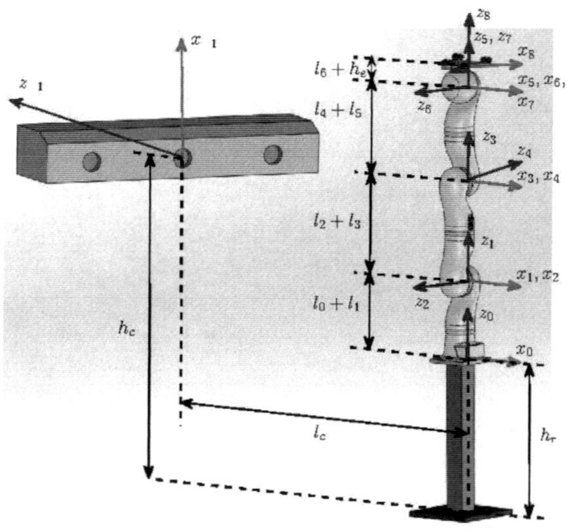

Figure 3: Coordinate frames for camera, robot, and LEDs fixture

Table 2: Constraints for Configurations Optimization

Parameter	**Constraint**
Joint angles, [rad]	$q_{1,3,5,7} \in [-\frac{17\pi}{18}, \frac{17\pi}{18}]$, $q_{2,4} \in [-\frac{2\pi}{3}, \frac{2\pi}{3}]$, $q_6 \in [-\frac{\pi}{2}, \frac{\pi}{2}]$
Fixture roll, [rad]	$\phi_{x,LED} \in [-\frac{4\pi}{9}, \frac{7\pi}{18}]$
Fixture pitch, [rad]	$\theta_{y,LED} \in [-\frac{\pi}{4}, \frac{\pi}{4}]$
Distance between LED TCP and robot base, [m]	$\sqrt{{}^0x_8^2 +{}^0 y_8^2} \geq 0.2$
z-coordinate of LED TCP, [m]	${}^0z_8 \geq -0.1$

visible. It gives us bounds on roll, pitch, and yaw angles of the LEDs fixture, expressed in the camera coordinate frame. Two more constraints were imposed to guarantee that the LEDs fixture will not collide with the robot base. All these constraints are summarized in Table 2.

4.2. Identification Results

As initial conditions for the optimization procedure we randomly selected M robot configurations, satisfying (4), such that it was uniformly distributed across the robot's workspace and provided distinguishable variation in joint angles as well.

As the result of the non-identifiable parameters elimination procedure, we determined that 4 parameters θ_0, r_0, θ_8, and r_8 are non-identifiable that defines a set of $4(n+2) - 4 = 32$ base parameters Φ_B to be calibrated and the base parametric Jacobian J_B of size $[6M \times 32]$.

Comparative results of the optimization for sets of 10 and 20 configurations for all three algorithms and two observability indices are given in Table 3.

Fig. 4 illustrates how joint angles change between initial and optimized calibration configurations.

Fig. 5 visualizes how robot end-effector positions (relative to the camera coordinate frame) change between nominal and optimized calibration configurations.

Table 3: Optimised Observability Indices

Number of Config.	Algorithm	Observability Index	
		O_1	O_2
$M = 10$	Non-optimized	0.000716	0.5685
	Conjugate-type deterministic	0.001535	0.8065
	Conjugate-type with multistart	0.001613	0.7955
	Randomized iterative with Tabu rule	0.001497	0.7904
$M = 20$	Non-optimized	0.000811	0.6435
	Conjugate-type deterministic	0.001618	0.8142
	Conjugate-type with multistart	0.001631	0.7973
	Randomized iterative with Tabu rule	0.001571	0.8091

Table 4: Calibrated mDH parameters

Frame	α_i, [rad]	d_i, [m]	θ_i, [rad]	r_i, [m]
0	0.01515	$1.07 \cdot 10^{-2}$	1.549	−4.407
1	1.52662	$2.7 \cdot 10^{-3}$	$q_1 - 0.815$	−0.147
2	1.57093	$-2.5 \cdot 10^{-4}$	$q_2 - 3.9 \cdot 10^{-3}$	$1.3 \cdot 10^{-4}$
3	−1.57029	$-2.5 \cdot 10^{-4}$	$q_3 - 9.7 \cdot 10^{-3}$	0.399
4	−1.57031	$1.0 \cdot 10^{-3}$	$q_4 - 6.5 \cdot 10^{-3}$	$2.5 \cdot 10^{-4}$
5	1.57025	$-9.1 \cdot 10^{-4}$	$q_5 + 8.3 \cdot 10^{-4}$	0.391
6	1.57274	$3.9 \cdot 10^{-4}$	$q_6 - 1.6 \cdot 10^{-3}$	$9.5 \cdot 10^{-5}$
7	−1.57261	$-5.7 \cdot 10^{-4}$	$q_7 - 0.267$	$1.1 \cdot 10^{-2}$
8	$6.5 \cdot 10^{-4}$	$-1.2 \cdot 10^{-4}$	1.053	0.079

And finally, Fig. 6 shows if the constraints imposed on the LEDs fixture position and orientation were matched.

For the calibration experiment we used a set of $M = 20$ optimised configurations found by the conjugate-type algorithm with the observability index O_1. Out of this set we picked $M = 17$ configurations, where we had good visibility of LEDs.

Estimated values for system's mDH parameters are in Table 4.

Fig. 7 illustrates the difference in calculated end-effector's Cartesian positions when nominal and calibrated (identified) parameters sets are used. It is clear that for estimated geometric parameters predicted values are much more closer to the measured ones.

5. Conclusions and Future Work

This work discusses the problem of robot kinematic calibration.

Open-loop geometric calibration of the KUKA LWR4+ redundant robotic arm, when a full 6D end-effector's pose is measured using Nikon K610 optical CMM considered as a particular example. To the best of our knowledge, such an experiment is a unique attempt on its own.

We compared 3 different algorithms and 2 observability indices solving the configurations optimization task. Comparative analysis of the obtained results leads to the following conclusions:

Journal of Physics: Conference Series **659** (2015) 012011 doi:10.1088/1742-6596/659/1/012011

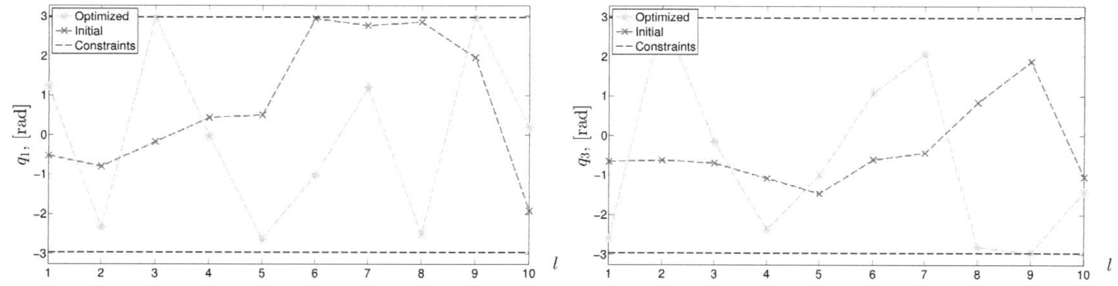

(a) Conjugate-type algorithm, O_1

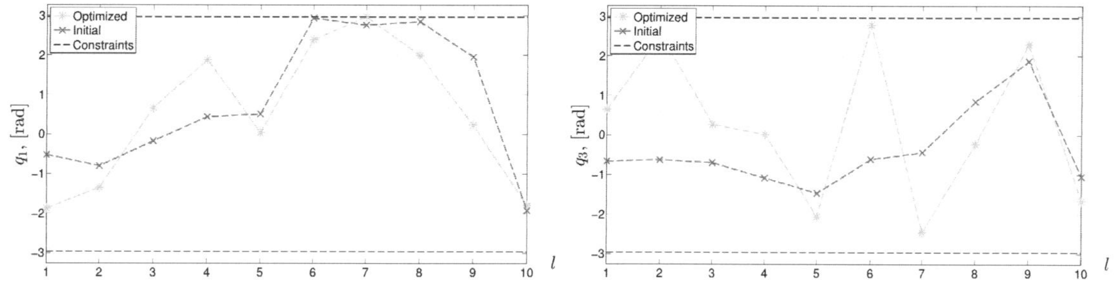

(b) Conjugate-type algorithm, O_2

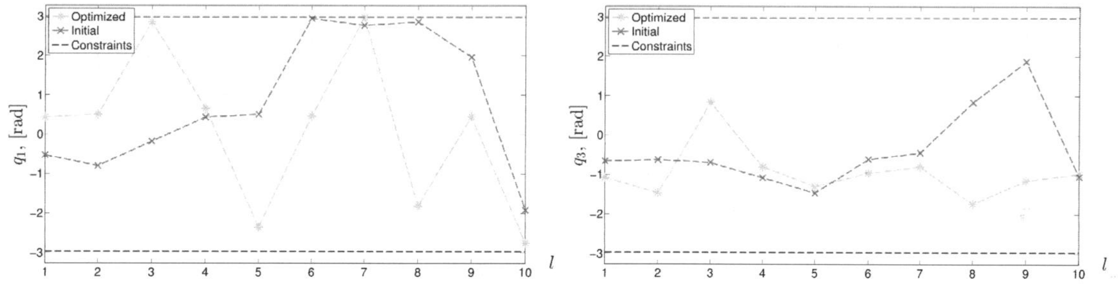

(c) Iterative algorithm, O_1

Figure 4: Joint coordinates for $M = 10$ selected calibration poses (only q_1 and q_3 are presented for illustration purposes)

- Optimization algorithms gave tangible difference in terms of joint configurations and end-effector's poses allocation, but provided quite similar resulting observability indices. It reveals features of the cost function, which is highly nonlinear and has many local minimums.
- The observability index O_1 associated with the Jacobian condition number shows higher relative improvement between initially selected and optimized configurations compared to the Born-Menq index O_2.
- In practice, the iterative meta-heuristic algorithm with Tabu rule didn't ensure better results, while required longer computation time.

Afterwards, based on the experimental data we estimated system's modified DH parameters that define geometric relations between consequent joint frames. Instead of the traditional approach we solved the original nonlinear identification task. Identification results show that there is a difference in range of [0.001 . . . 0.01] [rad] for angular displacements and in range of [0.3 . . . 3] [mm] for linear displacements between nominal parameters provided by robot's manufacturer and its estimated values.

Journal of Physics: Conference Series **659** (2015) 012011 doi:10.1088/1742-6596/659/1/012011

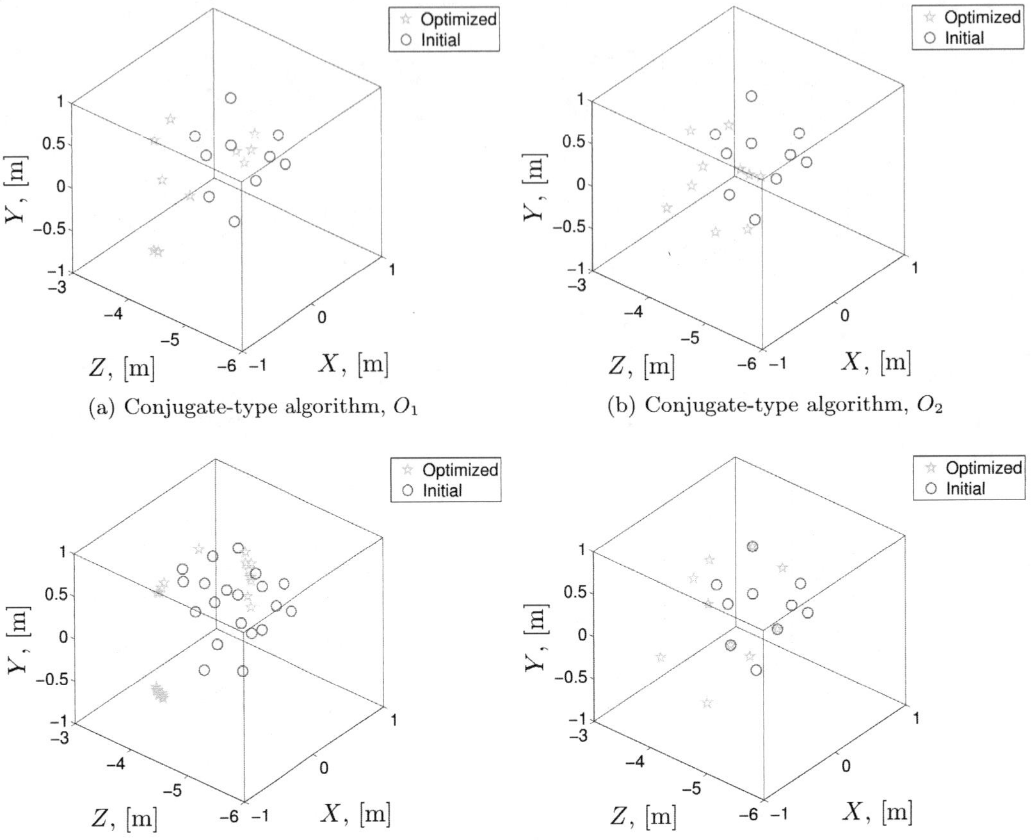

(a) Conjugate-type algorithm, O_1

(b) Conjugate-type algorithm, O_2

(c) Conjugate-type algorithm with multistart, O_1

(d) Iterative algorithm, O_1

Figure 5: LED frame Cartesian coordinates for $M = 10$ selected calibration poses

The proposed procedure is highly automated and can be implemented to improve manipulator's performance both right after the installation for initial calibration and on a periodic basis, since estimated parameters can deviate because of the tear and wear and other factors.

References

[1] ISO 9283:1998, *Manipulating industrial robots Performance criteria and related test methods.*

[2] Borm J H and Menq C H 1991 Determination of optimal measurement configurations for robot calibration based on observability measure, *J. of Robotic Systems* **10** 51-63

[3] van den Bossche A 2004 Procedure for determining the dynamic behavior of a vehicle on a test bench *US Patent 6,748,796*

[4] Daney D, Y. Papegay Y and Madeline B 2005 Choosing measurement poses for robot calibration with the local convergence method and Tabu search *The Int. J. of Robotics Research* **24** 501-18

[5] Driels M R and Pathre U S 1990 Significance of observation strategy on the design of robot calibration experiments *J. Robot. Sys.* **7** 197-223

[6] Hollerbach J M and Wampler C 1996 The calibration index and taxonomy for robot kinematic calibration methods *Int. J. Rob. Res.* **15(6)** 573-91

[7] Schreiber G, Stemmer A and Bischoff R 2010 The Fast Research Interface for the KUKA Lightweight Robot *Proc. IEEE ICRA 2010 Workshop on Innovative Robot Control Architectures*

[8] Khalil W, Besnard S and Lemoine P 2000 Comparison study of the geometric parameters calibration methods *Int. J. of Robotics and Automation* **15(2)** 56-67

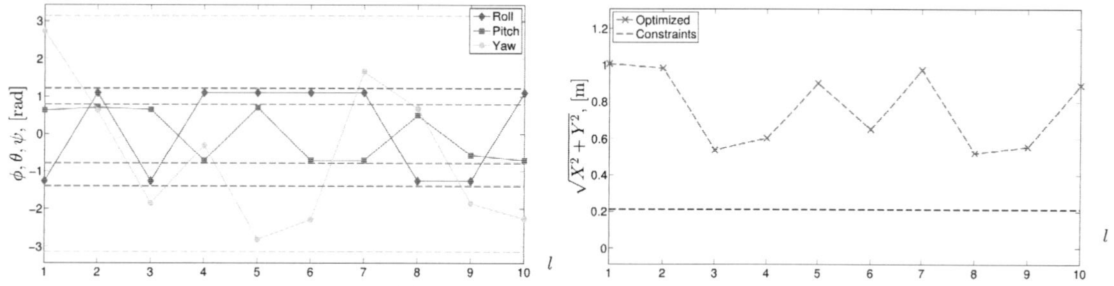

(a) Conjugate-type algorithm, O_1

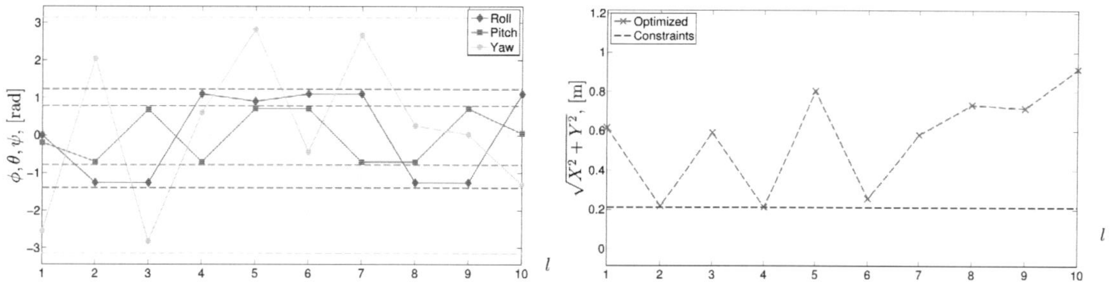

(b) Conjugate-type algorithm, O_2

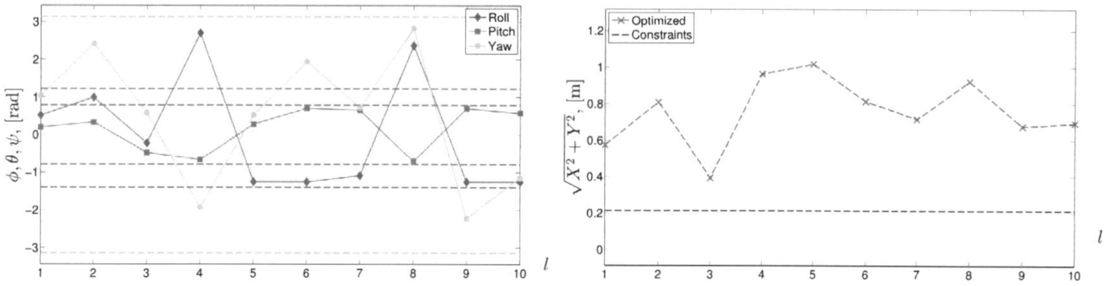

(c) Iterative algorithm, O_1

Figure 6: Constraints for the LED frame orientation and its distance from the robot base for $M = 10$ selected calibration poses

[9] Khalil W, Gautier M and Enguehard Ch 1991 dentifiable parameters and optimum configurations for robots calibration *Robotica* **9** 63-70

[10] Klimchik A, Pashkevich A, Wu Y, Caro S and Furet B 2012 Design of calibration experiments for identification of manipulator elastostatic parameters *J. of Mechanics Engineering and Automation* **2** 531-42

[11] Marie S, Courteille E and Maurine P 2013 Elasto-geometrical modeling and calibration of robot manipulators: Application to machining and forming applications *Mechanism and Machine Theory* **69** 13-43

[12] Renders J M, Rossignol E, Becquet M and Hanus R 1991 Kinematic calibration and geometrical parameter identification for robots *IEEE Transactions on Robotics and Automation* **7(6)** 721-32

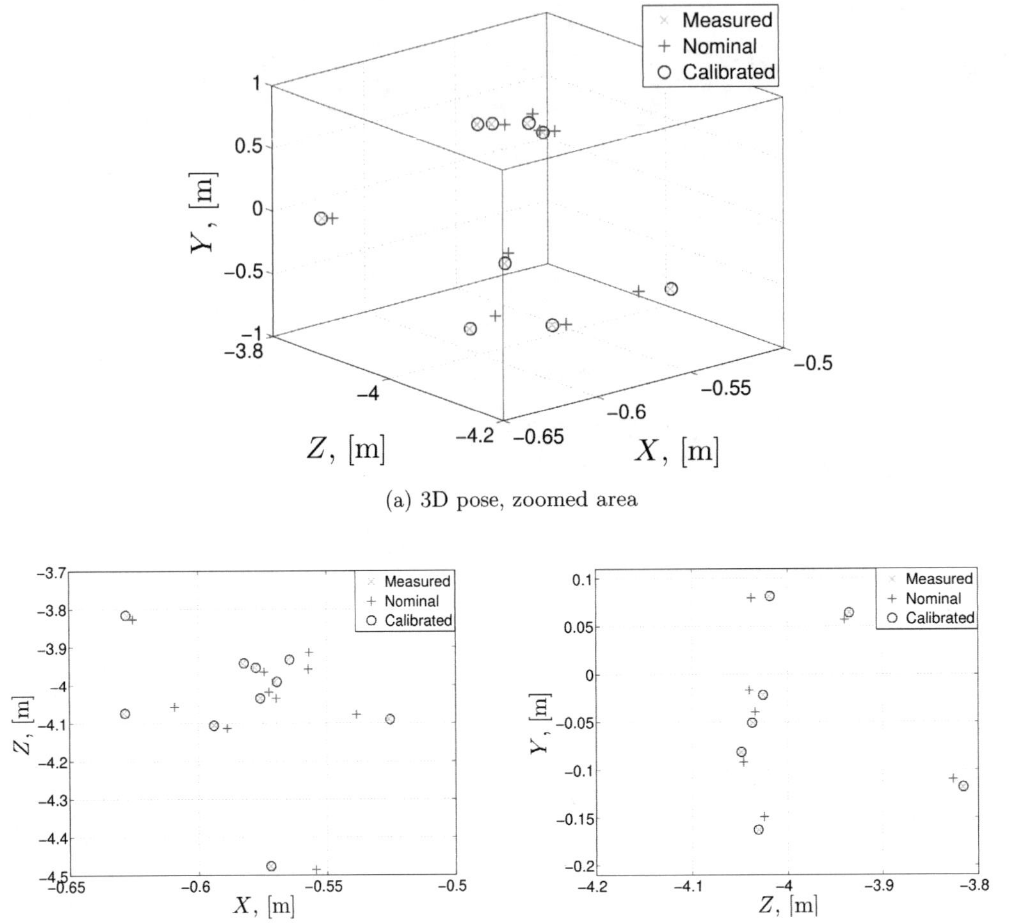

(a) 3D pose, zoomed area

(b) Projection on the XZ plane, zoomed area

(c) Projection on the YZ plane, zoomed area

Figure 7: Comparison of end-effector's Cartesian coordinates for nominal and calibrated geometric parameters

Time-frequency methods for signal analysis in wind turbines

Karel Kalista, Jindrich Liska

University of West Bohemia in Pilsen, NTIS, Technicka 8, 30614 Pilsen

kalistak@ntis.zcu.cz

Abstract. Since wind turbines became one of the most often source of renewable energy, appropriate health and condition monitoring systems are required. Especially proper monitoring of offshore plants is very significant because the accessibility is difficult and inspections are very costly. In comparison with conventional rotating machine vibration monitoring, where steady conditions and stationary signal are usually assumed, the wind turbines are characterized by unsteady conditions due to variable rotational speed. Hence the vibration signal is non-stationary and interpretation of signal signatures may be more complex. The common approach to analyze such non-stationary signals is the use of a time-frequency method, usually Short-Time Fourier Transform, which is the most popular one due to its simplicity. Nevertheless, there are other methods which can give a different view at the analyzed data and provide new information. This article investigates the potential use of some other time-frequency methods, namely Wavelet Transform, Wigner-Ville distribution and Hilbert-Huang transform in wind plants monitoring systems and apply these methods to real measured data with additional simulated bearing fault signal. Finally, the mentioned methods are compared based on computational complexity, readability and interpretability. Though the last two criteria are very subjective, Short-Time Fourier Transform was finally chosen as the most effective method followed by Wavelet Transform.

1. Introduction

Renewable energy is currently very popular topic, especially in relation to the global fear of nuclear energy. It helps to bring new investments and money into this branch to make it maximally effective. Currently, many wind turbines are under construction and many were already built offshore or onshore. Since the investments at the beginning are not small as well as cost of maintenance, a great demand on turbine reliability and availability exists because it is the only way to make wind power more competitive [4].

The condition monitoring systems (CMS) provided by the commercial vendors are usually based on time-domain signal analysis and tracking and bearing vibration diagnostic methods. The time-domain features are simple to calculate and may indicate some faults, however, they cannot provide any information on which component the fault occurred. Moreover, these features are usually sensitive to load and speed. Therefore, the time-domain features are used as initial or additional fault indicators.

Some of the more advanced system provides frequency-domain monitoring based on well known Fourier Transform. Nevertheless, due to the signal non-stationarity, the information may be

misleading and detection of fault occurrence is difficult. On the contrary, time-frequency representation (TFR) methods may enable tracking of signal components related to particular parts of the plant, thus significant information about time event caused by a defect can be provided for the HMS. The new trends in wind turbine condition monitoring are introduced in [1]. The most popular TFR is Short-Time Fourier Transform due to its simplicity and sufficient results. Nevertheless, the wavelet transform seems to be a proper alternative in CMS for specific fault detections [9],[10],[11]. The Wigner-Ville Distribution (WVD) was used in [12], but it is not common approach in wind turbine CMS as well as the Hilbert-Huang Transform (HHT).
Despite of the rich scientific literature, most of the sources apply the methods to fully simulated data with lack of relevant noise and other operating signal components [2] or the test is done with experimental test rig. It facilitates the demonstration of the pros and cons but without relation to reality. In this study, the mentioned methods are applied to real data measured in wind turbine. A reasonable fault signal was added to the real data to investigate the potential quality of fault detection of particular methods. Finally, the methods and theirs representations are compared from the viewpoint of computational complexity, readability, interpretability, resolution and the ability to detect the generated fault.

2. Wind Power Plants Vibration

Every rotating machine generates vibration, even in good condition. At least a rotating shaft generates a one-per-revolution frequency component caused by mass imbalance. It is demonstrated in Fig. 2, where the real signal (baseline) spectrum is shown. The vibration components are directly linked to periodic events in the machine operation, such as meshing gear teeth, irregularities on rolling elements in bearings, rotating electric fields and so on [3]. The specific frequencies give a direct indication of the source and enable its health evaluation. The most significant sources of vibration in wind power plant are rotor shaft, rolling elements bearings, gearbox, coupling and generator. Following figure shows the wind power plant schematic and sensor positions. The data used in this article were measured in main bearing by accelerometer (Sensor 08). Unfortunately, the details of measurement chain and data preprocessing were not provided.

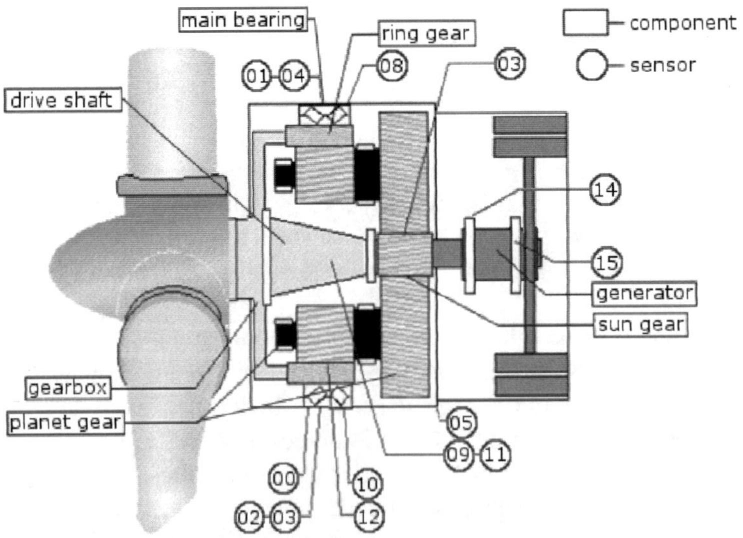

Figure 1: Wind power plant schematic and sensor positions.

2.1. Rotor vibration

Rotor shaft is the basic part of every rotating machine. The shaft is connected to all other parts and its vibrations are transmitted to them. The shaft imbalance component is always contained in the vibration signal because no shaft can be perfectly balanced. This imperfection is represented in spectrum by the peak in rotating speed frequency.
When two or more shafts are coupled together, a misalignment may occur. The dominant harmonics (order of rotating speed, X) are 1X for angular misalignment and 2X for axial misalignment.When misalignment becomes severe, it can generate high amplitude peaks at much higher harmonics.
Bent shaft generates high vibration and creates a lot of stress on other components. A permanent shafts bow produces similar response as a combination of unbalance and misalignment (especially angular one). Axial vibration measurement on one of the bearings usually reveals excited 1X when bow is near to the center of the shaft and 2X when bow is closer to coupling.The way to distinguish bent shaft from misalignment or unbalance is vibration phase measurement because the bent shaft is usually 180 ο out of phase in the axial direction.
Another common issue of a shaft are cracked shaft, eccentricity, looseness or rubbing.

2.2. Gearbox

Gearbox transmit power from one shaft to another, with change in speed and torque. Gearbox generates high vibration components event if not defect occurs. The vibration are caused by gear meshing when teeth meet together. Hence, the defect frequency is called gear mesh frequency (GMF). The spectrum can show the low harmonics as well as high harmonics. The reason is that the GMF is a product (multiplication) of the number of teeth and rotating speed. The common spectrum components are 1X and 2X of the speed, along with the GMF. Moreover, the GMF will have running speed sidebands relative to the shaft speed to which the gear is attached. These sidebands around GMF and its harmonics are quite common. Hence, the gearbox spectrum contains a range of frequencies due to different GMF and their harmonics. If the gearbox is in a good condition, all peaks have low amplitudes and no natural frequencies are excited.
Common defect of a gearbox are tooth wear and overload, gear eccentricity, backlash or misalignment, cracked or broken tooth or hunting tooth.

2.3. Bearings

A large bearing is used to support the rotor shaft and several smaller bearings support the generator shaft. Typical bearing defect frequencies are related to the bearing geometry and shaft speed and are associated with bearing component wear. The specific defect frequencies are related to cage, ball, outer and inner race wear and can be described by following equations.

$$Fundamental\ Train\ Frequency\ (FTF) = \frac{f}{2}\left[1 - \frac{d}{D}cos\varphi\right],$$
$$Ball\ Spin\ Frequency\ (BSF) = \frac{f}{2}\frac{d}{D}\left[1 - \left(\frac{d}{D}cos\varphi\right)^2\right],$$
$$Ball\ Passing\ Frequency\ Outer\ Race\ (BPFO) = \frac{nf}{2}\left[1 - \frac{d}{D}cos\varphi\right],$$
$$Ball\ Passing\ Frequency\ Inner\ Race\ (BPFI) = \frac{nf}{2}\left[1 + \frac{d}{D}cos\varphi\right],$$

where D is the pitch diameter, d is the ball diameter, n is the number of rolling elements and φ is the contact angle.

Moreover, bearing defects generates an impulse which excites the natural frequencies of bearing components and the structure as a whole. Therefore a signal resulting from these impulses appears as periodic burst of high frequency energy at intervals determined by the particular bearing defect.

2.4. *Electric generator*

Electric motors and generators vibrations are generated by mechanical forces such a e.g. unbalance, mentioned above and electromagnetic forces. In wind power plants, there are usually installed permanent magnet synchronous generators. Typical generator defect cause peaks in spectrum in multiples of the line frequency, pole pass frequency, stator winding faults or air gap eccentricity.

2.5. *Blade passing frequency*

Any rotating machine with blades generates increased amplitudes in spectra at frequency that is number of blades times running speed. In case of three-blade power plant, higher amplitudes can be by three times running speed and multiples of this frequency.

2.6. *Structural and other vibrations*

Besides kinematic frequencies of rotating parts, single components such as foundation, tower, rotor, blades or drive train can vibrate with their natural frequencies, The wind turbines are designed so that resonances are avoided.

.

3. Time-Frequency domain analysis

Time-Frequency domain methods are used to describe the distribution of amplitude (or energy) of signal in time and frequency domain simultaneously and allow to track changes of signal in time and frequency. The mentioned methods are described in more detail in [5],[6].

3.1. *Short Time Fourier Transform*

The fundamental method in time-frequency analysis is Short Time Fourier Transform (STFT) which naturally applies FT as it is shown in following equation

$$X(t,f) = \int_{-\infty}^{\infty} x(\tau)h(\tau - t)e^{-j2\pi f\tau}d\tau ,$$

where $h(t)$ is a window function with unit energy i.e.

$$\int_{-\infty}^{\infty} |h(t)|^2 dt = 1 ,$$

This window function allows to divide the non-stationary signal $x(t)$ into short parts which are assumed to be stationary. A long window leads to a representation coarsed in time but fine in the frequency domain and vice versa. The density of energy spectrogram is defined as

$$P_{SP}(t,f) = |X(t,f)|^2 = \left| \int_{-\infty}^{\infty} x(\tau)h(\tau - t)e^{-j2\pi f\tau}d\tau \right|^2$$

3.2. *Continuous Wavelet Transform*

Continuous Wavelet Transform (CWT) is another type of time-domain method defined as

$$CWT(a, b, \psi) = \int_{-\infty}^{\infty} x(\tau)\psi_{a,b}(\tau)d\tau$$

where ψ is mother wavelet and function $\psi_{a,b}$ denotes translation and scaling of mother wavelet

$$\psi_{a,b}(t) = \frac{1}{\sqrt{a}}\psi\left(\frac{t-b}{a}\right).$$

To obtain an admissible representation, $\psi(t)$ must have zero-mean i.e.

$$\int_{-\infty}^{\infty} \psi(t)dt = 0.$$

In comparison to STFT, which use complex exponentials as basis functions, wavelet transform basis is formed by orthonormal series generated by mother wavelet with use of scaling by a factor of a and translation by a factor of b. The change in scaling of wavelet function influences the time-frequency resolution because a window with variable length with only one wavelet is used. Hence, CWT enables to gather good frequency resolution for low frequencies at the expense of time accuracy. On the other hand, high frequencies can be analysed with fine time resolution at the expense of frequency one, which allows to reveal high-frequency time events. Therefore, CWT is so-called a multi-resolution technique.

In general, CWT is time-scale representation, since it displays the signal time-evolution at different scales. Nevertheless, there is a relation between scale and frequency. In case of Morlet wavelet used in this article, the scale-frequency relation is

$$f = \frac{f_0}{a}.$$

3.3. *Wigner-Ville Distribution*

Wigner-Ville Distribution (WVD) is an alternative method to STFT and CWT which decompose the signal into function basis. On the other hand, WVD is focused on decomposition of signal energy in the time-frequency domain.

By definition, it is a FT of central covariance function of a signal (instantaneous autocorrelation function)

$$WVD(t,f) = \int_{-\infty}^{\infty} x(t+\tau/2)x^*(t-\tau/2)e^{-j2\pi f\tau}d\tau,$$

where τ is the time lag and * represents the complex conjugate of the signal x.

In comparison to STFT and CWT, the WVD does not require any window, which is the limitation for resolution in time and frequency described by uncertainty principle (*Heisenberg–Gabor limit*) as

$$\sigma_f\sigma_t \geq \frac{1}{2}.$$

In other words, when frequency resolution is increased, the time resolution is decreased and vice versa.

Nevertheless, the real signal is not infinite, therefore Pseudo Wigner-Ville Distribution (PWVD) was defined as

$$PWVD(t,f) = \int_{-\infty}^{\infty} p(\tau)x(t+\tau/2)x^*(t-\tau/2)e^{-j2\pi f\tau}d\tau,$$

where

$$p(\tau) = h^*(\tau/2)h(-\tau/2).$$

Inconvenience of WVD is its nonlinearity which is responsible for the introduction of interference terms. This property of WVD causes existence of fake signal component between two real ones. This side effect can be suppressed by modifying the WVD to so-called Smoothed-Pseudo Wigner-Ville Distribution (SPWVD) defined as

$$SPWVD(t,f) = \int_{-\infty}^{\infty} h(\tau) \int_{-\infty}^{\infty} g(\varepsilon - t)x(\varepsilon + \tau/2)x^*(\varepsilon - \tau/2)\, d\varepsilon\, e^{-j2\pi f\tau} d\tau\, .$$

However, interference term suppressing deteriorates time-frequency resolution. Although SPWVD provides better resolution than STFT and CWT, the computation is time-consuming especially for long signals.

3.4. *Hilbert-Huang Transform*

Aforementioned methods use classic time-frequency analysis approach. It means that frequency determination is based on repetition of specific event with certain period. Thus, the event occurs in signal with certain frequency.

The approach in this section is based on knowledge of analytical signal (complex signal) which allows to define instantaneous phase and instantaneous frequency for each time sample of a signal. The analytical signal is a complex signal

$$z(t) = x_r + jx_i = A(t)e^{j\varphi(t)},$$

where its real part is the real measured signal. The imaginary part x_i can be determined directly by Hilbert Transform (HT)

$$H[x(t)] = x_i = \frac{j}{\pi}P \int_{-\infty}^{\infty} \frac{x(\tau)}{t-\tau}d\tau = x(t) * \frac{1}{\pi t},$$

where P is Cauchy principal value. Parameters x_r and x_i form a complex conjugate pair.

The instantaneous amplitude and phase of analytical signal is then

$$A(t) = \sqrt{x_r^2 + x_i^2}\ ;\ \varphi(t) = arctan\frac{x_r}{x_i}\, .$$

The instantaneous frequency is a derivation of instantaneous phase

$$\omega(t) = \frac{d\varphi}{dt} = \frac{x_i x_r - x_r x_i}{A^2}\, .$$

Unfortunately, the general signal is composed of many signal components, thus the theory mentioned above cannot be used directly. At first, the Empirical Modal Decomposition (EMD) must be carried out to decompose the signal to particular components. This algorithm is described in detail in [5].

EMD is an algorithm which is not covered by analytical definition [6] and is described by the following steps:

I. Identification of all extrema of the signal $x(t)$
II. Interpolation between minimal (maximal) ending up with some envelope e_{min} (e_{max}).
III. Computation of the mean value:
$$m(t) = \frac{e_{min}(t) + e_{max}(t)}{2}$$
IV. Extraction of the Intrinsic Mode Function (IMF) :
$$imf(t) = x(t) - m(t)$$
V. Iteration on the residual $m(t)$.

This algorithm is repeated till *imf(t)* is considered to be zero-mean or the stopping criteria is reached. Then, the set of IMFs is obtained and the instantaneous amplitude (IA) and frequency (IF) can be extracted with use of HT mentioned above. Finally, the time-frequency representation is obtained by displaying the evolution of IA and IF for each IMF.

4. Experimental Results

In this section, aforementioned methods and theirs representations are compared together. The data used for computation was measured by the accelerometer on the main bearing in a offshore wind

Journal of Physics: Conference Series **659** (2015) 012012 doi:10.1088/1742-6596/659/1/012012

turbine during no-fault state, therefore it can be considered as baseline data. The defect signal is added to the data to simulate damaged bearing. Since the main bearing parameters are not available due to intellectual property policy of the wind turbine owner, the defect properties were proposed by estimation. It was suggested, that the bearing defect signal frequency BPFO is 6,3 Hz, which produces impulses with related period 0,16s. These impulses excite natural frequency of the bearing housing which was proposed to be 300 Hz. Then the bursts can be located in peak-free area in the time-frequency representation. The amplitude of the burst was chosen in respect to the amplitude of the baseline signal so that the defect signal does not exceed the baseline signal. Hence, the final fault signal does not contain any issue at first glance. The sampling frequency is 5kHz and the time length is 2s.

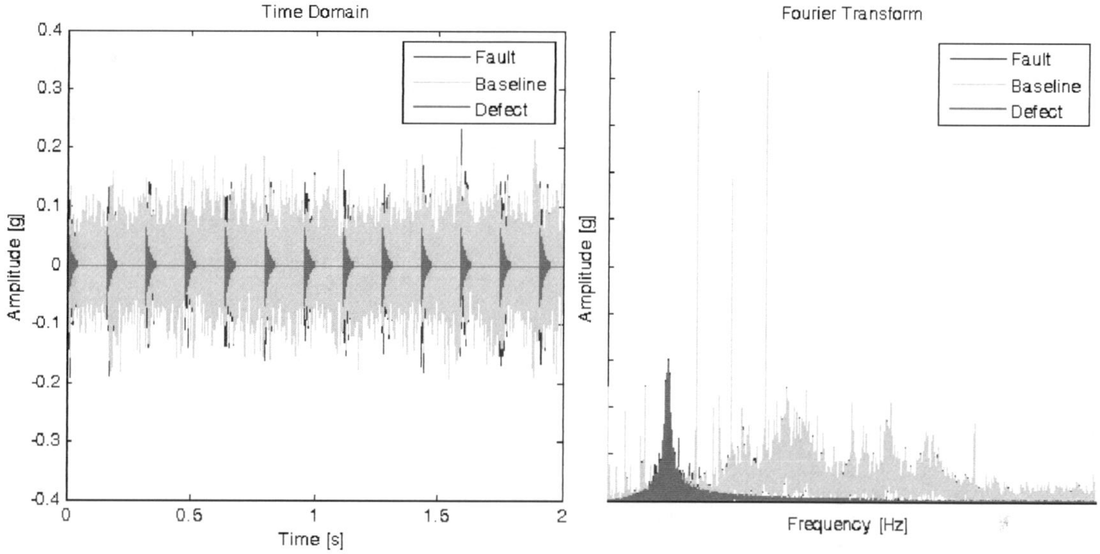

Figure 2: Time Domain and Fourier Transform Representation.

In the figure above, it is shown the signals in time domain. The fault signal (blue) represents the addition of baseline and defect signals. In the FT representation, an extra peak is apparent in the spectrum but it can be confused with regular signal component. Moreover, there is no time information about events.

Following figures show the time-frequency representation of baseline and fault signal acquired with use of aforementioned methods. Setting of particular methods was

- STFT - Hanning window of 512 samples with 99% overlap.
- CWT - Morlet Wavelet with central frequency $f_0 = 2Hz$.
- WVD, PWVD, SPWVD - Number of frequency bins is 5000.
- HHT

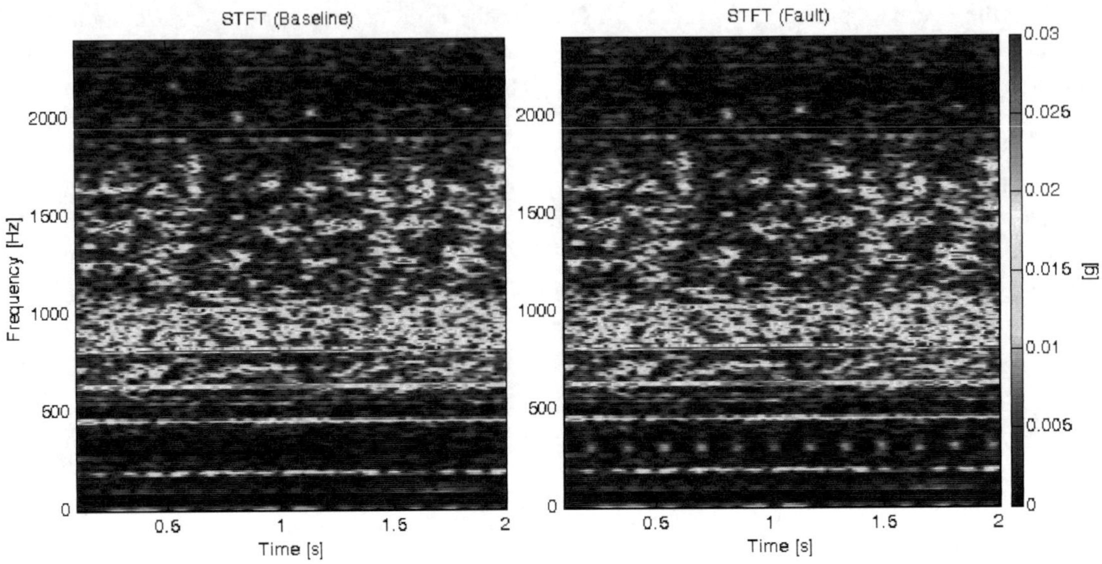

Figure 3: Short Time Fourier Transform of Baseline and Fault signal.

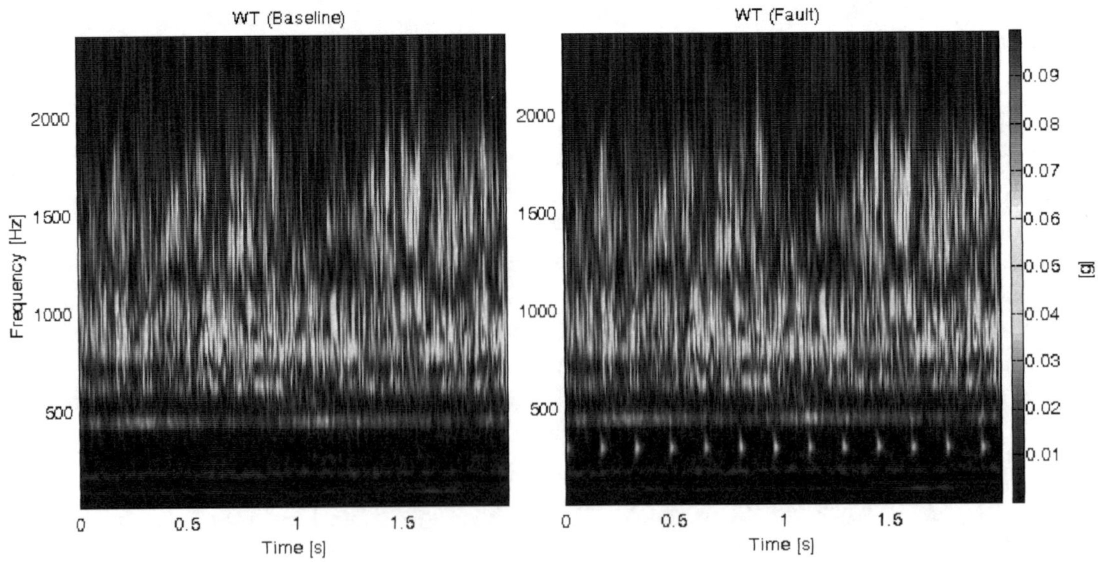

Figure 4: Wavelet Transform of Baseline and Fault signal.

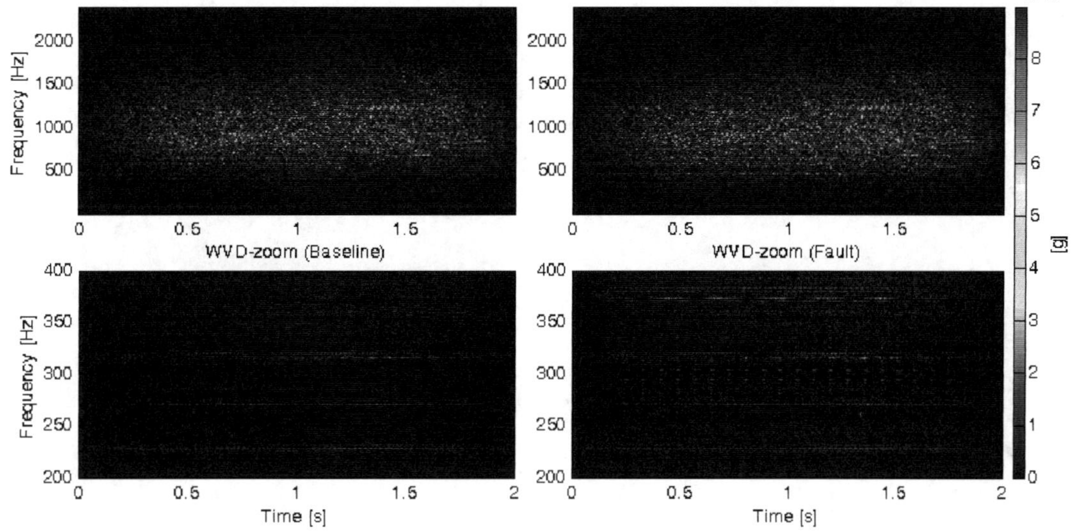

Figure 5: Wigner-Ville Distribution of Baseline and Fault signal.

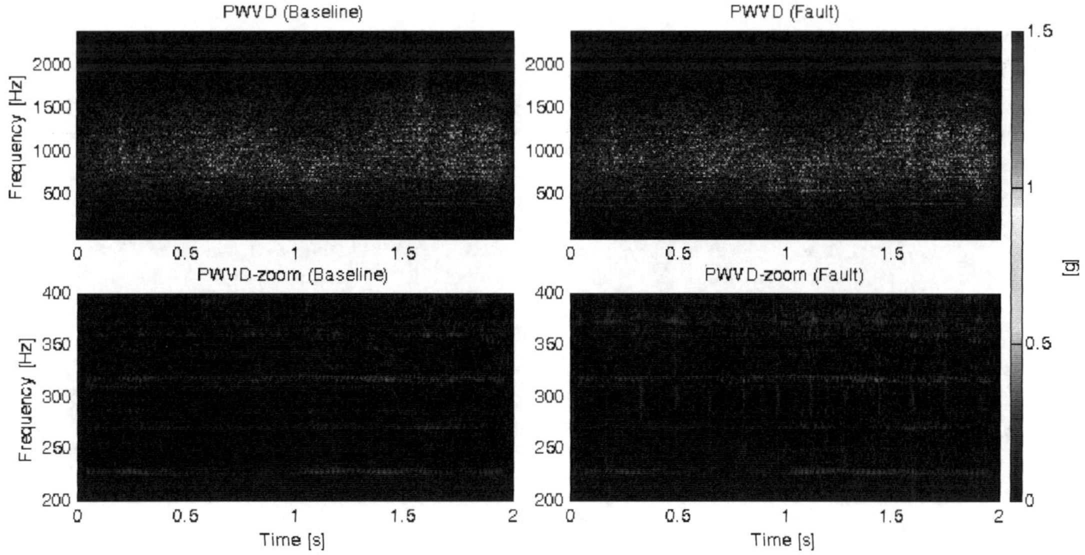

Figure 6: Pseudo Wigner-Ville Distribution of Baseline and Fault signal.

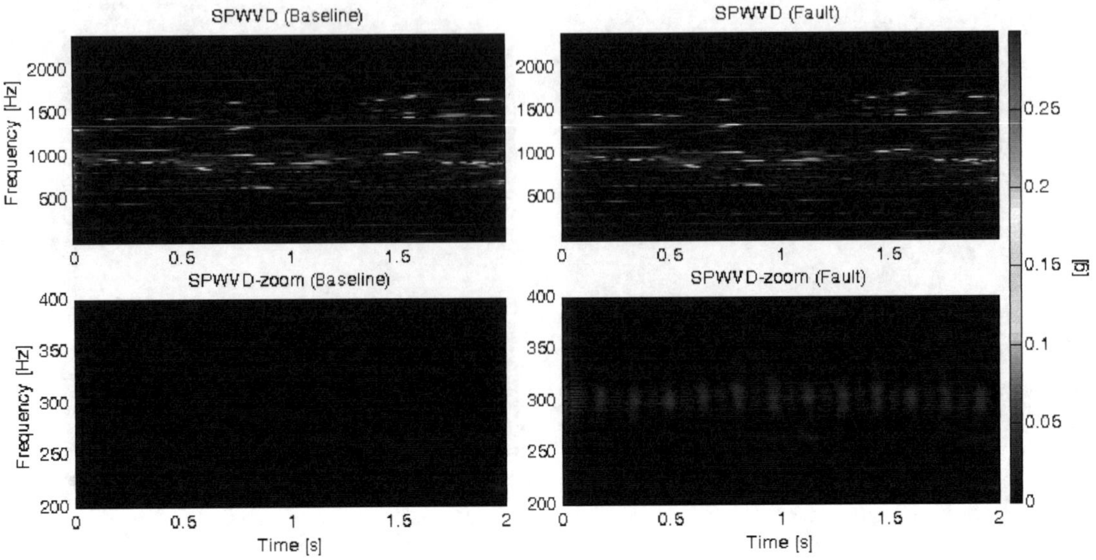

Figure 7: Smoothed Pseudo Wigner-Ville Distribution of Baseline and Fault signal.

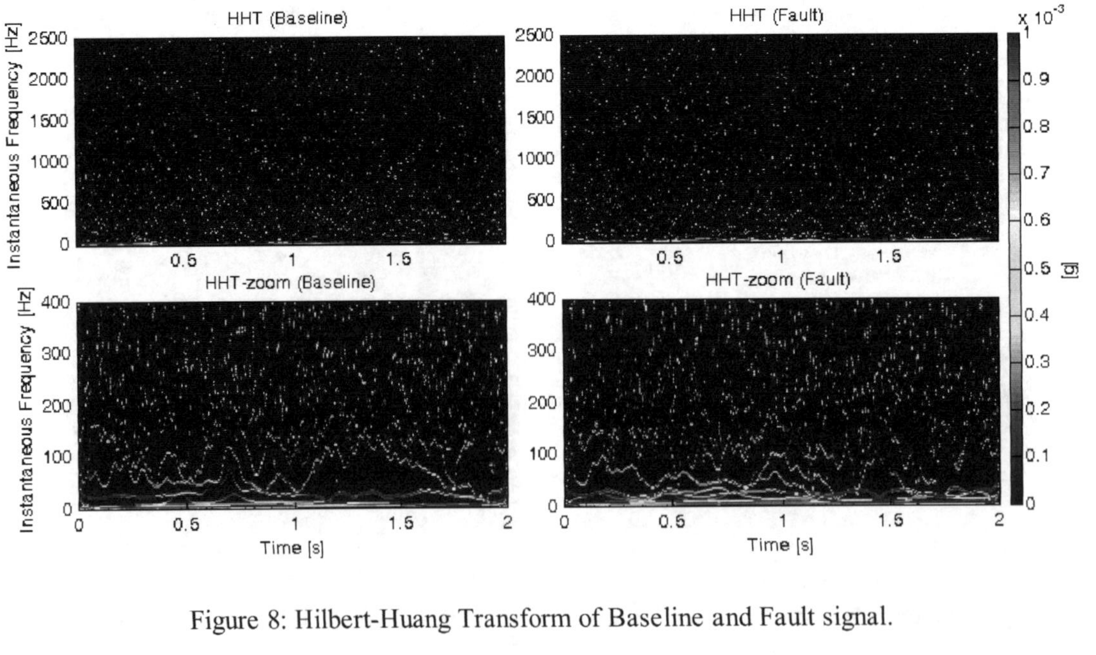

Figure 8: Hilbert-Huang Transform of Baseline and Fault signal.

4.1. Computational Complexity

The mentioned methods were programed in Matlab and were run offline in the same PC with configuration CPU Intel Core i5/2,7GHz and 8 GB RAM. The computational complexity results are reported in the following table.

Method	Time [s]
Short Time Fourier Transform	0,05
Continuous Wavelet Transform	3,2
Wigner-Ville Distribution	13,3
Pseudo Wigner-Ville Distribution	16,4
Smoothed Pseudo Wigner-Ville Distribution	440,0
Hilbert-Huang Transform	2,8

Table 1: Computational Complexity

It is obvious, that only STFT can be used in real-time systems, in practice.

4.2. Representation readability and interpretability

Figures 2-7 display the time-frequency representations for baseline and fault signals. It is apparent, that there are strong differences among particular representations in readability and interpretability of the results.

All of the methods except HHT revealed some disturbances in the frequency around 300 Hz, which was set in simulated defect signal as bearing natural frequency. The STFT and WT representations show the defect the most clearly followed by SPWVD. However, the SPWVD representation is affected by the cross-terms effect. The WVD and PWVD also display the defect, but the readability is very low and it can be easily confused with noise. Moreover, the presence of cross-terms is obvious.

The resolution of STFT and CWT is significantly lower compared to WVD, PWVD and HHT which confirms mentioned Heisenberg-Gabor uncertainty principle.

5. Conclusion

This article has dealt with time-frequency methods and their potential usage in analysis of wind turbine data to detect faults. The Short Time Fourier Transform, the Continuous Wavelet Transform, the Wigner-Ville Transform and Hilbert-Huang Transform have been applied to real data. The data was measured in main bearing in offshore wind power plant and a simulated bearing defect signal was added to it.

It has been shown, that all methods except Hilbert-Huang transform were able to reveal the defect, however with various resolution, readability and of the resultant representation, which is tricky to compare. Further, the computational complexity of the methods were also compared. The final results are summarized in the following table.

Method	Readability	Resolution	Computational Cost	Defect Revealed
STFT	Excellent	Good	Excellent	Yes
CWT	Excellent	Good	Good	Yes
WVD	Bad	Excellent	Bad	Yes
PWVD	Bad	Excellent	Bad	Yes
SPWVD	Good	Good	Very Bad	Yes
HHT	Good	Excellent	Good	No

Table 2: Time-frequency method comparison results.

References

This work was supported by the project PUNTIS - LO1506 and by the project SGS-2013-041.

[1] Tchakoua P, Wamkeue R, Ouhrouche M, Slaoui-Hasnaoui F, Tameghe TA, Ekemb G. Wind Turbine Condition Monitoring: State-of-the-Art Review, New Trends, and Future Challenges. *Energies*. 2014; 7(4):2595-2630.

[2] El.H. Bouchikhi et al., "A Comparative Study of Time-Frequency Representations for Fault Detection in Wind Turbine", IEEE Transactions on Industry Applications, 2011

[3] R. Gasch and J. Twele. Wind Power Plants: Fundamentals, Design, Construction and Operation. Electrical Engineering. Springer Berlin, Heidelberg, 2012.

[4] C. Scheer and P. Girdhar. Practical Machinery Vibration Analysis and Predictive Maintenance. Practical professional books from Elsevier. Elsevier Science, 2004.

[5] Robert Bond Randall. VIBRATION-BASED CONDITION MONITORING.ISBN: 978-0-470-97765-1. A John Wiley and Sons, Ltd., Publication, School of Mechanical and Manufacturing Engineering, University of New South Wales, Australia, 2011.

[6] Flandrin, P.: Time-Frequency/Time-Scale Analysis. Academic Press, London, 1999, ISBN 0-12-259870-9

[7] Huang, N. E. et al.: The empirical mode decomposition and the Hilbert spectrum for nonlinear and non-stationary time series analysis. Proc. Roy. Soc. Lond., 454, 903-993, 1998.

[8] Huang, N. E.... A confidence limit for the empirical mode decomposition and the Hilbert spectral analysis. Proc. Roy. Soc. Lond., 459, 2317-2345, 2003

[9] Tsai, C.-S.; Hsieh, C.-T.; Huang, S.-J. Enhancement of damage-detection of wind turbine blades via CWT-based approaches. IEEE Trans. Energy Convers. 2006, 21, 776–781.

[10] Watson, S.J.; Xiang, B.J.; Yang, W.; Tavner, P.J.; Crabtree, C.J. Condition monitoring of the power output of wind turbine generators using wavelets. IEEE Trans. Energy Convers. 2010, 25, 715–721.

[11] Meng, H. L., Leong M. S.: Detection of early faults in rotating machinery based on wavelet analysis, Hindawi Publishing Corporation, Advances in Mechanical Engineering, Vol. 2013

[12] Tang, Baoping & Liu, Wenyi & Song, Tao, 2010. Wind turbine fault diagnosis based on Morlet wavelet transformation and Wigner-Ville distribution, Renewable Energy, Elsevier, vol. 35(12), pages 2862-2866.

Parameter Estimation in Electrical Power Systems Using Prony's Method

Claudius Ortbandt[1], Cezary Dzienis[1], Robert Matussek[1] and Horst Schulte[2]

[1] Siemens AG, EM EA PRO D 1 3 Energy Automation, Berlin, Germany
[2] HTW Berlin, School of Engineering 1, Control Engineering, Berlin, Germany

E-mail: *cezary.dzienis@siemens.com, horst.schulte@htw-berlin.de, robert.matussek@siemens.com*

Abstract. This paper presents the basic principle of Prony's method, which is used for determining electrical power system parameters such as phasors, angular speeds as well as damping factors in a large frequency range. For fixed sampling frequencies, the approach to finding an optimal model order and to extracting transient parameters for further identification of different system effects is shown. Furthermore, it is shown that the method can be applied to the diverse plausibility checks for a confirmation of the result obtained using a conventional Fourier based approach. It will be presented that the significant advantage of the Prony algorithm in comparison to Fourier based methods is the possibility of computing an accurate frequency and in determining a damping factor.

1. Introduction

Power system protection technology has undergone large steps in development from electro-mechanics to digital devices. Thanks to that the power system protection became more sensitive, reliable and faster. This contributes to a high network stability. and security. Modern digital protection systems are highly complex and require a large degree of knowledge in measurement and signal processing techniques. In order for the protection devices to work accurately, the precise determination of the network state is necessary. This is done primarily through so called phasor quantities determined by Fourier-based algorithms. These provide good results for many applications, but in dynamic processes, they have a limited performance. Besides the FFT, other methods used for calculating network parameters and quantities can be applied successfully. Prony's method is one of these, which makes a prediction on the dynamics of a system using a mathematical model and a limited amount of measurement values. As such, it can be used to determine the phasors and frequency components of an electrical network. This publication examines whether Prony's method is applicable to modern protection systems and whether it can be used to identify different power system effects..

First examinations of Prony's method in the context of electrical power systems were published in 1990 [1]. The authors describe applications of the method and first results. Furthermore, they compare the methods against other known mathematical models. Moreover, Prony's method is primarily used for spectral analysis of distorted signals as in [6], as well as in the area of power electronics, to determine the harmonics and inter-harmonics in frequency inverters [7]. In [8] Prony's method is shortly explained and used for the analysis of fault currents. Statements are made about the model order, length of the measuring window, and the sampling frequency. In this context, [9] and [10] should also be mentioned. They also deal with the choice of the sampling frequency, in order to improve the efficiency of Prony's method. Fault currents were also analyzed in [11, 12, 13, 14]. In [11] the authors analyzed phase-to-ground faults using Prony's method in isolated and compensated power systems. Symmetrical fault-currents are examined in [12]. The authors use the presence of a decaying DC component in the currents as an indication of a symmetrical fault. Inrush currents are examined in [13] with Prony's method and are compared to the FFT-analysis. In [14] Prony's method is brought into conjecture with the reconstruction of currents, which are distorted by transformer saturation.

Prony's method is also examined in the context of fault-location algorithms. In [15, 16] the authors use the calculated frequencies and magnitudes of the currents during the process to calculate the fault location in DC networks. Other methods for determining the fault location in transmission and distribution networks are presented in [17, 18, 19, 20].

In connection with the identification of power swing effects, Prony's method appears, among others, in [21]. The author also describes the suppression of these effects using Power System Stabilizers [21].

The paper is organized as follows: Section 2 gives an introduction into Prony's method. Specifically, the classic variant of Prony's method and the least square Prony method are presented. In Section 3 Prony's method is used for the identification of different power system effects. Section 4 gives a conclusion and an outlook.

2. Prony's method

Prony's method is a parametric identification procedure which was developed in 1795 by Gaspard de Prony. In Figure 1 Prony's method is represented schematically with inputs and outputs.

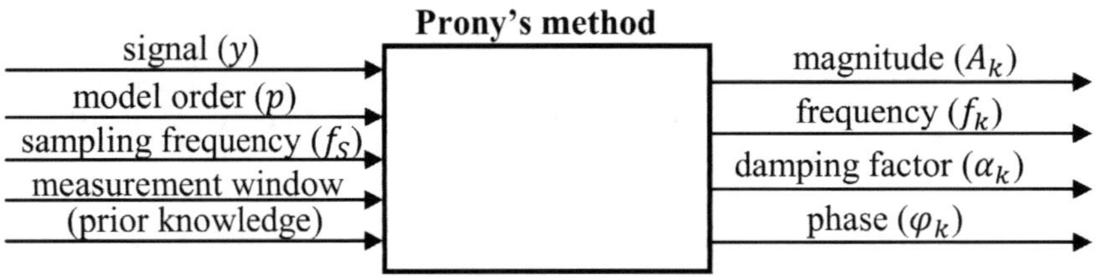

Figure 1. Input-output diagram of Prony's method

The method is based on the assumption that a discrete series of N measured values $y[1],..,y[N]$ of any sinusoidal or exponentially damped signal can be approximated by a linear combination of p exponential terms. The calculated signal components consist of positive, as well as negative frequencies. This can be interpreted as a phasor rotating one or the other way. The usage of Euler's formula relates the components to scalar and time-discrete values $\hat{y}[n]$.

$$\hat{y}[n] = \sum_{k=1}^{p} A_k \cdot \mathrm{e}^{(\alpha_k + j\omega_k)\cdot(n-1)\cdot T_s} \cdot \mathrm{e}^{j\varphi_k} \tag{1}$$

where:

n - 1,2,... N; N is the length of the signal
p - Number of exponentials / model order
T_s - Sampling period in [s]
A_k - Magnitude
α_k - Damping factor in [1/s]
ω_k - Angular velocity in [rad/s]
φ_k - Initial phase angle in [rad]

The discrete-time function in equation (1) can be simplified by separation into time-dependent and time-independent components:

$$\hat{y}[n] = \sum_{k=1}^{p} \underline{h}_k \cdot \underline{z}_k{}^{n-1} \tag{2}$$

The time-independent component is introduced as the complex magnitude $\underline{h}_k$, defined as:

$$\underline{h}_k = A_k \cdot e^{j\varphi_k} \tag{3}$$

The complex exponent $\underline{z}_k$ is a time-dependent parameter, defined as:

$$\underline{z}_k = e^{(\alpha_k + j\omega_k)T_s} \tag{4}$$

In order to represent the signal in a parametric form, the deviation of the model output from the original signal needs to be produced. The model error $e[n]$ is calculated for each data point.

$$e[n] = y[n] - \hat{y}[n] = y[n] - \sum_{k=1}^{p} \underline{h}_k \cdot \underline{z}_k{}^{n-1} \tag{5}$$

A better indicator is the root mean square error V over N measurement points. This will be the criterion for the model's quality.

$$V = \frac{1}{N} \sum_{n=1}^{N} \sqrt{e[n]^2} \tag{6}$$

The goal function minimization process leads to a nonlinear problem. The idea of Prony's method is to solve the difficult nonlinear problem by using linear equations. Therefore, Prony proposed a method which consideres the nonlinear aspect with the help of a polynomial factorization process.

2.1. *Classic variant of Prony's method*

The classic variant uses twice as many sampled data N as exponential components p to exactly model $2p$ complex parameters $\underline{h}_1, \ldots, \underline{h}_p$ and $\underline{z}_1, \ldots, \underline{z}_p$.

$$y[n] = \sum_{k=1}^{p} \underline{h}_k \cdot \underline{z}_k{}^{n-1} \tag{7}$$

Equation (7) shows the aim of Prony's method. A detailed expression for $1 \leq n \leq p$ is given in equation (8):

$$\begin{bmatrix} 1 & 1 & \cdots & 1 \\ \underline{z}_1^1 & \underline{z}_2^1 & \cdots & \underline{z}_p^1 \\ \vdots & \vdots & \ddots & \vdots \\ \underline{z}_1^{p-1} & \underline{z}_2^{p-1} & \cdots & \underline{z}_p^{p-1} \end{bmatrix} \cdot \begin{bmatrix} \underline{h}_1 \\ \underline{h}_2 \\ \vdots \\ \underline{h}_p \end{bmatrix} = \begin{bmatrix} y[1] \\ y[2] \\ \vdots \\ y[p] \end{bmatrix} \tag{8}$$

Assuming the knowledge of the elements $\underline{z}$, the complex magnitude vector $\underline{h}$ can be solved through the linear equations. Furthermore, Prony proposed that equation (14) is the solution of a homogeneous linear differential equation and defines a polynomial $F(\underline{z})$ of degree p that has the $\underline{z}_k$ exponents as its roots [2, 3]:

$$F_p(\underline{z}) = \prod_{k=1}^{p} (\underline{z} - \underline{z}_k) = (\underline{z} - \underline{z}_1) \cdot (\underline{z} - \underline{z}_2) \cdot \ldots \cdot (\underline{z} - \underline{z}_p) \tag{9}$$

The polynomial can be represented as a sum after utilizing the fundamental theorem of algebra:

$$F_p(\underline{z}) = \sum_{m=0}^{p} a[m] \cdot \underline{z}^{p-m} = a[0] \cdot \underline{z}^p + a[1] \cdot \underline{z}^{p-1} + \cdots + a[p-1] \cdot \underline{z} + a[p] \tag{10}$$

The coefficient $a[0] = 1$ is defined. A shift in the index in equation (7) from *n* to *n-m* and a multiplication by the parameter $a[m]$ results in equation (11) [2]:

$$a[m] \cdot y[n-m] = a[m] \cdot \sum_{k=1}^{p} \underline{h}_k \cdot \underline{z}_k{}^{n-m-1} \tag{11}$$

The sum of the products results in equation (12) for $p + 1 \leq n \leq 2p$:

$$\sum_{m=0}^{p} a[m] \cdot y[n-m] = \sum_{m=0}^{p} \left[a[m] \cdot \sum_{k=1}^{p} \underline{h}_k \cdot \underline{z}_k{}^{n-m-1} \right] = \sum_{k=1}^{p} \left[\underline{h}_k \cdot \sum_{m=0}^{p} a[m] \cdot \underline{z}_k{}^{n-m-1} \right] \tag{12}$$

The substitution of

$$\underline{z}_k{}^{n-m-1} = \underline{z}_k{}^{n-p-1} \cdot \underline{z}_k{}^{p-m} \tag{13}$$

results in:

$$\sum_{m=0}^{p} a[m] \cdot y[n-m] = \sum_{k=1}^{p} \left[\underline{h}_k \cdot \underline{z}_k{}^{n-p-1} \cdot \sum_{m=0}^{p} a[m] \cdot \underline{z}_k{}^{p-m} \right] = 0 \tag{14}$$

This equation is the desired linear differential equation. Looking at the summation in equation (14), the defined polynomial in equation (10) can be identified. By determining the roots $\underline{z}_k$ we obtain the sought zeros. Equation (14) can be solved exactly for the coefficients a[m] that are represented by *p* equations.

$$\begin{bmatrix} y[p] & y[p-1] & y[p-2] & \cdots & y[0] \\ y[p+1] & y[p] & y[p-1] & \cdots & y[1] \\ y[p+2] & y[p+1] & y[p] & \cdots & y[2] \\ \vdots & \vdots & \vdots & \ddots & \vdots \\ y[2p] & y[2p-1] & y[2p-2] & \cdots & y[p] \end{bmatrix} \cdot \begin{bmatrix} a[0] \\ a[1] \\ a[2] \\ \vdots \\ a[p] \end{bmatrix} = \begin{bmatrix} 0 \\ 0 \\ 0 \\ \vdots \\ 0 \end{bmatrix} \tag{15}$$

Equation (15) shows an over-determined, linear Toeplitz-type system. To determine a solution vector, the top line in the equation needs to be deleted and the first column needs to be substracted.

$$\underbrace{\begin{bmatrix} y[p] & y[p-1] & \cdots & y[1] \\ y[p+1] & y[p] & \cdots & y[2] \\ \vdots & \vdots & \ddots & \vdots \\ y[2p-1] & y[2p-2] & \cdots & y[p] \end{bmatrix}}_{Y} \cdot \begin{bmatrix} a[1] \\ a[2] \\ \vdots \\ a[p] \end{bmatrix} = - \begin{bmatrix} y[p+1] \\ y[p+2] \\ \vdots \\ y[2p] \end{bmatrix} \tag{16}$$

The basic procedure shown in Figure 2 can be separated into three steps:

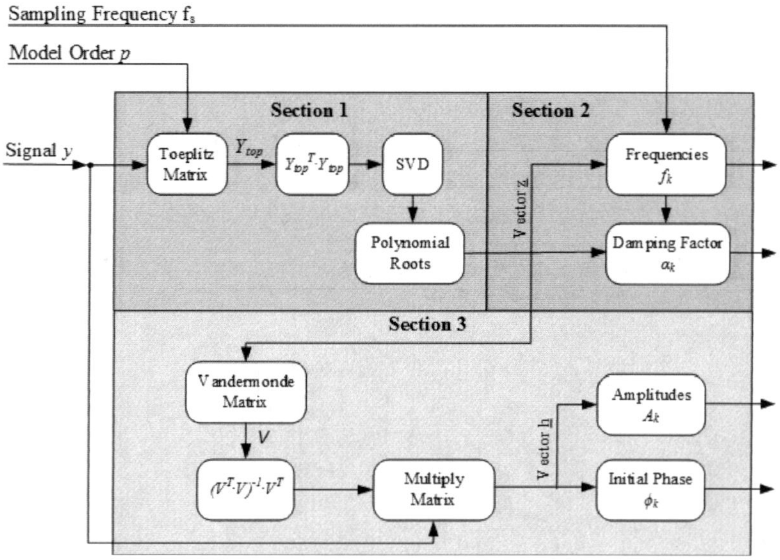

Figure 2. Simplified block diagram of Prony's method [5]

1. Step:
   Solve equation (16) and determine the parameters $a[0] \dots a[p]$, which are the polynomial's coefficients in equation (10).

2. Step:
   Find the roots of the characteristic polynomial in equation (10). These roots represent the time-dependent parameter vector $\underline{z}_k$ in equation (7). The damping factors and the frequency components can be determined with the aid of $\underline{z}_k$.

   The kth frequency is defined as:

$$f_k = \frac{\arctan\left(\frac{Im(\underline{z}_k)}{Re(\underline{z}_k)}\right)}{2 \cdot \pi \cdot T_s} \tag{17}$$

   Further the kth damping factor can be calculated:

$$\alpha_k = Re\left(\frac{\ln(\underline{z}_k)}{T_s}\right) \tag{18}$$

3. Step:
   Solve the system of linear equations (8) and determine the kth magnitude A_k and the kth initial phase angle φ_k.

$$A_k = |\underline{h}_k| \tag{19}$$

$$\varphi_k = \arctan\left(\frac{Im(\underline{h}_k)}{Re(\underline{h}_k)}\right) \tag{20}$$

2.2. Least squares Prony's method

If the amount of data points is greater than the number of parameters $= 2p$ then the sampled data can only be approximated but not interpolated by exponential functions. This approach is referred to as the minimization of the forward prediction.

$$e[n] = \sum_{m=1}^{p} a[m] \cdot y[n-m] \tag{21}$$

The identification of the relevant parameters $a[m]$ corresponds to the classic variant of Prony's method. The zeros of the characteristic polynomial with the coefficients $a[m]$ are $\underline{z}_1, \dots, \underline{z}_p$. This results in the following system of equations:

$$\underbrace{\begin{bmatrix} 1 & 1 & \cdots & 1 \\ \underline{z}_1^1 & \underline{z}_2^1 & \cdots & \underline{z}_p^1 \\ \vdots & \vdots & \ddots & \vdots \\ \underline{z}_1^{N-1} & \underline{z}_2^{N-1} & \cdots & \underline{z}_p^{N-1} \end{bmatrix}}_{Z} \cdot \begin{bmatrix} \underline{h}_1 \\ \underline{h}_2 \\ \vdots \\ \underline{h}_p \end{bmatrix} = \begin{bmatrix} y[1] \\ y[2] \\ \vdots \\ y[N] \end{bmatrix} \tag{22}$$

To obtain the complex magnitudes $\underline{h}_1, \dots, \underline{h}_p$, the following calculation is applied:

$$Z \cdot \underline{h} = y \tag{23}$$

$$Z' \cdot Z \cdot \underline{h} = Z' \cdot y \tag{24}$$

$$\underline{h} = (Z' \cdot Z)^{-1} \cdot Z' \cdot y \tag{25}$$

where $Z \in \mathbb{C}^{N \times p}$, $\underline{h} \in \mathbb{C}^{p \times 1}$ and $y \in \mathbb{R}^{N \times 1}$

The remaining calculation steps are analog to the classic variant of Prony's method.

3. Analysis of the power system quantities

Electrical Power Systems, in particular medium and high voltage systems, are naturally time variant due to event-driven switching of power lines and network areas. Thus, based on a local data acquisition in protection devices the structure and electrical parameters of the circuit diagram (of the electrical power system) change arbitrarily. For this reason, the model based approaches like state observer are not straightforward applicable to estimate states and system parameters like damping factors. During normal power system operation, currents and voltages are pure sinusoidal curves containing one fundamental component. However, it cannot be avoided that unforeseeable disturbances appear in power system. These contribute to significant changes in current and in voltage waves which include fundamental and periodic components. It can be noted that, during disturbances, the frequency, phase, magnitude, as well as the damping of undesired oscillations in current and voltage, are unknown. The commonly used Fourier-Transformation based methods do not deliver some signal parameters in such situations (they must be assumed and depend on the measurement window in which the analysis takes place) or their qualities are not acceptable. A very good alternative to the Fourier-Transformation seems to be the Prony-Algorithm. The description of the signal consists of its decomposition into exponential functions, including all periodic parameters (frequency, phase, magnitude and the damping factor) to be estimated. Based on the decomposed signal through Prony's method and analysis of the characteristic components, it can be concluded which effect is occurring in the network. The result of Prony's method does not depend on the measurement windows, which is the case for a Fourier-Transformation. The significant advantage of Prony's method over a Fourier-Transformation, is the possibility of online estimation of the damping factor of the periodic signal. It can be used in a plausibility check for a confirmation of the result. Especially the damping factor might show heavy fluctuations and still it corresponds to a realistic measurement.

The data analysis with Prony's method is performed in such way that a measurement window with a fixed length is chosen. The analysis is then subsequently carried out for every new data point. The number of frequency components delivered by Prony's method correlates with the data window length. Following this, only the more significant components arranged by magnitude will be considered.

$$magnitude\ (f_1) \geq magnitude\ (f_2) \geq magnitude\ (f_3) \geq magnitude\ (f_4) \qquad (26)$$

3.1. Short circuit analysis

In order to simplify the consideration, the simulated fault current is analyzing using Prony's method. The pre-fault condition is a current containing only a fundamental component with a frequency of 50 Hz and a magnitude of 1A. From fault inception, an increasing current in the 50 Hz component and an additional DC component can be seen. This DC component can be interpreted as unwanted noise in the conventional Fourier-based phasor calculation. The expectation is that Prony's method will estimate the fundamental component of 50 Hz with higher accuracy.

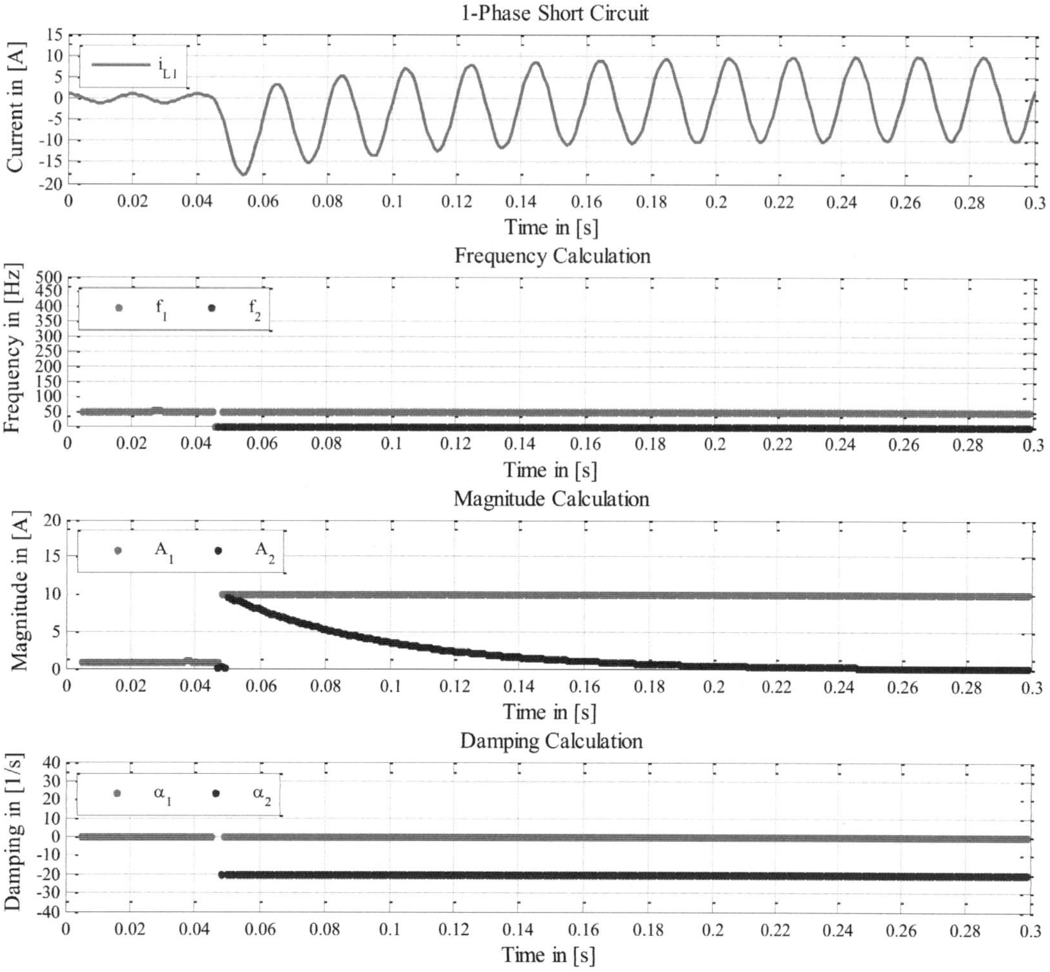

Figure 3. Prony parameter analysis of short circuit: $p = 3$; $N = 6$, $f_S = 1$ kHz

In Figure 3, the results of the fault analysis using Prony's method is presented. The measurement window length is 6 ms = 6 samples. The resulting model order is p = 3. This chosen model order allows for the estimation of two periodic components. During the power system fault, only the fundamental and DC component appears. Thus, the order of 3 is sufficient to accurately identify the signal parameters . This is a simulated case, therefore the signal does not include additional distortion or noise. All parameters obtained using Prony's method are presented with the appropriate time assignments. The presented results are correct and confirm the abilities of Prony's method. The analysis of the damping factor of the DC component shows that the power system time constant τ can be estimated. The calculation is only possible if there is a low fluctuation in this parameter. From this example the time constant can be extracted numerically.

$$e^{\alpha_2} = e^{\frac{-1}{\tau}} \tag{27}$$

$$\tau = \frac{-1}{\alpha_2} \tag{28}$$

$$\tau = \frac{-1}{\alpha_2} = \frac{-1}{-20\frac{1}{s}} = 50\ ms \tag{29}$$

Prony's method enables an identification of power system fault components and a description in form of phasor quantities. Also, the very short measurement window does not limit the application of Prony's method in the field of power system protection and monitoring.

3.2. Inrush analysis:

When energizing a power transformer, large inrush currents can occur. These currents vary in shape and magnitude with the voltage angle and with the transformer core's premagnetization. The voltage switching moment has an impact on the magnetic flux in every phase. In addition, the residual magnetism (remanence) affects the inrush current flow. Peak currents of 8 times the nominal current can be expected. Current differential protection is one of the most selective protection schemes. It is commonly used in medium and high voltage power systems. The principle of differential protection is based on Kirchhoff's first law. The protected zone is defined by the current transformer (CT) positions. All devices process distributed current measurements at the terminals. The complex current sum of all terminals is close to zero in a normal operating state. If the complex current sum is not zero, then an internal fault is detected and the protection devices trip the circuit breaker. The protected object is then disconnected from the power system. The occurrence of inrush currents within the protected zone might lead to a malfunction in the protection devices. The magnetizing current remains within the protected zone and the transformer is energized. A common method to cope with inrush events is to increase the restraint current or to block the differential protection's operation. The inrush detection can be triggered by a high proportion of the second harmonic in the current signal.

Selecting a measurement window length and model order:
Selecting the right measurement window length is of great importance in order to parametrically describe a signal sequence. The following figure shows the spectral decomposition of an inrush current signal. Short data windows lead to unstable and jittering frequency components but an increasing window length generates stable results. A window length of 40 ms (= 40 samples) provides the best results for that case. The frequency components contained in the signal can be extracted in a steady-state manner.

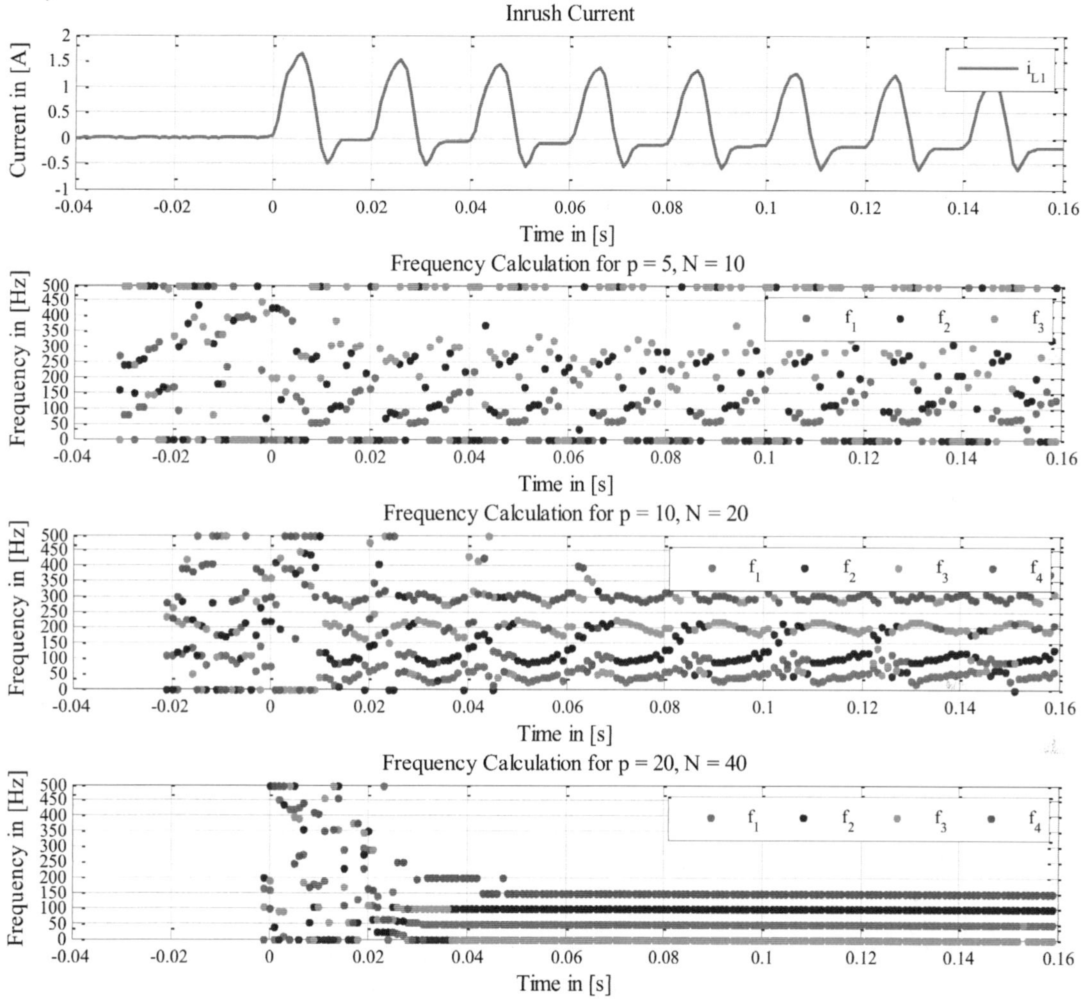

Figure 4. Selection of model order and measurement window

The analysis of the frequency components phase angle, magnitude and damping are depicted in figure 5. It can be shown that the signal contains a high second and third harmonic component. Also, it can be seen that the frequency components vary in time. The damping constant α at a model order of $p = 20$ yields values between 5 [1/s] and -15 [1/s]. This indicates an increasing oscillation or a decreasing oscillation, respectively. A data window at the length of approximately one period of the fundamental frequency yields results which are nearly constant over time.

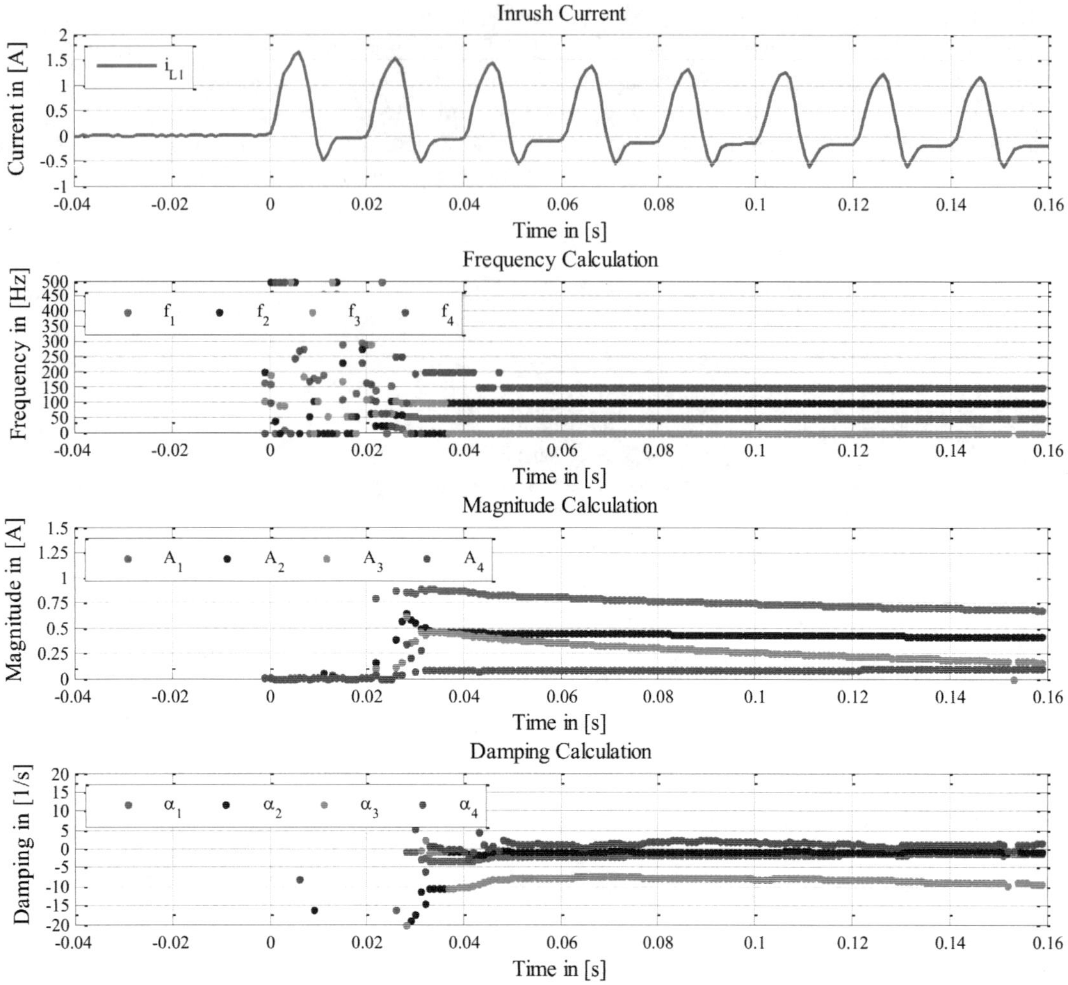

Figure 5. Analysis of a typical inrush current: $p = 20$, $N = 40$, $f_S = 1\ kHz$

The analysis of a power system fault and an inrush current utilizing Prony's method delivers substantial differences. In most cases, a power system fault consists of a large ratio of a fundamental frequency. Evaluating the second harmonic content can be an appropriate indicator to detect inrush events and to distinguish between other power system effects. For example, if the current is affected by CT saturation it can be complicated to discriminate these two effects. To achieve more solid results the measuring windows length could be increased. A windows length of 40 ms delivers a robust and reliable spectral decomposition. The inrush detection can be carried out for 30 ms after the energization. To achieve a better real-time performance it is possible to decrease the sampling rate (downsampling). The Nyquist-Shannon theorem must still be fulfilled.

4. Conclusion

The accuracy of the result using Prony's method strongly depends on the chosen model order, the sampling frequency, and the measurement window. Parameters that were chosen badly, lead to an inaccurate result, and thus to a large deviation from the original signal. Prony's method is not able to describe the input signal in spectral components in this case. Another effect which has to be considered is the algorithmic performance. Large data windows lead to a huge computational effort. An important advantage of Prony's method in comparison to the current method is that the frequencies being calculated are not limited in their bandwidth. The frequency components are the result of a minimization process and can take non-integer values. An application which benefits from this fact can be the power swing detection. Under optimal circumstances the results can be calculated within shorter measurement windows than it is possible with classic methods and is available only a few seconds after the start of an event. More complicated signal sequences will need more time for evaluation. Prony's method is able to create a new cluster of signal processing algorithms and to accelerate and improve current techniques. In the future, these methods can have a share in power system protection algorithms and the network stability and reliability can be further increased.

References

[1] J.F. Hauer, C.J. Demeure, I.L. Scharf. "*Initial Results in Prony Analysis of Power System Response Signals*", IEEE Transaction on Power Systems, vol. 5, no. 1, February 1990.

[2] T. Lobos, J. Rezmer, P. Schegner. "*Parameter Estimation of Distorted Signals Using Prony Method*", IEEE Bologna Power Tech Conference proceedings, vol. 4 p.5 pp. Vol.4-1, 2003.

[3] G. Pilz, P. Schegner. Method for predicting a future voltage and/or current curve, US Patent 7,692,336. 2010.

[4] L. Wenzel: Digitale Signalverarbeitung ist keine Hexerei, Teil 8. Datenfitting und Vorhersage, Elektronik 1999, H. 11, S. 62ff.

[5] J. Zygarlicki, M. Zygarlicka, J. Mroczka and K. J. Latawiec. "*A Reduced Prony's Method in Power-Quality Analysis – Parameters Selection*", IEEE Transaction on Power Delivery, vol. 25, no. 2, April 2010.

[6] T. Lobos, J. Rezmer: "*Spectral Estimation of Distorted Signals Using Prony Method*", Przeglad Elektrotechniczny (2003), vol. 79, no. 10, p. 735 – 738

[7] Z. Leonowicz, T. Lobos, J. Rezmer: "*Advanced Spectrum Estimation Methods for Signal Analysis in Power Electronics*", IEEE Transaction on Industrial electronics, vol. 50, no. 3, June 2003.

[8] Y. Jun: "*Prony Algorithm's Application on Extracting the Parameters of the Power System Transient Fault*", Applied Mechanics and Materials 2004, vols. 602-605, p. 3877-3880.

[9] O. Yu Bushuev, O. Ibryaeva: "*Choosing an Optimal Sampling Rate to Improve the Performance of Signal Analysis by Prony's Method*", 35th International Conference on Telecommunications and Signal Processing 2012, p. 634 – 638.

[10] J.-H. Lee, H.-T. Kim: "*Selection of Sampling Interval for Least Squares Prony Method*", Electronics Letters, vol. 41 no. 1, 6th January 2005

[11] O. Chaari, P. Bastard, M. Meunier: "*Prony's Method: An Efficient Tool for the Analysis of Earth Fault Currents in Petersen-Coil-Protected Networks*", IEEE Transactions on Power Delivery, vol. 10, no. 3, July 1995.

[12] S. Lotfifard, J. Faiz, M. Kezunovic:, "*Detection of Symmetrical Faults by Distance Relays during Power Swings*", IEEE Transaction on Power Delivery, vol. 25, no. 1, January 2010.

[13] S. Wylie, P. Tiwari: "*Harmonics Detection using Prony Frequency Analysis Algorithm in Transformer Inrush Current Generation after Fault Clearance*", International Journal of Innovative Research in Electrical, Electronics, Instrumentation and Control Engineering, vol. 2, issue 12, 2014.

[14] R. Schulze, P. Schegner: "*A Survey on Modern Fault Record Analysis*", IEEE Power and Energy Society General Meeting 2012, vol., p. 1 – 8.
[15] J. Rezmer, T. Sikorski: „*Signal Processing Method for Faults Location in DC Transmission System*“, 10th International Conference on Environment and Electrical Engineering 2011, p.1–4.
[16] M. Xu, L. Xiao, H. Wang: „A prony-based method of locating short-circuit fault in DC distribution system“, 2nd IET Renewable Power Generation Conference 2013, p. 1 – 4.
[17] M. Farshad, J. Sadeh: "*Transmission Line Fault Location using Hybrid Wavelet-Prony Method and Relief Algorithm*", International Journal of Electrical Power & Energy Systems 2014, vol. 61, p. 127 – 136.
[18] M. Tawfik, M. Morcos: "A F*ault Locator for Transmission Lines Based on Prony Method*", IEEE Power Engineering Society Summer Meeting. Conference Proceedings 1999, vol. 2, p. 943 – 947.
[19] A. Thakallapelli, R. Mehra, H.A. Mangalvedekar: "*Differentiation of Faults from Power Swings and Detection of High Impedance Faults by Distance Relays*“, IEEE 1st International Conference on Condition Assessment Techniques in Electrical Systems 2013, p. 374 – 377.
[20] T. JinRui, Y. XiangGen, Z. Zhe: "*Fault Location in Distribution Networks using Prony Analysis*", International Conference on Advanced Power System Automation and Protection 2011, vol. 1, p. 54 – 59.
[21] B. Qu, J. Wu, H. Ai, Y. Zhou: "*Online Identification and Suppression of Low Frequency Oscillation in Power System Based on WAMS*", Applied Mechanics and Materials, Vols. 513-517 (2014), pp. 2855-2861.
[22] W. Liang, H. Kang, L. Yao: „*Detection of Power System Oscillation Using Moving Window Prony Method*“, Interntional Conference on Power System Technology, 2010

Application of Advanced Process Control techniques to a pusher type reheating furnace

S M Zanoli[1,2], C Pepe[1,2] and L Barboni[3]
[2] DII, Università Politecnica delle Marche, Ancona (AN), Italy
[3] i.Process s.r.l., Falconara M.ma (AN), Italy

E-mail: s.zanoli@univpm.it, c.pepe@univpm.it

Abstract. In this paper an Advanced Process Control system aimed at controlling and optimizing a pusher type reheating furnace located in an Italian steel plant is proposed. The designed controller replaced the previous control system, based on PID controllers manually conducted by process operators. A two-layer Model Predictive Control architecture has been adopted that, exploiting a chemical, physical and economic modelling of the process, overcomes the limitations of plant operators' mental model and knowledge. In addition, an ad hoc decoupling strategy has been implemented, allowing the selection of the manipulated variables to be used for the control of each single process variable. Finally, in order to improve the system flexibility and resilience, the controller has been equipped with a supervision module. A profitable trade-off between conflicting specifications, e.g. safety, quality and production constraints, energy saving and pollution impact, has been guaranteed. Simulation tests and real plant results demonstrated the soundness and the reliability of the proposed system.

1. Introduction

Steel industry is aimed to the production of quality products, based on various types of raw material as input, e.g. billets or slabs. Complex physical and chemical reactions take place in the production chain, requiring specific control and monitoring strategies. Final products consist in steel bars, e.g. iron rods that then can be used in different fields [1], [2].

In the last decades, the considerable development of technology resulted in a rising level of automation in the process industries. Traditional control systems, e.g. PID controllers manually run by plant operators, cannot guarantee optimality in this new context. Optimality call for finding a profitable trade-off between opposing specifications: product quality increasing and safety constraints meeting versus energy saving and pollution impact decreasing. In order to balance these conflicting requirements, Advanced Process Control (APC) solutions have been proposed and applied [3].

This work proposes an APC system design for the control and the optimization of a pusher type reheating furnace located in an Italian steel industry. A two-layer Model Predictive Control (MPC) strategy has been adopted that overcomes the limitations of plant operators' mental model. A suited cooperation between the two layers, together with their collaboration with a supervision module, assures a flexible and resilient control, able to manage most critical situations. Moreover, an ad hoc decoupling strategy has been implemented, allowing the inhibition of specific manipulated variables

[1] To whom any correspondence should be addressed.

for the control of selected process variables. In particular, two equivalent mathematical approaches to obtain this feature are proposed and detailed.

The designed APC system has been tested on simulation and applied on the considered steel plant.

The present paper is organized as follows: a detailed description of the process, together with its control specifications, is given in section 2; section 3 explains the designed APC architecture, focusing on the adopted MPC solution. In section 4, the proposed decoupling strategy and its mathematical formulations are described while in section 5 simulations and practical results are resumed. Finally, conclusions are given in section 6.

2. Pusher type reheating furnace description and modelling

This section gives a detailed description of the considered unit of the Italian steel industry, i.e. its pusher type reheating furnace. In addition, control specifications and the consequent adopted modelling of the considered process are reported.

2.1. Focusing on the process features

The considered steel industry processes billets, that is small steel bars at an intermediate stage of manufacture. The production flow can be summarized as in figure 1.

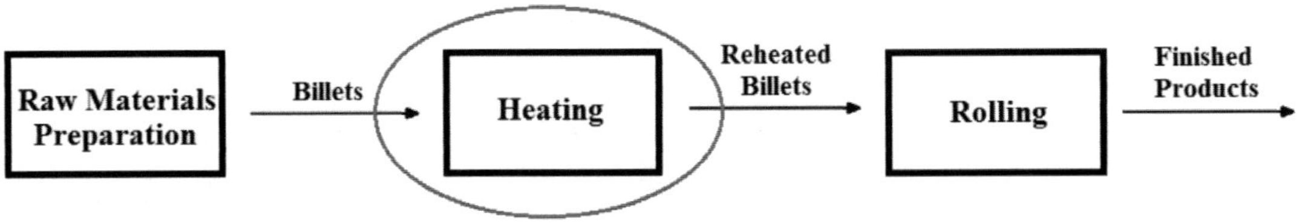

Figure 1. Schematic representation of the considered steel industry production chain.

In the initial phase, billets are prepared for their entry in a reheating furnace where the heating process takes place. In the furnace billets are reheated following a temperature profile guaranteeing a given exit temperature. After the exit from the furnace, the reheated billets are conveyed to the rolling mill stands. In this last stage of the production chain, the billets are subjected to a plastic deformation: in this way finished products, e.g. iron rods, are obtained.

This work is focused on the heating phase of the just described process (see figure 1). This production phase plays the most important role in steel industry, in terms of products quality, environmental impact and energy saving.

In figure 2, the reheating furnace is represented: the billets enter the furnace from the left side and they are moved along it by pusher tools, giving the name *pusher type* to the analyzed reheating furnace. Inside the furnace billets are subjected to a (re)heating process; conduction, convection and radiation heat exchanges occur in the furnace chambers where the billets transit. Oxy reactions are performed by fuel burners. The billets input temperature is variable, ranging from 50 °C to 300°C,

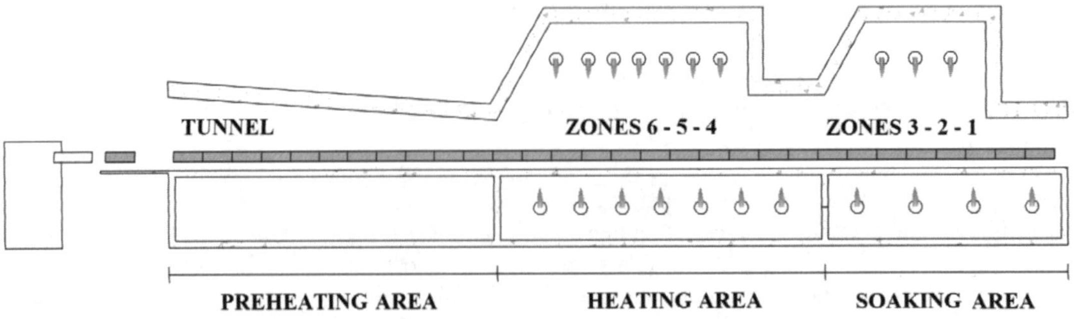

Figure 2. Schematic representation of the considered pusher type reheating furnace.

while their output temperature must be contained within the range 1000 °C – 1100 °C. On the basis of the local temperatures, the considered furnace can be divided into three different areas (see figure 2):

- *Preheating area*: this furnace area is the only one with no burners (see figure 2); hot gasses resulting from downstream areas are conveyed to this area, allowing the billets preheating. The chamber of this area, called *tunnel*, exhibits temperatures varying from 700 °C to 800 °C.
- *Heating area*: in this area, the most crucial phase of the billets reheating process takes place; here there are three chambers, named from the entry to the exit *zone 6*, *zone 5* and *zone 4*. Each of these chambers is equipped with an own burner; in this area, temperature may vary from 900°C to 1150 °C.
- *Soaking area*: it is the last furnace area, composed by three chambers, named from the entry to the exit *zone 3*, *zone 2* and *zone 1*. Each of these chambers is equipped with an own burner; in this area, temperature may vary from 1120°C to 1230 °C.

2.2. Process and control specifications

The most challenging requirement in a pusher type reheating furnace is the research of an optimal equilibrium between reheated billets quality and the binomial of energy saving and pollution impact decreasing, tied to fuel consumption minimization; billet quality requires homogeneous internal and external exit temperatures. Besides, safety constraints have to be fulfilled. Through several meetings with plant managers and engineers, these objectives have been converted into the following practical plant control specifications:

- minimization of air/fuel ratios, according to nominal stoichiometry; in this way a proper conduct of the process thermo-dynamical reactions is guaranteed;
- meeting of the assigned furnace temperature constraints, in order to assure a smooth billets heating process;
- compliance with safety constraints related to smoke-exchanger temperature, which is measured at beginning of the tunnel;
- minimization of fuel consumption.

For the regulation of the temperature of each zone of the furnace, an additional feature was required by plant engineers that consisted in the utilization of the zone own burner as sole control variable. This specification forces the neglecting of the physics of the process: in fact, the temperature of each zone is also influenced by the heat of the downstream zones. The only zone that is controlled acting on other zones burners is the tunnel, clearly due to the absence of power supplies: plant engineers required exploiting, as control variables, zone 6 and zone 5 burners. To fulfil this specific request, a decoupling strategy has been created, as will be detailed in section 4.

After this initial phase regarding control specifications, a key stage consisted in the selection of the input and output control variables together with measured disturbance input variables. These variables, in the industrial framework, are usually referred as Manipulated Variables (MVs), Controlled Variables (CVs) and Disturbance Variables (DVs), respectively. Fuel flow rate and stoichiometric ratio of each zone of the furnace have been selected as MVs. Temperature of each zone, fuel valves opening (per cent), temperature difference between adjacent zones and smoke-exchanger temperature have been chosen as CVs. Finally, furnace pressure, furnace production rate and air pressure have been selected as DVs. Based on the chosen MVs, CVs and DVs, an identification procedure with a black box approach has been fulfilled, in order to obtain reliable CVs-MVs and CVs-DVs dynamic models for the controller formulation: first order strictly proper linear models without delays have resulted from this phase [4], [5]. The obtained models consider deviations of MVs, DVs and CVs vectors from a vectors triplet (u_0, d_0, z_0 respectively), representing an operating point of the considered process. During the identification phase, the sampling time has been set to one minute; consistently, the APC cycle time has been set to one minute. The variables considered (see sections 4 - 5) for the proof of the proposed decoupling strategy are summarized in tables 1–2 –3.

Journal of Physics: Conference Series **659** (2015) 012014 doi:10.1088/1742-6596/659/1/012014

Table 1. The selected MVs.

TAG	Variable Name	Acronym [Units]
PTRF_F06	Zone 6 Fuel	*Fuel 6 [Nm3/h]*
PTRF_F05	Zone 5 Fuel	*Fuel 5 [Nm3/h]*
PTRF_F04	Zone 4 Fuel	*Fuel 4 [Nm3/h]*
PTRF_F03	Zone 3 Fuel	*Fuel 3 [Nm3/h]*
PTRF_F02	Zone 2 Fuel	*Fuel 2 [Nm3/h]*
PTRF_F01	Zone 1 Fuel	*Fuel 1 [Nm3/h]*

Table 2. The selected CVs.

TAG	Variable Name	Acronym [Units]
PTRF_TU	Tunnel Temperature	*Tunnel [°C]*
PTRF_ZT06	Zone 6 Temperature	*Temp 6 [°C]*
PTRF_ZT05	Zone 5 Temperature	*Temp 5 [°C]*
PTRF_ZT04	Zone 4 Temperature	*Temp 4 [°C]*
PTRF_ZT03	Zone 3 Temperature	*Temp 3 [°C]*
PTRF_ZT02	Zone 2 Temperature	*Temp 2 [°C]*
PTRF_ZT01	Zone 1 Temperature	*Temp 1 [°C]*

Table 3. The selected DVs.

TAG	Variable Name	Acronym [Units]
PTRF_FP	Furnace Pressure	*FP [mbar]*
PTRF_FPR	Production Rate	*FPR [t/h]*
PTRF_AP	Air Pressure	*AP [mbar]*

Tables 4 and 5 depict the relationships between the selected CVs and the selected MVs/DVs as resulting from the identification phase: denoting by m the number of CVs, by l the number of MVs and by p the number of DVs, the CVs-MVs transfer matrix is a $(m \times l)$ matrix, while the CVs-DVs transfer matrix is a $(m \times p)$ matrix. The structure of table 5 is in accordance with the physical behaviour of the process where hot gases from downstream zones influence upstream zone temperatures, but not vice versa.

Table 4. The CVs-DVs mapping matrix.

Acronym	*FP*	*FPR*	*AP*
Tunnel	X	X	X
Temp 6	X	X	X
Temp 5	X	X	X
Temp 4	X	X	X
Temp 3	X	X	X
Temp 2	X	X	X
Temp 1	X	X	X

Table 5. The CVs-MVs mapping matrix.

Acronym	*Fuel 6*	*Fuel 5*	*Fuel 4*	*Fuel 3*	*Fuel 2*	*Fuel 1*
Tunnel	X	X	X	X	X	X
Temp 6	X	X	X	X	X	X
Temp 5		X	X	X	X	X
Temp 4			X	X	X	X
Temp 3				X	X	X
Temp 2					X	X
Temp 1					X	X

3. Advanced Process Control system design

In this section, the architecture of the designed APC system is analysed: all its functional blocks are described, focusing on the adopted MPC strategy.

3.1. APC system architecture
The proposed APC system architecture is depicted in figure 3.

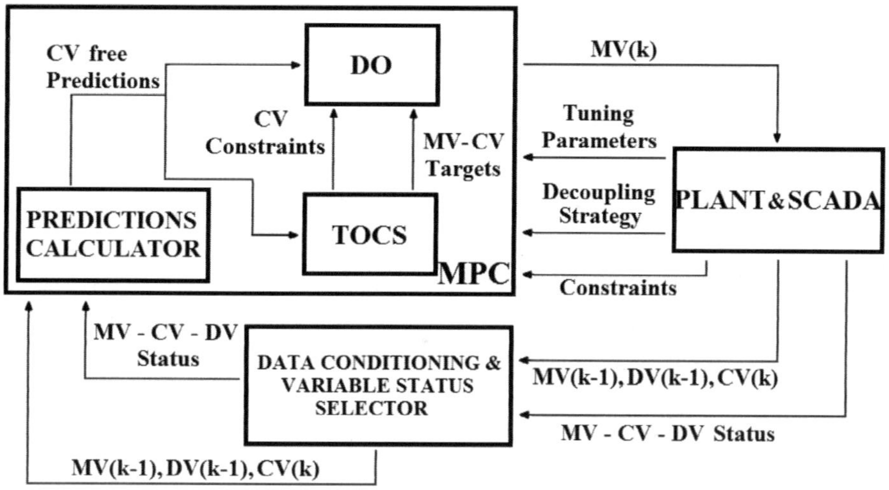

Figure 3. The proposed APC system architecture.

At a generic control instant k, updated MVs, DVs and CVs measurements are available from the plant: the communication between the process and the control system takes place through a Supervisory Control and Data Acquisition (*SCADA*) interface [6]. A *Data Conditioning & Variable Status Selector* module suitably processes new data, performing data conditioning if necessary (e.g. applying suitable filters): bad data (e.g. sensors spikes) or low level control loops faults are detected, contributing to the determination of the status of each MV, DV or CV, together with process management specifications. In this way, the selection of the variables subset to take into account in MPC formulation is performed. Together with this information, *MPC* block receives other fundamental data from the *SCADA* system: a set of tuning parameters, the settings for the adopted decoupling strategy implementation and the CVs and MVs constraints. In the following, these notions will be detailed.

MPC block, through an optimization strategy, computes a set of MVs moves: only the first of these control moves is forwarded to the plant and, at the next control instant, the entire procedure is repeated adding feedback from the process. In this way, a *receding horizon* strategy is performed [7]. *MPC* block is composed of three functional blocks: a two-layer scheme is supported by a *Predictions Calculator* module. The lower layer is constituted by a Dynamic Optimizer (*DO*) that dynamically calculates the MVs moves [7], [8]. The Targets Optimizing and Constraints Softening (*TOCS*) module lies on the upper layer: keeping into account economic and chemical objectives, it assures a coherency between DO steady state targets and constraints [9].

3.2. Predictions Calculator
The two-layer MPC architecture (see subsections 3.3 - 3.4) is assisted by a *Predictions Calculator* module. In the present case, based on a linear state space model, this block computes the CVs *free predictions* over a *prediction horizon* H_p, i.e. the CVs future trends keeping constant the MVs at the last value $MV(k-1)$ and exploiting all information available up to the time instant k. In the present case study, no future knowledge about DVs is available, so that in the CVs *free predictions* their value is kept constant to the last value $DV(k-1)$. According with the absence of delays in the considered case study, the CVs *free predictions* over the prediction horizon H_p can be formulated as follows [7]:

$$Z_{free}(k) = \begin{bmatrix} \hat{z}_{free}(k+1|k) \\ \hat{z}_{free}(k+2|k) \\ \vdots \\ \hat{z}_{free}(k+H_p|k) \end{bmatrix} = M \cdot \hat{x}(k|k) + N \cdot u_{past}(k) + T \cdot d_{past}(k) + Z_0 \tag{1}$$

$$Z_0 = ones(H_p, 1) \otimes z_0 \tag{2}$$

$$u_{past}(k) = ones(H_p, 1) \otimes (MV(k-1) - u_0) \tag{3}$$

$$d_{past}(k) = ones(H_p, 1) \otimes (DV(k-1) - d_0) \tag{4}$$

where $Z_{free}(k)$ is a $(m \cdot H_p \times 1)$ vector and $\hat{z}_{free}(k+i|k)$ is a $(m \times 1)$ vector, representing the CVs *free predictions* in the future control instant $k+i$. $\hat{x}(k|k)$ is the $(n \times 1)$ state estimation vector at the present instant k and M, a $(m \cdot H_p \times n)$ matrix, resumes the influence of this vector on CVs *free predictions*. Z_0 is a $(m \cdot H_p \times 1)$ vector containing the repetition of z_0 $(m \times 1)$ vector, $ones(H_p, 1)$ is a $(H_p \times 1)$ vector composed by elements all equal to 1 and $\otimes$ represents the Kronecker product [10]. $u_{past}(k)$ is a $(l \cdot H_p \times 1)$ vector containing the variation vectors of $MV(k-1)$ from u_0. $d_{past}(k)$ is a $(p \cdot H_p \times 1)$ vector containing the variation vectors of $DV(k-1)$ from d_0. N and T are $(m \cdot H_p \times l \cdot H_p)$ and $(m \cdot H_p \times p \cdot H_p)$ matrices, respectively: they represent the relationships between the deviations of $MV(k-1)$ and $DV(k-1)$ from their operating point and the CVs *free predictions* over the prediction horizon H_p.

3.3. Dynamic Optimizer

The Dynamic Optimizer, together with *Predictions Calculator* block, constitutes the basic scheme of a Model Predictive Controller. In the present work, the *DO* represents the lower layer of the proposed architecture. It is based on a Quadratic Programming (QP) problem, subject to linear inequality constraints. Consistently with the considered case study, no delays are assumed on MVs-CVs and DVs-CVs channels.

The *DO* objective function to minimize at each control instant k is:

$$V(k) = \sum_{i=0}^{H_p} \left\| \hat{u}(k+i|k) - u_{target}(k+i|k) \right\|_{S(i)}^2 + \sum_{i=0}^{H_u-1} \|\Delta\hat{u}(k+i|k)\|_{R(i)}^2 + \|\varepsilon(k)\|_\rho^2 \tag{5}$$

subject to

$$\begin{aligned} &\text{i. } lb_{du}(i) \le \Delta\hat{u}(k+i|k) \le ub_{du}(i),\ i = 0, \dots, H_u - 1 \\ &\text{ii. } lb_u(i) \le \hat{u}(k+i|k) \le ub_u(i),\ i = 0, \dots, H_u - 1 \\ &\text{iii. } lb_z(i) - ECR_{lbz}(i) \cdot \varepsilon(k) \le \hat{z}(k+i|k) \le ub_z(i) + ECR_{ubz}(i) \cdot \varepsilon(k),\ i = 1, \dots, H_p \\ &\text{iv. } \varepsilon(k) \ge 0 \end{aligned} \tag{6}$$

where $\hat{z}(k+i|k)$, $\hat{u}(k+i|k)$ are, respectively, the CVs and MVs prediction and $u_{target}(k+i|k)$ is the MVs reference trajectory at the future instant $k+i$. $\Delta\hat{u}(k+i|k)$ represents the future control moves that are assumed to be performed in the first H_u control steps. H_u is the *control horizon*, and it indicates the designed number of MVs moves available for the resolution of the *DO* problem. $S(i)$ is a positive (semi-)definite matrix, containing the tracking error weights for MVs at the future instant $k+i$. $R(i)$ is a positive definite matrix, that allows control efforts monitoring. The terms lb_{du} and ub_{du} are the lower and upper constraints for MVs moves while lb_u and ub_u indicate the lower and upper bounds for MVs. The terms lb_z and ub_z are the lower and upper constraints for CVs. MVs constraints can never be violated while CVs constraints may be possibly relaxed. In the proposed DO formulation, as a common MPC practice, these violations are allowed through the introduction of a non-negative slack variables vector $\varepsilon(k)$. $\varepsilon(k)$ is a $(m \times 1)$ vector, containing a slack variable for each CV. This vector has been introduced in the DO constraints through *Equal Concern for the Relaxation* (*ECR*) parameters, i.e. ECR_{lbz} and ECR_{ubz} in inequalities (6.iii) [11]. The importance of each single element of vector $\varepsilon(k)$ in the cost function (5) is tuned by a positive definite ρ matrix: this matrix, in cooperation with ECR parameters, defines a CVs priority ranking in DO constraints relaxation [12], [13]. According to CVs *free predictions* calculated by *Predictions Calculator* module (see equation (1)), the CVs (forced) predictions over the prediction horizon H_p are formulated as in the following:

$$Z(k) = \begin{bmatrix} \hat{z}(k+1|k) \\ \hat{z}(k+2|k) \\ \vdots \\ \hat{z}(k+H_p|k) \end{bmatrix} = Z_{free}(k) + F \cdot \Delta U(k) \tag{7}$$

where the $(l \cdot H_u \times 1)$ vector of MVs moves has the form:

$$\Delta U(k) = \begin{bmatrix} \Delta\hat{u}(k|k) \\ \Delta\hat{u}(k+1|k) \\ \vdots \\ \Delta\hat{u}(k+H_u-1|k) \end{bmatrix} \tag{8}$$

As can be noticed, CVs (forced) predictions can be expressed as a sum of two terms: CVs *free predictions* and an additional contribution, which depends on the dynamical relations between CVs and the MVs moves. These dynamical relations are explained by the F $(m \cdot H_p \times l \cdot H_u)$ matrix. Exploiting the information of equations (1) and (7), the CVs prediction in a future control instant $k+i$ becomes:

$$\hat{z}(k+i|k) = \hat{z}_{free}(k+i|k) + F((i-1)\cdot m + 1{:}\, i \cdot m, :) \cdot \Delta U(k) \tag{9}$$

where $\hat{z}_{free}(k+i|k)$ derives from equation (1), while the symbolism $F((i-1)\cdot m+1{:}\,i\cdot m,:)$ indicates a $(m \times l \cdot H_u)$ sub-matrix of F, composed by the same number of columns and including the F matrix rows between $((i-1)\cdot m+1)$th and $(i \cdot m)$th rows (extremes included): it details the dynamic relationship between $\hat{z}(k+i|k)$ and $\Delta U(k)$ [7].

In figure 3, the terms H_p, H_u, S, R, ECR_{lbz}, ECR_{ubz} and ρ are among the *Tuning Parameters*, while lb_{du}, ub_{du}, lb_u, ub_u, lb_{z_op} and ub_{z_op} represent the *Constraints*. An important remark is the possible modification of lb_{z_op} and ub_{z_op} by TOCS module determining lb_z and ub_z (see *CV Constraints* in figure 3), as will be shown in subsection 3.4.

3.4. Targets Optimizing and Constraints Softening

In the previous subsection DO module has been described: among its parameters, the definition of $u_{target}(k+i|k)$ terms has not been treated. A proper setting of these terms, representing the MVs reference trajectory, can guarantee the reaching of a profitable economic/chemical operating point for the process: in order to overcome the limitations of plant operators' mental models, an optimization approach for the computation of their end terms, i.e. $u_{target}(k+H_p|k)$, has been introduced. This is performed by TOCS module. In the present work, all $u_{target}(k+i|k)$ over the prediction horizon terms will be set equal to the target calculated by TOCS module.

During the driving of a process, in some abnormal situations, e.g. caused by improper operators' manoeuvres and settings, the CVs constraints set through *SCADA* interface (see figure 3) may result not coherent with the plant chemical principles [12], [13]. A protection against these unexpected events is provided by TOCS module that, as additional feature, relaxes DO constraints.

In the designed APC system, a Linear Programming (LP) problem, subject to linear constraints, constitutes the core of the TOCS module. The TOCS cost function to minimize at a generic control instant k is:

$$V_{ss}(k) = c_u{}^T \cdot \Delta u_{ss} + \rho_{ss\_min}{}^T \cdot \varepsilon_{ss\_min} + \rho_{ss\_max}{}^T \cdot \varepsilon_{ss\_max} \tag{10}$$

subject to

$$\begin{aligned}
&\text{i.} && lb_{du_ss} \le \Delta u_{ss} \le ub_{du_ss} \\
&\text{ii.} && lb_{u_ss} \le u_{ss} \le ub_{u_ss} \\
&\text{iii.} && u_{ss} = u_{target}(k + H_p|k) = MV(k-1) + \Delta u_{ss} \\
&\text{iv.} && z_{ss} = \hat{z}_{free}(k + H_p|k) + \Delta z_{ss} \\
&\text{v.} && lb_{z_ss} - ECR_{lb_ss} \cdot \varepsilon_{ss_min} \le z_{ss} \le ub_{z_ss} + ECR_{ub_ss} \cdot \varepsilon_{ss_max} \\
&\text{vi.} && \varepsilon_{ss_min} \ge 0; \ \varepsilon_{ss_max} \ge 0
\end{aligned} \tag{11}$$

In these expressions, Δu_{ss} and Δz_{ss} represent the steady state moves for MVs and CVs; making use of $MV(k-1)$ and $\hat{z}_{free}(k + H_p|k)$, they provide u_{ss} and z_{ss}, i.e. the MVs and CVs steady state targets. c_u is the MVs economic(/chemical) cost vector. The consistency between TOCS and DO modules is granted by the TOCS terms lb_{du_ss}, ub_{du_ss}, lb_{u_ss} and ub_{u_ss}, i.e. the steady state constraints for MVs steady state move and MVs steady state value, respectively:

$$\begin{aligned}
&\text{i.} && lb_{du_ss} = \textstyle\sum_{i=0}^{H_u-1} lb_{du}(i) \\
&\text{ii.} && ub_{du_ss} = \textstyle\sum_{i=0}^{H_u-1} ub_{du}(i) \\
&\text{iii.} && lb_{u_ss} = lb_u(H_u - 1) \\
&\text{iv.} && ub_{u_ss} = ub_u(H_u - 1)
\end{aligned} \tag{12}$$

Moreover, in equation (11.v) lb_{z_ss} and ub_{z_ss} terms correspond to the unchanged CVs bounds $lb_{z\_op}(H_p)$ and $ub_{z\_op}(H_p)$ (see *Constraints* in figure 3). According to DO module, violations on CVs steady state constraints are permitted in the proposed TOCS architecture: this feature is performed by the introduction of the non-negative slack variables vectors ε_{ss_min} and ε_{ss_max}. The interaction between TOCS-ECR parameters, i.e. ECR_{lb_ss} and ECR_{ub_ss}, and positive vectors ρ_{ss_min} and ρ_{ss_max} allows the creation of a CVs priority ranking in steady state constraints relaxation [12], [13]. An additional cooperation feature between TOCS and DO modules is represented by the relationship between lb_{z_ss}, ub_{z_ss} and $lb_z(i)$, $ub_z(i)$ $(i = 1, \dots, H_p)$:

$$\begin{aligned}
&\text{i.} && lb_z(i) = \min(lb_{z_op}(i), lb_{z_ss} - ECR_{lb_ss} \cdot \varepsilon_{ss_min}) \\
&\text{ii.} && ub_z(i) = \max(ub_{z_op}(i), ub_{z_ss} + ECR_{ub_ss} \cdot \varepsilon_{ss_max})
\end{aligned} \tag{13}$$

where min and max indicate minimum and maximum vector operators. According to CVs *free predictions* calculated by *Predictions Calculator* module (see equation (1)), the steady state CVs (forced) predictions exploited in the TOCS module have been formulated as in equation (11.iv). Exploiting the CVs-MVs $(m \times l)$ gain matrix (indicated with G), derived from the identification CVs-MVs transfer matrix, Δz_{ss} (the CVs steady state moves vector) can be expressed as:

$$\Delta z_{ss} = G \cdot \Delta u_{ss} \tag{14}$$

Substituting this relation in the equation (11.iv), CVs steady state predictions become:

$$z_{ss} = \hat{z}_{free}(k + H_p|k) + G \cdot \Delta u_{ss} \tag{15}$$

Terms c_u, ρ_{ss_min}, ρ_{ss_max}, ECR_{lb_ss} and ECR_{ub_ss} are part of *Tuning Parameters* of the figure 3.

4. MPC decoupling strategy

In this section, the proposed decoupling approach introduced in the MPC formulation is described, and two equivalent mathematical implementations are presented.

4.1. Mathematical approaches

As stated in the previous sections, to deal with a specific request of plant engineers, i.e. to regulate each zone temperature exploiting only its own burner, a decoupling strategy has been introduced in the MPC formulation. Typically, the control of a generic CV can be performed through the MVs related to that CV: this information is contained in the identification CVs-MVs transfer matrix; table 5 summarizes these relations for the case at issue. Thus, the decoupling strategy should enable a selection on which manipulated variables to move for controlling each of the controlled variables: in this way, an inhibition of the action of certain MVs on such CVs is obtained, that is not dependent on the tuning parameters. In the following, two equivalent approaches for the fulfilment of this objective are described.

The first approach modifies the setup of the MPC problem. Considering the $(m \times l)$ CVs-MVs mapping matrix, the number of MVs to be inhibited for the control of at least one CV is determined (each MV can be counted maximum once), denoted by b. An MV j can be inhibited if in the jth column of G matrix there is at least a row i (related to ith CV) for which both the following conditions hold:

- the (i, j) element of G matrix is nonzero;
- the control specification requires the inhibition of the jth MV moves for the ith CV.

After the determination of the b number, together with all pairs CV-MV to inhibit, the initial CVs-MVs transfer matrix (and consequently also G matrix) is modified by zeroing each (i, j) element to inhibit. Then, the b selected MVs are duplicated as DVs, thus resulting in a $(m \times (p + b))$ CVs-DVs transfer matrix and the removed MV-CVs transfer functions are restored in each of the added b columns. In this way, the original transfer matrices CVs-MVs and CVs-DVs are modified, and a new MPC problem is obtained. For the case at issue, starting from initial CVs-MVs (7×6) and CVs-DVs (7×3) transfer matrices, being b number equal to 5, the tables 4-5 are modified as follows:

Table 6. The CVs-MVs modified mapping matrix.

Acronym	*Fuel 6*	*Fuel 5*	*Fuel 4*	*Fuel 3*	*Fuel 2*	*Fuel 1*
Tunnel	X	X				
Temp 6	X					
Temp 5		X				
Temp 4			X			
Temp 3				X		
Temp 2					X	
Temp 1						X

Table 7. The CVs-DVs modified mapping matrix.

Acronym	*FP*	*FPR*	*AP*	***Fuel 5***	***Fuel 4***	***Fuel 3***	***Fuel 2***	***Fuel 1***
Tunnel	X	X	X		**X**	**X**	**X**	**X**
Temp 6	X	X	X	**X**	**X**	**X**	**X**	**X**
Temp 5	X	X	X		**X**	**X**	**X**	**X**
Temp 4	X	X	X			**X**	**X**	**X**
Temp 3	X	X	X				**X**	**X**
Temp 2	X	X	X					**X**
Temp 1	X	X	X				**X**	

The new setup of the MPC problem gives rise to a modified CVs-MVs (7×6) transfer matrix and a new CVs-DVs (7×8) transfer matrix.

Alternatively, a different approach that exploits the structure of the matrices F and G, not changing the original setup of the MPC problem, is proposed. The definition of a *Decoupling Matrix* D_E is performed: an (i, j) element of the D_E $(m \times l)$ matrix is set to 0 if both the following conditions hold:

- the (i, j) element of G matrix is nonzero;
- the control specification requires the inhibition of the jth MV moves for the ith CV.

The remaining elements of D_E matrix are set to 1. From D_E an additional $(m \times l \cdot H_u)$ matrix is constructed:

$$D_E^{ext} = ones(1, H_u) \otimes D_E \tag{16}$$

where $ones(1, H_u)$ indicates a row vector composed of elements all equal to 1. These matrices have been inserted in equations (9) - (14) - (15), resulting in:

$$\hat{z}(k+i|k) = \hat{z}_{free}(k+i|k) + (F((i-1)\cdot m + 1 : i \cdot m, :) \circ D_E^{ext}) \cdot \Delta U(k),\ i = 1, \dots, H_p \tag{17}$$

$$\Delta z_{ss} = (G \circ D_E) \cdot \Delta u_{ss} \tag{18}$$

$$z_{ss} = \hat{z}_{free}(k + H_p|k) + (G \circ D_E) \cdot \Delta u_{ss} \tag{19}$$

where $\circ$ indicates the element-wise multiplication.

For the case at issue, the D_E matrix has been reported in table 8.

Table 8. The decoupling matrix.

Acronym	*Fuel 6*	*Fuel 5*	*Fuel 4*	*Fuel 3*	*Fuel 2*	*Fuel 1*
Tunnel	1	1	0	0	0	0
Temp 6	1	0	0	0	0	0
Temp 5	1	1	0	0	0	0
Temp 4	1	1	1	0	0	0
Temp 3	1	1	1	1	0	0
Temp 2	1	1	1	1	1	0
Temp 1	1	1	1	1	0	1

The equivalence of the two mathematical approaches for the required decoupling strategy can be proved. This proof is not reported here, for brevity.

5. Simulation and field results

In this section, first a simulation example of the proposed decoupling approach is provided. Then, field results deriving from the implementation of the APC system on the real plant are depicted. Before showing the simulation example, a brief description of the design and tuning parameters is provided.

Tuning of S, R and ρ matrices of DO formulation has been performed in such a way that major importance to the CVs constraints meeting is assured, while considering MVs targets and monitoring control efforts. DO-ECR matrices and ρ matrix have been set so to guarantee the desired priority ranking on CVs constraints. For example, the meeting of smoke exchanger temperature bounds has been considered as a top priority objective compared to the compliance with zone temperatures constraints. The prediction horizon H_p has been set at 60 minutes, assuring steady state reaching for all CVs, while the control horizon H_u has been set at 8 control moves [14]. Tuning of TOCS module parameters has been performed according to DO one: c_u, ρ_{ss_min} and ρ_{ss_max} have been set in order to guarantee the fulfilment of CVs steady state constraints, while searching for economic and chemical objectives. The desired priority ranking in CVs steady state constraints relaxation is assured by the cooperation between ρ_{ss_min}, ρ_{ss_max} and ECR parameters, while for attaining fuel minimization c_u has been designed as positive vector.

In the simulation example, only zone 5 and zone 6 temperatures are considered in the APC control (no changes are made on the other MVs and DVs) and a typical operator's manoeuvre is simulated: a variation on both constraints of zone 6 temperature is performed, because of the entry of cold billets in the furnace. The effectiveness of the proposed decoupling strategy is proved by comparing the system performances without and with its introduction.

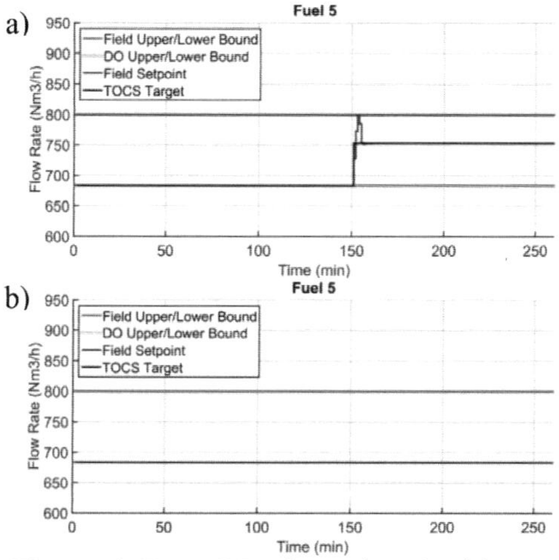

Figure 4. Zone 5 Fuel trends - a) without - b) with decoupling strategy.

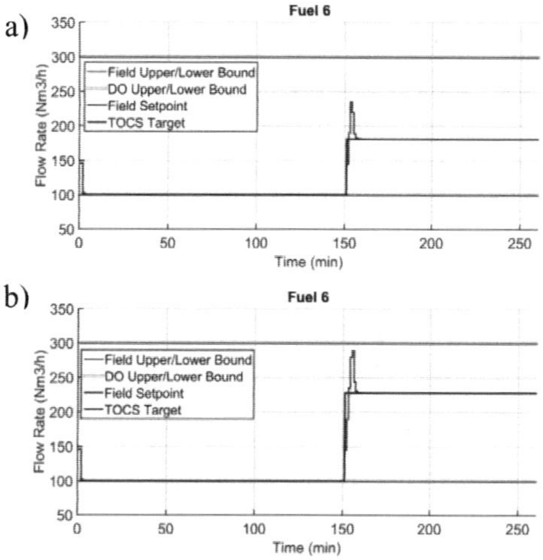

Figure 5. Zone 6 Fuel trends - a) without - b) with decoupling strategy.

In the presented simulation, as starting point a typical operating condition has been chosen: zone 6 fuel, zone 6 and zone 5 temperatures are inside their bounds, while zone 5 fuel lies on its lower limit. DO module, supported by TOCS module action, saves energy minimizing fuels flow rate. At time instant 150, as shown in figure 6, after the operator's move, the lower and upper constraints of zone 6 temperature are subjected to +20 °C and +30 °C change, respectively. Figures 4-5 show zone 5 and zone 6 fuel flow rates (zone 5 temperature has not be shown for brevity). In figures 4-6.a the control results without the decoupling strategy are reported, while figures 4-6.b refer to the decoupling approach. In the first control solution, when the operator performs the constraint change, TOCS module, in order to fulfil the request, modifies both zone 6 and zone 5 fuel targets (black dotted lines) and no steady state relaxation is required (cyan, red and green lines in figure 6 overlap). Consequently, DO module moves both fuel flow rates (blue line in figures 4-5), driving the entire subsystem towards a new operating point. In this way, zone 6 temperature (blue line in figure 6), after a brief transient, that causes a DO relaxation (magenta lines in figure 6), approaches its new lower constraint. In the second control solution, thanks to the decoupling strategy, when the constraints change is performed, TOCS module moves only zone 6 fuel target, leaving unchanged zone 5 fuel target. No steady state relaxation is required also in this second case. DO module, because of the inhibition of zone 5 fuel flow rate action for zone 6 temperature, acts on zone 6 fuel flow rate only, thus fulfilling the required management specification. Zone 6 temperature, after a transient (longer than in the first case but equally satisfactory), approaches its new lower constraint.

The designed APC system has been installed in an Italian steel plant, substituting the manual driving of PID control systems. In figure 7, comparisons of the system performances before and after the APC introduction are reported. In particular, zone 4 and zone 3 temperatures probability density functions (PDF) are shown: both situations refer to a three weeks period, with similar average boundary conditions, e.g. billets input temperature and furnace production rate.

After two years since the first start-up of the proposed APC system, the following results have been obtained:

- reduction of 3% average value and 8% variance of zones temperatures;
- reduction of 2.5% average specific consumption, computed considering fuel usage and productivity of the furnace.

a)
b)
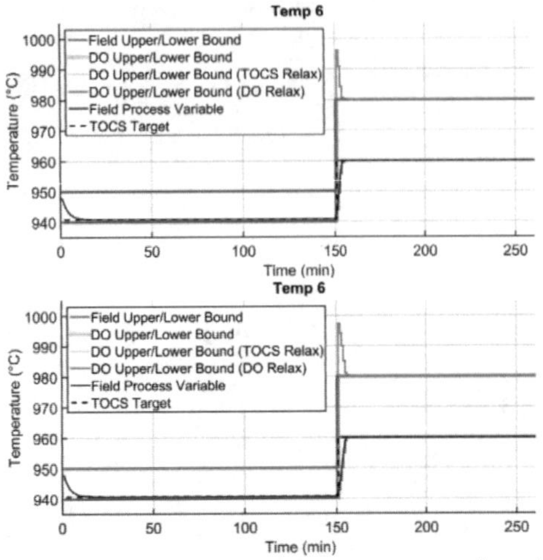

Figure 6. Zone 6 Temperature trends - a) without - b) with decoupling strategy

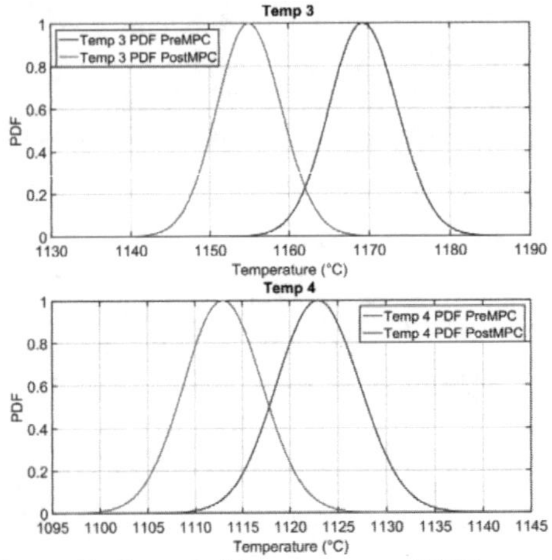

Figure 7. Zone 3-4 Temperature PDF pre and post MPC

6. Conclusion

In this work an APC system aimed at controlling a pusher type reheating furnace located in an Italian steel plant has been proposed. An architecture based on the interaction between a two-layer Model Predictive Controller and a supervisory block assures overcoming the limitations of plant operators' mental model in local PID control loops management. In addition, through a decoupling strategy, a selection on which MVs to use for controlling each CV has been introduced. In this way, a mathematical compromise between conflicting objectives, regarding quality, safety, pollution and energy aspects, has been guaranteed. With respect to the previous control system, a reduction of the variance of the main controlled variables has been observed, allowing to operate closer to the operating limits.

References

[1] Marino P, Pignotti A and Solis D 2002 *Latin American Applied Research* **32** 257-61
[2] Jaklic A, Vode F and Kolenko T 2007 *Applied Thermal Engineering* **27** 1105-14
[3] Bauer M and Craig I K 2008 *Journal of Process Control* **18**(1) 2-18
[4] Ljung L 1987 *System Identification: Theory for the User* (Englewood Cliffs, New Jersey, USA: Prentice-Hall)
[5] Liu T, Wang Q G and Huang H P 2013 *Journal of Process Control* **23**(10) 1597-623
[6] IEEE Standard for SCADA and Automation Systems 2007 *IEEE Std C37.1™*
[7] Maciejowski J M 2002 *Predictive Control with Constraints* (Harlow, UK: Prentice-Hall)
[8] Camacho E F and Bordons C 2005 *Model Predictive Control* (London, UK: Springer)
[9] Rawlings J B and Mayne D Q 2013 *Model Predictive Control: Theory and Design* (Madison, Wisconsin, USA: Nob Hill Publishing)
[10] http://www.siam.org/books/textbooks/OT91sample.pdf
[11] Bemporad A, Morari M and Ricker N L 2015 *Model Predictive Control Toolbox User's Guide*
[12] Zanoli S M, Pepe C and Rocchi M 2016 *Journal of Automation and Control Engineering* **4**(4) 262-267
[13] Zanoli S M, Pepe C, Rocchi M and Astolfi G 2015 *19th International Conference on System Theory, Control and Computing*
[14] Garriga J L and Soroush M 2010 *Industrial & Engineering Chemistry Research* **49**(8) 3505-15

Particle swarm optimization method for the control of a fleet of Unmanned Aerial Vehicles*

A. Belkadi[1,2], L. Ciarletta[3,4], and D. Theilliol[1,2]

[*] Research supported by Conseil Régional de Lorraine, France.
[1] Centre de Recherche en Automatique de Nancy (CRAN), Université de Lorraine, CNRS, 2 avenue de la forêt de Haye F-54516, Vandoeuvre-les-Nancy.
[2] CNRS, CRAN, UMR 7039, France.
[3] Permanent member the Madynes research team of Loria, UMR 7503, France.
[4] école des Mines Nancy, Université de Lorraine.

E-mail : adel.belkadi@univ-lorraine.fr, laurent.ciarletta@loria.fr, didier.theilliol@univ-lorraine.fr

Abstract. This paper concerns a control approach of a fleet of Unmanned Aerial Vehicles (UAV) based on virtual leader. Among others, optimization methods are used to develop the virtual leader control approach, particularly the particle swarm optimization method (PSO). The goal is to find optimal positions at each instant of each UAV to guarantee the best performance of a given task by minimizing a predefined objective function. The UAVs are able to organize themselves on a 2D plane in a predefined architecture, following a mission led by a virtual leader and simultaneously avoiding collisions between various vehicles of the group. The global proposed method is independent from the model or the control of a particular UAV. The method is tested in simulation on a group of UAVs whose model is treated as a double integrator. Test results for the different cases are presented.

Keywords : PSO algorithm, Fleet control, Virtual Leader, generating path.

1. Introduction

Recently, various applications of UAVs (Unmanned Aerial Vehicles) are emerging in the military, but also in the civilians areas. Using multiple vehicles fleet can cover large areas of research, surveillance[1], inspection[2] etc .. Recently, the attention of some scientists turned to imitate the behavior of birds in a fleet in order to make an autonomous flying formation [3][4].

Some solutions use leader, and in these approaches, there are various flight training control strategies : Leader-follower[5][6] (hierarchical approach), Virtual Leader [7][8][9] and control based on behavior [10][11] (decentralized approach). The method considered in our approach consists in replacing the head of the formation of the leader-follower approach by a virtual leader. All training entities receive the path of the mission which is considered as the virtual leader itself. A major disadvantage of the classical leader-follower strategy is that the risk of collision between agents increases[6], because

there is no return (feedback) training. In our approach, each agent also receives (and ises) information from its neighbors.
In this article, the objective is to develop a fleet control strategy for a group of UAVs capable of self-organizing and covering a surface in a predefined architecture/topology or to fly in formation, following a mission led by a virtual leader while ensuring collision avoidance between team members. In our case, we postulate that the gathering fleet is a reference generation problem rather than a control design problem. It is formulated as an optimization problem in order to find for each agent of the fleet reference to follow the trajectory at each instant t which minimizes a predefined objective function. Each UAV is assumed to be further controlled by a control law and it can measure the positions of all the agents in his neighborhood thanks to vision techniques and/or wireless communication capabilities. We consider in this paper, a simplified case of figure 2 without data loss communication delay or disruption.
This strategy was introduced in [12]. In our previous work, the objective function has been minimized in a simplified manner. The goal was to choose among a number of reachable points from a UAV, the position that minimized at best the objective function. In order to improve our approach we have chosen to use an optimization method that is more developed and responsive to the command of our fleet. This method is derived from meta-heuristics and is called Particle Swarm Optimization (PSO).
The control systems by Meta-heuristics, such as genetic algorithms[13], ant colony optimization algorithms[14] or by particle swarm optimization [15][16] have been the subject of several studies in the literature. Although genetic algorithms are very effective in finding the global minimum, their complexity and their execution time is their biggest disadvantage. The PSO algorithm has demonstrated very good performance by minimum search term with a minimum calculation time and relative ease of implementation [17]. These performances have oriented us to choose PSO
The paper is organized as follows. Section 2 is dedicated to the problem statement. Then, Section 3 is devoted to the optimization problem and the PSO algorithm description. In Section 4, the performance of the method based on various simulations has been illustrated. Finally, a conclusion and perspectives are presented.

2. Problem statement

The goal of our approach is to get a set of UAVs to fly on a 2D plane following a trajectory considered as a virtual leader while converging to a predefined spatial configuration (a flying pattern) and avoiding collisions.
Let us consider n mobile UAVs operating in space R^2 with $X_i(t)$ the robot position i at time $t > 0$. The system can be described by a weighted graph $G\,(N, \varepsilon(t), A(t))$ where $N = \{1,\ldots,n\}$ and $\varepsilon(t) \in N \times N$ are the set of nodes and the set of arcs respectively of the graph G and $A(t)$ a matrix of $M_n(R)$ of elements in R as $a_{ij}(t) > 0$ if : $(j,i) \in \varepsilon(t)$. At each sample i is an incoming neighbor of j and j is an outgoing neighbor of i.
G is assumed to be an undirected graph such as: $\forall (i,j) \in \varepsilon(t) \Rightarrow (j,i) \in \varepsilon(t)$. In this case the matrix $A(t)$ is symmetrical with $a_{ij}(t) = a_{ji}(t)$.
Let us define for each i node a set of neighbors $\Xi_i(t)$ such that :

$$\Xi_i(t) = \{ j \in N : || x_j(t) - x_i(t) || \leq l \} \quad (1)$$

where l defines the scope of the neighborhood.
Let's now consider for simplification, a fully actuated UAV i with dynamic model given by :

$$\dot{X}_i = v_i \quad (2)$$
$$\dot{v}_i = u_i \quad (3)$$

where $X_i = [x_i, y_i]^T$ is the configuration of UAV i, u_i is its control input and v_i its velocity vector. The control law is given by :

$$u_i = \alpha\left(X_i^d - X_i\right) + \beta\left(\dot{X}_i^d - \dot{X}_i\right) + \ddot{X}_i^d \quad (4)$$

The objective is to achieve a formation flight based on a virtual leader P. All training UAVs receive the trajectory of the mission. This trajectory is that of the virtual leader. The control strategy considered is completely independent from the underlying conception of the individual robot control.
The main idea is to find at each time t a desired reference input $X_i^d \in R^2$ for each UAV i of the fleet based on the desired reference trajectory $P_d(t)$ and the λ positions of the servants in its field of vision $x_j(t)$ with $j \in \Xi_i(t)$ such that $\lim A(t) = A_d$ while ensuring the anti-collision of the UAV with A_d being the desired adjacency matrix. This constraint can be rewritten as :

$$\forall n_i, n_j \in N : \ ||x_j(t) - x_i(t)|| > c \quad (5)$$

where c is a constant safety distance.
This objective is achieved by minimizing the cost function $\Lambda_i(t)$ for each one of the team member i.

3. Methodology

3.1. Assumptions and Optimization Problem

Given a team of n UAVs operating in space R^2. The entire system is described by a non-oriented weighted graph $G(N, \varepsilon(t), A(t))$, where $A(t)$ is a $n \times n$ matrix defined positive and symmetric, with $a_{i,j}(t)$ the distance between the node i and j at time t.
The neighbors of a robot i are defined in eq.(1) where $\Xi_i(t)$ is called the metrical neighborhood of the robot i. The control objective is to bring the fleet from an initial configuration $A(t_0)$ to a desired configuration A_d(i.e the desired formation). This problematic can be formulated as an optimization problem.
At first, let us define a cost function $\Lambda_i(t)$ for each UAV i such that:

$$\Lambda_i(t) = \rho\left(||(P_d(t) - (x_i(t) + h))|| - a_{ip}(t)\right) + \sum_{j=1}^{q} a_{ij}(t)\left(||\left(x_j(t) - (x_i(t) + h)\right)|| - a_{ij}(t)\right) \quad (6)$$

with $i \neq j$, $q = card(\Xi_i(t))$, $h \in R^2$, and $\rho \gg 1$
The main goal is to find the best vector h for each agent i minimizing the cost function $\Lambda_i(t)$ such that:

$$\forall i \in N, \lim_{t\to\infty} \Lambda_i(t) = 0 \Rightarrow \lim_{t\to\infty} A(t) = A_d \quad (7)$$

The reference trajectory at time $t+1$ for the agent i will then be:

$$x_{id}\,(t+1) = x_i\,(t) + h \quad (8)$$

The ρ coefficient in the cost function should be larger than 1. This choice is due to the fact that we favor each agent i position that tends to be in the direction of the virtual leader ***P*** affected by the reference trajectory $P_d(t)$.
During the evolution of the robots in R^2 space, non-collision of the agents should be ensured. This constraint, as shown in the previous section, can be written as follows: $\forall n_i \in N, \forall n_j \in h_i : \ ||x_j(t) - x_i(t)|| > c$, where $c \in R$ is the minimum distance between the two agents i and $j \in$ N do not cross.
On matrix $A(t)$, this means that :

$$\forall\, a_{ij}(t) \in A(t) :\ a_{ij}(t) > c, \text{ with } i \neq j \tag{9}$$

In order to introduce this condition in the cost function $\Lambda_i(t)$, we define new functions $\alpha_{ij}(t)$ between each agent i and its neighbors j such that :

$$\alpha_{ij}(t) = 1 + \exp(\frac{c - a_{ij}(t)}{\sigma}) \tag{10}$$

The value of the $\alpha_{ij}(t)$function depends on the difference between $a_{ij}(t)$ the distance between two Robots i and j with the safety distance c, it is all the greater when $a_{ij}(t) < c$ and converges to the value 1 when $a_{ij}(t) \gg c$.
So, each agent i is more likely to favor the values of h which avoids the need for the distances $a_{ij}(t) < c$ and minimizes at best the cost function $\Lambda_i(t)$, thanks to the PSO algorithm.

3.2. PSO Algorithm

The particle swarm optimization method appeared in 1995 in the U.S.A and was designed by Eberhart and Russel James Kennedy [20]. It is based on the notion of cooperation between agents called "particles", which manage to solve complex problems by exchanging information.
The PSO algorithm is initialized by a random population of potential solutions $x(t)$, interpreted as particles moving in the search space where $V(t)$ represents their speed. Each particle is attracted to its best position discovered in the past noted P_1 as well as to the best position of the particles discovered by its neighborhood P_2.
The algorithm includes several setting parameters to act on the compromise Exploration - Exploitation. Exploration is the ability to test various regions of space for "looking good" candidate solutions. The Exploitation is the ability to focus research around promising solutions to get as close as possible to the optimum.
The classic PSO algorithm is written as follows [18] :

$$V(t+1) = a\,V(t) + b_1 r_1\left(P_1(t) - x(t)\right) + b_2 r_2\left(P_2(t) - x(t)\right) \tag{11}$$
$$x(t+1) = x(t) + V(t+1) \tag{12}$$

Where, a is the coefficient of inertia, b_1 and b_2 two real representing the intensity of attraction and r_1, r_2 two random values between 0 and 1.
To simplify the study, we consider the deterministic version of the algorithm, where random numbers are replaced by their average values $1/2$.
With simplifications, the algorithm can be written as:

$$V(t+1) = a\,V(t) + b\left(P(t) - x(t)\right) \tag{13}$$
$$x(t+1) = x(t) + V(t+1) \tag{14}$$

Where :

$$P(t) = \frac{b_1 P_1(t) + b_2 P_2(t)}{b_1 + b_2} \quad \text{and} \quad b = \frac{b_1 + b_2}{2}$$

To make a dynamic analysis of the algorithm, the equations (13) and (14) are rewritten in the matrix form :

$$V(t+1) = -bx(t) + a\,V(t) + b\,P(t) \tag{15}$$
$$x(t+1) = x(t) + [-bx(t) + a\,V(t) + b\,P(t)] \tag{16}$$

Then :

$$V(t+1) = -bx(t) + a\,V(t) + b\,P(t) \tag{17}$$
$$x(t+1) = (1-b)x(t) + +a\,V(t) + b\,P(t) \tag{18}$$

The equations of the algorithm can then be written in the following matrix form :

$$\begin{pmatrix} x(t+1) \\ V(t+1) \end{pmatrix} = A \begin{pmatrix} x(t) \\ V(t) \end{pmatrix} + B\,P(t) \tag{19}$$

Where : $\begin{pmatrix} x(t) \\ V(t) \end{pmatrix}$ is the state of the system, consisting of the position of the particle and its velocity, P the system input, $A = \begin{bmatrix} 1-b & a \\ -b & a \end{bmatrix}$ is the dynamical matrix, and $B = \begin{bmatrix} b \\ b \end{bmatrix}$ the input matrix.

The equilibrium point of the system is such that the particle is positioned in P "$x(t) = P$" and has a zero speed $V(t) = 0$.

Behaviors of the particle depend on the eigenvalues of the matrix A are the solutions of :
$det(\lambda I - A) = 0$, i.e :

$$\lambda^2 - (a - b + 1)\lambda + a = 0 \tag{20}$$

The behavior and the convergence of the algorithm depend on the parameters a and b. The analysis of the equation (20) leads to determine the area of convergence of the PSO according to their values.

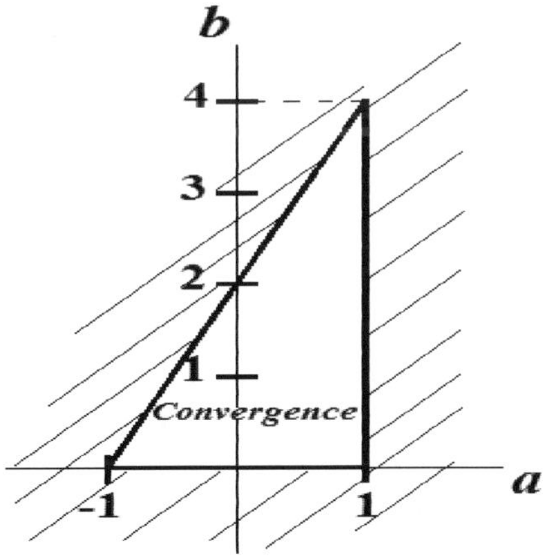

Figure 1. PSO convergence area

Figure 1 shows the values (plain area) that should have the parameters a and b for a convergence of the algorithm [19].

4. Simulation Results

The proposed method for the control of the fleet has been evaluated by the simulation of a group of 6 UAVs. The trajectory of the fictitious agent is defined initially in a fixed position, and secondly in motion. The results are compared with different initial conditions (positions of UAVs).

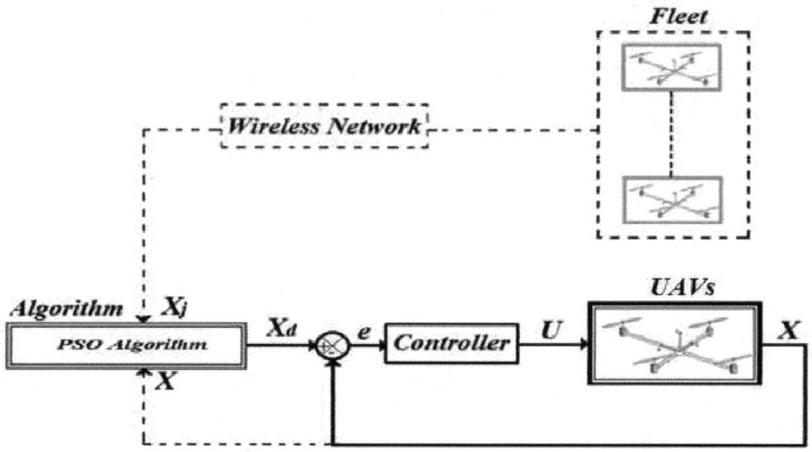

Figure 2. Control scheme of the fleet of UAVs

4.1. Stationary Leader Case

The constant sampling time is set to 10^{-2} seconds. The leader's position is stationary. The desired configuration for the UAVs is set from the desired A_d adjacency matrix.

In this case, the form of a circle is given as the desired configuration and we simulate the behavior of the fleet for different initial conditions.

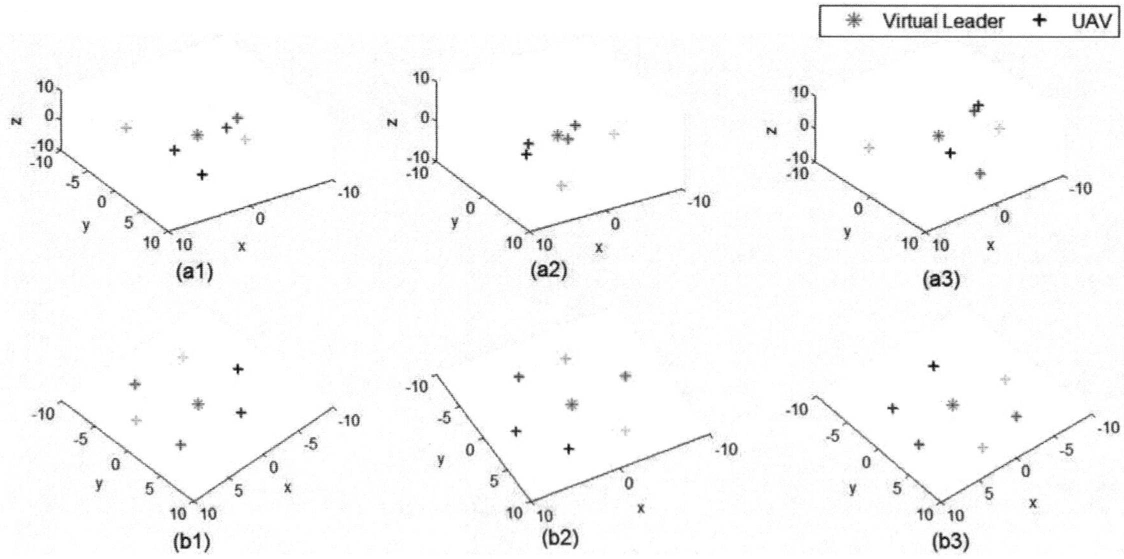

Figure 3. Circle Formation of the UAV Fleet with Stationnary Virtual Leader

Figure 3 shows the results obtained b1, b2, b3 with respectively the initial positions a1, a2 and a3, the UAVs fleet is able to reproduce the desired configuration i.e. the shape of a circle with the virtual leader at its center.

To evaluate the evolution of the distances between the UAVs and ensure that there is no collision between the flying robots, changes of the cost function $\Lambda_i(t)$ of each UAV with the initial positions (a1) is plotted in figure 4.

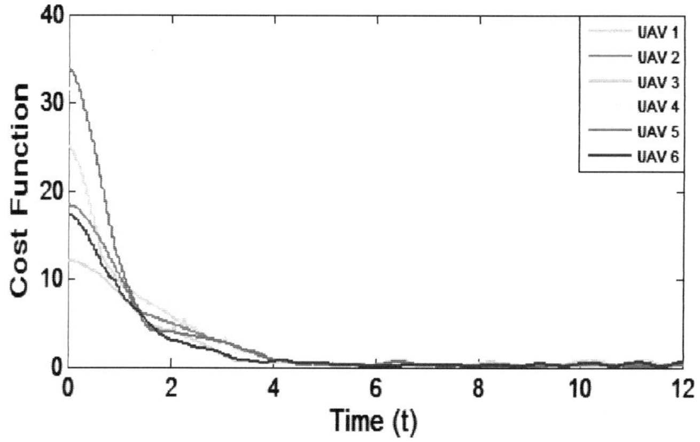

Figure 4. The variation of the cost function for Each UAVs with Stationnary Virtual Leader

The cost function decreases more and more as the fleet converges towards the desired configuration. A collision between two UAVs would appear on the cost function as a great increase in its value according to equation (10). As presented in Figure 4, it should be noticed that the safety distance and scope of the neighborhood is respected throughout the simulation. The performance of the PSO algorithm are good and the system converges after 4s. The simulations are repeated 100 times with different initial conditions. We have considered that the initial distance between two UAVs is greater than the safety distance. The results are equivalent to those obtained for the case a1 with average convergence time equal to 4s.

4.2. Mobile Leader Case

In the second part of the simulation, the leader is now in motion following a straight trajectory.

As in the previous example, various initial conditions a1, a2 and a3 are considered and a form of a circle is given as the desired configuration. The results obtained with the initial conditions a1 are illustrated in the following graphs (Figures 5 to 7).

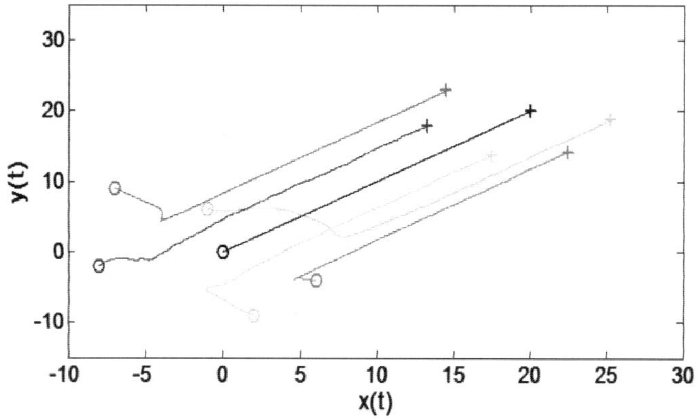

Figure 5. Circle Formation of the UAV Fleet with Mobile Virtual Leader

The figure 5 shows the evolution of the fleet on a 2D space. The desired configuration is a circle as in the previous example. The simulation demonstrates clearly that the fleet converges to the desired configuration while following the trajectory of the virtual leader.

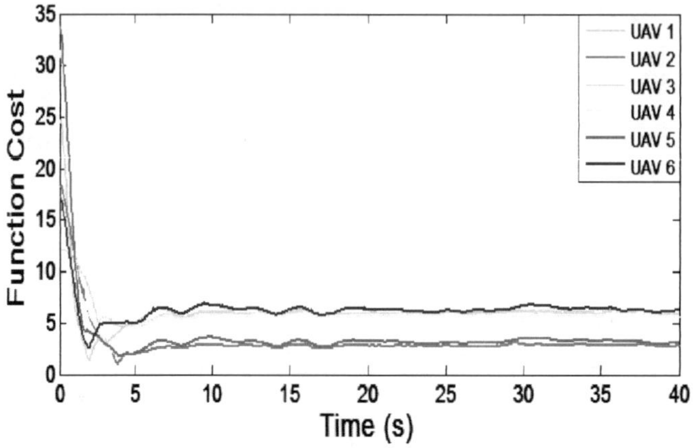

Figure 6. The variation of the cost function for Each UAVs with Mobile Virtual Leader

Figure 6 shows the performance of the PSO algorithm in the case of a moving leader. The convergence speed is reasonable and in the order of 5s. The simulations are also repeated 100 times with different initial conditions and the results are equivalent to the case illustrated in the above example. The approach considered in our simulations is able to reproduce without difficulty geometric configurations and flight training while ensuring the constraint of the anti-collision.

5. Conclusion

The paper presents a fleet control strategy based on a multi-agent system with a virtual leader. The use of the PSO algorithm that is part of optimization algorithms family called Meta-heuristic allowed us to have better results in terms of convergence speed and optimal solution.

This approach has been tested in simulation on a group of UAVs simply modeled as double integrators for our study. In our simulations, both the cases of the virtual leader standing and in motion has been considered.

The results clearly show that the proposed method is effective to cover the surface and the fleet command. The records obtained also show a robust approach against collisions, which was the main drawback of control by the virtual leader in the literature.

A future application of control strategy on non-linear complex systems and implementations on real systems are planned. An improvement would be to introduce the dynamics of controlled systems in the optimization algorithms to better find the most appropriate one for the studied systems solutions.

References

[1] CHAIMOWICZ, Luiz and KUMAR, Vijay. Aerial shepherds: Coordination among uavs and swarms of robots. In : Distributed Autonomous Robotic Systems 6. Springer Japan, 2007. p. 243-252.

[2] HERNANDEZ, Andres, COPOT, Cosmin, CERQUERA, Juan, and al. Formation Control of UGVs using an UAV as Remote Vision Sensor⋆. 2014, Cape Town, South Africa.

[3] HAFEZ, Ahmed, ISKANDARANI, Mohamad, GIVIGI, SidneyUAVs in Formation and Dynamic Encirclement via Model Predictive Control. In : IFAC World Congress. 2014, Cape Town, South Africa, pp. 1241-1246.

[4] SASKA, Martin, KASL, Zdenek, and PREUCIL, Libor. Motion planning and control of formations of micro aerial vehicles. In : IFAC World Congress. 2014, Cape Town, South Africa

[5] GUERRERO, Jose Alfredo and LOZANO, R. UAV Flight formation control. 2012.

[6] CHIARAMONTI, Martina. Formation control laws for autonomous flight vehicles. In : 2006 14th Mediterranean Conference on Control and Automation. 2006, ANCONA, ITALY. p. 1-5.

[7] SU, Housheng, WANG, Xiaofan, et LIN, Zongli. Flocking of multi-agents with a virtual

leader. Automatic Control, IEEE Transactions on, 2009, vol. 54, no 2, p. 293-307.

[8] LEONARD, Naomi Ehrich et FIORELLI, Edward. Virtual leaders, artificial potentials and coordinated control of groups. In : Decision and Control, 2001. Proceedings of the 40th IEEE Conference on. IEEE, 2001. p. 2968-2973.

[9] SHI, Hong, WANG, Long, et CHU, Tianguang. Virtual leader approach to coordinated control of multiple mobile agents with asymmetric interactions.Physica D: Nonlinear Phenomena, 2006, vol. 213, no 1, p. 51-65.

[10] OLFATI-SABER, Reza. Flocking for multi-agent dynamic systems: Algorithms and theory. IEEE Transactions on Automatic Control, ,2006, vol. 51, no 3, p. 401-420.

[11] PARK, Jaemann, KIM, H. Jin, and HA, Seung-Yeal. Cucker-Smale flocking with inter-particle bonding forces. IEEE Transactions onAutomatic Control, 2010, vol. 55, no 11, p. 2617-2623.

[12] BELKADI, A., CIARLETTA, L., and THEILLIOL, D. UAVs team flight training based on a virtual leader: Application to a fleet of Quadrirotors. In : Unmanned Aircraft Systems (ICUAS), 2015 International Conference on. IEEE, 2015, Denver, CO, USA. p. 1364-1369.

[13] DEB, Kalyanmoy, PRATAP, Amrit, AGARWAL, Sameer, et al. A fast and elitist multiobjective genetic algorithm: NSGA-II. Evolutionary Computation, IEEE Transactions on, 2002, vol. 6, no 2, p. 182-197.

[14] DORIGO, Marco, BIRATTARI, Mauro, BLUM, Christian, et al. (ed.). Ant Colony Optimization and Swarm Intelligence: 6th International Conference, ANTS 2008, Brussels, Belgium, September 22-24, 2008, Proceedings. Springer, 2008.

[15] KENNEDY, James. Particle swarm optimization. In : *Encyclopedia of Machine Learning*. Springer US, 2010. p. 760-766.

[16] ALFI, Alireza et MODARES, Hamidreza. System identification and control using adaptive particle swarm optimization. Applied Mathematical Modelling, 2011, vol. 35, no 3, p. 1210-1221.

[17] KAO, Chih-Cheng, CHUANG, Chin-Wen, et FUNG, Rong-Fong. The self-tuning PID control in a slider–crank mechanism system by applying particle swarm optimization approach. Mechatronics, 2006, vol. 16, no 8, p. 513-522.

[18] SHI, Yuhui et EBERHART, Russell C. Empirical study of particle swarm optimization. In : Evolutionary Computation, 1999. CEC 99. Proceedings of the 1999 Congress on. IEEE, 1999.

[19] TRELEA, Ioan Cristian. The particle swarm optimization algorithm: convergence analysis and parameter selection. *Information processing letters*, 2003, vol. 85, no 6, p. 317-325.

[20] EBERHART, Russ C. et KENNEDY, James. A new optimizer using particle swarm theory. In : Proceedings of the sixth international symposium on micro machine and human science. 1995. p. 39-43.

New virtual laboratories presenting advanced motion control concepts

Martin Goubej, Alois Krejčí, Jan Reitinger
University of West Bohemia, Univerzitni 8, 301 00, Pilsen, Czech Republic
E-mail: mgoubej@ntis.zcu.cz, krejcia@ntis.zcu.cz, reitinge@ntis.zcu.cz

Abstract.
The paper deals with development of software framework for rapid generation of remote virtual laboratories. Client-server architecture is chosen in order to employ real-time simulation core which is running on a dedicated server. Ordinary web browser is used as a final renderer to achieve hardware independent solution which can be run on different target platforms including laptops, tablets or mobile phones. The provided toolchain allows automatic generation of the virtual laboratory source code from the configuration file created in the open-source Inkscape graphic editor. Three virtual laboratories presenting advanced motion control algorithms have been developed showing the applicability of the proposed approach.

1. Introduction

Presently, virtual laboratories are recognized as a powerful tool which is often used for education in many technical areas such as power industry [1], [2] or automation and robotics [3–5]. These labs have many advantages like high efficiency in students or operators training or low cost maintenance. They are often more demonstrative than real experiments and can be very helpful for people with motor disabilities [5]. Virtual labs can be employed for simulation of many complex scenarios which cannot be simply reproduced using physical models or real plants (e.g. external disturbance injection or system failure). Unfortunately, many laboratories cannot be directly used for controlling physical systems, because their algorithms cannot be easily converted into a real-time control code [6, 7].

There are two principal groups of virtual laboratories: monolithic applications and labs based on server - client architecture. Both ways have their pros and cons. Monolithic applications do not require internet connection for operation. There is no need for any server running additional program, which is in most cases *Matlab* [8], [9] or *LabView* [10]. On the other hand, SW platforms which are usually used to develop monolithic (self-contained) labs cannot provide sufficient tools to simulate virtual systems. For example, *Easy Java/Javascript Simulations* (EjsS) is often used to develop more self-contained labs which do not need remote server. However, the model cannot be built from mature function blocks which is problematic especially when creating complex control schemes [11]. Technology of self-contained labs with Matlab/Simulink compatible algorithm based on Java SE applet technology was described in [12]. Unfortunately, employment of the Java applets has become problematic due to the recent security policy changes in modern web browsers. Therefore, an innovated solution which does not rely on the Java platform was developed. It provides a unified framework for the automatic generation of both

Journal of Physics: Conference Series **659** (2015) 012016 doi:10.1088/1742-6596/659/1/012016

virtual laboratories and human machine interfaces for real-time control of virtual models or real physical plants.

The virtual laboratory is based on the server - client architecture. A virtual model is developed in *Matlab/Simulink* environment using special block library called *RexLib* just as in [12]. This algorithm is executed on the server part in control system called REX [13]. The visualisation interface is assembled in open-source vector graphics editor *Inkscape* with a *RexHMI* extension, which is intended for easy creation of graphical Human-Machine Interfaces (HMI) [14]. The final vector graphic is converted into HTML5 and can be accessed on the client side via a web browser. This approach leads to platform independent solution which can be used on a wide range of target platforms.

The paper describes the individual components of the developed software framework. The proposed solution is employed for the generation of three virtual laboratories dedicated to various motion control problems including time-optimal position control, electronic camming and coordinated control of robotic arm. The laboratories present advanced features provided by motion control functional blocks implementing the industrial PLCopen Motion Control standard.

The outline of the paper is as follows: Section 2 presents the proposed toolchain for rapid HMI development. Section 3 deals with brief introduction to PLCopen Motion Control standard followed by description of three Virtual laboratories related to various control problems. Concluding remarks and ideas for future work are given in section 4.

2. Human Machine Interface

There are several ways how to design a webpage providing human machine interface for various virtual models or physical plants. First one is manual coding using HTML language. The unified framework contains a "low level" javascript API which deals with the interconnection between the simulation/control core and the HMI. However, this option is meant only for special adjustments. For rapid prototyping, several extensions for the Inkscape vector graphic editor were developed.

A basic principle for any HMI created by the framework is that there is only a single page which is scaled according to the resolution of the target device (desktop, tablet, mobile phone, etc.). This presumption allows to create multiplatform and multidevice HMIs. When one runs the Inkscape editor, there is only one main page. All the components inserted to the page will be automatically generated to the web page at the corresponding location. There are several extensions written in Python language supporting this behavior. The basic principle of the HMI generation is shown in Fig. 1.

The whole concept uses a set of SVG components and javascript files which control their behavior. These two parts are called *HMI module*. Each module contains configuration (SVG file) and control part (JavaScript file). The configuration is stored in SVG file inside the element `<description>` in JSON (Javascript Object Notation) format. When user inserts new component to the page, the configuration is cloned and then one can use the HMI extensions for Inkscape to alter it.

When the page is ready, it is exported to the final HTML file. There is a basic template which contains all the necessary libraries like jQuery, Three.js and also the HMI library from the framework. This template is extended with all the javascript controllers for the HMI modules inside the webpage. When the HTML file is loaded to the browser, the HMI library traverses all the components and starts their appropriate controllers. There are several HMI extensions which are described in the following section.

Element editor is the first extension (shown in Fig.2) which must be used. It serves for editing of individual HMI modules properties. The Element editor changes the JSON configuration

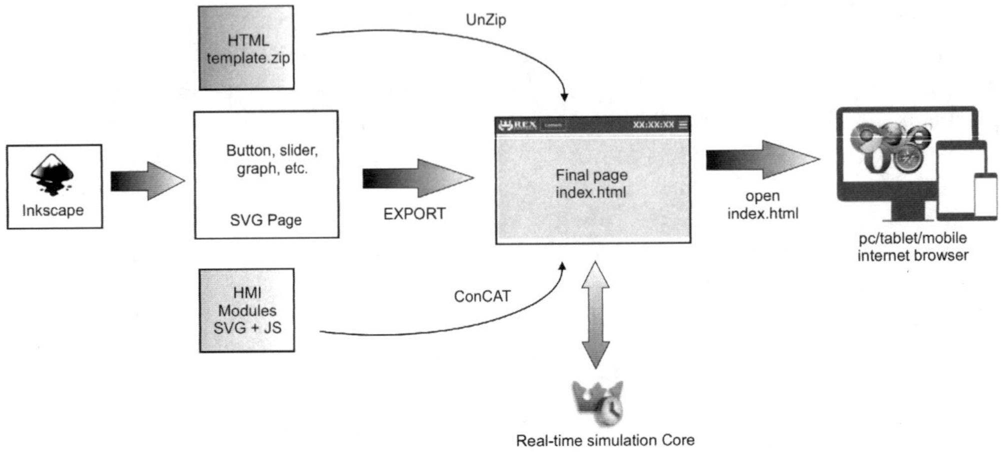

Figure 1: Inkscape toolchain for HMI/virtual laboratory generation

inside the `<description>` element inside the SVG group. One can edit Title, Module, Options and Connections using this tool.

Title is shown in the web browser after hovering the cursor above the element and is also used in Connections as shortening during the initialization.

Module uniquely identifies the type of the component and assigns the proper JavaScript functionality.

Connections tab contains names of signals that trigger the element behavior. There is a possibility to use an existing name after clicking on the Browse button.

Options tab edits some additional options of the component.

HMIConfig editor is similar to the Element editor (see Fig. 3). The options for the whole visualization project are edited here. HMIConfig editor is shown after calling the Element editor with no element selected. It contains three main tabs *Connections*, *Options* and *HTML export*. *Connections* are the signals used in the visualization which are assigned to the signals from the simulation/control core. If there is a proper scheme already running in the core, the signals might be browsed online.

Options tab contains an IP address of the target device and other options for the whole project.

HTML export tab contain configuration of the web page exporter.

Library Browser extension just opens the windows explorer with the selection of HMI modules.

HTML Builder extension compiles the web site code into the target folder, provided that a valid location has been set in the HMI Configuration menu. This extension unpacks the webpage template and inserts the content of the current SVG project. It also loads all the javascript controllers for the used modules and link them to the page. Such a exported project can be run from webserver or locally from generated file.

Group animation is the last extension (shown in Fig.4) which is different from the others. It creates a HMI module from any group by defining its behavior (translation, rotation, changes of color, size and opacity etc.). Such a module can be consequently edited by the Element editor.

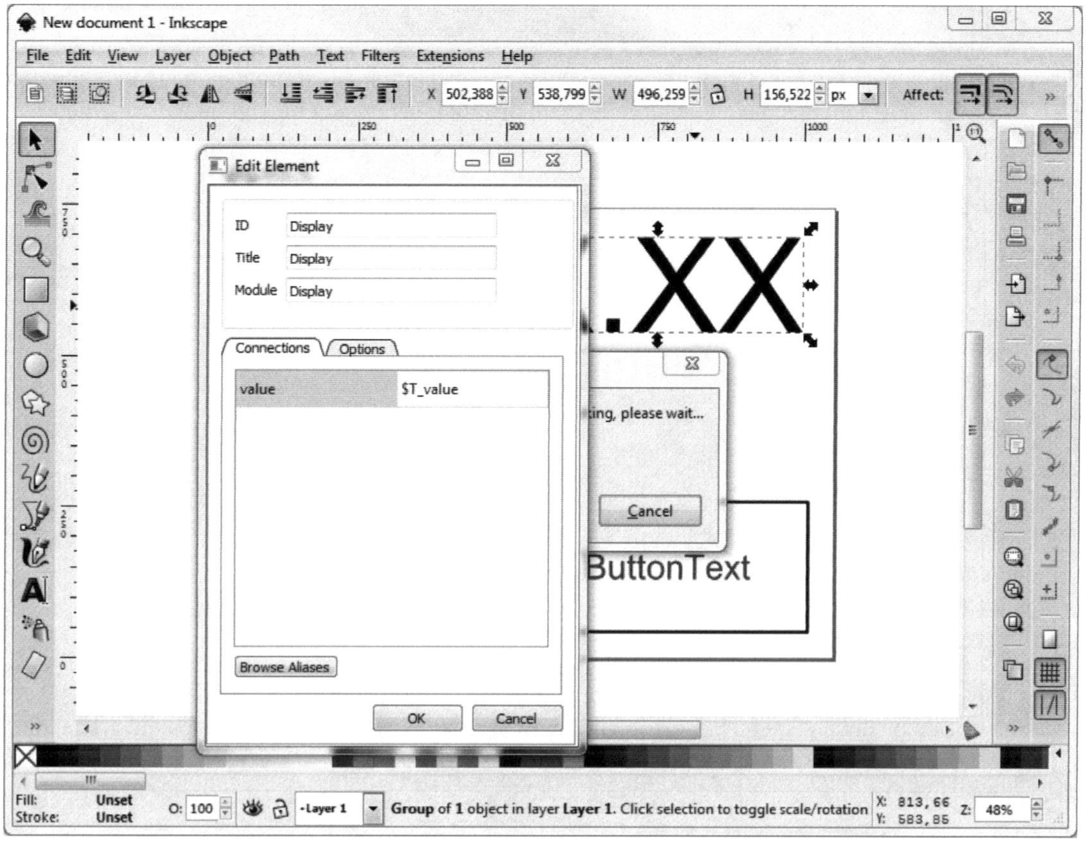

Figure 2: Element editor extension in Inkscape

3. Virtual Laboratories

This section describes three virtual laboratories which were created by means of the developed HMI framework. These labs are dedicated to specific motion control problems which emerge in many industrial applications. The goal is to present advanced functional blocks implementing the PLCopen standard which is briefly explained in the following section.

PLCopen is an independent organisation providing efficiency in industrial automation based on the needs of users. PLCopen members have concentrated on technical specifications around the IEC 61131-3 standard and its implementations in order to reduce cost in industrial engineering. The outcome includes standardized libraries for:

- Motion Control
- Communication
- Safety
- XML, etc.

PLCopen Motion Control: The motion control industry incorporates an extensive variety of incompatible solutions and systems in terms of various programming languages,

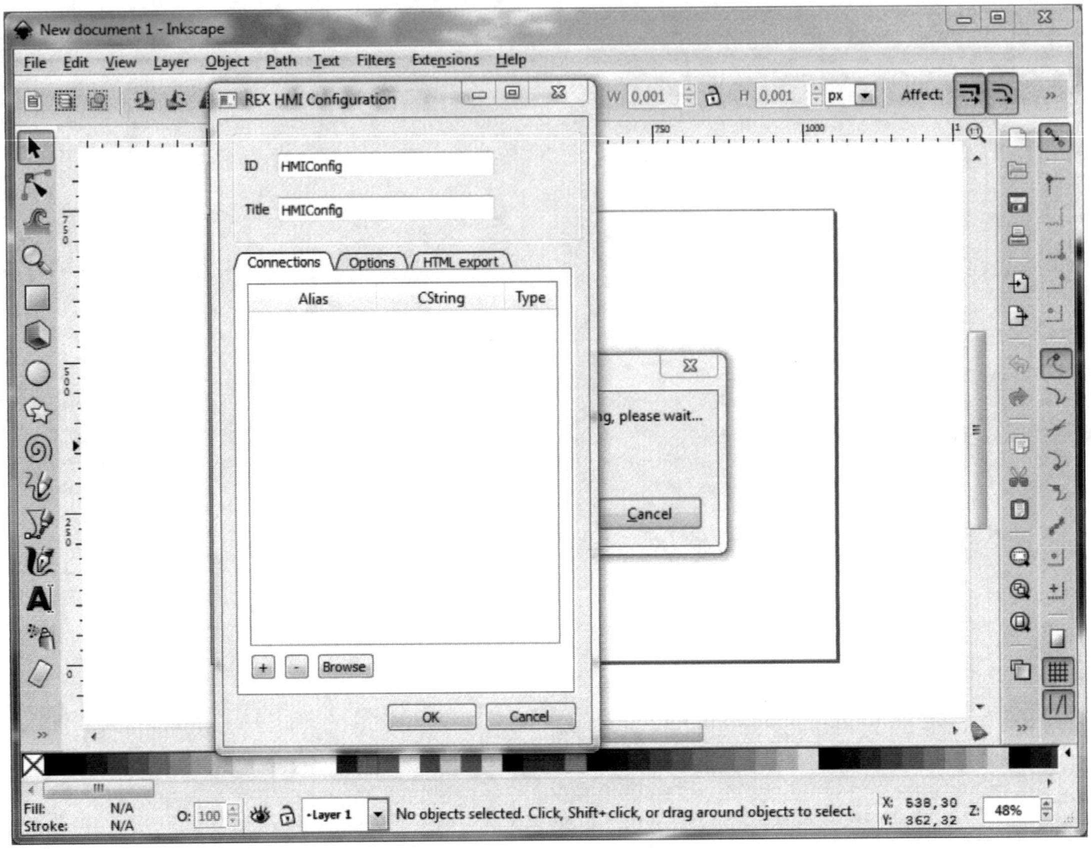

Figure 3: Configuration of Inkscape - REX HMI extension

communication and application interfaces and hardware platforms. This incompatibility induces considerable additional costs for the end users due to prolonged commissioning time, complicated staff training and difficult integration of vendor-specific functions. There was a strong demand for standardization which initiated the establishment of the Motion Control Task Force which was formed within the PLCopen organization. This task force released a set of standardized programming interfaces for motion control systems in the form of a functional block library which covers a wide range of industrial application domains [15].

Three pillars of the newly developed platform have been defined as: performance, functionality and standardization (see Fig. 5). These objectives are achieved by means of the specific functional properties:

- **Simplicity** - ease of use, towards the application program builder and installation & maintenance
- **Efficiency** - in the number of Function Blocks, directed to efficiency in design
- **Consistency** - conforming to IEC 61131-3 standard
- **Universality** - hardware independent
- **Flexibility** - future extensions / range of applications

Only the user interface and overall behavior of the individual functional blocks is defined by the PLCopen standard. Particular implementation issues are left for a motion control

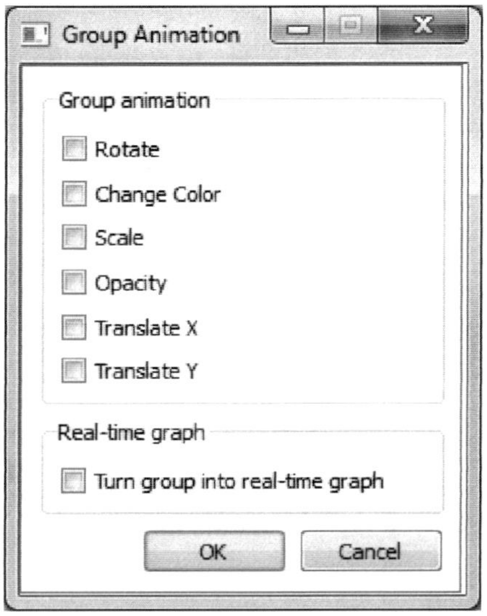

Figure 4: Group animation component in Inkscape

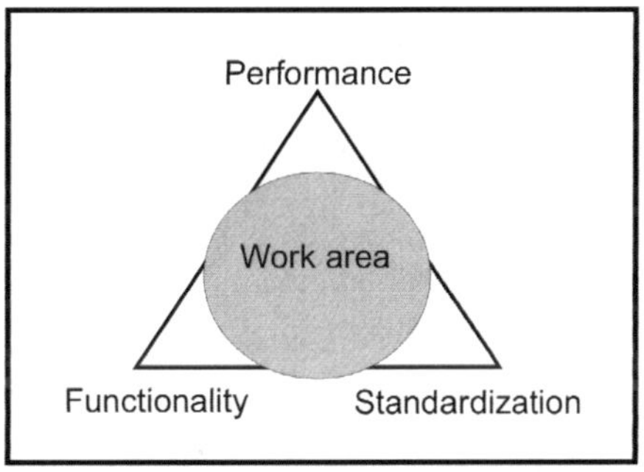

Figure 5: The triangle with user options

system supplier. The specification deals only with the trajectory generation layer of the motion control system providing setpoint values for the individual plant actuators. Hardware specific regulatory layer which tracks the commanded motion can be connected by means of a suitable communication interface (Fig. 6). Functional block library conforming the PLCopen standard has been developed at the University of West Bohemia. It is available as a set of C++ functions for various SW platforms including Matlab/Simulink and real-time control system REX. It consists of three subgroups of functional blocks (SingleLib, MultiLib and CoordLib) dedicated to different application requirements including single axis control, synchronized motion with electronic gear or cam and coordinated multi-axis control. Virtual laboratory presenting a fundamental functionality of the individual parts of the library were developed in the presented HMI framework.

3.1. SingleLib Time-optimal position control of single degree of freedom mechanism

This laboratory represents the task of point to point position control which is very frequent in various industrial applications. The goal is to perform a smooth transition of a working mechanism from an initial displacement to a final destination in the shortest time possible while respecting the physical constraints of the system given by a maximum velocity, acceleration and jerk (derivative of acceleration). This leads to a problem of state constrained time-optimal control of triple integrator plant. Most of the commercially available systems do not consider the jerk limit which is essential for a smooth shape of the resulting trajectory of motion. Either only acceleration limit is assumed which significantly reduces the complexity of the problem or a suboptimal heuristic solution is used. The presented functional blocks use a fully analytical time-optimal algorithm which was presented in [16].

FUNDAMNETAL BLOCKS:

- RM_Axis - Motion control axis
- MC_Power - Axis activation (power on/off)
- MC_Stop - Stopping a movement
- MC_Reset - Reset axis errors
- MC_MoveAbsolute - Move to position (absolute coordinates) - Block moves an axis to specified position as fast as possible. If no further action is pending, final velocity is zero, axis moves to position and stops.
- MC_MoveRelative - Move to position (relative to the execution point).

Case Study: This task represents point to point time-optimal control. One can activate the axis via the *MC_power* button. There are two ways how to set the end position: The first is by using the *MoveAbsolute* block. The user can change the parameters of this block (e.g. Position, Velocity, Acceleration). The motion can be started by clicking on the *Execute* button. The second way uses *MoveRelative* block. It has the same parameters as *MoveAbsolute* except that the final position is a relative distance to the actual execution (parameter Distance). User can switch to the trend or regulation scheme panel with a ComboBox. The trend shows the actual position, velocity and acceleration. Visualization of the motion is located on the bottom panel of the laboratory. The lab appearance is shown in Fig. 7.

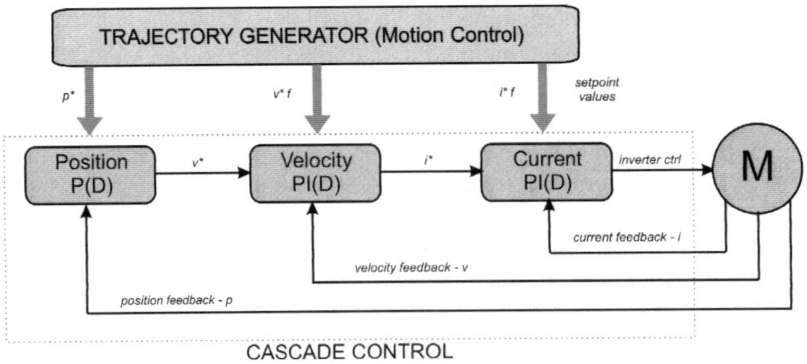

Figure 6: Single DoF Motion Control system - Interconnection of trajectory generator layer and cascade control scheme of a electrical drive

3.2. MultiLib - electronic cam and electronic gear

The second laboratory deals with electronic camming. Mechanical cams transform a rotary motion into a linear one and allow keeping mutual synchronization of multiple moving mechanisms. They are used in numerous applications including piston pumps, packaging machines, textile producing machines, machine tools or transport systems. The mechanical cam

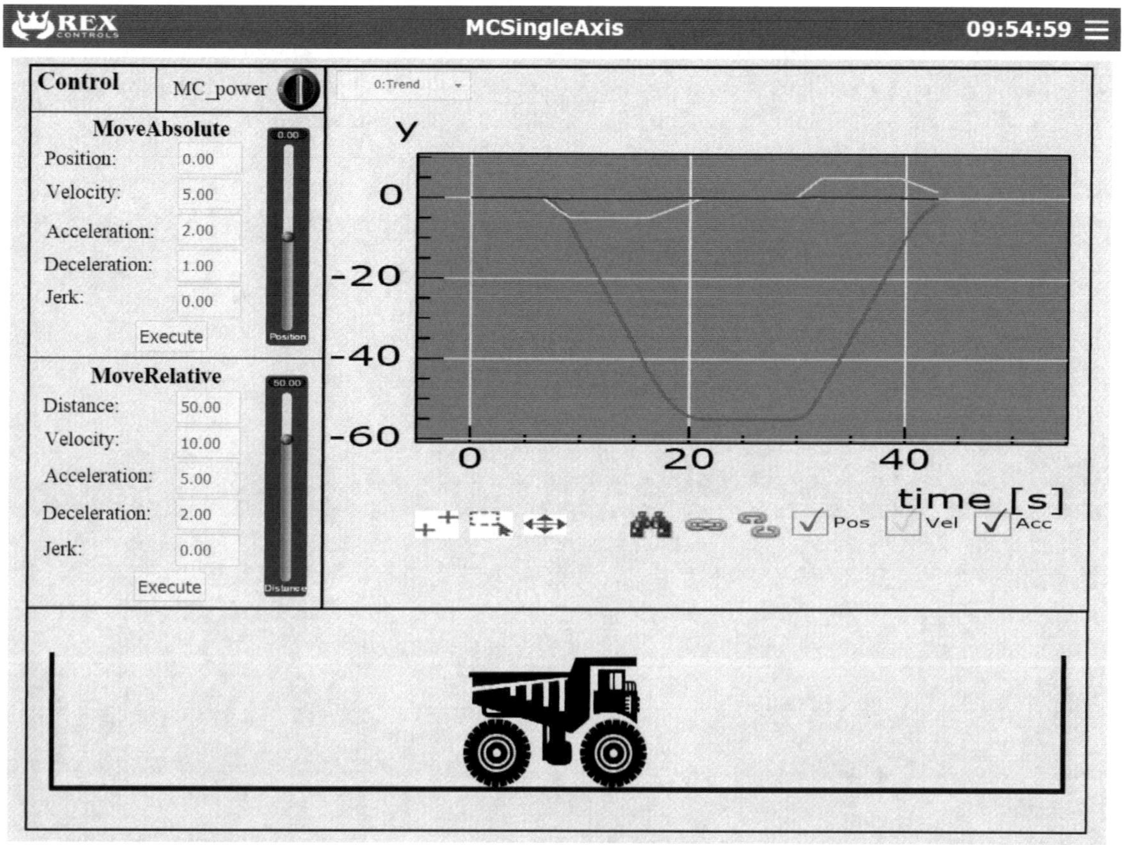

Figure 7: Virtual laboratory of time-optimal control of single degree of freedom system

can be replaced by its electronic counterpart. An electronic control system keeps synchronization between an independently controlled (master) axis and a dependent (slave) axis. Their relative motion is derived from a prescribed displacement diagram - cam profile, which determines the relation between the master and slave position. The master axis can be a real physical drive or a virtual motion generator inside the control system software. Multiple slave axes can be assigned to a single master axis to achieve precise synchronization of several actuators with a variable overall speed of operation. The main advantages of the electronic cam include fast and easy change of the cam profile, reliable high-speed and high accuracy operation and no mechanical wear of the cam-follower mechanism which would lead to performance deterioration. The electronic gear is a special case of the electronic cam with a linear displacement diagram which results in constant velocity and position ratio between the master and slave axis (Fig. 8).

FUNDAMNETAL BLOCKS:

- MC_CamTableSelect - Cam definition
- MC_CamIn - Engage the cam
- MC_GearIn - Engange the master/slave velocity ratio
- MC_MoveVelocity (SingleLib) - Move with constant velocity

Case Study: The laboratory demonstrates the motion control in multiple axes connected trough a cam and gear relation. User can switch on the axes (master and slaves) with the *MC_power* button. The cam profile is fixed and cannot be changed. Gear ratio and parameters of the master axis velocity (Velocity, Acceleration, Deceleration, Jerk and Direction) are specified in *GearIn* and *MoveVelocity* blocks. The motion can be started/stopped with *MP_start*/*MP_stop* buttons. The run of the simulation is displayed in the trend on the bottom side (see Fig. 9).

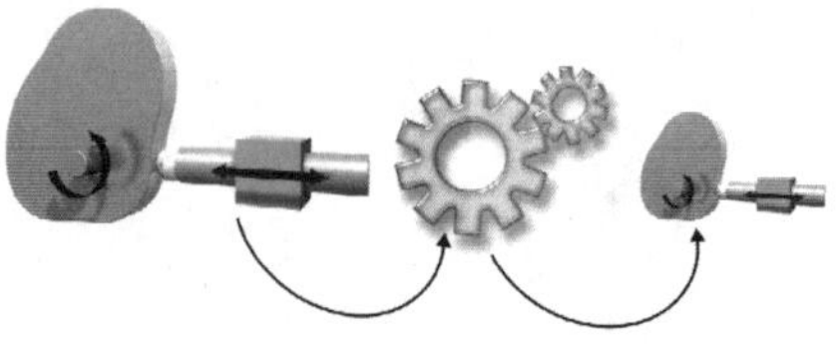

Figure 8: Cam and gear

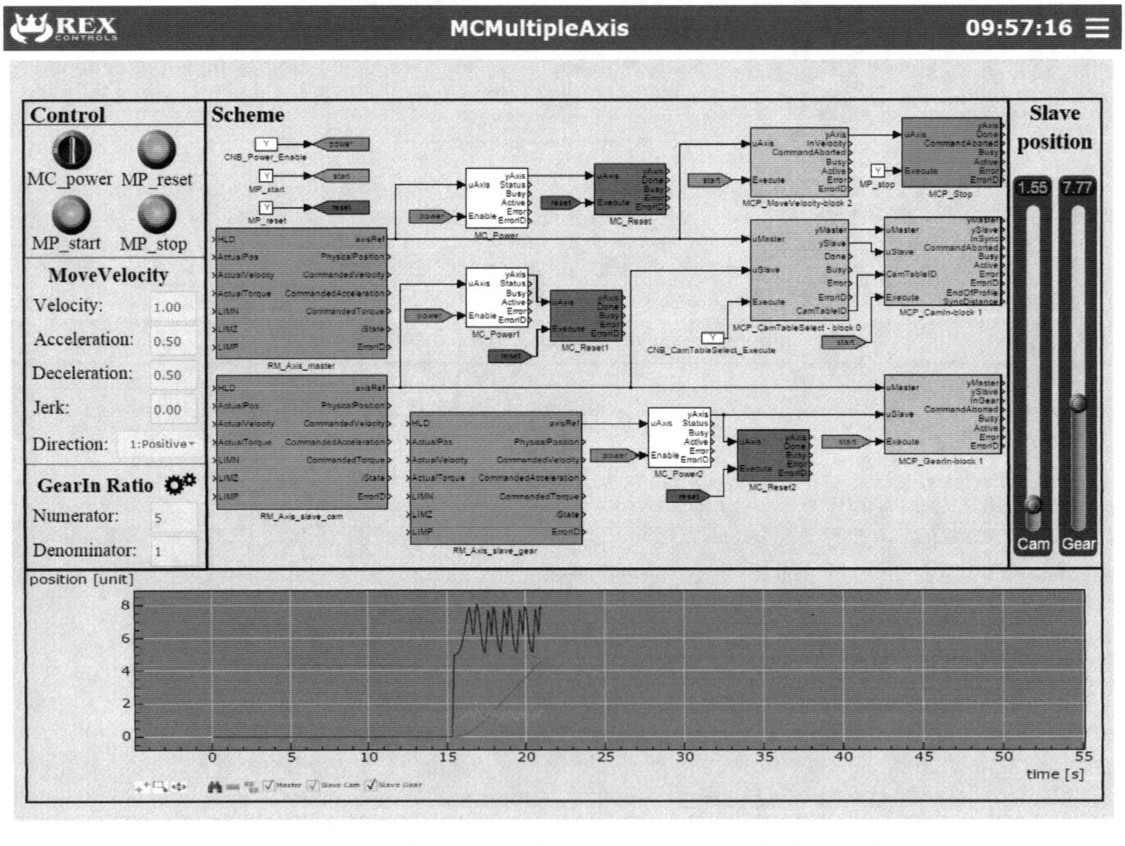

Figure 9: Virtual laboratory of electronic cam and electronic gear

3.3. CoordLib Coordinated control of robotic manipulator AGEBOT

Robotic manipulators are widely used in many industrial applications (e.g. welding, soldering, painting, material handling or machining). Commercial robotic arms manufacturers have developed several different programming languages and application interfaces. The PLCopen standard provides a unified programming standard in the form of functional blocks specifically designed for the coordinated control of multi-axes machines. The geometric shape of desired trajectory of motion is usually parameterized in the machine or tool/workspace coordinate system with the use of a proper interpolation algorithm (e.g. linear, circular or spline trajectory). Time path of the executed motion is generated based on the desired federate specified by the user. The resulting trajectory is consequently transformed into the axes coordinate system using a machine specific kinematic transform block which computes a contribution of the individual axes to the coordinated motion.

FUNDAMNETAL BLOCKS:

- RM_AxesGroup - Motion control axes group
- MC_MoveLinearAbsolute - linear movement on an axes group
- MCP_SetKinTransform_Agebot - sets a kinematic transforms between the machine and actuator coordinate systems

AGEBOT was designed as a special robot architecture which consist of two main parts [17], [18]. Serial manipulator (SM) and parallel manipulator (PM). SM ensures basic positioning of the end effector including the translations in x, y, z axes and orientation of the longitudinal axis of PM. This motion is used for handling of parts which are to be exposed of cleaning process in cleaning chambers. PM plays an important role in precise positioning of handled parts inside cleaning chambers because PM allows full orientating about x, y, z axes. Mechanical design of PM is allows separation of vulnerable components (motors, sensors, etc.) from an aggressive environment inside the cleaning chamber. Overall view of the AGEBOT is shown in Fig. 10.

Case Study: The laboratory presents basic principles of coordinate motion control on the serial part of the AGEBOT robotic manipulator. The robot axes can be activated by *MC_power* button. Target position of the effector in a 2D plane can be set with x, y and angle coordinates (tilt angle of the last arm). This motion is executed by *MoveLinear* block by clicking on the *Execute* button. There is possibility to manually operate the individual robot axes. User can switch between the visualisation and regulation scheme panel with a ComboBox. Trend chart is displayed on the bottom of the lab. The HMI is shown in Fig. 11.

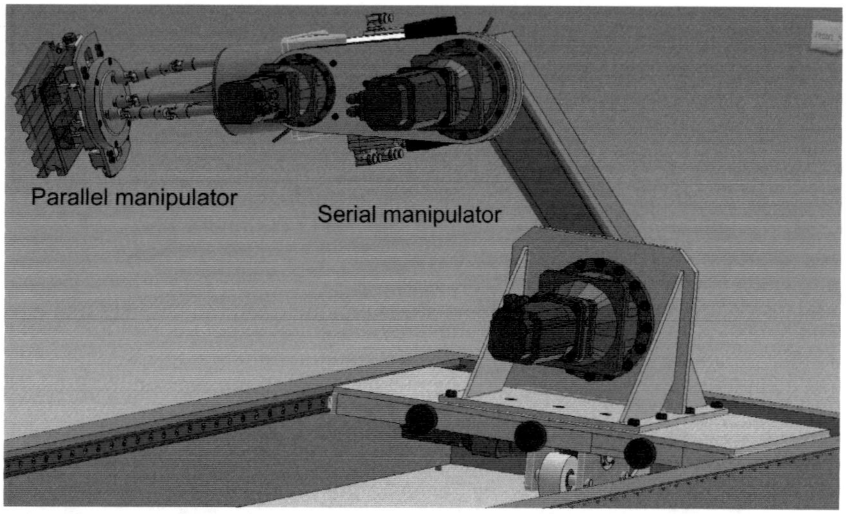

Figure 10: AGEBOT fom 3D CAD vizualization

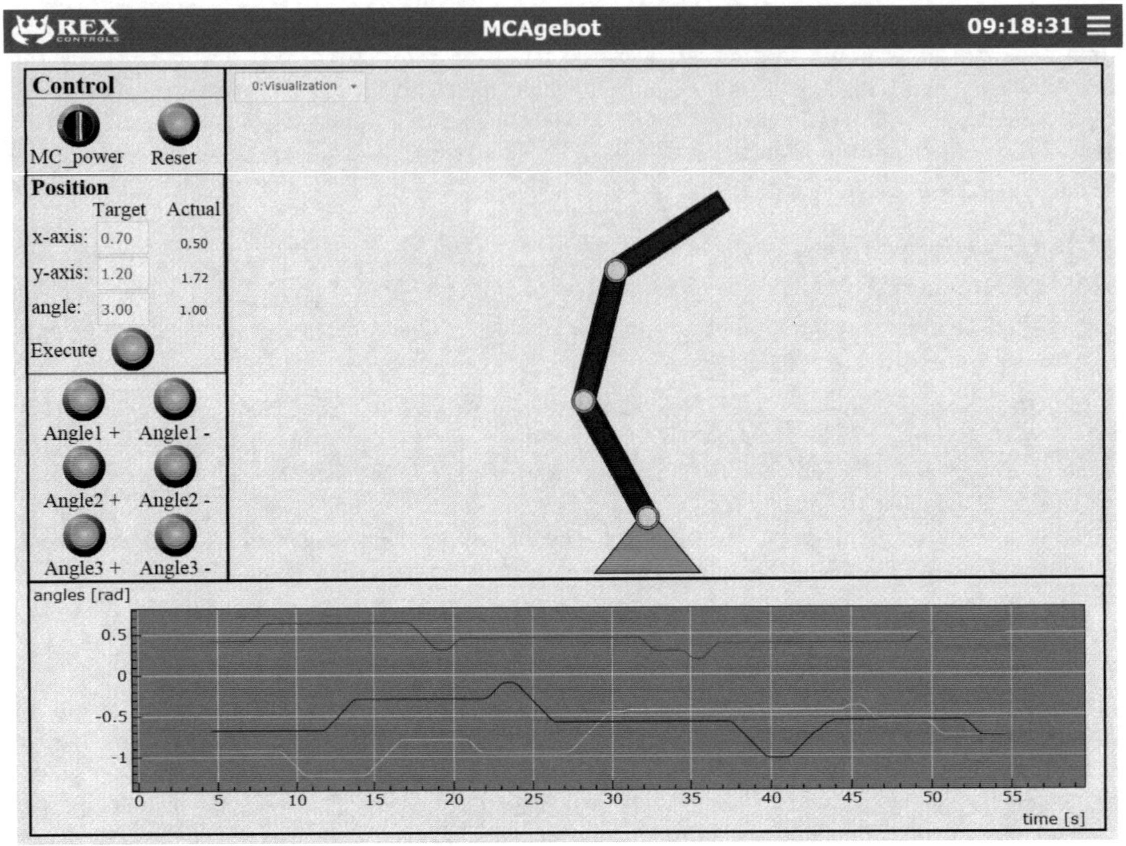

Figure 11: Virtual laboratory of robotic manipulator AGEBOT in 2D

4. Conclusion

The presented software tools allow simple creation of virtual laboratories with attractive visual appearance. Rapid development and automatic code generation is supported by a set of components integrated into the Inkscape graphic editor. The provided framework allows simple connection to REX control system which offers the functionality of real-time simulation using an extensive set of Matlab compatible functional block library. However, the visualization tools can be adjusted to another simulation core using a proper communication interface. Case study dealing with advanced motion control algorithms is provided as an example of use. THE VIRTUAL LABORATORIES ARE FREELY ACCESSIBLE AT WEBSITE: `http://lab.fav.zcu.cz/motioncontrol`. The developed tools can also serve for creation of human machine interfaces for model or hardware-in-the-loop simulators as well as for real-time control of a physical plant. Future work will be dedicated to extension of the advanced functions of the visualization framework, especially in the field of 3D graphics.

Acknowledgement

This work was supported by the Technology Agency of the Czech Republic – project No. TE02000103 and by Czech Ministry of Education, Youth and Sports – project No. LO1506.

References

[1] Redel-Macas M, Pinzi S, Martnez-Jimnez M, Dorado G and Dorado M 2015 Virtual laboratory on biomass for energy generation *Journal of Cleaner Production*
[2] Iovan M, Surianu F D and Molnar-Matei F 2015 Virtual laboratory for power quality study *Procedia - Social and Behavioral Sciences*
[3] Prieto-Blazquez J, Arnedo-Moreno J and Herrera-Joancomarti J 2008 An integrated structure for a virtual networking laboratory *IEEE Transactions on Industrial Electronics* ISBN 0278-0046
[4] Villar-Zafra A, Zarza-Sanchez S, Lazaro-Villa J A and Fernandez-Cant R M 2012 Multiplatform virtual laboratory for engineering education *9th International Conference on Remote Engineering and Virtual Instrumentation (REV)*
[5] Duarte M, Butz B P, Miller S M and Mahalingam A 2008 An intelligent universal virtual laboratory (uvl) *IEEE Transactions on Education* ISBN 0018-9359
[6] Koretsky M, Amatore D, Barnes C and Kimura S 2008 Enhancement of student learning in experimental design using a virtual laboratory *IEEE Transactions on Education*
[7] Uran S and Jezernik K 2008 Virtual laboratory for creative control design experiments *IEEE Transactions on Education*
[8] Elmagroud B and Ziyyat A 2012 Virtual laboratory for teaching electromagnetism using matlab *Proceedings of 2012 International Conference on Interactive Mobile and Computer Aided Learning (IMCL)*
[9] Sengupta D, Jain N and Kumar C S 2013 An educational website on kinematics of robots *IEEE Fifth International Conference on Technology for Education*
[10] Yang-Mei L and Bo C 2014 Electronic circuit virtual laboratory based on labview and multisim *2014 7th International Conference on Intelligent Computation Technology and Automation*
[11] Christian W and Esquembre F 2009 *Modeling Science: From Free Fall to Chaos*
[12] Reitinger J, Čech M, Schlegel M and Balda P 2014 New tools for teaching vibration damping concepts: Contlab.eu *IFAC Proceedings Volumes (IFAC-PapersOnline)*
[13] Severa O and Čech M 2012 Rex - rapid development tool for automation and robotics *Proceedings of 2012 IEEE/ASME 8th IEEE/ASME International Conference on Mechatronic and Embedded Systems and Applications*
[14] REX Controls 2015 Software downloads URL `http://www.rexcontrols.com/download`
[15] PLCopen 2015 Plcopen motion control Available online: http://www.plcopen.org/pages/tc2_motion_control/ accessed on: 2015-9-16 URL `http://www.plcopen.org/pages/tc2_motion_control/`
[16] Blaha L, Schlegel M and Mošna 2009 Optimal control of chain of integrators with constraints *17th International Conference on Process Control 2009*
[17] Švejda M and Goubej M 2012 Innovative design and control of robotic manipulator for chemically aggressive environments *13th International Carpathian Control Conference (ICCC)*
[18] Švejda M and Goubej M 2013 Dynamic analysis and control of robotic manipulator for chemically aggressive environments *IEEE International Conference on Mechatronics (ICM)*

Gas Bearing Control for Safe Operation in Critical Speed Regions - Experimental Verification

Lukas R. S. Theisen[1], Hans H. Niemann[1], Roberto Galeazzi[1], Ilmar F. Santos[2]

[1]Dept. of Electrical Engineering and [2]Dept. of Mechanical Engineering
Technical University of Denmark, DK 2800 Kgs. Lyngby, Denmark

E-mail: {lrst,hhn,rg}@elektro.dtu.dk, ifs@mek.dtu.dk

Abstract. Gas bearings are popular for their high speed capabilities, low friction and clean operation, but require low clearances and suffer from poor damping properties. The poor damping properties cause high disturbance amplification near the natural frequencies. These become critical when the rotation speed coincides with a natural frequency. In these regions, even low mass unbalances can cause rub and damage the machine. To prevent rubbing, the variation of the rotation speed of machines supported by gas bearings has to be carefully conducted during run-ups and run-downs, by acceleration and deceleration patterns and avoidance of operation near the critical speeds, which is a limiting factor during operation, specially during run-downs. An approach for reducing the vibrations is by feedback controlled lubrication. This paper addresses the challenge of reducing vibrations in rotating machines supported by gas bearings to extend their operating range. Using $\mathcal{H}_\infty$-design methods, active lubrication techniques are proposed to enhance the damping, which in turn reduces the vibrations to a desired safe level. The control design is validated experimentally on a laboratory test rig, and shown to allow safe shaft rotation speeds up to, in and above the two first critical speeds, which significantly extends the operating range.

1. Introduction

Gas bearings offer clean operation with low friction, but suffer from poor damping properties and require low clearances. Rotating machines supported by gas bearings are therefore very sensitive towards mass unbalance and disturbances. The natural frequencies especially become critical as they coincide with the rotation speed, where the mass unbalance response grows until the shaft rubs the bearing surface. Operation near the critical speeds is therefore avoided, which limits the usability of the machines. An ad-hoc approach to reduce vibrations when crossing the critical speeds, is to quickly accelerate rotation shaft across the critical speed before the full vibration amplitude is obtained. Such an approach however results in the "cat in the tree"-problem - it may be easy to get up, but difficult to get safely down again. Due to the low viscosity there are almost no friction losses, which in turn results in a slow deceleration of the shaft. This gives enough time for the undesired vibrations to build up.

For a given fixed machine design, the shaft vibrations can be reduced in two ways. Proper balancing of the shaft can drastically reduce mass unbalance, but not eliminate it. Further this obviously does not diminish sensitivity to external disturbances. Active lubrication techniques through feedback control represent a valid alternative approach, which can handle both mass unbalance and external disturbances. Feedback control has been widely applied to active magnetic bearing (AMB)-systems. Many authors have proposed various control designs, e.g. linear parameter-varying (LPV)-controllers to eliminate the mass unbalance response by placing closed-loop zeros in the sensitivity function at the shaft rotation speed. e.g. in [1, 3, 11]. Such controllers completely eliminate the mass unbalance at the cost of a high control effort. The use of phase shift filters has also been proposed in literature [12]. These mentioned approaches only reject disturbances at the rotational frequency.

Previous papers [5,9,10] on control of gas bearings have treated design of classical controllers, but none of these had sufficient robustness, and therefore were not able to reduce vibrations enough to allow a safe crossing of the critical speed. In [8] we proposed an $\mathcal{H}_\infty$ and an LPV controller to enhance the damping of the gas bearing. The controllers were found able to both reject the external disturbances and reduce the vibration amplitude within the considered operating range.

In continuation of that work, this paper explores further the capabilities of such designs. The same $\mathcal{H}_\infty$ setup is used to obtain a damping enhancing controller, and the vibration reduction capabilities are investigated to extend the region of operation. The available model relies on open loop identification, which poses a challenge, since open loop operation is not possible in these regions. A controller is therefore required, which should again be validated from a model of the system at that particular operating condition. Instead, an alternative must be sought. Our approach therefore relies on designing a controller from the LPV model inside the identified region, and investigating the controller performance outside the identified region. The upper limits for safe rotation are investigated both for closed loop experiments and open loop experiments.

The paper is structured as follows. Section 2 contains an overview of the test rig utilised for experimental validation and some highlights of the model of the gas bearing. The design of the $\mathcal{H}_\infty$ controller is detailed in Section 3, and the available LPV model is used in extrapolation to provide an expectation of the controller performance for higher rotation speeds. This performance is investigated experimentally in Section 4. Last, some conclusions are drawn and future aspects are discussed in Section 5.

Notation

The paper uses upper case bold letters for matrices $\mathbf{A}$, lower case bold letters for vectors $\mathbf{a}$, the Laplace variable is denoted s. Continuous time signals are addressed $a(t)$. Signals in the Laplace domain are addressed $a(s)$, and sampled signals at time instants kT_s are denoted $a(kT_s)$. State-space dynamics is formulated in shorthand as $\mathbf{G} = \left[\begin{array}{c|c}\mathbf{A} & \mathbf{B} \\ \hline \mathbf{C} & \mathbf{D}\end{array}\right]$, which defines the state-space relation $\begin{bmatrix}\dot{\mathbf{x}} \\ \mathbf{y}\end{bmatrix} = \mathbf{G}\begin{bmatrix}\mathbf{x} \\ \mathbf{u}\end{bmatrix} = \left[\begin{array}{c|c}\mathbf{A} & \mathbf{B} \\ \hline \mathbf{C} & \mathbf{D}\end{array}\right]\begin{bmatrix}\mathbf{x} \\ \mathbf{u}\end{bmatrix}$. The identity matrix of size n is denoted $\mathbf{I}_n$. Shaft rotation speed units are given in Hz, though the common in rotordynamics is revolutions per minute: $(1\ Hz = 60\ rpm)$.

2. Controllable Gas Bearing

2.1. Test Rig

The experimental controllable gas bearing setup at hand is shown in Fig. 1. It consists of a turbine (1) driving a flexible shaft (2) supported by both a ball bearing (3) and the controllable gas bearing (4), in which pressurised air is injected through four piezoactuated injectors numbered as shown. The injection pressure is constant $P_{\text{inj}} = 0.3\ MPa$, which is measured before splitting up to the four piezoactuators. The pressure is not controlled, but the variations are negligible. A disc (5) is mounted in one end to pre-load the journal. The horizontal and vertical disc movement $\mathbf{p} \triangleq [p_x, p_y]^{\text{T}}$ is measured at the disc location using eddy current sensors (6) in the coordinate frame specified in the figure. For zero input and when the

Figure 1. The experimental controllable gas bearing setup. A turbine (1) drives a flexible shaft (2), which is supported by both a ball bearing (3) and the controllable gas bearing (4) with four piezoactuated injectors. A disc (5) is mounted in one end to pre-load the journal and displacement sensors (6) measure the lateral movement of the disc in the shown reference frame. A quadrature encoder (7) measures the angular position.

rotor is at stand still, the position is $\mathbf{p} = \mathbf{0}$. The angular position of the rotor ϕ is measured by an optical quadrature encoder (7), from which the rotation speed Ω is calculated. The injectors are controlled in a pairwise differential mode. Thereby one piezoactuator reference r_x is sent to control the position of the horizontal injectors, and one reference r_y is sent to control the vertical ones. These references are in intervals $[-5, 5]$ V, which corresponds to full-span motion of the piezoactuator positions in the interval $[0, 45]$ μm. The nominal clearance of the gas bearing is 25 μm. Given the right conditions of sufficient injection pressure and sufficiently low rotational speed, the gas film generates restoring forces and thereby keeps the rotor levitating about a stable equilibrium. All measurements are sampled with period $T_s = 0.2\ ms$. A detailed description of the setup is available in [4].

For equipment safety, the vibrations must be within a safety region, in this case chosen as a circle:

$$x_s = r \cdot cos(\theta) - x_0, \quad y_s = r \cdot sin(\theta) - y_0, \quad \theta \in [0, 2\pi[\ rad \tag{1}$$

Manual tests were performed to assess safe limits of the circle parameters, and the following were found sufficiently conservative: radius $r = 20\mu m$, and centre $x_0 = 10\mu m$, $y_0 = -3\mu m$. These limits are deliberately chosen to be conservative for equipment safety.

2.2. Gas Bearing Model

The gas bearing test rig is modelled using the linear parameter-varying (LPV)-identification approach from [10], where local LTI-models are identified from data collected in a grid of injection pressures and shaft rotation speeds. For the present work, the injection pressure is kept constant, and the model is then only function of the rotation speed. The identification grid contained six uniformly spaced shaft rotation speeds in the interval $\Omega \in [0, 92]\ Hz$. The model can be decomposed to a cascade coupling of the actuator dynamics $\mathbf{G}_{act}$, time delays $\mathbf{G}_{\boldsymbol{\tau}}$ and the rotor-bearing dynamics $\mathbf{G}_{rb}$:

$$\mathbf{G}(t, \Omega) = \mathbf{G}_{rb}(t, \Omega)\mathbf{G}_{\boldsymbol{\tau}}(t, \Omega)\mathbf{G}_{act}(t) \tag{2}$$

The actuator dynamics are independent of the scheduling parameter and have the following diagonal second order form with two real poles and a gain:

$$\mathbf{G}_{act}(s) = \begin{bmatrix} G_{a,x}(s) & 0 \\ 0 & G_{a,y}(s) \end{bmatrix}, \qquad G_{a,j}(s) = \frac{\kappa_{a,j}}{\left(\frac{1}{p_{1,j}}s+1\right)\left(\frac{1}{p_{2,j}}s+1\right)}, \quad j \in \{x, y\} \tag{3}$$

The rotor-bearing dynamics is modelled as the interconnection of a parameter-varying delay and a second order parameter-varying mass-spring-damper system. The latter has a state-space realisation:

$$\mathbf{G}_{rb} = \left[\begin{array}{cc|c} \mathbf{0}_2 & \mathbf{I}_2 & \mathbf{0}_2 \\ \mathcal{K}(\Omega) & \mathcal{D}(\Omega) & \mathcal{B}(\Omega) \\ \hline \mathbf{I}_2 & \mathbf{0}_2 & \mathbf{0}_2 \end{array}\right] \tag{4}$$

in which the parameter-varying matrices $\{\mathcal{K}, \mathcal{D}, \mathcal{B}\}$ are second order polynomials in the rotation speed Ω. The delays are second order polynomials in rotation speed, and finite models are obtained by a first order Padé approximation:

$$\mathbf{G}_{\mathcal{T}}(t, \Omega) = \begin{bmatrix} G_{\mathcal{T}_x(\Omega)}(t) & 0 \\ 0 & G_{\mathcal{T}_y(\Omega)}(t) \end{bmatrix}, \qquad G_{\mathcal{T}_j(\Omega)}(t) = \left[\begin{array}{c|c} -2/\mathcal{T}_j(\Omega) & 1 \\ \hline 4/\mathcal{T}_j(\Omega) & -1 \end{array}\right], \quad j \in \{x, y\} \tag{5}$$

The natural frequencies of the LPV model change with rotation speed but are approximately $\omega_1 = 105\ Hz$ and $\omega_2 = 115\ Hz$. It is desired to extend the region of safe operation, which implies increasing the rotation speed to values outside the identification region. The actuator dynamics contains only real poles acting at frequencies higher than $400Hz$. Since actuation in this range of frequencies is not desired, the actuator dynamics is approximated by a static gain. The gas bearing model is then of order $n = 6$.

The nominal model is chosen as the LPV model evaluated at 91 Hz.

$$\mathbf{G} = \mathbf{G}(t, \Omega)\Big|_{\Omega=91\ Hz} \tag{6}$$

The mass unbalance is not included in the model, but acts as a force on the shaft given by:

$$\mathbf{f}_u = m_u e_u \Omega^2 \begin{bmatrix} \sin(\Omega t + \varphi) \\ \sin(\Omega t - \pi/2 + \varphi) \end{bmatrix} \tag{7}$$

in which Ω is the shaft rotation speed, m_u is the unknown unbalance mass, and e_u is the unknown distance between the mass unbalance and the geometrical shaft centre, and φ is the phase of the disturbance. This can be modelled as an input disturbance in the gas bearing. The force from mass unbalance therefore grows by Ω^2 as the rotation speed increases, and the response is greatly amplified near the resonance frequencies. At this point, the mass unbalance remains largely unknown except for its frequency, which suffices for control design.

3. $\mathcal{H}_\infty$ Control Design

This section details the damping enhancing $\mathcal{H}_\infty$ control design from [8] to allow the safe crossing of the first two critical speeds by reducing the vibrations to be within the desired safety region from Eq. (1). These requirements are not easily included directly in the $\mathcal{H}_\infty$ setup especially since the mass unbalance is largely unknown. Instead, the controller $\mathbf{K}$ should enhance the damping and thereby reduce the gain magnitude at the resonance frequencies without wearing the actuator out. These disturbance and noise rejection requirements are formulated using the mixed sensitivity setup [7], which seeks to minimise:

$$\mathbf{K}^* = \arg\min_{\mathbf{K}} \|\mathbf{N}\|_\infty, \quad \mathbf{N} = \begin{bmatrix} \mathbf{W}_p\mathbf{S} \\ \mathbf{W}_u\mathbf{K}\mathbf{S} \end{bmatrix}, \tag{8}$$

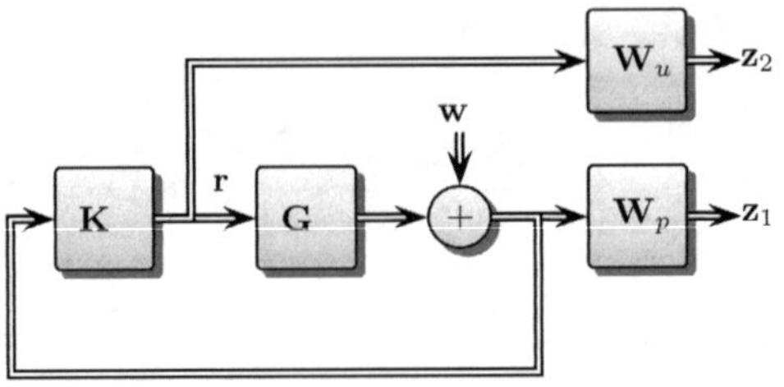

Figure 2. The augmented plant with controller for LPV controller design with performance weights $\mathbf{W}_p$ and controller sensitivity weight $\mathbf{W}_u$.

where the closed loop sensitivity functions for a given $\Omega = \bar{\Omega}$ are :

$$\mathbf{S}(s, \bar{\Omega}) \triangleq (\mathbf{I} + \mathbf{GK})^{-1} \tag{9}$$

and **KS** represents the control sensitivity. They are shaped by the weight functions $\mathbf{W}_p$ and $\mathbf{W}_u$. The external output disturbance $\mathbf{w}$ and external outputs $\mathbf{z} = [\mathbf{z}_1^{\mathrm{T}}, \mathbf{z}_2^{\mathrm{T}}]^{\mathrm{T}}$ are included into the system to obtain the augmented plant as shown in Fig. 2. The controller then satisfies $\|\mathbf{N}\|_\infty < \gamma$.

To enhance damping, the controller should have high performance in the frequency range around the under-damped eigenfrequencies of the rotor-bearing ω_x and ω_y. The performance filter is therefore chosen to contain inverse notch like filters:

$$\mathbf{W}_p(s) = \mathrm{diag}\left(w_{px}(s), w_{py}(s)\right) \tag{10}$$

in which w_{px} and w_{py} both have the form:

$$w_{px}(s) = \frac{s^2 + 2\zeta_1\omega_x s + \omega_x^2 k_0}{s^2 + 2\zeta_2\omega_x s + \omega_x^2}, \quad w_{py}(s) = \frac{s^2 + 2\zeta_1\omega_y s + \omega_y^2 k_0}{s^2 + 2\zeta_2\omega_y s + \omega_y^2} \tag{11}$$

The natural frequencies ω_x, ω_y are chosen as the under-damped natural frequencies of the gas bearing to obtain a high weight around these. The weight at the resonance frequencies is set to 19 dB, which is obtained by the damping factors $\zeta_1 = 0.3$ and $\zeta_2 = 0.05$. The low sensitivity around the natural frequencies must come at the cost of increased sensitivity in another frequency range due to Bode's sensitivity integral [2]. It was argued in [9] to place this sensitivity increase in the low frequency range, where an amplification of mass unbalance and disturbances is acceptable. The constant $k_0 = 1/3$ determines the low frequency weight, and the sensitivity at low frequency is guaranteed to be less than γk_0. The control signal sensitivity weight $\mathbf{W}_u$ is chosen to penalise control action at high frequency. This is achieved with the high-pass filter from [7, Sec. 2, Eq. (2.72)]:

$$\mathbf{W}_u(s) = \mathbf{I}w_u(s), \quad w_u(s) = \frac{s/M_b + \omega_b}{s + \omega_b A_b}, \tag{12}$$

where the low frequency gain is $1/A_b$, $A_b = 10$; the high frequency gain is $1/M_b = 15$, and the approximate crossover frequency is $\omega_b = 2000\pi$ rad/s. The weights are shown in Fig. 3. Since $|\mathbf{S}| > 1$ for low frequencies, the controller amplifies mass unbalance for low rotation speeds, which is affordable, whereas at high rotation speeds, the vibrations are attenuated. A 12-th order $\mathcal{H}_\infty$ controller is synthesised with $\gamma = 1.15$ and the performance shown in Fig. 4. At low

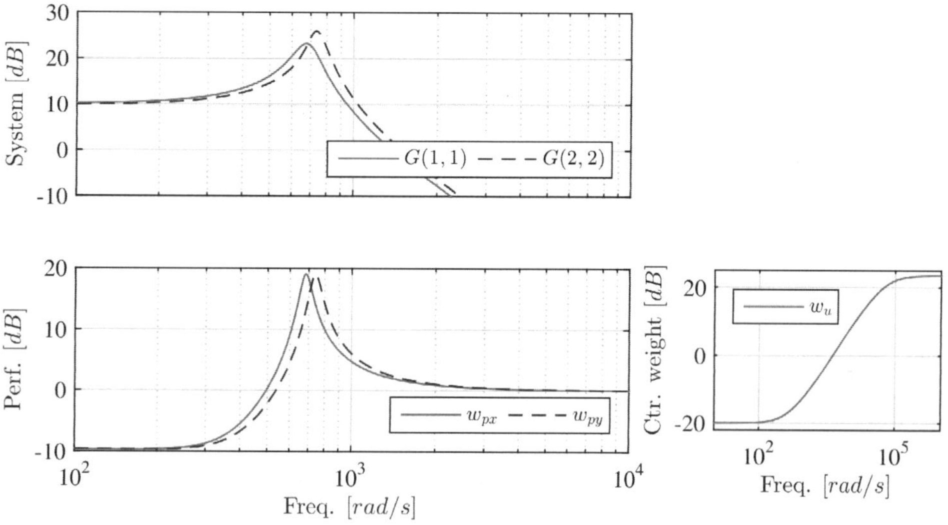

Figure 3. System gain, performance weights $w_{px}(s)$, $w_{py}(s)$ as defined in Eq. (11) and controller sensitivity weight $w_u(s)$ as defined in Eq. (12). The two left plots share the frequency axis to show the performance weight peaks coincide with the natural frequencies.

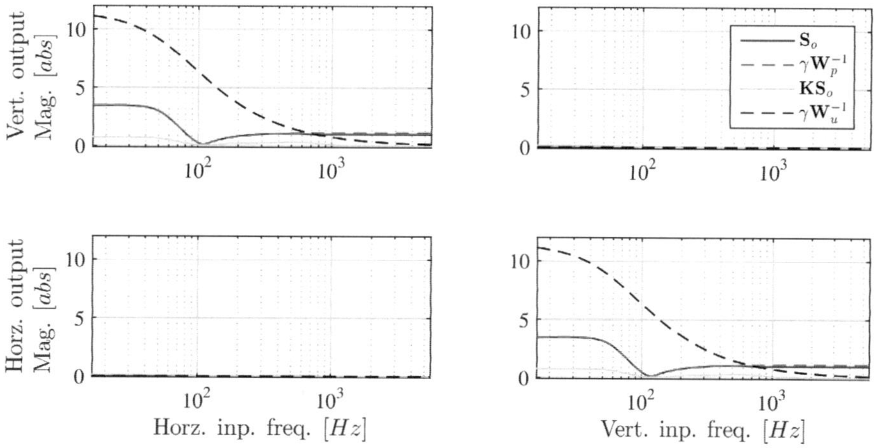

Figure 4. Resulting sensitivity functions **S** and **KS** and corresponding inverse scaled weights $\gamma \mathbf{W}_p^{-1}$ and $\gamma \mathbf{W}_u^{-1}$. At low frequency, the performance closely follows the performance weight.

frequencies up to above the first two critical speeds, the performance is limited by **S** as desired, whereas at high frequency, **KS** becomes limiting.

To allow implementation on the dSpace system, the controller is converted to discrete time using Tustin transformation and sampling frequency $f_s = 5\ kHz$.

3.1. Closed-Loop Performance Assessment Based on Extrapolated LPV Model
Since no gas bearing model is available for rotational speeds higher than 92 Hz, the performance of the controller is attempted to be assessed by means of model extrapolation from the identified LPV model. The performance assessment is strongly affected by the extrapolation, therefore the obtained results must be carefully interpreted. The LPV model Eq. (6) is extrapolated for various rotation speeds, and used to calculate the loop sensitivities using Eq. (9). The results are shown in Fig. 5. In open loop, the input disturbances are greatly amplified near the resonances, whereas the closed loop disturbance functions **SG** show a significant reduction in disturbance gain as desired. The controller has low sensitivity near the resonance frequencies, and obtains $|\mathbf{S}| < 1$ in an interval above the resonance frequencies. This indicates that the controller reduces the mass unbalance both before and after the critical speeds are crossed. Investigations also show, that the control effort **KS** is sufficiently low. The performance analysis based on the extrapolated models suggests that the designed controller may succeed in increasing the system damping at and about the critical frequencies. The validity of these predictions needs to be verified experimentally, and this is shown in the following section.

4. Experimental Results
The results of the previous section should be investigated experimentally, both the open loop and the closed loop.

This is done by applying the controller at low speed, and then slowly accelerating the shaft while monitoring the vibration amplitude. A standard runout filter is applied to remove artefacts from the measurements from mechanical imperfections in the disk using the procedure described in [10]. At standstill, the rotor equilibrium position is below the geometric bearing centre [6]. The given shaft positions **p** correspond to deviations from the equilibrium position. Two different experiment types where performed to assess the upper limit of safe rotation speed: an open loop and a closed loop. To avoid rotor-bearing rubbing during experiments, the shaft rotation speed is increased slowly until the vibrations exceed the safety region. The vibrations of the open loop experiment are shown in Fig. 6. At $\Omega = 94.8\ Hz$, the vibrations exceed the safety region, and the rotor reaches a speed of $\Omega = 97.5\ Hz$ before the experiment is stopped. The critical speeds can therefore not be crossed safely in open loop. During the closed loop experiment, the shaft is again accelerated slowly allow the vibrations to build up during the operation. A bias is applied to the control signal to allow a shift of the vibrations' centre. According to the model, the critical speeds are expected to be near 105 Hz and 115 Hz. It is evident, that the shape of the vibrations changes near these speeds. These speeds are crossed very slowly to allow the vibrations to build up and validate, the critical speeds pose no challenge in closed loop. The rotation speed can therefore be increased even further. During the experiment, the rotor reaches a speed of $\Omega = 153\ Hz$ where the control signals approaches the level of saturation. Such saturation in best case deteriorates the performance, but potentially destabilises the system. It is therefore decided to stop the experiment.

The vibrations are often investigated in rotor-dynamics as function of the rotation speed in a waterfall diagram. This is obtained as the FFT of smaller sections of the data. Such a diagram is shown for the vertical shaft direction in Fig. 7 for both the open loop and the closed loop case. The synchronous vibrations for $\Omega > 50Hz$ are significantly reduced in the closed loop case, and the critical speeds can therefore be crossed safely. It may be argued, that the approach of

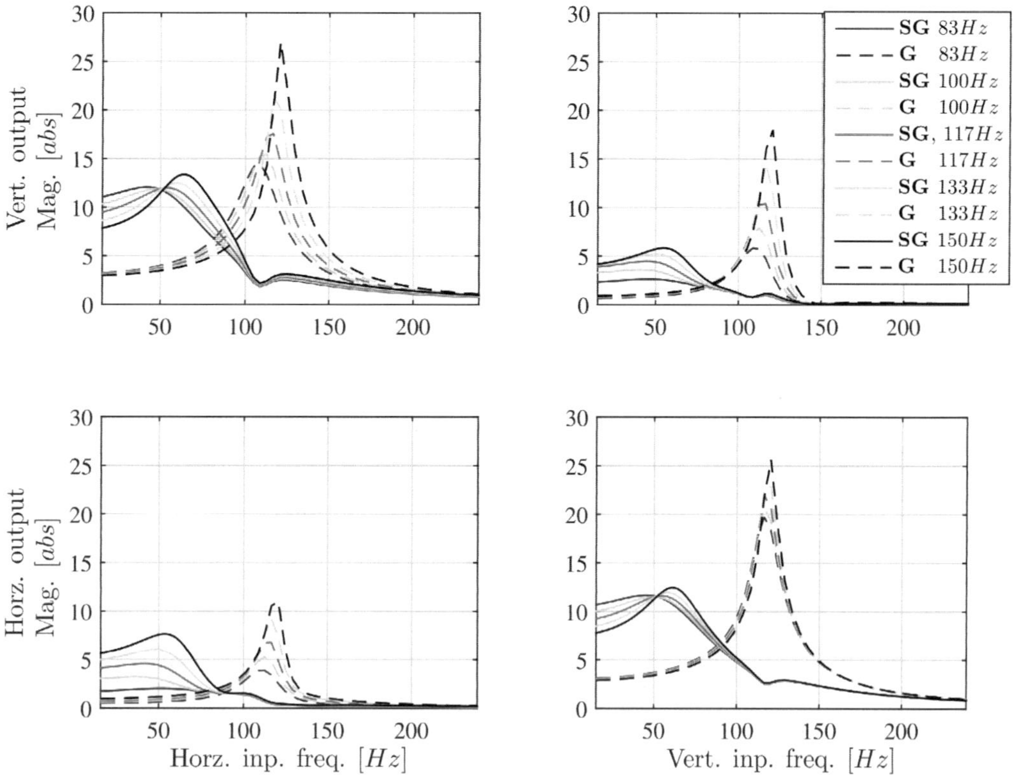

Figure 5. Evaluation of open and closed loop input disturbance responses (**G** , **SG**) for different rotation speeds extrapolated with the LPV model. The model shows, that the controller reduce the peak gain significantly.

applying an input bias should also have been used in the open loop experiment, but the vibration amplitude would still quickly have grown exceeded the allowed vibration level.

A later experiment was performed, where the rotor speed was increased even further up to $\Omega = 161.4\ Hz$, and the vibrations were within the desired level. This corresponds to an increase in operating range of approximately 70%.

5. Conclusions and Future Aspects

This paper investigated control designs to reduce the vibrations in rotating machinery supported by gas bearings. A controller was designed using the developed LPV model, and experimental results demonstrated the feasibility of using the controller to extend the operation range. The controller allowed rotation speeds up to, in and above the first two critical speeds, which extended the operation range by 70%. Future experiments will investigate the performance as the rotation speed is increased even further. In this region, sub-synchronous whirl dominates the response, and it is of interest to use controllers to postpone the onset of the whirl [4].

References

[1] H. M. N. K. Balini, J. Witte, and C. W. Scherer. Synthesis and implementation of gain-scheduling and lpv controllers for an amb system. *Automatica*, 48(3):521–527, 2012.

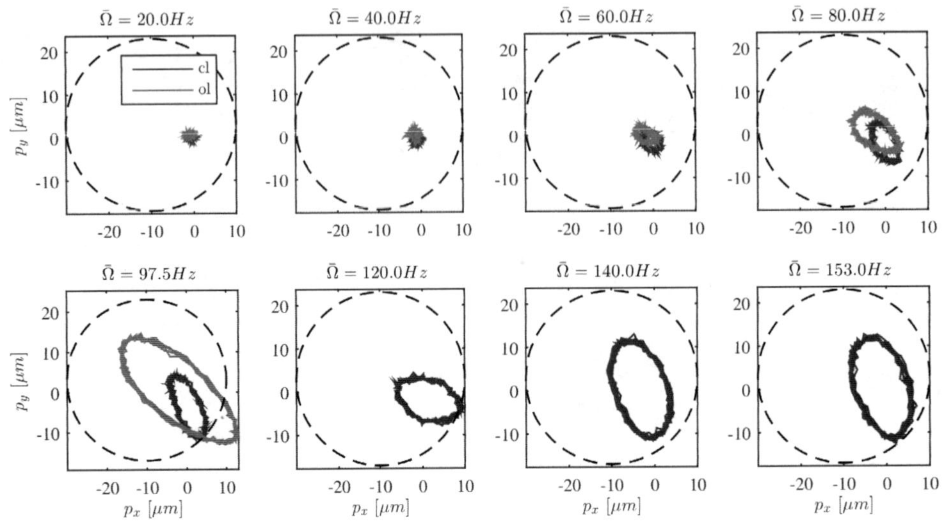

Figure 6. Experimental measurements for various shaft rotation speeds. In open loop (ol), the vibrations exceed the limit of safe operation at $\Omega = 94.8\ Hz$. In closed loop (cl), the control reduces the vibrations and significantly increases this limit.

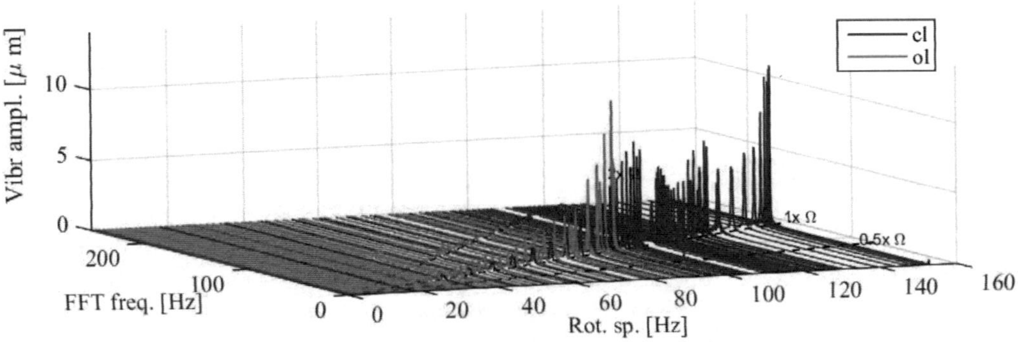

Figure 7. Waterfall diagram for vertical shaft direction to investigate the frequency content of the vibrations. In open loop (ol) the synchronous vibrations grow fast as the critical speed is approached. In closed loop (cl), the vibrations are sufficiently attenuated to allow faster rotation speeds.

[2] J. S. Freudenberg and D. P. Looze. Right half plane poles and zeros and design tradeoffs in feedback systems. *IEEE Transactions on Automatic Control*, 30(6):555–565, 1985.

[3] S. Mason, P. Tsiotras, and P. Allaire. Linear parameter varying controllers for flexible rotors supported on magnetic bearings. 2009.

[4] S. Morosi and I. F. Santos. *From Hybrid to Actively-Controlled Gas Lubricated Bearings Theory and Experiment.* PhD thesis, Technical University of Denmark, 2011.

[5] S. Morosi and I. F. Santos. Experimental investigations of active air bearings. In *Proceedings of ASME Turbo Expo 2012*, volume 7, pages 901–910. American Society of Mechanical Engineers, 2012.

[6] F. G. Pierart and Ilmar F. Santos. Steady state characteristics of an adjustable hybrid gas bearing - computational fluid dynamics, modified reynolds equation and experimental validation. *Proc. Inst. Mech. Eng. Part J.-j. Eng. Tribol*, 229(7):807–822, 2015.

[7] S. Skogestad and I. Postlethwaite. *Multivariable Feedback Control: Analysis and Design.* John Wiley & Sons, 2005.

[8] L. R. S. Theisen, H. H. Niemann, R. Galeazzi, and I. F. Santos. Enhancing damping of gas bearings: Comparison of $\mathcal{H}_\infty$ and lpv-control, theory & experiment. *Submitted to IEEE Trans. on Control Systems Technology*, 2015.

[9] L. R. S. Theisen, H. H. Niemann, I. F. Santos, and R. Galeazzi. Experimental investigations of decentralised control design for the stabilisation of rotor-gas bearings. In *Proceedings of the XVII International Symposium on Dynamic Problems of Mechanics*, 2015. In press.

[10] L. R. S. Theisen, H. H. Niemann, I. F. Santos, R. Galeazzi, and M. Blanke. Modelling and identification for control of gas bearings. *Mechanical Systems and Signal Processing*, pages –, 2015.

[11] J. Witte, H.M.N.K. Balini, and C.W. Scherer. Robust and LPV control of an AMB system. *American Control Conference (ACC), 2010*, pages 2194–2199, 2010.

[12] S. Zheng, B. Han, Y. Wang, and J. Zhou. Optimization of damping compensation for a flexible rotor system with active magnetic bearing considering gyroscopic effect. *Ieee-asme Transactions on Mechatronics*, 19(4):1130–1137, 2014.

Realization of a Complex Control & Diagnosis System on Simplified Hardware

R Stetter, M Swamy Prasad

Hochschule Ravensburg-Weingarten, 88241 Weingarten, Germany

E-mail: stetter@hs-weingarten.de

Abstract. Energy is an important factor in today's industrial environment. Pump systems account for about 20% of the total industrial electrical energy consumption. Several studies show that with proper monitoring, control and maintenance, the efficiency of pump systems can be increased. Controlling pump systems with intelligent systems can help to reduce a pump's energy consumption by up to one third of its original consumption. The research in this paper was carried out in the scope of a research project which involves modelling and simulation of pump systems. This paper focuses on the future implementation of modelling capabilities in PLCs (programmable logic controllers). The whole project aims to use a pump itself as the sensor rather than introducing external sensors into the system, which would increase the cost considerably. One promising approach for an economic and robust industrial implementation of this intelligence is the use of PLCs. PLCs can be simulated in multiple ways; in this project, Codesys was chosen for several reasons which are explained in this paper. The first part of this paper explains the modelling of a pump itself, the process load of the asynchronous motor with a control system, and the simulation possibilities of the motor in Codesys. The second part describes the simulation and testing of a system realized. The third part elaborates the Codesys system structure and interfacing of the system with external files. The final part consists of comparing the result with an earlier Matlab/SIMULINK model and original test data.

1. Introduction and background

For several centuries, human kind has depended mainly on agriculture for its well-being. The growth of agriculture in main (dry) regions depends on the efficiency of irrigation systems. Irrigation efficiency experienced tremendous growth when pumps were introduced. Due to their versatility, pump systems have many applications in industrial environments, from small water pumps to large petroleum industries. Even though the designing and modeling of pump systems has evolved significantly, the efficiency of certain kinds of pumps, however, is not in accordance with the needs of the 21st century. The efficient monitoring and controlling of pump systems can reduce their power consumption by up to one third [1]. A promising and economic option for efficient monitoring and controlling of (complete) pump systems (including driving means and consumer system) is the implementation of programmable logic controllers (PLCs). PLCs are very robust, developed to industrial standards, and relatively cost effective. This paper describes an approach for the future implementation of control and monitoring systems for pumps in PLCs, within which the complete system is modeled and simulated in controller-based software. Similar attempts can be observed in the field of realization of fractional models and fractional regulators into PLC controllers [2].

This research is part of a large project aiming to develop a well-founded prognosis of future control and diagnosis systems for pumps (compare e. g. Kleinmann et al. [3], [4]). For the development of smart and efficient pumps, the Hochschule Ravensburg-Weingarten, Germany, and the Hochschule für Technik Rapperswil, Switzerland, along with a leading German pump manufacturer formed a research team. The aim of the project team is to develop an advanced monitoring, diagnosis and control system for pumps. The whole project is mainly based on a burner application of a positive displacement pump – a spindle screw pump. Consequently, the statements in this paper are related to burner applications.

2. Modelling

In prior research, an elaborate model of the pump, the consumer system (burner) and the motor was developed in Matlab/SIMULINK (compare e. g. Kleinmann et al. [5]). The results are promising but the requirements for computer performance are rather large. Therefore, in ongoing research, the possibility of achieving similar results on less powerful computer hardware are investigated; this paper focuses on simulating PLCs with Codesys (an acronym for controller development system, previously stylised CoDeSys); Codesys is a development environment for programming controller applications according to the international industrial standard IEC 61131-3 [6].

Matlab/SIMULINK is software mainly designed for simulation, and hence provides great flexibility for modeling. Codesys is mainly designed for rather simple controlling tasks and has very limited built-in mathematical functions, and hence mathematical functions, such as differentiation, integration and matrix operations, are not possible.

The modeling of a pump can be divided into three major parts: the pump, consumer system or process load (burner application), and electrical motor. Such a model is built into separate blocks which are then integrated. This provides greater flexibility for analysis and testing. In this research, Codesys Control Win V3 is used as a PLC simulator. Figure 1 gives an overview of the blocks of the model and the sources for the formulae and parameters.

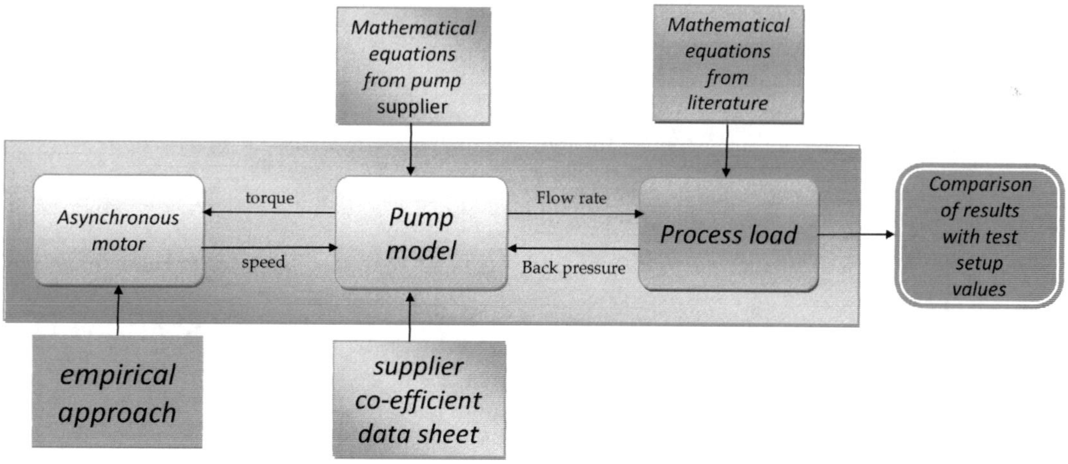

Figure 1. Complete system model.

2.1. Pump model

In this research, three screw spindle pumps (including different variants) are taken into account and modelled. The system is modelled in such a way that different variants can be tested and analyzed simply by changing the model number. The project aims to use devices with low computational power. Hence, the mathematical equations are kept as simple as possible. The mathematical equations are mostly based on prior research (compare Kleinmann et al. [4], [5]), but external references are also added.

For monitoring and controlling, certain parameters are measured and given as feedback to other blocks. The parameters include flow rate, speed, torque and differential pressure. As shown in Figure 2, torque (M) from the pump is given as a feedback to the motor in the on-line simulation so that the power of the motor can be controlled. For advanced control of whole system, the back pressure from the process load is also calculated. The inputs and outputs of the model are shown in Figure 2.

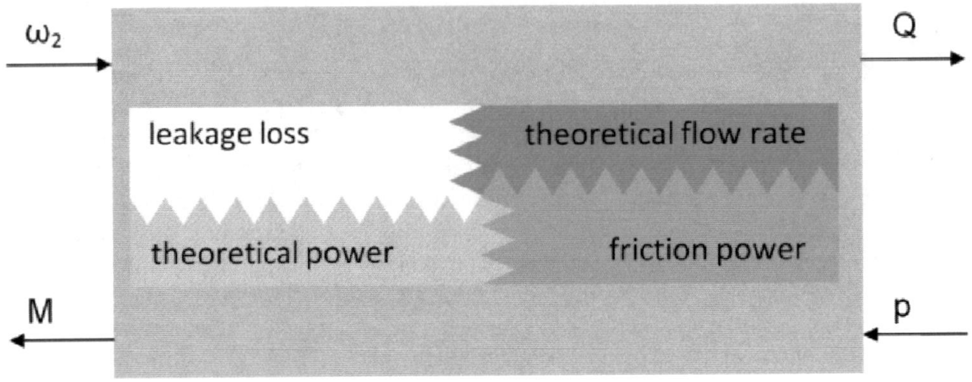

Figure 2. Pump model input and output.

In order to calculate the mentioned parameters, further parameters are needed, which include losses due to leakage and friction along with theoretical power and theoretical flow rate [5]. The following equations are used to model the pump. For the calculation of the flow rate, Equation 1 may be used:

$$Q_{th} = \left(\frac{3.D_a^{\ 3}.\pi.\tan\alpha.K_p}{2.10^7} \right).n_2 \tag{1}$$

Equation 2 may be used for the calculation of the theoretical power:

$$P_{th} = \frac{Q_{th2}.\Delta p}{600} \tag{2}$$

For the calculation of the pump torque load, Equation 3 may be used:

$$M_{load} = \frac{P_{act}}{\omega_2} \tag{3}$$

In these calculations, the value of the profile factor (K_p), diameter (D_a) of the pump and pitch angle (α) must be provided by the company which manufactures the pump. Figure 3 (left) shows a flowchart of the realization of this model in Codesys.

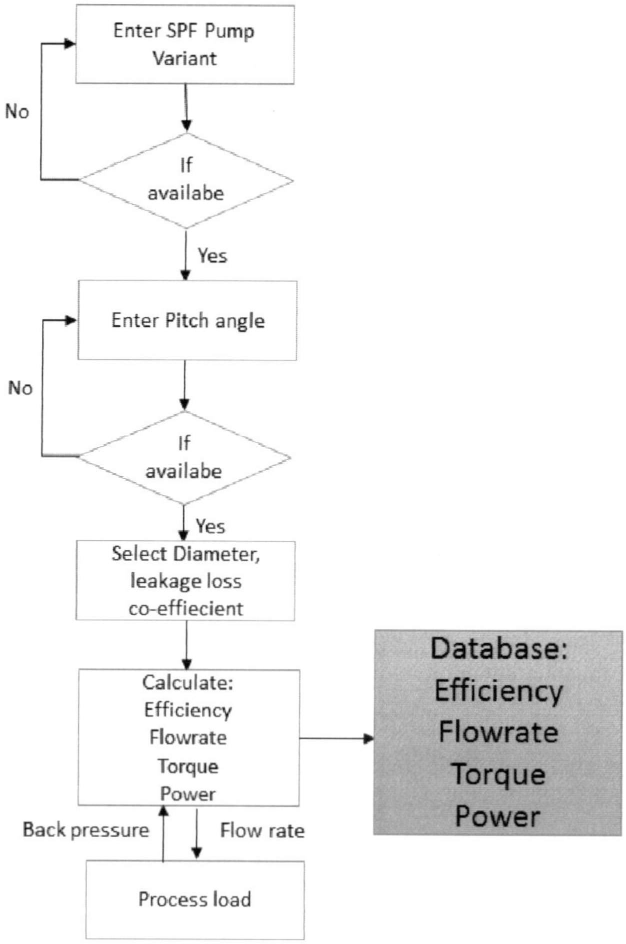

Figure 3. Pump model input and output.

2.2. Process load model

The process load is application specific. Since screw spindle pumps have a wide variety of applications, the requirements for each component can vary. The component includes different types of pipe, pipe bend, valve, nozzle, Y- pipe.

For modeling pipes, the mathematical formulae are taken from literature. For nozzle modelling, the direction of the nozzle was considered. Since nozzles can be both expanding and contracting, it was modelled in such a way that it will be decided whether it is expanding or contracting according to the user specifications. Additionally, pipe-bends were modelled similarly to prior research and the bend loss co-efficient values were refined.

The components are modeled into separate function blocks and put together to form a library. The advantage of creating a library for the process load is that different combination of components can be used. Every combination produces different results. Hence accuracy and efficiency can be improved. It also helps to avoid repetitive coding.

The output from each component is back pressure. The summation of pressure of each component will provide the total back pressure of the process load (Equation 4).

$$\Delta p_{tot} = \Delta p_n + \Delta p_{n+1} + \Delta p_{n+2} + \ldots \quad (4)$$

2.3. Electric Motor

The induction motor, or asynchronous motor, is generally used in industrial drive due to its robustness, reliability, and economical causes (in the test stand in the validation (compare section 5) a 2.2 KW nominal power asynchronous motor is used). As mentioned earlier, Codesys is controller based software and does not have the flexibility of using built-in library functions, and hence it was decided to model the system using an empirical approach. The advantage of an empirical approach is that simple mathematical equations, along with control systems, can be modelled.

Since the model runs continuously, the pump torque and speed of the motor, which can easily be gathered from a frequency converter for the asynchronous motor, are stored in a separate array. The current value of the pump torque is compared with the previous value of the pump torque and if the value exceeds defined limits (in both directions), then a change of the system state will be assumed and the speed of the motor will be increased or decreased proportionally according to the amount specified. Since the asynchronous motor is controlled by the frequency converter to a certain set (constant) speed motor, the speed will return to the set speed following a state change after a specified amount of time, which is necessary for the motor control to adapt to the new situation. In order to simulate that characteristic with rather simple calculations, a time delay function is used. If a defined delay time is over, the speed of motor will return to the set speed. The proportional factor and the time delay need to be found for each motor and each control system. The flowchart of this empirical motor model is shown in Figure 4.

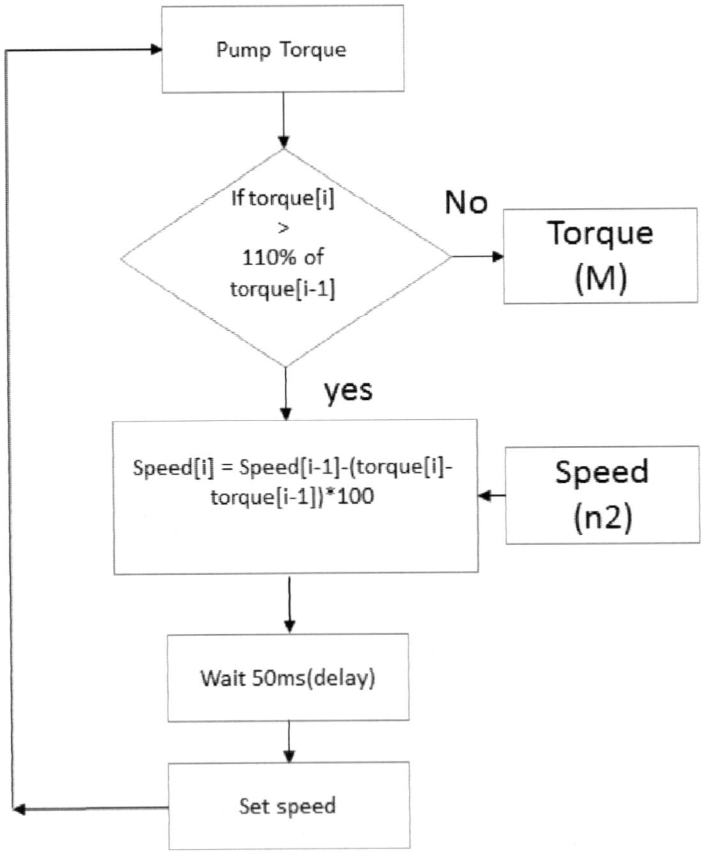

Figure 4. Flowchart of the motor model.

3. Programmable Logic Controller

Automation is the key word in today's industrial growth. In order to achieve maximum efficiency and reliability in production systems, industries are in need of automated production process. Programmable logic controllers (PLCs) help to achieve industrial requirements. The PLCs are available for small to large scale applications with varying input and output ports, as well as processing speed and built-in memory. PLCs are rigid and immune to electrical noise, which helps to achieve hard, real-time requirement. In this project, the advanced control system which is used in burner applications must be as rigid as possible. Therefore, it was decided to build an advanced control system for pumps with PLCs.

As the first step, the control system must be modelled in a simulated environment with hard, real-time requirements. Codesys helps to model such PLC-based systems in a simulated environment.

3.1. CoDeSys

CoDeSys stands for COntroller Development SYStem, which is a freeware provided by 3S-Smart Software Solutions. Codesys has an abundant library available to model a variety of systems. Codesys supports all five PLC programming languages defined by IEC 61131-3 to model and simulate the system. Here in this project, the system is modelled using "structure text".

The structure text (ST) is one of the PLC programming languages which is very similar to high level programming languages such as C, C++ or C#. Structure text has the advantage of modelling the system with conditional and looping statements.

Codesys allows the user to interface the model between different programming languages with the help of PLCopen XML. PLCopen XML is a special program which is only available in versions after V3.5 and created by PLCopen. PLCopen XML exports the program to a XML file, which is mostly hardware and software independent. The newly created XML file can be modified according to hardware requirements – this feature is quite unique.

3.2. Device

The output project file of Codesys is supported by most of the PLC hardware device providers. Codesys uses a device tree to know which devices are used in the project. For simulation, eCodesys provides a variety of PLC simulators. The simulators includes win V3, RTE V3, x64, HMI, and softmotion win V3. For simulating under windows OS, win V3 is suitable. Therefore, this project uses Codesys Control Win V3 as PLC simulator. The Win V3 device includes a API driver interface for customized I/O drivers and C functions, and also a SoftPLC-specific configuration. It supports standard field bus configurations, which are mentioned in IEC 61131-3, with no additional tool requirement.

3.3. Task configuration

The task configuration helps to manage the priority between applications. The type of process has to be mentioned in the task configuration. Since the model has to run in a loop for undetermined time, the type of process for this project is cyclic. The task configuration also helps to provide information about the maximum number of tasks and the maximum number of cyclic tasks.

3.4. Visualization

In order to realize a Human Machine Interface (HMI), Codesys provides visualization support. This support offers many possibilities for HMIs. In this project, the visualization part of Codesys helps to get values from the user without a lot of programming. For the customized monitoring of system, the visualization part of Codesys seems to be very helpful.

3.5. Advantages

Codesys is a free license software and can be installed legally without copy protection on further workstations. When compared to other PLC software, Codesys offers a wide variety of simulation possibilities. Codesys also provides object-oriented programming which is only available at higher programming levels. Object-oriented programming is widely used in modelling in this project. Codesys offers online as well as offline debugging capability.

3.6. Limitations

Since Codesys is automation software, mathematical functions are limited. While modelling the electrical motor, mathematical equations which involve differentiation and integration cannot be implemented.

4. Simulation

As mentioned earlier in Section 2, the project is divided into three subsystems: the pump, consumer system and electrical motor. The simulation of the subsystems is discussed in this section.

4.1. Program Organization Unit (POU)

The starting point of every project is PLC_PRG. From PLC_PRG, all POUs are executed.
The object-oriented programming part of Codesys can be realized by using POUs. The POUs are subcategorized into Program, Function and Function Block. The Program returns one or several output values. The values assigned to variables can be retained from the last run of the program. The Program can be called by other Programs, but not by Functions. The Function Block is also a POU with several output values. The variables in a Function Block can be used by other Functions or Function Blocks using the object-oriented concept of Codesys, which is only available in higher level programming languages. The major difference between a Program and a Function Block is that unlike a Program, the value of the last run cannot be retained in a Function Block.

As shown in Figure 5, the model of pump is subcategorized into efficiency, flow rate, power, speed and torque load.

The subcategory "intia" is the initialization of the system. All subcategories are defined under Program (PRG) since they return more than one value. The model is subcategorized in a way that essential parts of the model are built separately and then collectively form a folder in which all detailed information about the pump is given. The advantage of this separate modeling of subsystems is that they can be run and tested separately, which simplifies the debugging process.

In the process load, the system is subcategorized into pipe, pipe bend and nozzle. The major part of process load modeling is that the components should return the value of back pressure. Since back pressure is the only output value, the system will return only one value, and hence, they are defined under Function (FUN).

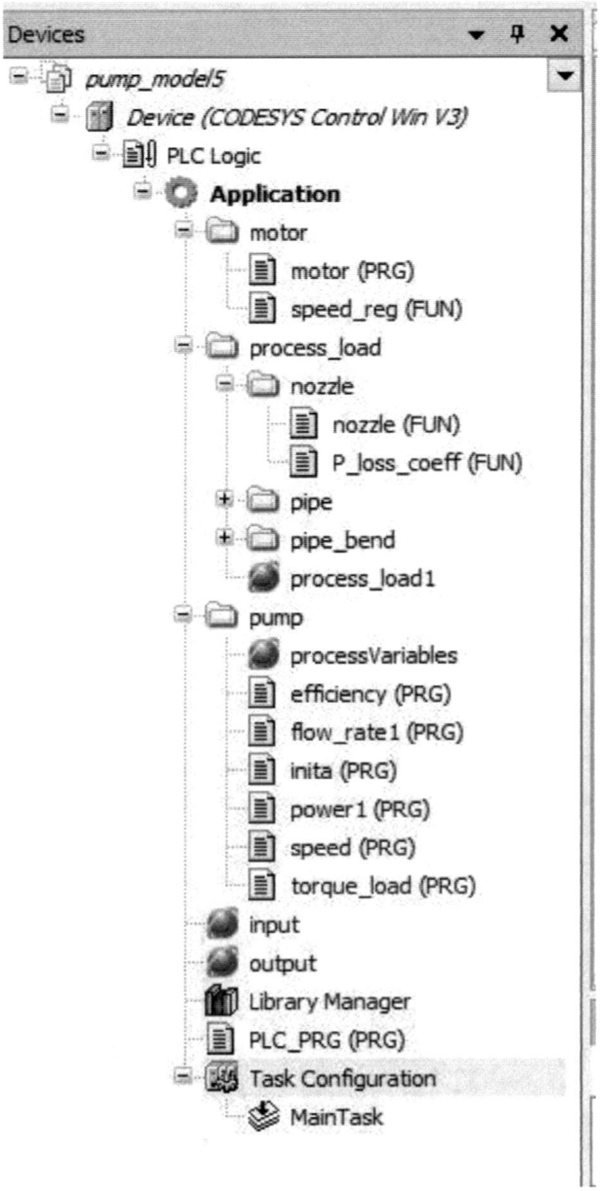

Figure 5. Structure of the project

4.2. Global Variables

Global variables provide a method to access variables throughout the system. A global variable (GVL) is marked with a separate symbol. There are three global variable lists used in this project.

There are separate global variable lists for input and output. The input global variable list consists of the pump model, the viscosity and the pitch angle. They are defined as global variables, since they are used all through the project. The output global variable list consists of the pump efficiency, the theoretical and the actual flow rate and of different losses in the system. There is another global variable list for process variables. The major disadvantage of GVL is that computation for GVL is much higher than normal variables, which are defined inside the POUs.

5. VALIDATION OF THE MODEL

To check the result obtained from simulation, results from the test setup used in prior research were compared (compare Kleinmann et al. [5] - Figure 6).

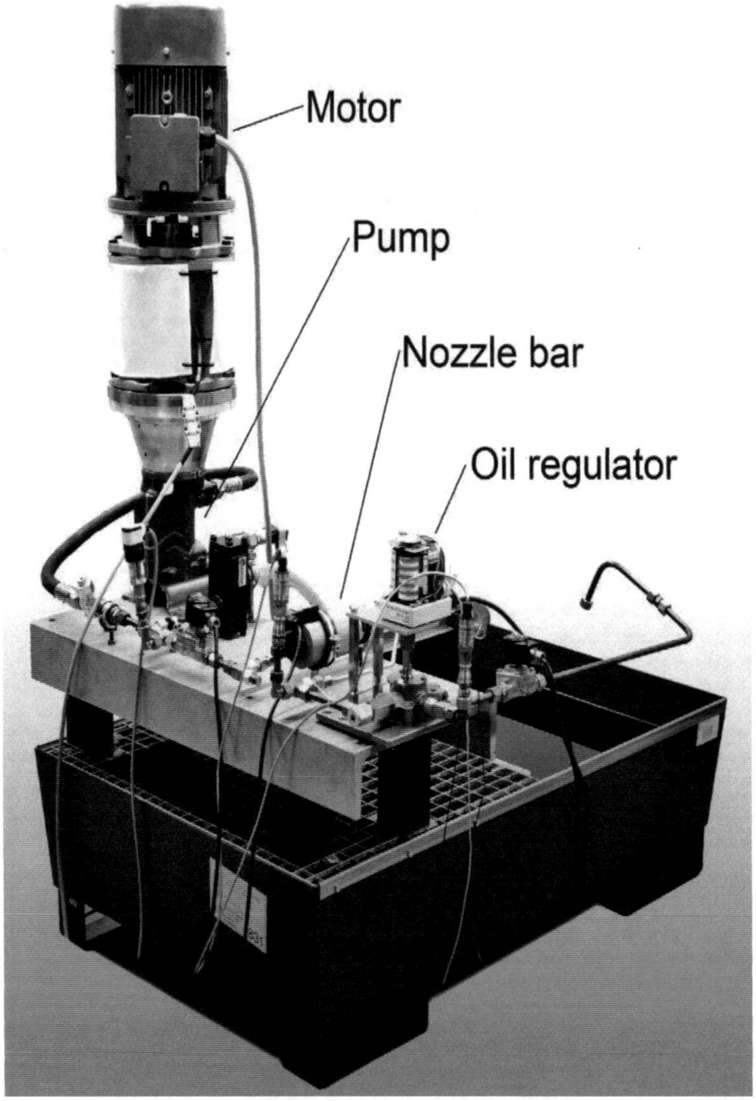

Figure 6. Test stand

The values obtained from the Codesys model were compared with the values from a prior model realized in Matlab/SIMULINK and also with test setup values. The results of a comparison of both models are plotted as graphs (see Figures 7 and 8).

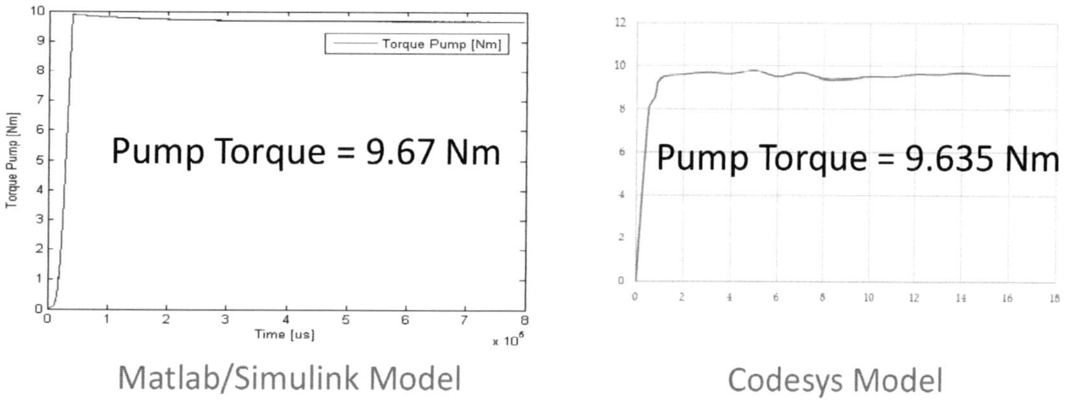

Figure 7. Comparison of models for torque: Matlab/SIMULINK (left) and Codesys (right)

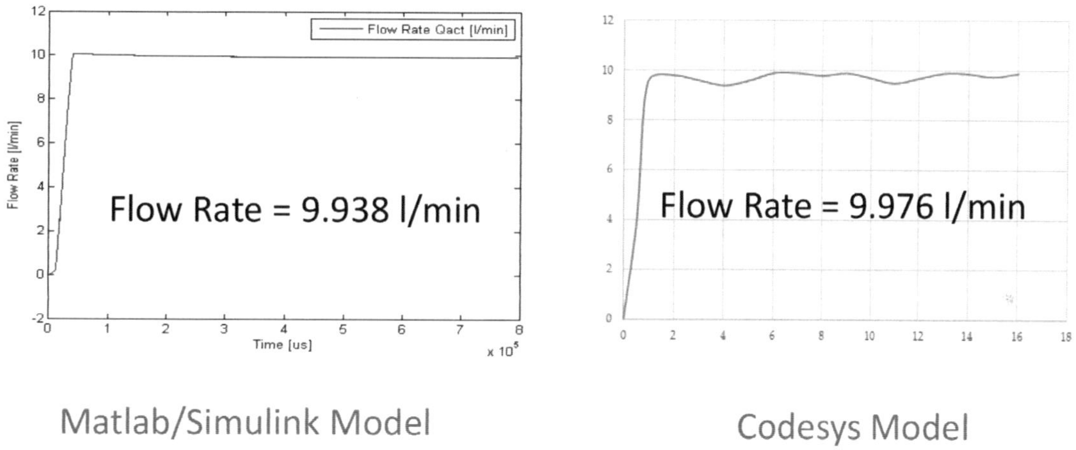

Figure 8. Comparison of models for flow rate: Matlab/SIMULINK (left) and Codesys (right)

6. Conclusion

Several survey show an astonishing contradiction between high a high energy saving potential of control systems and a relatively low application ratio in industrial practice. One root cause may be economic considerations – sometimes unstrung concentration on investment costs can be observed. Consequently, research projects which focus on the application of relatively cheap hardware for elaborate control activities such as the research of Kopka [2] are of paramount importance. This paper presents the modelling and simulation of smart pumps using the PLC simulation system Codesys. The results of this investigation show that PLC's can provide an economical and robust implementation possibility for industrial applications. It proved to be especially significant to model the asynchronous motor with its control system integrated into the frequency converter. For this purpose, a rather mundane empirical model was chosen, which relies on state changes and control delays. The simulation results obtained from the Codesys model correspond well with results from an elaborate Matlab/SIMULINK model from prior research which we require much more elaborate hardware. This comparison shows that the values obtained are within the tolerance limit. Since Codesys has general syntax defined by IEC 61131-3, the code can be implemented in any controller for further verification. Further steps will include this implementation and an expanded verification procedure.

References

[1] Friedl, C.: Energie-Optimierung von Pumpen rechnet sich. In: *Maschinenmarkt*, November 2007.

[2] Kopka, R.: Implementation of a Fractional Model–Based Fault Detection Algorithm into a PLC Controller. J. Phys.: Conf. Ser. 570 072005, 2014.

[3] Kleinmann, S., Dabrowska, A., Hoffmann, M., Kühn, H., Koller-Hodac, A., Stetter, R.: Advanced Control of Industrial Pump Systems. In: *Proceedings of the 7th Workshop on Advanced Control and Diagnosis*, ACD'2009, Zielona Góra, Poland.

[4] Kleinmann, S., Dabrowska, A., Zuniga Mera, A., Koller-Hodac, A., Stetter, R.: Diagnosis of Three Screw Spindle Pumps Based on Static and Dynamic Modelling. In: Bokor, J. (Ed.): *Proceedings of the 9th Workshop on Advanced Control and Diagnosis*, ACD'2011, 17. and 18. November 2011, Budapest, Ungarn.

[5] Kleinmann, S.; Stetter, R.; Abdul Jalal, M. F.: Praxisorientierte Modellierung von Pumpensystemen für Monitoring, Regelung und Diagnose. In *at – Automatisierungstechnik* 59 (2011) 5, pp. 309 – 322.

[6] Vogel-Heuser, B.; Wannagat A.: Modulares Engineering und Wiederverwendung mit CoDeSys V3, Oldenbourg Industrieverlag: 2008.

[7] Kleinmann, S., Leonardo, D., Koller-Hodac, A.: Intelligente Pumpenantriebe. Simulation, Condition Monitoring, Fehlerdiagnose und Energieeffizienz. *VDI-Konferenz Antriebssysteme 2011*, Stuttgart, September 13-14, 2011-08-10.

[8] Isermann, R.: Mechatronic Systems: Fundamentals. Berlin: Springer, 2005.

[9] Isermann, R.: Fault–Diagnosis Systems: An Introduction from Fault Detection to Fault Tolerance. Berlin: Springer 2005.

[10] Kallesöe, C.; Cocquempot, V.; Izadi-Zamanabadi, R.: Model Based Fault Detection in a Centrifugal Pump Application. In: *IEEE Transactions on Control Systems Technology*, Vol. 14, No. 2, 2006, p. 204-215.

[11] Miklosevic, K.; Spoljaric, Z.; Jerkovic, V.: Analysis of Electric DC Drive Using Matlab Simulink and Simpower System. J. J. Strossmayer University of Osijek. Osijek, Croatia

[12] Kleinmann, S., Dabrowska, Koller-Hodac, A., Leonardo, D., Stetter, R.: Model of a combined pump and drive system for advanced control and diagnosis. In: *The Archive of Mechanical Engineering*. Volume LVII, Number 4 / 2010, pp 405 - 414. 2010.

[13] Räbinger, K., Maksoud, T.M.A., Ward, J.: "Development of a finite volume model for the compressible gap flow inside a screw pump", University of Glamorgan School of Technology Wales, UK, 2005.

[14] Stetter, R.; Paczynski, A.; Dabrowska, A.: Industrial Application of Advanced Control and Diagnosis. In: Witczak, M.; Rozpra-Faygel, M. (Eds.): *Proceedings of the 1th European Workshop on Control Engineering for Industry, CEIND'2012*, 16. and 17. April 2012, Glogów, Polen.

12th European Workshop on Advanced Control and Diagnosis (ACD 2015) IOP Publishing
Journal of Physics: Conference Series **659** (2015) 012019 doi:10.1088/1742-6596/659/1/012019

Flatness-based embedded control in successive loops for spark-ignited engines

Gerasimos Rigatos
Unit of Industrial Automation, Industrial Systems Institute, 26504, Rion Patras, Greece
E-mail: grigat@ieee.org

Abstract. Embedded control units for transportation systems make use of advanced nonlinear control methods. In this research article a new nonlinear control method is applied to spark ignited (SI) engines. The proposed SI engine's control scheme is based on differential flatness theory The considered method succeeds the efficient control of the SI engine parameters such as intake pressure and turn speed. The method makes use of a state-space model of the SI-engine in the so-called triangular form. The controller design proceeds by showing that each row of the state-space model of the SI engine stands for a differentially flat system, where the flat output is chosen to be the associated state variable. Next, for each subsystem which is linked with a row of the state-space model, a virtual control input is computed, that can invert the subsystem's dynamics and can eliminate the subsystem's tracking error. From the last row of the state-space description, the control input that is actually applied to the SI engine is found. This control input contains recursively all virtual control inputs which were computed for the individual subsystems associated with the previous rows of the state-space equation. Thus, by tracing the rows of the state-space model backwards, at each iteration of the control algorithm, one can finally obtain the control input that should be applied to the SI-engine so as to assure that all its state vector elements will converge to the desirable setpoints.

1. Introduction

In this paper a new nonlinear control scheme is proposed for SI-engines making use of differential flatness theory [Rigatos, 2015],[Rigatos, 2011a]. The controller of the spark-ignited engines is structured in successive loops. The method makes use of a description of the engine's state-space model in the so-called triangular form. The controller design proceeds by showing that each row of the state-space model of the SI engine stands for a differentially flat system, where the flat output is chosen to be the associated state variable. Next, for each subsystem which is linked with a row of the state-space model a virtual control input is computed, that can invert the subsystem's dynamics and can eliminate the subsystem's tracking error. From the last row of the state-space description, the control input that is actually applied to the spark-ignited engine is found. This control input contains recursively all virtual control inputs which were computed for the individual subsystems associated with the previous rows of the state-space equation. Thus, by tracing the rows of the state-space model backwards, at each iteration of the control algorithm, one can finally obtain the control input that should be applied to the SI engine so as to assure that all its state vector elements will converge to the desirable setpoints.

The paper's results come to complement the significant research effort towards the development

Journal of Physics: Conference Series **659** (2015) 012019 doi:10.1088/1742-6596/659/1/012019

of embedded control systems for the automotive industry, aiming at improving the performance of vehicles' engines in terms of produced power, at reducing fuel consumption and at eliminating the emission of exhaust gases. In particular, the problem of control of the rotation speed of SI-engines as well as the problem of control of the engine's pressure manifolds has been approached with different methods [Sugihira, 2008],[Zhang, 2010]. In [Ali, 2006] a nonlinear state space controller for turn speed (and consequently for torque) of a spark-ignited engine is proposed. The controller's design is based on feedback linearization in combination with pole placement. In [Zhang, 2009] time-varying internal model-based design is applied to compensate for the time-varying but angle dependent pressure pulsations in the fuel injection system of SI-engines. In [Leroy, 2009] a control method for the air-path system of SI-engines is presented. The first part considers generation of the motion-planning trajectory of the intake manifold pressure from a torque set point. Then, feedforward and feedback control laws are presented. In [Moulin, 2011] a feedback linearization approach based on differential flatness theory is proposed for the control of the air system of a turbocharged gasoline engine. Finally, in [Flärdh, 2014] a model-based approach is pursued to maximize an SI-engine's torque through optimal control of the variable valve timing (VVT) and the variable gas turbine (VGT) model. Several other results on nonlinear control of spark ignited engines have been presented in [Sepuvelda, 2012], [Ballachi, 2013], [Blake Vance 2008],[Alipi, 2003],[Colin, 2007], [Nguyen, 2013]].

Moreover, the paper contributes a different perspective to the design and implementation of flatness-based controllers. Differential flatness theory is currently a main direction in nonlinear control systems analysis and synthesis [Rigatos, 2015],[Rigatos, 2011a]. A system is considered to be differentially flat if all its state variables and its control inputs can be expressed as functions of one single algebraic variable which is the so-called flat output, and also as functions of the flat-output's derivatives [Rudolph, 2003],[Sira-Ramirez, 2004],[Lévine, 2011],[Fliess, 1999], [Menhour, 2013]. The differential flatness property enables the transformation of the nonlinear system's dynamics into the linear canonical form and the design of a state feedback controller through the application of pole placement techniques in the linearized equivalent model of the system [Rouchon, 2005], [Martin, 1999],[Bououden, 2011], [Laroche, 2007],[Rigatos, 2014],[Rigatos, 2012]. In this paper, a different approach is developed for controller design in nonlinear dynamical systems which exhibit the differential flatness property. The method makes use of the initial nonlinear model of the system and of its decomposition into a set of nonlinear subsystems for which the differential flatness property holds. The proposed flatness-based controller consists finally of cascading interconnected loops.

The structure of the paper is as follows: in Section 2 the dynamic model of the SI-engine is analyzed and its state-space description is given. In Section 3 the implementation of flatness-based control in cascading loops is proposed for nonlinear dynamical systems having a state-space model in the triangular form. In Section 4 the dynamics of the closed loop of the spark-ignited engine is analyzed and asymptotic stability is proven. In Section 5 simulation experiments are carried out to assess the performance of the proposed nonlinear control scheme for the SI-engine. Finally, in Section 6 concluding remarks are stated.

2. Dynamic model of the Spark-Ignited Engine

It is possible to control the intake pressure p_m and the rotational speed of the engine's shaft ω by adjusting the angle of the air throttle. It is considered that the associated control loop is independent from the loops of the fuel injection control and spark timing control (Fig. 1).

The basic equations of the system are [Sugihira, 2008]:

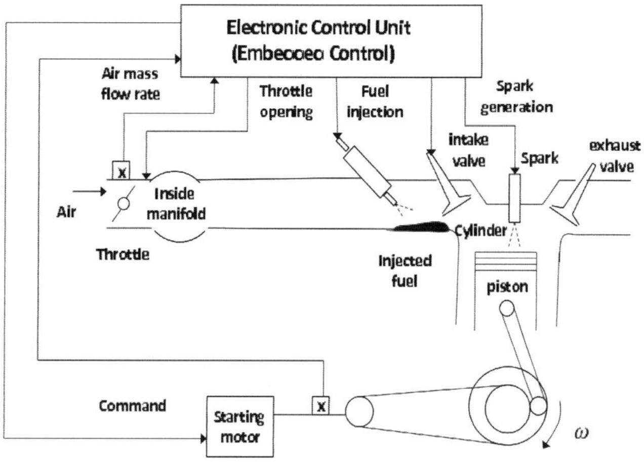

Figure 1. Diagram of the spark-ignited engine

$$\begin{aligned} \dot{\omega} &= k_{\omega_1} p_m(t-\tau_d) + k_{\omega_2} + k_{\omega_3} T_{f_m} \\ \dot{p}_m &= k_{p_1} \omega p_m + k_{p_2} \omega + k_{p_3} u \\ y_1 &= \omega \end{aligned} \tag{1}$$

The above model is easy to be confirmed by considering the rotational motion dynamics of the engine' shaft and the time delay in the measurement of the intake air pressure. The variable of the intake pressure appears with time delay in the equation of the turn speed in the second row of the model of the SI engine. Using that $p_{md} = p_m(t-\tau_d)$ and

$$p_m(t-\tau_d) = \frac{1}{\tau s+1} p_m \tag{2}$$

while $\tau = a_d/\omega$ and a_d is a parameter that is measured in radians. Denoting $k_d = -1/a_d$ one has about the dynamics of the delayed intake pressure variable

$$p_{md} = k_d \omega (p_{m_d} - p_m) \tag{3}$$

Using the previous formulation, and defining the state variables $x_1 = \omega$, $x_2 = p_{m_d}$ and $x_3 = p_m$, the dynamics of the SI engine is written as [Rigatos, 2015], [Sugihira, 2008]

$$\begin{aligned} \dot{x}_1 &= k_{\omega_1} x_2 + k_{\omega_2} + k_{\omega_3} T_{f_m} \\ \dot{x}_2 &= K_d x_1 (x_2 - x_3) \\ \dot{x}_3 &= k_{p_1} x_1 x_3 + k_{p_2} x_1 + k_{p_3} u \end{aligned} \tag{4}$$

where T_{f_m} are friction torques, which can be also perceived as disturbances. In the above equations coefficients k_{p_i}, $i = 1,2,3$, k_{ω_i}, $i = 1,2,3$ and K_d are associated with the combustion cycle of the SI-engine and are defined in [Rigatos, 2015], [Sugihira, 2008]. The model also takes the matrix form $\dot{x} = f(x) + g(x)u$ with

$$f(x) = \begin{pmatrix} k_{\omega_1} x_2 + k_{\omega_2} + k_{\omega_3} T_{f_m} \\ k_d x_1 (x_2 - x_3) \\ k_{p_1} x_1 x_3 + k_{p_2} x_1 \end{pmatrix} \quad g(x) = \begin{pmatrix} 0 \\ 0 \\ k_{p_3} \end{pmatrix} \tag{5}$$

The considered model of the spark-ignited engine can be also written in the so-called triangular form. Indeed one has

$$\begin{aligned} \dot{x}_1 &= f_1(x_1) + g_1(x_1)x_2 \\ \dot{x}_2 &= f_2(x_1,x_2) + g_2(x_1,x_2)x_3 \\ \dot{x}_3 &= f_3(x_1,x_2,x_3) + g_3(x_1,x_2,x_3)u \end{aligned} \tag{6}$$

where

$$\begin{aligned} f_1(x_1) &= k_{\omega_2} + k_{\omega_3}T_{fm} & g_1(x_1) &= k_{\omega 1} \\ f_2(x_1,x_2) &= k_d x_1 x_2 & g_2(x_1,x_2) &= k_d x_1 \\ f_3(x_1,x_2,x_3) &= k_{p_1}x_1x_3 + k_{p_2}x_1 & g_3(x_1,x_2,x_3) &= k_{p3} \end{aligned} \tag{7}$$

For the implementation of model-based control of the SI-engine it is important to use a reliable model of the engine's dynamics and to have precise knowledge of the engine's parameters. To this end, identification and model validation techniques are used, capable of providing estimates of the engine's parameters through the processing of sensor data [Rigatos, 2007],[Basseville, 1993],[Rigatos, 2009]].

3. A new approach to flatness-based control for nonlinear dynamical system

The control method to be applied to the SI-engine's model is also suitable for a wide class of nonlinear dynamical systems of the form:

$$\begin{aligned} \dot{x} &= f(x) + g(x)u \;\; x\in R^n \quad u\in R \\ y &= h(x) \end{aligned} \tag{8}$$

It is considered that such systems can be written in the so-called triangular form:

$$\begin{aligned} \dot{x}_1 &= f_1(x_1) + g_1(x_1)x_2 \\ \dot{x}_2 &= f_2(x_1,x_2) + g_2(x_1,x_2)x_3 \\ \dot{x}_3 &= f_3(x_1,x_2,x_3) + g_3(x_1,x_2,x_3)x_4 \\ &\cdots \\ \dot{x}_i &= f_i(x_1,x_2,\cdots,x_i) + g_i(x_1,x_2,\cdots,x_i)x_{i+1} \\ &\cdots \\ \dot{x}_{n-1} &= f_{n-1}(x_1,x_2,\cdots,x_{n-1}) + g_{n-1}(x_1,x_2,\cdots,x_{n-1})x_n \\ \dot{x}_n &= f_n(x_1,x_2,\cdots,x_n) + g_n(x_1,x_2,\cdots,x_n)u \end{aligned} \tag{9}$$

The following virtual control inputs $\alpha_i = x_{i+1}$ are defined per row of the state-space model of Eq. (9)

$$\begin{aligned} \dot{x}_1 &= f_1(x_1) + g_1(x_1)\alpha_1 \\ \dot{x}_2 &= f_2(x_1,x_2) + g_2(x_1,x_2)\alpha_2 \\ \dot{x}_3 &= f_3(x_1,x_2,x_3) + g_3(x_1,x_2,x_3)\alpha_3 \\ &\cdots \\ \dot{x}_i &= f_i(x_1,x_2,\cdots,x_i) + g_i(x_1,x_2,\cdots,x_i)\alpha_i \\ &\cdots \\ \dot{x}_{n-1} &= f_{n-1}(x_1,x_2,\cdots,x_{n-1}) + g_{n-1}(x_1,x_2,\cdots,x_{n-1})\alpha_{n-1} \\ \dot{x}_n &= f_n(x_1,x_2,\cdots,x_n) + g_n(x_1,x_2,\cdots,x_n)u \end{aligned} \tag{10}$$

The system of Eq. (10) is a differentially flat one. It is considered that $y = x_1$ is the flat output of system. It can be easily shown that each virtual control input $\alpha_i = x_{i+1}$, $i = 1,2,\cdots$ can be expressed as a function of the flat output and its derivatives, since it holds

$$\alpha_i = \frac{1}{g_i(x_1,x_2,\cdots,x_i)}(\dot{x}_i - f_(x_1,x_2,\cdots,x_i)) \tag{11}$$

For $i=1$ one has

$$\alpha_1 = \frac{1}{g_1(x_1}(\dot{x}_1 - f_(x_1)) \tag{12}$$

which means that α_1 is a function of the flat output and its derivative. For $i=2$ one has

$$\alpha_2 = \frac{1}{g_2(x_1,x_2)}(\dot{x}_2 - f_2(x_1,x_2)) \tag{13}$$

which means that $\alpha_2 = x_3$ is a function of the flat output $y = x_1$ and its derivatives. Continuing in a similar manner one has that $\alpha_{n-1} = x_n$ and consequently $\alpha_n = u$ is also a function of the flat output $y = x_1$ and its derivatives. According to the above, one has a nonlinear dynamical system in which, all its state variables and the control input can be written as functions of the flat output and its derivatives. Therefore, such a system is differentially flat.

Additionally, by considering each row of the model of Eq. (10), one has a set of n subsystems of the form

$$\dot{x}_i = f_i(x_1,x_2,\cdots,x_i) + g_i(x_1,x_2,\cdots,x_i)\alpha_i \tag{14}$$

where each subsystem describes the dynamics of the single state variable x_i. For each one of these subsystems one can consider the state variable x_i as the flat output. Obviously, the virtual control input α_i is a function of this flat output and its derivatives. Therefore, each local subsystem is also differentially flat.

Next, one can compute the virtual inputs which are applied to each subsystem. For the first subsystem, which is associated with the first row of Eq. (9), and by defining $z_i = x_i - x_i^* = x_1 - \alpha_{i-1}$, the virtual control input is given by

$$\begin{aligned} \alpha_1 = x_2^* = \frac{1}{g_1(x_1)}\cdot \\ \cdot(\dot{x}_1^d - f_(x_1) - K_1^1(x_1 - x_1^d)) \Rightarrow \\ \alpha_1 = x_2^* = \frac{1}{g_1(x_1)}\cdot \\ \cdot(\dot{x}_1^d - f_(x_1) - K_1^1 z_1) \end{aligned} \tag{15}$$

From the second row of Eq. (9), and using that $z_2 = x_2 - x_2^* = x_2 - \alpha_1$ one has

$$\begin{aligned} \alpha_2 = x_3^* = \frac{1}{g_2(x_1,x_2)}\cdot \\ \cdot(\dot{x}_2^* - f_2(x_1,x_2) - K_1^2(x_2 - x_2^*)) \Rightarrow \\ \alpha_2 = x_3^* = \frac{1}{g_2(x_1,x_2)}\cdot \\ \cdot(\dot{\alpha}_1 - f_2(x_1,x_2) - K_1^2 z_2) \end{aligned} \tag{16}$$

From the third row of Eq. (9), and using that $z_3 = x_3 - x_3^* = x_3 - \alpha_2$ one has

$$\begin{aligned} \alpha_3 = x_4^* = \frac{1}{g_3(x_1,x_2,x_3)}\cdot \\ \cdot(\dot{x}_3^* - f_3(x_1,x_2,x_3) - K_1^3(x_3 - x_3^*)) \Rightarrow \\ \alpha_3 = x_4^* = \frac{1}{g_3(x_1,x_2,x_3)}\cdot \\ \cdot(\dot{\alpha}_2 - f_3(x_1,x_2,x_3) - K_1^3 z_3) \end{aligned} \tag{17}$$

Continuing in a similar manner and from the i-th row of the state-space description of the system given in Eq. (9), and also using that $z_i = x_i - x_i^* = x_i - \alpha_{i-1}$ one obtains

$$\begin{array}{c} \alpha_i = x_{i+1}^* = \frac{1}{g_i(x_1,x_2,\cdots,x_i)} \cdot \\ \cdot(\dot{x}_i^* - f_i(x_1,x_2,\cdots,x_i) - K_1^i(x_i - x_i^*)) \Rightarrow \\ \alpha_i = x_{i+1}^* = \frac{1}{g_i(x_1,x_2,\cdots,x_i)} \cdot \\ \cdot(\dot{\alpha}_{i-1} - f_i(x_1,x_2,\cdots,x_3) - K_1^i z_i) \end{array} \tag{18}$$

Equivalently, from the $n-1$-th row of the state-space model of Eq. (9) and using that $z_{n-1} = x_{n-1} - x_{n-1}^* = x_{n-1} - \alpha_{n-2}$ one has

$$\begin{array}{c} \alpha_{n-1} = x_n^* = \frac{1}{g_{n-1}(x_1,x_2,\cdots,x_{n-1})} \cdot \\ (\dot{x}_{n-1}^* - f_{n-1}(x_1,x_2,\cdots,x_{n-1}) - K_1^{n-1}(x_{n-1} - x_{n-1}^*)) \Rightarrow \\ \alpha_{n-1} = x_n^* = \frac{1}{g_{n-1}(x_1,x_2,\cdots,x_{n-1})} \cdot \\ (\dot{\alpha}_{n-2} - f_{n-1}(x_1,x_2,\cdots,x_{n-1}) - K_1^{n-1} z_{n-1}) \end{array} \tag{19}$$

Finally, from the n-th row of the state-space model of Eq. (9) and using that $z_n = x_n - x_n^* = x_n - \alpha_{n-1}$ one has

$$\begin{array}{c} \alpha_n = u = \frac{1}{g_n(x_1,x_2,\cdots,x_n)} \cdot \\ \cdot(\dot{x}_n^* - f_n(x_1,x_2,\cdots,x_n) - K_1^n(x_n - x_n^*)) \Rightarrow \\ \alpha_n = x_n^* = \frac{1}{g_n(x_1,x_2,\cdots,x_n)} \cdot \\ \cdot(\dot{\alpha}_{n-1} - f_n(x_1,x_2,\cdots,x_n) - K_1^n z_n) \end{array} \tag{20}$$

The computation of the control input u that should be actually applied to the nonlinear system is performed in a recursive manner by processing backwards the virtual control inputs described in Eq. (15) to Eq. (20).

Thus, from the last row of the state-space description, the control input that is actually applied to the nonlinear system is found. This control input contains recursively all virtual control inputs which were computed for the individual subsystems associated with the previous rows of the state-space equation. Consequently, by tracing the rows of the state-space model backwards, at each iteration of the control algorithm, one can finally obtain the control input that should be applied to the nonlinear system so as to assure that all its state vector elements will converge to the desirable setpoints.

4. Closed-loop dynamics

By substituting Eq. (20) into the last row of the state space model of Eq. (9), and using the definition $\dot{x}_n - \dot{a}_{n-1} = z_n$, one obtains:

$$\begin{array}{c} \dot{x}_n = \dot{a}_{n-1} - K_1^n(x_n - \alpha_{n-1}) \Rightarrow \\ (\dot{x}_n - \dot{a}_{n-1}) + K_1^n(x_n - \alpha_{n-1}) = 0 \Rightarrow \\ \dot{z}_n + K_1^n z_n = 0 \end{array} \tag{21}$$

By substituting Eq. (19) into the last row of the state space model of Eq. (9), and using the definition $\dot{x}_{n-1} - \dot{a}_{n-2} = z_{n-1}$, one obtains:

$$\begin{gathered}\dot{x}_{n-1} = \dot{a}_{n-2} - K_1^{n-1}(x_{n-1} - \alpha_{n-2}) \Rightarrow \\ (\dot{x}_{n-1} - \dot{a}_{n-2}) + K_1^{n-1}(x_{n-1} - \alpha_{n-2}) = 0 \Rightarrow \\ \dot{z}_{n-1} + K_1^{n-1} z_{n-1} = 0\end{gathered} \tag{22}$$

By substituting Eq. (18) into the last row of the state space model of Eq. (9), and using the definition $\dot{x}_i - \dot{a}_{i-1} = z_i$, one obtains:

$$\begin{gathered}\dot{x}_i = \dot{a}_{i-1} - K_1^{n-1}(x_i - \alpha_{i-1}) \Rightarrow \\ (\dot{x}_i - \dot{a}_{i-1}) + K_1^i(x_i - \alpha_{i-1}) = 0 \Rightarrow \\ \dot{z}_i + K_1^i z_i = 0\end{gathered} \tag{23}$$

while continuing backwards and by substituting Eq. (16) into the second row of the state space model of Eq. (9), and using the definition $\dot{x}_2 - \dot{a}_1 = z_2$, one gets:

$$\begin{gathered}\dot{x}_2 = \dot{a}_1 - K_1^2(x_2 - \alpha_1) \Rightarrow \\ (\dot{x}_2 - \dot{a}_1) + K_1^2(x_2 - \alpha_1) = 0 \Rightarrow \\ \dot{z}_2 + K_1^2 z_2 = 0\end{gathered} \tag{24}$$

Finally, by substituting Eq. (15) into the first row of the state space model of Eq. (9), one has:

$$\begin{gathered}\dot{x}_1 = \dot{x}_1 - K_1^1(x_1 - x_1^d) \Rightarrow \\ (\dot{x}_1 - \dot{x}_1^d) + K_1^1(x_1 - x_1^d) = 0 \Rightarrow \\ \dot{z}_1 + K_1^1 z_1 = 0\end{gathered} \tag{25}$$

Therefore, after the application of the feedback control law, the closed-loop dynamics becomes $\dot{z}_1 + K_1^1 z_1 = 0$, $\dot{z}_2 + K_1^2 z_2 = 0$, $\dot{z}_3 + K_1^3 z_3 = 0$, $\cdots$, $\dot{z}_i + K_1^i z_i = 0$, $\cdots$, $\dot{z}_{n-1} + K_1^{n-1} z_{n-1} = 0$, $\dot{z}_n + K_1^n z_n = 0$.

In matrix form, the closed-loop dynamics is written as

$$\begin{pmatrix} \dot{z}_1 \\ \dot{z}_2 \\ \dot{z}_3 \\ \cdots \\ \dot{z}_i \\ \cdots \\ \dot{z}_{n-1} \\ \dot{z}_n \end{pmatrix} = \begin{pmatrix} -K_1^1 \; 0 \; 0 \; \cdots \; 0 \; \cdots \; 0 \; 0 \\ 0 \; -K_1^2 \; 0 \; \cdots \; 0 \; \cdots \; 0 \; 0 \\ 0 \; 0 \; -K_1^3 \; \cdots \; 0 \; \cdots \; 0 \; 0 \\ \cdots \; \cdots \; \cdots \; \cdots \; \cdots \; \cdots \\ 0 \; 0 \; 0 \; \cdots \; -K_1^i \; \cdots \; 0 \; 0 \\ \cdots \; \cdots \; \cdots \; \cdots \; \cdots \; \cdots \\ 0 \; 0 \; 0 \; \cdots 0 \cdots \; -K_1^{n-1} \; 0 \\ 0 \; 0 \; 0 \; \cdots \; 0 \; \cdots \; 0 \; -K_1^n \end{pmatrix} \begin{pmatrix} z_1 \\ z_2 \\ z_3 \\ \cdots \\ z_i \\ \cdots \\ z_{n-1} \\ z_n \end{pmatrix} \tag{26}$$

or equivalently

$$\dot{Z} = KZ \tag{27}$$

By selecting the eigenvalues of matrix K to be in the left complex semiplane, one has that

$$lim_{t\to\infty} Z = 0_{n\times 1} \tag{28}$$

which also implies that $lim_{t\to\infty} x_1 = x_1^*$, $lim_{t\to\infty} x_2 = \alpha_1 = x_2^*$, $lim_{t\to\infty} x_3 = \alpha_2 = x_3^*$, $\cdots$, $lim_{t\to\infty} x_i = \alpha_{i-1} = x_i^*$, $\cdots$, $lim_{t\to\infty} x_{n-1} = \alpha_{n-2} = x_{n-1}^*$, and $lim_{t\to\infty} x_n = \alpha_{n-1} = x_n^*$.

To prove asymptotic stability for the proposed control scheme the following Lyapunov function can be defined

$$V = \sum_{i=1}^{N} \frac{1}{2} z_i^2 \tag{29}$$

The time derivative of the aforementioned Lyapunov function is

$$\dot{V} = \sum_{i=1}^{N} z_i \dot{z}_i \Rightarrow \dot{V} = -\sum_{i=1}^{N} K_1^i z_i^2 \Rightarrow \dot{V} < 0 \tag{30}$$

By selecting the feedback control gains K_1^i, $i = 1, \cdots, n$ to be $K_1^i > 0$, the asymptotic stability of the control loop is assured.

5. Simulation tests

Simulation tests have been performed for the evaluation of the proposed nonlinear control scheme, that is based on implementation of a flatness-based controller in successive loops. Indicative results where obtained for the tracking of three different setpoints. It can be observed that the SI engine's state vector elements (that is engine's turn speed ω, input pressure subjected to time delay P_{m_d}, and input pressure P_m) converged fast and smoothly to the associated setpoints. The associated results in the case of tracking of setpoint 1 are depicted in Fig. 2 to Fig. 3. Additional results for the case of tracking of setpoint 2 are shown in Fig. 4 to Fig. 5. Finally, the case of tracking of setpoint 3 which exhibits more rapid variations is depicted in Fig. 6 to Fig. 7.

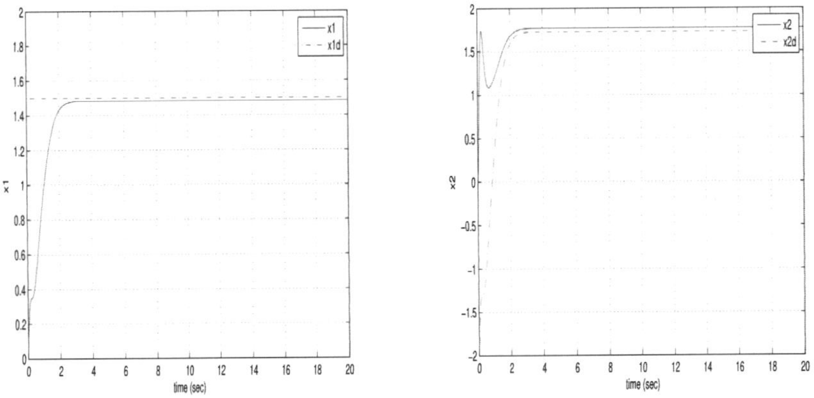

(a) (b)

Figure 2. Tracking of setpoint 1 from (a) state variables x_1, (b) state variables x_2

It can be also noted that the so-called backstepping control which is based on the recursive computation of the control signal of the system after applying virtual control inputs to the individual rows of the state-space model, can be completely substituted by the proposed flatness-based control method [Khalil, 1996], [Rigatos, 2011a],[Rigatos, 2011b],[Rigatos, 2006]. Moreover, it is noted that the proposed control method stands for a different implementation of flatness-based control in successive loops, in which there is no need for prior transformation of the system's dynamics in the canonical (trivial form) through the application of a change of variables (diffeomorphisms). Therefore, the control input is computed directly on the initial nonlinear state-space description of the system and not on an equivalent linearized form of it.

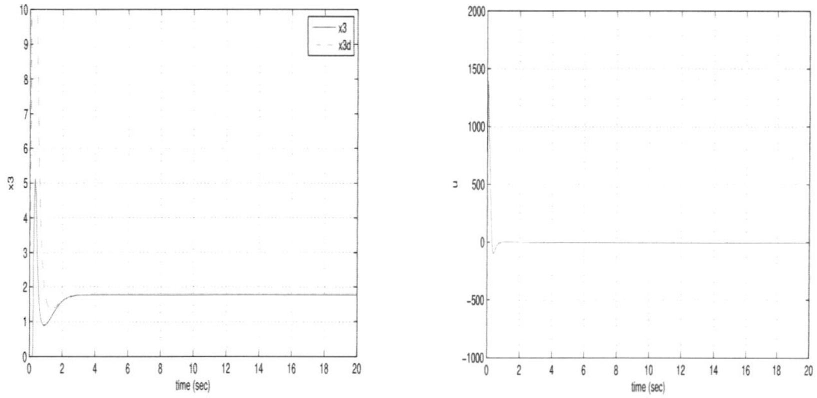

(a) **(b)**

Figure 3. (a) Tracking of setpoint 1 from state variables x_3, (b) control input u

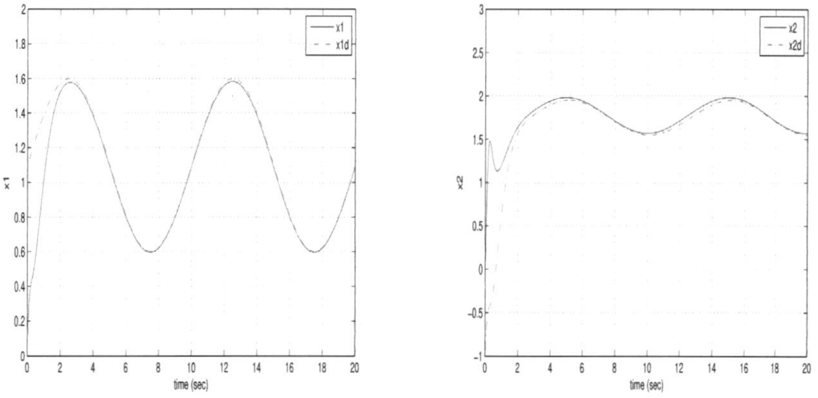

(a) **(b)**

Figure 4. Tracking of setpoint 2 from (a) state variables x_1, (b) state variables x_2

6. Conclusions

A nonlinear embedded control scheme for spark-ignited (SI) engines has been proposed in this article. The method is based on differential flatness theory. The method makes use of a description of the state-space model of the SI-engine in the so-called triangular form. In the proposed controller design method each row of the state-space model of the SI-engine is shown to be a differentially flat system, while the associated state variable is taken to be the flat output. Next, a virtual control input is computed for each subsystem which is obtained from a row of the state-space model. The virtual control input can invert the subsystem's dynamics while also eliminating the subsystem's tracking error. The control input that is actually applied to the SI engine is computed from the last row of the state-space description. This control input contains recursively all virtual control inputs which were found for the individual subsystems associated with the previous rows of the state-space equation. Thus, at each iteration of the

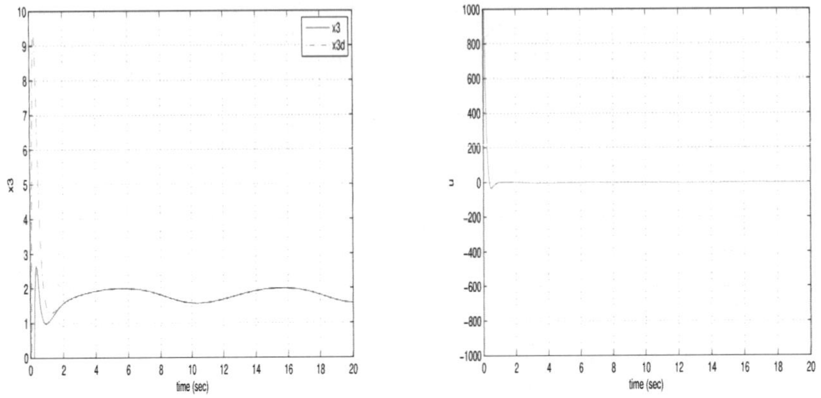

(a) **(b)**

Figure 5. (a) Tracking of setpoint 2 from state variables x_3, (b) control input u

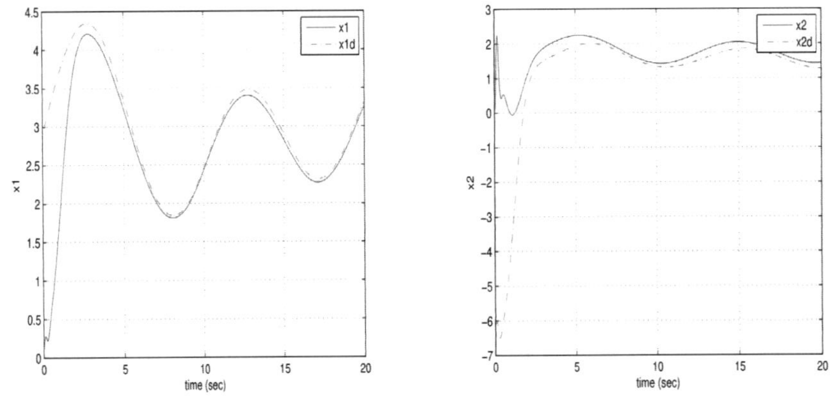

(a) **(b)**

Figure 6. Tracking of setpoint 3 from (a) state variables x_1, (b) state variables x_2

control algorithm and by tracing the rows of the state-space model backwards, one can finally obtain the control input that should be applied to the SI-engine so as to assure that all its state vector elements will converge to the desirable setpoints.

References

[Ali, 2006] Ali A. and Blath J.P. 2006 Nonlinear Torque Control of a Spark-Ignited Engine, *Proc. of the 2006 American Control Conference Minneapolis*, Minnesota, USA.

[Alipi, 2003] Alipi C. de Russion C. Piuri V. 2003 A neural network-based control solution to air-fuel ratio control for automotive fuel injection systems *IEEE Transactions on Systems, Man and Cybernetics - Part C: Applications and Reviews* **33** 259-268.

[Ballachi, 2013] Ballachi A. Benvenuti L. di Benedetto M.D. Sangiovanni-Vincentelli A. 2013 The design of dynamical observers for hybrid systems: theory and application of an automotive control problem *IEEE Transactions on Control Systems Technology* **49** 915-925.

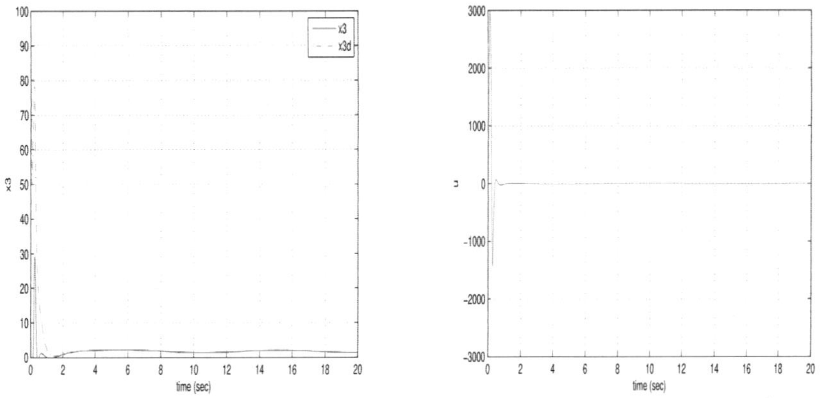

(a) **(b)**

Figure 7. (a) Tracking of setpoint 3 from state variables x_3, (b) control input u

[Basseville, 1993] Basseville M. Nikiforov I. 1993 *Detection of abrupt changes: Theory and Applications* Prentice-Hall.

[Blake Vance 2008] Blake Vance J. Kaul B.C. Jagannathan S. Drallmeier J.A. 2008 Output feedback controller for operation of spark ingition engines at lean conditions using neural networks *IEEE Transactions on Control Systems Technology* **16** 214-227.

[Bououden, 2011] Bououden S. Boutat D. Zheng G. Barbot J.P. Kratz F. 2011 A triangular canonical form for a class of 0-flat nonlinear systems *International Journal of Control, Taylor and Francis* **84** 261-269.

[Flärdh, 2014] Flärdh O. Ericsson G. Klingborg E. Martensson J. 2014 Optimal Air Path Control During Load Transients on a Spark Ignited Engine With Variable Geometry Turbine and Variable Valve Timing *IEEE Transactions on Control Systems Technology.*

[Fliess, 1999] Fliess M. Mounier H. 1999 Tracking control and π-freeness of infinite dimensional linear systems, In: G. Picci and D.S. Gilliam Eds. *Dynamical Systems, Control, Coding and Computer Vision* **258** 41-68 Birkhaüser.

[Khalil, 1996] Khalil H. 1996 *Nonlinear Systems* Prentice Hall.

[Leroy, 2009] Leroy T. Chauvin J. Petit N. 2009 Motion planning for experimental air path control of avariable-valve-timing spark ignition engine *Control Engineering Practice* Elsevier **17** 14321439.

[Colin, 2007] Colin G. Chamuillard Y. Bloch G. Corde G. 2007 Neural control of fast nonlinear systems: Application to a turbocharged SI engine with VCT *IEEE Transactions on Neural Networks* **18** 1101-1114.

[Laroche, 2007] Laroche B. Martin P. Petit N. 2007 Commande par platitude: Equations différentielles ordinaires et aux derivées partielles *Ecole Nationale Supérieure des Techniques Avancées* Paris.

[Lévine, 2011] Lévine J. 2011 On necessary and sufficient conditions for differential flatness, Applicable Algebra in Engineering, *Communications and Computing* Springer **22** 47-90.

[Martin, 1999] Martin P. Rouchon P 1999 Systèmes plats: planification et suivi des trajectoires *Journées X-UPS* École des Mines de Paris, Centre Automatique et Systèmes.

[Menhour, 2013] Menhour L. d'Andréa-Novel B. Fliess M. Mounier H. 2014 Coupled nonlinear vehicle control: Flatness-based setting with algebraic estimation techniques *Control Engineering Practice* **2** 135146.

[Moulin, 2011] Moulin P. Chauvin J. 2011 Modeling and control of the air system of a turbocharged gasoline engine, *Control Engineering Practice* **19** 287297.

[Nguyen, 2013] Nguyen A. Lauber J. Dambrine M. 2013 Multi-objective control design for turbo-charged spark ignited air system: a switching Takagi-Sugeno model approach *IEEE 2013 American Control Conference* Washington DC, USA.

[Rigatos, 2006] Rigatos G. Tzafestas S. 2006 Adaptive Fuzzy Control for the Ship Steering Problem *Journal of Mechatronics* Elsevier **16** 479-489.

[Rigatos, 2007] Rigatos G. Tzafestas S. 2007 Extended Kalman Filtering for Fuzzy Modelling and Multi-Sensor Fusion *Mathematical and Computer Modelling of Dynamical Systems* Taylor & Francis **13** 251-266.

[Rigatos, 2009] Rigatos G. Zhang Q. Fuzzy model validation using the local statistical approach *Fuzzy Sets and*

Systems, Elsevier **60** 882-904.

[Rigatos, 2011a] Rigatos G. (2011) *Modelling and control for intelligent industrial systems: Adaptive algorithms in robotics and industrial engineering*, Springer.

[Rigatos, 2011b] Rigatos G. 2011 Adaptive fuzzy control for field-oriented induction motor drives *Neural Computing and Applications* Springer **21** pp. 9-23.

[Rigatos, 2012] Rigatos G. 2012 A derivative-free Kalman Filtering approach to state estimation-based control of nonlinear dynamical systems *IEEE Transactions on Industrial Electronics* **59** 3987-3997.

[Rigatos, 2014] Rigatos G. Siano P. Zervos N. 2014 Sensorless Control of Distributed Power Generators With the Derivative-free Nonlinear Kalman Filter *IEEE Transactions on Industrial Electronics.*

[Rigatos, 2015] Rigatos G. 2015 *Nonlinear control and filtering using differential flatness approaches: applications to electromechanical systems*, Springer.

[Rigatos et al., 2015] Rigatos G. Siano P. Zervos N. 2015 A new concept on flatness-based control of nonlinear dynamical systems *IEEE INDIN 2015* Anglia Ruskin University, Cambridge, UK.

[Rouchon, 2005] Rouchon P 2005 Flatness-based control of oscillators *ZAMM - Journal of Applied Mathematics and Mechanics* **85** 411-421.

[Rudolph, 2003] Rudolph J. 2003 *Flatness Based Control of Distributed Parameter Systems, Examples and Computer Exercises from Various Technological Domains* Shaker Verlag Aachen.

[Sepuvelda, 2012] Sepulveda R. Montiel O. Castello O. Melin P. 2012 Embedding a high-speed interval type-2 fuzzy controller for a real plant into a FPGA *Applied Soft Computing*, Elsevier **12** 988-998.

[Sira-Ramirez, 2004] Sira-Ramirez H. Agrawal S. (2004) Differentially Flat Systems, *Marcel Dekker* New York.

[Sugihira, 2008] Sugihira S. Ohmori H. 2008 Model-based starting control of SI engines via adaptive feedback linearization, *SICE Annual Conference 2008* The University of Electro-Communications, Japan.

[Zhang, 2010] Zhang J. Shen T. Marino R. 2010 Model-based cold-start speed control scheme for spark ignition engines, *Control Engineering Practice*, Elsevier, **18** 1285-1294.

[Zhang, 2009] Zhang Z. and Sun Z. (2009) Rotational Angle Based Pressure Control of a Common Rail Fuel Injection System for Internal Combustion Engines, *2009 American Control Conference*, St. Louis, MO, USA.

A gain-scheduled observer under transmissions without delivery acknowledgment

Daniel Dolz, Ignacio Peñarrocha, and Roberto Sanchis

Department of Industrial System Engineering and Design, Universitat Jaume I of Castelló, Spain

E-mail: {ddolz,ipenarro,rsanchis}@uji.es

Abstract. This paper addresses the estimation problem for discrete-time systems where both measurements and control commands are sent to a central station through a lossy network without delivery acknowledgment. The central unit implements the estimation and control algorithms. We propose a jump observer that uses the expected value of the unknown control input at the actuator to run an open loop estimation. Then, the absence of acknowledgment in the control input transmission is dealt with through the introduction of an unknown disturbance. The observer update is performed by means of jumping gains when there are available measurements. We employ an statistic of the control error (new disturbance), i.e., the difference between the control inputs at the plant and at the observer, to schedule the observer gains in real time. The observer is designed to minimize the H_∞ norm from disturbances, measurement noises and control input errors, to estimation error. The infinite-dimensional design problem is turn into an optimization problem over polynomials using sum-of-squares decomposition techniques. Benefits of the proposal are shown in a simulation example.

1. Introduction

In networked control systems (NCS) the different elements of the control architecture are not collocated. The transfer of information (control inputs and measurements) between them is performed through a communication network, which increases flexibility and maintenance [1]. However, networks offer an unreliable communication link subject, e.g., to dropouts and time delays [2,3].

Focusing on the dropout case, communication protocols can be divided in two groups: 1) TCP-like, which implement a successful delivery acknowledgment and 2) UDP-like, where there is no reception acknowledgment. Clearly, this last protocol alleviates the network data traffic at the expense of introducing some new challenging issues.

Dealing with packet losses in the measurement channel (from sensors to controller/observer) does not depend on the use of delivery acknowledgment. Using jump observers relating their jumping modes to the measurement availability leads to better performances than implementing gain invariant approaches [4,5]. These approaches may obtain similar performances than the ones based on Kalman filters [6] with less computational effort, and can handle a larger class of disturbances.

However, the absence of a delivery acknowledgment is a critical issue when affecting control input transmission. In [7,8] the authors showed that under UDP-like transmission the separation principle does not hold when designing LQG controllers since the estimation algorithm depends on the control input reception, which is unknown. Several works have studied the jointly design of estimators and controller under this paradigm, e.g. [9–11]. Another approach to solve this problem is the one studied in [12, 13] where the authors proposed to estimate the transmission fate so the system fulfills the separation principle as in TCP-like communication.

In the present work, we introduce a different approach to tackle this problem in the remote state estimation problem under dropouts in UDP-like networks. We use the expected value of the received control input at the plant over a finite time window to run the open loop estimation. Then, we model the absence of control reception acknowledgment as the presence of a new unknown disturbance, which we call control error, that affects the estimation algorithm. With that, we obtain a control error statistic, calculated in real time, that can be described by a bounded time-varying parameter.

Some works as [14] started developing this idea but with invariant gains, which leads to conservative estimation performances. Here, we propose a gain-scheduled jump observer to estimate the system states. The observer gain jumping modes are related to the measurement reception scenario, modeled as a Markov chain, and depend on the control error statistic. The most significant contribution of this paper is the design of an observer whose gains are scheduled with the unknown disturbance introduced by the control input losses, i.e., with the control error statistic. Moreover, the presented results could allow to apply the separation principle when closing the loop if the controller is designed to satisfy the considered bounds on the control error statistic.

The gain-scheduled jump design is addressed by minimizing the H_∞ norm from external inputs to estimation error. To handle this optimization procedure, we impose the gain-scheduling function to be polynomial so we can exploit sum-of-squares (SOS) decomposition techniques (see [15]). Similar to [16] the conceptual originality introduced with respect previous works using SOS methods (e.g. [17,18]) resides in scheduling the observer with a time-varying parameter that depends on the behavior of the network.

Notation : Let $\mathbb{R}$ and $\mathbb{N}$ denote the real and positive integer numbers set. Let A and B be some matrices. $A \preceq 0$ means that matrix A is negative semidefinite. Similar applies to $\prec$, $\succ$ and $\succeq$. The direct sum is represented as $\bigoplus$, where $A \bigoplus B$ is a block diagonal matrix with A and B on its diagonal. Let $x_k \in \mathbb{R}^n$ be a stochastic process. $\mathbf{E}\{\cdot\}$ and $\mathbf{Pr}\{\}$ denotes expectation and probability. We denote the RMS norm of process x_k by $\|x\|_{\text{RMS}} = \lim_{K\to\infty} \sqrt{\frac{1}{K}\sum_{k=0}^{K-1} x_k^T x_k}$.

2. Problem setup

We consider the remote state estimation problem of a linear time-invariant discrete-time system where measurements and control inputs are transmitted through a communication network without successful delivery acknowledgment of received packets (e.g. UDP-like networks) subject to dropouts. We describe the system in its state-space form as

$$x_{k+1} = A\,x_k + B_u\,u_k + B_w\,w_k, \tag{1a}$$

$$y_k = C\,x_k + v_k \tag{1b}$$

where $x_k \in \mathbb{R}^n$ is the state at instant $k \in \mathbb{N}$, $u_k \in \mathbb{R}^{n_u}$ is the applied control input, $w_k \in \mathbb{R}^{n_w}$ is the state disturbance, $y_k \in \mathbb{R}^{n_y}$ is the measured output and $v_k \in \mathbb{R}^{n_y}$ is the measurement noise. We assume that both state disturbances and measurement noise are wide-sense stationary stochastic processes[1] with bounded variances. Then their RMS norms are such as

$$\|w\|_{\text{RMS}} \leq \bar{w}_{\text{rms}}, \quad \|v\|_{\text{RMS}} \leq \bar{v}_{\text{rms}}.$$

Let us also assume that the sent control input at each instant is known as the observer and controller are collocated in the central station.

[1] If x_k is wide sense stationary its RMS norm becomes $\|x\|_{\text{RMS}} = \sqrt{\mathbf{E}\{x_k^T x_k\}}$.

2.1. Measurement reception description

Each sensor s samples at instant k its output, synchronously with the control input update, and sends independently to the central station a time-tagged message with the correspondent measurement $y_{s,k}$. We characterize the individual measurement reception from sensor s at instant k with

$$\alpha_{s,k} = \begin{cases} 1 & \text{if } y_{s,k} \text{ is acquired at instant } k, \\ 0 & \text{otherwise.} \end{cases} \tag{2}$$

Aggregating the above binary variables such as $\alpha_k = \bigoplus_{s=1}^{n_m} \alpha_{s,k}$, we model the full measurement reception scenario at instant k. Let us consider that α_k follows a finite Markov chain having its states in the set

$$\alpha_k \in \Xi = \{\eta_0, \eta_1, \ldots, \eta_r\}, \ r = 2^{n_m} - 1, \tag{3}$$

where η_0 represents the case when $\alpha_k = 0$. We assume that the transition probability matrix $\Lambda = [p_{i,j}]$ with

$$p_{i,j} = \mathbf{Pr}\{\alpha_{k+1} = \eta_j | \alpha_k = \eta_i\}$$

is known.

Remark 1. *Considering mutually independent Markovian dropouts (see [19]), i.e.,*

$$\mathbf{Pr}\{\alpha_{s,k} = 0|\alpha_{s,k-1} = 0\} = q_s^y, \quad \mathbf{Pr}\{\alpha_{s,k} = 1|\alpha_{s,k-1} = 0\} = 1 - q_s,$$
$$\mathbf{Pr}\{\alpha_{s,k} = 1|\alpha_{s,k-1} = 1\} = p_s^y, \quad \mathbf{Pr}\{\alpha_{s,k} = 0|\alpha_{s,k-1} = 1\} = 1 - p_s,$$

for all $s = 1, \ldots, n_y$, the transition probabilities of α_k are calculated as

$$p_{i,j} = \prod_{s=1}^{n_m} \mathbf{Pr}\{\alpha_{s,k} = \eta_{s,j} | \alpha_{s,k-1} = \eta_{s,i}\}$$

where $\eta_{s,i}$ is the s-th diagonal element of η_i. ■

2.2. Control update description

The controller transmits (through the network) at each instant $k-1$ a single packet that aggregates all the control commands to be used at instant k (see [20]). Let us denote by u_k^t the known sent control input at $k-1$ to be applied at k. We describe the control input reception at instant $k-1$ with

$$\theta_{k-1} = \begin{cases} 1 & \text{if } u_k^t \text{ is received at instant } k-1, \\ 0 & \text{otherwise.} \end{cases} \tag{4}$$

We consider that the actuators implement a zero order hold strategy, i.e., the last received control input is applied if no new one is acquired.

Let us assume that we know the probability of employing at instant k the control input sent $\tau + 1$ instants before, i.e.

$$\varphi_\tau = P\{u_k = u_{k-\tau}^t\}, \quad \tau = 0, \ldots, N^u, \quad \sum_{\tau=0}^{N^u} \varphi_\tau = 1, \tag{5}$$

where N^u is the maximum number of consecutive control packet dropouts .

Remark 2. *Considering again mutually independent Markovian dropouts characterized by*

$$\mathbf{Pr}\{\theta_k = 0|\theta_{k-1} = 0\} = q^u, \quad \mathbf{Pr}\{\theta_k = 1|\theta_{k-1} = 0\} = 1 - q^u,$$
$$\mathbf{Pr}\{\theta_k = 1|\theta_{k-1} = 1\} = p^u, \quad \mathbf{Pr}\{\theta_k = 0|\theta_{k-1} = 1\} = 1 - p^u.$$

and assuming that the control link can assure that the maximum number of consecutive dropouts is below N^u*, the probabilities* (5) *can be computed as*

$$\varphi_0 = \frac{1}{\xi}\pi_1^u, \quad \varphi_{\tau>0} = \frac{1}{\xi}(q^u)^{\tau-1}(1-p^u)\,\pi_1^u$$

where $\pi_1^u = \mathbf{Pr}\{\theta_k = 1\}$,

$$\xi = \pi_1^u + \sum_{\tau=1}^{N^u}(q^u)^{\tau-1}(1-p^u)\,\pi_1^u, \quad \pi^u = \pi^u\Lambda^u,$$

$\pi^u = [\pi_0^u\ \pi_1^u]$ *(being* $\pi_0^u = \mathbf{Pr}\{\theta_k = 0\}$*) and* Λ_u *is the associated transition probability matrix of* θ_k*.* ■

Since the current control input value acting on the plant is unknown, we propose employing its expectation $\mathbf{E}\{u_k\}$ to run the open loop estimation. We denote $\mathbf{E}\{u_k\}$ by u_k^c where

$$u_k^c = \sum_{d=0}^{N^u}\varphi_d\, u_{k-d}^t. \tag{6}$$

The next lemma defines some statistics of the involved control error $\tilde{u}_k = u_k - u_k^c$ (between the applied control input in the plant and the one assumed in the observer).

Lemma 1. *The control error* $\tilde{u}_k$ *has a zero expected value and*

$$\mathbf{E}\{\tilde{u}_k^T\tilde{u}_k\} = \sum_{d=0}^{N^u}\varphi_d\left(u_{k-d}^t - \sum_{d=0}^{N^u}\varphi_d\, u_{k-d}^t\right)^T\left(u_{k-d}^t - \sum_{d=0}^{N^u}\varphi_d\, u_{k-d}^t\right). \tag{7}$$

Proof. First, the expected value of $\tilde{u}_k$ is zero by definition of u_k^c, see (6). Second, using the total probability law we obtain (7). □

Let us denote $\mathbf{E}\{\tilde{u}_k^T\tilde{u}_k\}$ by δ_k. The value of δ_k can be computed at each time instant with (7), and therefore is a known value. Let us assume that δ_k is a bounded time-varying parameter such that $\delta_k \in \mathcal{S}$ with

$$\mathcal{S} = \{\delta_k : 0 \leq \delta_k \leq \bar{\delta},\ \forall k\} \tag{8}$$

where $\bar{\delta}$ is a known value or, if not, can be used as a tuning parameter as we will see later.

2.3. Estimation algorithm

Considering the previous description, the estimation algorithm is:

$$\hat{x}_{k^-} = A\,\hat{x}_{k-1} + B_u\, u_{k-1}^c, \tag{9a}$$
$$\hat{x}_k = \hat{x}_{k^-} + L_k\,\alpha_k\,(y_k - C\,\hat{x}_{k^-}). \tag{9b}$$

At instant k, we compute an open loop estimation (9a) employing the expected value of the control command at the plant u_{k-1}^c. Then, when possible, we amend the estimation with the received measurements (nonzero diagonal elements of α_k) through the gain matrix L_k, see (9b).

The evolution on the state estimation error $\tilde{x}_k = x_k - \hat{x}_k$ is

$$\tilde{x}_k = (I - L_k\,\alpha_k\,C)\,(A\,\tilde{x}_{k-1} + B_{\mathcal{W}}\,\mathcal{W}_{k-1}) - L_k\,\alpha_k\,v_k \tag{10}$$

where

$$B_{\mathcal{W}} = [B_w \quad B_u], \quad \mathcal{W}_{k-1} = [w_{k-1}^T \quad \tilde{u}_{k-1}^T]^T.$$

Note that the control error is viewed by the observer as a new unknown disturbance.

In this work we aim to compute gain matrices L_k to minimize the estimation error under state disturbance, measurement noises and the uncertainty due to the control input dropouts. Previous works have shown that jump estimators whose jumping modes depend on the measurement reception scenario improve estimation performance [4]. Following that idea and in order to avoid conservativeness with respect to control errors, we propose a jump observer whose gains are scheduled in real time with δ_k. The gain-scheduling law is

$$L_k = \begin{cases} L_i(\Delta_{k-1}) & \text{if } \alpha_k = \eta_i \text{ for } i = 1, \dots, r, \\ 0 & \text{if } \alpha_k = \eta_0. \end{cases} \tag{11}$$

where

$$\Delta_{k-1} = \begin{bmatrix}\delta_{k-1} & \delta_{k-2}\end{bmatrix}^T. \tag{12}$$

Note that L_k is scheduled with both the current control error statistic δ_{k-1} (at instant k) and the previous one δ_{k-2} (at instant $k-1$). This allows us to include in the estimation algorithm the present control uncertainty δ_{k-1} and also its variation from δ_{k-2} to δ_{k-1}.

3. Observer design

Let us address in this section the design of the gain-scheduled jump observer $L(\alpha_k, \Delta_{k-1})$ using an H_∞-based method. First, we derive a sufficient condition for its existence based on parameter-dependent matrix inequalities (PDMI) related to Δ_k. Second, constraining those dependences to be polynomial, we present a SOS-based procedure to solve the design problem in polynomial time.

The next theorem shows the H_∞ observer design based on PDMI.

Theorem 1. *Consider the estimation algorithm (9) over system (1). If there exist positive definite symmetric matrices $P_j(\Delta_k) \in \mathbb{R}^{n\times n}$, full matrices $G_j(\Delta_k) \in \mathbb{R}^{n\times n}$ and $X_j(\Delta_k) \in \mathbb{R}^{n\times n_y}$, and positive real scalar functions $\gamma_w(\Delta_k)$, $\gamma_v(\Delta_k)$ and $\gamma_u(\Delta_k)$ for all $i, j = 0, \dots, r$ and $\delta_k \in \mathcal{S}$ fulfilling*

$$\Upsilon_i(\Delta_k, \Delta_{k-1}) = \begin{bmatrix} \Omega & \bar{M}_{A,i} & \bar{M}_{B,i} \\ \bar{M}_{A,i}^T & P_i(\Delta_{k-1}) - I & 0 \\ \bar{M}_{B,i}^T & 0 & \Gamma \end{bmatrix} \succ 0 \tag{13}$$

with

$$\begin{aligned} \Omega &= \bigoplus_{j=0}^{r} G_j(\Delta_k) + G_j(\Delta_k)^T - P_j(\Delta_k), \\ \bar{M}_{A,i} &= \begin{bmatrix}\sqrt{p_{i,0}}M_{A,0}^T & \cdots & \sqrt{p_{i,r}}M_{A,r}^T\end{bmatrix}^T, \\ M_{A,j} &= (G_j(\Delta_k) - X_j(\Delta_k)\,\eta_j\,C)\,A, \\ \bar{M}_{B,i} &= \begin{bmatrix}\sqrt{p_{i,0}}M_{B,0}^T & \cdots & \sqrt{p_{i,r}}M_{B,r}^T\end{bmatrix}^T, \\ M_{B,j} &= \begin{bmatrix}(G_j(\Delta_k) - X_j(\Delta_k)\,\eta_j\,C)B_{\mathcal{W}} & -X_j(\Delta_k)\eta_j\end{bmatrix}, \\ \Gamma &= \gamma_w(\Delta_k)\,I_{n_w} \oplus \gamma_u(\Delta_k)\,I_{n_u} \oplus \gamma_v(\Delta_k)\,I_{n_m} \end{aligned}$$

Journal of Physics: Conference Series **659** (2015) 012020 doi:10.1088/1742-6596/659/1/012020

and $B_{\mathcal{W}} = [B_w \;\; B_u]$, then, defining the observer gains as $L_i(\Delta_k) = G_i(\Delta_k)^{-1} X_i(\Delta_k)$, the next assertions hold for all $\alpha_k \in \Xi$, $\delta_k \in \mathcal{S}$, $\|w\|_{\mathrm{RMS}} \le \bar{w}_{\mathrm{rms}}$ and $\|v\|_{\mathrm{RMS}} \le \bar{v}_{\mathrm{rms}}$: (i) under no disturbances, measurement noises and control errors, system (10) is asymptotically mean square stable; (ii) under zero initial conditions, the estimation error is bounded by

$$\mathbf{E}\{\|\tilde{x}\|_{\mathrm{RMS}}^2\} < \bar{\gamma}_w \bar{w}_{\mathrm{rms}}^2 + \bar{\gamma}_v \bar{v}_{\mathrm{rms}}^2 + \bar{\gamma}_u, \tag{14}$$

where

$$\bar{\gamma}_u = \lim_{K\to\infty} \frac{1}{K} \sum_{k=0}^{K-1} \gamma_u(\Delta_k)\, \delta_k, \quad \bar{\gamma}_w = \lim_{K\to\infty} \frac{1}{K} \sum_{k=0}^{K-1} \gamma_w(\Delta_k), \quad \bar{\gamma}_v = \lim_{K\to\infty} \frac{1}{K} \sum_{k=0}^{K-1} \gamma_v(\Delta_k). \tag{15}$$

Proof. If (13) holds, then $G_j(\Delta_k)+G_j(\Delta_k)^T-P_j(\Delta_k) \succ 0$ and therefore, $G_j(\Delta_k)$ is nonsingular. Since $P_j(\Delta_k)$ is a symmetric positive definite matrix, then

$$(\, P_j(\Delta_k) - G_j(\Delta_k))\; P_j(\Delta_k)^{-1}\, (\, P_j(\Delta_k) - G_j(\Delta_k))^T \succeq 0,$$

which means that

$$G_j(\Delta_k) + G_j(\Delta_k)^T - P_j(\Delta_k) \preceq G_j(\Delta_k)\, P_j(\Delta_k)^{-1}\, G_j(\Delta_k)^T.$$

Now, let us define the Lyapunov function as

$$V_k = V(\tilde{x}_k, \alpha_k, \Delta_{k-1}) = \tilde{x}_k^T\, P_i(\Delta_{k-1})\, \tilde{x}_k$$

for $\alpha_k = \eta_i$ and $i = 0, \ldots, r$. Remembering that $X_j(\Delta_k) = G_j(\Delta_k)\, L_j(\Delta_k)$, applying a congruence transformation on (13) by matrix $\left(\bigoplus_{j=0}^{r} G_j(\Delta_k)^{-1}\right) \oplus I \oplus I \oplus I$, taking Schur's complements and premultiplying the result by $\begin{bmatrix}\tilde{x}_k^T & w_k^T & \tilde{u}_k^T & v_{k+1}^T\end{bmatrix}$ and postmultiplying by its transpose, we obtain that

$$\mathbf{E}\{V_{k+1}|\alpha_k = \eta_i\} - V_k + \tilde{x}_k^T \tilde{x}_k - \gamma_w(\Delta_k) w_k^T w_k - \gamma_u(\Delta_k) \tilde{u}_k^T \tilde{u}_k - \gamma_v(\Delta_k) v_{k+1}^T v_{k+1} < 0 \tag{16}$$

for all $i = 0, \ldots, r$ and $\delta_k \in \mathcal{S}$.

(i) Under no disturbances ($w_k = 0$), control errors ($\tilde{u}_k = 0$) and measurement noises ($v_k = 0$), (16) leads to $\mathbf{E}\{V_{k+1}|\alpha_k = \eta_i\} - V_k < 0$ for all $i = 0, \ldots, r$ and $\delta_k \in \mathcal{S}$, which guarantees asymptotically mean square stability. This proves the first statement.

(ii) Let us write $\mathbf{E}\{V_{k+1}|\alpha_k = \eta_i\}$ as $\mathbf{E}\{V_{k+1}|\alpha_k\}$. Taking conditional expectation given α_{k-1} over (16), remembering that α_k is known at instant k, we get

$$\begin{aligned} &\mathbf{E}\{V_{k+1}|\alpha_k\} - \mathbf{E}\{V_k\} + \mathbf{E}\{\tilde{x}_k^T \tilde{x}\} - \gamma_w(\Delta_k)\, \bar{w}_{\mathrm{rms}}^2 \\ &\quad - \gamma_u(\Delta_k)\, \delta_k - \gamma_f(\Delta_k)\, \Delta f_k^T\, \Delta f_k - \gamma_v(\Delta_k)\, \bar{v}_{\mathrm{rms}}^2 < 0 \end{aligned} \tag{17}$$

for all $\alpha_k \in \Xi$ and $\delta_k \in \mathcal{S}$, where we have considered the assumptions on w_k and v_k, that $\delta_k = \mathbf{E}\{\tilde{u}_k^T\, \tilde{u}_k\}$ and that $\|w\|_{\mathrm{RMS}}^2 \le \bar{w}_{\mathrm{rms}}^2$, $\|v\|_{\mathrm{RMS}}^2 \le \bar{v}_{\mathrm{rms}}^2$.

For brevity reasons, let us omit in the following the fact that the inequalities are fulfilled for all $\alpha_k \in \Xi$ and $\delta_k \in \mathcal{S}$. Under zero initial conditions ($V_0 = 0$), adding the above expression from $k = 0$ to $K - 1$, dividing the result by K and taking the limit when K tends to infinity we obtain that

$$\lim_{K\to\infty} \frac{1}{K} \sum_{k=0}^{K-1} \mathbf{E}\{\tilde{x}_k^T \tilde{x}_k\} < \bar{\gamma}_w \bar{w}_{\mathrm{rms}}^2 + \bar{\gamma}_u + \bar{\gamma}_v \bar{v}_{\mathrm{rms}}^2 \tag{18}$$

where we have considered that $\mathbf{E}\{V_{K+1}|\alpha_K\} > 0$, that $\bar{\gamma}_{\_}$ is as given in (15) and that

$$\lim_{K\to\infty} \frac{1}{K} \sum_{t=0}^{K-1} (\mathbf{E}\{V_k|\alpha_{k-1}\} - \mathbf{E}\{V_k\}) = 0.$$

Then, expression (18) leads to (14), which ends this proof. □

Note that gains $\gamma_{\_}(\Delta_k)$ (with $\gamma_{\_} = \{\gamma_w, \gamma_u, \gamma_v, \gamma_f\}$) depend on Δ_k to improve the characterization of the propagation of the state disturbances, control errors and measurements noises to the estate estimation error for all $\delta_k \in \mathcal{S}$.

Remark 3. *The proposed observer design method would allow to apply the separation principle when closing the loop with an observer-based controller whenever the controller is designed to assure a control error* $\delta_k \leq \bar{\delta}$. ■

3.1. SOS decomposition

Considering conditions of Theorem 1 leads to an infinite-dimensional problem. To work with a finite-dimensional one, we impose matrices and scalar functions to be polynomial functions of δ_k of fixed degree. Then, SOS decompositions offer a computationally tractable approach.

The following lemmas extracted from [15] define SOS polynomials and show that checking whether a polynomial matrix is nonnegative over a domain characterized by polynomial constraints can be restated with sufficient linear matrix inequality (LMI) conditions.

Lemma 2. *Let* $p(x)$ *be* 2*d-order polynomial in* $x \in \mathbb{R}^n$. *Let* $x^{\{d\}}$ *be a vector with all the monomials in* x *of degree* $\leq d$. *Then,* $p(x)$ *is said to be SOS if and only if there is a positive semidefinite matrix* Q *fulfilling* $p(x) = (x^{\{d\}})^T Q x^{\{d\}}$. *The set of SOS polynomials in* x *is denoted by* $\Sigma(x)$.

Lemma 3. *Let* $p(x)$ *be a polynomial in* $x \in \mathbb{R}^n$, *and let* $X = \{x \in \mathbb{R}^n : g_j(x) \geq 0, j = 1, \ldots, m\}$. *Suppose there exist SOS polynomials* $s_j(x) \in \Sigma(x)$ $(j = 1, \ldots, m,\ x \in \mathbb{R}^n)$ *fulfilling* $p(x) - \sum_{j=1}^{m} s_j(x)\, g_j(x) \in \Sigma(x)$, *then, the following condition holds:* $p(x) \geq 0,\ \forall\, x \in X$.

Lemma 4. *Let* $P(x) \in \mathbb{R}^{N\times N}$ *be a symmetric polynomial matrix in* $x \in \mathbb{R}^n$ *and let* $X = \{x \in \mathbb{R}^n : g_j(x) \geq 0, j = 1, \ldots, m\}$. *Suppose there exist SOS polynomials* $s_j(x,v) \in \Sigma(x,v)$ $(j = 1, \ldots, m)$ *fulfilling* $v^T P(x) v - \sum_{j=1}^{m} s_j(x,v)\, g_j(x) \in \Sigma(x,v)$ *with* $v \in \mathbb{R}^N$, *then, the following condition holds:* $P(x) \succeq 0,\ \forall\, x \in X$.

The next theorem derives sufficient conditions to obtain the parametric matrices and functions that guarantee the properties stated in Theorem 1. We use δ_1, δ_2, δ_3, $\Delta_1 = \begin{bmatrix}\delta_1 & \delta_2\end{bmatrix}^T$ and $\Delta_2 = \begin{bmatrix}\delta_2 & \delta_3\end{bmatrix}^T$ to denote independent SOS variables representing the possible values of δ_k, δ_{k-1}, δ_{k-2}, Δ_k and Δ_{k-1} respectively where $\delta_k \in \mathcal{S}$ for all k.

Theorem 2. *If there exist symmetric polynomial matrices*

$$P_i(\Delta_1) = \left(\Delta_1^{\{d_P\}} \otimes I_n\right)^T P_i \left(\Delta_1^{\{d_P\}} \otimes I_n\right),$$
$$P_i(\Delta_2) = \left(\Delta_2^{\{d_P\}} \otimes I_n\right)^T P_i \left(\Delta_2^{\{d_P\}} \otimes I_n\right)$$

with real symmetric matrices P_i *for* $i = 0, \ldots, r$, *polynomial matrices*

$$G_i(\Delta_1) = G_i \left(\Delta_1^{\{d_G\}} \otimes I_n\right), \quad X_i(\Delta_1) = X_i \left(\Delta_1^{\{d_P\}} \otimes I_{n_m}\right)$$

with real matrices G_i and X_i, and polynomial functions

$$\gamma_w(\Delta_1) = \gamma_w^T \Delta_1^{\{d_\gamma\}}, \quad \gamma_u(\Delta_1) = \gamma_u^T \Delta_1^{\{d_\gamma - 1\}}, \quad \gamma_v(\Delta_1) = \gamma_v^T \Delta_1^{\{d_\gamma\}}$$

with real vectors γ_w, γ_u and γ_v, where $2d_P$, d_G, d_X, d_γ, $d_\gamma - 1$ and d_f are the degrees of the involved polynomials, that fulfill the following constraints

$$\mu^T P_i(\Delta_1)\,\mu - s_{P1,i}\,h_1 - s_{P2,i}\,h_2 \in \Sigma(\Delta_1, \mu), \tag{19a}$$
$$s_{P1,i} \in \Sigma(\delta_1, \mu), \quad s_{P2,i} \in \Sigma(\delta_2, \mu), \tag{19b}$$
$$\nu^T \Upsilon_i(\Delta_1, \Delta_2)\,\nu - s_{\Upsilon 1,i}\,h_1 - s_{\Upsilon 2,i}\,h_2 - s_{\Upsilon 3,i}\,h_3 \in \Sigma(\Delta_1, \Delta_2, \nu) \tag{19c}$$
$$s_{\Upsilon 1,i} \in \Sigma(\delta_1, \nu), \ s_{\Upsilon 2,i} \in \Sigma(\delta_2, \nu), \ s_{\Upsilon 3,i} \in \Sigma(\delta_3, \nu), \tag{19d}$$
$$\gamma_j(\Delta_1) - s_{j1}\,h_1 - s_{j2}\,h_2 \in \Sigma(\Delta_1), \tag{19e}$$
$$s_{j1} \in \Sigma(\delta_1), \quad s_{j2} \in \Sigma(\delta_2), \quad j = \{w,\ u,\ f,\ v\}, \tag{19f}$$

for $i = 0, \ldots, r$ with $\Upsilon_i(\cdot)$ as in (13), μ and ν real vectors of corresponding dimensions, and

$$h_m = \delta_m\,(\bar{\delta} - \delta_m), \quad m \in \{1, 2, 3\}, \tag{20}$$

then constraints of Theorem 1 are fulfilled.

Proof. Let us note first that the set $\mathcal{S}$ is reformulated with its corresponding polynomial h as $\mathcal{S} = \{\delta_m : h_m \geq 0\}$, see (20). Second, by Lemma 3 and Lemma 4 constraints (19c) and (19d) assure the positive definiteness of $\Upsilon_i(\Delta_1, \Delta_2)$ for any $\delta_1, \delta_2, \delta_3 \in \mathcal{S}$, which guarantee (13) in Theorem 1. Similar apply to constraints (19a), (19b), (19e) and (19f) which guarantee the positive definiteness of $P_i(\Delta_1)$, $\gamma_w(\Delta_1)$, $\gamma_u(\Delta_1)$, and $\gamma_v(\Delta_1)$ for any $\delta_1, \delta_2 \in \mathcal{S}$, as required in Theorem 1. □

In the previous feasibility SOS problem μ and ν are scalarization vectors employed to turn polynomial matrices into polynomials (see Lemma 4). The decision variables are matrices P_i, G_i, X_i, γ_w, γ_u and γ_v; as well as the coefficients of SOS polynomials $s_{\_}$ in (19). We select the degree of the SOS polynomials to have the same degree on all the variables in the corresponding SOS expression. Choosing d_P, d_G, d_X, and d_γ , this can be obtained by imposing for all $j = 1, 2$ and $i = 0, \ldots, r$

$$\begin{aligned}
\deg\, s_{Pj,i} &= \deg\left\{\delta_j^{\max\{2d_P - 2,\, 0\}},\ \mu^2\right\}, \\
\deg\, s_{\Upsilon j,i} &= \deg\left\{\delta^{\max\{2d_P - 2,\, d_G - 2,\, d_X - 2, 0\}},\ \nu^2\right\}, \\
\deg\, s_{\Upsilon 3,i} &= \deg\left\{\delta^{\max\{2d_P - 2, 0\}},\ \nu^2\right\}, \\
\deg\, s_{wj} &= \deg\, s_{vj}(\delta_j) = \deg\, \delta_j^{\max\{d_\gamma - 2,\, 0\}}, \\
\deg\, s_{uj} &= \deg\, \delta_j^{\max\{d_\gamma - 3,\, 0\}}
\end{aligned}$$

where deg gives the maximum degree for each variable.

3.2. Optimization-based design

In order to minimize the estimation error, one possibility could be to minimize the right hand side of expression (14) for any k and $\delta_k \in \mathcal{S}$. This would lead to the worst case estimation performance. In order to obtain less conservative results, we propose to introduce some weighting function $g(\Delta_1)$ such that

$$\bar{\Gamma} = \int_{\mathcal{S}\times\mathcal{S}} g(\Delta_1)\Gamma_\chi \mathrm{d}\Delta_1 \tag{21}$$

with

$$\Gamma_\chi = \gamma_w(\Delta_1)\bar{w}_{\mathrm{rms}}^2 + \gamma_v(\Delta_1)v_{\mathrm{rms}}^2 + \gamma_u(\Delta_1)\delta_1.$$

Then, the minimization problem

$$\begin{aligned} \text{minimize} \quad & J \\ \text{subject to} \quad & (19), \\ & \bar{\Gamma} < J, \end{aligned} \tag{22}$$

minimizes the RMS norm of the state estimation error of the resulting observer under the weighting function $g(\Delta_1)$.

Remark 4. *A thoughtful choice for $g(\Delta_1)$ is the one that considers the scenario when δ_k is time invariant (constant or ramp-like transmitted control inputs, see (7)), i.e., $\delta_1 = \delta_2$, and none of the possible values of $0 \le \delta_k \le \bar{\delta}$ dominates over the others (i.e., assuming no knowledge about which value is more likely). Thus,*

$$g(\Delta_1) = \begin{cases} \frac{1}{\bar{\delta}} & \text{if } \delta_2 = \delta_1 \text{ and } 0 \le \delta_1 \le \bar{\delta}, \\ 0 & \text{otherwise.} \end{cases} \tag{23}$$

4. Examples

For ease of analysis, let us study a linear time invariant discrete time system described by matrices

$$A = \begin{bmatrix} 0.48 & 0.11 \\ 0.11 & 0.97 \end{bmatrix}, \quad B_u = \begin{bmatrix} -0.5 \\ 0.7 \end{bmatrix}, \quad B_w = \begin{bmatrix} 0.2 \\ -0.6 \end{bmatrix}, \quad C = \begin{bmatrix} 0.18 & 0.8 \end{bmatrix},$$

where there is only one measurable output. The state disturbances and measurement noises are Gaussian noises with zero mean and bounded RMS norms given by

$$w_{\mathrm{rms}} = 0.05, \quad v_{\mathrm{rms}} = 0.01.$$

The control input is assumed to be defined by

$$u_k = 2.7\sin(k/11).$$

The transition probability matrix of the possible measurement reception scenario is characterized by $q^y = 0.8$ and $p^y = 0.3$ (see Remark 1). The control input dropouts follow a modified version of a Markov chain with parameters $q^u = 0.5$ and $p^u = 0.4$ with a maximum number of consecutive dropouts of $N^u = 6$ (see Remark 2). Taking that into account, the probabilities of applying at the plant the control input sent $\tau - 1$ instants before are

$$\varphi = \begin{bmatrix} 0.668 & 0.2 & 0.08 & 0.032 & 0.0128 & 0.005 & 0.002 \end{bmatrix},$$

where $\varphi = [\varphi_0 \ \varphi_1 \ \varphi_2 \ \varphi_3 \ \varphi_4 \ \varphi_5 \ \varphi_6]$. Let us consider that in the worst case the control error statistic δ_k is bounded by $0 \le \delta_k \le 0.1$ for all k, i.e. $\bar{\delta} = 0.1$.

In this example, we analyze the performance of the proposed gain-scheduled jump observer obtained using second order polynomials ($d_P = 1$, $d_G = d_X = d_\gamma = 2$) with the weighting function in (23) (see Section 3) and the resulting one from imposing zero order polynomials ($d_P = d_G = d_X = d_\gamma = 0$, i.e. gains that do not depend on δ_k). Both observers are calculated using YALMIP parser [21].

Figure 1 shows how the observer gains depend on δ_k for the case when the control error statistic is time invariant, i.e., when $\delta_k = \delta_{k-1} = \delta_1$. Note that thanks to the scheduling approach the gains change up to a 30% while δ_1 changes from 0 to 0.1 .

Simulating the designed observers under the proposed scenario we get Figure 2 showing that at the instants when the control error is low, the gain-scheduling approach of degree 2 performs better than the one with gains not depending on δ_k. When the control error is near to its maximum both approaches have similar behavior. This could have been predicted from Figure 1 since the gains have similar values for control errors close to the upper bound.

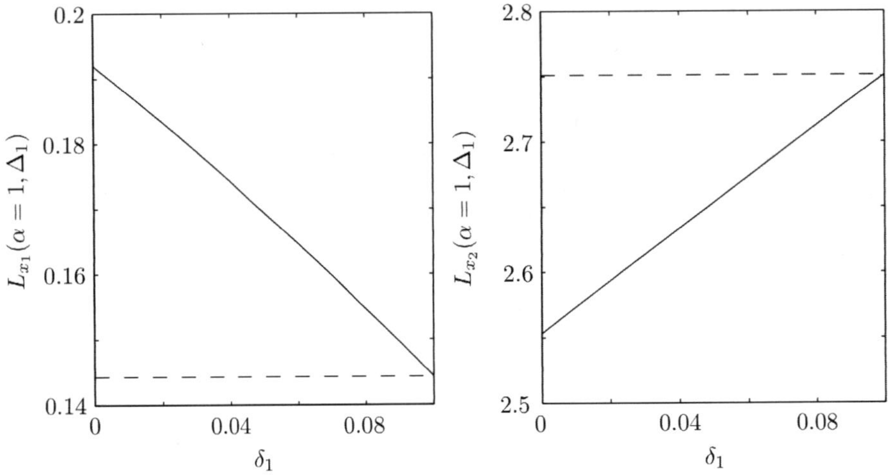

Figure 1. Observer gains as a function of δ_1, with polynomials of second order (continuous line) and zero order (dashed line).

5. Conclusion

In this work, we have studied the remote estimation problem under measurement and control input dropouts without successful delivery acknowledgment. We used a Markov chain to characterize the measurement reception scenario and described the control error between the control input at the plant and the one at the observer. We handle this control error as a new unknown disturbance input. Then, we related the observer gain to the measurement reception scenario and scheduled it with a control error statistic. We proved by simulation that our gain scheduling approach performs as well as the design with invariant gains (zero order polynomials) for the worst control error case, but improves the estimation performance whenever the control error statistic is not near its upper bound.

Further research directions may include extensions of the current results to observer-based control, where the separation principle holds if the controller can be designed to produce a control error that fulfills the imposed bounds, and uncertain transmission probabilities.

Acknowledgments

This work has been supported by MICINN project number DPI2011-27845-C02-0 and by grant PI15734 from *Universitat Jaume I.*

References

[1] R. A. Gupta and M.-Y. Chow, "Networked control system: overview and research trends," *IEEE Transactions on Industrial Electronics.*, vol. 57, no. 7, pp. 2527–2535, 2010.

[2] J. P. Hespanha and P. Naghshtabrizi, "A survey of recent results in networked control systems," *Proceedings of the IEEE*, vol. 95, no. 1, pp. 138–162, 2007.

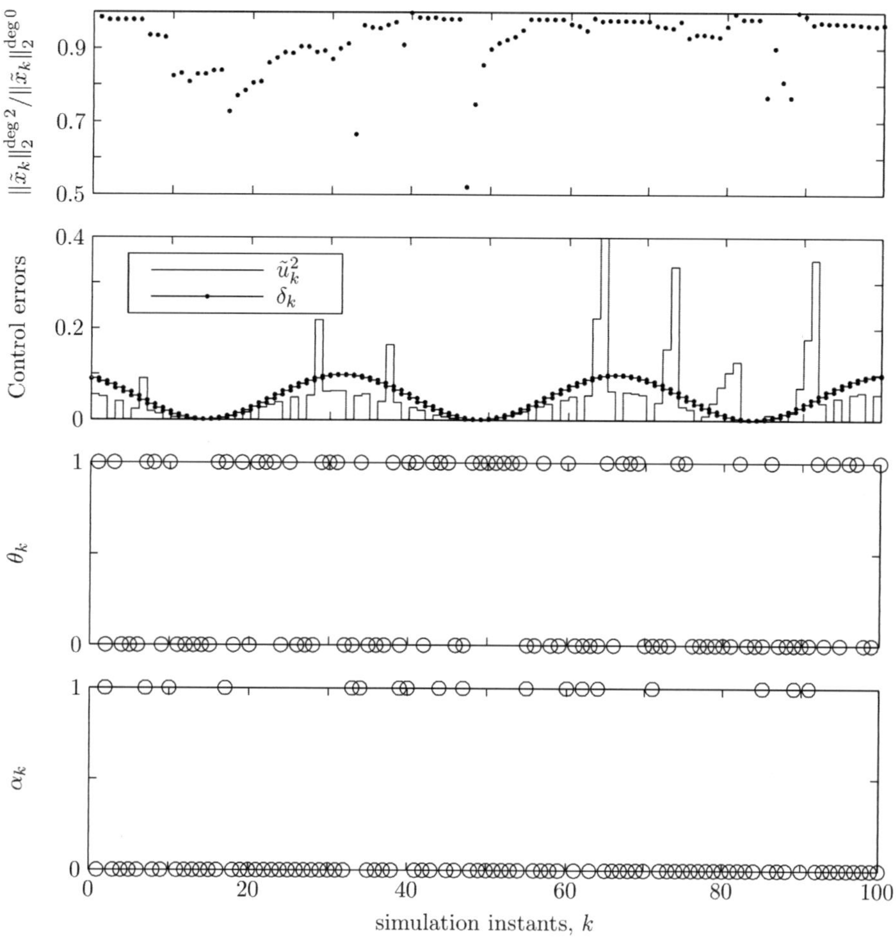

Figure 2. State estimation performances in simulation under control input and measurement dropouts.

[3] J. Chen, K. H. Johansson, S. Olariu, I. C. Paschalidis, and I. Stojmenovic, "Guest editorial special issue on wireless sensor and actuator networks," *IEEE Transactions on Automatic Control*, vol. 56, no. 10, pp. 2244–2246, 2011.

[4] S. C. Smith and P. Seiler, "Estimation with lossy measurements: jump estimators for jump systems," *IEEE Transactions on Automatic Control*, vol. 48, no. 12, pp. 2163–2171, 2003.

[5] D. Dolz, I. Peñarrocha, and R. Sanchis, "Jump state estimation with multiple sensors with packet dropping and delaying channels," *International Journal of Systems Science*, 2014.

[6] B. Sinopoli, L. Schenato, M. Franceschetti, K. Poolla, M. I. Jordan, and S. S. Sastry, "Kalman filtering with intermittent observations," *IEEE Transactions on Automatic Control*, vol. 49, no. 9, pp. 1453–1464, 2004.

[7] L. Schenato, B. Sinopoli, M. Franceschetti, K. Poolla, and S. S. Sastry, "Foundations of control and estimation over lossy networks," *Proceedings of the IEEE*, vol. 95, no. 1, pp. 163–187, 2007.

[8] B. Sinopoli, L. Schenato, M. Franceschetti, K. Poolla, and S. Sastry, "Optimal linear LQG control over lossy networks without packet acknowledgment," *Asian Journal of Control*, vol. 10, no. 1, pp. 3–13, 2008.

[9] M. Moayedi, Y. Foo, and Y. Soh, "Networked LQG control over unreliable channels," *International Journal of Robust and Nonlinear Control*, vol. 23, no. 2, pp. 167–189, 2013.

[10] E. Garone, B. Sinopoli, A. Goldsmith, and A. Casavola, "LQG control for mimo systems over multiple erasure channels with perfect acknowledgment," *IEEE Transactions on Automatic Control*, vol. 57, no. 2, pp. 450–456, 2012.

[11] J.-T. Yu and L.-C. Fu, "An optimal compensation framework for linear quadratic gaussian control over lossy networks," *IEEE Transactions on Automatic Control*, 2015.

[12] M. Epstein and R. M. Murray, "Estimation schemes for networked control systems using UDP-like communication," in *46th IEEE Conference on Decision and Control*, 2007, pp. 3945–3951.

[13] R. Blind and F. Allgöwer, "Estimating the fates of the control packets for Networked Control Systems with loss of control and measurement packets," in *48th IEEE Conference on Decision and Control*, 2009, pp. 2687–2692.

[14] Y. Wang, H. Ye, S. X. Ding, G. Wang, and D. Zhou, "Residual generation and evaluation of networked control systems subject to random packet dropout," *Automatica*, vol. 45, no. 10, pp. 2427–2434, 2009.

[15] G. Chesi, "LMI techniques for optimization over polynomials in control: a survey," *IEEE Transactions on Automatic Control*, vol. 55, no. 11, pp. 2500–2510, 2010.

[16] I. Peñarrocha, D. Dolz, and R. Sanchis, "A polynomial approach for observer design in networked control systems with unknown packet dropout rate." in *52nd IEEE Conference on Decision and Control*, 2013, pp. 5933–5938.

[17] F. Wu and S. Prajna, "SOS-based solution approach to polynomial LPV system analysis and synthesis problems," *International Journal of Control*, vol. 78, no. 8, pp. 600–611, 2005.

[18] I. Peñarrocha, D. Dolz, N. Aparicio, and R. Sanchis, "Synthesis of nonlinear controller for wind turbines stability when providing grid support," *International Journal of Robust and Nonlinear Control*, 2013.

[19] D. Wang, J. Wang, and W. Wang, "H_∞ controller design of networked control systems with markov packet dropouts," *IEEE Transactions on Systems, Man, and Cybernetics: Systems*, vol. 43, no. 3, pp. 689–697, 2013.

[20] I. Peñarrocha, D. Dolz, and R. Sanchis, "Inferential networked $\mathcal{H}_\infty$ control with accessibility constraints in both the sensor and actuator channels," *International Journal of Systems Science*, vol. 45, no. 5, pp. 1180–1195, 2014.

[21] J. Löfberg, "Pre-and post-processing sum-of-squares programs in practice," *IEEE Transactions on Automatic Control*, vol. 54, no. 5, pp. 1007–1011, 2009.

Implemented Wavelet Packet Tree based Denoising Algorithm in Bus Signals of a Wearable Sensorarray

M. Schimmack, S. Nguyen, P. Mercorelli

Institute of Product and Process Innovation
Leuphana University of Lueneburg
Volgershall 1, D-21339 Lueneburg, Germany

E-mail: `[schimmack, mercorelli]@uni.leuphana.de`

Abstract. This paper introduces a thermosensing embedded system with a sensor bus that uses wavelets for the purposes of noise location and denoising. From the principle of the filter bank the measured signal is separated in two bands, low and high frequency. The proposed algorithm identifies the defined noise in these two bands. With the Wavelet Packet Transform as a method of Discrete Wavelet Transform, it is able to decompose and reconstruct bus input signals of a sensor network. Using a seminorm, the noise of a sequence can be detected and located, so that the wavelet basis can be rearranged. This particularly allows for elimination of any incoherent parts that make up unavoidable measuring noise of bus signals. The proposed method was built based on wavelet algorithms from the WaveLab 850 library of the Stanford University (USA). This work gives an insight to the workings of Wavelet Transformation.

1. Introduction

Over the past decades, Wavelets have become another important tool that has established its industrial use in fields such as signal analysis, data compression and also noise reduction. They are mathematical functions, that essentially present an extension to the already primarily used Fourier Transform, however, with the properties to have not only time-based but also frequency-based results. The Fourier Transform is helpful for trigonometric functions but lacks in performance with other types. One reason to prefer the Wavelet Transform over the traditional Fourier Transform is its better resolution to give accurate information about frequencies at certain times, and it is also more suited in approximation of signals with discontinuities and sharp spikes.

The proposed wearable application can be used for thermosensing of the changes and fluctuations in temperature of the extremity surfaces of a patient. In a human body, the body temperature stays constant as long as its energy gain and energy loss are equal. If this condition is not met, the temperature is able to fluctuate, resulting in either heat generation or loss. Thereby, the body temperature can be increased by parameters such as metabolism and emotions, food or physical activity. Another important impact is conducted through the environment, where a human being is able to experience heat exchanges through the skin. One can differentiate into the body core temperature and the skin temperature, the latter being highly dependent on the environmental conditions. As humans are warm-blooded mammals, the body core temperature is maintained constant at a range of $36.8 - 37.7°C$. In contrast, the extremities and skin are the ones that adapt to outer changes to protect the body core.

Journal of Physics: Conference Series **659** (2015) 012021 doi:10.1088/1742-6596/659/1/012021

A thermoregulation process is performed by the body with help of the hypothalamus, where central sensors for the body core temperature are located. It also receives information about the skin temperature through thermal receptors under the skin, and also brain and spinal cord. When the body temperature strays from the set point its supposed to be, the hypothalamus is able to adjust the temperature through reflexes like sweating, shivering or changes in skin blood flow. With this, the body is able to regulate its own temperature. To measure the changes of the skin temperature, a temperature sensor network, which is implemented in a textile substrate is used. These sensors are connected to each other over conductive polymer-metal composite fibers to a sensor array and communicate over a bus system. Conductive polymer-metal composite fibers are the fundamental core for wearable technology and an integral part of smart textiles. There is a wide field of applications both in use and in development, which is analyzed in [1]. As traditional utilization, the yarn is used as a textile electrode as well as a transmission line for monitoring of physiological parameters as exhibited in [2]. Combined with an analyzing algorithm the fibers have been presented in context of a wearable orthosis-system after a bone fracture in [3] for monitoring during the predictive rehabilitation.

One of the pionier contributions of wavelet-denoising approach and thresholding can be found in [4] and also in [5]. Naturally measurement noises and disturbances occur during application, thus an adaptive filter is needed. Therefore a wavelet based denoising algorithm is built based on the WaveLab 850 of Stanford University [6], using the Haar wavelet family to accurately represent the form of a rectangular bus signal.

2. Wavelet Decomposition Method

The Haar Wavelet has a simple square wave structure, making it preferably used in analyzing compact discrete signals. For Wavelet Transformation, a wavelet function and a scaling function are needed. The Haar Wavelet's basis wavelet function, also known as the mother wavelet, can be described as follows

$$\psi_{(d,b,k)}(t) = \psi_b(2^d t - n), \tag{1}$$

with a support of size 2^{-d} of the Nyquist frequency and the indices (d, b, k). Herein, d is set as a scale index describing the tree level, b as a phase parameter for the frequency shift and k as a time related parameter indicating the shift in time. Also the mother wavelet's structure is shown by the dependencies

$$\psi(t) = \begin{cases} 1 & \text{if } 0 \leq t < \frac{1}{2} \\ -1 & \text{if } \frac{1}{2} \leq t < 1 \\ 0 & \text{for all other } t \end{cases} \tag{2}$$

As a decomposition method, this paper uses the Wavelet Packet Transform which belongs to the method of Discrete Wavelet Transform and creates a complete wavelet basis decomposition tree. Using this tree structure, the input signal that is going to be analyzed, will be disassembled into subspaces that hold either high frequency or low frequency sequences. Subsequently this procedure is done for different levels of the tree, repeating the filtration of Wavelet Father into two Wavelet Children, one of high and one of low frequency. Such a process enables the separation of the noise and the actual signal. The weighted coefficients $wp(d, b, k)$ are calculated in more detail as follows

$$wp_{(d,b,k)} = \int_I f(t)\psi^h_{(d,b,k)}(t)d(t). \tag{3}$$

For $b = 0$ it is possible to define the following coefficients

$$s_{(d,0,k)} = \int_I f(t)\psi^h_{(d,0,k)}(t)d(t), \tag{4}$$

where $f(t)$ is the required signal, I is the considered time interval and $\psi^h_{(d,0,k)}(t)$ is the well known mother function of a Haars wavelet [7]. To conclude

$$f(t) = \sum_k s_{(d,k)}\psi^h_{(d,0,k)}(t) + \sum_b \sum_k wp_{(d,b,k)}\psi^h_{(d,b,k)}, \tag{5}$$

where $s_{(d,k)} = wp_{(d,0,k)}$. It is shown that there are just two independent parameters and that parameter d characterizes the level of the tree. The wavelet packets under consideration are derived from the Haar mother wavelet. The Haar functions are identified using the parameter tuple, (d, b, k), here $d = 1$ represents the highest degree of refinement with respect to time. Fig. 1 shows the corresponding tree with the wavelet coefficients, described by $wp_{(d,b,k)}$. It represents the contribution of each of the wavelets to the signal based on n samples. The notation $wp_{(1,0,0\ldots\frac{n}{2}-1)}$ denotes the coefficients on the first level on the left with time shifts 0 through $\frac{n}{2} - 1$.

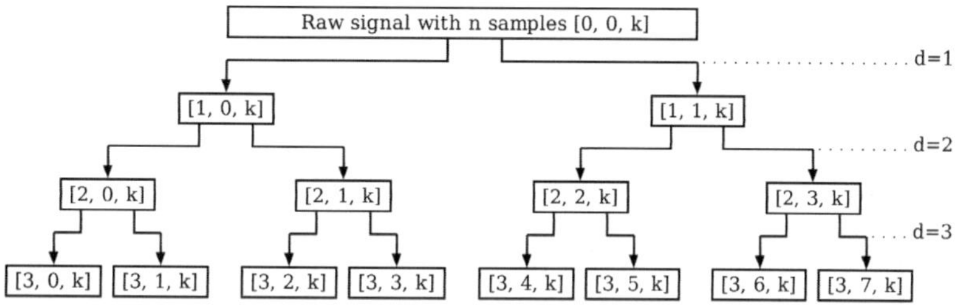

Figure 1. The Wavelet coefficients arranged in a tree-like structure based on WaveLab Version 850 with different scale index d

3. Wavelet Denoising Method using an axiomatic Seminorm

In order to detect the noise $e(t)$ in a measured signal $y(t)$ characterization through a possible definition is required. For the definition of the noise in a bus signal a seminorm will be used.

Definition 1

$$p(\mathbf{V}) = \left|\sum_{k=0}^{2^{D-d}} \{wp_{(d,b,k)}\}\right|, \tag{6}$$

where $\mathbf{V}$ *represents a vectorial field and in general* $p : \mathbf{V} \to \mathbb{R}^+$. □

Function $p(\mathbf{V})$ satisfies the positivity, absolute homogeneity and subadditivity properties, but it does not satisfy the separation property. In more detail, if $p(\mathbf{V}) = 0$, it does not necessarily imply that sequence (vector) $\mathbf{V} = 0$. A possible definition of a noise is proposed as follows.

Definition 2 *Given an observed sequence*

$$y(t) = x(t) + e(t). \tag{7}$$

Let $e(t)$ *be defined as the incoherent part of the sequence* $y(t)$ *at every level* d *of the packet tree as shown below*

$$e(t) = \sum_{(\mathbf{d}_{ml},\mathbf{b}_{ml})} \sum_{k=0}^{2^{D-d}} wp_{(\mathbf{d}_{ml},\mathbf{b}_{ml},k)}\psi_{(\mathbf{d}_{ml},\mathbf{b}_{ml},k)}(t) + \sum_{(\mathbf{d}_{mh},\mathbf{b}_{mh})} \sum_{k=0}^{2^{D-d}} wp_{(\mathbf{d}_{mh},\mathbf{b}_{mh},k)}\psi_{(\mathbf{d}_{mh},\mathbf{b}_{mh},k)}(t), \tag{8}$$

where if a specified length n (dyadic length) is considered to determine the height of the wavelet tree: $D = \log_2(n)$*. For characterization of the selected wavelets, the whole decomposition tree starting from level d=1 is split in half, dividing the tree into a low frequency domain on the left and a high frequency domain on the right. This is in analogy to wavelets working as a filter bank, as the tree is separated into two bands, low and high frequency. Here, the selected wavelets are defined by* $(\mathbf{d}_{ml}, \mathbf{b}_{ml})$ *for the low frequency and* $(\mathbf{d}_{mh}, \mathbf{b}_{mh})$ *for the high frequency subspaces indices such that*

$$\{(\mathbf{d}_{ml}, \mathbf{b}_{ml})\} = arg\Big(\min_{(d,b)} \big|\sum_{k=0}^{2^{D-d}} \{wp_{(d,b,k)}\}\big|\Big) : \{0 \leq k \leq 2^{D-d}, 0 \leq b \leq 2^{d-1} - 1, \forall d \in \mathbb{N}\}\Big), \tag{9}$$

$$\{(\mathbf{d}_{mh}, \mathbf{b}_{mh})\} = arg\Big(\min_{(d,b)} \big|\sum_{k=0}^{2^{D-d}} \{wp_{(d,b,k)}\}\big|\Big) : \{0 \leq k \leq 2^{D-d}, 2^{d-1} \leq b \leq 2^{d}-1, \forall d \in \mathbb{N}\}\Big). \tag{10}$$

□

In effect, the definition of the noise looks for the subspace characterized by the minimum of the components in both the low and the high frequency wavelet domain. Defenition 2 is used to sort the wavelet basis, which can illuminate the difference between the coherent and incoherent part of the sequence.

3.1. Description of the algorithm

The proposed algorithm works in the following way, starting from the second level of the wavelet packet tree (d=2). With the defined axiomatic seminorm, the minima of both frequency domains are calculated, resulting in one minimum subspace for the low and one for the high frequency domain. Once the minima at level d=2 are calculated with respect to b, the seminorm for each of their two children are calculated at level d=3 (d+1). Now when comparing the seminorm values for the minimum of a father to his descending children, if the sum of the children is less than that of the father, the noise is localized with the children. Otherwise the noise if located with the father subspace. In case of the noise being located with the children, both the subspaces are compared again, and the one with a smaller seminorm calculation represents the noise. This process is done for both high and low frequency domain, specifying one minimum subspace in each domain where the noise is located. For denoised reconstruction, those subspaces that are determined to represent noise, are left out. This way, the algorithm subsequently searches for a minimum down until the end of the tree.

3.2. Signal cleaning procedure

Given the sampled signal

- **Step 1**: Specify dyadic length, n, in order to determine the height of the wavelet tree: $D = \log_2(n)$.
- **Step 2**: Construct the wavelet coefficient tree $wp_{(d,b,k)}$ for every $d \geq 0$ and for every k.
- **Step 3a**: For the time-frequency intervals such that $b = 0, 1, ...2^{d-1} - 1$ with $d > 1$, calculate the absolute value of the $\sum_k$ for the low frequency intervals of the tree, that is, $W_{ml}sum(d, b) = |\sum_k W(d, b)|$.

Journal of Physics: Conference Series **659** (2015) 012021 doi:10.1088/1742-6596/659/1/012021

- **Step 3b**: For all time-frequency intervals such that $b = 2^{d-1}, 2^{d-1}+1, ...2^d - 1$ with $d > 1$, calculate the absolute value of the $\sum_k$ for the high frequency intervals of the tree, that is, $W_{mh}sum(d,b) = |\sum_k W(d,b)|$.
- **Step 4a**: For every "wavelet father" of the low frequency domain $W_{ml}(d,b)$ at the node (d,b) with $b = 0, 1, ...2^{d-1} - 1$, calculate its left child at the node $(d+1, bLeft)$ and its right child $(d+1, bRight)$. Then, calculate the absolute value of the sum with respect to k and denote them as

$$W_{ml}sumChildLeft = \left|\sum\nolimits_k W_{ml}sum(d, bLeft)\right| \tag{11}$$

and

$$W_{ml}sumChildRight = \left|\sum\nolimits_k W_{ml}sum(d, bRight)\right|. \tag{12}$$

While $(d < D)$
For $b = 0 : 2^{d-1} - 1$
Case

$$W_{ml}sum(d,b) \leq W_{ml}sumChildLeft + W_{ml}sumChildRight :$$
$$\mathbf{W}_{ml}b(d_{ml}, b_{ml}) = arg\big(\min_{(d,b)} W_{ml}sum(d,b)\big)$$

Case

$$W_{ml}sumChildLeft \leq W_{ml}sumChildRight :$$
$$\mathbf{W}_{ml}b(d_{ml}, b_{ml}) = arg\big(\min_{(d,b)} W_{ml}sumChildLeft\big)$$

Case

$$W_{ml}sumChildRight \leq W_{ml}sumChildLeft :$$
$$\mathbf{W}_{ml}b(d_{ml}, b_{ml}) = arg\big(\min_{(d,b)} W_{ml}sumChildRight\big)$$

End b loop

- **Step 4b**: In the same pattern as in the previous step, the procedure of noise location is done for the high frequency domain. For every "wavelet father" of the high frequency domain $W_{mh}(d,b)$ at the node (d,b) with $b = 2^{d-1}, 2^{d-1}+1, ...2^d - 1$, calculate its left child at the node $(d+1, bLeft)$ and its right child $(d+1, bRight)$. Then, calculate the absolute value of the sum with respect to k and denote them as

$$W_{mh}sumChildLeft = \left|\sum\nolimits_k W_{mh}sum(d, bLeft)\right| \tag{13}$$

and

$$W_{mh}sumChildRight = \left|\sum\nolimits_k W_{mh}sum(d, bRight)\right|. \tag{14}$$

For $b = 2^{d-1} : 2^d - 1$.
Case

$$W_{mh}sum(d,b) \leq W_{mh}sumChildLeft + W_{mh}sumChildRight :$$
$$\mathbf{W}_{mh}b(d_{mh}, b_{mh}) = arg\big(\min_{(d,b)} W_{mh}sum(d,b)\big)$$

Case

$$W_{ml}sumChildLeft \leq W_{mh}sumChildRight :$$
$$\mathbf{W}_{mh}b(d_{mh}, b_{mh}) = arg\big(\min_{(d,b)} W_{mh}sumChildLeft\big)$$

Case

$$W_{ml}sumChildRight \leq W_{mh}sumChildLeft$$
$$\mathbf{W}_{mh}b(d_{mh}, b_{mh}) = arg(\min_{(d,b)} W_{mh}sumChildRight)$$

End b loop

- **Step 5**: Reconstruct the noise

$$e(t) = \sum_{(\mathbf{d}_{ml},\mathbf{b}_{ml})} \sum_{k=0}^{2^{D-d}} wp_{(\mathbf{d}_{ml},\mathbf{b}_{ml},k)} \psi_{(\mathbf{d}_{ml},\mathbf{b}_{ml},k)}(t) + \sum_{(\mathbf{d}_{mh},\mathbf{b}_{mh})} \sum_{k=0}^{2^{D-d}} wp_{(\mathbf{d}_{mh},\mathbf{b}_{mh},k)} \psi_{(\mathbf{d}_{mh},\mathbf{b}_{mh},k)}(t) \quad (15)$$

and reconstruct the denoised signal

$$y_d(t) = y(t) - e(t). \quad (16)$$

4. Methodology and Embedded System

4.1. Sensor Bus

The DS18S20 digital thermosensor provides with a modification 12-bit Celsius temperature measurements and has an alarm function with nonvolatile user-programmable upper and lower trigger points. The sensor communicates over a 1-Wire bus which used one data line, power line and ground for communication with a embedded system. Fig. 2 shows that each sensor has a unique 64-bit serial code, which allows multiple sensors to function on the same 1-Wire bus. Thus, it is possible to use the embedded system to control many sensors distributed over a large area.

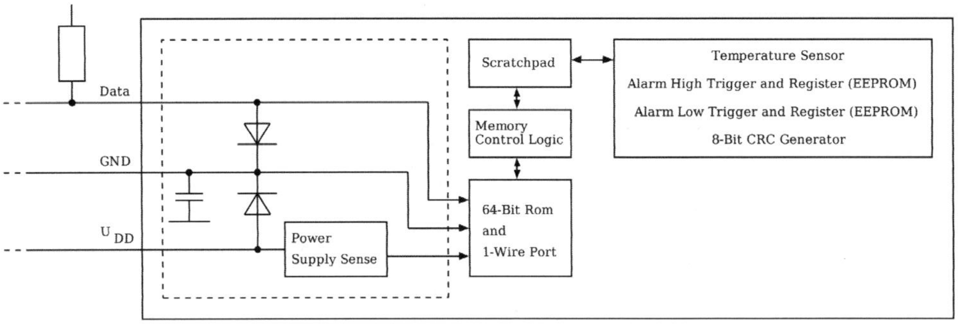

Figure 2. General block diagram of the high-precision 1-Wire digital thermosensor and its components.

4.2. Embedded System

A portable Linux based embedded system, which is depicted in Fig. 3 with the corresponding components, works like a signal processing module. The sensorarray them self is implemented in a textile substratum and connected over the 1-Wire bus to the sensorarray. The embedded system based on the Broadcom BCM2836 system on a chip (SoC) and has a quad-core ARM Cortex-A7 central processing unit (CPU) with a VideoCore IV dual-core graphics processing unit (GPU). The used Linux kernel for the operating system based on the ARM hard-float (armhf) Debian architecture and the proposed algorithm works with GNU Octave.

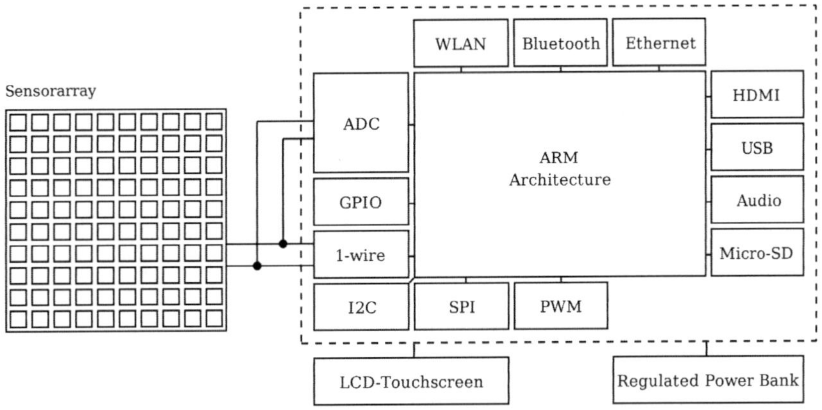

Figure 3. General block diagram of the measurement system with the Linux based embedded system.

5. Validation

To showcase the effectiveness of the algorithm, the denoising performance is observed for a synthetically generated noisy square wave signal, which can be analogously compared to that of a bus signal. With this, Fig. 4 shows the result of the algorithm for the reconstruction with the global minimum of the frequency domains as the noise component. On the other hand, Fig. 5 shows the result of the algorith for the reconstruction with both the minimum value of the low and minimum value of the high frequency domain as noise components. It is shown graphically, how the method in Fig. 5 is superior in suppressing noise, as the amplitude of the reconstruction is visibly lower than in Fig. 4. This proves that the here proposed algorithm method is indeed more effective in denoising than a method of only eliminating the global minimum of the frequency domains. Fig. 6 a shows the free upper limb with the implemented sensors in the textile substrate and a general visualization of the temperature surface [8]. Therefor the char of temperature values were transformed into the RGB-colours as

$$\lambda = \begin{cases} (0, k, 255), & k < 256 \\ (0, 255, 512 - k), & 256 \leq k < 768 \\ (255, 1024 - k, 0), & 768 \leq k < 1024 \end{cases} \tag{17}$$

The value k based on the interval $[0, 1024[$ to $\mathbb{R}$ and λ presents the colour of the RGB-sequence with $R \in [0, 255], G \in [0, 255], B \in [0, 255]$. The nearest neighbor interpolation represents in Fig. 6b a general algorithm for the visualization of the sensorarray values with

$$f(x, y) = \begin{cases} f(Q_{11}), x < x_2 - \frac{\Delta x}{2}, y < y_2 - \frac{\Delta y}{2} \\ f(Q_{12}), x > x_2 - \frac{\Delta x}{2}, y < y_2 - \frac{\Delta y}{2} \\ f(Q_{21}), x < x_2 - \frac{\Delta x}{2}, y > y_2 - \frac{\Delta y}{2} \\ f(Q_{22}), x > x_2 - \frac{\Delta x}{2}, y > y_2 - \frac{\Delta y}{2} \end{cases} \tag{18}$$

was used. Fig. 7 shows the graphical representation of the forearm with a) the bilinear interpolation like

$$\begin{aligned} f(x, y) \approx & \frac{f(Q_{11})}{(x_2 - x_1)(y_2 - y_1)}(x_2 - x)(y_2 - y) + \frac{f(Q_21}{(x_2 - x_1)(y_2 - y_1)}(x - x_1)(y_2 - y) \\ & + \frac{f(Q_12}{(x_2 - x_1)(y_2 - y_1)}(x_2 - x)(y - y_1) + \frac{f(Q_22}{(x_2 - x_1)(y_2 - y_1)}(x - x_1)(y - y_1) \end{aligned} \tag{19}$$

and b) a bicubic spline-interpolation of the sensorarray values was used.

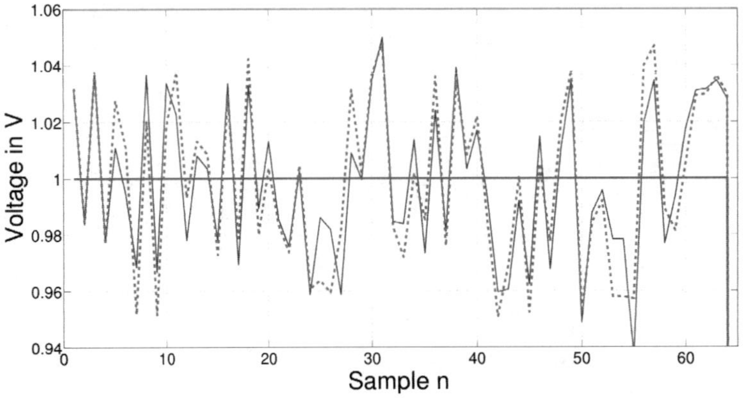

Figure 4. Graphical representation of the reconstruction without the global minimum of the frequency domains.

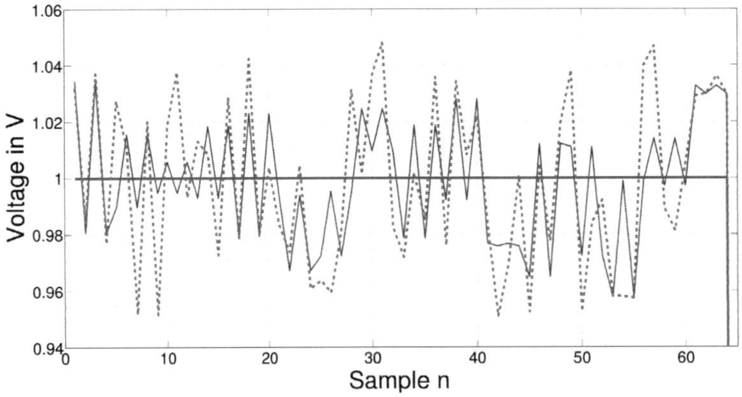

Figure 5. Graphical representation the reconstruction without both minimum values of the low and the high frequency domain.

6. Conclusion

The paper analyzes the properties of denoising 1-Wire bus signals of a temperature sensorarray for longterm monitoring of the skin surface based on Haar wavelets. With the Wavelet Packet Transform as a method of Discrete Wavelet Transform, it was able to decompose and reconstruct bus input signals of a sensor network. Using a seminorm, the noise of a sequence can be detected and located, so that the wavelet basis can be rearranged. In fact, the wavelets reduces the effect of the noise which can be applied directly without using model noise. This particularly allows for elimination of any incoherent parts that make up unavoidable measuring noise of bus signals. The proposed method was built based on wavelet algorithms from the WaveLab 850 library of the Stanford University (USA). This work gives an insight to the workings of Wavelet Tranformation. The results show the effectiveness of the proposed technique. Even though the

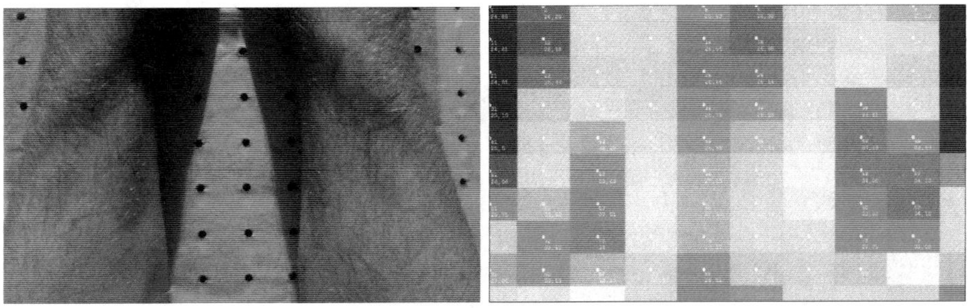

Figure 6. Free upper limb with the implemented sensors in the textile substrate and a general visualization of the temperature surface.

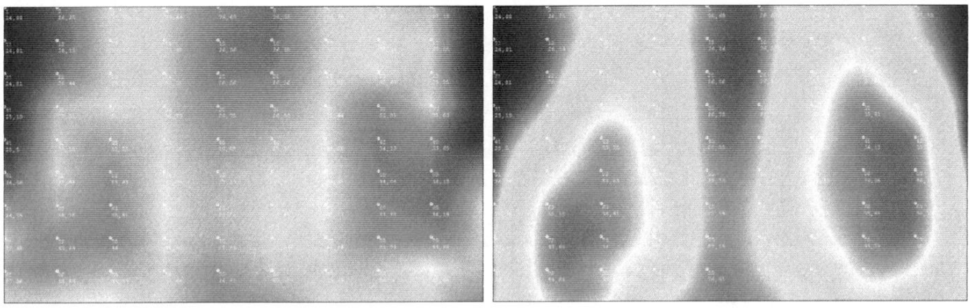

Figure 7. Graphical representation of the forearm with a) the bilinear interpolation and b) the bicubic spline-interpolation of the sensorarray values.

obtained results are quite general, results using GNU Octave and MATLAB® considering the experimental setup are shown and commented.

References

[1] A. Pantelopoulos and N. G. Bourbakis. A survey on wearable sensor-based systems for health monitoring and prognosis. *IEEE Transaction on Systems, Man, and Cybernetics Part C: Application and Review*, 40(1):1–12, 2010.

[2] N. Noury, A. Dittmar, C. Corroy, R. Baghai, J.L. Weber, D. Blanc, F. Klefstat, A. Blinovska, S. Vaysse, and B. Comet. A smart cloth for ambulatory telemonitoring of physiological parameters and activity: the vtamn project. In *IEEE HEALTHCOM - Proceedings of the 6th International Workshop on Enterprise Networking and Computing in Healthcare Industry*, pages 155–160, Japan, 2004.

[3] M. Schimmack, A. Hand, P. Mercorelli, and A. Georgiadis. Using a seminorm for wavelet denoising of sEMG signals for monitoring during rehabilitation with embedded orthosis system. In *IEEE MeMeA - International Symposium on Medical Measurements and Applications*, Italy, 2015.

[4] D. L. Donoho. Denoising and soft thresholding. *IEEE Transactions on Information Theory*, 41(3):613–627, 1995.

[5] S. Neville and N. Dimopoulos. Wavelet denoising of coarsely quantized signals. *IEEE Transactions on Instrumentation and Measurement*, 55(3):892–901, 2006.

[6] J. Buckheit, S. Chen, D. Donoho, I. Johnstone, and J. Scargle. About wavelab. *Handbook of WaveLab Version .850 by Standford University and NASA-Ames Research Center*, pages 1–37, 2005.

[7] P. Mercorelli and A. Frick. *Noise Level Estimation Using Haar Wavelet Packet Trees for Sensor Robust Outlier Detection.* Series: Lecture Note in Computer Sciences, Springer-Verlag publishers, 2006.

[8] M. Schimmack, R. Sukhikh, A. Georgiadis, and C. H. Siemsen. Drahtloses und bildgebendes Multi-Sensorsystem integriert in eine Handgelenksorthese für ein nichtinvasives Langzeitmonitoring. *Sensoren und Messsysteme*, 44(3):603–609, 2012.

Journal of Physics: Conference Series **659** (2015) 012022 doi:10.1088/1742-6596/659/1/012022

Performance evaluation of iterated extended Kalman filter with variable step-length

Jindřich Havlík and Ondřej Straka

Department of Cybernetics & European Center of Excellence: New Technologies for Information Society
University of West Bohemia, Univerzitní 8, 30614 Plzeň, Czech Republic

E-mail: havlikj@ntis.zcu.cz, straka30@ntis.zcu.cz

Abstract. The paper deals with state estimation of nonlinear stochastic dynamic systems. In particular, the iterated extended Kalman filter is studied. Three recently proposed iterated extended Kalman filter algorithms are analyzed in terms of their performance and specification of a user design parameter, more specifically the step-length size. The performance is compared using the root mean square error evaluating the state estimate and the noncredibility index assessing covariance matrix of the estimate error. The performance and influence of the design parameter, are analyzed in a numerical simulation.

1. Introduction

Nonlinear state estimation of discrete-time stochastic dynamic systems is an important field of study, which has undergone a rapid development especially in the last two decades. The importance of this field stems from its crucial role in many areas such as signal processing, target tracking, satellite navigation, fault detection, and adaptive and optimal control problems. It constitutes an essential part of any decision-making process. The state estimation methods are mostly based on the Bayesian approach or on the optimization approach.

Application of the Bayesian approach to the recursive state estimation problem leads to the Bayesian recursive relations (BRRs). The BRRs provide probability density functions (PDFs) of the state conditioned by the measurements. The conditional PDFs represent a full description of the state. A closed-form solution to the BRRs is available only for a few special cases such as a linear Gaussian system [1]. In other cases, an approximate method must be used. The approximate methods include the particle filtering [2, 3], the point-mass method [4], and the Gaussian sum method [5].

The optimization approach is largely based on minimizing the mean square error criterion and for linear Gaussian systems leads to the celebrated Kalman filter. As opposed to the Bayesian approach providing the conditional PDF of the state, the optimization approach provides a point estimate of the state and corresponding covariance matrix (CM) of the estimate error. An analytical solution to the state estimation problem based on the optimization approach can be found in a few special cases only, as is the case of the Bayesian approach. For other cases, the estimation methods usually follow the Kalman filter framework and utilize some approximation techniques such as linearization of the nonlinear functions. For example, the extended (EKF) and the second-order extended Kalman filters [6], [1] approximate the nonlinear functions by the Taylor series expansion around the current estimate up to the first or second order, respectively.

When the effects of the linearization errors tend to disrupt performance of the filter or its convergence, the relinearizing of the measurement equation around the updated state may alleviate the difficulties.

This method is known as the iterated extended Kalman filter (IEKF) [6]. The IEKF computes the state estimates not as an approximate conditional mean (as the EKF does), but as a maximum a posteriori (MAP) estimate [7]. In [8] it was shown that the IEKF measurement update is an application of the Gauss-Newton (GN) method and the EKF is a special case of the IEKF with only one GN method iteration. In [9] a new technique improving the EKF and the IEKF has been proposed based on a control of the step-length.

The paper presented a derivation of three alternative algorithms of the IEKF with the step-control, but a comprehensive performance analysis was not given. However, a complex performance analysis evaluating not only the state estimates, but also the CMs of the estimate error provided by filters is known to be important [10]. The reason is that the CM represents a self-assessment of the filter and as such it is used in the next filtering step to compute the estimate. Therefore, the goal of this paper is to deliver a comprehensive performance analysis of the filters proposed in [9] in terms of performance metrics utilizing the estimate error and the CM. More specifically, the root mean square error (RMSE), its distance from the Cramer-Rao lower bound (CRLB) [11, 3] and the noncredibility index (NCI) [12] will be analyzed. The comparison of the new filters with the EKF and the IEKF using the above mentioned performance metrics should provide better understanding of their capabilities.

The paper is organized as follows: Section 2 is devoted to a brief introduction to the nonlinear state estimation and its solution by the EKF and the IEKF. The recent development of the IEKFs is briefly presented in Section 3. Section 4 presents various performance metrics used in the analysis. Next, in Section 5 the filters are compared in two numerical examples; and the paper is concluded by Section 6.

2. Nonlinear state estimation by EKF and IEKF

2.1. System description

Let a discrete-time nonlinear stochastic system be considered in the following state-space form

$$\mathbf{x}_{k+1} = \mathbf{f}_k(\mathbf{x}_k) + \mathbf{w}_k, \qquad k = 0, 1, 2, \ldots, \tag{1}$$

$$\mathbf{z}_k = \mathbf{h}_k(\mathbf{x}_k) + \mathbf{v}_k, \qquad k = 0, 1, 2, \ldots, \tag{2}$$

where the vectors $\mathbf{x}_k \in \mathbb{R}^{n_x}$, and $\mathbf{z}_k \in \mathbb{R}^{n_z}$ represent the state of the system, and the measurement at time instant k, respectively, $\mathbf{f}_k : \mathbb{R}^{n_x} \to \mathbb{R}^{n_x}$ and $\mathbf{h}_k : \mathbb{R}^{n_x} \to \mathbb{R}^{n_z}$ are known vector functions, and $\mathbf{w}_k \in \mathbb{R}^{n_x}$, $\mathbf{v}_k \in \mathbb{R}^{n_z}$ are mutually independent state and measurement white noises. The PDFs of the noises are Gaussian with zero means and known CMs $\Sigma_k^{\mathbf{w}}$ and $\Sigma_k^{\mathbf{v}}$, respectively, i.e., $p(\mathbf{w}_k) = \mathcal{N}\{\mathbf{w}_k; \mathbf{0}_{n_x\times 1}, \Sigma_k^{\mathbf{w}}\}*$ and $p(\mathbf{v}_k) = \mathcal{N}\{\mathbf{v}_k; \mathbf{0}_{n_z\times 1}, \Sigma_k^{\mathbf{v}}\}$, respectively, where $\mathbf{0}_{a\times b}$ denotes an $a \times b$ matrix of zeros. The PDF of the initial state is Gaussian and known as well, i.e., $p(\mathbf{x}_0) = \mathcal{N}\{\mathbf{x}_0; \hat{\mathbf{x}}_0, \mathbf{P}_0\}$. The initial state is independent of the noises.

2.2. Extended Kalman filter

Originally, the Kalman filter was derived using the orthogonality principle [13] in 1960. It yields the minimum mean square error estimate for the system (1) and (2) with linear functions $\mathbf{f}_k$ and $\mathbf{h}_k$. For nonlinear functions $\mathbf{f}_k$ and $\mathbf{h}_k$ an approximation has to be used, such as that used in the EKF. The EKF is based on the first order Taylor series expansion (TE1). The TE1 of $\mathbf{h}_k$ under the assumption of known state predictive mean $\hat{\mathbf{x}}_{k|k-1} = \mathrm{E}[\mathbf{x}_k|\mathbf{z}^{k-1}]$ (defining the linearization point) is given by

$$\mathbf{h}_k(\mathbf{x}_k) \approx \mathbf{h}_k(\hat{\mathbf{x}}_{k|k-1}) + \mathbf{H}_k\left(\mathbf{x}_k - \hat{\mathbf{x}}_{k|k-1}\right), \tag{3}$$

where $\mathbf{H}_k = \frac{\partial \mathbf{h}(\mathbf{x}_k)}{\partial \mathbf{x}_k}\Big|_{\mathbf{x}_k=\hat{\mathbf{x}}_{k|k-1}}$ is the Jacobian matrix of $\mathbf{h}_k(\cdot)$ evaluated at $\hat{\mathbf{x}}_{k|k-1}$.

Use of the TE1 approximation for the prediction update step is analogous. The TE1 of $\mathbf{f}_k$ in (1) under assumption of known filtering mean $\hat{\mathbf{x}}_{k|k} = \mathrm{E}[\mathbf{x}_k|\mathbf{z}^k]$ is of the form

$$\mathbf{f}_k(\mathbf{x}_k) \approx \mathbf{f}_k(\hat{\mathbf{x}}_{k|k}) + \mathbf{F}_k\left(\mathbf{x}_k - \hat{\mathbf{x}}_{k|k}\right), \tag{4}$$

$*$ For the sake of simplicity all PDFs will be given by their argument, if not stated otherwise, i.e., $p(\mathbf{w}_k) = p_{\mathbf{w}_k}(\mathbf{w}_k)$.

where $\mathbf{F}_k = \frac{\partial \mathbf{f}(\mathbf{x}_k)}{\partial \mathbf{x}_k}\Big|_{\mathbf{x}_k=\hat{\mathbf{x}}_{k|k}}$ is the Jacobian matrix of $\mathbf{f}_k(\cdot)$ evaluated at $\hat{\mathbf{x}}_{k|k}$.

Approximations (3) and (4) are used in the algorithm of the EKF (see Algorithm 1).

Algorithm 1: Extended Kalman filter

Step 1: (*initialization*) Set the step $k = 0$ and define a priori initial condition by its first two moments

$$\hat{\mathbf{x}}_{0|-1} \triangleq \mathsf{E}[\mathbf{x}_0] = \hat{\mathbf{x}}_0, \tag{5}$$

$$\mathbf{P}_{0|-1} \triangleq \mathsf{E}[(\mathbf{x}_0 - \hat{\mathbf{x}}_{0|-1})(\mathbf{x}_0 - \hat{\mathbf{x}}_{0|-1})^T] = \mathbf{P}_0. \tag{6}$$

Step 2: (*filtering, measurement update*) The filtering mean $\hat{\mathbf{x}}_{k|k}$ and the CM of the estimate error $\mathbf{P}_{k|k} = \mathsf{E}[(\mathbf{x}_k - \hat{\mathbf{x}}_{k|k})(\mathbf{x}_k - \hat{\mathbf{x}}_{k|k})^T]$ are computed by means of

$$\hat{\mathbf{x}}_{k|k} = \hat{\mathbf{x}}_{k|k-1} + \mathbf{K}_k(\mathbf{z}_k - \mathbf{h}_k(\hat{\mathbf{x}}_{k|k-1})), \tag{7}$$

$$\mathbf{P}_{k|k} = \mathbf{P}_{k|k-1} - \mathbf{K}_k\mathbf{H}_k\mathbf{P}_{k|k-1}, \tag{8}$$

where

$$\mathbf{K}_k = \mathbf{P}_{k|k-1}\mathbf{H}_k^T(\mathbf{H}_k\mathbf{P}_{k|k-1}\mathbf{H}_k^T + \mathbf{\Sigma}_k^{\mathbf{v}})^{-1} \tag{9}$$

is the filter gain.

Step 3: (*prediction, time update*) The predictive mean $\hat{\mathbf{x}}_{k+1|k}$ and the CM of the estimate error $\mathbf{P}_{k+1|k} = \mathsf{E}[(\mathbf{x}_{k+1} - \hat{\mathbf{x}}_{k+1|k})(\mathbf{x}_{k+1} - \hat{\mathbf{x}}_{k+1|k})^T]$ are given by

$$\hat{\mathbf{x}}_{k+1|k} = \mathbf{f}_k(\hat{\mathbf{x}}_{k|k}), \tag{10}$$

$$\mathbf{P}_{k+1|k} = \mathbf{F}_k\mathbf{P}_{k|k}\mathbf{F}_k^T + \mathbf{\Sigma}_k^{\mathbf{w}}. \tag{11}$$

Let $k = k + 1$. The algorithm then continues by **Step 2**.

The EKF performs well for systems with mildly nonlinear functions $\mathbf{f}_k$ and $\mathbf{h}_k$, but if the measurement equation (2) is strongly nonlinear (such as in bearings-only tracking problems), the performance of the filter deteriorates. In such situations, the IEKF tends to provide more accurate estimates than the EKF.

2.3. *Iterated extended Kalman filter*

The idea of the IEKF [6] is to improve a reference trajectory, and consequently the estimate, in the presence of significant nonlinearities. Those improvements are achieved by local iterations of the EKF measurement update (see Algorithm 2). The iterations are usually stopped when there is no significant change in consecutive iterations or other criteria such as a maximum number of iterations are met.

Algorithm 2: Iterated extended Kalman filter measurement update

---

Step 1: (*initialization*) Set the iteration $i = 0$, and set the predictive estimate $\hat{\mathbf{x}}_k^0 = \hat{\mathbf{x}}_{k|k-1}$.

Step 2: (*measurement update iterations*) Compute the Jacobi matrix at the best state estimate available, the Kalman gain, and the next iteration of the state estimate as

$$\mathbf{H}_k^i = \frac{\partial \mathbf{h}_k(\mathbf{s})}{\partial \mathbf{s}}\bigg|_{\mathbf{s}=\hat{\mathbf{x}}_k^i} \tag{12}$$

$$\mathbf{K}_k^i = \mathbf{P}_{k|k-1}[\mathbf{H}_k^i]^T(\mathbf{H}_k^i\mathbf{P}_{k|k-1}[\mathbf{H}_k^i]^T + \Sigma_k^{\mathbf{v}})^{-1} \tag{13}$$

$$\hat{\mathbf{x}}_k^{i+1} = \hat{\mathbf{x}}_{k|k-1} + \mathbf{K}_k^i(\mathbf{z}_k - \mathbf{h}_k(\hat{\mathbf{x}}_k^i) - \mathbf{H}_k^i(\hat{\mathbf{x}}_{k|k-1} - \hat{\mathbf{x}}_k^i)) \tag{14}$$

Repeat **Step 2** with $i = i + 1$ until a stopping criterion is met†.

† The stopping-condition is usually represented by a maximum number of iterations or a negligible change between two consecutive iterations, e.g., $\|\hat{\mathbf{x}}_k^{i+1} - \hat{\mathbf{x}}_k^i\| \leq \epsilon$, where ϵ is a pre-specified threshold.

Step 3: (*updating the state and CM*) Save the last iteration as the new filtering mean and compute the CM based on the last iteration, i.e.,

$$\hat{\mathbf{x}}_{k|k} = \hat{\mathbf{x}}_k^i, \tag{15}$$

$$\mathbf{P}_{k|k} = (\mathbf{I} - \mathbf{K}_k^i \mathbf{H}_k^i)\mathbf{P}_{k|k-1}. \tag{16}$$

It is important to mention, that even the relinearization occurring in the IEKF algorithm does not ensure convergence of the filter nor the IEKF generally outperforms the EKF. Nevertheless, the IEKF measurement update possesses two very interesting properties:

- It can be viewed as an application of the Gauss-Newton method [8].
- It generates the MAP estimate $\hat{\mathbf{x}}_{k|k}^{\mathrm{MAP}} = \arg\max_{\mathbf{x}_k} p(\mathbf{x}_k|\mathbf{z}^k)$ [6].

3. Improvements of the iterated extended Kalman filter

As has been mentioned above, the IEKF can be seen as an application of the GN method used for solving the nonlinear least square problem. Hence, improvements of the GN method enhancing performance of the method and its convergence can be used in the IEKF to improve quality of the estimates. Changing length of a step is one of the frequently used improvements. First, the MAP criterion, which is minimized by the IEKF, is introduced.

3.1. The MAP criterion

If a probabilistic description of the system (1) and (2) is used, the posterior PDF $p(\mathbf{x}_k|\mathbf{z}^k)$ is sought using prior PDF $p(\mathbf{x}_k|\mathbf{z}^{k-1})$. The mean $\hat{\mathbf{x}}_{k|k-1}$ and the CM $\mathbf{P}_{k|k-1}$ computed in the time update of Algorithm 1 correspond to the prior PDF $p(\mathbf{x}_k|\mathbf{z}^{k-1})$, which is assumed to be Gaussian, and the mean $\hat{\mathbf{x}}_{k|k}$ and the CM $\mathbf{P}_{k|k}$ computed in the measurement update of Algorithm 2 correspond to the posterior PDF $p(\mathbf{x}_k|\mathbf{z}^k)$. The posterior PDF $p(\mathbf{x}_k|\mathbf{z}^k)$ is proportional to the product of the likelihood $p(\mathbf{z}_k|\mathbf{x}_k) = p_{\mathbf{v}_k}(\mathbf{z}_k - \mathbf{h}_k(\mathbf{x}_k))$ and the prior PDF $p(\mathbf{x}_k|\mathbf{z}^{k-1})$

$$\begin{aligned} p(\mathbf{x}_k|\mathbf{z}^k) &\propto p(\mathbf{z}_k|\mathbf{x}_k)p(\mathbf{x}_k|\mathbf{z}^{k-1}) \\ &\propto \exp -\frac{1}{2}\Big([\mathbf{z}_k - \mathbf{h}_k(\mathbf{x}_k)]^T (\Sigma_k^v)^{-1} [\mathbf{z}_k - \mathbf{h}_k(\mathbf{x}_k)] + \left[\hat{\mathbf{x}}_{k|k-1} - \mathbf{x}_k\right]^T \mathbf{P}_{k|k-1}^{-1} \left[\hat{\mathbf{x}}_{k|k-1} - \mathbf{x}_k\right] \Big), \end{aligned} \tag{17}$$

where the terms not depending on $\mathbf{x}_k$ have been dropped for convenient purposes. The MAP estimate is then given by

$$\begin{aligned} \hat{\mathbf{x}}_{k|k} &= \arg\min_{\mathbf{x}_k} \frac{1}{2}\Big([\mathbf{z}_k - \mathbf{h}_k(\mathbf{x}_k)]^T (\Sigma_k^v)^{-1} [\mathbf{z}_k - \mathbf{h}_k(\mathbf{x}_k)] + \left[\hat{\mathbf{x}}_{k|k-1} - \mathbf{x}_k\right]^T \mathbf{P}_{k|k-1}^{-1} \left[\hat{\mathbf{x}}_{k|k-1} - \mathbf{x}_k\right] \Big) \\ &= \arg\min_{\mathbf{x}_k} \mathbf{V}_k(\mathbf{x}_k). \end{aligned} \tag{18}$$

Now, having the MAP criterion specified using (18), the GN method can be used to find $\hat{\mathbf{x}}_{k|k}$ minimizing the criterion. Ideally, each GN method iteration (i.e. the IEKF measurement update iteration) should decrease the criterion $\mathbf{V}_k$.

The IEKF measurement update (16) can be rewritten in an implicit manner

$$\hat{\mathbf{x}}_k^{i+1} = \hat{\mathbf{x}}_{k|k-1} + \mathbf{K}_k^i(\mathbf{z}_k - \mathbf{h}_k(\hat{\mathbf{x}}_k^i) - \mathbf{H}_k^i(\hat{\mathbf{x}}_{k|k-1} - \hat{\mathbf{x}}_k^i)) = \hat{\mathbf{x}}_k^i + \Delta_k^i, \tag{19}$$

where the shift between the iterations Δ_k^i is given by

$$\Delta_k^i = \hat{\mathbf{x}}_{k|k-1} - \hat{\mathbf{x}}_k^i + \mathbf{K}_k^i(\mathbf{z}_k - \mathbf{h}_k(\hat{\mathbf{x}}_k^i) - \mathbf{H}_k^i(\hat{\mathbf{x}}_{k|k-1} - \hat{\mathbf{x}}_k^i)) \tag{20}$$

and $\hat{\mathbf{x}}_k^0 = \hat{\mathbf{x}}_{k|k-1}$. The relations (19) and (20) then match the relations of the GN method minimizing the MAP criterion (18).

3.2. IEKF with variable step-length

Ideally, the value of the criterion $\mathbf{V}_k$ should decrease with each iteration, but due to nonlinearity of $\mathbf{h}_k$ this behavior cannot be guaranteed. Thus, it is beneficial to check whether the criterion is decreasing, i.e.,

$$\mathbf{V}_k(\hat{\mathbf{x}}_k^{i+1}) < \mathbf{V}_k(\hat{\mathbf{x}}_k^{i}) \tag{21}$$

or rather to ensure it. This is done by introducing a variable step-length. The parameter $\alpha^i \in (0, 1\rangle$ is used to govern a step-length for a fixed direction Δ_k^i given by (20) in the following manner

$$\hat{\mathbf{x}}_k^{\mathrm{L},i+1} = \hat{\mathbf{x}}_k^{\mathrm{L},i} + \alpha^i \Delta_k^i, \tag{22}$$

where the notation $\hat{\mathbf{x}}_k^{\mathrm{L},i}$ was introduced to distinguish between $\hat{\mathbf{x}}_k^i$ for a full step given by (19).

There are several ways of determining the parameter α^i [9] to ensure the condition (21), but the exact line search given by

$$\alpha^i = \arg\min_{0<s\leq 1} \mathbf{V}_k\left(\hat{\mathbf{x}}_k^{\mathrm{L},i} + s\Delta_k^i\right) \tag{23}$$

provides the best results at the cots of increased computational demands.

The line search IEKF (IEKF-L) measurement update is then

$$\hat{\mathbf{x}}_k^{\mathrm{L},i+1} = \hat{\mathbf{x}}_k^{\mathrm{L},i} + \alpha^i \Delta_k^i = \hat{\mathbf{x}}_k^{\mathrm{L},i} + \alpha^i \left(\hat{\mathbf{x}}_{k|k-1} - \hat{\mathbf{x}}_k^{\mathrm{L},i} + \mathbf{K}_k^i\left(\mathbf{z}_k - \mathbf{h}_k(\hat{\mathbf{x}}_k^{\mathrm{L},i}) - \mathbf{H}_k^i(\hat{\mathbf{x}}_{k|k-1} - \hat{\mathbf{x}}_k^{\mathrm{L},i})\right)\right). \tag{24}$$

The IEKF-L algorithm is then given by substituting (14) by (24).

3.3. Quasi-Newton IEKF

Another approach improving the GN method is the Quasi-Newton technique (QN). The GN method approximates second order terms (Hessians) by first order terms (Jacobians) [14]. The QN attempts to compensate this approximation by introducing a correction term. The IEKF-QN equations, which were derived in [9], are as follows:

$$\hat{\mathbf{x}}_k^{\mathrm{QN},i+1} = \hat{\mathbf{x}}_k^{\mathrm{QN},i} + \alpha^i \left(\tilde{\mathbf{x}}_k^{\mathrm{QN},i} + \mathbf{K}_k^i(\mathbf{z}_k - \mathbf{h}_k(\hat{\mathbf{x}}_k^{\mathrm{QN},i}) - \mathbf{H}_k^i(\tilde{\mathbf{x}}_k^{\mathrm{QN},i})) - \mathbf{S}_k^i \mathbf{T}_k^i \tilde{\mathbf{x}}_k^{\mathrm{QN},i}\right), \tag{25}$$

where

$$\tilde{\mathbf{x}}_k^{\mathrm{QN},i} = \hat{\mathbf{x}}_{k|k-1} - \hat{\mathbf{x}}_k^{\mathrm{QN},i}, \tag{26}$$

$$\mathbf{S}_k^i = \left((\mathbf{H}_k^i)^T(\mathbf{\Sigma}_k^{\mathrm{v}})^{-1}\mathbf{H}_k^i + \mathbf{P}_{k|k-1}^{-1} + \mathbf{T}_k^i\right)^{-1}, \tag{27}$$

$$\mathbf{K}_k^i = \mathbf{S}_k^i(\mathbf{H}_k^i)^T(\mathbf{\Sigma}_k^{\mathrm{v}})^{-1} \tag{28}$$

and the matrix $\mathbf{T}_k^i$ is implicitly built using

$$\hat{\mathbf{T}}_k^i = \mathbf{T}_k^{i-1} + \frac{\boldsymbol{\upsilon}_i \boldsymbol{\nu}_i^T + \boldsymbol{\nu}_i \boldsymbol{\upsilon}_i^T}{\boldsymbol{\nu}_i^T \boldsymbol{\varsigma}_i} - \frac{\boldsymbol{\upsilon}_i \boldsymbol{\varsigma}_i}{(\boldsymbol{\nu}_i^T \boldsymbol{\varsigma}_i)^2} \boldsymbol{\nu}_i \boldsymbol{\nu}_i^T, \tag{29}$$

$$\mathbf{T}_k^i = \min\left(1, \frac{|\boldsymbol{\varsigma}_i^T \boldsymbol{\kappa}_i|}{|\boldsymbol{\varsigma}_i^T \hat{\mathbf{T}}_k^i \boldsymbol{\varsigma}_i|}\right) \hat{\mathbf{T}}_k^i, \tag{30}$$

where

$$\nu_i = -(\mathbf{H}_k^i)^T(\Sigma_k^v)^{-1}(\mathbf{z}_k - \mathbf{h}_k(\hat{\mathbf{x}}_k^{\mathrm{QN},i})) + (\mathbf{H}_k^{i-1})^T(\Sigma_k^v)^{-1}(\mathbf{z}_k - \mathbf{h}_k(\hat{\mathbf{x}}_k^{i-1})), \tag{31}$$

$$\kappa_i = (\mathbf{H}_k^{i-1} - \mathbf{H}_k^i)^T(\Sigma_k^v)^{-1}(\mathbf{z}_k - \mathbf{h}_k(\hat{\mathbf{x}}_k^{\mathrm{QN},i})), \tag{32}$$

$$\varsigma_i = \hat{\mathbf{x}}_k^{\mathrm{QN},i} - \hat{\mathbf{x}}_k^{\mathrm{QN},i-1}, \tag{33}$$

$$\upsilon_i = \kappa_i - \mathbf{T}_k^{i-1}\varsigma_i, \tag{34}$$

with initial condition $\mathbf{T}_k^0 = \mathbf{0}_{nx \times nx}$. The IEKF-QN algorithm is then given by substituting (14) and (13) by (25) - (34). And the only parameter α^i should be again chosen to satisfy criterion (21) and thus the exact line search should be preferred (23) with the same argumentation as for the IEKF-L.

3.4. Levenberg-Marquardt IEKF

The Levenberg-Marquardt (LM) method [15] (also known as the trust region method) adds a damping parameter μ_i to the GN method. The damping parameter controls interpolation between the GN and the steepest-descent. For a small μ_i the LM method behaves like the GN method and for a large μ_i like the steepest-descent method. The IEKF-LM relations are given by

$$\hat{\mathbf{x}}_k^{\mathrm{LM},i+1} = \hat{\mathbf{x}}_k^{\mathrm{LM},i} + \alpha^i \left(\tilde{\mathbf{x}}_k^{\mathrm{LM},i} + \mathbf{K}_k^i(\mathbf{z}_k - \mathbf{h}_k(\hat{\mathbf{x}}_k^{\mathrm{LM},i}) - \mathbf{H}_k^i(\tilde{\mathbf{x}}_k^{\mathrm{LM},i})) - \mu_i(\mathbf{I} - \mathbf{K}_k^i\mathbf{H}_k^i)\tilde{\mathbf{P}}_k^i\mathbf{B}_k^i\tilde{\mathbf{x}}_k^{\mathrm{LM},i}\right), \tag{35}$$

$$\tilde{\mathbf{P}}_k^i = \frac{\frac{1}{\mu_i \mathbf{B}_k^i}}{\mathbf{P}_{k|k-1} + \frac{1}{\mu_i \mathbf{B}_k^i}}\mathbf{P}_{k|k-1}, \tag{36}$$

$$\mathbf{K}_k^i = \tilde{\mathbf{P}}_k^i[\mathbf{H}_k^i]^T(\mathbf{H}_k^i\tilde{\mathbf{P}}_k^i[\mathbf{H}_k^i]^T + \Sigma_k^v)^{-1}, \tag{37}$$

$$\mathbf{B}_k^i = \mathrm{diag}(\mathbf{H}_k^i(\Sigma_k^v)^{-1}[\mathbf{H}_k^i]^T + (\tilde{\mathbf{P}}_k^i)^{-1}), \tag{38}$$

where $\tilde{\mathbf{x}}_k^{\mathrm{LM},i} = \hat{\mathbf{x}}_{k|k-1} - \hat{\mathbf{x}}_k^{\mathrm{LM},i}$. The IEKF-LM algorithm is then given substituting (14) and (13) by (35-38).

There are two parameters in the IEKF-LM algorithm, which have to be set. The α^i is again to be found using the exact line search (23) because it influences the results most. The selection of the damping parameter μ_i is rather tricky, because it influences the step-length too. Some discussion on this topic can be found in [9]. On the other hand, in our simulations an adaptation of both parameters was not observed to be of large effect, thus damping parameter was set constant.

4. Performance metrics

Behavior of a newly developed filter is frequently studied using benchmark problems where it is compared with the standard filters. As far as the criterion for the comparison is concerned, most authors rely only on the RMSE, which is related to the design criterion of most filters (minimum mean square error). However, the RMSE only evaluates quality of the first moment provided by the filter, the conditional mean of the state. In this paper besides the state estimate $\hat{\mathbf{x}}_{k|k}$ also comparison of the CM of the estimate error will be considered. The quality of the state estimate error CM provided by the filters will be compared using the NCI [12] as the credibility measure. Further, the RMSE of the state estimates will be supplemented with the CRLB [11], which represents an objective limit of cognizability of the system.

4.1. Root mean square error

The RMSE for the state estimate $\hat{\mathbf{x}}_{k|k}$ at time instant k calculated using M Monte Carlo simulations is defined as

$$\mathrm{RMSE}(\hat{\mathbf{x}}_{k|k}) \approx \sqrt{\frac{1}{M}\sum_{i=1}^{M}\tilde{\mathbf{x}}_{k|k}^T(i)\tilde{\mathbf{x}}_{k|k}(i)}, \tag{39}$$

where $\tilde{\mathbf{x}}_{k|k}(i) = \mathbf{x}_k(i) - \hat{\mathbf{x}}_{k|k}(i)$ is the estimate error and $\hat{\mathbf{x}}_{k|k}(i)$ is the estimate of $\mathbf{x}_k(i)$ at the i-th Monte Carlo simulation.

4.2. Cramer-Rao lower bound

The filtering CRLB denoted as $\mathbf{C}_{k|k}$ is an objective limit of cognizability of the system state, since it is a lower bound for the mean square error matrix (MSEM) $\mathbf{\Pi}_{k|k} = \mathrm{E}[\tilde{\mathbf{x}}_{k|k}\tilde{\mathbf{x}}_{k|k}^T]$ produced by a filter, i.e.,

$$\mathbf{\Pi}_{k|k} \geq \mathbf{C}_{k|k}. \tag{40}$$

Note that the inequality in (40) means that $\mathbf{\Pi}_{k|k} - \mathbf{C}_{k|k}$ is positive definite. The MSEM $\mathbf{\Pi}_{k|k}$ can be approximated using M Monte Carlo simulations by

$$\mathbf{\Pi}_{k|k} \approx \frac{1}{M}\sum_{i=1}^{M} \tilde{\mathbf{x}}_{k|k}(i)\tilde{\mathbf{x}}_{k|k}^T(i). \tag{41}$$

The comparison of the MSEM with the CRLB is beneficial in nonlinear filtering to show effects of approximations and linearization errors.

For the system given by (1) and (2) the filtering CRLB [11] can be recursively computed by

$$\mathbf{C}_{k|k} = \mathbf{C}_{k|k-1} - \mathbf{C}_{k|k-1}\mathcal{H}_k^T\left(\mathcal{H}_k\mathbf{C}_{k|k-1}\mathcal{H}_k^T + \Sigma_k^v\right)^{-1}\mathcal{H}_k\mathbf{C}_{k|k-1}, \tag{42}$$

$$\mathbf{C}_{k+1|k} = \mathcal{F}_k\mathbf{C}_{k|k}\mathcal{F}_k^T + \Sigma_k^w, \tag{43}$$

where the Jacobi matrices $\mathcal{F}_k = \left.\frac{\partial \mathbf{f}_k(\mathbf{x})}{\partial \mathbf{x}}\right|_{\mathbf{x}=\mathbf{x}_k}$, $\mathcal{H}_k = \left.\frac{\partial \mathbf{h}_k(\mathbf{x})}{\partial \mathbf{x}}\right|_{\mathbf{x}=\mathbf{x}_k}$ are functions of the true state $\mathbf{x}_k$ and the initial condition is $\mathbf{C}_{0|-1} = \mathbf{P}_{0|-1}$. It shall be noticed that the relations (42) and (43) are formally identical to those of the CM computation in the EKF (8) and (11). The only difference is that the Jacobi matrices in the CRLB are computed at the true state and in the EKF in the state estimate. The CRLB will later help to determine how significant the approximation errors are. The MSEM and CRLB are often compared using the matrix trace, because the square root of MSEM trace is the RMSE, i.e.,

$$\sqrt{\mathrm{tr}\,\mathbf{\Pi}_{k|k}} = \mathrm{RMSE}(\hat{\mathbf{x}}_{k|k}) \geq \sqrt{\mathrm{tr}\,\mathbf{C}_{k|k}}. \tag{44}$$

4.3. Noncredibility index

The NCI [12] compares the normalized estimation error squared (NEES) [7] of the estimator defined as $\epsilon_{k|k}(i) = \tilde{\mathbf{x}}_{k|k}^T(i)\mathbf{P}_{k|k}^{-1}(i)\tilde{\mathbf{x}}_{k|k}(i)$ with the NEES of a perfectly credible estimator given by $\epsilon_{k|k}^*(i) = \tilde{\mathbf{x}}_{k|k}^T(i)\mathbf{\Pi}_{k|k}^{-1}\tilde{\mathbf{x}}_{k|k}(i)$, where $\mathbf{\Pi}_{k|k}$ is the MSEM computed by (41). Finally, the NCI is defined as

$$\mathrm{NCI}(k) = \frac{10}{M}\sum_{i=1}^{M}\left|\log_{10}\frac{\epsilon_{k|k}(i)}{\epsilon_{k|k}^*(i)}\right|. \tag{45}$$

The NCI is capable of determining whether the estimator is credible, i.e., whether the CM provided by the estimator is close to the MSEM $\mathbf{\Pi}_{k|k}$. The closer it is, the smaller the NCI value is, i.e., the NCI of a perfectly credible estimator is zero.

The NCI is tightly coupled with the inclination index (II), which is defined in the same way as the NCI with the only exception of omitting the absolute value, i.e.,

$$\mathrm{II}(k) = \frac{10}{M}\sum_{i=1}^{M}\log_{10}\frac{\epsilon_{k|k}(i)}{\epsilon_{k|k}^*(i)}. \tag{46}$$

The II specifies whether the noncredibility of a filter is rather of the optimistic or pessimistic nature., i.e., whether the CM of the filter is smaller or larger than the MSEM, respectively.

5. Numerical illustrations

The analysis of the IEKFs with a variable step-length will be performed in this section. After specification of considered models, the appropriate setting of the step-length parameter will be discussed. Finally, the comprehensive analysis involving the RMSE, the CRLB, and the NCI will be given.

5.1. Specification of benchmark models

All numerical illustrations will be performed within the scope of two examples: the univariate non-stationary growth model [16] and the bearings-only tracking model [9]. Both examples utilize highly nonlinear measurement functions, where the IEKFs should outperform non-iterative Kalman filters.

The *Univariate non-stationary growth model* (UNGM) is specific for its high degree of nonlinearity and bi-modality of the measurement, which causes difficulties for filters. The model is specified as

$$x_k = 0.5x_{k-1} + 25\frac{x_{k-1}}{1+x_{k-1}^2} + 8\cos(1.2(k-1)) + w_k, \tag{47}$$

$$z_k = \frac{x_k^2}{20} + v_k, \tag{48}$$

where $p(w_k) \sim \mathcal{N}\{0, 1\}$, $p(v_k) \sim \mathcal{N}\{0, 1\}$ and the initial condition is $p(x_0) \sim \mathcal{N}\{0.1, 1\}$. All simulations were started with $x_0 = 0.1$. The initial condition is thus set within a sensitivity region, which places high demands on the estimators.

The *Bearings only tracking* (BOT) is represented by the motion model with measurements from two static radars providing bearing only. The state $\mathbf{x}_k = [x_{1,k}, x_{2,k}]^T$ represents position of a tracked object in the Cartesian coordinates. The model is given by

$$\mathbf{x}_{k+1} = \mathbf{x}_k + \mathbf{w}_k, \tag{49}$$

$$\mathbf{z}_k = \begin{bmatrix} \arctan\left(\frac{x_{2,k}-r_y^1}{x_{1,k}-r_x^1}\right) \\ \arctan\left(\frac{x_{2,k}-r_y^2}{x_{1,k}-r_x^2}\right) \end{bmatrix} + \mathbf{v}_k, \tag{50}$$

where the position coordinates of the first radar are $(r_x^1, r_y^1) = [0, 1.5]$ and the position coordinates of the second radar are $(r_x^2, r_y^2) = [0, 0]$. The statistics of random variables are $p(\mathbf{w}_k) \sim \mathcal{N}\{0, 0.1\mathbf{I}_2\}$, $p(\mathbf{v}_k) \sim \mathcal{N}\{0, \pi^2 10^{-5}\mathbf{I}_2\}$, i.e., the standard deviation of the measurement is approximately 0.59[°] and the initial condition is $p(\mathbf{x}_0) \sim \mathcal{N}\{[1.5,\ 1.5]^T, 0.1\mathbf{I}_2\}$. All simulations were started with $\mathbf{x}_0 = [1.5,\ 1.5]^T$.

5.2. Step-length choice discussion

For the recently proposed filters (the IEKF-L, the IEKF-QN, and the IEKF-LM) the parameter α has to be chosen at every iteration of each measurement update. The paper [9], where the filters were derived, contains a discussion about the step-length parameter. Nevertheless, in the simulations given in [9] only a fixed step-length $\alpha^i = 0.5$ was used. Hence, prior to the performance analysis of the IEKFs the attention will be focused on the step-length parameter setting itself as it is a prerequisite for the IEKFs.

First, a measurement update at time instant $k = 0$ in the UNGM example (denoted as UNGM-0) will be examined. Suppose the predictive statistics are $\hat{\mathbf{x}}_{0|-1} = 0.1$, $\mathbf{P}_{0|-1} = 1$ and the measurement is $\mathbf{z}_0 = 1.3$. The examined filters are: the EKF, the IEKF ($\alpha^i = 1$), the filters discussed in Section 3 with a fixed step-length $\alpha^i = 0.5$ denoted as IEKF-L 0.5, IEKF-QN 0.5, and IEKF-LM 0.5 and the filters discussed in Section 3 with exact line search, i.e. with α^i optimized, denoted as IEKF-L, IEKF-QN, and IEKF-LM. The filters IEKF-LM and IEKF-LM 0.5 need setting of the dumping parameter μ_i. A discussion can be found in [9], but to alleviate simultaneous optimization of two parameters, a fixed value $\mu_i = 0.01$ is set in this paper‡. The state estimate and the MAP criterion evolutions over iterations for

‡ To find a suboptimal value of μ_i, a small test was performed with following results: In case of the UNGM example, the best results appear for $\mu_i \in [0, 0.01]$ whilst for $\mu_i > 0.01$ the results (time, RMSE and NCI) deteriorate linearly. In case of the BOT example, the minimum of the RMSE and the NCI is reached for $\mu_i \in [0.01, 0.1]$.

$k = 0$ are depicted in Figures 1 and 2, respectively. In this particular example all filters except the EKF finally converge to the minimal criterion value and the only difference is a speed of convergence. The IEKF and the optimized variable step-length filters are superior to the IEKFs with a fixed step-length.

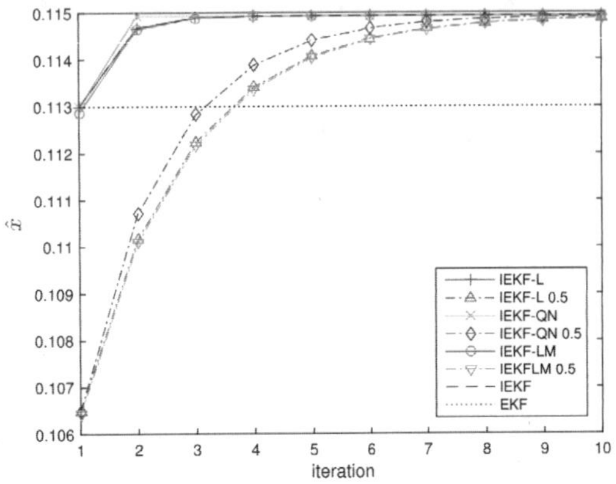

Figure 1. State estimate in a single measurement update over 10 iterations for UNGM-0.

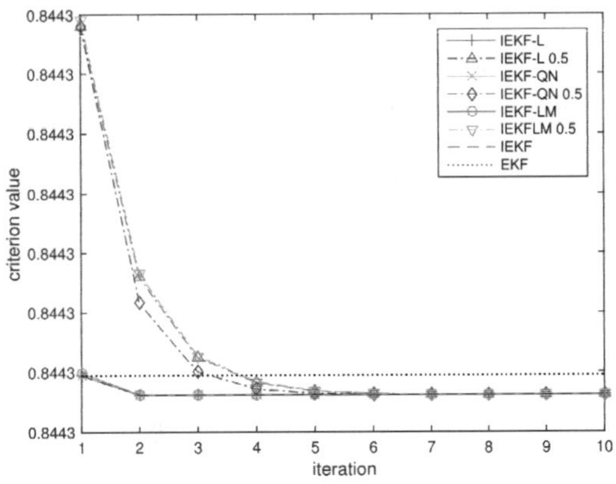

Figure 2. Value of the MAP criterion over 10 iterations for UNGM-0.

Second, a measurement update at time instant $k = 1$ will be examined in the UNGM example (denoted as UNGM-1). In the considered simulation the value of the predictive statistics were obtained as $\hat{\mathbf{x}}_{1|0} = 3.9$, $\mathbf{P}_{1|0} = 604$ and measurement was $\mathbf{z}_1 = -0.73$. The examined filters are the same as in the previous example, but their behavior is completely different. The results (evolution of the state estimate, the state error variance, the value of α^i and the value of the MAP criterion) are depicted in Figures 3-6. An interesting finding is the fact, that only the IEKFs with the exact line search and the IEKF-QN 0.5 converge. The EKF, the IEKF, the IEKF-L 0.5 and the IEKF-LM 0.5 fail to converge in terms of the state estimate, its variance and even the criterion, thus they should be used with extreme caution.

As far as the computational costs of the filters are concerned, it is clear that finding an optimal α^i in the IEKFs with exact line search is more costly than setting a fixed step-length ($\alpha^i = 0.5$ or $\alpha^i = 1$ for the IEKF). However, the filters with optimal α^i may not be slower than the filters with a constant α^i due to possible lower number of iterations performed. This little test served as a guideline for finding a robust parameter setting to the IEKFs, hence the comprehensive analysis can be performed now.

5.3. Comprehensive analysis of the IEKF-L, the IEKF-QN, and the IEKF-LM

The analysis was carried out for the UNGM and the BOT examples with 10^4 Monte Carlo simulations with 10 and 20 time instants, respectively. The IEKFs with the exact line search were used and compared to the standard filters - the EKF and the IEKF.

The RMSE development over time is depicted in Figures 7 (BOT) and 8 (UNGM). In the BOT example the IEKF-L and the IEKF-LM possess one degree of magnitude lower RMSE than the EKF and the IEKF. The RMSE of the IEKF-QN is slightly worse. In the UNGM example the results of IEKFs with the exact line search are even approaching to the CRLB, contrary to the EKF and IEKF, which are far from the CRLB. The estimate error CM analysis was performed using the NCI and its extension, the II. In the BOT example the course of the NCI (Figure 9) and the II (Figure 11) is similar to the RMSE development, i.e., the IEKF-L and the IEKF-LM outperform the EKF, the IEKF in terms of credibility of the results. Moreover all filters are too optimistic in their performance self-assessment, because the II is positive, indicating that the CM of the estimate error is too small in a perspective of the MSEM. Contrary to this, in the UNGM example the NCI (Figure 10) and the II (Figure 12) for the exact line search IEKFs tend to be perfectly credible from time instant $k = 3$ nay they are pessimistic in 2 time instants due to simulations starting from a point with zero variance contrary to the initial conditions for the filters.

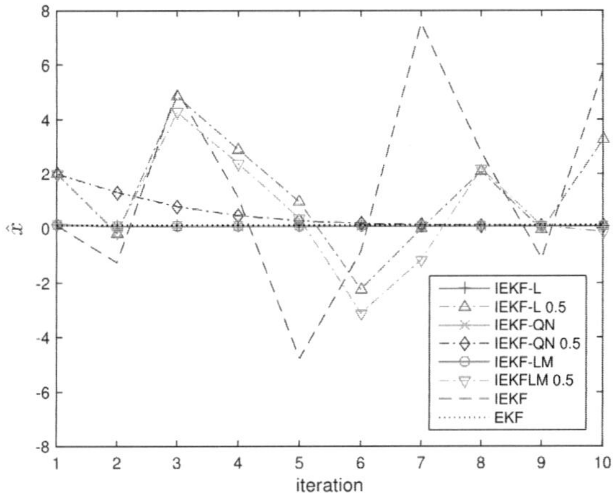

Figure 3. State estimate in a single measurement update over 10 iterations for UNGM-1.

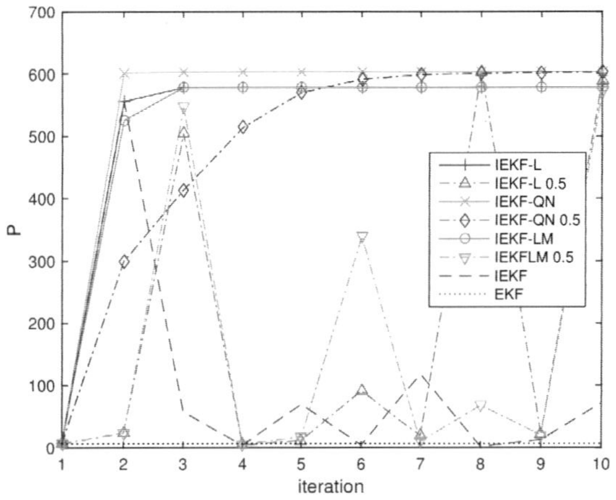

Figure 4. State estimate error variance development over 10 iterations for UNGM-1.

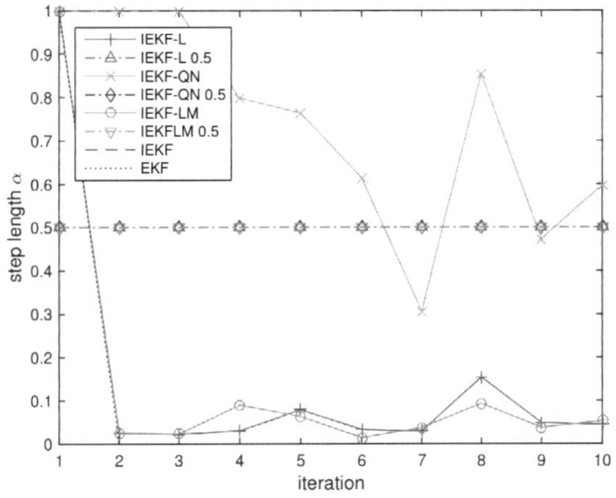

Figure 5. Value of α^i in a single measurement update over 10 iterations for UNGM-1.

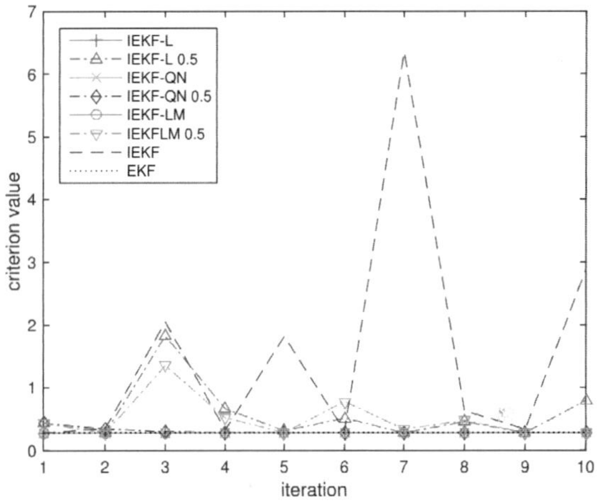

Figure 6. Value of the MAP criterion over 10 iterations for UNGM-1.

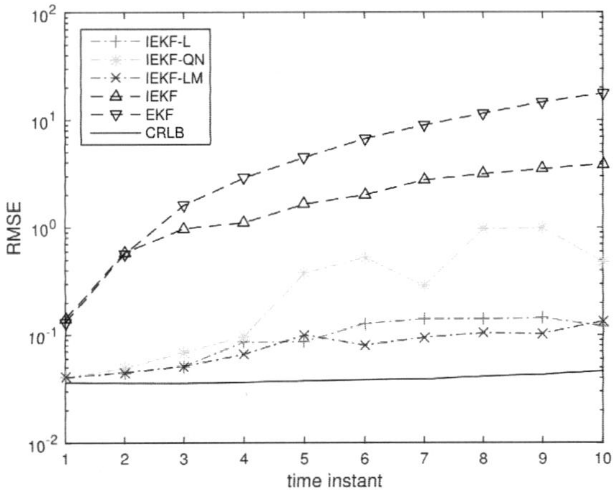

Figure 7. RMSE over time in the BOT example.

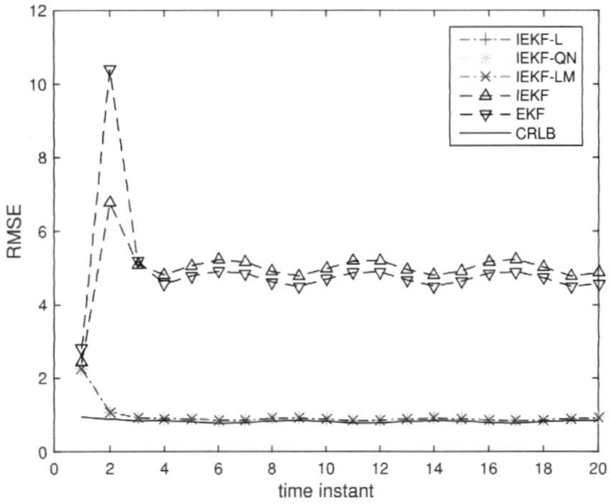

Figure 8. RMSE over time in the UNGM example.

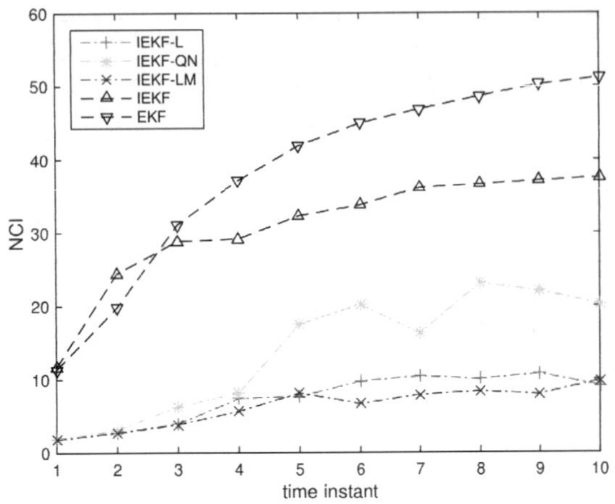

Figure 9. NCI over time in the BOT example.

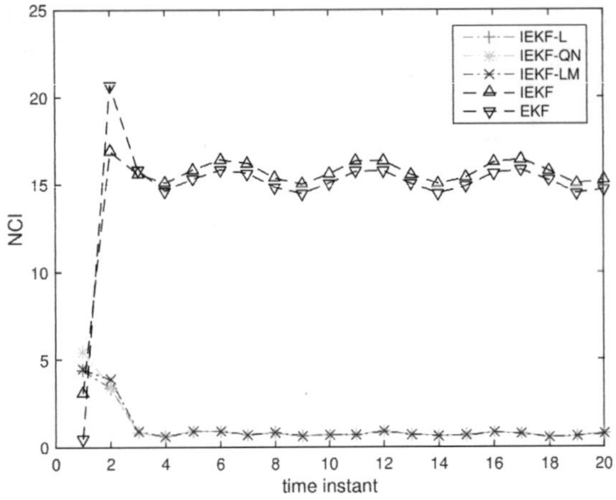

Figure 10. NCI over time in the UNGM example.

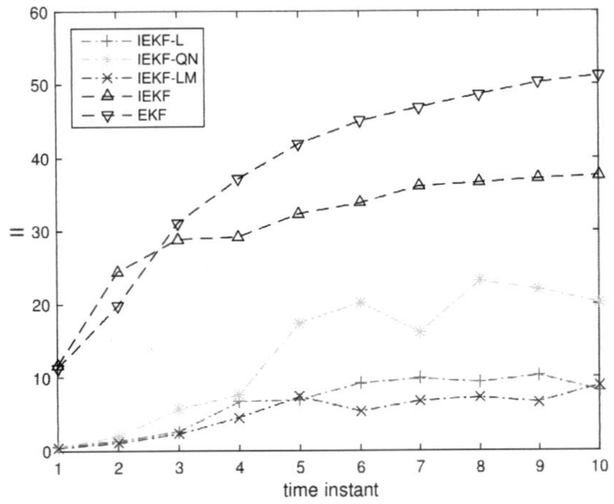

Figure 11. II over time in the BOT example.

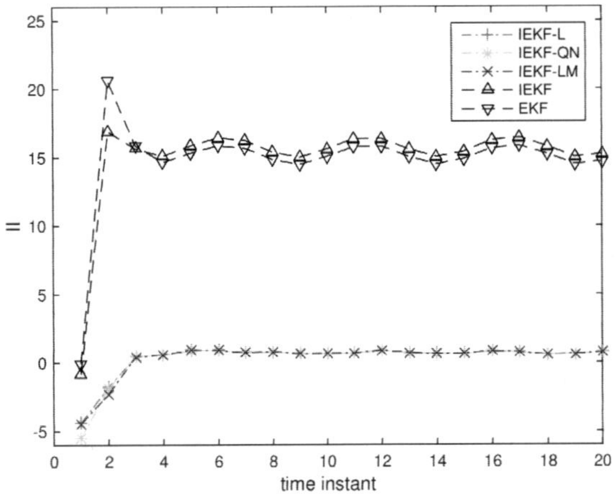

Figure 12. II over time in the UNGM example.

To complete the picture the time average of the RMSE and the NCI and computational time normalized with respect to the EKF are summed up in the Table 1 for the BOT example and in the Table 2 for the UNGM example, respectively. The measured time was normalized with respect to time of the EKF.

The results confirm the superiority of the IEKF-L and the IEKF-LM both in terms of the RMSE and the NCI. The results question the fixed parameter choice for the IEKFs and the IEKF-QN in general.

Note that the simulations were performed using the GaFT toolbox [17].

6. Conclusions

The paper dealt with state estimation of nonlinear stochastic dynamic systems. Particularly, a comprehensive analysis of the three iterated extended Kalman filters proposed in [9] was performed. The main goal was to verify that the filters provide not only better state estimates, but also the CM of the estimate errors. This was experimentally proved for the IEKF-L and the IEKF-LM using the noncredibility index, which compares the estimate error CM of the filter with the estimate error CM of a perfectly credible filter. Additionally, it was studied whether considering a fixed step-length leads to results comparable with optimized step-length. It was observed that the fixed step-length leads to

Table 1. BOT average results in 10^4 simulations.

Filter	time	RMSE	NCI
EKF	1.0	8.63	38.3
IEKF	2.2	2.22	30.8
IEKF-L	12.8	0.10	7.4
IEKF-L 0.5	6.2	0.31	13.7
IEKF-LM	16.8	0.08	6.3
IEKF-LM 0.5	7.1	0.69	13.8
IEKF-QN	43.0	0.49	13.9
IEKF-QN 0.5	17.5	86.09	25.3

Table 2. UNGM average results in 10^4 simulations.

filter	time	RMSE	NCI
EKF	1.0	4.98	2.2
IEKF	1.6	4.90	2.3
IEKF-L	7.4	0.98	0.7
IEKF-L 0.5	3.1	3.76	1.6
IEKF-LM	8.3	0.98	0.7
IEKF-LM 0.5	3.6	3.77	1.6
IEKF-QN	7.4	0.98	0.7
IEKF-QN 0.5	3.9	0.98	0.7

slightly better results than the IEKF can provide, however the results with the optimal step-length were better in order of magnitude. In fact, the results of the IEKFs with the optimized step-length in the UNGM problem were approaching to the objective limit of cognizability, i.e., to the Cramer-Rao lower bound.

Acknowledgments

This work was supported by the Czech Science Foundation, project no. GA 15-12068S.

References

[1] B. D. O. Anderson and J. B. Moore. *Optimal Filtering*. Englewood Cliffs, New Jersey: Prentice Hall Ins., 1979.

[2] A. Doucet, N. De Freitas, and N. Gordon, editors. *Sequential Monte Carlo Methods in Practice*. Springer, 2001. (Ed. Doucet A., de Freitas N., and Gordon N.).

[3] M. Šimandl and O. Straka. Nonlinear estimation by particle filters and Cramér-Rao bound. In *Proceedings of the 15th Triennial World Congress of the IFAC*, Barcelona, 2002.

[4] M. Šimandl, J. Královec, and T. Söderström. Advanced point – mass method for nonlinear state estimation. *Automatica*, 42(7):1133–1145, 2006.

[5] X Luo, I. M. Moroz, and I. Horeit. Scaled unscented transform Gaussian sum filter: Theory and application. *Physica D*, 239(10):684–701, 2010.

[6] A. H. Jazwinski. *Stochastic Processes and Filtering Theory*. Academic Press, 1970.

[7] Y. Bar-Shalom, X. R. Li, and T. Kirubarajan. *Estimation with Applications to Tracking and Navigation: Theory Algorithms and Software*. John Wiley & Sons, 2001.

[8] B. M. Bell and F. W. Cathey. The iterated Kalman filter update as a Gauss-Newton method. *IEEE Transactions on Automatic Control*, 38(3):294–297, 1993.

[9] M. A. Skoglund, G. Hendeby, and D. Axehill. Extended Kalman filter modifications based on an optimization view point. In *18th International Conference on Information Fusion*, 2015.

[10] X. R. Li and Z. Zhao. Evaluation of estimation algorithms part I: Incomprehensive measures of performance. *IEEE Transactions on Aerospace and Electronic Systems*, 42(4):1340–1358, 2006.

[11] M. Šimandl, J. Královec, and P. Tichavský. Filtering, predictive and smoothing Cramér-Rao bounds for discrete-time nonlinear dynamic filters. *Automatica*, 37(11):1703–1716, 2001.

[12] X. R. Li and Z. Zhao. Measuring estimators credibility: Noncredibility index. In *Proceedings of 2006 International Conference on Information Fusion*, Florence, 2006.

[13] R. E. Kalman. New results in linear filtering and prediction theory. *Journal of Basic Engineering*, pages 34–45, 1960. Transactions ASME.

[14] A. Björck. *Numerical Methods for Least Squares Problems*. Society for Industrial and Applied Mathematics, 1996.

[15] Jorge More. The Levenberg-Marquardt algorithm: Implementation and theory. In G.A. Watson, editor, *Numerical Analysis*, volume 630 of *Lecture Notes in Mathematics*, pages 105–116. Springer Berlin Heidelberg, 1978.

[16] Yuanxin Wu, Dewen Hu, Meiping Wu, and Xiaoping Hu. Unscented Kalman filtering for additive noise case: Augmented versus nonaugmented. *Signal Processing Letters, IEEE*, 12(5):357 – 360, 2005.

[17] J. Havlík, M. Šimandl and O. Straka. A New Practically Oriented Generation of Nonlinear Filtering Toolbox. In *Proceedings of the 17th IFAC Symposium on System Identification*, Beijing, 2015.

Interval observer for Linear Time Invariant (LTI) uncertain systems with state and unknown input estimations

N Ellero, D Gucik-Derigny, D Henry

University of Bordeaux, IMS-Lab, Automatic control group, 351 Cours de la libération,33405, Talence,France

E-mail: nicolas.ellero/david.gucik-derigny/david.henry@ims-bordeaux.fr

Abstract. In this paper, the problem of interval observer design for LTI uncertain systems is addressed. The aim is to estimate simultaneously the upper and lower bounds of unmeasurable states and unknown inputs. Based on a set membership approach, the proposed method interest consists in reducing the estimate conservatism assuming that all known and unknown inputs, disturbances, uncertainties and noises are bounded with a priori known bounds. For that, decoupling methods are used to reduce the latter propagation. The estimation problem is led back to a singular model representation. An example is given to illustrate the proposed design method.

Keywords : Uncertain systems, bounded uncertainties, state transformations, unknown input observers, interval observer.

1. INTRODUCTION

Interval observers appeared in the last two decades to estimate in a guaranteed way the admissible unmeasured states for uncertain systems. In the literature, many works has been dedicated to design interval observers assuming that all known and unknown inputs, disturbances, uncertainties and noises are bounded with a priori known bounds. There exists three main approaches to synthesize an interval observer. The first approach is based on the prediction/correction method as the well known Kalman filter approach [1, 2]. For that, the uncertain model knowledge is used to predict the state set at the next time instant and the incompatible state subset is deleted when new measured outputs are available. The second approach [3] performs the first one by considering the Müller theorem [4]. The uncertain system is transformed into two models without considering uncertainty effect, ensuring also the upper and the lower bound estimations of the state at any time. The last method is initially proposed by [5] dealing only with uncertain monotone systems. Two observers, based on the Luenberger observer structure, are computed to guarantee the upper and the lower bounds of the state at any time. This approach is based on the cooperative property [6] of observation errors which is one of the main challenge in interval observer design. Recent works propose new methods to relax this constraint for a larger class of systems [7, 8, 9, 10]. Thereafter, new approaches are proposed in [11, 12, 13, 14, 15] to design interval observers for more general class of uncertain systems .

Journal of Physics: Conference Series **659** (2015) 012023 doi:10.1088/1742-6596/659/1/012023

In presence of unknown inputs, conventional interval observer approach could be applied by considering unknown input as an additional disturbance with a priori known bounds. However, the state and unknown input estimates can be very large and then useless when disturbances, uncertainties and noises are propagated without decoupling a part of their effects. To overcome such problem, two methods were proposed in [16, 17] to design interval observers for LTI systems with noise and unknown input that reduce the jointly estimation conservatism for state and unknown input.

In this context, this paper proposes to extend the results of [16, 17]. The aim of this paper is to jointly estimate the unmeasurable states and the unknown inputs in presence of noises and uncertain parameters. For that, the proposed approach is based on a method developed for a class of nonlinear singular system [18] that can be applied on a large class of LTI systems. Nevertheless, the method proposed in [18] does not consider uncertain parameters and it is not applied in the context of interval observer. Inspired by the methods of [17] and [18], this paper proposes a new interval observer design for a large class of uncertain LTI systems.

This paper is organized as follows. In the section 2, some notations and background commonly used in interval observers theory are recalled. The problem statement and all considered assumptions useful for the interval observer design are introduced. Section 3 is devoted to the observer structure and demonstrations. Section 4 gives an example to illustrate the proposed method.

2. PRELIMINARIES

2.1. NOTATIONS AND BACKGROUND

For two matrices $A, B \in \mathbb{R}^{n\times m}$ (respectively two vectors $x_1, x_2 \in \mathbb{R}^m$), the relation $A \leq B$ (respectively $x_1 \leq x_2$) should be understood componentwise and define $A^+ = max(A, 0)$, $A^- = A^+ - A$. Given a matrix $A(\theta) \in \mathbb{R}^{n\times m}$ with $\theta \in \Theta \subset \mathbb{R}^r$ (respectively a vector x_1) and two matrices $\underline{A}, \overline{A} \in \mathbb{R}^{n\times m}$ (respectively two vectors $\underline{x_1}, \overline{x_1} \in \mathbb{R}^m$), we denote the set $\left[\begin{array}{cc}\underline{A} & \overline{A}\end{array}\right] = \left\{A(\theta) \in \mathbb{R}^{n\times m} \mid \underline{A} \leq A(\theta) \leq \overline{A},\ \forall\, \theta \in \Theta\right\}$ (respectively $\left[\begin{array}{cc}\underline{x_1} & \overline{x_1}\end{array}\right] = \left\{x_1 \in \mathbb{R}^m \mid \underline{x_1} \leq x_1 \leq \overline{x_1}\right\}$). Considering a locally essentially bounded signal u : $\mathbb{R}_+ \to \mathbb{R}$, we define the $\mathcal{L}_\infty$ norm of u by the symbol $\|u\|_{[t_0,t_1]}$:

$$\|u\|_{[t_0,t_1]} = ess\ sup|u(t)|, t \in [t_0, t_1]$$

Lemma 1. *[19] Consider a vector variable $x \in \mathbb{R}^n$ such that $\underline{x} \leq x \leq \overline{x}$ with $\underline{x}, \overline{x} \in \mathbb{R}^n$*
(1) Let $A \in \mathbb{R}^{n\times m}$ be a constant matrix, then

$$A^+\underline{x} - A^-\overline{x} \leq Ax \leq A^+\overline{x} - A^-\underline{x} \tag{1}$$

(2) Let $A \in \mathbb{R}^{n\times m}$ be a matrix variable, $\underline{A} \leq A \leq \overline{A}$ for some $\underline{A}, \overline{A} \in \mathbb{R}^{n\times m}$, then

$$\underline{A}^+\underline{x}^+ - \overline{A}^+\underline{x}^- - \underline{A}^-\overline{x}^+ + \overline{A}^-\overline{x}^- \leq Ax \leq \overline{A}^+\overline{x}^+ - \underline{A}^+\overline{x}^- - \overline{A}^-\underline{x}^+ + \underline{A}^-\underline{x}^- \tag{2}$$

Proof. See [19]. □

A square matrix $A \in \mathbb{R}^{n\times n}$ is called Hurwitz if the real part of all its eigenvalues is strictly negative definite and is said to be Metzler if all its off-diagonal elements are nonnegative.

Lemma 2. *[6] Given a non-autonomous system described by $x(t) = Ax(t) + B(t)$ where A is Metzler matrix and $B(t) \leq 0$ ($B(t) \geq 0$). Then $x(t) \leq 0$ ($x(t) \geq 0$) $\forall t \geq 0$ provided that $x(0) \leq 0$ ($x(0) \geq 0$)*

2.2. HOSM differentiator

Consider a measurement signal $y(t) = \psi(t) + \gamma(t)$ with $y(t), \psi(t), \gamma(t) \in \mathbb{R}^q$. $\psi(t)$ denotes the useful signal assumed to be continuously differentiable and $\gamma(t) \in \mathcal{L}_\infty$ represents the measurement noise assumed to be such that $|\gamma(t)| \leq \overline{\gamma}$. Then the first derivative of $\psi(t)$ can be estimated using a HOSM differentiator [20] described by :

$$\begin{cases} q_0(t) &= v_0(t) \\ v_0(t) &= q_1(t) - \lambda_0|q_0(t) - y(t)|^{\frac{1}{2}} sign(q_0(t) - y(t)) \\ q_1(t) &= -\lambda_1 sign(q_0(t) - y(t)) \end{cases} \tag{3}$$

where λ_k, $k = 0, 1$ are positive parameters.
The sufficient conditions allowing the convergence of the estimate $q_1(t)$ to $\psi(t)$ is related to the appropriate choice of the parameters λ_0 and λ_1, see [21].

Theorem 1. *[21] Let an input signal $\psi(t)$ be a continuously differentiable signal and $\gamma(t) \in \mathcal{L}_\infty$ then there exist $0 \leq T < \infty$ and a constant $\mu > 0$ such that for all $t \geq T$:*

$$|q_1(t) - \psi(t)| \leq \mu \,\|\gamma\|^{\frac{1}{2}} \tag{4}$$

Proof. see [21] □

Based on theorem 1 and the boundedness assumption on the noise, the first derivative error is defined as :

$$|q_1(t) - \psi(t)| \leq \mu\overline{\gamma}^{\frac{1}{2}} \tag{5}$$

The upper and lower bounds for the first derivative of $\psi(t)$ can be computed based on (5) :

$$\begin{aligned} \overline{\psi}(t) &= q_1(t) + \mu\overline{\gamma}^{\frac{1}{2}} \\ \underline{\psi}(t) &= q_1(t) - \mu\overline{\gamma}^{\frac{1}{2}} \end{aligned} \tag{6}$$

Remark (1). *In [21], the general case is considered in order to estimate the sth derivative of a noisy signal.*

2.3. PROBLEM STATEMENT

Consider an uncertain system described by :

$$\begin{cases} x(t) &= A(\theta)x(t) + B(\theta)u(t) + F_1(\theta)d_1(t) + H_1(\theta)v(t) \\ y(t) &= C(\theta)x(t) + D(\theta)u(t) + F_2(\theta)d_2(t) + H_2(\theta)v(t) \end{cases} \tag{7}$$

where $x \in \mathbb{R}^n$, $y \in \mathbb{R}^p$, $u \in \mathbb{R}^m$ represent respectively the state, the output and the known input. $d_1 \in \mathbb{R}^{k_1}$ and $d_2 \in \mathbb{R}^{k_2}$ denote the state and sensor faults, considered here as two unknown inputs. $\theta \in \Theta$ defines the unknown parameter vector where $\Theta \in [\underline{\theta}, \overline{\theta}]$ with $\underline{\theta}$,$\overline{\theta} \in \mathbb{R}^l$ assumed to be known a priori. $v \in \mathbb{R}^s$ describes the noise vector such that $v \in [\underline{v}, \overline{v}]$ where $\underline{v}, \overline{v} \in \mathbb{R}^s$ are known vectors with $s = n + p$. Matrices $A(\theta) \in \mathbb{R}^{n \times n}$, $B(\theta) \in \mathbb{R}^{n \times m}$, $F_1(\theta) \in \mathbb{R}^{n \times k_1}$, $H_1(\theta) \in \mathbb{R}^{n \times s}$, $C(\theta) \in \mathbb{R}^{p \times n}$, $D(\theta) \in \mathbb{R}^{p \times m}$, $F_2(\theta) \in \mathbb{R}^{p \times k_2}$ and $H_2(\theta) \in \mathbb{R}^{p \times s}$.

This section gives some useful assumptions to design an interval observer which estimates jointly the lower and the upper bounds of the state x and the unknown inputs d_1, d_2.

Assumption 1. $\forall \theta \in [\underline{\theta}, \overline{\theta}]$, *there exist three matrices* $\underline{\;X_i}$, $\overline{\;X_i} \in \mathbb{R}^{\alpha \times \beta}$ *and* $X_i \in \mathbb{R}^{\alpha \times \beta}$ *such that* $X_i(\theta) = X_i + \;\; X_i(\theta)$ *where* $\;\; X_i(\theta) \in [\;\underline{\;X_i} \;\; \overline{\;X_i}\;]$.

Remark (2). *In this assumption, all uncertainties are assumed to be bounded by known bounds and are assumed to have a linear additive form.*

Assumption 2. *There exist two vectors $w_1(t) \in \mathbb{R}^{q_1}$, $w_2(t) \in \mathbb{R}^{q_2}$ and three constant matrices $G_{11} \in \mathbb{R}^{n\times q_1}$, $G_{12} \in \mathbb{R}^{n\times q_2}$, $G_2 \in \mathbb{R}^{p\times q_2}$ such that :*

$$G_{11}w_1(\theta,t) + G_{12}w_2(\theta,t) = A(\theta)x(t) + F_1(\theta)d_1(t) \tag{8a}$$

$$G_2w_2(\theta,t) = C(\theta)x(t) + F_2(\theta)d_2(t) \tag{8b}$$

Remark (3). *The uncertain state distribution matrices $A(\theta)$, $C(\theta)$ provided by assumption 1 and all unknown inputs d_1, d_2 can be rewritten in a new form to create two unknown input vectors with constant known matrix distributions. This assumption is necessary to be able to decouple a part of unknown inputs and disturbance effects.*

Assumption 3.

$$rank\left(\begin{bmatrix} \underline{F_i}^+ & 0 & 0 & \overline{F_i}^- \\ 0 & \overline{F_i}^+ & \underline{F_i}^- & 0 \\ 0 & \underline{F_i}^- & \overline{F_i}^+ & 0 \\ \overline{F_i}^- & 0 & 0 & \underline{F_i}^+ \end{bmatrix}\right) = 4k_i, \ \forall i = 1,2 \tag{9}$$

Remark (4). *Based on [17, 22], this assumption is necessary in order to estimate the upper and lower bounds of the unknown inputs $d_1(t)$ and $d_2(t)$ from equations (8a-8b). An inversion method is applied to estimate the unknown inputs considering the uncertainties of $F_i(\theta)$ with $i = 1,2$. The lemma 1 allows to transform the uncertain equations (8a-8b) into a determinist form such that the inversion is feasible if assumption 3 holds .*

Under Assumption 1 and Assumption 2 and based on equalities (8a)-(8b), the system (7) can be rewritten as :

$$\left\{\begin{array}{ll} x(t) & = Ax(t) + B(\theta)u(t) + H_1(\theta)v(t) + G_{11}w_1(\theta,t) + G_{12}w_2(\theta,t) \\ y(t) & = Cx(t) + D(\theta)u(t) + H_2(\theta)v(t) + G_2w_2(\theta,t) \end{array}\right. \tag{10}$$

Assumption 4.

$$\begin{array}{ll} rank(CG_{11}) = rank(G_{11}) = q_1, & q_1 < n \\ rank\left(\begin{bmatrix} CG_{11} & G_2 \end{bmatrix}\right) = q_1 + q_2, & q_1 + q_2 < p \end{array} \tag{11}$$

Remark (5). *This assumption is standard in the literature [16, 17, 22] for observer design, especially for the unknown input estimation. From this assumption, a non-singular matrix can be computed to decouple the effects of w_1 and secondly, a joint estimation of the state x and the unknown input w_2 is achived by considering an augmented state vector.*

A non-singular state transformation is applied on (10) in order to decompose the state in two parts. One part is still affected by the unknown input w_1 whereas the second is decoupled from unknown input effect:

$$x(t) = M_0z(t), \quad M_0 = [G_{11}, G], \quad G \in \mathbb{R}^{n\times(n-q_1)} \tag{12}$$

Then the system (10) becomes:

$$\left\{\begin{array}{ll} z(t) & = A_0z(t) + B_0(\theta)u(t) + H_0(\theta)v(t) + G_{01}w_1(\theta,t) + G_{02}w_2(\theta,t) \\ y(t) & = C_0z(t) + D(\theta)u(t) + H_2(\theta)v(t) + G_2w_2(\theta,t) \end{array}\right. \tag{13}$$

where

$$A_0 = M_0^{-1}AM_0 = \begin{bmatrix} A_{01} \\ A_{02} \end{bmatrix}, B_0(\theta) = M_0^{-1}B(\theta) = \begin{bmatrix} B_{01}(\theta) \\ B_{02}(\theta) \end{bmatrix}, H_0(\theta) = M_0^{-1}H_1(\theta) = \begin{bmatrix} H_{01}(\theta) \\ H_{02}(\theta) \end{bmatrix},$$

$$G_{01} = M_0^{-1}G_{11} = \begin{bmatrix} I_{q_1} \\ 0_{(n-q_1)\times q_1} \end{bmatrix}, G_{02} = M_0^{-1}G_{12} = \begin{bmatrix} \tilde{G}_{01} \\ \tilde{G}_{02} \end{bmatrix}, C_0 = CM_0.$$

Now, keeping only the part of the dynamics from z decoupled from the unknown input w_1 effect, the system is now rewritten by considering the singular representation of (13) :

$$\begin{cases} Ez(t) &= A_{02}z(t) + B_{02}(\theta)u(t) + H_{02}(\theta)v(t) + \tilde{G}_{02}w_2(\theta,t) \\ y(t) &= C_0 z(t) + D(\theta)u(t) + H_2(\theta)v(t) + G_2 w_2(\theta,t) \end{cases} \tag{14}$$

where $E = \begin{bmatrix} 0_{(n-q_1)\times q_1} & I_{(n-q_1)} \end{bmatrix}$.
Finally, augmenting the state z with the unknown input w_2, the system (14) is rewritten :

$$\begin{cases} \tilde{E}\tilde{z}(t) &= \tilde{A}\tilde{z}(t) + \tilde{B}(\theta)u(t) + \tilde{H}_1(\theta)v(t) \\ y(t) &= \tilde{C}\tilde{z}(t) + D(\theta)u(t) + H_2(\theta)v(t) \end{cases} \tag{15}$$

where $\tilde{z}(t) = \begin{bmatrix} z(t) \\ w_2(\theta,t) \end{bmatrix}$, $\tilde{E} = \begin{bmatrix} E & 0_{(n-q_1)\times q_2} \end{bmatrix}$, $\tilde{A} = \begin{bmatrix} A_{02} & \tilde{G}_{02} \end{bmatrix}$, $\tilde{C} = \begin{bmatrix} C_0 & G_2 \end{bmatrix}$, $\tilde{B}(\theta) = B_{02}(\theta)$, $\tilde{H}_1(\theta) = H_{02}(\theta)$.

Assumption 5. *The pair ($T\tilde{A}$,$\tilde{C}$) is observable (or at least detectable) where $T \in \mathbb{R}^{(n+q_2)\times(n-q_1))}$ is a matrix chosen such that $\begin{bmatrix} T & F \end{bmatrix} \begin{bmatrix} \tilde{E} \\ \tilde{C} \end{bmatrix} = I_{n+q_2}$ with F a matrix of appropriate dimension.*

Remark (6). *Assumption 5 is necessary to guarantee the stability of the interval observer proposed in the sequel. The choice of matrices T and F is guaranteed if the matrix $\begin{bmatrix} \tilde{E} \\ \tilde{C} \end{bmatrix}$ is full column rank. This last condition holds if Assumption 4 holds also.*

3. MAIN RESULT

3.1. UNKNOWN INPUT INTERVAL OBSERVER

In this subsection, an interval observer based on Assumption 1-5 is proposed to estimate the state of system (15). For that, a non-singular state transformation $\tilde{z}(t) = M_1 r(t)$, $M_1 \in \mathbb{R}^{(n+q_2)\times(n+q_2)}$ is applied such that the system (15) is rewritten as :

$$\begin{cases} \tilde{E}M_1 r(t) &= \tilde{A}M_1 r(t) + (\tilde{B} + \tilde{B}(\theta))u(t) + \tilde{H}_1(\theta)v(t) \\ y(t) &= \tilde{C}M_1 r(t) + (D + D(\theta))u(t) + H_2(\theta)v(t) \end{cases} \tag{16}$$

Then, the interval observer for (16) is now introduced as follows:

$$\begin{cases} \xi_1(t) &= M_1^{-1}NM_1\xi_1(t) + M_1^{-1}Jy(t) + M_1^{-1}Ku(t) - \overline{W}_1 \\ \underline{r}(t) &= \xi_1(t) + M_1^{-1}Fy(t) - M_1^{-1}FDu(t) - \overline{W}_2 \end{cases} \tag{17a}$$

$$\begin{cases} \xi_2(t) &= M_1^{-1}NM_1\xi_2(t) + M_1^{-1}Jy(t) + M_1^{-1}Ku(t) + \overline{W}_1 \\ \overline{r}(t) &= \xi_2(t) + M_1^{-1}Fy(t) - M_1^{-1}FDu(t) + \overline{W}_2 \end{cases} \tag{17b}$$

where

$$\overline{W}_1 = \overline{\tilde{W}_1}^{+} \overline{\begin{bmatrix} u(t) \\ v(t) \end{bmatrix}}^{+} - \underline{\tilde{W}_1}^{+} \overline{\begin{bmatrix} u(t) \\ v(t) \end{bmatrix}}^{-} - \overline{\tilde{W}_1}^{-} \underline{\begin{bmatrix} u(t) \\ v(t) \end{bmatrix}}^{+} + \underline{\tilde{W}_1}^{-} \underline{\begin{bmatrix} u(t) \\ v(t) \end{bmatrix}}^{-}$$

$$\overline{W}_2 = \overline{\tilde{W}_2}^{+} \overline{\begin{bmatrix} u(t) \\ v(t) \end{bmatrix}}^{+} - \underline{\tilde{W}_2}^{+} \overline{\begin{bmatrix} u(t) \\ v(t) \end{bmatrix}}^{-} - \overline{\tilde{W}_2}^{-} \underline{\begin{bmatrix} u(t) \\ v(t) \end{bmatrix}}^{+} + \underline{\tilde{W}_2}^{-} \underline{\begin{bmatrix} u(t) \\ v(t) \end{bmatrix}}^{-}$$

with

$$\begin{array}{llll} \tilde{W}_1 & = \left[\begin{array}{cc} \tilde{W}_{11} & \tilde{W}_{12} \end{array} \right] & \tilde{W}_2 & = \left[\begin{array}{cc} \tilde{W}_{21} & \tilde{W}_{22} \end{array} \right] \\ \tilde{W}_{11} & = M_1^{-1}(J \quad D(\theta) - T \quad \tilde{B}(\theta)) & \tilde{W}_{21} & = M_1^{-1}F \quad D(\theta) \\ \tilde{W}_{12} & = M_1^{-1}(JH_2(\theta) - T\tilde{H}_1(\theta)) & \tilde{W}_{22} & = M_1^{-1}FH_2(\theta) \end{array}$$

Here, M_1, N, J, K, F, and T are unknown matrices of appropriate dimension such that the proposed methodology is able to be determined.

In the sequel, we denote the upper and lower bounds of observation errors by :

$$\mathcal{E}_1(t) = \xi_1(t) - M_1^{-1}T\tilde{E}M_1r(t) \tag{18a}$$

$$\mathcal{E}_2(t) = \xi_2(t) - M_1^{-1}T\tilde{E}M_1r(t) \tag{18b}$$

$$e_1(t) = \underline{r}(t) - r(t) \tag{18c}$$

$$e_2(t) = \overline{r}(t) - r(t) \tag{18d}$$

Theorem 2. *Consider $\underline{r}_0 \in \mathbb{R}^{n+q_2}, \overline{r}_0 \in \mathbb{R}^{n+q_2}$ such that $\underline{r}_0 \leq r(0) \leq \overline{r}_0$. Under Assumptions 1-5, there exist two non-singular constant matrices $N \in \mathbb{R}^{(n+q_2)\times(n+q_2)}$, $M_1 \in \mathbb{R}^{(n+q_2)\times(n+q_2)}$ such that $M_1^{-1}NM_1$ is Hurwitz and Metzler. If the following equalities are satisfied :*

$$\mathcal{E}_1(0) \leq 0, \ \mathcal{E}_2(0) \geq 0 \tag{19a}$$

$$NT\tilde{E} + J\tilde{C} - T\tilde{A} = 0 \tag{19b}$$

$$K = T\tilde{B} - JD \tag{19c}$$

$$\left[\begin{array}{cc} T & F \end{array} \right] \left[\begin{array}{c} \tilde{E} \\ \tilde{C} \end{array} \right] = I_{n+q_2} \tag{19d}$$

Then $\underline{r}(t) \leq r(t) \leq \overline{r}(t)$ $\forall t \geq 0$. The system (17) is an unknown input interval observer for (16).

Proof. The proof of the Theorem 2 is organized in two steps. The first step shows the stable cooperative dynamic property of (18a) and (18b). The second step ensures the negativity property of (18c) and the positivity property of (18d) for all $t \geq 0$.
Step 1 : Using the derivative of expressions (18a) and (18b) , we get :

$$\mathcal{E}_1(t) = \quad M_1^{-1}NM_1\mathcal{E}_1(t) + M_1^{-1}(NT\tilde{E} + J\tilde{C} - T\tilde{A})M_1r(t) + W_1 - \overline{W}_1 \tag{20a}$$

$$\mathcal{E}_2(t) = \quad M_1^{-1}NM_1\mathcal{E}_2(t) + M_1^{-1}(NT\tilde{E} + J\tilde{C} - T\tilde{A})M_1r(t) + W_1 + \overline{W}_1 \tag{20b}$$

where

$$\begin{array}{rl} W_1 = & M_1^{-1}(J(D + \quad D(\theta)) + K - T(\tilde{B} + \quad \tilde{B}(\theta))u(t) \\ + & M_1^{-1}(JH_2(\theta) - T\tilde{H}_1(\theta))v(t) \end{array}$$

It is trivial to prove with the Lemma 1 and the condition (19c) that $W_1 - \overline{W}_1 \leq 0$ and $W_1 + \overline{W}_1 \geq 0$, $\forall t \geq 0$. Using the same Lemma, there exist $\xi_1(0) \leq M_1^{-1}T\tilde{E}M_1r(0)$, $\xi_2(0) \geq M_1^{-1}T\tilde{E}M_1r(0)$ provided by $\underline{r}_0 \leq r(0) \leq \overline{r}_0$ and it follows that $\mathcal{E}_1(0) \leq 0$, $\mathcal{E}_2(0) \geq 0$. Considering the property (19b) satisfied then (20a) and (20b) are rewritten as:

$$\mathcal{E}_1(t) = \quad M_1^{-1}NM_1\mathcal{E}_1(t) + W_1 - \overline{W}_1 \tag{21a}$$

$$\mathcal{E}_2(t) = \quad M_1^{-1}NM_1\mathcal{E}_2(t) + W_1 + \overline{W}_1 \tag{21b}$$

From previous results and considering $M_1^{-1}NM_1$ Hurwitz and Metzler, all conditions for the Lemma 2 hold. Then, we deduce that (21a) and (21b) are stable cooperative dynamical systems.

Step 2 : To prove $e_1(t) \leq 0$ and $e_2(t) \geq 0 \ \forall t \geq 0$, we combine the second equation of (17a) with (18a) and the second equation of (17b) with (18b) :

$$\underline{r}(t) = \mathcal{E}_1(t) + M_1^{-1} \begin{bmatrix} T & F \end{bmatrix} \begin{bmatrix} \tilde{E} \\ \tilde{C} \end{bmatrix} M_1 r(t) + W_2 - \overline{W}_2 \tag{22a}$$

$$\overline{r}(t) = \mathcal{E}_2(t) + M_1^{-1} \begin{bmatrix} T & F \end{bmatrix} \begin{bmatrix} \tilde{E} \\ \tilde{C} \end{bmatrix} M_1 r(t) + W_2 + \overline{W}_2 \tag{22b}$$

where $W_2 = M_1^{-1}F \quad D(\theta)u(t) + M_1^{-1}FH_2(\theta)v(t)$

Under Assumption 4, there exist $T \in \mathbb{R}^{(n+q_2)\times(n-q_1)}$ and $F \in \mathbb{R}^{(n+q_2)\times p}$ such that $\begin{bmatrix} T & F \end{bmatrix} \begin{bmatrix} \tilde{E} \\ \tilde{C} \end{bmatrix} = I_{n+q_2}$. Equations (22a) and (22b) are now defined as :

$$e_1(t) = \mathcal{E}_1(t) + W_2 - \overline{W_2} \tag{23a}$$

$$e_2(t) = \mathcal{E}_2(t) + W_2 + \overline{W_2} \tag{23b}$$

Lemma 1 allows us to prove $(W_2 - \overline{W_2}) \leq 0$ and $(W_2 + \overline{W_2}) \geq 0$. Finally, with the previous results proposed in the first step, we obtain $e_1(t) \leq 0$ and $e_2(t) \geq 0$, $\forall t \geq 0$. □

Remark (7). *As mentioned in Theorem 2, an unknown input interval observer is designed if there exist two non-singular constant matrices $N \in \mathbb{R}^{(n+q_2)\times(n+q_2)}$, $M_1 \in \mathbb{R}^{(n+q_2)\times(n+q_2)}$ such that $M_1^{-1}NM_1$ is Hurwitz and Metzler. A practical method to compute N, M_1 and J is based on Assumption 5 and on the equations (19b),(19d). From (19d), we get $T\tilde{E} = I - F\tilde{C}$. Using the previous expression in (19b), an expression of N is chosen such that : $N = T\tilde{A} - L\tilde{C}$ with $L = J - NF$. Finally, from Assumption 5, it is shown that there exists an Hurwitz matrix N. Using a methodology that leads to get the Metzler and Hurwitz properties for the matrix $M_1^{-1}NM_1$ as described in [8], M_1 is computed and J is deduced from $J = NF + L$.*

3.2. STATES AND UNKNOWN INPUTS ESTIMATION

Corollary 1. *Assuming that Lemma 1 and Theorem 2 hold, there exist $\forall t \geq 0$, $\underline{x}(t), \overline{x}(t) \in \mathbb{R}^n$ and $\underline{w}_2(\theta,t), \overline{w}_2(\theta,t) \in \mathbb{R}^n$ such that :*

$$\begin{cases} \underline{x}(t) \leq x(t) \leq \overline{x}(t) \\ \underline{w}_2(\theta,t) \leq w_2(\theta,t) \leq \overline{w}_2(\theta,t) \end{cases} \tag{24}$$

Proof. If Theorem 2 holds then $\underline{r}(t) \leq r(t) \leq \overline{r}(t)$. Recalling that $\tilde{z}(t) = M_1 r(t)$ where $\tilde{z}(t) = \begin{bmatrix} z(t) \\ w_2(\theta,t) \end{bmatrix}$ and $x(t) = M_0 z(t)$, it is shown under Lemma 1 that:

$$\begin{cases} \underline{\tilde{x}}(t) = M^+ \underline{r}(t) - M^- \overline{r}(t) \\ \overline{\tilde{x}}(t) = M^+ \overline{r}(t) - M^- \underline{r}(t) \end{cases} \tag{25}$$

where $\tilde{x}(t) = \begin{bmatrix} x(t) \\ w_2(\theta,t) \end{bmatrix}$, $M = \tilde{M}_0 M_1$ with $\tilde{M}_0 = \begin{bmatrix} M_0 & 0 \\ 0 & I_{q_2} \end{bmatrix}$. □

Corollary 2. *Considering corollary 1 and Theorem 2 hold, there exist $\forall t \geq 0$, $\underline{d}_1(t), \overline{d}_1(t) \in \mathbb{R}^{k_1}$ and $\underline{d}_2(t), \overline{d}_2(t) \in \mathbb{R}^{k_2}$ such that:*

$$\begin{cases} \underline{d_1}(t) \leq d_1(t) \leq \overline{d_1}(t) \\ \underline{d_2}(t) \leq d_2(t) \leq \overline{d_2}(t) \end{cases} \tag{26}$$

Proof. **Step 1 :** This part is devoted to prove that $\underline{d_1}(t) \leq d_1(t) \leq \overline{d_1}(t)$. The differentiation of y in (10) leads to :

$$y(t) = Cx(t) + D(\theta)u(t) + H_2(\theta)v(t) + G_2 w_2(\theta, t) \tag{27}$$

Injecting the expression of x described in (10), we get :

$$\begin{aligned} y(t) = \; & CAx(t) + CB(\theta)u(t) + CH_1(\theta)v(t) + CG_{11}w_1(\theta,t) + CG_{12}w_2(\theta,t) \\ & + D(\theta)u(t) + H_2(\theta)v(t) + G_2 w_2(\theta,t) \end{aligned} \tag{28}$$

Under Assumption 4, there exists a non-singular output transformation $\tilde{M} = \begin{bmatrix} CG_{11} & G_2 & R \end{bmatrix}$, $R \in \mathbb{R}^{p\times(p-q_1-q_2)}$ such that $\tilde{M}^{-1} = \begin{bmatrix} \tilde{M_1} \\ \tilde{M_2} \end{bmatrix}$ with $\tilde{M_1} \in \mathbb{R}^{q_1 \times p}$, $\tilde{M_2} \in \mathbb{R}^{(p-q_1)\times p}$. Multiplying (28) by $\tilde{M_1}^{-1}$ and isolating the expression $w_1(t)$ in the latter, we get :

$$\begin{aligned} w_1(t) = \; & \psi(t) - \tilde{M_1}CAx(t) - \tilde{M_1}CB(\theta)u(t) - \tilde{M_1}CH_1(\theta)v(t) - \tilde{M_1}CG_{12}w_2(\theta,t) \\ & - \tilde{M_1}D(\theta)u(t) \end{aligned} \tag{29}$$

where $\psi(t) = \tilde{M_1}y(t) - \tilde{M_1}H_2(\theta)v(t)$.
To reduce the conservatism, the augmented state $\tilde{x}$ defined in (25) is used in (29) :

$$w_1(t) = \psi(t) - Sr(t) - \tilde{M_1}CB(\theta)u(t) - \tilde{M_1}CH_1(\theta)v(t) - \tilde{M_1}D(\theta)u(t) \tag{30}$$

where $S = \tilde{M_1}C \begin{bmatrix} A & G_{12} \end{bmatrix} M$ with M defined in (25).
From Lemma 1 and equation (30), the expressions of $\underline{w}_1$ $\overline{w}_1$ are deduced.

Remark (8). *For the brevity of presentation, the expressions of $\underline{w}_1$ $\overline{w}_1$ are not detailed but can be easily obtained.*

Remark (9). *ψ and $\tilde{M_1}H_2(\theta)v$ are respectively assumed to be the useful signal and the noise effect (see part 2.2 for further details). An HOSM differentiator is computed to estimate the first derivative of ψ and u in (30).*

Applying Lemma 1, the equation (8a) is rewritten as :

$$S_1 \tilde{d_1}(t) = P_1(t) \tag{31}$$

where

$$S_1 = \begin{bmatrix} \underline{F_1}^+ & 0 & 0 & \overline{\tilde{F_1}}^- \\ 0 & \overline{F_1}^+ & \underline{F_1}^- & 0 \\ 0 & \underline{F_1}^- & \overline{F_1}^+ & 0 \\ \overline{F_1}^- & 0 & 0 & \underline{F_1}^+ \end{bmatrix}, \quad \tilde{d_1}(t) = \begin{bmatrix} \underline{d_1}(t)^+ \\ \underline{d_1}(t)^- \\ \overline{d_1}(t)^+ \\ \overline{d_1}(t)^- \end{bmatrix}, \quad P_1(t) = \begin{bmatrix} P_{11}(t) \\ P_{12}(t) \\ P_{13}(t) \\ P_{14}(t) \end{bmatrix}$$

with

$$\begin{aligned} P_{11}(t) &= G_{11}^+ \underline{w}_1(t) + G_{12}^+ \underline{w}_2(t) + \underline{A}^+ \overline{x}(t)^- + \overline{A}^- \underline{x}(t)^+ \\ P_{12}(t) &= G_{11}^- \overline{w}_1(t) + G_{12}^- \overline{w}_2(t) + \overline{A}^+ \overline{x}(t)^+ + \underline{A}^- \underline{x}(t)^- \\ P_{13}(t) &= G_{11}^+ \overline{w}_1(t) + G_{12}^+ \overline{w}_2(t) + \overline{A}^+ \underline{x}(t)^- + \underline{A}^- \overline{x}(t)^+ \\ P_{14}(t) &= G_{11}^- \underline{w}_1(t) + G_{12}^- \underline{w}_2(t) + \underline{A}^+ \underline{x}(t)^+ + \overline{A}^- \overline{x}(t)^- \end{aligned}$$

Finally, under Assumption 3, the matrix S_1 is full column rank and it follows that $\tilde{d_1}(t) = (S_1^T S_1)^{-1} S_1^T P_1(t)$. The upper and the lower bounds for the unknown input d_1 are obtained by $\underline{d_1}(t) = \underline{d_1}(t)^+ - \underline{d_1}(t)^-$ and $\overline{d_1}(t) = \overline{d_1}(t)^+ - \overline{d_1}(t)^-$.

Step 2 : To prove that $\underline{d_2}(t) \leq d_2(t) \leq \overline{d_2}(t)$, we consider that step 1 holds. Using Lemma 1, the equation (8b) is rewritten as :

$$S_2\tilde{d}_2(t) = P_2(t) \tag{32}$$

where

$$S_2 = \begin{bmatrix} \underline{F_2}^+ & 0 & 0 & \overline{F_2}^- \\ 0 & \overline{F_2}^+ & \underline{F_2}^- & 0 \\ 0 & \underline{F_2}^- & \overline{F_2}^+ & 0 \\ \overline{F_2}^- & 0 & 0 & \underline{F_2}^+ \end{bmatrix}, \tilde{d}_2(t) = \begin{bmatrix} \underline{d_2}(t)^+ \\ \underline{d_2}(t)^- \\ \overline{d_2}(t)^+ \\ \overline{d_2}(t)^- \end{bmatrix},$$

$$P_2(t) = \begin{bmatrix} G_2^+\underline{w_2}(t) + \underline{\quad C}^+\overline{x}(t)^- + \overline{\quad C}^-\underline{x}(t)^+ \\ G_2^-\overline{w_2}(t) + \overline{\quad C}^+\overline{x}(t)^+ + \underline{\quad C}^-\underline{x}(t)^- \\ G_2^+\overline{w_2}(t) + \overline{\quad C}^+\underline{x}(t)^- + \underline{\quad C}^-\overline{x}(t)^+ \\ G_2^-\underline{w_2}(t) + \underline{\quad C}^+\underline{x}(t)^+ + \overline{\quad C}^-\overline{x}(t)^- \end{bmatrix}$$

Under Assumption 3, the matrix S_2 is full column rank and it follows that $\tilde{d}_2(t) = (S_2^TS_2)^{-1}S_2^TP_2(t)$. The upper and the lower bounds for the unknown input d_2 are obtained by $\underline{d_2}(t) = \underline{d_2}(t)^+ - \underline{d_2}(t)^-$ and $\overline{d_2}(t) = \overline{d_2}(t)^+ - \overline{d_2}(t)^-$. □

4. EXAMPLE

4.1. MODEL DESCRIPTION

To illustrate the proposed approach, consider the nonlinear model of a single-flexible joint robot coupled with a DC motor described by :

$$\begin{cases} L\frac{di(t)}{dt} + Ri(t) = V(t) - K_e\theta_m(t) \\ J_m\theta_m(t) + k(\theta_m(t) - \theta_l(t)) + B\theta_m(t) = K_ii(t) \\ J_l\theta_l(t) + k(\theta_l(t) - \theta_m(t)) + mgh\ sin(\theta_l(t)) = 0 \end{cases} \tag{33}$$

where i, V, θ_m, θ_m and θ_m are respectively the current, the tension, the angular position, the angular velocity and the angular acceleration of the motor. θ_l, θ_l and θ_l are respectively the angular position, the angular velocity and the angular acceleration of the link.
For the simulation, we consider k unknown such that $k = k_0 + \quad k$ with $\quad k \in [-\overline{\quad k}\quad \overline{\quad k}] = \theta$. The measurement output is defined as $y(t) = \psi(t) + v(t)$, where $\psi(t) = \begin{bmatrix} \theta_m(t) & (\theta_m(t) - \theta_l(t)) & i(t) & (\theta_m(t) - \theta_l(t)) & \theta_l(t) \end{bmatrix}^T$ and the measurement noise $v(t)$ belongs in $\begin{bmatrix} -\overline{v} & \overline{v} \end{bmatrix}$ with $\overline{v} = 0.03I_{5\times1}$. In the sequel, we consider $d_1(t) = sin(\theta_l(t))$ and $d_2(t) = f_s(t)$ where $f_s(t)$ is the fault located on the current sensor.

4.2. UKNOWN INPUT INTERVAL OBSERVER DESIGN

Under Assumption 1-2, the system (33) can be rewritten under the form described by (10) defining the state $x(t)$,the input u(t) and the unknown inputs $w_1(\theta, t)$, $w_2(\theta, t)$ by

$$\begin{array}{lcl} x(t) & = & \begin{bmatrix} \theta_m(t) & \theta_l & i(t) & \theta_m(t) & \theta_l(t) \end{bmatrix}^T \\ u(t) & = & V(t) \\ w_1(\theta, t) & = & \frac{mgh}{J_m}sin(\theta_l(t)) \\ w_2(\theta, t) & = & \begin{bmatrix} -\frac{\Delta k}{J_m}(\theta_l(t) - \theta_m(t)) - \frac{mgh}{J_m}sin(\theta_l(t)) \\ f_s(t) \end{bmatrix} \end{array} \tag{34}$$

Table 1. Simulation parameters table.

Parameter		Value	Parameter		Value
Inductance (H)	L	1	Rotor inertia $(kg.m^2)$	J_m	3.7×10^{-3}
Resistance (Ohm)	R	0.5	Link inertia $(kg.m^2)$	J_l	9.3×10^{-3}
Motor gain $(V.rad^{-1}.s^{-1})$	K_e	0.1	Stiffness $(M.m.rad^{-1})$	k	0.17
Motor gain $(N.m.A^{-1})$	K_i	0.1	Stiffness $(M.m.rad^{-1})$	k_0	0.2
Link mass (kg)	m	0.21	Stiffness $(M.m.rad^{-1})$	$\overline{k}$	5×10^{-2}
Earth gravity $(m.s^{-2})$	g	9.81	Center of mass (m)	h	0.31
Viscous friction $(M.m.s.rad^{-1})$	B	4.6×10^{-2}			

The associated distribution matrices to (10) are now given by:

$$A = \begin{bmatrix} 0 & 0 & 0 & 1 & 0 \\ 0 & 0 & 0 & 0 & 1 \\ 0 & 0 & -\frac{R}{L} & -\frac{K_e}{L} & 0 \\ -\frac{k_0}{J_m} & \frac{k_0}{J_m} & \frac{K_i}{J_m} & -\frac{B}{J_m} & 0 \\ \frac{k_0}{J_l} & -\frac{k_0}{J_l} & 0 & 0 & 0 \end{bmatrix}, B = \begin{bmatrix} 0 \\ 0 \\ \frac{1}{L} \\ 0 \\ 0 \end{bmatrix}, G_{11} = \begin{bmatrix} 0 \\ 0 \\ 0 \\ -1 \\ 0 \end{bmatrix}, G_{12} = \begin{bmatrix} 0 & 0 \\ 0 & 0 \\ 0 & 0 \\ -1 & 0 \\ \frac{J_m}{J_l} & 0 \end{bmatrix}$$

$$C = \begin{bmatrix} 1 & 0 & 0 & 0 & 0 \\ 1 & -1 & 0 & 0 & 0 \\ 0 & 0 & 1 & 0 & 0 \\ 0 & 0 & 0 & 1 & -1 \\ \frac{k_0}{J_l} & -\frac{k_0}{J_l} & 0 & 0 & 0 \end{bmatrix}, G_2 = \begin{bmatrix} 0 & 0 \\ 0 & 0 \\ 0 & 1 \\ 0 & 0 \\ \frac{J_m}{J_l} & 0 \end{bmatrix}, H_1 = 0_{n \times s}, H_2 = \begin{bmatrix} 0_{p \times n} & I_p \end{bmatrix}$$

The LTI model, proposed above, allows us to consider that Assumption 4 holds. Then, there exists a non-singular state transformation M_0 such that the system (10) is rewritten under the singular form presented in (15):

$$M_0 = \begin{bmatrix} 0 & 0.5883 & -0.2215 & -0.2243 & -0.9209 \\ 0 & 0.8645 & 0.7880 & -0.5945 & -0.8970 \\ 0 & -0.1439 & 0.9087 & 0.8102 & -0.8687 \\ -1 & 0.8784 & 0.0617 & -0.7832 & 0.2541 \\ 0 & 0.9596 & 0.0512 & 0.0907 & 0.5007 \end{bmatrix} \tag{35}$$

It follows that the matrix $\begin{bmatrix} \tilde{E} \\ \tilde{C} \end{bmatrix}$ is full column rank. Hence, T and F are computed from (19d) and it is deduced that the pair $(T\tilde{A},\tilde{C})$ is detectable. Hence, Assumption 5 holds and a pole placement $\{ -0.1 \quad -0.3 \quad -0.6 \quad -1 \quad -1.3 \quad -1.7 \quad -2 \}$ is used to compute the matrix N as in remark (7). From this result, the expressions of the matrices J and M_1 are determined. Finally, we compute K such that the theorem 2 holds.

For the estimation of the unknown input d_1, the output transformation $\tilde{M}$ is chosen as

$$\tilde{M} = \begin{bmatrix} 0 & 0 & 0 & -0.9538 & 0.9892 \\ 0 & 0 & 0 & 0.2093 & 0.7393 \\ 0 & 0 & 1.0000 & 0.3085 & 0.4540 \\ -1.0000 & 0 & 0 & 0.0078 & 0.0514 \\ 0 & 0.3978 & 0 & 0.5060 & -0.0508 \end{bmatrix} \tag{36}$$

Remark (10). *For the brevity of presentation, all matrices are not detailed but could be easily deduced from M_0. The matrix M_1 is computed here to get the cooperative property by the diagonalization of the matrix N.*

4.3. SIMULATIONS AND RESULTS

For the simulation, we consider the known input as :

$$u(t) = \begin{cases} -5sin(t) & \forall\, t < 10s \\ -5sin(t) - 3sin(3(t-10)) & \forall\, t \geq 10s \end{cases} \tag{37}$$

The fault sensor is assumed to be unknown. For the simulation, we consider

$$f_s(t) = \begin{cases} 0 & \forall\, t < 15s \\ 2sin(\frac{t-15}{2}) + 0.6cos(\frac{\pi}{2}(t-14)) & \forall\, t \geq 15s \end{cases} \tag{38}$$

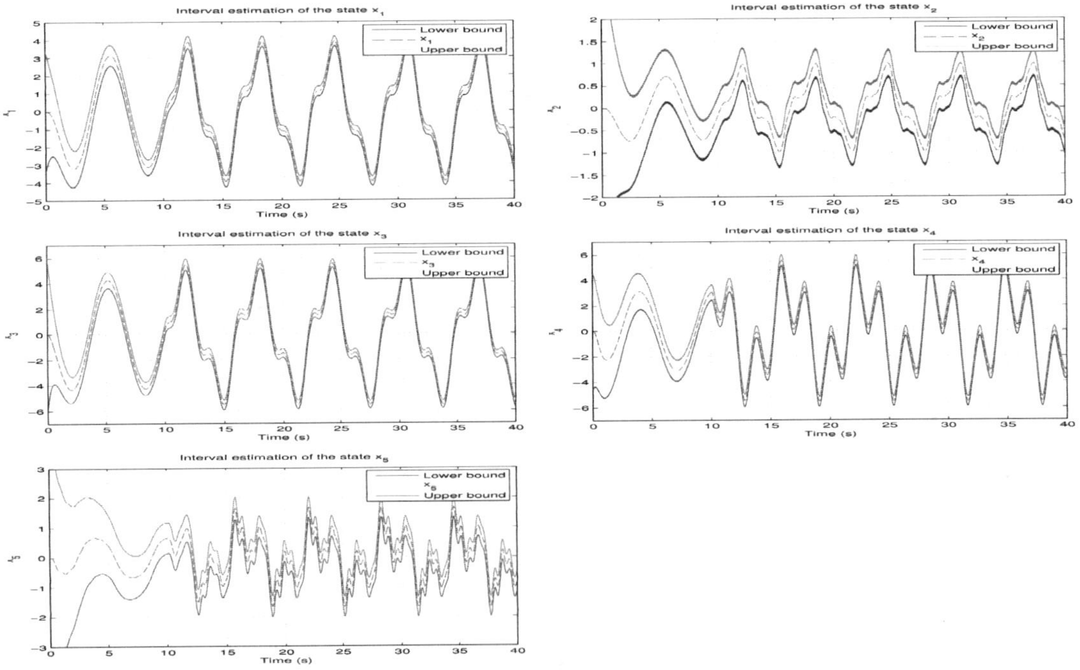

Figure 1. Interval estimation of the state x

It is considered that $\overline{x}(0) = \begin{bmatrix} \frac{\pi}{2} & \frac{\pi}{2} & 1 & \frac{\pi}{2} & \frac{\pi}{2} \end{bmatrix}^T$, $\underline{x}(0) = -\overline{x}(0)$, $\overline{w_2}(0) = \begin{bmatrix} 44 & 1 \end{bmatrix}^T$, $\underline{w_2}(0) = -\overline{w_2}(0)$. The initial conditions of the interval observer are given by $\underline{\xi_1}(0) = \tilde{W}^+\underline{\tilde{x}}(0) - \tilde{W}^-\overline{\tilde{x}}(0)$ and $\underline{\xi_2}(0) = \tilde{W}^+\overline{\tilde{x}}(0) - \tilde{W}^-\underline{\tilde{x}}(0)$ where $\tilde{W} = M_1^{-1}T\tilde{E}\tilde{M}_0^{-1}$ and $\tilde{M}_0$, $\tilde{x}$ defined in (25).

To estimate the unknown input $d_1(t)$, the upper and the lower bounds of $y(t)$ are determined by the first order HOSM differentiator. The parameters are chosen as $\lambda_0 = 100$, $\lambda_1 = 12$ and the parameter μ is consider as $\mu = 17.3$ for each output. The sample time for the simulation is $T_e = 1$ms. The upper and lower bounds of the state, presented in figure 1, converge asymptotically to a tight interval showing the efficiency of the decoupling method proposed in this paper to reduce the conservatism caused by noises, unknown inputs and other disturbances.

Journal of Physics: Conference Series **659** (2015) 012023 doi:10.1088/1742-6596/659/1/012023

The convergence speed is fixed by the interval observer design with the pole placement. The figure 2 is dedicated to the upper and lower bound estimations of unknown inputs d_1 and d_2. The interval estimation of the unknown input d_2 converge asymptotically to a tight interval given, in this case, by the interval observer properties. The width of the interval estimation for d_1 is proportional to state r, the noise v, the input u and the uncertain parameters θ.

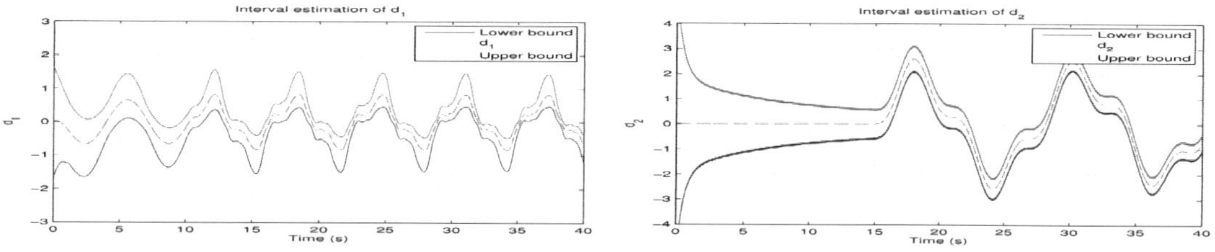

Figure 2. Interval estimation of the unknown inputs d_1 and d_2

5. CONCLUSION

A methodology to estimate jointly the unmeasurable states and unknown inputs estimation for uncertain LTI systems is proposed and is shown in this paper. The pertinence of the proposed approach has been illustrated on an academic example. The results obtained are significant to show the pertinence of the proposed approach. Indeed, the uncertainty, disturbance and noise propagation is reduced, decreasing significantly the conservatism of the proposed method.

REFERENCES

[1] Jaulin L 2002 *Automatica* **38** 1079–1082
[2] Raïssi T, Ramdani N and Candau Y 2004 *Automatica* **40** 1771–1777
[3] Kieffer M and Walter E 2006 *In Proc. 5th IFAC Symposium on Robust Control Design*
[4] Müller M 1927 *Mathematische Zeitschrift* **26** 619–645
[5] Gouzé J L, Rapaport A and Hadj-Sadok M Z 2000 *Ecological modelling* **133** 45–56
[6] Smith H L 1995 *Monotone dynamical systems: an introduction to the theory of competitive and cooperative systems* 41 (American Mathematical Soc.)
[7] Mazenc F and Bernard O 2011 *Automatica* **47** 140–147
[8] Raissi T, Efimov D and Zolghadri A 2012 *Automatic Control, IEEE Transactions on* **57** 260–265
[9] Combastel C 2013 *Automatic Control, IEEE Transactions on* **58** 481–487
[10] Thabet R E H, Raïssi T, Combastel C, Efimov D and Zolghadri A 2014 *Automatica* **50** 2677–2684
[11] Moisan M and Bernard O 2010 *Systems & Control Letters* **59** 687–694
[12] Bolajraf M, Rami M A and Tadeo F 2010 *18th Mediterranean Conference on Control and Automation, MED'10*
[13] Thabet R, Raissi T, Combastel C and Zolghadri A 2013 *Control Conference (ECC), 2013 European* pp 1126–1131
[14] Efimov D, Raïssi T, Chebotarev S and Zolghadri A 2013 *Automatica* **49** 200–205
[15] Chebotarev S, Efimov D, Raïssi T and Zolghadri A 2015 *Automatica* **58** 82–89
[16] Gucik-Derigny D, Raıssi T and Zolghadri A 2014 *IFAC World Congress 2014*
[17] Gucik-Derigny D, Raissi T and Zolghadri A 2015 *International Journal of Control* 1–21
[18] Darouach M, Boutat-Baddas L and Zerrougui M 2011 *Automatica* **47** 2517–2525
[19] Efimov D, Fridman L, Raissi T, Zolghadri A and Seydou R 2012 *Automatica* **48** 2365–2371
[20] Levant A 1998 *Automatica* **34** 379–384
[21] Levant A 2003 *International journal of Control* **76** 924–941
[22] Hou M and Muller P 1992 *Automatic Control, IEEE Transactions on* **37** 871–875

Journal of Physics: Conference Series **659** (2015) 012024 doi:10.1088/1742-6596/659/1/012024

Exponential Boundary Observers for Pressurized Water Pipe

Idellette Judith Hermine Som
CRIStAL: Research Center in Computer Science, Signal and Automatic Control of Lille, Lille 1 University. E-mail: hs.idellette-judith@etudiant.univ-lille1.fr

Vincent Cocquempot
CRIStAL: Research Center in Computer Science, Signal and Automatic Control of Lille, Lille 1 University. E-mail: vincent.cocquempot@univ-lille1.fr

Abdel Aitouche
CRIStAL: Research Center in Computer Science, Signal and Automatic Control of Lille, HEI: Hautes Etudes d'Ingénieur. E-mail: abdel.aitouche@hei.fr

Abstract. This paper deals with state estimation on a pressurized water pipe modeled by nonlinear coupled distributed hyperbolic equations for non-conservative laws with three known boundary measures. Our objective is to estimate the fourth boundary variable, which will be useful for leakage detection. Two approaches are studied. Firstly, the distributed hyperbolic equations are discretized through a finite-difference scheme. By using the Lipschitz property of the nonlinear term and a Lyapunov function, the exponential stability of the estimation error is proven by solving Linear Matrix Inequalities (LMIs). Secondly, the distributed hyperbolic system is preserved for state estimation. After state transformations, a Luenberger-like PDE boundary observer based on backstepping mathematical tools is proposed. An exponential Lyapunov function is used to prove the stability of the resulted estimation error. The performance of the two observers are shown on a water pipe prototype simulated example.

1. Introduction

Supervision of physical transport plant has been an active research topic in recent years and water distribution network (WDN) monitoring is a major concern [1], [2], [3]. Indeed a water distribution network is a set of interconnected physical components (pipes, pumps, tanks and valves) with the mission of supplying required water flow under sufficient pressure for various load conditions.

The real-time knowledge of the network state is essential to improve the efficiency rate of the network which is usually around 75% mainly due to leakages [4]. For that purpose, WDN are equipped with flow and pressure sensors at certain nodes. Due to cost reasons and physical constraints, it is not possible to implement sensors in every nodes. Thus, unmeasured quantities should be estimated to provide useful information for network supervision and management. Different kinds of estimators or state observers may be used depending on the quantities to be estimated, the available sensors and the desired performance of the estimation scheme.

The pressurized pipes of a WDN are relatively long to consider a one-dimensional flow movement modeled by nonlinear coupled distributed hyperbolic equations derived from the non-conservative laws [5]. Flow and pressure measures are available at certain boundaries in a WDN. Exact boundary observability for one dimensional first order quasilinear hyperbolic systems has been studied in [6]. The observability is based on the existence of a unique semi-global solution and leads to the possibility to design an exponential boundary observer.

Two approaches may be used to design a boundary observer. 1) The indirect approach where Partial Differential Equations (PDE) operating in a functional infinite dimensional space are approximated by Ordinary Differential Equations (ODE) in a finite-dimensional space through a differentiation operator (finite differences schemes are commonly used [7], [8]). The advantage of such approach is to give access to many techniques well developed for observers design in finite dimension. 2) The direct approach that preserves all the system information for state estimation [9], [10] by designing directly PDE observers. This approach should potentially lead to better estimation results since no model approximation is made, however it is much more complex and needs mathematical background.

This paper presents both observer design approaches. An exponential boundary observer for the discretized system is proposed. The Lipschitz property of the nonlinear term and Linear Matrix Inequalities (LMIs) techniques are used to prove the Lyapunov stability of the estimation error and the exponential convergence of the proposed observer. For the direct approach, a Luenberger-like PDE observer based on backstepping mathematical tools [11], [12] is proposed. An exponential Lyapunov function is used to prove the stability of the resulted dynamic estimation error. The exponential convergence of the PDE observer is guaranteed with the wise choice of a parameter μ (see section 4). Behaviors of observers based on ODE model and PDE model are shown with a length varying pipe.

This paper is structured as follows. Section 2 describes the mathematical model of a pressurized water pipe. The model is discretized and an exponential boundary observer is proposed in section 3. Section 4 deals with the PDE observer. Simulations of both observers are shown and discussed in section 5. Section 6 gives concluding remarks.

2. Mathematical model

We consider an isothermal flow of a non-compressible fluid in a non-deformable pipeline with constant cross-sectional area. The pipe wall friction is assumed to be constant along the pipe and is supposed to be the only contributor to pressure drop along the pipe according to the Darcy-Weisbach formula [5]. Further we assume that the wave speed is large compared with the flow velocity and the length of the pipe is sufficiently large to consider a uniform flow movement in the transversal direction which is described by a set of two nonlinear hyperbolic PDE obtained by applying the mass and momentum balance conditions on a control volume of length dx [5],

$$\begin{cases} \partial_t p(t,x) + \dfrac{\rho a^2}{A}\partial_x Q(t,x) = 0 \\ \partial_t Q(t,x) + \dfrac{A}{\rho}\partial_x p(t,x) + \dfrac{\zeta Q(t,x)\left|Q(t,x)\right|}{2DA} = 0 \end{cases} \tag{1}$$

where $t \in [0,+\infty)$ is the time, $x \in [0,L]$ is the coordinate along the pipe, $p(t,x)$ is the pressure drop (Pa) and $Q(t,x)$ is the volumetric fluid flow ($m^3.s^{-1}$). a is the wave speed ($m.s^{-1}$), A is a constant cross-section(m^2), D is the pipe diameter (m). ρ is the water density ($kg.m^{-3}$) and ζ the friction coefficient. ∂_t (resp. ∂_x) stands for derivative with respect to time (resp. to x axis).

We suppose that we measure three boundary conditions of (1): $p(t,L)$, $Q(t,L)$ and $p(t,0)$. In the following the inputs $u(t)$ and the output $y(t)$ of the system are

$$u(t)=\left[u_1(t)\; u_2(t)\right]^T=\left[p(t,L)\; Q(t,L)\right]^T \tag{2}$$

$$y(t)=p(t,0) \tag{3}$$

Our goal is to design in healthy conditions (without leakage) an exponential boundary observer for both discretized and original PDE system to estimate the non-measured state $Q(t,0)$ which will be noted $\hat{Q}(t,0)$ in order to compare the behavior of both observers.

3. Exponential boundary observer for the discretized system

Let us define,

$$\chi=\left[\chi_1\; \chi_2\right]^T=\left[p(t,x)\; Q(t,x)\right]^T \text{ and } \tau=\frac{\zeta}{2DA} \tag{4}$$

Spatial discretization by finite differences of matched states $p(t,x)$ and $Q(t,x)$ at x=0 gives,

$$\partial_x p(t,0)\approx\frac{p(t,L)-p(t,0)}{L};\; \partial_x Q(t,0)\approx\frac{Q(t,0)-Q(t,L)}{L} \tag{5}$$

Substituting (4)-(5) into (1)-(3) gives

$$\dot{\chi}(t,0)=\breve{C}\chi(t,0)+\phi(\chi(t,0),u(t)) \tag{6}$$

$$y(t)=E\chi(t,0) \tag{7}$$

where $$\breve{C}=\frac{1}{L}\begin{bmatrix}0 & -\dfrac{\rho a^2}{A}\\ +\dfrac{A}{\rho} & 0\end{bmatrix};\; \phi(\chi(t,0),u(t))=\begin{bmatrix}+\dfrac{\rho a^2}{AL}u_2(t)\\ -\dfrac{A}{\rho L}u_1(t)-\tau Q^2(t,0)\end{bmatrix};\; E=[1\; 0] \tag{8}$$

Since $\breve{c}_{12}=-\dfrac{\rho a^2}{A}\neq 0$, $(\breve{C},E)$ is an observable pair.

The corresponding exponential boundary observer is,

$$\dot{\hat{\chi}}(t,0)=\breve{C}\hat{\chi}(t,0)+\phi(\hat{\chi}(t,0),u(t))+\Pi(y(t)-\hat{y}(t))=0 \tag{9}$$

$$\hat{y}(t)=E\hat{\chi}(t,0) \tag{10}$$

This particular observer structure has already been studied and several design methods may be found in the literature. One of them is the non-constructive Thau's method [13]. Another one is an iterative constructive method proposed in [14]. Also a direct computation of observer gain by solving LMIs was developed in [15]. In the following described method, an exponential term is added to the aforementioned work and the Lyapunov stability of the estimation error is proven using LMIs.

Theorem 1

Consider system (6)-(7) and observer (9)-(10). If the observer gain Π is chosen such that the following inequalities holds

$$\begin{bmatrix} \breve{C}^T N - E^T \Pi^T N + N\breve{C} - N\Pi E + \alpha P - M & N \\ N & -\beta I_2 \end{bmatrix} \leq 0 \tag{11}$$

with β, M, N such that

$$\begin{cases} \beta\gamma_\phi I_2 + M \prec 0 \\ \beta > 0, \ \gamma_\phi > 0 \\ N > 0, \ N^T = N \\ M < 0, \ M^T = M \end{cases} \tag{12}$$

then the estimation error $e(t,0) = \chi(t,0) - \hat{\chi}(t,0)$ tends exponentially to 0 when t tends to infinity.

Proof of theorem 1

The dynamic state estimation error is

$$\dot{e}(t,0) = \dot{\chi}(t,0) - \dot{\hat{\chi}}(t,0) = (\breve{C} - \Pi E)e(t) + \phi(\chi(t,0), u(t)) - \phi(\hat{\chi}(t,0), u(t)) \tag{13}$$

Note $\Delta\phi = \phi(\chi(t,0), u(t)) - \phi(\hat{\chi}(t,0), u(t))$. $\Delta\phi$ is Lipchitz with respect to χ i.e. there exists a constant $\gamma_\phi > 0$ such that the following inequality is satisfied

$$\left\| \phi(\chi(t,0), u(t)) - \phi(\hat{\chi}(t,0), u(t)) \right\|^2 \leq \gamma_\phi \left\| \chi(t,0) - \hat{\chi}(t,0) \right\|^2 \tag{14}$$

$\| \ \|$ represents the 2-norm.

Let V be a Lyapunov function candidate,

$$V = e^T N e \ ; \left(N = N^T \succ 0 \right) \tag{15}$$

According to [16], the dynamic error is exponentially stable if $\dot{V} + \alpha V \leq 0$, where $\alpha > 0$,

$$\dot{V} + \alpha V = \dot{e}^T N e + e^T N \dot{e} + \alpha e^T N e = e^T \left(\breve{C}^T N - E^T \Pi^T N + N\breve{C} - N\Pi E + \alpha N \right) e + \Delta\phi^T N e + e^T N \Delta\phi \tag{16}$$

Consider the following inequality [17],

$$X^T Y + Y^T X \prec \beta X^T X + \beta^{-1} Y^T Y \quad (\beta > 0) \tag{17}$$

Arranging (16) with (17) we get

$$\dot{V} + \alpha V = e^T \underbrace{\left(\breve{C}^T N - E^T \Pi^T N + N\breve{C} - N\Pi E + \beta^{-1} N^T N + \alpha N\right)}_{R} e + \beta \Delta\phi^T \Delta\phi \le 0 \tag{18}$$

Knowing that

$$\Delta\phi^T \Delta\phi \le \left\|\Delta\phi\right\|^2 \text{ and} \left\|\Delta\phi\right\|^2 \le \gamma_\phi \left\|e\right\|^2, \tag{19}$$

equation (18) turns into

$$\dot{V} + \alpha V = e^T Re + \beta\gamma_\phi \left\|e\right\|^2 \le 0 \tag{20}$$

It can be easily proved [17] that there exists a matrix $M\left(M < 0\right);\ M^T = M$ such that,

$$e^T Me \le \left\|e\right\|^2 \cdot Max\left(eig\left(M\right)\right) < 0 \tag{21}$$

and

$$\beta\gamma_\phi \left\|e\right\|^2 + \left\|e\right\|^2 \cdot Max\left(eig\left(M\right)\right) < 0 \tag{22}$$

Applying (21) to (22), we get

$$e^T Me \le -\beta\gamma_\phi \left\|e\right\|^2 \tag{23}$$

From (20) and (23), we obtain:

$$e^T Re \le e^T Me \le -\beta\gamma_\phi \left\|e\right\|^2 \tag{24}$$

Π is chosen by solving the following Linear Matrix Inequalities (LMIs)

$$\begin{cases} R \le M \\ \beta\gamma_\phi I_2 + M < 0 \end{cases} \tag{25}$$

With Schur complement, we have (11)-(12). This proves theorem 1.

As we mentionned in our objectives, it would be interesting to design an observer using the original PDE system. This is the aim of the next section.

4. PDE exponential boundary observer

System (1) is a strictly hyperbolic system with two distinct real eigenvalues $\lambda_1 = +a$ and $\lambda_2 = -a$. The exact boundary observability of hyperbolic systems is studied in [6]. The observation is carried out from one-sided of the pipe on boundary $x = 0$. We first apply changes of coordinates to relax our observer's design task.

4.1 *State transformation*

Let us define the nonsingular transition matrix $\begin{bmatrix} l_1 & l_2 \end{bmatrix} = \begin{bmatrix} +\frac{A}{\rho a} & 1; & -\frac{A}{\rho a} & 1 \end{bmatrix}$ and the change of coordinates $\breve{w} = \begin{bmatrix} l_1 & l_2 \end{bmatrix} \times \begin{bmatrix} p(t,x) & Q(t,x) \end{bmatrix}^T$. $\breve{w}_1$ and $\breve{w}_2$ are called *Riemann invariants* [6]. $\breve{w}_1$ (resp. $\breve{w}_2$) is the constant solution of (1)-(2) along the trajectory $x = at$ (resp. $x = -at + L$).

It gives,

$$\breve{w}_1(t,x) = +\frac{A}{\rho a} p(t,x) + Q(t,x),\ \breve{w}_2(t,x) = -\frac{A}{\rho a} p(t,x) + Q(t,x) \tag{26}$$

System (1) may be rewritten

$$\begin{cases} \partial_t \breve{w}_1(t,x) + a\partial_x \breve{w}_1(t,x) + \frac{\tau}{4}\left(\breve{w}_1(t,x) + \breve{w}_2(t,x)\right)^2 = 0 \\ \partial_t \breve{w}_2(t,x) - a\partial_x \breve{w}_2(t,x) + \frac{\tau}{4}\left(\breve{w}_1(t,x) + \breve{w}_2(t,x)\right)^2 = 0 \end{cases} \tag{27}$$

Variables $\breve{w}_1(t,x)$ and $\breve{w}_2(t,x)$ remain partially matched. We first search for a change of coordinate to put (27) into the following form:

$$\partial_t w(t,z) + \Lambda \partial_z w(t,z) + H(z) w(t,z) = 0;\ (t,z) = [0,+\infty) \times [0,1] \tag{28}$$

$$\Lambda = \begin{bmatrix} +\Lambda_1 & 0 \\ 0 & -\Lambda_2 \end{bmatrix}, H(z) = \begin{bmatrix} 0 & H_1(z) \\ H_2(z) & 0 \end{bmatrix} \tag{29}$$

Consider the normalization $z = \frac{x}{L}$, and define new variables

$$w_i(t,z) = \breve{w}_i(t,zL) e^{\sigma_i \mu z}, \sigma_i = \{sign\lambda_i, i = 1,2\} \tag{30}$$

where μ is a constant to be fixed.

μ represents the transfer of a volumetric flow rate $Q(t,x)$, due to the pressure wave speed $+a$ propagation of from 0 to L, along a rough pipe with friction coefficient ζ.

Original variables $\begin{bmatrix} p(t,x) & Q(t,x) \end{bmatrix}^T$ are obtained through inverse transformations of (30) and (26)

$$p(t,x) = \frac{\rho a}{2A}\left(w_1\left(t,\frac{x}{L}\right) e^{-\frac{\mu x}{L}} - w_2\left(t,\frac{x}{L}\right) e^{+\frac{\mu z}{L}} \right),\ Q(t,x) = \frac{1}{2}\left(w_1\left(t,\frac{x}{L}\right) e^{-\frac{\mu x}{L}} + w_2\left(t,\frac{x}{L}\right) e^{+\frac{\mu x}{L}} \right) \tag{31}$$

Taking the spatial-derivative of (30), we get

$$\partial_z w_i(t,z) = L\partial_z \breve{w}_i(t,zL) e^{\sigma_i \mu z} + \mu \sigma_i w_i(t,z) \tag{32}$$

Taking its time derivative we obtain

$$\partial_t w_i(t,z) = \partial_t \breve{w}_i(t,zL)e^{\sigma_i \mu z} \tag{33}$$

For each i, introducing the corresponding expression of $\partial_t \breve{w}_i(t,zL)$ from (27) in (33) and then inserting (32) in this result we get:

$$\partial_t w_1(t,z) + \frac{a}{L}\partial_z w_1(t,z) - \left(\frac{a}{L}\mu - \breve{H}\right) w_1(t,z) + \breve{H} w_2(t,z) e^{+2\mu z} = 0 \tag{34}$$

$$\partial_t w_2(t,z) - \frac{a}{L}\partial_z w_2(t,z) - \left(\frac{a}{L}\mu - \breve{H}\right) w_2(t,z) + \breve{H} w_1(t,z) e^{-2\mu z} = 0 \tag{35}$$

where $\breve{H} = \frac{\tau}{4}\left(w_1(t,z)e^{-\mu z} + w_2(t,z)e^{+\mu z}\right)$. From (31), we get a simplified expression, $\breve{H} = \frac{\tau}{2}Q(t,zL)$

To get the form (28)-(29), let us choose

$$\mu = \frac{L}{a}\breve{H} = \frac{L\tau}{2a}Q(t,zL) \tag{36}$$

Then (28) is obtained with,

$$\Lambda = \frac{1}{L}\begin{bmatrix} +a & 0 \\ 0 & -a \end{bmatrix}; \quad H(z) = \frac{\tau}{2}Q(t,zL)\begin{bmatrix} 0 & e^{+2\mu z} \\ e^{-2\mu z} & 0 \end{bmatrix} \tag{37}$$

We get the corresponding output $p(t,0)$ by using (31),

$$y(t) = \frac{\rho a}{2A}\left(w_1(t,0) - w_2(t,0)\right) \tag{38}$$

4.2 *Observer setup and stability analysis*

A Luenberger-like observer is designed. This observer is built as a copy of the system with a correction function of the output estimation error

$$\partial_t \hat{w}(t,z) + \Lambda \partial_z \hat{w}(t,z) + H(z)\hat{w}(t,z) + \Pi(z)\left(y(t) - \hat{y}(t)\right) = 0 \tag{39}$$

$$y(t) = \frac{\rho a}{2A}\left(\hat{w}_1(t,0) - \hat{w}_2(t,0)\right) \tag{40}$$

where $\Pi(z)$ is a function to be designed to stabilize the state estimation error

$$\tilde{w}(t,z) = w(t,z) - \hat{w}(t,z) \tag{41}$$

From the system and observer equations, we obtain the error equation

$$\partial_t \tilde{w}(t,z) + \Lambda \partial_z \tilde{w}(t,z) + H(z)\tilde{w}(t,z) - \Pi(z)\left(y(t) - \hat{y}(t)\right) = 0 \tag{42}$$

To prove the stability of (42), the following backstepping transformation is used as in [11]

$$\tilde{w}(t,z) = \xi(t,z) - \int_0^1 k(z,\eta)\xi(t,\eta)d\eta \tag{43}$$

for $(z,\eta) \in D$; $D = \{z,\eta : 0 \le \eta \le z \le 1\}$ where $k(z,\eta)$ must be determined so that (43) is transformed into the following system

$$\partial_t \xi(t,z) + \Lambda \partial_z \xi(t,z) = 0 \tag{44}$$

$$\tilde{y}(t) = \frac{\rho a}{2A}\left(\xi_1(t,0) - \xi_2(t,0)\right) \tag{45}$$

Differentiating (43) with respect to t yields,

$$\begin{aligned}\partial_t \tilde{w}(t,z) &= \partial_t \xi(t,z) - \int_0^1 k(z,\eta)\partial_t \xi(t,\eta)d\eta = \partial_t \xi(t,z) + \int_{10}^1 k(z,\eta)\left(\Lambda \partial_z \xi(t,\eta)\right)d\eta \\ &= \partial_t \xi(t,z) + k(z,z)\Lambda \xi(t,z) - k(z,0)\Lambda \xi(t,0) - \int_0^1 \partial_\eta k(z,\eta)\Lambda \xi(t,\eta)d\eta\end{aligned} \tag{46}$$

and with respect to z,

$$\partial_z \tilde{w}(t,z) = \partial_z \xi(t,z) - k(z,z)\xi(t,z) - \int_0^1 \partial_z k(z,\eta)\xi(t,\eta)d\eta \tag{47}$$

Introducing resulted equations (46)-(47) into (42) yields

$$\begin{aligned}&\partial_t \xi(t,z) + k(z,z)\Lambda \xi(t,z) - k(z,0)\Lambda \xi(t,0) - \int_0^1 \partial_\eta k(z,\eta)\Lambda \xi(t,\eta)d\eta + \Lambda\left(\partial_z \xi(t,z) - k(z,z)\xi(t,z) - \int_0^1 \partial_z k(z,\eta)\xi(t,\eta)d\eta\right) \\ &+ H(z)\left(\xi(t,z) - \int_0^1 k(z,\eta)\xi(t,\eta)d\eta\right) - \Pi(z)\left(y(t) - \hat{y}(t)\right) = 0\end{aligned} \tag{48}$$

The equivalence of systems (42) and (44)-(45) is guaranteed if the following conditions are satisfied

$$\left(k(z,z)\Lambda - \Lambda k(z,z) + H(z)\right)\xi(t,z) = 0 \tag{49}$$

$$-k(z,0)\Lambda \xi(t,0) - \Pi(z)\left(y(t) - \hat{y}(t)\right) = 0 \tag{50}$$

$$\int_0^1 \left(\partial_\eta k(z,\eta)\Lambda + \Lambda \partial_z k(z,\eta) + k(z,\eta)H(z)\right)\xi(t,\eta)d\eta = 0 \tag{51}$$

Our goal is to estimate $Q(t,0)$. Conditions (49)-(50) are sufficient to compute the desired $\Pi(0)$.

Let $k(z,\eta) = \begin{bmatrix} k_1(z,\eta) & k_1(z,\eta) \\ k_2(z,\eta) & k_2(z,\eta) \end{bmatrix}$, developing (49) and rearranging the terms gives

$$k_1(z,z) = +\frac{L}{2a}H_1(z),\ k_2(z,z) = -\frac{L}{2a}H_2(z) \tag{52}$$

From equations (50) and (45), we obtain

$$\Pi_1(z) = -\frac{2A}{\rho} Lk_1(z,0),\ \Pi_2(z) = -\frac{2A}{\rho} Lk_2(z,0) \tag{53}$$

At $z = 0$, injecting (52) in (53) with (37), we get

$$\Pi_1(0) = -\frac{AL^2}{2\rho a}\tau\hat{Q}(t,0),\ \Pi_2(0) = +\frac{AL^2}{2\rho a}\tau\hat{Q}(t,0) \tag{54}$$

$\Pi(0)$ is a varying nonlinear gain function of the estimated $\hat{Q}(t,0)$. The exponential convergence of the proposed PDE observer depends on the parameter μ.

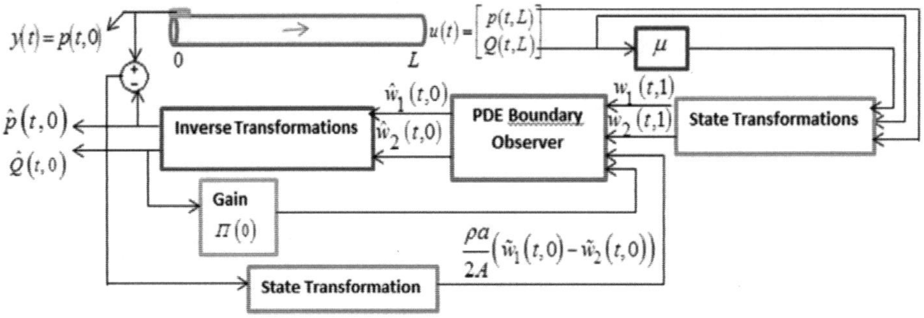

Fig. 1. PDE observer scheme for a pressurized water pipe

Furthermore transformation (43) is invertible [11] and the stability of the state estimation error (42) can be concluded from the stability of the target system (44)-(45).
The following Lyapunov function is introduced

$$V(t) = \int_0^1 \xi^T(t,z) D(z) \xi(t,z) dz \tag{55}$$

where $D(z) = diag\left[\theta_1 e^{+\mu z},\ \theta_2 e^{-\mu z}\right];\ \theta_1 > 0,\ \theta_2 > 0$. Then, $D(z)\Lambda = \Lambda D(z)$

Its time derivative along the trajectory is

$$\begin{aligned}\dot{V}(t) &= \int_0^1 \left(\partial_t \xi^T(t,z) D(z)\xi(t,z) + \xi^T(t,z) D(z)\partial_t\xi(t,z)\right) dz \\ &= -\xi^T(t,1)D(1)\Lambda\xi(t,1) + \xi^T(t,0)D(0)\Lambda\xi(t,0) - 2\mu\Lambda_1 \underbrace{\int_0^1 \xi^T(t,x)D(x)\xi(t,x)\,dx}_{V(t)>0} \\ &= -\xi^T(t,1)D(1)\Lambda\xi(t,1) + \xi^T(t,0)D(0)\Lambda\xi(t,0) - 2\mu\Lambda_1 \dot{V}(t)\end{aligned} \tag{56}$$

Boundary conditions are chosen such that $-\xi^T(t,L)D(L)\Lambda\xi(t,L) + \xi^T(t,0)D(0)\Lambda\xi(t,0) \le 0$ we get

$$\dot{V}(t) \le -2\mu\Lambda_1 V(t) \tag{57}$$

Thus, the exponential stability of the target system is established. Since transformation (43) is invertible, (42) remains stable if conditions (49)-(51) are fulfilled.

5. Simulations and results

Results are validated in simulation on a pipe with parameters presented in table 1. We have considered outlet pressure and flow rate respectively at $0.656\text{x}10^5$ *Pa* and *0.0082544* $m^3.s^{-1}$ as input of the system.

The observer gain for the discretized system is computed according to LMIs (11)-(12), we get $\Pi = \begin{bmatrix} +20363\times10^2 \\ -29\times10^2 \end{bmatrix}$. For the PDE boundary observer, $\Pi(0)$ is computed with (54) and illustrated in figure 3a. From equation (36) we get $\mu = 0.074181$.

Table 1. Simulation parameters

Parameter	Symbol	Value	Unit
Pipe length	L	86.49	(m)
Internal pipe diameter	D	6.54×10^{-2}	(m)
wave speed	a	375	(ms^{-1})
Friction coefficient	ζ	1.72×10^{-2}	/
Density	ρ	1000	$(kg.m^{-3})$

Source [7]

The estimated inlet pressure and flow rate with the two observers are represented in figures 2a and 2.b. Both observers converge exponentially; however, the convergence of the PDE observer is faster than the nonlinear observer based on the discretized system.

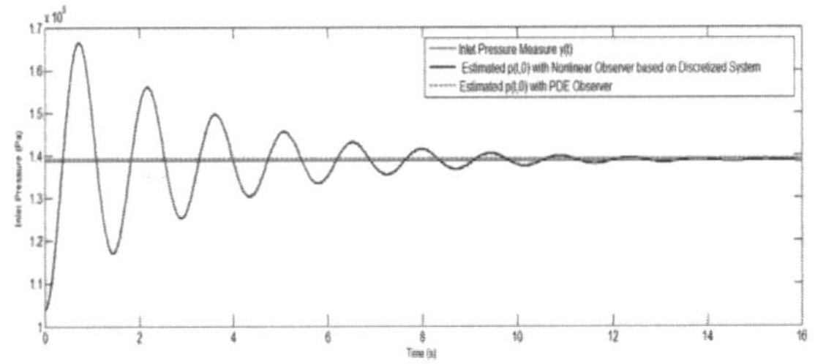

Fig. 2a. Estimated inlet pressure with L_1=L

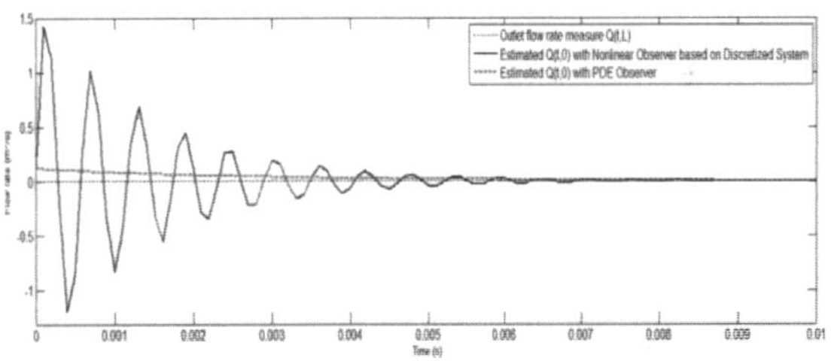

Fig. 2b. Estimated inlet flow rate with L_1=L

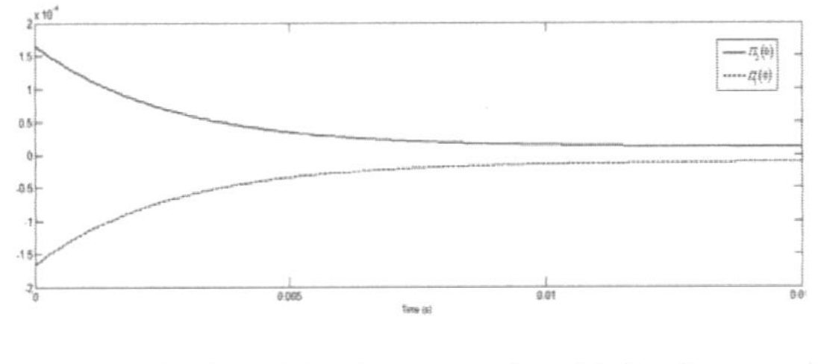

Fig. 3a. PDE observer gain with L_1=L

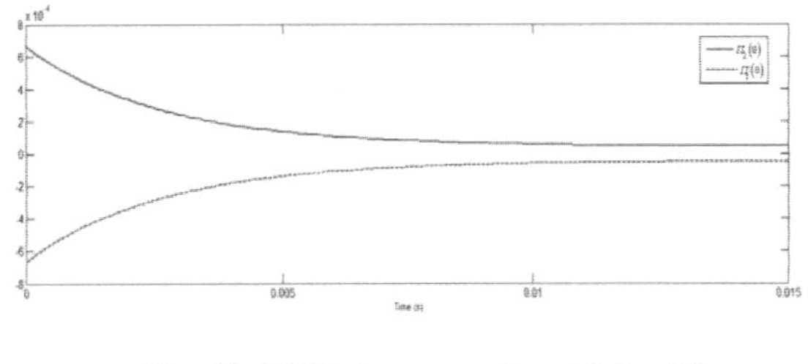

Fig. 3b. PDE observer gain with L_1=$2L$

The length of the pipe is changed to $L_1 = 2L$. We get $\mu = 0.1490478$ and the corresponding $\Pi(0)$ illustrated in figure 3b. In parallel, a new value $\Pi = \begin{bmatrix} +16461\times10^2 \\ -36\times10^2 \end{bmatrix}$ is obtained. The simulation results are shown in Figures 4a and 4b.

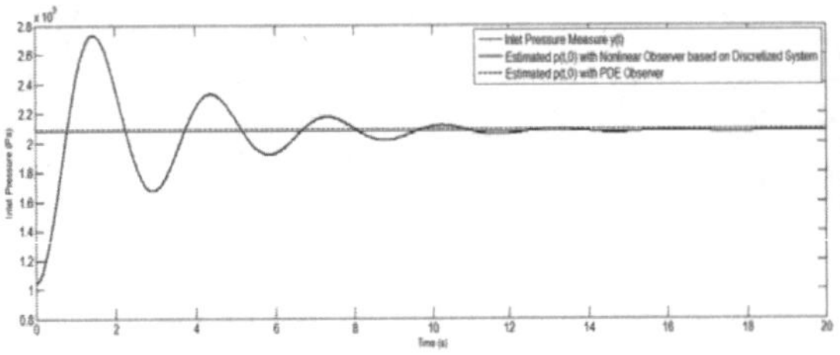

Fig. 4a Estimated inlet pressure with L_1=2L

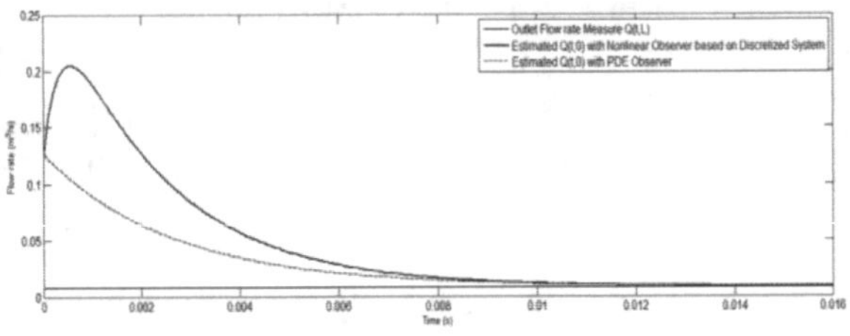

Fig. 4b Estimated inlet flow rate with L_1=2L

We note a relative long convergence time for the discretized observer. Figure 4b shows with a PDE observer, a stabilization of the inlet pressure estimation error around $+100Pa = 10^{-3}bar$ and an accuracy error of 0.48% (shown in figure 5b). Compared with the sensitivity of pressure sensors on distributed water network, we can conclude that both observers converge exponentially to the desired value. Furthermore, the implementation of PDE boundary observer seems to be suitable for relatively long pipes in order to alleviate the computation of observer gain and to guarantee a real time monitoring of distributed water network.

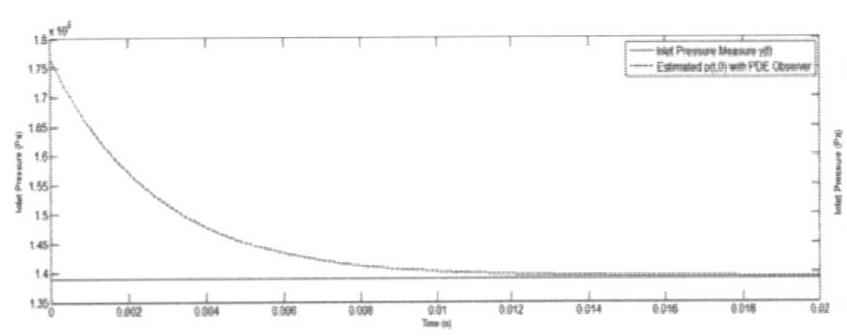

Fig. 5a PDE estimated inlet pressure with L_1=L

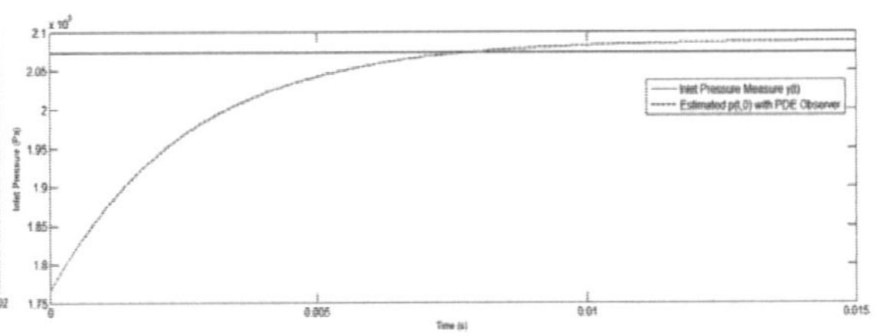

Fig. 5b PDE estimated inlet pressure with L_1=2L

6. Conclusion

State estimation on a pressurized water pipe with three known measures at boundary has been addressed. Two approaches have been explored. The indirect approach uses the discretized system; an exponential boundary nonlinear observer is designed by using the Lipschitz property of the nonlinear term and by solving LMIs to ensure the stability of estimation error. The direct approach, where some state transformations are used in order to design a Luenberger-like PDE boundary observer based on backstepping mathematical tools. An exponential Lyapunov function is used to prove the stability of the error equation.

Simulations have been carried out for a water pipe prototype. Results have shown the effectiveness of both approaches. The convergence time of the nonlinear observer increases with the length of pipe, The PDE observer seems more efficient. These two observers may be used to detect a leakage in the pipe using a reduced number of sensors. We will study in future work the efficiency of both observers for leakage detection. Early studies and simulations show that the leakage detection is more efficient using a PDE observer. Moreover, the PDE observer may also be easily extended to localize the leakage in the pipe if the four boundary conditions are measured.

References

[1] Andersen J. H. and Powell R.S., Implicit state estimation technique for water network monitoring. *J. Urban Water*, Vol. 2, no. 2, pp. 123-130, 2000.

[2] Preis A., Andrew J., Whittle M., Ostfeld A. and Perelman L. Efficient hydraulic state estimation technique using reduced models of urban water networks. *J. Water resources planning and management,* Vol. 137, no. 4, pp. 343-351, 2011.

[3] Leirens S., Zamora R., Negenborn R. and De Schutter B. Coordination in urban water supply networks using distributed model predictive control. American Control Conference, Baltimore, USA, pp. 3957-3962, 2010.

[4] Salvetti M. The network efficiency rate: a key performance indicator for water services asset management? 7th IWA International Conference on Efficient Use and Management of Water (Efficient 2013) Paris, France, 2013.

[5] Chaudry, M. H., 1979. Applied Hydraulic Transients. Springer, New York, 2014.

[6] Tatsien L. Exact boundary observability for quasilinear hyperbolic systems, *ESAIM: Control, Optimization and Calculus of Variations* Vol.14, no. 4, pp. 759-766, 2008.

[7] Navarro A., Begovich O., Besançon G. and Murillo J. Leak detection and isolation using an observer based on robust sliding mode differentiators. World Automation Congress (WAC), pp. 1-6, Puerto Vallarta, Mexico, 2012.

[8] Torres L., Besançon G., Navarro A. and Begovich O. Examples of pipeline monitoring with nonlinear observers and real-data validation. 8th International Multi-Conference on Systems, Signals & Devices, 2011.

[9] Hauge E., Aamo O.M. and Godhavn J.M. Application of an infinite-dimensional observer for drilling systems incorporating kick and loss detection. European Control Conference, Zurich, Switzerland, pp. 1605-1070, 2013.

[10] Castillo F., Witrant E., Prieur C. and Dugard L. Boundary observers for linear and quasi-linear hyperbolic systems with application to flow control. Automatica Vol. 49, no. 11, pp. 3180-3188, 2013.

[11] Coron J.M., Vazquez R., Miroslav K. and Bastin G. Local exponential H2 stabilization of 2× 2 quasilinear hyperbolic system using backstepping. SIAM J. Control and Optimization. Vol. 51, no. 3, pp. 2005-2035, 2013.

[12] Vazquez R. and Miroslav K. Marcum Q-functions and explicit feedback laws for stabilization of constant coefficient 2×2 linear hyperbolic systems. 52nd IEEE Conference on Decision and Control, Florence, Italy, 10-13 December, 2013.

[13] Thau F.E. Observing the state of non-linear dynamic systems. *Int. J. Control.* Vol. 17, no. 3, pp. 471–479, 1973.

[14] Raghavan S. and Hedrick J.K Observer design for a class of nonlinear systems. *International Journal of Control*, Vol. 59, no. 2, pp. 515-528, 1994.

[15] Rajamani R. Observers for Lipschitz Nonlinear Systems. *IEEE Transactions on Automatic,* Vol. 43, no. 3, pp. 397-401, 1998.

[16] Shauying K., Elliott D. and Jong Tarn T. Exponential observers for nonlinear dynamic systems *Information and control Journal* Vol .29 no.3, pp. 204-216, 1975.

[17] Boyd S., El Ghaoui L., Feron E. and Balakrishnan V. Linear Matrix Inequalities in system and control theory. SIAM *Studies in Applied Mathematics*, Vol. 15, Philadelphia, 1994.

Robust state estimation and control for nonlinear system with uncertain parameters

Mariusz Buciakowski[1], Marcin Witczak[1], Didier Theilliol[2]

[1]Institute of Control and Computation Engineering, University of Zielona Góra, ul. Podgórna 50, 65-246 Zielona Góra, Poland

[2]Centre de Recherche en Automatique de Nancy CRAN-UMR 7039, Nancy-Universite, CNRS, F-54506 Vandoeuvre-les-Nancy Cedex, France

E-mail: [1]{m.buciakowski, m.witczak}@issi.uz.zgora.pl,
[2]didier.theilliol@univ-lorraine.fr

Abstract. The paper focuses on robust state estimation and control for nonlinear system with uncertain parameters. In particular, the problem is oriented towards a practical application for a laboratory three-tank system. The proposed approach starts with a general description of the system and assumptions regarding uncertain parameters and a nonlinear function. The subsequent part of the paper is concerned with the design of the robust observer and controller for nonlinear systems. To confirm the performance of the proposed approaches, the final part presents results obtained for the laboratory three-tank system.

1. Introduction

The problem of designing a robust observer and a state feedback control to guarantee the global stability of a nonlinear system with uncertain parameters has received considerable attention during the last three decades. There are meany technique for state estimation and control for nonlinear system, for example: extended Kalman filter, adaptive technique, sliding mode, LPV technique, neural-networks, fuzzy rules base on Takagi–Sugeno etc. [10, 7, 11, 3, 16, 22, 20, 13, 9, 2]. Irrespective of the large spectrum of available methods non of them is universal to be suitable for arbitrary nonlinear systems. Thus, the complexity of the nonlinear systems, in particular multi–input multioutput system forces the development of new methods. It can be perceived as a generalization of the approaches proposed for Lipschitz systems as well as its recent one-sided counterpart. The proposed approach provides a novel output feedback control strategy that is divide into two traditional steps, i.e., observer design and controller design. Firstly, a robust observer design strategy is described that can tackle both parametric uncertainty and additive external disturbances. Subsequently, the robust control strategy is portrayed that follows general ideas employed for the observer design strategy.

Note that the proposed strategy is a preliminary step to designing robust fault-tolerant control scheme[22]. However, this is beyond the scope of this paper.

The paper is organised as follows. Section 2 introduces a description of the system along with assumptions regarding the nonlinearities. Whilst section 3 describes the concept of designing the robust observer. The subsequent section 4 presents the robust control framework. The final part of the paper is concerned with an illustrative example and conclusions.

2. A general description of system

Let us consider a non-linear system:

$$\boldsymbol{x}_{k+1} = (\boldsymbol{A} + \Delta\boldsymbol{A})\boldsymbol{x}_k + \boldsymbol{B}\boldsymbol{u}_k + \boldsymbol{g}(\boldsymbol{x}_k) + \boldsymbol{W}_1\boldsymbol{w}_k, \tag{1}$$

$$\boldsymbol{y}_k = \boldsymbol{C}\boldsymbol{x}_k + \boldsymbol{W}_2\boldsymbol{w}_k \tag{2}$$

where matrices $\boldsymbol{A}$, $\boldsymbol{B}$ define the nominal model and $\Delta\boldsymbol{A}$ represent admissible uncertainties which are assumed to be of the form

$$\left[\Delta\boldsymbol{A}\right] = \boldsymbol{H}\boldsymbol{\mathcal{F}}\left[\boldsymbol{E}_A\right] \tag{3}$$

where $\boldsymbol{H}$, $\boldsymbol{E}_A$ are now constant matrices of compatible dimensions, $\boldsymbol{\mathcal{F}}$ is an unknown matrix satisfies

$$\boldsymbol{\mathcal{F}}^T\boldsymbol{\mathcal{F}} \preceq \boldsymbol{I}, \tag{4}$$

where $\boldsymbol{x}_k \in \mathbb{X} \subset \mathbb{R}^n$ is the state, $\boldsymbol{u}_k \in \mathbb{U} \in \mathbb{R}^r$ stands for the input, $\boldsymbol{y}_k \in \mathbb{R}^m$ denotes the output, $\boldsymbol{g}(\boldsymbol{x}_k)$ is non-linear function with respect to $\boldsymbol{x}_k$ and $\boldsymbol{u}_k$ $\boldsymbol{w}_k \in l_2$ is a an exogenous disturbance vector and $\boldsymbol{W}_1 \in \mathbb{R}^{n\times n}$, $\boldsymbol{W}_2 \in \mathbb{R}^{m\times n}$ stand for its distribution matrices while

$$l_2 = \{\mathbf{w} \in \mathbb{R}^n | \; \|\mathbf{w}\|_{l_2} < +\infty\}, \|\mathbf{w}\|_{l_2} = \left(\sum_{k=0}^{\infty} \|\boldsymbol{w}_k\|^2\right)^{\frac{1}{2}}. \tag{5}$$

Moreover, the following set of assumptions are imposed:

Assumption 1:

$$\boldsymbol{g}(\boldsymbol{x})^T\boldsymbol{x} \leq \boldsymbol{x}^T\boldsymbol{M}\boldsymbol{x}, \quad \boldsymbol{M} \in \mathbb{M}. \tag{6}$$

Assumption 2:

$$\boldsymbol{g}(\boldsymbol{x})^T\boldsymbol{g}(\boldsymbol{x}) \leq \boldsymbol{x}^T\boldsymbol{M}^T\boldsymbol{M}\boldsymbol{x}, \quad \boldsymbol{M} \in \mathbb{M}. \tag{7}$$

Assumption 3: There exists a matrix $\boldsymbol{M}$ such that

$$\left(\boldsymbol{g}(\boldsymbol{a}) - \boldsymbol{g}(\boldsymbol{b})\right)^T(\boldsymbol{a} - \boldsymbol{b}) \leq (\boldsymbol{a} - \boldsymbol{b})^T\boldsymbol{M}(\boldsymbol{a} - \boldsymbol{b}), \quad \forall \boldsymbol{a}, \boldsymbol{b} \in \mathbb{X}. \tag{8}$$

Assumption 4: There exists a matrix $\boldsymbol{M}$ such that

$$\left(\boldsymbol{g}(\boldsymbol{a}) - \boldsymbol{g}(\boldsymbol{b})\right)^T\left(\boldsymbol{g}(\boldsymbol{a}, \boldsymbol{u}) - \boldsymbol{g}(\boldsymbol{b}, \boldsymbol{u})\right) \leq (\boldsymbol{a} - \boldsymbol{b})^T\boldsymbol{M}^T\boldsymbol{M}(\boldsymbol{a} - \boldsymbol{b}), \quad \forall \boldsymbol{a}, \boldsymbol{b} \in \mathbb{X}. \tag{9}$$

The paper extends the idea of one-sided Lipschitz condition [19, 25] by imposing *Assumption 3*. *Assumption 4* extends the usual Lipschitz condition. Using the Differential Mean Value Theorem (DMVT) [24], it is possible to describe Lipschitz constant matrices by supremum of the norm of Jacobian of the function.

$$\boldsymbol{g}(\boldsymbol{a}) - \boldsymbol{g}(\boldsymbol{b}) = \boldsymbol{M}_x(\boldsymbol{a} - \boldsymbol{b}), \tag{10}$$

with

$$\boldsymbol{M}_{x,u} = \left[\left(\frac{\partial g_1}{\partial x}(\boldsymbol{c_1}, \boldsymbol{u})\right)^T \; \ldots \; \left(\frac{\partial g_n}{\partial x}(\boldsymbol{c_n}, \boldsymbol{u})\right)^T\right]^T, \tag{11}$$

where $\boldsymbol{c}_1, \ldots, \boldsymbol{c}_n \in \mathrm{Co}(\boldsymbol{a}, \boldsymbol{b})$, $\boldsymbol{c}_i \neq \boldsymbol{a}$, $\boldsymbol{c}_i \neq \boldsymbol{b}$, $i = 1, \ldots, n$. Assuming that

$$\bar{a}_{i,j} \geq \frac{\partial g_i}{\partial x_j} \geq \underline{a}_{i,j}, \quad i = 1, \ldots, n, \quad j = 1, \ldots, n, \tag{12}$$

it is clear that:

$$\mathbb{M}_x = \left\{\boldsymbol{M} \in \mathbb{R}^{n\times n} | \bar{a}_{i,j} \geq m_{x,i,j} \geq \underline{a}_{i,j}, \; i, j = 1, \ldots, n, \right\} \tag{13}$$

It is worthwhile that, if $\boldsymbol{M}^T\boldsymbol{M} = \gamma^2\boldsymbol{I}$, then *Assumption 3* becomes a usual Lipschitz condition [1, 15, 18, 17], with γ being a Lipschitz constant. This property makes the proposed employed strategy more general than those presented in the literature [1, 15, 18, 17].

3. Design of the state observer

The main objective of this section is to present the design procedure of the robust observer that will be able to estimate all state vector for proposed nonlinear system. For that purpose, the following structure is proposed

$$\hat{\boldsymbol{x}}_{k+1} = \boldsymbol{A}\hat{\boldsymbol{x}}_k + \boldsymbol{B}\boldsymbol{u}_k + \boldsymbol{g}\left(\hat{\boldsymbol{x}}_k\right) + \boldsymbol{K}_o(\boldsymbol{y}_k - \boldsymbol{C}\hat{\boldsymbol{x}}_k), \tag{14}$$

while the state estimation error is given by

$$\boldsymbol{e}_{k+1} = \boldsymbol{x}_{k+1} - \hat{\boldsymbol{x}}_{k+1} = \tilde{\boldsymbol{A}}\boldsymbol{e}_k + \tilde{\boldsymbol{G}}\boldsymbol{x}_k + \boldsymbol{s}_k + \tilde{\boldsymbol{W}}\boldsymbol{w}_k, \tag{15}$$

where

$$\begin{aligned} &\tilde{\boldsymbol{A}} = (\boldsymbol{A} - \boldsymbol{K}_o\boldsymbol{C}), \quad \tilde{\boldsymbol{G}} = (\boldsymbol{H}\boldsymbol{\mathcal{F}}\boldsymbol{E}_A), \quad \boldsymbol{s}_k = \boldsymbol{g}\left(\boldsymbol{x}_k\right) - \boldsymbol{g}\left(\hat{\boldsymbol{x}}_k\right) \\ &\tilde{\boldsymbol{W}} = (\boldsymbol{W}_1 - \boldsymbol{K}_o\boldsymbol{W}_2). \end{aligned} \tag{16}$$

Theorem 1. *For a prescribed disturbance attenuation level $\mu > 0$, the observer design for the system* (15) *is solvable if there exist $\boldsymbol{N}$, $\boldsymbol{U}$, $\boldsymbol{P} \succ \boldsymbol{0}$, $\alpha > 0$, $\beta > 0$, $\epsilon > 0$ such that the following condition is satisfied:*

$$\left[\begin{array}{cccccccccccc}
\boldsymbol{I}-\boldsymbol{P}+\alpha(\boldsymbol{M}+\boldsymbol{M}^T) & \boldsymbol{0} & -\alpha\boldsymbol{I} & \boldsymbol{0} & \boldsymbol{A}^T\boldsymbol{U}^T-\boldsymbol{C}^T\boldsymbol{N}^T & \boldsymbol{M}^T\boldsymbol{U}^T & \boldsymbol{0} & \boldsymbol{0} & \boldsymbol{0} & \boldsymbol{0} & \boldsymbol{0} & \boldsymbol{0} \\
\boldsymbol{0} & -\boldsymbol{P}+\epsilon\boldsymbol{E}_A^T\boldsymbol{E}_A^T & \boldsymbol{0} & \boldsymbol{0} & \boldsymbol{0} & \boldsymbol{0} & \boldsymbol{0} & \boldsymbol{0} & \boldsymbol{0} & \boldsymbol{0} & \boldsymbol{H}^T\boldsymbol{U}^T & \boldsymbol{0} \\
-\alpha\boldsymbol{I} & \boldsymbol{0} & -\beta\boldsymbol{I} & \boldsymbol{0} & \boldsymbol{U}^T & \boldsymbol{0} & \boldsymbol{0} & \boldsymbol{0} & \boldsymbol{0} & \boldsymbol{0} & \boldsymbol{0} & \boldsymbol{0} \\
\boldsymbol{0} & \boldsymbol{0} & \boldsymbol{0} & -\mu^2\boldsymbol{I} & \boldsymbol{W}_1^T\boldsymbol{U}^T-\boldsymbol{W}_2^T\boldsymbol{N}^T & \boldsymbol{0} & \boldsymbol{0} & \boldsymbol{0} & \boldsymbol{0} & \boldsymbol{0} & \boldsymbol{0} & \boldsymbol{0} \\
\boldsymbol{U}\boldsymbol{A}-\boldsymbol{N}\boldsymbol{C} & \boldsymbol{0} & \boldsymbol{U} & \boldsymbol{U}\boldsymbol{W}_1-\boldsymbol{N}\boldsymbol{W}_2 & \boldsymbol{P}-\boldsymbol{U}-\boldsymbol{U}^T & \boldsymbol{0} & \boldsymbol{0} & \boldsymbol{0} & \boldsymbol{0} & \boldsymbol{0} & \boldsymbol{0} & \boldsymbol{0} \\
\boldsymbol{U}\boldsymbol{M} & \boldsymbol{0} & \boldsymbol{0} & \boldsymbol{0} & \boldsymbol{0} & \beta\boldsymbol{I}-\boldsymbol{U}-\boldsymbol{U}^T & \boldsymbol{0} & \boldsymbol{0} & \boldsymbol{0} & \boldsymbol{0} & \boldsymbol{0} & \boldsymbol{0} \\
\boldsymbol{0} & \boldsymbol{0} & \boldsymbol{0} & \boldsymbol{0} & \boldsymbol{0} & \boldsymbol{0} & -\epsilon\boldsymbol{I} & \boldsymbol{0} & \boldsymbol{0} & \boldsymbol{0} & \boldsymbol{0} & \boldsymbol{0} \\
\boldsymbol{0} & \boldsymbol{0} & \boldsymbol{0} & \boldsymbol{0} & \boldsymbol{0} & \boldsymbol{0} & \boldsymbol{0} & -\epsilon\boldsymbol{I} & \boldsymbol{0} & \boldsymbol{0} & \boldsymbol{0} & \boldsymbol{0} \\
\boldsymbol{0} & \boldsymbol{0} & \boldsymbol{0} & \boldsymbol{0} & \boldsymbol{0} & \boldsymbol{0} & \boldsymbol{0} & \boldsymbol{0} & -\epsilon\boldsymbol{I} & \boldsymbol{0} & \boldsymbol{0} & \boldsymbol{0} \\
\boldsymbol{0} & \boldsymbol{0} & \boldsymbol{0} & \boldsymbol{0} & \boldsymbol{0} & \boldsymbol{0} & \boldsymbol{0} & \boldsymbol{0} & \boldsymbol{0} & -\epsilon\boldsymbol{I} & \boldsymbol{0} & \boldsymbol{0} \\
\boldsymbol{0} & \boldsymbol{U}\boldsymbol{H} & \boldsymbol{0} & \boldsymbol{0} & \boldsymbol{0} & \boldsymbol{0} & \boldsymbol{0} & \boldsymbol{0} & \boldsymbol{0} & \boldsymbol{0} & -\epsilon\boldsymbol{I} & \boldsymbol{0} \\
\boldsymbol{0} & \boldsymbol{0} & \boldsymbol{0} & \boldsymbol{0} & \boldsymbol{0} & \boldsymbol{0} & \boldsymbol{0} & \boldsymbol{0} & \boldsymbol{0} & \boldsymbol{0} & \boldsymbol{0} & -\epsilon\boldsymbol{I}
\end{array}\right] \prec 0. \tag{17}$$

Proof. The problem of $\mathcal{H}_\infty$ observer design is to determine the matrix $\boldsymbol{K}_o$ such that[12, 14, 23]

$$\lim_{k\to\infty} \boldsymbol{e}_k = \boldsymbol{0} \quad \text{for } \boldsymbol{w}_k = \boldsymbol{0} \tag{18}$$

$$\|\boldsymbol{e}_k\|_{l_2} \leq \mu\|\boldsymbol{w}_k\|_{l_2} \quad \text{for } \boldsymbol{w}_k \neq \boldsymbol{0},\, \boldsymbol{e}_0 = \boldsymbol{0}. \tag{19}$$

In order to settle the above problem it is sufficient to find a Lyapunov function V_k such that:

$$\Delta V_k + \boldsymbol{e}_k^T \boldsymbol{e}_k - \mu^2 \boldsymbol{w}_k^T \boldsymbol{w}_k < 0,\ k = 0, \ldots \infty, \tag{20}$$

where

$$\Delta V_k = V_{k+1} - V_k, \tag{21}$$

$$V_k = \boldsymbol{e}_k^T \boldsymbol{P} \boldsymbol{e}_k + \boldsymbol{x}_k^T \boldsymbol{P} \boldsymbol{x}_k, \tag{22}$$

$$\Delta V_k = \boldsymbol{e}_{k+1}^T \boldsymbol{P} \boldsymbol{e}_{k+1} - \boldsymbol{e}_k^T \boldsymbol{P} \boldsymbol{e}_k - \boldsymbol{x}_k^T \boldsymbol{P} \boldsymbol{x}_k. \tag{23}$$

Consequently, using (15)

$$\begin{aligned}
&\Delta V_k + \boldsymbol{e}_k^T \boldsymbol{e}_k - \mu^2 \boldsymbol{w}_k^T \boldsymbol{w}_k = \\
&\boldsymbol{e}_k^T \left(\tilde{\boldsymbol{A}}^T \boldsymbol{P} \tilde{\boldsymbol{A}} + \boldsymbol{I} - \boldsymbol{P}\right) \boldsymbol{e}_k + \boldsymbol{e}_k^T \left(\tilde{\boldsymbol{A}}^T \boldsymbol{P} \tilde{\boldsymbol{G}}\right) \boldsymbol{x}_k + \boldsymbol{e}_k^T \left(\tilde{\boldsymbol{A}}^T \boldsymbol{P}\right) \boldsymbol{s}_k + \boldsymbol{e}_k^T \left(\tilde{\boldsymbol{A}}^T \boldsymbol{P} \tilde{\boldsymbol{W}}\right) \boldsymbol{w}_k + \\
&\boldsymbol{x}_k^T \left(\tilde{\boldsymbol{G}}^T \boldsymbol{P} \tilde{\boldsymbol{A}}\right) \boldsymbol{e}_k + \boldsymbol{x}_k^T \left(\tilde{\boldsymbol{G}}^T \boldsymbol{P} \tilde{\boldsymbol{G}} - \boldsymbol{P}\right) \boldsymbol{x}_k + \boldsymbol{x}_k^T \left(\tilde{\boldsymbol{G}}^T \boldsymbol{P}\right) \boldsymbol{s}_k + \boldsymbol{x}_k^T \left(\tilde{\boldsymbol{G}}^T \boldsymbol{P} \tilde{\boldsymbol{W}}\right) \boldsymbol{w}_k + \\
&\boldsymbol{s}_k^T \left(\boldsymbol{P} \tilde{\boldsymbol{A}}\right) \boldsymbol{e}_k + \boldsymbol{s}_k^T \left(\boldsymbol{P} \tilde{\boldsymbol{G}}\right) \boldsymbol{x}_k + \boldsymbol{s}_k^T \left(\boldsymbol{P}\right) \boldsymbol{s}_k + \boldsymbol{s}_k^T \left(\boldsymbol{P} \tilde{\boldsymbol{W}}\right) \boldsymbol{w}_k + \\
&\boldsymbol{w}_k^T \left(\tilde{\boldsymbol{W}} \boldsymbol{P} \tilde{\boldsymbol{A}}\right) \boldsymbol{e}_k + \boldsymbol{w}_k^T \left(\tilde{\boldsymbol{W}} \boldsymbol{P} \tilde{\boldsymbol{G}}\right) \boldsymbol{x}_k + \boldsymbol{w}_k^T \left(\tilde{\boldsymbol{W}} \boldsymbol{P}\right) \boldsymbol{s}_k + \boldsymbol{w}_k^T \left(\tilde{\boldsymbol{W}}^T \boldsymbol{P} \tilde{\boldsymbol{W}} - \mu^2 \boldsymbol{I}\right) \boldsymbol{w}_k < \boldsymbol{0}.
\end{aligned} \tag{24}$$

By defining

$$\tilde{\boldsymbol{v}}_k = [\boldsymbol{e}_k^T, \boldsymbol{x}_k^T, \boldsymbol{s}_k^T, \boldsymbol{w}_k^T]^T, \tag{25}$$

it can be shown that (25) is equivalent to

$$\tilde{\boldsymbol{v}}_k^T \begin{bmatrix} \tilde{\boldsymbol{A}}^T \boldsymbol{P} \tilde{\boldsymbol{A}} + \boldsymbol{I} - \boldsymbol{P} & \tilde{\boldsymbol{A}}^T \tilde{\boldsymbol{G}} & \tilde{\boldsymbol{A}}^T \boldsymbol{P} & \tilde{\boldsymbol{A}}^T \boldsymbol{P} \tilde{\boldsymbol{W}} \\ \tilde{\boldsymbol{G}}^T \boldsymbol{P} \tilde{\boldsymbol{A}} & \tilde{\boldsymbol{G}}^T \boldsymbol{P} \tilde{\boldsymbol{G}} - \boldsymbol{P} & \tilde{\boldsymbol{G}}^T \boldsymbol{P} & \tilde{\boldsymbol{G}}^T \boldsymbol{P} \tilde{\boldsymbol{W}} \\ \boldsymbol{P} \tilde{\boldsymbol{A}} & \boldsymbol{P} \tilde{\boldsymbol{G}} & \boldsymbol{P} & \boldsymbol{P} \tilde{\boldsymbol{W}} \\ \tilde{\boldsymbol{W}}^T \boldsymbol{P} \tilde{\boldsymbol{A}} & \tilde{\boldsymbol{W}} \boldsymbol{P} \tilde{\boldsymbol{G}} & \tilde{\boldsymbol{W}}^T \boldsymbol{P} & \tilde{\boldsymbol{W}}^T \boldsymbol{P} \tilde{\boldsymbol{W}} - \mu^2 \boldsymbol{I} \end{bmatrix} \tilde{\boldsymbol{v}}_k \prec \boldsymbol{0}. \tag{26}$$

Analyzing assumption of non-linear function, it can be show that (8) is equivalent to

$$\boldsymbol{s}_k^T \boldsymbol{e}_k \leq \boldsymbol{e}_k^T \boldsymbol{M} \boldsymbol{e}_k = \frac{1}{2} \boldsymbol{e}_k^T (\boldsymbol{M} + \boldsymbol{M}^T) \boldsymbol{e}_k. \tag{27}$$

Inequality (27) can be written as

$$\frac{1}{2} \boldsymbol{e}_k^T (\boldsymbol{M} + \boldsymbol{M}^T) \boldsymbol{e}_k - \boldsymbol{s}_k^T \boldsymbol{e}_k \leq 0,$$

which is equivalent to

$$\frac{1}{2} \boldsymbol{e}_k^T (\boldsymbol{M} + \boldsymbol{M}^T) \boldsymbol{e}_k - \frac{1}{2} \boldsymbol{s}_k^T \boldsymbol{e}_k - \frac{1}{2} \boldsymbol{e}_k^T \boldsymbol{s}_k \geq \boldsymbol{0}. \tag{28}$$

Thus, for any $\alpha > 0$

$$\alpha \tilde{\boldsymbol{v}}_k^T \begin{bmatrix} (\boldsymbol{M} + \boldsymbol{M}^T) & \boldsymbol{0} & -\boldsymbol{I} & \boldsymbol{0} \\ \boldsymbol{0} & \boldsymbol{0} & \boldsymbol{0} & \boldsymbol{0} \\ -\boldsymbol{I} & \boldsymbol{0} & \boldsymbol{0} & \boldsymbol{0} \\ \boldsymbol{0} & \boldsymbol{0} & \boldsymbol{0} & \boldsymbol{0} \end{bmatrix} \tilde{\boldsymbol{v}}_k \geq \boldsymbol{0}. \tag{29}$$

Journal of Physics: Conference Series **659** (2015) 012025 doi:10.1088/1742-6596/659/1/012025

Similarly, for some β inequality (9) can be written as

$$\beta \boldsymbol{e}_k^T \boldsymbol{M}^T \boldsymbol{M} \boldsymbol{e}_k - \beta \boldsymbol{s}_k^T \boldsymbol{s}_k \geq \boldsymbol{0}. \tag{30}$$

Thus, for any $\beta > 0$

$$\beta \bar{\boldsymbol{v}}_k^T \begin{bmatrix} \boldsymbol{M}^T\boldsymbol{M} & \boldsymbol{0} & \boldsymbol{0} & \boldsymbol{0} \\ \boldsymbol{0} & \boldsymbol{0} & \boldsymbol{0} & \boldsymbol{0} \\ \boldsymbol{0} & \boldsymbol{0} & -\boldsymbol{I} & \boldsymbol{0} \\ \boldsymbol{0} & \boldsymbol{0} & \boldsymbol{0} & \boldsymbol{0} \end{bmatrix} \bar{\boldsymbol{v}}_k \geq \boldsymbol{0}. \tag{31}$$

Combining together (26), (29) and (31) gives

$$\begin{bmatrix} \tilde{\boldsymbol{A}}^T\boldsymbol{P}\tilde{\boldsymbol{A}} + \boldsymbol{I} - \boldsymbol{P} + \alpha(\boldsymbol{M}+\boldsymbol{M}^T) + \beta\boldsymbol{M}^T\boldsymbol{M} & \tilde{\boldsymbol{A}}^T\tilde{\boldsymbol{G}} & \tilde{\boldsymbol{A}}^T\boldsymbol{P} - \alpha\boldsymbol{I} & \tilde{\boldsymbol{A}}^T\boldsymbol{P}\tilde{\boldsymbol{W}} \\ \tilde{\boldsymbol{G}}^T\boldsymbol{P}\tilde{\boldsymbol{A}} & \tilde{\boldsymbol{G}}^T\boldsymbol{P}\tilde{\boldsymbol{G}} - \boldsymbol{P} & \tilde{\boldsymbol{G}}^T\boldsymbol{P} & \tilde{\boldsymbol{G}}^T\boldsymbol{P}\tilde{\boldsymbol{W}} \\ \boldsymbol{P}\tilde{\boldsymbol{A}} - \alpha\boldsymbol{I} & \boldsymbol{P}\tilde{\boldsymbol{G}} & \boldsymbol{P} - \beta\boldsymbol{I} & \boldsymbol{P}\tilde{\boldsymbol{W}} \\ \tilde{\boldsymbol{W}}^T\boldsymbol{P}\tilde{\boldsymbol{A}} & \tilde{\boldsymbol{W}}\boldsymbol{P}\tilde{\boldsymbol{G}} & \tilde{\boldsymbol{W}}^T\boldsymbol{P} & \tilde{\boldsymbol{W}}^T\boldsymbol{P}\tilde{\boldsymbol{W}} - \mu^2\boldsymbol{I} \end{bmatrix} \prec \boldsymbol{0} \tag{32}$$

Lemma 1. *The following statements are equivalent[6]:*

(i) There exists $\boldsymbol{X} \succ 0$ *such that*

$$\boldsymbol{V}^T\boldsymbol{X}\boldsymbol{V} - \boldsymbol{W} \prec 0 \tag{33}$$

(ii) There exists $\boldsymbol{X} \succ 0$ *such that*

$$\begin{bmatrix} -\boldsymbol{W} & \boldsymbol{V}^T\boldsymbol{U}^T \\ \boldsymbol{U}\boldsymbol{V} & \boldsymbol{X} - \boldsymbol{U} - \boldsymbol{U}^T \end{bmatrix} \prec 0. \tag{34}$$

Rewriting (32) as

$$\begin{bmatrix} \tilde{\boldsymbol{A}}^T \\ \tilde{\boldsymbol{G}}^T \\ \boldsymbol{I}^T \\ \tilde{\boldsymbol{W}}^T \end{bmatrix} \boldsymbol{P} \begin{bmatrix} \tilde{\boldsymbol{A}} & \tilde{\boldsymbol{G}} & \boldsymbol{I} & \tilde{\boldsymbol{W}} \end{bmatrix} + \begin{bmatrix} \boldsymbol{I} - \boldsymbol{P} + \alpha(\boldsymbol{M}+\boldsymbol{M}^T) + \beta\boldsymbol{M}^T\boldsymbol{M} & \boldsymbol{0} & -\alpha\boldsymbol{I} & \boldsymbol{0} \\ \boldsymbol{0} & -\boldsymbol{P} & \boldsymbol{0} & \boldsymbol{0} \\ -\alpha\boldsymbol{I} & \boldsymbol{0} & -\beta\boldsymbol{I} & \boldsymbol{0} \\ \boldsymbol{0} & \boldsymbol{0} & \boldsymbol{0} & -\mu^2\boldsymbol{I} \end{bmatrix} \prec \boldsymbol{0}, \tag{35}$$

and using *Lemma 1* to (35) yields

$$\begin{bmatrix} \boldsymbol{I} - \boldsymbol{P} + \alpha(\boldsymbol{M}+\boldsymbol{M}^T) & \boldsymbol{0} & -\alpha\boldsymbol{I} & \boldsymbol{0} & \tilde{\boldsymbol{A}}^T\boldsymbol{U}^T & \boldsymbol{M}^T\boldsymbol{U}^T \\ \boldsymbol{0} & -\boldsymbol{P} & \boldsymbol{0} & \boldsymbol{0} & \tilde{\boldsymbol{G}}^T\boldsymbol{U}^T & \boldsymbol{0} \\ -\alpha\boldsymbol{I} & \boldsymbol{0} & -\beta\boldsymbol{I} & \boldsymbol{0} & \boldsymbol{U}^T & \boldsymbol{0} \\ \boldsymbol{0} & \boldsymbol{0} & \boldsymbol{0} & -\mu^2\boldsymbol{I} & \tilde{\boldsymbol{W}}^T\boldsymbol{U}^T & \boldsymbol{0} \\ \boldsymbol{U}\tilde{\boldsymbol{A}} & \boldsymbol{U}\tilde{\boldsymbol{G}} & \boldsymbol{U} & \boldsymbol{U}\tilde{\boldsymbol{W}} & \boldsymbol{P} - \boldsymbol{U} - \boldsymbol{U}^T & \boldsymbol{0} \\ \boldsymbol{U}\boldsymbol{M} & \boldsymbol{0} & \boldsymbol{0} & \boldsymbol{0} & \boldsymbol{0} & \beta\boldsymbol{I} - \boldsymbol{U} - \boldsymbol{U}^T \end{bmatrix} \prec \boldsymbol{0}, \tag{36}$$

with $\boldsymbol{U}\tilde{\boldsymbol{A}} = \boldsymbol{UA} - \boldsymbol{UK}_o\boldsymbol{C} = \boldsymbol{UA} - \boldsymbol{NC}$, $\boldsymbol{U}\tilde{\boldsymbol{W}} = \boldsymbol{UW}_1 - \boldsymbol{UK}_o\boldsymbol{W}_2 = \boldsymbol{UW}_1 - \boldsymbol{NW}_2$ and $\boldsymbol{U}\tilde{\boldsymbol{G}} = \boldsymbol{UH\mathcal{F}E}_A$. Inequality (36) can be written in equivalent form

$$\begin{bmatrix} \boldsymbol{I}-\boldsymbol{P}+\alpha(\boldsymbol{M}+\boldsymbol{M}^T) & \boldsymbol{0} & -\alpha\boldsymbol{I} & \boldsymbol{0} & \boldsymbol{A}^T\boldsymbol{U}^T-\boldsymbol{C}^T\boldsymbol{N}^T & \boldsymbol{M}^T\boldsymbol{U}^T \\ \boldsymbol{0} & -\boldsymbol{P} & \boldsymbol{0} & \boldsymbol{0} & \boldsymbol{0} & \boldsymbol{0} \\ -\alpha\boldsymbol{I} & \boldsymbol{0} & -\beta\boldsymbol{I} & \boldsymbol{0} & \boldsymbol{U}^T & \boldsymbol{0} \\ \boldsymbol{0} & \boldsymbol{0} & \boldsymbol{0} & -\mu^2\boldsymbol{I} & \boldsymbol{W}_1^T\boldsymbol{U}^T-\boldsymbol{W}_2^T\boldsymbol{N}^T & \boldsymbol{0} \\ \boldsymbol{UA}-\boldsymbol{NC} & \boldsymbol{0} & \boldsymbol{U} & \boldsymbol{UW}_1-\boldsymbol{NW}_2 & \boldsymbol{P}-\boldsymbol{U}-\boldsymbol{U}^T & \boldsymbol{0} \\ \boldsymbol{UM} & \boldsymbol{0} & \boldsymbol{0} & \boldsymbol{0} & \boldsymbol{0} & \beta\boldsymbol{I}-\boldsymbol{U}-\boldsymbol{U}^T \end{bmatrix} +$$

$$\begin{bmatrix} \boldsymbol{0} & \boldsymbol{0} & \boldsymbol{0} & \boldsymbol{0} & \boldsymbol{0} & \boldsymbol{0} \\ \boldsymbol{0} & \boldsymbol{E}_A^T & \boldsymbol{0} & \boldsymbol{0} & \boldsymbol{0} & \boldsymbol{0} \\ \boldsymbol{0} & \boldsymbol{0} & \boldsymbol{0} & \boldsymbol{0} & \boldsymbol{0} & \boldsymbol{0} \\ \boldsymbol{0} & \boldsymbol{0} & \boldsymbol{0} & \boldsymbol{0} & \boldsymbol{0} & \boldsymbol{0} \\ \boldsymbol{0} & \boldsymbol{0} & \boldsymbol{0} & \boldsymbol{0} & \boldsymbol{0} & \boldsymbol{0} \\ \boldsymbol{0} & \boldsymbol{0} & \boldsymbol{0} & \boldsymbol{0} & \boldsymbol{0} & \boldsymbol{0} \end{bmatrix} \begin{bmatrix} \boldsymbol{\mathcal{F}}^T & \boldsymbol{0} & \boldsymbol{0} & \boldsymbol{0} & \boldsymbol{0} & \boldsymbol{0} \\ \boldsymbol{0} & \boldsymbol{\mathcal{F}}^T & \boldsymbol{0} & \boldsymbol{0} & \boldsymbol{0} & \boldsymbol{0} \\ \boldsymbol{0} & \boldsymbol{0} & \boldsymbol{\mathcal{F}}^T & \boldsymbol{0} & \boldsymbol{0} & \boldsymbol{0} \\ \boldsymbol{0} & \boldsymbol{0} & \boldsymbol{0} & \boldsymbol{\mathcal{F}}^T & \boldsymbol{0} & \boldsymbol{0} \\ \boldsymbol{0} & \boldsymbol{0} & \boldsymbol{0} & \boldsymbol{0} & \boldsymbol{\mathcal{F}}^T & \boldsymbol{0} \\ \boldsymbol{0} & \boldsymbol{0} & \boldsymbol{0} & \boldsymbol{0} & \boldsymbol{0} & \boldsymbol{\mathcal{F}}^T \end{bmatrix} \begin{bmatrix} \boldsymbol{0} & \boldsymbol{0} & \boldsymbol{0} & \boldsymbol{0} & \boldsymbol{0} & \boldsymbol{0} \\ \boldsymbol{0} & \boldsymbol{0} & \boldsymbol{0} & \boldsymbol{0} & \boldsymbol{H}^T\boldsymbol{U}^T & \boldsymbol{0} \\ \boldsymbol{0} & \boldsymbol{0} & \boldsymbol{0} & \boldsymbol{0} & \boldsymbol{0} & \boldsymbol{0} \\ \boldsymbol{0} & \boldsymbol{0} & \boldsymbol{0} & \boldsymbol{0} & \boldsymbol{0} & \boldsymbol{0} \\ \boldsymbol{0} & \boldsymbol{0} & \boldsymbol{0} & \boldsymbol{0} & \boldsymbol{0} & \boldsymbol{0} \\ \boldsymbol{0} & \boldsymbol{0} & \boldsymbol{0} & \boldsymbol{0} & \boldsymbol{0} & \boldsymbol{0} \end{bmatrix} + \quad (37)$$

$$\begin{bmatrix} \boldsymbol{0} & \boldsymbol{0} & \boldsymbol{0} & \boldsymbol{0} & \boldsymbol{0} & \boldsymbol{0} \\ \boldsymbol{0} & \boldsymbol{0} & \boldsymbol{0} & \boldsymbol{0} & \boldsymbol{0} & \boldsymbol{0} \\ \boldsymbol{0} & \boldsymbol{0} & \boldsymbol{0} & \boldsymbol{0} & \boldsymbol{0} & \boldsymbol{0} \\ \boldsymbol{0} & \boldsymbol{0} & \boldsymbol{0} & \boldsymbol{0} & \boldsymbol{0} & \boldsymbol{0} \\ \boldsymbol{0} & \boldsymbol{UH} & \boldsymbol{0} & \boldsymbol{0} & \boldsymbol{0} & \boldsymbol{0} \\ \boldsymbol{0} & \boldsymbol{0} & \boldsymbol{0} & \boldsymbol{0} & \boldsymbol{0} & \boldsymbol{0} \end{bmatrix} \begin{bmatrix} \boldsymbol{\mathcal{F}} & \boldsymbol{0} & \boldsymbol{0} & \boldsymbol{0} & \boldsymbol{0} & \boldsymbol{0} \\ \boldsymbol{0} & \boldsymbol{\mathcal{F}} & \boldsymbol{0} & \boldsymbol{0} & \boldsymbol{0} & \boldsymbol{0} \\ \boldsymbol{0} & \boldsymbol{0} & \boldsymbol{\mathcal{F}} & \boldsymbol{0} & \boldsymbol{0} & \boldsymbol{0} \\ \boldsymbol{0} & \boldsymbol{0} & \boldsymbol{0} & \boldsymbol{\mathcal{F}} & \boldsymbol{0} & \boldsymbol{0} \\ \boldsymbol{0} & \boldsymbol{0} & \boldsymbol{0} & \boldsymbol{0} & \boldsymbol{\mathcal{F}} & \boldsymbol{0} \\ \boldsymbol{0} & \boldsymbol{0} & \boldsymbol{0} & \boldsymbol{0} & \boldsymbol{0} & \boldsymbol{\mathcal{F}} \end{bmatrix} \begin{bmatrix} \boldsymbol{0} & \boldsymbol{0} & \boldsymbol{0} & \boldsymbol{0} & \boldsymbol{0} & \boldsymbol{0} \\ \boldsymbol{0} & \boldsymbol{E}_A & \boldsymbol{0} & \boldsymbol{0} & \boldsymbol{0} & \boldsymbol{0} \\ \boldsymbol{0} & \boldsymbol{0} & \boldsymbol{0} & \boldsymbol{0} & \boldsymbol{0} & \boldsymbol{0} \\ \boldsymbol{0} & \boldsymbol{0} & \boldsymbol{0} & \boldsymbol{0} & \boldsymbol{0} & \boldsymbol{0} \\ \boldsymbol{0} & \boldsymbol{0} & \boldsymbol{0} & \boldsymbol{0} & \boldsymbol{0} & \boldsymbol{0} \\ \boldsymbol{0} & \boldsymbol{0} & \boldsymbol{0} & \boldsymbol{0} & \boldsymbol{0} & \boldsymbol{0} \end{bmatrix} \prec \boldsymbol{0},$$

Lemma 2. *Let $\boldsymbol{H}$, $\boldsymbol{E}$ be given real matrices of appropriate dimensions and $\boldsymbol{\mathcal{F}}$ satisfy $\boldsymbol{\mathcal{F}}^T\boldsymbol{\mathcal{F}} \preceq \boldsymbol{I}$. Then for any $\epsilon > 0$ the following holds*

$$\boldsymbol{H\mathcal{F}E} + \boldsymbol{E}^T\boldsymbol{\mathcal{F}}^T\boldsymbol{H} \preceq \epsilon\boldsymbol{HH}^T + \frac{1}{\epsilon}\boldsymbol{E}^T\boldsymbol{E}. \quad (38)$$

Applying *Lemma 2* to (37), we obtain

$$\begin{bmatrix} \boldsymbol{I}-\boldsymbol{P}+\alpha(\boldsymbol{M}+\boldsymbol{M}^T) & \boldsymbol{0} & -\alpha\boldsymbol{I} & \boldsymbol{0} & \boldsymbol{A}^T\boldsymbol{U}^T-\boldsymbol{C}^T\boldsymbol{N}^T & \boldsymbol{M}^T\boldsymbol{U}^T \\ \boldsymbol{0} & -\boldsymbol{P}+\epsilon\boldsymbol{E}_A^T\boldsymbol{E}_A^T & \boldsymbol{0} & \boldsymbol{0} & \boldsymbol{0} & \boldsymbol{0} \\ -\alpha\boldsymbol{I} & \boldsymbol{0} & -\beta\boldsymbol{I} & \boldsymbol{0} & \boldsymbol{U}^T & \boldsymbol{0} \\ \boldsymbol{0} & \boldsymbol{0} & \boldsymbol{0} & -\mu^2\boldsymbol{I} & \boldsymbol{W}_1^T\boldsymbol{U}^T-\boldsymbol{W}_2^T\boldsymbol{N}^T & \boldsymbol{0} \\ \boldsymbol{UA}-\boldsymbol{NC} & \boldsymbol{0} & \boldsymbol{U} & \boldsymbol{UW}_1-\boldsymbol{NW}_2 & \boldsymbol{P}-\boldsymbol{U}-\boldsymbol{U}^T & \boldsymbol{0} \\ \boldsymbol{UM} & \boldsymbol{0} & \boldsymbol{0} & \boldsymbol{0} & \boldsymbol{0} & \beta\boldsymbol{I}-\boldsymbol{U}-\boldsymbol{U}^T \end{bmatrix} + \quad (39)$$

$$\epsilon^{-1}\begin{bmatrix} \boldsymbol{0} & \boldsymbol{0} & \boldsymbol{0} & \boldsymbol{0} & \boldsymbol{0} & \boldsymbol{0} \\ \boldsymbol{0} & \boldsymbol{0} & \boldsymbol{0} & \boldsymbol{0} & \boldsymbol{0} & \boldsymbol{0} \\ \boldsymbol{0} & \boldsymbol{0} & \boldsymbol{0} & \boldsymbol{0} & \boldsymbol{0} & \boldsymbol{0} \\ \boldsymbol{0} & \boldsymbol{0} & \boldsymbol{0} & \boldsymbol{0} & \boldsymbol{0} & \boldsymbol{0} \\ \boldsymbol{0} & \boldsymbol{UH} & \boldsymbol{0} & \boldsymbol{0} & \boldsymbol{0} & \boldsymbol{0} \\ \boldsymbol{0} & \boldsymbol{0} & \boldsymbol{0} & \boldsymbol{0} & \boldsymbol{0} & \boldsymbol{0} \end{bmatrix} \begin{bmatrix} \boldsymbol{0} & \boldsymbol{0} & \boldsymbol{0} & \boldsymbol{0} & \boldsymbol{0} & \boldsymbol{0} \\ \boldsymbol{0} & \boldsymbol{0} & \boldsymbol{0} & \boldsymbol{0} & \boldsymbol{H}^T\boldsymbol{U}^T & \boldsymbol{0} \\ \boldsymbol{0} & \boldsymbol{0} & \boldsymbol{0} & \boldsymbol{0} & \boldsymbol{0} & \boldsymbol{0} \\ \boldsymbol{0} & \boldsymbol{0} & \boldsymbol{0} & \boldsymbol{0} & \boldsymbol{0} & \boldsymbol{0} \\ \boldsymbol{0} & \boldsymbol{0} & \boldsymbol{0} & \boldsymbol{0} & \boldsymbol{0} & \boldsymbol{0} \\ \boldsymbol{0} & \boldsymbol{0} & \boldsymbol{0} & \boldsymbol{0} & \boldsymbol{0} & \boldsymbol{0} \end{bmatrix} \prec \boldsymbol{0}.$$

Applying the Schur complements leads to (17), which completes the proof. □

4. Design of the state feedback controller

The main objective of this section is to present the design procedure of the robust controller for the proposed system. The controller will be designed in such a way that a predefined disturbance attenuation level with respect to the state of the system is achieved. To solve the above problem the following control scheme is proposed

$$\boldsymbol{u}_k = -\boldsymbol{K}_c \boldsymbol{x}_k. \tag{40}$$

Substituting (40) in to (1) gives

$$\boldsymbol{x}_{k+1} = \bar{\boldsymbol{A}}\boldsymbol{x}_k + \boldsymbol{g}\left(\boldsymbol{x}_k\right) + \bar{\boldsymbol{W}}\boldsymbol{w}_k \tag{41}$$

where

$$\bar{\boldsymbol{A}} = \left(\boldsymbol{A} - \boldsymbol{B}\boldsymbol{K}_c + \boldsymbol{H}\boldsymbol{\mathcal{F}}\boldsymbol{E}_A\right), \quad \bar{\boldsymbol{W}} = \boldsymbol{W}_1. \tag{42}$$

Theorem 2. *For a prescribed disturbance attenuation level $\mu > 0$, the controller design for the system* (41) *is solvable if there exist $\boldsymbol{N}$, $\boldsymbol{U}$, $\boldsymbol{P} \succ \boldsymbol{0}$, $\alpha > 0$, $\beta > 0$, $\epsilon > 0$ such that the following condition is satisfied:*

$$\begin{bmatrix}
\boldsymbol{I} - \boldsymbol{P} + \alpha(\boldsymbol{M} + \boldsymbol{M}^T) + \boldsymbol{H}\boldsymbol{H}^T & -\alpha\boldsymbol{I} & \boldsymbol{0} & \boldsymbol{A}\boldsymbol{U} - \boldsymbol{B}\boldsymbol{N} & \boldsymbol{M}\boldsymbol{U} & \boldsymbol{0} & \boldsymbol{0} & \boldsymbol{0} & \boldsymbol{0} & \boldsymbol{0} \\
-\alpha\boldsymbol{I} & -\beta\boldsymbol{I} & \boldsymbol{0} & \boldsymbol{U} & \boldsymbol{0} & \boldsymbol{0} & \boldsymbol{0} & \boldsymbol{0} & \boldsymbol{0} & \boldsymbol{0} \\
\boldsymbol{0} & \boldsymbol{0} & -\mu^2\boldsymbol{I} & \boldsymbol{W}_1\boldsymbol{U} & \boldsymbol{0} & \boldsymbol{0} & \boldsymbol{0} & \boldsymbol{0} & \boldsymbol{0} & \boldsymbol{0} \\
\boldsymbol{U}^T\boldsymbol{A}^T - \boldsymbol{N}^T\boldsymbol{B}^T & \boldsymbol{U}^T & \boldsymbol{U}^T\boldsymbol{W}_1^T & \boldsymbol{P} - \boldsymbol{U} - \boldsymbol{U}^T & \boldsymbol{0} & \boldsymbol{U}^T\boldsymbol{E}_A^T & \boldsymbol{0} & \boldsymbol{0} & \boldsymbol{0} & \boldsymbol{0} \\
\boldsymbol{U}^T\boldsymbol{M}^T & \boldsymbol{0} & \boldsymbol{0} & \boldsymbol{0} & \beta\boldsymbol{I} - \boldsymbol{U} - \boldsymbol{U}^T & \boldsymbol{0} & \boldsymbol{0} & \boldsymbol{0} & \boldsymbol{0} & \boldsymbol{0} \\
\boldsymbol{0} & \boldsymbol{0} & \boldsymbol{0} & \boldsymbol{E}_A\boldsymbol{U} & \boldsymbol{0} & -\epsilon\boldsymbol{I} & \boldsymbol{0} & \boldsymbol{0} & \boldsymbol{0} & \boldsymbol{0} \\
\boldsymbol{0} & \boldsymbol{0} & \boldsymbol{0} & \boldsymbol{0} & \boldsymbol{0} & \boldsymbol{0} & -\epsilon\boldsymbol{I} & \boldsymbol{0} & \boldsymbol{0} & \boldsymbol{0} \\
\boldsymbol{0} & \boldsymbol{0} & \boldsymbol{0} & \boldsymbol{0} & \boldsymbol{0} & \boldsymbol{0} & \boldsymbol{0} & -\epsilon\boldsymbol{I} & \boldsymbol{0} & \boldsymbol{0} \\
\boldsymbol{0} & \boldsymbol{0} & \boldsymbol{0} & \boldsymbol{0} & \boldsymbol{0} & \boldsymbol{0} & \boldsymbol{0} & \boldsymbol{0} & -\epsilon\boldsymbol{I} & \boldsymbol{0} \\
\boldsymbol{0} & \boldsymbol{0} & \boldsymbol{0} & \boldsymbol{0} & \boldsymbol{0} & \boldsymbol{0} & \boldsymbol{0} & \boldsymbol{0} & \boldsymbol{0} & -\epsilon\boldsymbol{I}
\end{bmatrix} \prec 0. \tag{43}$$

Proof. The problem of $\mathcal{H}_\infty$ controller design is to determine the matrix $\boldsymbol{K}_c$ such that[21, 4]

$$\lim_{k\to\infty} \boldsymbol{x}_k = \boldsymbol{0} \quad \text{for } \boldsymbol{w}_k = \boldsymbol{0} \tag{44}$$

$$\|\boldsymbol{x}_k\|_{l_2} \le \mu\|\boldsymbol{v}_k\|_{l_2} \quad \text{for } \boldsymbol{w}_k \ne \boldsymbol{0},\, \boldsymbol{x}_0 = \boldsymbol{0}. \tag{45}$$

In order to settle the above problem it is sufficient to find a Lyapunov function V_k such that:

$$\Delta V_k + \boldsymbol{x}_k^T\boldsymbol{x}_k - \mu^2\boldsymbol{w}_k^T\boldsymbol{w}_k < 0,\, k = 0, \ldots \infty, \tag{46}$$

where

$$\Delta V_k = V_{k+1} - V_k, \tag{47}$$

$$V_k = \boldsymbol{x}_k^T\boldsymbol{P}\boldsymbol{x}_k, \tag{48}$$

$$\Delta V_k = \boldsymbol{x}_{k+1}^T\boldsymbol{P}\boldsymbol{x}_{k+1} - \boldsymbol{x}_k^T\boldsymbol{P}\boldsymbol{x}_k, \tag{49}$$

Consequently, using (41)

$$\Delta V_k + \boldsymbol{x}_k^T\boldsymbol{x}_k - \mu^2\boldsymbol{w}_k^T\boldsymbol{w}_k =$$
$$\boldsymbol{x}_k^T\left(\bar{\boldsymbol{A}}^T\boldsymbol{P}\bar{\boldsymbol{A}} - \boldsymbol{P} + \boldsymbol{I}\right)\boldsymbol{x}_k + \boldsymbol{x}_k^T\left(\bar{\boldsymbol{A}}^T\boldsymbol{P}\right)\boldsymbol{g}(\boldsymbol{x}_k) + \boldsymbol{x}_k^T\left(\bar{\boldsymbol{A}}^T\boldsymbol{P}\bar{\boldsymbol{W}}\right)\boldsymbol{w}_k +$$
$$\boldsymbol{g}(\boldsymbol{x}_k)^T\left(\boldsymbol{P}\bar{\boldsymbol{A}}\right)\boldsymbol{x}_k + \boldsymbol{g}(\boldsymbol{x}_k)^T(\boldsymbol{P})\boldsymbol{g}(\boldsymbol{x}_k) + \boldsymbol{g}(\boldsymbol{x}_k)^T\left(\boldsymbol{P}\bar{\boldsymbol{W}}\right)\boldsymbol{w}_k +$$
$$\boldsymbol{w}_k^T\left(\bar{\boldsymbol{W}}^T\boldsymbol{P}\bar{\boldsymbol{A}}\right)\boldsymbol{x}_k + \boldsymbol{w}_k^T\left(\bar{\boldsymbol{W}}^T\boldsymbol{P}\right)\boldsymbol{g}(\boldsymbol{x}_k) + \boldsymbol{w}_k^T\left(\bar{\boldsymbol{W}}^T\boldsymbol{P}\bar{\boldsymbol{W}} - \mu^2\boldsymbol{I}\right)\boldsymbol{w}_k < \boldsymbol{0}.$$

By defining

$$\boldsymbol{v}_k = [\boldsymbol{x}_k^T, \boldsymbol{g}(\boldsymbol{x}_k)^T, \boldsymbol{w}_k^T]^T, \tag{50}$$

it can be shown that (50) is equivalent to

$$\boldsymbol{v}_k^T\begin{bmatrix}\bar{\boldsymbol{A}}^T\boldsymbol{P}\bar{\boldsymbol{A}} + \boldsymbol{I} - \boldsymbol{P} & \bar{\boldsymbol{A}}^T\boldsymbol{P} & \bar{\boldsymbol{A}}^T\boldsymbol{P}\bar{\boldsymbol{W}} \\ \boldsymbol{P}\bar{\boldsymbol{A}} & \boldsymbol{P} & \boldsymbol{P}\bar{\boldsymbol{W}} \\ \bar{\boldsymbol{W}}^T\boldsymbol{P}\bar{\boldsymbol{A}} & \bar{\boldsymbol{W}}^T\boldsymbol{P} & \bar{\boldsymbol{W}}^T\boldsymbol{P}\bar{\boldsymbol{W}} - \mu^2\boldsymbol{I}\end{bmatrix}\boldsymbol{v}_k \prec \boldsymbol{0}. \tag{51}$$

Analyzing assumption of non-linear function, it can be show that (6) is equivalent to

$$\boldsymbol{g}(\boldsymbol{x})^T\boldsymbol{x} \leq \boldsymbol{x}^T\boldsymbol{M}\boldsymbol{x} = \frac{1}{2}\boldsymbol{x}^T(\boldsymbol{M} + \boldsymbol{M}^T)\boldsymbol{x}, \quad \boldsymbol{M} \in \mathbb{M}. \tag{52}$$

Inequality (52) can be written as

$$\frac{1}{2}\boldsymbol{x}^T(\boldsymbol{M} + \boldsymbol{M}^T)\boldsymbol{x} - \boldsymbol{g}(\boldsymbol{x}_k)^T\boldsymbol{x}_k \succeq 0, \quad \boldsymbol{M} \in \mathbb{M}, \tag{53}$$

which is equivalent to

$$\frac{1}{2}\boldsymbol{x}_k^T(\boldsymbol{M} + \boldsymbol{M}^T)\boldsymbol{x}_k - \frac{1}{2}\boldsymbol{g}(\boldsymbol{x}_k)^T\boldsymbol{x}_k - \frac{1}{2}\boldsymbol{x}_k^T\boldsymbol{g}(\boldsymbol{x}_k) \geq \boldsymbol{0}. \tag{54}$$

Thus, for any $\alpha > 0$

$$\alpha\boldsymbol{v}_k^T\begin{bmatrix}(\boldsymbol{M} + \boldsymbol{M}^T) & -\boldsymbol{I} & \boldsymbol{0} \\ -\boldsymbol{I} & \boldsymbol{0} & \boldsymbol{0} \\ \boldsymbol{0} & \boldsymbol{0} & \boldsymbol{0}\end{bmatrix}\boldsymbol{v}_k \succeq \boldsymbol{0}, \quad \alpha > 0, \quad \boldsymbol{M} \in \mathbb{M}. \tag{55}$$

Similarly, for some β inequality (7) can be written as

$$\beta\boldsymbol{x}_k^T\boldsymbol{M}^T\boldsymbol{M}\boldsymbol{x}_k - \beta\boldsymbol{g}(\boldsymbol{x}_k)^T\boldsymbol{g}(\boldsymbol{x}_k) \succeq 0, \quad \beta > 0. \tag{56}$$

$$\beta\boldsymbol{v}_k^T\begin{bmatrix}\boldsymbol{M}^T\boldsymbol{M} & \boldsymbol{0} & \boldsymbol{0} \\ \boldsymbol{0} & -\boldsymbol{I} & \boldsymbol{0} \\ \boldsymbol{0} & \boldsymbol{0} & \boldsymbol{0}\end{bmatrix}\boldsymbol{v}_k \succeq \boldsymbol{0}. \tag{57}$$

Combining together (51), (55) and (57) gives

$$\begin{bmatrix}\bar{\boldsymbol{A}}^T\boldsymbol{P}\bar{\boldsymbol{A}} + \boldsymbol{I} - \boldsymbol{P} + \alpha(\boldsymbol{M} + \boldsymbol{M}^T) + \beta\boldsymbol{M}^T\boldsymbol{M} & \bar{\boldsymbol{A}}^T\boldsymbol{P} - \alpha\boldsymbol{I} & \bar{\boldsymbol{A}}^T\boldsymbol{P}\bar{\boldsymbol{W}} \\ \boldsymbol{P}\bar{\boldsymbol{A}} - \alpha\boldsymbol{I} & \boldsymbol{P} - \beta\boldsymbol{I} & \boldsymbol{P}\bar{\boldsymbol{W}} \\ \bar{\boldsymbol{W}}^T\boldsymbol{P}\bar{\boldsymbol{A}} & \bar{\boldsymbol{W}}^T\boldsymbol{P} & \bar{\boldsymbol{W}}^T\boldsymbol{P}\bar{\boldsymbol{W}} - \mu^2\boldsymbol{I}\end{bmatrix} \prec \boldsymbol{0} \tag{58}$$

Rewriting (58) as

$$\begin{bmatrix} \bar{\boldsymbol{A}}^T \\ \boldsymbol{I}^T \\ \bar{\boldsymbol{W}}^T \end{bmatrix} \boldsymbol{P} \begin{bmatrix} \bar{\boldsymbol{A}} & \boldsymbol{I} & \bar{\boldsymbol{W}} \end{bmatrix} \begin{bmatrix} \boldsymbol{I} - \boldsymbol{P} + \alpha(\boldsymbol{M} + \boldsymbol{M}^T) + \beta \boldsymbol{M}^T \boldsymbol{M} & -\alpha \boldsymbol{I} & \boldsymbol{0} \\ -\alpha \boldsymbol{I} & -\beta \boldsymbol{I} & \boldsymbol{0} \\ \boldsymbol{0} & \boldsymbol{0} & -\mu^2 \boldsymbol{I} \end{bmatrix} \prec \boldsymbol{0}, \tag{59}$$

and using a transposed version *Lemma 1* to (59) yields

$$\begin{bmatrix} \boldsymbol{I} - \boldsymbol{P} + \alpha(\boldsymbol{M} + \boldsymbol{M}^T) & -\alpha \boldsymbol{I} & \boldsymbol{0} & \bar{\boldsymbol{A}}\boldsymbol{U} & \boldsymbol{M}\boldsymbol{U} \\ -\alpha \boldsymbol{I} & -\beta \boldsymbol{I} & \boldsymbol{0} & \boldsymbol{U} & \boldsymbol{0} \\ \boldsymbol{0} & \boldsymbol{0} & -\mu^2 \boldsymbol{I} & \bar{\boldsymbol{W}}\boldsymbol{U} & \boldsymbol{0} \\ \boldsymbol{U}^T \bar{\boldsymbol{A}}^T & \boldsymbol{U}^T & \boldsymbol{U}^T \bar{\boldsymbol{W}}^T & \boldsymbol{P} - \boldsymbol{U} - \boldsymbol{U}^T & \boldsymbol{0} \\ \boldsymbol{U}^T \boldsymbol{M}^T & \boldsymbol{0} & \boldsymbol{0} & \boldsymbol{0} & \beta \boldsymbol{I} - \boldsymbol{U} - \boldsymbol{U}^T \end{bmatrix} \prec \boldsymbol{0}, \tag{60}$$

with $\bar{\boldsymbol{A}}\boldsymbol{U} = \boldsymbol{A}\boldsymbol{U} - \boldsymbol{B}\boldsymbol{K}_c\boldsymbol{U} + \boldsymbol{H}\boldsymbol{\mathcal{F}}\boldsymbol{E}_A\boldsymbol{U} = \boldsymbol{A}\boldsymbol{U} - \boldsymbol{B}\boldsymbol{N} + \boldsymbol{H}\boldsymbol{\mathcal{F}}\boldsymbol{E}_A\boldsymbol{U}$ and $\bar{\boldsymbol{W}}\boldsymbol{U} = \boldsymbol{W}_1\boldsymbol{U}$. Written (60) similarly as inequality (36) and then applying *Lemma 2* and Schur complements leads to (43),which completes the proof. □

5. Case study

To verify the proposed approach, it was implemented for real three-tank system portrayed in Fig. 1 [5, 14, 8]. The system is composed of three separate tanks with variable cross-area

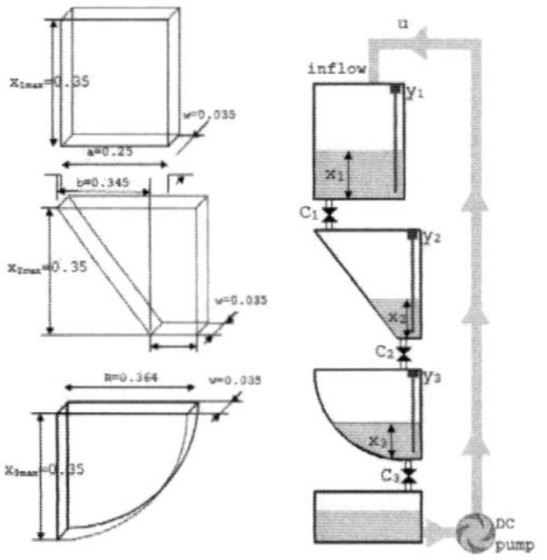

Figure 1. Three tank system

sections, where nonlinearities are imposed from shape of the second and third tank. The model was identified and following system matrices were obtained as well as nonlinear function of the system:

$$\boldsymbol{A} = \begin{bmatrix} 0.9982 & 0 & 0 \\ 0.0018 & 0.9973 & 0 \\ 0 & 0.0025 & 0.9972 \end{bmatrix}, \quad \boldsymbol{B} = \begin{bmatrix} 11.1400 \\ 0 \\ 0 \end{bmatrix}, \quad \boldsymbol{W}_1 = \begin{bmatrix} 0.0100 & 0 & 0 \\ 0 & 0.0100 & 0 \\ 0 & 0 & 0.0100 \end{bmatrix}$$

$$\boldsymbol{H} = \begin{bmatrix} 0.1 \\ 0.1 \\ 0 \end{bmatrix}, \quad \boldsymbol{E} = \begin{bmatrix} 0.2 \\ 0 \\ 0 \end{bmatrix}^T, \quad \boldsymbol{W}_1 = 0.05\boldsymbol{C}, \quad \boldsymbol{g}(\boldsymbol{x}_k) = \begin{bmatrix} \frac{1}{\beta_1(x_1)} C_1 x_{1,k}^{\alpha_1} \\ \frac{1}{\beta_2(x_{2,k})} C_1 x_{1,k}^{\alpha_1} - \frac{1}{\beta_2(x_{2,k})} C_2 x_{2,k}^{\alpha_2} \\ \frac{1}{\beta_3(x_{3,k})} C_2 x_{2,k}^{\alpha_2} - \frac{1}{\beta_3(x_{3,k})} C_3 x_{3,k}^{\alpha_3} \end{bmatrix}.$$

where: $\beta_i(x_i)$ is cross-sectional area of i-th tank, α_i denotes the flow coefficient parameter and C_i stands for i-th valve open level, the last two parameters were obtain through identification process. The numerical parameters obtained from identification along with operating ranges are: $0 \leq u \leq 1.5e^{-4}\left[\frac{m^3}{s}\right]$, $0.01 \leq x_{1-3} \leq 0.35[m]$ with $C_1 = 1.6721\mathrm{e}^{-4}$, $C_2 = 1.8804\mathrm{e}^{-4}$, $C_3 = 1.6637\mathrm{e}^{-4}$, $\alpha_1 = 0.4334$, $\alpha_2 = 0.4134$, $\alpha_2 = 0.3673$.

By varying each component of $\boldsymbol{x}_k$ within its possible domain, defined by the physical constraints of the three-tank system, it is possible to obtain the matrices

$$\boldsymbol{M}_{min} = \begin{bmatrix} -1.0094 & 0 & 0 \\ -0.0010 & -1.0457 & 0 \\ 0 & -0.0014 & -1.0593 \end{bmatrix}, \quad \boldsymbol{M}_{max} = \begin{bmatrix} -0.9997 & 0 & 0 \\ 0.0238 & -0.9968 & 0 \\ 0 & 0.0224 & -0.9979 \end{bmatrix}$$

which contain all elements $\bar{a}_{i,j}$ and $\underline{a}_{i,j}$, $i, j = 1, \ldots, n$ defining (13) respectively. Afterwards, taking into account all the possible combinations of elements $\boldsymbol{M}_{min}$ and $\boldsymbol{M}_{max}$, 32 matrices $\boldsymbol{M}_i$ are obtained.

For the purpose of further comparative study, two cases were employed:

Case SC1: It will be assumed that the whole state vector is directly measured

$$\boldsymbol{C} = \begin{bmatrix} 1 & 0 & 0 \\ 0 & 1 & 0 \\ 0 & 0 & 1 \end{bmatrix}, \quad \boldsymbol{K}_o = \begin{bmatrix} 0.7382 & -0.0008 & -0.0000 \\ 0.0012 & 0.7402 & 0.0005 \\ -0.0000 & 0.0030 & 0.7405 \end{bmatrix}, \quad \mu = 0.5623, \quad \boldsymbol{x}_0 = \boldsymbol{0}, \quad \boldsymbol{z}_0 = \boldsymbol{0}$$

Case SC2: It will be assumed that some state variables are not directly measured

$$\boldsymbol{C} = \begin{bmatrix} 1 & 0 & 0 \\ 0 & 0 & 1 \end{bmatrix}, \quad \boldsymbol{K}_o = \begin{bmatrix} 0.6977 & -0.0000 \\ 0.0002 & 0.0026 \\ -0.0000 & 0.7056 \end{bmatrix}, \quad \mu = 0.5623, \quad \boldsymbol{x}_0 = \boldsymbol{0}, \quad \boldsymbol{z}_0 = \boldsymbol{0}$$

Figure 2–3 present the results obtained for the proposed observer structure for case SC1. From these results, it is evident that the state estimation is performed with a good quality. Figure 3 also shows the state estimation for second state which is not directly measured. This corresponds to case SC2. From obtained results, it is evident that the state estimation is performed with a good quality. Of course there are differences between the quality of the estimation for the case SC1 and SC2 for state $\boldsymbol{x}_2$. Finally, Fig. 4 shows the system performance for the proposed control scheme. The goal is to control the liquid level in an upper tank. The reference signal (red solid line) is the target that has to be achieved by the controller. The robust controller was designed with (43) and $\mu = 0.5623$. The obtained robust controller is $K_c = [0.05731, 0.00076, 0.000001]$. The state vector is derived by the observer for SC2. From these results it can be seen that the state converges to the required setpoint.

6. Conclusions

The main objective of this paper was to propose problem of robust state estimation and robust control for nonlinear systems. The proposed design procedure for robust observer and robust controller is relatively simple and boils down to solving a set of linear matrix inequalities. The proposed approach can be efficiently implemented to the real-time system. The final part presents comprehensive case study of the three-tank system. The results clearly exhibit the high performance of the proposed scheme.

Acknowledgments

The work was supported by the National Science Centre of Poland under grant: 2013/11/B/ST7/01110

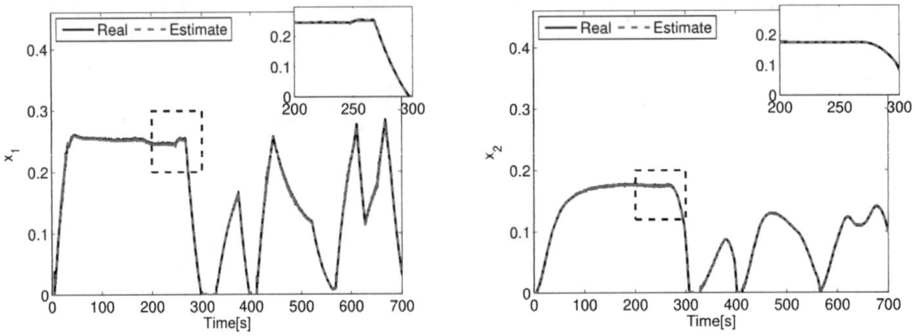

Figure 2. SC1: State variable $\boldsymbol{x}_1$ (left) $\boldsymbol{x}_2$ (right) (black line) and its estimates (red line) for $t = 0, \ldots, 700[s]$

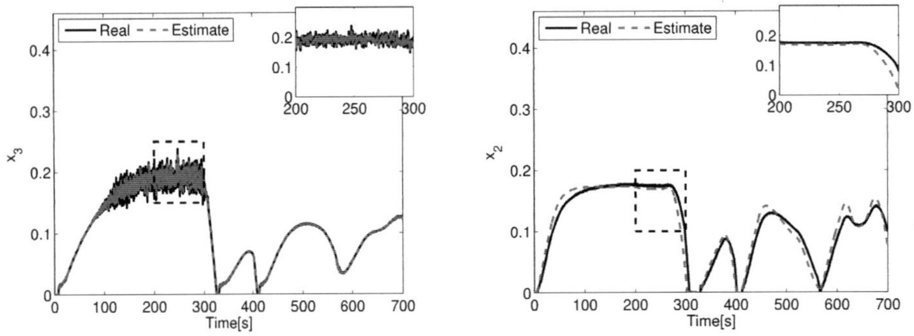

Figure 3. SC1(left): State variable $\boldsymbol{x}_3$ (black line) and its estimates (red line) for $t = 0, \ldots, 700[s]$. SC2(right): State variable $\boldsymbol{x}_2$ (black line) and its estimates (red line) for $t = 0, \ldots, 700[s]$

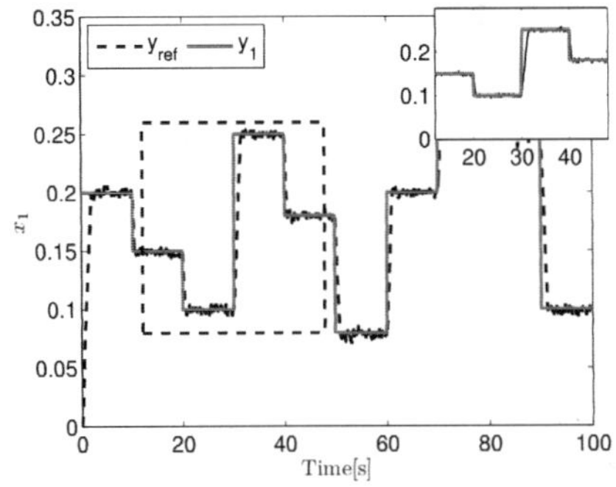

Figure 4. Performance of the system for $t = 0, \ldots, 100[s]$

[1] M. Abbaszadeh and H.J. Marquez. Lmi optimization approach to robust $h_infinity$ observer design and static output feedback stabilization for non-linear uncertain systems. *International Journal of Robust and Nonlinear Control*, 19(3):313–340, 2008.

[2] Stanisław Bańka, Paweł Dworak, and Krzysztof Jaroszewski. Design of a multivariable neural controller for control of a nonlinear mimo plant. *International Journal of Applied Mathematics and Computer Science*, 24(2):357–369, 2014.

[3] Ali Ben Brahim, Slim Dhahri, Fayçal Ben Hmida, and Anis Sellami. An h inf sliding mode observer for takagi–sugeno nonlinear systems with simultaneous actuator and sensor faults. *International Journal of Applied Mathematics and Computer Science*, 25(3):547–559, 2015.

[4] M. Buciakowski, M. de Rozprza Faygel, J. Ochalek, and M. Witczak. Actuator fault diagnosis and fault-tolerant control: application to the quadruple-tank process. In *11th European Workshop on Advanced Control and Diagnosis - ACD 2014*, pages 93–102, Berlin, Niemcy, 2014. [Berlin], [brak wydawcy].

[5] M. Buciakowski, M. Witczak, and M. Luzar. Robust fault-tolerant control for a multi-tank system. In *11th International Conference on Diagnostics of Processes and Systems - DPS 2013*, pages 1–12, Lagw Lubuski, Polska, 2013. [B. m.], –.

[6] M.C. de Oliveira, J. Bernussou, and J.C. Geromel. A new discrete-time robust stability condition. *Systems and Control Letters*, 37(4):261–265, 1999.

[7] S. Gillijns and B. De Moor. Unbiased minimum-variance input and state estimation for linear discrete-time systems. *Automatica*, 43:111–116, 2007.

[8] INTECO. *Multitank System - User's manual.* INTECO, www.inteco.com.pl, 2013.

[9] Cédric Join, J-C Ponsart, Dominique Sauter, and Didier Theilliol. Nonlinear filter design for fault diagnosis: application to the three-tank system. *IEE Proceedings-Control Theory and Applications*, 152(1):55–64, 2005.

[10] J. Y. Keller and M. Darouach. Two-stage Kalman estimator with unknown exogenous inputs. *Automatica*, 35(2):339–342, 1999.

[11] K. Kemir, F. Ben Hmida, J. Ragot, and M. Gossa. Novel optimal recursive filter for state and fault estimation of linear systems with unknown disturbances. *International Journal of Applied Mathematics and Computer Science*, 21(4):629–638, 2011.

[12] H. Li and M. Fu. A linear matrix inequality approach to robust h_∞ filtering. *IEEE Trans. Signal Processing*, 45(9):2338–2350, 1997.

[13] Francisco-Ronay López-Estrada, Jean-Christophe Ponsart, Carlos-Manuel Astorga-Zaragoza, Jorge-Luis Camas-Anzueto, and Didier Theilliol. Robust sensor fault estimation for descriptor–lpv systems with unmeasurable gain scheduling functions: Application to an anaerobic bioreactor. *International Journal of Applied Mathematics and Computer Science*, 25(2):233–244, 2015.

[14] M. Luzar, M. Witczak, M. Buciakowski, and V. Puig. Robust h infinity actuator fault diagnosis and fault-tolerant control for a multi-tank system. In *Conference on Control and Fault-Tolerant Systems - SysTol 2013*, pages 758–763, Nice, Francja, 2013. [B. m.].

[15] H.J. Marquez. *Nonlinear control systems. Analysis and Design.* Wiley, New Jersey, 2003.

[16] E.G. Nobrega, M.O. Abdalla, and K.M. Grigoriadis. Robust fault estimation of unceratain systems using an lmi-based approach. *International Journal of Robust and Nonlinear Control*, 18(7):1657–1680, 2008.

[17] A. M. Pertew, H. J. Marquez, and Q. Zhao. $\mathcal{H}_\infty$ synthesis of unknown input observers for non-linear lipschitz systems. *International Journal of Control*, 78(15):1155–1165, 2005.

[18] R. Rajamani. Observers for Lipschitz non-linear systems. *IEEE Transactions on Automatic Control*, 43(3):397–401, 1998.

[19] D.M. Stipanovic and D.D. Siljak. Robust stability and stabilization of discrete-time non-linear: the lmi approach. *International Journal of Control*, 74(5):873–879, 2001.

[20] Zhenhua Wang, Mickael Rodrigues, Didier Theilliol, and Yi Shen. Actuator fault estimation observer design for discrete-time linear parameter-varying descriptor systems. *International Journal of Adaptive Control and Signal Processing*, 29(2):242–258, 2015.

[21] M. Witczak, M. Buciakowski, and Ch. Aubrun. Predictive actuator fault-tolerant control under ellipsoidal bounding. *International Journal of Adaptive Control and Signal Processing*, 2015.

[22] M. Witczak, J. Korbicz, and R. Jozefowicz. Design of unknown input observers for non-linear stochastic systems and their application to robust fault diagnosis. *Control and Cybernetics*, 42(1):in print, 2013.

[23] Marcin Witczak, Mariusz Buciakowski, and Marcin Mrugalski. An h_∞ approach to fault estimation of non-linear systems: Application to one-link manipulator. In *Methods and Models in Automation and Robotics (MMAR), 2014 19th International Conference On*, pages 456–461. IEEE, 2014.

[24] A. Zemouche and M. Boutayeb. Observer design for Lipschitz non-linear systems: the discrete time case. *IEEE Trans. Circuits and Systems - II:Express Briefs*, 53(8):777–781, 2006.

[25] A. Zemouche, M. Boutayeb, and G. Iulia Bara. Observer for a class of Lipschitz systems with extension to $\mathcal{H}_\infty$ performance analysis. *Systems and Control Letters*, 57(1):18–27, 2008.

Microvibration Attenuation based on $\mathcal{H}_\infty$/LPV Theory for High Stability Space Missions

Valentin Preda[1], Jerome Cieslak[1], David Henry[1], Samir Bennani[2], Alexandre Falcoz[3]

[1]University of Bordeaux, IMS Laboratory, F-33400 Talence, France
{valentin.preda / jerome.cieslak / david.henry@ims-bordeaux.fr}
[2]European Space Agency, Noordwijk, 2201 AZ, The Netherlands {samir.bennani@esa.int}
[3]Airbus Defence and Space, 31 Rue Cosmonautes, F-31400 Toulouse, France
{alexandre.falcoz@astrium.eads.net}

Abstract. This paper presents a LPV (Linear Parameter Varying) solution for a mixed passive-active architecture used to mitigate the microvibrations generated by reaction wheels in satellites. In particular, $\mathcal{H}_\infty$/LPV theory is used to mitigate low frequency disturbances, current baseline for high frequency microvibration mitigation being based on elastomer materials. The issue of multiple harmonic microvibrations is also investigated. Simulation results from a test benchmark provided by Airbus Defence and Space demonstrate the potential of the proposed method.

1. Introduction

1.1. Motivations and learnt lessons from space experiments

As line-of-sight stability requirements get tighter with the increasing resolution, microvibrations become a key contributor to the performance of an increasing number of Earth observation and space science missions. The micro-dynamic behaviour of observation satellites has been studied as a potential contributor to the performance since the development of the SPOT/HELIOS satellite family. In the 80s and 90s, Airbus Defence and Space started gathering flight observations of microvibration effects, e.g. Doppler images on the SOHO satellite and the MICROMEDY experiment dedicated to the observation of SPOT 4 microvibrations, to name a few. Microvibrations degrade the image in two different ways: "high frequency" disturbances that cause image blurring, and "low frequency" vibrations that degrade the geometry inducing image distortions. In some cases, image distortions can be corrected on-ground by dedicated algorithms, but image blurring is definitely not rectifiable. This high/low frequency limit separation depends on the integration time; It is usually roughly equal to 0.1 times the inverse of the integration time. Hence, long integration time instruments like the ones used in scientific missions are more affected by high frequency microvibrations than short integration time ones as in LEO (Low Earth Orbit) observation missions. Any on-board equipment including one or several mobile devices shall then be considered as a potential source of microvibrations. This includes equipments such as Reaction Wheel Assemblies (RWAs) used for attitude control, gyroscopes, infrared-earth sensors, solar array drive mechanisms, thrusters, or payload elements such as pointing mechanisms, cryo-coolers, scan mechanisms, etc. However, the most critical sources are RWAs that generate multi-frequency and time-varying harmonic disturbances. Mass imbalances and ball bearing imperfections are considered as the main causes of microvibrations.

Two harmonic disturbances are generally considered as the main effects of microvibrations: The first harmonic, denoted H_1[1] by the space industry, is due to unbalance: the force and torques are proportional to the square of the wheel rate. The second harmonic, denoted $H_{0.6}$ generates forces and torques in the same order of magnitude as H_1 but at lower frequency (below 60 Hz for a wheel rate below 100 Hz).

Airbus Defence and Space flight proven isolation solutions consists of passive isolator which provides good rejection at high frequencies. A passive isolator can be roughly compared to a second order filter with amplification at a resonance frequency, and a rejection factor increasing with the disturbance frequency. Current baseline for micro vibration mitigation is based on elastomer design initially developed on Earth observation missions (Theos, Pleiades). Although apparently simple this concept requires a deep understanding of the elastomer materials and their physical properties when developing passive isolators taking into account all the space-related constraints (thermal, radiations, lifetime, launch stresses), the needs of compatibility with AOCS (Attitude and Orbit Control System), the desired isolation performance, etc. However, passive solutions are insufficient for future demanding Earth and Science Observation programmes. Usually, the order of magnitude of the best performances that can be reached when using passive isolators only set at the base of typically 3 wheels is about $1\mu rad$ for wheel rates varying from 0 to typically 6000 rpm; the major contributors are roughly in the $10 - 100$ Hz range; performances below 10 Hz are usually better than $0.1\mu rad$, thanks to the naturally low wheel disturbances at these frequencies; performances above 100 Hz are also generally below $0.1\mu rad$, thanks to the passive isolator filtering. 20 dB of disturbances rejection above 10 Hz is then required to meet the $0.1\mu rad$ objective and two solutions can be envisaged, i.e. *i)* the use of micro-propulsion system instead of wheels or *ii)* the use of mixed active/passive microvibration mitigation system. Micro-propulsion system allow to reach very high stability performances ($< 0.01\mu rad$) but is very expensive, massive ($>$ 140 kg without propellant) and complex whereas mixed active/passive microvibration mitigation provide lower cost/complexity and mass solution ($<$ 30 kg with the worst case assumption) while allowing using the wheels over the full operating range and during all the mission phases. Missions for which stability requirements evolve between 0.01 and $0.1\mu rad$ may be efficiently handled using such hybrid active/passive isolation solution.

The work presented in this paper should be understood in this context. It concerns the development of the active part of a mixed passive/active microvibrations mitigation solution.

1.2. Related work and antecedents

Mixed passive/active microvibrations attenuation solutions for satellites have been reported in a number of published works. In [1], a multi-purpose active isolation system based on six active struts (co-located piezo sensor and actuator), arranged in hexapod configuration, is considered. Active vibration isolation interface using piezoelectric ceramic actuators is discussed in [2]. A six-degree of freedom Stewart platform based on six local/decentralized integral force feedback controllers is proposed as a microvibration isolator in [3]. A complementary active control strategy is too discussed recently in [4].

Generally speaking, it could be argued that the aforementioned solutions suffer from a lack of formal proof, both in terms of stability and performances. More advanced control methodologies has to be investigated to provide formal stability and performance guarantees while taking into account the time-varying spectrum of microvibrations.

[1] The terminology $H_1/H_{0.6}$ is used by the space industry to outline that the frequency of the microvibration is equal to $1/0.6\Omega$, Ω being the speed of the reaction wheel.

Linear Parameter Varying (LPV) control strategies offer a solution to this problem. LPV framework is appealing because it provides stability and performance guarantees over a wide range of changing parameters, see for instance the none exhaustive list of papers [5–8]. In our case, noting that the frequencies of the microvibrations generated by RWAs have a known dependency on the angular velocity of the reaction wheel [9], it seems clear that it can be used as a scheduling parameter for the active control strategy, leading the philosophy of LPV (Linear Parameter Varying) theory to be a suitable framework.

Note that LPV solutions have been successfully used for active vibration control in a number of other applications such as active magnetic bearing systems [10], noise canceling headsets [11] or car engines [12], to name a few.

1.3. Contribution
The main contribution of this paper is the development of a LPV solution to mitigate low frequency microvibrations. The issue of multiple harmonic (i.e. H_1 and $H_{0.6}$) microvibrations is also investigated. The work is an extension of previous research results developed by the authors, see [13] and [14]. It is shown how the proposed LPV controller is capable of attenuating time-varying harmonic disturbances. This is achieved by introducing a parameter dependent weighting function within the $\mathcal{H}_\infty$/LPV synthesis procedure. This weighting function represents an extension of the notch filter used in [10]. This strategy is applied on a test benchmark provided by Airbus Defence and Space, that is representative of a flexible spacecraft structure.

2. Background on $\mathcal{H}_\infty$/LPV theory
2.1. Disturbance rejection for time-varying frequencies
Consider a LTI (Linear Time Invariant) plant described by the following system:

$$G(s) := \begin{cases} \dot{x}(t) = Ax(t) + Bu(t) + Ed(\theta, t) \\ y(t) = Cx(t) + v(t) \end{cases} \tag{1}$$

with the state vector $x \in \mathbb{R}^{n_x}$, control signal $u \in \mathbb{R}^{n_u}$, disturbance $d \in \mathbb{R}^{n_u}$ and measurement vector $y \in \mathbb{R}^{n_y}$ with measurement noise $v \in \mathbb{R}^{n_y}$. It is assumed that the spectrum of the disturbance d depends on a known θ parameter.

The aim is to design a controller $K(\theta)$ to attenuate as much as possible the disturbances d on the measurements y, in the L_2-norm sense. Emphasis is put around some known harmonics that are linearly dependent on the central frequency θ. This θ parameter is considered adjustable, bounded and time-varying i.e.

$$\theta(t) \in \mathcal{D}_\theta = \left[\underline{\theta},\ \overline{\theta}\right] \tag{2}$$

The control law is of course constrained by stability and robustness margins requirements. Let $T_{w \to z}$ represent the transfer between the signals w and z. We define $S_y = T_{r \to e}$, $KS_y = T_{r \to u}$, $S_yG = T_{d \to y}$ and $T_u = T_{d \to u}$ as the output, input, disturbance and input complementary sensibility functions respectively.

To carry out this problem, the mixed sensitivity approach of the $\mathcal{H}_\infty$ framework is used. To proceed, let us introduce the reference signal r and input disturbance d_n as illustrated on Fig. 1. It follows that the disturbance and performance channels are defined as $w = (r^T, d_n^T)^T$ and $z = (z_1^T, z_2^T)^T$ together with the weighting functions W_1, W_2 and W_3. It is assumed that the weight W_1 depends on a bounded and measured parameter θ so that $|\theta| \leq 1$. This can be done without loss of generality since the problem can always be scaled.

Remark 1 *In this work, it is assumed that θ is a scalar since, as it is explained later, θ plays the role of the central frequency of a notch filter. However, it is well known from the $\mathcal{H}_\infty$/LPV community that the LPV theory remains valid if θ is a vector, see for instance [6, 15].*

Then, the synthesis problem turns out to be the design of the controller $K(\theta)$ that (intrinsically) stabilizes the closed-loop and solves the following optimization problem

$$\min_{K(\theta)} \gamma \text{ s.t. } \left\| \mathcal{F}_l\Big(P(\theta), K(\theta)\Big) \right\|_\infty < \gamma, \ |\theta| \leq 1 \tag{3}$$

where $\mathcal{F}_l$ defines the *lower linear fractional transformation.* The performance z is given by (4).

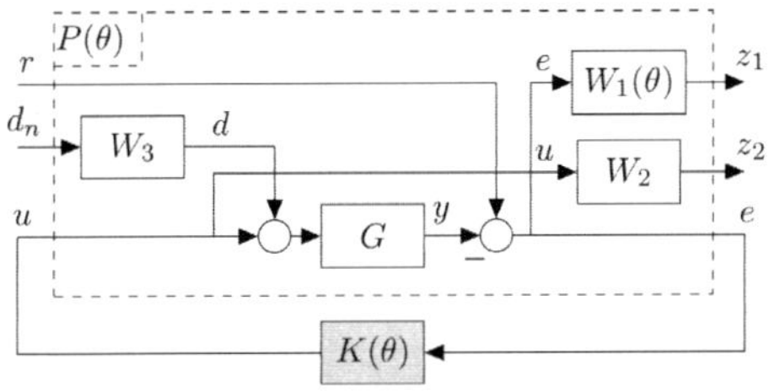

$$\underbrace{\begin{bmatrix} z_1 \\ z_2 \end{bmatrix}}_{z} = \underbrace{\begin{bmatrix} W_1(\theta)S_y & -W_1(\theta)S_yGW_3 \\ W_2KS_y & -W_2T_uW_3 \end{bmatrix}}_{\mathcal{F}_l\left(P(\theta), K(\theta)\right)} \underbrace{\begin{bmatrix} r \\ d_n \end{bmatrix}}_{w} \tag{4}$$

Figure 1. Augmented plant model

2.1.1. Problem formulation within the polytopic setting: By proper choice of $W_1(\theta), W_2, W_3$, the scheduled controller $K(\theta)$ can achieve the rejection of the disturbance signal d together with bounds on the different sensibility functions. Stability margin requirements and disturbance rejection performance can be managed through $W_1(\theta)$. Note that this filter shapes both the output S_y and disturbance S_yG sensitivity functions. The following expression of $W_1(\theta)$ is also retained:

$$W_1(\theta) = W_sW_p(\theta)I_{n_y} \tag{5}$$

In this equation, the function W_s is used to shape S_y and introduce a *modulus margin* stability requirement, whereas $W_p(\theta)$ is chosen as an inverted notch filter to provide attenuation around θ. Here, the formulation introduced in [10] is used for $W_p(\theta)$, i.e.

$$W_p(\theta) = \left(\begin{array}{cc|c} -2\theta\xi_2 & -\theta & 2\theta(\xi_1 - \xi_2) \\ \theta & 0 & 0 \\ \hline 1 & 0 & 1 \end{array}\right) = \left(\begin{array}{c|c} 0 & 0 \\ \hline C_p & 1 \end{array}\right) + \theta \left(\begin{array}{c|c} A_p & B_p \\ \hline 0 & 0 \end{array}\right) \tag{6}$$

where (6) is affine in θ. The damping ratio $\xi = \xi_2/\xi_1$ sets the notch depth corresponding to the disturbance attenuation requirement over Ω. For a fixed damping ratio, the ξ_2 parameter controls the width of the notch filter around the center frequency θ.

W_2 is also chosen to decrease the actuation effort and improve the controller roll-off. W_3 is used to bound the maximum value of the disturbance sensitivity function S_yG. Adjusting W_3 also influences the controller effort by shaping $T_u = T_{d \to u}$.

With these weight definitions, the augmented plant $P(\theta)$, has the following affine structure:

$$\begin{bmatrix} \dot{x} \\ \hline z \\ e \end{bmatrix} = \underbrace{\left[\begin{array}{c|cc} A(\theta) & B_1(\theta) & B_2 \\ \hline C_1(\theta) & D_{11} & D_{12} \\ C_2 & D_{12} & 0 \end{array}\right]}_{P(\theta)} \begin{bmatrix} x \\ \hline w \\ u \end{bmatrix} \text{ with } \begin{cases} A(\theta) = A_0 + \theta A_1 \\ B_1(\theta) = B_{10} + \theta B_{11} \\ C_1(\theta) = C_{10} + \theta C_{11} \end{cases} \tag{7}$$

2.1.2. Problem formulation within the LFR setting: Direct application of the LFR formalism enables the following alternative formulation for the $W_p(\theta)$ weight (6):

$$W_p(\theta) = \mathcal{F}_u\left(\begin{bmatrix} \frac{1}{s}A_p & B_p \\ \frac{1}{s}C_p & 1 \end{bmatrix}, \Delta\right), \quad \Delta = \begin{bmatrix} \theta & 0 \\ 0 & \theta \end{bmatrix} \tag{8}$$

where $\mathcal{F}_u$ represents the *upper linear fractional transformation.*

With this expression of $W_p(\theta)$ and incorporating $W_1(\theta)$, W_2 and W_3 into the augmented plant model (7), the following LFR can be derived,

$$P(\theta) = \mathcal{F}_u\left(\left(\begin{array}{c|ccc} A & B_\theta & B_1 & B_2 \\ \hline C_\theta & D_{\theta\theta} & D_{\theta 1} & D_{\theta 2} \\ C_1 & D_{1\theta} & D_{11} & D_{12} \\ C_2 & D_{2\theta} & D_{21} & 0 \end{array}\right), \Theta(t)\right) \quad \text{with} \quad \begin{array}{l} \Theta(t) = \widetilde{\theta}(t) I_{12} \;\; \text{and} \\ \widetilde{\theta}(t) = \dfrac{2(\theta(t) - \underline{\theta})}{\overline{\theta} - \underline{\theta}} - 1 \end{array} \tag{9}$$

which is then suitable for the LPV controller synthesis technique presented in [5,7]. This aspect is addressed in the next section.

2.2. Extension to multiple harmonics

The aforementioned formulated problem can be extended to take into account the first n harmonics of d, assuming that all harmonic frequencies denoted θ_i, depend linearly on θ, i.e.

$$\theta_i(t) = \alpha_i + \beta_i \theta(t), \quad i = 1...n \tag{10}$$

Using the affine property in θ of $W_p(\theta)$, see Eq. (6), a set of n weighting functions $W_{p_i}(\theta), i = 1...n$ can be defined according to

$$\begin{aligned} W_{p_i}(\theta) &= \left(\begin{array}{c|c} 0 & 0 \\ \hline C_{p_i} & 1 \end{array}\right) + \theta_i \left(\begin{array}{c|c} A_{p_i} & B_{p_i} \\ \hline 0 & 0 \end{array}\right) \\ &= \left(\begin{array}{c|c} \alpha_i A_{p_i} & \alpha_i B_{p_i} \\ \hline C_{p_i} & 1 \end{array}\right) + \theta(t) \left(\begin{array}{c|c} \beta_i A_{p_i} & \beta_i B_{p_i} \\ \hline 0 & 0 \end{array}\right), \quad i = 1...n \end{aligned} \tag{11}$$

so that each filter $W_{p_i}(\theta), i = 1...n$ can be independently parametrized to have a different width and depth. Combining all these filters leads to the following expression for $W_p(\theta)$:

$$W_p(\theta) = \left(\begin{array}{ccc|c} \alpha_1 A_{p1} & & & \alpha_1 B_{p1} \\ & \ddots & & \vdots \\ & & \alpha_n A_{pn} & \alpha_n B_{pn} \\ \hline C_{p1} & \cdots & C_{pn} & 1 \end{array}\right) + \theta \left(\begin{array}{ccc|c} \beta_1 A_{p1} & & & \beta_1 B_{p1} \\ & \ddots & & \vdots \\ & & \beta_n A_{pn} & \beta_n B_{pn} \\ \hline & 0 & & 0 \end{array}\right) \tag{12}$$

3. Design of the LPV controller $K(\theta)$

This section addresses different LPV synthesis techniques to design $K(\theta)$. Both the LFR and polytopic approaches are investigated. Due to space limitations, only brief descriptions will be provided. The reader can refer to the original papers for additional details.

For an LPV plant described either by (7) or by (9), the $\mathcal{H}_\infty$ LPV control problem consists of finding a dynamic output feedback controller $K(\theta)$ such that the closed-loop is internally stable and the $\mathcal{L}_2$-gain γ of the closed-loop operator between w and z is minimized, for all $\theta(t)$.

3.1. The LFR approach

The LPV synthesis approach introduced in [5] considers the LFR description (9) for the plant. A combination of symmetric and skew-symmetric matrices leads to a scaled version of the Bounded Real Lemma. The scalings commute with the Θ block and account for the realness of the scheduling parameter. The following theorem [5, 7] solves the problem:

Theorem 1 *Let the LPV plant be described by (9) with $\mathcal{N}_X = ker[C_2, D_{2\theta}, D_{21}, 0]$ and $\mathcal{N}_Y = ker[B_2^T, D_{\theta 2}^T, D_{12}^T, 0]$. There exists an LPV controller guaranteeing $\mathcal{L}_2$-performance level γ for the scaled Bounded Real Lemma, if and only if there exists the pairs of symmetric matrices $(\mathbf{X}, \mathbf{Y})$, $(\mathbf{S}, \mathbf{\Sigma})$ and a pair of skew-symmetric matrices $(\mathbf{T}, \mathbf{\Gamma})$ such that the following LMIs are feasible:*

$$\mathcal{N}_X^T \begin{bmatrix} A^T\mathbf{X}+\mathbf{X}A & \star & \star & \star & \star \\ B_\theta^T\mathbf{X}+\mathbf{T}C_\theta & -\mathbf{S}+\mathbf{T}D_{\theta\theta}+D_{\theta\theta}^T\mathbf{T}^T & \star & \star & \star \\ B_1^T\mathbf{X} & D_{\theta 1}^T\mathbf{T}^T & -\gamma I & \star & \star \\ \mathbf{S}C_\theta & \mathbf{S}D_{\theta\theta} & \mathbf{S}D_{\theta 1} & -\mathbf{S} & \star \\ C_1 & D_{1\theta} & D_{11} & 0 & -\gamma I \end{bmatrix} \mathcal{N}_X < 0, \quad \begin{bmatrix} \mathbf{S} & 0 \\ 0 & \mathbf{\Sigma} \end{bmatrix} > 0$$
$$\mathcal{N}_Y^T \begin{bmatrix} A\mathbf{Y}+\mathbf{Y}A^T & \star & \star & \star & \star \\ C_\theta\mathbf{Y}+\mathbf{\Gamma}B_\theta^T & -\mathbf{\Sigma}+\mathbf{\Gamma}D_{\theta\theta}^T+D_{\theta\theta}\mathbf{\Gamma}^T & \star & \star & \star \\ C_1\mathbf{Y} & D_{1\theta}\mathbf{\Gamma}^T & -\gamma I & \star & \star \\ \mathbf{\Sigma}B_\theta^T & \mathbf{\Sigma}D_{\theta\theta}^T & \mathbf{\Sigma}D_{\theta 1}^T & -\mathbf{\Sigma} & \star \\ B_1^T & D_{\theta 1}^T & D_{11}^T & 0 & -\gamma I \end{bmatrix} \mathcal{N}_Y < 0, \quad \begin{bmatrix} \mathbf{X} & I \\ I & \mathbf{Y} \end{bmatrix} > 0 \tag{13}$$

3.2. The polytopic approach
Another analysed LPV synthesis technique is the one used in [16]. This method applies to systems which admit a polytopic representation and uses the Projection Lemma to find a Single Quadratic Lyapunov Function (SQLF), valid for all scheduling domain (2). One issue with these techniques is that they sometimes lead to controllers with very fast poles. The poles in this case are calculated for fixed values of the scheduling parameter $\theta(t)$. To overcome this problem, one can use the concept of LMI regions introduced by [17]. Using a synthesis technique based on a change in the controller variables [18] allows the introduction of LMI regions [17]. These regions are new LMI constraints that restrict the closed-loop poles to subsets of the complex plane.

Parameter dependent Lyapunov Functions (PDLF) were used to reduce the conservatism associated with using a SQLF for the whole scheduling interval [6]. Considering bounds on $\dot{\theta}$ so that $\dot{\theta}(t) \in [\underline{v}, \overline{v}]$, leads to the following theorem:

Theorem 2 *For an LPV plant described by (7), there exists an LPV controller that guarantees closed-loop global asymptotic stability, $\mathcal{L}_2$-gain smaller than γ and closed-loop poles constrained to the circle of rayon r around the origin, whenever there exists parameter-dependent symmetric matrices $\mathbf{X}(\theta)$ and $\mathbf{Y}(\theta)$ and parameter-dependent matrices $(\widehat{\mathbf{A}}(\theta), \widehat{\mathbf{B}}(\theta), \widehat{\mathbf{C}}(\theta), \mathbf{D}(\theta))$ that satisfy for every $(\theta, \dot{\theta}) \in [\underline{\theta}, \overline{\theta}] \times [\underline{v}, \overline{v}]$ the following infinite dimensional matrix inequalities:*

$$\begin{bmatrix} \dot{\mathbf{X}}+\mathbf{X}A+\widehat{\mathbf{B}}C_2+(\star) & \star & \star & \star \\ \widehat{\mathbf{A}}^T+A+B_2\mathbf{D}C_2 & -\dot{\mathbf{Y}}+A\mathbf{Y}+B_2\widehat{\mathbf{C}}+(\star) & \star & \star \\ (\mathbf{X}B_1+\widehat{\mathbf{B}}D_{21})^T & (B_1+B_2\mathbf{D}D_{21})^T & -\gamma I & \star \\ C_1+D_{12}\mathbf{D}C_2 & C_1\mathbf{Y}+D_{12}\widehat{\mathbf{C}} & D_{11}+D_{12}\mathbf{D}D_{21} & -\gamma I \end{bmatrix} < 0, \quad \begin{bmatrix} \mathbf{X} & I \\ I & \mathbf{Y} \end{bmatrix} > 0$$
$$\begin{bmatrix} \mathcal{M}_{11} & \star \\ \mathcal{M}_{21} & \mathcal{M}_{11} \end{bmatrix} < 0, \quad \mathcal{M}_{11} = \begin{bmatrix} -r\mathbf{Y} & -rI \\ -rI & -r\mathbf{X} \end{bmatrix}, \quad \mathcal{M}_{21} = \begin{bmatrix} A\mathbf{Y}+B_2\widehat{\mathbf{C}} & \widehat{\mathbf{A}}^T \\ A+B_2\mathbf{D}C_2 & \mathbf{X}A+\widehat{\mathbf{B}}C_2 \end{bmatrix} \tag{14}$$

where the dependencies on θ are dropped for simplicity.

To reduce the optimization problem to a finite-dimensional case, the Lyapunov function variables $\mathbf{X}$, $\mathbf{Y}$ and the transformed controller variables $(\widehat{\mathbf{A}}, \widehat{\mathbf{B}}, \widehat{\mathbf{C}}, \mathbf{D})$ can been chosen to have the same affine dependence on the scheduling parameter θ as the plant model. This results in the following new expressions for the optimization variables:

$$\widehat{\mathbf{A}}(\theta) = \widehat{\mathbf{A}}_0 + \theta\widehat{\mathbf{A}}_1 \quad \widehat{\mathbf{B}}(\theta) = \widehat{\mathbf{B}}_0 + \theta\widehat{\mathbf{B}}_1 \quad \widehat{\mathbf{C}}(\theta) = \widehat{\mathbf{C}}_0 + \theta\widehat{\mathbf{C}}_1 \quad \mathbf{D}(\theta) = \mathbf{D}_0 + \theta\mathbf{D}_1$$
$$\text{and } \mathbf{X}(\theta) = \mathbf{X}_0 + \theta\mathbf{X}_1 \quad \mathbf{Y}(\theta) = \mathbf{Y}_0 + \theta\mathbf{Y}_1 \tag{15}$$

The problem is still not convex but can be approached by gridding the parameter space $[\underline{\theta},\overline{\theta}] \times [\underline{v},\overline{v}]$. The LMI constraints can be checked afterwards across a denser grid. However, stability and performances are not guaranteed between gridding points.

Another way to convexify the LMI problem (2), assuming (3.2), is by adding the following multiconvexity constraint, as in [19]:

$$\frac{\partial^2 F}{\partial \theta^2} \geq 0 \qquad \forall \theta \in [\underline{\theta},\overline{\theta}] \tag{16}$$

Thanks to the new convex nature of the problem, it is sufficient for the LMI constraints (2), (14) to hold at the bounding vertices of the domain $(\theta,\dot{\theta}) \in [\underline{\theta},\overline{\theta}] \times [\underline{v},\overline{v}]$.

The controller synthesis problem (3) is thus reduced to the following LMI optimization:

$$\underset{\substack{\widehat{\mathbf{A}}_i,\widehat{\mathbf{B}}_i,\widehat{\mathbf{C}}_i,\mathbf{D}_i,\mathbf{X}_i,\mathbf{Y}_i\\ i=0,1}}{\text{minimize}} \quad \gamma \quad \text{subject to } (2),(14) \text{ and } (16). \tag{17}$$

For a know $(\theta(t),\dot{\theta}(t))$ pair, the controller matrices can be reconstructed by undoing the initial controller transformation using the formulas (18), as described in [6].

$$\begin{aligned} A_K &= N^{-1}(\mathbf{X}\dot{\mathbf{Y}} + N\dot{M}^T + \widehat{\mathbf{A}} - \mathbf{X}(A - B_2\mathbf{D}C_2)\mathbf{Y} - \widehat{\mathbf{B}}C_2\mathbf{Y} - \mathbf{X}B_2\widehat{\mathbf{C}})M^{-T} \\ B_K &= N^{-1}(\widehat{\mathbf{B}} - \mathbf{X}B_2\mathbf{D}) \qquad C_K = (\widehat{\mathbf{C}} - \mathbf{D}C_2\mathbf{Y})M^{-T} \qquad D_K = \mathbf{D} \end{aligned} \tag{18}$$

where N, M solve the factorization problem, $I - \mathbf{XY} = NM^T$. To avoid the need for the scheduling parameter derivative $\dot{\theta}$ during controller reconstruction, either $\mathbf{X}$ or $\mathbf{Y}$ can be fixed to a constant value.

In the case of PDLF, the new LMI constraints have to be valid across the polytopic domain defined by $\theta \times \dot{\theta}$. This leads to an increase in the number of polytope vertices, compared to the synthesis based on SQLF. On-line computational burden is also increased because for every new value of the parameter $\theta(t)$, a new controller reconstruction is required based on the new values of $\mathbf{X}(\theta(t))$ or $\mathbf{Y}(\theta(t))$. This can be a critical problem to implement the LPV controller, which is the case here for micro-vibrations mitigation.

Fixing both $\mathbf{X}$ and $\mathbf{Y}$ to constant values (19), leads to a SQLF problem with additional pole constraints.

$$\mathbf{X}(\theta) = \mathbf{X}_0 \quad \text{and} \quad \mathbf{Y}(\theta) = \mathbf{Y}_0 \tag{19}$$

In this case, the derivatives $\dot{\mathbf{X}}$ and $\dot{\mathbf{Y}}$ are equal to zero and can be dropped from (2) and (18) together with the constraint (16). The controller can thus be scheduled, for a known θ, by performing the interpolation (20) between the controller matrices calculated at $\theta \in \{\underline{\theta},\overline{\theta}\}$.

$$K(\theta) = (1-\rho)K(\underline{\theta}) + \rho K(\overline{\theta}) \quad \text{with } \rho(t) = \frac{\theta(t) - \underline{\theta}}{\overline{\theta} - \underline{\theta}} \in [0,\ 1] \tag{20}$$

4. Application to microvibration mitigation

The method described in Section 3 is applied to the problem of microvibration mitigation. The application support is a physical system jointly developed by the European Space Agency (ESA) and Airbus Defence and Space. The final issue is to propose a solution for the active part of a mixed passive-active control architecture to mitigate the microvibrations generated by reaction wheels in satellites. In this sense, it is assumed that the application support is representative of a flexible satellite panel put in space.

4.1. System description
The hardware architecture of the system is presented in Fig. 2. The platform consists of a breadboard suspended by four springs. For this system, a shaker is installed on top of a dummy reaction wheel and is used to simulate the multi-harmonic microvibrations induced by a real reaction wheel.
The dummy wheel is connected to an active plate that can vibrate to compensate for the vibrations introduced by the shaker. Actuation of the active plate is provided using a set of six Proof Mass Actuators (PMA) mounted on three cubes; three along the vertical axis and three in the tangential direction. These actuators generate reactive forces by accelerating a small mass connected to a spring-damper system using a magnetic field. The field is produced by passing an electric current through a voice coil. The current in turn, is provided via amplifiers that convert the voltage control signal vector $u(t) \in \mathbb{R}^6$.
The active plate, in turn, is connected to an interface plate using four elastomer isolator modules. These passive isolators are used to attenuate the high frequency disturbances transmitted from the shaker to the interface plate. On the bottom of each isolator, a tri-axis force cell sensor is installed. When strained, piezoelectric elements inside these cells generate electric charges that are converted into voltages using charge amplifiers. The transmitted forces and moments are obtained by projecting the force cell measurements in the Force and Torques space and forming the output vector $y(t) \in \mathbb{R}^6$.

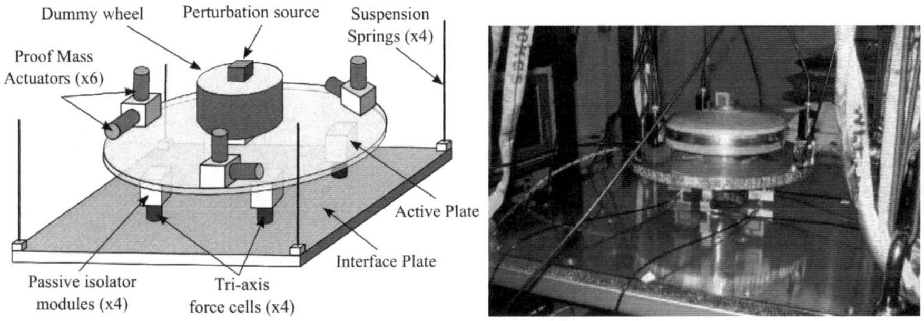

Figure 2. Hardware configuration.

4.2. System model
Model identification was performed by Airbus Defence and Space using real measurement data. The identification procedure was done by considering a null wheel rate and ignores gyroscopic effects. The resulting model is a 36 order nominal LTI model for $G(s)$, see Eq. (1). The estimated dynamics contains multiple bending modes with natural frequencies $\omega_n \in [1.92,\ 42.26]$ Hz and damping ratios $\xi_n \in [3.63 \cdot 10^{-2},\ 5.96 \cdot 10^{-2}]$. Additionally, the presence of strong coupling effect between axes is evident. Therefore, it is not possible to decouple the control laws for each axis and a full MIMO control law is judged necessary.[2].

4.3. Objectives and the $\mathcal{H}_\infty$ criterion
In this paper, the focus is put on mitigating two harmonics denoted H_1 and $H_{0.6} = 0.6H_1$, so that H_1 corresponds to the speed of the reaction wheel. Since the speed of the reaction wheel is measured, H_1 is chosen as the LPV scheduling parameter $\theta \in \mathcal{D}_\theta$. The interval $\mathcal{D}_\theta$ is fixed according to:

$$\theta(t) \in \mathcal{D}_\theta = [15,\ 40] \text{ Hz} \tag{21}$$

The harmonics H_1 and $H_{0.6}$ are chosen based on the fact that, for a wheel speed in the interval (21), most of the disturbance energy is concentrated in these two harmonics, see [20] if necessary.

[2] Additional details about the plant model cannot be supplied due to industrial confidentiality.

So the aim is to achieve around 20 dB disturbance attenuation for H_1 and 10 dB for $H_{0.6}$ on the force channels. Additionally, a $\|S_y\|_\infty < 6.5$ dB modulus margin is desired as a stability indicator. Robustness against additive and multiplicative output uncertainty is achieved by aiming for the bounds $\|KS_y\|_\infty < 7.5$ dB and $\|T_y\|_\infty < 6.5$ dB.

In practice, the reaction wheel acceleration is of small magnitude and thus the scheduling parameter θ is slow-varying. This fact motivated the use of the following method to translate the design requirements into different values for W_1, W_2 and W_3:

i) Choose $W_1(\theta)$ as in Eq. (5). To incorporate stability requirements, $W_s^{-1} = 6$ dB is set as the desired modulus margin. Two harmonics are to be attenuated therefore $W_p(\theta)$ contains two filters, W_{p_1} and W_{p_2}, defined for the H_1 and $H_{0.6}$ harmonics. The parameters used for these specifications are included in Table 4.3. The singular values of the $W_1^{-1}(\theta)$ weight are illustrated in Fig. 4.3 for 30 values of $\theta \in \mathcal{D}_\theta$.

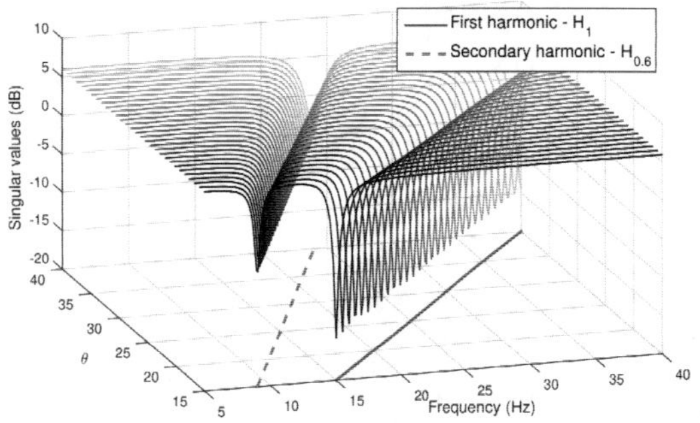

Figure 3. Singular values of $W_1(\theta)$ for $\theta \in [15, 40]$ Hz.

Table 1. Parameters for the $w_p(\theta)$ weights

	Frequency θ	Depth ξ_2/ξ_1	Width ξ_2
w_{p_1}	θ	20 dB	0.1
w_{p_2}	$0.6 \cdot \theta$	10 dB	0.1

ii) Parametrize W_2 and W_3 as in (22). The possible value intervals for w_2 and w_3 were chosen based on the fact that they are used to bound the KS_y and GS_y sensitivities.

$$W_2(s) = w_2^{-1} I_6 \qquad W_3(s) = w_3^{-1} I_6 \qquad \text{with} \qquad w_2 \in [0,\ 28] \text{ dB} \quad w_3 \in [-5,\ 22] \text{ dB} \tag{22}$$

iii) Grid the parameter space defined by w_2 and w_3 and construct a different controller at every grid point using the PDLF-LPV synthesis detailed in Section 3.2.

iv) For every controller, evaluate the frequency domain performance by computing the maximum bounds of $\|S_y(\theta)\|_\infty$, $\|T_y(\theta)\|_\infty$, $\|KS_y(\theta)\|_\infty$ and $\|GS_y(\theta)\|_\infty$ across multiple fixed $\theta \in \mathcal{D}_\theta$.

v) Reduce the possible values for w_2 and w_3 to subsets by only considering values such that $\|S_y(\theta)\|_\infty$, $\|T_y(\theta)\|_\infty$ and $\|KS_y(\theta)\|_\infty$ are below certain values.

Following this approach, the weights (22) are fixed according to:

$$w_2 = 0 \text{ dB} \quad w_3 = 16 \text{ dB} \tag{23}$$

For this choice, the optimal value of the optimisation problem (2) is found to be $\gamma = 5.77$.

This slightly high performance value can be explained due to the fact that some of the weights are too demanding and the $\mathcal{H}_\infty$ constraints can't be satisfied simultaneously. However, for this application, the rate of change of the scheduling variable θ is sufficiently low in practice. As such, the original $\mathcal{H}_\infty$ constraints can be validated *a posteriori* for pointwise values of θ. The bounds associated to the different closed-loop sensitivity functions, evaluated at 500 different fixed values of $\theta \in \mathcal{D}_\theta$, are found to be:

$$\begin{array}{ll} \|S_y(\theta)\|_\infty \le 6.02 \text{ dB} & \|T_y(\theta)\|_\infty \le 6.23 \text{ dB} \\ \|KS_y(\theta)\|_\infty \le 7.06 \text{ dB} & \|S_yG(\theta)\|_\infty \le 18.5 \text{ dB} \end{array} \tag{24}$$

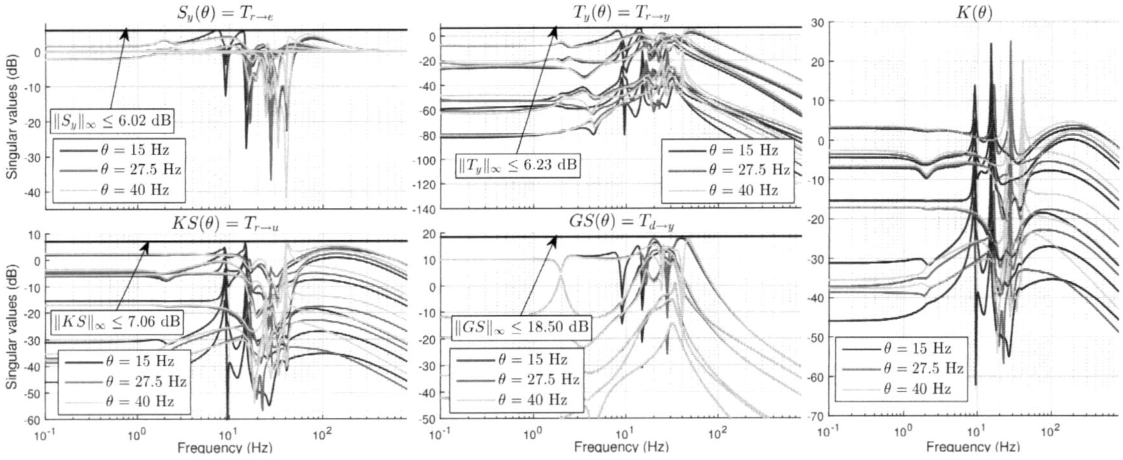

Figure 4. Singular value plots for the closed-loop sensitivity functions.

Figure 4 illustrates the singular value plots of these sensitivities together with the singular values of the controller $K(\theta)$, for three different $\theta \in \mathcal{D}_\theta$. The adaptive behaviour of the controller is evident, with the peaks in the singular values coinciding with harmonics to be attenuated. The harmonic attenuation performance was further analysed for multiple values of $\theta \in \mathcal{D}_\theta$ by measuring the gap between the singular values of the closed and open-loop transfer $T_{d\to y}$ (i.e. $|G| - |SyG|$) at the H_1 and $H_{0.6}$ frequencies. The attenuation levels in terms of statistics criteria (mean and standard deviations) are presented in Table 2 for each direction x, y, z. Based on this analysis, it can be concluded that the LPV controller achieves good pointwise multi-harmonic attenuation performances.

Table 2. Average attenuation performance

	F_x	F_y	F_z
H_1 harmonic	-30.32 ± 5.78 dB	-24.43 ± 2.84 dB	-21.57 ± 2.77 dB
$H_{0.6}$ harmonic	-12.31 ± 2.62 dB	-8.81 ± 3.43 dB	-7.52 ± 2.25 dB

4.4. Remarks on the choice of PDLF-LPV synthesis method

The weight tuning method uses a PDLF-Grid with pole constrains as the main LPV synthesis method. This was selected as the method of choice out of all other described in Section 3 after performing a comparison between the different methods as detailed in [13]. A summary of this comparison is presented in this section. The LPV comparison relies on the augmented plant model (4) using the following selection of weights:

$$W_1(\theta) = \frac{1}{6\text{ dB}} \left(\begin{array}{cc|c} -0.2\cdot\theta & -\theta & -0.19\cdot\theta \\ \theta & 0 & 0 \\ \hline 1 & 0 & 1 \end{array} \right) I_6 \qquad W_2(s) = 0.03\, I_6 \qquad W_3(s) = 0.5\, I_6 \quad (25)$$

After synthesising an LPV controller using each technique, the synthesis results are presented in Table 3. The methods are afterwards compared as follows: The controllers are reconstructed at 100 linearly spaced values for $\theta \in [15, 40]$. For every *a priori* fixed values, the achieved $\mathcal{H}_\infty$ criterion (3) performance criteria γ is computed, leading to a plot $\gamma(\theta)$. This function provides a pointwise indicator of the performance level for each LPV controller. A lower bound for the $\gamma(\theta)$ function can be obtained by performing a classical pointwise $\mathcal{H}_\infty$ synthesis for the same θ values. The $\gamma(\theta)$ plots for the LPV controllers results are presented in Figure 5 together with the pointwise $\mathcal{H}_\infty$ lower bound. As expected, all LPV synthesis methods introduce a degree of conservatism over the pointwise $\mathcal{H}_\infty$ synthesis.

Table 3. LPV controller synthesis results

Controller	γ	Fastest pole
LFR	1.43	19.2 kHz
SQLF	1.43	12.7 kHz
*SQLF	1.47	821 Hz
*PDLF	1.27	1392 Hz
*PDLF-Grid	1.00	531 Hz

*poles constrained via LMI regions [17]

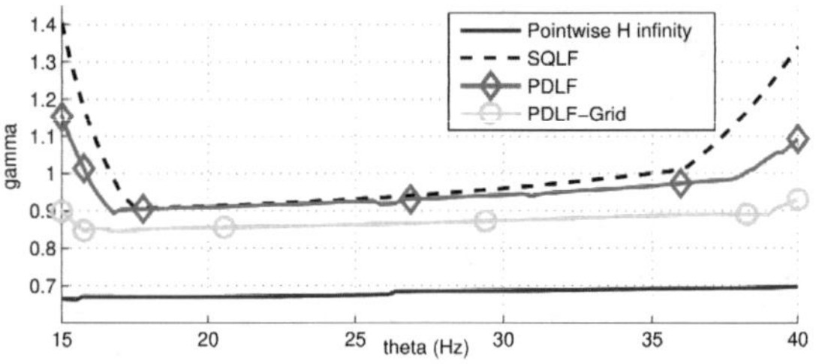

Figure 5. Pointwise $\gamma(\theta)$ performanace levels

The first employed synthesis methods are the LFR and Polytopic SQLF methods. As it can be seen from Table 3, these methods achieve the same performance level for this particular system and also lead to controllers with fast dynamics when evaluated for fixed values of θ. The next synthesis approach is based on Polytopic SQLF with pole constraints via LMI regions. This is performed by fixing $\mathbf{X}(\theta)$ and $\mathbf{Y}(\theta)$ as in (19). The introduction of pole constraints leads to a slight performance penalty ($\gamma = 1.47$ vs $\gamma = 1.43$). This small drop in performance was considered an acceptable compromise for implementability reasons and motivates the usage of pole constraints for all the synthesis methods further analysed.

Two other LPV synthesis methods based on PDLF have also been considered, considering bounds on the variations of the scheduling parameter θ: PDLF based on multiconvexity (16) and PDLF-Grid based on gridding the parameter space $[\underline{\theta}, \overline{\theta}] \times [\underline{v}, \overline{v}]$ over 20 grid points. As expected, both PDLF methods outperform the SQLF approaches. For the PDLF-Grid controller, the stability and performance levels defined by (2), (14) have been verified *a posteriori* using a dense grid defined by 10^4 points.

From Figure 5 and Table 2, it follows that the best performing LPV controller is the PDLF-Grid controller. This is why this method was selected during the weight optimization technique described in Section 4.3.

5. Simulations

The performances in the time-domain are evaluated using a Simulink-based simulator provided by Airbus Defence and Space. Noises acting on measurements ($\sigma^2_{n_m} = 10^{-5}$), plant inputs ($\sigma^2_{n_u} = 10^{-4}$) and on the scheduling variable θ ($\sigma^2_{n_\theta} = 10^{-3}$) are considered. The controller is rescheduled every 0.1s using the noisy measurement θ of the main disturbance frequency. The controller is discretized using a Tustin transformation with a $T_s = 10^{-3}$s sample time. Fig. 6

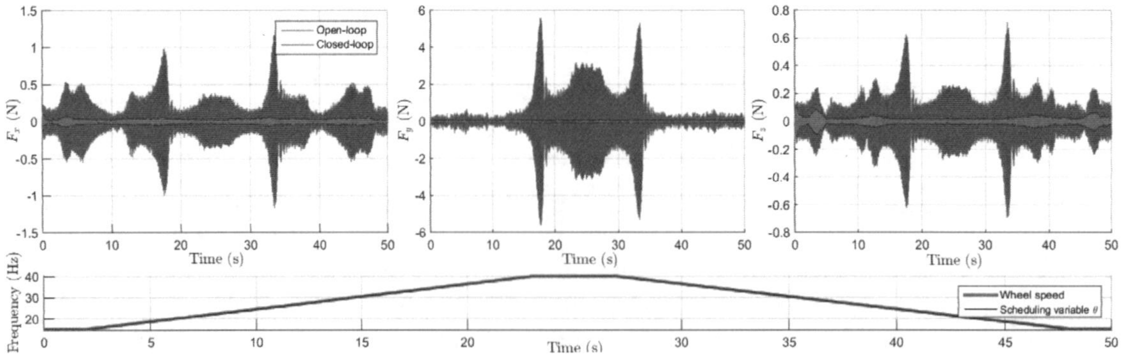

Figure 6. Transmitted forces on all three axes for a time-varying $\theta(t)$

presents the open and closed-loop transmitted force measurements (F_x, F_y, F_z) together with the wheel speed profile and the scheduling parameter θ. The wheel speed was selected to vary linearly between 15 and 40 Hz in 22 seconds. A side-by-side spectrogram comparison of these signals for the open and closed-loop case is shown in Fig. 7. The mitigation requirements are met for all $\theta \in [15, 40]$ Hz despite the noises affecting the system. The control signals $u \in \mathbb{R}^6$ are illustrated in Fig. 5 and are bounded by $|u(t)| < 0.3$ corresponding to reasonable actuator signals. These results suggest the fact that the proposed controller is a viable solution.

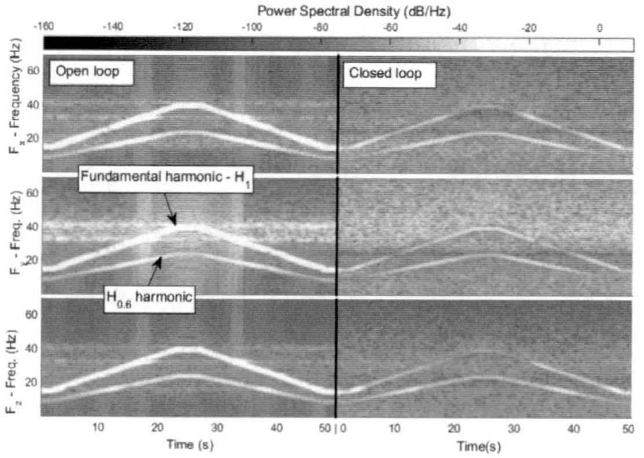

Figure 7. Spectrograms of F_x, F_y and F_z.

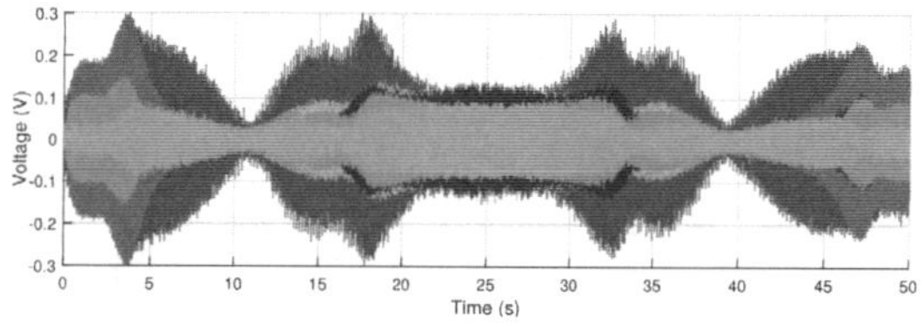

Figure 8. Control signals for the PMAs.

6. Conclusions

The study has shown how a mixed active-passive solution can be used for microvibration mitigation on-board satellite platforms. The control problem is formulated and managed in the $\mathcal{H}_\infty$/LPV framework. The performance and robustness of the proposed method was evaluated using simulations. Future works will concentrate on incorporating fault tolerant capabilities.

Acknowledgments

This research work was supported by Région Aquitaine and by the European Space Agency (ESA) in the framework of ESA's program NPI n. 4000109882/13/NL/GLC/al.

References

[1] Vaillon L and Philippe C 1999 *Smart Mater. Struct.* **8** 719
[2] Jalili N and Knowles IV D W 2004 *Smart Mater. Struct.* **13** 998
[3] Kamesh D, Pandiyan R and Ghosal A 2010 *J. Sound Vib.* **329** 3431–3450 ISSN 0022460X
[4] Liu C, Jing X, Daley S and Li F 2015 *Mech. Syst. Signal Process.* **56-57** 55–80 ISSN 08883270
[5] Scorletti G and El Ghaoui L 1995 *IEEE Conf. Decis. Control* vol 4 pp 3626–3631
[6] Apkarian P and Adams R J 1998 *IEEE Trans. Control Syst. Technol.* **6** 21–32
[7] Kajiwara H, Apkarian P and Gahinet P 1998 *AIAA Guid. Nav. Control Conf*
[8] Scherer C 2001 *Automatica* **37** 361–375
[9] Masterson R 2002 *J. Sound Vib.* **249** 575–598 ISSN 0022460X
[10] Witte J, Balini H and Scherer C W 2010 *IEEE Conf. Control Appl.* pp 950–955
[11] Shu X, Ballesteros P and Bohn C 2011 *Chinese Control Decis. Conf.* (IEEE) pp 728–733 ISBN 9781424487363
[12] Hoffmann C and Werner H 2015 *IEEE Trans. Control Syst. Technol.* **23** 416–433 ISSN 1063-6536
[13] Preda V, Cieslak J, Henry D, Bennani S and Falcoz A 2015 *8th IFAC Symp. Robust Control Des.*
[14] Preda V, Cieslak J, Henry D, Bennani S and Falcoz A 2015 *1st IFAC Work. Linear Param. Varying Syst.*
[15] Apkarian P and Gahinet P 1995 *Autom. Control. IEEE Trans.* **40** 853–864
[16] Apkarian P, Biannic J M and Gahinet P 1995 *Journ. Guid. Control. Dyn.* **18** 532–538
[17] Chilali M, Gahinet P and Apkarian P 1999 *IEEE Trans. Autom. Control* **44** 2257–2270
[18] Scherer C, Gahinet P and Chilali M 1997 *IEEE Trans. Autom. Control* **42** 896–911
[19] Apkarian P and Tuan H D 2000 *SIAM J. Control Optim.* **38** 1241–1264
[20] Taniwaki S and Ohkami Y 2003 *JSME Int. J. Ser. C* **46** 519–526

A Geometric Approach to Decouple Robotino Motions and its Functional Controllability

Daniel Straßberger, Paolo Mercorelli

Institute of Product and Process Innovation, Leuphana University of Lueneburg, Volgershall 1, D-21339 Lueneburg, Germany Phone: +49-(0)4131-677-5571, Fax: +49-(0)4131-677-5300.

E-mail: mercorelli@uni.leuphana.de

Oleg Sergiyenko

Engineering Institute of Autonomous University of Baja California, Blvd. Benito Juárez y Calle de La Normal, s n, Col. Insurgentes Este, C.P.21280, Mexicali, BC, Mexico, tel. fax. 01(52-686) 566-41-50.

E-mail: srgnk@iing.mxl.uabc.mx

Abstract. This paper analyses a functional control of the Robotino. The proposed control strategy considers a functional decoupling control strategy realized using a geometric approach and the invertibility property of the DC-drives with which the Robotino is equipped. For a given control structure the functional controllability is proven for motion trajectories of class C^3, continuous functions with third derivative also being continuous. Horizontal, Vertical and Angular motions are considered and the decoupling between these motions is obtained. Control simulation results using real data of the Robotino are shown. The used control which enables to produce the presented results is a standard Linear Model Predictive Control (LMPC), even though for sake of brevity the standard algorithm is not shown.

1. Introduction

This paper presents a systematic procedure for obtaining the decoupling controllability between horizontal, vertical and angular motions and their functional controllability. Here in this contribution the functional decoupling problem is investigated and roughly speaking, it consists of achieving motion tracking with no error variables transients. To achieve a decoupling effect a feedback control law is needed together with a feed-forward regulator. The functional controllability represents a structural property of the system which must be proven. In this paper a decoupled and functional controller is obtained. The relevance of the motion functional controllability of the Robotino control is ustified by the necessity of very fast and very precise loops of acceleration control counteracting errors in displacements caused by possible disturbance actions. In the past three decades, research on the geometric approach of dynamic systems theory and control has allowed this approach to become a powerful and a thorough tool for the analysis and synthesis of dynamic systems [1], [2], [3]. In [4], a robust decoupling controller using an algebraic state input feedback is presented, while this paper presents a robust decoupling controller using an algebraic output-input feedback. The goal of this paper is to propose a complete constructive procedure for the design of a decoupling controller. In Robotino case, to achieve a

Journal of Physics: Conference Series **659** (2015) 012027 doi:10.1088/1742-6596/659/1/012027

decoupling between Horizontal, Vertical and Angular motions a preselecting field law is needed together with a feed-forward regulator. The paper is organised in the following way: Section

Figure 1. Robotino in action on the pitch.

2 presents a possible model of the Robotino. Section 3 shows the decoupling strategy using the geometric approach. Section 4 presents the functional controllability problem in the case of Robotino. At the end, numerical computer simulations, considering real data of the Robotino, are shown using Model Predictive Control (MPC), even though per sake of brevity this standard algorithm is not discussed.

2. Mechanical and Electrical Model Description

In Fig. 2 a diagram with the representation of the forces in the considered system is shown. If

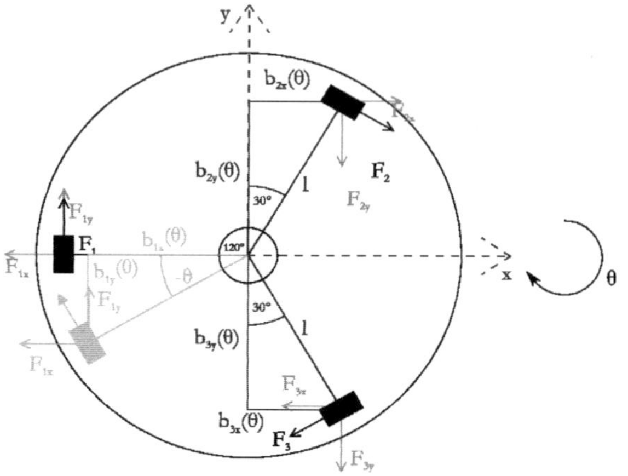

Figure 2. Mechanical schematic diagram of Robotino.

state vector $\mathbf{X}(t)$ is defined as, $x(t)$ and $y(t)$ positions of the center of mass of the system and its velocities, $x(t)$ and $y(t)$, moreover, considering the angular dynamics with its angular position $\theta(t)$ and its velocity $\theta(t)$, then the following system is derived:

$$\begin{cases} \dot{\mathbf{X}}(t) = \mathbf{A}\mathbf{X}(t) + \mathbf{g}(\theta)\mathbf{F}(t) \\ \mathbf{O}(t) = \mathbf{C}\mathbf{X}(t), \end{cases} \tag{1}$$

with

$$\mathbf{X}(t) = \begin{bmatrix} x(t) \\ x(t) \\ y(t) \\ y(t) \\ \theta(t) \\ \theta(t) \end{bmatrix}, \qquad \dot{\mathbf{X}}(t) = \begin{bmatrix} x(t) \\ x(t) \\ y(t) \\ y(t) \\ \theta(t) \\ \theta(t), \end{bmatrix} \tag{2}$$

matrix

$$\mathbf{A} = \begin{bmatrix} 0 & 1 & 0 & 0 & 0 & 0 \\ 0 & -\frac{k_v}{M} & 0 & 0 & 0 & 0 \\ 0 & 0 & 0 & 1 & 0 & 0 \\ 0 & 0 & 0 & -\frac{k_v}{M} & 0 & 0 \\ 0 & 0 & 0 & 0 & 0 & 1 \\ 0 & 0 & 0 & 0 & 0 & -\frac{k_\theta}{J} \end{bmatrix}, \tag{3}$$

which for sake of notation can be written as follows:

$$\mathbf{A} = \begin{bmatrix} \mathbf{A}_{x2\times2} & \mathbf{0}_{2\times2} & \mathbf{0}_{2\times2} \\ \mathbf{0}_{2\times2} & \mathbf{A}_{y2\times2} & \mathbf{0}_{2\times2} \\ \mathbf{0}_{2\times2} & \mathbf{0}_{2\times2} & \mathbf{A}_{\theta2\times2} \end{bmatrix}, \tag{4}$$

where $\mathbf{A}_{x2\times2}$, $\mathbf{A}_{y2\times2}$ and $\mathbf{A}_{\theta2\times2}$ are the minor matrices of $\mathbf{A}$ describing the x, y and θ dynamics of the system. Matrix $\mathbf{C}$ represents the output matrix and can be written in the following way:

$$\mathbf{C} = \begin{bmatrix} 1 & 0 & 0 & 0 & 0 & 0 \\ 0 & 0 & 1 & 0 & 0 & 0 \\ 0 & 0 & 0 & 0 & 1 & 0 \end{bmatrix}, \tag{5}$$

where the following notation is assumed:

$$\mathbf{C_x} = \begin{bmatrix} 1 & 0 & 0 & 0 & 0 & 0 \end{bmatrix}, \tag{6}$$
$$\mathbf{C_y} = \begin{bmatrix} 0 & 0 & 1 & 0 & 0 & 0 \end{bmatrix}, \tag{7}$$
$$\mathbf{C}_\theta = \begin{bmatrix} 0 & 0 & 0 & 0 & 1 & 0 \end{bmatrix}. \tag{8}$$

Field $\mathbf{g}(\theta)$ is as follows:

$$\mathbf{g}(\theta) = \begin{bmatrix} 0 & 0 & 0 \\ \frac{\sin(\theta)}{M} & \frac{\sin(\frac{2}{3}\pi+\theta)}{M} & -\frac{\sin(\frac{2}{3}\pi-\theta)}{M} \\ 0 & 0 & 0 \\ \frac{\cos(\theta)}{M} & \frac{\cos(\frac{2}{3}\pi+\theta)}{M} & \frac{\cos(\frac{2}{3}\pi-\theta)}{M} \\ 0 & 0 & 0 \\ a & b & c \end{bmatrix} \tag{9}$$

where

$$a = \frac{\sin(\theta)\cdot l\cdot\sin(\theta) + \cos(\theta)\cdot l\cdot\cos(\theta)}{J},$$

$$b = \frac{\sin(\frac{2}{3}\pi + \theta) \cdot l \cdot \cos(\frac{\pi}{6} + \theta)}{J} + \frac{\cos(\frac{2}{3}\pi + \theta) \cdot l \cdot \sin(\frac{\pi}{6} + \theta)}{J} \tag{10}$$

and

$$c = \frac{-\sin(\frac{2}{3}\pi - \theta) \cdot l \cdot -\cos(\frac{\pi}{6} - \theta)}{J} + \frac{\cos(\frac{2}{3}\pi - \theta) \cdot l \cdot \sin(\frac{\pi}{6} - \theta)}{J} \tag{11}$$

and force input vector signal $\mathbf{F}(t) = \begin{bmatrix} F_1(t) \\ F_2(t) \\ F_3(t) \end{bmatrix}$ in which $F_1(t) = K_m i_1(t)/r$, $F_2(t) = K_m i_2(t)/r$ and $F_3(t) = K_m i_3(t)/r$, where K_m represents the motor constant, r the radius of the wheels and variables $i_1(t)$, $i_2(t)$ and $i_3(t)$ are the currents of the three DC-electrical drives. In fact, the Robotino consists of 3 DC electrical drives that power the three omniwheels. The models of these three DC-drives are reported below.

$$\begin{bmatrix} \frac{di_1(t)}{dt} \\ \frac{d\omega_1(t)}{dt} \\ \frac{di_2(t)}{dt} \\ \frac{d\omega_2(t)}{dt} \\ \frac{di_3(t)}{dt} \\ \frac{d\omega_3(t)}{dt} \end{bmatrix} = \begin{bmatrix} -\frac{R}{L} & -\frac{K_m}{L} & 0 & 0 & 0 & 0 \\ \frac{K_m}{J} & -\frac{K}{J} & 0 & 0 & 0 & 0 \\ 0 & 0 & -\frac{R}{L} & -\frac{K_m}{L} & 0 & 0 \\ 0 & 0 & \frac{K_m}{J} & -\frac{K}{J} & 0 & 0 \\ 0 & 0 & 0 & 0 & -\frac{R}{L} & -\frac{K_m}{L} \\ 0 & 0 & 0 & 0 & \frac{K_m}{J} & -\frac{K}{J} \end{bmatrix} \cdot \begin{bmatrix} i_1(t) \\ \omega_1(t) \\ i_2(t) \\ \omega_2(t) \\ i_3(t) \\ \omega_3(t) \end{bmatrix} + \begin{bmatrix} \frac{1}{L} & 0 & 0 \\ 0 & 0 & 0 \\ 0 & \frac{1}{L} & 0 \\ 0 & 0 & 0 \\ 0 & 0 & \frac{1}{L} \\ 0 & 0 & 0 \end{bmatrix} \cdot \begin{bmatrix} U_{inp_1}(t) \\ U_{inp_2}(t) \\ U_{inp_3}(t) \end{bmatrix}, \tag{12}$$

where $U_{inp_1}(t)$, $U_{inp_2}(t)$ and $U_{inp_3}(t)$ represent the input voltage of the DC-drive, L is the inductance, R is the resistance, J represents the motor inertia factor, K_m is the motor moment and finally K can be seen as the friction factor. This model is important in realising the simulation and is a basis for the inverted drive used in the control strategy.

3. Design of a Decoupling Controller

This section describes the design of a decoupling controller with respect to the x-y and θ motions above defined. A geometric approach is used in this analysis. Since the geometric relations depend on the rotational position (θ), the forces actuated by the three DC drives are needed to move in a certain direction and they may be difficult to be obtained.

Definition 1 *A control law for the dynamic system (1) is* decoupling *with respect to the regulated outputs $x(t)$, $y(t)$, and $\theta(t)$, if there exist partitions $F_x(t), F_y(t)$, and $F_\theta(t)$ of the input vector $\mathbf{F}(t)$ such that for zero initial conditions, each input $F_{(\cdot)}(t)$ (with all other inputs, identically zero) only affects the corresponding output $x(t)$, $y(t)$, or $\theta(t)$.* □

It it to be shown that there exists a decoupling and stabilizing state feedback field $\mathbf{D}(\theta)$, along with three input partition fields $\mathbf{T}_x(\theta)$, $\mathbf{T}_y(\theta)$, and $\mathbf{T}_\theta(\theta)$ such that, for the dynamic triples

$$\begin{gathered} (\mathbf{C}_x, \mathbf{A} + \mathbf{g}(\theta)\mathbf{D}(\theta), \mathbf{g}(\theta)\mathbf{F}_1(t)), \\ (\mathbf{C}_y, \mathbf{A} + \mathbf{g}(\theta)\mathbf{D}(\theta), \mathbf{g}(\theta)\mathbf{F}_2), \\ (\mathbf{C}_\theta, \mathbf{A} + \mathbf{g}(\theta)\mathbf{D}(\theta), \mathbf{g}(\theta)\mathbf{F}_3(t)), \end{gathered} \tag{13}$$

it holds the following conditions:

$$\mathcal{R}_x(\theta) = \mathsf{min}\mathcal{I}\Big(\mathbf{A} + \mathbf{g}(\theta)\mathbf{D}(\theta),\, \mathbf{g}(\theta)\mathbf{T}_x(\theta)\Big) \subseteq \mathcal{C}_y \cap \mathcal{C}_\theta \;\; \forall\theta, \tag{14}$$

and

$$\mathbf{C}_x\mathcal{R}_x(\theta) = \mathsf{im}(\mathbf{C}_x), \;\; \forall\theta. \tag{15}$$

$$\mathcal{R}_y(\theta) = \mathsf{min}\mathcal{I}\Big(\mathbf{A} + \mathbf{g}(\theta)\mathbf{D}(\theta),\, \mathbf{g}(\theta)\mathbf{T}_y(\theta)\Big) \subseteq \mathcal{C}_x \cap \mathcal{C}_\theta \;\; \forall\theta, \tag{16}$$

and

$$\mathbf{C}_y\mathcal{R}_y(\theta) = \mathsf{im}(\mathbf{C}_y), \;\; \forall\theta. \tag{17}$$

$$\mathcal{R}_\theta(\theta) = \mathsf{min}\mathcal{I}\Big(\mathbf{A} + \mathbf{g}(\theta)\mathbf{D}(\theta),\, \mathbf{g}(\theta)\mathbf{T}_\theta(\theta)\Big) \subseteq \mathcal{C}_x \cap \mathcal{C}_y \;\; \forall\theta, \tag{18}$$

and

$$\mathbf{C}_\theta\mathcal{R}_\theta(\theta) = \mathsf{im}(\mathbf{C}_\theta), \;\; \forall\theta. \tag{19}$$

Here,

$$\mathsf{min}\mathcal{I}(\mathbf{A}, \mathsf{im}(\mathbf{g}(\theta))) = \sum_{i=0}^{n-1} \mathbf{A}^i \mathsf{im}(\mathbf{g}(\theta))$$

is a minimum $\mathbf{A}$–invariant subspace containing $\mathsf{im}(\mathbf{g}(\theta)) \; \forall \; \theta$. Moreover, the partition fields $\mathbf{T}_x(\theta)$, $\mathbf{T}_y(\theta)$ and $\mathbf{T}_\theta(\theta)$ satisfy the following relationships:

$$\begin{aligned} \mathsf{im}(\mathbf{g}(\theta) \cdot \mathbf{T}_x(\theta)) &= \mathsf{im}(\mathbf{g}(\theta)) \cap \mathcal{R}_x(\theta), \\ \mathsf{im}(\mathbf{g}(\theta) \cdot \mathbf{T}_y(\theta)) &= \mathsf{im}(\mathbf{g}(\theta)) \cap \mathcal{R}_y(\theta), \\ \mathsf{im}(\mathbf{g}(\theta) \cdot \mathbf{T}_\theta(\theta)) &= \mathsf{im}(\mathbf{g}(\theta)) \cap \mathcal{R}_\theta(\theta). \end{aligned} \tag{20}$$

The stabilizing field $\mathbf{D}(\theta)$ is such that:

$$(\mathbf{A} + \mathbf{g}(\theta)\mathbf{D}(\theta))\mathcal{R}_x(\theta) \subseteq \mathcal{R}_x(\theta), \tag{21}$$

$$(\mathbf{A} + \mathbf{g}(\theta)\mathbf{D}(\theta))\mathcal{R}_y(\theta) \subseteq \mathcal{R}_y(\theta), \tag{22}$$

and

$$(\mathbf{A} + \mathbf{g}(\theta)\mathbf{D}(\theta))\mathcal{R}_\theta(\theta) \subseteq \mathcal{R}_\theta(\theta). \tag{23}$$

Considering matrix $\mathbf{A}$ defined in (3) it is straightforward to see that this matrix is structurally already decoupled and intrinsically stable, this implies that field $\mathbf{D}(\theta) = \mathbf{0} \; \forall\theta$. Considering

$$\mathbf{T}(\theta) = [\mathbf{T}_x(\theta), \mathbf{T}_y(\theta), \mathbf{T}_\theta(\theta), \mathbf{T}_c(\theta)],$$

where $\mathbf{T}_c(\theta)$ is defined in a complementary fashion and it is straightforward to show that matrix $\mathbf{T}_c = \mathbf{0}$. In particular, matrix $\mathbf{T}_c$ represents the complementary field partition to the subspaces of x-position, y-position and angular position of the Robotino. The Robotino motion is described using ust three variables, therefore partitions $\mathbf{T}_x(\theta), \mathbf{T}_y(\theta), \mathbf{T}_\theta(\theta)$ complete the transformation and thus $\mathbf{T}_c = \mathbf{0}$.

$$\mathsf{im}\mathbf{T}(\theta) = \mathsf{im}[\mathbf{T}_x(\theta), \mathbf{T}_y(\theta), \mathbf{T}_\theta(\theta)] = \mathsf{im}\mathbf{T}_x(\theta) \oplus \mathsf{im}\mathbf{T}_y(\theta) \oplus \mathsf{im}\mathbf{T}_\theta(\theta). \tag{24}$$

Considering the output matrices (6), (7) and (8) corresponding to x-position, y-position and angular position their respective kernels are as follows:

$$\mathcal{C}_x = \text{im}\begin{bmatrix} 0 & 0 & 0 & 0 & 0 \\ 1 & 0 & 0 & 0 & 0 \\ 0 & 1 & 0 & 0 & 0 \\ 0 & 0 & 1 & 0 & 0 \\ 0 & 0 & 0 & 1 & 0 \\ 0 & 0 & 0 & 0 & 1 \end{bmatrix}, \quad \mathcal{C}_y = \text{im}\begin{bmatrix} 1 & 0 & 0 & 0 & 0 \\ 0 & -1 & 0 & 0 & 0 \\ 0 & 0 & 0 & 0 & 0 \\ 0 & 0 & -1 & 0 & 0 \\ 0 & 0 & 0 & -1 & 0 \\ 0 & 0 & 0 & 0 & 1 \end{bmatrix}, \quad \mathcal{C}_\theta = \text{im}\begin{bmatrix} 1 & 0 & 0 & 0 & 0 \\ 0 & 1 & 0 & 0 & 0 \\ 0 & 0 & 1 & 0 & 0 \\ 0 & 0 & 0 & -1 & 0 \\ 0 & 0 & 0 & 0 & 0 \\ 0 & 0 & 0 & 0 & 1 \end{bmatrix}. \tag{25}$$

According to relation (9) it is straightforward to observe that the following three equations hold $\forall\ \theta$;

$$\text{im}(\mathbf{g}(\theta)) \cap (\mathcal{C}_x \cap \mathcal{C}_y) \neq \mathbf{0}, \tag{26}$$

$$\text{im}(\mathbf{g}(\theta)) \cap (\mathcal{C}_x \cap \mathcal{C}_\theta) \neq \mathbf{0}, \tag{27}$$

$$\text{im}(\mathbf{g}(\theta)) \cap (\mathcal{C}_y) \cap \mathcal{C}_\theta) \neq \mathbf{0}. \tag{28}$$

Field $\mathbf{g}(\theta)$ is a function of θ without singularities if $\theta \neq \pi/2 + k\pi$ with $k \in \mathbb{N}$. For sake of brevity the following field is calculated considering ust $\theta = 0$:

$$(\mathbf{g}(\theta))^\dagger = \begin{bmatrix} 0 & 0 & 0 & a(\theta) & 0 & b(\theta) \\ 0 & c(\theta) & 0 & -a(\theta) & 0 & b(\theta) \\ 0 & -c(\theta) & 0 & -a(\theta) & 0 & b(\theta) \end{bmatrix}, \tag{29}$$

where with $(\mathbf{g}(\theta))^\dagger$ the pseudo inverse of field $\mathbf{g}(\theta)$ is indicated. Functions $a(\theta)$, $b(\theta)$ and $c(\theta)$ are functions of the variable θ with $\theta \approx 0$. The following calculation is obtained for $\theta \neq \pi/2 + k\pi$ with $k \in \mathbb{N}$:

$$\mathcal{C}_x \cap \mathcal{C}_y = \text{im}\begin{bmatrix} 0 & 0 & 0 & 0 \\ 1 & 0 & 0 & 0 \\ 0 & 0 & 0 & 0 \\ 0 & -1 & 0 & 0 \\ 0 & 0 & -1 & 0 \\ 0 & 0 & 0 & 1 \end{bmatrix}, \quad \mathcal{C}_\theta \cap \mathcal{C}_y = \text{im}\begin{bmatrix} 1 & 0 & 0 & 0 \\ 0 & 1 & 0 & 0 \\ 0 & 0 & 0 & 0 \\ 0 & 0 & 1 & 0 \\ 0 & 0 & 0 & 0 \\ 0 & 0 & 0 & 1 \end{bmatrix}, \quad \mathcal{C}_\theta \cap \mathcal{C}x = \text{im}\begin{bmatrix} 0 & 0 & 0 & 0 \\ 1 & 0 & 0 & 0 \\ 0 & 1 & 0 & 0 \\ 0 & 0 & 1 & 0 \\ 0 & 0 & 0 & 0 \\ 0 & 0 & 0 & 1 \end{bmatrix}. \tag{30}$$

The following calculation allows to get the required fields for the decoupling of the mechanical system:

$$\mathbf{T}_\theta(\theta) = (\mathbf{g}(\theta))^\dagger \cdot \text{im}(\mathbf{g}(\theta)) \cap \mathcal{C}_x \cap \mathcal{C}_y, \tag{31}$$

$$\mathbf{T}_x(\theta) = (\mathbf{g}(\theta))^\dagger \cdot \text{im}(\mathbf{g}(\theta)) \cap \mathcal{C}_\theta \cap \mathcal{C}_y, \tag{32}$$

$$\mathbf{T}_y(\theta) = (\mathbf{g}(\theta))^\dagger \cdot \text{im}(\mathbf{g}(\theta)) \cap \mathcal{C}_\theta) \cap \mathcal{C}_x. \tag{33}$$

Adding all 3 T-Fields together, we get a new field $\mathbf{T}(\theta)$:

$$\mathbf{T}(\theta) = \mathbf{T}_x(\theta) + \mathbf{T}_y(\theta) + \mathbf{T}_\theta(\theta). \tag{34}$$

Field $\mathbf{T}(\theta)$ can be seen as a preselecting field and the following product realises the mechanical decoupling:

$$\mathcal{B} = \text{im}(\mathbf{g}(\theta)\mathbf{T}(\theta)) = \text{im}\begin{bmatrix} 0 & 0 & 0 \\ 1 & 0 & 0 \\ 0 & 0 & 0 \\ 0 & 1 & 0 \\ 0 & 0 & 0 \\ 0 & 0 & 1 \end{bmatrix}, \tag{35}$$

in which matrix $\mathbf{B}$ can be seen as a resulting input matrix.

4. On the Functional Controllability of the Robotino

This section is aimed to the analysis of the output functional controllability of the Robotino system. As already pointed out, we are interested in controlling motion tracking without transients of error variables. In the whole, in the Robotino control, the exact tra ectory tracking is an ambitious ob ective and this is particularly emphasized by advanced control tasks recalled in Section 1. For solving the problem, the natural approach is to analyze the constrained output controllability idea, [5] and [1], formalized below.

Definition 2 (Perfect output controllability) *Given the triple* $(\mathbf{A}, \mathbf{B}, \mathbf{C})$*, the output subspace* $\mathcal{L}^i$ *is said to be perfect output functionally controllable with respect to i-th derivative and with respect to the subspace of states* $\mathcal{F}$ *if* $\mathcal{L}^i = \mathbf{C}\mathcal{F}$ *and, for every initial state* $\mathbf{x}_0 \in \mathcal{F}$*, it is possible, by means of proper bounded and measurable control functions, to follow in* $\mathcal{L}$ *any trajectory arbitrarily given in the class of functions which admit i-th derivative with respect to time, while the state evolves into* $\mathcal{F}$*.*

□

Recall, cf. [5], that the output functional controllability is strictly related to the geometric–type extension of the relative degree for multivariable systems and that each subspace $\mathcal{F}$ satisfying Definition 2 is $(\mathbf{A}, \mathbf{B})$–controlled invariant.
The following theorem which was demonstrated in [5] and in [1] is reported here below to be applied for the Robotino case.

Theorem 1 *(***Output Functional Controllability and Decoupling of the Robotino***) The output subspaces* $\mathsf{im}(\mathbf{C}_x)$*,* $\mathsf{im}(\mathbf{C}_y)$ *and* $\mathsf{im}(\mathbf{C}_\theta)$ *are output functionally controllable with respect to the* 3rd *derivative and with respect to the constrained reachable subspaces* $\mathcal{R}_x(\theta)$*,* $\mathcal{R}_y(\theta)$*, and* $\mathcal{R}_\theta(\theta)$ *respectively. Moreover, field* $\mathbf{T}(\theta)$ *makes the system, with outputs* $x(t)$*,* $y(t)$ *and* $\theta(t)$ *decoupled and output functionally controllable.*

□

Remark 1 *Regarding the functional controllability of rigid body as object motions, the* 3rd *order of derivative means that the outputs* $x(t)$*,* $y(t)$ *and* $\theta(t)$ *can perfectly track any desired trajectories* $x_d(t)$*,* $y_d(t)$ *and* $\theta_d(t)$ *which has a piecewise continuous* 3rd *derivative. This is true for all initial states* $\mathbf{x}_o$ *in* $\mathcal{R}_x(\theta)$*,* $\mathcal{R}_y(\theta)$*, and* $\mathcal{R}_\theta(\theta)$ *and with piecewise continuous control functions* $\mathbf{F}(t)$*. Furthermore, it could be easily shown that order 3 of the rigid body object motions is not due to the particular choice of the subspaces* $\mathcal{R}_x(\theta)$*,* $\mathcal{R}_y(\theta)$*, and* $\mathcal{R}_\theta(\theta)$ *but it is an inherent property of the system. It is related to the relative degree of the relationship between the rigid body object motion and the DC-Drive input voltage.*

□

In the following a fundamental condition is shown, further details in [5].

Projection Condition

Given the system represented as $(\mathbf{A}, \mathbf{B}, \mathbf{C})$*.* $\mathcal{F}^i$ *is a subspace of functional controllability of the output with respect to the i-derivative, if and only if:*

$$\mathcal{F}^i \subseteq \mathcal{F}^i \cap \mathcal{Z}_{i-1} + \mathcal{C}, \tag{36}$$

where $\mathcal{Z}^{i-1}$ *is defined in the following way:* $\mathcal{Z}_o = \mathcal{B}$ *and* $\mathcal{Z}_j = \mathcal{B} + \mathbf{A}(\mathcal{Z}_{j-1} \cap \mathcal{F}^i \cap \mathcal{C})$*.* □

Sketch of the proof considering the Robotino case
It is to be remarked that the Pro ection Condition states a necessary and sufficient condition for functional controllability of a system. If ust the mechanical part of the model is considered, it is possible to show that an output controllability with respect to the second derivative is obtained. In fact, in case $i = 2$, in a generic point $\overline{\mathbf{x}} \in \mathcal{F}^2$ of a generic tra ectory, exists, per definition, at least a velocity of state $\dot{\overline{\mathbf{x}}} \in \mathcal{F}^2$. All admissible velocities on $\mathcal{F}^2$ are obtained adding to $\dot{\overline{\mathbf{x}}}$ vectors belonging to subspace $\mathcal{F}^2 \cap \mathcal{B}$. Thus, such a subspace represents the set of all possible velocity variations and $\mathcal{F}^2 \cap \mathcal{B} \cap \mathcal{C}$ represents the set of all velocity variations which are not visible from the output. Being in general

$$\mathbf{x}(t) = \mathbf{A}\mathbf{x}(t) + \mathbf{B}\mathbf{u}(t), \tag{37}$$

then with a suitable choice of the control $u_1(t)+u_2(t)$, in which $u_1(t)$ is manipulable in its second derivative and such that $\mathbf{B}u_1(t) \in \mathcal{F}^2 \cap \mathcal{B} \cap \mathcal{C}$, and $u_2(t)$ is manipulable in its first derivative and such that $\mathbf{B}u_2(t) \in \mathcal{B}$. Thus, if the tra ectory evolves in the following subspace

$$\mathcal{Z}_1 = \mathcal{B} + \mathbf{A}\mathcal{F}^2 \cap \mathcal{B} \cap \mathcal{C}, \tag{38}$$

it is possible to variate the second derivative of it. A tra ectory belongs to $\mathcal{F}^2$ if its derivative belongs to $\mathcal{F}^2$ (invariant subspace definition according to [1]) and it is necessary that its variations belong to $\mathcal{Z}_1 \cap \mathcal{F}^2$. If $\mathcal{L}^2$ indicates an output subspace, then $\mathcal{L}^2 = \mathbf{C}\mathcal{F}^2$ is a subspace of output functional controllability with respect to the second derivative if and only if:

$$\mathbf{C}\mathcal{F}^2 = \mathbf{C}(\mathcal{F}^2 \cap \mathcal{Z}_1), \tag{39}$$

or

$$\mathcal{F}^2 + \mathcal{C} = \mathbf{C}(\mathcal{F}^2 \cap \mathcal{Z}_1) + \mathcal{C}. \tag{40}$$

□

Remark 2 *In other words, subspace $\mathcal{Z}_1$ represents the subspace of all the manipulable trajectories with continuous second derivative. If the trajectory belongs to $\mathcal{Z}_1$, then it is possible to manipulate the first and the second derivatives in $\mathcal{Z}_1$. This can be done by means of proper bounded and measurable control functions. In fact in so doing, it is possible to follow in $\mathcal{L}^2$ any trajectory arbitrarily given in the class of functions which admit 2nd derivative with respect to time, while the state evolves into $\mathcal{F}^2$.*

□

Considering again the mechanical system and choosing subspaces $\mathcal{F}_x$, $\mathcal{F}_y$ and $\mathcal{F}_\theta$ as candidate subspaces:

$$\mathcal{F}_x = \mathsf{im}\begin{bmatrix} 1 & 0 \\ 0 & 1 \\ 0 & 0 \\ 0 & 0 \\ 0 & 0 \\ 0 & 0 \end{bmatrix}, \quad \mathcal{F}_y = \mathsf{im}\begin{bmatrix} 0 & 0 \\ 0 & 0 \\ 1 & 0 \\ 0 & 1 \\ 0 & 0 \\ 0 & 0 \end{bmatrix}, \quad \text{and} \quad \mathcal{F}_\theta = \mathsf{im}\begin{bmatrix} 0 & 0 \\ 0 & 0 \\ 0 & 0 \\ 0 & 0 \\ 1 & 0 \\ 0 & 1 \end{bmatrix}. \tag{41}$$

It can be shown that the mechanical system with this choice of subspaces can be functionally controllable on tra ectories with continuous 2nd derivatives. In fact, considering that $\mathcal{F}^1 \not\subseteq \mathcal{F}^1 \cap \mathcal{Z}_0 + \mathcal{C}_x$, to be more precise

$$\mathcal{F}_x^1 \cap \mathcal{Z}_0 = \begin{bmatrix} 0 \\ 1 \\ 0 \\ 0 \\ 0 \\ 0 \end{bmatrix}, \tag{42}$$

so the mechanical system is not functionally controllable with respect to tra ectories of the 1st order. Similar results can be obtained for considering the output subspace $\mathsf{im}\mathbf{C}_y$ and $\mathsf{im}\mathbf{C}_\theta$. If the following relation is considered $\mathcal{F}_x^2 \subseteq \mathcal{F}_x^2 \cap \mathcal{Z}_1 + \mathcal{C}_x$, to be more precise

$$\mathcal{Z}_1 = \begin{bmatrix} 0 & 0 & 0 \\ 1 & 0 & 0 \\ 0 & 0 & 0 \\ 0 & 1 & 0 \\ 0 & 0 & 0 \\ 0 & 0 & 1 \end{bmatrix} + \begin{bmatrix} 0 & 1 & 0 & 0 & 0 & 0 \\ 0 & -\frac{k_v}{M} & 0 & 0 & 0 & 0 \\ 0 & 0 & 0 & 1 & 0 & 0 \\ 0 & 0 & 0 & -\frac{k_v}{M} & 0 & 0 \\ 0 & 0 & 0 & 0 & 0 & 1 \\ 0 & 0 & 0 & 0 & 0 & -\frac{k_\theta}{J} \end{bmatrix}.$$

$$\mathsf{im} \begin{bmatrix} 1 & 0 \\ 0 & 1 \\ 0 & 0 \\ 0 & 0 \\ 0 & 0 \\ 0 & 0 \end{bmatrix} \cap \begin{bmatrix} 0 & 0 & 0 \\ 1 & 0 & 0 \\ 0 & 0 & 0 \\ 0 & 1 & 0 \\ 0 & 0 & 0 \\ 0 & 0 & 1 \end{bmatrix} \cap \mathsf{im} \begin{bmatrix} 0 & 0 & 0 \\ 1 & 0 & 0 \\ 0 & 0 & 0 \\ 0 & -1 & 0 \\ 0 & 0 & 0 \\ 0 & 0 & 1 \end{bmatrix} = \mathcal{B}. \quad (43)$$

$$\mathcal{F}_x^2 \cap \mathcal{Z}_1 = \mathcal{F}_x^2. \quad (44)$$

It is possible to see then that equation (39) is satisfied thus also equation (40) is satisfied. Similar results can be obtained after considering the output subspace $\mathsf{im}\mathbf{C}_y$ and $\mathsf{im}\mathbf{C}_\theta$.

Remark 3 *After considering relation (40) it is possible to see that functional subspace $\mathcal{F}^2$ consists of the position subspace along trajectory x(t) and its velocity subspace. Moreover, the velocity subspace belongs to $\mathcal{C}_x$. Similar considerations can be obtained considering the output subspace* $\mathsf{im}\mathbf{C}_y$ *and* $\mathsf{im}\mathbf{C}_\theta$. □

Considering the electrical drive which is connected in input to the mechanical drive, it can be shown that this system is functionally controllable considering tra ectories with continuous 1st derivative. To conclude, considering the cascade structure of the whole system, the Robotino can be functionally controlled considering tra ectories with continuous 3rd derivative.

5. Simulation Results

Figure 3 shows the Simulink/Matlab block diagram used for simulation. Using a decentralized MPC-controller with the geometric decoupling approach acting as a pre-control, different tra ectories can be tracked. As such the MPC accepts tra ectories in horizontal, vertical and angular direction and attempts to direct the Robotino only in the specified direction. For sake of brevity the control strategy is not presented. Therefore, it is possible to move the Robotino along the x-axis without a rotation for example.

Figure 4 shows a very accurate movements along the desired tra ectory which is represented in the tested case by a circle and by an avoidance tra ectory path. In Fig. 5 the current in case of the circle and the avoidance tra ectory are represented. The circle is realized considering a continuous sine and a cosine function on the X and movements. These signals are visible in term of voltages of the X and drives in Fig. 6. It is possible to observe from Fig. 6 that the cosine function is not exactly reproducible because its not continuity function and this effect is also visible in the left part of Fig. 4 in which a considerable initial error occurs. Additionally it is vital, that the drives are not forced to output more than they can handle, so voltages and currents would have to remain in their respective boundaries. The right part of Fig. 4 shows a possible tra ectory avoidance which is a typical requirement in the Robotino tasks. From

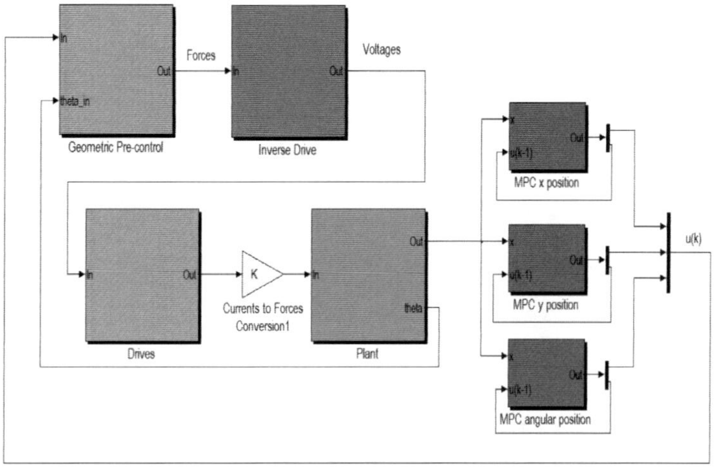

Figure 3. Block diagram of the control system structure

the right part of Fig. 4 it is possible to see an initial error in the tra ectory tracking. This is a structural error due, as before, to the desired tra ectory function which is a function of $\mathcal{C}^0$ (continuous function). Nevertheless, if the initial error represented in the left part of Fig. 4 is compared with that of the right part, it is possible to remark that in the case of obstacle avoidance the error seems to be smaller than that in the circle case. This aspect can be ustified considering that the X movement generating functions in the case of obstacle avoidance are a sine and a ramp, so they are continuous functions. Figure 5 shows in term of currents those two different loads. In particular, the case of obstacle avoidance presents a lower level of currents in the transient phase.

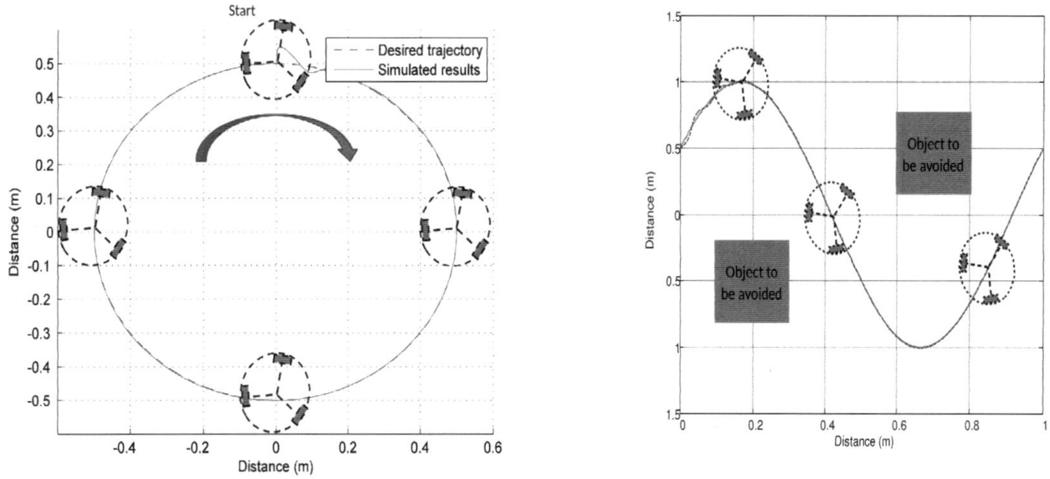

Figure 4. X- -Tra ectory of the Robotino (left). Tra ectory ob ect avoidance (right)

Journal of Physics: Conference Series **659** (2015) 012027 doi:10.1088/1742-6596/659/1/012027

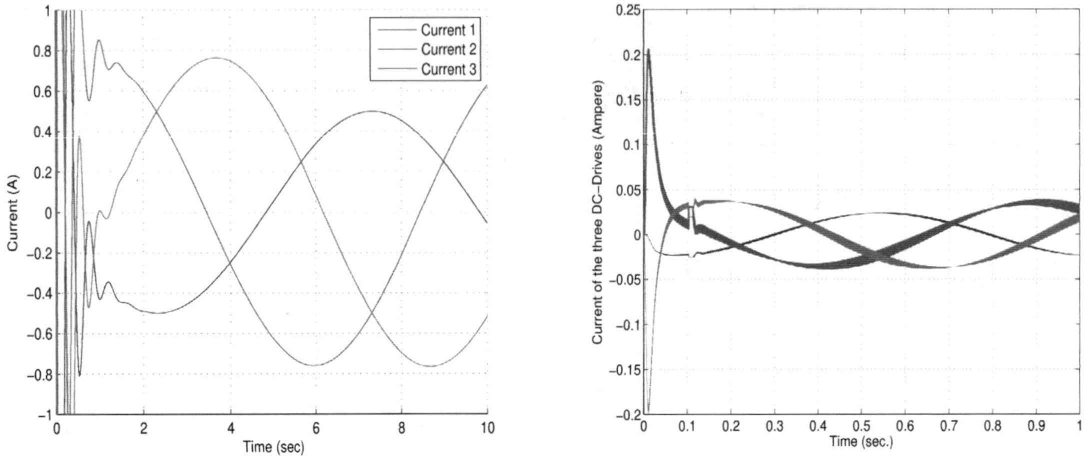

Figure 5. Currents of the three drives acting the three wheels in the case of tra ectory avoidance (left). Currents of the three drives acting the three wheels in the case of tra ectory avoidance (right)

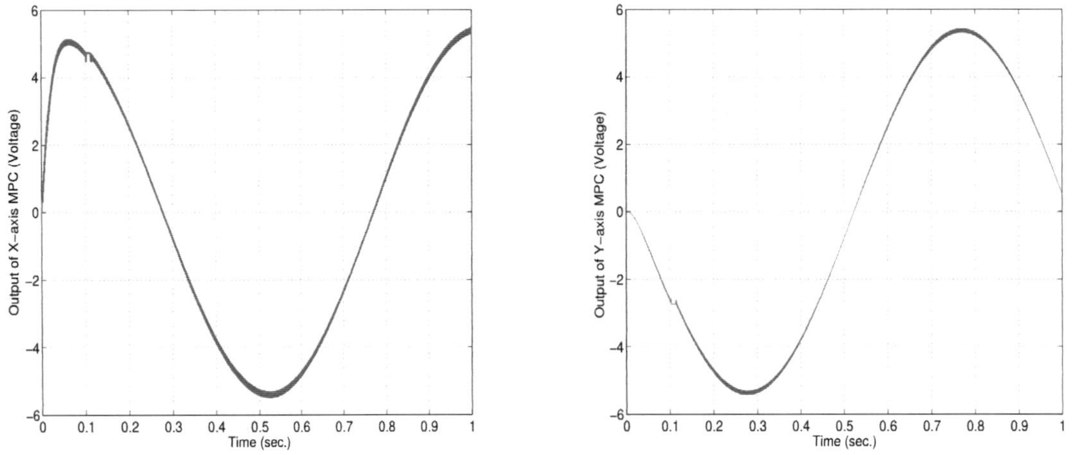

Figure 6. utput of X-axis and -axis MPC

6. Conclusion

This paper presents a procedure in order to obtain the decoupling functional controllability between horizontal, vertical and angular motions using an geometric approach. Here, the functional decoupling problem is investigated. To achieve a decoupling effect a feedback control law is needed together with a feed-forward regulator. The relevance of the motion functional controllability in the Robotino control is ustified by the necessity of very fast and very precise loops of acceleration control counteracting errors in displacements caused by possible disturbance actions. Simulation results are shown to indicate the potential of this approach.

References

[1] G. Basile and G. Marro. *Controlled and conditioned invariants in linear system theory.* Prentice Hall, New Jersey, 1992.

[2] A. Isidori. *Nonlinear control Systems: an introduction.* Springer, Berlin, 1989.

[3] W.M. Wonham. *Linear multivariable control: a geometric approach.* Springer, New York, 1979.

[4] P. Mercorelli and D. Prattichizzo. A geometric procedure for robust decoupling control of contact forces in robotic manipulation. *Kybernetika*, 39(4):433–455, 2003.

[5] G. Basile and G. Marro. A state space approach to non-interacting controls. *Ricerche di Automatica*, 1(1):68–77, 1970.

Journal of Physics: Conference Series **659** (2015) 012028 doi:10.1088/1742-6596/659/1/012028

Feedback control system based on a remote operated PID controller implemented using mbed NXP LPC1768 development board

Emil Pricop, Florin Zamfir, Nicolae Paraschiv

Petroleum-Gas University of Ploiesti
39, Bucuresti Blvd., Ploiesti, 100680, Prahova, Romania

E-mail: emil.pricop@gmail.com

Abstract. Process control is a challenging research topic for both academia and industry for a long time. Controllers evolved from the classical SISO approach to modern fuzzy or neuro-fuzzy embedded devices with networking capabilities, however PID algorithms are still used in the most industrial control loops. In this paper, we focus on the implementation of a PID controller using mbed NXP LPC1768 development board. This board integrates a powerful ARM Cortex-M3 core and has networking capabilities. The implemented controller can be remotely operated by using an Internet connection and a standard Web browser. The main advantages of the proposed embedded system are customizability, easy operation and very low power consumption. The experimental results obtained by using a simulated process are analysed and shows that the implementation can be done with success in industrial applications.

1. Introduction

Process control is a challenging topic of research for both academia and industry, being present in almost each industry and human activity from energy production to flying planes. Automatic process control refers to all the methods used to control the process variable in order to achieve an objective without any human intervention. The process control objectives are related to process efficiency and safety.

Industrial process control infrastructure evolved significantly in the last decades along with electronics and computer science. The analog devices used for implementation of basic controllers were replaced by digital devices with significant processing power such as PLCs and process computers. The control algorithms also evolved and the research shows how to use fuzzy or neuro-fuzzy approaches in various industrial facilities, but PID algorithm is still used in the majority of industrial control loops. Also the devices are now exchanging data using various protocols, starting from specific ones such as Modbus, Profibus or OPC and even on the global Internet, exposing the industrial infrastructure to various security threats [1].

In this paper we focus on the implementation of a PID controller using mbed NXP LPC1768 development board. This board integrates a powerful ARM Cortex-M3 core and has networking capabilities. The controller can be operated locally, from a PC, or remotely by using a TCP/IP or Internet connection and a simple Web browser. The development board characteristics are described in the second section of this paper.

The novelty of the implementation consists mainly in the development board used. This development board can be utilized in industrial applications due to its very low-cost and very low power consumption.

The controller implemented on the development board is very versatile, the control algorithm can be changed without any hardware modifications.

The control system was implemented in an experimental setup. The process and disturbances are simulated using Simulink software. The development board is connected to the simulated process using a National Instruments NI-6008 data acquisition equipment. In this way the controller can apply the command to the simulated process and receive the feedback in order to compute the next value of the command. This approach is presented in the third section of the paper.

The objective of the paper is to develop a networked controller. In order to reach this objective a RPC over HTTP server was implemented and a web based operational interface was designed. The server could be accessed from anywhere in the world where an Internet connection is available using a standard computer or a smartphone with a simple Web browser. The implementation and testing of the network components are presented in the fourth section of this paper.

The proposed networked controller was extensively tested using various process models in Simulink. The simulation results are presented in the last section of this paper.

2. Mbed NXP LPC1768 development board

Mbed NXP LPC1768 is a powerful and user-friendly development board based on an ARM Cortex-M3 microcontroller. It integrates the microcontroller unit together with peripherals such as SPI, I2C, USB and Ethernet interfaces. The development board has the main objective of rapid prototyping, supporting even an on-line compiler for writing code quickly and efficiently. In this section we will present the main features provided by mbed NXP LPC1768 development board.

Mbed board is built around an LPC1768 microcontroller produced by NXP semiconductor company. This microcontroller integrates an ARM Cortex-M3 core running at 100 MHz frequency with 63 KB of SRAM and 512 KB of Flash memory. The core architecture uses a multi-layer AHB bus that permit to use high-bandwidth peripherals such as Ethernet interface and USB to run simultaneously without having an impact on the board performance [2].

LPC1768 is a very versatile and powerful microcontroller. It can be used not only for implementing basic computing tasks but also it allows rapid prototyping for complex devices. The peripherals are divided in three categories: serial connected, analog and general purpose ones. The serial peripherals are:

- an 10/100 Ethernet controller offering IEEE 802.3 standard compatibility and full TCP/IP stack. The controller also supports WOL – Wake on LAN function, allowing the start-up of the device from network.
- USB 2.0 full-speed interface allowing a bandwidth of maximum 12 Mbps;
- three SPI controllers, full-duplex with a maximum bandwidth of 12,5 Mbps;
- three I2C bi-directional interfaces with a 1 Mbps bandwidth and the possibility to connect multiple equipment for master/slave communication;
- two CAN 2.0B controllers, with 1 Mbps bandwidth and 32-bit address space;
- one UART serial interface allowing 6,25 Mbps speed and full support of RS-232 and RS-485 communication standards [3].

The analog peripherals are represented by a 12-bit, 8 channels ADC (analog to digital converter) and a 10-bit DAC (digital to analog converter). The analog to digital converter has support for dynamic memory access. The conversion is made at 200 Hz multi-channel. Each converter channel has its own memory register for conversion result temporary storage. The conversion step is 0,806 mV with conversion limits between 0 and 3,3V [3].

In the general purposes category of peripherals, there are found a DMA controller with eight channels, PWM interfaces, 32-bit general timers or counters. For real-time application LPC1768 has a real time clock (RTC) with a 32kHz oscillator. In order to avoid microcontroller blockage there is implemented a Watch-dog timer that can reset the whole system in case of an error.

In order to interact easily with this great variety of peripherals mbed provides open-source libraries and APIs. In this way, the prototyping time is significantly reduced since the need to write low-level code is eliminated. If open-source stacks or third-party protocols are needed they can be added and reused in all the projects. All the mbed libraries comply to Cortex Microcontroller Software Interface Standard (CMSIS) allowing the user to migrate the project to other programming tools or to implement custom code for higher performance.

Mbed NXP LPC1768 is very easy to setup and use on any type of computer. The novelty of the device consists in the on-line compiler, available on http://developer.mbed.org. There is no need to install supplementary software apart form a standard web browser. Practically the user can login to mbed compiler from anywhere in the world, write a program in C++ using standard libraries and download the compiled firmware that can be copied on the development board. By using this approach, the user always has access to the latest version of compiler and to an update libraries set. Along with using the on-line compiler the user can program the device by using Keil MDK (Microcontroller Development Kit) that support ARM Cortex devices along with uVision Integrated Development Editor [4].

3. Feedback control system design

The feedback control system is based on a PID controller implemented on mbed NXP LPC1768 development board. The block diagram of the proposed system is presented in the figure 1. All the components listed in the block diagram are integrated on the mbed board.

The implemented PID controller has the standard inputs and output of an analogic PID controller. The set point, i, and the tuning parameters – Kp – proportional gain, Ti – integral gain and Td – derivative gain should be set by using the network interface via the web browser. The feedback from the process (r), also named process value is acquired using an analog input of the board and then is converted to a digital signal using the integrated ADC. The converted value is used to calculate the next control action (u) which is converted from digital to analog by using the internal 10-bit DAC of LPC1768. The limits for analog input/output are 0-3,3V.

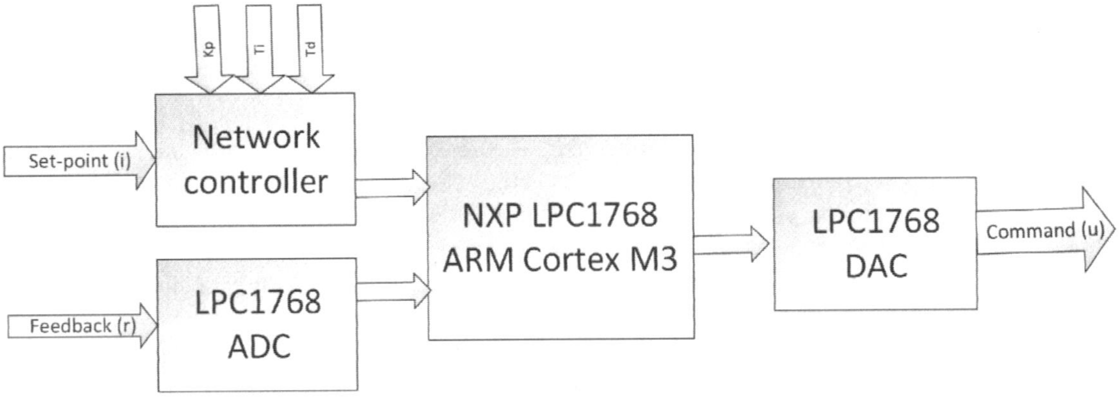

Figure 1. Proposed PID Controller system diagram

The controller is implemented in the mbed NXP LPC1768 board using the standard PID library from mbed repository. The PID loop functioning is not very complex. The controller read the process value and compare it to the desired value, which is also the set point. The following step is to compute an output value in order to adjust the process value to be closer or even to accede to the set-point. The algorithm is based on the following numerical form of PID command law:

$$u(t) = K_p\left(e(t) + \frac{1}{T_i}\int_0^t e(\tau)d\tau + T_d\frac{de(t)}{dt}\right) \tag{1}$$

The signification of the notations in (1) is presented below:

- u(t) is the control signal (command);

- e(t) is the deviation (error), computed as the difference between i– set point and r – process value;
- Kp is proportional gain;
- Ti is the integral time constant;
- Td is the derivative time constant [5].

The control signal is computed as a sum of three terms: the proportional value which is proportional to the current value of the error, an integral term that is an average of past error and an derivative value which linearly interpolate the future error values [5][1].

The PID library available on mbed repository implement the homonym class. The main object defined is the controller, which has a large number of specific methods. The programmer can set the input and output limits, the tuning parameters (Kp, Ti, Td), the set point, manual or automatic operation mode and can call the command computing routine. The control action is computed based on a numeric implementation of (1) [6].

The source code for the mbed program starts with preprocessor directives and variable declarations: the two pins used for data transfer from/to PC, the tuning parameters, the set point variable and the special variable PID controller as it is seen in figures 2 and 3. In the main function, all the variables are initialized, the controller limits are set and the web server is started. The controller enters an infinite loop where it listen for new tuning values or set point modification on a thread and on another thread calculates the control signal value periodically. For debugging purposes, the microcontroller writes the process values and corresponding control signal on serial port.

```
#include "PID.h"
#include "mbed.h"
PID controller(1.5, 1.5, 0, RATE);
AnalogIn pv(p15);
AnalogOut co(p18);
Serial pc(USBTX, USBRX); // tx, rx
```

Figure 2. PID controller declarations

```
float sp,kp,ti,td;
RPCVariable<float> rpc_sp(&sp, "sp");
RPCVariable<float> rpc_kp(&kp, "kp");
RPCVariable<float> rpc_ti(&ti, "ti");
RPCVariable<float> rpc_td(&td, "td");
```

Figure 3. RPC variable declarations

In order to obtain good precision, the output from Simulink should be filtered. The numerical filter is a nonlinear method used to reduce the noise of the input process variable. This can be done with a median filter that reorder the last *n* values acquired and return the value from the center of the string. The implementation is made with pointers and FIFO method, where the data is sorted in the acquisition order and the output is computed based of the value of element placed in the middle of the sorted buffer. The program contains two structures that force the values to be stored only if the data stays in certain limits.

4. Remote operator console

The PID controller designed in this paper should be remotely operated. This objective is accomplished by using the mbed TCP/IP stack in order to implement an HTTP server that can be accessed from anywhere in the world by using a simple web browser.

The HTTP implementation on mbed has mainly two components: one webserver engine programmed in the mbed firmware and HTML files that are located in the mbed file system. Practically the HTTP server is a process running on the mbed along with the PID routine, which represent another process running on the MCU, and the files are stored on a file system like a hard drive on a PC.

The server start when mbed board power on. The IP address can be configured to be static or dynamic, meaning that it can be obtained automatically form a DHCP server on the network if available. For debugging purposes, the IP address is printed on serial port. The server listens for connections on TCP port 80, allowing any user that has a standard browser like Firefox, Chrome, Safari or Internet Explorer to view or set parameters as it is shown in figure 6.

The web page programming uses RPC JavaScript library. RPC is a protocol that allows client and server communication by replacing network protocols and communication methods. This is an inter process communication where one program can uses services from a remote program. Once compiled, the RPC function can be used by any program with client/server capability [7].

The set point and the tuning parameters are defined as RPC variables, which can be read, written or deleted using RPCVariable class, included in RPCVariable.h library. The RPC variables are defined in two places – the microcontroller program (firmware) and also in the JavaScript stored in the mbed file system. The code section containing variable definition in the mbed program is presented in figure 4.

The JavaScript links standard variable to the RPC variables defined in the microcontroller program, using functions located in the RPCVariable.h library. For example, reading Set Point variable (Fig. 4) uses "readp" function that has two major roles. First task is to store in a variable "f" the content acquired through "sp" RPCVariable reading. The "sp" is declared in the HTML header along with the rest of the RPCVariable. Second task imply using the print function to display the "sp" variable in the location referred by "id" with the same name.

```
<script src="mbedRPC.js" language="javascript"></script>
<script type="text/javascript">
    mbed = new HTTPRPC();
    sp = new RPCVariable(mbed, "sp");
    ...
    function readp(){
        f = sp.read();
        ...
        print(f);
        ...
    }
    function print(str){
        document.getElementById("sp").innerHTML = str;
    }
    ...
        </script>
    ...
    <td>Set Point (sp): </td>
    <td><div id="sp"></div></td>
    ...
```

Figure 4. Source code used to read Set Point variable.

```
function writep()
{
var csp = document.getElementById("wsp").value;
    if(csp) sp.write(csp);
...
readp();
}
</script>
...
<th colspan="2">Tuning controller gains: </th>
<tr><td>sp:</td>
<td><input type="text" id="wsp"></td> </tr>
...
<th colspan="2">
<button type="button" onclick="writep()">Set</button>
</th>
...
```

Figure 5. Source code used to modify Set Point variable.

Modifying the Set Point can be done with the function "writep" (Fig. 5). The value placed by the operator in the text box identified by wsp id is sent to "csp" variable when Set button is used only. This time the RPC write function send the value to the set point RPCVariable stored in mbed memory.

The operator interface can be accessed using an URL in format http://x.y.z.t/index.htm, where x.y.z.t is the IP address allocated to the microcontroller unit. The operator can retrieve the set point, process value and tuning parameters by pressing the "View Values" button. If the operator wants to change any parameter or the set point, he should write the value in the corresponding text box and press the "Set" button. If the operator does not write a value in the text box corresponding to one parameter, its value does not change.

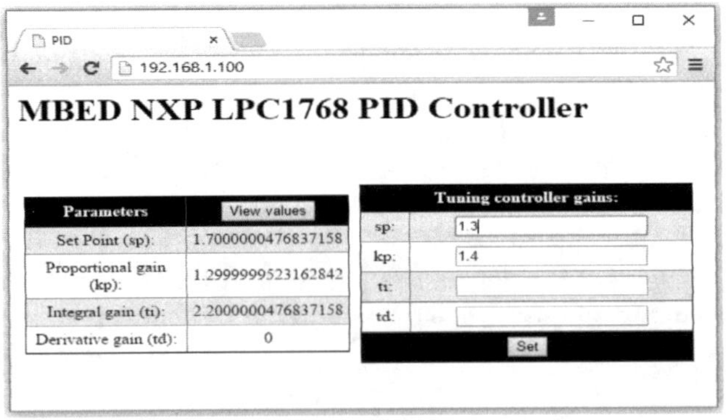

Figure 6. Web based operator interface for the PID controller

Journal of Physics: Conference Series **659** (2015) 012028 doi:10.1088/1742-6596/659/1/012028

5. Experimental results

Experimental testing of the proposed PID controller require the existence of a process to interact. The controller was tested with first and second order processes configured by the authors and simulated using Simulink.

In order to connect the physical PID controller implemented on the mbed NXP1768 development board it is necessary to acquire and transmit the signals from Simulink through a data acquisition board. We have used National Instruments USB-6008 data acquisition device for interconnecting the PID controller with the PC. NI USB-6008 has 8 analog inputs and 2 analog outputs. The board is supplied with NI DAQmx software drivers for operation on Microsoft Windows Operating System. The block diagram of the laboratory experimental setup is shown in figure 7.

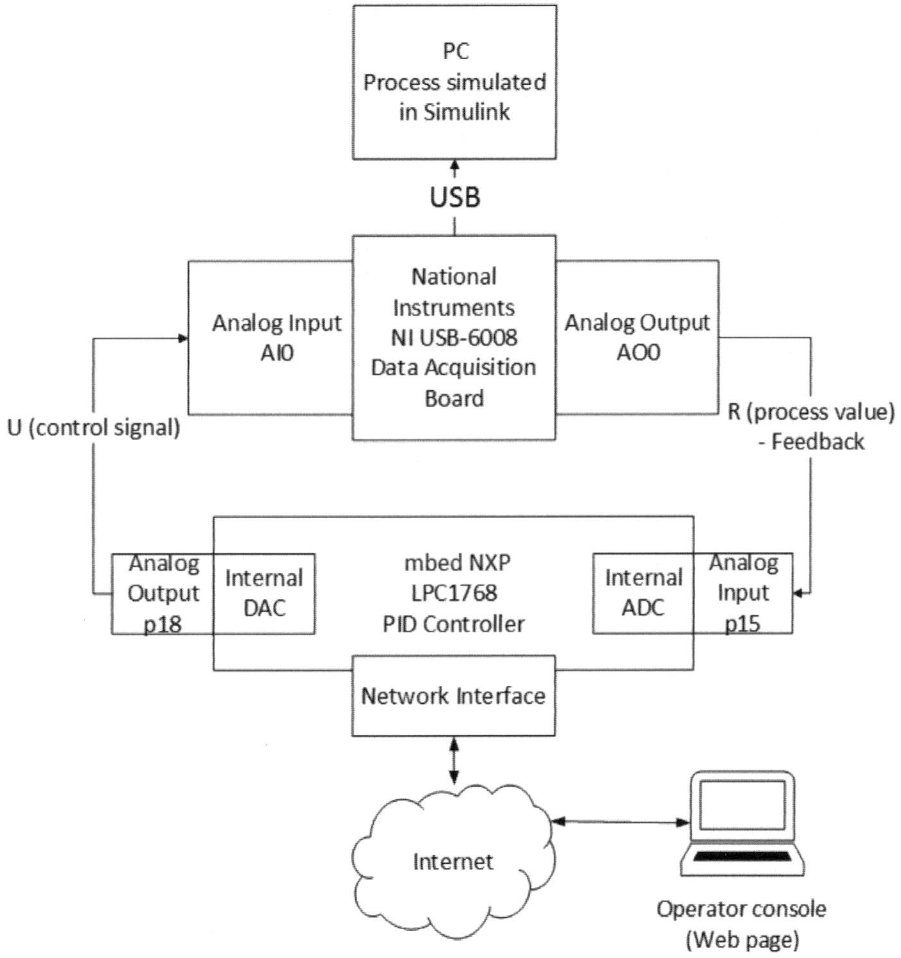

Figure 7. Block diagram of the experimental system

The experimental setup includes a PC where the process is simulated in Simulink. In order to transmit and acquire signals the Simulink model should integrate the NI DAQmx drivers for USB-6008 board. The board is connected to the PC by USB port. AI0 – analog input channel from USB-6008 is used for command acquisition from the PID controller and AO0 – analog output channel is designated for transmitting the feedback process value from simulator to the controller. The connections are made as shown in the following table:

Table 1. Pins assignment on USB-6008 and mbed NXP LPC1768.

USB-6008 pin	USB-6008 Simulink signal	mbed NXP LPC1768 pin	mbed Controller signal
AI0+ AI0-	Controller command signal for the process	P18 GND	Controller output (computed command signal)
AO0 GND	Process value – feedback signal	P15 GND	Controller input – feedback process value

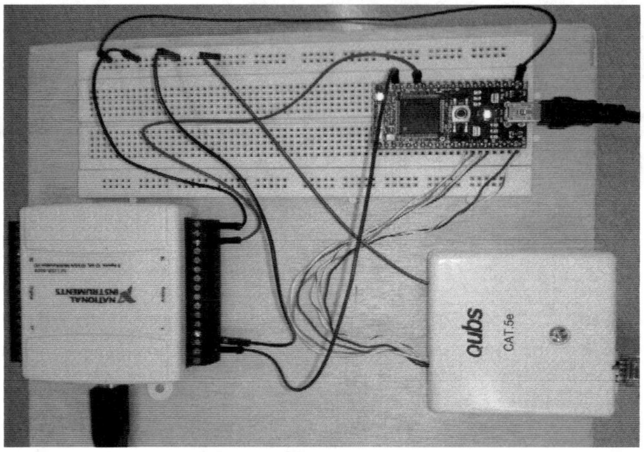

Figure 8. Experimental system for PID controller testing

The connection to NI USB-6008 board is made by an Analog Input block, from the Data Acquisition Simulink toolbox. The block is configured to acquire data from nidaq driver's channel 0, as shown in figure 9. The block used for analog output of the process value consists of a MATLAB function that uses the programming environment standard data acquisition and signal generation functions. The function define an analog output using nidaq driver, setup the desired output channel and write data at a desired sampling rate [8].

The process is simulated as a first order model with the transfer function

$$G(s) = \frac{1.5}{2s+1} \tag{2}$$

The disturbance is generated using the same model as shown in the block diagram in figure 9. There are two display block that show the current command (control signal) and the simulated process value (output). The time scope block has the role to generate a graphical representation of the two signals.

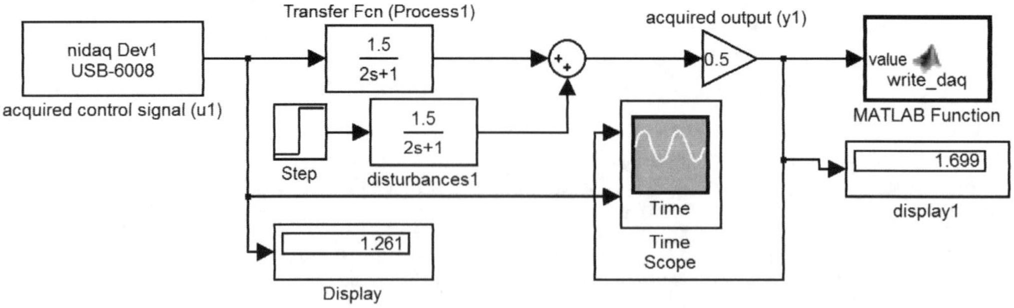

Figure 9. Block diagram of a simulated process with external PID controller

A second simulation was made in order to test the performances of the PID controller. The external controller was replaced by a simulated PID block included in Simulink libraries as shown in the block diagram from figure 10.

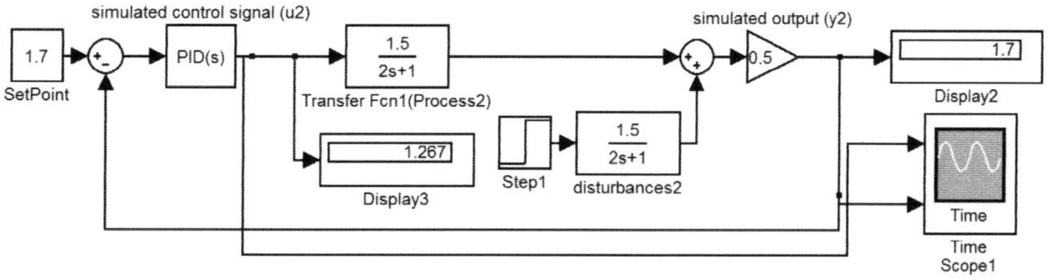

Figure 10. Block diagram of a simulated process with internal PID controller

The simulation results for the two systems are shown in figures 11 and 12. The PID controller parameters are set point sp=1.7, proportional gain kp=1.3, integral gain ti=2.2 and derivative gain td=0. The command signal is represented with blue and the process value is the red curve.

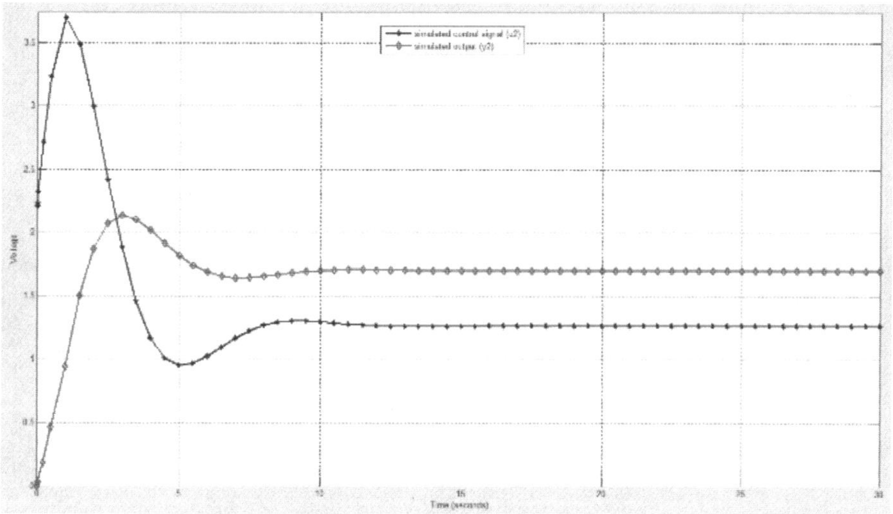

Figure 11. Control signal and process value for simulated PID controller

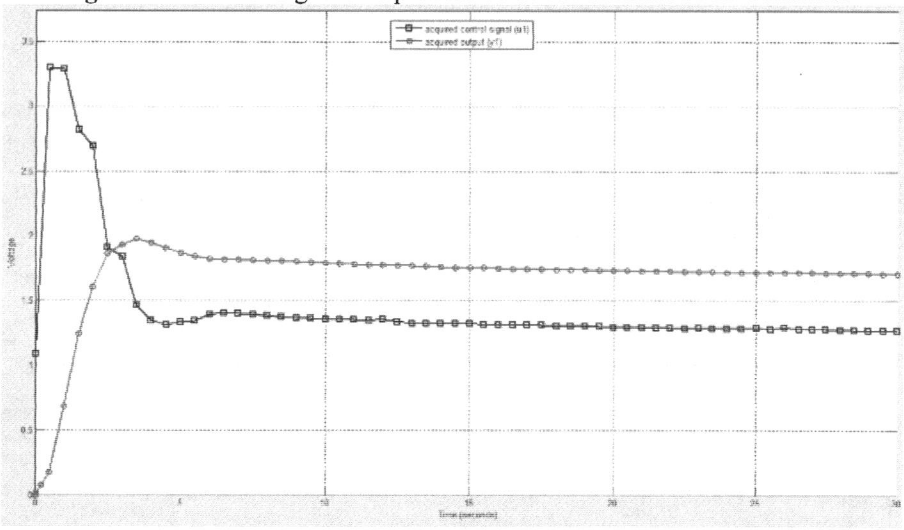

Figure 12. Control signal and process value for mbed NXP LPC1768 PID controller

A combined representation of the two systems can be seen in Fig.13. The number of signals on the time scope is increased to five: set point, acquired control signal from the microcontroller, output sent to microcontroller, simulated control signal and output from Simulink PID controller.

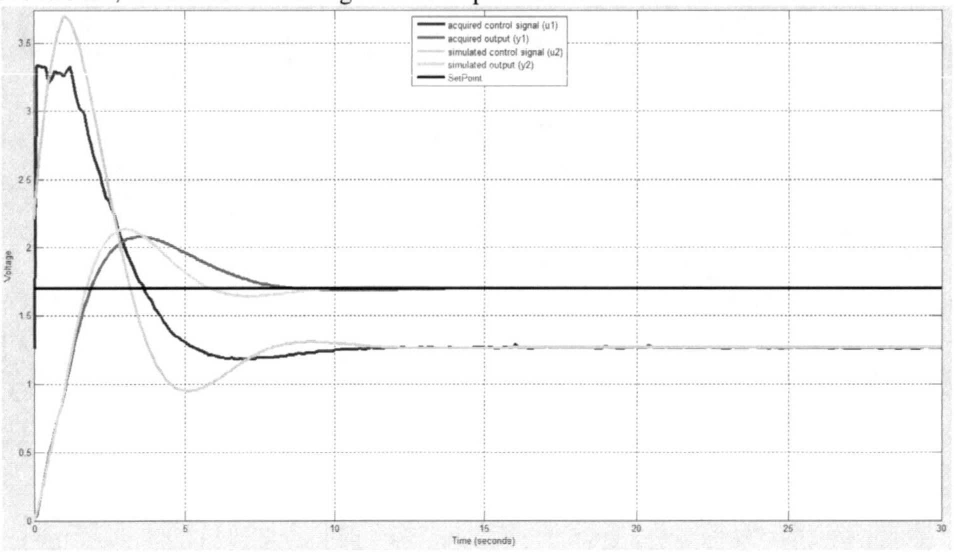

Figure 13. Simulation results of the simulated and experimental PID controllers

Analyzing the graphs presented in figures 11, 12 and 13 we can conclude that the controller work good on a first order process. The control signal (command) has approximately the same shape as the one generated by simulated Simulink controller.

The performances of the controller can be analyzed by using two parameters: overshoot and settling time. The results obtained in case of the first order process are shown in the table below.

Table 2. Overshoot and settling time values for the first order process – set point = 1,7

Parameter	**Simulated controller in MATLAB**	**Mbed implemented controller**
Overshoot	0,43	0,28
Maximum PV	2,13	1,98
Settling time (0.98 criteria)	8 sec.	16 sec.

The MATLAB simulated PID controller has a better settling time that the one implemented on mbed development board, being reduced by half, but the overshoot is better in the case of the implemented controller.

The experimental testing has been conducted also with a second order process as shown in the figures below.

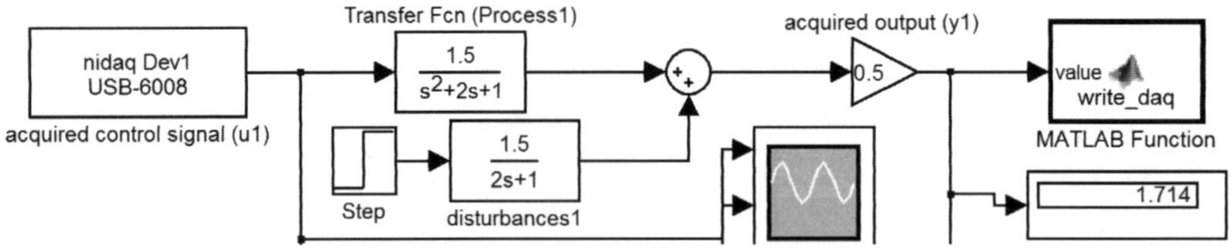

Figure 14. Block diagram of a second order process with external PID controller

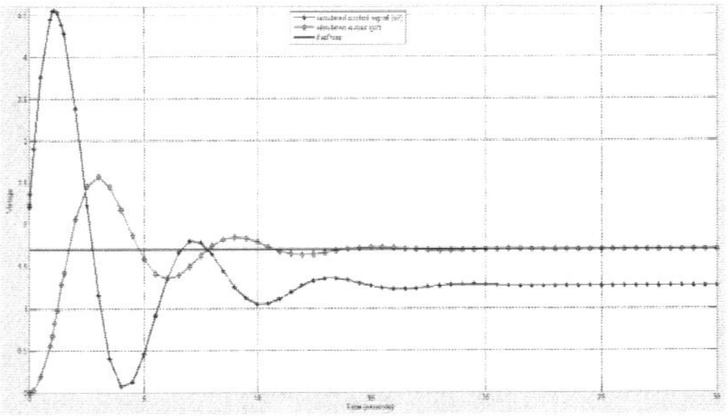

Figure 15. Command and process value for simulated controller for second order process

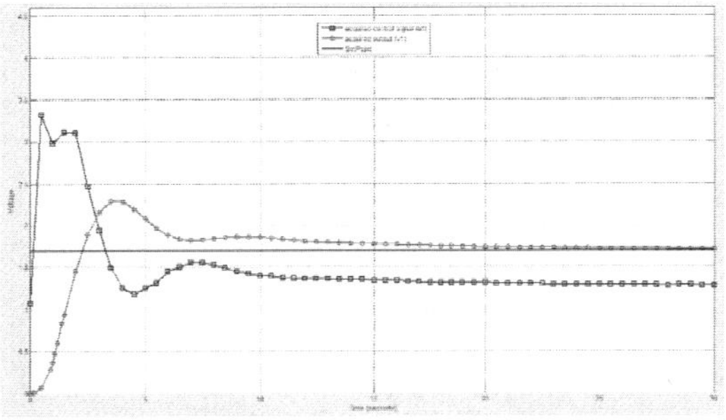

Figure 16. Control signal and process value for mbed controller for second order process

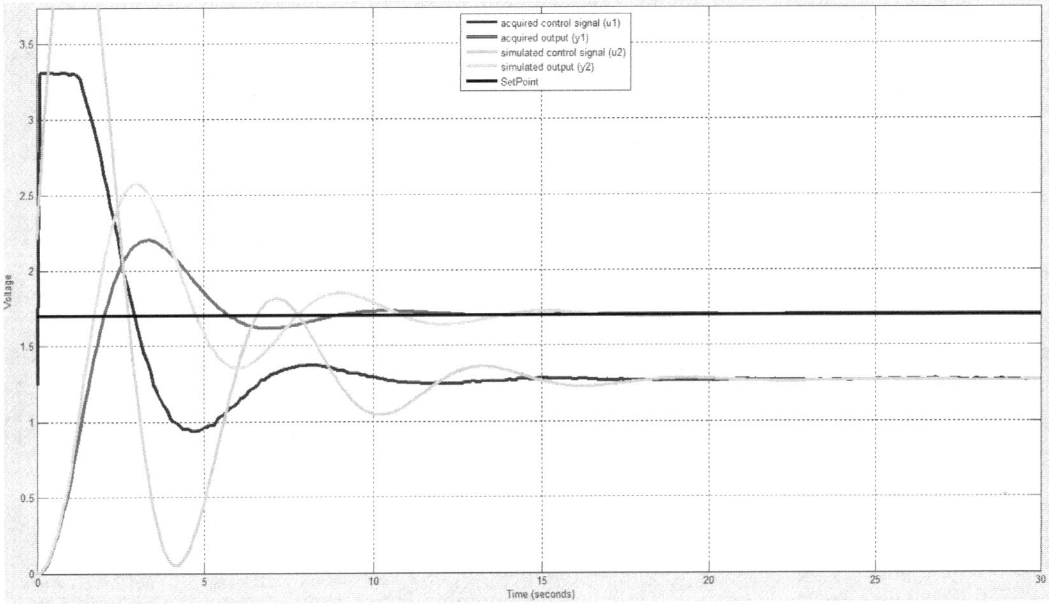

Figure 17. Results of the simulated and experimental PID controllers for second order process

The simulation results for a second order process show that the mbed NXP LPC1768 PID controller works very good. The results obtained with the two controllers (simulated and implemented) are presented in Table 3. The values were computed using graphical methods.

Table 3. Overshoot and settling time values for the second order process – set point = 1,7

Parameter	Simulated controller in MATLAB	Mbed implemented controller
Overshoot	0,84	0,59
Maximum PV	2,54	2,29
Settling time (0.98 criteria)	13 sec.	23 sec.

The overshoot value is better with the mbed implemented PID controller that with the PID controller simulated in MATLAB. The settling time is higher when the implemented controller is used.

We can conclude that the controller designed and implemented using the microcontroller has lower overshoot that a simulated controller with the same tuning parameters, but the settling time is approximatively double.

Journal of Physics: Conference Series **659** (2015) 012028 doi:10.1088/1742-6596/659/1/012028

6. Conclusion

In this paper is presented an innovative approach in implementing a networked PID controlled based on NXP LPC1768 development board. The controller main advantages are the versatility, easy remote operation, low cost and very low power consumption.

The implemented controller was laboratory tested by connecting it with a simulated process in Simulink. The communication was realized by using a National Instruments USB-6008 board for interfacing with the PC. The experimental results for first and second order processes are convincing and performances achieved by the PID controller are significant and comparable to the ones of a simulated controller with the same parameters. The experimental results show that the overshoot is lower when using the NXP LPC1768 implemented controller that the one obtained with the MATLAB simulated controller. Meanwhile the settling time is approximatively double in case our implementation than the MATLAB one. This result is good for industrial implementation since the controller does not action brutally on the controlled installation.

After securing the operator interface the controller can be used in any industrial setup. In case of the industrial implementation the cost of the development board solution is smaller than using a conventional PID controller with network interface.

The research should be continued in order to implement a remotely operated multi-channel controller using the same development board which is capable to run a real-time operating system – RTOS. Also the security of the remote interface should be improved in order to authenticate the operator and to log all the commands or changes issued.

References

[1] Ray Chaudhri U and Ray Chaudhri U 2013 *Fundamentals of Automatic Process Control* CRC Press (USA)
[2] http://www.nxp.com/documents/leaflet/LPC1768.pdf
[3] http://www.nxp.com/documents/data_sheet/LPC1769_68_67_66_65_64_63.pdf
[4] http://www.nxp.com/documents/user_manual/UM10360.pdf
[5] Aström K J and Murray R 2008 *Feedback Systems: An Introduction for Scientists and Engineers* Princeton University Press (New Jersey USA)
[6] https://developer.mbed.org/cookbook/PID
[7] https://developer.mbed.org/cookbook/Interfacing-Using-RPC
[8] http://www.mathworks.com/help/daq/ref/analog_input.html

12th European Workshop on Advanced Control and Diagnosis (ACD 2015) IOP Publishing
Journal of Physics: Conference Series **659** (2015) 012029 doi:10.1088/1742-6596/659/1/012029

A Decoupled MPC for Motion Control in Robotino Using a Geometric Approach

Daniel Straßberger, Paolo Mercorelli

Institute of Product and Process Innovation, Leuphana University of Lueneburg, Volgershall 1, D-21339 Lueneburg, Germany Phone: +49-(0)4131-677-5571, Fax: +49-(0)4131-677-5300.

E-mail: mercorelli@uni.leuphana.de

Oleg Sergiyenko

Engineering Institute of Autonomous University of Baja California, Blvd. Benito Juárez y Calle de La Normal, s n, Col. Insurgentes Este, C.P.21280, Mexicali, BC, Mexico, tel. fax. 01(52-686) 566-41-50.

E-mail: srgnk@iing.mxl.uabc.mx

Abstract. This paper proposes a controller for motion control of the Robotino. The proposed controller considers a functional decoupling control strategy realized using a geometric approach and the invertibility property of the DC-drives with which the Robotino is equipped. Horizontal, Vertical and Angular motions are considered and once the decoupling between these motions is obtained, a Model Predictive Control (MPC) strategy is used in combination with the inverse DC-drive model. Simulation results using real data of Robotino are shown.

1. Introduction

This paper presents a systematic procedure in order to obtain the decoupling controllability between horizontal, vertical and angular motions. Here, the decoupling problem is investigated. To achieve a decoupling effect a feedback control law is needed together with a feed-forward regulator. In the past three decades, research on the geometric approach to dynamic systems theory and control has allowed this approach to become a powerful and a thorough tool for the analysis and synthesis of dynamic systems [1], [2], [3]. ver the same time period, mechanical systems used in industry and developed in research labs have also evolved rapidly. Mobile robotics is a notable case of such evolution. The robotics community has developed sophisticated analysis and control techniques to meet increasing requirements on the control of motions of mechanical systems. These increasing requirements are motivated by higher performance specifications and an increasing number of degrees-of-freedom. References [4] and [5] mark progress in the analysis and synthesis of geometric controller for mechanical systems, and [6] proposes non-interacting force-motion control in robotic manipulation. Reference [7] reports the possibility of parameterising input controlled subspaces to guarantee non-interaction. In [4], a robust decoupling controller using an algebraic state input feedback is presented, while this paper presents a robust decoupling controller using an algebraic output-input feedback. The force/motion control problem has attracted significant attention of scientists over the last decade in the fields of robotic manipulation and mobile robotics. Approaches exploiting input-output decoupling controllers

Journal of Physics: Conference Series **659** (2015) 012029 doi:10.1088/1742-6596/659/1/012029

are found, for instance, in the work [8]. The geometric approach allows very elegant solutions of control problems. Nevertheless, robustness analysis using a linear geometric control offers answers through rank conditions of matrices that are necessary conditions. These conditions are often not constructive ones. Even though the rank conditions offer simple on-off conditions, it is also possible to measure the robustness. In [4], a robust decoupling controller is obtained using a state input feedback controller. The drawback of this approach consists of a wide sensing structure. In fact, the whole state space should be available and sensed. With the approach proposed in this paper, the robust decoupling is obtained with an output feedback from the contact forces and the oint positions. The goal of this paper is to propose a complete constructive procedure for the design of a decoupling controller. In Robotino's case, a preselecting field law is needed together with a feed-forward regulator for achieving a decoupling between Horizontal, Vertical and Angular motions. The paper is organised in the following way: Section 2 presents

Figure 1. Robotino in action on the pitch.

a possible model of the Robotino. Section 3 shows the decoupling strategy using the geometric approach. Section 4 is devoted to the derivation of the Model Predictive Control (MPC) strategy. At the end, numerical computer simulations, considering real data of the Robotino, are shown.

The main nomenclature
$\mathbf{A}$: state matrix of the mechanical model
$\mathbf{g}(\theta)$: input field of the mechanical model
$\mathbf{T}(\theta)$: decoupling input partition field
$\mathcal{B} = \mathbf{g}(\Theta) \cdot \mathbf{T}(\theta)$: decoupled input matrix of the mechanical model
$\mathcal{B} = \mathsf{im}\mathbf{B}$: image of matrix $\mathbf{B}$
(subspace spanned by the columns of matrix $\mathbf{B}$)
$\mathsf{min}\mathcal{I}(\mathbf{A}, \mathcal{B}) = \sum_{i=0}^{n-1} \mathbf{A}^i \mathsf{im}\mathbf{B}$: minimum $\mathbf{A}$–invariant subspace containing $\mathsf{im}(\mathbf{B})$
$\mathcal{C} = \mathsf{ker}\mathbf{C}$: kernel of matrix $\mathbf{C}$
(subspace spanned by the columns of matrix $\mathbf{C}$)

2. Mechanical and Electrical Model Description

In Fig. 2 a diagram with the representation of the forces in the considered system is shown. If state vector $\mathbf{X}(t)$ is defined as, $x(t)$ and $y(t)$ positions of the center of mass of the system and its velocities, $x(t)$ and $y(t)$, moreover, considering the angular dynamics with its angular position

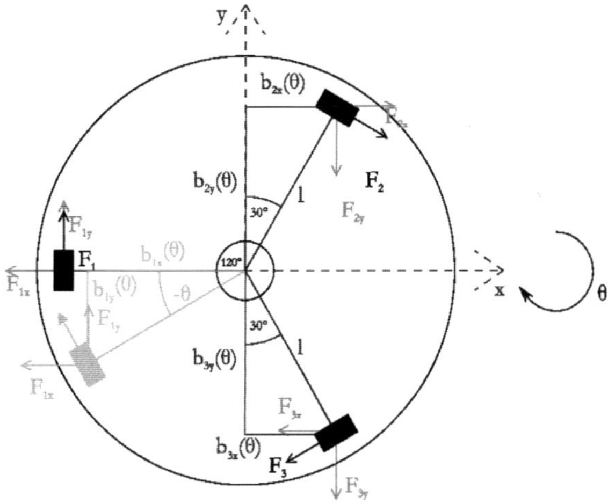

Figure 2. Mechanical schematic diagram of Robotino.

$\theta(t)$ and its velocity $\theta(t)$, then the following system is derived.

$$\begin{cases} \dot{\mathbf{X}}(t) = \mathbf{A}\mathbf{X}(t) + \mathbf{g}(\theta)\mathbf{F}(t) \\ \mathbf{O}(t) = \mathbf{C}\mathbf{X}(t), \end{cases} \tag{1}$$

with

$$\mathbf{X}(t) = \begin{bmatrix} x(t) \\ x(t) \\ y(t) \\ y(t) \\ \theta(t) \\ \theta(t) \end{bmatrix}, \qquad \dot{\mathbf{X}}(t) = \begin{bmatrix} x(t) \\ x(t) \\ y(t) \\ y(t) \\ \theta(t) \\ \theta(t), \end{bmatrix} \tag{2}$$

matrix

$$\mathbf{A} = \begin{bmatrix} 0 & 1 & 0 & 0 & 0 & 0 \\ 0 & -\frac{k_v}{M} & 0 & 0 & 0 & 0 \\ 0 & 0 & 0 & 1 & 0 & 0 \\ 0 & 0 & 0 & -\frac{k_v}{M} & 0 & 0 \\ 0 & 0 & 0 & 0 & 0 & 1 \\ 0 & 0 & 0 & 0 & 0 & -\frac{k_\theta}{J} \end{bmatrix}, \tag{3}$$

which for sake of notation can be written as follows:

$$\mathbf{A} = \begin{bmatrix} \mathbf{A}_{x2\times2} & \mathbf{0}_{2\times2} & \mathbf{0}_{2\times2} \\ \mathbf{0}_{2\times2} & \mathbf{A}_{y2\times2} & \mathbf{0}_{2\times2} \\ \mathbf{0}_{2\times2} & \mathbf{0}_{2\times2} & \mathbf{A}_{\theta2\times2} \end{bmatrix}, \tag{4}$$

where $\mathbf{A}_{x2\times2}$, $\mathbf{A}_{y2\times2}$ and $\mathbf{A}_{\theta2\times2}$ are the minor matrices of $\mathbf{A}$ describing the x, y and θ dynamics of the system. Matrix $\mathbf{C}$ represents the output matrix and can be written in the following way:

$$\mathbf{C} = \begin{bmatrix} 1 & 0 & 0 & 0 & 0 & 0 \\ 0 & 0 & 1 & 0 & 0 & 0 \\ 0 & 0 & 0 & 0 & 1 & 0 \end{bmatrix}, \tag{5}$$

where the following notation is assumed:

$$\mathbf{C_x} = \begin{bmatrix}1 & 0 & 0 & 0 & 0 & 0\end{bmatrix}, \tag{6}$$
$$\mathbf{C_y} = \begin{bmatrix}0 & 0 & 1 & 0 & 0 & 0\end{bmatrix}, \tag{7}$$
$$\mathbf{C}_\theta = \begin{bmatrix}0 & 0 & 0 & 0 & 1 & 0\end{bmatrix}. \tag{8}$$

Field $\mathbf{g}(\theta)$ is as follows:

$$\mathbf{g}(\theta) = \begin{bmatrix} 0 & 0 & 0 \\ \frac{\sin(\theta)}{M} & \frac{\sin(\frac{2}{3}\pi+\theta)}{M} & -\frac{\sin(\frac{2}{3}\pi-\theta)}{M} \\ 0 & 0 & 0 \\ \frac{\cos(\theta)}{M} & \frac{\cos(\frac{2}{3}\pi+\theta)}{M} & \frac{\cos(\frac{2}{3}\pi-\theta)}{M} \\ 0 & 0 & 0 \\ a & b & c \end{bmatrix} \tag{9}$$

where

$$a = \frac{\sin(\theta)\cdot l\cdot\sin(\theta)+\cos(\theta)\cdot l\cdot\cos(\theta)}{J},$$

$$b = \frac{\sin(\frac{2}{3}\pi+\theta)\cdot l\cdot\cos(\frac{\pi}{6}+\theta)+\cos(\frac{2}{3}\pi+\theta)\cdot l\cdot\sin(\frac{\pi}{6}+\theta)}{J}$$

and

$$c = \frac{-\sin(\frac{2}{3}\pi-\theta)\cdot l\cdot -\cos(\frac{\pi}{6}-\theta)}{J} + \frac{\cos(\frac{2}{3}\pi-\theta)\cdot l\cdot\sin(\frac{\pi}{6}-\theta)}{J} \tag{10}$$

and force input vector signal $\mathbf{F}(t) = \begin{bmatrix}F_1(t)\\F_2(t)\\F_3(t)\end{bmatrix}$ in which $F_1(t) = K_m i_1(t)/r$, $F_2(t) = K_m i_2(t)/r$ and $F_3(t) = K_m i_3(t)/r$, where K_m represents the motor constant, r the radius of the wheels and variables $i_1(t)$, $i_2(t)$ and $i_3(t)$ are the currents of the three DC-electrical drives. In fact, the Robotino consists of 3 DC electrical drives that power the three omniwheels. The models of these three DC-drive are reported below.

$$\begin{bmatrix}\frac{di_1(t)}{dt}\\ \frac{d\omega_1(t)}{dt}\\ \frac{di_2(t)}{dt}\\ \frac{d\omega_2(t)}{dt}\\ \frac{di_3(t)}{dt}\\ \frac{d\omega_3(t)}{dt}\end{bmatrix} = \begin{bmatrix} -\frac{R}{L} & -\frac{K_m}{L} & 0 & 0 & 0 & 0 \\ \frac{K_m}{J} & -\frac{K}{J} & 0 & 0 & 0 & 0 \\ 0 & 0 & -\frac{R}{L} & -\frac{K_m}{L} & 0 & 0 \\ 0 & 0 & \frac{K_m}{J} & -\frac{K}{J} & 0 & 0 \\ 0 & 0 & 0 & 0 & -\frac{R}{L} & -\frac{K_m}{L} \\ 0 & 0 & 0 & 0 & \frac{K_m}{J} & -\frac{K}{J} \end{bmatrix} \cdot \begin{bmatrix} i_1(t)\\ \omega_1(t)\\ i_2(t)\\ \omega_2(t)\\ i_3(t)\\ \omega_3(t)\end{bmatrix} + \begin{bmatrix} \frac{1}{L} & 0 & 0 \\ 0 & 0 & 0 \\ 0 & \frac{1}{L} & 0 \\ 0 & 0 & 0 \\ 0 & 0 & \frac{1}{L} \\ 0 & 0 & 0 \end{bmatrix} \cdot \begin{bmatrix} U_{inp_1}(t)\\ U_{inp_2}(t)\\ U_{inp_3}(t)\end{bmatrix}, \tag{11}$$

where $U_{inp_1}(t)$, $U_{inp_2}(t)$ and $U_{inp_3}(t)$ represent the input voltage of the DC-drive, L is the inductance, R is the resistance, J represents the motor inertia factor, K_m is the motor moment and finally K can be seen as the friction factor. This model is important in realising the simulation and is basis for the inverted drive used in the control strategy (see Fig. 3) but is otherwise not explicitly used in the MPC.

Journal of Physics: Conference Series **659** (2015) 012029 doi:10.1088/1742-6596/659/1/012029

3. Design of a Decoupling Controller

This section describes the design of a decoupling controller with respect to the x-y and θ motions above defined. A geometric approach is used in this analysis. The earliest geometric approaches to decoupling control were due to Basile and Marro ([9], [1]) and to Wonham and Morse ([10], [11], and [3]). Since the geometric relations depend on the rotational position (θ), the forces actuated by the three DC drives needed to move in a certain direction may be difficult to obtain.

Definition 1 *A control law for the dynamic system (1) is* decoupling *with respect to the regulated outputs $x(t)$, $y(t)$, and $\theta(t)$, if there exist partitions $F_x(t), F_y(t)$, and $F_\theta(t)$ of the input vector $\mathbf{F}(t)$ such that for zero initial conditions, each input $F_{(\cdot)}(t)$ (with all other inputs, identically zero) only affects the corresponding output $x(t)$, $y(t)$, or $\theta(t)$.* □

It it to be shown that there exists a decoupling and stabilizing state feedback field $\mathbf{D}(\theta)$, along with three input partition fields $\mathbf{T}_x(\theta)$, $\mathbf{T}_y(\theta)$, and $\mathbf{T}_\theta(\theta)$ such that, for the dynamic triples

$$\begin{array}{c}(\mathbf{C}_x,\ \mathbf{A}+\mathbf{g}(\theta)\mathbf{D}(\theta),\ \mathbf{g}(\theta)\mathbf{F}_1(t))\,,\\ (\mathbf{C}_y,\ \mathbf{A}+\mathbf{g}(\theta)\mathbf{D}(\theta),\ \mathbf{g}(\theta)\mathbf{F}_2)\,,\\ (\mathbf{C}_\theta,\ \mathbf{A}+\mathbf{g}(\theta)\mathbf{D}(\theta),\ \mathbf{g}(\theta)\mathbf{F}_3(t))\,,\end{array} \tag{12}$$

it holds the following conditions:

$$\mathcal{R}_x(\theta) = \mathsf{min}\mathcal{I}\Big(\mathbf{A}+\mathbf{g}(\theta)\mathbf{D}(\theta),\ \mathbf{g}(\theta)\mathbf{T}_x(\theta)\Big) \subseteq \mathcal{C}_y \cap \mathcal{C}_\theta \ \ \forall\theta, \tag{13}$$

and

$$\mathbf{C}_x\mathcal{R}_x(\theta) = \mathsf{im}(\mathbf{C}_x),\ \ \forall\theta. \tag{14}$$

$$\mathcal{R}_y(\theta) = \mathsf{min}\mathcal{I}\Big(\mathbf{A}+\mathbf{g}(\theta)\mathbf{D}(\theta),\ \mathbf{g}(\theta)\mathbf{T}_y(\theta)\Big) \subseteq \mathcal{C}_x \cap \mathcal{C}_\theta \ \ \forall\theta, \tag{15}$$

and

$$\mathbf{C}_y\mathcal{R}_y(\theta) = \mathsf{im}(\mathbf{C}_y),\ \ \forall\theta. \tag{16}$$

$$\mathcal{R}_\theta(\theta) = \mathsf{min}\mathcal{I}\Big(\mathbf{A}+\mathbf{g}(\theta)\mathbf{D}(\theta),\ \mathbf{g}(\theta)\mathbf{T}_\theta(\theta)\Big) \subseteq \mathcal{C}_x \cap \mathcal{C}_y \ \ \forall\theta, \tag{17}$$

and

$$\mathbf{C}_\theta\mathcal{R}_\theta(\theta) = \mathsf{im}(\mathbf{C}_\theta),\ \ \forall\theta. \tag{18}$$

Here,

$$\mathsf{min}\mathcal{I}(\mathbf{A}, \mathsf{im}(\mathbf{g}(\theta))) = \sum_{i=0}^{n-1} \mathbf{A}^i \mathsf{im}(\mathbf{g}(\theta))$$

is a minimum $\mathbf{A}$–invariant subspace containing $\mathsf{im}(\mathbf{g}(\theta))\ \forall\ \theta$. Moreover, the partition fields $\mathbf{T}_x(\theta)$, $\mathbf{T}_y(\theta)$ and $\mathbf{T}_\theta(\theta)$ satisfy the following relationships:

$$\begin{array}{l}\mathsf{im}(\mathbf{g}(\theta)\cdot\mathbf{T}_x(\theta)) = \mathsf{im}(\mathbf{g}(\theta)) \cap \mathcal{R}_x(\theta),\\ \mathsf{im}(\mathbf{g}(\theta)\cdot\mathbf{T}_y(\theta)) = \mathsf{im}(\mathbf{g}(\theta)) \cap \mathcal{R}_y(\theta),\\ \mathsf{im}(\mathbf{g}(\theta)\cdot\mathbf{T}_\theta(\theta)) = \mathsf{im}(\mathbf{g}(\theta)) \cap \mathcal{R}_\theta(\theta).\end{array} \tag{19}$$

The stabilizing field $\mathbf{D}(\theta)$ is such that:

$$(\mathbf{A}+\mathbf{g}(\theta)\mathbf{D}(\theta))\mathcal{R}_x(\theta) \subseteq \mathcal{R}_x(\theta), \tag{20}$$

$$(\mathbf{A} + \mathbf{g}(\theta)\mathbf{D}(\theta))\mathcal{R}_y(\theta) \subseteq \mathcal{R}_y(\theta), \tag{21}$$

and

$$(\mathbf{A} + \mathbf{g}(\theta)\mathbf{D}(\theta))\mathcal{R}_\theta(\theta) \subseteq \mathcal{R}_\theta(\theta). \tag{22}$$

Considering matrix $\mathbf{A}$ defined in (3) it is straightforward to see that this matrix is structurally already decoupled and intrinsically stable, this implies that field $\mathbf{D}(\theta) = \mathbf{0}\ \forall\theta$. Considering

$$\mathbf{T}(\theta) = [\mathbf{T}_x(\theta), \mathbf{T}_y(\theta), \mathbf{T}_\theta(\theta), \mathbf{T}_c(\theta)],$$

where $\mathbf{T}_c(\theta)$ is defined in a complementary fashion and it is straightforward to show that matrix $\mathbf{T}_c = \mathbf{0}$. In particular, matrix $\mathbf{T}_c$ represents the complementary field partition to the subspaces of x-position, y-position and angular position of the Robotino. The Robotino motion is described using ust three variables, therefore partitions $\mathbf{T}_x(\theta), \mathbf{T}_y(\theta), \mathbf{T}_\theta(\theta)$ complete the transformation and thus $\mathbf{T}_c = \mathbf{0}$.

$$\mathsf{im}\mathbf{T}(\theta) = \mathsf{im}[\mathbf{T}_x(\theta), \mathbf{T}_y(\theta), \mathbf{T}_\theta(\theta)] = \mathsf{im}\mathbf{T}_x(\theta) \oplus \mathsf{im}\mathbf{T}_y(\theta) \oplus \mathsf{im}\mathbf{T}_\theta(\theta). \tag{23}$$

Considering the output matrices (6), (7) and (8) corresponding to x-position, y-position and angular position their respective kernels are as follows:

$$\mathcal{C}_x = \mathsf{im}\begin{bmatrix} 0 & 0 & 0 & 0 & 0 \\ 1 & 0 & 0 & 0 & 0 \\ 0 & 1 & 0 & 0 & 0 \\ 0 & 0 & 1 & 0 & 0 \\ 0 & 0 & 0 & 1 & 0 \\ 0 & 0 & 0 & 0 & 1 \end{bmatrix}, \quad \mathcal{C}_y = \mathsf{im}\begin{bmatrix} 1 & 0 & 0 & 0 & 0 \\ 0 & -1 & 0 & 0 & 0 \\ 0 & 0 & 0 & 0 & 0 \\ 0 & 0 & -1 & 0 & 0 \\ 0 & 0 & 0 & -1 & 0 \\ 0 & 0 & 0 & 0 & 1 \end{bmatrix}, \quad \mathcal{C}_\theta = \mathsf{im}\begin{bmatrix} 1 & 0 & 0 & 0 & 0 \\ 0 & 1 & 0 & 0 & 0 \\ 0 & 0 & 1 & 0 & 0 \\ 0 & 0 & 0 & -1 & 0 \\ 0 & 0 & 0 & 0 & 0 \\ 0 & 0 & 0 & 0 & 1 \end{bmatrix}. \tag{24}$$

According to relation (9), it is straightforward to observe that the following three equations hold $\forall\ \theta$;

$$\mathsf{im}(\mathbf{g}(\theta)) \cap (\mathcal{C}_x \cap \mathcal{C}_y) \neq \mathbf{0}, \tag{25}$$

$$\mathsf{im}(\mathbf{g}(\theta)) \cap (\mathcal{C}_x \cap \mathcal{C}_\theta) \neq \mathbf{0}, \tag{26}$$

$$\mathsf{im}(\mathbf{g}(\theta)) \cap (\mathcal{C}_y) \cap \mathcal{C}_\theta) \neq \mathbf{0}. \tag{27}$$

Field $\mathbf{g}(\theta)$ is a function of θ without singularities if $\theta \neq \pi/2 + k\pi$ with $k \in \mathbb{N}$. For sake of brevity the following field is calculated considering ust $\theta = 0$:

$$(\mathbf{g}(\theta))^\dagger = \begin{bmatrix} 0 & 0 & 0 & a(\theta) & 0 & b(\theta) \\ 0 & c(\theta) & 0 & -a(\theta) & 0 & b(\theta) \\ 0 & -c(\theta) & 0 & -a(\theta) & 0 & b(\theta) \end{bmatrix}, \tag{28}$$

where with $(\mathbf{g}(\theta))^\dagger$ the pseudo inverse of field $\mathbf{g}(\theta)$ is indicated. Functions $a(\theta)$, $b(\theta)$ and $c(\theta)$ are functions of the variable θ with $\theta \approx 0$. The following calculation is obtained for $\theta \neq \pi/2 + k\pi$ with $k \in \mathbb{N}$:

$$\mathcal{C}_x \cap \mathcal{C}_y = \mathsf{im}\begin{bmatrix} 0 & 0 & 0 & 0 \\ 1 & 0 & 0 & 0 \\ 0 & 0 & 0 & 0 \\ 0 & -1 & 0 & 0 \\ 0 & 0 & -1 & 0 \\ 0 & 0 & 0 & 1 \end{bmatrix}, \quad \mathcal{C}_\theta \cap \mathcal{C}_y = \mathsf{im}\begin{bmatrix} 1 & 0 & 0 & 0 \\ 0 & 1 & 0 & 0 \\ 0 & 0 & 0 & 0 \\ 0 & 0 & 1 & 0 \\ 0 & 0 & 0 & 0 \\ 0 & 0 & 0 & 1 \end{bmatrix}, \quad \mathcal{C}_\theta \cap \mathcal{C}x = \mathsf{im}\begin{bmatrix} 0 & 0 & 0 & 0 \\ 1 & 0 & 0 & 0 \\ 0 & 1 & 0 & 0 \\ 0 & 0 & 1 & 0 \\ 0 & 0 & 0 & 0 \\ 0 & 0 & 0 & 1 \end{bmatrix}. \tag{29}$$

The following calculation allow to get the required fields for the decoupling of the mechanical system:

$$\mathbf{T}_\theta(\theta) = (\mathbf{g}(\theta))^\dagger \cdot \mathsf{im}(\mathbf{g}(\theta)) \cap \mathcal{C}_x \cap \mathcal{C}_y, \tag{30}$$

$$\mathbf{T}_x(\theta) = (\mathbf{g}(\theta))^\dagger \cdot \mathsf{im}(\mathbf{g}(\theta)) \cap \mathcal{C}_\theta \cap \mathcal{C}_y, \tag{31}$$

$$\mathbf{T}_y(\theta) = (\mathbf{g}(\theta))^\dagger \cdot \mathsf{im}(\mathbf{g}(\theta)) \cap \mathcal{C}_\theta) \cap \mathcal{C}_x. \tag{32}$$

Adding all 3 T-Fields together we get a new field $\mathbf{T}(\theta)$:

$$\mathbf{T}(\theta) = \mathbf{T}_x(\theta) + \mathbf{T}_y(\theta) + \mathbf{T}_\theta(\theta). \tag{33}$$

Field $\mathbf{T}(\theta)$ can be seen as a preselecting field and the following product realises the mechanical decoupling:

$$\mathcal{B} = \mathsf{im}(\mathbf{g}(\theta)\mathbf{T}(\theta)) = \mathsf{im} \begin{bmatrix} 0 & 0 & 0 \\ 1 & 0 & 0 \\ 0 & 0 & 0 \\ 0 & 1 & 0 \\ 0 & 0 & 0 \\ 0 & 0 & 1 \end{bmatrix}, \tag{34}$$

in which matrix $\mathbf{B}$ can be seen as a resulting input matrix.

4. Solving Three Decoupled Linear Position MPC Optimization Problems for the Mechanical System

If matrix (3) is observed, it is possible to notice that this matrix is a block matrix and thus represents, as already observed, an internally decoupled system. Considering that together with the preselecting matrix $\mathbf{T}_\theta$ above calculated, it is possible to consider a centralized decoupled model predictive controller as follows. Considering $k = nT_s$ and $k+1 = (n+1)T_s$, where T_s represents the sampling time and $n \in \mathbb{N}$, then:

$$u_{mpc}(k) = u_{mpc}(k-1) + \quad u_{mpc}(k), \tag{35}$$

then the following expression can be obtained:

$$\begin{aligned} \mathbf{x}(k+1) &= \mathbf{A}_{(\cdot)_k}\mathbf{x}(k) + \mathbf{B}_{(\cdot)_k}\Big(\quad u_{mpc}(k) + u_{mpc}(k-1)\Big) \\ y(k) &= \mathbf{H}_{(\cdot)_k}\mathbf{x}(k), \end{aligned} \tag{36}$$

where: $\mathbf{A}_{(\cdot)_k}$ represents the discretisation of each block matrix of definition (4) of matrix $\mathbf{A}$, matrix $\mathbf{B}_{(\cdot)_k}$ represents in the same way the discretisation of each block matrix of definition (34) of matrix $\mathbf{B}$, $\mathbf{x}(k)$ is the corresponding discrete part of the state variable and matrix $\mathbf{H}_{(\cdot)_k}$ is the output matrix which determines the position. A general MPC approach with control increments is considered. Using the recursive relation in (35) with a prediction horizon of four the following equation can be obtained:

$$\mathbf{Y}(k) = \mathbf{G}_p\mathbf{x}(k) + \mathbf{F}_{1p}(k)\mathbf{\Delta U}_{mpc}(k) + \mathbf{F}_{2p}u_{mpc}(k-1), \tag{37}$$

where:

$$\mathbf{Y}(k) = \begin{bmatrix} \hat{y}(k+1/k) \\ \hat{y}(k+2/k) \\ \hat{y}(k+3/k) \\ \hat{y}(k+4/k) \end{bmatrix}, \qquad \mathbf{\Delta U}_{mpc}(k) = \begin{bmatrix} u_{mpc}(k) \\ u_{mpc}(k+1) \\ u_{mpc}(k+2) \\ u_{mpc}(k+3) \end{bmatrix} \tag{38}$$

and matrices $\mathbf{G}_p$, $\mathbf{F}_{1p}$ and $\mathbf{F}_{2p}$ is given here below:

$$\mathbf{G}_p = \begin{bmatrix} \mathbf{H}_k\mathbf{A}_{(\cdot)_k} \\ \mathbf{H}_k\mathbf{A}^2_{(\cdot)_k} \\ \mathbf{H}_k\mathbf{A}^3_{(\cdot)_k} \\ \mathbf{H}_k\mathbf{A}^4_{(\cdot)_k} \end{bmatrix}, \tag{39}$$

$$\mathbf{F}_{1p} = \begin{bmatrix} \mathbf{H}_k\mathbf{B}_{(\cdot)_k} & \mathbf{0} & \mathbf{0} & \mathbf{0} \\ \mathbf{H}_k\mathbf{A}_{(\cdot)_k}\mathbf{B}_{(\cdot)_k} & \mathbf{H}_k\mathbf{B}_{(\cdot)_k} & \mathbf{0} & \mathbf{0} \\ \mathbf{H}_k\sum_{i=1}^{2}\left(\mathbf{A}^i_{(\cdot)_k}+\mathbf{I}\right)\mathbf{B}_{(\cdot)_k} & \mathbf{H}_k(\mathbf{A}_{(\cdot)_k}+\mathbf{I})\mathbf{B}_{(\cdot)_k} & \mathbf{H}_k\mathbf{B}_{(\cdot)_k} & \mathbf{0} \\ \mathbf{H}_k\sum_{i=1}^{3}\left(\mathbf{A}^i_{(\cdot)_k}+\mathbf{I}\right)\mathbf{B}_{(\cdot)_k} & \mathbf{H}_k\sum_{i=1}^{2}(\mathbf{A}^i_{(\cdot)_k}+\mathbf{I})\mathbf{B}_{(\cdot)_k} & \mathbf{H}_k(\mathbf{A}_{(\cdot)_k}+\mathbf{I})\mathbf{B}_{(\cdot)_k} & \mathbf{H}_k\mathbf{B}_{(\cdot)_k} \end{bmatrix}. \tag{40}$$

$$\mathbf{F}_{2p} = \begin{bmatrix} \mathbf{H}_k\mathbf{B}_{(\cdot)_k} \\ \mathbf{H}_k(\mathbf{A}_{(\cdot)_k}+\mathbf{I})\mathbf{B}_{(\cdot)_k} \\ \mathbf{H}_k\sum_{i=1}^{2}\left(\mathbf{A}^i_{(\cdot)_k}+\mathbf{I}\right)\mathbf{B}_{(\cdot)_k} \\ \mathbf{H}_k\sum_{i=1}^{3}\left(\mathbf{A}^i_{(\cdot)_k}+\mathbf{I}\right)\mathbf{B}_{(\cdot)_k} \end{bmatrix}, \tag{41}$$

If the following performance criterion is assumed,

$$J = \frac{1}{2}\sum_{j=1}^{N}\Big(y_d(k+j)-\hat{y}(k+j)\Big)^{\mathrm{T}}\mathbf{Q}_p\Big(y_d(k+j)-\hat{y}(k+j)\Big)+ \\ \sum_{j=1}^{N}\Big(\ u_{mpc}(k+j)\Big)^{\mathrm{T}}\cdot\mathbf{R}_p\ \ u_{mpc}(k+j), \tag{42}$$

where $x_d(k+j)$, $j=1,2,\ldots,N$ is the position reference tra ectory and N the prediction horizon, and $\mathbf{Q}_p$ and $\mathbf{R}_p$ are non-negative definite matrices, then the solution minimizing performance index (42) may be then obtained by solving

$$\frac{\partial J}{\partial\ \mathbf{U}_{mpc}} = 0. \tag{43}$$

A direct computation of the vectorial optimal solution may be obtained with

$$\tilde{\mathbf{U}}_{mpc} = (\mathbf{F}^{\mathrm{T}}_{1p}\mathbf{Q}_p\mathbf{F}_{1p}+\mathbf{R}_p)^{-1}\Big(\mathbf{F}^{\mathrm{T}}_{1p}\mathbf{Q}_p\big(\mathbf{Y}_{d_p}(k)-\mathbf{G}_p x(k)-\mathbf{F}_{2p}u_{mpc}(k-1)\big)\Big), \tag{44}$$

where $\mathbf{Y}_{d_p}(k)$ is the desired output column vector. The component at time k of the optimal vector of Eq. (44) is to be considered. Further details concerning this MPC can be found in [12].

5. Simulation Results

Fig. 3 shows the general setup used for simulation and validation. Using a centralized MPC-controller with the geometric decoupling approach acting as a pre-control, different tra ectories can be tracked. As such the MPC accepts tra ectories in horizontal, vertical and angular

direction and attempts to direct the Robotino only in the specified direction. Therefore, it is possible to move the Robotino along the x-axis without a rotation for example.
In the case presented here a circular movement was attempted while the angular position remained constantly zero. Figure 4 shows very accurate movements along the desired tra ectory which is represented in the tested case by a circle. The circle is realized considering a continuous sine and a cosine functions on the x and y movements. All the time the angular position of the robot remained unchanged at zero and this also has been achieved: As it can be observed in Figure 5, the simulated result is so small that it is considered to be a numerical error in the simulation. Additionally it is vital, that the drives are not forced to output more than they can handle, so voltages and currents would have to remain in their respective boundaries. Figure 5 shows that aside from the transients at the start, the nominal current of 0.9 Ampere was not exceeded. Further testings suggest that this could be avoided with better reference tra ectories.

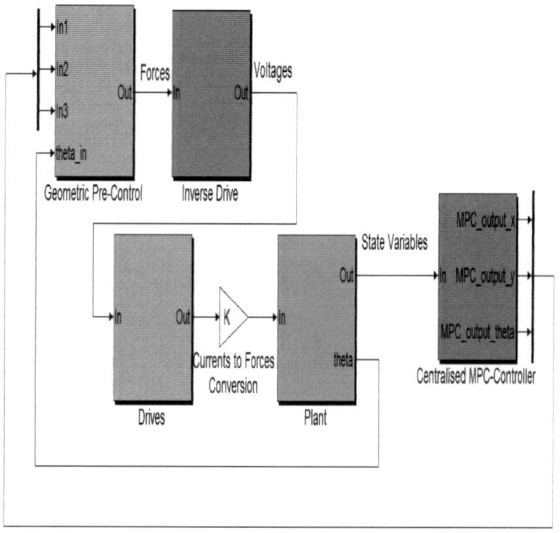

Figure 3. Block diagram of the control system structure

6. Conclusion

This paper presents a procedure in order to obtain the decoupling controllability between horizontal, vertical and angular motions using an geometric approach. A feedback control law is needed together with a feed-forward regulator to achieve a decoupling effect. Simulation results are shown to indicate the potential of this approach.

References

[1] G. Basile and G. Marro. *Controlled and conditioned invariants in linear system theory.* Prentice Hall, New Jersey, 1992.
[2] A. Isidori. *Nonlinear control Systems: an introduction.* Springer, Berlin, 1989.
[3] W.M. Wonham. *Linear multivariable control: a geometric approach.* Springer, New York, 1979.
[4] P. Mercorelli and D. Prattichizzo. A geometric procedure for robust decoupling control of contact forces in robotic manipulation. *Kybernetika*, 39(4):433–455, 2003.
[5] D. Prattichizzo and P. Mercorelli. On some geometric control properties of active suspension systems. *Kybernetika*, 36(5):549–570, 2000.

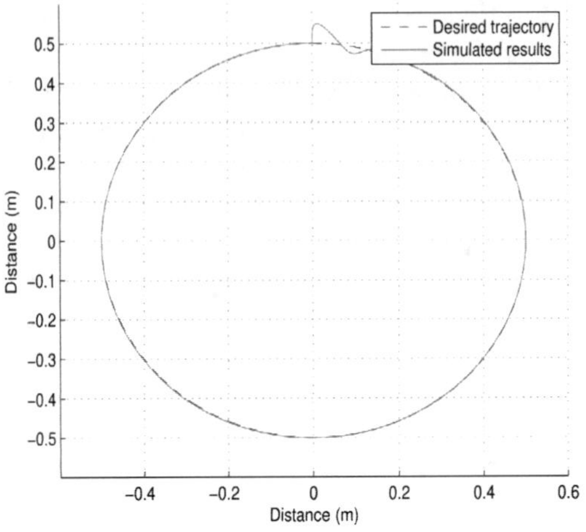

Figure 4. X- -Tra ectory of the Robotino

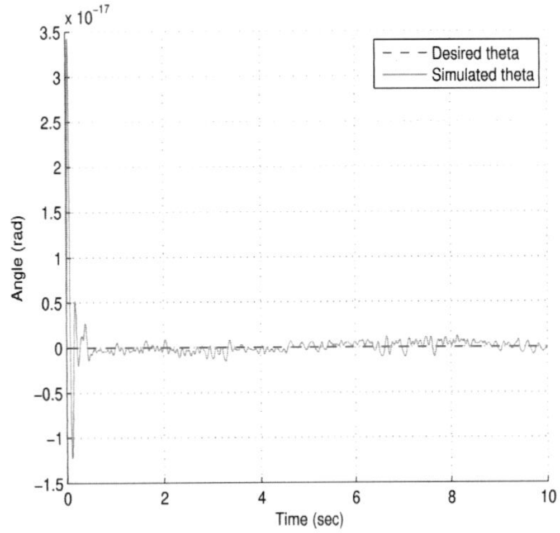

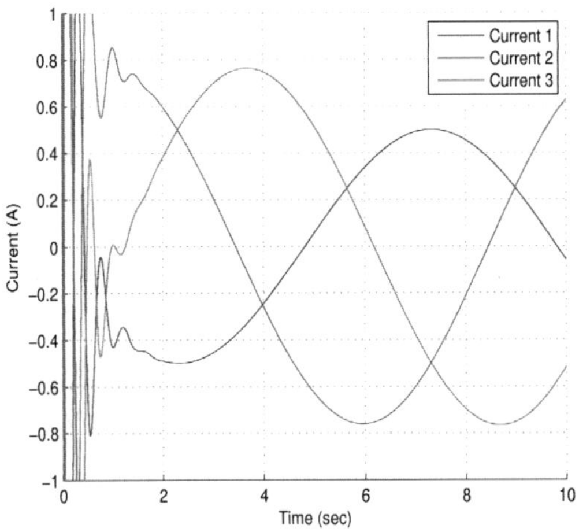

Figure 5. Change in angular position (left). Currents of the three drives acting the three wheels (right)

[6] P. Mercorelli. Invariant subspace for grasping internal forces and non-interacting force-motion control in robotic manipulation. *Kybernetika*, 48(6):12291249, 2012.

[7] P. Mercorelli. Geometric structures for the parameterization of non-interacting dynamics for multi-body mechanisms. *International Journal of Pure and Applied Mathematics-IJPAM*, 59(3):257–273, 2010.

[8] Y. Yamamoto and X. Yun. Effect of the dynamic interaction on coordinated control of mobile manipulators. *IEEE Transactions on Robotics and Automation*, 12(5):816–824, 1996.

[9] G. Basile and G. Marro. A state space approach to non-interacting controls. *Ricerche di Automatica*, 1(1):68–77, 1970.

[10] W.M. Wonham and A.S. Morse. Decoupling and pole assignment in linear multivariable systems: a geometric approach. *SIAM J. Control*, 8(1):1–18, 1970.

[11] A.S. Morse and W.M. Wonham. Decoupling and pole assignment by dynamic compensation. *SIAM J. Control*, (1):317–337, 1970.

[12] H. Sunan, T.K. Kiong, and L.T. Heng. *Applied Predictive Control*. Springer-Verlag London, Printed in Great Britain, 2002.

Towards energy efficient operation of Heating, Ventilation and Air Conditioning systems via advanced supervisory control design

A Oswiecinska[1,2], J Hibbs[2], I Zajic[1] and K J Burnham[1]

[1] Control Theory and Applications Centre, Futures Institute, 10 Innovation Village, Coventry University Technology Park, Cheetah Way, Coventry, CV1 2TL, UK
[2] Jet Environmental Systems Ltd, TS2, Pinewood Business Park, Coleshill Road, Solihull, B37 7HG, UK

E-mail: oswiecia@uni.coventry.ac.uk

Abstract. This paper presents conceptual control solution for reliable and energy efficient operation of heating, ventilation and air conditioning (HVAC) systems used in large volume building applications, e.g. warehouse facilities or exhibition centres. Advanced two-level scalable control solution, designed to extend capabilities of the existing low-level control strategies via remote internet connection, is presented. The high-level, supervisory controller is based on Model Predictive Control (MPC) architecture, which is the *state-of-the-art* for indoor climate control systems. The innovative approach benefits from using passive heating and cooling control strategies for reducing the HVAC system operational costs, while ensuring that required environmental conditions are met.

1. Introduction

Thermal systems are one of the main application domains of modern control engineering with the main focus on thermal comfort and efficient energy utilisation [1]. The research presented in this paper focuses on design of energy efficient control strategies for heating, ventilation and air conditioning (HVAC) systems used in large volume buildings. In this regard, the correct design and operation of the control system is essential to ensure that the HVAC system is achieving good thermal comfort at minimum energy use, operating cost and initial cost, see [2] and [3]. That is, at the design stage the HVAC system equipment should be accommodated to specific control needs together with the consideration for energy efficiency of chosen equipment and overall cost of the solution.

Since the use of HVAC systems is energy intensive and large amounts of energy amplify the peak demand for gas and electrical energy, the effective utilization of HVAC systems is essential. At present, the temperature control for large volume open plan industrial and commercial building structures such as warehouses, hangars and retail park stores, is niche research area, where considerable energy savings could be achieved. The thermal control of these structures is demanding because of their large space, small rate of change in indoor air temperature (compared to e.g. office buildings), and thermal mass within the building may change over the time. Some commercial sectors are imposed with legislative requirements creating a demand for accurate temperature regulation within the building, e.g. storage of pharmaceutical products and food.

In recognition of the fact that the computing capabilities of the existing local, low-level, controllers are limited as well as the fact that these controllers are already designed and installed it is proposed to deploy a supervisory control solution. This control solution is designed to operate remotely via internet connection, while eliminating the need to purchase additional HVAC control equipment. Such a solution can be offered on top of the existing, low-level, industrial standard control solution. The proposed control system makes use of adaptive and predictive control methods to provide optimal energy management without compromising HVAC system performance or indoor air quality. The overall control strategy has been implemented with use of MATLAB computational environment.

1.1. Motivation

Starting from simple heating optimizers, the development gradually progressed resulting in intelligent control systems, which are able to adjust themselves to upcoming changes in indoor conditions and control environment, e.g. outdoor climate, occupation scheme of the building, and control parameters [3]. Advantages resulting from predictive character of control have inspired the researchers to seek adoption of model predictive control (MPC) methods to improve energy performance and minimize operation costs of HVAC systems, see [4,5] for further discussion on MPC methods. Privara et al. [6] emphasize the contrast of MPC to conventional control strategies and its importance in building climate control, while Afram and Janabi-Sharifi [7] scrutinize theory and applications of MPC to HVAC systems and reflect on factors affecting performance of this technique.

The idea of optimal control was captured, among others, by Morris, Braun and Treado [8] which demonstrated use of thermal storage potential of building mass to reduce cooling costs. This study considered two optimal control strategies: minimum total energy cost and minimum peak electrical demand, showing significant energy reduction achieved by both control methods. The obtained results have been validated on experimental test facility showing that up to 51% of total cooling load can be shifted to electricity off-peak time. The peak cooling load was reduced by 15% using energy minimization control, and by 40% using peak minimization control, when compared to conventional night setback control.

2. Plant details

A typical setup of large volume building and HVAC system with heating and cooling capabilities is depicted in figure 1. In this case, the setup is based on an existing pharmaceutical warehouse in Midlands, UK, which resembles the facility in size and thermal response. The overall volume of conditioned space is 185,472m^3. In figure 1, the overall HVAC system comprises exterior air handling unit (AHU) with inlet air mixing damper, control unit and necessary ductwork.

To describe the functionality of the HVAC system, consider the air mixing damper shown in figure 1. The position of damper blades is regulated by the control unit, taking one of two possible states: closed, which is full recirculation of the return air from the controller area, and open, where only fresh air intake is assumed. The air leaving the mixing damper section passes through the AHU comprising a cooling unit, heating unit and main supply fan. The operation of the components within the AHU is regulated by the control unit, which continuously monitors the indoor and outdoor air temperatures. Subsequently, the controlled area is supplied with conditioned air, from the AHU, which is distributed evenly throughout the area via ductwork. By default, the air mixing damper is closed, i.e. full recirculation is assumed. Note that if the air mixing damper is closed, i.e. full recirculation is assumed, there is enough fresh air within the building due to air leakage (infiltration).

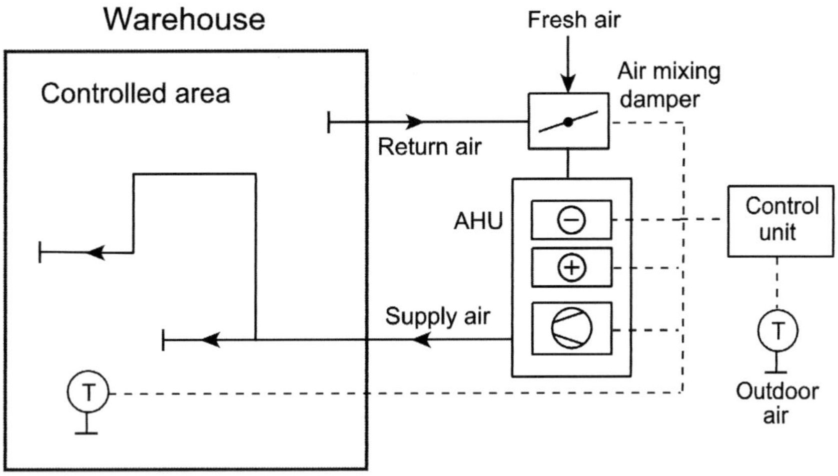

Figure 1. Air conditioning system of the pharmaceutical warehouse.

2.1. *Demanded environmental conditions*

The required indoor air temperature conditions are determined based on personnel comfort requirements, product storage requirements, and legislative requirements. Figure 2 shows an example of prescribed indoor air temperature set-point values of the warehouse facility. The space temperature is actively regulated to be within the control dead-band, denoted *db*. The control dead-band is defined by a cooling set-point from above, denoted r_c, and by a heating set-point from below, denoted r_h. In the case the indoor air temperature rises above the cooling set-point, i.e. is within the cooling band, the cooling unit is enabled. Similarly, if the air temperature decreases below the heating set-point, i.e. is within the heating band, the heating unit is enabled. Note, that no cooling or heating takes place within the control dead-band.

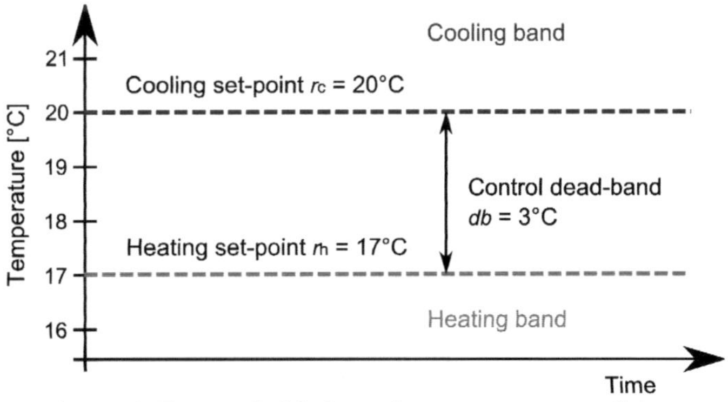

Figure 2. Demanded indoor air temperature conditions.

3. Control system structure

The proposed advanced control system has a two-level topology comprising low-level and high-level control unit. The concept of two-level control topology and its operation is depicted in figure 3. The low-level (or slave) controller represents the traditional standalone solution for control of HVAC systems, which is commonly located within the air conditioned building. The controller uses the local measurements of indoor and outdoor air temperatures, denoted *temperature* in figure 3, to compute the *control input* for actuators of the AHU. The control strategy of the low-level controller is designed to meet the demanded indoor air temperature conditions and contains basic, sub-optimal, strategy for energy efficient operation of the HVAC system.

Recognizing the fact that the computing capabilities of the local, low-level, controllers are limited as well as the fact that these controllers are already designed and installed on site it is proposed to deploy a supervisory control solution. This solution is designed to expand and build on the capabilities of the existing low-level control units via remote internet connection, while not compromising their safe operation. The supervisory controller is located in a remote command and control centre hosting the server room. The control solution, or service, is deployed via internet connection, where the controllers establish a communication link between each other in order to share information and achieve optimal, energy efficient, HVAC system operation. It should be emphasised, that the local control units work as a standalone HVAC system control solution, and hence if the remote connection is broken, the local controller continues to work as per industrial standard.

The high-level (or supervisory) controller is based on model predictive control architecture, which is in greater detail described in subsequent section 4. In general, the supervisory controller constantly monitors the performance of the existing low-level control units together with local environmental data, depicted as *warehouse env. data* in figure 3. Moreover, the controller is supplied with weather forecast data, current and anticipated energy prices, and anticipated heating and cooling loads. The supervisory controller uses a mathematical thermal model of the building together with the energy optimization algorithm to compute *optimal control settings* for the low-level controller. This process is continuously repeated in an adaptive, self-learning manner, to ensure continuous energy efficient HVAC system operation throughout its lifespan.

3.1. Low-level control design methodology

The control strategy for the low-level controller is designed and implemented using robust software design methods. The designed control strategy is validated by model-in-the-loop simulation techniques to achieve a rapid, low cost and, more importantly, reliable control software solution. For this purpose, the high-level control requirements are cascaded down to functional requirements, which then formulate the designed control strategy. Such control strategy is implemented in Matlab/Simulink programming environment and is tested, via simulation, against the functional requirements – this constitutes the model-in-the-loop testing and development methodology. Currently a prototype solution is under development and real-time trials are planned. The designed and validated low-level control strategy is implemented on the industrial programmable logic controller (PLC) of Trend Control Systems Ltd. The PLC belongs to a so called IQ line of products with enabled Ethernet connectivity, which enables the use of the proposed supervisory controller.

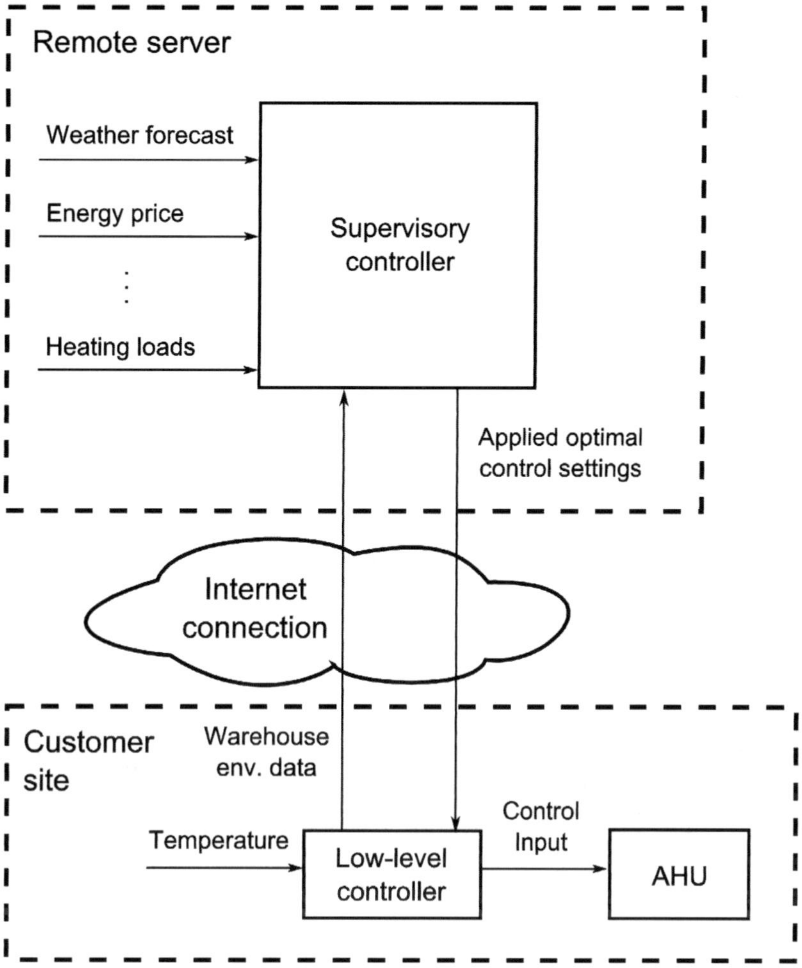

Figure 3. Proposed advanced control system architecture.

4. Supervisory control system

The supervisory controller is based on model predictive control architecture, which is the *state-of-the-art* for indoor climate control systems. The architecture is depicted in figure 4 and composes of three essential components, which are a mathematical model of the building (including AHU and low-level control logic), cost function and optimization algorithm. First, consider the mathematical thermal model of the building, which is used for computer simulation.

1) *Mathematical model*: Having the model it is possible to simulate and subsequently predict the indoor air temperature of the building, which is then utilized by the cost function. To obtain the prediction of the indoor air temperature the latest measurement of building environmental data (indoor and outdoor air temperature) is required as well as the prediction of future heating effects acting on the building. The heating effects are mainly defined by the outdoor weather conditions such as ambient air temperature, solar irradiation and wind speed. For this purpose the weather forecast data are required. Other possible cause of future heating load is the personnel and machinery activity within the building. The prediction of the indoor air temperature is obtained over a so called prediction horizon, denoted h_p.

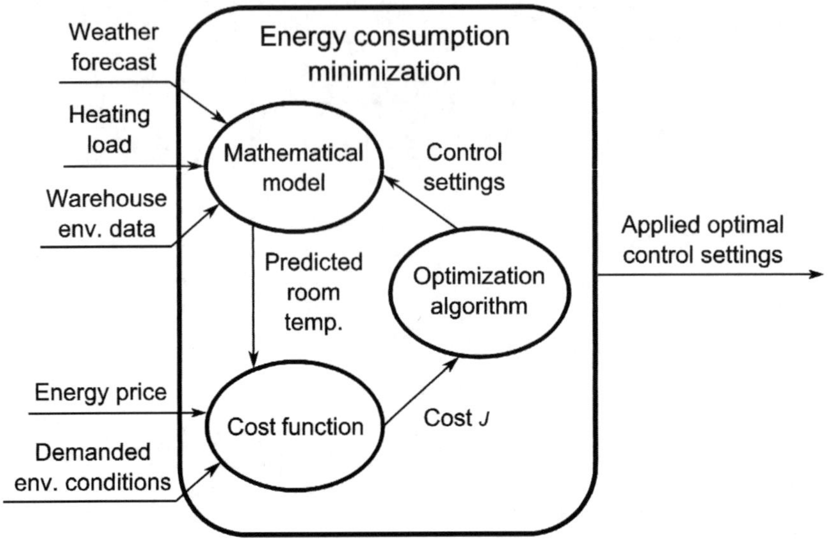

Figure 4. Model predictive control architecture of the supervisory controller designed for minimal energy utilization. The computed optimal control settings are then cascaded down and applied to the low-level, local, controller.

The indoor air temperature prediction is realized for a given low-level control setting (whose value is being optimized), i.e. specific configuration of the HVAC system such as cooling unit enabled or main fan disabled. Note, that different low-level control setting would cause indoor air temperature prediction to change despite using the same forecast of heating loads. The mathematical model representing the thermal response of the building to heating and cooling loads can be obtained through data analysis (black box modelling approach) or with the use of physical laws (white box modelling approach). The model derived using first principles is briefly introduced in appendix, see [9] for full reference.

Second, consider the cost function component in figure 4, which takes as an input the indoor air temperature prediction.

2) *Cost function*: The operating cost of the HVAC system (and with that the energy consumed) is minimized by means of minimizing the cost function, denoted J, defined in equation (1). This means that the cost function is defined such that its value directly relates to the energy consumed to achieve a particular prediction of indoor air temperature. The cost function arguments u_c and u_h are vectors comprising predicted cooling and heating loads over the prediction horizon h_p, respectively. The arguments u_c and u_h are weighted by respective future energy prices per unit of energy, p_c and p_h.

$$J = \sum_{t=1}^{h_p} p_c(t) \cdot u_c(t) + \sum_{t=1}^{h_p} p_h(t) \cdot u_h(t) \tag{1}$$

The achieved cost is a sum of products of HVAC unit activity and the price of the energy spent to run the unit over a selected prediction horizon.

Subsequently, the calculated cost, to achieve a given indoor air temperature over a given prediction horizon, is passed to the optimization algorithm. This is considered in the next third step.

3) *Optimization algorithm*: The optimization algorithm uses the latest computed cost J to compute a new set of low-level control settings, which will be used by the mathematical model in next iteration of the optimization procedure. The optimization algorithm seeks such a value of low-level control settings, which will cause the cost J to reduce. Note, that optimization methods are discussed in more detail in next subsection.

The steps 1), 2) and 3) are repeated until the optimization routine converges, i.e. a minimal value of the cost function J is achieved. Subsequently, the found optimal control setting is cascaded down and applied to a real low-level controller via the remote internet connection. It is required that the values of optimal low-level control setting are found within one sampling time instance, denoted Δt. This is the re-scheduling time of the controller at which the newly computed control inputs are applied to the AHU. Finally, the whole optimization routine repeats in the next sampling time interval, i.e. prediction and optimization steps until convergence is achieved. This gives an alternative name to MPC, namely receding horizon control.

Figure 5 shows an illustrative example of one realization of indoor air temperature prediction as part of one optimization step. The figure shows the past, measured, indoor and outdoor air temperatures, cooling and heating set-point values as well as the forecasted outdoor air temperature and predicted, corresponding, indoor air temperature. In this example, it is assumed that only the position of mixing damper blades needs to be optimized over the prediction horizon, where only open (fresh air supply only) or closed (full air recirculation) damper blades positions are considered. For this realization the optimization routine computes a particular combination of open/closed mixing damper blades positions over the prediction horizon $h_p = 7$, which is depicted by solid and empty circles in figure 5. After such a combination of mixing damper blades position is found, which results in the lowest value of the cost function J, then only the first damper blades position is actually applied to the low-level controller, while the rest is discarded. Subsequently, the whole optimization cycle repeats.

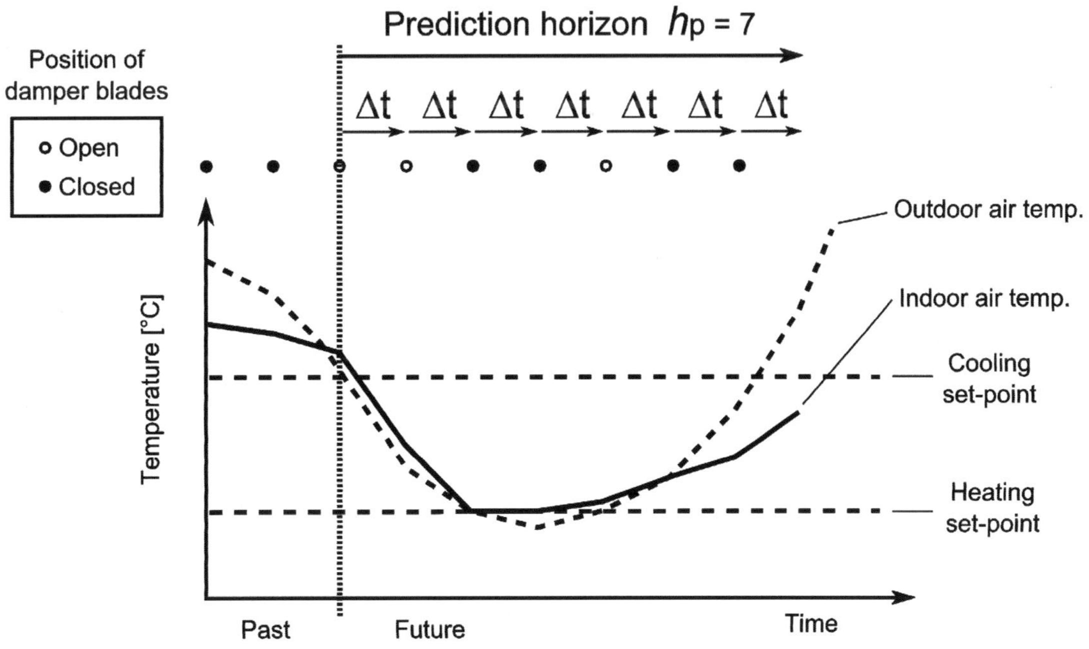

Figure 5. Conceptual Model Predictive Control of damper blades position. The predictive control strategy is either opening or closing the air mixing damper to force the indoor air temperature to remain in between cooling and heating set-point for maximum energy efficiency, i.e. no active heating cooling take place. Sampling time interval $\boldsymbol{\Delta t}$ is 15 min.

4.1. Optimization methods

The optimization problem is solved using a class of heuristic optimisation techniques. The solver proposed for this purpose is Genetic Algorithm (GA), which belongs to a group of evolutionary algorithms. This nature inspired method mirrors the process of natural selection by reproducing in each generation the members carrying the best solutions to create next offspring and rejecting the worst ones. New generations are created combining two methods, which are crossover and mutation method respectively. In the first method, being a fundamental way of procreation, each member of the new offspring inherits some part of genetic information from each of the two parents (solutions). The second method is used to introduce random changes for two reasons: first, this is important to avoid local optimal solution, second, the changes are made in hope that a desirable solution is introduced [10].

4.2. Weather forecasting

The performance of model predictive methods fundamentally rely on the knowledge of future inputs (heating gains), where the outdoor air temperature is of a high importance. The established air temperature forecasting algorithms are based on Kalman Filtering techniques using so called random walk models. These algorithms use the latest weather forecast and local measurements of the ambient air temperature of the building. The forecasted outdoor air temperature is extrapolated in a stochastic framework based on recorded past and present air temperature data and analysis of forecasted trends. The optimal predictive control algorithm can be improved by incorporating other measured variables into predictions, e.g. relative air humidity and solar irradiance, to achieve the highest rate of energy savings.

4.3. Exemplary case study

For the purpose of this exemplary case study a simplified mathematical thermal model of the facility described in section 2 has been derived using first principles (see appendix). For simplicity, it is assumed that the indoor space temperature is uniformly distributed implying perfect air mixing. Furthermore a perfect control is assumed such that whenever the indoor air temperature would rise above the cooling set-point or below heating set-point the air temperature will remain at these values, i.e. instantaneous cooling or heating response of the AHU is assumed. The indoor climate requirements hold as stated in section 2.1, where $r_c = 20°\text{C}$ and $r_h = 17°\text{C}$. The supervisory controller is designed to select optimal, energy efficient, position of mixing damper blades, where only open (fresh air supply only) or closed (full air recirculation) damper positions are considered.

The simulation results are shown in figure 6. The outdoor air temperature, depicted by a solid black line in the lower plot, is designed to be a sinusoidal signal being the main input to the thermal building model. Note, that it is also the case in practice that the outdoor air temperature naturally follows the night and day cycle. The upper plot shows the simulated indoor air temperature response subjected to the mixing damper blades position. In general, the indoor air temperature tends to follow the same trend as the ambient air temperature, but is limited from above by the cooling unit regulating the room temperature to remain at the cooling set-point value.

Considering the case of closed mixing damper (full air recalculation) the room temperature response, depicted by black solid line in figure 6, is very mild and prolonged. In comparison, the room temperature response obtained by having the mixing damper fully open (fresh/outdoor air intake only) follows the outdoor air temperature trend very closely and drops nearly as low as the ambient air temperature does, while rising sharply back. The room temperature response obtained using the optimal, energy efficient, combination of mixing damper blades positions is depicted by dashed dotted black line. The corresponding optimal damper blades position is depicted by grey solid line in the lower plot.

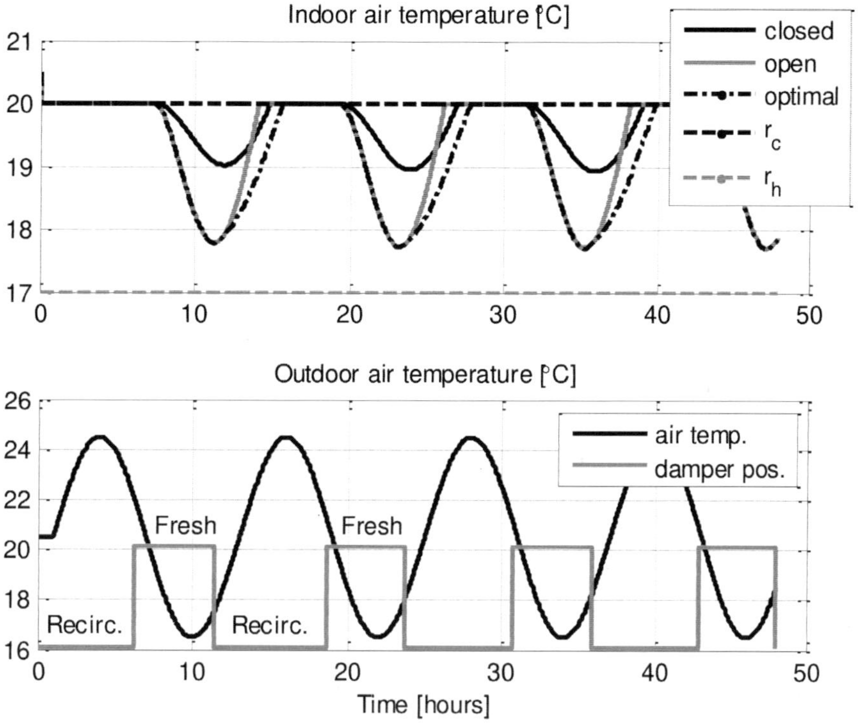

Figure 6. Simulation results of the simplified case study. The lower plot shows the selected outdoor air temperature, solid black line, and mixing damper position, solid grey line, where low state implies that the blades are closed, while the high state implies that blades are open. The upper plot shows the cooling and heating set-point values depicted by dashed black and grey lines, respectively. The black solid line shows the simulated room temperature for closed mixing damper (full recirculation), while the grey solid line shows the room temperature response for open mixing damper (fresh air supply only). The dashed dotted black line shows the simulated room temperature for the optimal combination of mixing damper blades positions depicted in the lower plot by solid grey line.

It can be observed that the room temperature response obtained by the optimal control strategy remains within the control dead-band the longest. This implies that minimal cooling is required, which, in return, reduces the HVAC system operating costs. The supervisory controller naturally selected such a strategy, which opens the mixing damper to cool down the room air temperature when ambient air temperature falls and closes the mixing damper when the ambient temperature rises. This strategy utilizes the natural cycling behavior of ambient air temperature to pre-cool the conditioned space when possible and prolong the periods where no cooling is required. In other words, the thermal storage capacity of the building is used to shift the heating loads for maximum energy efficiency.

Finally, assuming that the time for which cooling has been required directly relates to the energy consumed, then the energy savings can be numerically evaluated. The optimal control strategy reduced the cooling unit operating time by 10% when compared to the case of fully closed mixing damper. The cooling unit operating time has been reduced even more significantly, by 30%, when compared to the case of fully open mixing damper.

5. Case study: Pharmaceutical warehouse facility

In this case study, the potential for operating cost reduction of HVAC system installed in the existing pharmaceutical warehouse located in Midlands, UK, has been investigated. The results and detail description of the presented case study were originally presented in [9], while the general plant setup is introduced in section 2. The computer based simulation study has been chosen to compare the performance of the existing low-level controller with the proposed supervisory control strategy so that the HVAC system operation is not compromised. The thermal model of the building is similar in structure to the one used in subsection 4.3 and has been presented in detail in [9]. However, with the difference, that the model parameters have been correlated with measured local air temperature data collected in 2014. Additionally, the real measurements of ambient air temperature have been used during the simulation. This has resulted in more realistic case study, allowing for rapid changes in the ambient air temperature, which challenges the proposed control solution.

The supervisory controller has been designed to optimise the position of air mixing damper blades and to select energy efficient control set-point values. Adaptive set-point adjustment aims for temporary extension of the control dead-band envelope by reducing the heating set-point value down to $r_h = 16°C$ and increasing the value of cooling set-point up to $r_c = 22°C$, while not compromising the safe HVAC system operation. The maximization of the control dead-band envelope results in reduced heating and cooling demand. In order to see the full potential of proposed control solution the true value of future, forecasted, ambient air temperature has been made available to the MPC.

5.1. *Results & discussion*

The intelligent supervisory control system has been compared with the existing standard control system, which does not make use of predictive features and weather forecast data. The simulation study has shown that energy savings up to 7% in January, 57% in April and 76% in June can be theoretically achieved. Choosing representative data for cool, moderate and hot seasons, the obtained results indicate that energy reduction is possible at all times of the year. Nevertheless, visible differences in the level of energy savings are expected throughout the year as the typical seasonal temperatures vary. While almost continuous heating is required in a winter season, leaving little scope for reducing operational costs, the situation in summer season is more favourable and significant energy savings are possible.

6. Conclusions and further work

The conceptual control solution, which extends the capabilities of the existing low-level control strategies via remote internet connection, has been presented. A two-level scalable control topology has been proposed, where the high-level, supervisory controller, is based on Model Predictive Control (MPC) architecture feeding into existing low-level controller. The control design is targeted to work with established control solutions, e.g. provided by Trend Control Systems. The development of the supervisory controller goes alongside with the development of the control logic for the low-level controller. It is required that the control logic is implemented in a robust and reliable way and that this control logic is mated, by design, with the supervisory controller.

Results of simulation based case study have been presented. These results show potential for energy savings up to 7% in January, 57% in April and 76% in June when compared to equivalent simulation study where only existing, low-level, controller has been used.

In this work we are specifically targeting large volume buildings such as warehouse facilities, retail park stores or aircraft hangars. The control challenges include slow system dynamical response, air temperature stratification effects, large impact of solar irradiation and wind speed. Nevertheless, the applicability of the proposed control solutions is not limited to large volume buildings.

6.1. *Next steps*

The current work is focused on development and implementation of low-level controls and arrangement of the supervisory controller for its use with low-level controls. Then, the real-time field studies on the building facility will follow to investigate the performance of the HVAC system and the actual energy saving capabilities of proposed concepts.

Acknowledgements

The project was partially funded by the European Regional Development Fund under the Knowledge Exchange and Enterprise Network (KEEN). The research was a collaborative work between Jet Environmental Systems Ltd. and Coventry University.

Appendix

The simple air temperature model is based on first principles modelling methodology, see [3,11], where, in particular, a lumped parameter modelling approach has been considered. A schematic diagram of the room is shown in figure 7, where uniform air temperature distribution is assumed. It is also assumed that the air leaving the conditioned space, denoted $T_r(t)$ [K], is equal to the mean temperature of the whole room. The time index (t) is chosen to emphasize that the signals are not measured, but generated by physical based model. The final equations representing the physical system are reduced into a set of ordinary differential equations with a finite number of parameters.

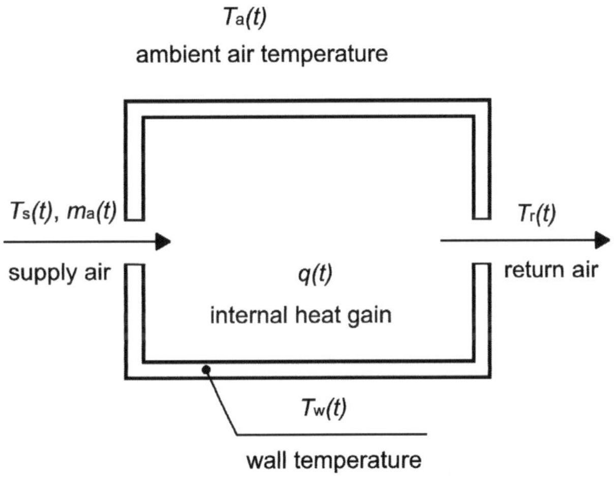

Figure 7. Schematic diagram of the room temperature model.

The continuous-time first principles room temperature dynamic model is based on energy balance equations, where two states are considered: the room air temperature, denoted $T_r(t)$, and the wall temperature, denoted $T_w(t)$. The energy balance equations for the temperature of the air within the conditioned space and adjacent walls, respectively, are

$$C_{\mathrm{r}}\frac{dT_{\mathrm{r}}(t)}{dt} = \mathrm{m_a}(t)\rho_{\mathrm{a}}c_{\mathrm{a}}[T_{\mathrm{s}}(t) - T_{\mathrm{r}}(t)] - (UA)_{\mathrm{int}}[T_{\mathrm{r}}(t) - T_{\mathrm{w}}(t)] + q(t) - n_{\mathrm{v}}\frac{\rho_{\mathrm{a}}V_{\mathrm{r}}c_{\mathrm{a}}}{3600}[T_{\mathrm{r}}(t) - T_{\mathrm{a}}(t)] \quad \text{(A.1a)}$$

$$C_{\mathrm{w}}\frac{dT_{\mathrm{w}}(t)}{dt} = (UA)_{\mathrm{int}}[T_{\mathrm{r}}(t) - T_{\mathrm{w}}(t)] - (UA)_{\mathrm{ext}}[T_{\mathrm{w}}(t) - T_{\mathrm{a}}(t)] \quad \text{(A.1b)}$$

where C_{r} [J/K] denotes the thermal air capacity of the room and C_{w} [J/K] the thermal capacity of the wall, c_{a} [J/(kg·K)] denotes the air specific heat capacity, $m_{\mathrm{a}}(t)$ [m^3/s] denotes the air volumetric-flow rate, U_{int} and U_{ext} [W/(m²·K)] are the heat transfer coefficients on the inner side and outer side of walls, respectively, A [m²] denotes the effective surface area of the walls, $T_{\mathrm{w}}(t)$ [K] denotes the mean wall temperature, $T_{\mathrm{a}}(t)$ [K] denotes the ambient (outdoor) temperature, $q(t)$ [W] denotes the heat load disturbance within the room and $T_{\mathrm{s}}(t)$ [K] denotes the supply air temperature. The thermal capacities C_{r} and C_{w} are calculated as follows:

$$C_{\mathrm{r}} = \rho_{\mathrm{a}} V_{\mathrm{r}} c_{\mathrm{a}} \tag{A.2a}$$

$$C_{\mathrm{w}} = \rho_{\mathrm{w}} V_{\mathrm{w}} c_{\mathrm{w}} \tag{A.2b}$$

where ρ [kg/m³] denotes the density of air, ρ_{a}, and wall material, ρ_{w}, c [J/(kg·K)] denotes the specific heat capacity of air, c_{a}, and wall material, c_{w}, V_{r} [m³] denotes the volume of the air within the room and V_{w} [m³] denotes the volume of the walls.

The left-hand side terms of equations (A.1a) and (A.1b) are regarded as the rate of change of energy in the room and the rate of change of energy within the walls, respectively. The rate of change must be in balance with the energy transfers placed on the right-hand side of the equations, as stated by the first law of thermodynamics. Analysing the first energy balance equation (A.1a), the first term represents the energy received from the HVAC system, the second term represents the transfer of internal energy into the wall, the third term introduces the heat loads, and the last term represents air infiltration. The ventilation air change rate is denoted n_{v} [1/h]. Note, that regulation of the position of damper blades will assume the supply temperature $T_{\mathrm{s}}(t) = T_{\mathrm{a}}(t)$ for damper open and $T_{\mathrm{s}}(t) = T_{\mathrm{r}}(t)$ for damper closed. Subsequently, the rate of change of energy in the walls is balanced with the heat transfer on the inner and outer side of walls.

References

[1] Ogata K. Modern Control Engineering. 5th ed. New Jersey: Prentice Hall; 2010.

[2] Chartered Institution of Building Services Engineers. CIBSE Guide H: Building Control Systems Oxford: Butterworth-Heinemann; 2000.

[3] Underwood CP. HVAC Control Systems: Modelling, analysing and design London, UK: E&FN SPON; 1999.

[4] Camacho EF, Bordons C. Model Predictive Control. 2nd ed. London, UK: Springer-Verlag; 1999.

[5] Wang L. Model Predictive Control System Design and Implementation Using MATLAB® London, UK: Springer-Verlang; 2009.

[6] Privara S, Vana Z, Cigler J, Oldewurtel F, Komarek J. Role of MPC in Building Climate Control. Computer Aided Chemical Engineering. 2011; 29: p. 728-732.

[7] Afram A, Janabi-Sharifi F. Theory and applications of HVAC control systems – A review of model predictive control (MPC). Building and Environment. 2014; 72: p. 343-355.

[8] Morris FB, Braun JE, Treado SJ. Experimental and Simulated Performance of Optimal Control of Building Thermal Storage. ASHRAE Transactions. 1994; 100(1): p. 402-414.

[9] Oswiecinska A. Energy efficient control strategies for HVAC systems. MSc Thesis. Wroclaw, Poland: Wroclaw University of Technology; 2014.

[10] Janiak A, Lichtenstein M. Advanced algorithms in combinatorial optimization Lodz, Poland: PRINTPAP; 2011.

[11] Cengel YA, Boles MA. Thermodynamics: An Engineering Approach. 5th ed. New York, NY, USA: McGraw Hill; 2006.

A Novel Control algorithm based DSTATCOM for Load Compensation

Sreejith R[1], Naran M. Pindoriya[1] and Babji Srinivasan[1]
[1] Department of Electrical Engineering, Indian Institute of Technology Gandhinagar, Ahmedabad, India -382424

Email: sreejith_r@mtech2012.iitgn.ac.in, naran@iitgn.ac.in, babji.srinivasan@iitgn.ac.in

Abstract. Distribution Static Compensator (DSTATCOM) has been used as a custom power device for voltage regulation and load compensation in the distribution system. Controlling the switching angle has been the biggest challenge in DSTATCOM. Till date, Proportional Integral (PI) controller is widely used in practice for load compensation due to its simplicity and ability. However, PI Controller fails to perform satisfactorily under parameters variations, nonlinearities, etc. making it very challenging to arrive at best/optimal tuning values for different operating conditions. Fuzzy logic and neural network based controllers require extensive training and perform better under limited perturbations. Model predictive control (MPC) is a powerful control strategy, used in the petrochemical industry and its application has been spread to different fields. MPC can handle various constraints, incorporate system nonlinearities and utilizes the multivariate/univariate model information to provide an optimal control strategy. Though it finds its application extensively in chemical engineering, its utility in power systems is limited due to the high computational effort which is incompatible with the high sampling frequency in these systems. In this paper, we propose a DSTATCOM based on Finite Control Set Model Predictive Control (FCS-MPC) with Instantaneous Symmetrical Component Theory (ISCT) based reference current extraction is proposed for load compensation and Unity Power Factor (UPF) action in current control mode. The proposed controller performance is evaluated for a 3 phase, 3 wire, 415 V, 50 Hz distribution system in MATLAB Simulink which demonstrates its applicability in real life situations.

1. Introduction

Nonlinear loads are increasing day by day with more and more use of power electronics equipments. Due to this, power quality is a major issue arisen in the power distribution system. Custom Power Devices (CPD) have been effectively used to mitigate the power quality issues such as voltage sags, swells, surges, poor power factor, unbalanced load currents and neutral current unbalance. In this regard, DSTATCOM is the most preferred option for load compensation. Load compensation involves management of reactive power for voltage regulation, power factor correction and load balancing.

Controlling the switching angle has been the biggest challenge in DSTATCOM. There are many control techniques adopted for DSTATCOM such as PI Controller and its variants [2], Fuzzy logic based Controller [3], Neural Network [4], Model free predictive Controller [5], Particle Swarm Optimization based control techniques [6] and a combination of these techniques [7,8] etc. Till date, PI controller is widely used in practice for load compensation due to its simplicity and ability. But since PI controller uses trial and error based tuning and Ziegler Nichols based tuning and is well suited for systems with large delays and large time constants, it is difficult to achieve efficient load compensation.

PI Controller also fails to perform satisfactorily under parameters variations, nonlinearities, etc. making it very challenging to arrive at best/optimal tuning values for different operating conditions [9]. Heuristic Fuzzy logic controllers rely on the expert knowledge database with its performance depending on the configuration used [10]. Neural network based techniques are good for tolerating only small perturbations; however they require extensive training (in case of supervised learning) and have been shown to perform poor when the range of the inputs deviate from those used in training conditions [11]. The hysteresis controller has several drawbacks such as variable switching frequency, tendency towards limit cycles, and generation of sub-harmonics etc. [12]. G. W. Moon [13] have introduced predictive current control of DSTATCOM, that uses the system model to predict the output voltages and current based on space vector PWM and uses coordinate transformation. But it does not incorporate any optimality framework.

Model predictive control is a powerful control strategy, used in the petrochemical industry and its application has been spread to different fields, especially hybrid systems [17, 18]. The main advantage of MPC over linear classical controllers is that it can handle various constraints, incorporate system nonlinearities and can be applied to Multiple Input Multiple Output (MIMO) systems [20]. MPC as such cannot be used in power systems, power electronics and drives applications since there is a huge amount of computational effort in order to solve the optimization problem online, which is incompatible with the small sampling times used in the converter control.

However, due to advent of fast processors like DSPs and FPGA, the problem of computational effort can be resolved and power electronics industry can look upon MPC as a good candidate. This work proposes a novel control algorithm, namely FCS-MPC for DSTATCOM application with appropriate modification in the algorithm. It incorporates the model information and has optimization criteria, which can help in optimization of the switching pulses, thus achieving an effective and excellent control of DSTATCOM. Easy implementation, control objective definition flexibility and superior transient performance are the key features of this method [14]. Even though variable switching frequency is considered as one of the drawbacks of this method, the average switching frequency is between 1/5th and 1/4th of the sampling frequency [14]. In this work, no co-ordinate transformation has been used unlike the conventional FCS-MPC used in the literature so far.

The paper is organised as follows. In Section 2, DSTATCOM Control strategy is described, followed by the formulation of the proposed control technique, namely, FCS-MPC for the compensator in Section 3. In Section 4, the simulation results and discussion on the performance of the DSTATCOM are presented. Here, the proposed current control technique is validated in SIMULINK. In Section 5, conclusions are presented.

2. DSTATCOM control strategy

DSTATCOM is operated in current control mode. The control system of DSTATCOM has to generate reference currents that compensate the harmonic, unbalanced and fundamental reactive components of non-linear and unbalanced load supply currents. The aim of current controller is to make the phase currents exactly follow the desired current reference with a minimum ripple current and phase delay. There are various reference current extraction methods available in the literature [16], in which Instantaneous Symmetrical Component theory (ISCT) based method is one of the most effective method compared to other methods.

2.1 Instantaneous Symmetrical Component Theory

ISCT based method is based on the theory of symmetrical components. This is mainly adopted due to the simplicity in formulation and the clarity in the definition of powers unlike the instantaneous p-q theory.

The main features of this algorithm are:

- It ensures that the source supplies only the load active power and the losses in the inverter.
- Source power factor can be set to any desired value.
- It can practically compensate any kind of unbalance and harmonics in the load, provided we have a high bandwidth current source to track the DSTATCOM current.

Here, the compensation reference current, i_f [15] is

$$i_{fa}^{*} = i_{La} - i_{sa}^{*} = i_{La} - \frac{v_{sa} + \beta(v_{sb} - v_{sc})}{\Delta s}(p_{lav} + p_{loss}) \quad (1)$$

$$i_{fb}^{*} = i_{Lb} - i_{sb}^{*} = i_{Lb} - \frac{v_{sb} + \beta(v_{sc} - v_{sa})}{\Delta s}(p_{lav} + p_{loss}) \quad (2)$$

$$i_{fc}^{*} = i_{Lc} - i_{sc}^{*} = i_{Lc} - \frac{v_{sc} + \beta(v_{sa} - v_{sb})}{\Delta s}(p_{lav} + p_{loss}) \quad (3)$$

where $\Delta s = v_{sa}^2 + v_{sb}^2 + v_{sc}^2$, $\beta = \frac{\tan\phi}{\sqrt{3}}$, p_{lav} is the average load power and p_{loss} is the losses in the inverter . Here, ϕ is the desired power factor angle.

It can be seen from the above equations that, for unity power factor action, $\beta = 0$. For voltage regulation, β is nonzero and source supplies or absorbs the reactive power corresponding to the β value. p_{lav} is calculated from the load power using the moving average filter. Here, p_{loss} is estimated by the PI Controller from the error difference between the actual DC bus voltage and the reference DC link voltage. The PI Controller loop maintains the dc link voltage value at reference value by drawing necessary active power from the ac source. For simulation, the positive sequence components of the PCC Voltage are extracted so that the algorithm is valid even for distorted and unbalanced voltage sources.

2.2 Finite Control Set MPC

Recently, Rodriguez et al. [14] came up with a novel method termed as FCS-MPC for power electronic drives application. The FCS-MPC control strategy is based on the fact that only a finite number of possible switching states can be generated by a static power converter and that models of the system can be used to predict the behavior of the variables for each switching state. This leads to drastic simplification of the optimization problem in MPC.

For the selection of the appropriate switching state to be applied, a selection criterion must be defined. This criterion consists of a cost function that will be evaluated for the predicted values of the variables to be controlled. Prediction of the future value of these variables is calculated for each possible switching state and then the state that minimizes the cost function is selected.

The procedure of switching state selection has been shown in Figure 1. Here, x(k) are the controlled variables. Based on the discrete model of system, the current values of the controlled variables, x(k) are used to predict their future values x(k+1) for all N possible switching states. The predicted values of all the controlled variables x(k + 1) are compared with their reference values x_{ref} (k + 1) in the cost function minimization block. Finally, the switching state (S) that minimizes the cost function will be selected as the next switching state and it will be applied to the converter. t = k is presenting the current state (now), t = k + 1 and t = k + 2 are the next time steps. The sampling time is T_s.

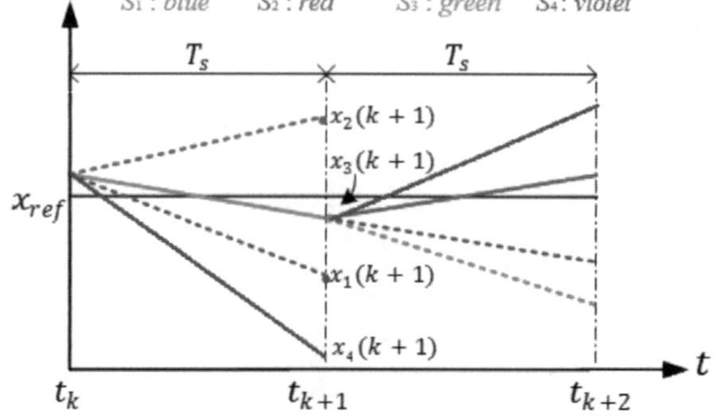

Figure 1. FCS-MPC Operating Principle

Assume that the FCS-MPC is applied to a converter with four possible switching states (S_1, S_2, S_3 and S_4) and the reference is constant in a short period of time. The distance between the controlled variable and its reference value is taken as the cost function that should be minimized in order to track the reference. The controlled variable at the next step time is predicted for all the switching states. However, choosing S_3 provides the least distance to the reference value x_{ref}; as a result, it will be applied to the converter at time t_{k+1}. Subsequently, all the process will be shifted one step forward. By repeating the procedure once again for t_{k+2}, S_2 will be selected due to its minimum distance with x_{ref}. Thus, the whole procedure will be repeated again.

3. FCS-MPC Based Proposed Methodology

In this paper, Finite Control Set MPC is applied to DSTATCOM as a current controller. This is a servo tracking problem from control point of view, where the actual DSTATCOM current should follow the reference compensation current. Figure 2 shows the basic block diagram for the predictive current control applied to the DSTATCOM.

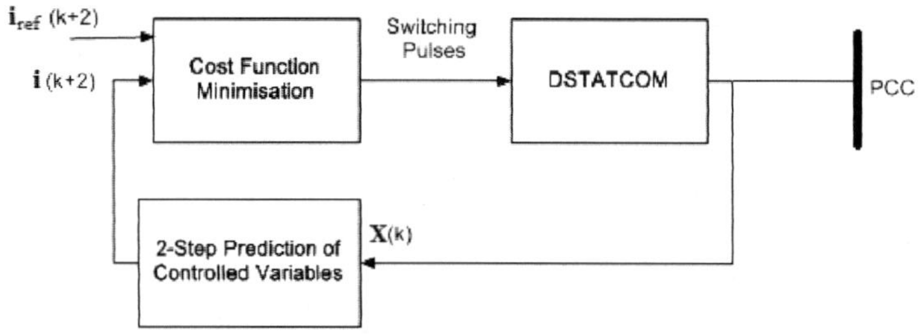

Figure 2. Block diagram for FCS-MPC based DSTATCOM

By striking a trade-off between accuracy and the computational efforts for the algorithm, 2 step ahead predictions has been considered.

The current control is performed in the following steps:

- The reference current $\mathbf{i}_{ref}$ (k) is obtained using ISCT based theory and the DSTATCOM current and other variables **X** (k) are measured or sensed.
- By two-step prediction of the system model, value of **i** (k +2) is predicted for each of the voltage vectors.

- The cost function is evaluated for each voltage vector and the one that minimizes the current error is selected and its corresponding switching state signals are applied to the Insulated Gate Bipolar Transistor (IGBT) Switches.

In short, the actual compensation DSTATCOM current is made to track the reference current (generated by ISCT based algorithm) using FCS-MPC as the current controller.

3.1. *DSTATCOM System Modeling for Prediction*

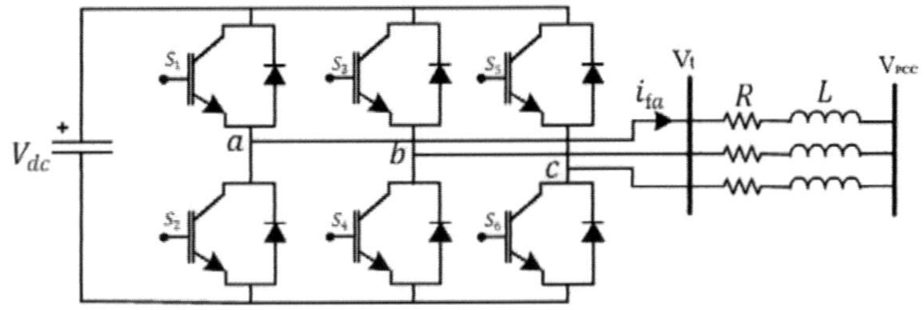

Figure 3. DSTATCOM Power Circuit

Here, we have considered a 3-phase 2-level voltage source inverter configuration with DC Link Capacitor as shown in Figure 3. R and L represent the internal losses and the coupling transformer inductance, respectively. i_{fa} is the DSTATCOM injected current in phase A. IGBT Switches with antiparallel diodes have been selected as the power switches. There shall be 8 switching states out of which two states are not possible:

- Both the switches of each legs of the inverter are simultaneously ON as it leads to short-circuiting of DC link capacitor
- Both the switches of each legs of the inverter are simultaneously OFF as it leads to turning off of the inverter, consequently no power is transferred.

Thus, avoiding the above two switching states and referring to the Figure 3, switching signals can be defined as follows :

$$S_a = \begin{cases} 1 & \text{if } S_1 \text{ on and } S_4 \text{ off} \\ 0 & \text{if } S_1 \text{ off and } S_4 \text{ on} \end{cases} \tag{5}$$

$$S_b = \begin{cases} 1 & \text{if } S_2 \text{ on and } S_5 \text{ off} \\ 0 & \text{if } S_2 \text{ off and } S_5 \text{ on} \end{cases} \tag{6}$$

$$S_c = \begin{cases} 1 & \text{if } S_3 \text{ on and } S_6 \text{ off} \\ 0 & \text{if } S_3 \text{ off and } S_6 \text{ on} \end{cases} \tag{7}$$

The relation between the DSTATCOM output terminal voltage and the DC Link voltage V_{dc} can then be written as

$$\begin{bmatrix} v_{fa} \\ v_{fb} \\ v_{fc} \end{bmatrix} = \frac{1}{3} \begin{bmatrix} 2S_a & -S_b & -S_c \\ -S_a & 2S_b & -S_c \\ -S_a & -S_b & 2S_c \end{bmatrix} [V_{dc}] \tag{8}$$

Using vector space analysis, the mathematical equation for the DSTATCOM power circuit can be written as

$$v_t = L\frac{di_f}{dt} + Ri_f + v_{pcc} \quad (9)$$

where v_t, i_f and v_{pcc} are the DSTATCOM terminal voltage, DSTATCOM current and point of common coupling voltage vectors, respectively.

The DSTATCOM current derivative can be discretised as follows using Eulers forward approximation method:

$$\frac{di_f}{dt} \approx \frac{i_f(k+1) - i_f(k)}{T_s} \quad (10)$$

which is substituted in the equation (9) to get the predicted output DSTATCOM current for each of the seven voltage vector values. Here, Ts is the sampling time. This expression is given by

$$i_f(k+1) = \left(1 - \frac{RT_s}{L}\right) i_f(k) + \frac{T_s}{L}(v_t(k) - v_{pcc}(k)) \quad (11)$$

Now, the expression for 2-step ahead predicted DSTATCOM output current can be written as follows:

$$i_f(k+2) = \left(1 - \frac{RT_s}{L}\right) i_f(k+1) + \frac{T_s}{L}(v_t(k+1) - v_{pcc}(k+1)) \quad (12)$$

3.2. Cost Function

Here, we need to track instantaneously the DSTATCOM reference current and the error between the reference and the measured values of the same should be reduced. The cost function for the algorithm is chosen as

$$G = \| i_{ref}(k+2) - i_{pred}(k+2) \| \quad (13)$$

Figure 4 shows the flow chart for the FCS-MPC based algorithm for DSTATCOM.

Table 1. System Parameters.

Parameters		3 Phase system (50μs)
Source		415V, 50Hz
Source Impedance		1Ω , 0.01H
Loads:		
1. RL Loads	(A)	(30 + j 22) Ω
	(B)	(120 + j125.6) Ω
	(C)	(60Ω + j31.4) Ω
2. Full Bridge rectifier with RL Load		150Ω , 300mH
RC Filter		5Ω , 10μF
DSTATCOM leakage filter RL		0.0476Ω , 0.6mH
DC Capacitor		500μF
V_{dcref}		750V

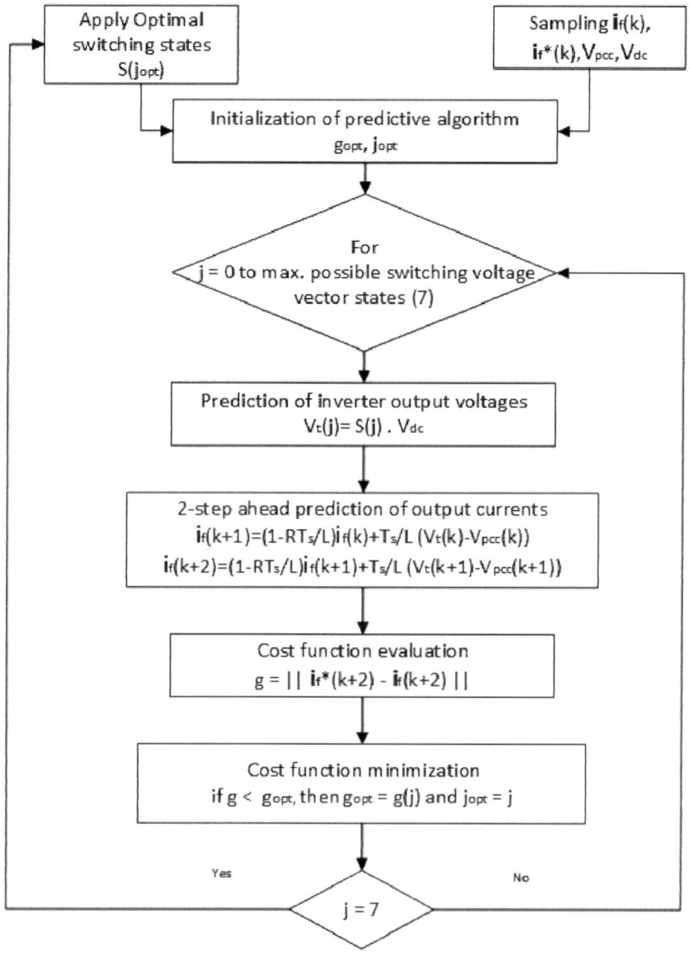

Figure 4. Flow chart for FCS-MPC based algorithm for DSTATCOM

4. Simulation Results

To verify the performance of the DSTATCOM, power distribution system compensated by DSTATCOM has been modeled in the SIMULINK using SimPowerSystem Block as shown in Figure 5. A 3-phase 3-wire 415V distribution system is taken as the initial test case with the parameters as shown in Table 1. A set of simulations were carried out for a base case without DSTATCOM but with RC Filter connected to the distribution network. RC Filter is used to reduce the higher order harmonics and the switching ripples in the PCC Voltage due to fast switching action of DSTATCOM.

The load is both unbalanced and nonlinear throughout the simulation. From 0.4 to 0.5 s, linear load of phase B at Node C is disconnected using a circuit breaker. From 0.4 to 0.6 s, linear load of phase A at Node C is disconnected using a circuit breaker.

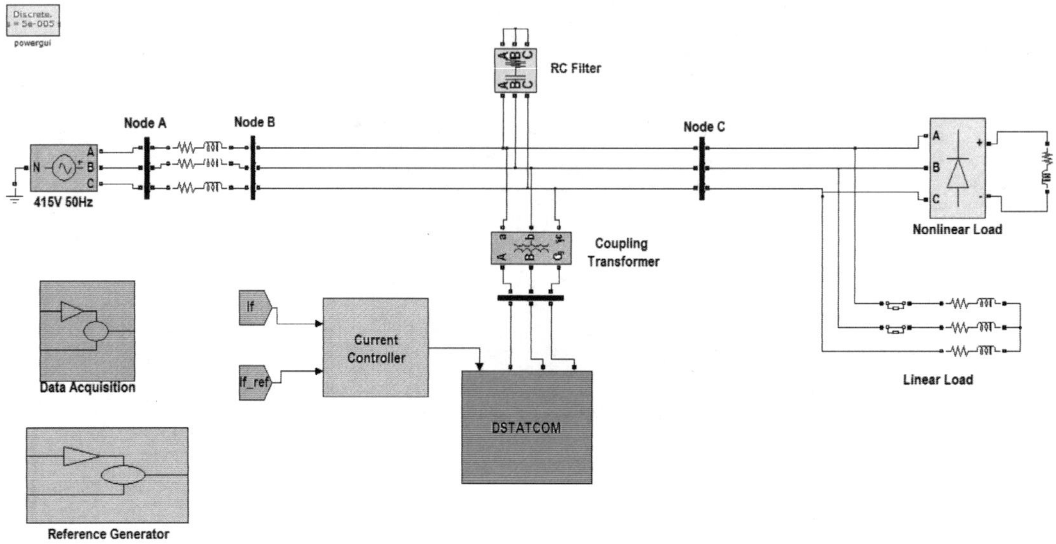

Figure 5. SIMULINK Model of the Test system.

Figure 6 shows that the PCC Voltage (V_{PCC}) is having harmonics due to nonlinear loads and the source current (i_S) is unbalanced due to the load current (i_L) characteristics for the base case scenario without DSTATCOM.

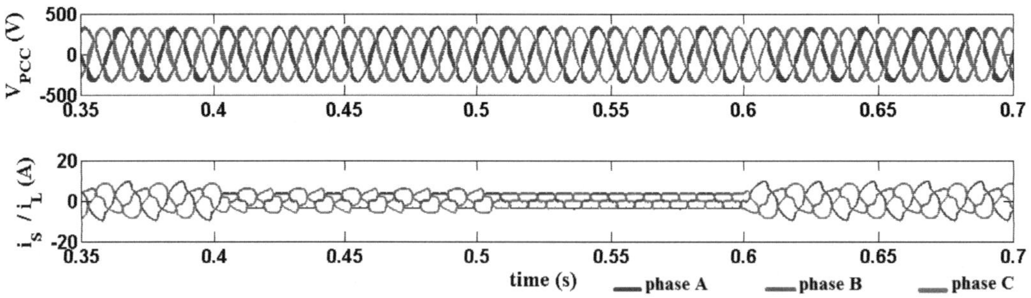

Figure 6. PCC voltage and load/source current before compensation.

A DSTATCOM was connected to the distribution system with the proposed control algorithm. K_p and K_i values for the DC voltage controller were obtained using Oscillation test based Ziegler Nichols tuning method. The simulation result showing the dynamic performance of DSTATCOM is shown in Figure 7. It was observed that the DSTATCOM was able to inject compensation current (i_D) in such a way that the source current becomes balanced and sinusoidal. Also, the PCC Voltage is almost maintained at the reference peak voltage value of 338V. Harmonics in PCC voltage has been considerably reduced. The DC link capacitor voltage (V_{dc}) of the DSTATCOM is maintained at the reference value of 750 V at the steady state following the disturbance period. Figure 8. shows the unity power factor action of DSTATCOM.

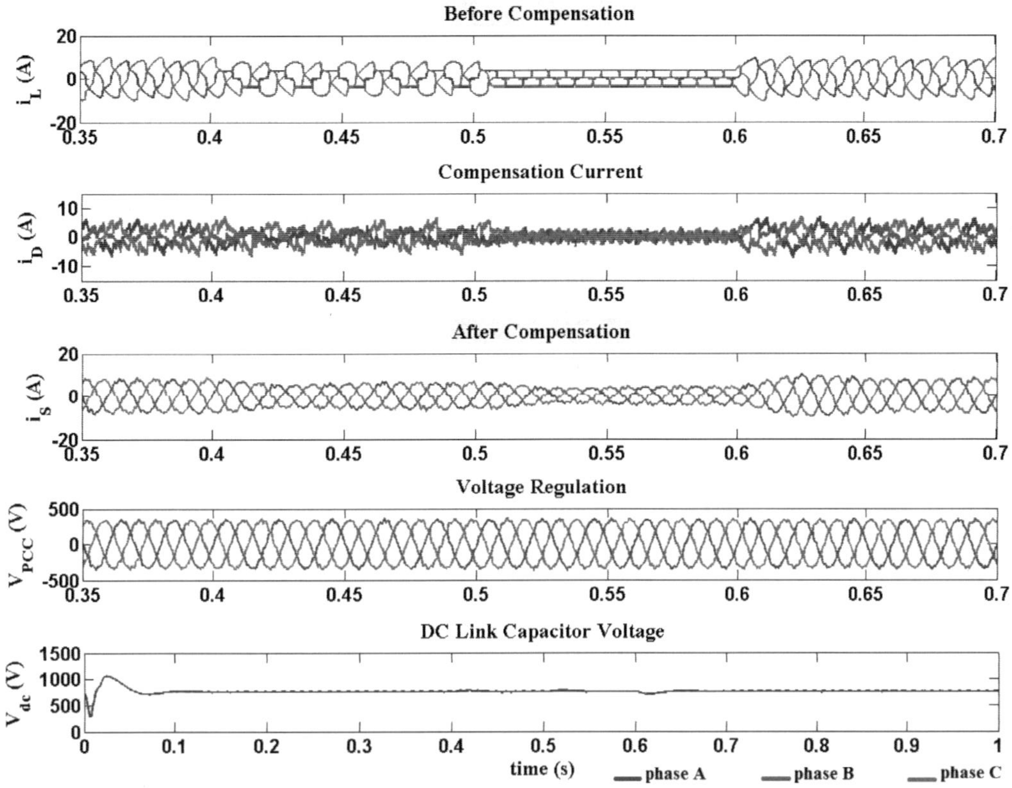

Figure 7. Performance of DSTATCOM with Proposed Algorithm.

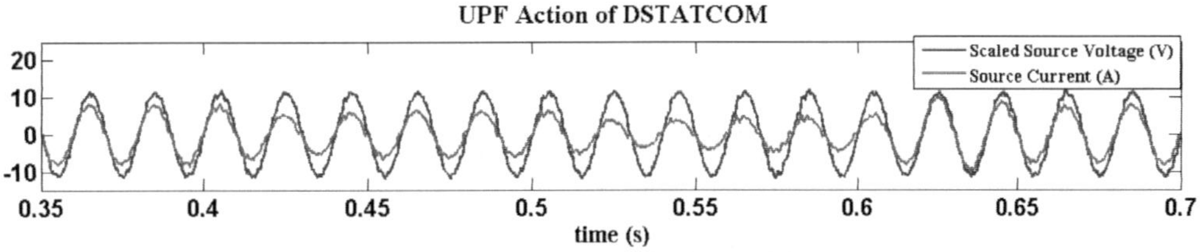

Figure 8. UPF action of DSTATCOM.

Table 2. %THD Comparison.

	Voltage %THD		Current %THD	
	Base Case	FCS-MPC	Base Case	FCS-MPC
Phase A	12.83	3.88	9.75	4.72
Phase B	9.42	3.74	15.81	4.48
Phase C	12.19	3.78	9.89	4.45

Comparison of the % Total Harmonic Distortion (%THD) values between FCS-MPC based DSTATCOM system and the base case scenario has been tabulated in the Table 2. Clearly, the %THD values has been considerably reduced and within the IEEE standard limit.

5. Conclusion

This paper proposes a new control technique, namely, Finite Control Set Model Predictive Controller (FCS-MPC) with ISCT based reference current extraction method for DSTATCOM. The proposed control technique is very simple and yet powerful, able to control the DSTATCOM without any additional modulation stage. This allows faster dynamic response compared to the classical controllers. It avoids additional coordinate transformation used in the conventional FCS-MPC method.

It was able to balance the source current under unbalanced load conditions as well as suppress the voltage harmonics along with maintaining the PCC voltage within the prescribed limit. In addition, unity power factor action is also verified. The secondary control objective to maintain the required reference voltage across the DC Link capacitor was also achieved. The %THD values below the IEEE-519:1992 validate the quality of the proposed control scheme. The V_{dc} control can be incorporated inside the FCS-MPC algorithm to make the control system much simpler and effective for future smart power distribution systems.

6. References

[1] T. J. E. Miller, *Reactive power control in electric systems*, Wiley, 1982.

[2] M. K. Mishra, K. Karthikeyan, A Fast-Acting DC-Link Voltage Controller for Three-Phase DSTATCOM to Compensate AC and DC Loads, *IEEE Transactions on Power Delivery* **24** (2009) 2291–2299, Oct. 2009.

[3] H. Suryanarayana, M. K. Mishra, Fuzzy logic based supervision of DC link PI control in a DSTATCOM, INDICON 2008. Annual IEEE India Conference, volume 2, IEEE, pp. 453–458, 11-13 Dec. 2008.

[4] B. Singh, P. Jayaprakash, S. Kumar, D. Kothari, Implementation of Neural Network Controlled Three-Leg VSC and a Transformer as Three-Phase Four-Wire DSTATCOM, *IEEE Transactions on Industry Applications* **47** (2011) 1892–1901, July-Aug. 2011.

[5] B. B. Bukata, Y. Li, A novel model-free prediction of power quality problems via DSTATCOM, in: Automation and Computing (ICAC), 2012 18th International Conference, IEEE, pp. 1 – 6, 7-8 Sept. 2012.

[6] S. B. Karanki, M. K. Mishra, B. Kumar, Particle swarm optimization based feedback controller for unified power-quality conditioner, *IEEE Transactions on Power Delivery* **25** (2010) 2814–2824.

[7] E. Deniz, R. Coteli, B. Dandil, S. Tuncer, F. Ata, M. T. Gencoglu, Neuro-fuzzy current controller for three-level cascade inverter based DSTATCOM, Universities Power Engineering Conference (UPEC), 2010 45th International Conference, IEEE, pp. 1 – 5, Aug. 31 2010-Sept. 3 2010.

[8] G. Escobar, A. M. Stankovic, P. Mattavelli, An adaptive controller in stationary reference frame for D-statcom in unbalanced operation, *IEEE Transactions on Industrial Electronics* **51** (2004) 401–409.

[9] A. Y. Sendjaja, V. Kariwala, Achievable PID performance using sums of squares programming, *Journal of Process Control* **19** (2009) 1061 – 1065.

[10] M. Sugeno, An introductory survey of fuzzy control, *Information Sciences* **36** (1985) 59 – 83.

[11] C. Wang, D. Hill, Learning from neural control, *IEEE Transactions on Neural Networks* **17** (2006) 130–146.

[12] J. Holtz, Pulsewidth modulation for electronic power conversion, Proceedings of the IEEE **82** (1994) 1194–1214.

[13] G. W. Moon, Predictive current control of distribution static compensator for reactive power compensation, Generation, Transmission and Distribution, IEE Proceedings 146 (1999) 515–520.

[14] J. Rodriguez, J. Pontt, C. Silva, P. Correa, P. Lezana, P. Cortes, U. Ammann, Predictive current control of a voltage source inverter, *IEEE Transactions on Industrial Electronics* **54** (2007) 495–503.

[15] A. Ghosh, G. F. Ledwich, *Power quality enhancement using custom power devices*, Kluwer academic publishers, 2002.

[16] R. Strzelecki, G. Benysek, *Power electronics in smart electrical energy networks*, Springer, 2008.

[17] F. Borrelli, A. Bemporad, M. Fodor, D. Hrovat, An MPC/hybrid system approach to traction control, *IEEE Transactions on Control Systems Technology* **14** (2006) pp. 541–552.

[18] S. Almer, H. Fujioka, U. Jonsson, C.-Y. Kao, D. Patino, P. Riedinger, T. Geyer, A. Beccuti, G. Papafotiou, M. Morari, A. Wernrud, A. Rantzer, Hybrid Control Techniques for Switched-Mode DC-DC Converters Part I: The Step-Down Topology, in: American Control Conference, 2007. ACC '07, pp. 5450–5457, 9-13 July 2007.

[19] M. Morari, J. H. Lee, Model predictive control: past, present and future, *Computers & Chemical Engineering* **23** (1999) 667 – 682.

[20] J. Rodriguez, P. Cortes, *Predictive control of power converters and electrical drives*, volume 37, John Wiley & Sons, 2012.

[21] J. Rodriguez, P. Cortes, R. Kennel, M. Kazrnierkowski, Model predictive control – a simple and powerful method to control power converters, in: Power Electronics and Motion Control, 2009. IPEMC '09. IEEE 6th International Conference, pp. 41–49.

[22] H. Abu-Rub, J. Guzinski, Z. Krzeminski, H. A. Toliyat, Predictive current control of voltage source inverters, in: Industrial Electronics Society, 2001. IECON '01. The 27th Annual Conference of the IEEE, volume 2, pp. 1195–1200.

[23] Young, H.; Rodriguez, J., "Comparison of finite-control-set model predictive control versus a SVM-based linear controller," Power Electronics and Applications (EPE), 2013 15th European Conference on , pp. 1 – 8, 2-6 Sept. 2013

Journal of Physics: Conference Series **659** (2015) 012032 doi:10.1088/1742-6596/659/1/012032

Detection, isolation and fault estimation of nonlinear systems using a directional study

Maya Kallas[1], Gilles Mourot[1], Didier Maquin[1], José Ragot[1]

[1]Centre de Recherche en Automatique de Nancy, CNRS, Université de Lorraine, 2 Avenue de la Forêt de Haye TSA 60604 54 518 Vandoeuvre-lès-Nancy

E-mail: {Maya.Kallas, Gilles.Mourot, Didier.Maquin, Jose.Ragot}@univ-lorraine.fr

Abstract. In terms of system diagnosis, several studies are generally performed. The diagnosis is composed of three different parts: detecting, isolating and estimating the value of the faults. If many results have been obtained for linear systems with a known model, the situation is quite different in the case of nonlinear systems behavior, especially when the model is not known *a priori*. This paper proposes to discuss the latter case using a study of the dissimilarities between data. The dissimilarities are evaluated by a nonlinear function of the Euclidean distances. To this end, a radial basis function is used, and a directional study is introduced for fault diagnosis. The relevance of the proposed technique is illustrated on simulated data.

1. Introduction

The supervision of systems covers various additional aspects, among which the diagnosis occupies a privileged place. Based on this statement, there exists a set of techniques used to know the status of a given system. In the state estimation step, the very important aspect is the detection of functioning changes due for example to sensors failure or actuators or even system components. Early detection of these defects can be used to react on the commands of the system so as to minimize the impact of the faults. Numerous techniques have been developed in recent decades, which differ depending on whether a model of the system is known *a priori* or not. This paper is concerned with the second situation, where only input and output measures of the system are available. Diagnosis strategies are thus based on the analysis of these data. The principal component analysis (PCA) is an example of a well-known technique for diagnosis. PCA and its variants have been the subject of many developments.

The PCA has been widely used for the detection and isolation of sensor failure [11], [13] and more generally for system supervision [2], [15]. As in the case of detection method using a model known *a priori*, one of the best performing PCA technique is based on the comparison between the value of a variable at a given time and the reconstructed value of this variable from the other variables. This reconstruction uses the eigenvectors of the covariance matrix of a dataset of observations, these eigenvectors belonging to the subspace called secondary or residual obtained by PCA. As indicated in [7], it is possible to extend this reconstruction to a system with n variables, subject to observability, where we can reconstruct simultaneously p variables out of the $n - p$ remaining ones, based on the fact that we have redundancy between some data in the whole dataset defined by the n variables.

Despite the proven performance of this technique, it is necessary to recall that it is applied on a system whose variables are mainly related linearly. Indeed, the PCA is a method of linear projection that cannot consider or reveal nonlinearities. The nonlinearity issue has motivated a number of work leading to the nonlinear PCA, or the use of kernel machines such as the kernel PCA (KPCA). In the field of diagnosis, the KPCA has been widely used to detect outliers from faulty sensors for example [6]. KPCA has been used to analyze partially observed systems, in applications of missing data, such as in image processing [3], [12]. The combination, for fault detection, of KPCA and discriminant analysis techniques, was proposed in data processing to increase the performance of each method [5], [10].

To address the difficult problem of isolating faults affecting the data, diagnosis techniques developed in a linear framework rely heavily on the concept of structuring fault indicators. The extension to the nonlinear case remains delicate due to the difficulty of the problem called pre-image (where the existence of a solution is not guaranteed *a priori*) but has been the subject of several works [1]. The principle used in [6] consists on applying KPCA technique to several subsets of data, each using a part of the original variables. The difficulty of such proposition is to build these subsets.

Based on the literature, it is clear that the detection of outliers is widely studied. However, only few papers propose new nonlinear technique for isolation and fault amplitude estimation affecting one or more variables. In [1], [11] and [8], the pre-image technique offers methods for isolation and estimation of faults.

As aforementioned, the PCA and its nonlinear extension KPCA have been studied in all their aspects in terms of fault detection, isolation and estimation. However, there exists a more general technique than PCA, the principal coordinate analysis (PCO) or the well-known multidimensional scaling (MDS) [16]. The main difference between the PCA and PCO can be given by three points. First, PCA seeks patterns in the variables, while PCO searches for similarities between cases. Second, PCA uses an eigen decomposition of a covariance matrix in order to reduce variable dimensionality, while PCO analyzes a distance matrix. Many different distance matrices can be used as long as they are metric. Third, the result of a PCO is a set of coordinates on a number of derived axes such that similar cases are close together. It is not possible to associate these axes with any variables. This technique was not used for diagnosis purposes. Therefore, we propose to study similarities between data. The aim of the scaling is to find the distances that well approximate the dissimilarities of data [14]. In [4], Critchley suggested a power transformation of the distances, if the dissimilarities are derived from Euclidean distances.

Based on the suggestion of Critchley, we propose in this paper a modified MDS study in order to evaluate the dissimilarities of data. We examine the possibility to detect and isolate the sensors providing faulty measures. Based on real non faulty dataset, the proposed technique is to compare the test dataset to these reference data. Because of the nonlinear behavior of systems that are of interest, this comparison will be performed in a space of suitable structure. More precisely, after selecting the appropriate structure of the kernel, the initial reference dataset are used to construct a Gram matrix, in order to represent the original data in a space of larger dimension. The new observations to be analyzed are then projected into this space and their comparison with the reference observations is done through their projections. From this comparison, we can discriminate new healthy observations from faulty ones (section 2.2). In the case where an observation is found abnormal, an isolation of the faulty variable is proposed in (section 2.3), and finally an estimate of the fault amplitude is examined in (section 2.4).

In this paper, the Gaussian kernel is applied. It is important to mention a difficulty when using this kind of kernels. It is a kernel with a decreasing bandwidth depending on the distance to its centers. Therefore, a faulty measure may no longer be in the area of influence of the kernel and thus may have near zero projection in the kernel space, making it difficult to locate the fault. To overcome this difficulty, a sequential search of one or more directions of fault is proposed.

As many fault detection methods, the proposed approach has a number of hyperparameters. It is necessary to adjust them depending on the problem to be treated. The hyperparameters, in this paper, are the number of principal components to retain from the decomposition of the Gram matrix, and finally the bandwidth of the kernel. The influence of these parameters is studied through the example in section 3, but in the present form of this paper, their choice is done experimentally.

2. Detection, isolation and estimation of the faults

2.1. Notations

Let us consider a set of healthy measures, denoted $\boldsymbol{x}_i$. They are taken from the system and they form the reference database:

$$X = \begin{bmatrix} \boldsymbol{x}_1^\top \\ \dots \\ \boldsymbol{x}_N^\top \end{bmatrix}, \quad \boldsymbol{x}_i \in \mathbb{R}^n, \quad X \in \mathbb{R}^{N\times n} \tag{1}$$

Based on using the idea of dissimilarities, and from the suggestion of Critchley [4], let us consider a nonlinear function κ, also called a kernel. By kernel, we mean a function from $X \times X$ into $\mathbb{R}$ with a Hermitian symmetry, that is $\kappa(\boldsymbol{x}_i, \boldsymbol{x}_j) = \kappa(\boldsymbol{x}_j, \boldsymbol{x}_i)$. Using this definition, we can mention the Radial Basis Function (RBF), and more precisely, the gaussian kernel, which is widely used. The form of this gaussian kernel is characterized by its bandwidth, or hyper-parameter c,

$$\kappa(\boldsymbol{x}_i, \boldsymbol{x}_j) = \exp\left(-\frac{\| \boldsymbol{x}_i - \boldsymbol{x}_j \|^2}{c}\right) \tag{2}$$

During an online diagnosis, we receive a new observation $\boldsymbol{x} \in \mathbb{R}^n$ to be tested. The analysis of this observation is done in order to detect the possible presence of errors in one or more variables of this measure. To perform this analysis, we need to evaluate the following expressions:

$$\begin{aligned} \boldsymbol{\kappa}(\boldsymbol{x}) &= \begin{bmatrix} \kappa(\boldsymbol{x}_1, \boldsymbol{x}) \\ \vdots \\ \kappa(\boldsymbol{x}_N, \boldsymbol{x}) \end{bmatrix}, \quad \boldsymbol{\kappa}(\boldsymbol{x}) \in \mathbb{R}^N \\ \kappa(\boldsymbol{x}_i, \boldsymbol{x}) &= \exp\left(-\frac{\| \boldsymbol{x}_i - \boldsymbol{x} \|^2}{c}\right), \quad \kappa(\boldsymbol{x}_i, \boldsymbol{x}) \in \mathcal{R} \end{aligned} \tag{3}$$

We already have the reference database. Thus, we can easily create the Gram matrix $K(X)$ using

$$K(X) = \begin{bmatrix} \kappa(\boldsymbol{x}_1, \boldsymbol{x}_1) \dots \kappa(\boldsymbol{x}_1, \boldsymbol{x}_N) \\ \vdots \\ \kappa(\boldsymbol{x}_N, \boldsymbol{x}_1) \dots \kappa(\boldsymbol{x}_N, \boldsymbol{x}_N) \end{bmatrix}, \quad K(X) \in \mathbb{R}^{N\times N} \tag{4}$$

As we can see from the expressions of $\kappa(\boldsymbol{x}_i, \boldsymbol{x}_j)$, the bandwidth c of the kernel influences the results. It is relative to the proximity between two observations $\boldsymbol{x}_i$ and $\boldsymbol{x}_j$.

If a fault affects a specific variable for a given observation $\boldsymbol{x}$, it can be defined by:

$$\boldsymbol{x} = \boldsymbol{x}^* + \boldsymbol{\xi}\, \boldsymbol{f} \tag{5}$$

where $\boldsymbol{x} \in \mathbb{R}^n$ is the faulty measure, $\boldsymbol{x}^* \in \mathbb{R}^n$ is the true value of the variable, $\boldsymbol{f} \in \mathbb{R}^{n_f}$ is a n_f-dimensional vector of faults amplitude, and $\boldsymbol{\xi} \in \mathbb{R}^{n\times n_f}$ their directions. For this paper, we take into consideration one fault, so $n_f = 1$, making it easier to explain. For example, $\boldsymbol{\xi} = \begin{pmatrix} 0 & 0 & 1 & 0 \end{pmatrix}^\top$ is a situation where we have a system with four variables, where a fault is occurring on the third one.

Diagnosis represents different tasks to be done. It is to decide whether the measure $\boldsymbol{x}$ is faulty, and if so, to determine the direction $\boldsymbol{\xi}$ of the detected fault, namely the faulty variable or component, and finally to estimate the fault amplitude f. In our case, where $n_f = 1$, f is considered to be a scalar.

2.2. Analysis of $\boldsymbol{\kappa}(\boldsymbol{x})$ for fault detection

The analysis of the validity of a new observation $\boldsymbol{x}$ is performed in the kernel space. We have evaluated the matrix $K(X)$ given by (4), which is the image of the matrix X in the kernel space. Therefore, it is quite natural to compare the gap between the new observation $\boldsymbol{x}$ and the database X. Working in the kernel space, we evaluate $\boldsymbol{\kappa}(\boldsymbol{x})$ and $K(X)$ representing the healthy dataset. To facilitate this comparison, we reduce the complexity of $K(X)$ by considering its spectral decomposition

$$K(X) = P\,\Lambda\,P^\top \tag{6}$$

where P and Λ contain respectively the eigenvectors and eigenvalues of the matrix $K(X)$, sorted by descending order of the magnitude of the eigenvalues. By neglecting the very small values of the eigenvalues, we take the ℓ eigenvectors, denoted P_ℓ, associated to the ℓ largest eigenvalues represented by Λ_ℓ. We propose an approximation of the matrix $K(X)$ using the following expression

$$K_\ell(X) = P_\ell\,\Lambda_\ell\,P_\ell^\top \tag{7}$$

We seek to decompose $\boldsymbol{\kappa}(\boldsymbol{x})$ given by (3) on the basis of the P_ℓ eigenvectors. To this end, we consider the criterion ψ defined by

$$\psi = \parallel \boldsymbol{\kappa}(\boldsymbol{x}) - P_\ell\,\boldsymbol{\theta} \parallel^2 \tag{8}$$

We seek its minimum, for a certain value of $\boldsymbol{\theta}$

$$\hat{\boldsymbol{\theta}} = \left(P_\ell^\top\,P_\ell\right)^{-1}\,P_\ell^\top\,\boldsymbol{\kappa}(\boldsymbol{x}) \tag{9}$$

Thus, the minimum is obtained by replacing (9) in (8), as follows

$$\hat{\psi} = \boldsymbol{\kappa}^\top(\boldsymbol{x})\left(I - P_\ell\left(P_\ell^\top\,P_\ell\right)^{-1}\,P_\ell^\top\right)\boldsymbol{\kappa}(\boldsymbol{x}) \tag{10}$$

The value of the criterion $\hat{\psi}$ is closely related to those of $\boldsymbol{\kappa}(\boldsymbol{x})$ and thus the measure $\boldsymbol{x}$ itself. One can study for an observation $\boldsymbol{x}$, on the one hand the value taken by $\boldsymbol{\kappa}(\boldsymbol{x}_i, \boldsymbol{x})$ and on the other hand, the value taken by $\hat{\psi}$. In particular, an observation $\boldsymbol{x}$ for which a fault affects one variable, will be revealed by the values close to zero of the $\hat{\psi}$ criterion and the component of the $\boldsymbol{\kappa}(\boldsymbol{x})$ vector. For the latter, a more comprehensive test is considered based on the sum of its parts

$$S = \sum_{i=1}^{N} \kappa(\boldsymbol{x}_i, \boldsymbol{x}) \tag{11}$$

Nevertheless, for a healthy observation $\boldsymbol{x}$, the values of S and $\hat{\psi}$ are significantly greater than zero. To decide the inconsistency of a measure, that is to say, the presence of a fault on one of its components, it is proposed to use the following two tests

$$\hat{\psi} < \delta_1 \tag{12a}$$
$$S < \delta_2 \tag{12b}$$

where δ_1 and δ_2 are two thresholds, having small values. They are chosen based on the original dataset during the learning stage. Theses thresholds are evaluated as the mean of different tests on many healthy observations.

2.3. Directional analysis of $\boldsymbol{\kappa}(\boldsymbol{x})$ for fault isolation

As indicated in the definition (3) of $\boldsymbol{\kappa}(\boldsymbol{x})$, the same weight is assigned to all components of the vectors involved in calculating the distances. In the case where one of the components of a test observation $\boldsymbol{x}$ is corrupted by a fault, the distance between this observation and those of reference $\boldsymbol{x}_i$ can become important. This can make $\kappa(\boldsymbol{x}_i, \boldsymbol{x})$ close to zero, which provides little information, except for the fact to say that this observation is faulty.

To overcome this problem, the solution would be to increase the bandwidth of the kernel function. However, this solution is hardly possible because it leads to a very particular Gram matrix, from a numerical point of view, because the obtained matrix would be nearly an identity matrix. Therefore, the main idea of the kernel methods is not represented, and it will be as we are in a linear case. However, using a rather restrictive assumption on the limited number of faulty variables on the observation $\boldsymbol{x}$, the evaluation of the distance can be performed with a particular metric by not taking into account some variables. To this end, $\kappa(\boldsymbol{x}_i, \boldsymbol{x})$ is changed to this form

$$\begin{aligned} \kappa_d(\boldsymbol{x}_i, \boldsymbol{x}) &= \exp\left(-\frac{(\boldsymbol{x}_i - \boldsymbol{x})^\top W_d\,(\boldsymbol{x}_i - \boldsymbol{x})}{c}\right) \\ W_d &= \mathtt{diag}\begin{pmatrix}1 & \ldots & 1 & 0 & 1 & \ldots & 1\end{pmatrix} \end{aligned} \tag{13}$$

where the diagonal matrix W_d contains 1 everywhere on its diagonal, except the component of rank d that is chosen to be zero. Thus, the distance between the observations $\boldsymbol{x}$ and $\boldsymbol{x}_i$ takes into account all the variables except the one of rank d. Therefore, if this variable is faulty, it will not be taken into account in the evaluation of the distance and the difficulty aforementioned about the null value of $\kappa(\boldsymbol{x}_i, \boldsymbol{x})$ is overlapped. Obviously, as the faulty variable is not known *a priori*, and in the case of just one faulty variable, it is necessary to examine all n matrices W_d, where $d = 1, \ldots, n$.

We also note that the weighting matrix W_d is directly related to the direction of a fault. Each direction, introduced in (5), corresponds in fact to a vector $\boldsymbol{\xi}_d$ of the canonical basis of $\mathbb{R}^n$

$$\boldsymbol{\xi}_d = \begin{pmatrix}\delta_{1,d} & \delta_{2,d} & \cdots & \delta_{n,d}\end{pmatrix}, \quad d = 1, \ldots, n \tag{14}$$

where $\delta_{i,d}$ is the Kronecker delta, defined by

$$\delta_{i,j} = \begin{cases} 0 & \text{if } i \neq j \\ 1 & \text{if } i = j \end{cases}$$

Therefore, the weighting matrix W_d, that can overcome the variable of rank d can be defined as

$$W_d = I_n - \boldsymbol{\xi}_d \boldsymbol{\xi}_d^\top \tag{15}$$

The rest of the procedure, proposed in section 2.2, is applied with these new definitions of kernel space given in (13). We can check for each test case $\boldsymbol{x}$, the values taken by $\kappa_d(\boldsymbol{x}_i, \boldsymbol{x})$ and $\hat{\psi}_d$ for different weighting matrices W_d where $d = 1, \ldots, n$. In particular, an observation $\boldsymbol{x}$ whose p rank variable is faulty, is identified by near-zero values of the components of $\boldsymbol{\kappa}_d(\boldsymbol{x})$ and a value close to zero of ψ_d for $d = 1, \ldots, n, d \neq p$. As before, a more general evaluation of $\boldsymbol{\kappa}_d(\boldsymbol{x})$ is suggested by defining the scalar indicator

$$S_d = \sum_{i=1}^{N} \kappa_d(\boldsymbol{x}_i, \boldsymbol{x}), \quad d = 1, \ldots, n \tag{16}$$

Given the different expressions of S_d according to the variables used, the truth table 1 summarizes the possible situations. In this table, the quantities $\delta\boldsymbol{x}_i$, for $i = 1, \ldots, n$ represent

Table 1. Theoretical fault signatures

	$\delta\boldsymbol{x}_1$	$\delta\boldsymbol{x}_2$	$\delta\boldsymbol{x}_3$	$\dots$	$\delta\boldsymbol{x}_n$
S_1	$\times$	0	0	$\dots$	0
S_2	0	$\times$	0	$\dots$	0
$\vdots$					
S_n	0	0	0	$\dots$	$\times$

the faults affecting the components of the observation $\boldsymbol{x}$, where the 0 symbol indicates that a fault makes the indicator S a very small value.

In summary, the steps of isolating a faulty variable for a given observation $\boldsymbol{x}$ can be summarized as

- Create the matrices W_d (15) for all possible directions $\boldsymbol{\xi}_d, d = 1, \dots, n$
- Evaluate $\kappa_d(\boldsymbol{x}_i, \boldsymbol{x})$ given by (13) for all fault directions and any reference observation $\boldsymbol{x}_i$
- Compute the indicator S_d for $d = 1, \dots, n$ given by (16) for fault presence of observation $\boldsymbol{x}$
- If one of the indicators S_d for $d = 1, \dots, n$, only the p rank is greater than the selected threshold, then the pth variable of the observation $\boldsymbol{x}$ is faulty

2.4. Estimation of the fault amplitude

This step finalizes the diagnosis since it completes the detection and fault isolation by estimating their amplitudes. The direction $\boldsymbol{\xi}_d$ of the fault is known, it is determined in the previous step. The objective is to estimate its amplitude f. To this end, we use the redundancy related to the kernel space. We know that any vector $\boldsymbol{\kappa}(\boldsymbol{x})$ can be formed from the column vectors of the matrix $K(X)$, or its approximate form $K_\ell(X)$ given by (7), where $\boldsymbol{x}$ is a vector of measures and X the reference healthy database. In reality, once the observation $\boldsymbol{x}$ is considered to be a faulty measure, we evaluate its estimation with:

$$\boldsymbol{x}^* = \boldsymbol{x} - \boldsymbol{\xi}\, f \tag{17}$$

Thus, we can evaluate the image of the corrected observation in the kernel space, using:

$$\begin{aligned} \boldsymbol{\kappa}(\boldsymbol{x}^*) &= \begin{bmatrix} \kappa(\boldsymbol{x}_1, \boldsymbol{x} - \boldsymbol{\xi}\, f \\ \vdots \\ \kappa(\boldsymbol{x}_N, \boldsymbol{x} - \boldsymbol{\xi}\, f) \end{bmatrix} \\ \kappa(\boldsymbol{x}_i, \boldsymbol{x} - \boldsymbol{\xi}\, f) &= \exp\left(-\frac{\| \boldsymbol{x}_i - \boldsymbol{x} + \boldsymbol{\xi}\, f \|^2}{c}\right) \end{aligned} \tag{18}$$

As aforementioned, the best value of f is the one that helps explaining $\boldsymbol{\kappa}(\boldsymbol{x}^*)$ from the column vectors of the matrix $K_\ell(X)$. From this statement, the value of f is evaluated by maximizing with respect to f the criterion:

$$\psi = \| K_\ell(X)\ \boldsymbol{\kappa}(\boldsymbol{x} - \boldsymbol{\xi}\, f)) \|^2 \tag{19}$$

The optimal value $\hat{f}$ of f is the solution to the following equation:

$$\frac{\partial \boldsymbol{\kappa}^\top(\boldsymbol{x} - \boldsymbol{\xi}\, f)}{\partial f} K_\ell^2(X)\, \boldsymbol{\kappa}(\boldsymbol{x} - \boldsymbol{\xi}\, f) = 0 \tag{20}$$

Since the derivative of κ with respect to f is given by

$$\frac{\partial \kappa(\boldsymbol{x}_i, \boldsymbol{x} - \boldsymbol{\xi}\, f)}{\partial f} = -\frac{2}{c}\, \kappa(\boldsymbol{x}_i, \boldsymbol{x} - \boldsymbol{\xi}\, f)\, \boldsymbol{\xi}^\top (\boldsymbol{x}_i - \boldsymbol{x} + \boldsymbol{\xi}\, f) \tag{21}$$

Therefore, we can write

$$\frac{\partial \boldsymbol{\kappa}(\boldsymbol{x} - \boldsymbol{\xi}\, f)}{\partial f} = -\frac{2}{c}\, (\boldsymbol{b}(\boldsymbol{x} - \boldsymbol{\xi}\, f) + \boldsymbol{\kappa}(\boldsymbol{x} - \boldsymbol{\xi}\, f)\, f) \tag{22}$$

where

$$\boldsymbol{b}(\boldsymbol{x} - \boldsymbol{\xi} f) = \begin{pmatrix} \kappa(\boldsymbol{x}_1, \boldsymbol{x} - \boldsymbol{\xi}\, f) \boldsymbol{\xi}^\top (\boldsymbol{x} - \boldsymbol{x}_1) \\ \vdots \\ \kappa(\boldsymbol{x}_N, \boldsymbol{x} - \boldsymbol{\xi}\, f) \boldsymbol{\xi}^\top (\boldsymbol{x} - \boldsymbol{x}_N) \end{pmatrix} \tag{23}$$

Using these expressions, we can write (20) as:

$$\boldsymbol{b}^\top (\boldsymbol{x} - \boldsymbol{\xi} f)\, K_\ell^2\, \boldsymbol{\kappa}(\boldsymbol{x} - \boldsymbol{\xi} f) = \boldsymbol{\kappa}^\top (\boldsymbol{x} - \boldsymbol{\xi} f)\, K_\ell(X)^2\, \boldsymbol{\kappa}(\boldsymbol{x} - \boldsymbol{\xi} f)\, f \tag{24}$$

Using (24), we can derive f with a fixed point iterative technique $f = g(f)$ starting with an initial point as mentioned in [1] and [9]. The final expression $\widehat{f}$ of the estimated value of f can be derived, and the estimation is given by:

$$\hat{f}^{(i+1)} = \frac{\boldsymbol{b}^\top (\boldsymbol{x} - \boldsymbol{\xi} f^{(i)})\, K_\ell^2(X)\, \boldsymbol{\kappa}(\boldsymbol{x} - \boldsymbol{\xi} f^{(i)})}{\boldsymbol{\kappa}^\top (\boldsymbol{x} - \boldsymbol{\xi} f^{(i)})\, K_\ell^2(X)\, \boldsymbol{\kappa}(\boldsymbol{x} - \boldsymbol{\xi} \hat{f}^{(i)})} \tag{25}$$

3. Experiments

In this section, we present the results obtained on an example deliberately chosen to be of small dimensions. It is used to illustrate the above principles of detection, isolation and fault estimation of outliers given by faulty sensors.

3.1. Simulated data

A set of 100 observations has been generated from a nonlinear system with $n = 3$ variables, where $\boldsymbol{x}_1$ are randomly selected in $[-7, 7]$. The system is defined by:

$$\begin{cases} \boldsymbol{x}_2 &= 0.15\, {\boldsymbol{x}_1}^2 - 0.5 \\ \boldsymbol{x}_3 &= \boldsymbol{x}_1 + \boldsymbol{x}_2 \end{cases} \tag{26}$$

However, the system is considered to be noisy. Therefore, the variables have been modified with additive noises generated randomly, as follows:

$$\begin{cases} \epsilon_i &\sim \mathbb{N}(0, 0.01), \quad i = 1, 2, 3 \\ \boldsymbol{x}_1 &= \boldsymbol{x}_1 + \epsilon_1 \\ \boldsymbol{x}_2 &= \boldsymbol{x}_2 + \epsilon_2 \\ \boldsymbol{x}_3 &= \boldsymbol{x}_3 + \epsilon_3 \end{cases} \tag{27}$$

Another set of $N_t = 9$ observations, denoted $x_j^t, j = 1, \ldots, N_t$, have been generated using the model (26) for testing purpose. The nine observations are divided into two groups: faulty and correct values. Figure 1 illustrates these observations and more precisely the red dot representing the faulty observations 1, 4, 7 and 9. The faulty ones have been corrupted as follows:

- x_1^t: adding 4 on the amplitude of the variable 1
- x_4^t: adding -4 on the amplitude of the variable 2
- x_7^t: adding 4 on the amplitude of the variable 3
- x_9^t: modifying the amplitude of the variables 1, 2 and 3

3.2. Detecting outliers

Figure 1 displays the N measures with the blue dots. It contains three subfigures in order to represent the three variables into three different $2D$ dimensions, namely, $\{\boldsymbol{x}_1, \boldsymbol{x}_2\}$, $\{\boldsymbol{x}_2, \boldsymbol{x}_3\}$ and $\{\boldsymbol{x}_3, \boldsymbol{x}_1\}$. This database is the reference dataset. It is used to test the proposed technique in order to detect the outliers.

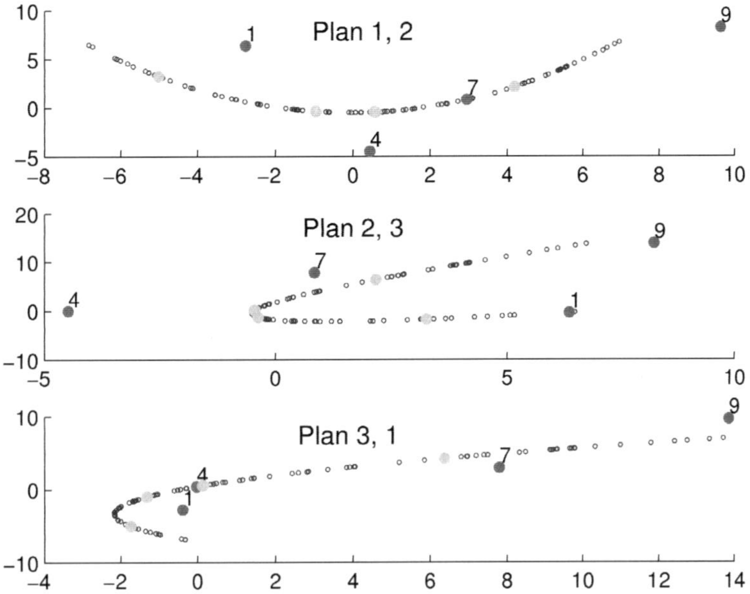

Figure 1. Representation of the simulated data in three subfigures. The blue dots represent the original database, green dots are the new non faulty observations, and the red dots are the new faulty ones. Each subfigure represents a different $2D$ representation of the $3D$ simulated data.

Of course, visual inspection of this figure allows direct isolation of the outliers from the real measures. For example, the observation 4 is abnormal and this is due to the variable 2 since in the $2D$ $\{\boldsymbol{x}_3, \boldsymbol{x}_1\}$ representation, the representative point of this observation can be considered as belonging to the field of true observations. However, in the $2D$ representations based on $\{\boldsymbol{x}_1, \boldsymbol{x}_2\}$ and $\{\boldsymbol{x}_2, \boldsymbol{x}_3\}$, the same point does not belong to the field of non faulty observations.

Figures 2 and 3 visualize changes in criterion $\hat{\psi}$ resulting from the approximation of $\boldsymbol{\kappa}(\boldsymbol{x})$ in the kernel space. This criterion was evaluated for two different values of the bandwidth of the kernel function $(0.5, 0.7)$, taking into account 30 principal components, for each of the N_t observations representing outliers. They are identified by numbers $1, 4, 7$ and 9. For very small values of the bandwidth less than 0.2, the criterion fails to indicate the real presence of fault by making almost all values near zero. For large values greater than 0.8, the threshold is difficult to estimate due to the extreme variations between values.

Given these figures, there is a clear separation between the values of the criteria related to the outliers $(1, 4, 7$ and $9)$ and the values of the non faulty observations. This statement is true for different numbers of principal components and bandwidth of the kernel function. This procedure is therefore quite relevant for the detection of outliers. However, no information can be given concerning the faulty variables.

Journal of Physics: Conference Series **659** (2015) 012032 doi:10.1088/1742-6596/659/1/012032

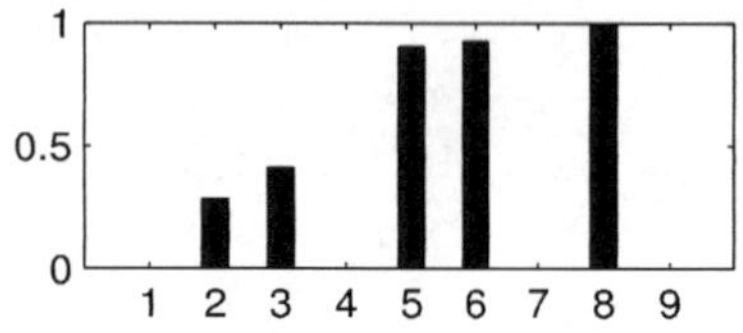

Figure 2. Criterion $\hat{\psi}$ with 30 PCs and bandwidth $c = 0.5$

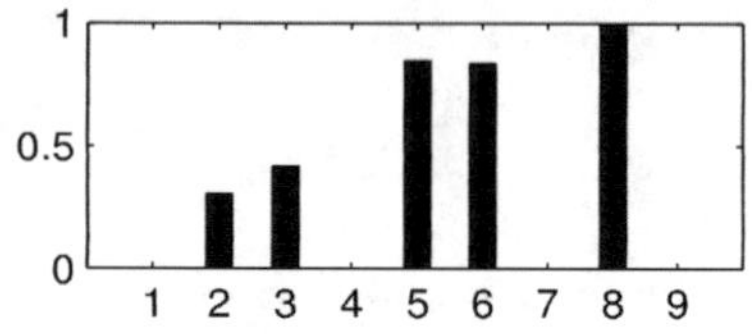

Figure 3. Criterion $\hat{\psi}$ with 30 PCs and bandwidth $c = 0.7$

3.3. Isolation of the faulty variables

In order to isolate the faulty variable, the indicator S_d defined by (16), was evaluated for all the directions $d = 1, \ldots, n$ on the test set. To identify the observation, the indicator is denoted $S_{d,j}$ with $j = 1, \ldots, N_t$. To make the interpretation and comparison of $S_{d,j}$ values easier, for the various observations and directions, the values of this indicator have been normalized. First, these values are written as $S_{d,j}$, for $d = 1, \ldots, n$, and $j = 1, \ldots, N_t$. The comparison is based on the normalized value of the indicator given by:

$$\overline{S}_{d,j} = \frac{S_{d,j}}{\max_j S_{d,j}} \tag{28}$$

Figures 4 and 5 concern the values of the normalized test indicator $\overline{S}_{d,j}$, defined in (28). It is evaluated according to the three directions of the data space. The comparison is based on a single kernel bandwidth with two different values of the number of principal components. For the case, where the number of principal components is correctly chosen, we can evaluate the indicator $\overline{S}_{d,j}$ in order to isolate the faulty direction. If the number of principal components is too small, the approximation of the initial dataset is wrong, and the information is lost. This number should be acceptable in order to maintain the main information needed to estimate the original dataset.

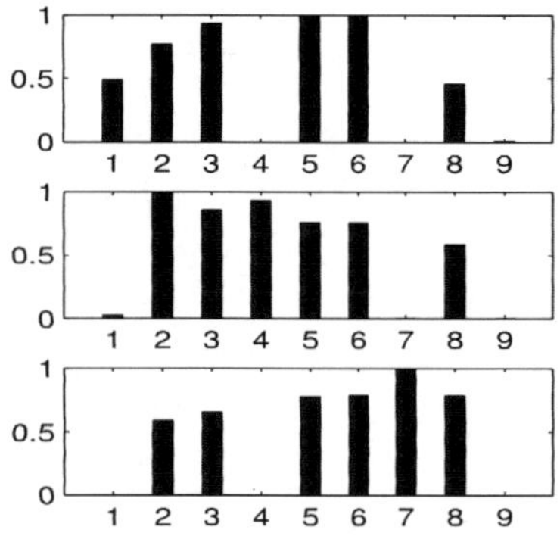

Figure 4. Values of $\overline{S}_{d,j}$ for 30 PCs with a bandwidth $c = 0.5$

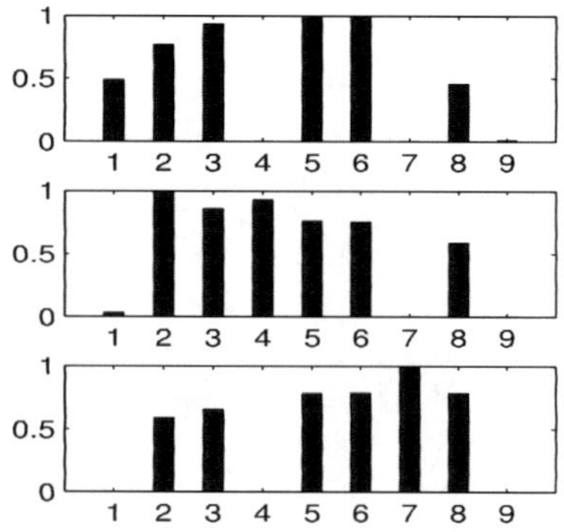

Figure 5. Values of $\overline{S}_{d,j}$ for 50 PCs with a bandwidth $c = 0.5$

Each figure contains three subfigures showing the values of the fault indicator $\overline{S}_{d,j}$ for the $N_t = 9$ test observations considering the $n = 3$ fault directions. More explicitly, the matrices W_d introduced aforementioned using (13) are defined by:

$$\begin{aligned} W_1 &= diag\begin{pmatrix}0 & 1 & 1\end{pmatrix} \\ W_2 &= diag\begin{pmatrix}1 & 0 & 1\end{pmatrix} \\ W_3 &= diag\begin{pmatrix}1 & 1 & 0\end{pmatrix} \end{aligned} \tag{29}$$

Therefore, for the metric W_1, the variable 1 is not involved in evaluating distances. Thus, the Gram matrix is insensitive to errors that can affect this variable. This perfectly explains the three values taken by the indicator $\overline{S}_{d,j}$. For example, for the first measure, $\overline{S}_{2,1}$ and $\overline{S}_{3,1}$ are evaluated from the variables $1,3$ and $1,2$ respectively. The fact that $\overline{S}_{2,1}$ and $\overline{S}_{1,1}$ provide practically zero values, whereas $\overline{S}_{3,1}$ is greater than zero, is due to the presence of the variable 1 that is common to the first two indicators, this variable being the faulty one. Moreover, for the observation x_9, we can detect the presence of fault, due to the zero values for each direction, however we can't isolate and indicate which direction is the faulty one. This can be explained by the fact that there exists different simultaneous faults affecting many directions.

Numerous tests were conducted with different values of the number of principal components and with different values of the kernel bandwidth. For a wide range of values for these parameters, the shapes of the preceding graphs are retained.

To analyze all the test observations, we construct the table 2 representing the theoretical signatures. This table shows the influence of an error impacting one of the three variables on the S_d indicators.

Table 2. Theoretical fault signatures: influence of a unique fault affecting each variable on the S_d indicator

	Variables		
	1	2	3
S_1	×	0	0
S_2	0	×	0
S_3	0	0	×

Then, from figures 4 and 5, we construct the table 3 which describes the experimental fault signatures. Two symbols have been used. The symbol 0 illustrates that a defect causes a near zero value of $\boldsymbol{\kappa}$ and thus the indicator S. However, the cross indicates that there is no faulty observation or measure. Once this table is constructed, we compare the experimental signature of each of the nine observations to the three theoretical signatures given in table 2, making it easier to detect in each case the variable presenting a faulty measure. Clearly, for observations $1, 4$, and 7, the variables $1, 2, 3$ are respectively detected as faulty.

Table 3. Experimental fault signatures

	observations								
	1	2	3	4	5	6	7	8	9
S_1	×	×	×	0	×	×	0	×	0
S_2	0	×	×	×	×	×	0	×	0
S_3	0	×	×	0	×	×	×	×	0

3.4. Fault amplitude estimation

Once the faulty observations have been detected (observations 1, 4, 7 and 9), and the variable isolation is performed, we, thus, proceed to estimate the amplitude of the fault. We apply the technique proposed in section 2.4. Table 4 illustrates the results, where we compare the true value of the fault and the estimated one.

Table 4. Real f and estimated $\widehat{f}$ fault amplitudes

	f	$\widehat{f}$
obs 1, var 1	4	3.98
obs 4, var 2	-4	-4.01
obs 7, var 3	4	3.95

4. Conclusion

In this paper, we propose to diagnose a nonlinear system using a directional study. To this end, the technique proposed in order to detect outliers derived from a nonlinear system is based on analysis of the observations in the kernel space. Three aspects are presented: the detection of faulty measures, the variable isolation of these faulty measures and the estimation of their fault amplitude. Therefore, we analyze the projections of the observations onto the kernel space. We take into consideration projections along directions favoring the recognition of healthy variables.

As for future work, a study of simultaneous many faulty variables will be derived, specifying the conditions for isolability. Moreover, an extension to nonlinear principal component regression can be proposed.

References

[1] C.F. Alcala and S.J. Qin. Reconstruction-based contribution for process monitoring with kernel principal component analysis. *Industrial & Engineering Chemistry Research*, 49(17):7849–7857, 2010.

[2] H.B. Aradhye, B.R. Bakshi, and R. Straus. Process monitoring by PCA, dynamic PCA, and multiscale PCA- theoretical analysis and disturbance detection in the tennessee eastman process. Technical report, AIChE Annual Meeting, 1999.

[3] M. Berar, M. Desvignes, G. Bailly, Y. Payan, and B. Romaniuk. Missing data estimation using polynomial kernels. In S. Singh, M. Singh, C. Apté, and P. Perner, editors, *International Conference on Advances in Pattern Recognition (1)*, volume 3686 of *Lecture Notes in Computer Science*, pages 390–399. Springer, 2005.

[4] F. Critchley. Multidimensional scaling: a short critique and a new method. In L.C.A. Corsten and J. Hermans, editors, *COMPSTAT*, pages 297–303. Physica, 1978.

[5] P. Cui and J. Fang. KPCA plus FDA for fault detection. In D. Liu, S. Fei, Z. Hou, H. Zhang, and C. Sun, editors, *4th International Symposium on Neural Networks : Advances in Neural Networks*, volume 4493 of *Lecture Notes in Computer Science*, pages 597–606. Springer Berlin Heidelberg, 2007.

[6] K. Fu, L. Dai, T. Wu, and M. Zhu. Sensor fault diagnosis of nonlinear processes based on structured kernel principal component analysis. *Journal of Control Theory and Applications*, 7(3):264–270, 2009.

[7] M.F. Harkat, G. Mourot, and J. Ragot. An improved PCA scheme for sensor FDI: Application to an air quality monitoring network. *Journal of Process Control*, 16(6):625–634, July 2006.

[8] M. Kallas, G. Mourot, D. Maquin, and J. Ragot. Diagnosis of nonlinear systems using kernel principal component analysis. In *11th European Workshop on Advanced Control and Diagnosis, ACD*, Berlin, Germany, 2014.

[9] M. Kallas, G. Mourot, D. Maquin, and J. Ragot. Fault estimation of nonlinear processes using kernel principal component analysis. In *14th European Control Conference*, Linz, Austria, 2015.

[10] Z. Li, G. Tan, and Y. Li. Fault diagnosis based on improved kernel fisher discriminant analysis. *Journal of software*, 7(12):2657–2662, 2012.

[11] S.J. Qin. Statistical process monitoring : basis and beyond. *Journal of Chemometrics*, 17(8-9):480–502, 2003.

[12] M. Scholz, F. Kaplan, C.L. Guy, J. Kopka, and J. Selbig. Non-linear PCA: a missing data approach. *Bioinformatics*, 21(20):3887–3895, 2005.

[13] M. Tamura and S. Tsujita. A study on the number of principal components and sensitivity of fault detection using PCA. *Computers & Chemical Engineering*, 31(9):1035–1046, 2007.

[14] C.K.I. Williams. On a connection between kernel pca and metric multidimensional scaling. In *Advances in Neural Information Processing Systems 13*, pages 675–681. MIT Press, 2001.

[15] J. Xu, J. Zhao, B. Ma, and S. Hu. Fault diagnosis of complex industrial process using KICA and sparse SVM. In *Mathematical Problems in Engineering*, 2013.

[16] A.F. Zuur. Principal coordinate analysis and non-metric multidimensional scaling. In *Analysing Ecological Data*, Statistics for Biology and Health, pages 259–264. Springer New York, 2007.

Fault Detection in the Blade and Pitch System of a Wind Turbine with $\mathcal{H}_2$ PI Observers

Ester Sales-Setién, Ignacio Peñarrocha, Daniel Dolz, Roberto Sanchis
Dept. of Industrial Systems Engineering and Design. Universitat Jaume I, Castelló, Spain.
E-mail: {esales,ipenarro,ddolz,rsanchis}@uji.es

Abstract. In this work, we present a fault detection strategy applicable to the blade and pitch system in offshore wind turbines. First, we model the system and possible faults and propose a PI observer to identify the faults. Then, the observer is designed accounting the sensors measurement noise, and addressing a trade off between the needs of false alarm rate, minimum detectable fault and detection time. By means of a well known benchmark, several simulations show the goodness of the approach and its flexibility to explicitly fix the fault detector performance.

1. Introduction

As a difference with respect to other energy production technologies, wind turbines are exposed to extremal and varying environmental conditions mainly due to wind nature. As a consequence, they are under permanently changing loads, what results in operation conditions with high variability hindering its correct operation. Recent studies, [1,2], show an increasing tendency on installing high power turbines, what results in a larger development of offshore wind farms. Its placement and size present a higher technological challenge than conventional wind turbines.

The periods of inactivity due to malfunctioning in wind turbines is in the order of a 3% of its life cycle, and the maintenance costs take between 10% and 30% of the total costs of a wind farm project [3]. A wind turbine consists of three main subsystems: the blade and pitch system, responsible of regulating the amount of wind energy that is transformed into mechanical one; the rotor, gathered by a drive train to the electrical generator, where the mechanical energy becomes electric; and the controller, who manages the operation of the previous systems. The authors in [4] explain that the pitch system is the main responsible of the downtimes of the machine.

In order to detect irregular operation of actuators and sensors, the control system can use fault detection techniques (FD) [5]. This methods will help the wind farm operator to detect operating problems and will help to avoid triggering critical faults. Thus, with FD not only downtimes can be reduced, but also the maintenance costs. The FD problem consists in the generation of residual signal that must be insensitive to noise measurements and disturbances, and a strategy to decide whether a fault has occurred by analysing the residual signal.

In this work, we assume that we have a sufficiently accurate model of the blade and pitch system and use a model-based FD technique [6–11]. We model the faults as additive signals, and the residual signal consists of an estimation of the fault signals. The design strategy of the fault detector consists in attenuating the $\mathcal{H}_2$ norm between both measurement noise and fault signal, and the fault estimation error. This norm is formalized via linear matrix inequalities. First, we define the required performance metrics as well as the detection algorithm. Then, using these requirements, we present a design strategy based on an optimization problem.

The structure of this work is as follows. First, we state the problem in Section 2, where we include a mathematical model of the blade and pitch system and the considered fault signals. In Section 2.2 we propose a strategy to detect these faults, and specify the requirements to account in the detector design in Section 3. In Section 4 we include some simulation results and, finally, in Section 5 we summarize the main conclusions.

2. Problem Description

The problem of fault detection of this work is focused on a real problem which deals with the blade and pitch system of an offshore wind turbine. The wind turbine's model is based on the one proposed by Odgaard et al. at [12], which describes a three-bladed variable-speed offshore wind turbine. The complexity of this model is sufficient enough to achieve the precision required to meet the FD requirements. The relevant faults that affect the blade and pitch system are also specified in [12], which is referred for further modeling details. Hereafter, we define the necessary elements for the design of the detector.

2.1. System

Due to its hydraulic character, the close-loop operation of each of the blade and pitch systems can be modeled as a second order system. For ease of notation and due to the similarity between all three systems, we will only work with one of them. The transfer function between the real pitch angle, β, and the corresponding reference angle, β_{ref}, given by the controller, results in

$$\frac{\beta(s)}{\beta_{ref}(s)} = \frac{\omega_n^2}{s^2 + 2\,\xi\,\omega_n\, s + \omega_n^2}. \tag{1}$$

This dynamics may vary due to either a drop in pressure in the hydraulic supply system or high air content in the oil. In both cases, these faults will result in a variation of parameters ω_n and ξ. Thus, we rewrite this parameters as $\omega_n = \omega_{n_0} + \Delta\omega_n$ and $\xi = \xi_0 + \Delta\xi$, where ω_{n_0} and ξ_0 refer to the non-faulty operation and $\Delta\omega_n$ and $\Delta\xi$ describe the respective variations. Thus, the relation between the pitch angle and its reference taking into account the possible dynamics variation (faulty situation) can be realized in state-space form as follows:

$$\dot{x} = \begin{bmatrix} 0 & 1 \\ -\omega_{n_0}^2 & -2\,\xi_0\,\omega_{n_0} \end{bmatrix} x + \begin{bmatrix} 0 \\ \omega_{n_0}^2 \end{bmatrix} u + \begin{bmatrix} 0 \\ 1 \end{bmatrix} f_a, \tag{2}$$

where $x = \begin{bmatrix} \beta & \dot{\beta} \end{bmatrix}^T$ represents the state vector and $u = \beta_{ref}$ is the input to the system. f_a is an additive signal that represents the faulty operation of the actuator (see [13]) and, in case of actuator faults, its value will be

$$f_{\dot{\beta}} = -\,2\,(\xi_0\,\Delta\omega_n + \omega_{n_0}\,\Delta\xi + \Delta\xi\,\Delta\omega_n)\,\dot{\beta} + (\Delta\omega_n^2 + 2\,\omega_{n_0}\,\Delta\omega_n)\,(\beta_{ref} - \beta). \tag{3}$$

In order to improve the system's reliability, each of the pitch angles is measured with two different sensors β_{m_1} and β_{m_2}. Each sensor is modeled by the sum of the actual variable measure, β_i, and a zero-mean gaussian noise of variance $\sigma_{\beta_m}^2$. This sensors can experiment either mechanical or electrical faults, what results, respectively, in a gain factor or an offset. An additive fault signal will take into account both situations and the measurements equation of the system results in

$$y = \begin{bmatrix} 1 & 0 \\ 1 & 0 \end{bmatrix} x + \begin{bmatrix} 1 \\ 0 \end{bmatrix} f_s + v, \tag{4}$$

with $y = \begin{bmatrix}\beta_{m1} & \beta_{m2}\end{bmatrix}^T$ the measurement vector and $v = \begin{bmatrix}v_1 & v_2\end{bmatrix}^T$ the white noises associated to each of the measurements. The additive signal f_s represents the malfunctioning of either of the sensors; so that, $f_s = f_{\beta_{m_1}}$ when the fault occurs in sensor β_{m_1} and $f_s = -f_{\beta_{m_2}}$ when the wrong measurements become from sensor β_{m_2}. If we assume that no simultaneous faults occur, this consideration will not influence the results.

Discretizing (2) and (4), the discrete state-space model results in

$$x_{k+1} = A_0\, x_k + B_0\, u_k + E_0\, f_{a,k}, \tag{5}$$

$$y_k = C_0\, x_k + H_0\, f_{s,k} + v_k. \tag{6}$$

In the current work, we model the faults as slow-varying signals, i.e.,

$$f_{k+1} = f_k + \Delta f_k, \tag{7}$$

where f_k represents the vector $\begin{bmatrix}f_{a,k} & f_{s,k}\end{bmatrix}^T$ and Δf_k the corresponding variation of the fault from instant k to $k+1$. Equation (7) allows modelling, for instance, step signals (Δf_k only takes a non-zero value at the fault appearance) or ramp signals (Δf_k takes a constant value). This kind of fault model have been widely used in the literature to analyze the behaviour of fault detection algorithms [13, 14]). Finally, we introduce an extended order model to include the fault dynamics as

$$z_{k+1} = A\, z_k + B\, u_k + B_f\, \Delta f_k, \tag{8}$$

$$y_k = C\, z_k + v_k, \tag{9}$$

where $z_k = \begin{bmatrix}x_k^T & f_k^T\end{bmatrix}^T$ is the extended state vector and

$$A = \begin{bmatrix} A_0 & \begin{bmatrix}E_0 & 0\end{bmatrix} \\ 0 & I \end{bmatrix},\quad B = \begin{bmatrix} B_0 \\ 0 \end{bmatrix},\quad B_f = \begin{bmatrix} 0 \\ I \end{bmatrix},\quad C = \begin{bmatrix} C_0 & \begin{bmatrix}0 & H_0\end{bmatrix} \end{bmatrix}.$$

2.2. Fault Detector

To reach fault detection objectives, we propose a scheme that relies on model-based observers. Specifically, we propose the following algorithm that estimates the extended state of the system,

$$\hat{z}_k = \hat{z}_{k-1} + L(y_k - C\,\hat{z}_{k-1}), \tag{10}$$

where L is the gain matrix that defines the relationship between the new measurements and the prediction obtained from the model. Defining the estimation error as $\tilde{z}_k = z_k - \hat{z}_k$, its dynamics is given by

$$\tilde{z}_{k+1} = (I - L\,C)\,A\,\tilde{z}_k - L\,v_k + (I - L\,C)\,B_f\,\Delta f_k \tag{11}$$

The estimation error of the fault signals is then $\tilde{f}_k = C_f\,\tilde{z}_k$. Following the approach developped at [15], we propose to use the estimated faults ($\hat{f}_k = C_f\,\hat{z}_k$ with $C_f = \begin{bmatrix}0 & I\end{bmatrix}$) to compute a quadratic residual at each instant k as

$$r_k = \hat{f}_k^T\, F^{-1}\, \hat{f}_k. \tag{12}$$

In all, the fault detection decision is given by

$$\begin{cases} \text{if } r_k \leq \delta & \text{No fault} \\ \text{if } r_k > \delta & \text{Fault} \end{cases} \tag{13}$$

where δ is a threshold to be defined. This means that fault detection is achieved by means of estimating the fault signals that affect the states and the measurements.

To evaluate the performance of a fault detector we should take into account different parameters. Firstly, according to [16, 17] the minimum detectable fault is defined as the fault that drives the residual to its threshold when no other fault occurs. We denote the minimum detectable fault of channel l as $f_{\min,l}$.

Note 1. *According to the previous definition, given (12) and (13) and assuming no estimation error ($\hat{f}_k = f_k$), the minimum detectable fault satisfies $f^2_{\min,l}\, F^{-1}(l,l) = \delta$, for $l = 1, ..., n_f$, where n_f is the number of faults that affect the system and $F^{-1}(l,l)$ is the l-th diagonal element of matrix F^{-1}.*

Secondly, the False Alarm Rate (FAR) is defined in [16] as the average probability of rising false alarms over an infinite-time window, i.e.,

$$\phi = \lim_{K\to\infty} \frac{1}{K} \sum_{k=0}^{K-1} \mathbf{Pr}\{r_k > \delta \mid f_k = 0\}. \tag{14}$$

Note 2. *Markov's inequality states that given x any positive random variable and $a > 0$, the inequality $\boldsymbol{Pr}\{x > a\} \le \frac{E\{x\}}{a}$ holds. Using this inequality and the definition of the* FAR *(14), we have that the* FAR *(14) of (12)-(13) is bounded by*

$$\phi \le \frac{1}{\delta}\, \boldsymbol{E}\{r_k \mid f_k = 0\}. \tag{15}$$

Thirdly, we define Integral of the Squared Error (ISE) of a fault estimation as

$$\varphi_l = \sum_{k=1}^{\infty} \tilde{f}_k^2(l). \tag{16}$$

The ISE characterizes the temporal behaviour of the detector against faults in accordance to the evaluation of disturbance rejection in control systems.

The design of a detector implies a trade-off between the FAR, the minimum detectable fault and the ISE. The objective of this work is to provide a method to compute the gain matrix L that satisfies this trade-off in a desirable and optimal way.

3. Design of the detector

To design matrix L, which regulates the estimator, and matrix F, which defines the residual, we will follow an approach based on the $\mathcal{H}_2$ norm of (11).

Theorem 1. *Consider the estimation algorithm (10) applied to system (8)-(9). If there exist any symmetric matrices P, F, Γ_f, Γ_v and a full matrix X fulfilling*

$$\begin{bmatrix} P & \mathcal{A} & 0 \\ \mathcal{A}^T & P & C_f^T \\ 0 & C_f & F \end{bmatrix} \succeq 0, \quad \begin{bmatrix} P & \mathcal{B}_\mu \\ \mathcal{B}_\mu^T & \Gamma_\mu \end{bmatrix} \succeq 0, \quad \mu = \{f, v\}, \tag{17}$$

with $\mathcal{A} = (P - X\,C)\,A$, $\mathcal{B}_f = (P - X\,C)\,B_f$ and $\mathcal{B}_v = -X$; then, defining the observer gain matrix as $L = P^{-1}\,X$, the following statements hold:

- *In the absence of faults and measurement noises the extended state estimation error (11) converges to zero.*

- *The residual signal (12) is bounded by*

$$\boldsymbol{E}\{\|r\|_{RMS}\} \leq tr(\Gamma_v V) + tr(\Gamma_f \Delta_f), \tag{18}$$

where $\Delta_f = \boldsymbol{E}\{\Delta f_k \Delta f_k^T\}$*; then, in the absence of faults, the residual signal satisfies* $\boldsymbol{E}\{\|r\|_{RMS}\} \leq tr(\Gamma_v V)$.

Proof. See Appendix A.1. □

The next theorem extends the results of the previous one to bound the FAR of the residual.

Theorem 2. *For a given threshold* $\delta > 0$ *and a positive number* $\bar{\phi} \in (0,1)$*, under the premisses of Theorem 1 if*

$$tr(\Gamma_v V) \leq \bar{\phi}\,\delta, \tag{19}$$

is fulfilled then, the following additional statement holds:

- *In the absence of faults and under null initial conditions, the fault detection algorithm assures a* FAR *bounded by* $\bar{\phi}$.

Proof. See Appendix A.2. □

Finally, the following theorem extends the previous results by bounding the ISE of the estimation of the faults.

Theorem 3. *Given a certain value* $\overline{\varphi}_l > 0$*, if the premisses of Theorem 1 are satisfied and the constraint*

$$F(l,l)\,\Gamma_f(l,l) \leq \overline{\varphi}_l, \tag{20}$$

is fulfilled, where $F(l,l)$ *and* $\Gamma_f(l,l)$ *denote the l-th diagonal element of matrices* F *and* Γ_f*, then, the following additional statement holds:*

- *In the absence of faults and under null initial conditions, the ISE of each element of the fault estimation vector,* $\hat{f}_k(l)$*, is bounded by* $\overline{\varphi}_l$.

Proof. See Appendix A.3. □

Note 3. *From the previous theorems we deduce that, given a certain threshold* δ*, the parameters* $\bar{\phi}$ *and* F *bound, respectively, the FAR and the minimum detectable faults. This determines the variability of the estimation signal caused by the noise in the absence of faults. Besides, the parameters* $\bar{\varphi}_l$ $(l = 1, \ldots, n_f)$ *bound the ISE after step faults in the absence of noise, i.e., they are proportional to the time of detection. For a certain value of* F*, the elements of matrix* Γ_f *allow to bound the ISE, i.e., to modify the time of detection.*

Depending on the desired performance of the detector, its design should be done following different strategies. Thereby, for a certain FAR, we may look either for the minimal estimation error ensuring certain minimum detectable faults or for the minimal value of the minimum detectable faults provided a certain estimation error. Since we do not require any determined estimation error in this case, we will fix the minimum detectable faults ($\bar{f}_{\min,l}$) and minimize the ISE, what we can write as:

$$\begin{aligned} &\text{minimize } \gamma \\ &\text{subject to } \mathcal{X}_1 = \begin{cases} (17),\ (19),\ tr(\Gamma_f) \leq \gamma, \\ F \preceq \mathcal{F} \end{cases} \end{aligned} \tag{21}$$

along the variables γ, Γ_f, F, Γ_v, P y X, where $\mathcal{F}$ is the matrix whose diagonal elements contain the minimum detectable faults, i.e., $\mathcal{F}(l,l) = \delta^{-1}\,\bar{f}^2_{\min,l}$.

It should be noticed that Note 2 uses Markov's inequality to bound the FAR. As it is well-known, this bound may be too conservative (see [18]) because it does not take into account the kind of probability distribution of the bounded signal. Under the assumption of gaussian noises, the estimation error of a fault signal will also follow a gaussian distribution. Thus, if we know this distribution we can recalculate a new threshold that will give us a tighten bound of the FAR.

Theorem 4. *Given a detector designed as shown in* (21)*, if the threshold in* (13) *is redefined as*

$$\underline{\delta} = \delta\, \delta_\Sigma\, \underline{\lambda}(F)/\overline{\lambda}(\Sigma_{\hat{f}}), \tag{22}$$

then $\bar{\phi}$ *is a tighten bound of the FAR. Here,* δ *is the parameter used in* (19) *and*

$$\delta_\Sigma = CDF^{-1}_{\chi^2_{n_f}}(1-\bar{\phi}), \tag{23}$$

with $CDF^{-1}_{\chi^2_{n_f}}$ *the inverse cumulative distribution function of a chi-squared random variable with* n_f *degrees of freedom.* $\underline{\lambda}(F)$ *and* $\overline{\lambda}(\Sigma_{\hat{f}})$ *represent, respectively, the minimum and maximum eigenvalues of* F *and matrix* $\Sigma_{\hat{f}}$ *defined as*

$$vec(\Sigma_{\hat{f}}) = \left(I - (I - LC)\, A \otimes (I - LC)\, A\right)^{-1} (L \otimes L)\, vec(V). \tag{24}$$

Proof. See Appendiix Appendix A.4. □

Note 4. *According to Theorem 1, if the detector* (12) *uses the threshold* $\underline{\delta}$ *instead of* δ*, the minimum detectable faults turn out to satisfy* $f_{\min,l}\, f^2_{\min,l}\, F^{-1}(l,l) = \underline{\delta}$*. These values will be smaller to those* $\bar{f}^2_{\min,l}$ *initially included in problem* (21)*. To achieve a minimum detectable fault close to* $\bar{f}^2_{\min,l}$ *when we use* $\underline{\delta}$ *as a threshold we can perform an iterative procedure by incrementing* $\mathcal{F}$ *in* (21)*.*

4. Simulations

In this section, we apply the FD strategy presented herein before to the blade and pitch system of a wind turbine. For this purpose, we use the parameters recommended in [6,12], i.e., $\xi_0 = 0.6$, $w_{n,0} = 11.11$ rad/s and $\sigma_{\beta_m} = 0.2°$. The sampling time used in the simulations is $T_s = 0.01$ s. We will simulate an operation of 35 s with a fault affecting the actuator and each of the senors. Thus, we can write the simulated fault signals as:

$$f_k = \begin{cases} \begin{bmatrix}\epsilon & 0\end{bmatrix}^T & \text{if } 2 \le t < 4 \\ \begin{bmatrix}0 & \epsilon\end{bmatrix}^T & \text{if } 14 \le t < 16 \\ \begin{bmatrix}0 & -\epsilon\end{bmatrix}^T & \text{if } 25 \le t < 27 \\ \begin{bmatrix}0 & 0\end{bmatrix}^T & \text{otherwise} \end{cases}, \tag{25}$$

where $\epsilon = 0.25$.

Figure 1 shows the simulation results of detectors with different guaranteed performances. Detector 1 has been designed with a FAR restricted to $\bar{\phi}_1 = 2.5 \cdot 10^{-6}$ and minimum detectable faults smaller than $\mathcal{F}_1(l,l) = 0.05$. For the same minimum detectable faults, $\mathcal{F}_2(l,l) = 0.05$, and a FAR limited to $\bar{\phi}_2 = 1 \cdot 10^{-6}$ we have obtained Detector 2. Finally, Detector 3 satisfies the same FAR as Detector 1, $\bar{\phi}_3 = 2.5 \cdot 10^{-6}$, and detects faults bigger than $\mathcal{F}_2(l,l) = 0.03$.

If we compare the performance of detectors 1 and 2 ($\mathcal{F}_1(l,l) = \mathcal{F}_2(l,l)$ and $\bar{\phi}_1 > \bar{\phi}_2$) we verify that the ISE increases when we try to estimate the same faults with a lower ISE ($\varphi_2(l) =$

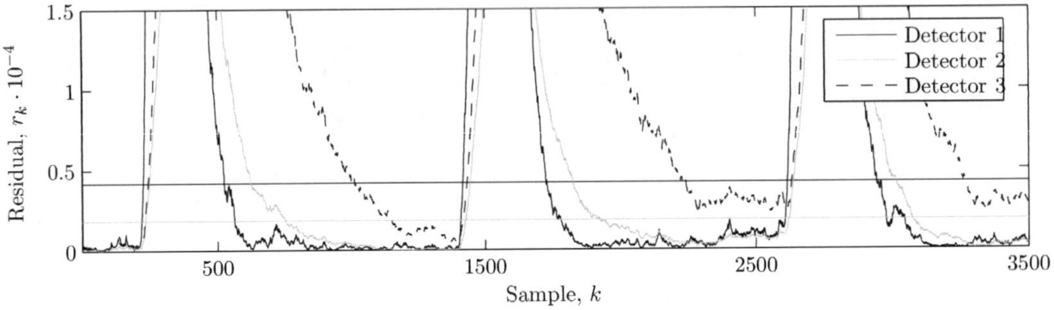

Figure 1. Residual signal and threshold of designs with different performances.

$2.31\,\varphi_1(l)$). This increment of the ISE with the same minimum detectable faults means a higher time of detection. Similarly, if we compare detectors 1 and 3 ($\mathcal{F}_1(l,l) > \mathcal{F}_3(l,l)$ and $\bar{\phi}_1 = \bar{\phi}_3$) we observe that the ISE increases when we try to detect smaller faults ($\varphi_3(l) = 4.55\,\varphi_1(l)$). This increment of ISE in the case of equal minimum detectable faults implies a larger time of detection.

5. Conclusion

In this work, we have developed a strategy to design a fault detection algorithm for the blade and pitch system of a wind turbine. The main goal of the approach is to provide a model-based observer that ensures an optimal detection while it guarantees certain performance requirements. Thus, we have proposed a design algorithm based on optimization techniques over the $\mathcal{H}_2$ norm that permits to fix, depending on the needs, the false alarm rate and the speed of detection, the accurateness of the estimations or the minimum detectable faults that will set off an alarm. The proposed technique has been validated on a international benchmark. Further research may extend this approach to other systems of the wind turbine such as the one that integrates the rotor, the drive train and the generator, including also uncertainties on the model.

Acknowledgments

This work has been supported by grant FPU14/01592 from *MECD*, grant PREDOC/2011/37 from *Universitat Jaume I de Castelló*, and by projects P11B2013-51 from *Universitat Jaume I de Castelló*, MICINN project number DPI2011-27845-C02-0 and *Generalitat Valenciana* project number GV/2015/099.

Appendix A. Proofs of Theorems

Appendix A.1. Proof of Theorem 1

Let us define a Lyapunov function at each instant k as $V_k = \tilde{z}_k^T\, P\, \tilde{z}_k$

- In the absence of noises or faults, after taking Schurs complements on the first linear matrix inequality presented in (17), premultiplying the result by $\tilde{z}_k^T$ and postmultipliying by its transpose, owe obtain that $V_{k+1} - V_k \leq 0$ what assures that the extended state estimation error (11) converges to zero.
- Performing similar steps on the other linear matrix inequalities in (17) (Schurs complements and operations with v_k^T y Δf_k^T), taking expected value on the results and adding the obtained constraints with the previous one, we get

$$\mathbf{E}\{V_{k+1}\} - \mathbf{E}\{V_k\} + \mathbf{E}\{\tilde{f}_k^T\, F^{-1}\, \tilde{f}_k\} - \mathbf{E}\{v_{k+1}^T\, \Gamma_v\, v_{k+1}\} - \Delta f_k^T\, \Gamma_f\, \Delta f_k \leq 0, \tag{A.1}$$

where we have considered the non correlation between $\tilde{x}_k$, v_{k+1} and Δf_k. Let w be a stochastic vector with mean μ and a covariance matrix W; if P is a symmetric matrix, then $\mathbf{E}\{w^T P w\} = \mu^T P \mu + tr(P W)$. Considering null initial conditions, applying the previous result to (A.1) and adding the result from $k = 0$ to $K - 1$, then

$$\sum_{k=0}^{K-1} \mathbf{E}\{\tilde{f}_k^T F^{-1} \tilde{f}_k\} \leq \sum_{k=0}^{K-1} \Delta x_k^T \Gamma_f \Delta x_k + K\, tr(\Gamma_v V)$$

where we have taken into account that $P \succeq 0$, $\mathbf{E}\{v_k\} = 0$ and $\sum_{k=0}^{K-1} tr(\Gamma_v V) = K\, tr(\Gamma_v V)$. Dividing the above expressions by K and taking the limit when $K \to \infty$, it leads to (18), which concludes this proof.

Appendix A.2. Proof of Theorem 2

Following Theoerm 1, if the system is fault-free, we have $\mathbf{E}\{\|r\|_{RMS}\} \leq tr(\Gamma_v V)$. Applying the results of Lemma 2, we have $\phi \leq \delta^{-1} tr(\Gamma_v V)$. Then, no matter the probabilistic distribution of r_k, the FAR (14) is bounded by $\bar{\phi}$ when constraint (19) is satisfied.

Appendix A.3. Proof of Theorem 3

As stated in Theorem 1, under null initial conditions we have

$$\sum_{k=0}^{\infty} \tilde{f}_k^T F^{-1} \tilde{f}_k \leq \sum_{k=0}^{\infty} \Delta f_k^T \Gamma_f \Delta f_k$$

when no noise exist. If a fault $f(l)$ appears and we can model it as a unitary step signal ($\Delta f_k(l) = 1$ at instant $k = 0$) then, $\sum_{k=0}^{\infty} \tilde{f}_k^2(l) \leq \gamma_{f_l} f_l^2 \Delta f_0(l)$ and the ISE is bounded by

$$\sum_{k=0}^{\infty} \tilde{f}_k^2(l) \leq \gamma_{f_l} f_l^2. \tag{A.2}$$

Then, bounding the product $\Gamma_f F$ we restrict the ISE of the fault estimations $f(l)$, what proves(20).

Appendix A.4. Proof of Theorem 4

We define $\Sigma_{\hat{f}}$ as the covariance matrix of the fault estimation signal, i.e., $\Sigma_{\hat{f}_k} = \mathbf{E}\{\hat{f}_k^T \hat{f}_k\}$. In the absence of faults, $\Sigma_{\hat{f}_k} = \mathbf{E}\{\tilde{f}_k^T \tilde{f}_k\}$, so that by the equation (11) we arrive to the expression included in (24).

In the absence of faults, a signal defined as $s_k = \hat{f}_k^T \Sigma_{\hat{f}}^{-1} \hat{f}_k$ follows a chi-squared distribution of n_f degrees of freedom, $\chi^2_{n_f}$; so, by definition,

$$\mathbf{Pr}\big(\hat{f}_k^T \Sigma_{\hat{f}}^{-1} \hat{f}_k > \delta_\Sigma\big) = \bar{\phi}.$$

Given a matrix M, we have $\underline{\lambda}(M)\, I \prec M \prec \overline{\lambda}(M)\, I$; thus we can bound s_k by

$$s_k = \hat{f}_k^T \Sigma_{\hat{f}}^{-1} \hat{f}_k > \hat{f}_k^T \hat{f}_k \underline{\lambda}(\Sigma_{\hat{f}}^{-1}),$$

and the residual r_k (defined in (12)) with

$$r_k = \hat{f}_k^T F^{-1} \hat{f}_k < \hat{f}_k^T \hat{f}_k \overline{\lambda}(F^{-1}).$$

From these two bounds we establish the relationship

$$s_k > \frac{\underline{\lambda}(\Sigma_f^{-1})}{\overline{\lambda}(F^{-1})} r_k.$$

For a signal $s'_k < s_k$ we have that $\mathbf{Pr}\big(s'_k > \delta_\Sigma\big) < \bar{\phi}$, what demonstrates that the threshold (22) bounds the FAR by $\bar{\phi}$.

References

[1] M. Bilgili, A. Yasar, and E. Simsek, "Offshore wind power development in europe and its comparison with onshore counterpart," *Renewable and Sustainable Energy Reviews*, vol. 15, no. 2, pp. 905–915, 2011.

[2] A. R. Henderson, C. Morgan, B. Smith, H. C. Sørensen, R. J. Barthelmie, and B. Boesmans, "Offshore wind energy in europea review of the state-of-the-art," *Wind energy*, vol. 6, no. 1, pp. 35–52, 2003.

[3] B. Hahn, M. Durstewitz, and K. Rohrig, "Reliability of wind turbines," in *Wind energy*. Springer, 2007, pp. 329–332.

[4] Z. Daneshi-Far, G. Capolino, and H. Henao, "Review of failures and condition monitoring in wind turbine generators," in *Electrical Machines (ICEM), 2010 XIX International Conference on*. IEEE, 2010, pp. 1–6.

[5] Z. Hameed, Y. Hong, Y. Cho, S. Ahn, and C. Song, "Condition monitoring and fault detection of wind turbines and related algorithms: A review," *Renewable and Sustainable energy reviews*, vol. 13, no. 1, pp. 1–39, 2009.

[6] P. F. Odgaard, J. Stoustrup, R. Nielsen, and C. Damgaard, "Observer based detection of sensor faults in wind turbines," in *Proceedings of European Wind Energy Conference*, 2009, pp. 4421–4430.

[7] X. Wei, M. Verhaegen, and T. van Engelen, "Sensor fault detection and isolation for wind turbines based on subspace identification and kalman filter techniques," *International Journal of Adaptive Control and Signal Processing*, vol. 24, no. 8, pp. 687–707, 2010.

[8] W. Chen, S. X. Ding, A. Sari, A. Naik, A. Q. Khan, and S. Yin, "Observer-based fdi schemes for wind turbine benchmark," in *Proceedings of IFAC world congress*, 2011, pp. 7073–7078.

[9] A. A. Ozdemir, P. Seiler, and G. J. Balas, "Wind turbine fault detection using counter-based residual thresholding," in *Proceedings of IFAC world congress*, vol. 2011, 2011, pp. 8289–8294.

[10] F. Shi and R. Patton, "An active fault tolerant control approach to an offshore wind turbine model," *Renewable Energy*, vol. 75, pp. 788–798, 2015.

[11] S. Georg and H. Schulte, "Takagi-sugeno sliding mode observer with a weighted switching action and application to fault diagnosis for wind turbines," in *Intelligent Systems in Technical and Medical Diagnostics*. Springer, 2014, pp. 41–52.

[12] P. F. Odgaard, J. Stoustrup, and M. Kinnaert, "Fault-tolerant control of wind turbines: A benchmark model," *Control Systems Technology, IEEE Transactions on*, vol. 21, no. 4, pp. 1168–1182, 2013.

[13] D. Koenig, "Unknown input proportional multiple-integral observer design for linear descriptor systems: application to state and fault estimation," *Automatic Control, IEEE Transactions on*, vol. 50, no. 2, pp. 212–217, 2005.

[14] A. Khedher, K. Benothman, D. Maquin, and M. Benrejeb, "State and sensor faults estimation via a proportional integral observer," in *Systems, Signals and Devices, 2009. SSD'09. 6th International Multi-Conference on*. IEEE, 2009, pp. 1–6.

[15] D. Dolz, I. Peñarrocha, and R. Sanchis, "Performance tradeoffs for networked jump observer-based fault diagnosis," *Signal Processing, IEEE Transactions on*, vol. 63, no. 10, pp. 2692–2703, 2015.

[16] J. Chen and R. Patton, *Robust model-based fault diagnosis for dynamic systems*. Springer Publishing Company, Incorporated, 2012.

[17] S. Ding, *Model-based fault diagnosis techniques: design schemes, algorithms, and tools*. Springer Science & Business Media, 2008.

[18] B. Ogunnaike, *Random phenomena: fundamentals of probability and statistics for engineers*. CRC Press, 2011.

Journal of Physics: Conference Series **659** (2015) 012034 doi:10.1088/1742-6596/659/1/012034

Design of sensor and actuator multi model fault detection and isolation system using state space neural networks

Andrzej Czajkowski

Institute of Control and Computation Engineering, University of Zielona Góra, ul. Podgórna 50, 65–246 Zielona Góra, Poland

E-mail: A.Czajkowski@issi.uz.zgora.pl

Abstract. This paper deals with the application of state space neural network model to design a Fault Detection and Isolation diagnostic system. The work describes approach based on multi-model solution where the SIMO process is decomposed into simple models (SISO and MISO). With such models it is possible to generate different residual signals which later can be evaluated with simple thresholding method into diagnostic signals. Further, such diagnostic signals with the application of Binary Diagnostic Table (BDT) can be used to fault isolation. All data used in experiments is obtain from the simulator of the real-time laboratory stand of Modular Servo under Matlab/Simulink environment.

1. Introduction

Recently, it has been observed an increasing development of the Fault Diagnosis (FD) methods for the Fault Tolerant Control (FTC) system desing purposes [3]. It is straightly connected to the advantages of the systems which can maintain current performance as close to the desirable one, and preserve stability conditions in the presence of faults [17]. Faults and equipment failures directly affect the performance of the control system and can result in large ecomomic losses and violation of the safety regulations. During the fault tolerant control system design, the basic problem is the early detection and isolation of possible faults. The proper isolation of the faults is crucial during FTC system design. The reason behind that statement is that the different control strategy needs to be applied in case of sensor faults and completely different behaviour is needed when actuator fault occurs. The main objective here is to have FDI methods as fast and as accurate as possible. Such assumptions are crucial to further design a control system which can provide the acceptable corrections of the control signal in case of fault. The paper focuses on two aspects of the FDI design, first is stage of system modelling and second, the application of the process models to residual signal generation and evaluation. One of the tools which are heavily exploited in this task are Artificial Neural Networks (ANN) [2, 14, 15]. In this work to construct the models of the system the so called State Space Neural Networks (SSNN) are applied. Such neural model of the system was successfully applied to Model Predictive Control of Twin Rotor Aerodynamical System in the previous work of the author [4]. The SSNN model is then used to carry out the fault detection by analysing the residual and fault isolation is done with the application of Binary Diagnostic Table. The methodology presented in the paper is tested on the example of a Modular Servo in case of both actuator and sensor faults.

The paper is organized as follows, Section 2 presents a general desciption of the Modular Servo System and provides information about faulty scenarios considered. The state space neural networks are described in Section 3. Section 4 presents a FDI algorithm, while experimental results are included in Section 5.

2. Modular Servo

In this work a Modular Servo System (MSS) laboratory stand is considered [7]. The main concept of the MSS is to create a rapid and direct path from control system design to hardware implementation. The MSS supports the real-time design and implementation of advanced control methods using MATLAB/Simulink enviroment. The MSS consists of the Inteco servomechanism and open-architecture software environment for real-time control experiments. A scheme of considered DC motor is presented in Fig. 1.

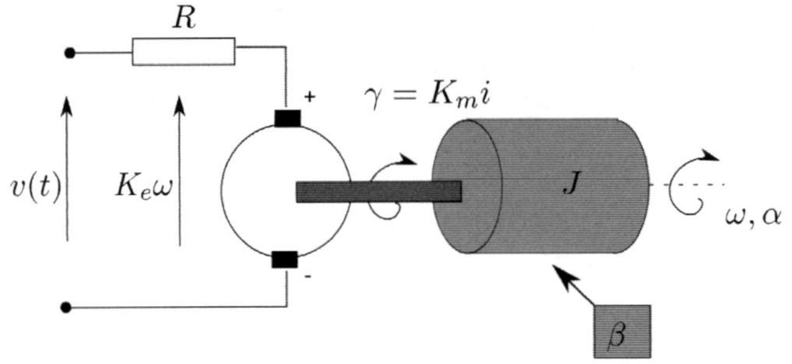

Figure 1. Diagram of DC motor.

The system is described by two classical equations, the electrical one:

$$v(t) = Ri(t) + K_e\omega(t) \tag{1}$$

and mechanical as follows:

$$J\dot{\omega}(t) = K_m i(t) - \beta\omega(t) \tag{2}$$

where $v(t)$ is the input voltage, $i(t)$ is the armature current, $\omega(t)$ is the angular velocity of the rotor, R is the resistance of armature winding, J is the moment of inertia of the moving parts, β is the damping coefficient due to viscous friction, $K_e\omega(t)$ is the back EMF and $K_M i(t)$ is the electromechanical torque.
The system used in this paper is described by the following nonlinear state space equations:

$$\begin{aligned} \dot{x_1} &= x_2 \\ \dot{x_2} &= c(u - g(x_2)) \end{aligned} \tag{3}$$

where state variable x_1 and x_2 are the angle α (position of the shaft) and angular velocity ω, respectively. The control applied is in PWM signal form so it is assumed that dimensionless control signal u is scaled input voltage $u(t) = v(t)/v_{max}$ and it satisfies $|u(t)| \leq 1$. The function g is the inverted steady state characteristics of the system, which needs to be determined experimentally.

The system can be classified as a multivariable (SIMO) because it has two measurable state variables and one control variable. This allows to simulate three different fault scenarios which are described in the Table 1.

Table 1. Specification of faulty scenarios.

Fault	Description	Type
f_1	DC motor power loss	multiplicative (0.7)
f_2	Tacho-Generator failure	additive (-30)
f_3	Angle positioner failure	additive (-30)

3. State Space Neural Network

The crucial part in FDI design process is the model of the plant which is used for residual generation. In this work the State Space Neural Network (SSNN) which are a very important class of dynamic neural networks are used [16, 12]. The structure of such network is depicted in Fig. 2. The output of the hidden layer is fed back to the input layer through a bank of unit delays. The number of unit delays determines the order of the model. In general, a user decides how many neurons are used to produce feedback.

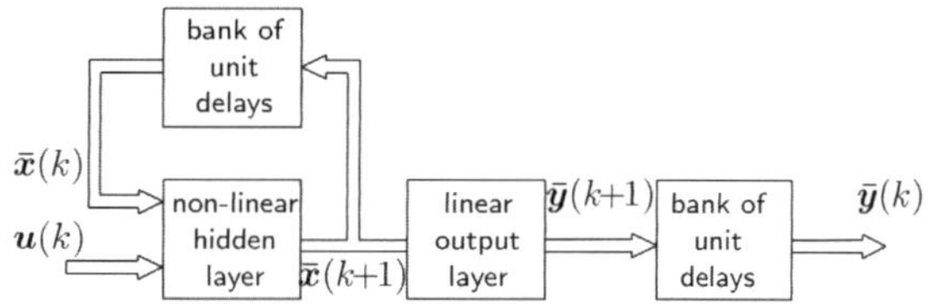

Figure 2. Block scheme of the state space neural network.

Let $\boldsymbol{u}(k) \in \mathbb{R}^n$ be the input vector, $\bar{\boldsymbol{x}}(k) \in \mathbb{R}^q$ - the output of the hidden layer at time k, and $\bar{\boldsymbol{y}}(k) \in \mathbb{R}^m$ - the output vector. The state space representation of the neural model is described by the equations

$$\begin{aligned} \bar{\boldsymbol{x}}(k+1) &= \bar{g}(\bar{\boldsymbol{x}}(k), \boldsymbol{u}(k)) \\ \bar{\boldsymbol{y}}(k) &= \boldsymbol{C}\bar{\boldsymbol{x}}(k) \end{aligned}, \tag{4}$$

where $\bar{g}(\cdot)$ is a nonlinear function characterizing the hidden layer, and $\boldsymbol{C}$ represents synaptic weights between hidden and output neurons. Introducing the weight matrix between input and hidden layers $\boldsymbol{W}^u$ and the matrix of recurrent links $\boldsymbol{W}^x$, the representation (4) can be rewritten in the following form:

$$\begin{aligned} \bar{\boldsymbol{x}}(k+1) &= h(\boldsymbol{W}^x\bar{\boldsymbol{x}}(k) + \boldsymbol{W}^u\boldsymbol{u}(k)) \\ \bar{\boldsymbol{y}}(k) &= \boldsymbol{C}\bar{\boldsymbol{x}}(k) \end{aligned}, \tag{5}$$

where $h(\cdot)$ stands for the activation function of the hidden neurons. In the most cases, the hyperbolic tangent activation function is selected giving pretty good modelling results. For the state space model the outputs of hidden neurons which constitute feedbacks are, in general, unknown during training. Therefore, state space neural models can be trained only by minimizing the simulation error. If state measurements are available, the training can be carried out much easier using the series-parallel identification scheme, similarly as for the external dynamic approach (the feedforward network with tapped delay lines). In spite of this inconvenience, state space models have a number of advantages, contrary to other recurrent neural networks ([13], [18]):

- The number of states (model order) can be selected independently from the number of hidden neurons. In the recurrent networks, e.g. Williams-Zipser, Elman, locally recurrent networks, the number of neurons directly influences the model order, which significantly impedes the modelling phase.
- Model states are easily accessible from the outside (they feed the network input). This property can be useful when state measurements are available at some time instants.
- State space neural models are useful in the fault tolerant control framework as they can be used to determine the approximation of a fault effect. As the fault effect can be represented in the state space one can handle different kind of faults including multiplicative and additive ones.

Presented advantages of State Space Neural Networks make such structure very interesting and promising in the solving the fault compensation problem. Also the class of nonlinear state space models is strongly evaluated in different scientific approaches as a nominal model (e.g [9]).

4. Fault Detection and Isolation

In this paper the FDI approach is based on the residual evaluation. To achieve the goal of early detection and proper localisation of faults it is needed to design a process model. In this paper both the sensor and actuator faults are taken into consideration. To distinguish one group from another it is proposed to use a multi-model solution. To solve this problem let us consider a nonlinear dynamic system governed by the following state equation:

$$\begin{aligned} \boldsymbol{x}(k+1) &= g(\boldsymbol{x}(k), \boldsymbol{u}(k)) \\ \boldsymbol{y}(k) &= \boldsymbol{C}\boldsymbol{x}(k), \end{aligned} \tag{6}$$

where $g(\cdot,\cdot)$ is a process working at the normal operating conditions, $\boldsymbol{x}(k)$ is a state vector and $\boldsymbol{u}(k)$ is the control. In case of the considered servomechanism system, the output $\boldsymbol{y}(k)$ consists of two variables – the position angle α and angular velocity ω. For diagnostic purposes it is proposed to decompose this SIMO system into the following SISO and MISO neural models:

$$\begin{aligned} \boldsymbol{x}_{ua}(k+1) &= g_{ua}(\boldsymbol{x}_{ua}(k), u(k)) \\ y_{ua}(k) &= \boldsymbol{C}\boldsymbol{x}_{ua}(k). \end{aligned} \tag{7}$$

which models the relation between voltage input u and angle position of the rotor shaft y_{ua}, through identified with SSNN function g_{ua}. This model further will be refereed as model 1 or input-angle model (uamodel).

$$\begin{aligned} \boldsymbol{x}_{uv}(k+1) &= g_{uv}(\boldsymbol{x}_{uv}(k), u(k)) \\ y_{uv}(k) &= \boldsymbol{C}\boldsymbol{x}_{uv}(k) \end{aligned} \tag{8}$$

which models the relation between voltage input u and angular velocity of the rotor y_{uv}, through identified with SSNN function g_{uv}. This model further will be refereed as model 2 or input-velocity model (uvmodel).

$$\begin{aligned} \boldsymbol{x}_{va}(k+1) &= g_{va}(\boldsymbol{x}_{va}(k), \omega(k)) \\ y_{va}(k) &= \boldsymbol{C}\boldsymbol{x}_{va}(k) \end{aligned} \tag{9}$$

which models the relation between measured angular velocity ω and angle position of the rotor shaft y_{va}, through identified with SSNN function g_{va}. This model further will be refereed as model 3 or velocity-angle model (vamodel).

$$\begin{aligned} \boldsymbol{x}_{uav}(k+1) &= g_{uav}(\boldsymbol{x}_{uav}(k), u(k), \alpha(k)) \\ y_{uav}(k) &= \boldsymbol{C}\boldsymbol{x}_{uav}(k) \end{aligned} \tag{10}$$

which models the relation between voltage input u together with measured angular velocity ω and angle position of the rotor shaft y_{uav}, through identified with SSNN function g_{uav}. This model further will be refereed as model 4 or input-angle-velocity model (uavmodel).

$$\begin{aligned} \boldsymbol{x}_{uva}(k+1) &= g_{uva}(\boldsymbol{x}_{uva}(k), u(k), \omega(k)) \\ y_{uva}(k) &= \boldsymbol{C}\boldsymbol{x}_{uva}(k) \end{aligned} \tag{11}$$

which models the relation between voltage input u together with measured angle position of the rotor shaft α and angular velocity of the rotor y_{uva}, through identified with SSNN function g_{uva}. This model further will be refereed as model 5 or input-velocity-angle model (uvamodel).

The models are realised by the means of the SSNN and are summarised in the Table 2.

Next, the angular velocity ω and angle position α can be measured as the system outputs and

name	inputs	outputs
model 1 (uamodel)	u	α
model 2 (uvmodel)	u	ω
model 3 (vamodel)	ω	α
model 4 (uavmodel)	u,α	ω
model 5 (uvamodel)	u,ω	α

Table 2. Diagnostic state space models

used for the residual calculation as follows:

$$r_1(k) = \alpha(k) - y_{ua}(k) \tag{12}$$

$$r_2(k) = \omega(k) - y_{uv}(k) \tag{13}$$

$$r_3(k) = \alpha(k) - y_{va}(k) \tag{14}$$

$$r_4(k) = \omega(k) - y_{uav}(k) \tag{15}$$

$$r_5(k) = \alpha(k) - y_{uva}(k) \tag{16}$$

To evaluate calculated residuals to fault detection and isolation, in this paper a simple thresholding method is applied [1]. The diagnostic signals s_i for each model are defined as follows:

$$s_i = \begin{cases} 0 : r_i > Tl_i \ \wedge \ r_i < Tu_i \\ 1 : r_i < Tl_i \ \vee r_i > Tu_i \end{cases}, \tag{17}$$

where $i \in \{1, 2, 3, 4, 5\}$ and Tu_i, Tl_i are upper and lower thresholds respectively. The thresholds are calculated using the so-called 3σ method using collected data during work of the system in nominal conditions.

Finally, the BDT can be created and applied for fault detection and isolation [8]. The designed BDT is presented in Table 3.

When all diagnostic signals are equal to the assumed level for the specific fault, then the fault alarm is raised. If any diagnostic signal is equal 1 then a unknown fault is signalled.

Table 3. Binary Diagnostic Table

	s_1	s_2	s_3	s_4	s_5
f_1	1	1	0	1	1
f_2	0	1	1	1	1
f_3	1	0	1	0	1

5. Experimental results

5.1. Neural Modelling

The first step in the design of fault diagnosis system is the process modelling. The proper model is crucial for residual generation and evaluation. As was shown in section 4 in this paper, the approach based on multiple models is used. The models described in section 4 were designed by the means of the SSNN. To design the training data multiple sinusoids were mixed with random signals. Such control signal gives a very rich data which characterise the whole operating range of the plant. The angular velocity operating range was equal to ± 200 and angle was assumed as ± 2000 rads. The collected outputs signals from encoder sensor and tacho-generator were normalised before the training process. The obtained data were divided into training and testing sets and resampled from 0.01s to 0.1s. During normal work of FDI system models outputs are linearly interpolated to match the sampling of the plant. The networks were trained for 100 epochs using Levenberg-Marquardt algorithm and the selection of the proper network structure is based on the criterion of the Sum of Squared Errors (SSE) index. The tested structures of specific models and training results are shown in table 4. The best structures and noted with frames.

Table 4. Selection of neural network structures.

		Model 1		Model 2		Model 3		Model 4		Model 5	
Model order	**No. neurons**	**Training SSE**	**Testing SSE**	**Training SSE**	**Testing SSE**	**Training SSE**	**Testing SSE**	**Training SSE**	**Testing SSE**	**Training SSE**	**Testing SSE**
1	2	126.2	279.4	1.58	1.80	0.18	61.21	1.58	1.80	0.1849	86.9
1	3	92.5	297.2	0.742	1.9e+04	0.0057	0.20	1.35	1.87	0.049	0.080
1	4	198.6	278.2	0.786	7.1e+03	0.0048	0.0256	0.26	1.54	0.1279	103.4
1	5	140.6	220.8	0.582	1.01	0.0037	0.11	0.108	0.220	0.0042	0.0593
1	6	124.3	198.0	0.652	3.3e+03	0.0041	0.12	0.107	0.221	0.2944	68.5
2	2	87.1	122.6	2.30	3.13	0.13	14.74	2.19	3.55	0.1525	19.5
2	3	0.77	1.47	0.0384	0.0729	0.0014	0.013	1.95	3.19	0.0103	6.55
2	4	0.64	0.72	0.0295	0.057	6.9e-04	0.0070	1.42	2.61	0.0017	0.0179
2	5	0.73	1.40	0.0299	0.0494	5.5e-04	6.7e-04	0.024	0.075	5.1e-04	0.0976
2	6	1.24	87.01	0.0397	0.0961	8.6e-04	9.6e-04	0.023	0.12	0.0010	0.0291
3	3	86.0	102.23	2.01	2.62	0.034	23.02	0.12	0.66	3.6993	216.1
3	4	1.12	14.40	0.28	0.402	11.35	252.13	1.64	3.36	1.4086	283.7
3	5	84.8	105.08	0.13	0.23	3.4e-04	0.0068	0.58	2.16	0.1877	95.24
3	6	8.3e+3	8.2e+03	3.34	2.2e+03	0.0018	0.0164	1.39	1.62	0.4683	162.05

Summarising the results, the best structures used in further experiments were shown in the Table 5.

5.2. Fault Management

The designed models of the plant allows to carry out the FDI experiments in which system is controlled by PID controller, tuned by the manufacturer of the plant. The control task was

Table 5. Selected neural network structures

	Model order	No neurons
model 1	2	4
model 2	2	5
model 3	2	5
model 4	2	5
model 5	2	4

to follow the reference angle signal, which changes between two values. The models are used for residual generation and simple thresholding method is used to evaluate residuals to obtain diagnostic signals. The calculated thresholds together with means and standard deviations are shown in Table 6. Three faults scenarios – one actuator and two sensor which are described in

Table 6. Thresholds calculation results.

	mean	σ	T_l	T_u
model 1	0.0132	7.8625	-23.5743	23.6006
model 2	0.2975	3.2198	-9.3619	9.9568
model 3	-0.0235	0.0165	-0.0730	0.0260
model 4	0.4268	3.2956	-9.4601	10.3137
model 5	-0.0563	0.2087	-0.6823	0.5697

Table 1, were considered. As shown in Figs. 3, 4 5 and 6 the actuator fault f_1 is very minimal and do not influence the work of the system as the PID controller compensate it easily (faults of greater magnitude should be diagnosed with even higher efficiency). In case of tacho-generator fault f_2 it does not influence the work of the control system but in the case of encoder fault f_3 which is used as the control feedback the fast detection and appropriate action is crucial.

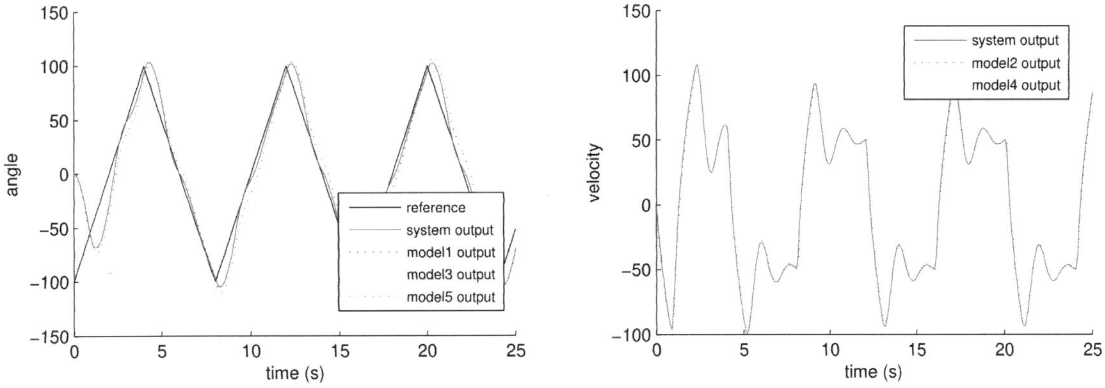

Figure 3. Outputs of the system and models in nominal conditions.

All of faults were introduced at the 14th second of simulation. The residuals were calculated and together with thresholds are presented in Figs. 7 and 8.

In all three scenarios as one can see the diagnostic system worked quite efficiently and it was easy to detect and isolate fault. The performance indexes in the form of false detection rate (r_{fd}), time of fault detection (t_{dt}), true fault detection rate (r_{td}), time of fault isolation (t_{ti}), false

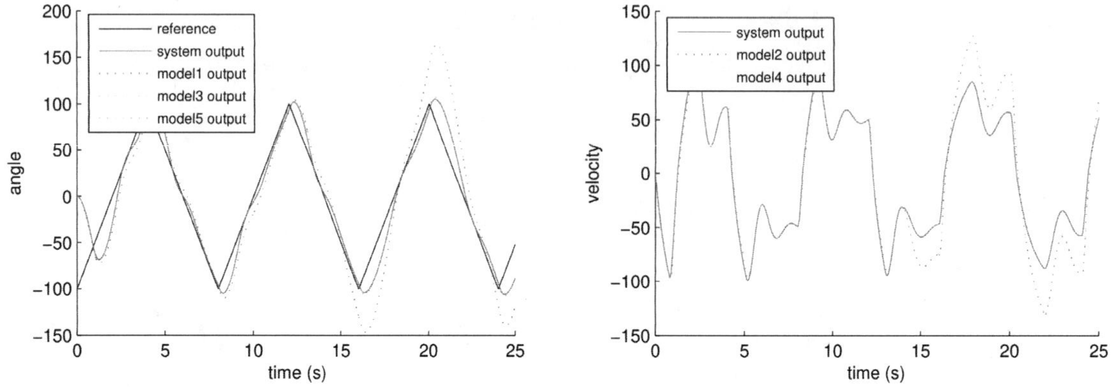

Figure 4. Outputs of the system and models in case of f_1.

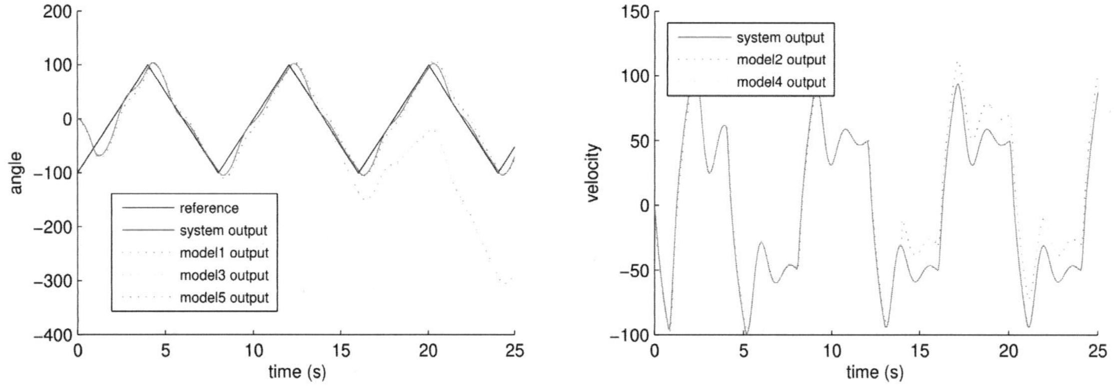

Figure 5. Outputs of the system and models in case of f_2.

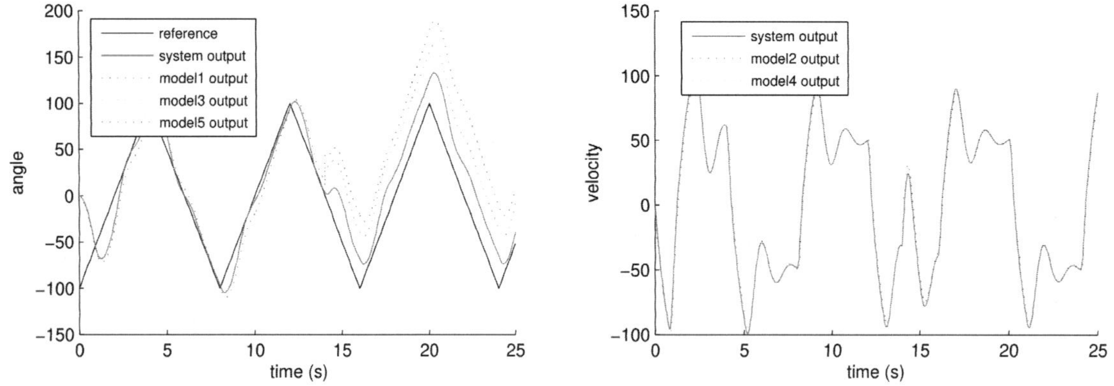

Figure 6. Outputs of the system and models in case of f_3.

isolation rate (r_{fi}) and true isolation rate (r_{ti}) were calculated according to formulas defined in [13]. Indexes are shown in Table 7. The obtained results are quite satisfying, especially in case of sensor faults where fault detection is instant and isolation rate is over 99%. Such high

Journal of Physics: Conference Series **659** (2015) 012034 doi:10.1088/1742-6596/659/1/012034

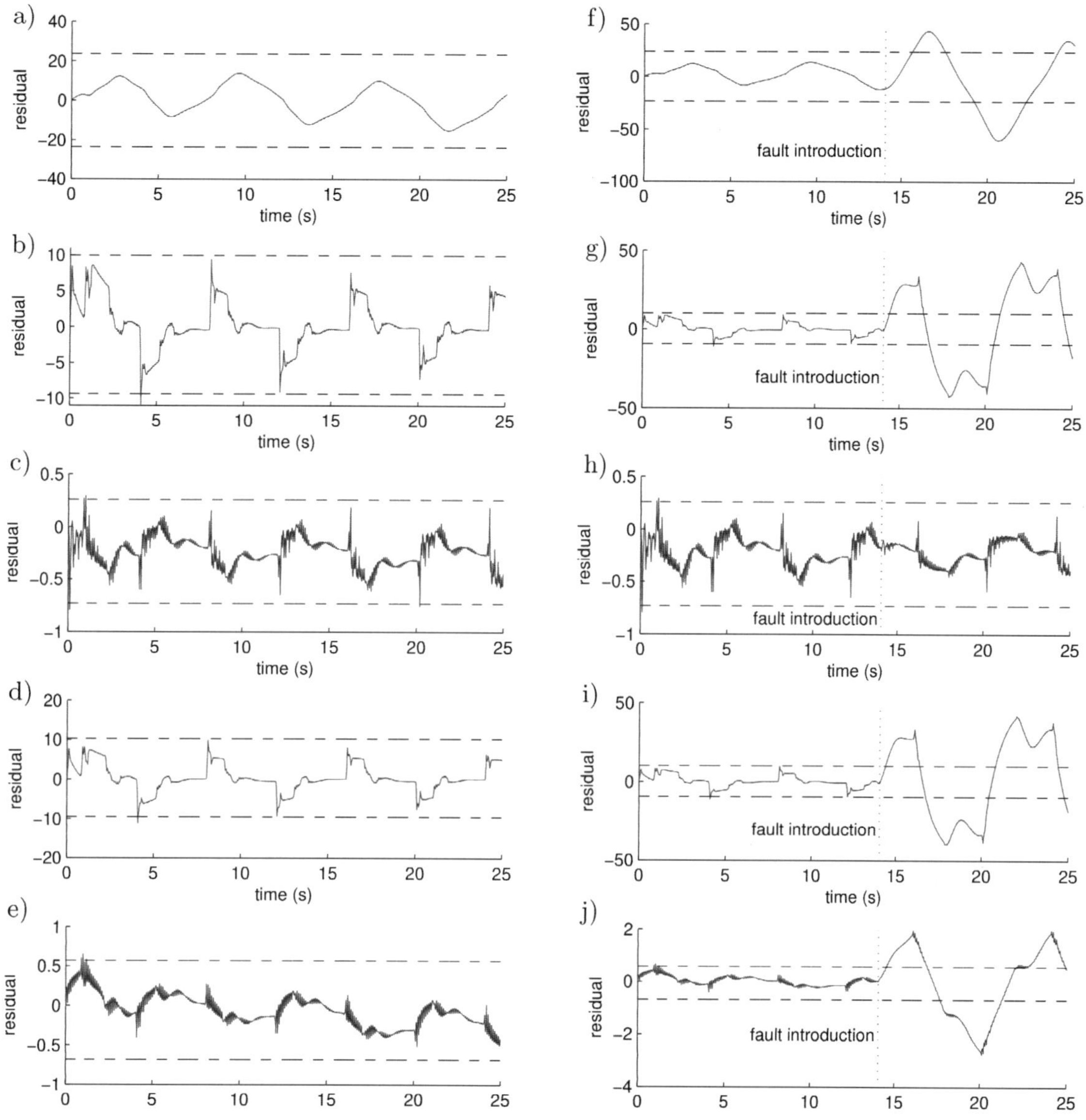

Figure 7. Residuals generated and thresholds with model 1 (a), model 2 (b), model 3 (c), model 4 (d) and model 5 (e) in nominal conditions and with model 1 (f), model 2 (g), model 3 (h), model 4 (i) and model 5 (j) in case of fault f_1.

Table 7. Diagnostic performance indexes

	r_{fd}	t_{dt}	r_{td}	t_{ti}	r_{fi}	r_{ti}
no fault	0.0032	-	-	-	-	-
f_1	0.0043	37	0.9673	164	0	0.2861
f_2	0.0043	1	1	4	0	0.9982
f_3	0.0043	1	1	2	0	0.9973

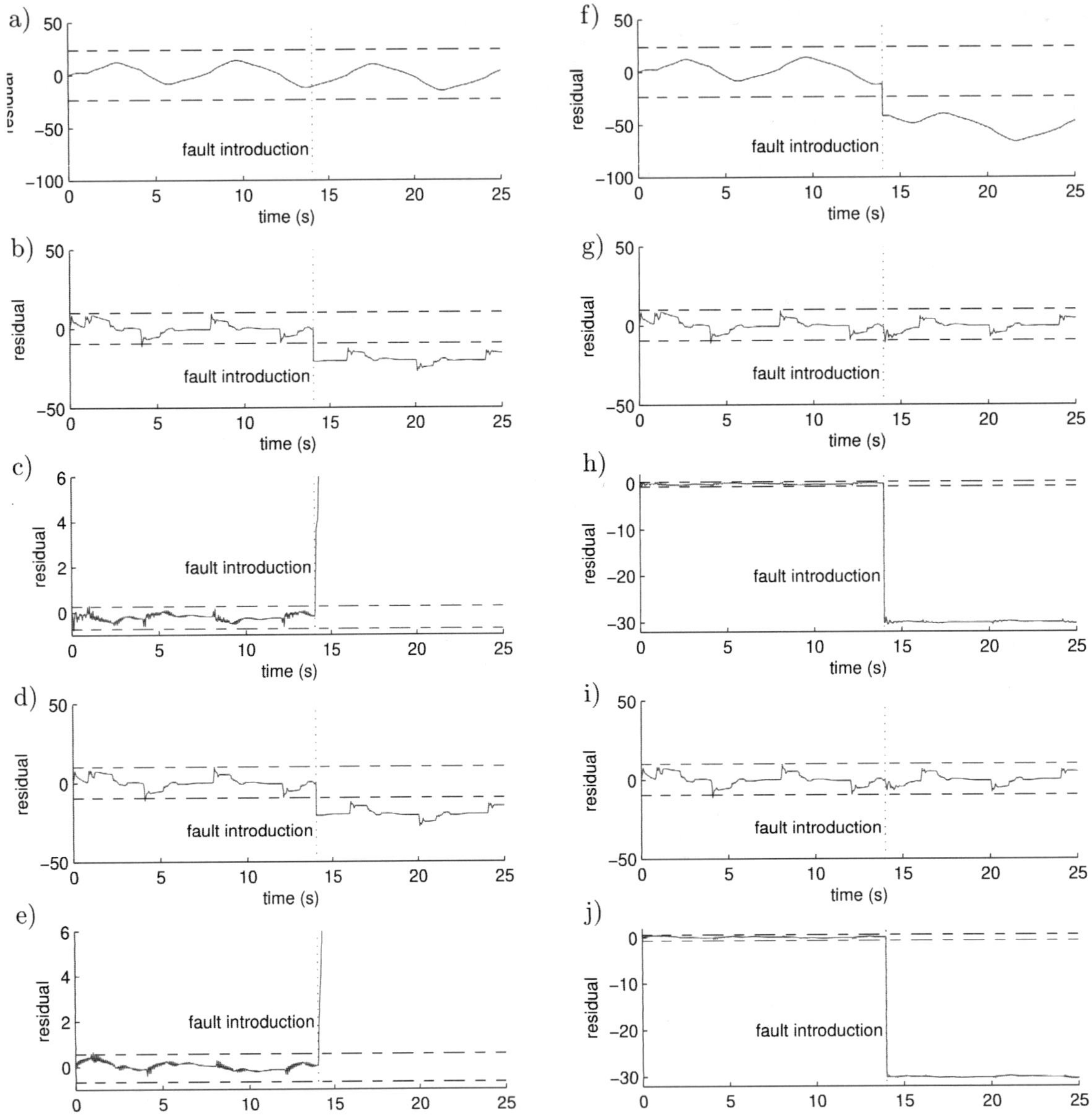

Figure 8. Residuals and thresholds generated with model 1 (a), model 2 (b), model 3 (c), model 4 (d) and model 5 (e) in case of fault f_2 and with model 1 (f), model 2 (g), model 3 (h), model 4 (i) and model 5 (j) in case of fault f_3.

performance is caused mostly by the model 3 (vamodel) which reacts very fast with great change of residual. The actuator fault is also immediately detected (after 0.3s) but the isolation is quite poor with only 28% rate. This problem can be addressed with the application of diagnostic submatrices or neural classifier and our future work will consider this approaches. The last conclusion concerns false alarms which rate is very low but such events can be disastrous in future design of fault tolerant control where inappropriate fault compensation action can occur. This will be addressed in future work with robust fault detection by application of the so-called MEM technique (model error modelling) used earlier by the authors in [6, 5]. What is most important the fault false isolation rate in each scenario is equal to 0 so that means the faults are no mistaken with each other.

6. Conclusions

In this work the complete sensor and actuator fault detection and isolation diagnostic system was designed. As it was shown the SSNN can be used very easily and successfully in application of such diagnostic system. The obtained residual signals allows for very fast and accurate fault detection. The so-called methods of simple thresholding and Binary Diagnostic Tables are very important tools in the process of residual evaluation and fault isolation.

Our future work will be focused on achieving the better results by the replacement of the simple thresholding method by robust fault detection with MEM in the diagnostic system. Also, more accurate fault isolation by applying diagnostic submatrices will be investigated.

Acknowledgments

This work was supported by the National Science Centre in Poland under the grant UMO-2012/07/N/ST7/03316.

References

[1] M. Basseville and I. V. Nikiforov. *Detection of Abrupt Changes: Theory and Application.* Prentice Hall, Englewood Cliffs, 1993.

[2] Piotr Bilski and Jacek Wojciechowski. Artificial intelligence methods in diagnostics of analog systems. *International Journal of Applied Mathematics and Computer Science*, 24(2):271–282, 2014.

[3] Mogens Blanke, Michel Kinnaert, Jan Lunze, Marcel Staroswiecki, and J. Schröder. *Diagnosis and Fault-Tolerant Control.* Springer-Verlag New York, Inc., Secaucus, NJ, USA, 2006.

[4] A. Czajkowski and K. Patan. Model predictive control of the two rotor aero-dynamical system using state space neural networks with delays. In M. Kowal eds. J. Korbicz, editor, *Intelligent systems in technical and medical diagnostics*, Advances in Intelligent Systems and Computing : Vol. 230, pages 113–124. Springer-Verlag, Berlin - Heidelberg, 2013.

[5] A. Czajkowski and K. Patan. Robust fault detection by means of echo state neural network. *Advanced and Intelligent Computations in Diagnosis and Control*, 386:341, 2015.

[6] Andrzej Czajkowski, Krzysztof Patan, and Mirosław Szymański. Application of the state space neural network to the fault tolerant control system of the plc-controlled laboratory stand. *Engineering Applications of Artificial Intelligence*, 30:168–178, 2014.

[7] INTECO. *Modular Servo System - User's Manual.* http://www.inteco.com.pl/Docs/Servo_um.pdf, 2014.

[8] J. Korbicz, J. M. Koscielny, Z. Kowalczuk, and W. (eds) Cholewa. *Fault Diagnosis. Models, Artificial Intelligence, Applications.* Springer-Verlag, Berlin, 2004.

[9] Pangao Kou, Jianzhong Zhou, Changqing Wang, Han Xiao, Huifeng Zhang, and Chaoshun Li. Parameters identification of nonlinear state space model of synchronous generator. *Engineering Applications of Artificial Intelligence*, 24(7):1227 – 1237, 2011.

[10] M. Luzar, M. Witczak, and C. Aubrun. Robust $\mathcal{H}_\infty$ sensor fault diagnosis with neural networks. In *Intelligent Systems in Technical and Medical Diagnosis.*, pages 125–136. Springer, Berlin, 2013.

[11] M. Luzar, M. Witczak, and P. Witczak. Neural-network based robust FTC: Application to wind turbines. In *Artificial Intelligence and Soft Computing)*, pages 97–108. Springer, Berlin, 2014.

[12] M. Norgaard, O. Ravn, N. K. Poulsen, and L. K. Hansen. *Neural Networks for Modelling and Control of Dynamic Systems.* Springer-Verlag, London, 2000.

[13] K. Patan. *Artificial Neural Networks for the Modelling and Fault Diagnosis of Technical Processes.* Springer-Verlag, Berlin, 2008.

[14] K. Patan. Neural network-based model predictive control: Fault tolerance and stability. *Control Systems Technology, IEEE Transactions on*, 23(3):1147–1155, May 2015.

[15] K. Patan and J. Korbicz. Nonlinear model predictive control of a boiler unit: a faul tolerant control study. *International Journal of Applied Mathematics and Computer Science*, Vol. 22(no 1):225–237, 2012.

[16] Ole Sørensen. *Neural networks in control applications.* PhD thesis, 1994.

[17] R.F. Stengel. Intelligent failure-tolerant control. In *Intelligent Control, 1990. Proceedings., 5th IEEE International Symposium on*, pages 548–557 vol.1, Sep 1990.

[18] Jesús M. Zamarreño Cosme and Pastora Vega. State space neural network. properties and application. *Neural Networks*, 11(6):1099–1112, 1998.

Journal of Physics: Conference Series **659** (2015) 012035 doi:10.1088/1742-6596/659/1/012035

Fault detection with principal component pursuit method

Yijun Pan, Chunjie Yang*, Youxian Sun, Ruqiao An and Lin Wang
Department of Control Science and Engineering, University of Zhejiang, Hangzhou Zhejiang, China

E-mail: cjyang@iipc.zju.edu.cn

Abstract. Data-driven approaches are widely applied for fault detection in industrial process. Recently, a new method for fault detection called principal component pursuit(PCP) is introduced. PCP is not only robust to outliers, but also can accomplish the objectives of model building, fault detection, fault isolation and process reconstruction simultaneously. PCP divides the data matrix into two parts: a fault-free low rank matrix and a sparse matrix with sensor noise and process fault. The statistics presented in this paper fully utilize the information in data matrix. Since the low rank matrix in PCP is similar to principal components matrix in PCA, a T^2 statistic is proposed for fault detection in low rank matrix. And this statistic can illustrate that PCP is more sensitive to small variations in variables than PCA. In addition, in sparse matrix, a new monitored statistic performing the online fault detection with PCP-based method is introduced. This statistic uses the mean and the correlation coefficient of variables. Monte Carlo simulation and Tennessee Eastman (TE) benchmark process are provided to illustrate the effectiveness of monitored statistics.

1. Introduction

Nowadays, the industrial process control system tends to be large-scale and complicated. In order to maintain the efficiency and reliability of the system, it's important to develop a fault detection method, which can monitor the process and detect the fault promptly. The fault detection method can be divided into three categories: model-based method, knowledge-based method and data-driven method [1]. Compared with model-based method and knowledge-based method, data-driven method can detect faults without the exactly models. Compared to other two methods, the model-based method is based on exact process models. The knowledge-based methods are based on the available knowledge of the process behavior and the experience of expert plant operators. The data-based process monitoring methods have no requirement of the process model and the associated expert knowledge. It has become more and more popular in recent years, especially in those complex industrial processes whose models and expert knowledge are difficult to build and obtain in practice. This method is suitable to complicated industrial process. The common applied data-driven methods are principal component analysis(PCA), partial least squares(PLS), support vector machine(SVM), artificial neural network(ANN) and some improved methods based on these basic methods. Process monitoring are divided into four steps: model building, fault detection, fault isolation and process reconstruction. An advantage of principal component pursuit(PCP) is that it can simultaneously accomplish the four steps of process monitoring. Besides, it's robust to outliers.

Candes first proposed the PCP method in detail in 2011[2]. A low rank matrix and a sparse matrix could be recovered from a data matrix by using a convex optimization algorithm. This method was able to realize the detection of objects in a cluttered background. Bouwmans reviewed the recent developments in the field of RPCA solved via PCP, and investigated how these methods were solved[3]. Then Isom indicated that PCA had several shortcomings compared with PCP, and PCP was robust to outliers[4]. In addition, PCP could simultaneously accomplish the objectives of model building, fault detection, fault isolation, and process reconstruction. This is the first time that the PCP-based method is used for process monitoring. Cheng proposed a new scaling preprocessing step that improved quality of data matrix in the sense of low coherence[5]. Moreover, a residual generator and a post-filter suitable for PCP generated process models were provided.

However, as far as we know, there are few appropriate monitored statistics to perform process monitoring with PCP-based method. The PCP is a useful matrix decomposition technique, and it can handle the data collected during some time. So its hard to perform online process monitoring in theory. In PCA-based method, a T^2 statistic is used for fault detection in principal component subspace. The principal component subspace is basically fault-free, which is similar to low rank matrix in PCP. In this paper, a T^2 statistic is introduced to be used for fault detection in low rank matrix. The process monitoring performance is better in PCP-based method than PCA-based method. PCA determines the number of principal components by calculating the cumulative variance proportion. But it's not sure that the total important information is put into the principal component subspace. In addition, the number of principal components must be an integer number, while the parameter λ can be any real number in PCP-based method. It can be adjusted little by little. This means that the adjustable range is flexible in PCP-based method. In PCA-based method, the important information is divided into principal component subspace and residual space. However, in PCP-based method, almost total important information is emerged in the low rank matrix. The low rank matrix is used for building the model. By using this monitored statistic, the fact that PCP is more sensitive to small variations in variables than PCA is illustrated. The reason why we compares the PCP with PCA is that they are both decomposition method. The PCA-based method divides the matrix into a principal component matrix and a residual matrix. The PCP-based method divides the matrix into a low rank matrix and a sparse matrix. Both principal component matrix and low rank matrix are fault-free.

In sparse matrix, the Q statistic in PCA is not suitable to fault detection. An online PCP-based process monitoring statistic that fully uses the information of noise and fault in sparse matrix is presented. The mean and correlation coefficient of variables in training matrix are used for building monitor statistic. In normal condition, the sparse matrix contains the noise. In abnormal condition, the sparse matrix contains sensor noises and process fault. Building a statistic by comparing the mean of variables in two sparse matrices. That can eliminate the influence of noises. So this statistic can increase the fault detection rate. This is the reason that this statistic is suitable to fault detection in sparse matrix. Both T^2 and new mean-coefficient statistics fully consider the advantage of PCP. The T^2 statistic uses the important process information, while the mean-coefficient statistic uses the noise and fault information. The total information in data matrix is used for fault detection. So these PCP-based statistics are appropriate to fault detection. The PCP-based method divides the data matrix into two parts including a low rank matrix and a sparse matrix. In actual industrial process, the variables are relevant. This means that the matrix is low rank only in the condition without sensor noise and process fault in process. This is the reason that low rank matrix is without fault data. The sparse matrix contains sensor noise and process fault. Finally, the monitored statistics are used for fault detection on numerical simulation and Tennessee Eastman(TE) process to prove the efficiency.

In section 2, a brief description of PCP is given. The proposed monitor statistic based on

PCP for process monitoring will be presented in section 3. Section 4 provides the simulation results on numerical simulation and TE process respectively. Section 5 gives the conclusion.

2. Principal component pursuit

Given a data matrix $X \in R^{n\times m}$ with n observations and m measurement variables. X can be decomposed by PCP as follows

$$X = A + E \tag{1}$$

where A is a low rank matrix and E is a sparse matrix. And $E_{ij} = 0$ means that the ith observation of jth variable is without sensor noise and process fault. The essence of PCP is to solve the optimization problem

$$\begin{aligned} &min \;\; \|A\|_* + \lambda\|E\|_1 \\ &s.t. \;\; X = A + E. \end{aligned} \tag{2}$$

where $\|A\|_*$ is the nuclear norm of matrix A, and it is equal to the sum of singular values. $\|E\|_1$ is the l_1 norm of matrix E, and it is equal to the sum of magnitudes of all entries[2]. λ is a parameter used for optimizing algorithm. The value of parameter λ represents the importance of $\|A\|_*$ and $\|E\|_1$ in (2). The larger value of parameter λ is, the more sensor noise and process fault the sparse matrix contains. But if the value of parameter λ is too large, the relationship of data in low rank matrix A will be broken. On the contrary, the sparse matrix has little sensor noise and process fault, which leads to a low fault detection rate. So choosing an appropriate λ is important. In this paper, the parameter λ is chosen by trial. The model in (2) is a simple convex optimization problem. In this work, we solve it by singular value thresholding method[6].

3. Fault detection based on PCP

3.1. A monitored statistic for fault detection in low rank matrix

As mentioned above, the low rank matrix contains the information without sensor noise and process fault. This is similar to the principal components space in PCA-based method. So the T^2 statistic is used for fault detection in low rank matrix in PCP-based method. And with this monitored statistic, the PCP-based method is illustrated to be sensitive to small variations in variables by simulation.

The steps of PCP-based method for fault detection are as follows:

- Normalization
  The normal training data matrix X is normalized by mean and variance of the corresponding variables.
- Matrix decomposition
  The PCP is used for dividing the normalized training data matrix X^* into a low rank matrix A and a sparse matrix E by solving the convex optimization problem.
- Eigenvalue decomposition
  Get the singular values diagonal matrix $\sum$ and loading vector P via eigenvalue decomposition of low rank matrix A.
- T^2 statistic threshold of normal condition
  The threshold of T^2 statistic in training data matrix can be determined by (3).

$$T_\alpha^2 = \frac{(n^2-1)m}{n(n-m)} \times [F_\alpha(m, n-m)] \tag{3}$$

where n, m are the dimension of matrix X^* and $F_\alpha(m, n-m)$ is the value of F-distribution at significance level with m and $(n-m)$ degrees of freedom [7].

- T^2 statistic of testing data matrix
  The T^2 statistic of testing data matrix can be computed by (4).

$$T^2 = \mathrm{x}^T P(\sum)^{-2} P^T \mathrm{x} \tag{4}$$

  where x is an observation vector.
- Fault detection
  If the T^2 statistic exceeds the normal threshold value, then a fault will happen.

3.2. A new monitored statistic for fault detection in sparse matrix
The data matrix is divided into two parts by PCP-based method: a low rank matrix and a sparse matrix. In training data matrix, the sparse matrix E_1 contains the sensor noise, while in testing data matrix, the sparse matrix E_2 contains the sensor noise and process fault. For the purpose of fault detection, the following assumptions is made. In the same operation condition, the magnitude and direction of noises are nearly the same in both training matrix and testing matrix. This statistic compares the mean of variables of two sparse matrices. This way eliminates the influence of noises, which improves the performance of fault detection. To get a normal condition threshold, the correlation coefficient of variables is calculated. This way realizes the online fault monitoring. Then the steps of building a new monitored statistic are as follows:

- Calculate the low rank matrix A_1 and sparse matrix E_1 of training matrix X_1 by PCP-based method.
- Calculate the correlation coefficient $c_i, i \in 1, 2, ..., m-1$ of every variables with the first variable in training matrix X.
- Calculate the proportion $p_i, i \in 1, 2, ..., m-1$ of every variables in correlation coefficient.

$$p_{i+1} = \frac{c_i}{1 + \sum_1^{m-1} c_i}, p_1 = \frac{1}{1 + \sum_1^{m-1} c_i} \tag{5}$$

- Calculate the normal threshold M. Multiply the proportion p_i by the mean of variables $d_i, i \in 1, 2, ..., m$ in training matrix.

$$M = p_i \times d_i \tag{6}$$

- Use the last $(n-i) \times m, i \in 1, 2, ..., n$ row data in training matrix and the first $i \times m$ row data in testing matrix to make up a new matrix D.
- Calculate the low rank matrix A_2 and sparse matrix E_2 of matrix D by PCP-based method.
- Calculate the monitored statistic G in testing matrix. Multiply the proportion p_i by sparse matrix E_2.

$$G = p_i \times E_2 \tag{7}$$

- Get the last value in G to make up the new monitored statistic MG .

$$MG(i) = G(n) \tag{8}$$

- Repeat n times to get the monitored statistic MG
- If the statistic MG exceeds the threshold value M, then a fault will happen.

4. Simulation result

In this section, the simulation results of PCP-based method and PCA-based method are discussed. The new monitored statistics are used for fault detection on numerical simulation and TE process.

4.1. Numerical simulation

The purpose of this Monte Carlo simulation is to verify that T^2 statistic and mean-coefficient statistic are suitable to the fault detection. Besides, PCP is more sensitive to small change in variables. Monte Carlo simulation results are provided for the case of modest fault magnitudes by randomly assigning fault sensors and fault magnitudes. In 2009, Alcala used the Monte Carlo simulation to build a numerical simulation example, which is suitable to test statistics[8]. The process model to be used is

$$\begin{pmatrix} x_1 \\ x_2 \\ x_3 \\ x_4 \\ x_5 \\ x_6 \end{pmatrix} = \begin{pmatrix} -0.2310 & -0.0816 & -0.2662 \\ -0.3241 & 0.7055 & -0.2158 \\ -0.217 & -0.3056 & -0.5207 \\ -0.4089 & -0.3442 & -0.4501 \\ -0.6408 & 0.2105 & -0.2372 \\ -0.5655 & -0.433 & -0.5938 \end{pmatrix} \begin{pmatrix} t_1 \\ t_2 \\ t_3 \end{pmatrix} + noise \tag{9}$$

where t_1 , t_2 and t_3 are zero-mean random variables with standard deviations of 1, 0.8 and 0.6, respectively. The noise included in the process is zero-mean with standard deviation of 0.2 and is normally distributed. The simulated testing data matrix is of the form

$$x_{test} = x^* + \varepsilon_i f \tag{10}$$

where x^* is generated according to the model given above and the fault magnitude f is a random number uniformly distributed between 0 and 0.9. The chosen fault magnitude is small. It not only can illustrate the effectiveness of T^2 statistic, but also illustrate that PCP is more sensitive to small variations. Also, the direction ε_i is uniformly random out of the six possible variable directions. In this section, total six variables are used for building the model.

4.1.1. The T^2 statistic for fault detection in low rank matrix The fault detection results of PCA-based method and PCP-based method for this process model are shown in Fig.1- Fig.2. The fault detection rate (FDR), fault alarm rate (FAR) of T^2 statistic are shown in Table 1. In PCA-based method, the T^2 and Q statistics are used for fault detection.

In this simulation, a small change in the variables which works as a process fault is introduced in testing data matrix. The testing data matrix consists of 100 normal data and 500 fault data. In PCA-based method, the number of principal components is three by calculating the accumulative variance contribution rate. The T^2 statistic threshold of normal condition is equal to 0.3468, and the Q statistic threshold of normal condition is equal to 8.4063×10^{-30}. In PCP-based method the parameter λ is equal 0.05. The T^2 statistic threshold of normal condition is equal to 0.6986.

From Fig.1-Fig.2 and Table 1, the PCP-based method gets higher FDR and lower FAR than PCA-based method. This shows that the T^2 statistic is suitable to fault detection in low rank matrix based on PCP. And PCP-based method is more sensitive to small variations.

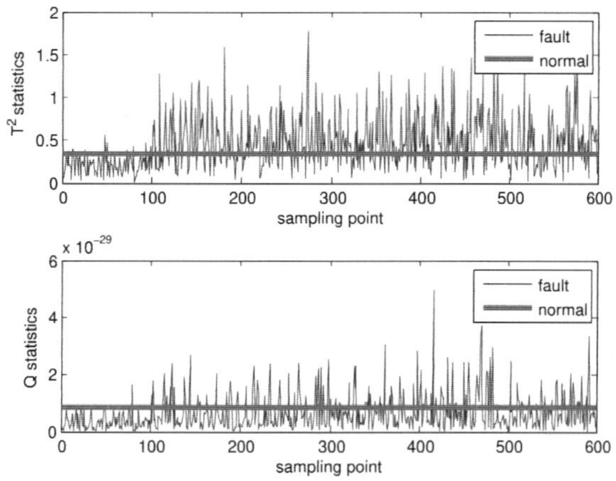

Figure 1. The T^2 statistic of PCA-based method for fault detection in numerical simulation

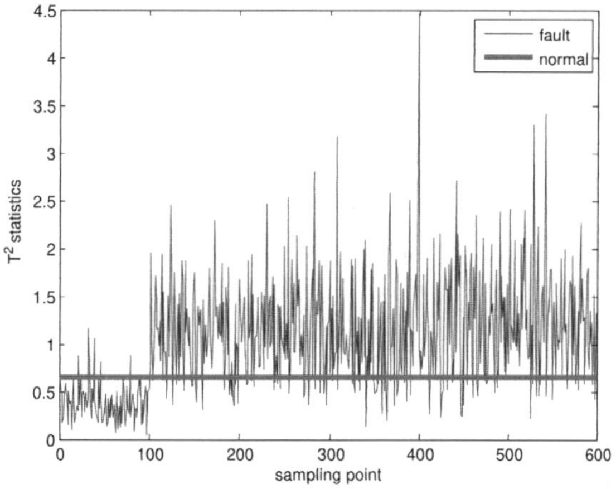

Figure 2. The T^2 statistic of PCP-based method for fault detection in numerical simulation

Table 1. The FAR and FDR of PCP and PCA.

Item	FDR	FAR
PCA	FDR	FAR
PCP	81.20	8.00

4.1.2. The new monitored statistic in sparse matrix In this part, an online monitored statistic is presented. The fault detection results with mean-coefficient monitored statistic are shown in Fig.3, and fault detection rate (FDR), fault alarm rate(FAR) are shown in Table 2. In this part, the fault magnitude f is a random number uniformly distributed between 0 and 5, which is the same as the original condition in (8). And the parameter λ is equal to 1.5. The threshold of normal condition is equal to 5.3434×10^{-4}. This new statistic can deal with online fault detection, which overcomes the main shortcoming of PCP-based method. However, this new statistic can't simultaneously deal with the four steps of process monitoring.

From Fig.3 and Table 2, we can obtain that the new statistic in sparse matrix performances better than the T^2 statistic in low rank matrix in numerical simulation. This means that the new PCP-based statistic is suitable to fault detection.

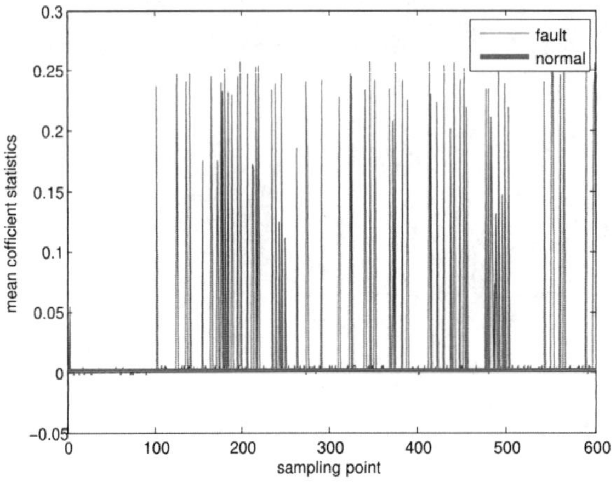

Figure 3. The mean-coefficient statistic of PCP-based method for fault detection in numerical simulation.

Table 2. The FAR and FDR of PCP and PCA.

Item	PCP
FDR	98.60
FAR	2.30

4.2. TE process simulation

4.2.1. TE process In this section, due to the actual industrial process has few fault data, TE process simulated the actual industrial process is used. This process contains 41 measured variables and 12 manipulated variables. In these 53 variables, reactor agitator speed remains constant. Therefore, reactor agitator speed is removed from the variables, remaining 52 variables.

There are 21 process faults in the TE process, and Chiang described it in detail[9]. The 21 faults in TE are introduced in Table 3.Two sets of data matrices are collected from the TE process - training and testing data matrix. To collect the data matrix, 22 running condition (1 normal condition and 21 abnormal conditions) are performed. In data matrix, the fault is introduced at 8th hour with a sampling interval of 3 minutes. Hence, the data matrix is composed of 22 separate data matrices containing 1 normal condition data matrix and 21 fault data matrices, each having dimension of 960×52 (160 normal data/800 fault data). In this part, the total 52 measured and manipulated variables in TE process are used for building the model in PCA-based and PCP-based method.

Table 3. Description of Faults in the TE Process.

no.	description	type
1	A/C feed ratio, B composition constant(stream 4)	step
2	D feed temperature (stream 2)	step
3	B composition, A/C ratio constant (stream 4)	step
4	reactor cooling water inlet temperature	step
5	condenser cooling water inlet temperature	step
6	A feed loss (stream 1)	step
7	C header pressure loss?reduced availability(stream 4)	step
8	A, B, and C feed composition (stream 4)	random variation
9	D feed temperature (stream 2)	random variation
10	C feed temperature (stream 4)	random variation
11	reactor cooling water inlet temperature	random variation
12	condenser cooling water inlet temperature	random variation
13	reaction kinetics	slow drift
14	reactor cooling water valve	sticking
15	condenser cooling water valve	sticking
16-20	unknown	unknown
21	valve for stream 4 was fixed at steady-state position	constant position

4.2.2. Simulation results in low rank matrix The fault detection results of PCA and PCP-based method for fault 1, 5 are shown in Fig.4-Fig.7. The parameters of simulation are shown below. In Fig.4, the T^2 statistic threshold of normal condition is equal to 4.8430. The parameter λ is equal to 0.05. In Fig.5, the T^2 statistic threshold of normal condition is equal to 2.1172, and the Q statistic threshold of normal condition is equal to 17.3456. The number of principal components is 24. In Fig.6, the T^2 statistic threshold of normal condition is equal to 7.8914. The parameter λ is equal to 0.05. In Fig.7, the T^2 statistic threshold of normal condition is equal to 2.1172, and the Q statistic threshold of normal condition is equal to 17.3456. The number of principal components is 24.

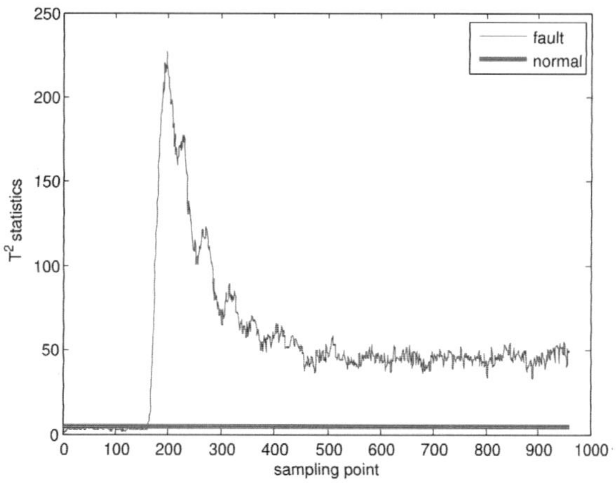

Figure 4. The T^2 statistic in low rank matrix of PCP-based method for fault detection of fault 1 in TE process.

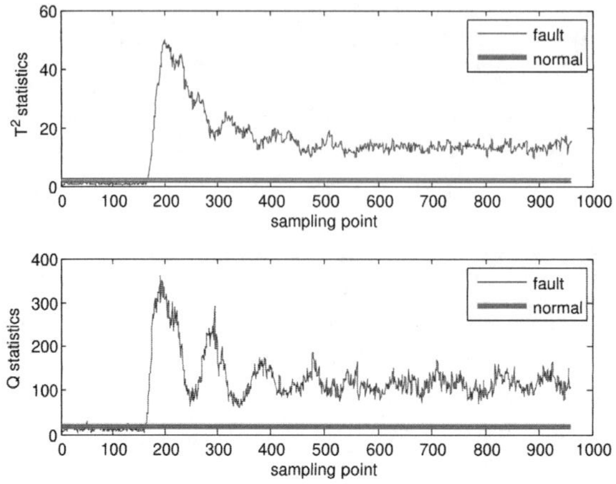

Figure 5. The T^2 and Q statistic of PCA-based method for fault detection of fault 1 in TE process.

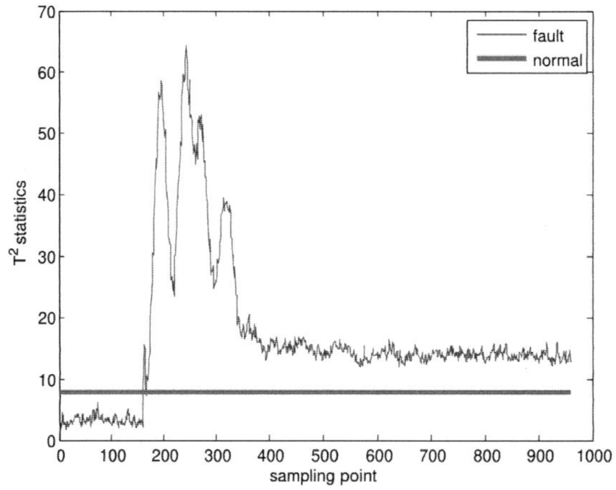

Figure 6. The T^2 statistic in low rank matrix of PCP-based method for fault detection of fault 5 in TE process.

Fault 5 is associated with step change in the condenser cooling water inlet temperature. The PCA-based method for fault 5 in Fig.7 only detects the fault from 160th to 350th. Since the control loops compensated some of the changes in the variables and only small variations remained, the PCA-based method fails to detect the small variations. PCP-based method for fault 5 in Fig.6 can detect the fault from 161th to 960th, which means that the PCP method is more sensitive to small variations in industrial process. And the T^2 statistic is suitable to fault detection with PCP-based method. The FDR and FAR of faults by using T^2 statistic in TE process are shown in Table 4. In most faults, the PCP-based method gets better results than PCA-based method. The average values of FAR and FDR of two methods also illustrate that PCP-based method is better.

According to the numerical simulation and TE process simulation, the PCP-based method gets higher FDR and lower FAR than PCA-based method. This means that the T^2 statistic is suitable to fault detection in PCP-based method than PCA-based method. PCA-based method determines the number of principal components by calculating the cumulative variance

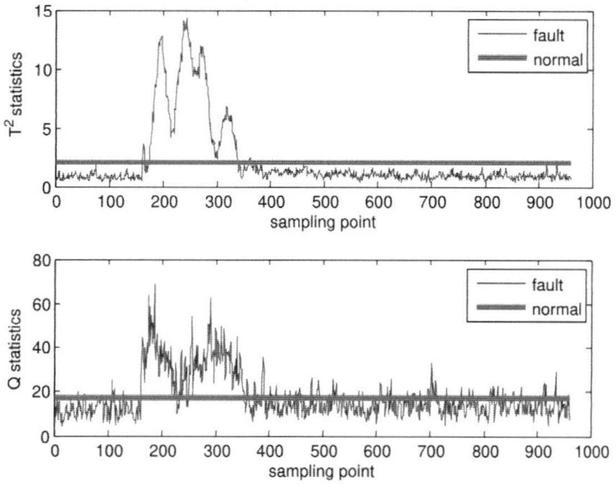

Figure 7. The T^2 and Q statistic of PCA-based method for fault detection of fault 5 in TE process.

proportion. The number of principal components must be an integer number. This means that the adjustable range is inflexible. So the important information is put into principal component subspace and residual space. However, in PCP-based method the parameter λ is a real number. It can be adjusted little by little, so the noises and faults are almost isolated from low rank data matrix.

Table 4. The FAR and FDR of PCP and PCA

Fault	FAR-PCP	FAR-PCA	FDR-PCP	FDR-PCA
1	0.00	0.00	98.75	99.25
2	1.25	0.00	98.38	98.25
3	6.25	97.25	16.75	11.85
4	1.25	2.50	91.25	56.50
5	0.00	2.50	99.50	33.00
6	0.00	1.87	99.50	99.12
7	5.63	0.00	86.50	99.75
8	10.00	2.50	87.88	97.88
9	23.13	15.00	16.63	11.13
10	3.13	1.87	83.13	50.75
11	5.63	11.25	76.38	70.50
12	5.63	16.25	99.00	99.25
13	1.97	4.37	94.50	95.87
14	5.63	8.75	99.88	99.75
15	10.63	6.25	23.13	26.12
16	5.00	41.25	85.12	48.25
17	6.88	6.88	84.25	88.50
18	5.00	8.13	89.88	91.87
19	11.87	3.76	51.38	36.25
20	5.00	4.37	77.50	60.75
21	11.87	16.25	53.13	50.75

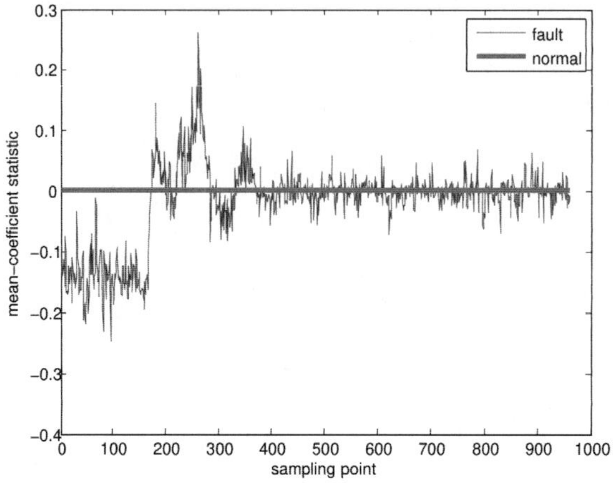

Figure 8. The mean-coefficient statistic in sparse matrix of PCP-based method for fault detection of fault 1 in TE process.

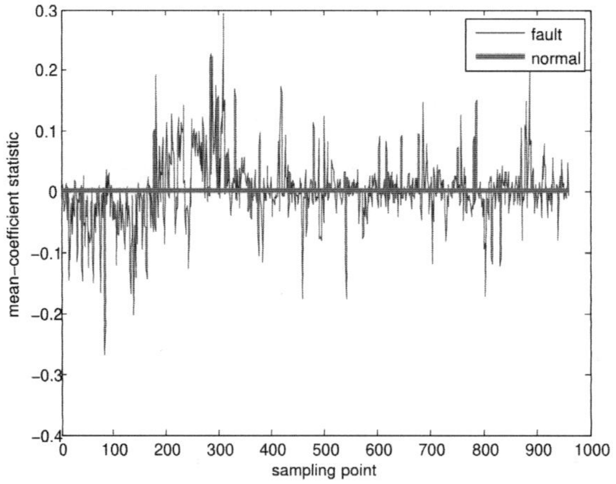

Figure 9. The mean-coefficient statistic in sparse matrix of PCP-based method for fault detection of fault 2 in TE process.

4.2.3. Simulation results in sparse matrix In this part, the fault detection results with new monitored statistic in sparse matrix for fault 1 and 2 in TE process are also presented in Fig.7-Fig.8. In Fig.7-Fig.8, the threshold of normal condition is equal to 0.0014. The parameter λ is equal to 0.5. The fault detection performance in low rank matrix is better than that in sparse matrix. Since the performance of fault detection is not well and the amount of calculation is huge, the FAR and FDR are not presented. There are two reasons to explain this poor performance. First, the values of parameter λ is not appropriate. Second, the assumption made in part 'A monitored statistic in sparse matrix' is hard to realize. Because in actual process, the noise is uncontrollable. Although the fault detection performance is not well, this method realizes the online monitoring in sparse matrix, which is more important.

5. Conclusion

In this paper, the monitored statistics based on PCP used for fault detection are introduced. In low rank matrix, a T^2 statistic is introduced. This statistic gets good performance. And by this statistic, the fact that PCP-based method is more sensitive to small variations is illustrated. In sparse matrix, an online mean-coefficient statistic is presented. The two statistics fully use the information in data matrix. The numerical simulation and TE process illustrate the effectiveness of these two monitored statistics. And there still exits large room to improve. While using the PCP to get two matrices, the parameter λ is chosen by trial. The value of λ is important to simulation results. In further work, the algorithm of how to get the optimal parameter will be discussed. And how to improve fault detection performance in sparse matrix is should also be considered.

References

[1] Z. Q. Ge, Z. H. Song, and F. R. Gao, Review of recent research on data-based process monitoring, Industrial and Engineering Chemistry Research, vol. 52, no. 10, pp. 3543-3562, 2013

[2] E. J. Candes, X. Li, Y. Ma, and J.Wright, Robust principal component analysis? Journal of ACM, vol. 58, no. 3, 2011.

[3] T. Bouwmans and E. H. Zahzah, Robust PCA via principal component pursuit: A review for a comparative evaluation in video surveillance, Computer Vision and Image Understanding, vol. 122, pp. 22-34, 2014.

[4] J. D. Isom and R. E. LaBarre, Process fault detection, isolation, and reconstruction by principal component pursuit, in 2011 American Control Conference, San Francisco, CA, USA, 2011, pp. 238-243.

[5] Y. Cheng and T. W. Chen, Application of principal component pursuit to process fault detection and diagnosis, in 2013 American Control Conference, Washington, DC, USA, 2013, pp. 3535-3540.

[6] J. F. Cai, E. J. Candes, and Z. W. Shen, A singular value thresholding algorithm for matrix completion, SIAM Journal on Optimization, vol. 20, no. 4, pp. 1956-1982, 2010.

[7] H. L. Chiang, L. E. Russell, and D. R. Braatz, Fault diagnosis in chemical processes using fisher discriminant analysis, discriminant partial least squares, and principal component analysis, Chemometrics and Intelligent Laboratory Systems, vol. 50, pp. 243-252, 2000.

[8] C. Alcala and S. Qin, Reconstruction-based contribution for process monitoring, Automatica, vol. 45, no. 7, pp. 1593-1600, 2009.

[9] L. Chiang, E. Russell, and R. Braatz, Fault detection and diagnosis in industrial systems. Springer Verlag, 2001.

Journal of Physics: Conference Series **659** (2015) 012036 doi:10.1088/1742-6596/659/1/012036

On fault detection filters design with unitary transfer function matrices

Dušan Krokavec, Anna Filasová and Pavol Liščinský

Department of Cybernetics and Artificial Intelligence, Faculty of Electrical Engineering and Informatics, Technical University of Košice, Letná 9, 042 00 Košice, Slovakia

E-mail: dusan.krokavec@tuke.sk; anna.filasova@tuke.sk; pavol.liscinsky@tuke.sk

Abstract. An adaptation of unitary system principle in fault detection filter design for continuous-time linear MIMO systems is presented in the paper. The conformation is based on an enhanced fault residual transfer function matrix with unitary construction and offers the key advantages on providing high residual sensitivity with respect to faults. Re ecting the emplacement of singular values in unitary construction, an associated structure of linear matrix inequalities with built-in structured constraints is outlined to verify the filter stability. The proposed design conditions are verified by the numerical illustrative example.

1. Introduction

A conventional control for complex systems may results in unsatisfactory performances in the event of system component malfunctions. To overcome these weaknesses, different approaches to control system design are developed to tolerate component malfunctions and to maintain acceptable performances of the system with faults. These control structures are known as fault-tolerant control (FTC) systems and force the ability to accommodate component failures. In that sense, research in FTC is sub ect of a wide range of publications, reflecting faults effect on control structure reconfiguration [4], [14], fault estimation [16], [20], and fault residuals generation and analysis [7], [25]. The ideal approach is to construct disturbance-decoupled residuals, with responsiveness to the faults, as shown in [6].

To scale up accuracy of fault detection, it is eligible to craft residuals with high sensitivity to faults under robustness to disturbances. ne of the options is the use of H_∞/H_- optimization principle [8], [9]. The restriction of these methods is mainly the necessity of existence of a full rank direct-feed external gain matrix from faults to residuals [21], which limits them to be used only to residuals revealing actuator faults. Another approach, based on an unitary system property, is proposed in [22], [23] where optimization is realized inherently in the sense that if the singular values of a unitary system are assigned as the magnitude frequency response of a first-order transfer function then H_∞ is the maximum and H_- is the minimum of the magnitude of the transfer function. To apply this, Riccati equation based formulation was proposed to design a stable unitary fault transfer function approximation.

Reflecting this approach in unitary systems design, the issue of this paper is to simplify the design conditions, mainly because solutions of the introduced Riccati equation form for this singular task often fail. Searching for accurate parameter matrices of the fault residual state-space description and to ensure the constructed unitary model closely approximates the actual fault residual transfer function matrix, the proposed extended adaptation of the results given in [13] presents design task which is non-singular with unitary conditions reaching approximately the theoretical limits for the prescribed set of singular values. To analyze stability of the observer-based residual filter, the proposed conditions use standard arguments and require to solve only LMIs with one built-in prescribed constraint, depending on the system output and fault input matrix structures.

The paper is organized as follows. Placed immediately after Introduction, Sec. 2 presents the problem statement and Sec. 3 summaries in basic preliminaries the auxiliary lemmas on the issue of the design task. The enhanced structure of unitary fault residual transfer function matrix, as well as the set of LMIs describing its stability, is theoretically explained in Sec 4. An example is provided to demonstrate the proposed approach in Sec. 5 and, finally, Sec. 6 draws some concluding remarks.

Used notations are conventional so that $\boldsymbol{x}^T$, $\boldsymbol{X}^T$ denote transpose of the vector $\boldsymbol{x}$ and matrix $\boldsymbol{X}$, respectively, $\boldsymbol{X} = \boldsymbol{X}^T > 0$ means that $\boldsymbol{X}$ is a symmetric positive definite matrix, the symbol $\boldsymbol{I}_n$ marks the n-th order unit matrix, $\rho(\boldsymbol{X})$ and $rank(\boldsymbol{X})$ indicate the eigenvalue spectrum and rank of a square matrix $\boldsymbol{X}$, $\boldsymbol{Y}^{\perp}$ designates the orthogonal complement to a rank-deficient matrix $\boldsymbol{Y}$, $\sigma_i(\boldsymbol{Z})$ labels the i-th singular value of matrix $\boldsymbol{Z}$, $I\!R$ denotes the set of real numbers and $I\!R^n$, $I\!R^{n\times r}$ refer to the set of all n-dimensional real vectors and $n \times r$ real matrices, respectively.

2. The Problem Statement

The systems under consideration are linear MIM continuous-time dynamic systems represented as follows

$$\boldsymbol{q}(t) = \boldsymbol{A}\boldsymbol{q}(t) + \boldsymbol{B}\boldsymbol{u}(t) + \boldsymbol{F}\boldsymbol{f}(t) + \boldsymbol{E}\boldsymbol{d}(t), \tag{1}$$

$$\boldsymbol{y}(t) = \boldsymbol{C}\boldsymbol{q}(t), \tag{2}$$

where $\boldsymbol{q}(t) \in I\!R^n$, $\boldsymbol{u}(t) \in I\!R^r$, $\boldsymbol{y}(t) \in I\!R^m$ are vectors of the state, input and output variables, respectively, $\boldsymbol{f}(t) \in I\!R^p$ is fault vector and $\boldsymbol{d}(t) \in I\!R^{r_d}$ is vector of disturbance. The real matrices $\boldsymbol{A} \in I\!R^{n\times n}$, $\boldsymbol{B} \in I\!R^{n\times r}$, $\boldsymbol{C} \in I\!R^{m\times n}$, $\boldsymbol{F} \in I\!R^{n\times p}$, $\boldsymbol{E} \in I\!R^{n\times r_d}$ are finite values, satisfying the rank conditions $rank(\boldsymbol{F}) = p$, $rank(\boldsymbol{C}) = m$, $p = m$, $p < n$. Moreover, it is supposed that the matrix $\boldsymbol{V} = \boldsymbol{C}\boldsymbol{F}$ is regular matrix such that $\boldsymbol{V} \in I\!R^{m\times m}$.

Problem of the interest is an unitary representation of the fault residual filter for the system with the square transfer function matrix of unknown fault input and the residuals. Note, such construction of an unitary systems to given linear system, with respect to the singular values of the system transfer function matrix, is non unique task also for square linear systems [13], [22].

3. Basic Preliminaries

3.1. Unitary Linear Systems

If $\mathcal{H}$ and $\mathcal{E}$ be Krein spaces [2], a continuous linear transformation

$$\begin{pmatrix} \boldsymbol{A} & \boldsymbol{B} \\ \boldsymbol{C} & \boldsymbol{D} \end{pmatrix} : \begin{matrix} \mathcal{H} \\ \oplus \\ \mathcal{E} \end{matrix} \rightarrow \begin{matrix} \mathcal{H} \\ \oplus \\ \mathcal{E} \end{matrix} \tag{3}$$

is called a linear system. The underlying Krein space $\mathcal{H}$ is called the state space and the auxiliary Krein space $\mathcal{E}$ is called the coefficient space or the external space [19]. The transformation $\boldsymbol{A}$, $\boldsymbol{B}$, $\boldsymbol{C}$ is the main, input and output transformation, respectively, the operator $\boldsymbol{D}$ is called the external operator. A linear system is said to be unitary if the above matrix is unitary, and the transfer function $\boldsymbol{G}(s)$ of the linear system is defined by

$$\boldsymbol{G}(s) = \boldsymbol{C}(s\boldsymbol{I}_n - \boldsymbol{A})^{-1}\boldsymbol{B} + \boldsymbol{D}\,. \tag{4}$$

If $\boldsymbol{A}$ has no imaginary eigenvalues then $\boldsymbol{G}(j\omega)$ is defined for all $\omega \in I\!R$ and the singular values of the transfer function matrix $\boldsymbol{G}(s)$, evaluated on the imaginary axis, are $\sigma_i(\boldsymbol{G}(j\omega))$, where the i-th singular value of the complex matrix $\boldsymbol{G}(j\omega)$ is the nonnegative square-root of the i-th largest eigenvalue of $\boldsymbol{G}(j\omega)\boldsymbol{G}^*(j\omega)$. The H_∞ norm of the transfer matrix $\boldsymbol{G}(s)$ is [5]

$$\|\boldsymbol{G}\|_\infty = \sup_{\omega \in I\!R} \sigma_1(\boldsymbol{G}(j\omega)) \,. \tag{5}$$

Some reflections can be found, e.g., in [3], [10] and [24].

Definition 1 *[22] A stable linear time-invariant system of m-inputs and m-outputs is defined as a unitary system if the singular values of its transfer function (matrix) $\boldsymbol{G}(s)$ satisfies*

$$\sigma_1 = \sigma_2 = \cdots = \sigma_m, \tag{6}$$

where σ_i is the i-th singular value of $\boldsymbol{G}(s)$.

Defining the unitary system gain matrix equal to $\boldsymbol{V} = \boldsymbol{C}\boldsymbol{F}$, the following system state coordinate transformation can be done.

Lemma 1 *If for the matrix parameters $\boldsymbol{C}$ and $\boldsymbol{F}$ of the system (1), (2)*

$$\boldsymbol{C} = \begin{bmatrix} \boldsymbol{V} & \boldsymbol{0} \end{bmatrix} \boldsymbol{T}, \quad \boldsymbol{T}\boldsymbol{F} = \begin{bmatrix} \boldsymbol{I}_p \\ \boldsymbol{0} \end{bmatrix}, \quad \boldsymbol{V} = \boldsymbol{C}\boldsymbol{F}, \tag{7}$$

then for $m = p$ the transform matrix $\boldsymbol{T} \in I\!R^{n \times n}$ takes the form

$$\boldsymbol{T} = \begin{bmatrix} \boldsymbol{V}^{-1}\boldsymbol{C} \\ \boldsymbol{F}^{\perp} \end{bmatrix}, \tag{8}$$

where $\boldsymbol{V}^{-1}\boldsymbol{C} \in I\!R^{p \times n}$, $\boldsymbol{F}^{\perp} \in I\!R^{(n-p) \times n}$, respectively, and $\boldsymbol{F}^{\perp}$ is the left orthogonal complement to $\boldsymbol{F}$ [1], [11].

Proof: Writing the first term of (7) as

$$\boldsymbol{C} = \begin{bmatrix} \boldsymbol{V} & \boldsymbol{0} \end{bmatrix} \begin{bmatrix} \boldsymbol{T}_1 \\ \boldsymbol{T}_2 \end{bmatrix} = \boldsymbol{C}, \tag{9}$$

it is evident that

$$\boldsymbol{V}\boldsymbol{T}_1 = \boldsymbol{C}, \quad \boldsymbol{T}_1 = \boldsymbol{V}^{-1}\boldsymbol{C}. \tag{10}$$

Analyzing the second term of (7), i.e.,

$$\begin{bmatrix} \boldsymbol{T}_1 \\ \boldsymbol{T}_2 \end{bmatrix} \boldsymbol{F} = \begin{bmatrix} \boldsymbol{I}_p \\ \boldsymbol{0} \end{bmatrix}, \tag{11}$$

the following condition results

$$\boldsymbol{T}_2\boldsymbol{F} = \boldsymbol{0}, \quad \boldsymbol{T}_2 = \boldsymbol{F}^{\perp}. \tag{12}$$

Thus, (10), (12) imply (8).

It is easily verified using (9), (12) that

$$\boldsymbol{T}_1\boldsymbol{F} = \boldsymbol{V}^{-1}\boldsymbol{C}\boldsymbol{F} = \boldsymbol{V}^{-1}\boldsymbol{V} = \boldsymbol{I}_p\,, \tag{13}$$

$$\boldsymbol{C}\boldsymbol{T}^{-1} = \begin{bmatrix} \boldsymbol{V} & \boldsymbol{0} \end{bmatrix}, \tag{14}$$

respectively.

Note, the last equality imply that, in general, $\boldsymbol{T}^{-1}$ can be computed only numerically. This concludes the proof. ■

3.2. Structures of Unitary Fault Transfer Matrices

The basic structures of unitary system transfer matrices are introduced by the following lemmas.

Lemma 2 *For the system (1), (2) with $m = p$ there exist, for a prescribed positive scalar $s_o \in I\!R$, the matrix $\boldsymbol{L}^o \in I\!R^{n\times m}$ such that the fault transfer function matrix takes the form*

$$\boldsymbol{G}_f(s) = \boldsymbol{G}_\Delta(s)\boldsymbol{G}_{f\Delta}(s)\,, \tag{15}$$

where

$$\boldsymbol{G}_{f\Delta}(s) = \boldsymbol{C}\left(s\boldsymbol{I}_n - (\boldsymbol{A} - \boldsymbol{L}^o\boldsymbol{C})\right)^{-1} = \frac{\boldsymbol{V}}{s+s_o}\,, \tag{16}$$

$$\boldsymbol{G}_\Delta(s) = \boldsymbol{I}_m + \boldsymbol{C}(s\boldsymbol{I}_n - \boldsymbol{A})^{-1}\boldsymbol{M}, \tag{17}$$

$$\boldsymbol{L}^o = \begin{bmatrix} s_o\boldsymbol{I}_m + \boldsymbol{A}_{o11} \\ \boldsymbol{A}_{o21} \end{bmatrix}\boldsymbol{V}^{-1}, \quad \boldsymbol{A}_o = \boldsymbol{T}\boldsymbol{A}\boldsymbol{T}^{-1} = \begin{bmatrix} \boldsymbol{A}_{o11} & \boldsymbol{A}_{o12} \\ \boldsymbol{A}_{o21} & \boldsymbol{A}_{o22} \end{bmatrix}, \quad \boldsymbol{M} = \boldsymbol{T}^{-1}\boldsymbol{L}^o \tag{18}$$

and $\boldsymbol{T}$ is given by (8).

Proof: (compare [13], [22]) Since

$$\boldsymbol{G}_f(s) = \boldsymbol{C}(s\boldsymbol{I}_n - \boldsymbol{A})^{-1}\boldsymbol{F} \tag{19}$$

is the fault transfer function matrix of dimension $m\times p$, then by using (7), (14), (18) the transfer function matrix (19) can be rewritten as

$$\begin{aligned} \boldsymbol{G}_f(s) &= \boldsymbol{C}\boldsymbol{T}^{-1}\boldsymbol{T}(s\boldsymbol{I}_n - \boldsymbol{A})^{-1}\boldsymbol{T}^{-1}\boldsymbol{T}\boldsymbol{F} \\ &= \boldsymbol{C}\boldsymbol{T}^{-1}(\boldsymbol{T}(s\boldsymbol{I}_n - \boldsymbol{A})\boldsymbol{T}^{-1})^{-1}\boldsymbol{T}\boldsymbol{F} \\ &= \begin{bmatrix} \boldsymbol{V} & \boldsymbol{0} \end{bmatrix}(s\boldsymbol{I}_n - \boldsymbol{A}_o)^{-1}\begin{bmatrix} \boldsymbol{I}_p \\ \boldsymbol{0} \end{bmatrix}. \end{aligned} \tag{20}$$

Defining the matrix product $\boldsymbol{A}^o = \boldsymbol{T}\boldsymbol{M}\boldsymbol{C}\boldsymbol{T}^{-1}$, where $\boldsymbol{M} \in I\!R^{n\times m}$ is a real matrix, then by exploiting (8), (14) it yields

$$\boldsymbol{A}^o = \boldsymbol{T}\boldsymbol{M}\boldsymbol{C}\boldsymbol{T}^{-1} = \begin{bmatrix} \boldsymbol{V}^{-1}\boldsymbol{C} \\ \boldsymbol{F}^{\perp} \end{bmatrix}\boldsymbol{M}\begin{bmatrix} \boldsymbol{V} & \boldsymbol{0} \end{bmatrix} = \begin{bmatrix} \boldsymbol{V}^{-1}\boldsymbol{C}\boldsymbol{M}\boldsymbol{V} & \boldsymbol{0} \\ \boldsymbol{F}^{\perp}\boldsymbol{M}\boldsymbol{V} & \boldsymbol{0} \end{bmatrix} \tag{21}$$

and, accepting the analogy between (18) and (21), it can define

$$\boldsymbol{A}_o = \boldsymbol{A}_o - \boldsymbol{A}^o = \begin{bmatrix} \boldsymbol{A}_{o11} - \boldsymbol{V}^{-1}\boldsymbol{C}\boldsymbol{M}\boldsymbol{V} & \boldsymbol{A}_{o12} \\ \boldsymbol{A}_{o21} - \boldsymbol{F}^{\perp}\boldsymbol{M}\boldsymbol{V} & \boldsymbol{A}_{o22} \end{bmatrix}. \tag{22}$$

Setting

$$\boldsymbol{A}_{o11} - \boldsymbol{V}^{-1}\boldsymbol{C}\boldsymbol{M}\boldsymbol{V} = -s_o\boldsymbol{I}_m\,, \quad \boldsymbol{A}_{o21} - \boldsymbol{F}^{\perp}\boldsymbol{M}\boldsymbol{V} = \boldsymbol{0}\,, \tag{23}$$

where $s_o \in I\!R$ is a prescribed positive real value, and writing (23) as

$$\begin{bmatrix} s_o\boldsymbol{I}_m + \boldsymbol{A}_{o11} \\ \boldsymbol{A}_{o21} \end{bmatrix} = \begin{bmatrix} \boldsymbol{V}^{-1}\boldsymbol{C} \\ \boldsymbol{F}^{\perp} \end{bmatrix}\boldsymbol{M}\boldsymbol{V} = \boldsymbol{T}\boldsymbol{M}\boldsymbol{V} = \boldsymbol{T}\boldsymbol{T}^{-1}\boldsymbol{L}^o\boldsymbol{V} = \boldsymbol{L}^o\boldsymbol{V}, \tag{24}$$

then, with

$$\boldsymbol{M} = \boldsymbol{T}^{-1}\boldsymbol{L}^o, \tag{25}$$

the following yields

$$\boldsymbol{A}^o = \boldsymbol{T}\boldsymbol{M}\boldsymbol{C}\boldsymbol{T}^{-1} = \boldsymbol{L}^o\boldsymbol{C}\boldsymbol{T}^{-1}, \tag{26}$$

$$\boldsymbol{A}_o = \boldsymbol{A}_o - \boldsymbol{A}^o = \boldsymbol{T}(\boldsymbol{A} - \boldsymbol{M}\boldsymbol{C})\boldsymbol{T}^{-1} = \boldsymbol{T} \quad \boldsymbol{A}\boldsymbol{T}^{-1}, \tag{27}$$

where

$$\boldsymbol{A} = \boldsymbol{A} - \boldsymbol{M}\boldsymbol{C} = \boldsymbol{A} - \boldsymbol{T}^{-1}\boldsymbol{L}^o\boldsymbol{C} \tag{28}$$

and (22) takes the form

$$\boldsymbol{A}_o = \begin{bmatrix} -s_o\boldsymbol{I}_m & \boldsymbol{A}_{o12} \\ \boldsymbol{0} & \boldsymbol{A}_{o22} \end{bmatrix}. \tag{29}$$

Defining the transfer function matrix $\boldsymbol{G}_{f\Delta}(s)$ as follows

$$\boldsymbol{G}_{f\Delta}(s) = \boldsymbol{C}(s\boldsymbol{I}_n - \quad \boldsymbol{A})^{-1}\boldsymbol{F}, \tag{30}$$

then with (27) it is

$$\boldsymbol{G}_{f\Delta}(s) = \boldsymbol{C}\boldsymbol{T}^{-1}(s\boldsymbol{I}_n - \boldsymbol{T} \quad \boldsymbol{A}\boldsymbol{T}^{-1})^{-1}\boldsymbol{T}\boldsymbol{F} = \begin{bmatrix} \boldsymbol{V} & \boldsymbol{0} \end{bmatrix}(s\boldsymbol{I}_n - \quad \boldsymbol{A}_o)^{-1}\begin{bmatrix} \boldsymbol{I}_p \\ \boldsymbol{0} \end{bmatrix}. \tag{31}$$

Since

$$s\boldsymbol{I}_n - \quad \boldsymbol{A}_o = \begin{bmatrix} (s+s_o)\boldsymbol{I}_m & -\boldsymbol{A}_{o12} \\ \boldsymbol{0} & s\boldsymbol{I}_{n-m} - \boldsymbol{A}_{o22} \end{bmatrix} \tag{32}$$

and

$$(s\boldsymbol{I}_n - \quad \boldsymbol{A}_o)^{-1} = \begin{bmatrix} (s+s_o)^{-1}\boldsymbol{I}_m & (s+s_o)^{-1}\boldsymbol{A}_{o12}(s\boldsymbol{I}_{n-m} - \boldsymbol{A}_{o22})^{-1} \\ \boldsymbol{0} & (s\boldsymbol{I}_{n-m} - \boldsymbol{A}_{o22})^{-1} \end{bmatrix}, \tag{33}$$

then, substituting (33) into (31), it can obtain

$$\boldsymbol{G}_{f\Delta}(s) = \begin{bmatrix} \boldsymbol{V} & \boldsymbol{0} \end{bmatrix}(s\boldsymbol{I}_n - \quad \boldsymbol{A}_o)^{-1}\begin{bmatrix} \boldsymbol{I}_p \\ \boldsymbol{0} \end{bmatrix} = \frac{\boldsymbol{V}}{s+s_o}, \tag{34}$$

which implies (16).

The transfer function (30) together with (28) and (19) can be rewritten as

$$\begin{aligned} \boldsymbol{G}_{f\Delta}(s) &= \boldsymbol{C}(s\boldsymbol{I}_n - \quad \boldsymbol{A})^{-1}\boldsymbol{F} \\ &= \boldsymbol{C}\left((s\boldsymbol{I}_n - \boldsymbol{A})(\boldsymbol{I}_n + (s\boldsymbol{I}_n - \boldsymbol{A})^{-1}\boldsymbol{M}\boldsymbol{C})\right)^{-1}\boldsymbol{F} \\ &= \boldsymbol{C}(\boldsymbol{I}_n + (s\boldsymbol{I}_n - \boldsymbol{A})^{-1}\boldsymbol{M}\boldsymbol{C})^{-1}(s\boldsymbol{I}_n - \boldsymbol{A})^{-1}\boldsymbol{F}, \end{aligned} \tag{35}$$

which gives, by using the equality

$$(\boldsymbol{F}\boldsymbol{C})^{-1}\boldsymbol{F}\boldsymbol{C} = \boldsymbol{I}_n, \tag{36}$$

$$\begin{aligned} \boldsymbol{G}_{f\Delta}(s) &= \boldsymbol{C}(\boldsymbol{I}_n + (s\boldsymbol{I}_n - \boldsymbol{A})^{-1}\boldsymbol{M}\boldsymbol{C})^{-1}(\boldsymbol{F}\boldsymbol{C})^{-1}\boldsymbol{F}\boldsymbol{C}(s\boldsymbol{I}_n - \boldsymbol{A})^{-1}\boldsymbol{F} \\ &= \boldsymbol{C}(\boldsymbol{I}_n + (s\boldsymbol{I}_n - \boldsymbol{A})^{-1}\boldsymbol{M}\boldsymbol{C})^{-1}(\boldsymbol{F}\boldsymbol{C})^{-1}\boldsymbol{F}\,\boldsymbol{G}_f(s), \end{aligned} \tag{37}$$

as well as, after some manipulations,

$$\begin{aligned} \boldsymbol{G}_{f\Delta}(s) &= \boldsymbol{C}\left((\boldsymbol{F}\boldsymbol{C})(\boldsymbol{I}_n + (s\boldsymbol{I}_n - \boldsymbol{A})^{-1}\boldsymbol{M}\boldsymbol{C})\right)^{-1}\boldsymbol{F}\,\boldsymbol{G}_f(s) \\ &= \boldsymbol{C}\left(\boldsymbol{F}(\boldsymbol{I}_m + \boldsymbol{C}(s\boldsymbol{I}_n - \boldsymbol{A})^{-1}\boldsymbol{M})\boldsymbol{C}\right)^{-1}\boldsymbol{F}\,\boldsymbol{G}_f(s) \\ &= \left(\boldsymbol{I}_m + \boldsymbol{C}(s\boldsymbol{I}_n - \boldsymbol{A})^{-1}\boldsymbol{M}\right)^{-1}\boldsymbol{G}_f(s). \end{aligned} \tag{38}$$

Thus, with the notation (17), then (38) implies (15). This concludes the proof. ■

With existence of such transformation the structure of (20) really means that there exists the subset of transformed state variables whose dynamics is explicitly affected by the fault $\boldsymbol{f}(t)$ and the second one, whose dynamics is not affected explicitly by the fault $\boldsymbol{f}(t)$.

Defined by (6), a linear time-invariant system is considered as unitary if all singular values of its transfer function matrix are equal. Because the construction given in Lemma 2 is not unique, equivalent structures can be used.

Lemma 3 *An equivalent structure of the fault transfer function matrix of the system (1), (2) takes the form*

$$\boldsymbol{G}_f(s) = \boldsymbol{G}^{\circ}_{\Delta}(s)(\boldsymbol{G}^{\circ}_{f\Delta}(s) - \boldsymbol{V}), \tag{39}$$

where

$$\boldsymbol{G}^{\circ}_{f\Delta}(s) = \boldsymbol{C}(s\boldsymbol{I}_n - \boldsymbol{A} + \boldsymbol{N}\boldsymbol{C})^{-1}(\boldsymbol{F} + \boldsymbol{N}\boldsymbol{V}) + \boldsymbol{V}, \tag{40}$$

$$\boldsymbol{G}^{\circ}_{\Delta}(s) = \boldsymbol{I}_m + \boldsymbol{C}(s\boldsymbol{I}_n - \boldsymbol{A})^{-1}\boldsymbol{N}. \tag{41}$$

If $\boldsymbol{M} = \boldsymbol{N}$ *then*

$$\frac{\boldsymbol{G}^{\circ}_{f\Delta}(s)}{s + s_o + 1} = \frac{\boldsymbol{V}}{s + s_o} = \boldsymbol{G}_{f\Delta}(s). \tag{42}$$

Proof: Considering the associated system (40), it can be written

$$(s\boldsymbol{I}_n - \boldsymbol{A} + \boldsymbol{N}\boldsymbol{C})^{-1} = ((s\boldsymbol{I}_n - \boldsymbol{A})(\boldsymbol{I}_n + (s\boldsymbol{I}_n - \boldsymbol{A})^{-1}\boldsymbol{N}\boldsymbol{C}))^{-1} = \boldsymbol{G}^{\diamond}_{\Delta}(s)(s\boldsymbol{I}_n - \boldsymbol{A})^{-1}, \tag{43}$$

where

$$\boldsymbol{G}^{\diamond}_{\Delta}(s) = (\boldsymbol{I}_n + (s\boldsymbol{I}_n - \boldsymbol{A})^{-1}\boldsymbol{N}\boldsymbol{C})^{-1}. \tag{44}$$

Therefore, substituting (43) in (40) leads to

$$\boldsymbol{G}^{\circ}_{f\Delta}(s) = \boldsymbol{C}\left(\boldsymbol{G}^{\diamond}_{\Delta}(s)(s\boldsymbol{I}_n - \boldsymbol{A})^{-1}(\boldsymbol{I}_n + \boldsymbol{N}\boldsymbol{C}) + \boldsymbol{I}_n\right)\boldsymbol{F} \tag{45}$$

and it yields

$$\begin{aligned} \boldsymbol{G}^{\circ}_{f\Delta}(s) &= \boldsymbol{C}\left(\boldsymbol{G}^{\diamond}_{\Delta}(s)((s\boldsymbol{I}_n - \boldsymbol{A})^{-1} + (s\boldsymbol{I}_n - \boldsymbol{A})^{-1}\boldsymbol{N}\boldsymbol{C}) + \boldsymbol{I}_n\right)\boldsymbol{F} \\ &= \boldsymbol{C}\left(\boldsymbol{G}^{\diamond}_{\Delta}(s)((s\boldsymbol{I}_n - \boldsymbol{A})^{-1} + \boldsymbol{G}^{\diamond -1}_{\Delta}(s)\right)\boldsymbol{F} \\ &= \boldsymbol{C}\boldsymbol{G}^{\diamond}_{\Delta}(s)(s\boldsymbol{I}_n - \boldsymbol{A})^{-1}\boldsymbol{F} + \boldsymbol{V}. \end{aligned} \tag{46}$$

Since, using the equality (36), it can obtain

$$\begin{aligned} \boldsymbol{C}\boldsymbol{G}^{\diamond}_{\Delta}(s) &= \boldsymbol{C}\boldsymbol{G}^{\diamond}_{\Delta}(s)(\boldsymbol{F}\boldsymbol{C})^{-1}\boldsymbol{F}\boldsymbol{C} \\ &= \boldsymbol{C}\left(\boldsymbol{F}\boldsymbol{C}(\boldsymbol{I}_n + (s\boldsymbol{I}_n - \boldsymbol{A})^{-1}\boldsymbol{N}\boldsymbol{C})\right)^{-1}\boldsymbol{F}\boldsymbol{C} \\ &= \boldsymbol{C}\left(\boldsymbol{F}(\boldsymbol{I}_m + \boldsymbol{C}(s\boldsymbol{I}_n - \boldsymbol{A})^{-1}\boldsymbol{N})\boldsymbol{C}\right)^{-1}\boldsymbol{F}\boldsymbol{C} \\ &= \boldsymbol{G}^{\circ -1}_{\Delta}(s)\boldsymbol{C}, \end{aligned} \tag{47}$$

then the substitution of (47) into (46) gives

$$\boldsymbol{G}^{\circ}_{f\Delta}(s) = \boldsymbol{G}^{\circ -1}_{\Delta}(s)\boldsymbol{C}(s\boldsymbol{I}_n - \boldsymbol{A})^{-1}\boldsymbol{F} + \boldsymbol{V} = \boldsymbol{G}^{\circ -1}_{\Delta}(s)\boldsymbol{G}_f(s) + \boldsymbol{V}. \tag{48}$$

Thus, (48) implies (39).

Considering that $\boldsymbol{M} = \boldsymbol{N}$ then (17) and (41) implies $\boldsymbol{G}^{\circ}_{\Delta}(s) = \boldsymbol{G}_{\Delta}(s)$ and (15) defines $\boldsymbol{G}_f(s) = \boldsymbol{G}_{\Delta}(s)\boldsymbol{G}_{f\Delta}(s)$. Thus, (39) can be written as

$$\boldsymbol{G}_f(s) = \boldsymbol{G}_{\Delta}(s)(\boldsymbol{G}^{\circ}_{f\Delta}(s) - \boldsymbol{V}) = \boldsymbol{G}_{\Delta}(s)\boldsymbol{G}_{f\Delta}(s), \tag{49}$$

which gives, with respect to (16),

$$\boldsymbol{G}^{\circ}_{f\Delta}(s) = \boldsymbol{G}_{f\Delta}(s) + \boldsymbol{V} = \boldsymbol{V}\left(\frac{1}{s + s_o} + 1\right) = \frac{\boldsymbol{V}}{s + s_o}(s + s_o + 1), \tag{50}$$

which gives (42). This concludes the proof. ■

Note, the relations (42) implies that both transfer functions have the same gain but with different pole. This property gives the possibility to combine two specific gains in the design of unitary fault transfer function matrix by the way specified in the following section. Singular values related properties, such as H_2 norm, H_∞ norm, as well as H_- index, can be determined based on this pole in the FTC structures where the optimization of singular values related properties are of key importance.

Journal of Physics: Conference Series **659** (2015) 012036 doi:10.1088/1742-6596/659/1/012036

4. Enhanced Structure of Unitary Fault Transfer Function Matrix

To exploit the properties all structures presented above, the enhanced form of unitary fault transfer matrix is proposed in the form

$$\boldsymbol{G}^{\circ}_{f}(s) = \boldsymbol{C}(s\boldsymbol{I}_n - \boldsymbol{A} + (\boldsymbol{M} + \boldsymbol{N})\boldsymbol{C})^{-1}\boldsymbol{V}\,, \tag{51}$$

where $\boldsymbol{M}$ is introduced in (18) and $\boldsymbol{N}$ is designed in such way that $\boldsymbol{F} + \boldsymbol{N}\boldsymbol{V} = \boldsymbol{0}$.

To formulate the stability condition of the unitary system, approximated by the equivalent transform function matrix (40), the following theorems are given.

Lemma 4 *The state-space representation of the enhanced structure of transfer function matrix (51) in the form of a closed-loop system is*

$$\dot{\boldsymbol{q}}^{\circ}_{d}(t) = \boldsymbol{A}^{\circ}\boldsymbol{q}^{\circ}_{d}(t) + \boldsymbol{B}^{\circ}\boldsymbol{u}^{\circ}(t)\,, \tag{52}$$

$$\boldsymbol{y}^{\circ}(t) = \boldsymbol{C}^{\circ}\boldsymbol{q}^{\circ}_{d}(t)\,, \tag{53}$$

$$\boldsymbol{u}^{\circ}(s) = -\boldsymbol{K}^{\circ}\boldsymbol{q}^{\circ}(t) = -(\boldsymbol{M}^T + \boldsymbol{N}^T)\boldsymbol{q}^{\circ}(t)\,, \tag{54}$$

where the system constraint is

$$\boldsymbol{R}^{\circ} = \begin{bmatrix} \boldsymbol{0} & \boldsymbol{S} \\ \boldsymbol{S}^T & -\boldsymbol{R} \end{bmatrix} = \begin{bmatrix} \boldsymbol{0} & \boldsymbol{F}\boldsymbol{V}^T \\ \boldsymbol{V}\boldsymbol{F}^T & -\boldsymbol{V}\boldsymbol{V}^T \end{bmatrix} \geq 0 \tag{55}$$

and the matrix parameters are

$$\boldsymbol{A}^{\circ} = \boldsymbol{A}^T, \quad \boldsymbol{B}^{\circ} = \boldsymbol{C}^T, \quad \boldsymbol{C}^{\circ} = \boldsymbol{F}^T, \tag{56}$$

$$\boldsymbol{M}^T = \boldsymbol{L}^{oT}\boldsymbol{T}^{-T}, \quad \boldsymbol{N}^T = -\boldsymbol{V}^{-T}\boldsymbol{F}^T. \tag{57}$$

Thus,

$$\dot{\boldsymbol{q}}^{\circ}_{d}(t) = \boldsymbol{A}^{\circ}_{dc}\boldsymbol{q}^{\circ}_{d}(t), \quad \boldsymbol{A}^{\circ}_{dc} = \boldsymbol{A}^{\circ} - \boldsymbol{B}^{\circ}\boldsymbol{K}^{\circ}\,. \tag{58}$$

$$\boldsymbol{y}^{\circ}(t) = \boldsymbol{C}^{\circ}\boldsymbol{q}^{\circ}_{d}(t)\,. \tag{59}$$

Proof: Considering that

$$\boldsymbol{F} + \boldsymbol{N}\boldsymbol{V} = \boldsymbol{0} \tag{60}$$

then

$$\boldsymbol{N} = -\boldsymbol{F}\boldsymbol{V}^{-1} \tag{61}$$

and using the Laplace transform property [15], then (51) conditioned by the design constraint (60) implies

$$\tilde{\boldsymbol{q}}_d(s) = (s\boldsymbol{I}_n - \boldsymbol{A} + \boldsymbol{M}\boldsymbol{C} + \boldsymbol{N}\boldsymbol{C})^{-1}\boldsymbol{V}\tilde{\boldsymbol{f}}(s) = (s\boldsymbol{I}_n - \boldsymbol{A} + \boldsymbol{M}\boldsymbol{C} - \boldsymbol{F}\boldsymbol{V}^{-1}\boldsymbol{C})^{-1}\boldsymbol{V}\tilde{\boldsymbol{f}}(s)\,, \tag{62}$$

Writing in the dual state-space form [17], the response (62) reflects a free-response of the system

$$\dot{\boldsymbol{q}}^{\circ}_{d}(t) = (\boldsymbol{A}^T - \boldsymbol{C}^T\boldsymbol{M} + \boldsymbol{C}^T\boldsymbol{V}^{-T}\boldsymbol{F}^T)\boldsymbol{V}\boldsymbol{f}(t)\,, \tag{63}$$

$$\boldsymbol{y}^{\circ}(t) = \boldsymbol{F}^T\boldsymbol{q}^{\circ}_{d}(t) \tag{64}$$

and can be interpreted as the response of the closed-loop system (52)-(54), (56), (57). It is evident that this system should be stable under constraint (64) for $\boldsymbol{N}$ of structure given in (61).

Multiplying the right side of (60) by $\boldsymbol{V}^T$ leads to

$$\boldsymbol{F}\boldsymbol{V}^T + \boldsymbol{N}\boldsymbol{V}\boldsymbol{V}^T = \boldsymbol{0}\,, \tag{65}$$

which implies

$$\boldsymbol{N} = -\boldsymbol{F}\boldsymbol{V}^T(\boldsymbol{V}\boldsymbol{V}^T)^{-1}. \tag{66}$$

Since pre-multiplying the right side of (66) by $\boldsymbol{V}\boldsymbol{F}^T$ and then using (60) gives

$$\boldsymbol{F}\boldsymbol{V}^T(\boldsymbol{V}\boldsymbol{V}^T)^{-1}\boldsymbol{V}\boldsymbol{F}^T = -\boldsymbol{N}\boldsymbol{V}\boldsymbol{F}^T = \boldsymbol{F}\boldsymbol{F}^T \geq 0 \tag{67}$$

and because a dual state-space description (53), (56) implies

$$\boldsymbol{y}^{\circ T}(t)\boldsymbol{y}^{\circ}(t) = \boldsymbol{q}^{\circ T}(t)\boldsymbol{C}^{\circ T}\boldsymbol{C}^{\circ}\boldsymbol{q}^{\circ}(t) = \boldsymbol{q}^{\circ T}(t)\boldsymbol{F}\boldsymbol{F}^T\boldsymbol{q}^{\circ}(t), \tag{68}$$

it is evident that (67) is the constraint given on $\boldsymbol{q}^{\circ}(t)$.

Thus, using the Schur complement property, (67) implies the quadratic constraint (55). This concludes the proof. ■

Theorem 1 *The equivalent system (52)-(57) is stable if there exists a symmetric positive definite matrix $\boldsymbol{P} \in I\!R^{n\times n}$ such that*

$$\boldsymbol{P} = \boldsymbol{P}^T > 0, \tag{69}$$

$$\begin{bmatrix} (\boldsymbol{A}-\boldsymbol{M}\boldsymbol{C}+\boldsymbol{F}\boldsymbol{V}^{-1}\boldsymbol{C})\boldsymbol{P}+\boldsymbol{P}(\boldsymbol{A}-\boldsymbol{M}\boldsymbol{C}+\boldsymbol{F}\boldsymbol{V}^{-1}\boldsymbol{C})^T & \boldsymbol{P}\boldsymbol{C}^T+\boldsymbol{F}\boldsymbol{V}^T \\ \boldsymbol{C}\boldsymbol{P}+\boldsymbol{V}\boldsymbol{F}^T & -\boldsymbol{V}\boldsymbol{V}^T \end{bmatrix} < 0 \tag{70}$$

and the common gain matrix is

$$\boldsymbol{K}^{\circ} = -\boldsymbol{M} - \boldsymbol{N} = -\boldsymbol{M} + \boldsymbol{F}\boldsymbol{V}^{-1}. \tag{71}$$

Proof: Introducing the notation

$$\boldsymbol{q}^{\bullet T}(t) = \begin{bmatrix} \boldsymbol{q}_d^{\circ T}(t) & \boldsymbol{u}^{\circ T}(t) \end{bmatrix}, \tag{72}$$

the Lyapunov function candidate can be considered in the form

$$v(\boldsymbol{q}^{\bullet}(t)) = \boldsymbol{q}_d^{\circ T}(t)\boldsymbol{P}\boldsymbol{q}_d^{\circ}(t) + \int_0^t \boldsymbol{q}^{\bullet T}(v)\boldsymbol{R}^{\circ}\boldsymbol{q}^{\bullet}(v)\mathrm{d}v > 0, \tag{73}$$

where $\boldsymbol{P} \in I\!R^{n\times n}$ is a symmetric, positive definite matrix and $\boldsymbol{R}^{\circ} \in I\!R^{(n+m)\times(n+m)}$ is given in (55).

Therefore, the time derivative of (73) can be written as

$$\dot{v}(\boldsymbol{q}^{\bullet}(t)) = \dot{\boldsymbol{q}}^{\circ T}(t)\boldsymbol{P}\boldsymbol{q}^{\circ}(t) + \boldsymbol{q}^{\circ T}(t)\boldsymbol{P}\dot{\boldsymbol{q}}^{\circ}(t) + \boldsymbol{q}^{\bullet T}(t)\boldsymbol{R}^{\circ}\boldsymbol{q}^{\bullet}(t) < 0, \tag{74}$$

which, by substituting (52), takes the form

$$\dot{v}(\boldsymbol{q}^{\bullet}(t)) = \boldsymbol{q}^{\bullet T}(t)\boldsymbol{R}^{\circ}\boldsymbol{q}^{\bullet}(t) + (\boldsymbol{A}^{\circ}\boldsymbol{q}_d^{\circ}(t)+\boldsymbol{B}^{\circ}\boldsymbol{u}^{\circ}(t))^T\boldsymbol{P}\boldsymbol{q}_d^{\circ}(t) + \boldsymbol{q}_d^{\circ T}(t)\boldsymbol{P}(\boldsymbol{A}^{\circ}\boldsymbol{q}_d^{\circ}(t)+\boldsymbol{B}^{\circ}\boldsymbol{u}^{\circ}(t)) < 0. \tag{75}$$

Then, using (72), the inequality (75) can be written as

$$\dot{v}(\boldsymbol{q}^{\bullet}(t)) = \boldsymbol{q}^{\bullet T}(t)\boldsymbol{P}^{\bullet}\boldsymbol{q}^{\bullet}(t) < 0, \tag{76}$$

where

$$\boldsymbol{P}^{\bullet} = \begin{bmatrix} \boldsymbol{P}\boldsymbol{A}^{\circ}+\boldsymbol{A}^{\circ T}\boldsymbol{P} & \boldsymbol{P}\boldsymbol{B}^{\circ}+\boldsymbol{F}\boldsymbol{V}^T \\ \boldsymbol{B}^{\circ T}\boldsymbol{P}+\boldsymbol{V}\boldsymbol{F}^T & -\boldsymbol{V}\boldsymbol{V}^T \end{bmatrix} < 0. \tag{77}$$

Constructing the closed-loop system matrix

$$\boldsymbol{A}^{\circ}_{dc} = \boldsymbol{A}^{\circ} - \boldsymbol{B}^{\circ}\boldsymbol{K}^{\circ} = \boldsymbol{A}^{T} - \boldsymbol{C}^{T}(\boldsymbol{M}^{T} + \boldsymbol{N}^{T}), \tag{78}$$

which, by using (57), is of the form

$$\boldsymbol{A}^{\circ}_{dc} = \boldsymbol{A}^{T} - \boldsymbol{C}^{T}\boldsymbol{M}^{T} + \boldsymbol{C}^{T}\boldsymbol{V}^{-T}\boldsymbol{F}^{T}, \tag{79}$$

it is evident that

$$\boldsymbol{A}^{\circ T}_{dc}\boldsymbol{P} = (\boldsymbol{A} - \boldsymbol{M}\boldsymbol{C} + \boldsymbol{F}\boldsymbol{V}^{-1}\boldsymbol{C})\boldsymbol{P} \tag{80}$$

and

$$\boldsymbol{P}\boldsymbol{B}^{\circ} = \boldsymbol{P}\boldsymbol{C}^{T}. \tag{81}$$

Thus, replacing $\boldsymbol{A}^{\circ}$ in (77) by $\boldsymbol{A}^{\circ}_{c}$ and using (81), the matrix inequality (77) takes the form (70). This concludes the proof. ■

5. Unitary Fault Detection Filter

Writing the dual form of the state-space description (58), (78) in the direct form, it is obtained

$$\dot{\boldsymbol{q}}^{\circ}(t) = \boldsymbol{A}^{\circ T}_{dc}\boldsymbol{q}^{\circ}(t), \tag{82}$$

$$\boldsymbol{A}^{\circ T}_{dc} = \boldsymbol{A}^{\circ T} - \boldsymbol{K}^{\circ T}\boldsymbol{B}^{\circ T} = \boldsymbol{A} - (\boldsymbol{M} + \boldsymbol{N})\boldsymbol{C}. \tag{83}$$

Denoting

$$\boldsymbol{e}(t) = \boldsymbol{q}^{\circ}(t), \quad \boldsymbol{J} = \boldsymbol{M} + \boldsymbol{N}, \tag{84}$$

it yields

$$\dot{\boldsymbol{e}}(t) = (\boldsymbol{A} - \boldsymbol{J}\boldsymbol{C})\boldsymbol{e}(t), \tag{85}$$

which is the autonomous equation of the estimation error of Luenberger observer to the nominal noise-free system (1), (2), defined in the form [11]

$$\dot{\boldsymbol{q}}_e(t) = \boldsymbol{A}\boldsymbol{q}_e(t) + \boldsymbol{B}\boldsymbol{u}(t) + \boldsymbol{J}\boldsymbol{C}(\boldsymbol{q}(t) - \boldsymbol{q}_e(t)), \tag{86}$$

Introducing $\boldsymbol{e}(t) = \boldsymbol{q}(t) - \boldsymbol{q}_e(t)$, then with (1), (2) and (86) it yields

$$\begin{aligned} \dot{\boldsymbol{e}}(t) &= \boldsymbol{A}\boldsymbol{q}(t) + \boldsymbol{B}\boldsymbol{u}(t) + \boldsymbol{F}\boldsymbol{f}(t) + \boldsymbol{E}\boldsymbol{d}(t) - \boldsymbol{A}\boldsymbol{q}_e(t) - \boldsymbol{B}\boldsymbol{u}(t) - \boldsymbol{J}\boldsymbol{C}(\boldsymbol{q}(t) - \boldsymbol{q}_e(t)) \\ &= (\boldsymbol{A} - \boldsymbol{J}\boldsymbol{C})\boldsymbol{e}(t) + \boldsymbol{F}\boldsymbol{f}(t) + \boldsymbol{E}\boldsymbol{d}(t) \end{aligned} \tag{87}$$

$$\tilde{\boldsymbol{e}}(s) = (s\boldsymbol{I}_n - (\boldsymbol{A} - \boldsymbol{J}\boldsymbol{C}))^{-1}(\boldsymbol{F}\tilde{\boldsymbol{f}}(s) + \boldsymbol{E}\tilde{\boldsymbol{d}}(s) \tag{88}$$

respectively, where $\tilde{\boldsymbol{e}}(s)$, $\tilde{\boldsymbol{f}}(s)$, $\tilde{\boldsymbol{d}}(s)$ stand for the Laplace transform of the n dimensional observer error vector, the m dimensional fault vector and the p dimensional disturbance vector.

Designing the fault residuals as

$$\boldsymbol{r}(t) = \boldsymbol{V}^{-1}\boldsymbol{C}\boldsymbol{e}(t), \quad \tilde{\boldsymbol{r}}(s) = \boldsymbol{V}^{-1}\boldsymbol{C}\tilde{\boldsymbol{e}}(s), \tag{89}$$

then the residual transfer function matrices of the fault and the disturbance are

$$\boldsymbol{G}_f(s) = \boldsymbol{V}^{-1}\boldsymbol{C}(s\boldsymbol{I}_n - (\boldsymbol{A} - \boldsymbol{J}\boldsymbol{C}))^{-1}\boldsymbol{F}, \tag{90}$$

$$\boldsymbol{G}_d(s) = \boldsymbol{V}^{-1}\boldsymbol{C}(s\boldsymbol{I}_n - (\boldsymbol{A} - \boldsymbol{J}\boldsymbol{C}))^{-1}\boldsymbol{E}. \tag{91}$$

It is evident that with $\boldsymbol{J}$ of the structure (84), and designed by (57), $\boldsymbol{G}_f(s)$ is an unitary transfer function matrix with optimized singular values related properties.

6. Illustrative Example

The considered unstable system is represented by the model (1), (2) with the parameters

$$\boldsymbol{A}=\begin{bmatrix} 1.380 & -0.208 & 6.175 & -1.576 \\ 0.581 & -4.290 & 0.000 & 0.675 \\ 1.067 & 4.273 & -6.654 & 5.893 \\ 0.048 & 4.273 & 1.343 & -2.104 \end{bmatrix},\quad \boldsymbol{B}=\begin{bmatrix} 0.000 & 0.000 & 0.000 \\ 5.679 & -1.000 & 0.000 \\ 1.136 & -3.146 & 1.324 \\ 1.136 & 0.000 & 3.496 \end{bmatrix},\quad \boldsymbol{E}=\begin{bmatrix} 1.400 \\ 1.504 \\ 2.233 \\ 0.610 \end{bmatrix},$$

$$\boldsymbol{B}=\boldsymbol{F},\ \boldsymbol{C}=\begin{bmatrix} 4 & 0 & 1 & 0 \\ 0 & 0 & 0 & 1 \\ 0 & 1 & 0 & 0 \end{bmatrix},\quad \boldsymbol{V}=\boldsymbol{C}\boldsymbol{F}=\begin{bmatrix} 1.136 & -3.146 & 1.324 \\ 1.136 & 0.000 & 3.496 \\ 5.679 & -1.000 & 0.000 \end{bmatrix},\quad \det(\boldsymbol{V})\neq 0.$$

The parameters of the matrix $\boldsymbol{T}$ were computed as follows

$$\boldsymbol{V}^{-1}\boldsymbol{C}=\begin{bmatrix} -0.2331 & 0.1833 & -0.0583 & 0.0221 \\ -1.3237 & 0.0411 & -0.3309 & 0.1253 \\ 0.0757 & -0.0596 & 0.0189 & 0.2789 \end{bmatrix},$$

$$\boldsymbol{F}^{\perp}=\begin{bmatrix} -1.0000 & 0.0000 & 0.0000 & 0.0000 \end{bmatrix},$$

that is

$$\boldsymbol{T}=\begin{bmatrix} -0.2331 & 0.1833 & -0.0583 & 0.0221 \\ -1.3237 & 0.0411 & -0.3309 & 0.1253 \\ 0.0757 & -0.0596 & 0.0189 & 0.2789 \\ -1.0000 & 0.0000 & 0.0000 & 0.0000 \end{bmatrix}.$$

Computing (18) and separating the blocks of the matrix $\boldsymbol{A}^{\circ}$ give the results

$$\boldsymbol{A}_{o11}=\begin{bmatrix} -6.1156 & 4.1078 & -0.9991 \\ -11.1342 & 19.0383 & -8.0335 \\ 8.6811 & -3.7656 & -1.2707 \end{bmatrix},\quad \boldsymbol{A}_{o12}=\begin{bmatrix} -3.8116 \\ -21.0651 \\ 2.7614 \end{bmatrix},$$

$$\boldsymbol{A}_{o21}=\begin{bmatrix} -4.0432 & 19.2186 & -2.6660 \end{bmatrix},\quad \boldsymbol{A}_{o22}=\begin{bmatrix} -23.3200 \end{bmatrix}$$

and so, choosing $s_o=22$, it was obtained using (18) and (61) that

$$\boldsymbol{L}^{o}=\begin{bmatrix} -2.3040 & 0.5868 & 3.1405 \\ -13.0844 & 2.6574 & 0.1252 \\ 1.1328 & 5.5004 & 0.2018 \\ -6.1750 & 1.5760 & 0.2080 \end{bmatrix},\quad \boldsymbol{M}=\begin{bmatrix} 6.1750 & -1.5760 & -0.2080 \\ 0.0000 & 0.6750 & 17.7100 \\ 15.3460 & 5.8930 & 4.2730 \\ 1.3430 & 19.8960 & 4.2730 \end{bmatrix},$$

$$\boldsymbol{N}=\begin{bmatrix} 0 & 0 & 0 \\ 0 & 0 & -1 \\ -1 & 0 & 0 \\ 0 & -1 & 0 \end{bmatrix},\quad \boldsymbol{J}=\begin{bmatrix} -6.1750 & 1.5760 & 0.2080 \\ 0.0000 & -0.6750 & -18.7100 \\ -16.3460 & -5.8930 & -4.2730 \\ -1.3430 & -20.8960 & -4.2730 \end{bmatrix},$$

respectively. Thus, constructing the observer system matrix

$$\boldsymbol{A}_e=\boldsymbol{A}-\boldsymbol{J}\boldsymbol{C}=\begin{bmatrix} -23.3200 & 0.0000 & 0.0000 & 0.0000 \\ 0.5810 & -23.0000 & 0.0000 & 0.0000 \\ -64.3170 & 0.0000 & -23.0000 & 0.0000 \\ -5.3240 & 0.0000 & 0.0000 & -23.0000 \end{bmatrix},$$

it is evident that the eigenvalues spectrum of $\boldsymbol{A}_e$ is

$$\rho(\boldsymbol{A}_e)=\{\ -23.32 \quad -23.00 \quad -23.00 \quad -23.00\ \}$$

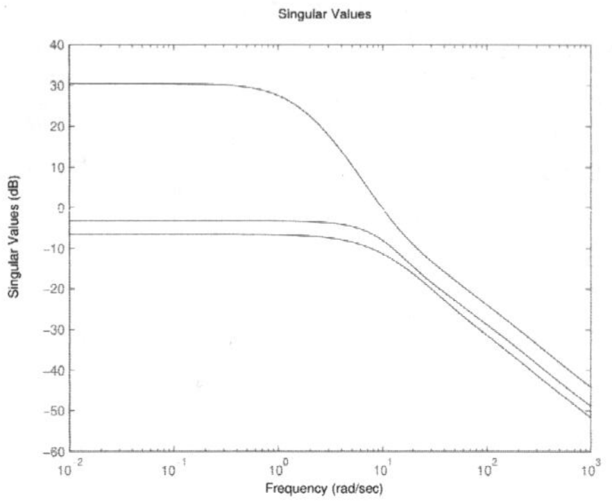

Figure 1. Singular values of the original system.

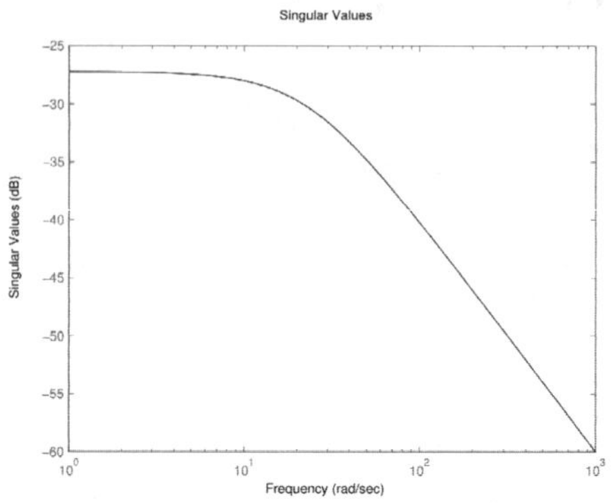

Figure 2. Singular values of the unitary approximation of the system.

and the steady-state value of the residual transfer function matrices are

$$\boldsymbol{G}_f(0) = \boldsymbol{V}^{-1}\boldsymbol{C}\boldsymbol{A}_e^{-1}\boldsymbol{F} = -\begin{bmatrix} 0.0435 & 0.0000 & 0.0000 \\ 0.0000 & 0.0435 & 0.0000 \\ 0.0000 & 0.0000 & 0.0435 \end{bmatrix},$$

$$\boldsymbol{G}_d(0) = \boldsymbol{V}^{-1}\boldsymbol{C}\boldsymbol{A}_e^{-1}\boldsymbol{E} = -\begin{bmatrix} 0.0027 \\ 0.0027 \\ -0.0517 \end{bmatrix}.$$

For completeness it was verified that in sense of Lyapunov stability there exist the positive definite matrix

$$\boldsymbol{P} = \begin{bmatrix} 3.5370 & 0.2172 & -1.1139 & 0.1226 \\ 0.2172 & 0.5332 & -0.1550 & -0.0333 \\ -1.1139 & -0.1550 & 0.8587 & -0.1443 \\ 0.1226 & -0.0333 & -0.1443 & 0.6646 \end{bmatrix}.$$

such that (69)-(71) are affirmative.

Note, using the coordinate transformation defined by the transform matrix (8), the block matrix $\boldsymbol{A}_{o22}$ in the matrix structure (18) is uncontrollable in the closed-loop structure (52)-(54) which describe the equivalent unitary systems. That means, the eigenvalues of $\boldsymbol{A}_{o22}$ determine the unprescribed subset of singular eigenvalues of $\rho(\boldsymbol{A}_e)$ (compare the value $\boldsymbol{A}_{o22} = -23.3200$ and the first singular value of $\rho(\boldsymbol{A}_e)$ in this example).

The results are closer to the theoretical expectations, and are simpler than those that have been published in [22].

7. Concluding Remarks

Based on the singular value approach a simpler method, but of the same precision, for unitary approximation of unknown fault transfer function matrix for continuous-time linear systems is introduced in the paper. Presented version is derived in terms numerical procedures to manipulate dynamics of the residual fault transfer function matrix. Formulated in sense of the second Lyapunov method and expressed through LMIs, stability conditions guaranteing the asymptotic convergence of unitary system observer state are derived over and above. The numerical simulation results show very good approximation performances.

Journal of Physics: Conference Series **659** (2015) 012036 doi:10.1088/1742-6596/659/1/012036

Acknowledgment
The work presented in this paper was supported by VEGA, the Grant Agency of the Ministry of Education and the Academy of Science of Slovak Republic, under Grant No. 1/0348/14. This support is very gratefully acknowledged.

References

[1] Adkins W A and Weintraub S H 1992 *Algebra. An Approach via Module Theory* (Berlin:Springer)
[2] Azizov T Y and Iokhvidov I S 1989 *Linear Operators in Spaces with an Indefinite Metric* (Chichester:John Wiley & Sons)
[3] Bernstein D S 2009 *Matrix Mathematics. Theory, Facts, and Formulas*, (Princeton:Princeton University Press)
[4] Blanke M Kinnaert M Lunze J and Staroswiecki M 2006 *Diagnosis and Fault-Tolerant Control* (Berlin:Springer-Verlag)
[5] Boyd D and Balakrishnan V 1990 A regularity result for the singular values of a transfer matrix and a qudratically convergent algorithm for computing its L_∞-norm *Systems & Control Letters*, **15** 1-7
[6] Chen J and Patton R J 1999 *Robust Model-Based Fault Diagnosis for Dynamic System* (Norwel:Kluwer Academic Publishers)
[7] Ding S 2008 *Model-Based Fault Diagnosis Techniques. Design Schemes, Algorithms, and Tools* (Berlin:Springer-Verlag)
[8] Ding S X Jeinsch T Frank P M and Ding E L 2000 A unified approach to the optimization of fault detection systems *Int. J. Adaptive Control and Signal Processing* **14** 725-45
[9] Hou M and Patton R J 1996 An LMI approach to H_- H_∞ fault detection observers *Proc. UKACC Int. Conf. Control CONTROL'96* Exeter England 305-10
[10] Gasparyan O N 2008 *Linear and Nonlinear Multivariable Feedback Control. A Classical Approach* (Chichester:John Wiley & Sons)
[11] Krokavec D and Filasová A 2013 Actuator faults reconstruction using reduced-order fuzzy observer structures *Proc. 12th European Control Conference ECC'13* Zurich Switzerland 4299-4304
[12] Krokavec D and Filasová A 2014 Robust residual generator design for systems with distributed time delay *Intelligent Systems in Technical and Medical Diagnostics* eds J Korbicz and M Koval (Berlin:Springer-Verlag) 417-28
[13] Krokavec D and Filasová A 2015 Construction process of unitary systems using LMI based approach *IFAC-PapersOnLine* **48**(14) 174-179
[14] Mahmoud M S and Xia Y 2014 *Analysis and Synthesis of Fault-Tolerant Control Systems*, (Chichester:John Wiley & Sons)
[15] Miles J W 1971 *Integral Transforms in Applied Mathematics* (New York:Cambridge University Press)
[16] Noura H Theilliol D Ponsart J.C. and Chamseddine A 2009 *Fault-tolerant Control Systems: Design and Practical Applications* (London:Springer-Verlag)
[17] Schultz D G and Melsa J L 1967 *State Functions and Linear Control Systems* (New York:McGraw-Hill)
[18] VanAntwerp J G and Braatz R D 2000 A tutorial on linear and bilinear matrix inequalities *J. Process Control* **10** 363-385
[19] Yang M H 1995 The state space of a canonical linear system *J. Korean Mathematical Society* **32** 447-459
[20] Zhang K Jiang B and Shi P 2013 *Observer-Based Fault Estimation and Accomodation for Dynamic Systems* (Berlin:Springer-Verlag)
[21] Zhang Y Wu L. Li J and Chen X 2012 LMI approach to mixed H_- H_∞ fault detection observer design *Tran. Tianjin University* **8** 343-349
[22] Zhao Z Xie W F Hong H and Zhang Y 2011 Unitary system I. Constructing a unitary fault detection observer *Prep. 18th IFAC World Congress* Milano Italy 7725-7730
[23] Zhao Z Xie W F Hong H and Zhang Y 2011 Unitary system II. Application to H_∞ H_- optimization of strictly-proper systems *Prep. 18th IFAC World Congress* Milano Italy 7731-7736
[24] Zhou K Doyle J C and Glover K 1996 *Robust and Optimal Control* (Englewood Cliffs:Prentice Hall)
[25] Zolghadri A Henry D Cieslak J Efimov D and Goupil P 2014 *Fault Diagnosis and Fault-Tolerant Control and Guidance for Aerospace Vehicles. From Theory to Application* (London:Springer-Verlag)

Fault Detection for Automotive Shock Absorber

Diana Hernandez-Alcantara, Ruben Morales-Menendez
Tecnológico de Monterrey, School of Engineering and Sciences
Av Eugenio Garza Sada # 2501, Col. Tecnológico
64,849 Monterrey, Nuevo León, México
E-mail: { A00469139, rmm }@itesm.mx

Luis Amezquita-Brooks
Universidad Autonoma de Nuevo León
Av. Universidad S/N
66,451 San Nicolas de los Garza, Nuevo León, México
E-mail: amezquita-brooks@ieee.org

Abstract. Fault detection for automotive semi-active shock absorbers is a challenge due to the non-linear dynamics and the strong influence of the disturbances such as the road profile. First obstacle for this task, is the modeling of the fault, which has been shown to be of multiplicative nature. Many of the most widespread fault detection schemes consider additive faults. Two model-based fault algorithms for semi-active shock absorber are compared: an observer-based approach and a parameter identification approach. The performance of these schemes is validated and compared using a commercial vehicle model that was experimentally validated. Early results shows that a parameter identification approach is more accurate, whereas an observer-based approach is less sensible to parametric uncertainty.

1. Motivation

Safety processes and technologies typically improve the overall equipment effectiveness and plant efficiency, with benefits that drop to the bottom line. One study also showed the following benefits from embedded safety systems: 5% productivity increase, 3% reduction in production expenses, 5% reduction in maintenance costs, 1% savings in capital expenses and 20% reduction in insurance costs, [4]. Considering that an automobile is integrated by many systems, such as, transmission, steering, exhaust, cooling, electrical, gasoline engine, fuel supply, electrical, braking and suspension system; where each system is a set of interdependent parts fulfilling a specific function and capable of functioning on its own. These systems can fail due to any number of possible failures that may occur in the sensors, actuators and devices that integrate them.

Although the automobile is not a safety-critical system, several driving conditions can lead to lose the vehicle stability with fatal consequences. The World Health Organization [19] estimates that 2.2 % of the global mortality (around 1.24 million deaths) is related to vehicle collisions caused by multiple factors, e.g. human factors, road quality, vehicle design and its maintenance, etc. The economic cost of motor vehicle collisions is estimated at over US $ 100 billion per year in middle-income countries (i.e. 1 - 2 % of their gross national product). Therefore, the preservation of the safety and comfort of passengers are the most fundamental goals for vehicle manufacturers and automotive control engineers. This research will focus on the controlled shock absorbers of the *Semi-Active* (*SA*) suspension system.

The primary functions of an automotive suspension system are to [9]: (1) isolate the chassis motion from road irregularities, (2) keep the tire-road contact with minimal load variations, (3) resist roll of the chassis, (4) react to the control forces (longitudinal and lateral) produced by the tires and braking and driving torques, and (5) maintain the wheels in the proper steer and camber attitudes to the road. The last two objectives are related to the kinematic behavior of the suspension geometry which can be solid axle or independent suspension; while, the first three objectives directly depend on the vertical force that it must transmit from the tires to the chassis.

According to the capability to adjust the damping force, the automobile suspensions can be classified as: *passive*, *active* or *SA*. *Passive* suspensions are only able to dissipate the energy and their damping characteristics are time invariant, while active ones are able to store, dissipate and generate energy through a variable damping coefficient but they are very expensive to apply because require an external power supply. *SA* suspensions represent a good solution to control the vertical vehicle dynamics using lower energy consumption and lower cost.

A suspension is integrated by 3 main elements: (1) an elastic element (i.e. a coil spring) which delivers a force proportional and opposite to the suspension elongation, this part carries all the static load; (2) a damping element (i.e. shock absorber), which provides a dissipative force proportional and opposite to the elongation speed; this part delivers a negligible force at steady-state, but plays a crucial role in the dynamic behavior of the suspension and (3) a set of mechanical elements which links the suspended (sprung) body to the unsprung mass. From the dynamic point of view, the spring and the damper are the two key elements, [21].

A *SA* suspension has an adaptable shock absorber where the continuous variable damping coefficient adjusted by external control signals along a road surface offers much better comfort and road holding performance with respect to *Passive* suspensions. The most fundamental advantage is its cost-benefit in comparison with *Active* suspensions. Although, the variable absorbing force may only affect transient behavior of the suspension deflection and does not affect its steady state as in an *Active* damper, *SA* shock absorbers are efficient to manage the balance between comfort and road holding. During past decades, different *SA* damper technologies have been examined, whose bandwidth oscillates from 0 to 30 Hz. There are four main technologies of *SA* shock absorbers in the market: *Electro-Hydraulic*, *Pneumatic*, *Magneto-Rheological* (*MR*), and *Electro-Rheological*.

An extensive work on *SA* suspension control systems emphasizes that *Magneto-Rheological* shock absorbers are one of the best solutions to manage the compromise between comfort and road holding. The transient response time of *MR* shock absorbers is around 12 ms, the power requirement is lower, they are robust to extreme temperature variations, and they are easier to manufacture and implement [5]. However, the use of these *MR* shock absorbers increases the necessity of fault detection. It has been noted that *SA* shock absorbers are more susceptible to faults. This is specially relevant since the effect of damper faults could vary from the increase of chassis vibrations to the loss of stability.

This paper, an extended version of [12], is organized as follows: Section 2 briefly reviews previous works on Fault Detection and Isolation for *SA* suspensions Section 3 defines the vehicle model. Section 4 describes the proposal. Section 5 discusses the results. Section 6 concludes the research project.

2. State of the art

Failures are inevitable and unpredictable in vehicle suspensions. The considered fault as case of study is represented as a loss of oil in an automobile shock absorber. There are different aspects that cause a damper leakage [20], which are related to a bad installation, an excessive use, a damage by external devices or a bad design of control system that saturates the manipulation. The direct consequence of a faulty shock absorber is the loss of damping force that causes a performance reduction in the suspension capabilities (comfort and safety for passengers) at any driving situation.

There are few papers dealing with *Fault Detection and Isolation* (*FDI*) for *SA* shock absorbers, and most of them are based on methods that inherently handle additive faults. These methods have the advantage of low computational requirements and do not need permanent excitation.

FDI modules based on analytical redundancy are the most accepted strategies for fault diagnosis and estimation [28]; especially parity-space approaches, because of its applicability to nonlinear systems, speed of detection, isolability property, robustness and computational complexity. Some FDI approaches, based on parity space theory, have been proposed for automotive suspensions [8], [16] to diagnose faults in actuators or sensors. However, these approaches have been designed for *Linear Time Invariant* (*LTI*) systems without consideration of model uncertainties or noisy measurements.

[18] explored an optimized parity space approach for actuator *FDI*. The parity spaces spans all the parity relations that quantify the analytical redundancies available between the sensor outputs and the actuator inputs. A transformation matrix is then optimized to transform these parity relations into residuals that are especially sensitive to specific actuator faults. Actuator faults cause the variance of parity space residuals to increase. A *CUmulative SUMmation* (*CUSUM*) procedure is used to determine when residual variance has changed sufficiently to indicate an actuator fault. The added advantage of this design procedure is that it uses less computational power than some other *FDI* systems such as a bank of extended *Kalman* filters.

In [24] the parity space method is used to detect the damper fault. However, the *SA* damper nonlinearities are not considered nor model uncertainties. An extension of the *FDI* module based on parity space theory by adding robustness to exogenous unknown disturbances is proposed by [23]. A novel Fault-Tolerant Controller is proposed for an automotive suspension system based on a*Quarter of Vehicle QoV* model. The design is divided in a robust *Linear Parameter Varying* (*LPV*) controller used to isolate vibrations from external disturbances and in a compensation mechanism used to accommodate actuator faults. The compensation mechanism is based on a robust fault detection and estimation scheme that reconstructs a fault on the *SA* damper; this information is used to reduce the failure effect into the vertical dynamics to achieve good control performances. Results show the effectiveness of the fault-tolerant *SA* damper versus an uncontrolled damper.

[1, 2] presented a fault detection method for suspension system using continuous wavelet transform in which Morlet wave function is used to determine the natural frequencies and amplitudes of signal components and include maximum contribution to the signal energy at the specific scale. Fault diagnosis has been investigated for dampers and upper damper bushings. After determination of faulty parts of suspension system by using signal energy distribution, by observing the changes in the natural frequencies and corresponding energy amplitudes, damper fault is distinguished from damper bushing fault.

[27] presents an application of a diagnosis and a fault tolerant control method for an active suspension system. The method detects and identifies an actuator fault and reconfigures the controller to maintain the ride comfort and vehicle performance in the presence of road disturbances. The proposed scheme consists of a diagnosis module and a controller reconfiguration module. The diagnosis module is based on the *Unknown Input Observer* (*UIO*)approach to detect and identify actuator faults. A bank of observers is designed to generate residual signals such that each of them matches the system in a defined actuator fault mode. The controller reconfiguration module consists of a bank of controllers, each corresponding to a fault mode of the system. The module changes the control law after the detection and the identification of an actuator fault. In [25] a Lyapunov based restoring force observer that leads to the design of a robust fault detection system is proposed. The fault estimation method provides a direct estimate of fault (i.e. stiffness and damping variations).

[29] designed a fault detection filter in finite frequency domain. A sufficient condition for residual system with the prescribed H_∞ performance index is derived based on the generalized *Kalman-Yakubovich-Popov* lemma. The fault detection filter is designed in middle-frequency domain such that the residual systems are asymptotically stable. Results are validated with a numerical example.

Even more focus on control, [6] demonstrated the efficiency of the *LPV* framework for vehicle dynamics control, when some actuators may be in failure (e.g. suspension and braking actuators failures). A LPV/H_∞ fault tolerant gain-scheduled vehicle dynamic control involving the steering actuators, rear brakes and four suspension systems is proposed for yaw stability, lateral and vertical vehicle performances. The *LPV* control system can handle such failures through online adaptation of the control input distribution. Simulation results on a model experimentally validated (e.g. Renault Mégane Coupé) show the proposed methodology is effective and robust. Based on same LPV/H_∞ framework, [17] introduced a new *Fault Tolerant Control* (*FTC*) for a faulty semi-active suspension system (e.g. oil leakages damper). A fast adaptive fault algorithm is used to estimate the fault. Then, the *LPV-FTC* is designed to ensure the damper dissipativity constraint and to achieve the designed performance objectives in both comfort and road holding. The controller solution is based on the *LMI* solution for polytopic systems

Some *FDI* approaches based on parametric identification have been also proposed. These approaches have the advantage of providing deeper insight into the process; however, most online estimation methods need excitation and other operating requirements which may be restrictive when the road profile is unknown. [26] presented two simple models for the vertical dynamics. A two-step scheme for fault diagnosis is proposed, which consists of a detection of the fault in the first step and a fault diagnosis in the second. Results with measured data, drawn from a test rig showed that fault detection with parameter estimation leads to good results. Several different faults can be distinguished due to their unique patterns of the estimated parameter changes. Also, parity equations enable detection of some faults and require only little computational effort. Weak dampers can be detected.

Artificial Intelligence approaches had been exploited too. For automatic fault detection, estimated parameters were mapped to the different faults, [3]. Neuronal networks (e.g. feedforward perceptron networks) were proven for such tasks, the generalization error was below 4 %. Even a *FDI* system was implemented with parameter estimation, the process has to be excited all the time to guarantee convergence of the estimation algorithm. Then, parity equations were combined because they do not need permanent excitation and require less computational effort. Also, [7] designed the *FDI* system by developing the parity equations based on simple kinematical equations and on models of the lateral vehicle dynamics. The calculated residuals are then used for fuzzy logic fault diagnosis. The detection and diagnosis of sensor and process faults, as well as sensor reconfiguration, were presented for lateral and vertical vehicle dynamics. Small faults could be detected with a good isolation performance.

Two different approaches are applied to *SA* shock absorbers to estimate faults: the observer-based approach and the parameter estimation approach. The advantages and disadvantages of each method are analysed in order to determine which one is more suitable to estimate actuator faults in a *SA* suspension systems.

3. Quarter of Vehicle model

A *QoV* model is the most basic automotive suspension, Fig. 1. Its use assumes an equivalent load distribution among the four corners and a linear dependency with respect to the translational and rotational chassis motions. The lateral and longitudinal wheel dynamics is not considered, while the wheel road contact is ensured.

This very simple one-dimensional model consists of a spring-damper-system representing the suspension strut and a single spring replacing the tire. This friction element is added to estimate offset values to detect offset faults, or increased friction. The damper contributes the drive safety and the drive quality to the same extent. Its tasks are the prevention of an amplification of the body and the prevention of a skipping wheel. A non-skipping wheel is the condition for a good road-contact. The task of the spring is to carry the body-mass and to isolate the body from road disturbances maximally. A *QoV* model is a well-known simplification for vertical dynamics analysis of vehicle, [3]. Faults occurring in the damping system can be as a result of one or more of the following factors (numbered in Fig. 1): (1) Damage of leak of the damper, (2) Weak springs, (3) Worn bushing and broken of support member assembly, (4) the looseness joins, (5) under-inflation or over-inflation of tyres and (6) wear of the tyre.

The dynamic behavior of a *QoV* model with a *SA* suspension is described by:

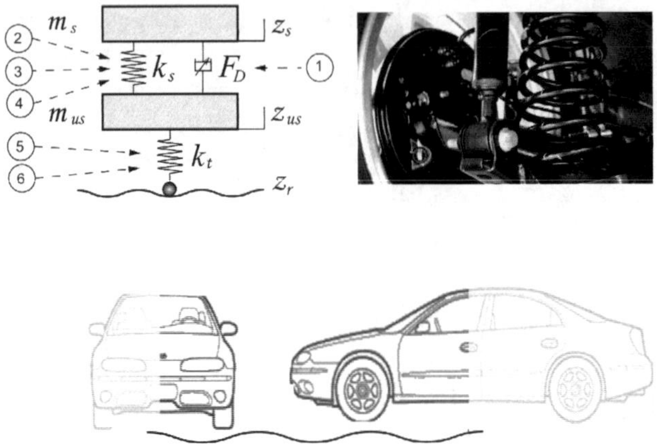

Figure 1. *QoV* model

$$\begin{aligned} m_s\ddot{z}_s &= -F_D - k_s(z_s - z_{us}) \\ m_{us}\ddot{z}_{us} &= k_s(z_s - z_{us}) + F_D - k_t(z_{us} - z_r) \end{aligned} \tag{1}$$

where z_s, z_{us} and z_r are the sprung mass, unsprung mass and road profile vertical positions, respectively; m_s is the sprung mass which represents the chassis, m_{us} is the unsprung mass which represents the wheel, tire, etc.; k_s and k_t are the suspension and the tire stiffness, and F_D represents the damping force.

The *SA* damper force behavior may be represented according to, [10]:

$$F_{SA} = k_p z_{def} + c_p \dot{z}_{def} + f_c \upsilon \tanh(a_v \dot{z}_{def} + a_d z_{def}) \tag{2}$$

where $z_{def} = z_s - z_{us}$ is the suspension deflection, and υ is the damper control input (*i.e* electric current for a *MR* damper), f_c is the force due to the input, c_p is the viscous damping coefficient, k_p is the stiffness coefficient, a_v and a_d are coefficients which characterize the hysteretic behavior, Table 1.

The considered model for multiplicative fault is :

$$F_D = \alpha[k_p z_{def} + c_p \dot{z}_{def} + f_c \upsilon \tanh\left(a_v \dot{z}_{def} + a_d z_{def}\right)] \tag{3}$$

where α represents the effectiveness of the damper, *i.e.* $\alpha = 1$ is a healthy damper, $\alpha = 0$ represents a 100 % damper failure. The aim is to estimate α using the masses accelerations and the suspension deflection, while the road profile z_r is an unknown perturbation input, because it is expensive to measure.

4. Approaches

Two complementary model-based fault estimation schemes for *SA* shock absorbers are proposed: an observer-based approach, which is intended to estimate additive faults; and a parameter identification approach, which is intended to estimate multiplicative faults.

4.1. Observer-based Fault Estimation

A *FDI* system based on an *UIO* and the *QoV* model (1) is proposed. The idea is to synthesize a residual, which is sensitive to the damper fault and insensitive to disturbances (i.e. road profile). The damper fault is first modelled as an additive fault:

$$F_D = \bar{F}_D - F_\delta \tag{4}$$

where $\bar{F}_D$ is the nominal force (*i.e.* the force of a healthy damper) and F_δ is the loss of force due to the fault. By using (1), (3) and (4), a state space system is:

$$\begin{array}{rcl} \dot{x}(t) & = & Ax(t)+B_u u(t)+B_z z_r(t)+B_F F_\delta(t) \\ y(t) & = & Cx(t)+D_u u(t)+D_z z_r(t)+D_f F_\delta(t) \end{array} \tag{5}$$

with

$$x=\begin{bmatrix} z_s & \dot{z}_s & z_{us} & \dot{z}_{us} \end{bmatrix}^\top, y=\begin{bmatrix} \ddot{z}_{us} & \ddot{z}_s & z_{def} \end{bmatrix}^\top$$

$$A=\begin{bmatrix} 0 & 1 & 0 & 0 \\ -\frac{k}{m_s} & -\frac{c_p}{m_s} & \frac{k}{m_s} & \frac{c_p}{m_s} \\ 0 & 0 & 0 & 1 \\ \frac{k}{m_{us}} & \frac{c_p}{m_{us}} & -\frac{k+k_t}{m_{us}} & -\frac{c_p}{m_{us}} \end{bmatrix} B_u=\begin{bmatrix} 0 \\ -\frac{f_c}{m_s} \\ 0 \\ \frac{f_c}{m_{us}} \end{bmatrix} B_z=\begin{bmatrix} 0 \\ 0 \\ 0 \\ \frac{k_t}{m_{us}} \end{bmatrix} B_F=\begin{bmatrix} 0 \\ -\frac{1}{m_s} \\ 0 \\ \frac{1}{m_{us}} \end{bmatrix}$$

$$C=\begin{bmatrix} \frac{k}{m_{us}} & \frac{c_p}{m_{us}} & -\frac{k+k_t}{m_{us}} & -\frac{c_p}{m_{us}} \\ -\frac{k}{m_s} & -\frac{c_p}{m_s} & \frac{k}{m_s} & \\ 1 & 0 & -1 & 0 \end{bmatrix} D_u=\begin{bmatrix} \frac{f_c}{m_{us}} \\ \frac{-f_c}{m_s} \\ 0 \end{bmatrix} D_z=\begin{bmatrix} \frac{k_t}{m_{us}} \\ 0 \\ 0 \end{bmatrix} D_f=\begin{bmatrix} \frac{1}{m_{us}} \\ \frac{-1}{m_s} \\ 0 \end{bmatrix}$$

where $k=k_s+k_p$, $u(t)=\upsilon\rho(t)$ and $\rho(t)$ is the nonlinear term of the *SA* damper model:

$$\rho(t)=\tanh\left(a_v\dot{z}_{def}+a_d z_{def}\right) \tag{6}$$

Note that system (5) is comprised by a linear subsystem in a closed loop with a memoryless nonlinear function $\upsilon\rho(t)$.

An important result of applying the *UIO* theory to this domain is, without road information, the system is non-observable and non-detectable [11]. This will affect the state estimations.

An approximated *UIO* observer can be obtained by decoupling the system (5) from the road surface using the procedure reported in [14]. The decoupled system can be written as:

$$\begin{array}{rcl} \dot{x}(t) & = & \bar{A}x(t)+\bar{B}_u\rho(t)u(t)+\bar{B}_F F_\delta(t)+Gy_1(t) \\ y_2(t) & = & \bar{C}x(t)+\bar{D}_u\rho(t)u(t)+\bar{D}_f F_\delta(t) \end{array} \tag{7}$$

where

$$\bar{A}=\begin{bmatrix} 0 & 1 & 0 & 0 \\ -\frac{k}{m_s} & -\frac{c_p}{m_s} & \frac{k}{m_s} & \frac{c_p}{m_s} \\ 0 & 0 & 0 & 1 \\ 0 & 0 & 0 & 0 \end{bmatrix} \bar{B}_u=\begin{bmatrix} 0 \\ -\frac{f_c}{m_s} \\ 0 \\ 0 \end{bmatrix} \bar{B}_f=\begin{bmatrix} 0 \\ -\frac{1}{m_s} \\ 0 \\ 0 \end{bmatrix} G=\begin{bmatrix} 0 \\ 0 \\ 0 \\ 1 \end{bmatrix} \bar{C}=\begin{bmatrix} \bar{C}_1 \\ \bar{C}_2 \end{bmatrix}=\begin{bmatrix} 1 & 0 & -1 & 0 \\ -\frac{k}{m_s} & -\frac{c_p}{m_s} & \frac{k}{m_s} & \frac{c_p}{m_s} \end{bmatrix}$$

$$\bar{D}_u=\begin{bmatrix} \bar{D}_{u_1} \\ \bar{D}_{u_2} \end{bmatrix}=\begin{bmatrix} 0 \\ \frac{-f_c}{m_s} \end{bmatrix} \bar{D}_f=\begin{bmatrix} \bar{D}_{f_1} \\ \bar{D}_{f_2} \end{bmatrix}=\begin{bmatrix} 0 \\ \frac{-1}{m_s} \end{bmatrix}, y_1=\ddot{z}_{us}, y_2=z_{def}, y_3=\ddot{z}_s, \bar{y}=\begin{bmatrix} y_2 \\ y_3 \end{bmatrix}$$

For the state estimation the decoupled system (7) with the output y_2 will be used since this output does not depend on the fault. For the residual generation, y_3, which is affected by the fault, will be used.

If $(\bar{A},\bar{C}_1)$ are observable or detectable, an observer can be designed for the decoupled system to estimate the states. The observability matrix of system (7), has rank 2; thus, two of the observer poles cannot be located arbitrarily. For the system (7) the non-observable subset is related to two poles in $s=0$; therefore, the system is non-observable and non-detectable. A possible solution consists on generating a stable system which approximates the frequency response of the *UIO* observer, [13].

An observer is a dynamical system whose inputs are the process input and output vectors. For the decoupled system (7), the observer poles in $s=0$ are related to the input contribution to the state estimation. Since these poles are not stable regardless the observer gain, the best estimation effort is the open-loop simulation of these states. In general the state estimation based only in the simulation of the process (which is equivalent to an observer with observer gain equal to zero) is not advised. However, the input related with these integrators is the unsprung-mass acceleration, which is indeed a measured output, but it was considered as an input to achieve the decoupling of the system from the road profile. Thus, the non-observable states are estimated via the output, like it is done by using high-gain observers.

The observer gain matrix L is designed to place the two observable poles at $(-20.38 \pm 1.09i)$. The resulting observer combines the closed-loop estimation dynamics and the integration of acceleration measurements. For the states, which are estimated by numerical integration, it is well known that low frequency noise in the measurements induces a divergence called drift [22]. Without additional measurements, it is not possible to eliminate the drift by modifying the observer gain, [11, 15]. Since the suspension deflection is limited and the masses velocities converge to zero, it is possible to use high-pass filters to eliminate the drift. Then, the observer is complemented by adding a pair of high-pass filters $g_f(s) = s/(s+\omega_f)$ for each of the integrators. Only the non-detectable states are filtered, while the observable ones are estimated by a closed-loop observer. The cut-frequency ω_f of the filter is set at 0.14 Hz.

By using the estimated states with the *UIO* observer and the measurement of the sprung mass acceleration, the residual can be computed to detect a fault (loss of damping force):

$$r(t) = y_3 - \left[\bar{C}_2\hat{x}(t) + \bar{D}_{u_2}u(t)\right] = \bar{D}_{f_2}F_\delta(t) \tag{8}$$

then the magnitude of the fault can be estimated as $\hat{F}_\delta = \left[\bar{D}_{f_2}\right]^{\dagger} r(t)$ where $\dagger$ stands for the *Moore Penrose* pseudo inverse.

4.2. Parameter Estimation Approach

By considering the multiplicative model of the fault in (3), the system can be represented in state-space form as:

$$\begin{aligned} \dot{x}(t) &= A_n x(t) + g(x,u)\alpha + B_z z_r(t) \\ y(t) &= C_n(\alpha)x(t) + h(x,u)\alpha + D_z z_r(t) \end{aligned} \tag{9}$$

with

$$A_n = \begin{bmatrix} 0 & 1 & 0 & 0 \\ -\frac{ks}{m_s} & 0 & \frac{ks}{m_s} & 0 \\ 0 & 0 & 0 & 1 \\ \frac{ks}{m_{us}} & 0 & -\frac{ks+k_t}{m_{us}} & 0 \end{bmatrix} \quad g(x,u) = \begin{bmatrix} 0 \\ -\frac{f_c\rho u + c_p(x_2-x_4) + k_p(x_1-x_3)}{m_s} \\ 0 \\ \frac{f_c\rho u + c_p(x_2-x_4) + k_p(x_1-x_3)}{m_{us}} \end{bmatrix}$$

$$C_n = \begin{bmatrix} \frac{ks}{m_{us}} & 0 & -\frac{ks+k_t}{m_{us}} & 0 \\ -\frac{ks}{m_s} & 0 & \frac{ks}{m_s} & 0 \\ 1 & 0 & -1 & 0 \end{bmatrix} \quad h(x,u) = \begin{bmatrix} \frac{f_c\rho u + c_p(x_2-x_4) + k_p(x_1-x_3)}{m_{us}} \\ -\frac{f_c\rho u + c_p(x_2-x_4) + k_p(x_1-x_3)}{m_s} \\ 0 \end{bmatrix}$$

Although the fault affects both the sprung and unsprung mass dynamics, for this estimator only the sprung mass dynamic will be considered since it is independent of the unmeasured road profile. Two measurements are used: $\ddot{z}_s$ and z_{def}. From (9), the discrete-time representation of the sprung mass dynamic can be written in the regression form as $\zeta = \psi^\top\theta$ where $\zeta = -k_s(z_{def}) - m_s(\ddot{z}_s)$, $\psi^\top = -\frac{f_c\rho u + c_p(\dot{z}_{def}) + k_p(z_{def})}{m_s}$ and $\theta = \alpha$. The damper fault can be online estimated by using the *Recursive Least Squares* (*RLS*) with forgetting factor λ :

$$\begin{aligned} e(k) = \zeta - \psi^\top(k)\hat{\theta}(k-1) \quad \gamma(k) = \frac{P(k-1)\psi(k)}{\lambda + \psi^\top(k)P(k-1)\psi(k)} \\ P(k) = \frac{1}{\lambda}(I - \gamma(k)\psi^\top(k))P(k-1) \quad \hat{\theta}(k) = \hat{\theta}(k-1) + \gamma(k)e(k) \end{aligned} \tag{10}$$

where $\hat{\theta}$ is the estimation of θ. The measurement of $\dot{z}_{def}$ is not available, therefore an estimation is required to implement the parameter estimation. $\dot{z}_{def}$ may be estimated either by the numerical derivative approximation of z_{def} or by an state-observer. In this case, an approximation based on the bandwidth-limited derivative is used.

For a sufficiently high sampling rate, the discrete estimator will approximate the continuous time model parameters. For this system the simulations show that 500 Hz was a sufficiently high sampling rate to have a good approximation.

Table 1. Commercial vehicle model parameters

Parameter	Value	Units	Parameter	Value	Units
m_s	315	Kg	m_{us}	37.5	Kg
k_s	29,500	N/m	k_t	230,000	N/m
c_p	1,500	Ns/m	k_p	-10,239	N/m
a_v	7.89	s/m	a_d	-13.8	1/m
f_c	441	Ns/A			

5. Results

Data from a commercial vehicle was used, Table 1.

Six tests were used to assess the performance of the *FDI* proposals. These tests consider a) an *ISO* Type *D* at a vehicle speed of 30 Km/h, and b) a series of bumps as road profile signals, which represents a normal driving conditions. Table 2 summarizes the tests characteristics.

Table 2. Tests used to evaluate the *FDI* methods.

Test	Description		Objective analysis
	α	Current	
#1	$[0.5,1]$	$u = 2\ A$	Estimation of abrupt fault at constant damper input.
#2	$[0.1,1]$	$u = 2\ A$	Estimation of gradual fault at constant damper input.
#3	$[0.1,1]$	$u \in [0.5,3]\ A$	Estimation of gradual fault at varying damper input.
#4	$[0.5,1]$	steps in z_r	Sensitivity to road disturbances.
#5	$[0.5,1]$	$1.2\ m_s$	Robustness to uncertainty in m_s.
#6	$[0.2,1]$	$0.8\ K_t$	Robustness to uncertainty in k_t.

Although both strategies are used to estimate faults of different nature (*i.e.* additive or multiplicative), a comparison can be made by calculating the equivalence between them. From (4) and (3) the relation between the multiplicative and the additive faults can be obtained as $F_\delta = F_D\left(\frac{1}{\alpha} - 1\right)$.

Figure 2 (left plot) shows the estimation of the loss of force F_δ in the Test #1. An abrupt damper fault is modelled as a step change in α; however, the equivalent additive fault also depends on F_D, therefore even if α is constant, F_δ is not. This figure shows the *Least Squares Estimator* (*LEE*) error converges to zero for a step change in α, while the *Observer-Based Estimator* (*OBE*) error does not.

Figure 2 (right plot) shows the multiplicative equivalent fault of the Test #1, this confirms the last discussion. The *LEE* is better posed for constant multiplicative faults than the *OBE*. Since real damper faults are more similar to multiplicative faults, only this type of faults will be discussed.

Figure 3 (left plot) shows that both approaches are able to estimate gradual variations for multiplicative faults, Test # 2. When the multiplicative fault α has a constant slope, neither of the estimator errors converge to zero. The *LEE* has a lower overall error. In Fig. 3 (right plot) the estimation of α is compared in the same fault conditions as in Test #2, but, in this case a varying damper input ($[1,3]\ Amp$) is considered. This is an important observation because this characteristic, is essential for *SA* suspension systems. This figure also shows that the *LEE* has lower sensitivity to damper input variation than the *OBE*.

Test #4 is used to asses the sensitivity of the fault estimators to sudden road profile changes, which can represent road abnormalities such as potholes. The road decoupling of the *OBE* is not perfect since a perfect *UIO* was not possible to design due to observability issues. This is more evident when the road presents sudden changes, inducing a greater level of error. The *LEE* is not visibly affected by these road characteristics, Fig. 4 (left plot).

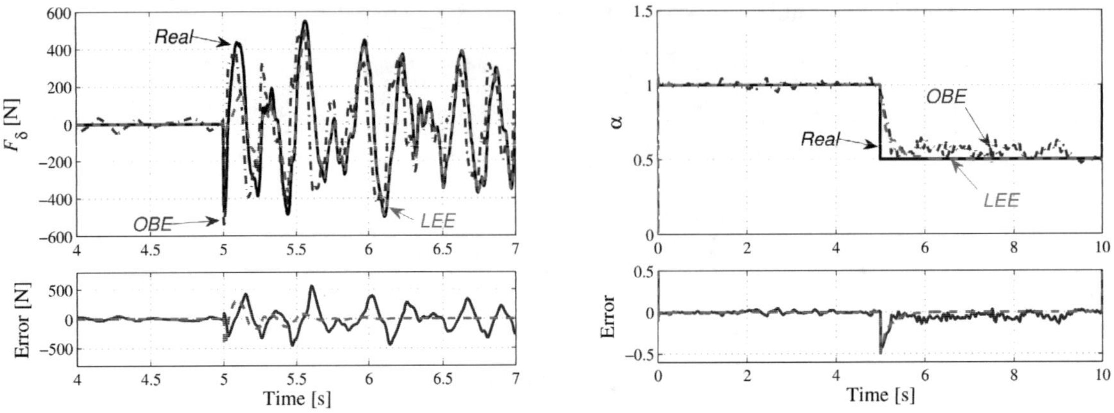

Figure 2. Comparison of estimated force loss and fault α in Test #1 (*ISO* Type *D*).

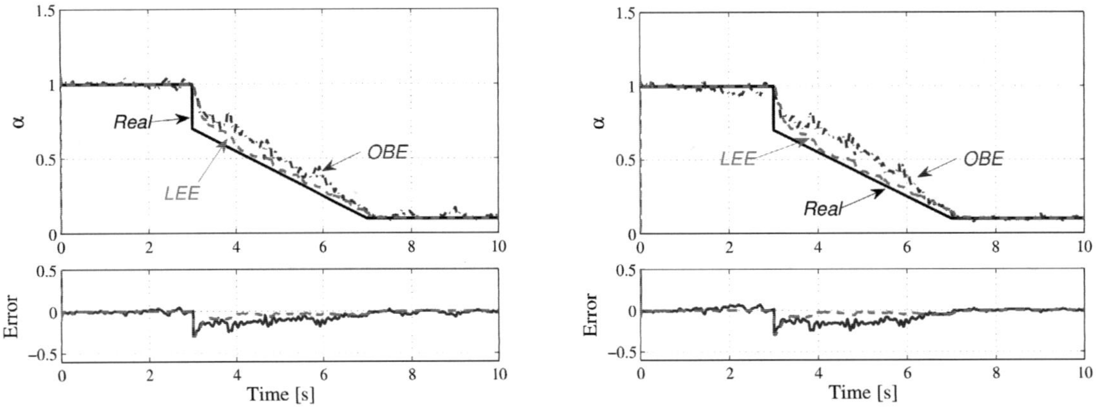

Figure 3. Comparison of estimated fault α in Tests #2 and #3 (*ISO* Type *D*).

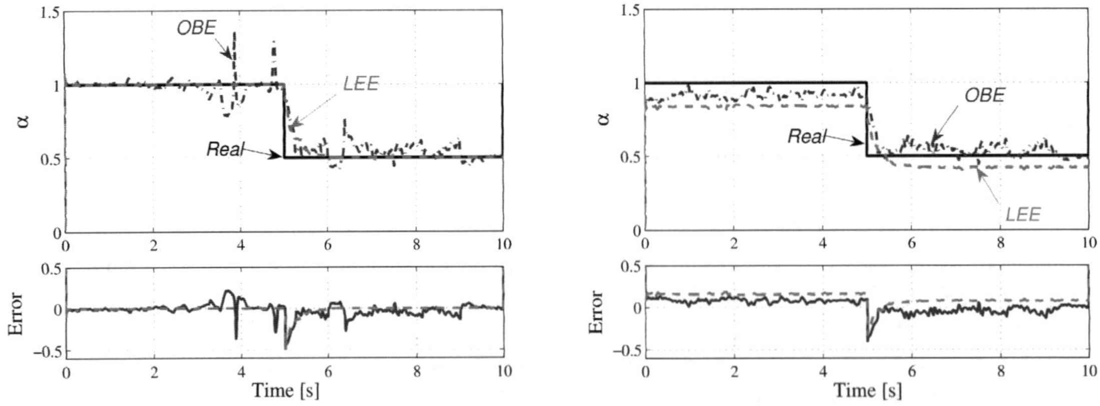

Figure 4. Comparison of estimated fault α in Tests #4 and #5 (*ISO* Type *D*).

Figure 4 (right plot) is used to assess the effect of sprung mass uncertainty on the estimated fault, a sprung mass 20% heavier was considered. The results show, that the mass uncertainty induces steady state error in both algorithms. The *LEE* has a greater error.

Figure 5 (left plot) shows the estimated fault in the Test #1 under other road condition, the series of bumps. As observed in this figure, the *LEE* is less sensitive to the road profile and it presents a lower error level. The *LEE* is better posed for constant multiplicative faults than the *OBE*. In Fig. 5 (right plot), it can be observed that he *LEE* has lower sensitivity to damper input variation than the *OBE*.

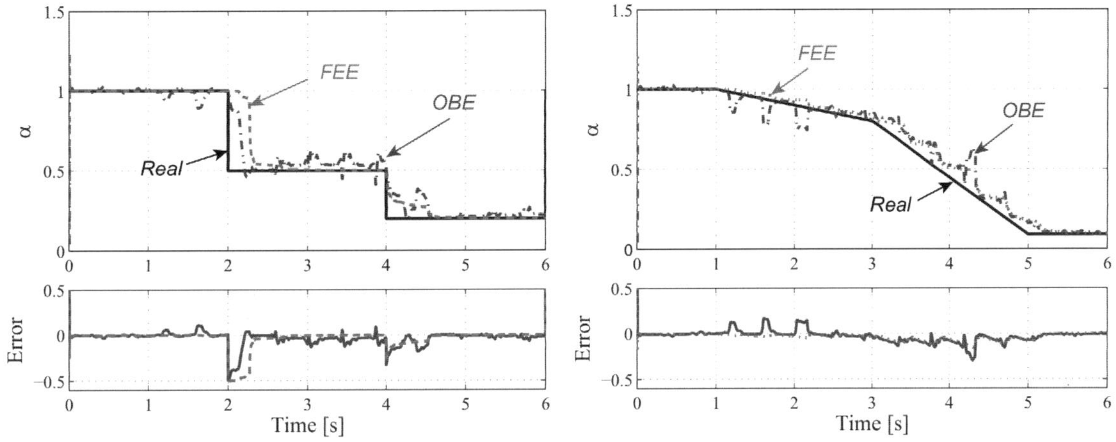

Figure 5. Comparison of estimated fault α in Test #1a and #3a (Series of bumps).

Test #6 is used to evaluate the effect of tire stiffness uncertainty on the estimated fault. This variation may occur due to variations in the tire inflation pressure. A 20% lower tire stiffness was considered, which corresponds to a variation of 5 psi in the inflation pressure. The results in Fig. 6 show that both methods are robust to the uncertainty in this parameter. The *OBE* error is slightly higher, 10 % with respect to the unperturbed case, whereas the estimation of the *LEE* is not affected since it is independent of this parameter.

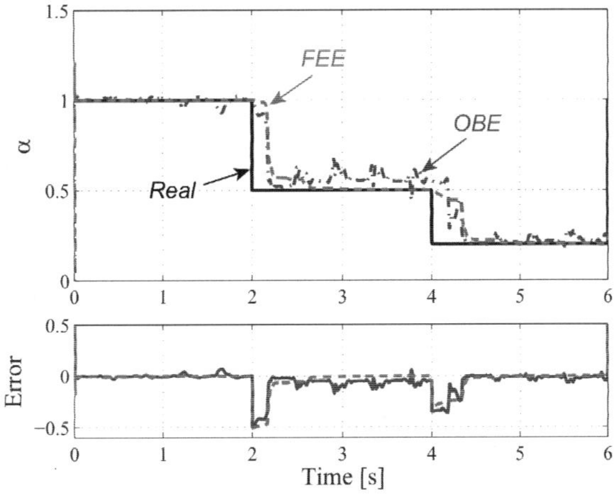

Figure 6. Comparison of estimated fault α in Test #6 (Series of bumps).

Finally, a quantitative comparison of the approaches was made using the *Root Mean Square Error*

(*RMSE*) index. Table 3 summarizes the results. It can be observed that the *LEE* presents an overall lower level of error; for instance, in Test #2 the *LEE* has 493% lower *RMSE* in the estimation of the loss of force F_δ. If the fault is represented in the multiplicative form, the *LEE* also yields a lower *RMSE*, but the differences are smaller. For example, in test #2 the *LEE* is also lower but 57%. Only when sprung-mass uncertainty is considered the *OBE* has a lower index, in this case 55% lower for F_δ.

Table 3. *RMSE* of fault estimation approaches under different scenarios.

Test	Road Profile	*OBE*		*LEE*	
		F_δ	α	F_δ	α
1	*ISO* Type *D*	1.46221	0.00062	0.36250	0.00050
2	*ISO* Type *D*	2.40880	0.00071	0.40596	0.00045
3	*ISO* Type *D*	1.78609	0.00080	0.41011	0.00044
4	*ISO* Type *D*	1.66067	0.00073	0.34299	0.00049
5	*ISO* Type *D*	1.51313	0.00077	0.91140	0.00120
1a	Series of bumps	1.74457	0.00106	0.97300	0.00121
3a	Series of bumps	1.75683	0.00077	1.16826	0.00065
6	Series of bumps	1.71662	0.00119	0.89106	0.00121

6. Conclusions

A comparison of two fault detection algorithms for the estimation of *semi-active* shock absorber faults is presented. One method is based on classical least square parametric identification to estimate a multiplicative fault. The second method is based on an *Unknown Input Observer*, which is used to calculate a residual and to estimate the loss of force due to an additive fault.

It has been reported that shock absorber faults are normally of multiplicative nature; nonetheless, additive fault estimation methods are more widely reported. Since it is possible to derive the multiplicative equivalent from the additive estimation, both methods were compared through different scenarios, including mass uncertainty, shock absorber input sensitivity and road profile sensitivity. The results show that the *Least Squares Estimator* achieves better qualitative and quantitative performance in all cases, with the exception of mass uncertainty. This also suggests that, for this system, identification algorithms based on regression models are highly sensible to process parameters changes. It can be concluded that the two estimators present complementary characteristics which could be exploited depending on the operating conditions to improve the overall system performance. The results also show that for this application a simple fault estimation algorithm, such as *Least Squares Estimator*, is enough as long as a multiplicative fault structure is used and a low level of parameter uncertainty is present.

Acknowledgment

Authors thank *Tecnológico de Monterrey* and *CONACyT* because the *Automotive Consortium* and bilateral PCP 06/13 (México-France) project.

References

[1] S. Azadi and A. Soltani. Application of Wavelet Analysis to the Suspension System Fault Detection of a Vehicle. In *SAE 2007 Noise and Vibration Conf. and Exhibition*, USA, 2007. Technical paper 2007-01-2370.

[2] S. Azadi and A. Soltani. Fault Detection for Passive Suspension Systems using Wavelet Transform. In *SAE 2007 Noise and Vibration Conf. and Exhibition*, USA, 2007. Technical paper 2007-01-2368.

[3] M. Börner, R. Isermann, and M. Schmitt. A Sensor and Process Fault Detection System for Vehicle Suspension Systems. In *SAE 2002 World Congress & Exhibition*, USA, 2002. Technical paper 2002-01-0135.

[4] ABB Consulting. 5 Ways to Improve Safety and Profitability, without Disrupting Operations. *White paper*, pages 1–5, 2015.
[5] J.C. Dixon. *The Shock Absorber Handbook*. Wiley Ed., 2nd Ed., 2007.
[6] S. Fergani, O. Sename, L. Dugard, and P Gáspár. Further Development on the LPV Fault Tolerant Control for Vehicle Dynamics. In *Preprints of* 9th *IFAC-Safeprocess*, pages 24–29, France, Sept 2015.
[7] D. Fischer, M. Börner, J. Schmitt, and R. Isermann. Fault Detection for Lateral and Vertical Vehicle Dynamics. *Control Eng. Practice*, 15:315–324, 2007.
[8] D. Fischer and R. Isermann. Mechatronic Semi-Active and Active Vehicle Suspensions. *Control Eng. Practice*, 12:1353–1367, 2004.
[9] T.D. Gillespie. *Fundamentals of Vehicle Dynamics*. Society of Automotive Engineers, Inc., 1992.
[10] S. Guo, S. Yang, and C. Pan. Dynamical Modeling of Magneto-Rheological Damper Behaviors. *J. of Intell. Mater., Syst. and Struct.*, 17:3–14, 2006.
[11] J.K. Hedrick, R. Rajamani, and K. Yi. Observer Design for Electronic Suspension Applications. *Vehicle Sys Dyn*, 23(1):413–440, 1994.
[12] D. Hernández-Alcantara, R. Morales-Menendez, and L. Amezquita-Brooks. Fault Estimation Methods for Semi-Active Suspension Systems. In *to appear in XVII Int. Autum Meeting on Power, Electronics and Computing*, México, Nov 2015.
[13] D. Hernández-Alcantara, J.C. Tudón-Martínez, L. Amezquita-Brooks, C. Vivas-Lopez, and R. Morales-Menendez. State Observers for Semi-Active Suspensions: Experimental Results. In *IEEE Multi-Conf on Systems and Control*, France, 2014.
[14] M. Hou and P.C. Muller. Design of Observers for Linear Systems with Unknown Inputs. *IEEE Trans Autom Control*, 37(6):871–875, 1992.
[15] Ling-Yuan Hsu and Tsung-Lin Chen. Vehicle Full-State Estimation and Prediction System Using State Observers. *IEEE Trans on Vehicular Tech*, 58(6):2651–2662, July 2009.
[16] H. Kim and H. Lee. Fault-Tolerant Control Algorithm for a Four-Corner Closed-Loop Air Suspension System. *IEEE Transactions on Industrial Electronics*, 58(10):4866–4879, 2011.
[17] M.Q. Nguyen, O. Sename, and L. Dugard. An LPV Fault Tolerant Control for Semi-Active Suspension - Scheduled by Fault Estimation. In *Preprints of* 9th *IFAC-Safeprocess*, pages 42–47, France, Sept 2015.
[18] Hendrik M. Odendaal and Thomas Jones. Actuator Fault Detection and Isolation: An Optimised Parity Space Approach. *Control Eng Practice*, 26(0):222 – 232, 2014.
[19] World Health Organization. Global Status Report on Road Safety, 2013.
[20] Sachs. Shock Absorbers Leak, Potential Causes. Technical report, Sachs, www.zfsachs.com, 2008.
[21] S.M. Savaresi, C. Poussot-Vassal, C. Spelta, O. Sename, and L. Dugard. *Semi-Active Suspension Control for Vehicles*. Elsevier - Butterworth Heinemann, 2010.
[22] U-X Tan et al. Estimating Displacement of Periodic Motion with Inertial Sensors. *IEEE Sensors J*, 8(8):1385–1388, 2008.
[23] J.C. Tudón-Martínez, S. Varrier, R. Morales-Menendez, R. Ramírez-Mendoza, D. Koenig, J-J. Martínez, and O. Sename. Fault Tolerant Control with Additive Compensation for Faults in an Automotive Damper. In *Proc. of the* 10th *IEEE Int Conf on Networking, Sensing and Control*, pages 810–814, France, 2013.
[24] S. Varrier, J.J. Lozoya-Santos, D. Hernandez, D. Koenig, J-J. Martínez, and R. Morales-Menendez. Fault Detection in Automotive Semi-active Suspensions: Experimental Results. In *Proc. of SAE World Congress 2013*, USA, 2013. Technical paper: 2013-01-1234.
[25] Y. Vidal, L. Acho, F. Pozo, and J. Rodellar. Fault Detection in Base-Isolation Systems via a Restoring Force Observer. In *Conf. on Control and Fault Tolerant Systems*, pages 777–782, France, 2010.
[26] T. Weispfenning. Fault Detection and Diagnosis of Components of the Vehicle Vertical Dynamics. *Meccanica*, 32(5):459–472, 1997.
[27] A. Yetendje, M. Seron, and J. De Dona. Diagnosis and Actuator Fault Tolerant Control in Vehicle Active Suspension. In *Int Conf on Information and Automation for Sustainability*, pages 153–158, Australia, 2007.
[28] Y. Zhang, J. Sheng, S. J. Qin, and T. Hesketh. Dynamic Process Monitoring using Multiscale PCA . In *IEEE Conf on Electrical and Computer Eng*, pages 1579–1584, May 1999.
[29] X. Zhu, Y. Xia, S. Chai, and P. Shi. A Finite-Frequency Domain Approach to Fault Detection Filter Design for Vehicle Active Suspension Systems. In *Proc of the* 11th *World Congress on Intelligent Control and Automation*, pages 485–490, China, Jun 2014.

An improved method for estimating the domain of attraction of nonlinear systems containing rational functions

Péter Polcz[1], Gábor Szederkényi[1,2], and Tamás Péni[2]

[1]Faculty of Information Technology and Bionics, Pázmány Péter Catholic University, Práter u. 50/a, H-1083 Budapest, Hungary

[2]Systems and Control Laboratory, Institute for Computer Science and Control (MTA SZTAKI), Hungarian Academy of Sciences, Kende u. 13-17, H-1111 Budapest, Hungary

E-mail: polpe@digitus.itk.ppke.hu, szederkenyi@itk.ppke.hu, peni.tamas@sztaki.mta.hu

Abstract. An optimization based method is proposed in this paper for the computation of Lyapunov functions and regions of attractions for nonlinear systems containing polynomial and rational terms. The Lyapunov function is given in a special quadratic form, and the negativity of its derivative is ensured using appropriate LMI conditions. The conservatism of the solution is reduced by utilizing Finsler's lemma. The number of monomial and rational terms in the computational problem is kept as low as possible using linear fractional transformation (LFT) and automatic model simplification steps. The operation of the method is illustrated on two examples taken from the literature.

1. Introduction

Finding or at least approximating the region of attraction of nonlinear dynamical systems is an important task in model analysis and controller design/evaluation, and numerous works have been devoted to this issue (see, e.g. [1]). An important early result in this field is the existence of so-called maximal Lyapunov functions for a wide class of nonlinear systems and the corresponding iterative procedure to approximate them [2]. In [3], maximal Lyapunov functions were defined and computed for hybrid (piecewise nonlinear) systems. At the same time, the use of linear matrix inequalities (LMI) and semidefinite programming (SDP) techniques for nonlinear systems has become very popular due to their advantageous properties and the availability of efficient numerical tools to solve LMI problems. These new techniques provide a powerful framework for stability analysis, robust control, and filtering problems. Ghaoui et.al. [4] used quadratic Lyapunov functions and linear fractional transformations (LFT) to represent a rational nonlinear system and defined convex conditions for stability analysis and state feedback design. The application of sum of squares (SOS) programming to maximize the estimate of the region of attraction can be found in [1, 5]. Stability conditions in both references are converted into LMIs using SOS relaxations and the generalization of the S-procedure. Topcu et.al. [6] utilized a further branch-and-bound type refinement in the parameter space to reduce the solution's conservatism. According to Trofino et.al. [7], LMI conditions can be obtained by using the Finsler's Lemma and the notion of annihilators. The newly introduced sufficient conditions for the stability are affine parameter dependent LMIs because they are characterized by affine functions of the state (x) and uncertain parameters (δ). Affine parameter dependent LMIs can be computationally handled by checking their feasibility at the corner points (vertices) of a polytopic region, on which the uncertain parameters are defined. In [7] it is shown that with some additional conservatism, the use of the vertices can be avoided by modifying the LMIs with the S-Procedure.

Based on the results of [7], in this work we present improved sufficient linear matrix inequality (LMI) conditions for local and regional asymptotic stability of polynomial and rational nonlinear systems. These LMI conditions are given through a Lyapunov function containing monomial and rational terms with the prescribed properties. We also present our computational results on illustrative examples taken from the literature.

2. Background

In this section we present the basic notions and known results on which our computational results are based.

2.1. System class, Lyapunov functions and domain of attraction

We consider nonlinear systems of the form

$$\dot{x} = f(x, \delta), \;\; x \in \mathbb{R}^n, \;\; x_0 \in \mathcal{X}, \;\; \delta \in \mathcal{D}, \;\; \dot{\delta} \in \mathcal{D}_d, \tag{1}$$

where x is the state vector, x_0 is the initial condition, and δ is the vector of (possibly time-dependent) uncertain parameters. Furthermore, it is assumed that $\mathcal{X}$, $\mathcal{D}$, and $\mathcal{D}_d$ are known polytopic regions. We assume that $x^* = 0 \in \mathbb{R}^n$ is a locally asymptotically stable equilibrium point of (1) for all $\delta \in \mathcal{D}$. The set of all initial conditions from which the solutions converge to x^* is called the *domain of attraction* (DOA). We search for Lyapunov functions in the form

$$V(x, \delta) = x^T P(x, \delta) x, \;\; v_l(x) \leq V(x, \delta) \leq v_u(x) \tag{2}$$

with the property

$$\dot{V}(x) \leq -v_d(x), \;\; \forall (x, \delta, \dot{\delta}) \in \mathcal{X} \times \mathcal{D} \times \mathcal{D}_d, \tag{3}$$

where P is a positive definite symmetric matrix function, and v_l, v_u and v_d are continuous positive functions on $\mathcal{X}$. Clearly, if (3) is fulfilled, then any closed level set of V completely inside $\mathcal{X}$ bounds an invariant region of the state space that is part of the domain of attraction.

The measure of *conservatism* is a property, which describes an invariant region. Let us consider two domains bounded by two different ε and γ level sets. We call ε less *conservative* than γ if ε is a better estimate of the actual region of attraction in the sense that the area/volume inside ε is larger than that of γ.

Roughly speaking, our main objective is to find a $V(x)$ function having a level set, which bounds the least conservative invariant region. In fact, introducing higher degree monomials into $V(x)$ generally results in better estimates, although a small increment in the number of monomials generates a huge increase in the dimension of the problem. Therefore, the rapidly growing computational burden must be taken into consideration through considering the possibilities of dimension reduction.

2.2. Two relevant approaches in the literature

The main differences between the methods proposed by Topcu et.al. [5] and by Trofino et.al. [7] are related to the definition of the sets, on which the conditions are stated, and the objective function, with which the size of the invariant region will be maximized.

Topcu prescribed the Lyapunov function to be positive definite on the whole $\mathbb{R}^n$ but the function's time derivative should be negative definite only in the inside of a level set, which would be the final invariant level set as it reached its maximal size[1]. In order to maximize the size of the level set, Topcu defined a variable sized region $\mathcal{P}_\beta = \{x \in \mathbb{R}^n \mid p(x) \leq \beta\}$, which should lie inside the invariant level set, while the variable β is maximized. The polynomial $p(x)$ is a design factor, which determines the shape and the orientation of the inner region.

On the other hand, the authors in [7] do not prescribe any constraint outside a given $\mathcal{X} \subset \mathbb{R}^n$ polytopic set, but for every $x \in \mathcal{X}$ the Lyapunov function and its derivative are required to be positive and negative definite, respectively. Furthermore, an additional constraint was introduced, more specifically, that the level set $\varepsilon_1 = \{x \in \mathbb{R}^n \mid V(x) = 1\}$ should be located inside the $\mathcal{X}$ polytope. In this case, the objective function to be minimized constitutes the sum of values of the Lyapunov function in some $x \in \mathcal{X}$ points, which are strategically chosen. Such an objective function can enforce that the level set $V(x) = 1$ is as close to the boundary of $\mathcal{X}$ as possible.

2.3. Further notations

In the paper, we will use the following additional notations:

- $\text{Co}(\mathcal{X})$ denotes the convex expansion of the set $\mathcal{X}$, more specifically, $\text{Co}(v_i, i = 1 \ldots n)$ is the convex hull of the set of vertices $\{v_i \mid i = \overline{1, n}\}$.
- $\vartheta(\mathcal{X})$ denotes the set of all vertices (corner points) of the polytope $\mathcal{X}$ (i.e. $\text{Co}(\vartheta(\mathcal{X})) = \mathcal{X}$).
- $\nabla V(x)$ is the gradient of the scalar function $V(x)$.
- $O_{n\times m}$ and I_n denote the $n \times m$ zero matrix and $n \times n$ unit matrix, respectively.
- A *basis function* is an element of a particular basis for a function space, e.g. monomials of the form $x_1^{p1} x_2^{p2} \cdots x_k^{pk}$ are a special class of basis functions. In this paper, we have considered every rational function with a monomial numerator as a basis function.

2.4. LMI stability conditions and Finsler's lemma

Let us suppose that we have the $\dot{x} = Ax$ linear time invariant (LTI) dynamical system, then the origin is globally asymptotically stable if there exists a Lyapunov function $V(x) = x^T Px$ such that the following conditions are satisfied:

$$\begin{aligned} &\forall x: \quad V(x) = x^T P x > 0 && \iff P > 0 \quad \text{(P is positive definite)} \\ &\forall x: \quad \dot{V}(x) = \nabla V(x)\dot{x} = x^T(PA + A^T P)x < 0 && \iff PA + A^T P < 0 \end{aligned} \tag{4}$$

This small example shows how can the Lyapunov conditions converted into a system of LMIs. As it is presented in [8] (Definition 1.37), every system of LMIs can be easily converted into a single LMI condition.

In the case of an uncertain linear system $\dot{x} = A(\delta)x$ with an affine $A(\delta)$ state transition matrix function, the second LMI of (4) will become parameter (δ) dependent LMI. According to Proposition 5.4 in [8], if we can specify a bounded polytopic domain in which the $\delta \in \mathcal{D}$ uncertain parameter operates, then it is enough to test the inequality at the corner points of the $\mathcal{D}$ polytope: $PA(\delta) + A(\delta)^T P < 0$, $\forall \delta \in \vartheta(\mathcal{D})$, being still *equivalent* to the original Lyapunov condition.

Consider the nonlinear system $\dot{x} = A(x)x$, where $A(x) \in \mathbb{R}^{n\times n}$. Let us choose again a simple quadratic Lyapunov function $V(x) = x^T Px$. Then the negative definiteness of $\dot{V}(x)$ constitutes $x^T K(x)x < 0 \; \forall x$, where $K(x) = PA(x) + A^T(x)P$. In this case, prescribing that $K(x) < 0$ for

[1] for more details, see Lemma 1. in [5]

all x introduces conservatism, because of its being only a sufficient but not a necessary condition for $x^T K(x)\, x < 0$.

Alternatively, we can define a non-quadratic polynomial Lyapunov function of the form $V(x) = \pi^T P\,\pi$, where π is a column vector of monomials in x, eg. $\pi^T = \begin{bmatrix} x_1 & x_2 & x_1^2 & x_1x_2 \end{bmatrix}$. Now we can see again that the condition $P > 0$ is only a sufficient condition for $\pi^T P\pi > 0$, because it is equivalent to $z^T Pz > 0 \;\; \forall z \in \mathbb{R}^p$, which is definitely a more strict condition for P than the original one.

From the above examples it is clear that it would be advantageous to decrease the conservatism of the inequalities corresponding to the stability condition. It is shown in [7] that Finsler's lemma and the notion of annihilators will help us to bring more freedom into our LMIs.

Lemma 2.1 (Finsler's lemma) *Let $\mathcal{X} \subseteq \mathbb{R}^n$ be given a polytopic set, $P : \mathcal{X} \mapsto \mathbb{R}^{p\times p}$, $N : \mathcal{X} \mapsto \mathbb{R}^{r\times p}$ be given matrix functions, with P symmetric. Let $Q(x)$ be a basis for the null space of $N(x)$. Then, the following are equivalent:*

(i) $\forall x \in \mathcal{X}$: $\pi^T P(x)\pi > 0$ *is satisfied* $\forall \pi \in \mathbb{R}^p, N(x)\pi = 0$

(ii) $\forall x \in \mathcal{X}$: $\exists L : \mathcal{X} \mapsto \mathbb{R}^{q\times r}$ *matrix function such that* $P(x) + L(x)N(x) + N^T(x)L^T(x) > 0$

(iii) $\forall x \in \mathcal{X}$: $Q^T(x)P(x)Q(x) > 0$ *is satisfied.*

Assuming that $P(x)$ and $N(x)$ are affine functions and L is constant matrix, we obtain a special case of the Finsler's lemma. Henceforth, the conditions (i) and (ii) are no longer equivalent, but (ii) shall continue to be a sufficient condition for (i), which in our case is a satisfactory result. Furthermore, (ii) will became a polytopic LMI (i.e. an affine parameter dependent LMI) of the following form

$$\forall x \in \mathcal{X} \;\; \exists L \in \mathbb{R}^{p\times r} : \quad P(x) + LN(x) + N^T(x)L^T > 0, \tag{5}$$

which again can be transformed into a single parameter independent LMI with a higher dimension. We have to design an $N(x)$ affine matrix function in such a way that $N(x)\pi = 0$ for all π. Therefore, $N(x)$ is called an annihilator of π. It is obvious that (5) is a less conservative sufficient condition for $\pi^T P\,\pi > 0$, than $P > 0$. In fact, increasing the size of the annihillator introduces more freedom into the computational problem.

2.5. Dynamical system representation

As a starting point, we use the same differential-algebraic representation of nonlinear models that was introduced in [7], namely:

$$\begin{aligned} \dot{x} &= f(x,\delta) = A\,x + B\,\pi \qquad && x_0 \in \mathcal{X} \\ 0 &= G(x,\delta)\,x + F(x,\delta)\,\pi \qquad && \delta \in \mathcal{D}, \dot{\delta} \in \mathcal{D}_d \end{aligned} \tag{6}$$

where $G : \mathcal{X} \times \mathcal{D} \mapsto \mathbb{R}^{p\times n}$ and $F : \mathcal{X} \times \mathcal{D} \mapsto \mathbb{R}^{p\times p}$ are affine matrix functions of (x, δ). This form separates the linear part of the system (Ax) from its nonlinear part $(B\pi)$. The second equation introduces a constraint representing the relation between x and π, where $\pi(x)$ is a vector of monomials in x. In [7], the authors propose that π contains all basic monomials of degree less than or equal to the maximal degree term in the system equation. This clearly causes a combinatorial explosion as the number of variables and their degrees increase. Therefore, in this work, we propose the application of LFT to decrease the number of monomials in π, hoping that the solutions conservatism will not increase significantly.

Journal of Physics: Conference Series **659** (2015) 012038 doi:10.1088/1742-6596/659/1/012038

2.6. Linear fractional transformation (LFT)

The main objective of LFT is to separate the linear part of a system from the nonlinear and uncertain parts. The LFT requires the system model in the form $\dot{x} = \mathcal{A}(x)x$, therefore, in case of a system represented in a more general $\dot{x} = f(x)$ form, we have to find a matrix function $\mathcal{A} : \mathbb{R}^n \mapsto \mathbb{R}^{n\times n}$, such that $\mathcal{A}(x)x = f(x)$. This computation step can be handled e.g. with the built-in functions of Matlab's Symbolic Math Toolbox. Let us consider the following uncertain (parameter dependent) nonlinear system:

$$\dot{x} = \mathcal{A}(x,\delta)\,x, \quad \delta \in \mathcal{D} \tag{7}$$

The linear fractional representation (LFR) of the system is the following:

$$\begin{aligned} \dot{x} &= M_{11}x + M_{12}w \\ z &= M_{21}x + M_{22}w \end{aligned} \Longrightarrow \begin{bmatrix}\dot{x}\\ z\end{bmatrix} = \left(\begin{array}{c|c} M_{11} & M_{12} \\ \hline M_{21} & M_{22}\end{array}\right) \cdot \begin{bmatrix} x \\ w\end{bmatrix} \tag{8}$$

$$w = \Delta(x,\delta)\,z \tag{9}$$

Equation (8) can be considered as a linear time invariant (LTI) system equipped with an input defined by the nonlinear uncertain relation (9). The block diagram of an LFR can be seen in Figure 1. Eliminating z and w from (8) and (9) we get the following:

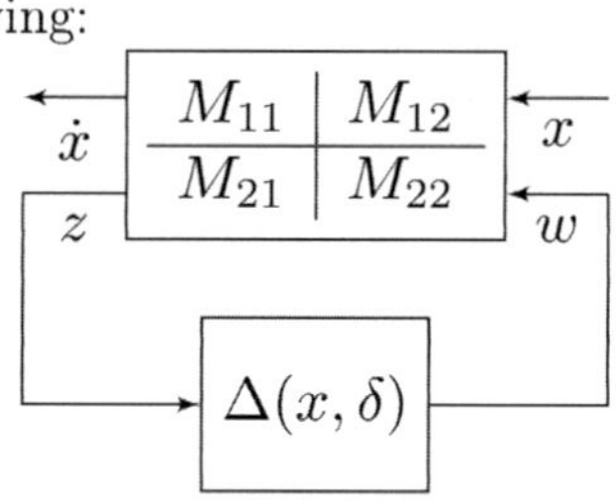

Figure 1: Block diagram of an LFR system

$$\mathcal{A}(x,\delta) = \underbrace{M_{21}(I_n - \Delta M_{11})^{-1}\Delta M_{12}}_{\text{nonlinear part}} + \underbrace{M_{22}}_{\text{linear part}}, \tag{10}$$

where M_{ij} are constant matrices and Δ is a diagonal matrix containing monomials of the state variables and uncertain parameters.

3. Estimating the domain of attraction

In this section we describe the main steps of the improved method for estimating the domain of attraction.

3.1. Preliminary transformations of the model

Due to the fact that the LFR is a special case of (6), it can be easily converted into that form by introducing the following notations:

$$\begin{aligned} A &= M_{22} \\ B &= M_{21} \end{aligned}, \quad \begin{aligned} G(x) &= -\Delta M_{12} \\ F(x) &= I_n + \Delta M_{11} \end{aligned}, \quad \pi = (I_n - \Delta M_{11})^{-1}\Delta M_{12}\,x \tag{11}$$

In this form, π contains only nonlinear elements. The LFR may result in such M_{ij} and Δ matrices that π will contain polynomials and rational terms with polynomial numerator and denominator. In this case, these elements of π should be split into monomials and rational terms with monomial numerators, respectively. At the same time, B and $F(x)$ should be modified appropriately, to satisfy the model equations (6) with the modified π vector. The same steps should be performed on the matrix B.

Additionally, the LFT and the previous transformations may generate several linearly dependent and thus redundant entries in π (the same term can appear multiple times, optionally with different constant multipliers). This may bring a significant unnecessary increase in the dimensions of the representation. Therefore, we designed an algorithm which eliminates the repetitive terms from π. The basic principle is to merge two rows in π into a single row, consequently, we also have to merge two columns in B and $F(x)$. Let us denote with $p(x)$ the monomial appearing at least twice and with $a(x)$, $b(x)$ those two elements in an arbitrary row of B or $F(x)$, which will be multiplied by the two identical monomials in π. Using these notations, we can define the following transformation:

$$\underbrace{\begin{bmatrix} \cdots & \cdots & \cdots & \cdots & \cdots \\ \cdots & a(x) & \cdots & b(x) & \cdots \\ \cdots & \cdots & \cdots & \cdots & \cdots \end{bmatrix}}_{B \text{ or } F(x)} \underbrace{\begin{bmatrix} \cdots \\ \alpha p(x) \\ \cdots \\ \beta p(x) \\ \cdots \end{bmatrix}}_{\pi} = \begin{bmatrix} \cdots & \cdots\cdots\cdots & \cdots \\ \cdots & \alpha a(x) + \beta b(x) & \cdots \\ \cdots & \cdots\cdots\cdots & \cdots \end{bmatrix} \begin{bmatrix} \cdots \\ p(x) \\ \cdots \end{bmatrix} \tag{12}$$

We have to note here that it is comfortable to keep $F(x)$ a square matrix, because in a few computation steps the left inverse of $F(x)$ is used. However, all multiplications by the left inverse can be avoided. We remark that the authors in [7] assumed $F(x)$ to be square matrix, i.e. $F(x) \in \mathbb{R}^{p\times p}$ and $\pi \in \mathbb{R}^p$. In order to ease this strict condition on $F(x)$, we will give a more general formula for the LMIs in Section 3.3. In that formula, we assume that $F(x) \in \mathbb{R}^{q\times p}$ is a general rectangular matrix. This relaxation does not affect the size of the LMIs, they still remain semidefinite problems (SDP). Although it is not necessary for $F(x)$ to be of maximum rank, it is advisable to remove the redundant rows from $C_b(x) = \begin{bmatrix} G(x) & F(x) \end{bmatrix}$ (i.e. to keep linearly independent rows only). The following pseudocode shows the entire simplification procedure:

Data:
  π, indices(π) = (first) $1 \dots p$ (last), having repetitive monomials
  B, $F(x)$, $G(x)$: system matrices
Result:
  π without repetitive monomials,
  modified B, $F(x)$, $G(x)$ model matrices corresponding to the new π
for $j = p$ *down to 2* **do**
  for $i =$ *1 to j-1* **do**
    $\frac{\alpha\, p(x)}{r_i(x)} \leftarrow \pi[i]$, $\frac{\beta\, q(x)}{r_j(x)} \leftarrow \pi[j]$
    if $p(x) = q(x)$ *and* $r_i(x) = r_j(x)$ **then**
      Generate the new column in B and $F(x)$ (element-wise):
      B[column i] = α B[column i] + β B[column j]
      $F(x)$[column i] = α $F(x)$[column i] + β $F(x)$[column j]
      $\pi[i] = p(x)/r_i(x)$
      Remove column j from B and $F(x)$.
      Remove the jth element from π
    end
  end
end
$p \leftarrow$ size(π)
$C_b(x) \leftarrow \begin{bmatrix} G(x) & F(x) \end{bmatrix}$
Clear linearly dependent rows from $C_b(x)$

3.2. Annihillator generation

In the processed form, π contains only non-repetitive nonlinear basis functions (monomials and rational terms with monomial numerator). Let us introduce the notations $\pi_b = [x^T\ \pi^T]^T$ and $C_b(x) = \begin{bmatrix} G(x) & F(x) \end{bmatrix}$. Thus, the second equation in (6) can be modified to $0 = C_b(x)\pi_b$. It is clear that $C_b(x)$ is a linear annihilator of π_b, which represents the dependence of π upon x. However, the size of $C_b(x)$ as an annihilator is far from being maximal, therefore, finding a second annihilator with rows not appearing in $C_b(x)$ is essential. Using Matlab's symbolic toolbox, we have written an algorithm, which generates an affine annihilator $N_{\pi_b}(x)$ having a special form. Since π_b contains only basis functions (optionally multiplied by an arbitrary constant), it is enough if in each row of the matrix there appear only two nonzero items[2], which will eliminate the two corresponding elements in π_b. In our algorithm, we search for an adequate $N_{\pi_b}(x)$ in the following form:

$$N_{\pi_b}(x) = \begin{bmatrix} \cdots & \alpha x_i & \cdots & \beta x_j & \cdots \\ \cdots & \alpha x_i & \cdots & \beta & \cdots \\ \cdots & \alpha & \cdots & \beta x_j & \cdots \end{bmatrix} \tag{13}$$

We have chosen two elements from π, let them be $a(x), c(x)$. If the numerator $\mathrm{num}(x)$ and the denominator $\mathrm{den}(x)$ of their simplified fraction $a(x)/c(x)$ are monomials of degree 0 or 1, then there exists $b(x) = \mathrm{den}(x)$, $d(x) = -\mathrm{num}(x)$ affine monomials such that $a(x)b(x)+c(x)d(x) = 0$. Finally, a new row can be appended to the annihilator matrix:

$$\begin{bmatrix} \cdots & b(x) & \cdots & d(x) & \cdots \end{bmatrix} \tag{14}$$

This procedure is evaluated on each pair of the elements of π. Possibly, $N_{\pi_b}(x)$ and $C_b(x)$ will have common rows, which can be skipped from $N_{\pi_b}(x)$. For this we used the built-in Matlab function $\left[\begin{smallmatrix} C_b(x) \\ N_{\pi_b}(x) \end{smallmatrix}\right]$=`unique(`$\left[\begin{smallmatrix} C_b(x) \\ N_{\pi_b}(x) \end{smallmatrix}\right]$`,'rows','static')`.

3.3. Finding an appropriate Lyapunov function

After the transformations presented in Section 3.1, we have the following model:

$$\begin{cases} \dot{x} = Ax + B\pi, & x \in \mathbb{R}^n, \quad \pi \in \mathbb{R}^p \\ 0 = G(x)x + F(x)\pi, & G(x) \in \mathbb{R}^{q\times n}, \quad F(x) \in \mathbb{R}^{q\times p} \end{cases} \tag{15}$$

Furthermore, $C_b(x) = \begin{bmatrix} G(x) & F(x) \end{bmatrix} \in \mathbb{R}^{q\times(n+p)}$ and $N_{\pi_b}(x) \in \mathbb{R}^{s_b\times(n+p)}$ are annihilators of π_b. A Lyapunov function candidate for this system will be given as $V(x) = x^T\mathbf{P}(x)x = \pi_b^T P \pi_b$, where $P \in \mathbb{R}^{(n+p)\times(n+p)}$ is a constant symmetric matrix. As it has been mentioned in Section 2.4, the positive definiteness of $V(x)$ can be ensured by a stricter inequality:

$$\forall x \in \mathcal{X} \;\; \exists L \in \mathbb{R}^{(n+p)\times(q+s_b)}: \quad P + L\begin{bmatrix} C_b(x) \\ N_{\pi_b}(x) \end{bmatrix} + \begin{bmatrix} C_b^T(x) & N_{\pi_b}^T(x) \end{bmatrix} L^T > 0 \tag{16}$$

According to Theorem 4.1 in [7], the negative definiteness of $\dot{V}(x) = \nabla V(x)\,\dot{x}$ in ensured by the following sufficient LMI condition:

$$\forall x \in \mathcal{X} \;\; \exists L_a \in \mathbb{R}^{(n^2+n+2p+np)\times s_b}: \quad P_a + P_a^T + L_a\begin{bmatrix} C_a(x) \\ N_{\pi_a}(x) \end{bmatrix} + \begin{bmatrix} C_a^T(x) & N_{\pi_a}^T(x) \end{bmatrix} L_a^T < 0 \tag{17}$$

In contrast to [7], we assume a general rectangular $F(x) \in \mathbb{R}^{q\times p}$ matrix (not necessarily square), therefore, the following variable is redefined as:

$$C_a(x) = \begin{bmatrix} G(x) & F(x) & O_{q\times p} & O_{q\times n^2} & O_{q\times np} \\ W_1(x) & W_2(x) & F(x) & O_{q\times n^2} & \bar{F}_a \\ W_3(x) & W_4(x) & O_{n^2\times p} & I_{n^2} & O_{n^2\times np} \\ O_{nq\times n} & O_{nq\times p} & O_{nq\times p} & -G_b(x) & F_b(x) \end{bmatrix} \tag{18}$$

where the matrices P_a, $N_{\pi_a}(x)$, $W_1(x)$, $W_2(x)$, $W_3(x)$, $W_4(x)$, $\bar{F}_a$, $G_b(x)$, $F_b(x)$ remain the same as they are defined in [7] in equations (39-42).

[2] In case of monomials, if an element $a(x)$ of π_b can be eliminated by two other elements $b(x), c(x)$, than $a(x)$ surely can be eliminated by using only one from $b(x)$ and $c(x)$. This statement is no longer valid when having monomials and basis rational terms, too. Consider the following simple example: $\begin{bmatrix} -1 & x & x \end{bmatrix}\begin{bmatrix} x & \frac{1}{x^2+1} & \frac{x^2}{x^2+1} \end{bmatrix}^T = 0$.

3.4. Finding the maximal level set, for a given $\mathcal{X}$

In order to find the maximal invariant level set, we adopted a combined method of the two approaches presented in Section 2.2. First of all, we defined a small $\mathcal{Y}$ polytope around the locally stable origin inside $\mathcal{X}$. With reference to [5], $\mathcal{Y}$ could be a variable size polytope, the size of which can be maximized during the optimization producing a maximal level set around $\mathcal{Y}$. The challenge is that the matrix inequality conditions are no longer linear using this approach, hence the need arises for further relaxations. In comparison, we set $\mathcal{Y}$ to have a small but constant size. In short, we have two polytopes around the origin: a larger one ($\mathcal{X}$) and a smaller one ($\mathcal{Y}$) in the inside of $\mathcal{X}$. We are looking for a level set $\varepsilon_\alpha = \{x \in \mathcal{X} \mid V(x) = \alpha,\ 1 \le \alpha\}$ that is outside of $\mathcal{Y}$ but inside of $\mathcal{X}$, subject to the 1-level set $\varepsilon_1 = \{x \in \mathcal{X} \mid V(x) = 1\}$ is in the same region. To rephrase, we have to maximize α, s.t.

(1) ε_α is located in the inside of $\mathcal{X}$,

(2) $\mathcal{Y}$ is inside ε_1 (without this condition, $V(x)$ can be scaled arbitrarily, resulting in an unbounded optimum).

Condition (C1) means that for each facet $\mathcal{F}_k^{(\mathcal{X})}$ of the $\mathcal{X}$ polytope the following condition must be satisfied:

$$V(x) = \pi_b^T P \pi_b \ge \alpha \Longleftrightarrow \begin{bmatrix}\pi_b^T & 1\end{bmatrix}\begin{bmatrix}P & 0\\ 0 & -\alpha\end{bmatrix}\begin{bmatrix}\pi_b\\ 1\end{bmatrix} \ge 0, \quad \forall x \in \vartheta(\mathcal{F}_k^{(\mathcal{X})}),\ \forall k = \overline{1, M_\mathcal{X}}, \tag{19}$$

where $M_\mathcal{X}$ is the number of facets of $\mathcal{X}$. According to [7] and using the equivalence between (i), (iii) of the Finsler's lemma, we can rewrite the condition (19) into

$$Q_k^T\left(P_{c_k}(x) + \Gamma_{c_k}(x)\right)Q_k > 0, \quad \forall x \in \vartheta(\mathcal{F}_k^{(\mathcal{X})}),\ \forall k = \overline{1, M_\mathcal{X}}, \tag{20}$$

where Q_k, $P_{c_k}(x)$, $\Gamma_{c_k}(x)$ are described in detail by equations (82-89) of [7]. With the same consideration, condition (C2) can be written as

$$V(x) = \pi_b^T P \pi_b \le 1 \Longleftrightarrow \begin{bmatrix}\pi_b^T & 1\end{bmatrix}\begin{bmatrix}P & 0\\ 0 & -1\end{bmatrix}\begin{bmatrix}\pi_b\\ 1\end{bmatrix} \le 0, \quad \forall x \in \vartheta(\mathcal{F}_k^{(\mathcal{Y})}),\ \forall k = \overline{1, M_\mathcal{Y}}, \tag{21}$$

which can be converted into the same form as (20). Finally, the optimization problem is traced back to a minimization of $-\alpha$ subject to eqs. (16), (17), (20) and (21).

3.5. Finding the most appropriate outer polytope

Finding the most suitable $\mathcal{X}$, in which the Lyapunov conditions could be fulfilled is an iterative problem. The basic concept is to choose an $\mathcal{X}^{(0)}$ inital polytope, which surely satisfies the LMI conditions, than in each step ($\mathcal{X}^{(k)}$) find out the $\varepsilon^{(k)}$ maximal level set and give a new, larger $\mathcal{X}^{(k+1)}$ polytope considering the shape of $\varepsilon^{(k)}$. One possible solution can be choosing some uniformly distributed discrete points placed on the $\varepsilon^{(k)}$ level set. These points will span a polytope, which should be enlarged with a given increment, such that its shape is not changed (practically, the coordinates of every corner are multiplied by an $1 < \gamma \ll 2$ scalar factor, since the stable equilibrium point is assumed to be at the origin).

Journal of Physics: Conference Series **659** (2015) 012038 doi:10.1088/1742-6596/659/1/012038

4. Numerical examples and results

The results presented in this section have been computed in the Matlab environment. For symbolic computations we used Matlab's built-in Symbolic Math Toolbox based on Mupad. For LFT we used the Enhanced LFR-toolbox [9, 10]. To model and solve semidefinite optimization (SDP) problems we used the SeDuMi and Mosek solvers with YALMIP [11].

4.1. Van der Pol dynamics

The equation

$$\begin{aligned}\dot{x}_1 &= -x_2 \\ \dot{x}_2 &= x_1 - \varepsilon(1 - x_1^2)x_2\end{aligned} \quad \text{with } \varepsilon = 1 \tag{22}$$

describes a time-inverted oscillator introduced by the Dutch electrical engineer and physicist Balthasar van der Pol. This system has an asymptotically stable equilibrium point at the origin, and it has a limit cycle, which defines the boundary of the region of attraction (ROA). This limit cycle and some trajectories of the system are illustrated in Figure 2 (dashed red line and black trajectories, respectively). In Figure 2, you can also see the maximal invariant level set (green line) generated by the method described in Section 3. The nonlinear basis functions of π used by Trofino et.al. were

$$\pi^T = \begin{bmatrix} x_1{}^2 & x_1 x_2 & x_2^2 & x_1^3 & x_1^2 x_2 & x_1 x_2^2 & x_2^3 \end{bmatrix} \tag{23}$$

In comparison, our method by applying the LFT generates a smaller number of monomials:

$$\pi^T = \begin{bmatrix} x_1^2 x_2 & x_1 x_2 \end{bmatrix}. \tag{24}$$

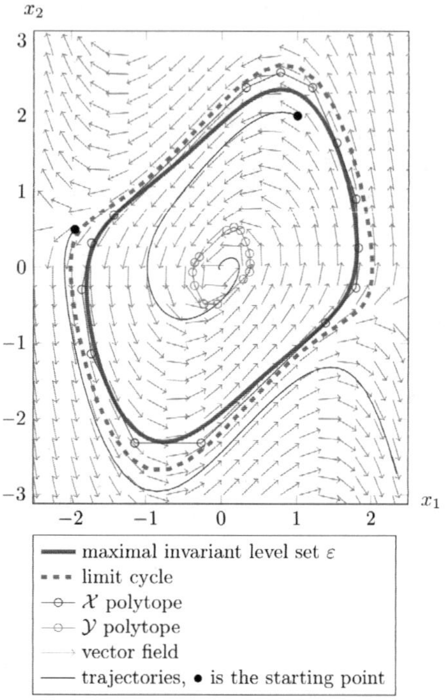

Figure 2: Van der Pol system, phase diagram. The ε level set (green line) approximates the limit cycle (dashed red line) in a quite acceptable manner. The blue and the red polygons constitutes the $\mathcal{X}$ and $\mathcal{Y}$, respectively

However, the area of the estimated invariant region is close enough to that of the true DOA. The matrices generated by the LFR Toolbox are the following:

$$M_{11} = \begin{bmatrix} 0 & 1 \\ 0 & 0 \end{bmatrix}, \quad M_{12} = \begin{bmatrix} 0 & 0 \\ 1 & 0 \end{bmatrix}, \quad M_{21} = \begin{bmatrix} 0 & 0 \\ 1 & 0 \end{bmatrix}, \quad M_{22} = \begin{bmatrix} 0 & -1 \\ 1 & -1 \end{bmatrix}, \quad \Delta = \begin{bmatrix} x_1 & 0 \\ 0 & x_2 \end{bmatrix},$$

The final Lyapunov function and invariant level set we obtained is the following:

$$\begin{aligned} V(x) = &-0.05746\, x_1^4\, x_2^2 + 0.03872\, x_1^3\, x_2^2 + 2.235\, x_1^3\, x_2 + 0.2525\, x_1^2\, x_2^2 - 0.06664\, x_1^2\, x_2 \\ &+ 5.961\, x_1^2 - 0.01476\, x_1\, x_2^2 - 8.719\, x_1\, x_2 + 5.05\, x_2^2 \\ \varepsilon = &\left\{ x \in \mathcal{X} \;\middle|\; V(x) = \alpha = 17.4086 \right\} \end{aligned} \tag{25}$$

4.2. Continuous fermentation process

Bioreactors often show strongly nonlinear dynamical behaviour, therefore, they can be interesting subjects for stability analysis. In our work, we have analysed the stability region of a widely used model presented in e.g. [12], which is a rational nonlinear system having a locally asymptotically stable equilibrium point. The equations of the normalized system are the following:

$$\dot{x} = \begin{bmatrix} \dot{\bar{X}} \\ \dot{\bar{S}} \end{bmatrix} = \begin{bmatrix} (\bar{X} + X_0) \cdot \mu(\bar{S} + S_0) - \dfrac{(\bar{X} + X_0)F_0}{V} \\ -\dfrac{(\bar{X} + X_0) \cdot \mu(\bar{S} + S_0)}{Y} + \dfrac{(S_F - (\bar{S} + S_0))F_0}{V} \end{bmatrix} \tag{26}$$

$$\mu(S) = \mu_{max} \frac{S}{K_2 S^2 + S + K_1}, \tag{27}$$

where the variables and parameters are explained in the following table:

Variables and parameters of the process				Steady-state operating point			
X	Biomass concentration		$[g/l]$	X_0	equilibrium point of X	4.8907	[g/l]
S	Substrate concentration		$[g/l]$	S_0	equilibrium point of S	0.2187	[g/l]
F	Feed flow rate		$[l/h]$	F_0	Inlet feed flow rate	3.2089	[l/h]
V	Volume	4	$[l]$				
S_F	Substrate feed concentration	10	$[g/l]$				
Y	Yield coefficient	0.5					
μ_{max}	maximal growth rate	1	$[l/h]$				
K_1	Saturation parameter	0.03	$[g/l]$				
K_2	Inhibition parameter	0.5	$[l/g]$				

We cannot apply the LFT in this form of the system, because there appear constant terms as well: if $\bar{X} = 0$ and $\bar{S} = 0$ then $\dot{x} = \begin{bmatrix} X_0 \cdot \mu(S_0) - X_0 F_0/V \\ -X_0 \cdot \mu(S_0)/Y - (S_F - S_0)F_0/V \end{bmatrix}$. Knowing that $F_0 = V\mu(S_0)$ and $X_0 = (S_0 - S_F)Y$, these constant terms can be eliminated. After the factorization ($f(x) = A(x)x$) we obtain the desired form:

$$\dot{x} = A(x)x, \quad \text{where} \quad A(x) = \begin{bmatrix} -\frac{c_2 F_0 \bar{S}^2 + c_1 F_0 \bar{S}}{q(\bar{S})V} & \frac{\mu_{max} V(\bar{X}+X_0) - X_0 F_0 (c_2 \bar{S} + c_1)}{q(\bar{S})V} \\ -\frac{\mu_{max} S_0}{q(\bar{S})Y} & -\frac{F_0}{V} - \frac{\mu_{max} V(\bar{X}+X_0) + Y F_0 (S_0 - S_F)(c_2 \bar{S} + c_1)}{q(\bar{S})VY} \end{bmatrix} \tag{28}$$

$$c_2 = K_2$$
$$c_1 = 2\,K_2\,S_0 + 1$$
$$c_0 = K_2\,S_0^2 + S_0 + K_1 \qquad q(\bar{S}) = c_2\,\bar{S}^2 + c_1\,\bar{S} + c_0$$

Using the LFT we obtain a π with rational nonlinear elements having $q(\bar{S})$ as their denominator. The final $\mathcal{X}$ polytope with the corresponding invariant level set can be seen in Figure 3, as a gray and red line, respectively. We can see that the final level set adequately approximates the boundary of the DOA (dashed line). The final Lyapunov functions was the following:

$$V(x) = \frac{75232(-22.278\,x_1^2\,x_2^4 + 349.79\,x_1^2\,x_2^3 + 43.209\,x_1^2\,x_2^2 + 1.6758\,x_1^2\,x_2 + 0.093181\,x_1^2 + 123.37\,x_1\,x_2^5 + 577.76\,x_1\,x_2^4)}{1.3657\,x_2^4 + 6.6573\,x_2^3 + 9.6019\,x_2^2 + 3.6291\,x_2 + 0.40585}$$
$$+ \frac{75232(46.861\,x_1\,x_2^3 + 1.3259\,x_1\,x_2^2 + 0.038426\,x_1\,x_2 - 43.571\,x_2^6 - 120.6\,x_2^5 + 23.602\,x_2^4 + 1.3087\,x_2^3 + 0.031804\,x_2^2))}{1.3657\,x_2^4 + 6.6573\,x_2^3 + 9.6019\,x_2^2 + 3.6291\,x_2 + 0.40585}$$

4.3. Continuous fermentation process with a simple linear feedback

In the previous normalized process, we constrained the inlet feed flow rate to be constant (F_0). Here, we introduce a centered input u, and the actual feed flow rate will be $F = F_0 + u$. Consequently, the model should be modified with an input term as follows.

$$\dot{x} = A(x)x + g(x)u, \qquad g(x) = \begin{bmatrix} -\frac{X_0 + \bar{X}}{V} \\ -\frac{\bar{S} + S_0 - S_F}{V} \end{bmatrix} \tag{29}$$

In [12] it is shown that the zero dynamics of the model is globally stable if the output is the substrate concentration. Therefore, we chose a simple feedback of the form $u = k\bar{S}$ with $k \in \mathbb{R}$ ($k = -1$ was used for the computations). Then we can write:

$$g(x)u = \begin{bmatrix} -\frac{k\bar{S}}{V} & -\frac{kX_0}{V} \\ 0 & -\frac{k(\bar{S}+S_0-S_F)}{V} \end{bmatrix} \begin{bmatrix} \bar{X} \\ \bar{S} \end{bmatrix} \tag{30}$$

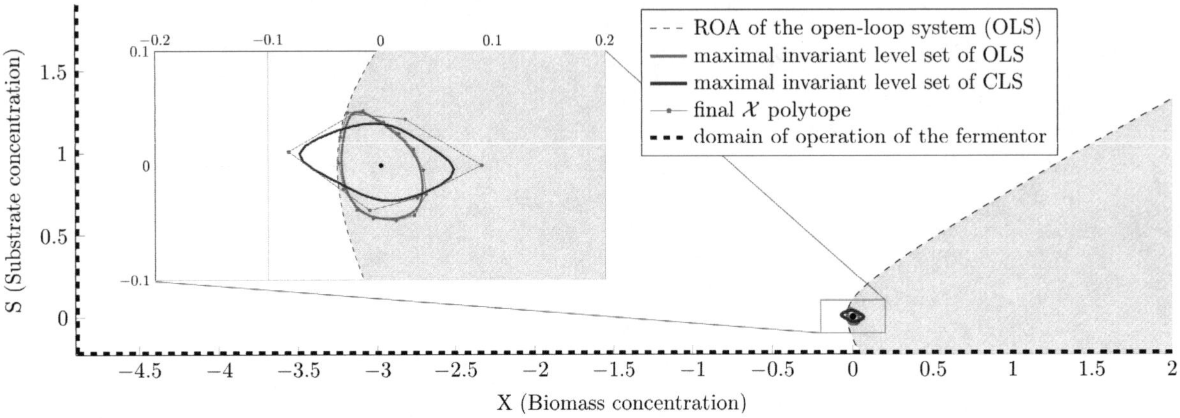

Figure 3: Region of attraction (ROA) of the continuous fermentation process without feedback (green area). The final invariant level set for the open-loop system (OLS) and for the closed-loop system (CLS) with $k = -1$ can be seen as a red and blue line, respectively. The most appropriate outer polytopes are illustrated by dotted gray lines.

The equation of the closed loop system can be written as $\dot{x} = \mathfrak{A}(x)x$, where

$$\mathfrak{A}(x) = \begin{bmatrix} -\frac{c_2 F_0 \bar{S}^2 + c_1 F_0 \bar{S}}{r(\bar{S})V} - \frac{k\bar{S}}{V} & \frac{\mu_{max} V(\bar{X}+X_0) - X_0 F_0 (c_2 \bar{S} + c_1)}{r(\bar{S})V} - \frac{kX_0}{V} \\ -\frac{\mu_{max} S_0}{r(\bar{S})Y} & \frac{Y F_0 (S_F - S_0)(c_2 \bar{S} + c_1) - \mu_{max} V(\bar{X}+X_0)}{r(\bar{S})VY} - \frac{k(\bar{S}+S_0-S_F)+F_0}{V} \end{bmatrix}$$

Due to numerical difficulties, we used an outer polytope ($\mathcal{X}$) with a reduced number of corner points with the purpose of reducing the SDP problem's dimension. Furthermore, the corners' position were chosen manually, according to the following procedure. First of all, we defined an initial polytope, which surely satisfies the LMI conditions. Then, in each step, we enlarged the polytope with small increments in a quasi-random manner ('mutation' of the polytope). In one step, only a few corners were modified, and they were strategically chosen by considering the distance of the facets to the maximal invariant level set of the previous iteration. The position of the chosen corner points were altered randomly by small amount with the intention of increasing the distance of the neighbouring facets to the level set. If the problem with the new outer polytope is feasible producing no numerical failures, than we continue with the next step, in the other case we try another mutation of the previous polytope. This procedure is based on quasi-random modifications of an initially given feasible polytope, therefore, it can be automated. The final Lyapunov function we achieved was $V(x) = p(x)/q(x)$, and its maximal invariant level set was $\varepsilon = \{x \in \mathcal{X} \mid V(x) = 12.886\}$, where:

$$\begin{aligned} p(x) = {} & 7893.1\, x_1^2 x_2^{12} + 47168\, x_1^2 x_2^{11} + 1.5464e6\, x_1^2 x_2^{10} + 1.2694e7\, x_1^2 x_2^9 + 4.3955e7\, x_1^2 x_2^8 + 7.9184e7\, x_1^2 x_2^7 + 8.0827e7\, x_1^2 x_2^6 \\ & + 4.8265e7\, x_1^2 x_2^5 + 1.5898e7\, x_1^2 x_2^4 + 2.6443e6\, x_1^2 x_2^3 + 1.7643e5\, x_1^2 x_2^2 + 1392\, x_1^2 x_2 + 147.6\, x_1^2 + 13824\, x_1 x_2^1 3 + 94230\, x_1 x_2^{12} \\ & - 6.131e5\, x_1 x_2^{11} - 8.0351e6\, x_1 x_2^{10} - 3.2513e7\, x_1 x_2^9 - 6.5635e7\, x_1 x_2^8 - 7.123e7\, x_1 x_2^7 - 4.1178e7\, x_1 x_2^6 - 1.261e7\, x_1 x_2^5 \\ & - 1.8956e6\, x_1 x_2^4 - 93978\, x_1 x_2^3 + 4561\, x_1 x_2^2 + 291.41\, x_1 x_2 + 46964\, x_2^1 4 + 5.953e5\, x_2^1 3 + 3.2591e6\, x_2^1 2 + 1.0158e7\, x_2^{11} \\ & + 1.9889e7\, x_2^{10} + 2.5173e7\, x_2^9 + 2.0184e7\, x_2^8 + 9.7704e6\, x_2^7 + 2.8587e6\, x_2^6 + 5.3869e5\, x_2^5 + 76424\, x_2^4 + 8727.2\, x_2^3 + 537.97\, x_2^2 \\ q(x) = {} & x_2^{10} + 12.187\, x_2^9 + 62.131\, x_2^8 + 171.36\, x_2^7 + 276.58\, x_2^6 + 265.61\, x_2^5 + 150.77\, x_2^4 + 50.925\, x_2^3 + 10.065\, x_2^2 + 1.0762\, x_2 \\ & + 0.048143 \end{aligned}$$

5. Conclusions

In this work, we presented an optimization-based computational method for determining Lyapunov functions and invariant regions for nonlinear dynamical systems. The starting point of the method is the approach presented in [7]. The improvements and new contributions can be summarized as follows: 1) The model transformation to the required form for optimization is done automatically using LFT with auxiliary algorithmic simplifications. This technique results in the dimension reduction of the problem compared to known solutions in the literature. 2) An algorithm was given for the generation of appropriate annihilators for the vector π. 3) An improved method was proposed for determining the possible largest invariant set for the dynamics using the computed Lyapunov function. 4) A generalized formula was given for the case when the system matrix $F(x)$ is not regular in the model (6). The operation of the approach was illustrated through examples taken from the literature. Although the developed method itself is capable of handling uncertain models, it will be the target of future work to test it on examples containing uncertainties.

6. Acknowledgement

The authors gratefully acknowledge the support of the National Research, Development and Innovation Office – NKFIH through grant no. NF104706, as well as the support of Pázmány Péter Catholic University through the grant KAP15-052-1.1-ITK.

References

[1] Graziano Chesi. *Domain of attraction: analysis and control via SOS programming*, volume 415. Springer Science & Business Media, 2011. 1
[2] A. Vannelli and M. Vidyasagar. Maximal Lyapunov functions and domains of attraction for autonomous nonlinear systems. *Automatica*, 21:69–80, 1985. 1
[3] Sz. Rozgonyi, K. M. Hangos, and G. Szederkényi. Determining the domain of attraction of hybrid non-linear systems using maximal Lyapunov functions. *Kybernetika*, 46:19–37, 2010. 1
[4] Laurent El Ghaoui and Gérard Scorletti. Control of rational systems using linear-fractional representations and linear matrix inequalities. *Automatica*, 32(9):1273 – 1284, 1996. 1
[5] U. Topcu, A. K. Packard, and P. Seiler. Local stability analysis using simulations and sum-of-squares programming. *Automatica*, 44(10):2669–2675, 2008. 1, 2, 3, 8
[6] U. Topcu, A. Packard, P. Seiler, and G. Balas. Robust region-of-attraction estimation. *IEEE Transactions on Automatic Control*, 55(1):137–142, 2010. 1
[7] A. Trofino and T. J. M. Dezuo. LMI stability conditions for uncertain rational nonlinear systems. *International Journal of Robust and Nonlinear Control*, 2013. 1, 2, 3, 4, 6, 7, 8, 12
[8] Carsten Scherer and Siep Weiland. Linear matrix inequalities in control. *Lecture Notes, Dutch Institute for Systems and Control, Delft, The Netherlands*, 2000. 3
[9] *Enhanced LFR-toolbox for MATLAB*, 2004. 9
[10] JF Magni. Linear fractional representation toolbox (version 2.0) for use with matlab. *Free Web publication http://www. cert. fr/dcsd/idco/perso/Magni*, 2006. 9
[11] J. Löfberg. YALMIP : A toolbox for modeling and optimization in MATLAB. In *Proceedings of the CACSD Conference*, Taipei, Taiwan, 2004. 9
[12] G. Szederkényi, N. R. Kristensen, K. M. Hangos, and S. B. Jorgensen. Nonlinear analysis and control of a continuous fermentation process. *Computers and Chemical Engineering*, 26:659–670, 2002. 9, 10

Journal of Physics: Conference Series **659** (2015) 012039 doi:10.1088/1742-6596/659/1/012039

Digraph-Based Algorithm for Finding Minimal Positive Realisation of Two-Dimensional Linear System with Delays

Konrad Andrzej MARKOWSKI

Warsaw University of Technology, Faculty of Electrical Engineering,
Institute of Control and Industrial Electronics,
Koszykowa 75, 00-662 Warsaw, POLAND

E-mail: Konrad.Markowski@ee.pw.edu.pl

Abstract. In this paper, the new method of the positive minimal realisation two-dimensional linear system with delays described by general model using multi-dimensional digraphs theory $\mathfrak{D}^{(n)}$ has been presented. For the proposed method, a digraph-based algorithm was constructed. The algorithm is based on a parallel computing method to gain needed speed and computational power for such a solution. The proposed solution allows minimal digraphs construction for any positive two-dimensional linear system with delays. The proposed method was discussed and illustrated with some numerical examples.

1. Introduction

The most popular models of two-dimensional linear system have been introduced by Roesser [1], Fornasini and Marchesini [2], [3] and Kurek [4]. In recent years, many researchers have been interested in positive linear systems. In this type of the system, state variables and outputs take only non-negative values. Analysis of positive one-dimensional (1D) systems is more difficult than of standard systems. Examples of positive systems include industrial processes involving chemical reactors, heat exchangers and distillation columns, storage systems, compartmental systems, water and atmospheric pollution models. An overview of the state of the art in the positive systems theory is given in [5], [6] and [7].

The realisation problem is a very difficult task. In many research studies, we can find the canonical form of the system, i.e. constant matrix form, which satisfies the system described by the transfer function. With the use of this form, we are able to write only one realisation (or some by the transformation matrices) of the system, while there exist many sets of matrices which fit into the system transfer function. The realisation problem for positive discrete-time systems without and with delays was considered in [6], [8], [9], [10], while in [11], [12] a solution for finding a set of possible realisations of the characteristic polynomial was proposed, that allows for finding many sets of matrices. In paper [13], [14], [15], [16], [17] the proposed method for finding a minimal positive realisations is an extension of the method for finding realisation of the characteristic polynomial. The optimisation of the proposed algorithm is presented in the paper [18] and [19].

The digraphs theory was applied to the analysis of dynamical systems. The use of the multi-dimensional theory was proposed for the first time in the paper [20], [21], [22] for analysis of positive two-dimensional systems.

In this paper, a new method of determination of a positive minimal realisation for the two-dimensional linear system with delay will be proposed, and the procedure for computation of the minimal realisation will be given. The proposed method is based on multi-dimensional digraphs theory, and it will be illustrated with a numerical example.

This work has been organised as follows: Section 2 presents some notations, defines a positive two-dimensional linear system with delays as the state-space representation, presents basic definitions of the digraph theory and introduces the concept of a multi-dimensional digraph and formulates the problem statement. In Section 3, we construct an algorithm for the determination of a positive minimal realisation of the linear two-dimensional system with delays described by the general model. Finally, we demonstrate the working of the algorithm on a numerical example in Section 4, and at the end we present some concluding remarks, open problems and bibliography positions.

2. Preliminaries and Problem Formulation

2.1. Notion

In this paper, the following notion will be used. The matrices will be denoted by a bold font (for example $\mathbf{A}$, $\mathbf{B}$, ...), the sets by the double line (for example $\mathbb{A}$, $\mathbb{B}$, ...), lower/upper indices and polynomial coefficients will be written using a small font (for example a, b, ...) and the digraph will be denoted using a mathfrak font $\mathfrak{D}$. The set $n \times m$ real matrices will be denoted by $\mathbb{R}^{n\times m}$ and $\mathbb{R}^n = \mathbb{R}^{n\times 1}$. If $\mathbf{G} = [g_{ij}]$ is a matrix, we write $\mathbf{G} \gg 0$ (matrix $\mathbf{G}$ is called strictly positive), if $g_{ij} > 0$ for all i, j; $\mathbf{G} > 0$ (matrix $\mathbf{G}$ is called positive), if $g_{ij} > 0$ for all i, j; $\mathbf{G} \geqslant 0$ (matrix $\mathbf{G}$ is called non-negative), if $g_{ij} \geqslant 0$ for all i, j. The set of $n \times m$ real matrices with non-negative entries wll be denoted by $\mathbb{R}_+^{n\times m}$ and $\mathbb{R}_+^n = \mathbb{R}_+^{n\times 1}$. The $n \times n$ identity matrix will be denoted by $\mathbf{I}_n$. For more information about matrix theory, an interested reader is referred, for instance, to: [23], [24].

2.2. Positive Two-Dimensional Linear System with Delays

Let be given the model described by the equations:

$$x_{i+1,j+1} = \sum_{k_1=0}^{q_1}\sum_{l_1=0}^{q_2}\left(\mathbf{A}^0_{k_1l_1}x_{i-k_1,j-l_1} + \mathbf{A}^1_{k_1l_1}x_{i-k_1+1,j-l_1} + \mathbf{A}^2_{k_1l_1}x_{i-k_1,j-l_1+1}\right) + \tag{1a}$$

$$+\sum_{k_2=0}^{p_1}\sum_{l_2=0}^{p_2}\left(\mathbf{B}^0_{k_2l_2}u_{i-k_2,j-l_2} + \mathbf{B}^1_{k_2l_2}u_{i-k_2+1,j-l_2} + \mathbf{B}^2_{k_2l_2}u_{i-k_2,j-l_2+1}\right),$$

$$y_{ij} = \mathbf{C}x_{ij} + \mathbf{D}u_{ij}, \qquad i,j \in \mathbb{Z}_+ = \{0,1,\ldots\}. \tag{1b}$$

The model described by the equations (1a)–(1b) is called the two-dimensional general model with delays, where $x_{ij} \in \mathbb{R}^n$, $u_{ij} \in \mathbb{R}^m$, $y_{ij} \in \mathbb{R}^p$ are state, input and output vectors respectively and $\mathbf{A}^t_{k_1l_1} \in \mathbb{R}^{n\times n}$, $\mathbf{B}^t_{k_2l_2} \in \mathbb{R}^{n\times m}$, $t = 0,1,2$; $k_1 = 0,1,\ldots,q_1$, $l_1 = 0,1,\ldots,q_2$,$k_2 = 0,1,\ldots,p_1$, $l_2 = 0,1,\ldots,p_2$, $\mathbf{C} \in \mathbb{R}^{p\times n}$, $\mathbf{D} \in \mathbb{R}^{p\times m}$.

Boundary conditions for model described by the equations (1a)–(1b) have the form

$$x_{i-k_1,-l_1},\ x_{-k_1,j-l_1},\ k_1 = 0,1,\ldots,q_1;\ l_1 = 0,1,\ldots,q_2;\ i,j \in \mathbb{Z}_+. \tag{2}$$

The model described by the equations (1a)–(1b) is called a two-dimensional linear system with (q_1, q_2)–delays in state vector and (p_1, p_2)–delays in input vector.

Using to (1a) $\mathcal{Z}$–transform with zero boundary conditions (2) and taking into account that $Y(z_1,z_2) = \mathbf{C}X(z_1,z_2) + \mathbf{D}U(z_1,z_2)$ we can determine the transfer matrix in the following form:

$$\mathbf{T}(z_1,z_2) = \mathbf{C}\left[\mathbf{I}z_1z_2 - \sum_{k_1=0}^{q_1}\sum_{l_1=0}^{q_2} z_1^{-k_1}z_2^{-l_1}\left(\mathbf{A}^0_{k_1l_1} + \mathbf{A}^1_{k_1l_1}z_1 + \mathbf{A}^2_{k_1l_1}z_2\right)\right]^{-1} \times \left[\sum_{k_2=0}^{p_1}\sum_{l_2=0}^{p_2} z_1^{-k_2}z_2^{-l_2}\left(\mathbf{B}^0_{k_2l_2} + \mathbf{B}^1_{k_2l_2}z_1 + \mathbf{B}^2_{k_2l_2}z_2\right)\right] + \mathbf{D} = \frac{\mathbf{N}(z_1,z_2)}{d(z_1,z_2)}, \tag{3}$$

where

$$\mathbf{N}(z_1,z_2) = \begin{bmatrix} n_{11}(z_1,z_2) & \dots & n_{1m}(z_1,z_2) \\ \vdots & \ddots & \vdots \\ n_{p1}(z_1,z_2) & \dots & n_{pm}(z_1,z_2) \end{bmatrix},$$

$$n_{ij}(z_1,z_2) = \sum_{\substack{k=0 \\ k+l\le N_1+N_2}}^{N_1}\sum_{l=0}^{N_2} n^{ij}_{kl}z_1^kz_2^l = n^{ij}_{N_1,N_2}z_1^{N_1}z_2^{N_2} + n^{ij}_{N_1,N_2-1}z_1^{N_1}z_2^{N_2-1} + + n^{ij}_{N_1-1,N_2}z_1^{N_1-1}z_2^{N_2} + \ldots + n^{ij}_{10}z_1 + n^{ij}_{01}z_2 + n^{ij}_{00}, \tag{4}$$

$$d(z_1,z_2) = z_1^{nq_1}z_2^{nq_2} - \sum_{\substack{k=0 \\ k+l<n(q_1+q_2)}}^{nq_1}\sum_{l=0}^{nq_2} d_{kl}z_1^kz_2^l = z_1^{nq_1}z_2^{nq_2} - d_{nq_1,nq_2-1}z_1^{nq_1}z_2^{nq_2-1} - - d_{nq_1-1,nq_2}z_1^{nq_1-1}z_2^{nq_2} - \ldots - d_{10}z_1 - d_{01}z_2 - d_{00}. \tag{5}$$

Definition 1 *The system described by the equations* (1a)–(1b) *is called internally positive if and only if* $x(i,j) \in \mathbb{R}^n_+$ *and* $y(i,j) \in \mathbb{R}^p_+$, $i,j \in \mathbb{Z}_+$ *for any boundary conditions* $x_{i-k_1,-l_1} \in \mathbb{R}^n_+$, $x_{-k_1,j-l_1} \in \mathbb{R}^n_+$, $k_1 = 0,1,\ldots,q_1$, $l_1 = 0,1,\ldots,q_2$, $i,j \in \mathbb{Z}_+$ *and all input sequence* $u(i,j) \in \mathbb{R}^m_+$, $i = -p_1, -p_1+1, \ldots;\ j = -p_2, -p_2+1, \ldots$.

Theorem 1 *The two-dimensional linear system with delay described by the equations* (1a)–(1b) *is positive if and only if*

$$\begin{aligned} &\mathbf{A}^k_{k_1l_1} \in \mathbb{R}^{n\times n}_+, \quad k_1 = 0,1,\cdots,q_1;\ \ l_1 = 0,1,\cdots,q_2;\ k = 0,1,2 \\ &\mathbf{B}^k_{k_2l_2} \in \mathbb{R}^{n\times m}_+, \quad k_2 = 0,1,\cdots,p_1;\ \ l_2 = 0,1,\cdots,p_2;\ k = 0,1,2, \\ &\mathbf{C} \in \mathbb{R}^{p\times n}_+, \quad \mathbf{D} \in \mathbb{R}^{p\times m}_+. \end{aligned} \tag{6}$$

The proof of the Theorem 1 is given in [25].

Matrices (6) are called positive realisations of the transfer matrix if they satisfy the equality (3). The realisation is called minimal if the dimension of the state matrix $\mathbf{A}^k_{k_1l_1}$, $k_1 = 0,1,\ldots,q_1$; $l_1 = 1,2,\ldots,q_2$; $k = 0,1,2$ is minimal among all possible realisations of $\mathbf{T}(z_1,z_2)$. In [26] the basic definitions of the positive realisation problem have been presented in detail.

Journal of Physics: Conference Series **659** (2015) 012039 doi:10.1088/1742-6596/659/1/012039

2.3. Multi-Dimensional Digraphs

A directed graph (or just digraph) $\mathfrak{D}$ consists of a non-empty finite set $\mathbb{V}(\mathfrak{D})$ of elements called vertices and a finite set $\mathbb{A}(\mathfrak{D})$ of ordered pairs of distinct vertices called arcs [27]. We call $\mathbb{V}(\mathfrak{D})$ the vertex set and $\mathbb{A}(\mathfrak{D})$ the arc set of $\mathfrak{D}$. We will often write $\mathfrak{D} = (\mathbb{V}, \mathbb{A})$ which means that $\mathbb{V}$ and $\mathbb{A}$ are the vertex set and arc set of $\mathfrak{D}$, respectively. The order of $\mathfrak{D}$ is the number of vertices in $\mathfrak{D}$. The size of $\mathfrak{D}$ is the number of arc in $\mathfrak{D}$. For an arc (v_1, v_2) the first vertex v_1 is its tail and the second vertex v_2 is its head.

A two-dimensional digraph $\mathfrak{D}^{(2)}$ is a directed graph with two types of arcs and input flows. For the first time, this type of digraph was presented in papers [20], [21] and [22]. When we generalise this approach, we can define multi-dimensional digraphs $\mathfrak{D}^{(n)}$ in the following form.

Definition 2 *A n-dimensional digraph $\mathfrak{D}^{(n)}$ is a directed graph with q types of arcs and input flows. In detail, it is $(\mathbb{S}, \mathbb{V}, \mathfrak{A}_1, \mathfrak{A}_2, \ldots, \mathfrak{A}_q, \mathfrak{B}_1, \mathfrak{B}_2, \ldots, \mathfrak{B}_q)$, where $\mathbb{S} = \{s_1, s_2, \ldots, s_m\}$ is the set of sources, $\mathbb{V} = \{v_1, v_2, \ldots, v_n\}$ is the set of vertices, $\mathfrak{A}_1$, $\mathfrak{A}_2$, ..., $\mathfrak{A}_q$ are the subsets of $\mathbb{V} \times \mathbb{V}$ whose elements are called $\mathfrak{A}_1$-arcs, $\mathfrak{A}_2$-arcs, ..., $\mathfrak{A}_q$-arcs respectively, $\mathfrak{B}_1$, $\mathfrak{B}_2$, ..., $\mathfrak{B}_q$ are the subsets of $\mathbb{S} \times \mathbb{V}$ whose elements are called $\mathfrak{B}_1$-arcs, $\mathfrak{B}_2$-arcs, ..., $\mathfrak{B}_q$-arcs respectively.*

When we consider multi-dimensional digraphs $\mathfrak{D}^{(n)}$, we write arc in the following form: $(v_i, v_j)_{\mathfrak{A}_q}$ or $(v_i, v_j)_{\mathfrak{B}_q}$, where the first vertex v_i is its tail, the second vertex v_j is its head and $\mathfrak{A}_q$ or $\mathfrak{B}_q$ corresponding to the matrix $\mathbf{A}_q$ and $\mathbf{B}_q$ respectively.

Remark 1 *$\mathfrak{A}_q$ -arcs and $\mathfrak{B}_q$ -arcs, are drawn by the other colour or line style. In this paper, $\mathfrak{A}_1$ -arc and $\mathfrak{B}_1$-arc is drawn by the solid line, $\mathfrak{A}_2$-arc and $\mathfrak{B}_2$ -arc is drawn by the dashed line and $\mathfrak{A}_3$ -arc and $\mathfrak{B}_3$ -arc is drawn by dotted line.*

Example 1 *For the system described by the matrices*

$$\underbrace{\begin{bmatrix} 0 & 0 & 1 \\ 1 & 0 & 0 \\ 0 & 1 & 0 \end{bmatrix}}_{\mathbf{A}_1}, \underbrace{\begin{bmatrix} 1 & 0 & 0 \\ 0 & 0 & 1 \\ 1 & 0 & 0 \end{bmatrix}}_{\mathbf{A}_2}, \underbrace{\begin{bmatrix} 0 & 0 & 0 \\ 1 & 0 & 1 \\ 0 & 0 & 1 \end{bmatrix}}_{\mathbf{A}_3}, \underbrace{\begin{bmatrix} 1 & 0 \\ 0 & 0 \\ 0 & 0 \end{bmatrix}}_{\mathbf{B}_1}, \underbrace{\begin{bmatrix} 0 & 0 \\ 0 & 0 \\ 0 & 1 \end{bmatrix}}_{\mathbf{B}_2}, \underbrace{\begin{bmatrix} 0 & 0 \\ 0 & 0 \\ 0 & 0 \end{bmatrix}}_{\mathbf{B}_3} \tag{7}$$

we can draw three-dimensional digraphs $\mathfrak{D}^{(3)}$ consisting of vertices v_1, v_2, v_3 and source s_1, s_2. A three-dimensional digraph corresponding to system (7) is presented in Figure 1.

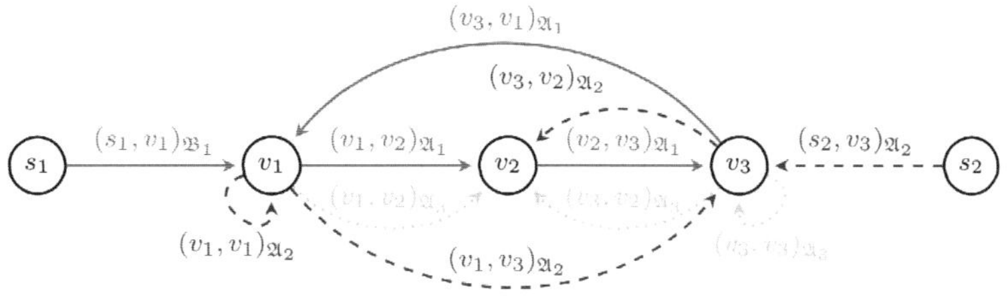

Figure 1. Three-dimensional digraph $\mathfrak{D}^{(3)}$ corresponding to matrices (7)

We present below some basic notions from the graph theory which are used in farther considerations [27], [28], [29], [30]. A walk in a multi-dimensional digraph $\mathfrak{D}^{(n)}$ is a nite sequence of arcs in which every two vertices v_i and v_j are adjacent or identical. A walk in which all of the arcs are distinct is called a path. The path, that goes through all vertices, is called a nite path. If the initial and the terminal vertices of the path are the same, then the path is called a cycle.

2.4. Problem Formulation

For the given transfer matrix (3), determine a minimal positive realisation of the system (1a)–(1b) using the multi-dimensional $\mathfrak{D}^{(n)}$ digraphs theory. The dimension of the system must be the minimal among possible.

3. Problem Solution

The essence of the proposed method for determining a minimal realisation for a positive linear two-dimensional general model with delays described by the equations (1a)–(1b) will be presented in single-input single-output (SISO) systems. The transfer matrix (3) can be written in the following form:

$$\begin{aligned}\mathbf{T}(z_1,z_2) = & \quad \mathbf{C}\left[\mathbf{I}z_1z_2 - \sum_{\substack{k_1=-1\\k_1+l_1>-2}}^{q_1}\sum_{l_1=-1}^{q_2}\mathbf{A}_{k_1l_1}z_1^{-k_1}z_2^{-l_1}\right]^{-1}\left[\sum_{\substack{k_2=-1\\k_2+l_2>-2}}^{p_1}\sum_{l_2=-1}^{p_2}\mathbf{B}_{k_2l_2}z_1^{-k_2}z_2^{-l_2}\right]+\mathbf{D} = \\ = & \quad \frac{\mathbf{N}(z_1,z_2)}{d(z_1,z_2)},\end{aligned} \tag{8}$$

where

$$\begin{aligned}&\mathbf{A}_{-1,0}=\mathbf{A}_{00}^1,\ \mathbf{A}_{0,-1}=\mathbf{A}_{00}^2,\ \mathbf{A}_{-1,1}=\mathbf{A}_{01}^1,\ \mathbf{A}_{1,-1}=\mathbf{A}_{10}^2,\\ &\mathbf{A}_{k_1l_1}=\mathbf{A}_{k_1l_1}^0+\mathbf{A}_{k_1+1,l_1}^1+\mathbf{A}_{k_1,l_1+1}^2,\ k_1=0,1,\dots,q_1-1;\ l_1=0,1,\dots,q_2-1,\\ &\mathbf{A}_{q_1-1,q_2}=\mathbf{A}_{q_1-1,q_2}^0+\mathbf{A}_{q_1,q_2}^1,\ \mathbf{A}_{q_1,q_2-1}=\mathbf{A}_{q_1,q_2-1}^0+\mathbf{A}_{q_1,q_2}^2,\ \mathbf{A}_{q_1,q_2}=\mathbf{A}_{q_1,q_2}^0,\\ &\mathbf{B}_{-1,0}=\mathbf{B}_{00}^1,\ \mathbf{B}_{0,-1}=\mathbf{B}_{00}^2,\ \mathbf{B}_{-1,1}=\mathbf{B}_{01}^1,\ \mathbf{B}_{1,-1}=\mathbf{B}_{10}^2,\\ &\mathbf{B}_{k_2l_2}=\mathbf{B}_{k_2l_2}^0+\mathbf{B}_{k_2+1,l_2}^1+\mathbf{B}_{k_2,l_2+1}^2,\ k_2=0,1,\dots,q_1-1;\ l_2=0,1,\dots,q_2-1,\\ &\mathbf{B}_{q_1-1,q_2}=\mathbf{B}_{q_1-1,q_2}^0+\mathbf{B}_{q_1,q_2}^1,\ \mathbf{B}_{q_1,q_2-1}=\mathbf{B}_{q_1,q_2-1}^0+\mathbf{B}_{q_1,q_2}^2,\ \mathbf{B}_{q_1,q_2}=\mathbf{B}_{q_1,q_2}^0.\end{aligned} \tag{9}$$

From (8) we have

$$\mathbf{D} = \lim_{z_1\to\infty,z_2\to\infty}\mathbf{T}(z_1,z_2) = \begin{bmatrix} n_{N_1,N_2}^{ij} & \cdots & n_{N_1,N_2}^{1m} \\ \vdots & \ddots & \vdots \\ n_{N_1,N_2}^{p1} & \cdots & n_{N_1,N_2}^{pm}\end{bmatrix}, \tag{10}$$

since $\lim_{z_1\to\infty,\ z_2\to\infty}[\mathbf{I}z_1z_2 - \sum_{k_1=-1}^{q_1}\sum_{l_1=-1}^{q_2} z_1^{-k_1}z_2^{-l_1}\mathbf{A}_{k_1l_1}z_1^{-k_1}z_2^{-l_1}]^{-1} = 0$. The strictly proper transfer function is given by the equation

$$\mathbf{T}_{sp}(z_1,z_2) = \mathbf{T}(z_1,z_2)-\mathbf{D} = \frac{\widetilde{\mathbf{N}}(z_1,z_2)}{d(z_1,z_2)} = \frac{1}{d(z_1,z_2)}\begin{bmatrix}\widetilde{n}_{11}(z_1,z_2) & \dots & \widetilde{n}_{1m}(z_1,z_2)\\ \vdots & \ddots & \vdots \\ \widetilde{n}_{p1}(z_1,z_2) & \dots & \widetilde{n}_{pm}(z_1,z_2)\end{bmatrix}, \tag{11}$$

where

$$\begin{aligned}\widetilde{n}_{ij}(z_1,z_2) = & \sum_{\substack{k=0\\k+l<N_1+N_2}}^{N_1}\sum_{l=0}^{N_2}\widetilde{n}_{kl}^{ij}z_1^kz_2^l = \\ = & \ \widetilde{n}_{N_1,N_2-1}^{ij}z_1^{N_1}z_2^{N_2-1}+\widetilde{n}_{N_1-1,N_2}^{ij}z_1^{N_1-1}z_2^{N_2}+\ldots+\widetilde{n}_{10}^{ij}z_1+\widetilde{n}_{01}^{ij}z_2+\widetilde{n}_{00}^{ij},\end{aligned} \tag{12}$$

and a characteristic polynomial is described by the equation (5).

After multiplying the nominator and denominator of the transfer function (3) by $z_1^{-nq_1} z_2^{-nq_2}$, we obtain:

$$\begin{aligned} \widetilde{n}_{ij}(z_1, z_2) = \quad & \widetilde{n}^{ij}_{N_1,N_2-1} z_1^{N_1-nq_1} z_2^{N_2-nq_2-1} + \widetilde{n}^{ij}_{N_1-1,N_2} z_1^{N_1-nq_1-1} z_2^{N_2-nq_2} + \dots \\ & \dots + \widetilde{n}^{ij}_{10} z^{1-nq_1} z_2^{-nq_2} + \widetilde{n}^{ij}_{01} z_1^{-nq_1} z_2^{1-nq_2} + \widetilde{n}^{ij}_{00} z_1^{-nq_1} z_2^{-nq_2}, \end{aligned} \tag{13}$$

$$\begin{aligned} d(z_1^{-1}, z_2^{-1}) = \quad & 1 - d_{nq_1,nq_2-1} z_2^{-1} - d_{nq_1-1,nq_2} z_1^{-1} - \dots \\ & \dots - d_{10} z_1^{1-nq_1} z_2^{-nq_2} - d_{01} z_1^{-nq_1} z_2^{1-nq_2} - d_{00} z_1^{-nq_1} z_2^{-nq_2}. \end{aligned} \tag{14}$$

In the first step, we must find matrices $\mathbf{A}_{k_1 l_1} \in \mathbb{R}_+^{n\times n}$, $k_1 = -1, 0, 1, \cdots, q_1$; $l_1 = -1, 0, 1, \cdots, q_2$; using a decomposition characteristic polynomial (14). We decompose the polynomial into a set of simple monomials

$$d(z_1, z_2) = (1 - d_{M_1}(z_1, z_2)) \cup (1 - d_{M_2}(z_1, z_2)) \cup \dots \cup \big(1 - d_{M_p}(z_1, z_2)\big) \tag{15}$$

where p is a number of simple monomials in the characteristic polynomial (14). For each simple monomial, we create digraphs representations. Then we can determine all possible characteristic polynomial realisations using all combinations of the digraphs monomial representations. Finally, we combine received digraphs in one digraph which is corresponding to a characteristic polynomial $d(z_1, z_2)$.

Theorem 2 *There exists positive matrices* $\mathbf{A}_{k_1 l_1} \in \mathbb{R}_+^{n\times n}$, $k_1 = -1, 0, 1, \cdots, q_1$; $l_1 = -1, 0, 1, \cdots, q_2$ *of the linear two-dimensional positive system with delays described by the equations* (1a)–(1b) *corresponding to the characteristic polynomial* $d(z_1, z_2)$ *if*

- ***(C1)*** *the coefficients of the polynomial* $d(z_1, z_2)$ *are non-negative.*
- ***(C2)*** *the sets* $\mathbb{D}_{M_1} \cap \mathbb{D}_{M_2} \cap \cdots \cap \mathbb{D}_{M_p}$, *where* p *is a number of simple monomials in polynomial* $d(z_1, z_2)$ *corresponding to multi-dimensional digraphs are not disjoint.*
- ***(C3)*** *the obtained multi-dimensional digraph does not have additional cycles.*

Proof: *Condition (C1):* This condition came from an internal positivity of the system (Definition 1) and must be satisfied if we consider positive systems. If coefficients of the characteristic polynomial are negative, then in state matrices negative elements appear. The proof is given in paper [25]. *Condition (C2):* The sets $\mathbb{D}_{M_1} \cap \mathbb{D}_{M_2} \cap \cdots \cap \mathbb{D}_{M_p}$, where p is number of simple monomials in polynomial $d(z_1, z_2)$ are disjoint if $\mathbb{D}_{M_1} \cap \mathbb{D}_{M_2} \cap \cdots \cap \mathbb{D}_{M_p} = \emptyset$ then we have a digraph whose vertices can be divided into two disjoint sets. It means that we obtain an additional simple monomial in a characteristic polynomial $d(z_1, z_2)$. In this situation, we obtain a new characteristic polynomial $\bar{d}(z_1, z_2)$, $d(z_1, z_2) \neq \bar{d}(z_1, z_2)$. *Condition (C3):* Each monomial is represented by one cycle. If after combining all digraphs (each corresponding to one monomial) we obtain an additional cycle, this means that in the polynomial an additional simple monomial appears. □

Using the Theorem 2 we can construct the Algorithm 1.

Let us assume that the matrix $\mathbf{B}_{k_2 l_2}$, $k_2 = -1, 0, 1, \cdots, p_1$, $l_2 = -1, 0, 1, \cdots, p_2$ and matrix $\mathbf{C}$ have the following form:

$$\mathbf{B}_{k_2 l_2}|_{k_2=-1,0,1,\cdots p_1\ l_2=-1,0,1,\cdots p_2;} = \begin{bmatrix} b_{11}^{k_2 l_2} & b_{12}^{k_2 l_2} & \dots & b_{1,m}^{k_2 l_2} \\ b_{21}^{k_2 l_2} & b_{22}^{k_2 l_2} & \dots & b_{2,m}^{k_2 l_2} \\ \vdots & \vdots & \ddots & \vdots \\ b_{n,1}^{k_2 l_2} & b_{n,2}^{k_2 l_2} & \dots & b_{n,m}^{k_2 l_2} \end{bmatrix}, \mathbf{C} = \begin{bmatrix} c_{11} & c_{12} & \dots & c_{1,n} \\ c_{21} & c_{22} & \dots & c_{2,n} \\ \vdots & \vdots & \ddots & \vdots \\ c_{p,1} & c_{p,2} & \dots & c_{p,n} \end{bmatrix}. \tag{16}$$

After determining the state matrix $\mathbf{A}_{k_1 l_1} \in \mathbb{R}_{+}^{n\times n}$, $k_1 = -1, 0, 1, \cdots, q_1$; $l_1 = -1, 0, 1, \cdots, q_2$ from the Algorithm 1 and inserting matrices (16) to the equation (8) we obtain the polynomial $\overline{n}_{ij}(z_1, z_2)$. After comparing variables with the same power of z_1 and z_2 polynomials $\widetilde{n}_{ij}(z_1, z_2) = \overline{n}_{ij}(z_1, z_2)$ we receive the set of equations. After solving the equation and inserting to (9), we obtain matrices (6).

Algorithm 1 *DetermineMinimalRealisation*()

Determine matrix $\mathbf{D}$ using (10);
Determine strictly proper transfer function/matrix using (11);
if $\mathbf{D} => 0$ **then**
    monomial $= 1$;
    Determine number of *cycles* in characteristic polynomial (14);
    for *monomial* $= 1$ **to** *cycles* **do**
        Determine multi-dimensional digraph $\mathfrak{D}^{(n)}$ for all monomial;
        MonomialRealisation(*monomial*);
    end for
    for *monomial* $= 1$ **to** *cycles* **do**
        Determine digraph as a combination of the digraph monomial representation;
        PolynomialRealisation(*monomial*);
        if *PolynomialRealisation* ! = *cycles* **then**
            Digraph contains additional cycles or digraph contains disjoint union;
            BREAK
        else if *PolynomialRealisation* == *cycles* **then**
            Digraph satisfies characteristic polynomial;
            Determine weights of the arcs in digraph;
            Write state matrix $\mathbf{A}_{k_1 l_1}|_{k_1=-1,0,1,\cdots,q_1;\ l_1=-1,0,1,\cdots,q_2}$;
            return (*PolynomialRealisation*, $\mathbf{A}_{k_1 l_1}|_{k_1=-1,0,1,\cdots,q_1,\ l_1=-1,0,1,\cdots,q_2}$);
        end if
    end for
    for *PolynomialRealisation* $= 1$ **to** j **do**
        Input - state matrix $\mathbf{A}_{k_1 l_1}|_{k_1=-1,0,1,\cdots,q_1;\ l_1=-1,0,1,\cdots,q_2}$;
        Determine polynomial $\widetilde{n}(z_1, z_2)$;
        Compare variables with the same power of the $n(z_1, z_2)$;
        Solve non-linear set of the equations;
        if Matrix $\mathbf{B}_{k_2 l_2}|_{k_2=-1,0,1,\cdots p_1\ l_2=-1,0,1,\cdots p_2} \geqslant 0$ AND matrix $\mathbf{C} \geqslant 0$ **then**
            return (*MinimalRealisation*, $\mathbf{A}, \mathbf{B}, \mathbf{C}$);
        else if Matrix $\mathbf{B}_{k_2 l_2}|_{k_2=-1,0,1,\cdots p_1\ l_2=-1,0,1,\cdots p_2} < 0$ OR matrix $\mathbf{C} < 0$ **then**
            Realisation is not positive;
            BREAK
        end if
    end for
    BREAK
else if $\mathbf{D} < 0$ **then**
    Realisation is not positive;
    BREAK
end if

4. Numerical Example

For the given transfer function

$$T(z_1, z_2) = \tag{17}$$
$$= \frac{2z_1^3 z_2^6 + 5z_1^3 z_2^5 + z_1^3 z_2^4 + 4z_1^2 z_2^6 + 4z_1^2 z_2^5 + z_1^2 z_2^4 + 3z_1^2 z_2^3 + 3z_1 z_2^6 + z_1 z_2^3 + 2z_1 z_2^2 + 2z_1 + 2}{z_1^3 z_2^6 - 2z_1^2 z_2^5 - z_1^2 z_2^3 - z_1 z_2^2 - z_1 z_2 - 1},$$

find all possible realisations of the two-dimensional positive linear system with $(q_1 = 1, q_2 = 1)$ delays in state vector and $(p_1 = 1, p_2 = 0)$ delays in input vector using multi-dimensional digraph theory.

Solution. In the first step, after using formula (10) and transfer function (17), we compute

$$\mathbf{D} = \lim_{z_1 \to \infty,\, z_2 \to \infty} T(z_1, z_2) = 2 \tag{18}$$

and a strictly proper transfer function (using (11) and (18)):

$$\begin{aligned} T_{sp}(z_1, z_2) &= T(z_1, z_2) - \mathbf{D} = \frac{n(z_1, z_2) - 2 \cdot d(z_1, z_2)}{d(z_1, z_2)} = \\ &= \frac{5z_1^3 z_2^5 + 4z_1^2 z_2^6 + 3z_1 z_2^6 + z_1^3 z_2^4 + z_1^2 z_2^4 + z_1^2 z_2^3 + z_1 z_2^3}{z_1^3 z_2^6 - 2z_1^2 z_2^5 - z_1^2 z_2^3 - z_1 z_2^2 - z_1 z_2 - 1} = \frac{\widetilde{n}(z_1, z_2)}{d(z_1, z_2)}. \end{aligned} \tag{19}$$

In the next step, after using (8), we write the transfer function in the following form:

$$\begin{aligned} \mathbf{T}_{sp}(z_1, z_2) = {} & \mathbf{C}\left[\mathbf{I} - \mathbf{A}_{-1,0} z_2^{-1} - \mathbf{A}_{-1,1} z_2^{-2} - \mathbf{A}_{0,-1} z_1^{-1} - \mathbf{A}_{0,0} z_1^{-1} z_2^{-1} - \mathbf{A}_{0,1} z_1^{-1} z_2^{-2} - \right. \\ & \left. \mathbf{A}_{1,-1} z_1^{-2} - \mathbf{A}_{1,0} z_1^{-2} z_2^{-1} - \mathbf{A}_{1,1} z_1^{-2} z_2^{-2}\right] \times \\ & \times \left(\mathbf{B}_{-1,0} z_2^{-1} + \mathbf{B}_{0,-1} z_1^{-1} + \mathbf{B}_{0,0} z_1^{-1} z_2^{-1} + \mathbf{B}_{1,-1} z_1^{-2} + \mathbf{B}_{1,0} z_1^{-2} z_2^{-1}\right) + \mathbf{D} \end{aligned} \tag{20}$$

After using (20), we determine possible weights from which we will build digraphs:

$$z_1^{-1},\quad z_1^{-2},\quad z_2^{-1},\quad z_2^{-2},\quad z_1^{-1} z_2^{-1},\quad z_1^{-1} z_2^{-2},\quad z_1^{-2} z_2^{-1},\quad z_1^{-2} z_2^{-2}. \tag{21}$$

After multiplying the nominator and denominator of the strictly proper transfer function (19) by $z_1^{-3} z_2^{-6}$, we obtain:

$$T_{sp}(z_1, z_2) = \frac{5z_2^{-1} + 4z_1^{-1} + 3z_1^{-2} + z_2^{-2} + z_1^{-1} z_2^{-2} + z_1^{-1} z_2^{-3} + z_1^{-2} z_2^{-3}}{1 - 2z_1^{-1} z_2^{-1} - z_1^{-1} z_2^{-3} - z_1^{-2} z_2^{-4} - z_1^{-2} z_2^{-5} - z_1^{-3} z_2^{-6}} = \frac{\widetilde{n}(z_1, z_2)}{d(z_1, z_2)}, \tag{22}$$

where $d(z_1, z_2)$ is the characteristic polynomial. The proposed method finds state matrices $\mathbf{A}_{k_1 l_1}$, $k_1 = -1, 0, 1$; $l_1 = -1, 0, 1$ using the decomposing characteristic polynomial (15).

Using (21) and characteristic polynomial $d(z_1, z_2)$, we write initial conditions:

- number of vertices in digraph: $vertices = 3$;
- number of colours and line style in digraph: $colour = 4$, $line\_style = 4$;
- possible weights from which we will build digraphs: $(v_i, v_j)_{\mathfrak{A}_{-1,0}} z_2^{-1}$ (blue line), $(v_i, v_j)_{\mathfrak{A}_{-1,1}} z_2^{-2}$ (blue dashed line), $(v_i, v_j)_{\mathfrak{A}_{0,-1}} z_1^{-1}$ (red line), $(v_i, v_j)_{\mathfrak{A}_{0,0}} z_1^{-1} z_2^{-1}$ (black line), $(v_i, v_j)_{\mathfrak{A}_{0,1}} z_1^{-1} z_2^{-2}$ (blue dotted line), $(v_i, v_j)_{\mathfrak{A}_{1,-1}} z_1^{-2}$ (red dashed line), $(v_i, v_j)_{\mathfrak{A}_{1,0}} z_1^{-2} z_2^{-1}$ (red dotted line), $(v_i, v_j)_{\mathfrak{A}_{1,1}} z_1^{-2} z_2^{-2}$ (black dashed line);
- monomials: $M_1 = 1 - 2z_1^{-1} z_2^{-1}$, $M_2 = 1 - z_1^{-1} z_2^{-3}$, $M_3 = 1 - z_1^{-2} z_2^{-4}$, $M_4 = 1 - z_1^{-2} z_2^{-5}$, $M_5 = z_1^{-3} z_2^{-6}$.

Then, we determine all possible realisations of the:

- monomial M_1 presented in: Figure 2(a), for $i = 1, 2, 3$; Figure 2(b), for $i, j = 1, 2, 3$, $i \neq j$;,
- monomial M_2 presented in: Figure 2(c), for $i, j = 1, 2, 3$; $i \neq j$; Figure 2(d), for $i, j = 1, 2, 3$; $i \neq j$; Figure 2(e), for $i, j, k = 1, 2, 3$; $i \neq j$, $j \neq k$, $i \neq k$;
- monomial M_3 presented in Figure 2(f), for $i, j, k = 1, 2, 3$; $i \neq j$, $j \neq k$, $i \neq k$; Figure 2(g), for $i, j = 1, 2, 3$, $i \neq j$;

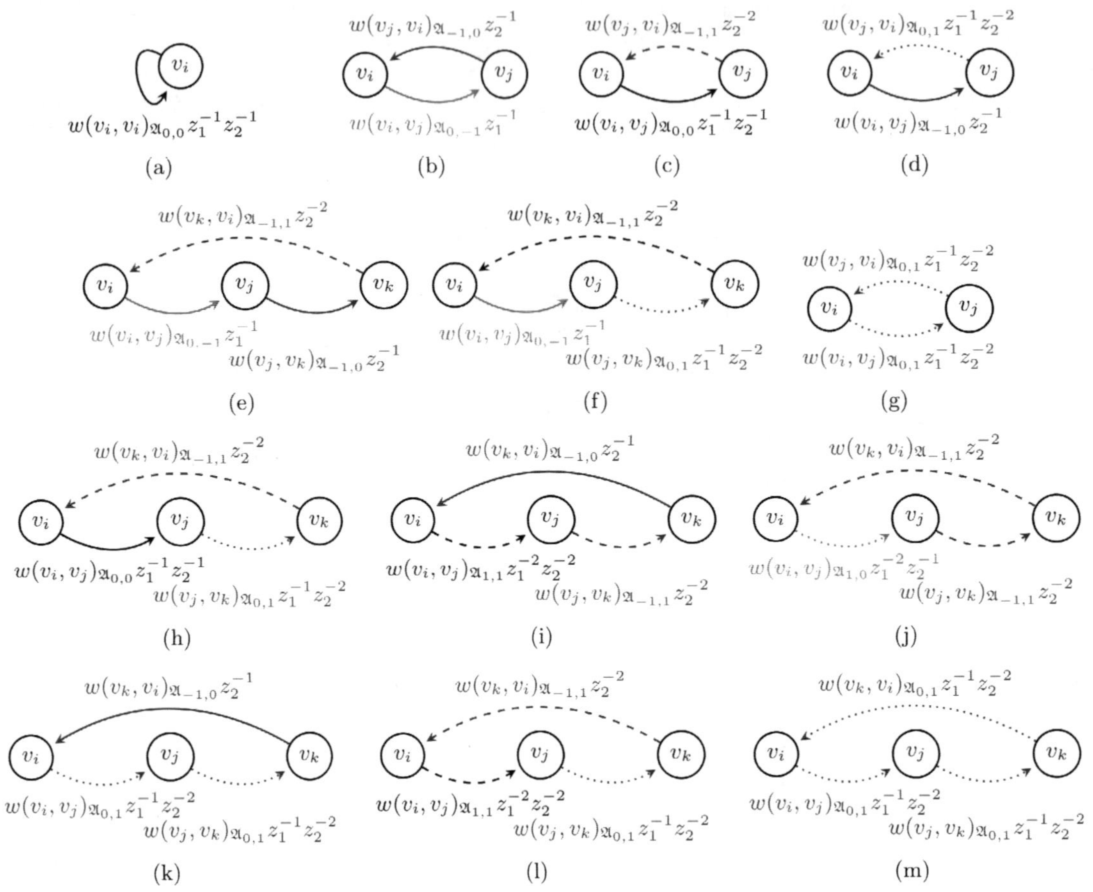

Figure 2. Multi-dimensional digraphs corresponding to the monomials: M_1:2(a), 2(b); M_2: 2(c), 2(d), 2(e); M_3: 2(f), 2(g); M_4: 2(h), 2(i), 2(j), 2(k); M_5: 2(l), 2(m).

- monomial M_4 presented in: Figure 2(h), Figure 2(i), Figure 2(j) and Figure 2(k) for $i, j, k = 1, 2, 3$; $i \neq j$, $j \neq k$, $i \neq k$;
- monomial M_5 presented in: Figure 2(l) and Figure 2(m) for $i, j, k = 1, 2, 3$; $i \neq j$, $j \neq k$, $i \neq k$;

In the defined problem, we assume that dimension of the state matrices must be the minimal among possible. Taking into account this condition, we have 23 328 variants of possible minimal polynomial realisations consisting of three vertices.

Consider the following two possible realisations:

Realisation 1 presented in Figure 3 consists of the digraphs presented in: Figure 2(a) for $i = 3$; Figure 2(d) for $i = 2$, $j = 1$; Figure 2(g) for $i = 1$, $j = 2$; Figure 2(k) for $i = 1$, $j = 2$, $k = 3$; Figure 2(m) for $i = 1$, $j = 2$, $k = 3$. After using Theorem 2, we can check the conditions. The coefficients of the characteristic polynomial satisfy the condition (C1). To verify the second condition, we must compare sets $\mathbb{D}_1$, $\mathbb{D}_2$, $\mathbb{D}_3$, $\mathbb{D}_4$ and $\mathbb{D}_5$ corresponding to representation of all simple monomial digraphs. The sets are disjoint $\mathbb{D}_1 \cap \mathbb{D}_2 \cap \mathbb{D}_3 \cap \mathbb{D}_4 \cap \mathbb{D}_5 = \emptyset$. Described realisation does not satisfy the condition (C2). The obtained digraphs presented in Figure 4 do not appear in additional cycles. The Condition (C3) is satisfied. The realisation does not satisfy all conditions and is rejected.

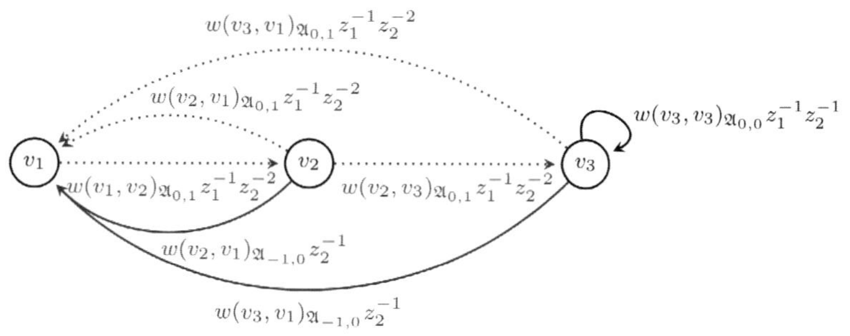

Figure 3. Multi-dimensional digraph corresponding to characteristic polynomial $d(z_1, z_2)$

Realisation 2 presented in Figure 4 consists of the digraphs presented in: Figure 2(a) for $i = 1$; Figure 2(d) for $i = 2$, $j = 1$; Figure 2(g) for $i = 1$, $j = 2$; Figure 2(k) for $i = 1$, $j = 2$, $k = 3$; Figure 2(m) for $i = 1$, $j = 2$, $k = 3$. After using Theorem 2, we can check the conditions. The coefficients of the characteristic polynomial satisfy the condition (C1). To verify the second condition, we must compare sets $\mathbb{D}_1$, $\mathbb{D}_2$, $\mathbb{D}_3$, $\mathbb{D}_4$ and $\mathbb{D}_5$ corresponding to representation of all simple monomial digraphs. The sets are not disjoint $\mathbb{D}_1 \cap \mathbb{D}_2 \cap \mathbb{D}_3 \cap \mathbb{D}_4 \cap \mathbb{D}_5 = \{v_1\}$ (vertex v_1 in the Figure 3). Described realisation satisfies the condition (C2). The obtained digraphs presented in Figure 3 do not appear in additional cycles. The condition (C3) is satisfied. The realisation does satisfy all conditions and is correct.

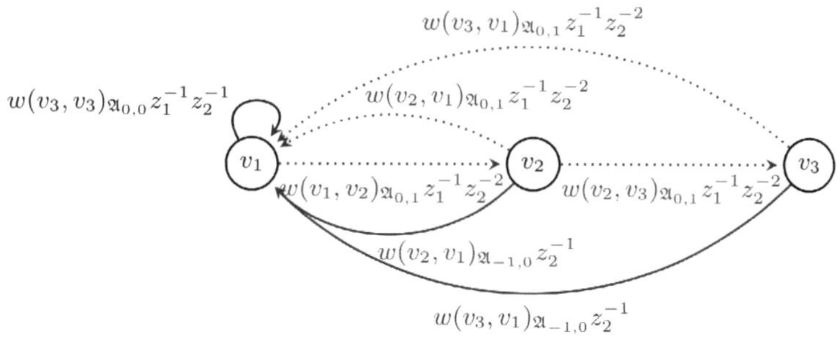

Figure 4. Multi-dimensional digraph corresponding to characteristic polynomial $d(z_1, z_2)$

From the obtained digraphs, we can write state matrices $\mathbf{A}_{q_1,q_2}$, $q_1 = -1, 0, 1$ and $q_2 = -1, 0, 1$ in the form:

$$\begin{aligned}
\mathbf{A}_{0,0} &= \begin{bmatrix} w(v_1, v_1)\mathfrak{a}_{0,0} & 0 & 0 \\ 0 & 0 & 0 \\ 0 & 0 & 0 \end{bmatrix} = \begin{bmatrix} 2 & 0 & 0 \\ 0 & 0 & 0 \\ 0 & 0 & 0 \end{bmatrix}, \\
\mathbf{A}_{-1,0} &= \begin{bmatrix} 0 & w(v_2, v_1)\mathfrak{a}_{-1,0} & w(v_3, v_1)\mathfrak{a}_{-1,0} \\ 0 & 0 & 0 \\ 0 & 0 & 0 \end{bmatrix} = \begin{bmatrix} 0 & 1 & 1 \\ 0 & 0 & 0 \\ 0 & 0 & 0 \end{bmatrix}, \\
\mathbf{A}_{0,1} &= \begin{bmatrix} 0 & w(v_2, v_1)\mathfrak{a}_{0,1} & w(v_3, v_1)\mathfrak{a}_{0,1} \\ w(v_1, v_2)\mathfrak{a}_{0,1} & 0 & 0 \\ 0 & w(v_2, v_3)\mathfrak{a}_{0,1} & 0 \end{bmatrix} = \begin{bmatrix} 0 & 1 & 1 \\ 1 & 0 & 0 \\ 0 & 1 & 0 \end{bmatrix}.
\end{aligned} \tag{23}$$

After inserting matrices (16) and (23) to the equation (20), we obtain the polynomial $\overline{n}(z_1, z_2)$. After comparison of the coefficients of the same powers of z_1 and z_2 polynomials $\widetilde{n}(z_1, z_2) = \overline{n}(z_{1,2})$, we receive the set of the equations. After solving them, we obtain the following matrices:

$$\begin{aligned}
\mathbf{C} &= \begin{bmatrix} c_{11} & 0 & 0 \end{bmatrix} = \begin{bmatrix} 1 & 0 & 0 \end{bmatrix}; \\
\mathbf{B}_{-1,0} &= \begin{bmatrix} b_{11}^{-1,0} \\ 0 \\ b_{31}^{-1,0} \end{bmatrix} = \begin{bmatrix} 5 \\ 0 \\ 1 \end{bmatrix}; \quad \mathbf{B}_{0,-1} = \begin{bmatrix} b_{11}^{0,-1} \\ 0 \\ 0 \end{bmatrix} = \begin{bmatrix} 4 \\ 0 \\ 0 \end{bmatrix}; \\
\mathbf{B}_{0,0} &= \begin{bmatrix} 0 \\ 0 \\ b_{31}^{0,0} \end{bmatrix} = \begin{bmatrix} 0 \\ 0 \\ 1 \end{bmatrix}; \quad \mathbf{B}_{1,-1} = \begin{bmatrix} b_{11}^{1,-1} \\ 0 \\ 0 \end{bmatrix} = \begin{bmatrix} 3 \\ 0 \\ 0 \end{bmatrix}; \quad \mathbf{B}_{1,0} = \begin{bmatrix} 0 \\ 0 \\ 0 \end{bmatrix}
\end{aligned} \tag{24}$$

The desired positive realisation of the (17) is given by (18), (23) and (24).

5. Concluding Remarks

The paper presents a method, based on the multi-dimensional digraph theory, for finding the complete set of multi-dimensional characteristic polynomial realisations, that can be used to solve the minimal positive realisation problem of a two-dimensional linear system with delays described by the general model which includes single-input and single-output (SISO). The difference between the proposed algorithm in this paper and currently used methods based on canonical forms of the system (i.e. constant matrix forms) is the creation of not one or few minimal realisations, but a set of possible minimal realisations. A number of colours used in the digraphs, increases with the number of delays (q_1, q_2) in the state vector, is a huge problem. It should be noted, that a large number of colours generates an additional number of potential solutions that must be analysed by the algorithm. This leads to a problem of allocating a cube on the graphics card.

Further work includes extension of the algorithm to find all possible solutions, solving the realisation problem, reachability and controllability of systems using the fast graph-based method. There is also a very difficult open problem of the analysis of systems dynamics for realisations on a different number of nodes in digraphs (not only minimal number of nodes). Currently, the method of determining a positive realisation using a parallel computing method and digraphs methods is optimised, and in the next step it will be implemented in a memory-efficient way.

Acknowledgments

Research has been financed with the funds of the Statutory Research of 2015.

References

[1] Roesser R B 1975 *IEEE Trans. Austom. Contr.* 1 – 10
[2] Fornasini E and Marchesini G 1976 *IEEE Trans, Autom. Contr.* **21**(4) 481–491
[3] Fornasini E and Marchesini G 1978 *Math. Sys. Theory* **12**(1) 59–72
[4] Kurek J 1985 *IEEE Trans. Austom. Contr.* **30**(6) 600 – 602
[5] Luenberger D G 1979 *Introduction to Dynamic Systems: Theory, Models, and Applications* (New York: Wiley) chap Positive linear systems
[6] Farina L and Rinaldi S 2000 *Positive linear systems: theory and applications* (New York: Wiley-Interscience, Series on Pure and Applied Mathematics)
[7] Benvenuti L, De Santis A and Farina L 2003 *Positive Systems* Lecture Notes on Control and Information Sciences 294 (Berlin: Springer-Verlag)
[8] Benvenuti L and Farina L 2004 *IEEE Transactions on Automatic Control* **49** 651–664
[9] Kaczorek T and Busłowicz M 2004 *Int. J. Appl. Math. Comput. Sci.* 181 – 187

[10] Kaczorek T 2007 *Polynomial and Rational Matrices* (London: Springer Verlag)
[11] Hryniów K and Markowski K A 2014 *Proceedings of 2014 15th International Carpathian Control Conference (ICCC)* pp 174–179 URL http://dx.doi.org/10.1109/CarpathianCC.2014.6843592
[12] Hryniów K and Markowski K A 2015 *Progress in Automation, Robotics and Measuring Techniques* (*Advances in Intelligent Systems and Computing* vol 350) ed Szewczyk R, Zieliski C and Kaliczyska M (Springer International Publishing) pp 63–72 ISBN 978-3-319-15795-5 URL http://dx.doi.org/10.1007/978-3-319-15796-2_7
[13] Markowski K A and Hryniów K 2015 *Proceedings of 7th Conference on Non-integer Order Calculus and its Applications7th Conference on Non-integer Order Calculus and its Applications* pp 105–118 ISBN 978-3-319-23038-2 URL http://dx.doi.org/10.1007/978-3-319-23039-9_9
[14] Hryniów K and Markowski K A 2015 *Proceedings of 20th International Conference on Methods and Models in Automation and Robotics, MMAR 2015, Miedzyzdroje, Poland, August 24–27* pp 110–115 URL http://dx.doi.org/10.1109/MMAR.2015.7283856
[15] Markowski K A and Hryniów K 2015 *Proceedings of 2nd IEEE International Conference on Cybernetics, CD-ROM, Gdynia, Poland, June 24-26, 2015* pp 172–177 URL http://dx.doi.org/10.1109/CYBConf.2015.7175927
[16] Markowski K A 2015 *Proceedings of 19th International Conference on System Theory, Control and Computing* pp 545–550 ISBN 978-1-4799-8481-7
[17] Markowski K A 2015 *Procedings of the International Symposium on Fractional Signal and Systems* ed Both-Rusu R (U.T. Press) pp 7–12 ISBN 978-606-737-084-3
[18] Hryniów K and Markowski K 2015 *Progress in Automation, Robotics and Measuring Techniques* (*Advances in Intelligent Systems and Computing* vol 350) ed Szewczyk R, Zieliński C and Kaliczyńska M (Springer International Publishing) pp 73–83 ISBN 978-3-319-15795-5 URL http://dx.doi.org/10.1007/978-3-319-15796-2_8
[19] Hryniów K and Markowski K A Accepted, 2015 *Proceedings of 20th International Conference on Methods and Models in Automation and Robotics, MMAR 2015, Miedzyzdroje, Poland, August 24–27* pp 1139–1144 URL http://dx.doi.org/10.1109/MMAR.2015.7284039
[20] Fornasini E and Valcher M E 1997 *Linear Algebra and Its Applications* **263** 275–310
[21] Fornasini E and Valcher M E 2003 *LCNIS* 297–304
[22] Fornasini E and Valcher M E 2005 *IEEE Transaction on Circuits and Systems I* 576–585
[23] Berman A, Neumann M and Stern R J 1989 *Nonnegative Matrices in Dynamic Systems* (New York: Wiley)
[24] Horn R A and Johnson C R 1991 *Topics in Matrix Analysis* (Cambridge Univ. Press)
[25] Kaczorek T 2008 *COMPEL* **28** 341–352
[26] Kaczorek T and Sajewski L 2014 *The Realization Problem for Positive and Fractional Systems* (Berlin: Springer International Publishing) URL http://dx.doi.org/10.1007/978-3-319-04834-5
[27] Bang-Jensen J and Gutin G 2009 *Digraphs: Theory, Algorithms and Applications* (London: Springer-Verlag)
[28] Godsil C and Royle G 2001 *Algebraic Graph Theory* (Springer Verlag)
[29] Foulds L R 1991 *Graph Theory Applications* (Springer Verlag)
[30] Wallis W D 2007 *A Beginner's Guide to Graph Theory* (Biiokhäuser)

On the Separation Theorem of Stochastic Systems in the Case Of Continuous Observation Channels with Memory

S V Rozhkova[1], V I Rozhkova[2], V V Lasukov[3] and E V Devyashina[4]
[1]Professor, Department of Higher Mathematics, Institute of Physics and Technology, National Research Tomsk Polytechnic University, Lenina 30, Tomsk 634050, Russia
[2]Senior Teacher, Department of Higher Mathematics, Institute of Physics and Technology,
National Research Tomsk Polytechnic University, Lenina 30, Tomsk 634050, Russia
[3]Assistant Professor, Department of Higher Mathematics, Institute of Physics and Technology,
National Research Tomsk Polytechnic University, Lenina 30, Tomsk 634050, Russia
[4]Student, Elite Technical Education,
National Research Tomsk Polytechnic University, Lenina 30, Tomsk 634050, Russia

E-mail: rozhkova@tpu.ru

Abstract. The article proves the separation theorem for optimal control of stochastic systems in the case when an observed continuous-time process possess memory of arbitrary ration relating to a state vector.

1. Introduction

The separation theorem [1] is the basis for a theory of optimal control by incompletely observable stochastic systems. Being fundamental theoretical results, it ensured solving of a number of important practical problems in a wide range of areas [2, 3]. With the results [4, 5], we provided a generalized separation theorem in the case when observations have memory of arbitrary ratio, i. e. they depend not only on current values but also on an arbitrary number of previous values of the system state vector.

Used notations: $P\{\cdot\}$ is event probability; $E\{\cdot\}$ is the mathematical expectation; normal (Gaussian) density with parameters a and b by $N\,\{a;\,b\}$; $|\cdot|$ and $tr[\cdot]$ are determinant and trace of matrix; D^{-1} is the inversion matrix of D; D^T denotes transpose of a matrix or a vector; $D>0$ $(D\geq 0)$ is positive (non-negative) definite matrix.

2. The problem statement

On a certain probability space $(\Omega, F, \mathbf{F}=(F_t)_{t\geq 0}, \mathbf{P})$ [6] the unobservable n-dimensional process x_t (state vector of system) and the observable l- dimensional process z_t with continuous time are determined by the stochastic differential equations (Ito's differential rule)

$$dx_t = f(t, x_t, u_t)dt + \Phi_1(t, x_t)dw_t, \quad t\in[0, \mathrm{T}], \tag{1}$$

$$dz_t = h(t, x_t, x_{\tau_1}, \cdots, x_{\tau_N}, z)dt + \Phi_2(t)dv_t, \tag{2}$$

where u_t is $r-$dimensional control vector, $0<\tau_N<\tau_{N-1}<\cdots<\tau_1<t_m\le t$, and $\tau_k=const$, $k=\overline{1;N}$. It is assumed: 1) w_t and v_t are r_1-dimensional and r_2-dimensional standard Wiener processes; 2) x_0, w_t, v_t are jointly independent; 3) $f(\cdot)$, $h(\cdot)$, $\Phi_1(\cdot)$, $\Phi_2(\cdot)$ are continuous functions for all arguments; 4) $Q(\cdot)=\Phi_1(\cdot)\Phi_1^T(\cdot)>0$, $R(\cdot)=\Phi_2(\cdot)\Phi_2^T(\cdot)>0$; 5) original density is set $p_0(x)=\partial P\{x_0\le x\}/\partial x$.

Task: to find control u_t^0 that provides optimum condition, on the class of $F_t^{z,\eta}$-measurable functional, where $u_t=u_t\left[z_0^t\right]$, $z_0^t=\{z_s;0\le s\le t\}$

$$J=E\left\{b(\mathrm{T},x_\mathrm{T})+\int_\tau^\mathrm{T}\Lambda(s,x_s,u_s)ds\middle|p_0(x)\right\}\to\min_{\{u_{t_0}^\mathrm{T}\}},\qquad(3)$$

where $\tau>\tau_1$, $u_{t_0}^T=\{u_s;\tau\le s\le \mathrm{T}\}$, $b(\cdot)\ge 0$, $\Lambda(\cdot)>0$.

To solve the set task, let us apply the method of sufficient coordinates [7], assuming that there is F_t^z - measurable process $\lambda_t\left[z_0^t\right]$ that fully characterizes posterior density

$$p_t(x)=\partial P\left\{x_t\le x\middle|z_0^t\right\}/\partial x\qquad(4)$$

of x_t system state vector, on the one hand. On the other hand, it can be found by means of $p_t(x)$.

Remark 1. We consider that the optimal control dates from moment $t_0>\tau_1$. An arbitrary F_t^z-measurable process is used as u_t on the interval $t\in[0,\tau]$.

3. Preliminary results

In accordance with the sufficient coordinates method, we introduce Bellman function

$$S(t,\lambda)=\min_{\{u_t^T\}}E\left\{b(\mathrm{T},x_\mathrm{T})+\int_t^\mathrm{T}\Lambda(t,x_{t'},u_{t'})dt'\middle|\lambda_t=\lambda\right\}.\qquad(5)$$

Theorem 1. Let the following conditions be met:

1) process λ_t is Markov diffusion process with characteristics

$$a(t,\lambda)=\lim_{\Delta t\to 0}\frac{1}{\Delta t}E\{\Delta\lambda_t|\lambda_t=\lambda\},\qquad(6)$$

$$D(t,\lambda)=\lim_{\Delta t\to 0}\frac{1}{\Delta t}E\left\{[\Delta\lambda_t][\Delta\lambda_t]^T\middle|\lambda_t=\lambda\right\},\qquad(7)$$

where $\Delta\lambda_t=\lambda_{t+\Delta t}-\lambda_t$.

2) Let process $\{x_t;\lambda_t\}$ be Markov process with transition probability density

$$p\left(t',x',\lambda'\middle|t,x,\lambda\right)=\partial^2P\left\{x_{t'}\le x',\lambda_{t'}\le\lambda'\middle|x_t=x,\lambda_t=\lambda\right\}/\partial x'\partial\lambda'.\qquad(8)$$

Then, Bellman equation for $S(t,\lambda)$ has the form

$$\min_{\{u\}}\left\{\frac{\partial S(t,\lambda)}{\partial t}+L_{t,\lambda}^*[S(t,\lambda)]+E\{\Lambda(t,x_t,u_t)|\lambda_t=\lambda\}\right\}=0,\qquad(9)$$

$$S(t,\lambda)|_{t=\mathrm{T}}=E\{b(\mathrm{T},x_\mathrm{T})|\lambda_\mathrm{T}=\lambda\},\qquad(10)$$

where $L^*_{t,\lambda}[\cdot]$ denotes inverse Kolmogorov operator that corresponds to process λ_t [6], i.e.

$$L^*_{t,\lambda}[S(t,\lambda)] = a^T(t,\lambda)\frac{\partial S(t,\lambda)}{\partial \lambda} + \frac{1}{2}tr\left[D(t,\lambda)\frac{\partial^2 S(t,\lambda)}{\partial \lambda^2}\right], \tag{11}$$

and the smallest value of performance criterion has the form $J^0 = S(\tau,\lambda)$.

Proof. Let

$$p_t(x|\lambda) = \partial P\{x_t \le x|\lambda_t = \lambda\}/\partial x. \tag{12}$$

Then, by expanding operator $E\{\cdot\}$ into (), and considering condition2), we obtain

$$S(t,\lambda) =$$

$$= \min_{\{u_t^T\}}\left\{\int p_t(x|\lambda)\left\{\int_t^T \Lambda(t',x',u)p(t',x',\lambda'|t,x,\lambda)dx'd\lambda'dt' + \int b(T,x')p(t',x',\lambda'|t,x,\lambda)dx'd\lambda'\right\}dx\right\}, \tag{13}$$

where $u_t^T = \{u_s; t \le s \le T\}$.

Lemma 1. Accurate to $o(\Delta t)$, the function $S(t,\lambda)$ satisfies Δt - recurrence equation

$$S(t,\lambda) = \min_{\{u\}}\left\{\int S(t+\Delta t,\lambda'')p(t+\Delta t,\lambda''|t,\lambda)d\lambda'' + (\Delta t)\int \Lambda(t,x,u)p_t(x|\lambda)dx + o(\Delta t)\right\}, \tag{14}$$

where

$$p(t+\Delta t,\lambda''|t,\lambda) = \partial P\{\lambda_{t+\Delta t} \le \lambda''|\lambda_t = \lambda\}/\partial\lambda'' \tag{15}$$

is transition probability density of Markov process λ_t.

Proof. Expressing interval $[t,T]$ as $[t,T] = [t,t+\Delta t] + [t+\Delta t,T]$, we obtain from (13)

$$S(t,\lambda) = \min_{\{u_t^{t+\Delta t}\}}\min_{\{u_{t+\Delta t}^T\}}\{S_1(t,t+\Delta t) + S_2(t+\Delta t,T)\}, \tag{16}$$

$$S_1(t,t+\Delta t) = \int p_t(x|\lambda)\left\{\int_t^{t+\Delta t}\Lambda(t',x',u)p(t',x',\lambda'|t,x,\lambda)dx'd\lambda'dt'\right\}dx, \tag{17}$$

$$S_2(t+\Delta t,T) =$$

$$= \int p_t(x|\lambda)\left\{\int_{t+\Delta t}^T \Lambda(t',x',u)p(t',x',\lambda'|t,x,\lambda)dx'd\lambda'dt' + \int b(T,x')p(t',x',\lambda'|t,x,\lambda)dx'd\lambda'\right\}dx, \tag{18}$$

So far as $p(t',x',\lambda'|t,x,\lambda) = \delta(x'-x,\lambda'-\lambda)$, when $\Delta t \downarrow 0$, it follows from (17), that

$$S_1(t,t+\Delta t) = (\Delta t)\int \Lambda(t,x,u)p_t(x|\lambda)dx + o(\Delta t). \tag{19}$$

Markov process transition probability density $\{x_t,\lambda_t\}$ (see condition 2) satisfies the Chapman-Kolmogorov equation [8]

$$p(t',x',\lambda'|t,x,\lambda) = \int p(t',x',\lambda'|t+\Delta t,x'',\lambda'')p(t+\Delta t,x'',\lambda''|t,x,\lambda)dx''d\lambda''. \tag{20}$$

With regard to (12), (15)

$$\int p(t+\Delta t,x'',\lambda''|t,x,\lambda)p_t(x|\lambda)dx = p_{t+\Delta t}(x''|\lambda'')p(t+\Delta t,\lambda''|t,\lambda), \tag{21}$$

Then, it follows from (18), (20), (21) that

$$S_2(t+\Delta t,\mathrm{T})=$$
$$=\int p(t+\Delta t,\lambda''|t,\lambda)\left\{\int p_{t+\Delta t}(x''|\lambda'')\left\{\int_{t+\Delta t}^{\mathrm{T}}\Lambda(t',x',u)p(t',x',\lambda'|t,x,\lambda)dx'd\lambda'dt'+\right.\right.$$
$$\left.\left.+\int b(\mathrm{T},x')p(t',x',\lambda'|t+\Delta t,x'',\lambda'')dx'd\lambda'\right\}dx''\right\}d\lambda''. \tag{22}$$

With regard to (19), it follows from (16), that

$$S(t,\lambda)=\min_{\{u\}}\left\{\min_{\{u_{t+\Delta t}^{\mathrm{T}}\}}S_2(t+\Delta t,\mathrm{T})+(\Delta t)\int\Lambda(t,x,u)p_t(x|\lambda)dx+o(\Delta t)\right\}. \tag{23}$$

Substituting (22) in (23) with regard to (13) leads to (14). Lemma 1 is proved.
We expand

$$S(t+\Delta t,\lambda'')=S(t+\Delta t,\lambda)+\left(\frac{\partial S(t+\Delta t,\lambda)}{\partial\lambda}\right)^T\Delta\lambda+\frac{1}{2}\Delta\lambda^T\frac{\partial^2 S(t+\Delta t,\lambda)}{\partial\lambda^2}\Delta\lambda+o\left([\Delta\lambda]^2\right) \tag{24}$$

Then, with regard to (6), (7), (15) it follows, that

$$\int S(t+\Delta t,\lambda)p(t+\Delta t,\lambda''|t,\lambda)d\lambda''=S(t+\Delta t,\lambda), \tag{25}$$

$$\int\left(\frac{\partial S(t+\Delta t,\lambda)}{\partial\lambda}\right)^T\Delta\lambda p(t+\Delta t,\lambda''|t,\lambda)d\lambda''=\left(\frac{\partial S(t+\Delta t,\lambda)}{\partial\lambda}\right)^T E\{\Delta\lambda_t|\lambda_t=\lambda\}=$$
$$=(\Delta t)\int\left(\frac{\partial S(t+\Delta t,\lambda)}{\partial\lambda}\right)^T a(t|\lambda)+o(\Delta t), \tag{26}$$

$$\int\Delta\lambda^T\frac{\partial^2 S(t+\Delta t,\lambda)}{\partial\lambda^2}\Delta\lambda p(t+\Delta t,\lambda''|t,\lambda)d\lambda''=tr\left[\frac{\partial^2 S(t+\Delta t,\lambda)}{\partial\lambda^2}E\left\{[\Delta\lambda_t][\Delta\lambda_t]^T|\lambda(t)=\lambda\right\}\right]=$$
$$=(\Delta t)tr\left[\int\frac{\partial^2 S(t+\Delta t,\lambda)}{\partial\lambda^2}D(t|\lambda)\right]+o(\Delta t), \tag{27}$$

$$\int o\left([\Delta\lambda]^2\right)p(t+\Delta t,\lambda''|t,\lambda)d\lambda''=E\left\{o\left([\Delta\lambda]^2\right)|\lambda_t=\lambda\right\}=o(\Delta t). \tag{28}$$

Substitution of (24) in (14) with regard to the latter formulae and (11) leads to the relation ($b^T g=g^T b$, $tr[BD]=tr[DB]$)

$$S(t,\lambda)=\min_u\left\{S(t+\Delta t,\lambda)+(\Delta t)L_{t,x}^*[S(t+\Delta t,\lambda)]+(\Delta t)\int\Lambda(t,x,u)p_t(x|\lambda)dx+o(\Delta t)\right\} \tag{29}$$

In (29) proceeding to the limit with $\Delta t\to 0$ and considering (12), we obtain equation (9). Boundary condition (10) and the expression for J^0 arise from (3), (5). Theorem 1 is proved.

Affirmation 1. Let

$$f(\cdot)=F(t)x_t+B(t)u_t,\quad \Phi_1(\cdot)=\Phi_1(t),\ p_0(x)=N\{x;\mu_0,\Gamma_0\}, \tag{30}$$

$$h(\cdot)=H_0(t)x_t+\sum_{k=1}^{N}H_k(t)x_{\tau_k}, \tag{31}$$

$$\mu(t) = E\{x_t \mid z_0^t\}, \tag{32}$$

$$\mu(\tau_k, t) = E\{x_{\tau_k} \mid z_0^t\}, \quad k = \overline{1;N}, \tag{33}$$

$$\Gamma(t) = E\{[x_t - \mu(t)][x_t - \mu(t)]^T \mid z_0^t\}, \tag{34}$$

$$\Gamma_{kk}(\tau_k, t) = E\{[x_{\tau_k} - \mu(\tau_k, t)][x_{\tau_k} - \mu(\tau_k, t)]^T \mid z_0^t\}, \tag{35}$$

$$\Gamma_{0k}(\tau_k, t) = E\{[x_t - \mu(t)][x_{\tau_k} - \mu(\tau_k, t)]^T \mid z_0^t\} \tag{36}$$

$$\Gamma_{ki}(\tau_k, \tau_i, t) = E\{[x_{\tau_k} - \mu(\tau_k, t)][x_{\tau_i} - \mu(\tau_i, t)]^T \mid z_0^t\} \tag{37}$$

$$k = \overline{1;N}, \quad i = \overline{2;N}, \quad i > k. \tag{38}$$

Then, for a posteriori density (4) property

$$p_t(x) = \mathrm{N}\{x; \mu(t), \Gamma(t)\} \tag{39}$$

The density parameters are specified by the equations

$$d\mu(t) = [F(t)\mu(t) + B(t)u_t]dt + \tilde{H}_0^T(t)R^{-1}(t)d\tilde{z}_t, \tag{40}$$

$$d\mu(\tau_k, t) = \tilde{H}_k^T(t)R^{-1}(t)d\tilde{z}_t, \tag{41}$$

$$d\Gamma(t)/dt = F(t)\Gamma(t) + \Gamma(t)F^T(t) - \tilde{H}_0^T(t)R^{-1}(t)\tilde{H}_0(t) + Q(t), \tag{42}$$

$$d\Gamma_{kk}(\tau_k, t)/dt = -\tilde{H}_k^T(t)R^{-1}(t)\tilde{H}_k(t), \tag{43}$$

$$d\Gamma_{0k}(\tau_k, t)/dt = F(t)\Gamma_{0k}(\tau_k, t) - \tilde{H}_0^T(t)R^{-1}(t)\tilde{H}_k(t), \tag{44}$$

$$d\Gamma_{ki}(\tau_k, \tau_i, t)/dt = -\tilde{H}_k^T(t)R^{-1}(t)\tilde{H}_i(t) \tag{45}$$

where

$$d\tilde{z}(t) = dz(t) - \left[H_0(t)\mu(t) - \sum_{j=1}^{N} H_j(t)\mu(\tau_j, t)\right]dt, \tag{46}$$

$$\tilde{H}_0(t) = H_0(t)\Gamma(t) + \sum_{j=1}^{N} H_j(t)\Gamma_{0j}^T(\tau_j, t), \tag{47}$$

$$\tilde{H}_k(t) = H_k(t)\Gamma_{kk}(\tau_k, t) + \sum_{j \neq k}^{N} H_j(t)\Gamma_{kj}^T(\tau_k, \tau_j, t). \tag{48}$$

This relation arises from [4].

Lemma 2. A sufficient coordinates vector is an optimal in a mean-root-square sense estimate $\mu(t)$ of the process x_t, i.e. $\lambda_t[z_0^t] = \mu(t)$, which is a Markov diffusion process with the characteristics (see (9))

$$a(t, \mu) = F(t)\mu + B(t)u, \quad D(t, \mu) = \tilde{H}_0^T(t)R^{-1}(t)\tilde{H}_0(t), \tag{49}$$

Proof. Since $\Gamma(t)$, according to (42), does not depend on z_t, then it follows from (39), that $\lambda_t = \mu(t)$. According to [6], process $\tilde{z}_t$, a differential of which is given by (46), and for which $\tilde{Z}_t = (\tilde{z}_t, F_t^z)$, is the Wiener process with

$$E\{\tilde{z}_t \tilde{z}_t^T | F_t^z\} = \int_0^t R(s)ds. \tag{50}$$

Then, Markov property $\mu(t)$ and formulae (49) arise from (41), (50).

Affirmation 2. The coupled processes $\{x_t; \mu(t)\}$ are a Markov diffusion process.

Proof. The statement is proved by (1), (31), (41) with regard to Lemma 2.

Remark 2. Since the conditions of Theorem 1 are satisfied for $\lambda_t = \mu(t)$, then $S(t,\lambda) = S(t,\mu)$, and it follows from (3)-(5), that

$$\min_u \left\{ \frac{\partial S(t,\mu)}{\partial t} + L_{t,\mu}^*[S(t,\mu)] + E\{\Lambda(t, x_t, u_t) | z_0^t\} \right\} = 0, \tag{51}$$

$$S(t,\mu)|_{t=\mathrm{T}} = E\{b(\mathrm{T}, x_\mathrm{T}) | z_0^T\}, \tag{52}$$

$$L_{t,\mu}^*[S(t,\mu)] = a^T(t,\mu) \frac{\partial S(t,\mu)}{\partial \mu} + \frac{1}{2} tr\left[D(t,\mu) \frac{\partial^2 S(t,\mu)}{\partial \mu^2} \right], \tag{53}$$

where $a(t,\mu)$ and $D(t,\mu)$ have the form (49).

Substitution of $E\{\cdot | z_0^t\}$ for $E\{\cdot | \lambda_t = \lambda\}$ results from F_t^z measurability of the process $\mu(t)$ (see (41), (46), (50)).

4. Main results

Theorem 2.

Let

$$\Lambda(\cdot) = x_t^T L(t) x_t + u_t^T N(t) u_t, \quad b(\cdot) = x_\mathrm{T}^T S_\mathrm{T} x_\mathrm{T}, \tag{54}$$

where $L(t)$, $N(t)$, S_T are symmetric matrices, and $L(t) \geq 0$, $S_\mathrm{T} \geq 0$, $N(t) > 0$. Hence, the Bellman equation (51) and boundary condition (52) take the following form

$$\begin{aligned} \min_u \Bigg\{ & \frac{\partial S(t,\mu)}{\partial t} + [F(t)\mu + B(t)u]^T \frac{\partial S(t,\mu)}{\partial \mu} + \frac{1}{2} tr\left[\tilde{H}_0^T(t) R^{-1}(t) \tilde{H}_0(t) \frac{\partial^2 S(t,\mu)}{\partial \mu^2} \right] + \\ & + \mu^T L(t) \mu + u^T N(t) u + tr[L(t)\Gamma(t)] \Bigg\} = 0, \end{aligned} \tag{55}$$

$$S(t,\mu)|_{t=\mathrm{T}} = tr[S_T \Gamma(\mathrm{T})] + \mu^T S_\mathrm{T} \mu. \tag{56}$$

Proof. From (54), with regard to (37) and F_t^z-measurability u_t, we obtain

$$\begin{aligned} E\{\Lambda(t, x_t, u_t) | z_0^t\} &= E\{x_t^T L(t) x_t + u_t^T L(t) u_t | z_0^t\} = E\{x_t^T L(t) x_t | z_0^t\} + u_t^T L(t) u_t = \\ &= \mu_t^T L(t) \mu_t + tr[L(t)\Gamma(t)] + u_t^T L(t) u_t. \end{aligned} \tag{57}$$

Analogous

$$E\{b(\mathrm{T},x_\mathrm{T})|z_0^T\}=E\{x_\mathrm{T}^T S_\mathrm{T} x_\mathrm{T}|z_0^T\}=\mu^T(\mathrm{T})S_\mathrm{T}\mu(\mathrm{T})+tr[S_\mathrm{T}\Gamma(\mathrm{T})] \tag{58}$$

Substitution of (49), (53), (57), in (51) results in (55), and substitution of (58) in (52) leads to (56). The theorem is proved.

From now on, the t-derivative will be identified by the point at the top.

Theorem 3 (The separation theorem). The optimal control u_t^0 has the form

$$u_t^0=-N^{-1}(t)B^T(t)S(t)\mu(t), \tag{59}$$

where an optimal in a mean-root-square sense estimate $\mu(t)$ of the state vector x_t is defined by filter equations (40)-(48), where $u_t=u_t^0$, the matrix $S(t)$ is defined by matrix differential Riccati equation

$$\dot{S}(t)=-F^T(t)S(t)-S(t)F(t)+S(t)B(t)N^{-1}(t)B^T(t)S(t)-L(t), \tag{60}$$

with boundary equation

$$S(\mathrm{T})=S_\mathrm{T}, \tag{61}$$

and the smallest value J^0 of quality criterion has the following form

$$J^0=\mu^T(t_0)S(t_0)\mu(t_0)+tr[S_\mathrm{T}\Gamma(\mathrm{T})]+\int_{t_0}^{\mathrm{T}}tr[L(t)\Gamma(t)]dt+\int_{t_0}^{t}tr\left[\tilde{H}_0(t)R^{-1}(t)\tilde{H}_0(t)S(t)\right]dt \tag{62}$$

Proof. Taking u-derivative from the left-hand side of (55), we obtain the equation to compute the optimal control

$$B^T(t)\frac{\partial S(t,\mu)}{\partial\mu}+2N(t)u=0. \tag{63}$$

Hence, we obtain the expression for optimal control by means of Bellman function in the following form

$$u_t^0=-(1/2)N^{-1}(t)B^T(t)[\partial S(t,\mu)/\partial\mu]. \tag{64}$$

Substituting (64) in (55), we obtain a second-order partial equation for Bellman function in the following form

$$\begin{aligned}\frac{\partial S(t,\mu)}{\partial t}+\mu^T F(t)\frac{\partial S(t,\mu)}{\partial\mu}-\frac{1}{4}\left(\frac{\partial S(t,\mu)}{\partial\mu}\right)^T B(t)N^{-1}(t)B^T(t)\frac{\partial S(t,\mu)}{\partial\mu}+\\+\mu^T L(t)\mu+tr[L(t)\Gamma(t)]+\frac{1}{2}tr\left[\tilde{H}_0^T(t)R^{-1}(t)\tilde{H}_0(t)\frac{\partial^2 S(t,\mu)}{\partial\mu^2}\right]=0.\end{aligned} \tag{65}$$

We solve the equation (65) by using the separation of variables method in the form

$$S(t,\mu)=l(t)+\mu^T S(t)\mu, \tag{66}$$

where $l(t)$ -unknown scalar function, and $S(t)$ - unknown matrix $(n\times n)$-function with imposed symmetry condition. Thereafter

$$\frac{\partial S(t,\mu)}{\partial t}=l(t)+\mu^T\dot{S}(t)\mu,\quad \frac{\partial S(t,\mu)}{\partial\mu}=2S(t)\mu,\quad \frac{\partial^2 S(t,\mu)}{\partial\mu^2}=2S(t). \tag{67}$$

As long as $S(t)$ is imposed with symmetry condition, then, with regard to (67) and the fact that $b=b^T=(1/2)(b+b^T)$ is true for scalar b, we obtain

$$\mu^T F^T(t)\frac{\partial S(t,\mu)}{\partial \mu}=2\mu^T F^T(t)S(t)\mu=\mu^T F^T(t)S(t)\mu+\mu^T S(t)F(t)\mu. \tag{68}$$

Substituting (67), (68) in (65), we have the following formula

$$\begin{aligned}\dot{l}(t)+\mu^T\dot{S}(t)\mu+\mu^T F^T(t)S(t)\mu+\mu^T S(t)F(t)\mu-\mu^T S(t)B(t)N^{-1}(t)B^T(t)S(t)\mu+\\+\mu^T L(t)\mu+tr[L(t)\Gamma(t)]+tr\left[\tilde{H}_0^T(t)R^{-1}(t)\tilde{H}_0(t)S(t)\right]=0\end{aligned} \tag{69}$$

Next, according to the method of separation of variables, we set the coefficients in (69) equal, provided powers μ are equal. Then, we have the equation (60) for $S(t)$, and for $l(t)$ we obtain the following equation

$$\dot{l}(t)=-tr[L(t)\Gamma(t)]-tr\left[\tilde{H}_0^T(t)R^{-1}(t)\tilde{H}_0(t)S(t)\right]. \tag{70}$$

According to (66)

$$S(t,\mu)\big|_{t=\mathrm{T}}=l(\mathrm{T})+\mu^T S(\mathrm{T})\mu. \tag{71}$$

The boundary condition (61) for equation (60) follows from comparison (56) and (71), while the boundary condition for (70) has the form

$$l(t)\big|_{t=\mathrm{T}}=tr[S_{\mathrm{T}}\Gamma(\mathrm{T})]. \tag{72}$$

According to Theorem 1, $J^0=S(t_0,\lambda(t_0))$, and $\lambda=\mu$, thus, it follows from (66), that

$$J^0=l(t_0)+\mu^T(t_0)S(t_0)\mu(t_0). \tag{73}$$

The solution to equation (70) with boundary condition (72) has the form

$$l(t)=tr[S_{\mathrm{T}}\Gamma(\mathrm{T})]+\int_t^{\mathrm{T}}tr[L(\tau)\Gamma(\tau)]d\tau+\int_t^{\mathrm{T}}tr\left[\tilde{H}_0^T(\tau)R^{-1}(\tau)\tilde{H}_0(\tau)S(\tau)\right]d\tau. \tag{74}$$

When $t=t_0$, substitution of (74) in (73) results in (62). Applying (67) in (64), we get (59).

5. Conclusions

Comparing results of Theorem 3 with separation Theorem for memoryless sources in the classical case [1], we come to the conclusion that the expression of optimal control has the same form (59). The difference is the following: in the classical case, Kalman filter estimates the state vector, while in the case considered above, it is the filter (40)-(48), that estimates not only filtering parameters $\mu(t)$ for the current value of state vector x_t, but also estimates interpolation $\mu(\tau_k,t)$ for previous values of the state vector x_{τ_k}, $k=\overline{1;N}$. Thus, the expression for smallest values of quality criterion J^0 changes.

Acknowledgments
All comments and suggestions by critical readers and editorial assistance are acknowledged with gratitude.

References

[1] Wonham W M 1968 On the separation theorem of stochastic control *SIAM J. Control* **6** 312-326

[2] Bryson A E, Yu-Chi-Ho 1969 *Applied optimal control* (London: Blaisdell publishing)

[3] Boguslavskii A I 1970 *Navigation and control methods by incomplete statistical information* (Moscow: Mashinostroenie)

[4] Abakumova O L, Demin N S, Sushko T V 1995 Filtration of stochastic processes for continuous and discrete observations with memory. II. Synthesis of filters *Automation and Remote Control* **56**:10, 1383–1393

[5] Demin N S, Rozhkova S V 2000 Continuously discrete estimation of stochastic processes for observations with memory under anomalous noise: Synthesis *Journal of Computer and Systems Sciences International* **39**:3, 335-346

[6] Liptser R S, Shiryaev A N 2001 *Statistics of Random Processes* (Berlin: Springer-Verlag Berlin Heidelberg)

[7] Stratonovich R L 1968 *Conditional Markov processes and their application to the theory of optimal control* (New York: American Elsevier Publishing Company)

[8] Gikhman I I, Skorokhod AV 2004 *The Theory of Stochastic Processes* (Berlin: Springer-Verlag Berlin Heidelberg)

Positive Stable Realisation of Fractional Electrical Circuits Consisting of n Subsystem

Konrad Andrzej MARKOWSKI
Warsaw University of Technology, Faculty of Electrical Engineering,
Institute of Control and Industrial Electronics,
Koszykowa 75, 00-662 Warsaw, POLAND

E-mail: Konrad.Markowski@ee.pw.edu.pl

Abstract. This paper presents a method of the determination of a positive stable realisation of the fractional continuous-time positive system consisting of n subsystems with one fractional order and with different fractional orders. For the proposed method, a digraph-based algorithm was constructed. In this paper, we have shown how we can realise the transfer matrix using electrical circuits consisting of resistances, inductances, capacitances and source voltages. The proposed method was discussed and illustrated with some numerical examples.

1. Introduction

In recent years, many researchers have been interested in positive linear systems. In this type of the system, state variables and outputs take only non-negative values. Analysis of positive one-dimensional (1D) systems is more difficult than of standard systems. Examples of positive systems include industrial processes involving chemical reactors, heat exchangers and distillation columns, storage systems, compartmental systems, water and atmospheric pollution models. An overview of the state of the art in the positive systems theory is given in [1], [2], [3], [4].

The first definition of the fractional derivative was introduced by Liouville and Riemann at the end of the 19th century. Mathematical fundamentals of fractional calculus are given in the monographs [5], [6], [7], [8], [9], [10]. The notion of positive fractional discrete-time linear systems have been introduced in [6].

The realisation problem is a very difficult task. In many research studies, we can find the canonical form of the system, i.e. constant matrix form, which satisfies the system described by the transfer function. With the use of this form, we are able to write only one realisation (or some by the transformation matrices) of the system, while there exist many sets of matrices which fit into the system transfer function. The realisation problem for positive discrete-time systems without and with delays was considered in [2], [11], [12], [13], while in [14], [15] a solution for finding a set of possible realisations of the characteristic polynomial was proposed, that allows for finding many sets of matrices. In paper [16], [17], [18], [19], [20] the proposed method for finding minimal positive realisations is an extension of the method for finding a realisation of the characteristic polynomial. The optimisation of the proposed algorithm is presented in the paper [21] and [22].

The digraphs theory was applied to the analysis of dynamical systems. The use of the multi-dimensional theory was proposed for the first time in the paper [23], [24], [25] for analysis of positive two-dimensional systems.

In this paper, a new method of determination positive minimal realisation for the fractional continuous one-dimensional system in the form of the electrical circuit will be proposed and the procedure for computation of the minimal realisation will be given. The procedure will be illustrated with a numerical example.

This work has been organised as follows: Section 1 presents some notations and basic definitions of one-dimensional digraph theory. Section 2 presents basics of fractional order systems theory. In this section fractional continuous-time system is defined as the state-space representation of the electrical circuit consisting of n subsystems with one and different fractional order. In Section 3, we construct an algorithm for the determination of a positive minimal realisation of the fractional continuous-time system as the electrical circuit. Finally, we demonstrate the workings of the algorithm on two numerical examples in Section 4, and at the end we present some concluding remarks, open problems and bibliography positions.

Notion: In this paper, the following notion will be used. The matrices will be denoted by a bold font (for example $\mathbf{A}$, $\mathbf{B}$, ...), the sets by a double line (for example $\mathbb{A}$, $\mathbb{B}$, ...), lower/upper indices and polynomial coefficients will be written as a small font (for example a, b, ...) and digraphs will be denoted using a mathfrak font $\mathfrak{D}$. The set $n \times m$ real matrices will be denoted by $\mathbb{R}^{n\times m}$ and $\mathbb{R}^n = \mathbb{R}^{n\times 1}$. If $\mathbf{G} = [g_{ij}]$ is a matrix, we write $\mathbf{G} \gg 0$ (matrix $\mathbf{G}$ is called strictly positive), if $g_{ij} > 0$ for all i, j; $\mathbf{G} > 0$ (matrix $\mathbf{G}$ is called positive), if $g_{ij} > 0$ for all i, j; $\mathbf{G} \geqslant 0$ (matrix $\mathbf{G}$ is called non-negative), if $g_{ij} \geqslant 0$ for all i, j. The set of $n \times m$ real matrices with non-negative entries will be denoted by $\mathbb{R}_+^{n\times m}$ and $\mathbb{R}_+^n = \mathbb{R}_+^{n\times 1}$. The set of $n \times n$ real matrices with off-diagonal non-negative entries will be denoted by $\mathbb{M}^{n\times n}$. The $n \times n$ identity matrix will be denoted by $\mathbf{I}_n$. For more information about matrix theory, an interested reader is referred, for instance, to: [26], [27].

Digraphs: A directed graph (or just digraph) $\mathfrak{D}$ consists of a non-empty finite set $\mathbb{V}(\mathfrak{D})$ of elements called vertices and a finite set $\mathbb{A}(\mathfrak{D})$ of ordered pairs of distinct vertices called arcs ([28], [29]). We call $\mathbb{V}(\mathfrak{D})$ the vertex set and $\mathbb{A}(\mathfrak{D})$ the arc set of $\mathfrak{D}$. We will often write $\mathfrak{D} = (\mathbb{V}, \mathbb{A})$ which means that $\mathbb{V}$ and $\mathbb{A}$ are the vertex set and arc set of $\mathfrak{D}$, respectively. The order of $\mathfrak{D}$ is the number of vertices in $\mathfrak{D}$. The size of $\mathfrak{D}$ is the number of arc in $\mathfrak{D}$. For an arc (v_1, v_2) the first vertex v_1 is its tail and the second vertex v_2 is its head.

There exists an $\mathfrak{A}$-arc from vertex v_j to vertex v_i if and only if the (i, j)-th entry of the matrix $\mathbf{A}$ is non-zero. There exists a $\mathfrak{B}$-arc from source s to vertex v_j if and only if the l-th entry of the matrix $\mathbf{B}$ is non-zero.

Example 1 *Let us be given the positive system single input described by the following matrices*

$$\underbrace{\begin{bmatrix} 1 & 0 & 1 \\ 1 & 0 & 1 \\ 0 & 1 & 0 \end{bmatrix}}_{\mathbf{A}}, \underbrace{\begin{bmatrix} 1 \\ 0 \\ 0 \end{bmatrix}}_{\mathbf{B}} \tag{1}$$

we can draw a one-dimensional digraph $\mathfrak{D}^{(1)}$ consisting of vertices v_1, v_2, v_3 and source s_1. A one-dimensional digraph corresponding to the system (1) is presented in Figure 1.

We present below some basic notions from the graph theory which are used in further considerations. A walk in a digraph $\mathfrak{D}^{(1)}$ is a finite sequence of arcs in which every two vertices v_i and v_j are adjacent or identical. A walk in which all of the arcs are distinct is called a path. The path that goes through all vertices is called a finite path. If the initial and the terminal vertices of the path are the same, then the path is called a cycle.

More information about use digraph theory in positive system is given in [23], [25].

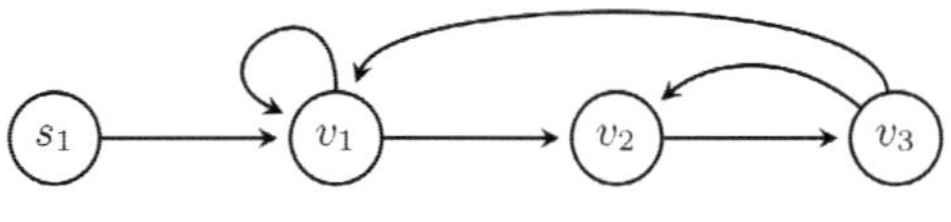

Figure 1. One-dimensional digraph

2. Fractional order system

2.1. Model and Representation

The equation for a continuous-time dynamic system of the fractional-order can be written as follows:

$$H(\mathfrak{D}^{\alpha_0\alpha_1\ldots\alpha_m})(y_1, y_2, \ldots, y_l) = G(\mathfrak{D}^{\beta_0\beta_1\ldots\beta_n})(u_1, u_2, \ldots, y_k),$$

where y_i, u_i are the function of time and $H(\cdot)$, $G(\cdot)$ are the combinations of the fractional-order derivative operator. If we describe the linear time-invariant single-variable case we obtain the following equation:

$$H(\mathfrak{D}^{\alpha_0\alpha_1\ldots\alpha_m})(y_t) = G(\mathfrak{D}^{\beta_0\beta_1\ldots\beta_n})(u_t) \tag{2}$$

with

$$H(\mathfrak{D}^{\alpha_0\alpha_1\ldots\alpha_n}) = \sum_{k=0}^{n} a_k\mathfrak{D}^{\alpha_k}, \quad a_k \in \mathbb{R}; \quad G(\mathfrak{D}^{\beta_0\beta_1\ldots\beta_m}) = \sum_{k=0}^{m} b_k\mathfrak{D}^{\beta_k}, \quad b_k \in \mathbb{R}.$$

or

$$a_n\mathfrak{D}^{\alpha_n}y(t) + a_{n-1}\mathfrak{D}^{\alpha_{n-1}}y(t) + \cdots + a_0\mathfrak{D}^{\alpha_0}y(t) = b_m\mathfrak{D}^{\beta_m}u(t) + b_{m-1}\mathfrak{D}^{\beta_{m-1}}u(t) + \cdots + b_0\mathfrak{D}^{\beta_0}u(t). \tag{3}$$

After applying the Laplace transform to (3) with zero initial conditions, the input-output representation of fractional-order system can be obtained. The fractional-order system as the transfer function has the following form:

$$G(s) = \frac{b_m s^{\beta_m} + b_{m-1}s^{\beta_{m-1}} + \cdots + b_0 s^{\beta_0}}{a_n s^{\alpha_n} + a_{n-1}s^{\alpha_{n-1}} + \cdots + a_0 s^{\alpha_0}}. \tag{4}$$

2.2. State-Space Representation

In this paper, the following Caputo definition of the fractional derivative will be used:

$$_a^C\mathfrak{D}_t^\alpha = \frac{d^\alpha}{dt^\alpha} = \frac{1}{\Gamma(n-\alpha)}\int_a^t \frac{f^{(n)}(\tau)}{(t-\tau)^{\alpha+1-n}}d\tau, \tag{5}$$

where $\alpha \in \mathbb{R}$ is the order of fractional derivative, $f^{(n)}(\tau) = \frac{d^n f(\tau)}{d\tau^n}$ and $\Gamma(x) = \int_0^\infty e^{-t}t^{x-1}dt$ is the gamma function.

The Laplace transform [6] of the derivative-integral (5) has the following form:

$$\mathcal{L}\left[\ _0^C\mathfrak{D}_t^\alpha\ \right] = s^\alpha F(s) - \sum_{k=1}^{n} s^{\alpha-k} f^{(k-1)}(0^+). \tag{6}$$

Let the current $i_C(t)$ in a supercondensator with the capacity C be the α–order derivative:

$$i_C(t) = C\frac{d^\alpha u_C(t)}{dt^\alpha}, \tag{7}$$

where $\frac{d^\alpha}{dt^\alpha}$ is the α–order derivative defined by (6) and $u_C(t)$ is the voltage on the condensator. Similarly, let the voltage $u_L(t)$ on the coil with the inductance L be the β–order derivative

$$u_L(t) = L\frac{d^\beta i_L(t)}{dt^\beta}, \tag{8}$$

where $\frac{d^\beta}{dt^\beta}$ is the β–order derivative defined by (6) and $i_L(t)$ is the current in the coil.

Let us consider electrical circuit composed of resistance, capacitances, coils and voltage sources. As the state variable, we choose the voltages on the capacitors and the currents in the coils. Using (7) and (8) and Kirhoff's laws we can write

$$\begin{bmatrix} \frac{d^\alpha x_C(t)}{dt^\alpha} \\ \frac{d^\beta x_L(t)}{dt^\beta} \end{bmatrix} = \begin{bmatrix} \mathbf{A}_{11} & \mathbf{A}_{12} \\ \mathbf{A}_{22} & \mathbf{A}_{22} \end{bmatrix} \begin{bmatrix} x_C(t) \\ x_L(t) \end{bmatrix} + \begin{bmatrix} \mathbf{B}_1 \\ \mathbf{B}_2 \end{bmatrix} u(t), \tag{9a}$$

$$y(t) = \mathbf{C}x(t) + \mathbf{D}u(t), \tag{9b}$$

where $x_C(t) \in \mathbb{R}^{n_1}$ are voltages on the condensators, $x_L(t) \in \mathbb{R}^{n_2}$ are currents in the coil, $u(t) \in \mathbb{R}^m$ are the voltage or current source and $\mathbf{A}_{ij} \in \mathbb{R}^{n_i \times n_j}$, $\mathbf{B}_i \in \mathbb{R}^{n_i \times m}$ for $i, j = 1, 2$.

After using the Laplace transform to (9a)–(9b) and taking into account

$$\begin{aligned} \mathcal{L}\left[\frac{d^\alpha x_C}{dt^\alpha}\right] &= s^\alpha X_C(s) - s^{\alpha-1}x_C(0), \\ \mathcal{L}\left[\frac{d^\beta x_L}{dt^\beta}\right] &= s^\beta X_L(s) - s^{\beta-1}x_L(0), \\ U(s) &= \mathcal{L}\left[u(t)\right], \end{aligned} \tag{10}$$

we obtain:

$$\begin{aligned} \begin{bmatrix} X_C(s) \\ X_L(s) \end{bmatrix} &= \begin{bmatrix} \mathbf{I}_{n_1}s^\alpha - \mathbf{A}_{11} & -\mathbf{A}_{12} \\ -\mathbf{A}_{21} & \mathbf{I}_{n_2}s^\beta - \mathbf{A}_{22} \end{bmatrix}^{-1} \left\{ \begin{bmatrix} s^{\alpha-1}x_C(0) \\ s^{\beta-1}x_L(0) \end{bmatrix} + \begin{bmatrix} \mathbf{B}_1 \\ \mathbf{B}_2 \end{bmatrix} U(s) \right\}, \\ Y(s) &= \mathbf{C}X(s) + \mathbf{D}U(s). \end{aligned} \tag{11}$$

After using (11) with zero initial conditions, we can determine transfer matrix of the system in the following form:

$$\mathbf{T}(s) = \mathbf{C}\begin{bmatrix} \mathbf{I}_{n_1}s^\alpha - \mathbf{A}_{11} & -\mathbf{A}_{12} \\ -\mathbf{A}_{21} & \mathbf{I}_{n_2}s^\beta - \mathbf{A}_{22} \end{bmatrix}^{-1} \begin{bmatrix} \mathbf{B}_1 \\ \mathbf{B}_2 \end{bmatrix} + \mathbf{D}. \tag{12}$$

Definition 1 *The fractional system* (9a)–(9b) *is called the internally positive fractional system if and only if* $x(t) \in \mathbb{R}^n_+$ *and* $y(t) \in \mathbb{R}^p_+$ *for* $t \geqslant 0$ *for any initial conditions* $x_0 \in \mathbb{R}^n_+$ *and all inputs* $u(t) \in \mathbb{R}^m_+$ *for* $t \geqslant 0$.

Definition 2 *A square real matrix* $\mathbf{A} = [a_{ij}]$ *is called the Metzler matrix if its off-diagonal entries are non-negative, i.e.* $a_{ij} \geqslant 0$ *for* $i \neq j$.

Definition 3 *The fractional system* (9a)–(9b) *is positive if and only if*

$$\mathbf{A} \in \mathbb{M}^{n \times n},\ \mathbf{B} \in \mathbb{R}^{n \times m}_+,\ \mathbf{C} \in \mathbb{R}^{p \times n}_+,\ \mathbf{D} \in \mathbb{R}^{p \times m}_+. \tag{13}$$

Matrices (13) are called a positive realisation of the transfer function $\mathbf{T}(s)$ if they satisfy the equality (12). The realisation is called minimal if the dimension of the state matrix $\mathbf{A}$ is minimal among all possible realisation of (12).

Let us consider single-input single-output multi-order fractional continuous-time linear system:

$$\begin{bmatrix} \mathfrak{D}^{\alpha_1}x_1(t) \\ \vdots \\ \mathfrak{D}^{\alpha_n}x_n(t) \end{bmatrix} = \begin{bmatrix} \mathbf{A}_{11} & \cdots & \mathbf{A}_{1n} \\ \vdots & \ddots & \vdots \\ \mathbf{A}_{n1} & \cdots & \mathbf{A}_{nn} \end{bmatrix} \begin{bmatrix} x_1(t) \\ \vdots \\ x_n(t) \end{bmatrix} + \begin{bmatrix} \mathbf{B}_1 \\ \vdots \\ \mathbf{B}_n \end{bmatrix} u(t),$$

$$y(t) = \begin{bmatrix} \mathbf{C}_1 & \cdots & \mathbf{C}_n \end{bmatrix} \begin{bmatrix} x_1(t) \\ \vdots \\ x_n(t) \end{bmatrix} + \mathbf{D}u(t), \tag{14}$$

where $0 < \alpha_i < 1$, $i = 1, \ldots, n$ $x_i(t) \in \mathbb{R}^{n_i}$, $i = 1, \ldots, n$ is the state vector, $u(t) \in \mathbb{R}$ and $y(t) \in \mathbb{R}$ are the input and output of the system, $\mathbf{A}_{ij} \in \mathbb{R}^{n_i \times n_j}$, $\mathbf{B} \in \mathbb{R}^{n_i}$, $\mathbf{C} \in \mathbb{R}^{1 \times n_j}$, $i, j = 1, \ldots, n$ and $\mathbf{D} \in \mathbb{R}$. The transfer function of the system (14) has the form

$$\mathbf{T}(s) = \begin{bmatrix} \mathbf{C}_1 & \mathbf{C}_2 & \cdots & \mathbf{C}_n \end{bmatrix} \begin{bmatrix} \mathbf{I}_{n_1}s^{\alpha_1} - \mathbf{A}_{11} & -\mathbf{A}_{12} & \cdots & -\mathbf{A}_{1n} \\ -\mathbf{A}_{21} & \mathbf{I}_{n_2}s^{\alpha_2} - \mathbf{A}_{22} & \cdots & -\mathbf{A}_{2n} \\ \vdots & \vdots & \ddots & \vdots \\ -\mathbf{A}_{n1} & -\mathbf{A}_{n2} & \cdots & \mathbf{I}_{n_n}s^{\alpha_n} - \mathbf{A}_{nn} \end{bmatrix}^{-1} \begin{bmatrix} \mathbf{B}_1 \\ \mathbf{B}_2 \\ \vdots \\ \mathbf{B}_n \end{bmatrix} + \mathbf{D}, \tag{15}$$

where:

$$\mathbf{A} = \begin{bmatrix} \mathbf{A}_{11} & \mathbf{A}_{12} & \cdots & \mathbf{A}_{1n} \\ \mathbf{A}_{21} & \mathbf{A}_{22} & \cdots & \mathbf{A}_{2n} \\ \vdots & \vdots & \ddots & \vdots \\ \mathbf{A}_{n1} & \mathbf{A}_{n2} & \cdots & \mathbf{A}_{nn} \end{bmatrix}, \quad \mathbf{B} = \begin{bmatrix} \mathbf{B}_1 \\ \mathbf{B}_2 \\ \vdots \\ \mathbf{B}_n \end{bmatrix}, \quad \mathbf{C} = \begin{bmatrix} \mathbf{C}_1 & \mathbf{C}_2 & \cdots & \mathbf{C}_n \end{bmatrix}, \quad \mathbf{D}. \tag{16}$$

Matrices (16) are called a realisation of the transfer matrix $\mathbf{T}(s)$ if they satisfy the equality (15).

Task: For the given transfer matrix (12) (or (15)), determine a minimal positive realisation as a electrical circuit of the system (9a)–(9a) using the one-dimensional $\mathfrak{D}^{(1)}$ digraphs theory. The dimension of the system must be the minimal among possible.

3. Problem Solution

Let us be given the matrix $\mathbf{A}$ in the following form

$$\mathbf{A} = \begin{bmatrix} \mathbf{A}_{11} & 0 & \cdots & 0 \\ 0 & \mathbf{A}_{22} & \cdots & 0 \\ \vdots & \vdots & \ddots & \vdots \\ 0 & 0 & \cdots & \mathbf{A}_{nn} \end{bmatrix} \tag{17}$$

then the strictly proper transfer matrix can be written in the following form:

$$\mathbf{T}_{sp}(s) = \begin{bmatrix} \mathbf{C}_1 & \mathbf{C}_2 & \cdots & \mathbf{C}_n \end{bmatrix} \begin{bmatrix} \mathbf{I}_{n_1}s^{\alpha_1} - \mathbf{A}_{11} & 0 & \cdots & 0 \\ 0 & \mathbf{I}_{n_2}s^{\alpha_2} - \mathbf{A}_{22} & \cdots & 0 \\ \vdots & \vdots & \ddots & \vdots \\ 0 & 0 & \cdots & \mathbf{I}_{n_n}s^{\alpha_n} - \mathbf{A}_{nn} \end{bmatrix}^{-1} \begin{bmatrix} \mathbf{B}_1 \\ \mathbf{B}_2 \\ \vdots \\ \mathbf{B}_n \end{bmatrix}. \tag{18}$$

The strictly proper transfer matrix (18) can be considered as a pseudo-rational function of the variable $\lambda_1 = s^{\alpha_1}$, $\lambda_2 = s^{\alpha_2}, \cdots$, $\lambda_n = s^{\alpha_n}$ in the form:

$$\mathbf{T}_{sp}(\lambda) = \begin{bmatrix} \mathbf{C}_1 \; \mathbf{C}_2 \cdots \mathbf{C}_n \end{bmatrix} \begin{bmatrix} \mathbf{I}_{n_1}\lambda_1 - \mathbf{A}_{11} & 0 & \cdots & 0 \\ 0 & \mathbf{I}_{n_2}\lambda_2 - \mathbf{A}_{22} & \cdots & 0 \\ \vdots & \vdots & \ddots & \vdots \\ 0 & 0 & \cdots & \mathbf{I}_{n_n}\lambda_n - \mathbf{A}_{nn} \end{bmatrix}^{-1} \begin{bmatrix} \mathbf{B}_1 \\ \mathbf{B}_2 \\ \vdots \\ \mathbf{B}_n \end{bmatrix} = \tag{19}$$

$$= \frac{1}{\det\left[(\mathbf{I}_{n_1}\lambda_1 - \mathbf{A}_{11})(\mathbf{I}_{n_2}\lambda_2 - \mathbf{A}_{22})\dots(\mathbf{I}_{n_n}\lambda_n - \mathbf{A}_{nn})\right]} \begin{bmatrix} n_{11}(\lambda) & n_{12}(\lambda) & \cdots & n_{1n}(\lambda) \\ n_{21}(\lambda) & n_{22}(\lambda) & \cdots & n_{2n}(\lambda) \\ \vdots & \vdots & \ddots & \vdots \\ n_{n1}(\lambda) & n_{n2}(\lambda) & \cdots & n_{nn}(\lambda) \end{bmatrix},$$

where

$$\begin{aligned} d(\lambda) &= \det\underbrace{(\mathbf{I}_{n_1}\lambda_1 - \mathbf{A}_{11})}_{d_{11}(\lambda)} \det\underbrace{(\mathbf{I}_{n_2}\lambda_2 - \mathbf{A}_{22})}_{d_{22}(\lambda)} \cdots \det\underbrace{(\mathbf{I}_{n_n}\lambda_n - \mathbf{A}_{nn})}_{d_{kk}(\lambda)} = \\ &= d_{11}(\lambda)d_{22}(\lambda)\cdots d_{kk}(\lambda), \qquad k = 1, 2\dots, n. \end{aligned} \tag{20}$$

In the first step, we must find matrices $\mathbf{A}_{kk} \in \mathbb{R}_+^{n\times n}$, $k = 1, 2, \cdots, n$; using decomposition characteristic polynomial (20). We decompose each simple polynomial into a set of simple monomials

$$d_{kk}(s) = \left(1 - d_{kk_{M_1}}(\lambda)\right) \cup \left(1 - d_{kk_{M_2}}(\lambda)\right) \cup \cdots \cup \left(1 - d_{kk_{M_p}}(\lambda)\right), \tag{21}$$

where p is number of simple monomials $d_{kk_{M_p}}$ in the simple polynomial $d_{kk}(\lambda)$. For each simple monomial, we create digraphs representations. Then we can determine all possible simple polynomial realisations using all combinations of the digraphs monomial representations. Finally, we combine received digraphs in one digraph which is corresponding to a characteristic polynomial $d(s)$.

Theorem 1 *There exists a stable positive fractional continuous-time linear system described by the equation (14) if a multi-dimensional digraph corresponding to matrix $\mathbf{A} \in \mathbb{R}_+^{n\times n}$, satisfies the following conditions:*

- ***(C1)*** *the sets $\mathbb{D}_{kk_{M_1}} \cap \mathbb{D}_{kk_{M_2}} \cap \cdots \cap \mathbb{D}_{kk_{M_p}}$, where $k = 1, 2, \dots, n$; and p is a number of simple monomials in simple polynomial $d_{kk}(s)$ corresponding to multi-dimensional digraphs are not disjoint.*
- ***(C2)*** *the obtained multi-dimensional digraph does not have additional cycles.*
- ***(C3)*** *the poles of the characteristic polynomial are distinct real and negative.*

Proof: *Condition (C1):* The sets $\mathbb{D}_{kk_{M_1}} \cap \mathbb{D}_{M_{kk_{M_2}}} \cap \cdots \cap \mathbb{D}_{kk_{M_p}}$, $k = 1, 2, \dots, n$; where p is a number of simple monomials in a simple polynomial $d(s)$ are disjoint if $\mathbb{D}_{kk_{M_1}} \cap \mathbb{D}_{kk_{M_2}} \cap \cdots \cap \mathbb{D}_{kk_{M_p}} = \emptyset$ then we have a digraph whose vertices can be divided into two disjoint sets. It means that we obtain an additional simple monomial in a simple characteristic polynomial $d_{kk}(s)$. In this situation, we obtain a new characteristic polynomial $\overline{d}(s)$, $d(s) \neq \overline{d}(s)$. *Condition (C2):* Each monomial is represented by one cycle. If after combining all digraphs (each corresponding to one simple monomial in every simple polynomial) we obtain an additional cycle, this means that in the polynomial an additional simple monomial appears. *Condition (C3):* If the state matrix **A** has the structure as (17) then the roots of the characteristic polynomial are real negative and the matrix **A** is stable and is a Metzler matrix. □

Using the Theorem 1 we can construct the Algorithm 1.

Algorithm 1 *DetermineMinimalRealisation*()

```
monomial = 1;
Determine number of simple_polynomial in characteristic polynomial (20);
for simple_polynomial = 1 to k do
    Determine number of cycles in simple polynomial;
    for monomial = 1 to cycles do
        Determine one-dimensional digraph D^(1) for all monomial;
        MonomialRealisation(monomial);
    end for
    for monomial = 1 to cycles do
        Determine digraph as a combination of the digraph monomial representation;
        SimplePolynomialRealisation(monomial);
        if SimplePolynomialRealisation != cycles then
            Digraph contains additional cycles or digraph contains disjoint union;
            BREAK
        else if SimplePolynomialRealisation == cycles then
            Digraph satisfies characteristic polynomial;
            Determine weights of the arcs in digraph;
            Write state matrix A_kk|k=1,...,n;
            return (SimplePolynomialRealisation, A_kk|k=1,...,n);
        end if
    end for
    return (PolynomialRealisation, A = diag(A_kk|k=1,...,n));
end for
```

Let us assume that the matrix $\mathbf{B}_k$, $k = 1, 2, \cdots, n$ and matrix $\mathbf{C}$ have the following form:

$$\mathbf{B}_k|_{k=1,2,\cdots n} = \begin{bmatrix} b_{11}^k & b_{12}^k & \cdots & b_{1,m}^k \\ b_{21}^k & b_{22}^k & \cdots & b_{2,m}^k \\ \vdots & \vdots & \ddots & \vdots \\ b_{n,1}^k & b_{n,2}^k & \cdots & b_{n,m}^k \end{bmatrix}, \mathbf{C} = \begin{bmatrix} c_{11} & c_{12} & \cdots & c_{1,n} \\ c_{21} & c_{22} & \cdots & c_{2,n} \\ \vdots & \vdots & \ddots & \vdots \\ c_{p,1} & c_{p,2} & \cdots & c_{p,n} \end{bmatrix}. \quad (22)$$

After determining the state matrix $\mathbf{A}_{kk} \in \mathbb{R}_+^{n \times n}$, $k = 1, 2, \cdots, n$ from the Algorithm 1 and inserting matrices (22) to the equation (18) we obtain the polynomial $\widetilde{n}_{ij}(s)$. After comparing variables with the same power of s polynomials $\widetilde{n}_{ij}(s) = n_{ij}(s)$ we receive the set of equations. After solving the equation, we obtain matrices (13).

Remark 1 *If there exists the characteristic polynomial in the form* (20)*, then we can always draw an electrical circuit consist from: resistances, inductances and source voltages; resistances and capacitances; resistances, inductances and capacitances.*

4. Numerical Example

4.1. Example 1 (one fractional order).

For the given strictly proper transfer matrix

$$\mathbf{T}_{sp}(s) = \frac{1}{s^{3\alpha} + 34s^{2\alpha} + 245s^{\alpha} + 500} \begin{bmatrix} 0 & 25s^{2\alpha} + 225s^{\alpha} + 500 \\ 4s^{2\alpha} + 120s^{\alpha} + 500 & 0 \end{bmatrix}, \quad (23)$$

find a minimal positive stable realisation as an electrical circuit using the one-dimensional digraph theory for $0 < \alpha < 1$.

Solution. In the first step, using a pseudo-rational function of the variable $\lambda = s^\alpha$ write transfer matrix (23) in the following form:

$$\mathbf{T}_{sp}(\lambda) = \frac{1}{\lambda^3 + 34\lambda^2 + 245\lambda + 500}\begin{bmatrix} 0 & 25\lambda^2 + 225\lambda + 500 \\ 4\lambda^2 + 120\lambda + 500 & 0 \end{bmatrix}. \quad (24)$$

After multiplying the nominator and denominator of the transfer function (24) by λ^{-3}, we obtain:

$$\mathbf{T}_{sp}(\lambda) = \frac{1}{d(\lambda)}\begin{bmatrix} 0 & n_{12}(\lambda) \\ n_{21}(\lambda) & 0 \end{bmatrix} = \quad (25)$$
$$= \frac{1}{1 + 34\lambda^{-1} + 245\lambda^{-2} + 500\lambda^{-3}}\begin{bmatrix} 0 & 25\lambda^{-1} + 225\lambda^{-2} + 500\lambda^{-3} \\ 4\lambda^{-1} + 120\lambda^{-2} + 500\lambda^{-3} & 0 \end{bmatrix}.$$

We can write the characteristic polynomial in the following form

$$d(\lambda) = 1 + 34\lambda^{-1} + 245\lambda^{-2} + 500\lambda^{-3} = \underbrace{\left(1 + 4\lambda^{-1}\right)}_{d_{11}(\lambda^{-1})}\underbrace{\left(1 + 5\lambda^{-1}\right)}_{d_{22}(\lambda^{-1})}\underbrace{\left(1 + 25\lambda^{-1}\right)}_{d_{33}(\lambda^{-1})}. \quad (26)$$

In the next step, we a decompose characteristic polynomial (26) into a set of simple polynomials $d_{11}(\lambda) = 1 + 4\lambda^{-1}$, $d_{22}(\lambda) = 1 + 5\lambda^{-1}$ and $d_{33}(\lambda) = 1 + 25\lambda^{-1}$. In the next step, for each simple monomial we write initial conditions. For a simple polynomial $d_{11}(\lambda)$, we have following conditions: number of vertices in digraph: $vertices = 1$; possible weights from which we will build digraphs: $(v_i, v_j)_{\mathfrak{A}_{11}}\lambda^{-1}$; monomials: $M_1 = 1 + 4\lambda^{-1}$.

Then, we determine all possible realisations of the simple polynomial $d_{11}(\lambda)$. In the considered example, we have only one realisation presented in Figure 2(a). The realisation meets conditions (C1) and (C2) of the Theorem 1. Finally, we must verify the third condition. In the considered simple polynomial, the poles are real and negative. Described realisation satisfies the condition (C3). **The realisation does satisfy all conditions and is correct.** In this same way, we can determine realisations of the simple polynomial $d_{22}(\lambda)$ presented in Figure 2(b) and $d_{33}(\lambda)$ presented in Figure 2(c).

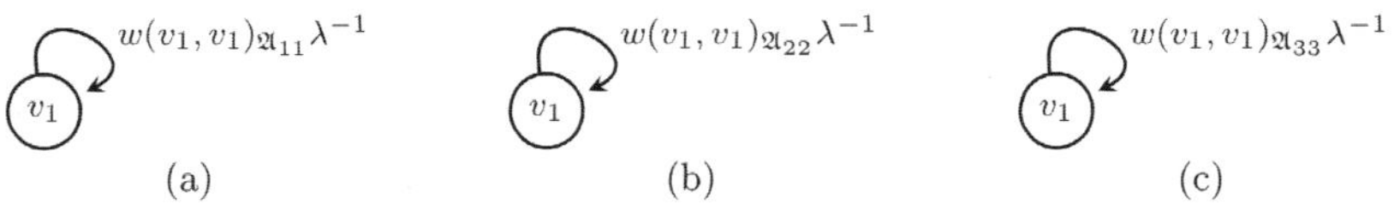

Figure 2. Multi-dimensional digraphs corresponding to a simple polynomial: 2(a)–$d_{11}(\lambda)$; 2(b)–$d_{22}(\lambda)$; 2(c)–$d_{33}(\lambda)$.

From the obtained digraphs, we can write a state matrix $\mathbf{A}$ in the form:

$$\mathbf{A} = \begin{bmatrix} \mathbf{A}_{11} & 0 & 0 \\ 0 & \mathbf{A}_{22} & 0 \\ 0 & 0 & \mathbf{A}_{33} \end{bmatrix} = \begin{bmatrix} w(v_1, v_1)_{\mathfrak{A}_{11}} & 0 & 0 \\ 0 & w(v_1, v_1)_{\mathfrak{A}_{22}} & 0 \\ 0 & 0 & w(v_1, v_1)_{\mathfrak{A}_{33}} \end{bmatrix}. \quad (27)$$

After using a characteristic polynomial (26) and obtaining matrix (27), we can draw an electrical circuit, presented in Figure 3(a), which consists of three loops: Loop L_1 corresponding

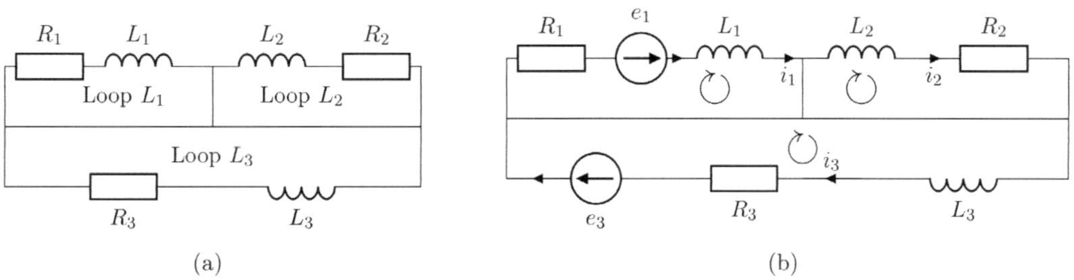

Figure 3. Electrical circuit consisting of resistances and inductances

to matrix $\mathbf{A}_{11}$ and polynomial $d_{11}(\lambda)$; Loop L_2 corresponding to matrix $\mathbf{A}_{22}$ and polynomial $d_2(\lambda)$; and Loop L_3 corresponding to matrix $\mathbf{A}_{33}$ and polynomial $d_{33}(\lambda)$. In every loop, we have resistances and inductances.

After using (27), Figure 3(a) and the value of the electrical circuit: $R_1 = 12\Omega$, $R_2 = 20\Omega$, $R_3 = 25\Omega$, $L_1 = 3H$, $L_2 = 4H$ and $L_3 = 1H$, we can write a state matrix in the following form:

$$\mathbf{A} = \begin{bmatrix} \mathbf{A}_{11} & 0 & 0 \\ 0 & \mathbf{A}_{22} & 0 \\ 0 & 0 & \mathbf{A}_{33} \end{bmatrix} = \begin{bmatrix} -\frac{R_1}{L_1} & 0 & 0 \\ 0 & -\frac{R_2}{L_2} & 0 \\ 0 & 0 & -\frac{R_3}{L_3} \end{bmatrix} = \begin{bmatrix} -4 & 0 & 0 \\ 0 & -5 & 0 \\ 0 & 0 & -25 \end{bmatrix} \tag{28}$$

In the next step, we must add source voltages for example to the Loop L_1 and Loop L_3 (Figure 3(a)). After this operation, we will obtain the electrical circuit presented in Figure 3(b).

Now, we can write matrix $\mathbf{B}$ in the following form:

$$\mathbf{B} = \begin{bmatrix} \frac{1}{L_1} & 0 \\ 0 & 0 \\ 0 & \frac{1}{L_3} \end{bmatrix} = \begin{bmatrix} 0.333 & 0 \\ 0 & 0 \\ 0 & 1 \end{bmatrix}. \tag{29}$$

After inserting matrices (28), (29) and (22) to the equation (20), we obtain the polynomial $\widetilde{n}_{ii}(\lambda)$, $i = 1, 2$. In the next step, we compare one of the coefficients of the same power of λ polynomials $\widetilde{n}_{ii}(\lambda) = n_{ii}(\lambda)$ for $i = 1, 2$, and we receive the set of the equations. After solving them, we obtain the following matrix:

$$\mathbf{C} = \begin{bmatrix} 0 & 0 & R_3 \\ R_1 & 0 & 0 \end{bmatrix} = \begin{bmatrix} 0 & 0 & 25 \\ 12 & 0 & 0 \end{bmatrix}. \tag{30}$$

The obtained realisation is stable, as can be seen in Figure 4, showing the step response of the system for α equal to: 0.3, 0.5, 0.7 and 0.9. In Figure 5(a), we can see the Nyquist characteristic for: $\alpha = 0.3$ (Figure 5(a)); $\alpha = 0.5$ (Figure 5(b)); $\alpha = 0.7$ (Figure 5(c)); $\alpha = 0.9$ (Figure 5(d)). Simulations were performed using the Matlab toolbox FOMCON (Fractional-order Modeling and Control) presented in the paper [30] and [31].

4.2. Example 2 (three fractional order).
For the given transfer matrix

$$\mathbf{T}_{sp}(s) = \frac{\begin{bmatrix} 0 & 25s^{1.2} + 100s^{0.9} + 125s^{0.3} + 500 \\ 4s^{1.3} + 100s^{0.9} + 20s^{0.4} + 500 & 0 \end{bmatrix}}{s^{1.6} + 4s^{1.3} + 25s^{1.2} + 100s^{0.9} + 5s^{0.7} + 20s^{0.4} + 125s^{0.3} + 500}, \tag{31}$$

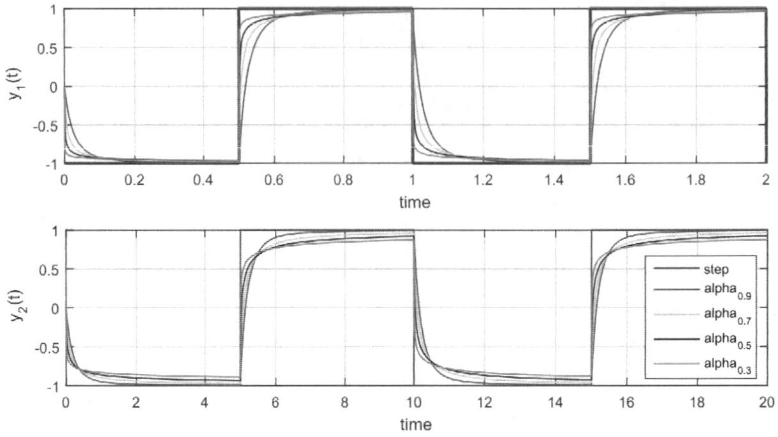

Figure 4. Step response of the system (23) for $\alpha = \{0.3; 0.5; 0.7; 0.9\}$.

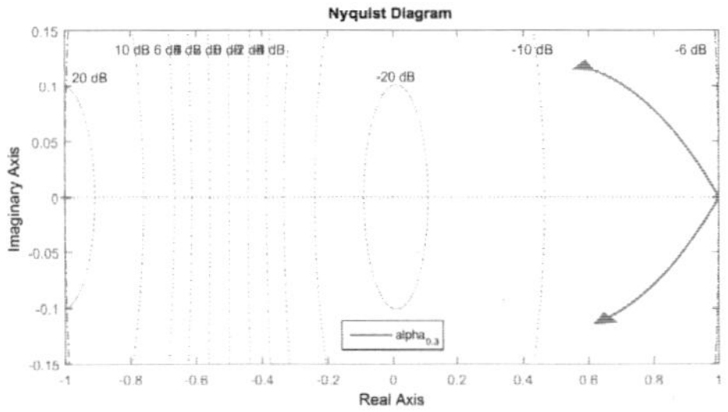

(a)

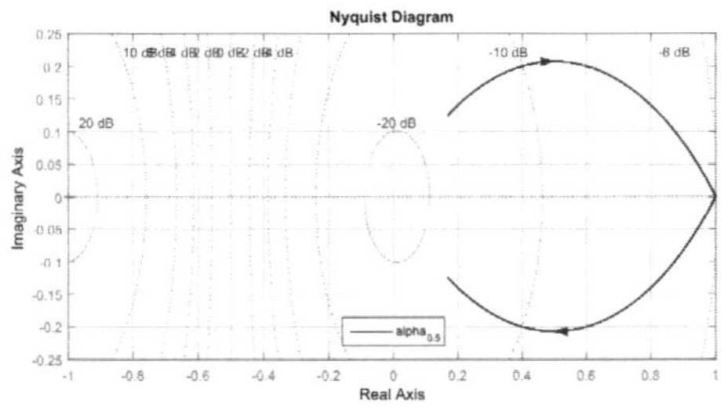

(b)

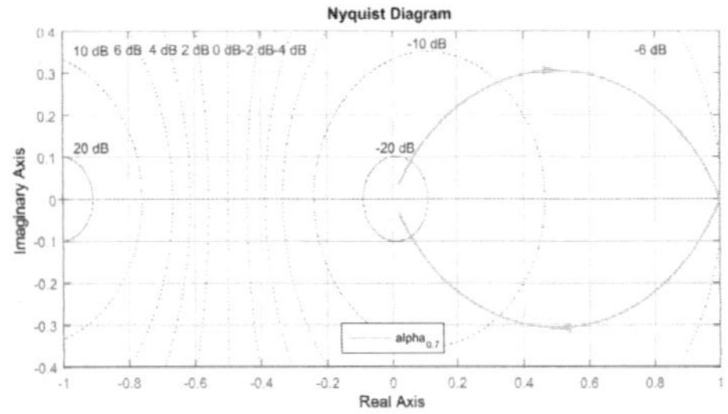

(c)

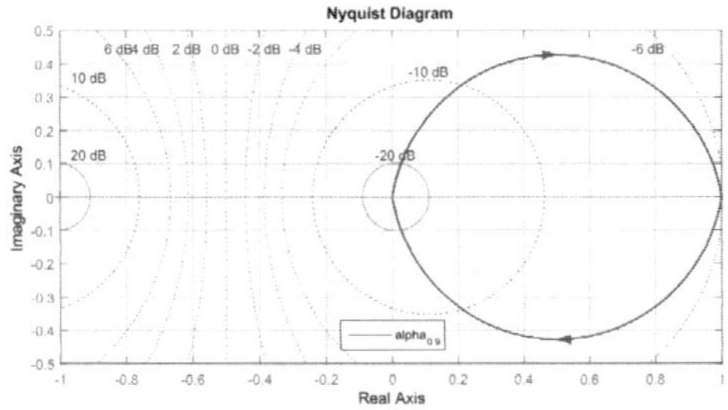

(d)

Figure 5. Nyquist characteristic of the system (23) for: 5(a) $\alpha = 0.3$; 5(b) $\alpha = 0.5$; 5(c) $\alpha = 0.7$; 5(d) $\alpha = 0.9$.

find a minimal positive stable realisation as an electrical circuit using the one-dimensional digraph theory for $0 < \alpha < 1$.

Solution. In the first step using pseudo-rational function of the variable $\lambda_1 = s^{0.3}$, $\lambda_2 = s^{0.9}$, $\lambda_3 = s^{0.4}$ we can write transfer matrix (31) in the following form:

$$\mathbf{T}_{sp}(\lambda) = \frac{\begin{bmatrix} 0 & 25\lambda_1\lambda_2 + 100\lambda_2 + 125\lambda_1 + 500 \\ 4\lambda_2\lambda_3 + 100\lambda_2 + 20\lambda_3 + 500 & 0 \end{bmatrix}}{\lambda_1\lambda_2\lambda_3 + 4\lambda_2\lambda_3 + 25\lambda_1\lambda_2 + 5\lambda_1\lambda_3 + 100\lambda_2 + 20\lambda_3 + 125\lambda_1 + 500}. \tag{32}$$

After multiplying the nominator and denominator of the transfer function (24) by $\lambda_1^{-1}\lambda_2^{-1}\lambda_3^{-1}$, we obtain:

$$\mathbf{T}_{sp}(\lambda^{-1}) = \frac{1}{d(\lambda)}\begin{bmatrix} 0 & n_{12}(\lambda) \\ n_{21}(\lambda) & 0 \end{bmatrix} = \tag{33}$$

$$= \frac{\begin{bmatrix} 0 & 25\lambda_3^{-1} + 100\lambda_1^{-1}\lambda_3^{-1} + 125\lambda_2^{-1}\lambda_3^{-1} + 500\lambda_1^{-1}\lambda_2^{-1}\lambda_3^{-1} \\ 4\lambda_1^{-1} + 100\lambda_1^{-1}\lambda_3^{-1} + 20\lambda_1^{-1}\lambda_2^{-1} + 500\lambda_1^{-1}\lambda_2^{-1}\lambda_3^{-1} & 0 \end{bmatrix}}{1 + 4\lambda_1^{-1} + 25\lambda_3^{-1} + 5\lambda_2^{-1} + 100\lambda_1^{-1}\lambda_3^{-1} + 20\lambda_1^{-1}\lambda_2^{-1} + 125\lambda_2^{-1}\lambda_3^{-1} + 500\lambda_1^{-1}\lambda_2^{-1}\lambda_3^{-1}}$$

The characteristic polynomial can be written in the following form:

$$\begin{aligned} d(\lambda) &= 1 + 4\lambda_1^{-1} + 25\lambda_3^{-1} + 5\lambda_2^{-1} + 100\lambda_1^{-1}\lambda_3^{-1} + 20\lambda_1^{-1}\lambda_2^{-1} + 125\lambda_2^{-1}\lambda_3^{-1} + 500\lambda_1^{-1}\lambda_2^{-1}\lambda_3^{-1} = \\ &= \underbrace{\left(1 + 4\lambda_1^{-1}\right)}_{d_{11}(\lambda)}\underbrace{\left(1 + 5\lambda_2^{-1}\right)}_{d_{22}(\lambda)}\underbrace{\left(1 + 25\lambda_3^{-1}\right)}_{d_{33}(\lambda)} \end{aligned} \tag{34}$$

After using Algorithm 1, we determine the matrices **A**, **B**, **C** in the form (28), (29), (30) and electrical circuit presented in Figure 3(b). The obtained realisation is stable, as can be seen

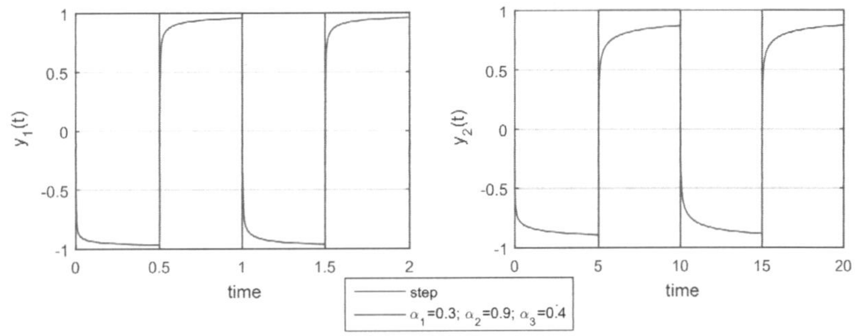

Figure 6. Step response of the system (23) for $\alpha_1 = 0.34$; $\alpha_2 = 0.9$; $\alpha_3 = 0.4$.

in Figure 6, showing the step response of the system for $\alpha_1 = 0.3$; $\alpha_2 = 0.9$ and $\alpha_3 = 0.4$. Simulations were performed using the Matlab toolbox FOMCON (Fractional-order Modeling and Control) presented in the paper [30] and [31].

5. Concluding Remarks

The paper presents a method, based on the one-dimensional digraph theory, for finding the realisations of the one-dimensional continuous-time fractional system. The difference between the proposed algorithm in this paper and currently used methods based on canonical forms of the system (i.e. constant matrix forms) is the creation of not one (or few) minimal realisations, but a set of every possible minimal realisation. In this paper, we have shown how we can realise the transfer matrix using electrical circuits. Also a method for determining a realisation using a system consisting of n subsystems with one fractional order and with many different fractional orders has been shown.

Further work includes extension of the algorithm to find the broader class of electrical circuits corresponding to the transfer matrix.

Acknowledgments

Research has been financed with the funds of the Statutory Research of 2015.

References

[1] Luenberger D G 1979 *Introduction to Dynamic Systems: Theory, Models, and Applications* (New York: Wiley) chap Positive linear systems

[2] Farina L and Rinaldi S 2000 *Positive linear systems: theory and applications* (New York: Wiley-Interscience, Series on Pure and Applied Mathematics)

[3] Kaczorek T 2001 *Positive 1D and 2D systems* (London: Springer Verlag)

[4] Benvenuti L, De Santis A and Farina L 2003 *Positive Systems* Lecture Notes on Control and Information Sciences 294 (Berlin: Springer-Verlag)

[5] Das S 2011 *Functional Fractional Calculus* (New York: Springer Berlin Heidelberg) URL `http://dx.doi.org/10.1007/978-3-642-20545-3`

[6] Kaczorek T and Sajewski L 2014 *The Realization Problem for Positive and Fractional Systems* (Berlin: Springer International Publishing) URL `http://dx.doi.org/10.1007/978-3-319-04834-5`

[7] Nishimoto K 1984 *Fractional Calculus* (Koriama: Decartess Press)

[8] Monje C A, Chen Y, Vinagre B M, Xue D and Feliu V 2010 *Fractional-Order systems and Control: Fundamentals and Applications* (Springer) ISBN 978-1-84996-334-3 URL `http://dx.doi.org/10.1007/978-1-84996-335-0`

[9] Ortigueira M D 2011 *Fractional Calculus for Scientists and Engineers* (Springer Netherlands: Academic Press) URL `10.1007/978-94-007-0747-4`

[10] Podlubny I 1999 *Fractional Differential Equations* (San Diego: Academic Press)

[11] Benvenuti L and Farina L 2004 *IEEE Transactions on Automatic Control* **49** 651–664

[12] Kaczorek T and Busłowicz M 2004 *Int. J. Appl. Math. Comput. Sci.* 181 – 187

[13] Kaczorek T 2007 *Polynomial and Rational Matrices* (London: Springer Verlag)

[14] Hryniów K and Markowski K A 2014 *Proceedings of 2014 15th International Carpathian Control Conference (ICCC)* pp 174–179 URL `http://dx.doi.org/10.1109/CarpathianCC.2014.6843592`

[15] Hryniów K and Markowski K A 2015 *Progress in Automation, Robotics and Measuring Techniques* (*Advances in Intelligent Systems and Computing* vol 350) ed Szewczyk R, Zieliski C and Kaliczyska M (Springer International Publishing) pp 63–72 ISBN 978-3-319-15795-5 URL `http://dx.doi.org/10.1007/978-3-319-15796-2_7`

[16] Hryniów K and Markowski K A 2015 *Proceedings of 20th International Conference on Methods and Models in Automation and Robotics, MMAR 2015, Miedzyzdroje, Poland, August 24–27* pp 110–115 URL `http://dx.doi.org/10.1109/MMAR.2015.7283856`

[17] Markowski K A and Hryniów K 2015 *Proceedings of 2nd IEEE International Conference on Cybernetics, CD-ROM, Gdynia, Poland, June 24-26, 2015* pp 172–177 URL `http://dx.doi.org/10.1109/CYBConf.2015.7175927`

[18] Markowski K A and Hryniów K 2015 *Proceedings of 7th Conference on Non-integer Order Calculus and its Applications7th Conference on Non-integer Order Calculus and its Applications* pp 105–118 ISBN 978-3-319-23038-2 URL `http://dx.doi.org/10.1007/978-3-319-23039-9_9`

[19] Markowski K A 2015 *Proceedings of 19th International Conference on System Theory, Control and Computing* pp 545–550 ISBN 978-1-4799-8481-7

[20] Markowski K A 2015 *Procedings of the International Symposium on Fractional Signal and Systems* ed Both-Rusu R (U.T. Press) pp 7–12 ISBN 978-606-737-084-3

[21] Hryniów K and Markowski K 2015 *Progress in Automation, Robotics and Measuring Techniques* (*Advances in Intelligent Systems and Computing* vol 350) ed Szewczyk R, Zieliński C and Kaliczyńska M (Springer International Publishing) pp 73–83 ISBN 978-3-319-15795-5 URL `http://dx.doi.org/10.1007/978-3-319-15796-2_8`

[22] Hryniów K and Markowski K A Accepted, 2015 *Proceedings of 20th International Conference on Methods and Models in Automation and Robotics, MMAR 2015, Miedzyzdroje, Poland, August 24–27* pp 1139–1144 URL `http://dx.doi.org/10.1109/MMAR.2015.7284039`

[23] Fornasini E and Valcher M E 1997 *Linear Algebra and Its Applications* **263** 275–310

[24] Fornasini E and Valcher M E 2003 *LCNIS* 297–304

[25] Fornasini E and Valcher M E 2005 *IEEE Transaction on Circuits and Systems I* 576–585

[26] Berman A, Neumann M and Stern R J 1989 *Nonnegative Matrices in Dynamic Systems* (New York: Wiley)

[27] Horn R A and Johnson C R 1991 *Topics in Matrix Analysis* (Cambridge Univ. Press)

[28] Bang-Jensen J and Gutin G 2009 *Digraphs: Theory, Algorithms and Applications* (London: Springer-Verlag)

[29] Foulds L R 1991 *Graph Theory Applications* (Springer Verlag)

[30] Tepljakov A, Petlenkov E and Belikov J 2011 *International Journal of Microelectronics and Computer Science* **2** 51–62

[31] Tepljakov A, Petlenkov E and Belikov J 2011 *The 18th International Conference Mixed Design of Integrated Circuits and Systems* (Gliwice, Poland) p 684–689

On the Separation Theorem of Stochastic Systems in the Case Of Continuous- Discrete Observation Channels with Memory

S V Rozhkova[1], V I Rozhkova[2], V V Lasukov[3] and E V Devyashina[4]
[1]Professor, Department of Higher Mathematics, Institute of Physics and Technology, National Research Tomsk Polytechnic University, Lenina 30, Tomsk 634050, Russia
[2]Senior Teacher, Department of Higher Mathematics, Institute of Physics and Technology,
National Research Tomsk Polytechnic University, Lenina 30, Tomsk 634050, Russia
[3]Assistant Professor, Department of Higher Mathematics, Institute of Physics and Technology,
National Research Tomsk Polytechnic University, Lenina 30, Tomsk 634050, Russia
[4]Student, Elite Technical Education,
National Research Tomsk Polytechnic University, Lenina 30, Tomsk 634050, Russia

E-mail: rozhkova@tpu.ru

Abstract. The article proves the separation theorem for optimal control of stochastic systems in the case when an observed continuous-discrete-time process possess memory of arbitrary ration relating to a state vector.

1. Introduction

A principle of dynamic programming [1], and separation theorem [2] are the basis for a theory of optimal control by incompletely observable stochastic systems. Being fundamental theoretical results, they ensured solving of a number of important practical problems in various application domains [3, 4]. Availability of time delays [5, 6] and a continuous-discrete character of the observable information [7-11] require study the problems, when: 1) observable information is a totality of realizations of continuous and discrete time-processes; 2) current observable process values depend both on current and arbitrary number of the past values of a state vector, i.e. observation channels have a memory of arbitrary multiplicity. The main result of this article is extension of the separation theorem [2] in the case of continuous-discrete observation channels with a memory of arbitrary multiplicity.

Used notations: $P\{\cdot\}$ is event probability; $E\{\cdot\}$ is the mathematical expectation; normal (Gaussian) density with parameters a and b by $N\ \{a;\ b\}$; $|\cdot|$ and $tr[\cdot]$ are determinant and trace of matrix; D^{-1} is the inversion matrix of D; D^T denotes transpose of a matrix or a vector; $D > 0$ $(D \geq 0)$ is positive (non-negative) definite matrix.

2. The problem statement

On a certain probability space $(\Omega, F, \mathbf{F} = (F_t)_{t\geq 0}, \mathbf{P})$ [12] the unobservable n-dimensional process x_t (state vector of system) and the observable l- dimensional process z_t with continuous time are determined by the stochastic differential equations (Ito's differential rule)

$$dx_t = f(t, x_t, u_t)dt + \Phi_1(t, x_t)dw_t, \quad t \in [0, \mathrm{T}], \tag{1}$$

$$dz_t = h(t, x_t, x_{\tau_1}, \cdots, x_{\tau_N}, z)dt + \Phi_2(t)dv_t, \tag{2}$$

and the observable q -dimensional process with discrete time $\eta(t_m)$ has the form

$$\eta(t_m) = g(t_m, x_{t_m}, x_{\tau_1}, \cdots, x_{\tau_N}) + \Phi_3(t_m)\xi(t_m), \quad m = 1,2\cdots k, \tag{3}$$

where u_t is $r-$dimensional control vector, $0 < \tau_N < \tau_{N-1} < \cdots < \tau_1 < t_m \le t$, and $\tau_k = const$, $k = \overline{1;N}$. It is assumed: 1) w_t and v_t are r_1-dimensional and r_2-dimensional standard Wiener processes; 2) x_0, w_t, v_t and $\xi(t_m)$ are jointly independent; 3) $f(\cdot)$, $h(\cdot)$, $\Phi_1(\cdot)$, $\Phi_2(\cdot)$, $\Phi_3(\cdot)$ are continuous functions for all arguments; 4) $Q(\cdot) = \Phi_1(\cdot)\Phi_1^T(\cdot) > 0$, $R(\cdot) = \Phi_2(\cdot)\Phi_2^T(\cdot) > 0$, $V(\cdot) = \Phi_3(\cdot)\Phi_3^T(\cdot) > 0$; 5) original density is set $p_0(x) = \partial P\{x_0 \le x\}/\partial x$.

Task: to find control u_t^0 that provides optimum condition, on the class of $F_t^{z,\eta}$-measurable functional, where $u_t = u_t[z_0^t; \eta_0^m]$, $z_0^t = \{z_s; 0 \le s \le t\}$, $\eta_0^m = \{\eta(t_0), \eta(t_1), \cdots, \eta(t_m); t_m \le t\}$

$$J = E\left\{ b(\mathrm{T}, x_\mathrm{T}) + \int_\tau^\mathrm{T} \Lambda(s, x_s, u_s)ds \middle| p_\tau(x) \right\} \to \min_{\{u_\tau^\mathrm{T}\}}, \tag{4}$$

where $\tau > \tau_1$, $u_\tau^T = \{u_s; \tau \le s \le \mathrm{T}\}$, $b(\cdot) \ge 0$, $\Lambda(\cdot) > 0$.

Remark 1. Suppose that on the time interval $t \in [0, \tau]$ as a control vector u_t to be used arbitrary $F_t^{z,\eta}$ - measurable functional.

Remark 2. To solve this problem, we use sufficient coordinates method [13], under the assumption that $F_t^{z,\eta}$ - measurable process $\lambda_t = \lambda_t[z_0^t; \eta_0^m]$ characterizing a posteriori density

$$p_t(x) = \partial P\{x_t \le x \mid z_0^t, \eta_0^m\}/\partial x \tag{5}$$

exists. Then, we have $u_t = u_t[\lambda_t]$.

3. Preliminary results

In accordance with the sufficient coordinates method, we introduce Bellman function

$$S(t, \lambda) = \min_{\{u_\tau^T\}} E\left\{ b(\mathrm{T}, x_\mathrm{T}) + \int_t^\mathrm{T} \Lambda(t, x_s, u_s)ds \middle| \lambda_t = \lambda \right\}. \tag{6}$$

Theorem 1. Let the process λ_t is Markov diffusion process with characteristics

$$a(t, \lambda) = \lim_{\Delta t \to 0} \frac{1}{\Delta t} E\{\Delta\lambda_t | \lambda_t = \lambda\},$$

$$D(t, \lambda) = \lim_{\Delta t \to 0} \frac{1}{\Delta t} E\{[\Delta\lambda_t][\Delta\lambda_t]^T | \lambda_t = \lambda\}, \tag{7}$$

where $\Delta\lambda_t = \lambda_{t+\Delta t} - \lambda_t$.

Then, Bellman equation for $S(t, \lambda)$ has the form

$$\min_{\{u\}}\left\{\frac{\partial S(t,\lambda)}{\partial t}+L^*_{t,\lambda}[S(t,\lambda)]+E\{\Lambda(t,x_t,u_t)|\lambda_t=\lambda\}\right\}=0, \tag{8}$$

$$S(t,\lambda)|_{t=\mathrm{T}}=E\{b(\mathrm{T},x_\mathrm{T})|\lambda_\mathrm{T}=\lambda\}, \tag{9}$$

where $L^*_{t,\lambda}[\cdot]$ denotes inverse Kolmogorov operator that corresponds to process λ_t [13], i.e.

$$L^*_{t,\lambda}[S(t,\lambda)]=a^T(t,\lambda)\frac{\partial S(t,\lambda)}{\partial\lambda}+\frac{1}{2}tr\left[D(t,\lambda)\frac{\partial^2 S(t,\lambda)}{\partial\lambda^2}\right], \tag{10}$$

and the smallest value of performance criterion has the form $J^0=S(\tau,\lambda)$.

Proof. From (6), taking into account $E\{\cdot|\lambda_t=\lambda\}=E\{E\{\cdot|\lambda_{t+\Delta t}=\lambda'\}|\lambda_t=\lambda\}$, we obtain

$$S(t,\lambda)=\min_{\{u_t^{t+\Delta t}\}}\min_{\{u_{t+\Delta t}^{\mathrm{T}}\}}E\left\{E\left\{b(\mathrm{T},x_\mathrm{T})+\int_{t+\Delta t}^{\mathrm{T}}\Lambda(s,x_s,u_s)ds|\lambda_{t+\Delta t}=\lambda'\right\}++\int_t^{t+\Delta t}\Lambda(\tau,x_\tau,u_\tau)d\tau|\lambda_t=\lambda\right\}. \tag{11}$$

Then, from (6), (11) it follows that $S(t,\lambda)$ is determined by recurrence equation

$$S(t,\lambda)=\min_{\{u\}}\{E\{S(t+\Delta t,\lambda_{t+\Delta t})|\lambda_t=\lambda\}+\Delta tE\{\Lambda(t,x_t,u_t)|\lambda_t=\lambda\}+o(\Delta t)\}. \tag{12}$$

Since

$$S(t+\Delta t,\lambda_{t+\Delta t})=S(t+\Delta t,\lambda_t)+\Delta\lambda_t^T\frac{\partial S(t+\Delta t,\lambda_t)}{\partial\lambda_t}+\frac{1}{2}\Delta\lambda_t^T\frac{\partial^2 S(t+\Delta t,\lambda_t)}{\partial\lambda_t^2}\Delta\lambda_t+o([\Delta\lambda]^2), \tag{13}$$

taking into account (7), (10), (13), we have

$$E\{S(t+\Delta t,\lambda_{t+\Delta t})|\lambda_t=\lambda\}=S(t+\Delta t,\lambda)+(\Delta t)L^*_{t,x}[S(t+\Delta t,\lambda)]+o(\Delta t). \tag{14}$$

Substituting (14) into (12) and proceeding to limit $\Delta t\to 0$ gives (8). Boundary condition (9) and expression for J^0 arise from (4), (6).

Proposition 1. Let

$$\Phi_1(\cdot)=\Phi_1(t),\ p_0(x)=N\{x;\mu_0,\Gamma_0\},$$

$$h(\cdot)=H_0(t)x_t+\sum_{k=1}^N H_k(t)x_{\tau_k},\quad g(\cdot)=G_0(t_m)x_{t_m}+\sum_{k=1}^N G_k(t_m)x_{t_m}, \tag{15}$$

$$\mu(t)=E\{x_t\,|\,z_0^t,\eta_0^m\},\quad \mu(\tau_k,t)=E\{x_{\tau_k}\,|\,z_0^t,\eta_0^m\},\quad k=\overline{1;N},$$

$$\Gamma(t)=E\{[x_t-\mu(t)][\cdot]^T|z_0^t,\eta_0^m\},$$

$$\Gamma_{kk}(\tau_k,t)=E\{[x_{\tau_k}-\mu(\tau_k,t)][\cdot]^T|z_0^t,\eta_0^m\},$$

$$\Gamma_{0k}(\tau_k,t)=E\{[x_t-\mu(t)][x_{\tau_k}-\mu(\tau_k,t)]^T|z_0^t,\eta_0^m\} \tag{16}$$

$$\Gamma_{ki}(\tau_k,\tau_i,t)=E\{[x_{\tau_k}-\mu(\tau_k,t)][x_{\tau_i}-\mu(\tau_i,t)]^T|z_0^t,\eta_0^m\}$$

$$k=\overline{1;N},\quad i=\overline{2;N},\quad i>k.$$

Journal of Physics: Conference Series **659** (2015) 012042 doi:10.1088/1742-6596/659/1/012042

Then, for a posteriori density (5) property

$$p_t(x) = \mathrm{N}\{x;\ \mu(t), \Gamma(t)\} \tag{17}$$

takes place, and parameters of this density on the time intervals $t_m \le t < t_{m+1}$ are determined by the equations

$$d_t\mu(t) = [F(t)\mu(t) + B(t)u_t]dt + \tilde{H}_0^T(t)R^{-1}(t)d\tilde{z}_t, \tag{18}$$

$$d\mu(\tau_k, t) = \tilde{H}_k^T(t)R^{-1}(t)d\tilde{z}_t, \tag{19}$$

$$d\Gamma(t)/dt = F(t)\Gamma(t) + \Gamma(t)F^T(t) - \tilde{H}_0^T(t)R^{-1}(t)\tilde{H}_0(t) + Q(t), \tag{20}$$

$$d\Gamma_{kk}(\tau_k, t)/dt = -\tilde{H}_k^T(t)R^{-1}(t)\tilde{H}_k(t), \tag{21}$$

$$d\Gamma_{0k}(\tau_k, t)/dt = F(t)\Gamma_{0k}(\tau_k, t) - \tilde{H}_0^T(t)R^{-1}(t)\tilde{H}_k(t), \tag{22}$$

$$d\Gamma_{ki}(\tau_k, \tau_i, t)/dt = -\tilde{H}_k^T(t)R^{-1}(t)\tilde{H}_i(t, z) \tag{23}$$

with the initial conditions

$$\mu(t_m) = \mu(t_m - 0) + \tilde{G}_0^T(t_m)W^{-1}(t_m)\tilde{\eta}(t_m), \tag{24}$$

$$\mu(\tau_k, t_m) = \mu(\tau_k, t_m - 0) + \tilde{G}_k^T(t_m)W^{-1}(t_m)\tilde{\eta}(t_m), \tag{25}$$

$$\Gamma(t_m) = \Gamma(t_m - 0) - \tilde{G}_0^T(t_m)W^{-1}(t_m)\tilde{G}_0(t_m), \tag{26}$$

$$\Gamma_{kk}(\tau_k, t_m) = \Gamma_{kk}(\tau_k, t_m - 0) - \tilde{G}_k^T(t_m)W^{-1}(t_m)\tilde{G}_k(t_m), \tag{27}$$

$$\Gamma_{0k}(\tau_k, t_m) = \Gamma_{0k}(\tau_k, t_m - 0) - \tilde{G}_0^T(t_m)W^{-1}(t_m)\tilde{G}_k(t_m), \tag{28}$$

$$\Gamma_{ki}(\tau_k, \tau_i, t_m) = \Gamma_{ki}(\tau_k, \tau_i, t_m - 0) - \tilde{G}_k^T(t_m)W^{-1}(t_m)\tilde{G}_i(t_m), \tag{29}$$

where $\varphi(\cdot, t_m - 0) = \varphi(\cdot, t)$ as $t \uparrow t_m$,

$$d\tilde{z}(t) = dz(t) - \left[H_0(t)\mu(t) + \sum_{j=1}^{N} H_j(t)\mu(\tau_j, t)\right]dt, \tag{30}$$

$$\tilde{H}_0(t) = H_0(t)\Gamma(t) + \sum_{j=1}^{N} H_j(t)\Gamma_{0j}^T(\tau_j, t), \tag{31}$$

$$\tilde{H}_k(t) = H_k(t)\Gamma_{kk}(\tau_k, t) + \sum_{j \ne k}^{N} H_j(t)\Gamma_{kj}^T(\tau_k, \tau_j, t), \tag{32}$$

$$\tilde{\eta}(t_m) = \eta(t_m) - G_0(t_m)\mu(t_m - 0) - \sum_{j=1}^{N} G_j(t_m)\mu(\tau_j, t_m - 0), \tag{33}$$

$$\tilde{G}_0(t_m) = G_0(t_m)\Gamma(t_m - 0) + \sum_{j=1}^{N} G_j(t_m)\Gamma_{0j}^T(\tau_j, t_m - 0), \tag{34}$$

$$\tilde{G}_k(t_m) = G_k(t_m)\Gamma_{kk}(\tau_k, t_m - 0) + \sum_{j \ne k}^{N} G_j(t_m)\Gamma_{kj}^T(\tau_k, \tau_j, t_m - 0), \tag{35}$$

$$W(t_m) = V(t_m) + G_{0,N}(t_m)\tilde{\Gamma}_{N+1}(\tilde{\tau}_N, t_m - 0)G_{0,N}^T(t_m), \tag{36}$$

$$G_{0,N}(t_m) = [G_0(t_m) \mid G_1(t_m) \mid \cdots \mid G_N(t_m)], \tag{37}$$

and $\tilde{\Gamma}_{N+1}(\tilde{\tau}_N, t)$ is $[n(N+1)] \times [n(N+1)]$ matrix, which consist of $\Gamma(t)$, $\Gamma_{kk}(\cdot)$, $\Gamma_{0k}(\cdot)$, $\Gamma_{ki}(\cdot)$ (see (16)). This Proposition arises from [7].

Proposition 2. A sufficient coordinates vector is an optimal in a mean-root-square sense estimate $\mu(t)$ of the process x_t, i.e. $\lambda_t[z_0^t, \eta_0^m] = \mu(t)$, which is a Markov diffusion process with the characteristics (see (7))

$$a(t,\mu) = F(t)\mu + B(t)u, \quad D(t,\mu) = \tilde{H}_0^T(t)R^{-1}(t)\tilde{H}_0(t), \tag{38}$$

$$D(t_m,\mu) = D(t_m - 0, \mu) + \Delta D(t_m, \mu), \quad \Delta D(t_m,\mu) = \tilde{G}_0^T(t_m)W^{-1}(t_m)\tilde{G}_0(t_m). \tag{39}$$

Proof. Since $\Gamma(t)$, according to (20), (26), does not depend on z_t and $\eta(t_m)$, then from (17), taking into consideration Remark 2, it follows that $\lambda_t = \mu(t)$. Process $\tilde{z}_t$, a differential of which is given by (30), is F_t^z-Wiener process with $E\{\tilde{z}_t\tilde{z}_t^T \mid F_t^z\} = \int_\tau^t R(s)ds$. Then, since $u_t = u_t[\mu(t)]$ (see Remark 2), so from (18) follows that process $\mu(t)$ is a diffusion Markov process with the characteristics (38). Formulae (39) follows from (24) taking into consideration $E\{\tilde{\eta}(t_m)\tilde{\eta}^T(t_m) \mid z_0^{t_m}, \eta_0^{m-1}\} = W(t_m)$.

4. Main results

Theorem 2 (The separation theorem). Then optimal control u_t^0 is given by

$$u_t^0 = -N^{-1}(t)B^T(t)S(t)\mu(t), \tag{41}$$

$$u_{t_m}^0 = u_{t_m - 0}^0 + \Delta u_{t_m}^0, \tag{42}$$

$$u_{t_m - 0}^0 = -N^{-1}(t_m)B^T(t_m)S(t_m)\mu(t_m - 0),$$

$$\Delta u_{t_m}^0 = -N^{-1}(t_m)B^T(t_m)S(t_m)\tilde{G}_0^T(t_m)W^{-1}(t_m)\tilde{\eta}(t_m), \tag{43}$$

an optimal, in a mean-root-square sense, estimate $\mu(t)$ of the state vector x_t is defined by expressions of Proposition 1 , where $u_t = u_t^0$, matrix $S(t)$ is defined by matrix differential Riccati equation

$$\dot{S}(t) = -F^T(t)S(t) - S(t)F(t) + S(t)B(t)N^{-1}(t)B^T(t)S(t) - L(t), \quad S(\mathrm{T}) = S_\mathrm{T}, \tag{44}$$

and a minimal value of the performance criterion J^0 is given by

$$\begin{aligned} J^0 = {} & \mu^T(\tau)S(\tau)\mu(\tau) + tr[S_\mathrm{T}\Gamma(\mathrm{T})] + \int_\tau^\mathrm{T} tr[L(t)\Gamma(t)]dt + \int_\tau^t tr[\tilde{H}_0(t)R^{-1}(t)\tilde{H}_0(t)S(t)]dt + \\ & + \sum_{m=1}^k tr[\tilde{G}_0^T(t_m)W^{-1}(t_m)\tilde{G}_0(t_m)S(t_m)] + \sum_{m=1}^n tr[L(t_m)\Gamma(t_m)]. \end{aligned} \tag{45}$$

Proof. Since $\lambda_t = \mu(t)$ and $u_t = u_t(\mu(t))$, then, taking into consideration F_t^z-measurability $\mu(t)$ from (40) follows

$$E\{\Lambda(t, x_t, u_t) \mid z_0^t, \eta_0^m\} = \mu^T(t)L(t)\mu(t) + tr[L(t)\Gamma(t)] + u_t^T N(t)u_t,$$

$$E\{b(\mathrm{T}, x_{\mathrm{T}}) | z_0^{\mathrm{T}}, \eta_0^k\} = \mu^T(\mathrm{T}) S_{\mathrm{T}} \mu(\mathrm{T}) + tr[S_{\mathrm{T}} \Gamma(\mathrm{T})]. \tag{46}$$

Use of (46) in Theorem 1 taking into account Proposition 2 and (26) we introduce Bellman equation, given by

$$\min_u \left\{ \frac{\partial S(t,\mu)}{\partial t} + [F(t)\mu + B(t)u]^T \frac{\partial S(t,\mu)}{\partial \mu} + \frac{1}{2} tr\left[\tilde{H}_0^T(t) R^{-1}(t) \tilde{H}_0(t) \frac{\partial^2 S(t,\mu)}{\partial \mu^2} \right] + \right.$$
$$+ \frac{1}{2} \sum_{m=1}^{k} \delta(t - t_m) tr\left[\tilde{G}_0^T(t_m) W^{-1}(t_m) \tilde{G}_0(t_m) \frac{\partial^2 S(t,\mu)}{\partial \mu^2} \right] + \mu^T L(t)\mu + u^T N(t) u + \tag{47}$$
$$\left. + tr[L(t)\Gamma(t)] + \sum_{m=1}^{k} \delta(t - t_m) tr[L(t)\Gamma(t)] \right\} = 0,$$

$$S(\mathrm{T}, \mu) = tr[S_{\mathrm{T}} \Gamma(\mathrm{T})] + \mu^T S_{\mathrm{T}} \mu. \tag{48}$$

Differentiating the left side (47) with respect to u, we obtain

$$u_t^0 = -(1/2) N^{-1}(t) B^T(t) [\partial S(t,\mu)/\partial \mu]. \tag{49}$$

Substitution of (49) into (47) gives the equation

$$\frac{\partial S(t,\mu)}{\partial t} + \mu^T F(t) \frac{\partial S(t,\mu)}{\partial \mu} - \frac{1}{4} \left(\frac{\partial S(t,\mu)}{\partial \mu} \right)^T B(t) N^{-1}(t) B^T(t) \frac{\partial S(t,\mu)}{\partial \mu} +$$
$$+ \mu^T L(t)\mu + tr[L(t)\Gamma(t)] + \frac{1}{2} tr\left[\tilde{H}_0^T(t) R^{-1}(t) \tilde{H}_0(t) \frac{\partial^2 S(t,\mu)}{\partial \mu^2} \right] + \tag{50}$$
$$+ \frac{1}{2} \sum_{m=1}^{k} \delta(t - t_m) tr\left[\tilde{G}_0^T(t_m) W^{-1}(t_m) \tilde{G}_0(t_m) \frac{\partial^2 S(t,\mu)}{\partial \mu^2} \right] + \sum_{m=1}^{k} \delta(t - t_m tr[L(t)\Gamma(t)]) = 0$$

we can solve in the form

$$S(t,\mu) = l(t) + \mu^T S(t) \mu. \tag{51}$$

Substitution (51) into (50) gives the expression

$$\dot{l}(t) + \mu^T \dot{S}(t)\mu + \mu^T F^T(t) S(t)\mu + \mu^T S(t) F(t)\mu - \mu^T S(t) B(t) N^{-1}(t) B^T(t) S(t)\mu +$$
$$+ \mu^T L(t)\mu + tr[L(t)\Gamma(t)] + tr[\tilde{H}_0^T(t) R^{-1}(t) \tilde{H}_0(t) S(t)] + \sum_{m=1}^{k} \delta(t - t_m) tr[L(t)\Gamma(t)] + \tag{52}$$
$$+ \sum_{m=1}^{k} \delta(t - t_m) tr[\tilde{G}_0^T(t_m) W^{-1}(t_m) \tilde{G}_0(t_m) S(t)] = 0.$$

Equating coefficients in (52) at equal degrees μ gives equation (44) for $S(t)$ and an equation for $l(t)$

$$\dot{l}(t) = -tr[L(t)\Gamma(t)] - tr[\tilde{H}_0^T(t) R^{-1}(t) \tilde{H}_0(t) S(t)] - \sum_{m=1}^{k} \delta(t - t_m) tr[\tilde{G}_0^T(t_m) W^{-1}(t_m) \tilde{G}_0(t_m) S(t)] -$$
$$- \sum_{m=1}^{k} \delta(t - t_m) tr[L(t)\Gamma(t)],$$

(53)

$$l(\mathrm{T}) = tr[S_{\mathrm{T}} \Gamma(\mathrm{T})],$$

and boundary conditions follow from (46), (51). Solution (53) for $t = \tau$ is given by

$$l(\tau)=tr[S_{\mathrm{T}}\Gamma(\mathrm{T})]+\int_{\tau}^{\mathrm{T}}tr[L(t)\Gamma(t)]dt+\int_{\tau}^{\mathrm{T}}tr\left[\tilde{H}_0^T(t)R^{-1}(t)\tilde{H}_0(t)S(t)\right]dt+ \\ +\sum_{m=1}^{k}tr\left[\tilde{G}_0^T(t_m)W^{-1}(t_m)\tilde{G}_0(t_m)S(t_m)\right]+\sum_{m=1}^{k}tr[L(t_m)\Gamma(t_m)]. \tag{54}$$

Use of (51) in (49) taking into account (24) gives (41)-(43). Since $J^0=S(\tau,\mu)$, then (45) follows from (51), (54).

5. Conclusions

1. Matrix $S(t)$, defining a regulator, is the same matrix as in the case of a classical separation theorem [2]. Discrepancy is explained by the fact that in a classical case estimate of state vector is produced by the Kalman filter, while in the considered case – by continuous-discrete filter-interpolation (*Proposition 1*), which produces not only filtering estimate $\mu(t)$ for a current value x_t of the sate vector, but also interpolation estimates $\mu(\tau_k,t)$ for the past values x_{τ_k}, $k=\overline{1;N}$ of the state vector. Expression for J^0 changes respectively.

2. From Theorem 2 follows solutions of problems for the following special cases: 1) only continuous observations with memory, or with delay ($H_0(t)=0$); 2) only discrete observations with memory, or with delay ($G_0(t_m)=0$); 3) continuous observations with memory, discrete observations, without memory ($G_k(t_m)=0$, $k=\overline{1;N}$), with delay ($G_0(t_m)=0$); 4) discrete observation with memory, continuous observation, without memory ($H_k(t)=0$, $k=\overline{1;N}$), with delay ($H_0(t)=0$); 5) various combinations of the cases 3) and 4).

Acknowledgments

All comments and suggestions by critical readers and editorial assistance are acknowledged with gratitude.

References

[1] Bellman R 1957 *Dynamic programming* (New-York: Princeton Univ. Press, Prinseton)

[2] Wonham W M 1968 On the separation theorem of stochastic control *SIAM J. Control* **6** 312-326

[3] Bryson A E, Yu-Chi-Ho 1969 *Applied optimal control* (London: Blaisdell publishing)

[4] Boguslavskii A I 1970 *Navigation and control methods by incomplete statistical information* (Moscow: Mashinostroenie)

[5] Dion J.M., Dugard J.L. and Fliss M. Linear Time-Delay Systems. Pergamon, London, 1999.

[6] Kolmanovskii V B, Shaikhet L E 1996 *Control of systems with aftereffect* (American Math. Soc. Providence)

[7] Abakumova O L, Demin N S, Sushko T V 1995 Filtration of stochastic processes for continuous and discrete observations with memory. II. Synthesis of filters *Automation and Remote Control* **56**:10, 1383–1393

[8] Dyomin N S, Rozhkova S V and Rozhkova O V 2000 Generalized moving extrapolation of stochastic processes based on the totality of continuous and discrete observations with memory *Journal of Computer and Systems Sciences International* **39**:4, 523-535

[9] Dyomin N S, Rozhkova S V and Safronova I E 2003 Information amount determination for joint problem of filtering and generalized extrapolation of stochastic processes with respect to the set of continuous and discrete memory *Informatica* **14**:3, 295-322

[10] Dyomin N S, Rozhkova S V and Safronova I E 2004 About structure of Shannon information amount for joint filtering and extrapolation problem by continuous-discrete memory observations *Informatica* **15**:2, 171-202

[11] Dyomin N S, Rozhkova S V 2006 Optimal control of stochastic systems in the case of continuous-discrete observation channels with memory *Avtomatika i Vychislitel'naya Tekhnika* **6**, 3-15
[12] Liptser R S, Shiryaev A N 2001 *Statistics of Random Processes* (Berlin: Springer-Verlag Berlin Heidelberg)
[13] Stratonovich R L 1968 *Conditional Markov processes and their application to the theory of optimal control* (New York: American Elsevier Publishing Company)

Statistical process control for AR(1) or non-Gaussian processes using wavelets coefficients

A Cohen, T Tiplica and A Kobi

L'UNAM, LARIS Systems Engineering Research Laboratory, ISTIA Engineering School, 62 Avenue Notre Dame du Lac 49000 Angers, France

E-mail: achraf.cohen@univ-angers.fr

Abstract. Autocorrelation and non-normality of process characteristic variables are two main difficulties that industrial engineers must face when they should implement control charting techniques. This paper presents new issues regarding the probability distribution of wavelets coefficients. Firstly, we highlight that wavelets coefficients have capacities to strongly decrease autocorrelation degree of original data and are normally-like distributed, especially in the case of Haar wavelet. We used AR(1) model with positive autoregressive parameters to simulate autocorrelated data. Illustrative examples are presented to show wavelets coefficients properties. Secondly, the distributional parameters of wavelets coefficients are derived, it shows that wavelets coefficients reflect an interesting statistical properties for SPC purposes.

1. Introduction

Statistical process control (SPC) gathers several statistical techniques in order to monitor manufacturing processes and service operations. SPC mainly involves the implementation of control charts. They are among the most widely techniques used to detect changes in Gaussian models parameters. Based on statistics and exploiting the data collected from processes, control charts are particularly designed to detect assignable causes. Various procedures and control schemes have been proposed for detecting mean and/or variance change. The Xbar-R and Xbar-S are the firsts control schemes proposed [1]. Afterwards, EWMA [2] and CUSUM [3] are introduced, which are more effective than Xbar chart, especially for small shifts. Nowadays, ideas to vary control charts parameters (sample size, sampling interval, control limit coefficient) have been developed considerably by proposing several charts, called adaptive control charts. For more details, see [4–6]. Recently, a comparison study was done to evaluate the detection performance of nine control charts, for monitoring process mean [7]. It shows that CUSUM/EWMA are the best charts in terms of FSSI (Fixed Sample Size and Sampling Intervals) control charts.

In the last decade the multi-scale SPC techniques, which combining SPC classical models ($\overline{X}$, EWMA, CUSUM, PCA, etc.) and multi-scale decomposition using wavelets analysis [8–11], are taking a big part in the SPC literature, especially in the multivariate context [12–20]. Most of the published papers that investigated this combination focused on the use of wavelets as data pre-processing tool, such as de-noising [21–23], data reduction [24; 25], or feature extraction [26; 27], in order to improve the detection performance. These research works are mainly motivated by the ability of wavelets analysis to extract signals time-frequency/scale components. For

Journal of Physics: Conference Series **659** (2015) 012043 doi:10.1088/1742-6596/659/1/012043

example to identify the corrosion intensity the time-frequency plan was used to localize the corrosion [28; 29].

Independence is one of the two hypotheses that most SPC procedures suppose valid. However, data collected from processes, mostly via sensors, are generally autocorrelated. This aspect can also exist in quality characteristic data, especially when short sampling interval are used. The main effect of autocorrelation in control charts is the inflation of the false alarms rate, consequently the Average Run Length (ARL) is improperly calculated. To face this issue two approaches exist: 1) adjust control limits to take into account the autocorrelation, it's done by estimating the true variance of the process and 2) construct a times series model that fits the process data, then monitor the residues using classical SPC tools. A good literature review regarding SPC procedures for monitoring autocorrelated processes is done here [30].

On the other hand, violation of normality hypothesis can also generate dramatic consequences in terms false alarms rate, for more details see [31–33]. Three solutions are proposed to deal with non-normality: 1) use data transformation techniques in order to make data normal distribution-like, such as Box-Cox [34] and Jonhson [35] transformations; 2) adjust control limits by taking into account data statistical characteristic (kurtosis and skewness) and 3) create control chart that is adapted to the distribution of the data. For instance, CUSUM for observations exponentially distributed are proposed [36].

In these references [13; 26] authors noted that wavelets analysis has the ability to decrease strongly autocorrelation degree, also wavelets coefficients are normally distributed even if original data are not Gaussian. Nevertheless, there is no study in the literature regarding the evaluation of autocorrelation and normality of wavelets coefficients. We note also that there is a lack of case studies in order to highlight the benefit of using wavelets coefficients to face autocorrelation and normality issues for SPC procedures.

The goal of this paper is to propose a result regarding wavelets coefficients probability distribution, using orthogonal wavelets families. Therefore we shed light on the use of wavelets coefficients for SPC purpose. We present briefly the result regarding the evaluation of normality and autocorrelation of wavelets coefficients.

This paper is organized as follows: the second section introduces the wavelets coefficients and their distributional characteristics. We present the findings related to study of autocorrelation and normality of wavelets coefficients. In the third section we are presenting the results and discussions for SPC issues. Finally, in the last section conclusions and perspectives are presented.

2. Statistical properties of wavelets coefficients

MultiResolution Analysis (MRA) [37] provides multiscale decomposition using orthogonal wavelets families across filter banks. MRA has led to create Fast Wavelet Transform (FWT) and then multiply applications of wavelets, especially in image processing. For more details see [8; 10; 37]. Wavelets coefficients, approximations $a_j(k)$ and details $d_j(k)$, are given as follows:

$$a_j(k) = \sum_{i=0}^{l} h[i]a_{j-1}[2k - i] \tag{1}$$

$$d_j(k) = \sum_{i=0}^{l} g[i]a_{j-1}[2k - i] \tag{2}$$

Where $a_0 = x$ the original signal, j represents the decomposition scale; $k \in Z$; l is the filter length; h and g are scaling and wavelets filters respectively.

2.1. Autocorrelation & wavelets coefficients

This part aims to evaluate the wavelets coefficients Autocorrelation Function (AF). We used an autoregressive model AR(1) with positive autocorrelation in order to simulate autocorrelated data. This model is used by several authors to evaluate control charts performance in presence of autocorrelation [38–40]. AR(1) model is defined as follows:

$$X_t = c + \varepsilon_t + \varphi_1 X_{t-1} \tag{3}$$

Where $\varepsilon_t \rightsquigarrow \mathcal{N}(0, \sigma_0^2)$ and we used a stationary model ($c = 0$ and $|\varphi_1| < 1$). Simulation study was done for different window size $N \in \{4, 8, 16, 32, 64, 128\}$, $\varphi_1 \in \{0.1, 0.2, ..., 0.7, 0.8, 0.9, 0.95\}$ and wavelets families (Haar, Daubechies, Symlets, Coiflets). Results for $\varphi_1 \in \{0.7, 0.8, 0.9, 0.95\}$ are presented in tables 1 and 2. One can remark that globally wavelets analysis decreases strongly the autocorrelation present in the original data. Moreover, in the case of Haar and *db*2 wavelets one can remark that serial correlation (AF($T = 2$)) is equal to zero. For instance, when $\varphi_1 = 0.95$ and using $N = 64$ we obtained for Haar a serial correlation that equal to $-0.12(sd = 0.21)$ (sd: standard-deviation) and for *db*2 we got 0.18(0.28). But for *db*3 one can see from table 2 that autocorrelation is evaluated as 0.52(0.22). Regarding results in tables 1 and 2, we have concluded that the autocorrelation can be removed using Haar, *db*2 especially for large signals $N > 32$, and by using others wavelets autocorrelation can be strongly debased.

2.1.1. An illustrative example

In this example we show the autocorrelation functions behavior of wavelets coefficients using *db*2 wavelet and $N = 128$. The signal has an autoregressive parameter equal to $\varphi_1 = 0.95$, see figure 1(a)-(b). The figure 1(c) shows the autocorrelation function of approximation coefficients (6 coefficients in this case), one can remark that there is no serial correlation (0.03(0.30), see the table 2), the same conclusion can be achieved in the case of wavelets coefficients (App. + Det.). By the same way, the figure 1(d) shows the autocorrelation function of details coefficients. It is clear that details coefficients are sightly autocorrelated (-0.31(0.16), see the table 2), but it could be very acceptable compared to the autocorrelation of original data $\varphi_1 = 0.95$. For more details see tables 1 and 2.

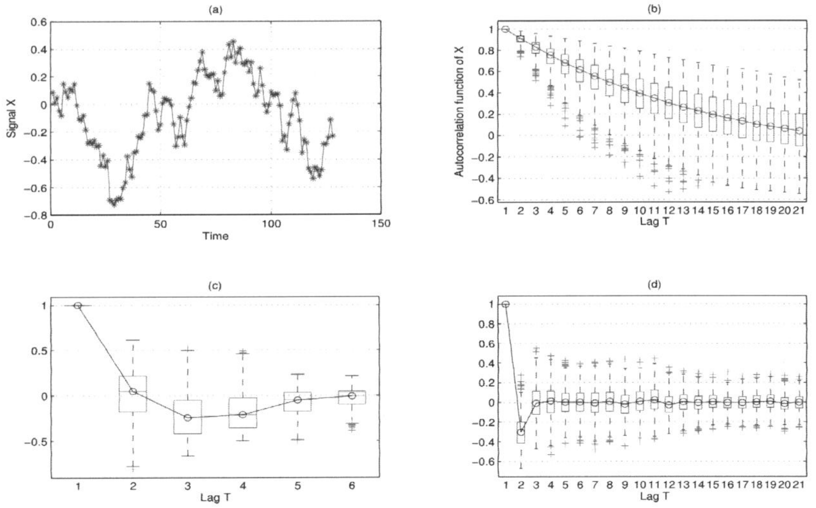

Figure 1. Wavelets coefficients autocorrelation functions; (a): original data X with $\varphi_1 = 0.95$; (b): autocorrelation function of X; (c): autocorrelation function of approximation coefficients of X using *db*2; (d): autocorrelation function of details coefficients of X using *db*2

							φ_1 = 0.7								
						Estimation of correlation coefficient- Mean (standard-deviation)									
$N \downarrow$	Coefficients	Haar	db2	db3	db4	db5	sym2	sym3	sym4	sym5	coif1	coif2	coif3	coif4	coif5
4	App.+Det.	-0.24(0.27)	—	—	—	—	—	—	—	—	—	—	—	—	—
	App.	—	—	—	—	—	—	—	—	—	—	—	—	—	—
	Det.	-0.34(0.24)	—	—	—	—	—	—	—	—	—	—	—	—	—
8	App.+Det.	-0.17(0.25)	0.16(0.30)	—	—	—	0.14(0.30)	—	—	—	—	—	—	—	—
	App.	—	0.06(0.28)	—	—	—	0.05(0.30)	—	—	—	—	—	—	—	—
	Det.	-0.20(0.28)	-0.34(0.27)	—	—	—	-0.36(0.26)	—	—	—	—	—	—	—	—
16	App.+Det.	-0.14(0.24)	0.15(0.25)	0.28(0.24)	0.28(0.23)	—	0.17(0.25)	0.29(0.24)	0.32(0.24)	—	0.30(0.24)	—	—	—	—
	App.	—	0.12(0.29)	0.24(0.27)	0.22(0.28)	—	0.14(0.28)	0.24(0.27)	0.34(0.27)	—	0.31(0.26)	—	—	—	—
	Det.	-0.16(0.24)	-0.26(0.21)	-0.20(0.21)	-0.12(0.28)	—	-0.24(0.21)	-0.21(0.27)	-0.30(0.25)	—	-0.29(0.23)	—	—	—	—
32	App.+Det.	-0.12(0.20)	-0.01(0.23)	0.18(0.24)	0.19(0.25)	0.37(0.18)	-0.00(0.23)	0.17(0.24)	0.23(0.23)	0.39(0.19)	0.21(0.21)	0.37(0.20)	—	—	—
	App.	—	0.01(0.30)	0.19(0.29)	0.23(0.29)	0.39(0.20)	0.01(0.31)	0.18(0.29)	0.30(0.29)	0.45(0.20)	0.29(0.25)	0.42(0.21)	—	—	—
	Det.	-0.13(0.20)	-0.25(0.18)	-0.20(0.18)	-0.23(0.21)	-0.17(0.21)	-0.25(0.18)	-0.21(0.18)	-0.27(0.17)	-0.26(0.22)	-0.26(0.17)	-0.23(0.21)	—	—	—
64	App.+Det.	-0.13(0.17)	-0.08(0.18)	0.08(0.19)	0.13(0.21)	0.18(0.18)	-0.07(0.19)	0.08(0.20)	0.17(0.21)	0.21(0.19)	0.10(0.19)	0.21(0.19)	0.43(0.14)	0.43(0.14)	0.43(0.14)
	App.	—	-0.01(0.31)	0.22(0.27)	0.26(0.27)	0.26(0.21)	-0.01(0.32)	0.21(0.26)	0.28(0.28)	0.38(0.20)	0.22(0.27)	0.33(0.22)	0.50(0.14)	0.49(0.14)	0.49(0.14)
	Det.	-0.14(0.16)	-0.25(0.15)	-0.22(0.15)	-0.17(0.12)	-0.21(0.15)	-0.23(0.14)	-0.23(0.16)	-0.20(0.14)	-0.34(0.15)	-0.23(0.14)	-0.28(0.14)	-0.19(0.16)	-0.18(0.17)	-0.16(0.17)
128	App.+Det.	-0.12(0.12)	-0.09(0.16)	-0.00(0.17)	0.06(0.18)	0.04(0.18)	-0.10(0.16)	0.00(0.18)	0.11(0.18)	0.10(0.17)	0.03(0.15)	0.11(0.18)	0.22(0.14)	0.22(0.15)	0.22(0.14)
	App.	—	0.05(0.31)	0.12(0.20)	0.18(0.28)	0.17(0.24)	0.03(0.31)	0.14(0.29)	0.31(0.27)	0.31(0.22)	0.21(0.25)	0.30(0.23)	0.34(0.16)	0.34(0.16)	0.32(0.16)
	Det.	-0.13(0.12)	-0.22(0.12)	-0.19(0.12)	-0.17(0.12)	-0.23(0.12)	-0.23(0.12)	-0.20(0.12)	-0.17(0.11)	-0.27(0.11)	-0.19(0.11)	-0.25(0.10)	-0.23(0.10)	-0.23(0.10)	-0.22(0.11)

							φ_1 = 0.8								
						Estimation of correlation coefficient - Mean (standard-deviation)									
$N \downarrow$	Coefficients	Haar	db2	db3	db4	db5	sym2	sym3	sym4	sym5	coif1	coif2	coif3	coif4	coif5
4	App.+Det.	-0.23(0.25)	—	—	—	—	—	—	—	—	—	—	—	—	—
	App.	—	—	—	—	—	—	—	—	—	—	—	—	—	—
	Det.	-0.32(0.24)	—	—	—	—	—	—	—	—	—	—	—	—	—
8	App.+Det.	-0.17(0.26)	0.2(0.29)	—	—	—	0.19(0.29)	—	—	—	—	—	—	—	—
	App.	—	0.08(0.28)	—	—	—	0.07(0.29)	—	—	—	—	—	—	—	—
	Det.	-0.19(0.30)	-0.33(0.26)	—	—	—	-0.34(0.26)	—	—	—	—	—	—	—	—
16	App.+Det.	-0.14(0.23)	0.22(0.25)	0.36(0.24)	0.39(0.23)	—	0.24(0.26)	0.38(0.23)	0.40(0.24)	—	0.40(0.23)	—	—	—	—
	App.	—	0.16(0.27)	0.29(0.28)	0.30(0.27)	—	0.16(0.28)	0.30(0.27)	0.40(0.25)	—	0.37(0.25)	—	—	—	—
	Det.	-0.14(0.24)	-0.26(0.21)	-0.20(0.25)	-0.11(0.28)	—	-0.26(0.21)	-0.20(0.26)	-0.31(0.24)	—	-0.29(0.23)	—	—	—	—
32	App.+Det.	-0.12(0.21)	0.09(0.25)	0.29(0.23)	0.31(0.23)	0.48(0.18)	0.06(0.25)	0.27(0.23)	0.35(0.24)	0.50(0.19)	0.32(0.22)	0.49(0.17)	—	—	—
	App.	—	0.06(0.31)	0.28(0.26)	0.31(0.27)	0.49(0.19)	0.04(0.30)	0.26(0.27)	0.37(0.27)	0.53(0.20)	0.37(0.23)	0.51(0.18)	—	—	—
	Det.	-0.13(0.21)	-0.27(0.18)	-0.21(0.18)	-0.23(0.21)	-0.17(0.22)	-0.27(0.18)	-0.22(0.18)	-0.28(0.18)	-0.27(0.22)	-0.26(0.17)	-0.22(0.21)	—	—	—
64	App.+Det.	-0.11(0.18)	-0.05(0.21)	0.17(0.21)	0.24(0.21)	0.32(0.18)	-0.05(0.20)	0.17(0.21)	0.24(0.23)	0.35(0.19)	0.18(0.20)	0.36(0.19)	0.55(0.13)	0.55(0.14)	0.56(0.13)
	App.	—	-0.00(0.32)	0.28(0.26)	0.33(0.26)	0.36(0.20)	0.00(0.32)	0.28(0.25)	0.32(0.28)	0.48(0.19)	0.27(0.25)	0.45(0.20)	0.59(0.13)	0.59(0.14)	0.60(0.13)
	Det.	-0.13(0.18)	-0.27(0.16)	-0.24(0.16)	-0.22(0.15)	-0.22(0.15)	-0.26(0.16)	-0.26(0.16)	-0.21(0.15)	-0.37(0.15)	-0.25(0.14)	-0.29(0.14)	-0.20(0.16)	-0.19(0.16)	-0.18(0.17)
128	App.+Det.	-0.12(0.14)	-0.11(0.17)	0.02(0.18)	0.09(0.20)	0.14(0.19)	-0.10(0.17)	0.02(0.19)	0.15(0.21)	0.21(0.18)	0.06(0.17)	0.22(0.19)	0.37(0.14)	0.37(0.14)	0.36(0.14)
	App.	—	-0.02(0.31)	0.14(0.28)	0.20(0.28)	0.25(0.23)	-0.01(0.31)	0.13(0.29)	0.31(0.27)	0.37(0.20)	0.21(0.26)	0.37(0.22)	0.47(0.15)	0.45(0.15)	0.45(0.15)
	Det.	-0.12(0.14)	-0.25(0.13)	-0.23(0.13)	-0.20(0.13)	-0.26(0.12)	-0.25(0.14)	-0.24(0.12)	-0.20(0.13)	-0.30(0.11)	-0.23(0.12)	-0.27(0.11)	-0.25(0.10)	-0.25(0.11)	-0.24(0.10)

Table 1. Evaluation of wavelets coefficients autocorrelation (App. : approximation, Det. : detail)

								$\varphi_1 = 0.9$							
						Estimation of correlation coefficient- Mean (standard-deviation)									
$N\downarrow$	Coefficients	Haar	db2	db3	db4	db5	sym2	sym3	sym4	sym5	coif1	coif2	coif3	coif4	coif5
	App.+Det.	-0.23(0.24)	—	—	—	—	—	—	—	—	—	—	—	—	—
4	App.	—	—	—	—	—	—	—	—	—	—	—	—	—	—
	Det.	-0.33(0.24)	—	—	—	—	—	—	—	—	—	—	—	—	—
	App.+Det.	-0.19(0.23)	0.25(0.28)	—	—	—	0.25(0.28)	—	—	—	—	—	—	—	—
8	App.	—	0.11(0.28)	—	—	—	0.10(0.29)	—	—	—	—	—	—	—	—
	Det.	-0.17(0.30)	-0.35(0.27)	—	—	—	-0.34(0.27)	—	—	—	—	—	—	—	—
	App.+Det.	-0.16(0.22)	0.36(0.25)	0.45(0.24)	0.46(0.22)	—	0.34(0.27)	0.46(0.24)	0.51(0.22)	—	0.48(0.23)	—	—	—	—
16	App.	—	0.25(0.27)	0.37(0.27)	0.34(0.27)	—	0.21(0.27)	0.37(0.26)	0.46(0.24)	—	0.42(0.24)	—	—	—	—
	Det.	-0.12(0.26)	-0.26(0.21)	-0.22(0.26)	-0.11(0.28)	—	-0.24(0.22)	-0.21(0.26)	-0.31(0.25)	—	-0.29(0.24)	—	—	—	—
	App.+Det.	-0.12(0.21)	0.21(0.27)	0.43(0.23)	0.45(0.22)	0.61(0.17)	0.20(0.28)	0.44(0.22)	0.49(0.24)	0.63(0.17)	0.47(0.22)	0.63(0.17)	—	—	—
32	App.	—	0.12(0.28)	0.38(0.24)	0.41(0.26)	0.58(0.18)	0.12(0.27)	0.37(0.26)	0.46(0.26)	0.62(0.17)	0.46(0.23)	0.61(0.18)	—	—	—
	Det.	-0.13(0.23)	-0.27(0.19)	-0.23(0.18)	-0.26(0.20)	-0.16(0.22)	-0.27(0.19)	-0.22(0.26)	-0.31(0.17)	-0.26(0.22)	-0.26(0.17)	-0.22(0.21)	—	—	—
	App.+Det.	-0.11(0.20)	0.05(0.25)	0.34(0.22)	0.38(0.21)	0.51(0.18)	0.05(0.24)	0.34(0.23)	0.42(0.22)	0.55(0.18)	0.36(0.22)	0.55(0.17)	0.71(0.11)	0.71(0.12)	0.71(0.12)
64	App.	—	0.04(0.30)	0.37(0.24)	0.42(0.23)	0.52(0.19)	0.04(0.29)	0.38(0.24)	0.45(0.24)	0.61(0.16)	0.40(0.23)	0.58(0.18)	0.73(0.11)	0.72(0.12)	0.72(0.12)
	Det.	-0.12(0.20)	-0.29(0.17)	-0.28(0.15)	-0.24(0.15)	-0.23(0.15)	-0.29(0.17)	-0.27(0.16)	-0.22(0.15)	-0.38(0.15)	-0.26(0.14)	-0.30(0.13)	-0.19(0.16)	-0.17(0.17)	-0.18(0.16)
	App.+Det.	-0.12(0.18)	-0.08(0.20)	0.13(0.21)	0.20(0.21)	0.36(0.19)	-0.08(0.21)	0.15(0.21)	0.28(0.22)	0.43(0.18)	0.19(0.20)	0.42(0.17)	0.58(0.13)	0.58(0.13)	0.59(0.14)
128	App.	—	-0.01(0.31)	0.20(0.26)	0.25(0.25)	0.44(0.20)	-0.03(0.31)	0.22(0.26)	0.37(0.26)	0.52(0.18)	0.29(0.24)	0.52(0.17)	0.63(0.13)	0.63(0.13)	0.63(0.13)
	Det.	-0.13(0.17)	-0.29(0.15)	-0.27(0.14)	-0.23(0.14)	-0.28(0.13)	-0.29(0.15)	-0.27(0.14)	-0.24(0.14)	-0.32(0.12)	-0.26(0.13)	-0.30(0.12)	-0.26(0.10)	-0.26(0.11)	-0.24(0.11)
								$\varphi_1 = 0.95$							
						Estimation of correlation coefficient- Mean (standard-deviation)									
$N\downarrow$	Coefficients	Haar	db2	db3	db4	db5	sym2	sym3	sym4	sym5	coif1	coif2	coif3	coif4	coif5
	App.+Det.	-0.23(0.24)	—	—	—	—	—	—	—	—	—	—	—	—	—
4	App.	—	—	—	—	—	—	—	—	—	—	—	—	—	—
	Det.	-0.32(0.24)	—	—	—	—	—	—	—	—	—	—	—	—	—
	App.+Det.	-0.20(0.23)	0.26(0.26)	—	—	—	0.27(0.26)	—	—	—	—	—	—	—	—
8	App.	—	0.14(0.28)	—	—	—	0.15(0.28)	—	—	—	—	—	—	—	—
	Det.	-0.13(0.30)	-0.32(0.27)	—	—	—	-0.33(0.27)	—	—	—	—	—	—	—	—
	App.+Det.	-0.16(0.21)	0.38(0.26)	0.50(0.23)	0.51(0.22)	—	0.38(0.25)	0.50(0.23)	0.56(0.22)	—	0.54(0.22)	—	—	—	—
16	App.	—	0.26(0.27)	0.39(0.27)	0.40(0.28)	—	0.25(0.26)	0.39(0.27)	0.50(0.24)	—	0.47(0.24)	—	—	—	—
	Det.	-0.09(0.27)	-0.25(0.22)	-0.19(0.27)	-0.14(0.27)	—	-0.27(0.22)	-0.21(0.26)	-0.31(0.24)	—	-0.28(0.24)	—	—	—	—
	App.+Det.	-0.15(0.20)	0.32(0.28)	0.52(0.22)	0.54(0.23)	0.68(0.15)	0.32(0.27)	0.53(0.23)	0.60(0.22)	0.56(0.19)	0.58(0.19)	0.70(0.16)	—	—	—
32	App.	—	0.18(0.28)	0.43(0.25)	0.47(0.26)	0.63(0.18)	0.18(0.28)	0.44(0.25)	0.53(0.24)	0.53(0.19)	0.53(0.19)	0.66(0.17)	—	—	—
	Det.	-0.10(0.25)	-0.26(0.17)	-0.23(0.18)	-0.24(0.20)	-0.17(0.22)	-0.25(0.19)	-0.22(0.18)	-0.30(0.18)	-0.25(0.18)	-0.25(0.18)	-0.22(0.21)	—	—	—
	App.+Det.	-0.12(0.21)	0.18(0.28)	0.48(0.22)	0.51(0.21)	0.63(0.17)	0.18(0.28)	0.48(0.22)	0.56(0.22)	0.50(0.21)	0.50(0.21)	0.67(0.19)	0.79(0.10)	0.79(0.11)	0.78(0.11)
64	App.	—	0.12(0.29)	0.48(0.22)	0.51(0.21)	0.60(0.18)	0.11(0.29)	0.48(0.21)	0.54(0.23)	0.49(0.22)	0.49(0.22)	0.67(0.16)	0.78(0.11)	0.78(0.11)	0.78(0.11)
	Det.	-0.11(0.23)	-0.30(0.17)	-0.28(0.16)	-0.25(0.16)	-0.23(0.15)	-0.30(0.17)	-0.28(0.16)	-0.24(0.15)	-0.25(0.14)	-0.25(0.14)	-0.30(0.13)	-0.20(0.16)	-0.19(0.17)	-0.18(0.17)
	App.+Det.	-0.11(0.20)	0.03(0.25)	0.31(0.22)	0.33(0.22)	0.54(0.17)	0.06(0.25)	0.30(0.23)	0.44(0.22)	0.37(0.21)	0.37(0.22)	0.59(0.16)	0.73(0.11)	0.73(0.11)	0.72(0.12)
128	App.	—	0.03(0.30)	0.33(0.24)	0.36(0.24)	0.57(0.17)	0.00(0.30)	0.32(0.24)	0.47(0.24)	0.41(0.23)	0.41(0.23)	0.63(0.16)	0.74(0.192)	0.74(0.12)	0.74(0.11)
	Det.	-0.12(0.21)	-0.31(0.16)	-0.28(0.14)	-0.26(0.14)	-0.30(0.13)	-0.32(0.16)	-0.28(0.14)	-0.26(0.15)	-0.27(0.12)	-0.27(0.13)	-0.31(0.12)	-0.25(0.10)	-0.26(0.11)	-0.24(0.11)

Table 2. Evaluation of wavelets coefficients autocorrelation (App. : approximation, Det. : detail)

2.2. Normality & wavelets coefficients

The evaluation of the normality of wavelets coefficients was carried out by using 5 normality tests (Jarque-Bera [41], Shapiro-Wilk [42; 43], Anderson-Darling [44], Lilliefors [45] and Kolmogorov-Smirnov [46; 47]). Simulations show that wavelets details coefficients are normally-like distributed as demonstrated in the figure 3. On the other hand, wavelets approximations are normally distributed except in some cases (the complete study is not presented here, because of limited number of pages). The decision was taken in each simulation by majority strategy (for example if there are 3 tests that do not reject normality then the normality is decided). The table 3 presents summary results regarding the most used distributions (Exponential, Uniform, Weibull, Rayleigh, Gamma). It shows the parameters formula of Gaussian distribution of wavelets coefficients (approximation and details), if normality is decided.

Distribution	Wavelets coefficients	Normality	Gaussian parameters
Exponential (λ)	App. a_j	☺	$\mathcal{N}(\mu = 2^{j/2}\lambda; \sigma^2 = \lambda^2)$
	Det. d_j	☺	$\mathcal{N}(0; \sigma^2 = \lambda^2)$
Uniform (a,b)	App. a_j	☺	$\mathcal{N}(\mu = 2^{j/2}\frac{a+b}{2}; \sigma^2 = \frac{(b-a)^2}{12})$
	Det. d_j	☺	$\mathcal{N}(0; \sigma^2 = \frac{(b-a)^2}{12})$
Weibull (α,β)	App. a_j	☺	$\mathcal{N}(\mu = 2^{j/2}\beta\Gamma\left(1 + \frac{1}{\alpha}\right); \sigma^2 = \beta^2\Gamma\left(1 + \frac{2}{\alpha}\right) - \mu^2)$
	Det. d_j	☺	$\mathcal{N}(0; \sigma^2 = \beta^2\Gamma\left(1 + \frac{2}{\alpha}\right) - \mu^2)$
Rayleigh (σ_R)	App. a_j	☺	$\mathcal{N}(\mu = 2^{j/2}\sigma_R\sqrt{\frac{\pi}{2}}; \sigma^2 = \frac{4-\pi}{2}\sigma_R^2)$
	Det. d_j	☺	$\mathcal{N}(0; \sigma^2 = \frac{4-\pi}{2}\sigma_R^2)$
Gamma (α,β)	App. a_j	☺	$\mathcal{N}(\mu = 2^{j/2}\alpha\beta; \sigma^2 = \alpha\beta^2)$
	Det. d_j	☺	$\mathcal{N}(0; \sigma^2 = \alpha\beta^2)$

Table 3. Distributional parameters of wavelets coefficients when they are normally distributed

2.2.1. An illustrative example

The aim of this example is to display the distribution of wavelets coefficients when original data are exponentially distributed. In the figure 2, data was generated from exponential distribution with $\lambda = 5$. One can see from the figure 3 that wavelets coefficients have the characteristic symmetric "bell curve" shape.

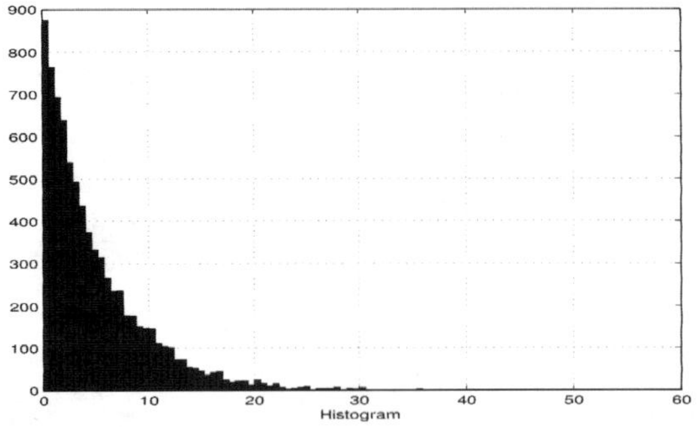

Figure 2. Exponentially distributed data with $\lambda = 5$

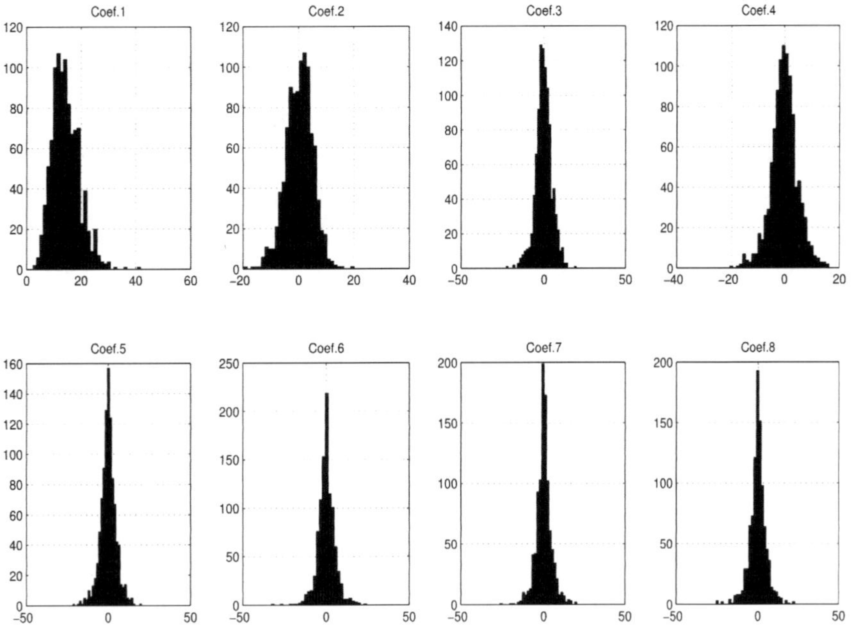

Figure 3. Waveletes coefficients histograms using Haar wavelet and $N = 8$; Coef.1 corresponds to the approximation and Coef.2-8 correspond to the details

3. Wavelets coefficients and SPC purposes

SPC aims to monitor process characteristics in order to detect changes in the mean and/or variance of the processes. Based on statistics (indicators) control charts are designed and control limits are derived from the statistic probability distribution. In the following theorem, we present a result regarding the probability distribution of wavelets coefficients (details and approximations).

Theorem 1. *Assume* $X = [x_1, x_2, .., x_n]$ *is a signal, where* x_i *are identically normally distributed random variables* ($x_i \rightsquigarrow \mathcal{N}(\mu, \sigma^2)$)*. Consider Orthonormal compactly supported wavelets (Haar, Daubechies, Symlets, Coiflets). The multiresolution analysis of* X *provides wavelets coefficients as follows:*

$$a_j(k) \rightsquigarrow \mathcal{N}(2^{j/2}\mu, \sigma^2)$$
$$d_j(k) \rightsquigarrow \mathcal{N}(0, \sigma^2)$$

Which are normally distributed random variables (μ *and* σ^2 *are defined in the table 3 for non-Gaussian distributions). Moreover, if the Haar wavelet is used then they are independent (even if* x_i *are autocorrelated), else autocorrelation is strongly debased (see tables 1 et 2).*

One can remark, from the proof of the theorem 1 (see appendix A), that wavelets coefficients are summation of normally distributed variables consequently they follow the normal distribution.

From the theorem 1, we show that wavelets coefficients present some interesting distributional characteristics that reveal the original data features. One can remark that the approximation coefficients amplify the mean of the original data. Regarding details coefficients, they have a mean equal to zero (see window2 in Fig. 4b). Therefore, this characteristic could be useful to detect change in variance, because the mean change does not affect the details coefficients, as we shown in previous work [20].

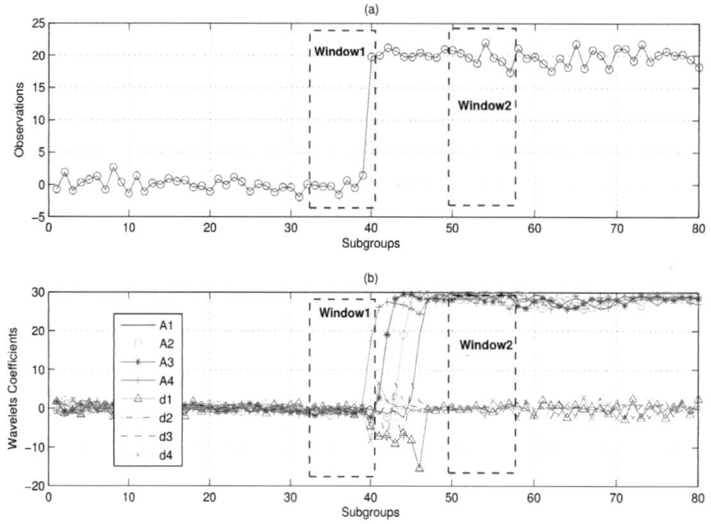

Figure 4. (a): Observations with mean change (positive shift) at 40 ; (b): Wavelets coefficients (approximations A_i and details d_i) using db2 wavelet

In the figure 4, we display the wavelets coefficients behavior when change in the mean is occurring. We consider a window size of length L=8 observations/subgroups, as the smallest one that can be used rationally with the *db*2 wavelet. Consequently we get eight wavelets coefficients at the scale one: four approximations coefficients and four details coefficients. The first 39 observations plotted in the figure 4(a) consist of observations randomly generated from a normal distribution $\mathcal{N}(0,1)$ (an in-control process), and the last 41 observations, consist of observations randomly generated from a normal distribution $\mathcal{N}(20,1)$ (out-of-control process with mean change). One can conclude that wavelets coefficients reflect clearly the mean change.

- Note, from the theorem 1, that $a_j(k)$ coefficients have the same distribution behavior as the average of the data $\overline{X}$ (Xbar). In fact, if $X \rightsquigarrow N(\mu_0, \sigma_0^2)$ then $\overline{X} \rightsquigarrow N(\mu_0, \frac{\sigma_0^2}{n})$. A mean change δ is defined as : $\mu_1 = \mu_0 + \delta * \sigma_0$ and can be detected in $\overline{X}$ chart as $\delta\sqrt{n}$ shift. When the sample size is $n = 2^J$, then the shift is equal to $\delta 2^{J/2}$. By the same way δ can be detected in wavelets approximations coefficients A_J (statistic based on wavelets approximations coefficients, in the case of the Haar wavelet the A_J is the approximation coefficient), at the highest scale J, as $\delta 2^{J/2}$ (see theorem 1). Furthermore, we note that $\overline{X}$ statistic reduces the variance of the data and keeps the mean equal to the mean of the data. However, in terms of detectability the A_J and $Xbar$ charts are totally equivalent (see the figure 5(b)). Regarding wavelets details coefficients $d_j(k)$, they follow normal distribution with mean equal to zero and variance reflecting the variance of the data.
- DeWave chart are proposed based on wavelet details in order to detect variance change. It shows that DeWave performs better than S-chart when large samples are used [20] .

- In practice, DeWave and A_J control charts can be can implemented by doing a Discrete Wavelet Transform. Using the Haar wavelet and a sample of size $n = 2^J$; we obtain one approximation coefficient that will be used to monitor the process mean, and n-1 details coefficients that will be used to control the process variability. A special case where n=2, the wavelets details coefficients are corresponding to the Range (R) chart to monitor the process variability.

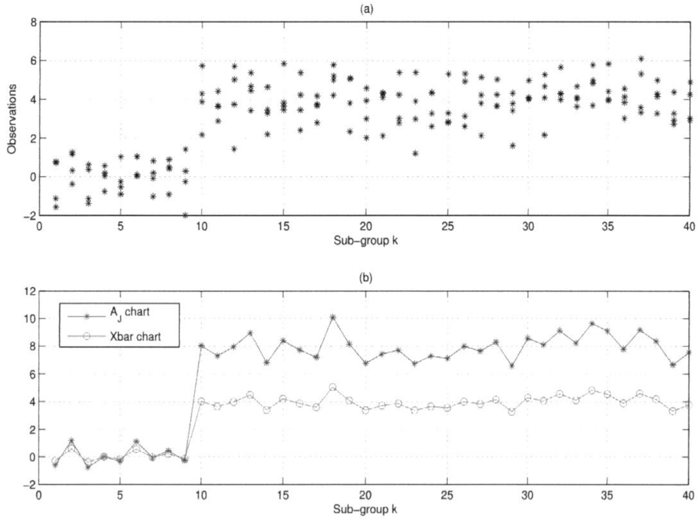

Figure 5. (a): Observations with mean change at 10th sub-group of size 4 ; (b): Xbar control chart and A_J statistic (approximation coefficient) using Haar wavelet.

4. Conclusion

In this paper we expose the muliresolution analysis benefits in terms of handling two challenges that face SPC procedures. Firstly we propose a result regarding distributional parameters of wavelets coefficients. Secondly, we demonstrate that wavelets coefficients could be normally-like distributed even if original data have non-Gaussian distributions and we drive their distribution parameters. Moreover, we show the ability of wavelets analysis to debase the autocorrelation degree. Our study on AR(1) processes shows that the Haar wavelet is very recommended in order to decorrelate data. Illustrative examples and discussions are also proposed. Next research will concern the publication of the complete studies with examples of industrial applications. Futur works should concern more applications of wavelets coefficients to monitor processes that are autocorrelated or characterised by non-Gaussian data. More studies are also needed to highlight the assets of wavelets coefficients to improve SPC performances.

Appendix A

Wavelets and scaling filters must satisfy some conditions [8; 27], in order to provide a multi-resolution analysis. We introduce the following conditions, which will be useful for the proof.

Journal of Physics: Conference Series **659** (2015) 012043 doi:10.1088/1742-6596/659/1/012043

We consider $x_i \rightsquigarrow \mathcal{N}(\mu, \sigma^2)$:

$$\sum_n h_n = \sqrt{2} \tag{4}$$

$$\sum_n g_n = 0 \tag{5}$$

Proof. The approximation wavelets coefficients, at the first scale:

$$a_1(k) = \sum_{i=1}^{l} h[i]x[2k-i]$$

The expectation and the variance give that:

$$\begin{aligned} E(a_1) &= E(\sum_{i=1}^{l} h[i]x[2k-i]) = \sum_{i=1}^{l} h_i * E(x) \\ &= 2^{1/2} * \mu \\ V(a_1) &= V(\sum_{i=1}^{l} h[i]x[2k-i]) = \sum_{i=1}^{l} h_i^2 * V(x) = \sigma^2 \end{aligned}$$

At the second scale:

$$a_2(k) = \sum_{i=1}^{l} h[i]a_1[2k-i]$$

The expectation and the variance give that:

$$\begin{aligned} E(a_2(k)) &= E(\sum_{i=1}^{l} h[i]a_1[2k-i]) = \sum_{i=1}^{l} h_i * E(a_1(i)) \\ &= 2^{2/2} * \mu \\ V(a_2(k)) &= \sigma^2 \end{aligned}$$

By the same way, at the higher scales, we conclude at scale j:

$$\begin{aligned} E(a_j(k)) &= 2^{j/2} * \mu \\ V(a_j(k)) &= \sigma^2 \end{aligned}$$

Proceeding by the same previous proof, one can show that details wavelets coefficients, at scale j:

$$\begin{aligned} E(d_j(k)) &= 0 \\ V(d_j(k)) &= \sigma^2 \end{aligned}$$

□

References

[1] Shewhart W A 1925 *Journal of the American Statistical Association* **20** pp. 546–548 ISSN 01621459

[2] Roberts S W 1959 *Technometrics* **1** pp. 239–250 ISSN 00401706

[3] Page E S 1954 *Biometrika* **41** pp. 100–115 ISSN 00063444

[4] Reynolds M R, Amin R W, Arnold J C and Nachlas J A 1988 *Technometrics* **30** 181–192

[5] Bai D and Lee K 1998 *International Journal of Production Economics* **54** 57–64

[6] De Magalhães M S, Epprecht E K and Costa A F 2001 *International Journal of Production Economics* **74** 191–200

[7] Ou Y, Wu Z and Tsung F 2012 *International journal of production economics* **135** 479–490

[8] Daubechies I 1992 *Ten lectures on wavelets* (Philadelphia, Pa: Society for Industrial and Applied Mathematics)

[9] Cohen A, Daubechies I and Feauveau J C 1992 *Communications on pure and applied mathematics* **45** 485–560

[10] Meyer Y 1993 *Wavelets-algorithms and applications* vol 1 (Society for Industrial and Applied Mathematics Translation)

[11] Mallat S 1999 *A wavelet tour of signal processing* (Academic press)

[12] Bakshi B R 1998 *AIChE journal* **44** 1596–1610

[13] Bakshi B R 1999 *Journal of chemometrics* **13** 415–434

[14] Misra M, Yue H H, Qin S J and Ling C 2002 *Computers and Chemical Engineering* **26** 1281–1293

[15] Lin J and Zhang A 2005 *NDT and E International* **38** 421–427

[16] Maulud A, Wang D and Romagnoli J 2006 *Journal of Process Control* **16** 671 – 683 ISSN 0959-1524

[17] Choi S W, Morris J and Lee I B 2008 *Chemical engineering science* **63** 2252–2266

[18] Konar P and Chattopadhyay P 2011 *Applied Soft Computing* **11** 4203–4211

[19] Alexandridis A K and Zapranis A D 2013 *Neural Networks* **42** 1 – 27 ISSN 0893-6080 URL `http://www.sciencedirect.com/science/article/pii/S0893608013000129`

[20] Cohen A, Tiplica T and Kobi A 2015 *Control Engineering Practice* ISSN 0967-0661

[21] Donoho D L and Johnstone J M 1994 *Biometrika* **81** 425–455

[22] Donoho D L and Johnstone J M 1995 *Jouranl of the American Statistical Association* 1200–1224

[23] Donoho D 1995 *IEEE Transactions on Information Theory* **41** 613–627 ISSN 0018-9448

[24] Jeong M K, Lu J C, Huo X, Vidakovic B and Chen D 2006 *Technometrics* **48**

[25] Jeong M K, Chen D and Lu J C 2003 *Applied Stochastic Models in Business and Industry* **19** 231–244 URL `http://dx.doi.org/10.1002/asmb.495`

[26] Ganesan R, Das T K and Venkataraman V 2004 *IIE transactions* **36** 787–806

[27] Gao R X and Yan R 2010 *Wavelets: Theory and Applications for manufacturing* (Springer)

[28] Dai X D, Motard R L, Joseph B and Silverman D C 2000 *Industrial and engineering chemistry research* **39** 1256–1263 ISSN 0888-5885

[29] Wharton J, Wood R and Mellor B 2003 *Corrosion science* **45** 97–122

[30] Psarakis S and Papaleonida G 2007 *Quality Technology and Quantitative Management* **4** 501–540

[31] Burr I W 1967 *Industrial Quality Control* **23** 563–569

[32] Yourstone S A and Zimmer W J 1992 *Decision Sciences* **23** 1099–1099

[33] Schilling E G and Nelson P R 1976 *Journal of Quality Technology* **8**

[34] Box G E and Cox D R 1964 *Journal of the Royal Statistical Society. Series B (Methodological)* 211–252

[35] Johnson N L, Kotz S and Balakrishnan N 1994 *Continuous univariate distributions* vol 1 (John Wiley and Sons)

[36] Vardeman S and Ray D o 1985 *Technometrics* **27** 145–150

[37] Mallat S G 1989 *IEEE Transactions on Pattern Analysis and Machine Intelligence* **11** 674–693

[38] Knoth S and Schmid W 2004 *Control charts for time series: a review* (Springer)

[39] Timmer D H, Pignatiello Jr J and Longnecker M 1998 *IIE transactions* **30** 525–534

[40] Iwersen J 1997 *Frontiers in Statistical Quality Control* (Springer) pp 138–155

[41] Jarque C M and Bera A K 1980 *Economics letters* **6** 255–259

[42] Shapiro S S and Wilk M B 1965 *Biometrika* 591–611

[43] Royston P 1992 *Statistics and Computing* **2** 117–119

[44] Anderson T W and Darling D A 1952 *The annals of mathematical statistics* 193–212

[45] Lilliefors H W 1969 *Journal of the American Statistical Association* **64** 387–389

[46] Massey Jr F J 1951 *Journal of the American statistical Association* **46** 68–78

[47] Miller L H 1956 *Journal of the American Statistical Association* **51** 111–121

Rotor thermal stress monitoring in steam turbines

Bouberle Antonín, Jakl Jan, Liška Jindřich

NTIS – New Technologies for the Information Society, University of West Bohemia, Czech Republic

bouban@ntis.zcu.cz

Abstract. One of the issues of steam turbines diagnostics is monitoring of rotor thermal stress that arises from nonuniform temperature field. The effort of steam turbine operator is to operate steam turbine in such conditions, that rotor thermal stress doesn't exceed the specified limits. If rotor thermal stress limits are exceeded for a long time during machine operation, the rotor fatigue life is shortened and this may lead to unexpected machine failure. Thermal stress plays important role during turbine cold startup, when occur the most significant differences of temperatures through rotor cross section. The temperature field can't be measured directly in the entire rotor cross section and standardly the temperature is measured by thermocouple mounted in stator part. From this reason method for numerical solution of partial differential equation of heat propagation through rotor cross section must be combined with method for calculation of temperature on rotor surface. In the first part of this article, the application of finite volume method for calculation of rotor thermal stress is described. The second part of article deals with optimal trend generation of thermal flux, that could be used for optimal machine loading.

Notation

a	thermal diffusivity
c_r	specific heat capacity
n_j, n_i	normal vector
m	mass
$q(t), q^k$	heat flux
Q	heat
dQ	amount of obtained/lost heat
r	radius from rotor (cylinder) center
$R[i]$	radius from rotor center to node i
S	area
S_i, S_{i-1}	i-th control volume surfaces
t	time
Δt	sampling period
T	temperature
$dT, \Delta T$	temperature change
T_i^k	temperature in layer (node) i in time point k
T_n^k	measured temperature (in time point k)
$T\_filt_n^k$	filtered inlet temperature
TS^k	thermal stress of rotor
TS_{max}^k	upper limit of maximum of permissible TS^k
TMI^k	mean integral temperature
V	volume
$V[i]$	volume of element i
z	coordinate corresponding to rotor length
ρ	density of material
φ	rotation angle
τ	time constant of filter

1. Introduction

Currently, the power industry development is divided into two main areas. The first area is aimed to the problem of depletion of non-renewable resources associated with gradually replacement by renewable resources. The second area is aimed to achieving of higher efficiency, durability and reliability of machines producing the electricity power. Both areas are associated with requirement of monitoring and diagnostics of machines, because early detection of possible faults can prevent machine fault and minimize associated economic losses. Some serious problems of steam turbine operation, for example high vibrations due to bend rotor or rotor/stator rub are related to inappropriate heating of rotor during machine startup. The non-uniform thermal field around rotor surface is also associated with rotor thermal stress. Generally, the largest thermal stress occurs during the turbine startup from the cold state – i.e. turbine run-up on the operating speed and especially on the respective nominal power. In ideal situation the machine startup is archived as fast as possible, without threat of machine operation, but thermal stress limits must be taken account. The first aim of this article is development of method for evaluation of thermal stress of the rotor based on temperature measurement in stator part of machine, because it is not possible to measure rotor surface temperature directly. The assumption for evaluation of rotor surface temperature from measurement of temperature in stator part is the same thermal conductivity of rotor and stator part - effects of flowing steam are in this article neglected. Because the only one thermocouple was used for stator temperature measurement, therefore the analysis is made only for 2D case (i.e. rotor cross section). The second aim of this article is development of algorithm for calculation of optimal thermal flux which can be used for evaluation machine run-up RPM and power trend.

2. Calculation of thermal field and thermal stress of the rotor

2.1. Heat equation in cylindrical coordinates

Differential equation of the heat conduction in cylindrical coordinates (for rigid homogeneous substance without internal heat resources), which describes a temperature distribution in the space (i.e. in the cylinder – in the rotor) and across time is as follows [2], [4], [11]:

$$\frac{\partial T}{\partial t} = a\left(\frac{\partial^2 T}{\partial r^2} + \frac{1}{r}\frac{\partial T}{\partial r} + \frac{1}{r^2}\frac{\partial^2 T}{\partial \varphi^2} + \frac{\partial^2 T}{\partial z^2}\right), \tag{1}$$

where T is a temperature, t is a time, a is a thermal diffusivity, r is a radius from the cylinder center (rotor), φ is a rotation angle and coordinate z corresponds to the rotor length. Because the subject of interest in our case is only temperature distribution in the cross section of the rotor, the term $\frac{\partial^2 T}{\partial z^2}$ in equation (1) is zero. For 3D case (also in the z-direction) it would be required to use an additional thermocouple. Assumption of rotor thermal symmetry leads to independency of temperature T on angle φ, and equation (1) can be rewritten in following form:

$$\frac{\partial T}{\partial t} = a\left(\frac{\partial^2 T}{\partial r^2} + \frac{1}{r}\frac{\partial T}{\partial r}\right). \tag{2}$$

Analytical solution of partial differential equations of heat conduction can be easily obtained only for simple cases and therefore we will use the numerical solution. By a spatial discretization – cross section of the rotor on each layer (i.e. annulus, see fig. 1, i-th layer), by a time discretization and by an application of finite volume method the system of algebraic equations for the calculation of temperatures in the individual layers in a given time can be obtained. For completeness the boundary conditions must be defined. The first boundary condition leads from rotor thermal field symmetry:

$$\frac{\partial T}{\partial r} = 0. \tag{3}$$

The second boundary condition is given by information from thermocouple in stator part. For equal rotor and stator thermal conductivity one can expect the same temperature in appropriate rotor layer.

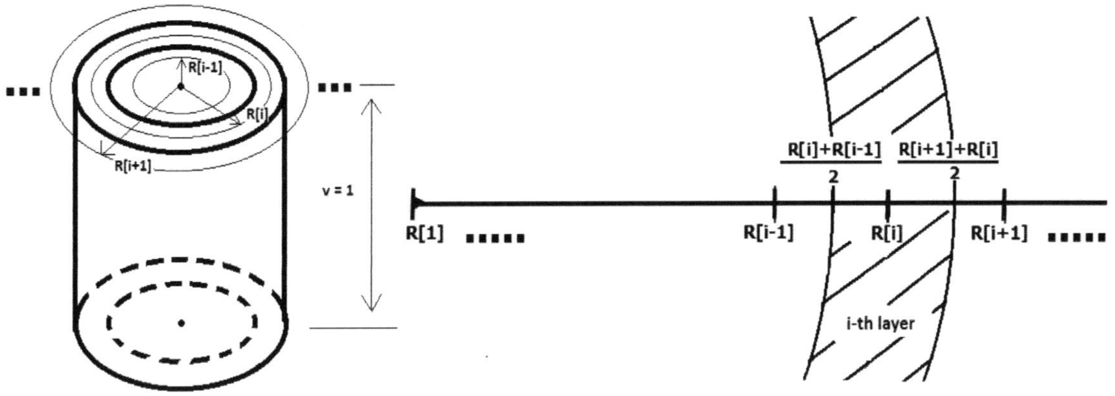

fig. 1: Spatial (rotor) discretization, *i*-th layer

2.2. Finite volume method

In this section will be derived equations for the temperature calculation in the *i*-th layer resp. in the *i*-th node (i=2,..,(n-1)) [5], [6], [9]. Relations for calculation of temperatures in nodes *1* and (n+1) will be derived later. Consider rotor like cylinder with unit height. Let divide this cylinder is into (n+1) elements – called control volumes (i.e. annulus with unit height, see fig. 1), which don't overlap each other. In the geometric center of each element (control volume) is located computing node (except for layer (n+1)). The temperature at any point of control volume is equal to the temperature in the computing node. Temperature calculation in the rotor cross section (at given time) runs across (n+1) nodes, where node *1* corresponds to the rotor geometric center, node n corresponds to the thermocouple location and node (n+1) corresponds to the rotor surface. Equation (2) can be written in the following form:

$$\frac{1}{a}\frac{\partial T}{\partial t} = \frac{1}{r}\frac{\partial}{\partial r}\left(r\frac{\partial T}{\partial r}\right). \tag{4}$$

By integration over the volume V we get:

$$\int_V \frac{1}{a}\frac{\partial T}{\partial t} dV = \int_V \frac{1}{r}\frac{\partial}{\partial r}\left(r\frac{\partial T}{\partial r}\right) dV. \tag{5}$$

For further derivations the Gauss theorem will be used:

$$\int_V \frac{\partial \phi_j}{\partial x_j} dV = \int_S \phi_j n_j dS, \quad n_j = \left[n_x\, n_y\, n_z\right]^T, \tag{6}$$

where S is the area and n_j is the normal to the area. Term $\frac{\partial T}{\partial t}$ in equation (5) is approximated by the backward difference and on the right side the Gauss theorem is applied:

$$\frac{1}{a}\frac{T_i^k - T_i^{k-1}}{\Delta t}\int_V dV = \int_S \frac{1}{r} r \frac{\partial T}{\partial r} n_r\, dS, \quad n_r = [-n_{i-1}\, n_i\,]^T \tag{7}$$

Flow through the control volume is given by the sum over the surfaces (through one surface heat enters into the control volume, through another surface heat leaves the element). Previous equation can be rewriten into form:

$$\frac{1}{a}\frac{T_i^k-T_i^{k-1}}{\Delta t}V[i] = -\int_{S_{i-1}}\frac{\partial T}{\partial r}n_{i-1}\,dS + \int_{S_i}\frac{\partial T}{\partial r}n_i\,dS, \tag{8}$$

where Δt is the sampling period, $V[i]$ is the volume of element i (i=2,..,(n-1)) defined by:

$$V[i] = \pi\left(\left(\frac{R[i+1]+R[i]}{2}\right)^2 - \left(\frac{R[i]+R[i-1]}{2}\right)^2\right), \tag{9}$$

and S_i resp. S_{i-1} is the i-th control volume surface, through which heat into the element (for example) enters resp. leaves, i.e. it is a surface of the respective cylinder.

$$S_i = \underbrace{2\pi\frac{R[i+1]+R[i]}{2}}_{2\pi r}\underbrace{1}_{v},\ S_{i-1} = \underbrace{2\pi\frac{R[i]+R[i-1]}{2}}_{2\pi r}\underbrace{1}_{v}. \tag{10}$$

Surface integral of partial derivative of temperature in equation (8) can be approximated by a mean value of this function on a given surface:

$$\frac{1}{a}\frac{T_i^k-T_i^{k-1}}{\Delta t}V[i] = -\frac{\partial T_{S[i-1]}}{\partial r}S[i-1] + \frac{\partial T_{S[i]}}{\partial r}S[i] \tag{11}$$

By approximation of partial derivatives in (11) by backward differencies the final relation for calculation of temperature in layer i for time step k can be obtained:

$$T_i^k = T_i^{k-1} + \frac{a.\Delta t.\pi}{V[i]}\left[\frac{R[i+1]+R[i]}{R[i+1]-R[i]}\left(T_{i+1}^{k-1} - T_i^{k-1}\right) - \frac{R[i]+R[i-1]}{R[i]-R[i-1]}\left(T_i^{k-1} - T_{i-1}^{k-1}\right)\right] \tag{12}$$

2.2.1. *Temperature calculation in the rotor geometrical center*

In following part of the article, the relation for evaluation of temperature in rotor geometric center will be derived. From the assumption rotor thermal symmetry follows that for any time step the temperature over all layers has local or global extrema in the rotor center. The mechanism for derivation of temperature in the rotor center is the same as for inner rotor layers, but now we use simplified partial differential equation:

$$\frac{\partial T}{\partial \tau} = a\frac{\partial^2 T}{\partial r^2}. \tag{13}$$

Let us consider that heat only enters in the first control volume (i.e. cylinder), respectively doesn't flow from it to another element as in the previous case (i=2,..,(n-1)). Flow through the first control volume is given by the flow through cylinder surface. Let us defined the first element volume:

$$V[1] = \pi\left(\left(\frac{R[2]+R[1]}{2}\right)^2 - R[1]^2\right) \tag{14}$$

and surface through which heat enters/leaves the element:

$$S[1] = \underbrace{2\pi\frac{R[2]+R[1]}{2}}_{2\pi r}\underbrace{1}_{v}. \tag{15}$$

Then is possible to derive the equation for calculation of temperature in the first node (in the geometric rotor center):

$$T_1^k = T_1^{k-1} + \frac{a.\Delta t.\pi}{V[1]}\left[\frac{R[2]+R[1]}{R[2]-R[1]}\left(T_2^{k-1} - T_1^{k-1}\right)\right] \tag{16}$$

2.2.2. *Temperature calculation on the rotor surface*

Finally, it remains to determine the relationship for calculation of temperature in node *(n+1)*. Temperature of this node corresponds to the temperature of rotor surface. It is worthy to mention that the temperature of node *(n+1)* doesn't belong to the solution of partial differential equation, because the boundary condition is the temperature in node *n* (value measured by thermocouple). The first manner of temperature calculation of the rotor surface is based on a formally rewritten equation (12) for the layer *n*, i.e. $i = n$:

$$T_n^k = T_n^{k-1} + \frac{a.\Delta t.\pi}{V[n]}\left[\frac{R[n+1]+R[n]}{R[n+1]-R[n]}\left(T_{n+1}^{k-1} - T_n^{k-1}\right) - \frac{R[n]+R[n-1]}{R[n]-R[n-1]}\left(T_n^{k-1} - T_{n-1}^{k-1}\right)\right], \tag{17}$$

where the temperature T_n^k is measured, term T_{n+1}^{k-1} is to be determined and $V[n]$ is given by expression (9). By modifying of previous equation we get a required relationship to the calculation of temperature in the layer *(n+1)*:

$$T_{n+1}^{k-1} = T_n^{k-1} + \left(\frac{R[n+1]-R[n]}{R[n+1]+R[n]}\right)\left[\frac{V[n]}{a.\Delta t.\pi}\left(T_n^k - T_n^{k-1}\right) + \frac{R[n]+R[n-1]}{R[n]-R[n-1]}\left(T_n^{k-1} - T_{n-1}^{k-1}\right)\right] \tag{18}$$

Calculation of temperature in the layer *(n+1)* according to (18) may come across the following problem. Thermal processes are generally slow processes. In the case of rough quantisation level of measured temperature, the values of T_n in two consecutive time steps may equal. If such situation occurs, the resulting evolution of calculated variable T_{n+1} over time shows significant fluctuations (high variance of signal), see the gray curve on fig. 2. This behaviour does not correspond to slow thermal processes.

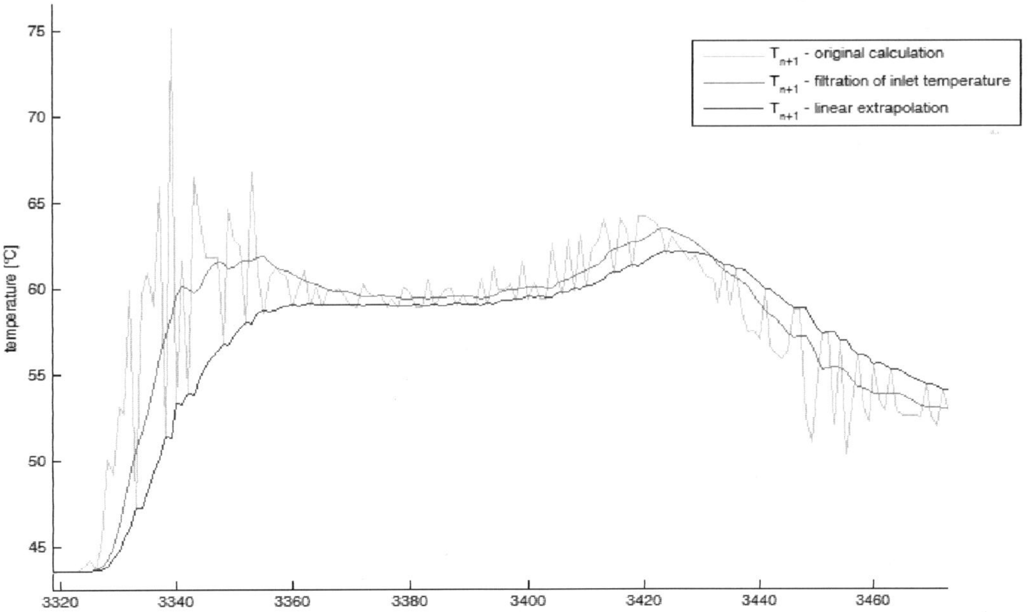

fig. 2: Temperature progress in the *(n+1)*. layer

One option for elimination of this problem consists of filtration of measured temperature T_n by appropriate filter. This filtering smooths T_n and avoid the temperature equality in two consecutive time steps. In our work we used second order filter:

$$F(p) = \frac{1}{(\tau p+1)^2}, \tag{19}$$

where τ = 1s. After discretization by Tustin approximation the resulting equation for filtered temperature calculation in time step k is as follows:

$$T\_filt_n^k = \frac{-2(\Delta t-2\tau)T\_filt_n^{k-1}}{(2\tau+\Delta t)} - \frac{(\Delta t-2\tau)^2 T\_filt_n^{k-2}}{(2\tau+\Delta t)^2} + \frac{\Delta t^2 T_n^k}{(2\tau+\Delta t)^2} + \frac{2\Delta t^2 T_n^{k-1}}{(2\tau+\Delta t)^2} + \frac{\Delta t^2 T_n^{k-2}}{(2\tau+\Delta t)^2} \tag{20}$$

where T_n^k is the measured temperature, $T\_filt_n^k$ is the filtered inlet temperature and Δt is the sampling period. From fig. 2 it is evident that the resulting temperature T_{n+1} has a significantly smoother character than temperature waveform T_{n+1} in case without filtered T_n, see red curve.

As mentioned, the node *(n + 1)* is located at the border of control volume (not in the geometric center). In the case of finite volume method values are on the border of control volumes obtained by interpolation. Because the node *(n + 1)* is boundary node of the edge control volume, it is necessary to use the extrapolation. Extrapolation was chosen as linear – line passing in the points $R[n]$ and $R[n-1]$ through values T_n and T_{n-1}. The equation for calculating of temperature in the layer *(n+1)* by linear extrapolation is as follows [8]:

$$\underbrace{T_{n+1}^{k-1}}_{y} = \underbrace{\frac{(T_n^{k-1}-T_{n-1}^{k-1})}{(R[n]-R[n-1])}}_{a} \underbrace{R[n+1]}_{x} + \underbrace{T_n^k - \frac{(T_n^{k-1}-T_{n-1}^{k-1})}{(R[n]-R[n-1])} R[n]}_{b} \tag{21}$$

The resulting temperature waveform is shown in fig. 2 (blue curve). From the figure it is evident that the linear extrapolation eliminates the significant fluctuations problem. Trends of both curves (according to (18) and linear extrapolation) are similar. Therefore the linear extrapolation can be also used for calculating of temperature in the layer *(n+1)*.

2.3. Explicit and implicit version of finite volume method

In this section the major difference between implicit and explicit version of the above method will be marginally mentioned. Note, that the above method is an explicit version. During deriving of formula for temperature calculation in the layer i in the time point k, the partial derivative $\frac{\partial T_{S[i-1]}}{\partial r}$ resp. $\frac{\partial T_{S[i]}}{\partial r}$ was approximated by difference, which containes the temperatures T_{i-1}^{k-1}, T_i^{k-1}a T_{i+1}^{k-1} in time *(k-1)*, see (12). If the temperatures T_{i-1}^k, T_i^k and T_{i+1}^k (in the time point k) were inserted into relevant expression then the calculation of temperature distribution in the rotor cross section would lead to the system of equations (vs. a simpler aforementioned recursive relationship). Both versions of the method have advantages and disadvantages. Of course, solving of the system of equations is more memory and computationally demanding, but on the other hand the stability of the explicit version isn't always guaranteed.

2.4. Thermal stress and mean integral temperature

As mentioned, the thermal stress occurs due to nonuniform temperature field. Thus, let's define thermal stress as a deviation of highest temperature in the rotor cross section from the average of the temperature field [1], [3], [7]. Because the thermal stress is particularly significant during the turbine startup (external rotor heating), it is often assumed that the highest temperature occurs just in the layer *(n+1)*. Thermal stress can be written as a following relationship:

$$TS^k = T_{n+1}^k - TMI^k \tag{22}$$

where TMI^k is the mean integral temperature representing the average temperature across all layers (control volumes). The every control element has different volume – different weight, therefore the calculation of TMI^k is carried out as a weighted average. Mean integral temperature is given by:

$$TMI^k = \frac{\prod_{i=1}^{n+1} T_i^k V[i]}{\sum_{i=1}^{n+1} V[i]} = \frac{T_1^k V[1]+T_2^k V[2]+..+T_n^k V[n]+T_{n+1}^k V[n+1]}{V[1]+V[2]+..+V[n]+V[n+1]} \quad (23)$$

where the volume $V[n+1]$ is:

$$V[n+1] = \pi\left(R[n+1]^2 - \left(\frac{R[n+1]+R[n]}{2}\right)^2\right) \quad (24)$$

2.5. Application and evaluation on the 60 MW steam turbine

Firstly, the application and evaluation is based on the above knowledge of simulated heat propagation in the rotor cross section. This simulation was compared with performed simulation in software Ansys. Note that the results of both simulations were perfectly coincided – the deviations of calculated temperatures (in space and in time) were in the order of hundredths of a °C or less. Then the above knowledges were applied for analysis of signals measured on the 60 MW steam turbine. The application results are shown in data comprising startup of the turbine from the cold state. Note that the rotor cross section was discretized into 26 layers (i.e. $n = 25$) and the thermocouple was located 8 mm under the inner turbine body surface – stator. Calculation of temperature field together with thermal stress was cyclically carried with a period of 0.5 sec (i.e. $\Delta t = 0.5$). In fig. 3 the time progresses of temperature in the various layers using color scale (in °C) are shown, where the progress of layer 25 corresponds to the measured values of the thermocouple. Red curve shows the waveform of mean integral temperature *TMI* and the blue one shows the thermal stress *TS*. Due to external rotor heating the *TMI* is gradually increasing and then stabilizes at around 428 °C. Thermal stress increases till 8000 sec., then returns to zero by heating the rotor gradually.

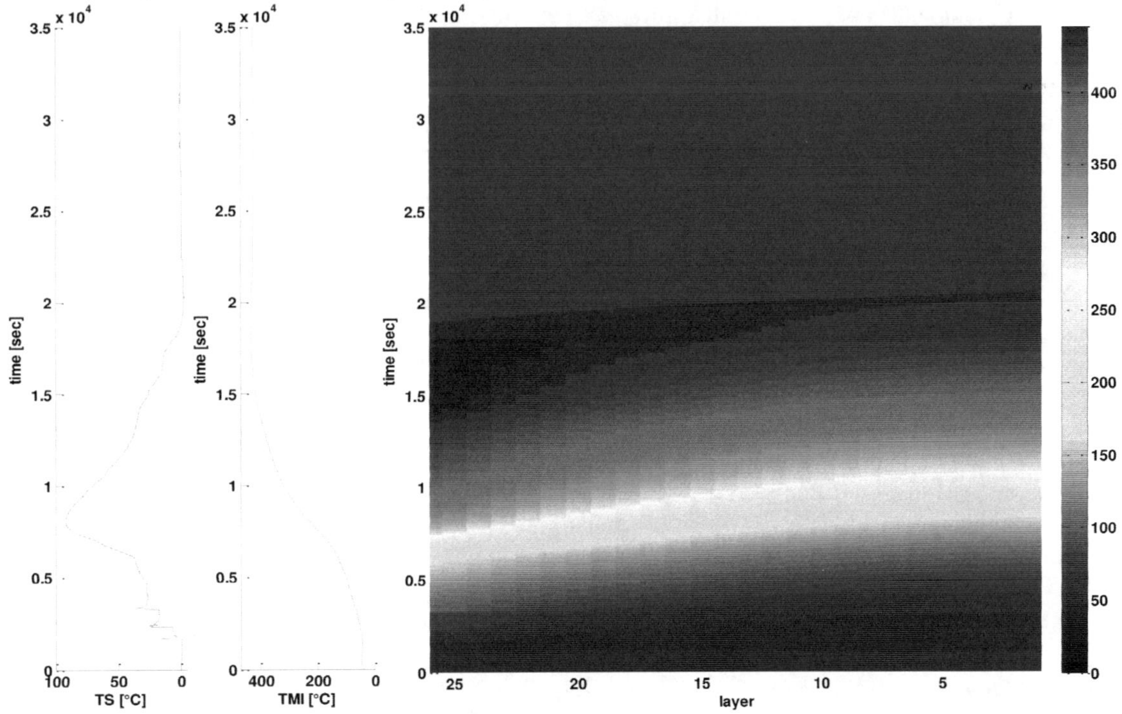

fig. 3: Progress of the rotor temperarture field and *TMI* and *TS* waveforms

3. Generating an optimal trend of steam turbine loading

One of the basic requirements for the turbine startup is achieving of the operating speed and corresponding nominal power in the shortest possible time without exceeding the specified thermal stress limits. Thermal stress is closely related to the amount of obtained/lost heat per time (i.e. change of temperature field). This change describes a heat flux variable [4], [10]:

$$q(t) = \frac{dQ}{dt} \tag{25}$$

Previous formula can be modified and applied to our case (dQ - unknown). From equation for calculating of the specific heat capacity, [11]:

$$c_r = \frac{1}{m}\frac{dQ}{dT} \tag{26}$$

the dQ can be expressed and put into the (25) where c_r is the specific heat capacity and dQ is the amount of heat that the body (in our case rotor) obtain/lost with mass m to its temperature changed by dT. Then we can get:

$$q(t) = \frac{c_r.m.dT}{dt} \tag{27}$$

Previous equation can be discreetly rewritten as follows:

$$q^k = \frac{c_r.m.\Delta T}{\Delta t} = \frac{c_r.\rho.V.(T^k - T^{k-1})}{\Delta t} \tag{28}$$

where Δt corresponds to the sampling period ($\Delta t = 0.5$), ρ is the density of material and V represents the volume which is given as a sum of all the elementary volumes $V[i]$. Because the temperature field of the rotor is generally inhomogeneous in the time point k resp. k-1, we can replace the temperature of the body (rotor) T^k resp. T^{k-1} with the average temperature – mean integral temperature TMI^k resp. TMI^{k-1}. Equation (28) can be rewritten into the following form:

$$q^k = \frac{c_r.\rho.\sum_{i=1}^{n+1} V[i].\left(\frac{\prod_{i=1}^{n+1} T_i^k V[i]}{\sum_{i=1}^{n+1} V[i]} - \frac{\prod_{i=1}^{n+1} T_i^{k-1} V[i]}{\sum_{i=1}^{n+1} V[i]}\right)}{\Delta t} = \frac{c_r.\rho.(\prod_{i=1}^{n+1} T_i^k V[i] - \prod_{i=1}^{n+1} T_i^{k-1} V[i])}{\Delta t} \tag{29}$$

According to equation (29) the heat flux into the rotor based on the known temperature field in the time k can be calculated, thus the real increase (or decrease) of heat in the rotor per time Δt. Note that the volume V (resp. partial volumes $V[i]$) is defined as a cylinder with unit height in our case. Then the variable q^k (see (29)) can also be called as the linear heat flux density.

As mentioned, the thermal stress is related to the time change of heat in the rotor. The formula (29) will be utilized to generate such a trend of heat flux (resp. linear heat flux density), that would be optimal from the turbine startup point of view – i.e. the actual rotor thermal stress reaches the upper boundary of the established limit (without exceeding this limit). Because there must be satisfied a number of operational rules during the turbine startup, then the optimal value of heat flux can be generated just one time step Δt forward. In the each time point k it is therefore necessary to calculate the value of the heat flux q^{k+1}, which will lead in the following time point *(k+1)* to equality of actual thermal stress TS^{k+1} with the desired value TS_{max}^{k+1} (e.g. the upper limit of thermal stress):

$$q^{k+1} = \frac{c_r.\rho.(\prod_{i=1}^{n+1} T_i^{k+1} V[i] - \prod_{i=1}^{n+1} T_i^k V[i])}{\Delta t} \tag{30}$$

The temperatures T_i^k *(i=1,..,n+1)* in the above relationship are known. However, the problem occurs with temperatures T_i^{k+1} *(i=1,..,n+1)*, because the future progress of temperature field leading to equality $TS^{k+1} = TS_{max}^{k+1}$ is unknown. Let's start from the equation for calculation of thermal stress in the time *(k+1)*.

$$TS^{k+1} = T_{n+1}^{k+1} - TMI^{k+1} \tag{31}$$

Using TMI^{k+1} according to (23) and $TS^{k+1} = TS_{max}^{k+1}$ in the previous equation we will get:

$$TS_{max}^{k+1} = T_{n+1}^{k+1} - \frac{\prod_{i=1}^{n+1} T_i^{k+1} V[i]}{\sum_{i=1}^{n+1} V[i]} \tag{32}$$

The unknown variables in equation (32) are again the temperatures T_i^{k+1} *(i=1,..,n+1)*. Notice that the future progress of temperatures $T_1^{k+1},..,T_{n-1}^{k+1}$ can be forward counted in the time point k in the case of explicit version, see equation (12). The same can be performed in the case of implicit version, but the calculation again will lead to solve the equation system. In formula (32) only variables T_n^{k+1} and T_{n+1}^{k+1} are remaining as unknown. If T_{n+1} is also function of T_n:

$$T_{n+1}^{k+1} = f(T_n^{k+1}), \tag{33}$$

where *f(.)* is a general function, e.g linear extrapolation (other expression than (21)):

$$T_{n+1}^{k+1} = T_n^{k+1} \frac{R[n+1]-R[n-1]}{R[n]-R[n-1]} - T_{n-1}^{k+1} \frac{R[n+1]-R[n]}{R[n]-R[n-1]} \tag{34}$$

then it is possible to substitute the T_{n+1}^{k+1} in equation (32) and express remaining unknown T_n^{k+1}. Specifically, in the case of linear extrapolation we will get:

$$T_n^{k+1} = \frac{TS_{max}^{k+1}(R[n]-R[n-1])(\sum_{i=1}^{n+1} V[i]) - T_{n-1}^{k+1}(R[n]-R[n+1])(\sum_{i=1}^{n+1} V[i])}{(R[n+1]-R[n-1])(\sum_{i=1}^{n+1} V[i]) - V[n](R[n]-R[n-1]) - V[n+1](R[n+1]-R[n-1])} + \frac{(\prod_{i=1}^{n-1} T_i^{k+1} V[i])(R[n]-R[n-1]) - V[n+1].T[n-1](R[n]-R[n+1])}{(R[n+1]-R[n-1])(\sum_{i=1}^{n+1} V[i]) - V[n](R[n]-R[n-1]) - V[n+1](R[n+1]-R[n-1])} \tag{35}$$

Finally, it remains to calculate the optimal value of the heat flux q^{k+1} in the time point k, see (30), because all temperatures T_i^{k+1} *(i=1,..,n+1)* are already known (T_{n+1}^{k+1} was calculate according to (33) resp.(34)). If the actual heat flux into the rotor is successively equal to the generated optimum value during the turbine startup (by regulation of amount and temperature of steam), then the actual rotor thermal stress moves around the desired value. This situation is demonstrated in fig. 4, where the blue line shows the actual waveform of the rotor thermal stress and the red one shows the desired waveform of thermal stress (gray line represents the waveform of temperature T_n). For the desired waveform of thermal stress the upper limit of the maximum of permissible thermal stress was elected. Note that the maximum of steam temperature is limited, therefore the temperature T_n also stabilizes at the appropriate value, see fig. 4. Of course, the thermal stress returns to zero after the entire rotor is heated. The heat flux into the rotor is influenced by the temperature and amount of steam – regulation of temperature and pressure.

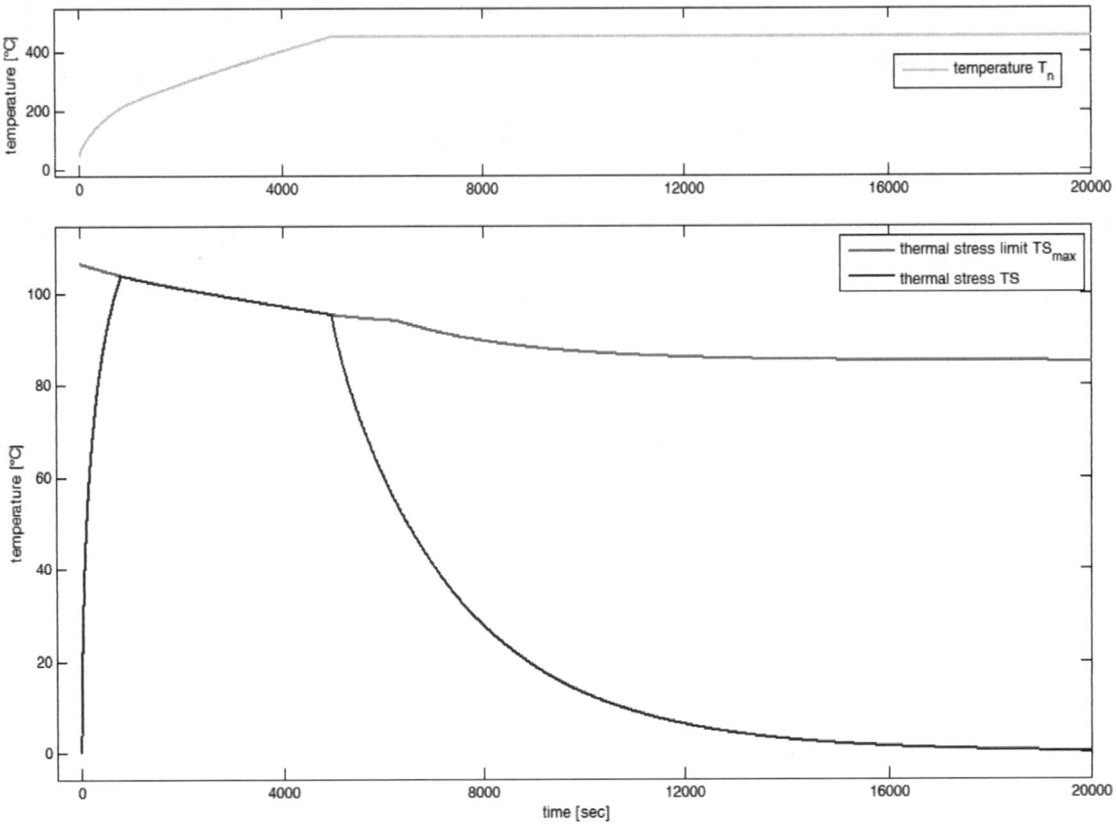

fig. 4: Optimal startup of turbine

4. Conclusion

This paper deals with the calculation of thermal stress of the rotor and generation of optimal steam turbine loading trend leading to equality of the actual thermal stress with the desired value. The entire problem is based on the calculation of the temperature field in the rotor cross section based on the temperature measurement in the inner turbine body – i.e. in the stator. First, using the finite volume method the partial differential equations of heat conduction is solved. The formula for calculation of temperatures in the inner layers of the rotor was obtained in its geometric center and on the rotor surface. Then there was also described the formula for calculation of thermal stress. The obtained calculation methods were afterthat applied to analyse signals from the 60 MW steam turbine. Finally, a method for generation of the optimum heat flux trend or linear heat flux density into the rotor, that leads to the desired behavior of actual thermal stress during the turbine startup, was proposed. Heat flux into the rotor is influenced by temperature and pressure of steam in the turbine.

Acknowledgment

This work was supported by the project PUNTIS - LO1506 and by the project SGS-2013-041.

References

[1] Akimasa Nakai, Masashi Nakamoto, Atsuyuki Kakehi, Shinji Hayashi, *Turbine Start-up Algorithm Based on Prediction of Rotor Thermal Stress,* 1995, Japan

[2] Shuang Bian, Wenyao Li, Calculation of Thermal Stress and Fatigue Life of 1000 MW Steam Turbine Rotor, 2013, 5, Energy and Power Engineering, 1484-1489

[3] Yunchun XIA, *Simulation System on the Thermal Stress and Fatigue Life Loss of Startup and Shutdown for a Domestic 600 MW Steam Turbo generator Unit,* 2009, International Conference on Energy and Environment Technology

[4] Özisik, M., N, *Heat conduction,* 1993, Department of Mechanical and Aerospace Engineering, North Carolina

[5] Eymard, R., Gallouët, T., Herbin, R., *Finite Volume Methods,* 2003, Marseille, n0 97-19 du LATP, UMR 6632

[6] Schäfer, M., *Computional Engineering – Introduction to Numerical Methods*, 2006, X, 321 p. 204 illus., Softcover, ISSBN: 978-3-540-30685-6

[7] Banaszkiewicz, M., *Steam turbines start-ups,* 2014, Transactions of the institute of fluid-flow machinery, No. 126, 2014, 196-198

[8] Press, W. H., Teukolsky, S. A., Vetterling, W. T., Flannery, B. P., *Numerical Recipes in Fortran 77: The Art of scientific computing,* 1986-1992, Cambridge, ISBN 0-521-43064-X

[9] Blomberg, T., Heat Conduction in Two and Three Dimensions: Computr Modelling of Building Physics Applications, 1996, Lund Universitry, Sweden, ISBN 91-88722-05-8

[10] Boelter, L. M. K., *Heat transfer notes,* 1964, New York

[11] Jiji, L. M., *Heat condiction,* 2009, New York, ISBN 978-3-642-01266-2

Data-Driven Assistance Functions for Industrial Automation Systems

Stefan Windmann, Oliver Niggemann
Fraunhofer Application Center Industrial Automation (IOSB-INA), Lemgo
E-mail: {stefan.windmann, oliver.niggemann}@iosb-ina.fraunhofer.de

Abstract. The increasing amount of data in industrial automation systems overburdens the user in process control and diagnosis tasks. One possibility to cope with these challenges consists of using smart assistance systems that automatically monitor and optimize processes. This article deals with aspects of data-driven assistance systems such as assistance functions, process models and data acquisition. The paper describes novel approaches for self-diagnosis and self-optimization, and shows how these assistance functions can be integrated in different industrial environments. The considered assistance functions are based on process models that are automatically learned from process data. Fault detection and isolation is based on the comparison of observations of the real system with predictions obtained by application of the process models. The process models are further employed for energy efficiency optimization of industrial processes. Experimental results are presented for fault detection and energy efficiency optimization of a drive system.

1. Introduction

The growing capacity of cost-effective equipment and the requirements of production technology are resulting in increasingly complex distributed automation systems. A consequence of this trend is an ever-increasing data volume in industrial processes [1], which must be analyzed and taken into account in process control. This leads to ever growing demands on data acquisition, process monitoring, fault recognition and analysis of failure causes. A possibility to cope with these challenges is the application of assistance systems. As a matter of course, application and configuration of such assistance systems is required to be as simple as possible. A step into this direction is the usage of process models that are automatically learned from process data. Self-learning of process models from data leads to considerable simplification of model creation and configuration. Process models can be applied to several model-based assistance functions such as visualization, diagnosis or optimization. An example is the energy efficiency optimization and fault detection for a conveying system. The purpose of the assistance system is to detect deviations in system behavior from the normal state (for example too much or not enough energy consumption of a drive). In addition, it diagnoses the causes for such anomalies (such as wear or wrong load). Furthermore, automatic optimization of system behavior regarding a given cost function can be accomplished. An example is the calculation of energy efficient movements of the conveying system.

This article deals with aspects of smart assistance systems such as process models, assistance functions and system interfaces. Novel research results with respect to self-diagnosis and self-optimization of industrial processes are presented. In doing so, the focus is set on components of industrial automation systems such as conveying systems or processing modules. The paper highlights how to integrate approaches to data-driven model learning, fault detection and energy efficiency optimization in an assistance system for such application cases.

The remaining part of this paper is structured as follows. First, the state of the art for industrial assistance system is resumed (section 2). In the main part, assistance functions, such as model learning (section 3.1), self-diagnosis (section 3.2) and self-optimization (section 3.3) are investigated. The interaction between the assistance functions is outlined in section 3.4. In section 4, architecture and integration of smart assistance systems are considered. Furthermore, experimental results for the proposed assistance system are presented (section 5).

2. State of the art

Todays assistance systems can be categorized as follows:

- Assistance systems based on universally usable statistics or data mining tools
- Complete solutions for data acquisition and analysis
- Sector-specific special-purpose solutions.

Examples of data mining tools are WEKA (Waikato Environment for Knowledge Analysis, [2]) and RapidMiner [3]. These tools offer numerous classification possibilities in addition to preprocessing methods and methods for cluster and association analysis. Examples of statistics packages are R [4] and the Matlab Statistics Toolbox. Both of these packages offer many types of statistical process analysis, such as variance analyses (ANOVA), regression methods and statistical time series analyses. Common data acquisition and analysis solutions in an industrial context are SAS Quality Lifecycle Analysis and SPSS from IBM. With their powerful prediction methods, these solutions allow an early detection and user notification of quality defects and arising problems in production processes. They also provide comprehensive means of merging and correcting large data volumes. The described approaches offer a wide range of functions, but are not always easy to operate for inexperienced users, since they require many parameters such as threshold values for automatic warning functions to be configured manually. Also available are special-purpose solutions, tailored to specific fields of application (e.g. Siemens Siplus CMS for condition monitoring of motors, generators, pumps, etc., CM4810 from B&R for condition monitoring at control system level or PUMon [5] for process monitoring in the chemical industry). Regarding the algorithms used in PUMon, which have been developed at Fraunhofer IOSB as part of the ProDaMi suite, there are already approaches for a generalization to other fields of application, such as monitoring of wind turbines.

In the described assistance systems, a conflict between the bandwidth of application possibilities and the required configuration effort can be observed. The price for a high level of flexibility in possible application areas is usually costly and time-consuming manual engineering. An approach to close this gap is the development of self-learning assistance functions, which, because of their learning capability, can be integrated into existing automation systems without extensive configuration.

3. Assistance functions

In the following section, particular assistance functions and the interaction between these assistance functions are described.

3.1. Data-driven process modelling

Data-driven process modelling allows the flexible use of smart assistance functions in complex automation systems. With this approach, process models are learned from process data and are used for diagnosis (section 3.2) and process optimization (section 3.3). With increasing complexity of industrial processes and the resulting data, manual generation of such process models is no longer practical and, furthermore, very time-consuming and prone to errors. The development of robust methods for automatic model learning from process data is becoming increasingly important. In these approaches, the interaction between continuous and discrete process variables must be considered. In many applications, distinct system states and modes can be observed, which alter as discrete process variables (such as switching signals) change. Especially the automatic learning of process models for discrete system elements has not, to date, been researched in great detail. Approaches to this task are ALERGIA [6], MDI [7], RTI+ [8], HyBUTLA [9] and OTALA [10]. In general, temporal dependencies must be considered in addition to the static dependencies between process variables. A further challenge consists in automatic learning of the structure of process models in addition to the model parameters.

Furthermore, continuous process behaviour such as energy consumption has to be taken into account. The continuous process behaviour is modeled with, among others, differential and difference equations, equivalent state space models, neural networks [11], self-organizing maps [12] or stochastic models such as Kalman filters (e.g. [13], [14]), particle filters (e.g. [15]) or Hidden Markov Models (e.g. [16]). In order to describe hybrid processes, such models are learned for each system state (the states of the machine shown in Fig. 2) from continuous process data.

The learning of energy consumption in the modes of a hybrid automaton is considered in section 3.1.1. The MoSDA algorithm [17] allows automatic learning of separated component models of hybrid industrial processes that can be used for both anomaly detection and the identification of causes for anomalies (see section 3.1.2). MapReduce technology is applied to learn the proposed process models from large amounts of data (see section 3.1.3).

3.1.1. Learning the energy consumption in the modes of hybrid automata: The learning of energy consumption in the modes of a hybrid system has been considered in [14]. The underlying system behavior is modelled in this approach with an automaton. The automaton consists of discrete system states which determine the system behavior. Transitions between system states are triggered by discrete events, e.g. control signals such as switch on/off or open/close valve. Model learning algorithms such as OTALA [10] can be applied to learn the automaton. The principle of the OTALA algorithm consists in creating a new state for each combination of discrete signals which occurs in the data. Transitions are inserted between states which correspond to subsequent combinations of discrete signals in the input data. The energy consumption in each particular mode of a hybrid automaton can be modelled as a sequence of linear state space models. Switches between the linear state space models account for non-linearities or switches in the energy consumption. The learning of state space models $M_{E1}, M_{E2}, \dots$ consists of two steps (see Fig. 1):

In step 1, splitting points are detected (vertical dotted lines in Fig. 1) where the state space models are assumed to switch. In this work, a time series analysis based on a moving average (MA) method [18] is employed to obtain predictions $\widehat{y}(k)$ for the measured energy consumption $y(k)$. Splitting point candidates are detected for $|\widehat{y}(k) - y(k) > y_S|$ with the given threshold y_S. In order to reduce the number of splitting points, adjacent splitting point candidates are removed. More precisely, splitting point candidates with a distance less than a given parameter δ_S are merged. In doing so, the first and the last splitting point in a time series of adjacent candidates is preserved in each case.

In step 2, linear state space models are learned for each time segment between two splitting points. The energy consumption in each time segment s is described with a linear state space model M_{Es} (a standard random walk model). The linear models are learned from a sequence of training data $y(k)$ which is observed in the respective time segments. Expectation-maximization (EM) is employed for parameter estimation (see [19]).

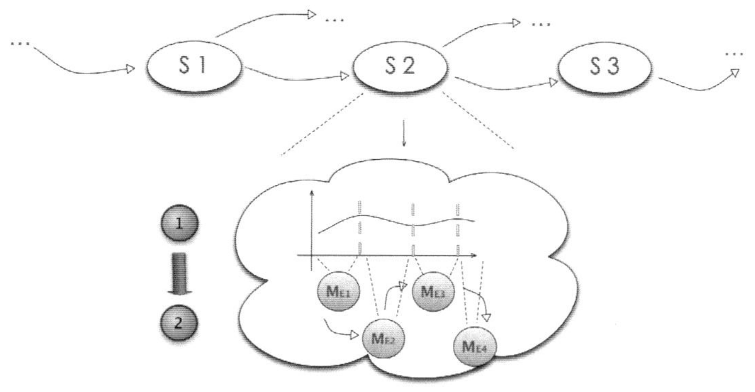

Figure 1. Model-based fault detection

3.1.2. Component-based process models: The algorithm MoSDA [20] exploits a typical situation for cyber-physical production systems: Anomalies often refer to anomalous observations on the system-level, e.g., an anomalous energy consumption of the production system as a whole. MoSDA leverages on the fact that these observations create an over-determined system and MoSDA can therefore break down the overall system model to component models. Fig. 2

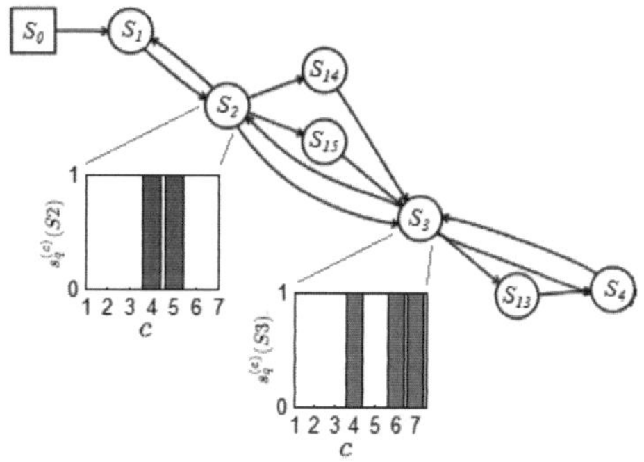

Figure 2. Automaton of the storage system with separation variables $s_{q,c}(S2)$ and $s_{q,c}(S3)$

illustrates the functional principle. The underlying system behavior is modelled with an automaton that can be learned from data as described in section 3.1.1. It is assumed that state $q = S1, S2, \ldots$ determines whether the output $y_c(k)$ of component c contributes at time instance k to the overall output signal

$$y(k) = \sum_{c=0}^{n_c-1} s_c(k) y_c(k). \tag{1}$$

This is expressed with a separation variable $s_{q,c}(q)$ for each component c in state q which can be either 0 or 1. In many practical cases, $s_{q,c}(q)$ is known: System mode q determines the active control signals, the dependency between control signals and components is in many cases given. However, if $s_{q,c}(q)$ is not known a-priori, $s_{q,c}(q)$ can be learnt from training data, e.g. by exploitation of correlation as accomplished in [21]. In this paper, we consider components c where the output signal $y_c(k)$ of a component is a function of two input signals $\mathbf{x}_c(k) = (x_{1,c}, x_{2,c})$ at each time instance k. This approach can be extended straightforwardly to an arbitrary amount of input and output signals. Output $y_c(k)$ is approximated by the quadratic function

$$y_c(k) = \theta_0^c + \theta_1^c x_{1,c}(k) + \theta_2^c x_{2,c}(k) + \theta_{1,1}^c x_{1,c}(k)^2 + \theta_{2,2}^c x_{2,c}(k)^2 + \theta_{1,2}^c x_{1,c}(k) x_{2,c}(k) \quad (2)$$

with coefficients θ_0^c, θ_1^c, θ_2^c, $\theta_{1,1}^c$, $\theta_{2,2}^c$ and $\theta_{1,2}^c$ which can be efficiently computed by solution of a linear over-determined optimization problem (see [17]).

3.1.3. Efficient model learning with MapReduce technology: MapReduce technology allows efficient model learning from large amounts of historical data (see e.g. [22]). The principle is illustrated in Fig. 3. Data is stored in a distributed filesystem in order to process splits of the input data on several host machines. A user-defined map function processes the records of each split and assigns the output to a key. Data is shuffled according to the keys and processed with a user-defined reduce function. Application of MapReduce technology to model learning for the OTALA algorithm described above is straightforward. Pairs of subsequent observations are processed in the map function in order to extract information for each transition (e.g. transition times) which is assigned to unique transition keys. The keys are computed from the discrete signals which change between the two observations. The transitions are shuffled according to the transition keys and are processed in the reduce function (e.g. distributions of the transition times are calculated) in order to build the final automaton. A great deal of model learning algorithms allow for parallelization in a similar fashion.

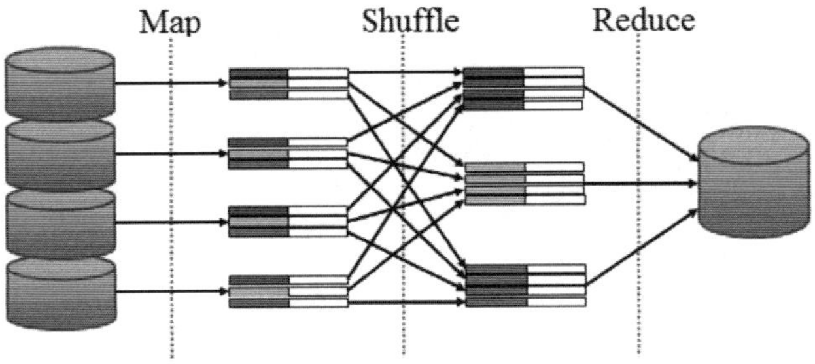

Figure 3. MapReduce technology

3.2. Fault Detection

Reliable fault detection allows costs and risks to be reduced through the early detection of faults and problems in the process and by preventing component failures and, in extreme cases, a production stop of the entire plant.

The principle of model-based fault detection in the operating phase (see Fig. 4) consists of comparing the plants actual behavior with its behavior as predicted from the process models. The process models can be learned from process data collected about the plant and its components in normal, fault-free operation, as described in section 3.1. This allows the detection of, on one hand, discrete events that are incompatible with the learned process models [23], and, on the other, deviations of the continuous behavior from the predicted process behavior.

For the latter, stochastic models can be used (see section 3.1.1). Measurements $y(k)$ are predicted in this approach according to a stochastic process model with input $u(k)$. A common approach to fault detection is the evaluation of the residuals $r(k) = y(k) - \hat{y}(k)$ between measurements $y(k)$ and predictions $\hat{y}(k)$. Methods such as log-likelihood ratios, generalized log-likelihood ratios, sequential probability ratio tests, or cumulative sum test statistics are used

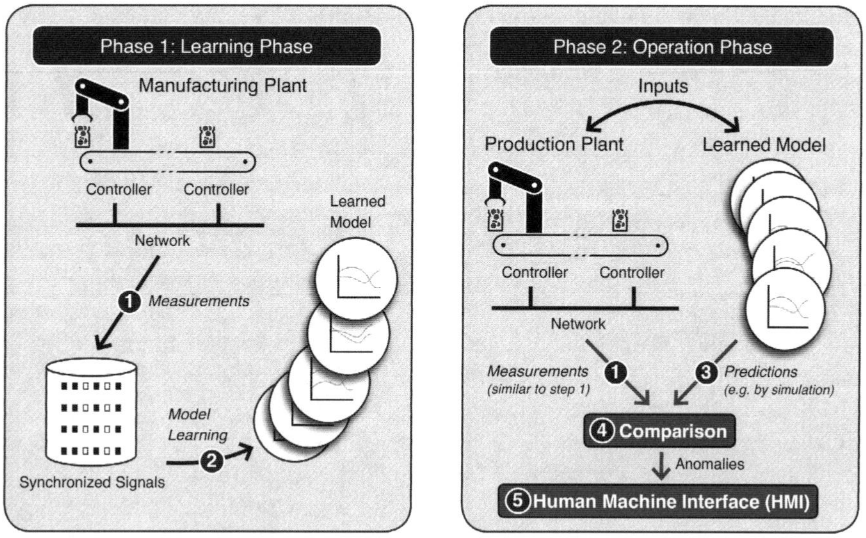

Figure 4. Model-based fault detection

in literature to detect changes in the residuums (see [24]) for an overview). A related approach is the evaluation of $y(k)$ with respect to the conditional probability density $p(y(k)|y(0)\ldots y(k-1))$ (see e.g. [14], [17]). Observations are assumed to be improbable once measurement $y(k)$ at discrete time k drops out of a given confidence interval. This is the case for [14]

$$F(y(k)) < \frac{1}{2} - \frac{C}{2} \text{ or } F(y(k)) > \frac{1}{2} + \frac{C}{2} \tag{3}$$

$$\text{where} \quad F(y(k)) = \int_{-\infty}^{y(k)} p(\tilde{y}(k)|y(0)\ldots y(k-1))d\tilde{y}(k)$$

denotes the distribution function of $p(y(k)|y(0)\ldots y(k-1))$ and C the width of the confidence interval ($C \in [0,1]$). Measurements beyond a confidence interval of width $C = C_1$ are detected as potential faults. In this case, anomaly $f_1(k)$ is reported for time instance k. Furthermore, anomalies with a duration of more than one time instance are taken into account as suggested in [8]. Measurements beyond a confidence interval of width $C = C_2$ are considered as candidates for such anomalies. If at least a given portion p_d of the last n_d measurements are beyond the confidence interval of C_2, anomaly $f_2(k)$ is reported. Evaluation of (3) requires an adequate model for the process behaviour that essentially determines in which way the probability density $p(y(k)|y(0)\ldots y(k-1)$ is obtained. A general approach is the application of discrete time nonlinear state space models (see e.g. [14], [15], [25]) that are specific for particular states or components of the hybrid automaton:

$$\begin{aligned} y(k) &= h(\mathbf{x}(k), u(k)) + v(k) \\ \mathbf{x}(k) &= f(\mathbf{x}(k-1), u(k)) + w(k) \end{aligned} \tag{4}$$

where f and h are the state and measurement dynamic functions, respectively. The two equations consider uncertainties in the form of additive terms for measurement noise $v(k)$ and state transition noise $w(k)$. The continuous system behaviour is modelled with a time-variant state vector $\mathbf{x}(k)$. The random walk model introduced in section 3.1.1 and the system model introduced in (1) are special cases of this general state space model. The component models which are learnt by the MoSDA algorithm (see section 3.1) enable beside fault detection the location of faults by relating the faults to particular components (see section 3.1.2). The application of component models for model-based fault detection is evaluated in section 5.1.

3.3. Energy Efficiency Optimization

Another field of application for smart assistance systems is the self-optimization of industrial processes. The goal here is to analyze and improve the performance and efficiency of a plant to optimize its operation. In particular the energy efficiency optimization of industrial automation systems is becoming increasingly relevant, driven by rising energy prices and political targets [1]. In existing plants, interventions in the process usually consist of manual stopping, energy-efficient time planning of the production steps in the MES system, or control actions at 15-minute intervals [26]. To date, little research has been carried out regarding active methods to implement real-time optimization at second or millisecond intervals, even though, due to their process dynamics, many applications require process parameters to be rapidly adapted to changing operating conditions. First approaches have been developed for stopping the plant based on energy data [27], for reducing reactive power and for controlling energy in Smart Grids [28]. Another approach is based on the use of predictive process models that can be learned with the methods described in section 3.1. The process models allow the future process behavior to be predicted for various parameter combinations in order to determine the optimum process parameters. Optimization is accomplished with respect to the energy and load management of electrical drive systems, which are the most voracious energy consumers, especially in transportation and logistics applications.

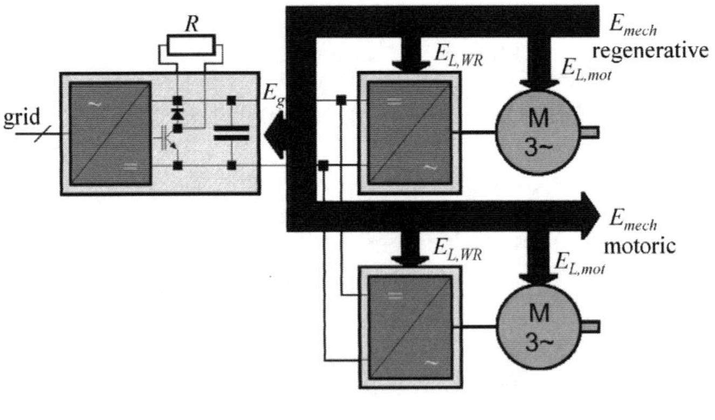

Figure 5. Energy flow between two drives [29]

The basic optimization problem is illustrated in Fig. 5 for two drives. The main optimization aim in this application is to minimize the overall energy consumption $y_0(k)$ and $y_1(k)$ of the two converters ($E_{L,WR}$) and motors ($E_{L,mot}$) for time instances $k = 0 \dots n-1$. In addition, the energy fed back $E_g(k)$ into the internal DC link or the grid (i.e. the difference between regenerative and motive power) is to be minimized. To this end, energy consumption and energy fed back are weighted with factor λ according to the primary usage of power fed back (consumption at braking resistances, regenerative feedback to the mains, and temporary storage in capacitors). Altogether, optimal driving speeds $\mathbf{v}$ of the conveying system can be obtained as solution of the following optimization problem [20]:

$$\mathbf{v} = \operatorname{argmin} \left\{ \sum_{k=0}^{n-1} \sum_{c=0}^{n_c-1} y_c(k) + \lambda \sum_{k=0}^{n-1} E_g(k) \right\} \tag{5}$$

where $y_c(k)$, $c = 0 \dots n_c - 1, n_c = 2$, denote the output variables of the component models of the particular drives (see (2)) which are functions of velocities $x_{1,c}(k)$ and accelerations $x_{2,c}(k)$ for the given application case. Optimization constraints regarding energy flows and the sequences of movements (starting position, end position, speed limits, and positioning time) are detailed in [20]. The overall optimization problem is a mixed integer quadratically constrained problem (MIQP) that can be efficiently solved with standard methods. Evaluation results for energy-efficiency optimization of movements in conveying systems by solution of the exposed optimization problem are given in section 5.2.

3.4. Interaction of the assistance functions
In this section, the interaction of the introduced assistance functions is outlined for the example of the drive system shown in Fig. 5. The overall system model (see (1)) with output variable

$$y(k) = \sum_{c=0}^{n_c-1} s_c(k) y_c(k) \tag{6}$$

is composed of n_c ($n_c = 2$ for the example) component models with output variables $y_c(k)$ according to (2). Switching variables $s_c(k)$ have been introduced in section 3.1.2. The system model is obtained from training data by application of the MoSDA algorithm. Energy efficient movements of the drives are calculated by solution of optimization problem (5) with respect to optimal velocities $\mathbf{v}$ and constraints detailed in [20]. Furthermore, anomalous energy consumption of the drive system is detected by evaluation of

$$F(y(k)) = \int_{-\infty}^{y(k)} p(\tilde{y}(k)|y(0) \dots y(k-1)) d\tilde{y}(k) \tag{7}$$

in (3) for probability density

$$p(y(k)|y(0) \dots y(k-1)) = \mathcal{N}\left(y(k); \sum_{c=0}^{n_c-1} s_c(k) y_c(k), \sigma\right) \tag{8}$$

where standard deviation σ of the prediction error is obtained from training data.

4. Architecture and integration
The flexible use of an assistance system in different industrial environments requires the simple integration of the system in various automation systems. Furthermore, an uncomplicated connection of application-specific user interfaces is required. Besides, aspects such as configurability, modularization capability and scalability have to be taken into account. In proKNOWS - an assistance system developed at Fraunhofer IOSB-INA - these requirements are met through the software architecture shown in Fig. 6.

A unified system interface based on OPC UA allows an abstraction of the assistance functions from heterogeneous networks and data acquisition methods. This, in turn, allows the assistance system to be connected to a wide range of plants and automation systems. The assistance system interfaces with the particular automation systems through OPC UA, with OPC DA servers, or particular instruments and controls or data loggers being integrated through corresponding interfaces. Interfaces for CSV and SQL are also implemented.

The assistance functions implemented in the modeling engine can be run on a central server in real time. In the modeling engine, the data-driven process modeling methods described in the previous sections (HyButla, OTALA, MoSDa, switching state-space models, and electrical drive models) are implemented for fault detection (see e.g. [23] and [14]), fault localisation [17])

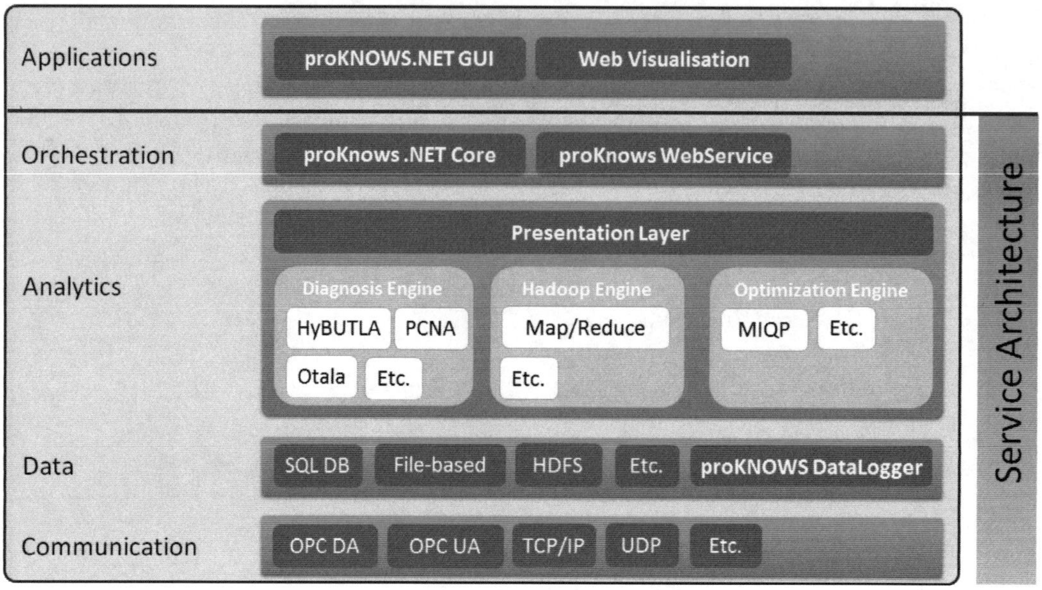

Figure 6. Architecture of proKNOWS

and for process visualization. The modeling engine uses a simple structure model of the plant, which can be configured manually in XML with little effort. The plant structure is split into separate modules. The behavior of these modules is described with process models, on which the represented assistance functions are executed. The process models can be learned automatically from process data, as described in section 3.1. These approaches allow the assistance system to be flexibly adapted to various processes.

The presentation layer converts representations of the process models and results of analysis into a simple exchange format which essentially consists of vector graphics and status information. This provides a convenient way of displaying the relevant information on various user interfaces. In addition to the control panel, these could include mobile apps for particular plant modules, which can be selected specifically for application cases such as commissioning or troubleshooting. A Hadoop engine allows for efficient model learning from large historical databases and provides a framework for efficient analysis of big data. The proKNOWS tool box also contains an optimization engine, in which the methods described in section 3.3 are implemented.

5. Evaluation Results

The introduced assistance functions have been evaluated in several research projects on demonstrators of the Lemgo Smart Factory (LSF), such as a bulk goods processing module (Fig. 7 a) and a conveying system (Fig. 7 b).

The structures of the learned process models for each of the demonstrators differ in terms of system states, transitions and characteristics of the continuous system variables. Typical faults in the represented application cases are errors in the discrete event system, for example delayed switching processes or signal failures, and deviations of continuous signals from their expected trend, such as unexpectedly high or low power consumption by drives or heater fans. Evaluation results for the assistance functions outlined in section 3.4 are presented in the subsequent sections.

Figure 7. a) Module for bulk goods processing b) Conveying system

5.1. Fault Detection

The separated component models, which have been learned in section 3.1.2, have been employed for both fault detection and diagnosis. The proposed methods have been evaluated in the research conveying system depicted in in Fig. 7 b). The system comprises five conveyor belts and seven drives, which are used to move objects between four storage positions on two levels (five drives for horizontal movements and two drives for vertical movements). A power meter continuously captures the overall power consumption. Further sensor information is collected via the programmable logic controller (PLC) of the storage system (Lenze Control 3200 C). For evaluation, process cycles have been recorded, in which an object has been moved between two positions of the conveying system. Faults have been artificially added to 10 process cycles of the conveying system. The created faults differ with respect to position (start time), duration (temporal length) and offset (signal jump or signal drop). The offset is assumed to be constant for each fault.

Fault detection has been accomplished with the method outlined in section 3.4. Results for 10 different durations are averaged with respect to the 10 test cycles, five different offsets and 20 randomly selected positions for each test cycle. The F-measure, i.e. the harmonic mean of sensitivity and precision, was calculated as performance metric.

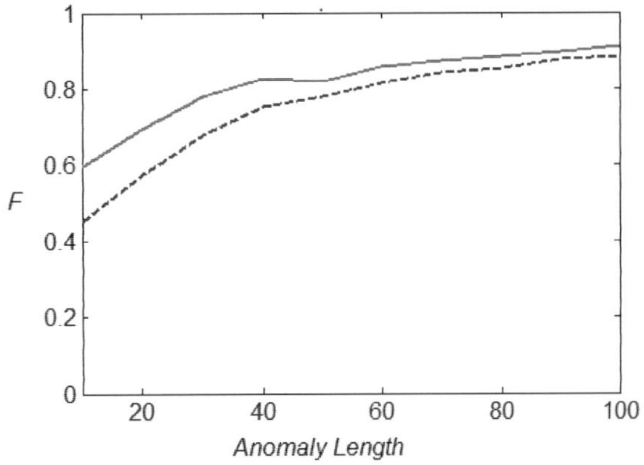

Figure 8. Fault detections results for different fault lengths

Fig. 8 shows the fault detection results for the novel separation approach (straight line). Further, the results for the baseline method without separation (see [11]) are depicted (dashed line). In the test case, significantly improved F-measures are achieved with the separation method. In total, the F-measure improved from 0.75 to 0.81. The introduced component models further allow for diagnosis by relating faults to particular components.

5.2. Optimization

Initial experimental results for the optimization approach introduced in section 3.3, which is summarized in section 3.4, are shown in Fig. 9. The graphs show the speed $v(t)$ and position $x(t)$ for a belt conveyor with a rated power of $470W$ as a function of time t. The movement sequence of the belt conveyor was optimized for given workpiece start and end positions and for given constraints regarding speeds $v(t)$. As can be seen in Fig. 9, the optimization results

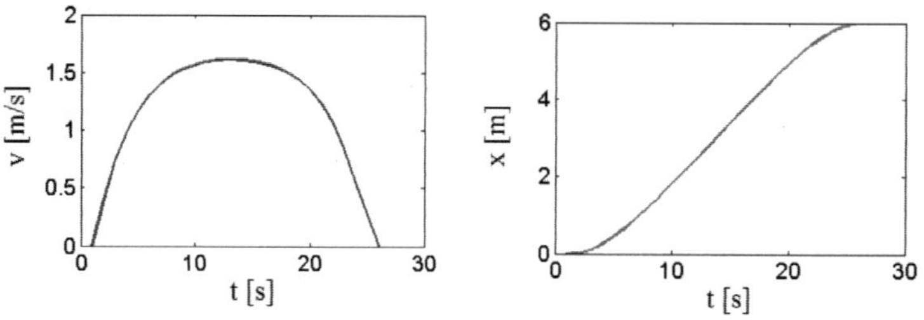

Figure 9. Optimization results

in a movement profile in which acceleration of the conveyor decreases with velocity. For the shown application case, energy consumption was reduced from $5988Ws$ to $5557Ws$ compared to a movement profile with maximum acceleration and deceleration, i.e. energy savings of 7.2% were obtained. Joint optimization of two drives results in a motion profile in which the first drive is decelerated exactly during the second drives acceleration phase. In this application case, the complete regenerative energy of the first drive could be recovered as input for the second drive. Hence, the automatic coordination of motor speeds leads to cancellation of undesired feedback of energy.

6. Conclusion

Because of the increasing complexity of industrial automation systems, process control and monitoring is becoming increasingly challenging. It is therefore desirable to support users in these tasks with smart assistance systems that can take over some of the key tasks. To date there are many partial, but not yet any integrated solutions available. To close this gap, various aspects of smart assistance systems have been integrated, including the system interface and the learning of process models as basis for various assistance functions, such as process monitoring and optimization. The described approaches have proven to make the complexity in automation better manageable, thereby sustainably increasing the productivity of industrial processes.

References

[1] MANUFUTURE-EU, "Factories of the Future PPP Strategic Multi-annual Roadmap," 2010.
[2] I. Witten, E. Frank, and M. Hall, *Data Mining: Practical Machine Learning Tools and Techniques*. Morgan Kaufmann, 2011.
[3] M. Hofmann and R. Klinkenberg, Eds., *RapidMiner: Data Mining Use Case and Business Analytics Applications*. Chapman & Hall/CRC Data Mining and Knowledge Discovery Series, 2013.
[4] R. Ihaka and R. Gentleman, "R: A language for data analysis and graphics," *Journal of Computational and Graphical Statistics*, vol. 5, no. 3, pp. 299–314, 1996.

[5] C. Frey, "Diagnosis and monitoring of complex industrial processes based on self-organizing maps and watershed transformations," in *Computational Intelligence for Measurement Systems and Applications, 2008. CIMSA 2008. 2008 IEEE International Conference on*, july 2008, pp. 87 –92.

[6] R. C. Carrasco and J. Oncina, "Learning deterministic regular grammars from stochastic samples in polynomial time," in *RAIRO (Theoretical Informatics and Applications)*, vol. 33, 1999, pp. 1–20.

[7] F. Thollard, P. Dupont, and C. de la Higuera, "Probabilistic DFA inference using Kullback-Leibler divergence and minimality," in *Proc. of the 17th International Conf. on Machine Learning*. Morgan Kaufmann, 2000, pp. 975–982.

[8] S. Verwer, "Efficient identification of timed automata: Theory and practice," Ph.D. dissertation, Delft University of Technology, 2010.

[9] O. Niggemann, B. Stein, A. Vodenčarević, A. Maier, and H. Kleine Buning, "Learning behavior models for hybrid timed systems," in *Twenty-Sixth Conference on Artificial Intelligence (AAAI-12)*, Jul 2012.

[10] A. Maier, "Online passive learning of timed automata for cyber-physical production systems," in *The 12th IEEE International Conference on Industrial Informatics (INDIN 2014)*. Porto Alegre, Brazil, Jul 2014.

[11] A. Vodenčarević, H. Kleine Büning, O. Niggemann, and A. Maier, "Using behavior models for anomaly detection in hybrid systems," in *Information, Communication and Automation Technologies (ICAT), 2011 XXIII International Symposium on*, 2011, pp. 1–8.

[12] C. Frey and T. Ross, "Anwendung datengetriebener Modelle zur Zustandsanalyse bei verfahrenstechnischen Anlageneinheiten: Theoretische Grundlagen und Beispiele aus der Industriellen Umsetzung," *Chemie Ingenieur Technik*, vol. 84, no. 8, p. 1378, 2012.

[13] S. Narasimhan and G. Biswas, "Model-based diagnosis of hybrid systems," in *Proceedings of the 18th international joint conference on Artificial intelligence*. San Francisco, CA, USA: Morgan Kaufmann Publishers Inc., 2003, pp. 376–381.

[14] S. Windmann, S. Jiao, O. Niggemann, and H. Borcherding, "A Stochastic Method for the Detection of Anomalous Energy Consumption in Hybrid Industrial Systems," in *INDIN*, 2013.

[15] M. Wang and R. Dearden, "Detecting and Learning Unknown Fault States in Hybrid Diagnosis," in *Proceedings of the 20th International Workshop on Principles of Diagnosis, DX09*, Stockholm, Sweden, 2009, pp. 19–26.

[16] M. Jager, C. Knoll, and F. Hamprecht, "Weakly supervised learning of a classifier for unusual event detection," *Image Processing, IEEE Transactions on*, vol. 17, pp. 1700–1708, 2008.

[17] S. Windmann and O. Niggemann, "Automatic model separation and application to diagnosis in industrial automation systems," in *IEEE International Conference on Industrial Technology (ICIT 2015)*, 2015.

[18] T. Mingyu and C. Jin, "The time series of the double moving average forecast," *China Academic Journal Electronic House*, vol. 1, pp. 33–37, 2010.

[19] Z. Ghahramani and G. E. Hinton, "Parameter estimation for linear dynamical systems," Univ. Toronto, Dept. Comput. Sci., Tech. Rep., 1996.

[20] S. Windmann, O. Niggemann, and H. Stichweh, "Energy efficiency optimization by automatic coordination of motor speeds in conveying systems." in *IEEE International Conference on Industrial Technology (ICIT 2015)*, 2015.

[21] O. Niggemann, S. Windmann, S. Volgmann, A. Bunte, and B. Stein, "Using learned models for the root cause analysis of cyber-physical production systems," in *DX*, 2014.

[22] T. White, *Hadoop: The Definitive Guide*, 2nd ed., M. Loukides, Ed. O'Reilly Media, Inc., 2011.

[23] S. Faltinski, H. Flatt, F. Pethig, B. Kroll, A. Vodenčarević, A. Maier, and O. Niggemann, "Detecting anomalous energy consumptions in distributed manufacturing systems," in *Industrial Informatics (INDIN), 2012 9th IEEE International Conference on*, 2012, pp. 358 – 363.

[24] W. Liu and I. Hwang, "Robust estimation and fault detection and isolation algorithms for stochastic linear hybrid systems with unknown fault input," *Control Theory Applications, IET*, vol. 5, no. 12, pp. 1353 –1368, august 2011.

[25] L. Christiansen, A. Fay, B. Opgenoorth, and J. Neidig, "Improved diagnosis by combining structural and process knowledge," in *Emerging Technologies Factory Automation (ETFA), 2011 IEEE 16th Conference on*, 2011, pp. 1–8.

[26] P. Palensky and D. Dietrich, "Demand Side Management: Demand Response, Intelligent Energy Systems, and Smart Loads," in *IEEE Transactions on Industrial Informatics*, vol. 7, 2011.

[27] A. Cannata, S. Karnoukos, and M. Taisch, "Energy efficient driven process analysis and optimization in discrete manufacturing," in *IECON*. IEEE, 2009, pp. 4449–4454.

[28] G. Reed, "Sample Survey of Smart Grid Approaches and Technology Gap Analysis," in *Innovative Smart Grid Technologies Conference Europe*, 2010, pp. 1–10.

[29] Lenze SE, "Internal Report," 2014.

Active fault detection: A comparison of probabilistic methods

Jan Škach, Ivo Punčochář

Department of Cybernetics and NTIS - New Technologies for the Information Society, Faculty of Applied Sciences, University of West Bohemia, Technická 8, 30614 Plzeň, Czech Republic

E-mail: `janskach@kky.zcu.cz`, `ivop@kky.zcu.cz`

Abstract. The paper deals with probabilistic methods for designing the active fault detectors that improve the quality of detection using an auxiliary input signal. Two probabilistic methods that assume a similar stochastic model of a monitored system are considered and compared with a special attention to various difficulties associated with active fault detector designs. The active fault detector design based on a general detection cost function is compared with the model sequence selection error minimization design in terms of assumptions and theoretical properties. Practical aspects of both methods are also considered and demonstrated through a numerical example.

1. Introduction

Over the last decades fault diagnosis has received an increased attention [1]. With growing demands on the quantity and quality of production, automated systems have become more complex and less predictable and protected at the same time. Fault diagnosis allows a monitored system to be supervised in order to achieve a required safety integrity level.

There are two approaches to fault detector design: passive approach and active approach. The difference between these two approaches is illustrated in the block diagrams in Figure 1. In the case of the passive approach [2, 3] the input-output data of the monitored system are measured, analyzed for faulty behavior, and consequently a decision about a fault is made based on this analysis. Passive detectors are designed using various methods ranging from signal based methods to model based methods or knowledge based methods [4, 5, 6]. Contrary to the passive approach that is mature and commonly used in many today's applications, the active approach [7, 8, 9, 10, 11] is historically younger, generaly more complex, and still under development. In addition to generating decisions, an active fault detector interacts with the monitored system by means of a suitably designed auxiliary input signal that is injected into the system to probe it for faults. This auxiliary input signal can then increase the quality of fault diagnosis.

Several methods with different levels of generality have been developed in the area of active fault detection (AFD). They can be classified into two groups based on the description of uncertainties in the monitored system. Methods in the first group can be called probabilistic as the uncertainties are described by stochastic processes with known properties. A common practice is to assume that the initial state, state noise, and measurement noise are white random processes with known probability density functions (pdfs). More information on the probabilistic

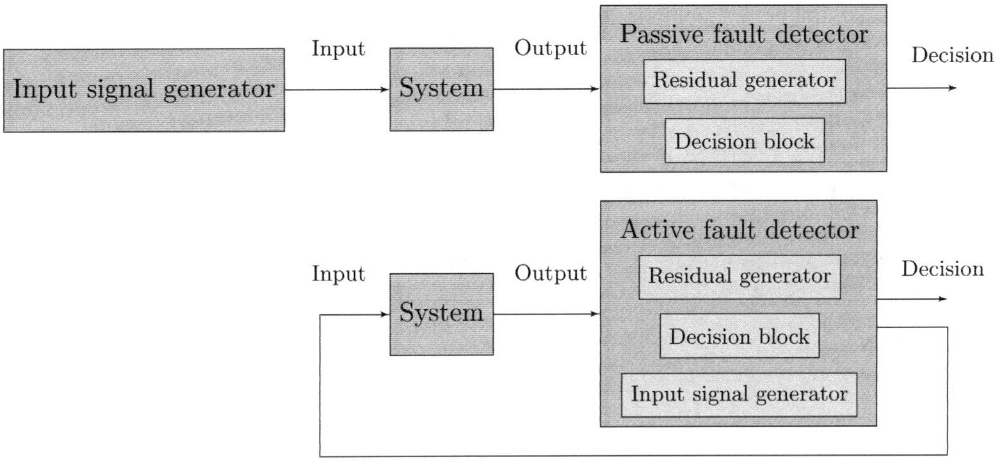

Figure 1. General schemes for passive and active fault detection.

methods can be found in [8, 12, 13, 11, 14]. The second group consists of deterministic methods which assume that the uncertainties are described as bounded signals. More specifically, the initial state, state noise, and measurement noise belong to particular bounded sets. Contrary to the probabilistic methods a perfect diagnosis can be achieved on a finite time horizon if input-output trajectories can be separated by a suitably designed auxiliary input signal. Deterministic methods based on an integrated design of a controller with four degrees of freedom using the $\mathcal{H}_2$ and $\mathcal{H}_\infty$ theory can be found in [15, 10, 16]. The other group of deterministic methods based on an idea of bounding system uncertainties is investigated, e.g., in [9, 17, 18, 19].

The AFD methods are generally more complex than the passive methods and finding a suitable auxiliary input signal can be a very challenging task. The methods usually struggle with similar design problems and throughout the solution may employ various practices from different technical fields such as filtering, estimation, optimization, artificial intelligence, etc. In this paper, two probabilistic methods will be discussed in order to show common features as well as different characteristics and difficulties of the probabilistic methods. This can be useful when closely studying the AFD. Moreover, a numerical example will be given to demonstrate the performance and to provide a better insight into the theoretical background of finding the auxiliary input signal. To the best of authors' knowledge, no detailed comparison of probabilistic AFD methods accompanied by a numerical example has been done.

In the first method, the design of an active fault detector is formulated as an optimization problem where a criterion penalizing wrong decisions is minimized. The method was presented both for a finite time horizon [11, 20] and an infinite time horizon [21, 22]. Note that the formulation contains three special cases based on the form of the criterion, i.e. AFD, active fault detection and control (AFDC), and dual control (DC). In this paper, only the AFD case is considered. A solution can be found by approximate dynamic programming (ADP) techniques [23]. The second method considered in this paper is based on finding an auxiliary input signal that minimizes the probability of the model selection error [13]. A solution to the finite horizon optimization problem can be found using sequential quadratic programming (SQP) techniques [24]. Moreover, a given control task can be achieved when considering constraints on admissible system state behavior.

The paper is organized as follows. In Section 2, a general formulation of the AFD problem for linear discrete-time stochastic systems and a state estimation problem is presented. Active fault detector design formulations and corresponding problem solutions of the two probabilistic

Journal of Physics: Conference Series **659** (2015) 012046 doi:10.1088/1742-6596/659/1/012046

methods are presented and comprehensively compared in Section 3. A numerical example in Section 4 demonstrates performance of designed active fault detectors.

2. Problem formulation

Since two considered probabilistic methods make a slightly different assumptions about the system and the aim of fault detection, this section presents only the part of the problem formulation that is common to both methods. At first, the multiple-model description of the system is presented. Then an approximate solution to the state estimation problem for such a model is shown. The other assumptions are discussed in Section 3.

2.1. System description

The AFD methods usually employ the multiple-model approach to represent the behavior of the monitored system under fault-free and faulty conditions. Therefore, it is considered that a monitored system can be described at each time step k by one of the following linear discrete-time stochastic dynamic models

$$\mathbf{x}_{k+1} = \mathbf{A}_{\mu_k}\mathbf{x}_k + \mathbf{B}_{\mu_k}\mathbf{u}_k + \mathbf{w}_k, \tag{1a}$$

$$\mathbf{y}_k = \mathbf{C}_{\mu_k}\mathbf{x}_k + \mathbf{v}_k, \tag{1b}$$

where $\mathbf{x}_k \in \mathbb{R}^{n_x}$ is an unknown continuous common part of the hybrid state $\mathbf{x}_k^{\mathrm{a}} = [\mathbf{x}_k^{\mathrm{T}}, \mu_k]^{\mathrm{T}} \in \mathbb{R}^{n_x} \times \mathcal{M}$, $\mu_k \in \mathcal{M} = \{1, 2, \ldots, N_\mu\}$ is an unknown index of the model, N_μ is the number of models with one fault-free model and $N_\mu - 1$ faulty models, $\mathbf{u}_k \in \mathcal{U} \subset \mathbb{R}^{n_u}$ is the input, $\mathbf{y}_k \in \mathbb{R}^{n_y}$ is the output measurement, $\mathbf{w}_k \in \mathbb{R}^{n_x}$ and $\mathbf{v}_k \in \mathbb{R}^{n_y}$ are corresponding mutually independent zero-mean Gaussian white noises with known covariances $\mathbf{Q}_{\mu_k}$ and $\mathbf{R}_{\mu_k}$, respectively. The matrices $\mathbf{A}_{\mu_k}$, $\mathbf{B}_{\mu_k}$, $\mathbf{C}_{\mu_k}$ are known and have matching dimensions. The model index μ_k is described by a finite-state Markov chain with a known transition probability matrix $\mathbf{P}_{j,i} = P(\mu_{k+1} = j|\mu_k = i)$. The initial hybrid-state distribution is described by probability $P(\mu_0)$ of the initial model index μ_k and initial condition on continuous part $\mathbf{x}_0$ such that it corresponds to the Gaussian distribution with a given mean value $\hat{\mathbf{x}}_{0|-1}$ and a symmetric positive definite covariance matrix $\mathbf{\Sigma}^{\mathbf{x}}_{0|-1}$.

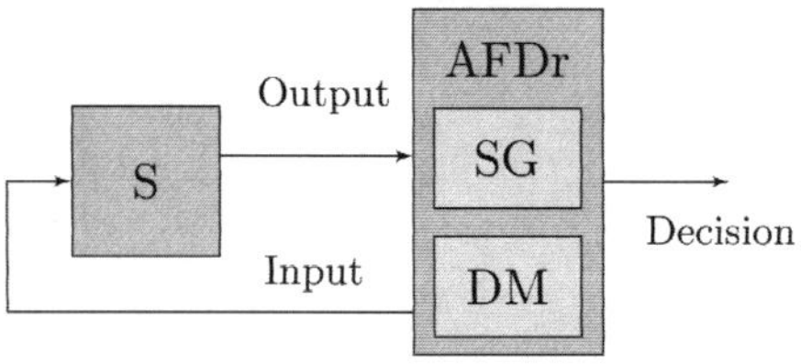

Figure 2. A block diagram of active fault detection.

Active fault detector (AFDr) consists of a decision making block (DM) that generates a decision about faults in the system (S) and an auxiliary input signal generator (SG) that probes the system, as shown in Figure 2. A goal of the AFD problem is to design AFDr such that a certain criterion will be optimized. In Section 3, an overview of two different design methods of the probabilistic approach to AFD will be covered.

2.2. State estimation

The multiple model approach brings a problem of state estimation which is described by a stochastic linear Markovian switching system. The question can be either to determine a conditional probability of model $P(\mu_k|\mathbf{y}_0^k, \mathbf{u}_0^{k-1})$ or a conditional probability of hybrid-state $P(\mathbf{x}_k^{\mathrm{a}}|\mathbf{y}_0^k, \mathbf{u}_0^{k-1})$. Both probabilities can be expressed using a sum over all of $(N_\mu)^k$ possible model sequences μ_0^{k-1}, i.e.

$$P(\mu_k|\mathbf{y}_0^k,\mathbf{u}_0^{k-1}) = \sum_{\mu_0^{k-1}} P(\mu_0^k|\mathbf{y}_0^k,\mathbf{u}_0^{k-1}), \tag{2a}$$

$$P(\mathbf{x}_k^{\mathrm{a}}|\mathbf{y}_0^k,\mathbf{u}_0^{k-1}) = \sum_{\mu_0^{k-1}} P(\mu_0^k|\mathbf{y}_0^k,\mathbf{u}_0^{k-1})P(\mathbf{x}_k|\mu_0^k,\mathbf{y}_0^k,\mathbf{u}_0^{k-1}). \tag{2b}$$

Note that the conditional probability of the model sequence μ_0^k can be determined by

$$P(\mu_0^k|\mathbf{y}_0^k,\mathbf{u}_0^{k-1}) = \frac{p(\mathbf{y}_k|\mu_0^k,\mathbf{y}_0^{k-1},\mathbf{u}_0^{k-1})P(\mu_0^k|\mathbf{y}_0^{k-1},\mathbf{u}_0^{k-2})}{\sum_{\mu_0^k} p(\mathbf{y}_k|\mu_0^k,\mathbf{y}_0^{k-1},\mathbf{u}_0^{k-1})P(\mu_0^k|\mathbf{y}_0^{k-1},\mathbf{u}_0^{k-2})}, \tag{3}$$

where the predictive probability of the model sequence is given by

$$P(\mu_0^k|\mathbf{y}_0^{k-1},\mathbf{u}_0^{k-2}) = \begin{cases} P(\mu_0) & \text{if } k=0, \\ P(\mu_k|\mu_{k-1})P(\mu_0^{k-1}|\mathbf{y}_0^{k-1},\mathbf{u}_0^{k-2}) & \text{if } k>0. \end{cases} \tag{4}$$

For a given model sequence μ_0^k the optimal state estimate can be obtained by the Kalman filter. The conditional pdf $p(\mathbf{y}_k|\mu_0^k,\mathbf{y}_0^{k-1},\mathbf{u}_0^{k-1})$ has the Gaussian distribution with the mean value $\hat{\mathbf{y}}_{k|k-1}$ and the covariance matrix $\boldsymbol{\Sigma}_{k|k-1}^{\mathbf{y}}$ that are given by

$$\hat{\mathbf{y}}_{k|k-1} = \mathbf{C}_{\mu_k}\hat{\mathbf{x}}_{k|k-1}(\mu_0^{k-1}), \tag{5a}$$

$$\boldsymbol{\Sigma}_{k|k-1}^{\mathbf{y}} = \mathbf{C}_{\mu_k}\boldsymbol{\Sigma}_{k|k-1}^{\mathbf{x}}\mathbf{C}_{\mu_k}^{\mathrm{T}} + \mathbf{R}_{\mu_k}. \tag{5b}$$

Note that the initial values $\hat{\mathbf{x}}_{0|-1}$, $\boldsymbol{\Sigma}_{0|-1}^{\mathbf{x}}$ are known. The filtering conditional pdf $p(\mathbf{x}_k|\mu_0^k,\mathbf{y}_0^k,\mathbf{u}_0^{k-1})$ has the Gaussian distribution with the mean value $\hat{\mathbf{x}}_{k|k}(\mu_0^k)$ and the covariance $\boldsymbol{\Sigma}_{k|k}^{\mathbf{x}}(\mu_0^k)$ given by

$$\hat{\mathbf{x}}_{k|k}(\mu_0^k) = \hat{\mathbf{x}}_{k|k-1}(\mu_0^{k-1}) + \mathbf{K}(\mu_0^k)[\mathbf{y}_k - \hat{\mathbf{y}}_{k|k-1}(\mu_0^k)], \tag{6a}$$

$$\boldsymbol{\Sigma}_{k|k}^{\mathbf{x}}(\mu_0^k) = \boldsymbol{\Sigma}_{k|k-1}^{\mathbf{x}}(\mu_k^{k-1}) - \mathbf{K}(\mu_0^k)\mathbf{C}_{\mu_k}\boldsymbol{\Sigma}_{k|k-1}^{\mathbf{x}}(\mu_0^{k-1}), \tag{6b}$$

where the Kalman gain $\mathbf{K}(\mu_0^k)$ can be found solving

$$\mathbf{K}(\mu_0^k) = \boldsymbol{\Sigma}_{k|k-1}^{\mathbf{x}}(\mu_0^{k-1})\mathbf{C}_{\mu_k}^{\mathrm{T}}\boldsymbol{\Sigma}_{k|k-1}^{\mathbf{y}}(\mu_0^k)^{-1}. \tag{7}$$

Finally, the predictive state mean value $\hat{\mathbf{x}}_{k+1|k}(\mu_0^k)$ and the predictive state covariance matrix $\boldsymbol{\Sigma}_{k+1|k}^{\mathbf{x}}(\mu_0^k)$ are given by

$$\hat{\mathbf{x}}_{k+1|k}(\mu_0^k) = \mathbf{A}_{\mu_k}\hat{\mathbf{x}}_{k|k}(\mu_0^k) + \mathbf{B}_{\mu_k}\mathbf{u}_k, \tag{8a}$$

$$\boldsymbol{\Sigma}_{k+1|k}^{\mathbf{x}}(\mu_0^k) = \mathbf{A}_{\mu_k}\boldsymbol{\Sigma}_{k|k}^{\mathbf{x}}(\mu_0^k)\mathbf{A}_{\mu_k}^{\mathrm{T}} + \mathbf{Q}_{\mu_k}. \tag{8b}$$

It can be seen that the conditional probabilities (2) can be computed using the equations above. However, with the higher k, the number of model sequences grows exponentially and it would be computationally intractable to compute the statistics for all the sequences. One of the common suboptimal solutions that prevents the problem of increasing number of all possible model sequences with time consists of tracking only a h-step history of the model sequences.

It means that when $k \geq h$, the conditional probability of the model sequence μ_{k-h+1}^{k} can be computed as

$$P(\mu_{k-h+1}^{k}|\mathbf{y}_0^k, \mathbf{u}_0^{k-1}) = \sum_{\mu_{k-h}} P(\mu_{k-h}^{k}|\mathbf{y}_0^k, \mathbf{u}_0^{k-1}) \tag{9}$$

and the conditional pdf $p(\mathbf{x}_k|\mu_{k-h+1}^{k}, \mathbf{y}_0^k, \mathbf{u}_0^{k-1})$ has a form of Gaussian sum replaced by a single Gaussian pdf with the first and second moments computed by a moment matching technique. This h-step history tracking corresponds to the Generalized Pseudo Bayes algorithm [25].

3. Active fault detector design

Two methods for designing an active fault detector will be briefly summarized and then compared in this section. Both methods assume the same model of the system described in Section 2.1. Also the approximate solution to the state estimation problem presented in Section 2.2 is employed by these methods. Besides some slight differences between methods, the key difference is the way the available information is utilized in those two methods. The first method was proposed in [22] and it uses so called closed loop information processing strategy. The second method was presented in [13] and it basically applies the open loop feedback information processing strategy. Without use of any approximations the closed loop information processing strategy will always result in a lower or the same value of the criterion compared to the open loop information processing strategy. Unfortunately, when approximations are used through the solution, it is almost impossible to tell in advance whether the use of the closed loop information processing strategy will be beneficial and to what extent. In such a case, a numerical comparison for a particular example can provide some insight into benefits of the closed loop information processing strategy.

3.1. Optimal active fault detector design

The design of an optimal active fault detector that minimizes a discounted sum of decision costs over an infinite time horizon assuming that the common continuous part of the hybrid state is known was considered in [22]. However, this part of the hybrid state is rather observed through noisy measurements in most of the real-world systems. A method for designing an active fault detector for such systems was proposed in [26]. Although the infinite time horizon formulation is theoretically more challenging, it has two advantages when compared to the finite time horizon formulation. First, the optimal active fault detector is represented by a time-invariant function that maps the pdf of the hybrid state into decisions and auxiliary inputs. Second, the actual time horizon need not be known in advance and the obtained active fault detector can be used even for a finite time horizon as an approximation of the optimal solution. The h-step history tracking is employed to limit the increasing number of sequences with time. In the following part a problem formulation of finding the active fault detector will be given.

The first step of the design procedure is the problem reformulation using a hyper-state that includes the results produced by the approximate state estimation procedure. The hyper-state allows the original model to be rewritten into a new model that basically includes the original model together with the approximate state estimator. Thus the monitored system coupled with the approximate state estimator can be described at each time step k by the following model

$$\boldsymbol{\xi}_{k+1} = \boldsymbol{\phi}(\boldsymbol{\xi}_k, \mathbf{u}_k, \mathbf{y}_{k+1}), \tag{10}$$

where $\boldsymbol{\xi}_k \in \mathcal{G} \subset \mathbb{R}^{n_\xi}$ is the hyper-state that consists of a finite number of sufficient statistics for the unknown hybrid state $\mathbf{x}_k^{\mathrm{a}}$, $\boldsymbol{\phi} : \mathcal{G} \times \mathcal{U} \times \mathbb{R}^{n_y} \mapsto \mathcal{G}$ is a time-invariant function that describes the monitored system coupled with the approximate state estimator, and $\mathbf{y}_{k+1}$ is a random

Journal of Physics: Conference Series **659** (2015) 012046 doi:10.1088/1742-6596/659/1/012046

variable with the known predictive conditional pdf $p(\mathbf{y}_{k+1}|\boldsymbol{\xi}_k, \mathbf{u}_k)$ that has the form of a Gaussian sum. The hyper-state $\boldsymbol{\xi}_k$ consists of the filtering estimates of the hybrid state, i.e., the mean values $\hat{\mathbf{x}}_{k|k}(\mu_{k-h+1}^k)$, covariance matrices $\boldsymbol{\Sigma}_{k|k}^{\mathbf{x}}(\mu_{k-h+1}^k)$, and probabilities $P(\mu_{k-h+1}^k|\mathbf{y}_0^k, \mathbf{u}_0^{k-1})$ for all possible model sequences μ_{k-h+1}^k. Note that the hyper-state dimension can be reduced simply by considering only the lower or upper triangular part of the covariance matrix $\boldsymbol{\Sigma}_{k|k}^{\mathbf{x}}(\mu_{k-h+1}^k)$ and only $(N_\mu)^h - 1$ conditional probabilities of model sequences since they sum up to one. Then the hyper-state dimension yields in $n_{\boldsymbol{\xi}} = (N_\mu)^h(n_{\mathbf{x}}^2 + 3n_{\mathbf{x}} + 2)/2 - 1$.

This reformulation of the original problem enables the active fault detector that depends only on the hyper-state $\boldsymbol{\xi}_k$ to be designed. The active fault detector is assumed to be a time-invariant mapping of the hyper-state into a decision and auxiliary input

$$\begin{bmatrix} d_k \\ \mathbf{u}_k \end{bmatrix} = \boldsymbol{\rho}(\boldsymbol{\xi}_k) = \begin{bmatrix} \sigma(\boldsymbol{\xi}_k) \\ \boldsymbol{\gamma}(\boldsymbol{\xi}_k) \end{bmatrix}, \tag{11}$$

where $\boldsymbol{\rho} : \mathcal{G} \mapsto \mathcal{M} \times \mathcal{U}$ is an unknown function that represents the active fault detector with components $\sigma : \mathcal{G} \mapsto \mathcal{M}$ generating the decisions $d_k \in \mathcal{M}$ and $\boldsymbol{\gamma} : \mathcal{G} \mapsto \mathcal{U}$ generating the auxiliary input signal $\mathbf{u}_k$. The active fault detector is designed such that the design criterion

$$J^{\mathrm{OAFD}}(\boldsymbol{\rho}) = \lim_{F\to\infty} \mathbb{E}\left\{\sum_{k=0}^{F} \lambda^k L^{\mathrm{d}}(\boldsymbol{\xi}_k, d_k)\right\} \tag{12}$$

is minimized. Note that $L^{\mathrm{d}} : \mathcal{G} \times \mathcal{M} \mapsto \mathbb{R}^+$ refers to a general detection cost function that penalizes wrong decisions about faults in the monitored system and $\lambda \in (0, 1)$ is a discount factor that influences the future detection costs. Since the detection function L^{d} is chosen in advance the designer must understand the system behavior. In case of the AFDC problem the criterion (12) must include a component that penalizes system state behavior.

The introduced problem reformulation leads to a dynamic optimization problem with perfect state information that can be solved by dynamic programming (DP) [27]. It follows from DP that the design of an optimal active fault detector is reduced to solving the following discrete-time Bellman functional equation

$$V^*(\boldsymbol{\xi}_k) = \min_{d_k\in\mathcal{M}} \mathbb{E}\{L^{\mathrm{d}}(\boldsymbol{\xi}_k, d_k)|\boldsymbol{\xi}_k, d_k\} + \min_{\mathbf{u}_k\in\mathcal{U}} \mathbb{E}\{\lambda V^*(\boldsymbol{\phi}(\boldsymbol{\xi}_k, \mathbf{u}_k, \mathbf{y}_{k+1}))|\boldsymbol{\xi}_k, \mathbf{u}_k\}, \tag{13}$$

where $V^* : \mathcal{G} \mapsto \mathbb{R}$ is the unknown Bellman function that represents the expected detection costs over the infinite time horizon $[k, \infty)$ discounted by λ when starting at $\boldsymbol{\xi}_k$ and using the optimal policy $\boldsymbol{\rho}^*$. Note that the term "policy" is commonly used in DP and refers to a control law. Then the optimal detector and the optimal auxiliary input signal generator are given by

$$d_k^* = \sigma^*(\boldsymbol{\xi}_k) = \arg\min_{d_k\in\mathcal{M}} \mathbb{E}\{L^{\mathrm{d}}(\boldsymbol{\xi}_k, d_k)|\boldsymbol{\xi}_k, d_k\} \tag{14a}$$

$$\mathbf{u}_k^* = \boldsymbol{\gamma}^*(\boldsymbol{\xi}_k) = \arg\min_{\mathbf{u}_k\in\mathcal{U}} \mathbb{E}\{V^*(\boldsymbol{\phi}(\boldsymbol{\xi}_k, \mathbf{u}_k, \mathbf{y}_{k+1}))|\boldsymbol{\xi}_k, \mathbf{u}_k\}. \tag{14b}$$

Although the formal solution described by (13) and (14) is obtained straightforwardly by DP, the design problem was only recast into solving the discrete-time Bellman functional equation. As an analytical form of the Bellman function is hard to find in general case, a suitable approximation of the Bellman function needs to be found using ADP [28, 23]. A common approximations to Bellman function include parametric approximations, nonparametric approximations, and other variants. Furthermore, the approximate Bellman function is found by an iterative algorithm such as the value iteration algorithm or the policy iteration algorithm. Finally, the conditional expected value in the discrete-time Bellman functional equation (13) must be approximated since the system function (10) is nonlinear. To approximate the expectation a sufficient number of Monte Carlo simulations or an unscented transformation can be used.

3.2. Model selection error minimization design

The design of an auxiliary input signal generator that minimizes the probability of the model sequence selection error at the end of a finite time horizon was presented in [13]. Although the time horizon is assumed to be finite, it can be very large. The exponentially growing number of model sequences must be then approximated using a pruning technique. Since the key idea of the auxiliary input signal design is to minimize the probability that the true model sequence will be pruned at the end of the finite horizon, a different approach is used to tackle this issue. The auxiliary input signal is designed in the open loop by considering only a reasonable number of the model sequences that are a priori the most probable. Thus, the model sequences with a low a priori probability are discarded and only a finite number of promising model sequences is tracked.

The auxiliary input signal is designed such that the following criterion

$$J^{\mathrm{MSEM}}\left(\mathbf{u}_0^{F-1}\right) = P\left(\mathrm{prune}|\mathbf{u}_0^{F-1}\right) = \sum_{\mu_0^{*F}} \int_{\mathbf{y}_0^F} P\left(\mathrm{prune}|\mathbf{y}_0^F, \mathbf{u}_0^{F-1}\right) p\left(\mathbf{y}_0^F|\mu_0^{*F}\right) d\mathbf{y}_0^F P\left(\mu_0^{*F}\right) \quad (15)$$

is minimized. The criterion J^{MSEM} denotes the probability of pruning the true model sequence marginalized over the measurements and true model sequences. The event of pruning true model sequence can be expressed as

$$\exists i \geq K : P\left(\mu_0^{*F}|\mathbf{y}_0^F, \mathbf{u}_0^{F-1}\right) < P\left(\left(\mu_0^F\right)_i|\mathbf{y}_0^F, \mathbf{u}_0^{F-1}\right), \quad (16)$$

where μ_0^{*F} is the true model sequence, K is a known number of tracked model sequences, the model sequences are sorted in descending order of probability, and F denotes the final time step of the finite time horizon.

However, the exact value of the probability of pruning the true model sequence cannot be determined analytically. Therefore, the method employs an approximation by a tractable upper bound of this probability. Then for the K tracked model sequences, the criterion (15) approximated by the upper bound has the following form

$$\bar{J}^{\mathrm{MSEM}}\left(\mathbf{u}_0^{F-1}\right) = \sum_{i=1}^{K}\sum_{j>i}^{K} P\left(\left(\mu_0^F\right)_i\right)^{\frac{1}{2}} P\left(\left(\mu_0^F\right)_j\right)^{\frac{1}{2}} e^{-\kappa_{ij}}, \quad (17)$$

where κ_{ij} is a quadratic function of $\mathbf{u}_0^{F-1}$ parametrized by predictive mean values and covariance matrices of the measurements conditioned by the corresponding tracked sequence $(\mu_0^F)_i$. The problem of finding K a priori most probable model sequences is not trivial, but it fits to a tree search problems known from computer science and artificial intelligence [29]. The tree search algorithms evaluate and expand tree nodes, i.e. the probabilities of model sequences. A^* search is a best-first search algorithm that features a memory and expands the tree based on a cost function of the known costs and a heuristic estimate. The three search algorithm such as A^* should find the K most probable sequences efficiently.

The problem formulation leads to an optimization problem suitable to be solved using SQP [24]. When extending the AFD problem to AFDC, the demands on system state behavior can be expressed in form of linear constraints. Then the SQP methods find a solution to the equality-constrained problem

$$\mathbf{u}_0^{*F-1} = \arg\min_{\mathbf{u}_0^{F-1}} \bar{J}^{\mathrm{MSEM}}\left(\mathbf{u}_0^{F-1}\right), \quad (18a)$$

$$\text{subject to } c_{\mathrm{eq}}\left(\mathbf{u}_0^{F-1}\right) = 0, \quad c_{\mathrm{iq}}\left(\mathbf{u}_0^{F-1}\right) \leq 0, \quad (18b)$$

where c_{eq} and c_{iq} are functions of equality and inequality constraints, respectively, e.g. constraints on the control inputs $\mathbf{u}_{\text{min}} \leq \mathbf{u}_i \leq \mathbf{u}_{\text{max}}$ for all $i = 0, 1, \ldots, F-1$ or expected system state $\mathbf{x}_{\text{min}} \leq \mathbb{E}[\mathbf{x}_i | \mathbf{y}_0^i, \mathbf{u}_0^{i-1}] \leq \mathbf{x}_{\text{max}}$ in case of the AFDC problem. Note that $\mathbf{u}_{\text{min}}$, $\mathbf{u}_{\text{max}}$, $\mathbf{x}_{\text{min}}$, and $\mathbf{x}_{\text{max}}$ are corresponding extreme values of the input and system state, respectively.

3.3. Comparison

A theoretical comparison of the two probabilistic methods is covered in Table 1. Both methods consider analogous system description and state estimation techniques that encouraged their joint comparison. However, one of the substantial difference lies in a type of the information processing strategy used. The closed loop information processing strategy should be advantageous because all available and the future information is consider in the design. Another significant difference can be seen in the criteria defining the auxiliary input signals. Therefore, this should be considered when evaluating a quality of the detectors.

Since the analytical solution to the formulated active fault detector designs is impossible to find, various approximations are used. A reformulation as a perfect state information problem is employed in case of OAFD. An analytical solution of the reformulated problem is found and approximated afterwards. In [22], the continuous state space is discretized using an uniform grid. The Bellman function is evaluated in the grid points. Then, the auxiliary input signal is generated for an arbitrary continuous state using an aggregation function that projects the state to the grid. The state-space quantization is used in [30] to transform the original problem into a Markov decision problem that can be solved using DP. A different approach is considered in [26] where a nonparametric local Bellman function is first trained using sampled data along the system trajectory and then employed to approximate the active fault detector. On the other hand, a problem reformulation in the MSEM design is used to propose an approximate solution. Such a reformulated problem is then solved by SQP [31, 13]. These approximations can largely influence the quality of suboptimal solutions as well as computational demands. As a consequence, different structures of auxiliary input signal generators can largely influence selection of the method for online detection.

4. Numerical example

Both probabilistic AFD methods follow a similar framework for the auxiliary input signal design as shown in Section 3. In this section, particular features of the methods are highlighted by means of a numerical example. Basically, three information processing strategies are used in the numerical example: the MSEM designed using open loop and open loop feedback information processing strategy and OAFD designed using the closed loop information processing strategy.

To simplify the design of the active fault detectors, it is assumed that the monitored system can be described by one of three first-order models. The matrices of these models are the following

$$\begin{aligned}
&\mathbf{A}_1 = 0.8187, \mathbf{A}_2 = 0.8106, \mathbf{A}_3 = 0.8959, \mathbf{B}_1 = 0.1813, \mathbf{B}_2 = 0.1894, \mathbf{B}_3 = 0.1042, \\
&\mathbf{C}_1 = \mathbf{C}_2 = \mathbf{C}_3 = 1, \mathbf{Q}_1 = \mathbf{Q}_2 = \mathbf{Q}_3 = 1 \cdot 10^{-6}, \mathbf{R}_1 = \mathbf{R}_2 = \mathbf{R}_3 = 1 \cdot 10^{-4}.
\end{aligned} \tag{19}$$

Two different lengths of the finite time horizon $F = 3$ and $F = 20$ are considered. The switching between models over time is given by the following transition probability matrix

$$\mathbf{P} = \begin{bmatrix} 0.98 & 0.2 & 0.1 \\ 0.01 & 0.6 & 0.3 \\ 0.01 & 0.2 & 0.6 \end{bmatrix}. \tag{20}$$

The probability of the initial model is $P(\mu_0) = [0.1,\ 0.5,\ 0.4]^{\mathrm{T}}$. The initial state mean value is $\hat{\mathbf{x}}_{0|-1} = 0$ and the covariance matrix is $\mathbf{\Sigma}^{\mathbf{x}}_{0|-1} = 2 \cdot 10^{-4}$. The set of admissible inputs is

	OAFD	MSEM
System	Linear discrete-time stochastic systems described by state-space models are considered. System uncertainties are described by known pdfs.	Linear discrete-time stochastic systems described by state-space models are considered. System uncertainties are described by known pdfs.
Statistics	The Kalman filter and Bayesian theory are used for hyper-state estimation.	The Kalman filter and Bayesian theory are used for hyper-state estimation.
Information processing strategy	The method uses the closed loop information processing strategy.	The method can either employ the open loop or open loop feedback information processing strategy.
Model sequences tracking	A finite history of model sequences are tracked.	The aim is to track a finite number of model sequences that contain the true model sequence.
Time horizon	The infinite time horizon formulation is presented due to its universal use. The finite time horizon case is presented in [11].	The auxiliary input signal is designed for a finite time horizon. A rolling horizon technique can be used for the infinite time horizon.
Fault detection	A general detection aim can be expressed using the detection cost function.	Penalization of true and false decisions is done only by zero and one, respectively.
System control	A control aim can be included into the criterion.	A control aim can be expressed using state constraints.
Criterion	The auxiliary input signal minimizes the given general design criterion.	The auxiliary input signal minimizes the upper bound of losing track of the true model sequence.
Approximations in formulation	The original problem is reformulated using hyper-state.	Upper bound on the probability of model selection error is considered.
Solution	The optimization problem is solved by ADP. The Bellman function is replaced by a local non-parametric function approximation.	The optimization problem is solved by SQP.
Auxiliary signal	The auxiliary input signal generator is a time invariant function of the hyper-state.	The auxiliary input signal is a given sequence independent of system state and measurements.

Table 1. Comparison of the optimal active fault detector design (OAFD) method and the model selection error minimization design (MSEM) method in terms of their features.

$\mathcal{U} = \{-1, 0, 1\}$. The GPB algorithm is used with parameter $h = 1$ to compute the approximate state estimate that consists of three mean values $\hat{\mathbf{x}}_{k|k}(\mu_k)$, three covariances $\mathbf{\Sigma}^{\mathbf{x}}_{k|k}(\mu_k)$ and two probabilities $P(\mu_k|\mathbf{y}_0^k, \mathbf{u}_0^{k-1})$. As a result, the hyper-state $\boldsymbol{\xi}_k$ is an eight dimensional vector

$$\boldsymbol{\xi}_k = [\hat{\mathbf{x}}_{k|k}(1), \hat{\mathbf{x}}_{k|k}(2), \hat{\mathbf{x}}_{k|k}(3), \mathbf{\Sigma}^{\mathbf{x}}_{k|k}(1), \mathbf{\Sigma}^{\mathbf{x}}_{k|k}(2), \mathbf{\Sigma}^{\mathbf{x}}_{k|k}(3), P(\mu_k = 1|\mathbf{y}_0^k, \mathbf{u}_0^{k-1}), P(\mu_k = 2|\mathbf{y}_0^k, \mathbf{u}_0^{k-1})].$$

Note that $P(\mu_k = 3|\mathbf{y}_0^k, \mathbf{u}_0^{k-1})$ can be simply computed from the hyper-state $\boldsymbol{\xi}_k$.

In order to make the detection aim as similar as possible in both methods, the detection cost function for OAFD method is chosen to be

$$L^{\mathrm{d}}(\boldsymbol{\xi}_k, d_k) = \mathrm{E}\{L(\mu_k, d_k)|\mathbf{y}_0^k, \mathbf{u}_0^{k-1}\} = \begin{cases} 1 - \boldsymbol{\xi}_{k,7} & \text{if } d_k = 1, \\ 1 - \boldsymbol{\xi}_{k,8} & \text{if } d_k = 2, \\ \boldsymbol{\xi}_{k,7} + \boldsymbol{\xi}_{k,8} & \text{if } d_k = 3, \end{cases} \tag{21}$$

where $L : \mathcal{M} \times \mathcal{M} \mapsto \{0, 1\}$ is the zero-one cost function defined as

$$L(\mu_k, d_k) = \begin{cases} 0 & \text{if } \mu_k = d_k, \\ 1 & \text{otherwise.} \end{cases} \tag{22}$$

Since the OAFD, which assumes the infinite time horzion, is used as an approximate solution to a finite time horizon, the discount factor is computed as

$$\lambda = 10^{\frac{\log(z/100)}{F+1}}, \tag{23}$$

where $z = 1\%$ denotes the desired value of the tail costs with respect to the overall costs in percent. The Bellman function was approximated by a nonparametric approximation analogously to [26]. The samples in the hyper-state space that were used to solve the discrete-time Bellman functional equation approximately using the value iteration algorithm were obtained by simulating the monitored system with the following input signal generators:

- zero input signal generator,
- sine input signal generator, and
- MSEM OL input signal generator.

These simulations resulted into 378 sample points in the hyper-state space. The value iteration algorithm convergence error threshold was set to $1 \cdot 10^{-2}$ and the algorithm converged in 5 iterations for $F = 3$ and 16 iterations for $F = 20$.

The following setup was used for the MSEM. The number of the most probable sequences was chosen to be $K = 10$ and the optimization problem with a relaxed constraint $u_k \in [-1, 1]$ was solved using a trust region method. The starting auxiliary input sequence for the trust region method was chosen randomly such that it satisfies the relaxed constraint on the input. The simulation results show that the optimal solution is on the boundary of the constraint set and thus the applied inputs belong to the original constraint set $\mathcal{U}$.

The results of 1000 Monte Carlo (MC) simulations together with the time and memory requirements are presented in Table 2 and Table 3 for the length of the horizon $F = 3$ and $F = 20$, respectively. The first two columns show the estimate of the criterion and the variance of this estimate computed using a bootstrap technique. As the estimates of the criterion value for the considered methods are close to each other, the methods seem to have a comparable quality of detection in this particular example. The third column shows the off-line time required to design the active fault detector. Note that there is zero off-line time for the MSEM OLF because all computations are performed on-line. The increase in the off-line computation time for the OAFD and $F = 20$ is mainly caused by the change in the discount factor λ and thus the number of iterations of the value iteration algorithm. The fourth column presents the average on-line time requirements during simulations. In all cases it also includes time of simulating the system and performing the state estimation. It can be seen that for a short time horizon the OAFD is the most demanding. However, the MSEM OLF becomes the most time expensive when the time horizon gets longer because of an increased size of the optimization problems that are solved on-line. Moreover, contrary to the MSEM OL and OAFD that have steady on-line

Table 2. Results of MC simulations, computational and memory requirements of the methods for $F = 3$.

method	$\hat{J}$	var $\hat{J}$	off-line time [s]	on-line time [s]	memory [bytes]
MSEM OL	0.7116	$0.2730 \cdot 10^{-6}$	0.0334	0.0025	24
MSEM OLF	0.7144	$0.2730 \cdot 10^{-6}$	0	0.0513	$\leq$7840
OAFD	0.7103	$0.2724 \cdot 10^{-6}$	91.0127	0.1900	27216

Table 3. Results of MC simulations, computational and memory requirements of the methods for F=20.

method	$\hat{J}$	var $\hat{J}$	off-line time [s]	on-line time [s]	memory [bytes]
MSEM OL	1.8865	$0.7285 \cdot 10^{-6}$	0.1295	0.0132	160
MSEM OLF	1.8496	$0.6970 \cdot 10^{-6}$	0	1.4449	$\leq$255360
OAFD	1.8921	$0.7527 \cdot 10^{-6}$	291.2339	1.0298	27216

time requirements, the on-line time requirements of the MSEM OLF for a longer time horizon fluctuate significantly between 1 [s] and 5.5 [s] depending on the actual state estimate. The last column provides rough estimates of the memory requirements. The value for the MSEM OL corresponds to the memory requirements of storing the whole auxiliary input sequence. In case of the MSEM OLF, the value represents an estimate of the memory requirements when the SQP optimization problem is set up at the first time step. Finally, the memory requirements for the OAFD accounts for storing the sampled hyper-states and corresponding values of the Bellman function. The MSEM OL has the lowest memory requirements. The memory requirements of the MSEM OLF increases rapidly for a longer time horizon whereas it remains the same for the OAFD because the Bellman function approximation is independent of the time horizon length.

5. Conclusion

The paper deals with a comparison of two probabilistic methods for designing an active fault detector. The methods are intentionally chosen for a comparison because they share the same assumptions on the system description to a great extent. Nevertheless, the detection aims of the methods are different and the provided active fault detectors that solve the problem are designed only approximately. Therefore, the comparison of the detection quality is not definite and other factors need to be considered as well. The preliminary results of Monte Carlo simulations indicate that, on a finite time horizon, the OAFD can provide a similar detection quality when compared to the MSEM and may become advantageous for longer time horizons when the on-line time requirements need to be kept reasonably small.

Acknowledgments

This work was supported by the project LO1506 of the Czech Ministry of Education, Youth and Sports and by the project GA15-12068S of the Czech Science Foundation.

References

[1] Severson K, Chaiwatanodom P and Braatz R D 2015 Perspectives on Process Monitoring of Industrial Systems *Proceedings of the 9th IFAC Symposium on Fault Detection, Supervision and Safety of Technical Processes* (Paris, France) pp 931–939

[2] Basseville M and Nikiforov I V 1993 *Detection of Abrupt Changes - Theory and Application* (New Jersey, USA: Prentice Hall)
[3] Isermann R 2011 *Fault-Diagnosis Applications* (Heidelberg, Germany: Springer)
[4] Miljković D 2011 Fault Detection Methods: A Literature Survey *Proceedings of the 34th International Convention MIPRO* (Opatija, Croatia) pp 750–755
[5] Gao Z, Cecati C and Ding S X 2015 *IEEE Transactions on Industrial Electronics* **62** 3757–3767
[6] Gao Z, Cecati C and Ding S X 2015 *IEEE Transactions on Industrial Electronics* **62** 3768–3774
[7] Nett C N, Jacobson C A and Miller A T 1988 An Integrated Approach to Controls and Diagnostics: The 4-Parameter Controller *1988 American Control Conference* (Atlanta, GA) pp 824–835 ISSN 07431619
[8] Zhang X J 1989 *Auxiliary Signal Design in Fault Detection and Diagnosis* (Heidelberg, Germany: Springer Verlag)
[9] Campbell S L and Nikoukhah R 2004 *Auxiliary Signal Design for Failure Detection* (Princeton University Press) ISBN 978-0691099873
[10] Niemann H 2006 *IEEE Transactions on Automatic Control* **51** 1572–1578 ISSN 0018-9286
[11] Šimandl M and Punčochář I 2009 *Automatica* **45** 2052–2059
[12] Poulsen N K l and Niemann H H 2008 *International Journal of Applied Mathematics and Computer Science* **18** 487–496 ISSN 1641-876X
[13] Blackmore L, Rajamanoharan S and Williams B C 2008 *IEEE Transactions on Automatic Control* **53** 2223–2236 ISSN 0018-9286
[14] Mesbah A, Streif S, Findeisen R and Braatz R D 2014 Active Fault Diagnosis for Nonlinear Systems with Probabilistic Uncertainties *Proceedings of the 19th IFAC World Congress* (Cape Town, South Africa) pp 7079–7084
[15] Khosrowjerdi M, Nikoukhah R and Safari-Shad N 2004 *Automatica* **40** 261–267 ISSN 00051098
[16] Niemann H and Poulsen N K 2014 Active Fault Detection in MIMO Systems *Proceedings of the 2014 American Control Conference* (Oregon, USA) pp 1975–1980 ISBN 9781479932740
[17] Ashari A E, Nikoukhah R and Campbell S L 2011 *Automatica* **47** 1887–1895 ISSN 00051098
[18] Ashari A E, Nikoukhah R and Campbell S L 2012 *Automatica* **48** 866–872 ISSN 00051098
[19] Scott J K, Findeisen R, Braatz R D and Raimondo D M 2014 *Automatica* **50** 1580–1589 ISSN 00051098
[20] Punčochář I, Široký J and Šimandl M 2015 *EEE Transactions on Automatic Control* **60** 253–258
[21] Škach J, Punčochář I and Šimandl M 2014 *Journal of Physics: Conference Series* **570** 1–9 ISSN 1742-6588
[22] Punčochář I and Šmandl M 2014 *International Journal of Applied Mathematics and Computer Science* **24** 795–807 ISSN 2083-8492
[23] Vrabie D, Vamvoudakis K G and Lewis F L 2013 *Optimal Adaptive Control and Differential Games by Reinforcement Learning Principles* (London, United Kingdom: IET) ISBN 9781849194891
[24] Nocedal J and Wright S J 2006 *Numerical Optimization* 2nd ed (New York, USA: Springer Verlag) ISBN 9780387303031
[25] Watanabe K and Tzafestas S G 1993 *Journal of Intelligent and Robotic Systems* **7** 95–112 ISSN 0921-0296
[26] Punčochář I, Škach J and Šimandl M 2015 Infinite Time Horizon Active Fault Diagnosis based on Approximate Dynamic Programming *Accepted for the Proceedings of the 54th IEEE Conference on Decision and Control* (Osaka, Japan)
[27] Bertsekas D P 2000 *Dynamic Programming and Optimal Control (Volume I)* 2nd ed (Belmont, Massachusetts: Athena Scientific) ISBN 1-886529-09-4
[28] Powell W B 2007 *Approximate Dynamic Programming: Solving the Curses of Dimensionality* (Hoboken, NJ, USA: Wiley-Interscience) ISBN 978-0-470-17155-4
[29] Russell S J and Norvig P 1995 *Artificial Intelligence: A Modern Approach* (Englewood Cliffs, New Jersey: Prentice Hall) ISBN 0-13-103805-2
[30] Punčochář I, Škach J and Šimandl M 2015 Adaptive Generalized Policy Iteration in Active Fault Detection and Control *Proceedings of the 9th IFAC Symposium on Fault Detection, Supervision and Safety of Technical Processes* (Paris, France) pp 505–510
[31] Blackmore L and Williams B 2006 Finite Horizon Control Design for Optimal Discrimination between Several Models *Proceedings of the 45th IEEE Conference on Decision and Control* (San Diefo, California: Ieee) pp 1147–1152 ISBN 1-4244-0171-2

Switched Fault Diagnosis Approach for Industrial Processes based on Hidden Markov Model

Lin Wang, Chunjie Yang*, Youxian Sun, Yijun Pan, and Ruqiao An
State Key Laboratory of Industrial Control Technology, Institute of Industrial processes Control, Zhejiang University, Hangzhou, Zhejiang, China

E-mail: cjyang@iipc.zju.edu.cn

Abstract. Traditional fault diagnosis methods based on hidden Markov model (HMM) use a unified method for feature extraction, such as principal component analysis (PCA), kernel principal component analysis (KPCA) and independent component analysis (ICA). However, every method has its own limitations. For example, PCA cannot extract nonlinear relationships among process variables. So it is inappropriate to extract all features of variables by only one method, especially when data characteristics are very complex. This article proposes a switched feature extraction procedure using PCA and KPCA based on nonlinearity measure. By the proposed method, we are able to choose the most suitable feature extraction method, which could improve the accuracy of fault diagnosis. A simulation from the Tennessee Eastman (TE) process demonstrates that the proposed approach is superior to the traditional one based on HMM and could achieve more accurate classification of various process faults.

1. Introduction

With the growth of complexity of industrial processes, effective monitoring and diagnosis play a significant role in ensuring operational safety of chemical equipment, maintaining product quality, optimizing product profit, and improving environmental sustainability [1].

In process industry, more and more data have been collected with a wide range utilization of sensors. However, it will lead to low efficiency if original data are stored and processed directly. Besides, they do not need all the features of objects in some particular applications. This is because many features cannot reflect the nature of the objects. Sometimes redundant features will even become obstacles to the subsequent processing. Based on the views above, feature extraction is needed to gain essential characteristics of data. It will reduce the interference information like noise and highlight the useful information. Feature extraction is the key step in fault diagnosis. Because of this, the accuracy of the fault diagnosis will be improved by enhancing the effectiveness of feature extraction.

There are a lot of traditional feature extraction methods, such as PCA、KPCA and ICA [2]. PCA can deal with linear relationships among variables effectively, but it cannot extract the nonlinear relationships, which are very common in the process industry [3]. KPCA was proposed to deal with it. It could extract nonlinear relationships among variables. Because of this, it is widely utilized in image recognition, feature extraction, fault detection and other fields [4]. However, there is no effective way to select the kernel function and parameters of KPCA at present. If KPCA was used to deal with process variables with linear relationships, not only the choice of kernel function and its parameters is cumbersome and time-consuming, but also it will greatly decrease accuracy of fault diagnosis with the inappropriate parameters. So it is inappropriate to use KPCA for feature extraction blindly. So far, when KPCA is utilized, nonlinear relationships among variables are a prior condition in most researches. That is to say, it is necessary to analyse the nonlinear relationships among process variables. This article proposes a switched feature extraction method to deal with the proposed problem. As we all know, a better feature extraction would lead to a better classification. So the proposed method has a great research value. A nonlinearity measure proposed by Kruger to analyse the nonlinearity among variables is used to choose the appropriate method for feature extraction [5]. The extracted features will be used as the observation sequences of HMM, which is used as a classifier.

HMM is a valuable pattern recognition tool for sequential data. Its application for industrial processes monitoring is gradually increasing. It is an intelligent diagnosis technology with some unique characteristics. It is a doubly stochastic process which establishes a statistical model of time series by Markov chains and random functions [6]. So HMM has a strong ability to set up model of the dynamic process and classify sequential patterns. Due to the characteristics above, HMM has a good application prospect in the field of fault diagnosis and detection in process industry. So we choose it to clarify the efficiency of the proposed method.

In this article, the proposed method properly takes the characteristics of process data into consideration. By analysing data characteristics, we could overcome the insufficient of the traditional fault diagnosis methods which sometimes extract features of process variables by inappropriate method. By this way, we could improve the accuracy of the fault diagnosis.

The remainder of this article is organized as follows. In section II, PCA, KPCA and HMM are briefly described. In section III, switched fault diagnosis approach for industrial processes based on HMM is developed in detail. In section IV, TE process is used to illustrate the effectiveness of the proposed method. Section V gives the conclusions.

2. Preliminaries

2.1. Principal component analysis
PCA is a linear dimension reduction technique. It transforms high-dimensional correlated variables into low-dimensional uncorrelated ones. PCA determines a series of orthogonal vectors which are called loading vectors. Consider a data matrix $X \in R^{n\times m}$, for n observations with m measurement variables. The loading vectors could be derived by Singular Value Decomposition (SVD) as follows:

$$\frac{1}{\sqrt{n-1}}X = U\Sigma V^T \tag{1}$$

where $U \in R^{n\times n}$ and $V \in R^{m\times m}$ are both unitary matrices. The matrix $\Sigma \in R^{n\times m}$ contains real and decreasing singular values ($\sigma_1 \geq \sigma_2 \geq \cdots \geq \sigma_m$) along the main diagonal, with other off-diagonal elements to be zero. Loading vectors are the orthogonal columns of the matrix V and the variance of the data set projecting along the kth column of V is equal to σ_k^2 . $\sigma_k^2 (i=1,2,\cdots,m)$ is arranged in decreasing order to determine the principal components. The first a PCs are selected to build the PCA model, so score vectors for kth sample can be represented by loading matrix P and sample vectors x_k .

$$t_k = P^T x_k \tag{2}$$

2.2. Kernel principal component analysis
PCA performs well when the relationships among variables are linear, but it can't deal with nonlinearity among variables. KPCA projects observation matrix to a high dimensional feature space F by using kernel function Φ and then implements PCA in F, which can be expressed as

$$\Phi : R^m \to F \tag{3}$$

Assume mapping data for each variable have been scaled to zero mean, just $\sum_{k=1}^{n}\Phi(x_k)=0$, and n denotes the data size. The corresponding covariance matrix can be defined as

$$S^F=\frac{1}{n}\sum_{k=1}^{n}\Phi(x_k)\Phi(x_k)^T \tag{4}$$

In essence, KPCA has realized the nonlinear transformation between data space and feature space. x_i and x_j are sample points in data space and the mapping function is Φ. The fundamental of KPCA is inner product transformation.

$$K_{ij}=\langle\Phi(x_i),\Phi(x_j)\rangle=K(x_i,x_j) \tag{5}$$

Data points $\Phi(x_k),(k=1,\cdots,n)$ need to be processed in the feature space F, the process is as follows

$$\tilde{K}=K-1_nK-K1_n+1_nK1_n \tag{6}$$

where

$$1_n=\frac{1}{n}\begin{bmatrix}1 & \cdots & 1\\ \vdots & \ddots & \vdots\\ 1 & \cdots & 1\end{bmatrix}\in R^{n\times n}$$

By eigenvalue decomposition to $\tilde{K}$ as shown

$$\lambda\alpha=\tilde{K}\alpha \tag{7}$$

We can obtain orthogonal eigenvectors $\alpha_1,\alpha_2,\cdots,\alpha_n$ and corresponding eigenvalues $\lambda_1\ge\lambda_2\ge\cdots\ge\lambda_n$. The dimension reduction can be obtained by mapping $\Phi(x)$ to the eigenvectors v_k in F, where $(k=1,2,\cdots,a)$, which can be expressed as

$$t_k=\langle v_k,\Phi(x)\rangle=\sum_{i=1}^{n}\alpha_i^k\langle\Phi(x_i),\Phi(x)\rangle \tag{8}$$

In the KPCA method, we just need to calculate matrix K to deal with the problem of eigenvalues by ignoring the specific nonlinear mapping relation Φ. In this article, we choose radial basis kernel, that is

$$K(x,y)=exp(-\frac{\|x-y\|^2}{\sigma}) \tag{9}$$

with $\sigma=rm$, where r is a constant to be selected and m is the dimension of the input space.

2.3. Hidden Markov model

Generally speaking, HMM is a valuable pattern recognition tool for sequential data. Its application for pro-cess monitoring is gradually increasing. A continuous HMM (CHMM) is used since the TE data are continuous. It usually can be described by the following parameters:

- N: the number of hid den states, the states are expressed as $S=\{S_1,S_2,\cdots,S_N\}$
- $B=\{b_j(O),j=1,2,\cdots N\}$: observation probability matrix

  where $b_j(O)=\sum_{m=1}^{M}C_{jm}H[O,\mu_{jm},U_{jm}],\quad 1\le j\le N$

- $A = \{a_{ij}\}$: state transition probability matrix

  $$a_{ij} = P(q_{t+1} = S_j \mid q_t = S_i), 1 \le i, j \le N$$

  where $\sum_{j=1}^{N} a_{i,j} = 1, \quad 1 \le i \le N$

- O is the observation vector. C_{jm} is a mixed coefficient of the mth mixing element of the hidden state j. H is Gaussian mixture probability density. μ_{jm} is mean vector of the mth mixing element of the hidden state S_j. U_{jm} is the covariance matrix of the mth mixing element of the hidden state S_j.

  where $\sum_{m=1}^{M} C_{jm} = 1, \quad 1 \le j \le N$

  $$C_{jm} \ge 0, \quad 1 \le j \le N, \quad 1 \le m \le M$$

- $\pi = \{\pi_i\}$:the initial state probability matrix

  $$\pi_i = P(q_1 = S_i), 1 \le i \le N$$

  where $\sum_{i=1}^{N} \pi_i = 1$

3. Switched fault diagnosis approach based on HMM

Industrial processes are usually sophisticated, so the process data have various characteristics, such as non-Gaussianity, dynamic correlations, nonlinear relations and uncertainties in the process measurement, which makes feature extraction significant [3]. Only one feature extraction method would be used in traditional fault diagnosis approach based on HMM, such as PCA [7]. But data have their own characteristics in different operation conditions, which makes it unable to perform well by using a unified feature extraction method. Here, we just consider linear and nonlinear relationships among process variables. This article proposes a switched feature extraction method based on PCA and KPCA. Nonlinearity measure is used to analyse the nonlinear relationships among process variables and choose the appropriate method for feature extraction. The method will be introduced below.

3.1. Nonlinear measure

The original data can be divided into several regions based on a priori knowledge of the process, and the accuracy boundaries are determined according to the confidence intervals of the correlation matrix of one region. By using the principle of cross-validation [8], we can deal with the problem of choosing region, which helps us to determine the accuracy boundaries. The relationships among variables are decided by comparing the residuals in each region with their corresponding accuracy boundaries. Specific steps are shown as follows.

Suppose a data matrix $Z \in R^{n \times m}$ having m variables and n samples is divided into l regions, each of which includes $\tilde{n} = n / l$ observations.

The correlation matrix for the original data is defined as:

$$S_{zz} = \begin{bmatrix} s_{11} & s_{12} & \cdots & s_{1m} \\ s_{21} & s_{22} & \cdots & s_{2m} \\ \vdots & \vdots & \ddots & \vdots \\ s_{m1} & s_{m2} & \cdots & s_{mm} \end{bmatrix} \tag{10}$$

The element in matrix is expressed as:

$$s_{ij} = \frac{1}{\tilde{n}-1}\sum_{k=1}^{\tilde{n}} \frac{z_{ki}-\tilde{z}_i}{\sigma_i}\frac{z_{kj}-\tilde{z}_j}{\sigma_j} \tag{11}$$

where $\tilde{z}_i$, $\tilde{z}_i$ are the mean values and σ_i, σ_j are the standard deviations of ith, jth variables.

Since the variance and mean values of variables follow χ^2 and t distribution, respectively. The confidence intervals of the mean value and variance of ith variable, ${}_z CONF_\alpha^{(i)}$, for a confidence level

Table 1. Calculation of confidence limits for mean value.

Step	Description	Equation
1	Determine solution of c_i for t distribution	$c_i = f_1^{-1}(\frac{1+\alpha}{2})$
2	Calculate mean $\tilde{z}_i$ and variance s_i	$\tilde{z}_i = \frac{1}{\tilde{n}}\sum_{k=1}^{\tilde{n}} z_{ki}, s_i = \frac{1}{\tilde{n}-1}\sum_{k=1}^{\tilde{n}}(z_{ki}-\tilde{z}_i)^2$
3	Calculate μ_i	$\mu_i = \frac{s_i c_i}{\sqrt{\tilde{n}}}$
4	Define confidence limit	${}_z CONF_\alpha^{(i)}\{\tilde{z}_i - \mu_i \le \hat{z}_i \le \tilde{z}_i + \mu_i\}$

Table 2. Calculation of confidence limits for variance.

Step	Description	Equation
1	Determine solution of c_{1i} and c_{2i} for χ^2 distribution	$c_{1i} = f_2^{-1}(\frac{1-\alpha}{2}), c_{2i} = f_2^{-1}(\frac{1+\alpha}{2})$
2	Calculate $(\tilde{n}-1)s_i$	
3	Calculate μ_{1i} and μ_{2i}	$\mu_{1i} = \frac{(\tilde{n}-1)s_i}{c_{1i}}, \mu_{2i} = \frac{(\tilde{n}-1)s_i}{c_{2i}}$
4	Define confidence limit	${}_s CONF_\alpha^{(i)}\{\mu_{2i} \le \hat{s}_i \le \mu_{1i}\}$

of $\alpha = 95\%$ or $\alpha = 99\%$ can be obtained as expressed in Table 1 and Table 2, respectively. According to them, we can obtain the upper and lower threshold for each element of matrix S_{zz}. It can be shown as:

$$S_{zz} = \begin{bmatrix} s_{11_L} \le s_{11} \le s_{11_U} & s_{12_L} \le s_{12} \le s_{12_U} & \cdots & s_{1m_L} \le s_{1m} \le s_{1m_U} \\ s_{21_L} \le s_{21} \le s_{21_U} & s_{22_L} \le s_{22} \le s_{22_U} & \cdots & s_{2m_L} \le s_{2m} \le s_{2m_U} \\ \vdots & \vdots & \ddots & \vdots \\ s_{m1_L} \le s_{m1} \le s_{m1_U} & s_{m2_L} \le s_{m2} \le s_{m2_U} & \cdots & s_{mm_L} \le s_{mm} \le s_{mm_U} \end{bmatrix} \tag{12}$$

where the subscripts U and L represent the upper and lower limit. The more details of calculation are presented in a paper authored by Kruger [5]. The formula above can be simplified as follows

$$S_{zz_L} \le S_{zz} \le S_{zz_U} \tag{13}$$

The accuracy boundaries can be determined according to the confidence interval of correlation matrix. In terms of frobenius norm, the sum of the standard deviation σ is equal to the sum of the discarded eigenvalues of a PCA model.

$$\sigma = \sum_{j=1}^{m} \sigma_j = \frac{1}{\tilde{n}-1} \sum_{i=1}^{\tilde{n}} \sum_{j=1}^{m} e_{ij}^2 = \sum_{k=a+1}^{m} \lambda_k \tag{14}$$

Because the elements in correlation matrix S_{zz} decide the eigenvalues $\lambda_{a+1}, \cdots, \lambda_m$. The estimates of accuracy boundaries can be converted into the following optimization problem.

$$\begin{aligned} \lambda_{k_{MAX}} &= \underset{\Delta S_{zz_{MAX}}}{argmax} \lambda_k (S_{ZZ} + \Delta S_{ZZ_{MAX}}) \\ \lambda_{k_{MIN}} &= \underset{\Delta S_{zz_{MIN}}}{argmin} \lambda_k (S_{ZZ} + \Delta S_{ZZ_{MIN}}) \end{aligned} \tag{15}$$

where $\Delta S_{ZZ_{MAX}}$ and $\Delta S_{ZZ_{MIN}}$ are volatilities of S_{ZZ} to determine $\lambda_{k_{MAX}}$ and $\lambda_{k_{MIN}}$. Finally, we get the accuracy boundaries as follows

$$\sigma_{MAX} = \sum_{k=a+1}^{m} \lambda_{k_{MAX}}, \sigma_{MIN} = \sum_{k=a+1}^{m} \lambda_{k_{MIN}} \tag{16}$$

If all the residual variances fall inside the accuracy boundaries, the process is considered to be linear and then we choose PCA for feature extraction, vice versa.

3.2. Fault diagnosis approach

In this article, HMM is integrated with the switched feature extraction method for fault diagnosis. The method mentioned above contains three procedures: nonlinearity measure, feature extraction and fault classification.

Firstly, measure the nonlinearity among original process variables for every mode. The original data are divided into several regions and the accuracy boundaries are determined. If all the residual variances lie inside the accuracy boundaries for each region, PCA is much more appropriate than KPCA for feature extraction and vice versa.

Secondly, the data from every mode are processed respectively by the feature extraction method above selectively. For example, if PCA was selected, the PCs extracted are utilized as observation sequences of HMM.

Thirdly, set up a model library of HMMs by Baum-Welch algorithm, including one HMM for normal operating condition and other HMMs for abnormal conditions. When an unknown fault needs to be classified, it is processed by the suitable method for feature extraction. $P(O \mid \lambda_i)(i = 1, \cdots, N)$ would be calculated, where λ_1 represents HMM for normal operating condition, $\lambda_2, \cdots, \lambda_N$ represent HMMs for all abnormal conditions, and O represents observation sequence. $P(O \mid \lambda_i)$ represents the probability of the appearance of the observation sequence o with given λ_i. The maximum $P(O \mid \lambda_i)$ is gained by calculating all the probability, which shows that fault i $(i = 1, \cdots, N)$ occurred. A flow diagram of the proposed approach is shown in Fig.1.

4. Application of examples

TE process was created to provide a realistic industrial process to test the performance of various monitoring and diagnosis approaches [9]. This process consists of five major units: reactor, condenser, separator, compressor and stripper. It has 41 measured variables (22 continuous process variables and 19 composition variables) and 12 manipulated variables. It includes 21 preset faults which are denoted

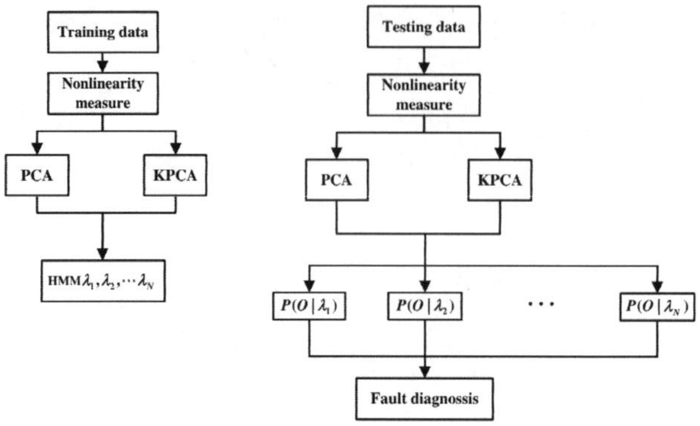

Figure 1. Flow diagram of the proposed approach.

as IDV(i) ($i=1,\cdots,21$).

In this article, training set consists of three modes, R_1, R_2 and R_3. R_1 represents the normal operating condition, whereas R_2 and R_3 represent two abnormal conditions, IDV(1) and IDV(7). There are 33 variables that have been selected: 22 process measurements and 11 manipulated variables. The agitation speed is not involved because it is not manipulated in simulation process. In addition, 19 composition measurements are excluded [10]. More details are shown in Table 3.

Table 3. Training set.

Pattern	Fault Number	Variable	Type	Time(minute)
R_1	Normal	------	------	480
R_2	IDV(1)	A/C feed ratio, B composition constant	step	480
R_3	IDV(7)	C header pressure loss-reduced availability	step	480

Table 4. Nonlinearity measure applied to the process using three sets.

Region	UCL99	UCL95	LCL99	LCL95	Region1	Region2	Region3
1	5.9241	5.8930	5.1061	4.8307	**6.0042**	**3.9945**	**3.5391**
2	5.8769	5.8439	5.0108	4.7111	**11.6015**	5.8740	**6.4194**
3	12.3295	11.1726	10.6114	9.8594	**17.4527**	12.1941	12.1379

Firstly, we analyse the nonlinear relationships among variables in the training set. R_1 is divided into three regions of 160 samples each. The results of nonlinearity measure of R_1 for each region are shown in Table 4. The accurate boundaries and residuals of each region are presented. The bold figures represent the residuals fall outside of the accurate boundaries. Fig.2 is a graphical representation of the case where the accuracy boundaries were obtained from the first region. According to it, the discarded eigenvalues of PCA model for all data sets fall outside the accuracy boundary. It shows the strong nonlinear relationships among variables. So it is necessary to use KPCA for feature extraction. The other modes are also processed in this way. Because of length limitations of paper, the results of the other faults are not presented here.

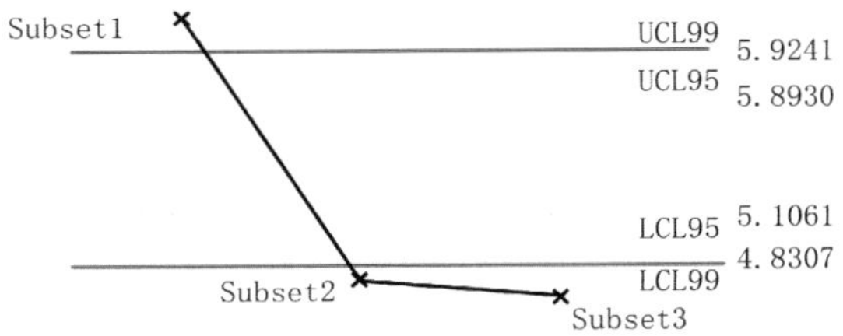

Figure 2. Graphical representation of nonlinearity measure for accuracy bounds for first region.

Secondly, there are three HMMs which need to be trained in this article, one for normal operating condition and the other two for corresponding faults. For each training mode R_i (i=1, 2, 3), the extracted features are used as observation sequences to train parameters λ_i (i =1, 2, 3) of HMM by

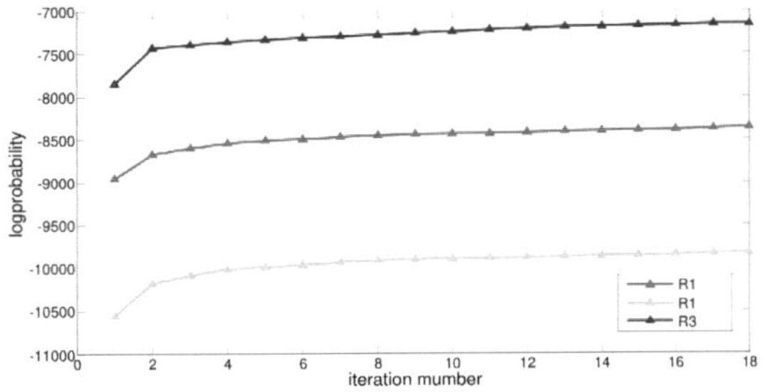

Figure 3. Iteration process for training three HMMs.

Table 5. Testing set.

Pattern	Fault Number	Fault Type	Fault occurs from(minute)
T_1	Normal	------	------
T_2	IDV(1)	step	160
T_3	IDV(7)	step	160

Baum-Welch method. The results are shown in Fig.3. It is clear that three training processes are quite different from each other, which indicates the modes of operation present distinct characteristics.

There are also three modes in the testing set, T_1,T_2 and T_3 . T_1 represents the normal operating condition, whereas T_2 and T_3 represent two abnormal conditions, IDV (1) and IDV (7). All faults in the testing set are same with the faults of the training set, and each begins with normal operation and introduces a fault after 160 minutes. The more details are shown in Table 5.

The classifying results of three modes in the testing set are shown in Fig. 4, Fig. 5, Fig. 6. The ordinates represent $\log P(O\,|\,\lambda_i)(i=1,2,3)$, the log probabilities of the appearance of the observation sequence sequence O with given λ_i in the model library, and abscissa represents time. From all figures, we can see that different faults can be identified correctly by the proposed fault diagnosis method. In Fig. 4, we can see that all the method can identify the fault correctly, but it is important to note that $P(O\,|\,\lambda_2)$ and $P(O\,|\,\lambda_3)$ are close in (b) and (c). This means that R_2 and R_3 are not distinguished well in the traditional method. In Fig. 5, either HMM based on PCA or KPCA cannot identify the fault accurately, while the switched method has a better effect. According to the results shown in Fig. 6, the proposed method performs better although HMM based on PCA could identify the fault.

5. Conclusion

Switched fault diagnosis for industrial processes based on hidden Markov model is proposed. Generally speaking, the swiched fault diagnosis approach can enhance the diagnostic capacity of TE process because the method properly takes the characteristics of process data into consideration. By analysing characteristic of data, we overcome the insufficient of the traditional methods which sometimes extracts data features by inappropriate method. By implementing the simulation on TE, we are able to get the results which exhibit that the proposed diagnosis approach possesses a superior performance when it is compared to the traditional diagnosis based on HMM.

This study shows the superiority of the switched fault diagnosis approach for industrial processes based on HMM. However, there are some issues to be further addressed. The effectiveness of the proposed method is demonstrated by using the simulated process data. Future work can implement the proposed method with the real world industrial data. In this article we just consider linearity and non-linearity by using PCA and KPCA. In the future, we can take more characteristic of data into consideration, such as Gaussianity and non-Gaussianity by enhancing the accuracy of fault diagnosis for different modes.

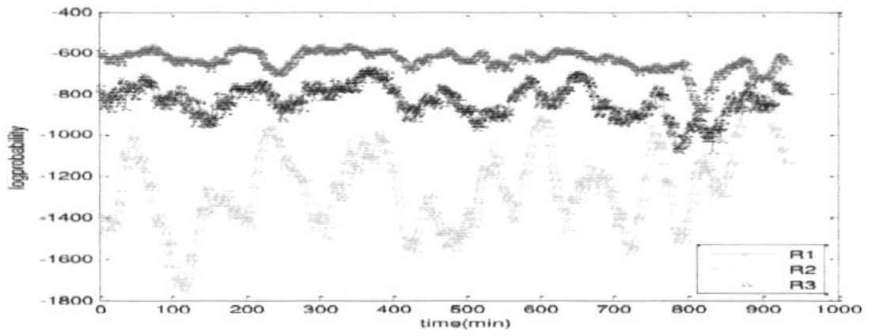

(a)

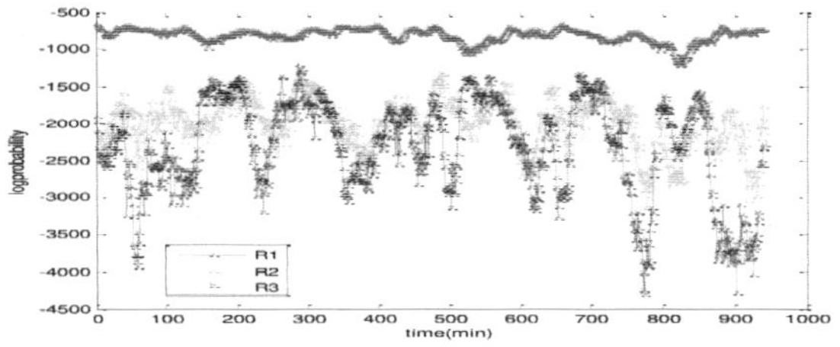

(b)

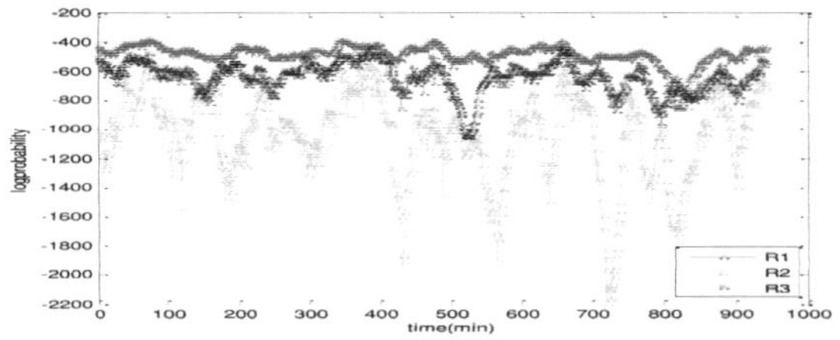

(c)

Figure 4. Diagnostic result for T1: (a) result based on switched method (b) result based on KPCA (c) result based on PCA.

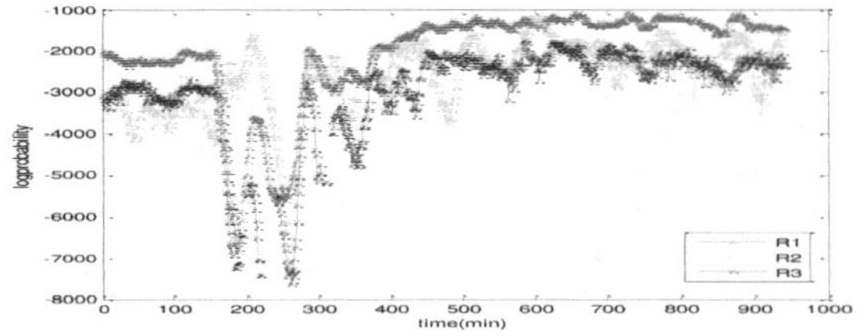

(a)

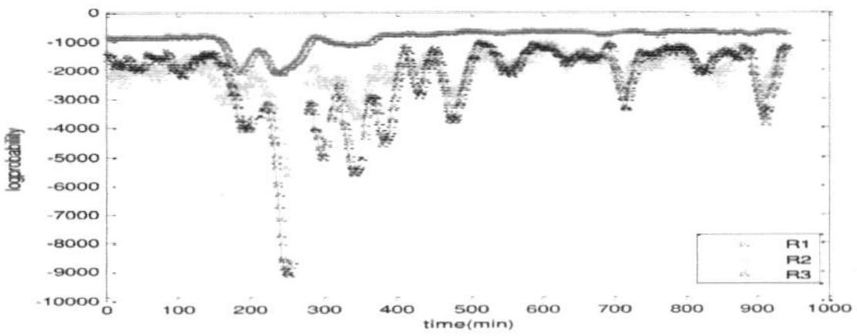

(b)

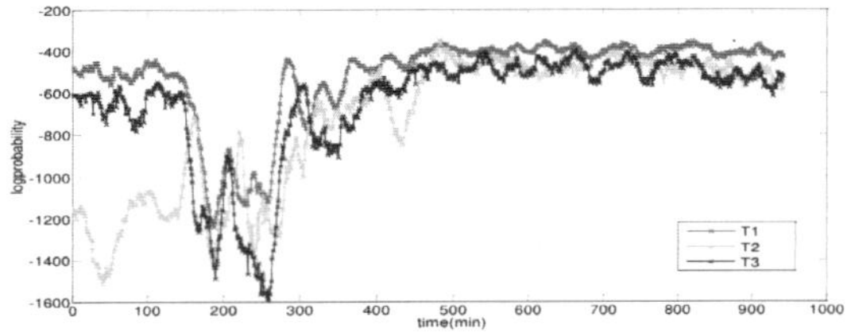

(c)

Figure 5. Diagnostic result for T2: (a) result based on switched method (b) result based on KPCA (c) result based on PCA.

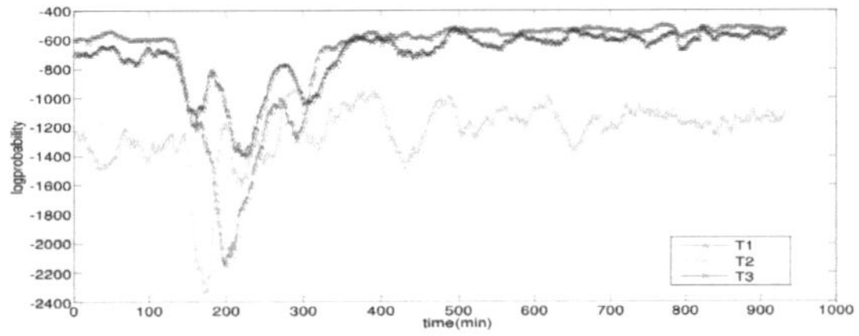

(a)

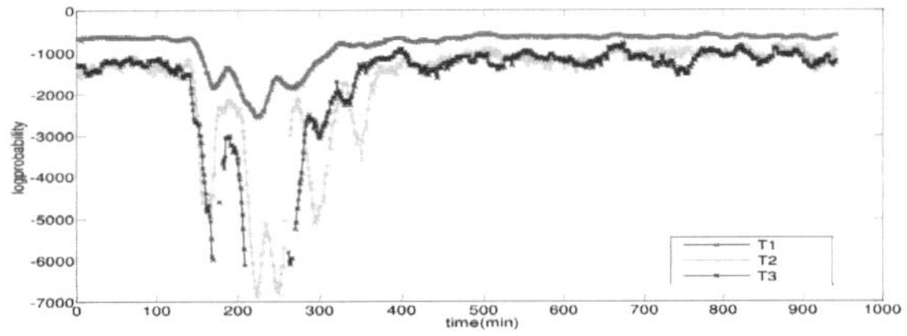

(b)

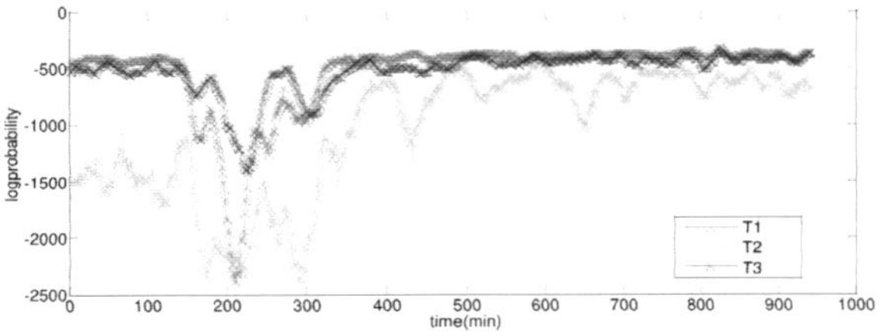

(c)

Figure 6. Diagnostic result for T3: (a) result based on switched method (b) result based on KPCA (c) result based on PCA.

Acknowledgments

This work was supported by the 863 Program (Grant 2012AA041709) and the National Natural Science Foundation of China (Grant No. 61290321)

References

[1] M. R. Rashid and Y. Jie, Hidden Markov model based adaptive independent component analysis approach for complex chemical process monitoring and fault detection, *Ind. Eng. Chem. Res*, vol. 51, pp. 5506–5514, Mar. 2012

[2] L. J. Cao, K. S. Chua, W. K. Chong, H. P. Lee and Q. M. Gu, A comparison of PCA, KPCA and ICA for dimensionality reduction in support vector machine, *Neurocomputing*, vol. 55, pp. 321–336, Mar. 2003

[3] Z. Q. Ge, Z. H. Song and F. R. Gao, Review of recent research on data-based process monitoring, *Ind. Eng. Chem. Res*, vol. 52, pp. 3544–3562, Feb. 2013.

[4] C. Y. Cheng, C. C. Hsu and M. C. Chen, Adaptive kernel principal component analysis (KPCA) for monitoring small disturbances of nonlinear processes, *Ind. Eng. Chem. Res*, vol. 46, pp. 2054–2262, Jan. 2010.

[5] U. Kruger, D. Antory, J. Hahn, G. W. Irwin and G. M. Cullough, Introduction of a nonlinearity measure for principal component models, *Computers & Chem. Eng*, vol. 29, pp. 2355–2362, July 2005.

[6] L. R. Rashid, A Tutorial on the hidden Markov models and selected applications in the speech recognition, *Proc. IEEE*, vol. 77, no. 2, pp. 257–286, Feb. 1989.

[7] S. Y. Zhou, J. M. Zhang and S. Q. Wang, Fault diagnosis in industrial processes using principal component analysis and hidden Markov model, in *Conf. Rec. 2004 American Control Conference*, pp. 5680–5685.

[8] M. Stone. Cross-validatory choice and assessment of statistical prediction (with discussion) , *Journal of the Royal Statistical Society, Series B Methodologyical*, vol. 36, pp. 111–133.

[9] A. Singhai and D. E. Seborg, Evaluation of a pattern matching method for the Tennessee Eastman challenge process, *Journal of Process Control*, vol. 16, pp. 601–613, Mar. 2006

[10] Z. Q. Ge and Z. H. Song, Process monitoring based on independent component analysis-principal component analysis (ICA-PCA) and similarity factors, *Ind. Eng. Chem. Res*, vol. 46, pp. 2054–2063, Mar. 2007

Journal of Physics: Conference Series **659** (2015) 012048 doi:10.1088/1742-6596/659/1/012048

Sliding mode control of continuous time systems with reaching law based on exponential function

Piotr Gamorski

Institute of Automatic Control Technical University of Łódź, Łódź, Poland

E-mail: pgamorski@wp.pl

Abstract. In this paper a pseudo-sliding mode control is proposed by introducing a continuous and smooth input signal in order to guarantee both chattering elimination and boundedness of sliding variable derivative in the presence of non-zero external disturbance. For this purpose, having fixed a suitable sliding manifold, a homogeneous differential equation describing the sliding variable evolution is considered. It is discussed later in this paper that the input signal formed on the basis of this equation provides asymptotic convergence of the sliding variable and its derivative to zero as well as the asymptotic stability of the non-linear system in the absence of external disturbance. The dynamics of the system affected by non-zero external disturbance make the state vector converge to domains in a vicinity of the origin at the exponential rate, as the norm of arbitrary trajectory is limited to decreasing exponential function. In order to expand the variety of controllers based on a reaching law and providing the above-mentioned properties, a certain class of functions is presented.

1. Introduction

In many practical applications it is very difficult to obtain an accurate model reflecting a physical phenomenon that needs to be controlled. Discrepancy between the actual dynamics and its mathematical model induced by complexity and non-linearity of the system, time-varying or unknown parameters as well as unmodelled structures, forced engineers to introduce a new control technique combating aforementioned modelling difficulties. One widely adopted approach to both robust and computationally effective control of high-order non-linear plants is the sliding mode control (SMC). The idea of SMC is based on the application of a control signal always moving the state vector toward a region adjacent to a differentiable manifold [10, 11] and allowing the representative point to remain on the sliding surface in its close vicinity thereafter. Once the system state hits a manifold, the dynamics of the system becomes insensitive to some disturbances and uncertainties of the model. The development of these concepts began in the late 1950s in the Soviet Union - a significant contribution in the progress of SMC is owed to Utkin [1, 2] and Itkis [3]. Since the sliding mode control law is usually not a continuous function of time, thus one needs to take into consideration control chattering problem impeding SMC implementations. Although chattering attenuation received a major boost in the early 1990s due to second-order concepts [6] and then in 2000s, when higher-order ideas appeared [7], it is still one of the most important aspects in practical applications of SMC.

In this paper we consider an alternative control strategy based on a reaching law [8, 9]. The designed controller is both continuous and smooth in order to provide complete chattering elimination in the presence of non-zero disturbance. Furthermore, proposed algorithm guarantees boundedness of the sliding mode derivative. The price we pay for obtaining such an input signal is a loss of accuracy, as the system trajectories converge to a vicinity of the origin (instead of the origin) in the presence of non-zero external disturbance.

2. Problem statement

We consider the time-varying, nonlinear n-th order dynamic system

$$\dot{x}_1 = x_2, \dot{x}_2 = x_3, ..., \dot{x}_n = f(t, x_1, ..., x_n) + b \cdot u + d(t) \quad (\dot{x} = dx/dt, t \geq 0), \tag{1}$$

where $x_1, x_2, ..., x_n$ are the state variables of the system and $\mathbf{x}(t) = [x_1(t), x_2(t), ..., x_n(t)]^T$ is the state vector, t denotes time, $u \in \mathbb{R}$ is the input signal. Function f is a priori known and maps $[t_0; \infty) \times \mathbb{R}^n$ into $\mathbb{R}$. Besides, $f = f(t, \mathbf{x})$ is assumed to be bounded and Lipschitz continuous with respect to the second coordinate . Moreover, b is a non-zero constant, while scalar-valued function d defined on $[t_0; \infty)$ is unknown and represents the system external disturbance. Further in this paper, it is assumed that d is continuous almost everywhere (discontinuity set is a measure-zero set) [16] and there exists a positive constant D such that

$$|d(t)| \leq D, \quad \text{for all} \quad t \in [t_0; \infty). \tag{2}$$

Switching function $\sigma : \mathbb{R}^n \to \mathbb{R}$ is introduced as follows

$$\sigma(\mathbf{x}) := c_1 x_1 + ... + c_n x_n, \tag{3}$$

where $c_1, ..., c_n$ are real constants. Furthermore, it is assumed that the constants $c_1, ..., c_n$ are chosen so that the eigenvalues of the homogeneous differential equation

$$c_1 x_1 + ... + c_n x_n = 0 \tag{4}$$

are identical, real and less than zero.

Until the end of the next chapter we will be analyzing dynamic system (1), wherein $d \equiv 0$ $(t_0 \leq t < \infty)$. The differential equation describing the sliding variable evolution is proposed in this brief in the following form

$$\dot{\sigma} = -k \left(1 - e^{-|\sigma|/\sigma_0}\right) sign(\sigma), \tag{5}$$

where k and σ_0 are certain constants greater than zero. Note that the latter equation is satisfied when the following controller is applied

$$u(t, \mathbf{x}) = -\frac{k}{b} \left(1 - e^{-|\sigma|/\sigma_0}\right) sign(\sigma) - \frac{1}{b}(c_1 x_2 + ... + c_{n-1} x_n) - \frac{1}{b} f(t, \mathbf{x}). \tag{6}$$

The solution of Eq. 5 is determined by the equations

$$\sigma(t) = \sigma_0 ln \left(1 + e^{-k(t+C)/\sigma_0}\right) sign(\sigma), \quad C = -\frac{\sigma_0}{k} ln \left(e^{\sigma(t_0)/\sigma_0} - 1\right) sign[\sigma(t_0)]. \tag{7}$$

Taking σ as a measure of the distance between a representative point $\mathbf{x}$ and a hyperplane $H = \{\mathbf{x} \in \mathbb{R}^n : \sigma(\mathbf{x}) = 0\}$ we obtain, that while representative point is reaching the hyperplane, its velocity of convergence $\dot{\sigma}$ is decreasing exponentially. Since the function $\dot{\sigma}$ is bounded upper and below by $\pm k$, thus the absolute value of $\dot{\sigma}$ never exceeds k.

Note that the control function (6) is both continuous and smooth despite discontinuity of *sign* function. Although simple calculation shows that

$$\lim_{t \to \infty} \sigma(t) = 0, \quad \lim_{t \to \infty} \dot{\sigma}(t) = 0, \tag{8}$$

the smooth feedback controller (6) cannot provide finite-time convergence of the sliding variable to zero. Convergence of the system trajectories is discussed later in this paper.

The use of the controller (6) for the system carries some mathematical advantage. Indeed, since one can prove that function (6) has Lipschitz property in the absence of external disturbance d, thus there is no need to understand the solutions of the system in the Filippov sense [5, 12] any longer. Thanks to Picard - Lindeløf theorem [13] the existence and uniqueness of the solution is guaranteed.

3. Asymptotic stability of the system in the absence of external disturbance

In this section asymptotic stability of the system including feedback controller (6) is proven. Let us consider once more dynamic system (1) in the absence of external disturbance d. The idea is to show that every solution of non-homogeneous differential equation

$$c_1x_1 + ... + c_nx_n = \sigma(t), \tag{9}$$

where σ is determined by (7), is asymptotically stable. To do so, one can select a proper Lyapunov - candidate function, however in this very situation it is not an easy task. Therefore, a general law in the theory of ordinary differential equations is proven in this chapter.

Definition 1: Consider differential equation

$$\dot{\mathbf{x}} = \mathbf{A}(t) \cdot \mathbf{x} + \mathbf{r}(t), \tag{10}$$

where $t \in (a; b)$, $\mathbf{A} : (a; b) \to \mathrm{L}(\mathbb{R}^n)$ is a continuous function, $\mathrm{L}(\mathbb{R}^n)$ denotes a space of bounded linear operators $\mathbb{R}^n \to \mathbb{R}^n$ and $\mathbf{r} : (a, b) \to \mathbb{R}^n$ is continuous almost everywhere. Function $\mathbf{U} : (a; b)^2 \to \mathrm{L}(\mathbb{R}^n)$ defined by the formula

$$\mathbf{U}(t; t_0) := \sum_{m=0}^{\infty} \mathbf{K}_m(t; t_0), \tag{11}$$

where $\mathbf{K}_0(t, t_0) = \mathbf{I}$ and

$$\mathbf{K}_{m+1}(t; t_0) = \int_{t_0}^{t} \int_{t_0}^{s} ... \int_{t_0}^{s_{m-1}} \mathbf{A}(s_m)...\mathbf{A}(s_1)\mathbf{A}(s)ds_m...ds_1ds = \int_{t_0}^{t} \mathbf{K}_m(s; t_0)ds \tag{12}$$

is called a resolvent of Eq. 10 [14, 15].

Let $\phi = \phi(t)$ be a solution of Eq. 10 determined by initial condition $\phi_0 = \phi(t_0)$. Note that the function $g(t, \mathbf{x}) = \mathbf{A}(t) \cdot \mathbf{x}$ is Lipschitz continuous with respect to the second coordinate, thus every solution can be extended to infinity. Taking into account that $\mathbf{U}(t_0; t_0) = \mathbf{I}$ and $\dot{\mathbf{U}}(t; t_0) = \mathbf{A}(t)\mathbf{U}(t; t_0)$, we obtain that the resolvent is a normalized fundamental matrix of Eq. 10, hence

$$\phi(t) = \mathbf{U}(t; t_0) \cdot \phi_0. \tag{13}$$

Furhermore, differentiating the following equation

$$\phi(t) = \mathbf{U}(t; t_0)\phi_0 + \int_{t_0}^{t} \mathbf{U}(t; s)\mathbf{r}(s)ds, \tag{14}$$

one can easily prove that (14) represents a unique solution of (10).

Lemma 1: If the trivial solution $\phi \equiv 0$ of homogeneous equation associated with Eq. 10 is asymptotically stable, then every solution of Eq. 10 is asymptotically stable.

Proof: Let $\psi = \psi(t)$ and $\eta = \eta(t)$ be arbitrary solutions of (10) determined respectively by initial conditions $\psi_0 = \psi(t_0)$, $\eta_0 = \eta(t_0)$. From (14) we get

$$\psi(t) - \eta(t) = \mathbf{U}(t;t_0)\psi_0 + \int_{t_0}^{t} \mathbf{U}(t;s)\mathbf{r}(s)ds + \quad (15)$$

$$= -\mathbf{U}(t;t_0)\eta_0 - \int_{t_0}^{t} \mathbf{U}(t;s)\mathbf{r}(s)ds = \quad (16)$$

$$= \mathbf{U}(t;t_0)(\psi_0 - \eta_0). \quad (17)$$

The latter equation shows that the difference between any two non-homogeneous system solutions is a solution of a homogeneous system, thus taking into account that the trivial solution of homogeneous equation is asymptotically stable, we obtain the following implication

$$\forall_{\varepsilon>0}\ \exists_{\delta>0}\ ||\psi(t_0) - \eta(t_0)|| < \delta \Rightarrow ||\psi(t) - \eta(t)|| < \varepsilon, \quad \text{for all } \ t_0 \leq t \leq \infty. \quad (18)$$

Obviously $\lim_{t\to\infty} ||\psi(t) - \eta(t)|| = 0$. This concludes the proof. Now we are in a position to formulate the main result of this chapter.

Theorem 1: If $d \equiv 0$ $(t_0 \leq t < \infty)$, then system (1) including control signal given by (6) is asymptotically stable.

Proof: Simple calculation shows that Eq. 9 might be represented in the following form

$$\dot{\mathbf{x}} = \mathbf{A} \cdot \mathbf{x} + \mathbf{r}(t), \quad (19)$$

where $\mathbf{x}$, $\mathbf{r}(t)$ are vectors of dimension $(n-1) \times 1$, wherein $\mathbf{x} = [x_1, ..., x_{n-1}]^T$ and $\mathbf{r}(t) = [0, ...0, \sigma(t)]^T$, $\mathbf{A}$ is in turn matrix of dimension $(n-1) \times (n-1)$ given by the formula

$$\mathbf{A} = \begin{bmatrix} 0 & 1 & 0 & \cdots & 0 & 0 \\ 0 & 0 & 1 & \cdots & 0 & 0 \\ \vdots & \vdots & \vdots & \ddots & \vdots & \vdots \\ 0 & 0 & 0 & \cdots & 0 & 1 \\ -c_1 & -c_2 & -c_3 & \cdots & -c_{n-2} & -c_{n-1} \end{bmatrix}. \quad (20)$$

Since homogeneous equation corresponding to (9) is asymptotically stable, therefore in particular trivial solution is asymptotically stable. Relying on lemma 1, we obtain asymptotic stability of (9). The theorem is proven.

4. Uniform and asymptotic boundedness of the system in the presence of non-zero external disturbance

Until now, we have been considering system (1) assuming the absence of disturbance function. Suppose now that $d \neq 0$, $k > D$ and control law is designed as in (6). Under above assumptions, one can easily notice that sliding variable satisfies the following differential equation

$$\dot{\sigma} = -k\left(1 - e^{-|\sigma|/\sigma_0}\right) sign(\sigma) + d(t). \quad (21)$$

Define a set

$$\Sigma_D := \{\sigma : |\sigma| \leq -\sigma_0 ln\,(1 - D/k)\}\,. \quad (22)$$

Lemma 2: If there exists time $t_1 \in [t_0; \infty)$ such that $|\sigma(t_1)| = -\sigma_0 ln\,(1 - D/k)$, then $\sigma(t) \in \Sigma_D$ for all $t \geq t_1$.

Proof: Suppose that there exists time $t_2 > t_1$ such that $\sigma(t_2) > -\sigma_0 ln\,(1 - D/k)$. Since disturbance function d is continuous almost everywhere, thus there must exist an interval

$(a;b) \subset (t_1;t_2)$ and for all $t \in (a;b)$ the switching function σ is increasing. Hence, $\dot{\sigma} > 0$ if $t \in (a;b)$. On the other hand, Eq. 21 yields $\dot{\sigma} < 0$ for all t and $\sigma(t) > -\sigma_0 ln\,(1 - D/k)$, which ultimately leads to a contradiction. Due to the symmetry of the problem, it is enough to consider the case of $\sigma > 0$.

The latter lemma means that if sliding variable reaches boundary set Σ_D, then it remains there thereafter. The following Lyapunov - candidate function

$$V(\sigma) = \frac{1}{2}\left[\sigma + sign(\sigma)\sigma_0 ln\,(1 - D/k)\right]^2 \tag{23}$$

guarantees the asymptotic convergence of σ to Σ_D. Considering $V(\sigma) = \frac{1}{2}\sigma^2$ as a Lyapunov - candidate function, one can easily prove finite-time convergence. Now we are in a position to formulate the main result of the paper.

Theorem 2: Let $\phi = \phi(t)$ be an arbitrary solution of system (1) in the presence of non-zero external disturbance, determined by initial condition $\phi_0 = \phi(t_0)$. Let λ denotes the eigenvalue of $\mathbf{A}$. Then

$$||\phi(t)|| \leq \frac{\sigma_0}{\lambda} ln\,(1 - D/k)\,, \quad \text{if} \quad t \to \infty. \tag{24}$$

Proof: The idea is to show, that every solution of non-homogeneous differential equation

$$c_1 x_1 + ... + c_n x_n = \sigma(t), \tag{25}$$

where σ is is related to Eq. 21, satisfies condition (24). As we have already seen, Eq. 25 is is equivalent to (19). Taking into account the Ważewski's inequality [15, 17]

$$||U(t;t_0)|| = ||e^{\mathbf{A}(t-t_0)}|| \leq e^{\lambda(t-t_0)} \quad \text{for all} \quad t \in [t_0;\infty), \tag{26}$$

we obtain that

$$||\phi(t)|| \;\leq\; ||\mathbf{U}(t;t_0)|| \cdot ||\phi(t_0)|| + \int_{t_0}^{t} ||\mathbf{U}(t;s)|| \cdot ||\mathbf{r}(s)|| ds \leq \tag{27}$$

$$\leq\; e^{\lambda(t-t_0)}||\phi(t_0)|| + \int_{t_0}^{t} e^{\lambda(t-s)}|\sigma(s)|ds = \tag{28}$$

$$=\; e^{\lambda(t-t_0)}||\phi(t_0)|| + e^{\lambda t}\int_{t_0}^{t} e^{-\lambda s}|\sigma(s)|ds. \tag{29}$$

Obviously, $\lim_{t\to\infty} e^{\lambda(t-t_0)}||\phi(t_0)|| = 0$. Relying on the generalized de l'Hospital theorem in the form of Stolz [15], one can easily prove that

$$\lim_{t\to\infty} e^{\lambda t}\int_{t_0}^{t} e^{-\lambda s}|\sigma(s)|ds = \frac{\sigma_0}{\lambda} ln\,(1 - D/k)\,. \tag{30}$$

Note that the existence of the latter limit is guaranteed only when $\sigma \notin \Sigma_D$. Suppose that there exists time $t_1 \in [t_0;\infty)$ and $|\sigma(t_1)| = -\sigma_0 ln\,(1 - D/k)$. Simple calculation shows that if $t \to \infty$, then

$$||\phi(t)|| \;\leq\; \lim_{t\to\infty} e^{\lambda(t-t_0)}||\phi(t_0)|| + \lim_{t\to\infty} e^{\lambda t}\int_{t_0}^{t_1} e^{-\lambda s}|\sigma(s)|ds + \tag{31}$$

$$+\; \lim_{t\to\infty} e^{\lambda t}\int_{t_1}^{t} e^{-\lambda s}|\sigma(s)|ds = \lim_{t\to\infty} e^{\lambda t}\int_{t_1}^{t} e^{-\lambda s}|\sigma(s)|ds \leq \tag{32}$$

$$\leq\; \lim_{t\to\infty} e^{\lambda t}\int_{t_1}^{t} e^{-\lambda s}\left[-\sigma_0 ln\,(1 - D/k)\right] ds = \frac{\sigma_0}{\lambda} ln\,(1 - D/k)\,. \tag{33}$$

The last equality follows from the generalized de l'Hospital theorem in the form of Stolz. The proof of the theorem is complete.

It is worth noting that the record $lim_{t\to\infty}||\phi(t)||$ is formally incorrect as the considered limit does not need to exist, though the statement (26) remains valid.

Corollary 1: Let $\phi = \phi(t)$ be an arbitrary solution of system (1) in the presence of non-zero external disturbance, determined by initial condition $\phi_0 = \phi(t_0)$. Let λ denotes the eigenvalue of $\mathbf{A}$. Then

$$||\phi(t)|| \leq ||\phi(t_0)||e^{\lambda(t-t_0)} + \frac{\sigma_0}{\lambda} ln\left(1 - D/k\right) \quad \text{for all} \quad t \in [t_0; \infty). \tag{34}$$

Proof: Since

$$\frac{d}{dt}\int_{t_0}^{t} e^{\lambda(t-s)}|\sigma(s)|ds = |\sigma(t)| \geq 0 \tag{35}$$

and the function of the upper limit is increasing, hence

$$||\phi(t)|| \leq ||\phi(t_0)||e^{\lambda(t-t_0)} + \lim_{t\to\infty}\int_{t_0}^{t} e^{\lambda(t-s)}|\sigma(s)|ds. \tag{36}$$

Computing the latter limit, we obtain the thesis.

Corollary 2: Let $\phi = \phi(t)$ be an arbitrary solution of system (1) in the presence of non-zero external disturbance, determined by initial condition $\phi_0 = \phi(t_0)$. Let λ denotes the eigenvalue of $\mathbf{A}$. If $det(\mathbf{A}) \neq 0$, then

$$||\phi(t)|| \leq ||\phi(t_0)||e^{\lambda(t-t_0)} + |\sigma(t_0)| \cdot ||A^{-1}|| \cdot \sup_{t\in[t_0;\infty)} (||\mathbf{U}(t;t_0) - \mathbf{I}||) \leq \tag{37}$$

$$\leq ||\phi(t_0)||e^{\lambda(t-t_0)} + |\sigma(t_0)| \cdot ||\mathbf{A}^{-1}|| \cdot n, \tag{38}$$

for all $t \in [t_0; \infty)$.

Proof: Note that

$$||\phi(t)|| \leq ||\phi(t_0)||e^{\lambda(t-t_0)} + ||\int_{t_0}^{t} e^{\mathbf{A}(t-s)}|\sigma(t_0)|ds|| = \tag{39}$$

$$= ||\phi(t_0)||e^{\lambda(t-t_0)} + ||\int_{t_0}^{t}\sum_{k=0}^{\infty}\frac{\mathbf{A}^k(t-s)^k}{k!}|\sigma(t_0)|ds||. \tag{40}$$

As the assumptions of the Fubini's theorem [16] are satisfied, thus

$$||\sum_{k=0}^{\infty}\frac{\mathbf{A}^{k+1}(t-t_0)^{k+1}}{(k+1)!}|| = ||\sum_{k=0}^{\infty}\frac{\mathbf{A}^k(t-t_0)^k}{k!} - \mathbf{I}|| \leq \sup_{t\in[t_0;\infty)} (||\mathbf{U}(t;t_0) - \mathbf{I}||) \leq \tag{41}$$

$$\leq \sup_{t\in[t_0;\infty)}\left(e^{\lambda(t-t_0)} + ||\mathbf{I}||\right) \leq n, \tag{42}$$

which concludes the proof.

5. Generalization on the control law

In this section a generalization on the differential equation (5) is proposed. Consider function $z : \mathbb{R} \to \mathbb{R}$ of variable σ such that

- $z \geq 0$, z is continuous and even,
- z is differentiable or differentiable except 0,

- z'_σ is continuous or continuous except 0,
- z reaches global minimum in 0 and $z(0) = 0$,
- $z\mid_{[0;\infty)}$ is increasing,
- $\lim_{\sigma\to\infty} z(\sigma) = \infty$, $\lim_{\sigma\to\infty} a^{-z(\sigma)/\sigma_0} \cdot z'_\sigma = g < \infty$, where $a > 1$.

Let $C^{0,1}_{a,L}$ denotes the class of all functions satisfying the upper conditions. The following controller can be used instead of the conventional control law (6)

$$u(t, \mathbf{x}) = -\frac{k}{b}\left(1 - a^{z(\sigma)/\sigma_0}\right) sign(\sigma) - \frac{1}{b}(c_1x_2 + ... + c_{n-1}x_n) - \frac{1}{b}f(t, \mathbf{x}). \tag{43}$$

It is easy to observe that appliance of controller (43) to the system makes the sliding variable satisfy the following differential equation

$$\dot{\sigma} = -k\left(1 - a^{-z(\sigma)/\sigma_0}\right) sign(\sigma) + d(t). \tag{44}$$

Depending on the choice of z, homogeneous equation associated with the Eq. 44 may be impossible to solve analytically, yet the existence and uniqueness is guaranteed by Picard - Lindeløf theorem. Indeed, if $\sigma > 0$, then

$$P'_\sigma = \left[-k\left(1 - a^{-z(\sigma)/\sigma_0}\right) sign(\sigma)\right]'_\sigma = -\frac{k \cdot ln(a)}{\sigma_0} \cdot a^{-z(\sigma)/\sigma_0} \cdot z'_\sigma \le 0 \tag{45}$$

and $z \in C^{0,1}_{a,L}$, thus P'_σ is continuous on $[0;\infty)$, which implies that $\lim_{\sigma\to 0} P'_\sigma = 0$, $\lim_{\sigma\to\infty} P'_\sigma = -k \cdot ln(a) \cdot g/\sigma_0$. Hence P'_σ is bounded and as a result Lipschitz continuous. The same is true for $\sigma < 0$.

Selecting a proper Lyapunov - candidate function, one can easily prove that in the absence of external disturbance

$$\lim_{t\to\infty} \sigma(t) = 0, \quad \lim_{t\to\infty} \dot{\sigma}(t) = 0. \tag{46}$$

From the foregoing considerations, we obtain that controller (43) makes the sliding variable and its derivative possess the characteristics and properties of the variable that satisfies (5). Therefore, all the lemmas and theorems discussed so far may be prescribed to an accuracy of some constants. Indeed, consider a set

$$\sigma_{min} := \inf\left\{\sigma \in (0;\infty) : z(\sigma) > -\sigma_0 log_a\left(1 - D/k\right)\right\}, \tag{47}$$

where $k > D$. Note that $z \in C^{0,1}_{a,L}$ implies the existence of $\sigma_1 \in [0,\infty)$ such that function $z_{min} = z|_{[\sigma_1;\, \sigma_{min}]}$ is strictly increasing and as a result injective. Hence,

$$\Sigma^a_D := \left\{\sigma \in \mathbb{R} : |\sigma| \le z^{-1}_{min}\left(-\sigma_0 log_a\left(1 - D/k\right)\right)\right\} \tag{48}$$

is used instead of Σ_D. Obviously, if the function z is injection on $[0;\infty)$, then it is enough to consider z^{-1} instead of z^{-1}_{min}.

5.1. *Example*

For the mass-spring-damper system with friction [18] modeled by

$$m\ddot{x} + b_0\dot{x} + b_1 sign(\dot{x}) + k_1 x = u + d(t), \tag{49}$$

where x is the displacement from equilibrium, u is the external force applied to system, m is the mass of the block, b_0 is damping constant, b_1 is friction, k_1 is force constant of spring, d denotes unknown external disturbance, the control law u

$$u(x) = -km\left(1 - e^{-\sigma^2}\right) sign(\sigma) - m\dot{x} + b_0\dot{x} + b_1 sign(\dot{x}) - k_1 x \tag{50}$$

applied to (49) makes the sliding variable satisfy (44) (for the purposes of this example, we will call (50) non-conventional control law), if the switching function is given by $\sigma(x) = x + \dot{x}$ and $a = exp$, $z(\sigma) = \sigma^2$. The control system is simulated with $m = 1$ [kg], $b_0 = 0.1$ [kg/s], $b_1 = 0.05$ [kg/s^2], $k_1 = 0.5$ [kg/s], $k = 2$, $\sigma_0 = 1$, $d(t) = sin(t)$, $D = 1$, $x(0) = 7$ [m] and $\dot{x}(0) = 10$ [m/s]. Furthermore, we demonstrate the results of the simulation with the same control system simulated with conventional sliding mode control law given by

$$u(x) = -x - 3 \cdot sign(x + \dot{x}). \tag{51}$$

Lines located above and below the hyperplane $\sigma(x) = 0$ (see Fig. 1) are given respectively by

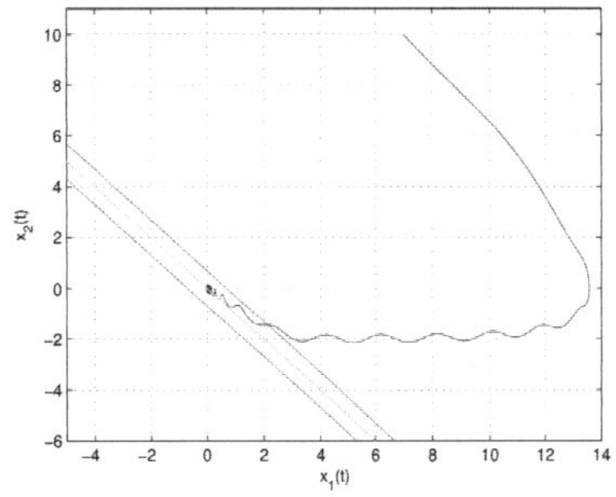

Figure 1: Phase portrait obtained using non-conventional control law

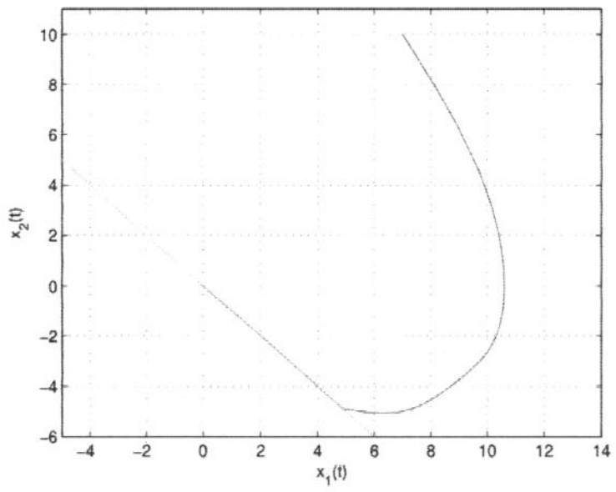

Figure 2: Phase portrait obtained using conventional control law

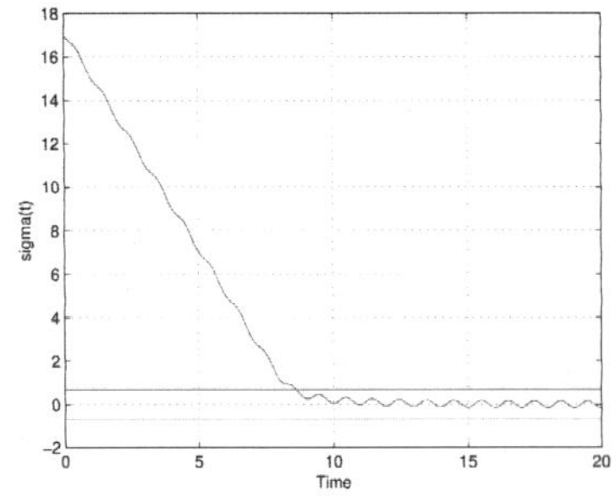

Figure 3: Sliding variable obtained by using non-conventional control law

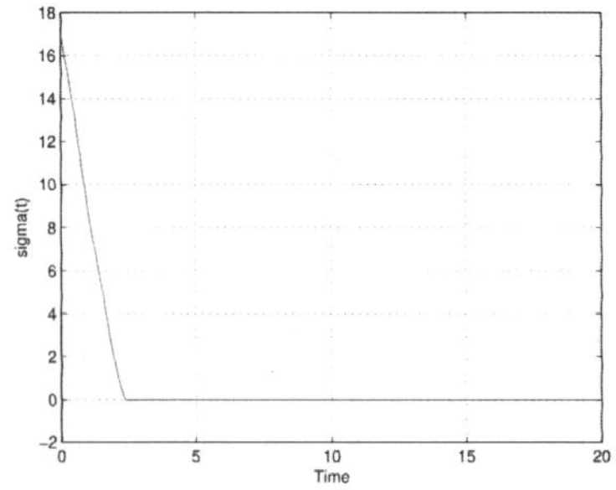

Figure 4: Sliding variable obtained by using conventional control law

$$x_2 = -x_1 \mp \sigma_0 ln\left(1 - D/k\right) = -x_1 \mp ln\left(0.5\right), \tag{52}$$

wherein $x_1 = x$ and $x_2 = \dot{x}$. Lines located in Fig. 3 are determined respectively by

$$\sigma = \mp ln\left(1 - D/k\right) = \mp ln\left(0.5\right). \tag{53}$$

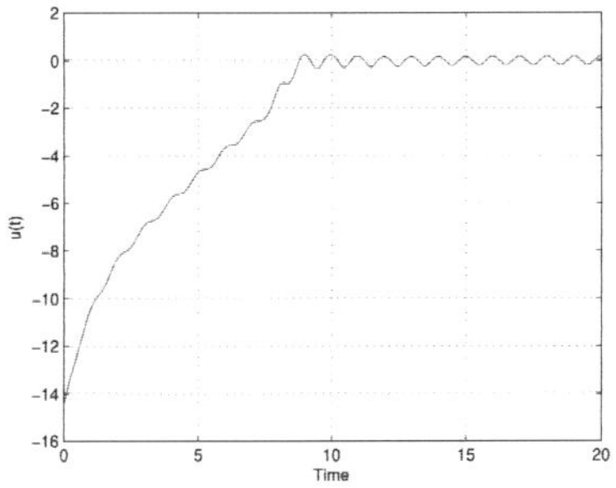

Figure 5: Control signal generated according to non-conventional sliding mode control

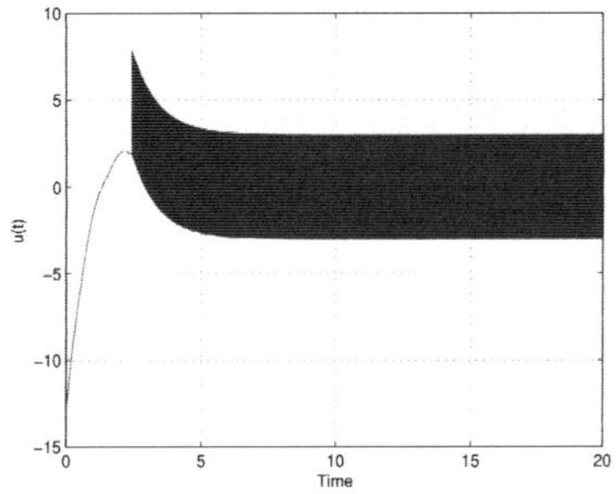

Figure 6: Control signal generated according to conventional sliding mode control

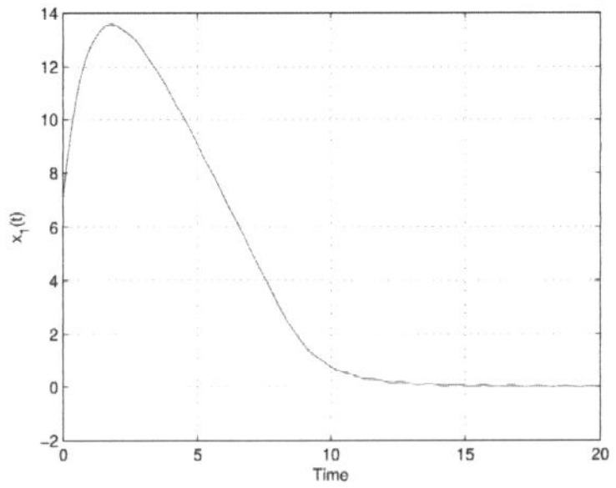

Figure 7: System output $x_1(t)$ obtained using non-conventional control law

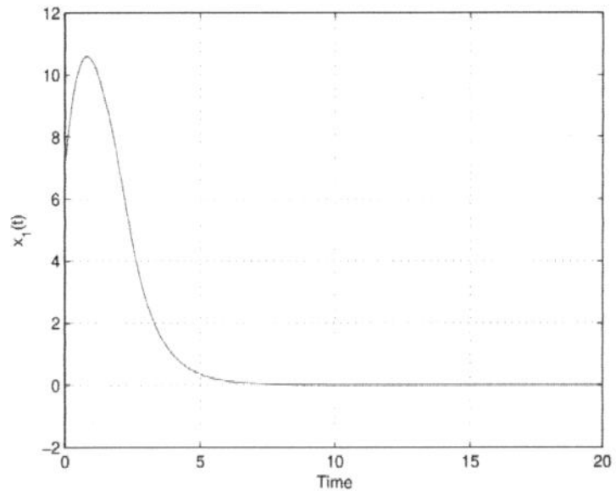

Figure 8: System output $x_1(t)$ obtained using conventional control law

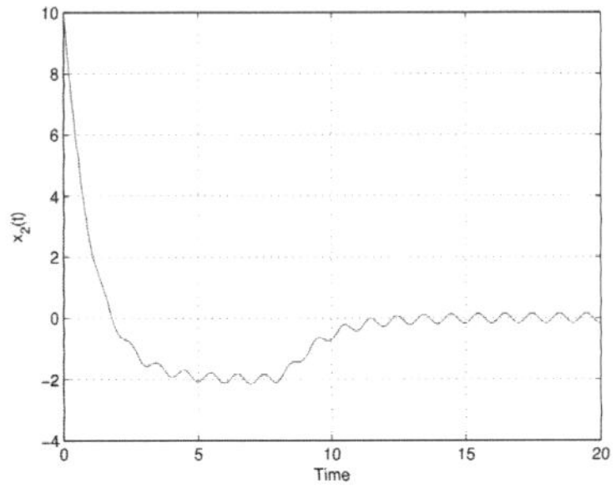

Figure 9: System output $x_2(t)$ obtained using non-conventional control law

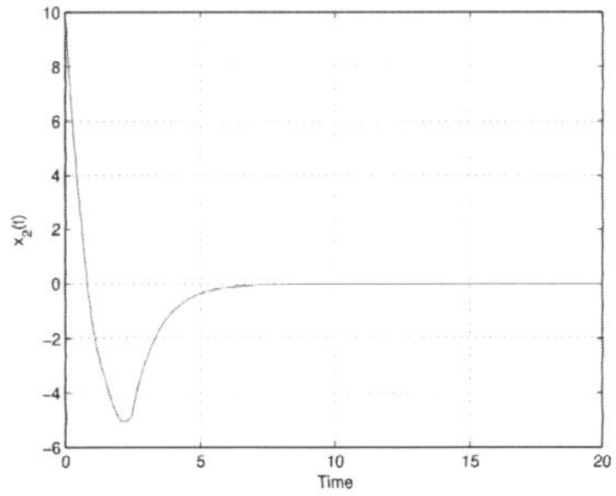

Figure 10: System output $x_2(t)$ obtained using conventional control law

6. Conclusions

In this paper an alternative sliding mode control for continuous time systems has been proposed. The method employs a reaching law - the strategy has been designed so that the rate of change of the sliding variable is always bounded and decreases exponentially with the sliding variable decrease. On the one hand, presented algorithm enforces the convergence of the representative point to the origin while considering unperturbed systems. On the other hand, for systems affected by external disturbance it has been proven that sliding variable enters in finite time a specified set and never leaves it. As a result, every trajectory starting from any initial position converges to a vicinity of the origin and the system becomes uniformly and asymptotically bounded. Furthermore, application of the proposed reaching law has resulted in chattering elimination. Techniques demonstrated in this article can be adapted for discrete time systems.

References

[1] Utkin V Variable structure systems with sliding modes 1977 *IEEE Trans. on Automatic Control*, vol. 22, pp. 212-222
[2] Utkin V Sliding mode and their applications in variable structure control 1978 *MIR*
[3] Itkis U Control systems of variable structure 1978 *John Wiley & Sons*
[4] DeCarlo R S, Żak S and Mathews G Variable structure control of nonlinear multivariable systems 1988 *Proc. of IEEE*, vol. 76, pp. 212-232
[5] Shtessel Y, Edwards C, Fridman L and Levant Arie Sliding mode control and observation 2014 *Springer*
[6] Levant A and Levantowsky L V Sliding order and sliding accuracy in sliding mode control 1993 *Int. J. Control*, pp. 1247-1263
[7] Levant A Universal SISO sliding mode controllers with finite-time convergence 2001 *IEEE Trans. on Automatic Control*, vol. 46, pp. 1447-1451
[8] Gao W and Hung J C Variable structure control of nonlinear systems: a new approach 1993 *IEEE Trans. on Industrial Electronics*, vol. 40, pp. 45-55
[9] Bartoszewicz A A new reaching law for sliding mode control of continuous time systems with constraints 2015 *Trans. of the Institute of Measurement and Control*, vol. 37, pp. 515-521
[10] Lee J M Introduction to topological manifolds 2000 *Springer*
[11] Hirsch M W Differential topology 1976 *Springer*
[12] Joó I and Tallos P The Filipov-Ważewski relaxation theorem revisited 1999 *Acta Math* pp. 171-177
[13] Gerald T Ordinary Differential equations and dynamical systems 2012 *American Mathematical Society*
[14] Grimmer R C Resolvent operators for integral equations in a Banach space 1982 *American Mathematical Society* vol. 273 pp. 333-336
[15] Demidovich B P Lectures on mathematical stability theory 1967 Moscow
[16] Bruckner A M, Bruckner J B and Thomson B S Real analysis 1997 *Prentice-Hall*
[17] Gil M I On Generalized Wazewski and Lozinski inequalities for semilinear abstract differential-delay equations 1998 *J. ofInequal. & Appl.*, vol. 2, pp.255-256
[18] Halliday D, Resnick Robert, Walker J Fundamentals of physics extended 2013 *Wiley*

Stability of distributed MPC in an intersection scenario

T Sprodowski[1] **and J Pannek**[1,2]

[1] Department of Production Engineering, University of Bremen

[2] BIBA Bremer Institut für Produktion und Logistik GmbH, 28359 Bremen, Germany

E-mail: {spr[1], pan[2]}@biba.uni-bremen.de

Abstract. The research topic of autonomous cars and the communication among them has attained much attention in the last years and is developing quickly. Among others, this research area spans fields such as image recognition, mathematical control theory, communication networks, and sensor fusion. We consider an intersection scenario where we divide the shared road space in different cells. These cells form a grid. The cars are modelled as an autonomous multi-agent system based on the Distributed Model Predictive Control algorithm (DMPC). We prove that the overall system reaches stability using Optimal Control for each multi-agent and demonstrate that by numerical results.

Today the development of autonomous cars and the insurance of their correct behavior are still a great challenge in both academic research and the automotive industry. To ensure that networked cars are reliable in shared-space scenarios, the car-to-car communication implemented as a cooperative strategy is a possibility for getting a solution in the intersection resource problem. The ideal situation will be if all participants support them. Yet, non-communicating participants should also be included within such scenarios. Here, we assume that even if communication data is missing, a networked car is able to estimate the future route of a respective participant sufficiently accurate to guarantee collision avoidance.

To incorporate the behavior of the other cars and of disturbance factors, we propose to utilize a Distributed Nonlinear Model Predictive Control Scheme to formulate our problem as an optimal control problem. Nonlinear Model Predictive control (NMPC) is nowadays a widespread method, which is used in many feedback control system applications [1]. A survey of the nonlinear version is given by [2] and [3]. NMPC itself is a three step procedure: First, the current state of the system is estimated. Then, based on that estimate, an internal model of the system dynamics is used to predict the state trajectory over a finite time horizon and to compute a control u minimizing a given cost functional subject to possible constraints. Last, only the initial segment of the computed control u is applied, which renders the algorithm to be iteratively applicable and repeat the procedure until the equilibrium is achieved. The goal of NMPC is to (practically) asymptotically stabilize the system to a target x^*. To guarantee this property, endpoint constraints [4] or terminal constraints and costs [5] as well as a relaxed Lyapunov inequality [3] have been proposed. Since terminal constraints are inadequate for a networked car scenario due to the size of the problem and also limit the feasible set [6], we follow the latter approach.

For our networked car scenario, a distributed control setting appears naturally. Stemming

Journal of Physics: Conference Series **659** (2015) 012049 doi:10.1088/1742-6596/659/1/012049

from a centralized idea, Richards and How [7], among others, proposed to split up the centralized optimal control problem into subproblems connected by a communication network, which lead to a Distributed Model Predictive Controller (DMPC). The general idea is to reduce the complexity of the problem induced by the number of constraints and control variables, which are needed by the centralized problem. While the total number of constraints and control variables remains unchanged, each subproblem is much smaller and can be solved more easily. We model these subproblems as an independent NMPC problems, and connect them to obtain all information about the chosen solution of the other subproblems. Here, we follow the approach to a fixed order, in which subproblems are solved.

The paper is organized as follows: In the following section, we describe the intersection scenario, which we consider throughout our work. In Section 2, we formally introduce the DMPC setting for our problem scenario. Respective stability conditions are shown to hold in Section 3 and illustrated using numerical experiments in Section 4. Last, we draw some conclusions in Section 5 and give a short outlook on future work.

1. Intersection Scenario

We consider a scenario based on an intersection without traffic lights similar to [8]. In this scenario, the authors proposed to divide the shared space of an intersection into multiple cells. These cells form a two-dimensional grid, which is administrated centrally. This scenario illustrates a simple form of a shared resource allocation problem. The cars are managed by one multi-agent each, which aims to cross the intersection in two directions straight ahead. The resulting problem is solved by a centralized approach utilizing the Organic Computing Principle [9]. An observer/controller-architecture is introduced to calculate a path through the intersection for each incoming networked car and to detect deviations using the A^* Search Algorithm [10]. The path is formulated as a sequence of tuples. Each tuple contains time (n) and space coordinates (k, m) to clarify when the cells will be occupied ($p = (n_1, k_1, m_1), \ldots (n_i, k_i, m_i)$). Here, n represents the nth discrete time step, k and m are the coordinates within the grid. Since the cars are modelled as multi-agents which can re-plan their original path, deviations are possible. The observer recognizes the latter and influences the affected cars including the causer by renewing the paths and spreading the information. The communication is centralized and controlled by the observer-/controller-system. Hence, only car-to-infrastructure communication is considered.

We propose to modify this scenario and implement a distributed control. Different from [11] where the paths are predefined and the optimization function considers speed and acceleration only, our aim is to model both the car and the scenario more generally. The collision avoidance in [11] is introduced as a constraint, which is imposed to a car following the priority assignment. We consider the reservation of grid cells to avoid any collisions. An illustration of our scenario is illustrated in Figure 1. To this end, we allow traffic from each direction within our extended

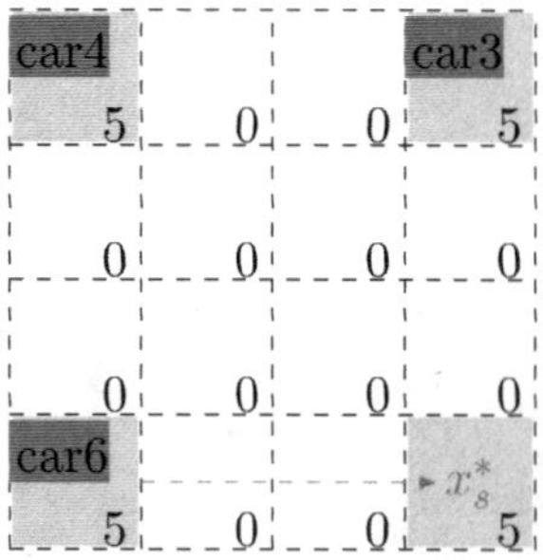

Figure 1. Example of a 4 × 4 - Intersection scenario divided in grid cells which shows the requested reservations made by cars, dashed line illustrates the planned path for car 6 to reach its target x_s^*

scenario. This raises the risk of a deadlock, which, for instance, may occur if two cars stand opposite to each other and cannot give way. Here, we avoid this possibility by introducing lanes, which are a necessity for conventional cars during the introduction phase of networked cars. The lanes determine the direction in which a car can move forward. Additionally, we introduce one multi-agent per car, which performs the path calculation via a DMPC algorithm to compute an optimal collision-free solution through the grid. To follow the approach of a fixed order, each car is set up with an index by initialization. The simulation procedure is executed as described in Algorithm 1.

Algorithm 1 Simulation procedure

Given: Initial time and set of active cars S.
While $S \neq \emptyset$ do

(i) **State estimation:**
Get current state for each car $x_s \in S$.

(ii) **Control computation:**
For each car $x_s \in S$ do

- (a) If current position is target position, then remove car x_s from S and break. Else set completed state to false.
- (b) While completed state is false
  1. Compute optimal path over finite horizon for given constraints.
  2. Try to reserve time/space tuples. If successful, set completed state to true. Else, increment time value and try again.

(iii) **Control implementation:**
Implement all first control elements and shift horizon forward in time.

We model the target distance as a cost function and let the path be calculated and optimized by each vehicle separately. The actual path optimization in every time step n is necessary as we have multiple lanes for each direction so the dynamics is in the path calculation. We consider the reservation of grid cells to avoid any collisions.

2. Formulation in DMPC

2.1. System definition

We suppose the cars to be modeled as multi-agents and denote the overall state of the system by x, where there can be S subsystems

$$x = \{x_1, \cdots, x_S\}.$$

where s denotes the sth car. From the physical point of view x describes the position of the car in the intersection scenario. The path p through the intersection scenario is given as a sequence of tuples $(n, k, m) \in \mathbb{N}_0 \times \{1, \ldots, L_k\} \times \{1, \ldots, L_m\}$ with $L_k, L_m \in \mathbb{N}$, where the latter set denotes the time-space grid. The term (n, k, m) denotes the coordinates, at which a car can be situated, i.e.

$$x_s(n) = (n, k, m) \in \mathbb{N}_0 \times \{1, \ldots, L_k\} \times \{1, \ldots, L_m\}$$

where n denotes the discrete time, and k and m are the coordinates within the grid. The dynamics of the sth subsystem

$$x_s(n+1) = f_s(x_s(n), u_s(n)) \tag{1}$$

connects these tupels to generate a path, where the control $u_s(n)$ implements the path item for the next time step. Hence, the path can be written as

$$\text{Path } p_s = (x_s(n))_{n\in\mathbb{N}_0}$$

The target for the overall system is denoted by x^* containing the targets

$$x^* = (x_s^*)_{s=\{1,\cdots,S\}}$$

for each subsystem, which we consider to be time invariant for simplicity of exposition. The control $u_s(n)$ of a subsystem s is given also as absolute coordinates of the grid similar to $x_s(n)$ which contains also the path tuple (n, k, m) for each system change.

2.2. Constraints of Subsystems

Each subsystem is contemplated with constraints that have to be satisfied. Here, we introduce three different kinds of constraints, the first to incorporate restrictions solely based on the local state and control, the second to include collision avoidance, and the third to enforce stability of each subsystem. The latter is modeled via

$$\#x_s = |\left((n_1, k_1, m_1), \cdots, (n_i, k_i, m_i)\right)| < M, \tag{2}$$

where M defines an upper bound on the length of a path sequence. As a consequence, each subsystem has to reach its target in at most M steps.
To avoid collisions, we do not allow two subsystems x_r and x_v, $r \neq v$, to take identical time-space coordinates. Hence, a tuple (n_i, k_i, m_i) can only appear at most once in all paths p_s. More formally, we obtain

$$\forall (n_i, k_i, m_i) \in x_r \quad \Rightarrow \quad (n_i, k_i, m_i) \notin x_v \tag{3}$$

for all $r, v \in \{1, \cdots, S\}$ with $r \neq v$.
Last, using local restrictions we restrict the motion of each car by either staying at the current position, i.e. $x_s(n+1) = x_s(n)$, or moving to a neighbouring cell, that is from $x_s(n) = (n, k_1, m_1)$ to $x_s(n+1) = (n+1, k_2, m_2)$, where

$$(|k_2 - k_1|, |m_2 - m_1|) \leq (1, 1) \tag{4}$$

2.3. Costs for each Subsystem

Incorporating the dynamics (1) and the constraints (2), (3) and (4), each multi-agent optimizes its subsystem with respect to the stage cost

$$\ell(x_s(k), u_s(k)) = \|x_s(k) - x_s^*\|^2 + \lambda\|u_s(k)\|^2,$$

where $\lambda > 0$ is a regularization factor to avoid bang-bang behavior of optimal solutions. Utilizing the DMPC approach, we define the cost function for each subsystem via

$$J_s^N(x_s^0, u_s) := \sum_{k=0}^{N-1} \ell_s(x_s(k), u_s(k)) \tag{5}$$

where x_s^0 denotes the initial value of the path to be optimized, i.e. $x_s(0) := x_s^0$, and J_s^N the cumulated costs for the each system state k over an horizon of length N.

2.4. Costs of Overall System

The cost function can also be interpreted as energy, i.e. the overall system loose potential energy if the cars are moving towards the target. Here, we assume that after each time step of the DMPC procedure, the overall system looses potential energy. While appearing restrictive, this assumption always holds due to our fixed order of computations for subsystems. Hence, at least one car x_s can move forward and reduces its own costs, and thereby also the cumulative system costs. The DMPC procedure implements a minimizer of (5), which we denote by

$$u_s^*(\cdot, x_s) = \underset{u_s}{\operatorname{argmin}}\, J_s^N\left(x_s, u_s\right).$$

Hence, the cumulative costs of the overall system for the DMPC setting are given by

$$V^N(x^0) = \sum_{i=1}^{S} J_s^N\left(x_i^0, u_s^*(\cdot, x_i^0)\right) \tag{6}$$

2.5. Feedback of the subsystem

Last, we obtain the feedback μ_s^N for each subsystem as the first element of the minimizer $u_s^*(\cdot, x_s)$ from (6), i.e.

$$\mu_s^N(x_s^0) = u_s^*(0, x_s^0) \tag{7}$$

Note that the feedback is implicitly coupled with the success of reserving a path item (n, k, m) of the grid in the optimization and coordination step. If more than one system attempt to get a reservation for the same cell at the same time n, only the highest priority path item is accepted. This fulfills the constraint formulated in (3). The other inquires are rejected and the systems have to attempt a new reservation at a later time or consider a different field to attempt a reservation.

3. Stability conditions

Our goal is to show asymptotic stability of the overall system. We use [3, Def. 2.14] to derive an upper bound for our system, which will guarantee the asymptotic stability of the system. To define the required bounding function $\beta(r, n) \in \mathcal{KL}$, we introduce the set $\mathcal{K}$ describing the functions $\alpha : \mathbb{R}_0^+ \to \mathbb{R}_0^+$, which are strictly increasing and satisfy $\alpha(0) = 0$. If additionally α is unbounded, then it called a class $\mathcal{K}_\infty$ function. Secondly, we call functions $\delta : \mathbb{R}_0^+ \to \mathbb{R}_0^+$ to be of class $\mathcal{L}$, if δ is continuous and strictly decreasing with $\lim_{n\to\infty} \delta(t) = 0$. Combining these properties, we call $\beta(r, n)$ to be a class $\mathcal{KL}$ function if $\beta(r, n)$ is continuous and $\beta(\cdot, n) \in \mathcal{K}$ for all $n \geq 0$ and $\beta(r, \cdot) \in \mathcal{L}$ for all $r \geq 0$. Based on the asymptotically stability we can show that our system is optimal controllable. This is presented in the numerical results 4 where the suboptimality α is used to evaluate the performance of the controller.

Note that for our intersection scenario, we cannot guarantee that a minimizer u_s^* to exist such that costs are decreasing for each car in each time step. Indeed, the constraints (2) and (3) allow cars to hold their position for a number of time steps to let other cars pass. Yet, some cars move closer to their target and the respective costs of these subsystems are decreasing. Hence, only an overall consideration is promising.

In the following, we prove stability of the presented intersection scenario by showing that for $r := |x_s - x_s^*|$

$$\beta(r, n) = S \max\left\{\overline{M} - \underline{M} - n, 0\right\}\underline{M}^2 + \lambda S \underline{M} + \sum_{j=1}^{\underline{M} - \max\left\{0, n - \overline{M} + \underline{M}\right\}} j^2$$

is a class $\mathcal{KL}$ function and by constructing an upper bound for the overall costs. To this end, we first derive a bound $\beta(r,0)$ for the initial conditions, which we extend in a second step to an upper bound for all time instances n. Here, we abbreviate the lower bound of the path length (2) as $\underline{M} = \min(\#x_s)$ and the upper bound as $\overline{M} = \max(\#x_s)$.

First, we ignore (3) and define $\beta(r,0)$ as a function, where all S cars stay in their initial position until they have to leave to get to the target without violating $\overline{M}$. Due to ignoring the collision constraint (3), $\beta(r,0)$ represent an upper bound on the costs. We split this expression into two subsums:

$$\beta(r,0) = \sum_{s=1}^{S} \left(\sum_{k=0}^{\overline{M}-\underline{M}-1} \ell(x_s, 0) + \sum_{k=\overline{M}-\underline{M}}^{\overline{M}-1} \ell(x_s(k)), u_s(k) \right).$$

Since by assumption cars remain at their initial position, we have $u_s(k) = 0, k = 0, \cdots, \overline{M} - \underline{M} - 1$ and can dissolve the inner sums to

$$\begin{aligned}\beta(r,0) &= \sum_{s=1}^{S} \left(\left(\overline{M} - \underline{M}\right) ||x_s^0 - x_s^*||^2 + \sum_{k=\overline{M}-\underline{M}}^{\overline{M}-1} ||x_s(k) - x_s^*||^2 + \lambda ||u_s||^2 \right) \\ &= \sum_{s=1}^{S} \left(\left(\overline{M} - \underline{M}\right) ||x_s^0 - x_s^*||^2 + \sum_{k=0}^{\underline{M}-1} ||x_s(k) - x_s^*||^2 + \lambda ||u_s||^2 \right),\end{aligned}$$

which allows us to shift the interval for the second sum to $k = 0, \cdots, \underline{M} - 1$. By our model definition, we obtain $||x_s^0 - x_s^*|| = \underline{M}$ for the initial position and $||x_s(k) - x_s^*|| = \underline{M} - k$ as a car moves towards its target with $||u_s||^2 = 1$. Hence, we obtain

$$\beta(r,0) = \sum_{s=1}^{S} \left(\left(\overline{M} - \underline{M}\right) \underline{M}^2 + \sum_{k=0}^{\underline{M}-1} \left(\underline{M} - k\right)^2 + \lambda \underline{M} \right).$$

Dissolving now the inner sum over the systems we get

$$\beta(r,0) = S \left(-\underline{M}^3 + \underline{M}^2 \overline{M} + \underline{M}\lambda + \sum_{k=1}^{\underline{M}} k^2 \right). \tag{8}$$

With the application of square pyramidal number for the last remaining sum in (8)

$$\sum_{k=1}^{\underline{M}} k^2 = \frac{1}{6} \left(\underline{M} + 1\right) \underline{M} \left(2\underline{M} + 1\right)$$

we can simplify (8) to

$$\beta(r,0) = S \left(-\frac{2}{3} \underline{M}^3 + \underline{M}^2 \left(\overline{M} + \frac{1}{2} \right) + \underline{M} \left(\lambda + \frac{1}{6} \right) \right), \tag{9}$$

which reveals an upper bound on the initial cost. To construct the decrease of the bound over time, we consider the sum over the subsystems S

$$\beta(r,0) = \sum_{s=1}^{S} \left(\frac{1}{3} \|x_s(k) - x_s^*\|^3 + \|x_s(k) - x_s^*\|^2 \left(k + \frac{1}{2} \right) + \|x_s(k) - x_s^*\| \left(\lambda + \frac{1}{6} \right) \right).$$

Now, we can use (9) to incorporate the time n in $\beta(r,n)$. The upper bound over time resembles the sequential steps of all cars. Hence, as time progresses, we get

$$\beta(r,n) = \sum_{s=1}^{S} \left(\sum_{k=0}^{\max\{\overline{M}-1-n,0\}} \ell(x_s(0),0) + \sum_{k=\max\{\overline{M}-\underline{M}-n,0\}}^{\max\{\overline{M}-1-n,0\}} \ell(x_s(k+n),u_s(k+n)) \right).$$

Again, the first inner sum describes those cars waiting at their position until they must start to reach their target. The second sum represents the costs over the interval during which the cars move towards their target. Dissolving the first sum reveals

$$\beta(r,n) = S \max\left\{\overline{M} - \underline{M} - n, 0\right\} \underline{M}^2 + S \sum_{k=\max\{\overline{M}-\underline{M}-n,0\}}^{\max\{\overline{M}-1-n,0\}} \ell(x_s(k),u_s(k)).$$

By shifting the index of the remaining sum to $k = 0$, we get

$$\begin{aligned}\beta(r,n) = {} & S \max\left\{\overline{M} - \underline{M} - n, 0\right\} \underline{M}^2 \\ & + S \sum_{k=0}^{\substack{\max\{\overline{M}-1-n,0\} \\ -\max\{\overline{M}-\underline{M}-n,0\}}} \ell\left(x_s\left(k + \max\left\{\overline{M} - \underline{M} - n, 0\right\}\right), u_s\left(k + \max\left\{\overline{M} - \underline{M} - n, 0\right\}\right)\right).\end{aligned}$$

As we combine the difference of the maxima of the upper limit, we get

$$\begin{aligned}\beta(r,n) = {} & S \max\left\{\overline{M} - \underline{M} - n, 0\right\} \underline{M}^2 \\ & + S \sum_{k=0}^{\underline{M}-1} \ell\left(x_s\left(k + \max\left\{\overline{M} - \underline{M} - n, 0\right\}\right), u_s\left(k + \max\left\{\overline{M} - \underline{M} - n, 0\right\}\right)\right).\end{aligned}$$

Resetting $k := k - n$, the latter expression simplifies to

$$\begin{aligned}\beta(r,n) = {} & S \max\left\{\overline{M} - \underline{M} - n, 0\right\} \underline{M}^2 \\ & + S \sum_{k=0}^{\underline{M}-1} \ell\left(x_s\left(k - n + \max\left\{\overline{M} - \underline{M}, n\right\}\right), u_s\left(k - n + \max\left\{\overline{M} - \underline{M}, n\right\}\right)\right).\end{aligned}$$

As $\overline{M} - \underline{M} \geq n$, we can eliminate the maximum terms in the index and get therefore

$$\begin{aligned}\beta(r,n) = {} & S \max\left\{\overline{M} - \underline{M} - n, 0\right\} \underline{M}^2 \\ & + S \sum_{k=0}^{\underline{M}-1} \ell\left(x_s\left(k - n + \overline{M} - \underline{M}\right), u_s\left(k - n + \overline{M} - \underline{M}\right)\right).\end{aligned} \tag{10}$$

and for $\overline{M} - \underline{M} \leq n$

$$\beta(r,n) = S \max\left\{\overline{M} - \underline{M} - n, 0\right\} \underline{M}^2 + S \sum_{k=0}^{\underline{M}-1} \ell(x_s(k),u_s(k)). \tag{11}$$

as distinction of both cases.

As $k+\overline{M}-\underline{M}-n \leq k+\overline{M}-\underline{M} \leq k+\overline{M}$ for case (10) we obtain

$$\beta(r,n) = S\max\left\{\overline{M}-\underline{M}-n,0\right\}\underline{M}^2 + S\sum_{k=0}^{\underline{M}-1}\left(k+\overline{M}\right)^2 + \lambda S1$$

and for the second case (11)

$$\beta(r,n) = S\max\left\{\overline{M}-\underline{M}-n,0\right\}\underline{M}^2 + S\sum_{k=0}^{\underline{M}-1}k^2 + \lambda S1$$

For the first case (10) we can assume that for the index $k+\overline{M}-\underline{M}-n$ the inequality $k \leq n-\overline{M}+\underline{M}$ and for the second case (11) $k \leq \underline{M}-1$ as given by the interval.

Therefore we can dissolve both cases by substitute with $j := \underline{M}+k$ and $\max\left\{0, n-\overline{M}+\underline{M}\right\} = \underline{M}-j$ to get $\underline{M}-1 = \underline{M}-j$ which leads to $j=1$ for the lower bound and we get

$$\beta(r,n) = S\max\left\{\overline{M}-\underline{M}-n,0\right\}\underline{M}^2 + \lambda S\underline{M} + \sum_{j=1}^{\underline{M}-\max\left\{0,n-\overline{M}+\underline{M}\right\}} j^2$$

Hence, by construction we have that $\beta(r,n) \in \mathcal{KL}$ and an upper bound of the overall costs satisfying the conditions of [3, Def. 2.14]. Utilizing the properties of the system and of $\underline{M}$, $\overline{M}$, the function $\beta(r,n)$ may largely overestimate the costs of the overall system. These properties allowed us to exclude (3) and avoid a collision detection as required in the closed loop. Thereby, we can conclude asymptotic stability of the overall closed loop system.

Within the analysis of the constructed feedback law, we are not only interested in stability of the closed loop, but also in the performance of the control. Since the bound $\beta(r,n)$ is a worst case bound, the performance estimate that can be derived from it is rather poor. Additionally, the values $\underline{M}$, $\overline{M}$ may be unknown for example in that case as we cannot predict the maximum wait time of one car which can prolong the path length. To circumvent these issues, we make use of the concept of relaxed Lyapunov arguments in the trajectory based setting. In particular, [3, Proposition 7.6] shows that if there exist a constant $\alpha \in (0,1]$ such that the relaxed Lyapunov inequality

$$V^N(x(n)) \geq \alpha\ell\left(x(n), \mu^N(x(n))\right) + V^N(x(n+1)) \tag{12}$$

holds for all $n \in \mathbb{N}_0$ and additionally there exist functions $\alpha_1, \alpha_2, \alpha_3 \in \mathcal{K}_\infty$ such that

$$\alpha_1(\|x(n)\|) \leq V^N(x(n)) \leq \alpha_2(\|x(n)\|), \tag{13}$$

$$\alpha_3(\|x(n)\|) \leq \ell\left(x(n), \mu^N(x(n))\right), \tag{14}$$

then the closed loop solution $x(n)$ behaves like a trajectory of an asymptotically stable system and the performance bound

$$J^\infty\left(x(0), \mu^N(x_s(0))\right) \leq V^N(x(0))/\alpha \leq V^\infty(x(0))/\alpha$$

holds.

Since we have shown the existence of a bound $\beta(r,n)$, we can apply converse Lyapunov results to guarantee the existence of α_1, α_2, $\alpha_3 \in \mathcal{K}_\infty$ such that (13), (14) hold, cf. [12]. This allows us to concentrate on the performance evaluation of the proposed control scheme via (12).

4. Experiments

For the implementation we use the discrete event simulator framework adevs [13] along with the Qt framework [14]. The cars are modelled as multi-agents objects, which include a controller instance, dynamical properties, constraints and a cost function with individual targets. The grid is implemented as an array containing all grid cells. Upon initialization, each car reserves a time and grid cell tuple in first-come-first-serve fashion. Hence, if a car enters the scenario at a later time instant, its initial time will also be later. To retain a unique order for the reservation process, the cars are indexed. For the optimization step within the MPC, we apply the COBYLA Algorithm (**C**onstrained **O**ptimization **BY** **L**inear **A**pproximations) developed by Powell [15] provided by the NLOpt library [16]. Minimizing the cost function including the individual target of the car implicitly defines the direction and route of the car. The model of the middle lane is given as a set of constraints which are defined for each car depending on its start position. Both the path and control values x_s, u_s are mapped from tuples to absolute values of the cell grid. In particular, we map the cell coordinate tuples (n, k, m) to global scalars w via

$$\vec{w} = (w_1, \ldots, w_N), w_i \in \{0, \ldots, L_k L_m - 1\}$$

with

$$w_i = k_i + m_i L_k.$$

If a car reaches its target, the car is removed from the scenario and is not considered in further calculations. We set up a standard intersection scenario with a 4×4-intersection grid and 20 cars starting in the corners. To guarantee a connected path, we imposed a penalty term η. Here, a path is connected if each pair of consecutive path items are neighbors in the grid. The penalty is defined as

$$\eta_s = Bv,$$

where v is the number of not connected pairs of consecutive path items and B a constant.

To analyze the performance of the controller, we utilize the minimal and maximal path lengths

$$p_{\min} = \min_s \#x_s \geq \underline{M}, \quad p_{\max} = \max_s \#x_s \leq \overline{M},$$

the mean path length

$$\bar{p} = \frac{1}{S} \sum_{s=0}^{S} \#x_s.$$

and the suboptimality bound α from (12). We evaluate the bound α as the minimum over all time steps $n \in \mathbb{N}_0$ and compute the mean over all systems $x_1, \ldots, x_S$ via

$$\alpha_n = \frac{\sum_{i=1}^{S} \min\left(\alpha_{n-1}, \frac{V_{n-1} - V_n}{\ell_{n-1}}\right)}{S} \quad \text{with } \alpha_{-1} = 1.$$

For α we used the value in the last step named here α_n and $\overline{\alpha}$ as the mean over all time steps.

For our experiments, we considered the parameters $\lambda = 0.2$ and $B = 100$ and a horizon length of $N = 3$.

We set up configurations with different quantities of cars and the intersection size. Pursuing a realistic approach we consider a 4×4-intersection scenario, i.e. two lanes for each direction,

Table 1. Experimental results for test scenarios, α_n as last time step, $\overline{\alpha}$ as mean value over all time steps in the simulation

$\#S$	Grid	N	$p_{\min}$	$p_{\max}$	$\bar{p}$	α_n	$\overline{\alpha}$
20	4×4	3	4	11	6.65	0.25	0.2155
30	4×4	3	4	10	6.2	0.27	0.2707
20	6×6	3	5	18	7.25	0.0019	0.1005
30	6×6	3	5	13	6.76	0.00013	0.1639
20	6×6	4	5	11	6.55	0.00587	0.43
30	6×6	4	5	16	6.8	0.00588	0.3587

and a 6×6-intersection scenario reflecting 3 lanes for each direction. In Table 1, we present our numerical results for both scenarios with 20 and 30 cars for each direction.

In the 4×4-intersection we observe a slightly larger mean path length $\bar{p}$ for 30 cars than for 20 cars. This due to higher reservation times for the direct way from the start to the target cell of a vehicle as more cars try to reserve the direct way. With the growing reservation time values for the direct lane, the minimum of cost function is attained using the second lane. Hence, cars entering the intersection at a later time step choose the second lane. Reflecting this situation we can argue that a certain amount of cars is necessary that the cost function gives an advantage over distance for a shorter time delay. For α we used the value in the last step named here mentioned α_n and $\overline{\alpha}$ as the mean over all time steps.

For the 6×6-scenario, we similarly observe that the mean path length $\bar{p}$ is smaller for the configuration of 30 cars than with 20 cars. The effect is also increased with the greater intersection dimensions. Hence, we can conclude that the influence of the higher distances for the cars leads to the greater gap in $\bar{p}$. Additionally, we considered different horizon lengths and observed an decrease in $\bar{p}$ for larger horizon for 20 cars. For the scenario with 30 cars $\bar{p}$ changes only slightly. Again, the reservation of more cells in the optimization has a large impact on the difference of $\bar{p}$ for 20 cars with greater horizon length. The time slots which are reserved by the cars reveal later times for late arriving cars. Hence, the cost function leads to an earlier change to the second lane. Yet, we also observed an improvement in the suboptimality index for both scenarios which improves greatly for the larger horizon length.

In the following figures, we give some graphs for the overall cost function according to Table 1. We included $\beta(r, n)$ as a monotonously decreasing function representing the upper bound for our cost function. For longer horizons illustrated in Figure 6 and 7, we observe that the cost function tends to zero more slowly. In this case, the control impact is less aggressive compared to a shorter horizon, which is a known typical effect of using MPC. In the figures 4-7 a slightly increase of the costs occurs in one step. As mentioned in the 4x4-evaluation, this is caused by changing to the second lane of the later following cars. As the cost function returns lower costs due to earlier available reservation times for the second lane over the optimization horizon for one car than reserving later reservation times in the crowded cells, the applied optimal control $u(0)$ increases the costs for the changing cars slightly. Still, the costs are decreasing and remain bounded by $\beta(r, n)$, which indicates that the solutions show stable behavior.

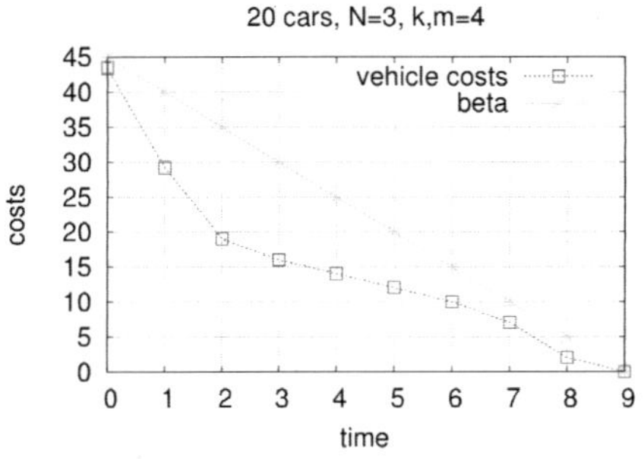

Figure 2. Overall costs for 20 cars, N=3, k,m=4

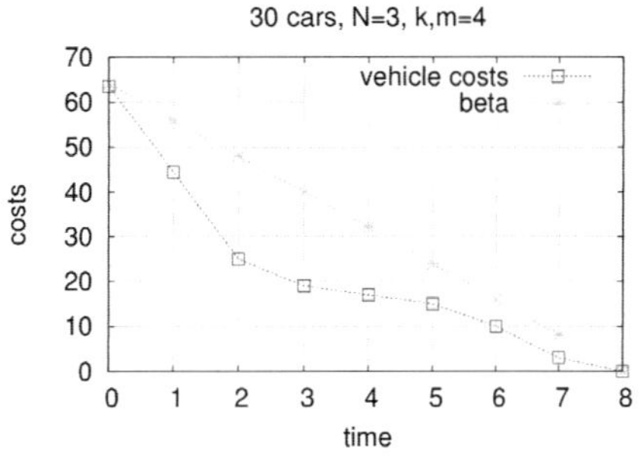

Figure 3. Overall costs for 30 cars, N=3, k,m=4

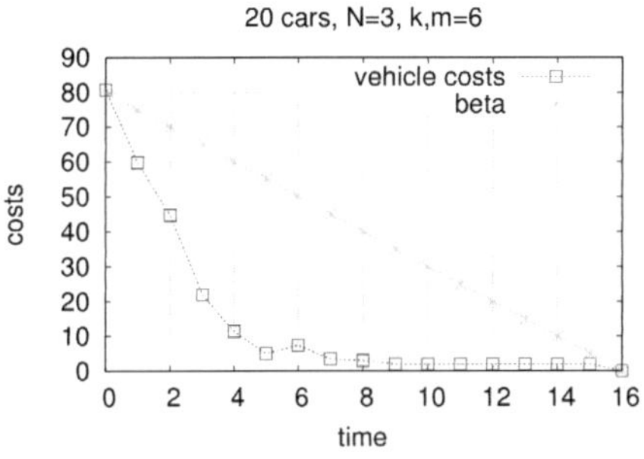

Figure 4. Overall costs for 20 cars, N=3, k,m=6

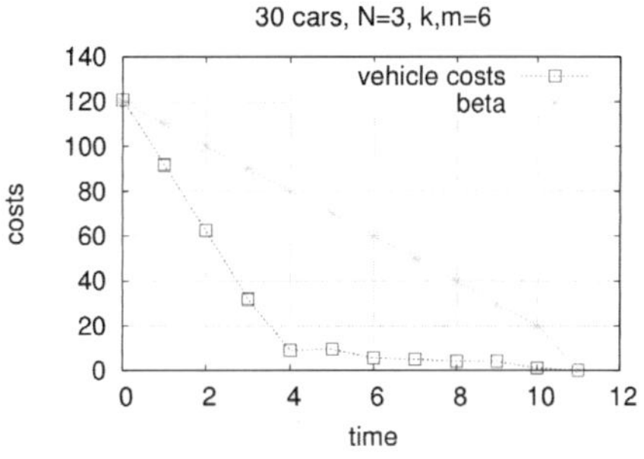

Figure 5. Overall costs for 30 cars, N=3, k,m=6

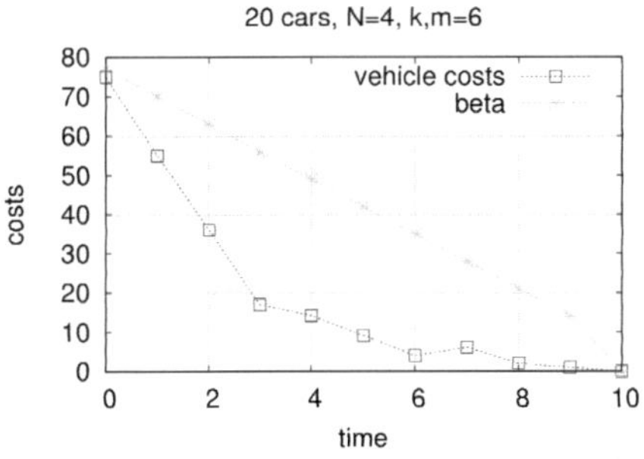

Figure 6. Overall costs for 20 cars, N=4, k,m=6

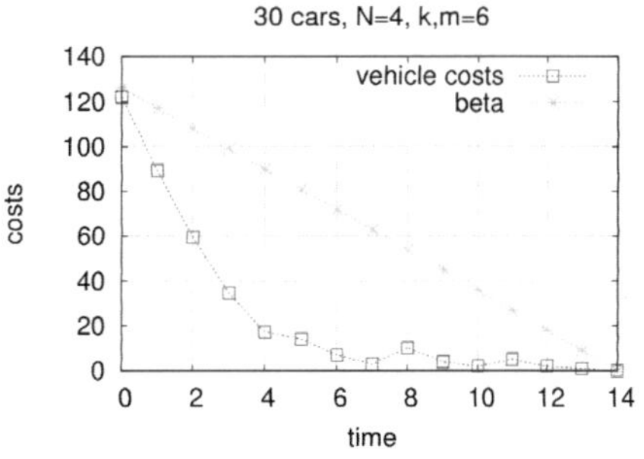

Figure 7. Overall costs for 30 cars, N=4, k,m=6

5. Conclusion

In this paper we used an intersection scenario as a discrete shared space grid. We formulated this scenario as a Distributed Model Predictive Control Problem in Section 2 where each car modelled as an multi-agent optimizes its own path. In Section 3 we have proven the stability for the overall system by deriving a class $\mathcal{KL}$ boundary function. Our numerical results showed that the suboptimality index α known from relaxed Lyapunov theory remains positive. Additionally, the cumulated costs are decreasing in every time step and satisfy our computed bound, which indicates stable behavior of the system.

In further research we will consider an exponential term depending on time to privilege long waiting cars. Converting the cost function into a continuous one will also extend the suitability for continuous optimization algorithms. Additionally, we plan to incorporate the risk of deadlocks by allowing any direction for the lanes to increase the throughput for the overall traffic. Last, by including path predictions of non-communicating participants we will improve practicability of the coordination.

Acknowledgment
We thank Prof. C. Müller-Schloer from the Leibniz University of Hannover for a fruitful discussion on this topic.

References

[1] Qin S J and Badgwell T A 2003 A survey of industrial model predictive control technology *Control Engineering Practice* **11** 733–764
[2] Rawlings J B and Mayne D Q 2009 *Model Predictive Control: Theory and Design* (Nob Hill Pub.)
[3] Grüne L and Pannek J 2011 *Nonlinear Model Predictive Control: Theory and Algorithms* Communications and Control Engineering (London: Springer)
[4] Keerthi S S and Gilbert E G 1988 Optimal Infinite Horizon Feedback Laws for a General Class of Constrained Discrete-Time Systems: Stability and Moving Horizon Approximations *Journal of Optimization Theory and Applications* **57** 265–293
[5] Chen H and Allgöwer F 1998 A Quasi-Infinite Horizon Nonlinear Model Predictive Control Scheme with Guaranteed Stability *Automatica* **34** 1205–1217
[6] Grüne L 2012 NMPC without terminal constraints *IFAC Proceedings Volumes (IFAC-PapersOnline)* **4** 1–13
[7] Richards A and How J 2003 Model predictive control of vehicle maneuvers with guaranteed completion time and robust feasibility *Proceedings of the 2003 American Control Conference, 2003.* **5** 4034–40
[8] Chaaban Y, Hähner J and Müller-Schloer C 2009 Towards Fault-Tolerant Robust Self-Organizing Multi-agent Systems in Intersections without Traffic Lights *2009 Computation World: Future Computing, Service Computation, Cognitive, Adaptive, Content, Patterns* 467–475
[9] Richter U and Mnif M 2006 Towards a generic observer / controller architecture for Organic Computing *GI Jahrestagung* **1** 112–119
[10] Hart P E, Nilsson N J and Raphael B 1968 A Formal Basis for the Heuristic Determination of Minimum Cost Paths *IEEE Transactions of systems science and cybernetics* 100–107
[11] Makarem L and Gillet D 2013 Model predictive coordination of autonomous vehicles crossing intersections *IEEE Conference on Intelligent Transportation Systems, Proceedings, ITSC* 1799–04
[12] Jiang Z P and Wang Y 2002 A converse Lyapunov theorem for discrete-time systems with disturbances *Systems and Control Letters* **45** 49–58
[13] Nutaro J 2013 adevs (A Discrete Event Simulator) library http://web.ornl.gov/˜1qn/adevs/
[14] Digia Qt Framework
[15] Powell M J D 1998 Direct search algorithms for optimization calculations *Acta Numerica* **7** 287–336
[16] Johnson S The {NLopt} nonlinear-optimization package

Journal of Physics: Conference Series **659** (2015) 012050 doi:10.1088/1742-6596/659/1/012050

Event-based and Multi Agent Control of an Innovative Wheelchair

U Diomin[1], P Witczak[2] and R Stetter[3]

[1] Brest State Technical University, Brest, Belarus
[2] University of Zielona Gora, Zielona Gora, Poland
[3] Hochschule Ravensburg-Weingarten, Weingarten, Germany

E-mail: stetter@hs-weingarten.de

Abstract. Due to the aging population more and more people require mobility assistance in form of a wheelchair. Generally it would be desirable that such wheelchairs would be easy to use and would allow their users the possibility to move in any direction at any time. Concepts which allow such movements are existing since many years but have for several reasons not found their way to the market. Additionally for semi-autonomous (assisted) operation and fully autonomous operation (e. g. an empty wheelchair driving to its charging station) the control task is much less challenging for such drive system, because no complex manoeuvres needs to be considered and planned. In an ongoing research a drive system for a wheelchair was developed which offers such possibilities employing a relatively simple mechanical design. This drive system is based on a certain steering principle which is based on torque differences between different wheels. This allows a relatively simple mechanical design but poses challenges on the control of the vehicle. This paper describes two possible approaches to address this challenge – the use of an event based control and the application of multiple software agents. Both approaches can solve the control problem individually but can also complement each other for better system performance. The paper stars with a description of the wheelchair drive system. Then the asynchronous event based control software is described as well the multi agent based approach. The next sections report the results of the experiments and discuss the further improvements.

1. Semi-autonomous Wheelchair

Wheelchairs are currently provided mainly for handicapped persons however, the rapid growth in the elderly population suggests that the numbers of electric wheelchair users will soon increase dramatically ([1], [2]). In general, electric wheelchairs are difficult to maneuver especially for elderly people who have little experience using a joystick to operate a driven wheel system. Current wheelchairs need a complex series of movements resembling parallel automobile parking when he or she wants to move sideways.

A major achievement for any vehicle or mobile robot is the ability to move freely in any direction and performing rotation at any arbitrary position and orientation. This can be realized by obtaining a locomotion system with the ability to perform a holonomic and omnidirectional motion [3]. Both

Journal of Physics: Conference Series **659** (2015) 012050 doi:10.1088/1742-6596/659/1/012050

abilities are necessary for realizing better maneuver in a crowded area or restricted space. A non-holonomic omnidirectional mechanism requires preliminary maneuvering for reorientation of steering wheel which may affect the time-travel cost and the spaces needed. The holonomic omnidirectional mechanism has better mobility and also less complexity in design compared with a non-holonomic omnidirectional mechanism.

During the last decades several approaches towards better maneuverability, usually in the form of omnidirectional mobile system were presented; Wang provides a good overview [1]. Notable is additionally the ongoing long-term research project “Omni” concerning similar systems [4]. In spite of the fact that several systems were proposed none are today available on the market. One of the causes of this can be the high complexity of the solutions. This paper presents are rather simple solution for a holonomic omnidirectional drive train for a wheelchair. The simplicity of the design is possible, because the steering is based on torque differences.

1.1. Torque steering system

Autonomous vehicles such as mobile robots and the respective steering systems have been successfully developed and built for some years (Anderson&Jones [5], Ashmore&Barnes [6], Dillard [7]). The distinctive quality of the steering system used is the dynamic behaviour. This innovative steering system that is already registered as a patent ([8], [9]) is based on the concept to use the torque of drive motors (more exactly the torque differences between wheels) to steer four independent axes of a vehicle. This system can be used for production vehicles [10], cars [11] and wheelchairs, as well. The principal steering system is shown in Figure 1.

In this example a vehicle consists of four drive motors which are fastened on arms that may freely rotate. These arms have no drive or brake, only an angle encoder is attached at the end of each axle. These angle encoders measure the angle of the motor and the wheel with regard to the vehicle platform. The distinct characteristic of the innovative drive system is the absence of dedicated steering motors. By means of angle encoders applied at the four steering axes and highly dynamic control algorithms it is possible to steer such vehicles only by means of the four drive motors (compare Figure 1).

Each of the wheels on the short axle can be directed into the desired position by means of the torque applied on the wheel. This could take place sequentially for each individual wheel but also simultaneously, if the control allows different torque on all wheels. This characteristic allows simpler and simultaneously more robust vehicle concepts. It is also a main advantage of this concept that the resulting vehicle is able to drive directly in any direction without time and space consuming turning maneuvers.

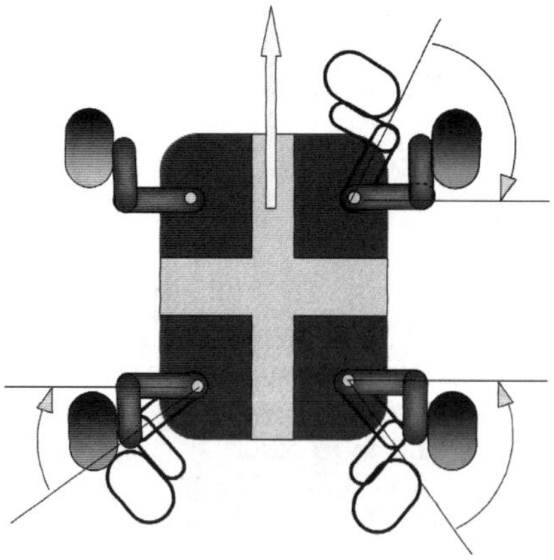

Figure 1. Principal steering system.

Furthermore, a vehicle based on the dynamic drive system is able to turn around its own centre. This characteristic is very important if cameras or other equipment are mounted on such vehicles which can only be used in a certain orientation. The innovative steering system shares these advantages with Omni drive systems (Ashmore&Barnes [6]), but has reduced friction as well as easier controllability and offers the possibility to determine an exact position and orientation from an analysis of the angles of the steering axes and the angles of the drive wheels (odometry).

1.2. Realization of Wheelchair

The wheelchair was designed, developed and realised in several steps. The drive system firstly consists of two front wheels which are each driven by an individual drive motor. The centre of these wheels is outside the vertical rotation axis, therefore torque steering is possible. On this axis a rotary encoder is fixed thus allowing a closed loop control of the steering angle of each front wheel. Additionally, a brake is present on both axes, which allows to brake the respective axis and by this to maintain a constant steering angle. This is advantageous for situations with higher driving speeds and if some obstacles as well as rough or slippery surfaces are present. These brakes are engaged (the angle is maintained) if there is no electrical current; they are powered only when steering angles are desired. The drive system secondly disposes of a third wheel in the back, which can be actively steered by means of a dedicated steering motor (the back wheel is driven by a second designated drive motor). The steering motor also disposes of an encoder functionality thus allowing a closed loop control of the back steering angle. Between the motor and the axis a self-retaining gear system is placed, which ensures a safe behavior also under dynamic operation conditions. The arrangement of the drive train is visible in Figure 2.

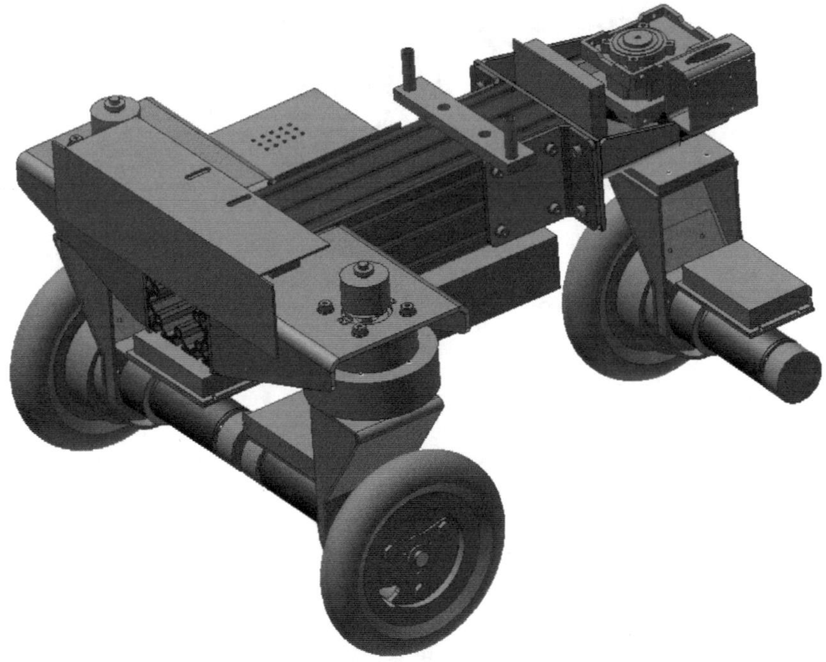

Figure 2. Arrangement of the drive train.

Generally, two driving modes can be realized. The first driving mode, which can be used for fast driving maintains a straight driving angle for the two front wheels and realizes the steering only through steering of the back wheel and speed differences between the front wheels. This mode allows fast driving, but does not allow, for instance, sideway driving. A second driving mode also relies on the torque steering capabilities of the two front wheels and allows unlimited maneuvering. Figure 3 shows the assembled wheelchair with all equipment such as main computer and battery.

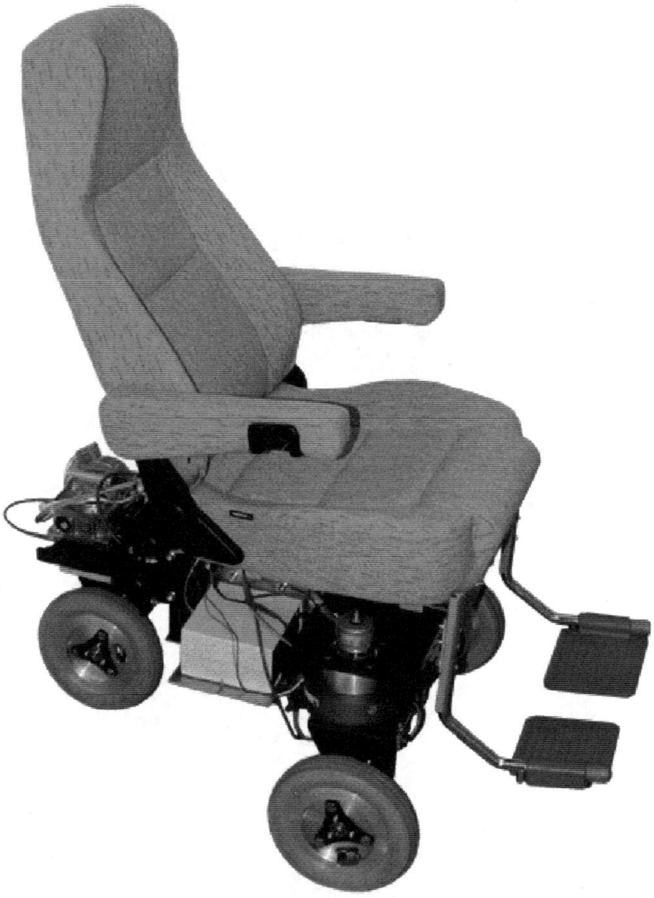

Figure 3. Assembled wheelchair.

2. Asynchronous event-based control software

In the course of the project a mathematical model of the wheelchair was derived. On the basis of this model, the control paradigms can be implemented. In order to maintain the real-time behavior of the system as well as safety of the driver and surroundings, a proper software needs to be done. The modules can be treated as separate agents with own control procedures, i.e. PID control of the wheel velocity and angular module position. Thus the main unit should handle data acquisition, control law reconfiguration and subsequently update respective agents inputs. To achieve this with relatively low computational cost, and to prevent misuse of power, the event-based control was implemented. Its idea bases on the discrete events broadcasted through the system entities (agents). Each event consists of unique identifier, creation time and the necessary data, e.g. new reference velocity, reference angle or display data. All agents are represented virtually within the main framework and as such can control the expiration time of events they handle. Each agent is spawned as thread (or threads).Thus the system emulates real-time environment, i.e. if time lapses the new event is required or buffered value is used, delays between actions are negligible as they are much smaller than the overall inertia of the system.). Due to the fact that events are broadcasted each agent is notified when events occurs but does not need to handle it. In other words - in the given time interval each agent works in the closed-loop until the event occurs, when they accept the input and follow new control law, until the reference point is reached or changed by the precedent controller.

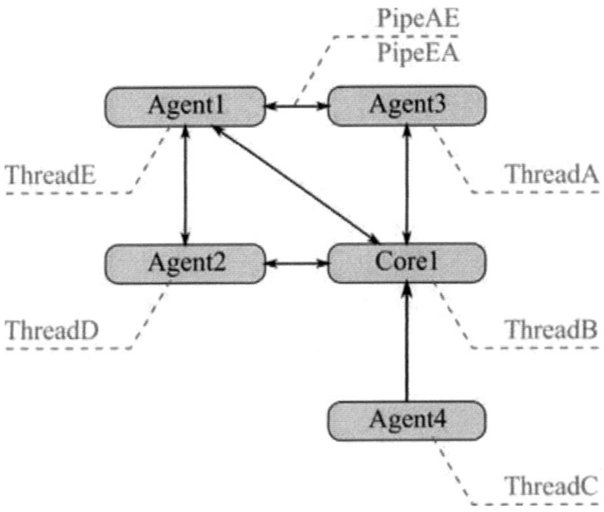

Figure 4. Thread communication scheme.

The on-board CPU and operating system handles priorities of threads, however the quality and computational power of industrial pc used is more than enough to respect the real-time schedule of the system itself. Additionally the computational unit is equipped with the software and hardware watchdogs (software faults detectors) that can restart particular threads or the entire program in order to avoid wheelchair failures due to CPU-related issues. The overall construction of software enables fast and reliable prototyping and changes both in mechanical design and control scheme.

Figure 4 presents the communication scheme between the agents represented by threads. Each Agent in depicted case is represented by a single thread. Events are send and received by unidirectional pipes(e.g. one from ThreadA to ThreadE, and second vice versa). Some agents can exchange the data, but some works only as sensor, collecting and broadcasting data (Agent4, ThreadC). Pipes collect the events that are an object of interest of the represented agent, and send them to the buffer. First-in first-out queue is then resolved internally in each thread. Obvious advantage of the solution appears in the multi-core architecture. The unavoidable disadvantage is performance of the hardware, which limits the real-time appearance of the solution. At the same time due to the inertia of the system and operator real-time can be perceived as fast-enough.

Note that in this case real-time system should not be perceived as defined in i.e. [12][13], so not like a synchronous system that requires quantitative expression of time to describe the behavior of the system. Instead it is asynchronous, event-driven with buffers and variables expiration times. Therefore a task can be performed faster than assumed, and in case of data unavailability the closest sample (i.e. from buffer) is used to preserve functioning of the system. Thus making it possibly faster, and not-less reliable than the conventional real-time systems.

3. Improving multi-agent control technique

The next sections describe the improved control technique based on the multi-agent control approach. In previous researches the multi-agent approach mixed with reinforcement learning (RL) was shown. The approach improves the control of the mobile robot based on complicated steering scheme [14, 15].

The control was done either for simulation or for real robot hardware. The main problem of such software transfer is that real hardware is not ready to be controlled by agents and needs a lot of development to use it.

To describe multi-agent systems one needs to start from the agent definition. An **agent** is an instance that can be viewed as one that perceives its **environment** through the **sensors** and acts upon that environment through the **actuators** [16]. In multi-agent systems the most important receive/perceive ability of the agent is the communication ability. The agents of the system should send messages and answer to each other. This is the main criterion to create multi-agency.

The second important criterion is the intelligence of the agent. An **intelligent** or **rational** agent for each possible percept sequence should select an action that is expected to maximize its performance measure, given the evidence provided by the percept sequence and whatever built-in knowledge the agent has [16]. Our intelligent agents contain reinforcement learning decision-making system. In reinforcement learning agents improve their behavior by exploration of the environment. To become really intelligent agent should make a lot of *epochs* (couple of iterations). Each epoch the agent starts with a random state and explores the world by making actions. Every action of the agent returns a feedback from the environment as a reward. In this case the evaluation of the reward will be: "the bigger the reward value the better". The reward value is saved by the agent to the knowledge base using one of the reinforcement learning rules. This knowledge base is used to help in making choice of the next action in the next epochs. The goal of the agent not to only to use maximum valuable steps from the base but also to investigate not visited ones. In our approach we used the improved multi-agent reinforcement learning algorithm with *sharing state*. The description and the framework will be revealed at the end of this chapter.

As it was mentioned above the wheelchair provides two driving modes: *fast mode* as common driving one and the *unlimited maneuvering mode* used when the wheelchair needs to drive along complicated trajectories or in the limited space. So the multi-agent system was developed for working in these two modes.

The asynchronous event-based control framework provides the *control system* of high quality and moreover it can be used as the abstraction layer for control system based on multi-agent approach. The designed multi-agent architecture consists of four agents where three of them are the representation of each of driving modules and the fourth agent is the *virtual head agent*. The *virtual head agent* is the agent that achieves the desired movements of the wheelchair user and sends the cooperating commands to the agents. The decomposition of the wheelchair is shown on the Figure 5.

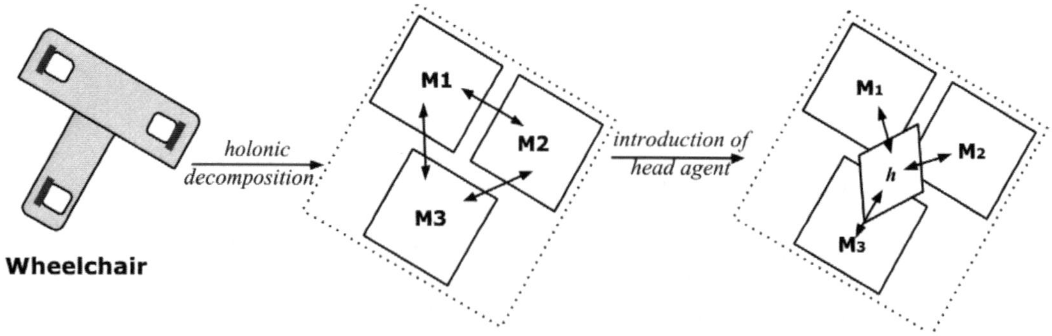

Figure 5. Wheelchair decomposition to multi-agent system with head agent.

The general architecture after decomposition step for multi-agent control system with head agent is shown on Figure 6. Every module provides control $P = \{p_i | i = 1..n\}$ of n input parameters. For every control parameter the agent is using policy π_i and after every step achieves the reward R_i from environment. The Environment/Planning subsystem provides information about the desired speed of the platform v^d and the global state of the agents $S = \{U_{i=1}^{N} s_i\} \cup \{s_h\}$, where $s_i \in s$ is the state of the i-th module and s_h is the state of the head agent.

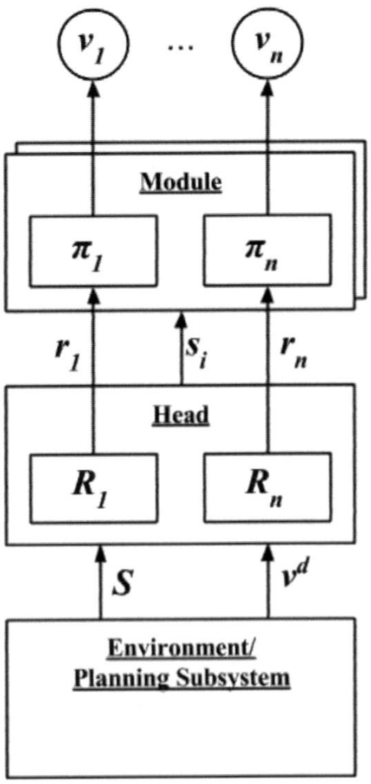

Figure 6. The architecture of the control system.

The intelligence of the agents is made by reinforcement learning approach that is extended to the multi-agent system. The standard reinforcement learning approaches are not suitable for the multi-agent systems. The curse of dimension becomes a huge problem when one needs to deal with more than one agent. In the presented approach an alternative way to learn many agents in one time was applied. The approach uses sharing of the agents' state. The sharing procedure is provided by head agent that sends synchronization command to the other agents. During the process of synchronization agents update their knowledge base due to the Q-Learning reinforcement learning rule. Q-Learning is an off-policy temporal-difference control algorithm [17]. Q-learning is a model-free reinforcement learning technique, which relies on learning an action-value function that ultimately gives the expected utility of taking a given action in a given state and following the optimal policy thereafter.

While reinforcement learning the ith agent executes an action a_i at the current state s_i. Then it goes to the next state s'_i and receives a numerical reward r as the feedback for the recent action [17], where $s_i, s_i' \in s$, $a_i' \in a$, $r \in R$. Ideally agents should explore state space (interact with environment) to build an optimal policy π^*.

Let $Q(s, a)$ – represent a Q-function that reflects the quality of the specified action a in state s. The optimal policy can be expressed in the terms of optimal Q-function Q^*:

$$\pi^*(s) = \text{argmax}_{a\in\alpha(s)} Q^*(s,a) \tag{1}$$

The initial values of Q-functions are unknown and equal to zero. The learning goal is to approximate the Q-function, (e.g. to find true Q-values for each action in every state using received sequences of rewards). The RL framework with *shared state* is shown on Figure 7.

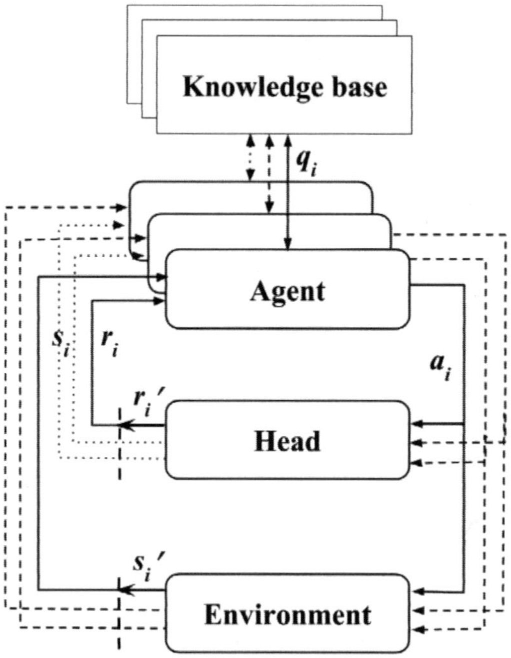

Figure 7. Reinforcement learning framework with shared state.

4. Simulation results

This section describes the results of simulations done with the wheelchair model. The experiments were divided into two parts: *the learning part* and *the driving part* that will be described sequentially.

4.1. The learning part

At this part agents learn how to work in two provided modes. Every driving agent has the driving scheme that is shown at Figure 8. Two agents at the front of the wheelchair have only one control parameter v – the velocity of the wheel. According to that we learn the agent with one output parameter. The third agent that is placed at the back of the wheelchair has two output parameters: wheel velocity v and the angle relative to the wheelchair φ. The full state of the agents and their possible actions are described in Table 1. The error of the wheelchair trajectory is an important parameter that is used as reward for the agents. The agents are learnt in team to drive in two modes.

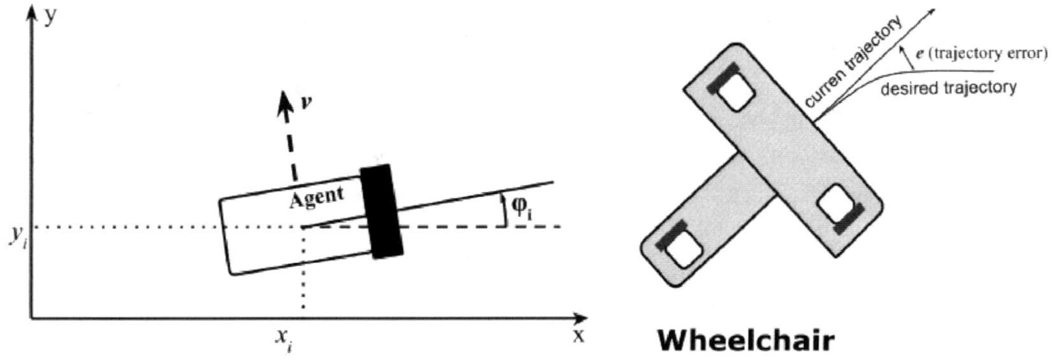

a) b)

Figure 8. a) The graphical representation of the agent-module. b) The wheelchair trajectory error calculation

Table 1. Reinforcement learning parameters of the agents

Agent	Controlled parameters	Actions	The state parameters
Front Agent	v – wheel velocity	$+v$, $-v$: increase and decrease the velocity	v – current velocity d – desired trajectory of the wheelchair
Back Agent	v – wheel velocity φ – rotation angle	$+v$, $-v$: increase and decrease the velocity; $+\varphi$, $-\varphi$: increase and decrease the rotation angle;	v – current velocity $+\varphi$, $-\varphi$: increase and decrease the rotation angle; d – trajectory error of the wheelchair
Head agent (virtual)	k – set the knowledge base of the agents accordingly to the mode d – set the desired trajectory for the agents	k_1, k_2 : set one of the driving modes	m – the current driving mode that set by user t – trajectory from the user

4.2. The driving part

To test driving abilities of the wheelchair model we create an artificial maze environment by Robotics Operation System. The Gazebo 2.0 was used as a simulator [18]. The simulation environment is shown in Figure 9.

Journal of Physics: Conference Series **659** (2015) 012050 doi:10.1088/1742-6596/659/1/012050

Figure 9. The simulation environment with the base of the wheelchair

For the virtual wheelchair three tasks were defined in order to allow testing. Figure 10 shows the prepared trajectories for the model. The tasks descriptions are:

a) Driving along the straight line in mode one.
b) Driving along the curve line in mode one.
c) Driving along the straight line in mode two,

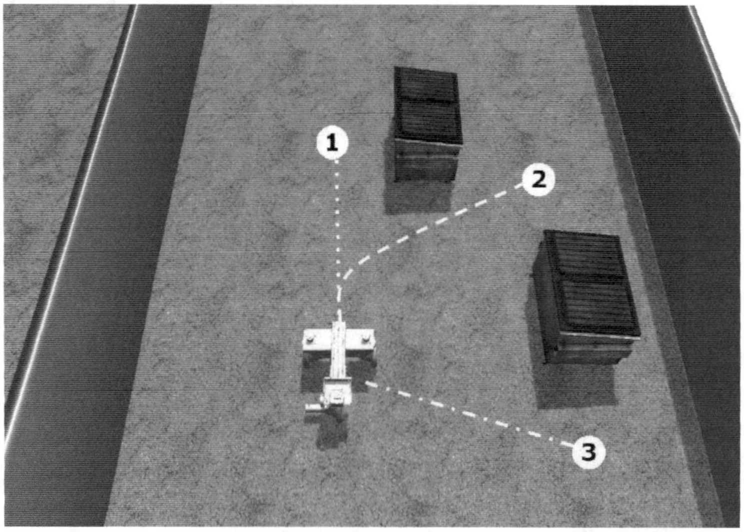

Figure 10. The trajectories of the tasks in driving part of experiments.

The *driving experiments* results are represented in Table 2. This table consists of the median results that were collected from 20 experiments for each of the tasks.

Table 2. The median of the experiments results in driving task

Task	Velocity, *seconds*	Trajectory error, %	Goal error, %
Task 1	1,43	0,8	0,3
Task 2	2,11	1,54	0,3
Task 3	1,89	1,1	0,3

This results are the proof of the quality of the control system, based on multi-agent reinforcement learning architecture. The advantages of this method are *adaptability* – using reinforcement learning the multi-agent system adapts to control the wheelchair platform through the commands of the head agent; *sharing state* of the agents *decreases the learning time and increases the provided quality* of the already learnt system.

In further experiments it is planned to use the real hardware of the robot to provide full comparison of the two control approaches:

- Asynchronous event-based control software.
- Combination of the asynchronous event-based control software and reinforcement learning multi-agent approach.

5. Summary and Outlook

The control approaches described in this paper were developed for and tested on a holonomic, omnidirectional wheelchair. Such wheelchairs are easier to use for elderly people but also beneficial for semi-autonomous and autonomous operations. The application of an event based control lead to promising results. It is important to note that not a synchronous real-time system was realized, but an asynchronous, event-driven system with buffers and variables expiration times. Therefore a task can be performed faster than assumed, and in case of data unavailability the closest sample (i.e. from buffer) is used to preserve functioning of the system. The use of a multi agent based approach consisted of a learning part with two different modes and a simulation in an artificial maze environment. The simulation was aimed at testing the driving abilities of the wheelchair model. The results of the simulation proofed of the quality of the control system, which is based on multi-agent reinforcement learning architecture. As mentioned above, both approaches can solve the control problem individually but can also complement each other for better system performance. Obviously up to know the results were only verified on one single drive train configuration. More research is needed to underline the findings of the research.

References

[1] Wada M: Motion Control of a Four-wheel-drive Omnidirectional Wheelchair with High Step Climbing Capability. In: Behnam Miripour (Ed.): Climbing and Walking Robots, 2010.

[2] M. Wada (2007), Holonomic and Omnidirectional Wheelchairs with synchronized 4WD Mechanism, Proceedings of the 2007 IEEE/RSJ International Conference on Intelligent Robots and Systems (IROS2007), pp. 1196-1202.

[3] Bin Safar, M. J. A.: Tip-over Stability Analysis and Enhancement for Omnidirectional Personal Mobility Robots with Active Dual-wheel Caster Assemblies. Ph. D. Thesis, Okayama University, 2014.

[4] Borgolte, U.; Hoyer, H.; Bühler, C.; Heck, H.; Hoelper, R.: Architectural Concepts of a Semi-autonomous Wheelchair. Journal of Intelligent and Robotic Systems 22. Kluwer Academic Publishers, 1998; pp. 233–253.

[5] Anderson, K. Jones, C.: “Autonomous Robotic Vehicle Design”, Proceedings of the Southwest Regional ASEE Conference, ASEE 2005.

[6] Ashmore, M., Barnes, N.: “Omni-drive robot motion on curved paths: The fastest path between two points is not a straight line", Proceedings of the Australian Joint Conference on Artificial Intelligence, Dec., 2002, pp. 225-236.

[7] Dillard, W.: "Revisting the Autonomous Robot. Finding the Engineer within the Student", Proceedings of the 2004 AEE Conf. , Salt Lake City, 2004.

[8] Patent „Fahrzeug" Aktenzeichen DE 10 2005 057 635 A1, Deutsches Patent- und Markenamt, München.

[9] Patentanmeldung „Fahrzeug" Aktenzeichen EP 1 954 516 A1 und WO 2007/062858 A1, Europäisches Patentamt, München und World Intellectual Property Organization, Genf, Schweiz.

[10] Seybold, L., Krokowicz, J., Pieczyński, A., Paczynski, A, Stetter R.: Prinzipien für das Monitoring, die Planung, Regelung und Diagnose von fahrerlosen Transportsystemen. In: Automation 2011. VDI-Berichte 2143. Düsseldorf: VDI, 2011 (S. 137 – 140).

[11] Stetter, R., Bertsch, S.; Eckart, P.; Paczynski, A.: Torque Steering System for Electrical and Hybrid Power Trains. In: Proceedings of the Fisita EAEC Conference 2011 in Valencia, Spain, 2011.

[12] Albert M. K. Cheng; Real-Time Systems: Scheduling, Analysis, and Verification, ohn Wiley & Sons, 2003

[13] Rajib Mall; Real-Time Systems: Theory and Practice. Pearson Education India, 2009

[14] Dziomin, U., Kabysh, A., Golovko, V., & Stetter, R. (2013, September). A multi-agent reinforcement learning approach for the efficient control of mobile robot. In Intelligent Data Acquisition and Advanced Computing Systems (IDAACS), 2013 IEEE 7th International Conference on (Vol. 2, pp. 867-873). IEEE.

[15] Dziomin, U., Kabysh, A., Golovko, V., & Stetter, R. (2014). A Multi-agent Efficient Control System for a Production Mobile Robot. In Neural Networks and Artificial Intelligence (pp. 171-181). Springer International Publishing.

[16] Russell, S., & Norvig, P. (1995). Artificial intelligence: a modern approach.Asgd

[17] Sutton, R.; Barto, A.G.: Reinforcement learning: An introduction. MIT press, 1998.

[18] Koenig, Nathan, and Andrew Howard.: Design and use paradigms for gazebo, an open-source multi-robot simulator. Intelligent Robots and Systems, 2004.(IROS 2004). Proceedings. 2004 IEEE/RSJ International Conference on. Vol. 3. IEEE, 2004.

Journal of Physics: Conference Series **659** (2015) 012051 doi:10.1088/1742-6596/659/1/012051

Diagnostic methods of a bladed disc mode shape evaluation used for shrouded blades in steam turbines

Jaromir Strnad, Jindrich Liska

University of West Bohemia, NTIS - European Centre of Excellence, Pilsen, Univerzitni 8, 30614 Czech Republic

E-mail: `jastrnad@ntis.zcu.cz`, `jinliska@ntis.zcu.cz`

Abstract. This paper deals with advanced methods for the evaluation of a bladed disc behavior in terms of the wheel vibration and blade service time consumption. These methods are developed as parts of the noncontact vibration monitoring system of the steam turbine shrouded blades. The proposed methods utilize the time-frequency processing (cross spectra) and the method using least squares to analyse the data from the optical and magnetoresistive sensors, which are mounted in the stator radially above the rotor blades. Fundamentally, the blade vibrations are detected during the blade passages under the sensors and the following signal processing, which covers also the proposed methods, leads to the estimation of the blade residual service life. The prototype system implementing above mentioned techniques was installed into the last stage of the new steam turbine (LP part). The methods for bladed disc mode shape evaluation were successfully verified on the signals, which were obtained during the commission operation of the turbine.

1. Introduction

Any mechanical stress of rotating blades reduces their service life. A damage of the blades causes changes in their frequency characteristics, frequency and intensity of the excitation forces, which are measured using contact or contactless method. The contact measurement by strain gauges gives the direct information about the blades stress during a whole revolution [1]. This method is unsuitable for long-term measurements, because a sensor service life is short in a corrosive environment. Furthermore, it is very difficult and expensive to monitor all blades of the bladed disc. The contactless method BTT (blade tip-timing) is based on the measurements with the sensors that are mounted radially on the circumference of the stator above the rotor blades [2]. The times of blade tip passages under the sensors are analyzed. Changes of the frequency characteristics, which are associated with their normal wear and tear, are reflected very slowly. We can not measure these changes using conventional methods. In contrast, the immediate blade damages are manifested by the rapid changes, which enforce the immediate shutdown and repair of the turbine.

It is important to monitor the blades frequency characteristics, during the turbine runup, rundown and power changes. The measurement of vibrations gives us the information about the blade deflection and blade residual service life. Based on these informations, we monitor the state of the blades and other parts of the turbine. The long-term measurement of the blade deflection is important for the planning shutdowns and optimization of the cost of maintenance and operation of the turbine.

The blade deflections are determined by the difference between the real and expected passage times of the blades under the sensors. The deflection of the specific blade is sampled once per revolution. It can not be used the antialiasing filter in the measurement chain, because the turbine rotation frequency is in fact the blade vibration sampling frequency and furtermore, the rotation speed changes. For these reasons, the vibrations of the specific blade are measured in the limited frequency bands with lot images of higher frequencies and the low frequency resolution. An increasing of the sensor amount, and thus a broadening of the frequency bands, has little impact for the expended resources.

Instead of one blade behavior analysis, the whole bladed disc could be analyzed. The deflections are sampled by the passages of all the blades. Then, the width of the frequency band is sufficient and the aliasing manifests itself not so much. The allblade spectrum, which is spectrum involving all blades of one disc, is calculated for each sensor from deflections measured for all the blades passed under the sensor using the short-time Fourier transform. Consequently, cross spectra are calculated from the couples of the allblade spectra from different sensors.

For the estimation of the nodal diameter, a new progresive method is presented; this method is based on the analysis of data from multiple sensors using the method of least squares.

2. Blade vibration manifestation

The blades interact with each other and their behavior affects the behavior of the whole bladed disc. A mode shape of a bladed disc describes the disc behavior, because the blade vibrations are not random. There is the specific relative position of the blades, which is periodically repeated. In such cases, there are points on the disc, which are not in a motion in one time. These points lie on the straight lines, which pass through the disc centre, or on the concentric circles around the disc center. These points are known as the nodal diameter and nodal circle. The nodal diameters and circles characterize the vibrations shapes. In practice, the rotating and standing shapes (travelling or standing waves) are monitored in term of the disc vibration. The following figure 1 illustrates the selected nodal diameters including the special case - zeroth nodal diameter.

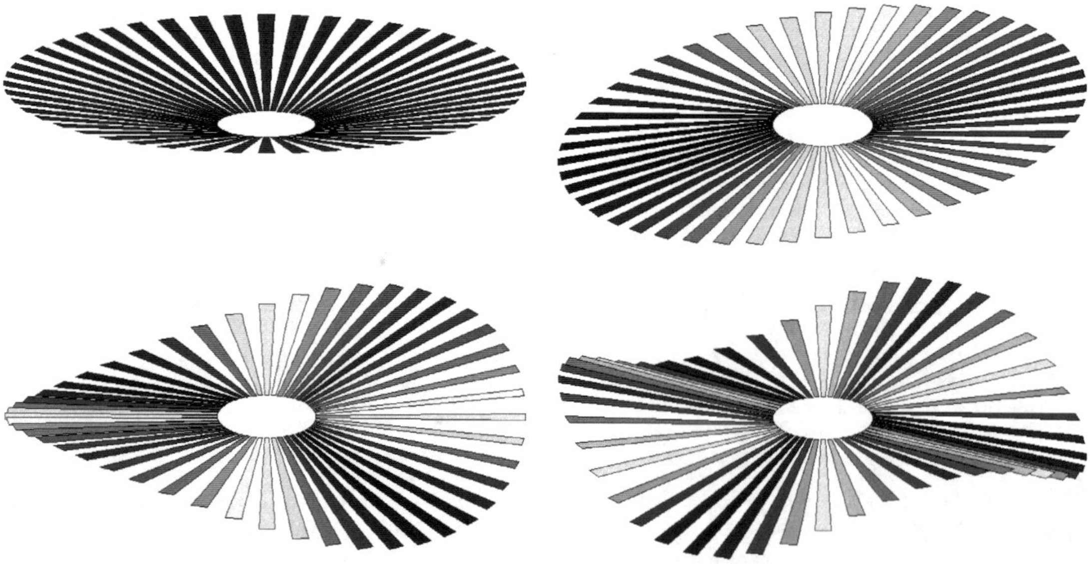

Figure 1. An example of several nodal diameters.

3. Standing wave

The bladed disc vibrates with ND^{th} nodal diameter with the frequency f_{osc}. This modulation is monitored using the timing of the blades with the stationary sensor.

$$f_{mod} = ND \cdot f_{rot} \pm f_{osc} \tag{1}$$

There are two frequencies due to the modulation.

$$f_k = f_{osc} \tag{2}$$
$$f_d = ND \cdot f_{rot} \tag{3}$$

Using two sensors s_1 and s_2, we obtain following signals

$$s_1 = \cos(2\pi f_k t_1 + \varphi_k) \cdot \cos(2\pi f_d t_1 + \varphi_{d1}) \tag{4}$$
$$s_2 = \cos(2\pi f_k t_1 + \varphi_k) \cdot \cos(2\pi f_d t_1 + \varphi_{d2}). \tag{5}$$

The wave with the frequency f_{osc} has the same phase shift for both signals because the sensors measure synchronously. We analyze different blades at the same time. The wave with the frequency f_d corresponds to the bladed disc shape. Different blades have different deflections at the same time. If we consider only the bladed disc shape, then the individual blade have always the same deflection. However, different blades pass under different sensors at the same time. Therefore, the vibrations on the frequency f_d have a different phase shift at the same time

$$\varphi_{d2} - \varphi_{d1} = \rho \cdot ND, \tag{6}$$

where ρ is the angle formed by sensors. According to the goniometric equality

$$\cos\alpha \cdot \cos\beta = \frac{1}{2}\left[\cos(\alpha - \beta) + \cos(\alpha - \beta)\right], \tag{7}$$

we monitor the signals according to equations (4) and (5) at each sensor.

$$s_1 = \frac{1}{2}\{\cos[2\pi(f_k - f_d)t_1 + \varphi_k - \varphi_{d1}] + \cos[2\pi(f_k + f_d)t_1 + \varphi_k + \varphi_{d1}]\} \tag{8}$$
$$s_2 = \frac{1}{2}\{\cos[2\pi(f_k - f_d)t_1 + \varphi_k - \varphi_{d2}] + \cos[2\pi(f_k + f_d)t_1 + \varphi_k + \varphi_{d2}]\} \tag{9}$$

After the discrete Fourier transform calculation, the complex frequency signals are obtained from the measured time signals. From these complex signals, we can calculate the amplitude and angle on the given frequency. Further, we focus on the determination of phase. From the equations (8) and (9), we determine the phases (φ) of both frequency components. For the first sensor, we apply the equation (8)

$$\Phi(S_1(f_k - f_d)) = \varphi_k - \varphi_{d1} \tag{10}$$
$$\Phi(S_1(f_k + f_d)) = \varphi_k + \varphi_{d1} \tag{11}$$

and for the second sensor the equation (9)

$$\Phi(S_2(f_k - f_d)) = \varphi_k - \varphi_{d2} \tag{12}$$
$$\Phi(S_2(f_k + f_d)) = \varphi_k + \varphi_{d2}. \tag{13}$$

It holds true for the multiply of two complex numbers $c1$ and $c2$, that the angle determined by the product is equal to the sum of the angles determined from the individual factors. If we will consider complex conjugate number $c1$, we will get the difference of the angles.

$$\Phi(c_1^* \cdot c_2) = \Phi(c_2) - \Phi(c_1) \tag{14}$$

We obtain the cross-correlation spectrum by the multiplying of the complex conjugate Fourier image of one measured signal with the Fourier image of the other measured signal. The phase of the frequency components corresponds to the following formulas.

$$\Phi(S_1(f_k - f_d)^* \cdot S_2(f_k - f_d)) = -(\varphi_k - \varphi_{d1}) + (\varphi_k - \varphi_{d2}) = -ND \cdot \rho \tag{15}$$

$$\Phi(S_1(f_k + f_d)^* \cdot S_2(f_k + f_d)) = -(\varphi_k + \varphi_{d1}) + (\varphi_k + \varphi_{d2}) = ND \cdot \rho \tag{16}$$

If we find two integer multiples of the angle between the sensors (one positive, the other negative) in the cross-spectrum, we can identify the founded nodal diameter.

4. Travelling wave

The travelling wave is constituted by the deflections of the circumferentially blades. All blades vibrate with the same amplitude and the adjacent blades vibrate with different phase. If the total circumferentially phase delay achieves 2π, we can see a rotating diameter, on which the blades have zero deflection at the moment. If phase delay achieves 4π, we can see two diameters, etc. The deflection of one blade is described as follows

$$\sin(2\pi f_{osc} t + \varphi). \tag{17}$$

The deflection of each blade for any time is described by the expression (18), where b is the blade number, N the blade quantity and D the diameters number.

$$\sin(2\pi f_{osc} t + 2\pi \frac{b}{N} D + \varphi_{begin}) \tag{18}$$

If we bring the disc rotation to the expression(18), we will detect the position of only one blade at one time. Assuming the uniform blades distribution around the disc circumference and constant speed, we can write

$$\sin(2\pi f_{osc}(t + b\Delta t) + 2\pi \frac{b}{N} D + \varphi_{begin}). \tag{19}$$

Because the initial time is constant, it is expressed as the part of the initial phase delay without loss of a generality. Correspondingly, we take the initial time as zero.

$$\sin(2\pi f_{osc}(b\Delta t) + 2\pi \frac{b}{N} D + \varphi_{begin}) \tag{20}$$

Due to the mentioned assumptions, the time differences are replaced

$$\Delta t = \frac{1}{N \cdot f_{rot}} \tag{21}$$

$$\sin(2\pi f_{osc}(\frac{b}{N \cdot f_{rot}}) + 2\pi \frac{b}{N} D + \varphi_{begin}). \tag{22}$$

After the modifications, we achieve

$$\sin(2\pi f_{osc} \frac{b}{N} (\frac{f_{osc}}{f_{rot}} + D) + \varphi_{begin}). \tag{23}$$

The sequentially coming blades, are replaced with time T

$$\frac{p}{N \cdot f_{rot}} = T \tag{24}$$

$$\sin(2\pi T(f_{osc} + D \cdot f_{rot}) + \varphi_{begin}). \tag{25}$$

In the spectrograms (whether allblade or cross), we monitor the travelling wave on the frequency

$$f_{obs} = f_{osc} + D \cdot f_{rot}. \tag{26}$$

The monitored wave has the form

$$\sin(2\pi T f_{obs} + \varphi_{begin}). \tag{27}$$

From the cross spectrogram, we find out with which diameter D and on what frequency f_{obs} the travelling wave seemingly vibrates. The real frequency of the blades movement is

$$f_{obs} = f_{osc} - D \cdot f_{rot}. \tag{28}$$

So far, we consider the travelling wave in one yet unspecified direction. The travelling waves are forward or backward, that depends on the direction, in which the wave propagates according to the disc rotation direction.

Let us take into account the rotation direction such that the blades sequentially pass under the sensor with increasing number p. Then, the forward wave propagates against the direction of the blades numbering. The maximum deflection passes from the blades with a higher number to the blades with a lower number p. This means, that the phase φ is the higher the more blade numbers increase.

The backward wave propagates in the direction of the blades numbering. It follows, that the phase is the more downward the more blade numbers increase. The opposite wave propagation is achieved by the negative phase φ. The directions are shown in figure 2. Let's assume that φ is positive, we can rewrite experssion (17) into forms

$$\sin(2\pi f_{osc} t + \varphi) \tag{29}$$

$$\sin(2\pi f_{osc} t - \varphi) \quad . \tag{30}$$

Expression (29) or (30) corresponds with forward or backward wave. In expression (18) the quantity of the rotating diameters D was defined. Further, to the sign of φ from the previous expression, we extend the value D to the negative numbers, which maintain the validity of the expression (18) for the forward and backward wave.

5. Evaluation of nodal diameters

Based on the angle ρ between the sensors, the phases are defined for each of the searched nodal diameter.

$$\pm \rho \cdot ND \tag{31}$$

The phases on the individual frequency components of the cross spectrum, which was discussed in the section 3, are compared with the searched phases. In the conformity with the phase tolerance, the calculated nodal diameter is assigned to the corresponding coefficient. When we process the signals from more than two sensors, we compare the coefficients for the individual frequency components of all cross spectra, which combine the signals from each sensor. The coefficient for each frequency component is assigned to the corresponding color that will be used for plotting.

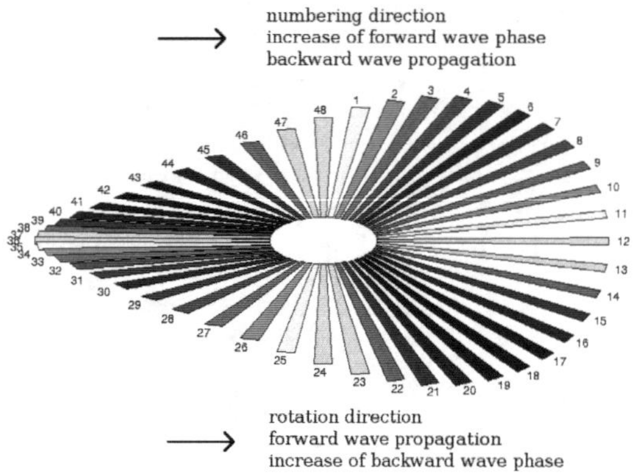

Figure 2. Definition of the directions on bladed disc.

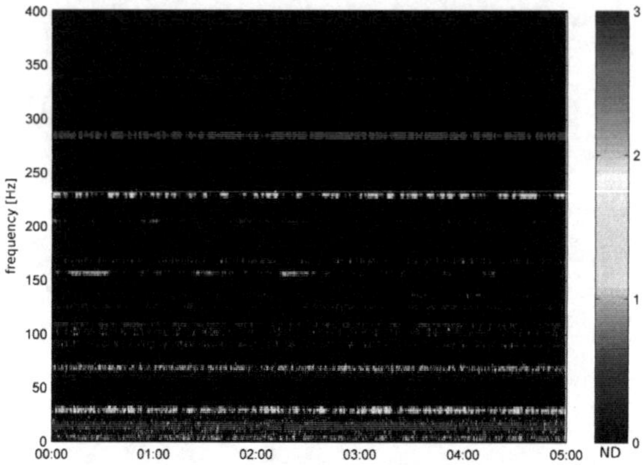

Figure 3. An example of cross spectrogram.

The colors of the frequency components with low amplitudes are suppressed, because of the differentiation of the nodal diameters from the amplitude insignificant frequency components and noise. From these cross spectra, we obtain the cross spectrogram in figure 3, which gives us the information about the substractional and additional components (same color) of individual nodal diameters.

The figure 3 shows strong second nodal diameter 130 Hz in yellow (additional and substractional component of 230 Hz and 30 Hz) or third nodal diameter 135 Hz in red (additional and substractional component 285 Hz and 15 Hz). Less apparent is the first nodal diameter 120Hz in light blue (additional and substractional component 170 Hz and 70 Hz). The zeroth nodal diameter in dark blue (not observed with modulation, so without additional and substractional components) is visible in the spectrogram at several frequencies.

6. Amplitude evaluation

At the first, let's assume standing wave with amplitude A vibrating on the disc. Using basic goniometric relation we can write

$$A \cdot \cos\alpha \cdot \cos\beta = A \cdot \frac{1}{2}[\cos(\alpha - \beta) + \cos(\alpha + \beta)]. \tag{32}$$

The standing wave can be expressed as linear combination of two traveling waves with opposite direction of propagation. Lets define amplitudes of both traveling waves

$$A_{\oplus} = A_{\ominus} = \frac{A}{2}. \tag{33}$$

These amplitudes can be calculated directly from allblade spectrum. For standing wave the following equation holds.

$$A \cdot \cos\alpha \cdot \cos\beta = A_{\ominus} \cdot \cos(\alpha - \beta) + A_{\oplus} \cdot \cos(\alpha + \beta) \tag{34}$$

The amplitude of standing wave is monitored on the blades in the antinodes. This is the maximum amplitude across all blades of the disc. The minimal amplitude is monitored on the blades in the nodes of the standing wave. In the ideal case, minimal amplitude is zero.

In the case of traveling wave, the amplitudes $A_{\oplus}$ and $A_{\ominus}$ are different, $A_{\oplus} \neq A_{\ominus}$. This is interpreted as the superposition of the standing and travelling wave. This behavior of the disc

Journal of Physics: Conference Series **659** (2015) 012051 doi:10.1088/1742-6596/659/1/012051

can not be described by the equation (34). It is necessary to expand the equation with the description of the traveling wave. Superposition of waves travelling in opposite direction can be expressed as superposition of standing wave and single travelling wave. The travelling wave propagates in the direction with the higher amplitude of the frequency component. In following example holds $A_{\oplus} > A_{\oplus}$.

$$A_{\ominus} \cdot \cos(\alpha - \beta) + A_{\oplus} \cdot \cos(\alpha + \beta) = \underbrace{2 \cdot A_{\ominus} \cdot \cos\alpha \cdot \cos\beta}_{standing} + \underbrace{(A_{\oplus} - A_{\ominus}) \cdot \cos(\alpha + \beta)}_{travelling} \tag{35}$$

In the case of $A_{\oplus} > A_{\ominus}$, the amplitude of the traveling wave is equal to $A_{\oplus} - A_{\ominus}$. In the same case, the amplitude of the standing wave is equal to

$$A = 2 \cdot A_{\ominus}. \tag{36}$$

In the comparison to the standing wave, the blade amplitudes are increased by the amplitude of the travelling wave. The blade amplitudes in the antinodes of the standing waves (maximum amplitude) are equal to $A_{\oplus} + A_{\ominus}$ and the blade amplitudes in the nodes of the standing wave (minimum amplitude) are equal to $|A_{\oplus} - A_{\ominus}|$.

7. Evaluation from multiple sensors

An another evaluation method of the nodal diameter provides using the method of least squares. This method, compared to the method of cross spectrum, is based on an evaluation of data from multiple sensor simultaneously. For a complex value for a given frequency in the allblade spectrum, the phase information is the superposition of the initial phase of vibration of the modulated wave, which is the same for all sensors, and phase of modulating wave, which is different for each sensor. When using the method of least squares, we use the regressive equation

$$a \cdot e^{iD\rho} = e^{i\varphi}. \tag{37}$$

D is a chosen diameter, which is integer, ρ is an angle of sensor placement, φ is a phase obtained from the allblade spectrum at a given frequency at a particular sensor and a is a complex number. The angle of this complex number is a phase information, which includes the sum of the initial phase of vibration of the modulated wave and the initial phase of modulating wave relative to the phase marker. The absolute value of the number a indicates the equivalent of the probability that the chosen diameter D corresponds to the true diameter. The higher the absolute value of a closer to one, the higher probability it was chosen the correct value D.

The sensor placement plays a major role in this method too. The placement has a great impact on the degree of credibility of the calculation. An inappropriate placement of sensors may produce falsely correct results.

8. Sensors placement

The sensors should be placed to each other at the angle, which is the integer multiple of the angle formed by the blades. In this arrangement, the blades with the corresponding angle pass under the sensor approximately at the same time. Therefore, there is no increase of the phase delay.

It is also good to choose the angle depending on the expected nodal diameter. Higher nodal diameter represents larger monitored phase. It is necessary to choose the angle between the sensors such that $ND\rho$ lies in the interval 0 to π rad. This prevents an overlap of the detected angles of different nodal diameters, for example the first and third nodal diameter (sensors form angle $\pi/2$ rad). It is recommended to choose the angle between the sensors on the value of prime multiples of the angle, which encloses the blades.

9. Conclusions

Any mechanical stress of blades reduces their service time. The blade vibration measurement provides the information about the long-lasting and rapid changes in the blades parameters. This allows early detection of the blade faults. With the blade tip timing method using the magnetoresistive sensors, the oscillations of the longest blades of the low-pressure part of the steam turbine 270 MW were measured. The nodal diameters were identified and the amplitudes of their oscillation were calculated. The frequencies of the nodal diameters were verified by the model, which were provided by the manufacturer of the turbine. The calculated data will be used for the determination of the blade residual service times.

Acknowledgments

The authors would like to thank to the company Doosan Skoda Power for the possibility to do the research in the area of industrial interest and to have an opportunity to solve the actual problems on the field of steam turbine troubleshooting.

This work was supported by the European Regional Development Fund (ERDF), project "NTIS New Technologies for the Information Society", European Centre of Excellence, CZ.1.05/1.1.00/02.0090 and by the project PUNTIS - LO1506 and by the project SGS-2013-041.

References

[1] Zielinski M, Ziller G 2005 *Noncontact Blade Vibration Measurement System for Aero Engine Application American Institute of Aeronautics and Astronautics, ISABE*

[2] Kadoya Y, Mase M, Kaneko Y, Umemura S, ODA T, Johnson MC 1995 *Noncontact Vibrational Measurement Technology of Steam Turbine Blade JSME International Journal* **38** 486–493

[3] Gallego-Garrido J, Dimitriadis G, Wright JR 2007 *A Class of Methods for the Analysis of Blade Tip Timing Data from Bladed Assemblies Undergoing Simultaneous Resonances–Part I:Theoretical Development International Journal of Rotating Machinery*

[4] Dimitriadis G, Carrington IB, Wright JR, Cooper JE 2002 *Blade-tip timing measurement of synchronous vibrations of rotating bladed assemblies Mechanical Systems and Signal Processing*

[5] Ye D, Duan F, Guo H, Li Y, Wang K 2012 *Turbine blade tip clearance measurement using a skewed dual-beam fiber optic sensor SPIE*

[6] Mercadal M, von Flotow A, Tappert P 2000 *Damage Identification By NSMS Blade Resonance Tracking in Mistuned Rotors IEEE*

[7] Ivey PC, Grant KR, Lawson C 2002 *Tip timing techniques for turbomachinery hcf condition monitoring The 16th Symposium on Measuring Techniques in Transonic and Supersonic Flow in Cascades and Turbomachines*

[8] Heath S, Imregun M 1998 *A Survey of Blade Tip-Timing Measurement Techniques for Turbomachinery Vibration ASME*

[9] Kawasima T, Iinuma H, Minagawa N 1994 *Optical Semiconductor Blade Vibration Monitoring System for Gas Turbine Engine IEEE*

[10] Allport JM, Jupp ML, Pezouvanis A, Janicki GW, Pieronczyk AI, Day AJ, Olley P, Mason B, Ebrahimi MK *Turbocharger blade vibration: Measurement and validation through laser tip-timing*

[11] Heath S, Imregun M 1996 *An improved single-parameter tip-timing method for turbomachinery blade vibration measurements using optical laser probes Elsevier Science Ltd*

[12] Hockaday BD 2011 *Quantifying optical tip-timing probe error with laboratory apparatus Proceedings of ASME Turbo Expo*

[13] Singh MP, Vargo JJ, Schiffer DM, Dello JD *Safe diagram – a design and reliability tool for turbine blading*

An LMI approach for the Integral Sliding Mode and H_∞ State Feedback Control Problem

Souad Bezzaoucha[1], **David Henry**[2]

[1] Interdisciplinary Centre for Security, Reliability and Trust, Campus Kirchberg, University of Luxembourg, L-1359, Luxembourg. *souad.bezzaoucha@uni.lu*
[2]University of Bordeaux, IMS Laboratory, F-33400 Talence, France.
david.henry@ims-bordeaux.fr

Abstract. This paper deals with the state feedback control problem for linear uncertain systems subject to both matched and unmatched perturbations. The proposed control law is based on an the Integral Sliding Mode Control (*ISMC*) approach to tackle matched perturbations as well as the H_∞ paradigm for robustness against unmatched perturbations. The proposed method also parallels the work presented in [1] which addressed the same problem and proposed a solution involving an Algebraic Riccati Equation (ARE)-based formulation. The contribution of this paper is concerned by the establishment of a Linear Matrix Inequality (LMI)-based solution which offers the possibility to consider other types of constraints such as $\mathcal{D}$-stability constraints (pole assignment-like constraints). The proposed methodology is applied to a pilot three-tank system and experiment results illustrate the feasibility. Note that only a few real experiments have been rarely considered using SMC in the past. This is due to the high energetic behaviour of the control signal.
It is important to outline that the paper does not aim at proposing a LMI formulation of an ARE. This is done since 1971 [2] and further discussed in [3] where the link between AREs and ARIs (algebraic Riccati inequality) is established for the H_∞ control problem. The main contribution of this paper is to establish the adequate LMI-based methodology (changes of matrix variables) so that the ARE that corresponds to the particular structure of the mixed ISMC/H_∞ structure proposed by [1] can be re-formulated within the LMI paradigm.

1. Motivations

Recent works on Sliding Mode Control (SMC) demonstrate that SMC approaches are able to withstand external disturbances and model uncertainties satisfying the matching condition, that is, perturbations that enter the state equation at the same point as the control input, see [4, 5] for instance. Perturbations not satisfying the matching condition are obviously called "unmatched perturbations". The matching condition is a sufficient condition for the existence of a SMC law able to reject matched perturbations (exact compensation or insensitivity) providing they are bounded by a known function $\in \mathcal{L}_\infty$. However, if the matching condition is not satisfied then the system behaviour depends on the influence of the disturbances [6, 5]. This motivates the work reported in this paper.

A sliding mode controller is composed of two modes: The first one called the *reaching mode* that drives the system onto the so-called sliding surface in a finite time. The second mode, known as the *sliding mode*, aims at maintaining the system on the sliding surface. Note that during the reaching phase, the system is sensitive, even to perturbations satisfying the matching condition.

Different solutions have been proposed to carry out the unmatched perturbations, see [7, 8, 9, 10, 11, 12, 13, 14] to name a few. The main idea is to define the control law as the sum of a continuous control (say u_0) designed by means of robust control techniques and a SMC signal (say u_1). u_0 is responsible of the attenuation of the unmatched disturbances on the system states and u_1, the discontinuous control, is used to reject the matched perturbations. The same philosophy is used in [15, 16] by adding an integral term in the definition of the sliding surface. This enables the system always to start at the sliding manifold. The technique is called Integral Sliding Mode Control (ISMC). The method proposed in [1] follows the same idea. The continuous signal u_0 is designed using H_∞ technique whereas u_1 is a ISMC-based law. Conditions are given for the existence of a stabilizing control law u_0 in terms of an algebraic Riccati equation (ARE), which leads the approach numerically suitable since it can be solved using matrix algebra theory. Note that, as outlined by the authors, the integral sliding mode controller, if improperly designed, while eliminating the matched perturbations, could lead to amplification of the unmatched ones. This further motivates the use of H_∞ techniques.

In this paper, a Semi Definite Programming (SDP) formulation involving Linear Matrix Inequality (LMI) constraints, is proposed. The major reason of using the LMI paradigm is that it is now well known that it offers the possibility to consider other types of constraints such as $\mathcal{D}$ stability constraints (pole assignment-like constraints) [17, 18, 19]. The proposed method also parallels the work presented in [1] which as already mentioned, proposed a ARE-based formulation.

It is important to outline that the paper does not aim at proposing a LMI formulation of an ARE. This is done since 1971 [2] and further discussed in [3] where the link between AREs and ARIs (algebraic Riccati inequality) is established for the H_∞ control problem. The main contribution of this paper is to establish the adequate LMI-based methodology (changes of matrix variables) so that the ARE that corresponds to the particular structure of the mixed ISMC/H_∞ structure proposed by [1] can be re-formulated as a SDP problem under LMI constraints. In fact, results from [3] cannot be used since some degrees of freedom in the control gain are used to ensure matched perturbation compensation by the SMC part. This will be highlighted in the paper, see remark 2. Establishing a LMI formulation of the ISMC/H_∞ control has not been done right now, see the literature in both the SMC and H_∞ communities.

The paper's structure is the following. In section 2, the Integral Sliding Mode Control (ISMC) combined with the H_∞ control is considered. The mathematical problem is stated formally in terms of LMI framework in section 3. In section 4, an application on a pilot three-tank system is given. The plant model is presented and the proposed control design is applied to the system. Experimental results are given to illustrate the developed approach efficiency. Finally, in section 5, conclusions and perspectives for future works are given.

2. Problem statement and material background

Similarly to the approach addressed in [1], the solution proposed in this paper to the state feedback control problem for linear uncertain systems subject to both matched and unmatched perturbations, consists in defining the control law as the sum of a continuous control signal u_0 and a discontinuous ISMC signal u_1, i.e.

$$u(t) = u_0(t) + u_1(t) \tag{1}$$

u_0 is in charge to manage the unmatched perturbations (this part is addressed through the H_∞ formalism) while u_1 is in charge to manage the matched perturbations (this part is addressed using sliding mode control). An integral term is included in the sliding manifold to guarantee that the system starts at the sliding manifold.

Thus, in the interest of brevity, throughout this section an earnest attempt will be made to avoid duplicating material presented in [1]. Towards this end, the focus of this section will lie wholly with the main results established in [1]. The reader is invited to refer to this paper for necessary backgrounds and proofs.

Consider the following state-feedback control problem

$$\begin{cases} \dot{x}(t) &= Ax(t) + B_1 w(t) + B_2 u(t) \\ z(t) &= C_1 x(t) + D_{12} u(t) \\ y(t) &= x(t) \end{cases} \tag{2}$$

where $x \in \mathbb{R}^n$ is the system's state vector, $u \in \mathbb{R}^m$ the control input vector, $w \in \mathbb{R}^q$ is an external perturbation due to model uncertainties or external disturbances (including here both matched and unmatched disturbances), $z \in \mathbb{R}^r$ is the objective vector (including tracking signals, regulation signals and actuator's output) and $y \in \mathbb{R}^n$ is the measurement signal. The following assumptions are made:

Assumption 1. $rank(B_1) = m$

Assumption 2. *$w(t)$ is assumed to be bounded by a known function $v(t) \in \mathcal{L}_\infty$, i.e. $||w(t)|| \leq v(t)$ for all t.*

Assumption 3. *The pair (A, B_2) is stabilisable, the pair (C_1, A) is detectable and $D_{12}^T[C_1\ D_{12}] = [0\ I]$.*

Assumption 1 is a necessary and sufficient condition to separate the matched from the unmatched part of w. Assumption 2 is common in the sliding mode theory and assumption 3 is common in the H_∞ theory. The interested reader can refer to [4, 5, 20].

From sliding mode control theory, the discontinuous control u_1 is selected as

$$u_1(x,t) = -\rho(x,t)\frac{(GB_2)^T s(x,t)}{||(GB_2)^T s(x,t)||} \tag{3}$$

where $\rho(x,t)$ is a matrix of adequate dimension so that its values are chosen high enough to enforce the sliding motion. The term $\dfrac{(GB_2)^T s(x,t)}{||(GB_2)^T s(x,t)||}$ plays the role of the "*sign*" function so that Eq. (3) is an alternative to the well known SM equation $u_1 = -\rho.sign(s)$. The sliding manifold s is defined by the set $\{x|s(x,t) = 0\}$ with:

$$s(x,t) = G\left[x(t) - x(t_0) - \int_{t_0}^{t}(Ax(t) + B_2 u_0(x,\tau))d\tau\right] \tag{4}$$

$G \in \mathbb{R}^{m\times n}$ is a projection matrix such that GB_2 is invertible. It can be noticed that at $t = t_0$, $s(x, t_0) = 0$ so the system always starts at the sliding manifold.

Following [1], the H_∞ norm of the transfer that goes from $w(t)$ to $z(t)$ satisfies $||T_{zw}||_\infty < \gamma$ if G is set to B_2^T and

$$u_0(t) = -B_2^T X x(t) \tag{5}$$

where X is a positive semi-definite matrix, solution of the ARE

$$XA + A^T X - X(B_2 B_2^T - \gamma^{-2}\overline{B}_w \overline{B}_w^T)X + C_1^T C_1 = 0 \tag{6}$$

$$\overline{B}_w = B_2^\perp (B_2^\perp)^+ B_1 \tag{7}$$

where the columns of $B_2^\perp$ span of null space of B_2^T. $(B_2^\perp)^+$ is the left inverse of $B_2^\perp$, that is $(B_2^\perp)^+ = ((B_2^\perp)^T B_2^\perp)^{-1}(B_2^\perp)^T$.

It follows that the sliding manifold is given by

$$s(x,t) = B_2^T \left[x(t) - x(t_0) - \int_{t_0}^{t} (A - B_2 B_2^T X)x(\tau)d\tau \right] \tag{8}$$

and the control law is defined according to

$$u(x,t) = -B_2^T X x(t) - \rho \frac{s}{||s||}, \ \rho > ||B_2^+ B_1 w|| \tag{9}$$

The reader can refer to the paper [1] for additional explanations and proofs.

Remark 1. *It should be outlined that, even if the stability for the control laws* $u_1 = -\rho\frac{s}{||s||}$ *and* $u_0 = -B_2^T X x$ *has been proven separately, there does not exist any formal proof of stability for the overall control law (9). This is a known problem from the SMC community. Fortunately, different authors have demonstrated that the stability is practically preserved [12, 13, 14]. Thus, since the method we proposed in this paper follows the same principle, the same problem arises. This is a known problem which is under current research investigation.*

Remark 2. *Discussion on AREs and LMIs for ISMC:*
The particular structure of the control law u_0 *given by (5) follows the requirements for* u_1 *to mitigate the matched perturbations (exact compensation), see the theoretical developments in [1]. In a pure* H_∞ *problem, the control signal has the general form* $u(t) = -X_\infty x(t)$[1] *where* X_∞ *is the solution of an ARE. The consequence of the presence of* B_2^T *in the definition of* u_0 *is that one can not directly apply, e.g. the bounded and projection lemmas technique [3, 21], the changes of LMI variables approaches [17, 18] or [2, 22, 23] for the state-feedback case, to derive a LMI formulation. In other words, even if it seems trivial to derive a LMI formulation from Eq. (6), this is in fact not the case due to the particular structure of* u_0 *given by Eq (5). The next section provides a solution to this problem.*

3. Main result: a LMI formulation

Consider the disturbances vector w in the system model (2). Then, applying the theoretical developments proposed by [1], the following proposition yields

Proposition 1. *For any matrix satisfying* B_1 *satisfying assumption 1, the disturbances* w *can be decomposed as*

$$B_1 w(t) = B_w w(t) + \overline{B}_w w(t), \quad B_w = B_2 B_2^+ B_1, \quad \overline{B}_w = B_2^\perp (B_2^\perp)^+ B_1 w(t) \tag{10}$$

[1] The minus sign is kept there for similarity reason with eq. (5) but it can obviously be removed.

□

The first part of (10) represents the matched disturbances and will be eliminated by the discontinuous control u_1; the second one refers to the unmatched disturbances and will be attenuated using the continuous control u_0. The following proposition gives the solution to the problem:

Theorem 1. *Consider the system model given by Eq. (2) subject to both matched and unmatched perturbations. The state-feedback controller given by Eq. (1) minimizing the H_∞ norm of the transfer that goes from $w_u(t) = B_2^\perp (B_2^\perp)^+ B_1 w(t)$ (unmatched perturbations) to $z(t)$ such that $||T_{zw_u}||_\infty < \gamma$ and annihilating the effect of $w_m(t) = B_2 B_2^+ B_1 w(t)$ (matched perturbations) is defined according to*

$$u(x,t) = -B_2^T Q^{-1} x(t) - \rho \frac{s}{||s||}, \; \rho > ||B_2^+ B_1 w|| \tag{11}$$

where $Q = Q^T > 0$ is a positive definite matrix solution of the following LMI constraint

$$\begin{pmatrix} AQ + QA^T - 2B_2 B_2^T & -QC_1^T & B_2 & \overline{B}_w \\ & -I & 0 & 0 \\ & * & -I & 0 \\ & * & * & -\gamma^2 I \end{pmatrix} \leq 0 \tag{12}$$

□

Proof 1. *Let us first consider the following well know lemma that associates a so-called algebraic Riccati inequality (ARI) to an ARE:*

Lemma 1. *Suppose matrices $A, W = W^T > 0$ and Q are given. If the Algebric Riccati Equation (ARE)*

$$PA + A^T P + Q - PWP = 0 \tag{13}$$

has a positive definite symmetric solution P, then for any $0 < W_1 \leq W$ and $Q_1 \geq Q$, the equation

$$P_1 A + A^T P_1 + Q_1 - P_1 W_1 P_1 \leq 0 \tag{14}$$

has a positive definite symmetric solution $P_1 \geq P$. □

Using this lemma and the ARE (6), we know that there exists a matrix $X_1 = X_1^T$ which satisfies the following inequality:

$$X_1 A + A^T X_1 + C_1^T C_1 + \gamma^{-2} X_1 \overline{B}_w \overline{B}_w^T X_1 - X_1 B_2 B_2^T X_1 \leq 0 \tag{15}$$

Let us choose $K = -B_2^T X_1$, we get:

$$X_1 A + A^T X_1 + C_1^T C_1 + \gamma^{-2} X_1 \overline{B}_w \overline{B}_w^T X_1 + K^T B_2^T X_1 + X_1 B_2 K + K^T K \leq 0 \tag{16}$$

Let us note $Q = X_1^{-1}$ and $W = KQ$. From the congruence principle and applying the Schur's complement, the LMI (12) is then deduced where $X_1 = Q^{-1}$ and $K = WQ^{-1} = -B_2^T X_1$ which terminates the the proof. □

This theorem also gives a numerical tractable solution to the ISMC/H_∞ design problem, since noting that γ^2 enters linearly in (12), the H_∞ norm $||T_{zw_u}||$ can be minimized by minimizing γ.

3.1. ISMC, H_∞ and poles assignment

As it is known, a good time response specifications can be achieved by forcing the closed-loop poles into a specific regions. In this case, the H_∞ synthesis being formulated as a convex optimization problem involving LMIs will give wide range of flexibility in combining several constraints on the closed-loop system. This flexible nature of LMI schemes can be used to handle H_∞ controller through the LMI regions concept [24].

Direct application of the developments proposed in [24] to the definition of the control law $u_0(t) = -B_2^T Q^{-1} x(t)$ shows that the eigenvalues of the evolution matrix of the closed loop $A_{cl} = A - B_2 B_2^T Q^{-1}$ can be assigned into a prescribed $\mathcal{D} = \cap_{k=1}^{n_s} \mathcal{D}_k$ if there exist a common Lyapunov matrix $Y = Y^T > 0$ such that the set of n_s LMIs

$$\alpha_k \otimes Y + \beta_k \otimes A_{cl} Y + \beta_k^T \otimes Y A_{cl}^T < 0 \quad k = 1, 2, ..., n_s \tag{17}$$

is simultaneously satisfied.

In this expression, " $\otimes$ " denotes the Kronecker product of matrices, α_k and β_k are matrices of appropriate dimension defining each region $\mathcal{D}_k = \{\chi \in \mathbb{C} : f_{\mathcal{D}_k}(\chi) < 0\}$, where $f_{\mathcal{D}_k}(\chi) = \alpha_k + \chi\beta + \chi^*\beta^T$ is the characteristic function of the region $\mathcal{D}_k$. Here $\chi = x + yj$ is the complex variable and χ^* denotes the conjugate of χ.

Setting $Y = Q$, it is easy to see that solving the following optimization problem gives the solution of the ISMC problem with jointly H_∞ attenuation performance against unmatched perturbations and pole assignment.

$$\min \gamma^2 \quad s.t. \, Eq.(12) \text{ and } Eq.(17) \tag{18}$$

Remark 3. *Note that since theorem 1 provides only a sufficient condition for a solution of the ISMC/H_∞ problem, the solution of Eq. (12) may be more conservative than the solution of the ARE-based formulation. Thus, if it is not required, e.g. $\mathcal{D}$-stability, the solution described in section 2 may be preferred to the one proposed in section 3.*

4. Application to a pilot three-tank system

4.1. Three-tank system description and modelling

The proposed state-feedback controller is now applied to an experimental study. The experimental study is based on a pilot three-tank system, manufactured by "AMIRA" industry (fig 1). The plant consists of three cylinders T_1, T_2, T_3 with cross-section S. These are connected serially with one another cylindrical pipes with a cross-section S_n. The out-flowing liquid (usually distilled water) is collected in a reservoir, which supplies the pumps 1 and 2. Here, the circle is closed. The three water levels (in m) are denoted h_1, h_2 and h_3. They are measured via piezo-resistive pressure sensors. Q_1 and Q_2 (in m^3/s) are the flow rates of the pumps 1 and 2. The pumps are rotational-speed controlled such that a well-defined incoming mass flow corresponds to the reference input introduced by the pump controller.

To model the dynamics of this multi-input/multi-output (MIMO) system, the *Torricielli* rule is applied to each tank. Since modelling this system has already done by many authors, the detailed are omitted here. The interested reader can refer to, e.g. [25].

Performing a first-order approximation around an equilibrium point $h^* = (h_1^* \; h_2^* \; h_3^*)^T$ of the Torricielli equations leads to

$$\begin{cases} \dot{x}(t) &= A(\theta)x(t) + Kf(t) + B_2 u(t) \\ y(t) &= x(t) \end{cases} \tag{19}$$

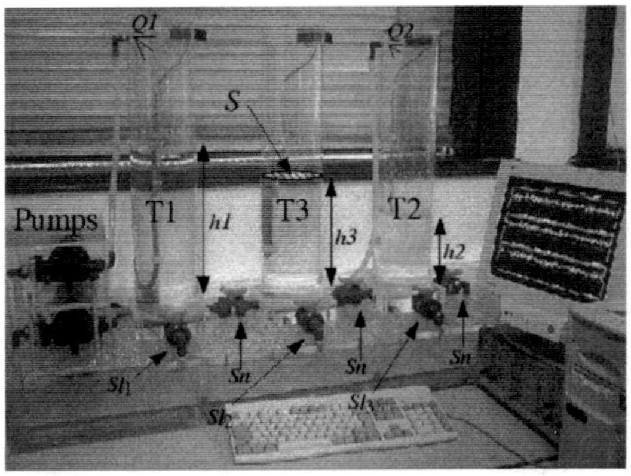

Figure 1. Physical structure of the three-tank system

where $u = (Q_1\ Q_2)^T$ is the controlled inputs and $y = (h_1\ h_2\ h_3)^T$ represents the measured outputs. $f(t)$ models a leakage in the first tank. $\theta = (a_1\ a_2\ a_3)^T$ denotes the vector of outflow coefficients that are uncertain but assumed to be bounded. The system matrices are the following:

$$A(\theta) = \begin{pmatrix} -k_1a_1 & 0 & k_1a_1 \\ 0 & -k_3a_3 - k_2a_2 & k_3a_3 \\ k_1a_1 & k_3a_3 & -k_1a_1k_3a_3 \end{pmatrix}, \; K = \begin{pmatrix} \frac{S_n\sqrt{2g}a_{10}}{2S\sqrt{h_1}} \\ 0 \\ 0 \end{pmatrix}, \; B_2 = \frac{1}{S}\begin{pmatrix} -1 & 0 \\ 0 & 1 \\ 0 & 0 \end{pmatrix} \tag{20}$$

The parameters $k_i, i = 1, 2, 3$ are defined as follows:

$$k_1 = \frac{S_n\sqrt{2g}}{2S\sqrt{h_1^* - h_3^*}}, \; k_2 = \frac{S_n\sqrt{2g}}{2S\sqrt{h_2^*}}, \; k_1 = \frac{S_n\sqrt{2g}}{2S\sqrt{h_3^* - h_2^*}} \tag{21}$$

where g the constant of the gravity. All numerical values are given in the appendix.

From the definition of K and B_2, it is easy to see that f satisfies the matching condition. Thus, following the methodology proposed in section 3, a state-feedback control law $u(t) = u_0(t) + u_1(t) = -B_2^T Q^{-1}x(t) - \rho\frac{s}{||s||}$ is proposed so that u_1 compensate exactly f, u_0 being in charge to attenuate, in the H_∞ sense, the effect of the uncertainties θ on the system state. u_1 being deduced from the ISMC theory according to section 2, the main drawback is concerned by the H_∞ part. So, f (the matched perturbation) is omitted in the following developments.

To proceed, the LFR (Linear Fractionnal Representation) formalism is used here. All parameters θ_i, $i = 1...3$ entering in $A(\theta)$ are "pulled out" so that the model appears as a LTI nominal model P subject to an artificial block diagonal operator $\Delta = blockdiag\left(\delta_1 I_{k_1}, ..., \delta_q I_{k_q}\right)$ specifying how δ_i enters P where $k_i > 1$ whenever the parameter δ_i is repeated, see [26, 27, 28, 29, 30] for more details. In this formalism, $|\delta_i| \leq 1 \Leftrightarrow ||\Delta||_\infty \leq 1$. This can be assumed without loss of generality since the model P can always be scaled. This boils down to the following model

$$\left\{ \begin{array}{lcl} \dot{x}(t) & = & Ax(t) + E_1\eta(t) + B_2u(t) \\ \varepsilon(t) & = & E_2x(t) \\ y(t) & = & x(t) \end{array} \right. , \quad \begin{array}{l} \eta(t) = \Delta\varepsilon(t), \\ \Delta = diag(\delta_1, \delta_2, \delta_3) \\ A = A(\theta_0) \end{array} \tag{22}$$

$$E_1 = \begin{pmatrix} w_1 & 0 & 0 \\ 0 & w_2 & w_3 \\ -w_1 & 0 & -w_3 \end{pmatrix}, \quad E_2 = \begin{pmatrix} -k_1 & 0 & k_1 \\ 0 & -k_2 & 0 \\ 0 & -k_3 & k_3 \end{pmatrix} \tag{23}$$

$w_i, i = 1,2,3$ are weights introduced to scale the uncertainties, that is $a_i = a_{i0} + w_i\delta_i, i = 1,2,3$. Then, by virtue of the small gain theorem, there exists a stabilizing H_∞ control law $u_0(t) = -\mathcal{K}x(t) = -B_2B_2^TQ^{-1}x(t)$ robust to Δ if $\mathcal{K}$ stabilizes the following model

$$\left\{ \begin{array}{rcl} \dot{x}(t) & = & Ax(t) + E_1\eta(t) + B_2u(t) \\ \varepsilon(t) & = & E_2x(t) \\ y(t) & = & x(t) \end{array} \right. \quad \text{so that } ||T_{\eta\varepsilon}||_\infty < 1 \tag{24}$$

where $||T_{\eta\varepsilon}||_\infty$ denotes the transfer between η and ε. The dimension of the problem is $x \in \mathbb{R}^3$, $u \in \mathbb{R}^2$, $\eta \in \mathbb{R}^3$ and $\varepsilon \in \mathbb{R}^3$ which clearly exhibits a MIMO problem.

4.2. Design of the control law

In terms of H_∞ objectives for u_0, it is only required attenuation of the unmatched perturbations on the system state, so the control penalty function is fixed to identity. Now, considering the above statement, it follows the following definition of the different matrices of the state-feedback problem (2)

$$B_w = K, \quad \overline{B}_w = E_1, \quad C_1 = \begin{pmatrix} E_2 \\ I_3 \end{pmatrix} \tag{25}$$

Theorem 1 is next used to derive the control law given by Eq. (11). In order to avoid fast dynamics of the control law (this prevents, e.g. measurement noise amplification), a pole clustering constraint is considered following the developments proposed in section 3.1. Especially, it is required pole clustering in the LMI region $\mathcal{D} = \{\lambda : \mathbb{R}\left(\lambda_i(A - B_2B_2^TQ^{-1})\right) > -\beta\}$ with $\beta = 70$. This leads to the following LMI:

$$2\beta Q + AQ + QA^T - 2B_2B_2^T > 0, \quad \beta = 70 \tag{26}$$

Noticed that the H_∞ LMI (12) enforces stability, it doesn't matter if this LMI contains the right-half complex plan. Finally, the SDP problem given by Eq. (18) is solved using the SDPT3 solver, leading to the following optimal solution.

$$\gamma \approx 0.847, \quad \lambda_i\left(A - B_2B_2^TQ^{-1}\right) \approx \{-0.0409, \ -67.3141, \ -67.3620\}, \quad ||B_2^TQ^{-1}||_2 \approx 1.27$$

Since $\gamma < 1$, robustness to the considered uncertainties (outflow coefficients) is guaranteed, thanks to the small gain theorem. Note that the numerical value of $||B_2^TQ^{-1}||_2$ indicates a noise amplification of $\approx 27\%$ on the control signal u_0. This may cause serious problems from a practical point of view. A solution to manage such a difficulty is proposed later using a Kalman filter, see remark 4.

With regards to the ISMC part, the main concern is the definition of the matrix ρ to enforce the sliding motion. Here, ρ has been chosen to $\rho = 5.10^{-5}I_2$. This numerical value has been fixed considering the saturation level of the pumps, i.e. it has been chosen to be close to the saturation level of the actuators $Q_{max} = 9.24.10^{-5}m^3/s$.

4.3. Simulation results

The proposed control law combining the *ISMC* and the H_∞ approach, is applied to the three-tank system presented above. In order to appreciate the effect of the *ISMC* and H_∞ parts, simulation results from a nonlinear simulator is first presented, ignoring the measurement noises and the effect of the DAC-DCA. The numerical method used to solve the integrations and differential equations is based on the Euler method with a sampling period equal to $T_s = 10ms$. This configuration is retained since they correspond to the one used for the real experiments, see section 4.4.

The considered experience corresponds to the following simulation: The system starts at the equilibrium point h_1^*, h_2^* and h_3^* respectively equal to $0.4m$, $0.2m$ and $0.3m$, a change of reference occurs at $t = 150s$, the new references to attain being fixed to $10cm$ more, i.e. $0.5m$, $0.3m$ and $0.4m$ for the first, second and third outputs. A simple second order polynomial path planing is also implemented for each output to track. Then, a sinusoidal variation of the outflow coefficients is simulated for $t \geq 200s$. Finally, a leak occurs in the first tank for $t \geq 300s$.

Because we consider the problem of output tracking, the definition of the control signals u_0 and u_1 defined in section 3, needs to be updated. The major difficulty is concerned by the sliding mode part since it is well known that the new form of u_0 is $u_0(t) = \mathcal{K}e_x(t) = B_2B_2^TQ^{-1}e_x(t)$ where $e_x(t) = x_c(t) - x(t)$ is the tracking error, x_c being the reference.
Based on Eq. (8), the following sliding function is chosen, the definition of u_1 being unchanged:

$$s(x,t) = B_2^T\left[e_x(t) - e_x(t_0) + \int_{t_0}^{t}(A - B_2B_2^TX)e_x(\tau)d\tau\right] \tag{27}$$

Figures 2 give the obtained simulation results. For comparison, the solution proposed by [1], i.e. the ARE-based solution described in section 2, is too tested. The obtained results are given in the figures 3. The following remarks can be made from these figures:

- Even if it can be observed that the behaviour of u_0 and u_1 differs between the LMI and the ARE solutions, the tracking errors exhibit very similar behaviour. In other words, the two solutions exhibit very similar control performance, which is reassuring. Note however, that the H_∞ performance of the ARE-based solution is found to $\gamma \approx 0.098$ with $||B_2^TQ^{-1}|| \approx 1.0734$ which illustrates that the LMI-based solution may be more conservative than the ARE-based solution and thus that, if it is not required poles assignment, the ARE-based solution should be preferred, see remark 3.
- The sliding surface is reached at the beginning of the simulation and is never left, thanks to the definition of the sliding function $s(x,t)$ given by Eq. (27) where it can be noticed that at $t = t_0 = 0s$, $s(x, t_0) = 0$, so the system always starts at the sliding manifold.
- An immediate consequence is that at $t = 300s$, when a leak occurs in the first tank (matched perturbation), it is immediately mitigated by the SM part of the control since the sliding motion is not left. This can be easily observed on the tracking errors. This can be too noticed on the behaviour of $u_1(t)$ and the sliding function $s(t)$, especially at $t = 300s$ where a change can be observed.
- Finally, it can be observed that at $t = 200s$, when the uncertainties manifest themselves, the H_∞ part of the control, i.e. $u_0(t)$, reacts whereas the SMC part seems to not react. This highlights the benefit of the proposed control scheme.

4.4. Experimental results
Finally, the LMI solution is implement within the control software of the three-tanks system. The considered experience is close to those considered previously, except for the outflow coefficients since it is obvious that they are physically inaccessible. The obtained results are given on figures 4. Clearly, due to the presence of noise, the quantization effect of the DAC-DCA, transient saturations of the control signals and probably many other unknown perturbations

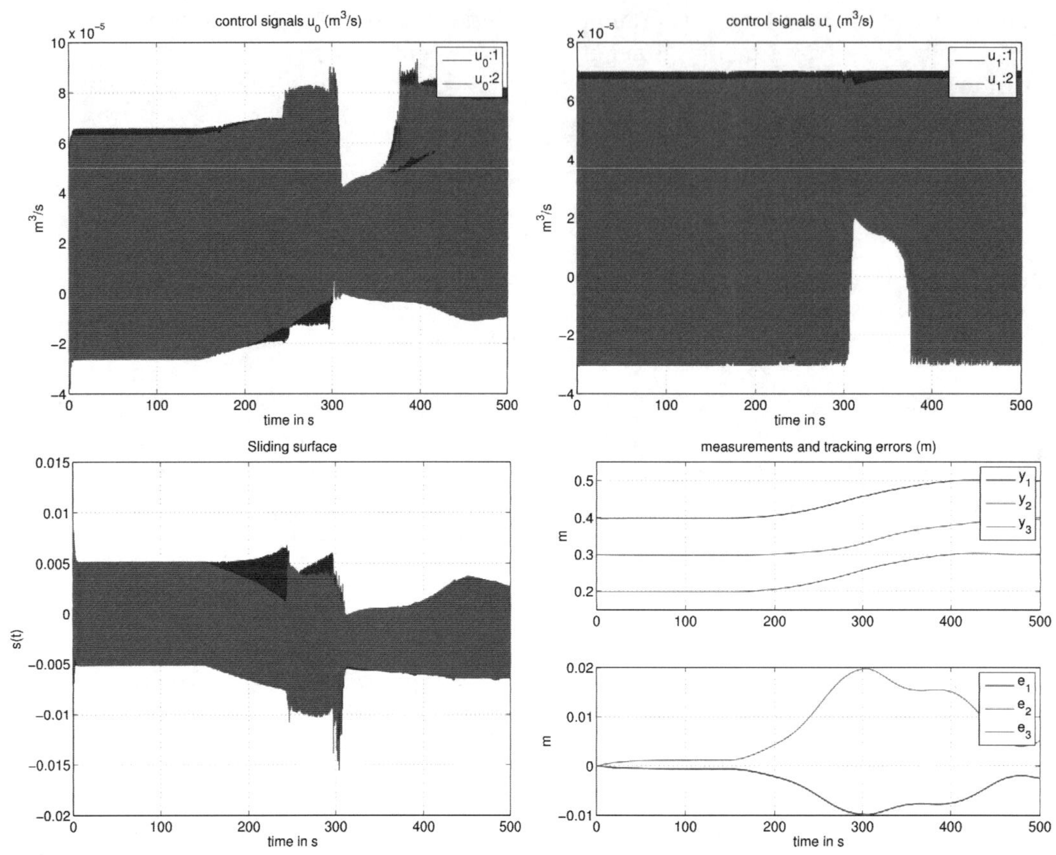

Figure 2. From top left to bottom right: control signals u_0, u_1, sliding function s and output signals $y = x$ and tracking error e_x. LMI solution

and disturbances, it is very hard to identify from the behaviour of $u0(t)$ and $u_1(t)$, how the H_∞ and SM parts of the control co-operate. However, it can be noticed the major benefits of both the techniques, i.e. the tracking errors are small, the leak is mitigated (but not instantaneously even if it seems that the sliding motion is not left), and the overall control system is stable despite the presence of the uncertainties. This tends to demonstrate that the proposed solution operates well. Note that the third output is correctly tracked as opposed to the simulations. A deeper investigation into the situation reveals that the outflow coefficients didn't really vary during the real experiment.

Remark 4. *It is well know that SM controllers are very sensitive to measurement noise. Thus, a Kalman filter is used to attenuate the measurement noise of the piezo-resistive pressure sensors, thus providing a "smooth" estimate* $\hat{h}_i, i = 1, 2, 3$ *of the three water levels* $h_i, i = 1, 2, 3$. *The model used correspond to Eq. (19) with* $\theta = \theta_0$ *and* $f = 0$ *and with covariance matrices fixed to* $Q = 0.5I_3$ *and* $R = I_3$.

5. Conclusions

In the proposed manuscript, a state feedback control based on *ISMC* and H_∞ approaches is proposed. The control gain (for H_∞ part) is calculated based on the LMI framework combined with poles placement constraints in order to improve the controller performances. An application on a three-tank pilot system is considered with satisfying results. Future work is concerned by

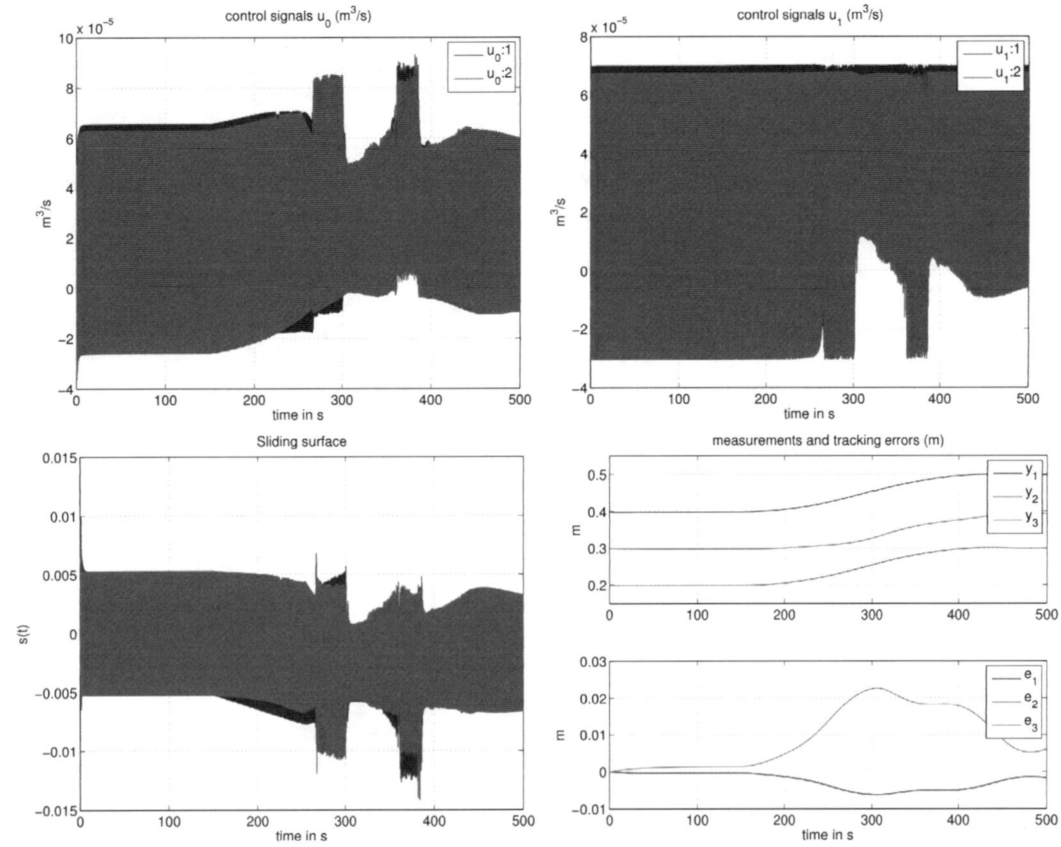

Figure 3. From top left to bottom right: control signals u_0, u_1, sliding function s and output signals $y = x$ and tracking error e_x. ARE solution

the establishment of a formal proof of stability of the overall control law, see remark 1.

Appendix A. Parameters of the three-tank system

$S = 0.0154m^2$, $Sn = 5.10^{-5}m^2$, $h_{max} = 60cm$, $Q_{max} = 9.24.10^{-5}m^3/s$, $0.58 \leq a_1 \leq 0.62$, $0.78 \leq a_2 \leq 0.82$, $0.58 \leq a_3 \leq 0.62$, $Sl = Sn$, $h_1^* = 40cm$, $h_2^* = 20cm$, $h_3^* = 30cm$.

References

[1] Castanos F and Fridman L 2006 *IEEE Transactions on Automatic Control* **51** 853–858
[2] Willems J 1971 *IEEE Transactions on Automatic Control* **16** 621–634
[3] Apkarian P, Gahinet P and Becker G 1995 *Automatica* **31** 1251–1261
[4] Utkin V, Guldner J and Shi J 1999 *Sliding Modes in Electromechanical Systems* (London, U.K.: Taylor & Francis)
[5] Shtessel Y, Edwards C, Fridman L and Levant A 2014 *Sliding Mode Control and Observation* (Springer New York Heidelberg Dordrecht London)
[6] Cao W J and Xu J X 2004 *IEEE Transaction on Automatic Control* **49** 1355–1360
[7] Edwards C and Spurgeon S K 1995 *International Journal of Control* **62** 1129–1144
[8] Bag K, Spurgeon S K and Edwards C 1997 *IEE Proc.-Control Theory Appl.* **144** 209–216
[9] Edwards C, Akoachere A and Spurgeon S K 2001 *IEEE Trans. Automat. Contr* **46** 115–119
[10] Levant A 2003 *International Journal of Control* **76** 924–941
[11] Stessel Y 1997 *IEEE Trans. Automat. Contr.* **42** 1282–1286
[12] Poznyak A, Stessel Y, and Jimenez C 2003 *IEEE Trans. Automat. Contr.* **48** 2141–2150

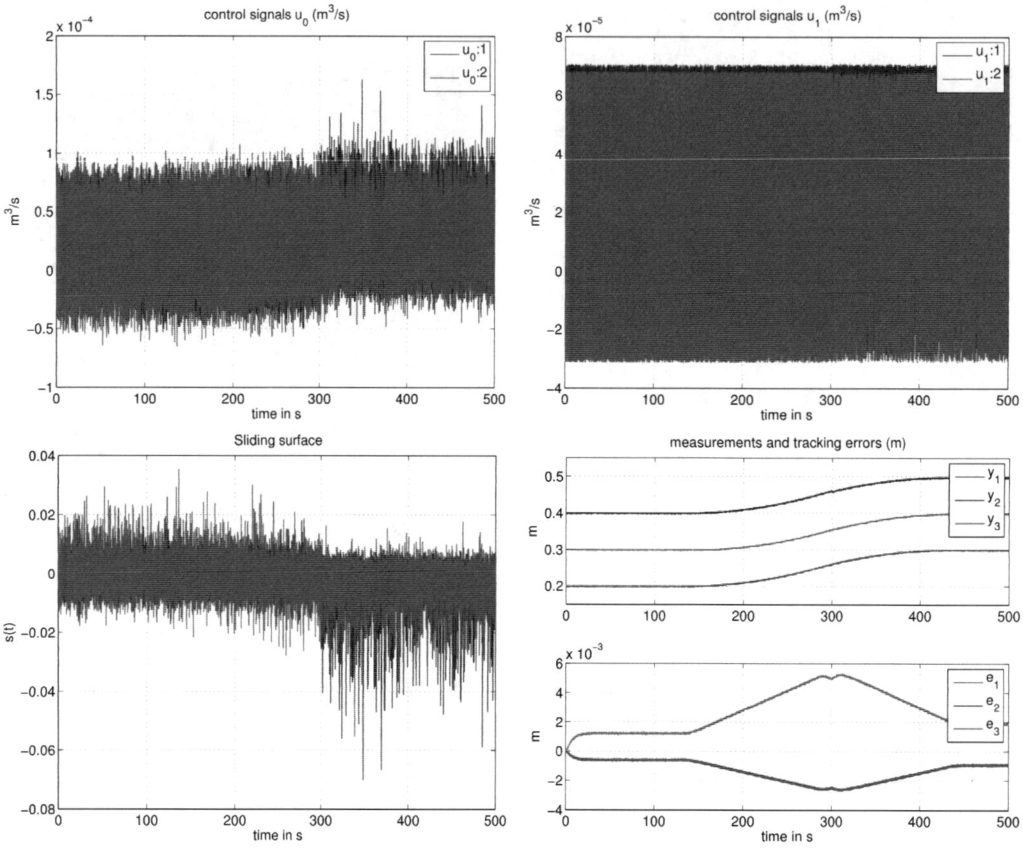

Figure 4. From top left to bottom right: control signals u_0, u_1, sliding function s and output signals $y = x$ and tracking error e_x. LMI solution

[13] Orlov Y, Perruquetti W and Richard J 2003 *Asian Journal of Control* **5** 568–577
[14] Choi H 2003 *IEEE Trans. Automat. Contr.* **48** 1634–1638
[15] Matthews G and DeCarlo R 1988 *Automatica* **24** 187–193
[16] Utkin V and Shi J 1996 *Proc. Conference on Decision and Control, Kobe, Japan* **Dec.** 45914596
[17] Chilali M and Gahinet P 1996 *IEEE Transactions on Automatic Control* **41** 358–367
[18] Scherer C, Gahinet P and Chilali M 1997 *IEEE Transactions on Automatic Control* **42** 896–911
[19] Chilali M, Gahinet P and Apkarian P 1999 *IEEE Transactions on Automatic Control* **44** 2257–2270
[20] Doyle J, Glover K, Khargonekar P and Francis B 1989 *IEEE Transaction on Automatic Control* **3** 831–847
[21] Apkarian P and Gahinet P 1995 *IEEE Transactions on Automatic Control* **40** 853–864
[22] Nemirovski A and Gahinet P 1994 The projective method for solving linear matrix inequalities *American Control Conference* (Baltimore, Mariland, USA)
[23] Xie R, Wang X and Li Y 2008 H_∞ state feedback control for the stabilization of the three Euler angles of helicopter based on LMI *International Conference on Intelligent Computation Technology and Automation*
[24] Chilali M and Gahinet P 1996 *IEEE Transactions Automatic Control* **41** 358–367
[25] Henry D and Zolghadri A 2006 *Control Engineering Practice* **14** 1081–1097
[26] Beck C, Doyle J and Glover K 1996 *IEEE Transactions on Automatic Control* **41** 1466–1477
[27] Cockburn J and Morton B 1997 *Automatica* **33** 1263–1271
[28] Varga A, Looye G, Moormann D and G G 1998 *Mathematical Modelling of Dynamical Systems* **4** 249–274
[29] Beck C and Doyle J 1999 *IEEE Transactions on Automatic Control* **44** 1802–1813
[30] Hecker S, Varga A and Magni J 2005 *Aerospace Science and Technology* **9** 173–180

Unified framework for generation of 3D web visualization for mechatronic systems

O. Severa, M. Goubej and J. Konigsmarkova
NTIS, University of West Bohemia, Pilsen, Czech Republic
E-mail: {osevera, mgoubej, jkonig}@ntis.zcu.cz

Abstract. The paper deals with development of a unified framework for generation of 3D visualizations of complex mechatronic systems. It provides a high-fidelity representation of executed motion by allowing direct employment of a machine geometry model acquired from a CAD system. Open-architecture multi-platform solution based on latest web standards is achieved by utilizing a web browser as a final 3D renderer. The results are applicable both for simulations and development of real-time human machine interfaces. Case study of autonomous underwater vehicle control is provided to demonstrate the applicability of the proposed approach.

1. Introduction

Computer aided development (CAD) has become a standard in many industrial automation domains including mechatronics, robotics and motion control systems. Various supporting software tools are used in the individual stages of the prototype development cycle. This process typically involves a set of subsequent steps:

(i) Problem formulation
(ii) Mathematical model synthesis
(iii) Mechanical design
(iv) Control system design
(v) *Employment of visualization framework*
(vi) Model and hardware-in-the-loop (MIL, HIL) simulation (preferably with *3D visualization*)
(vii) Prototype manufacturing
(viii) Implementation of control system and human-machine interface (HMI) (preferably with *3D visualization*)

The geometric modelling is carried out in some of the available CAD systems. Several alternatives exist for the mathematical modeling part such as Matlab/Simulink/Simscape by MathWorks, MapleSoft Maple, Wolfram Mathematica and Mathmodelica, OpenModelica or Dymola and special plugin toolboxes e.g. [1], [2], [3]. These tools can be used in the stages of model, software, processor and hardware-in-the-loop simulations (see [4] for detailed description of the individual simulation types). Validation of the model and control system is followed by manufacturing of the prototype and implementation of the control algorithms into a target hardware platform.

An important part of the development cycle is proper visualization of controlled plant behavior. This is essential both for simulations and real machine operation when dealing with complex mechatronic systems. High fidelity three dimensional representation of the executed motion using a geometric model acquired from the CAD system can simplify the process of simulation or real machine diagnostics by revealing possible bottlenecks in mechanical and control design. The above mentioned modeling and simulation tools offer several possibilities of connection of the mathematical model with the 3D machine geometry. However, the visualization tools cannot be simply used for both real-time HIL simulations and development of the HMI for the final prototype system. On the other hand, most of the commercial HMI tools for industrial use (InTouch by Wonderware, GENESIS by Iconics, WinCC by Siemens, etc.) are focused on the data processing. One can create extensive HMIs with multiple nodes, but usually the HMI is bound to some vendor specific client application and there is a limited support for direct employment of the CAD geometric model. Moreover, commercial software tools are burdened with high license fees. Therefore, there is a need for a cost-effective open-architecture system which

- is able to import models from various CAD systems,
- provides a high-fidelity real-time 3D visualization,
- is multiplatform,
- can be connected both to simulation model and real plant (preferably without significant changes in the HMI configuration)

The paper presents a unified framework based on latest web technologies which is intended for generation of 3D visualizations for mechatronic systems. Its development is built on the previous results described in [5] and [6]. The main purpose of the framework is to support rapid prototype development. Platform-independent solution is achieved by employing web browser as the final 3D HMI renderer. This allows to run the HMI on desktops with various operation systems, on tablets or even mobile phones. It can be connected both to a simulation model and real-time control system using a proper communication interface. A case-study of HMI development for autonomous underwater vehicle is provided to demonstrate the applicability of the proposed approach.

The paper is organized as follows: section 2 describes the individual components of the tool chain, section 3 shows the employment of the developed framework on the example of underwater vehicle. Finally, some concluding remarks and possible directions for a future research are given.

2. Framework for 3D web visualization of mechatronic systems

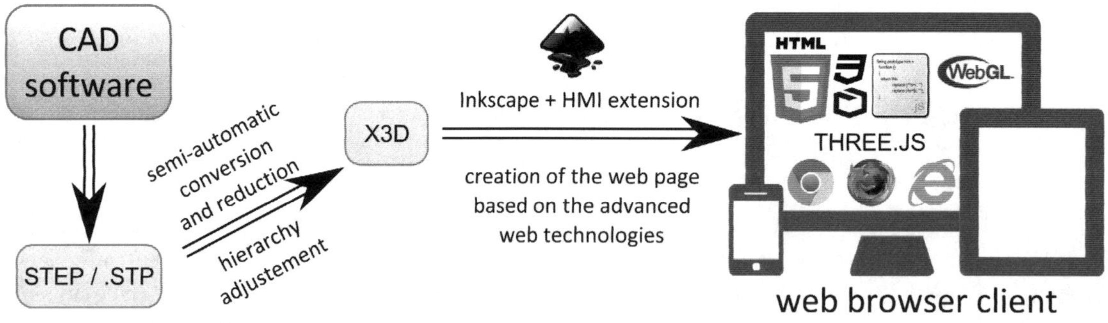

Figure 1: From CAD to 3D visualization

The framework extends the described prototype development steps with step 5 (*Employment of visualization framework*) and adds 3D HMI both for MIL and HIL simulations (step 6). Finally the same 3D HMI can be used for the final product (step 8). The overall view on the whole procedure of 3D HMI creation is depicted in Fig. 1. At the beginning, there is a CAD drawing of the mechanical part of the machine/robot. Then, the model is exported to a 3D exchange format. There are plenty of exchange formats like STEP, IGES, DXF, Parasolid XT, DRG, etc. but only some of them are suitable for 3D HMI because the format must preserve materials and hierarchy of the model. For the framework, the STEP format was chosen. Closer description of reasons will follow.

When the model is exported from CAD system, it must be converted to a web supported format, simplified and tagged for animation. Some of the actions are done automatically, some of them must be done by user / developer. Once again, there are several web supported 3D formats like X3D, VRML, COLLADA, etc. For the unified framework, the X3D was chosen since it is very similar to SVG, further explanation will follow.

When the 3D model is ready, it must be bound with a simulation / control system core. The Inkscape editor with HMI extension is used for this purpose. One just connects the chosen parts of the 3D model with signals from the core and sets appropriate animations rules. After the configuration is done, finally the 3D model is embedded to a HTML5 enabled web page automatically. Such a page can be served via a web server or run locally from the file system.

STEP - 3D exchange format The whole visualization is based on the 3D model which was exported from CAD (Computer aided design) system. There are several 3D exchange formats, where STEP (Standard for the Exchange of Product model data) is one of the most supported. This format is standardized under ISO 10303-21. Description of the 3D scene like volumes, hierarchy, materials, etc. is written in the text form using EXPRESS data modeling language. The main advantages of the STEP format are wide support from various CAD systems and the fact that the export preserves model hierarchy and materials (colors, opacity, shininess, etc.). The hierarchy of components of the model is very important especially for animation. The main disadvantage of the STEP format is that there is no suitable 3D web renderer which supports this format directly. Thus, further modifications are necessary for preparation of the model for visualization purposes.

Three.js - WebGL renderer There are several ways how to render 3D objects inside the webpage. Few years ago, the only way how to render 3D was to use third-party plugins such as Java applets [6], Adobe Flash plugins or various ActiveX components. Nowadays, the WebGL [7] is available. This technology allows to render 3D graphics directly in a webpage without any plugin. WebGL can be directly used from JavaScript, but there are also libraries built on top of the low level WebGL API. Project *Three.js* was used in this case. This open-source library is still in development process by wide community. It allows developer to easily render scenes, control cameras and animations, control lights, shadows, etc. This library supports also scene import in several formats.

The whole web rendering segment is under extensive development, thus described technologies are not fully supported in every web browser. It means that the visualization can be used only in advanced browsers i.e. Google Chrome / Chromium project, Mozilla Firefox or latest versions of Internet Explorer and Microsoft Edge browser.

Blender / Vivaty Studio - 3D editor During the 3D HMI development process it is necessary to simplify and adjust the hierarchy of the 3D model. This can be done manually or using available 3D editor. In the previous version of the framework, the Vivaty Studio was used to adjust hierarchy of the model. Since the software is no longer supported the Blender (free and open source 3D creation suite) is used. Both programs support hierarchy rearrangements, model simplification and overall editing functions. The final 3D model is then exported back to the X3D format.

Inkscape - the vector graphic editor The last tool in the row is Inkscape which is free and open source professional vector graphics editor for Windows, Mac OS X and Linux. It support creation and editing of vector graphic stored in Scalable Vector Graphics (SVG) format. The SVG is a XML-based vector image format for two-dimensional graphics with support for interactivity and animation. The SVG specification is an open standard developed by the World Wide Web Consortium (W3C) since 1999. Current version of the SVG specification is 1.1 standardized on 16th of August 2011 [8]. SVG is supported in all modern browsers (IE, Chrome, Firefox, Opera, etc.). The Inkscape in the 3D framework is used only as a supporting tool. It is extended with several plugins which help user / developer to configure and design the final 3D HMI. One can create the 3D HMI without the extended Inkscape but it requires more manual coding in HTML and Javascript. The description of extension is given in subsection 2.1.2.

HTML5 and Javascript libraries Final 3D HMI is running inside web browser thanks to the HTML5. HTML5 is specification for web pages which supports e.g. semantic elements like `<header>`, `<footer>`, `<article>`, `<section>`, form control attributes, multimedia elements like `<audio>`, `<video>` and finally graphic elements `<svg>` and `<canvas>`. The `<canvas>` element in this case uses WebGL (Web Graphics Library) which is a JavaScript API for rendering interactive 3D and 2D graphics within any compatible web browser without the use of plug-ins. WebGL does so by introducing an API that closely conforms to OpenGL ES 2.0.

Except plain HTML, the client application uses scripts in JavaScript. These scripts are responsible for interaction of the webpage with the user. There are several libraries in use. For example jQuery [9] is an extensive library for DOM (Document object model) manipulation. This library allows easy manipulation with all nodes of the DOM.

The next section describes how the whole development from user/developer point-of-view is done.

2.1. Development of the 3D HMI

In the Fig. 1 is shown that the first step for 3D HMI is a 3D model of the visualized machine or robot designed using CAD tool. Usually the designer creates the whole system with all the necessary components including gears, screws, nuts, cables, etc. Thus, a semi-automatic conversion is needed.

2.1.1. Preparation of the 3D model When the model is exported to the STEP file format, it must be converted to the X3D. Such a conversion can be done by several tools. One of them is a CAD Exchanger which we use in this case. Unfortunately, it is not open source software, so there is a room for improvements. The next step is adjustment of the hierarchy of the model in 3D editing software. It is a semi-automatic process in Vivaty Studio / Blender, where one needs to create group of the components for the animation. Each group need to contain all the elements which animates together.

For example the underwater vehicle has 8 motors. Each of them has a propeller, so each propeller represents one group which will rotate, consequently each motor is inside of a machine group which will rotate and translate simultaneously, see the use case section for more details.

In 3D HMI, the selected nodes/elements/groups are animated according to data from the simulation/control core. These selected nodes must be tagged. This is done via Inkscape extension which allows configuration of the animation and interconnection of the signals (see the following subsection). Finally, the 3D model is loaded to the WebGL renderer inside the webpage. The library Three.js do not directly supports X3D format, thus the framework contains special loader for X3D files. This loader allow users to load large X3D files (more than 10 MB) into the browser DOM and to convert them into the inner Three.js 3D representation as a source for the WebGL renderer. During the loading process, the connection to the simulation/control core is established and the selected nodes are animated based on the HMI configuration.

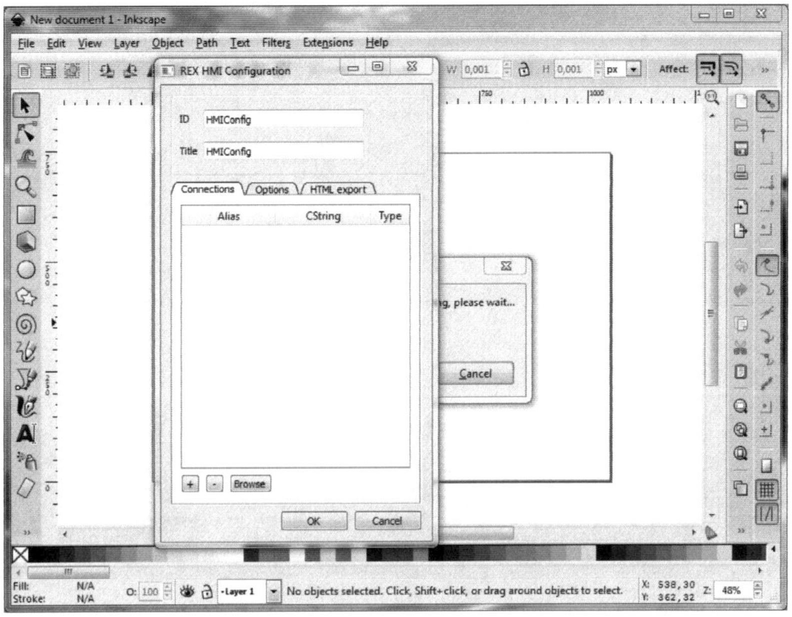

Figure 2: Inkscape with HMI configuration extension

2.1.2. Web-page design One has several ways how to design a webpage (HMI) for various virtual models or real devices. First one is to implement everything by manual coding using HTML language. The unified framework contains "low level" javascript API which allows interconnection between the simulation/control core and the HMI. However, such an option is meant only for special adjustments. For fast prototyping of the simulator, several extensions for Inkscape vector graphic editor were developed.

Basic principle for any HMI created by the framework is a single page which scales according to the size of the screen on the device(desktop, tablet, mobile phone, etc.). This presumption allows to create multiplatform and multidevice HMIs. When one run the Inkscape editor there is only one main page. All the components inserted to the page will be automatically generated to the web page in the corresponding positions. There are several extensions written in Python language which support such a behavior.

The whole concept uses SVG components with javascript files which control the final HMI behavior. These two files are called *HMI module*. Each module contains configuration (inside SVG file) and control part (JavaScript file). The configuration is stored in SVG file inside the element `<description>` in JSON (Javascript Object Notation) format. When user inserts new component to the page, the configuration is cloned and then one can use the HMI extension for Inkscape to alter it.

When the page is ready, it is exported to the final HTML file. There is a basic template which contains all the necessary libraries like jQuery, Three.js and at last the HMI library for the framework. This template is extended with all the javascript controllers for HMI modules inside the webpage. When the HTML file is loaded to the browser, the HMI library traverses all components and starts appropriate controllers.

There are several HMI extensions for Inkscape, the most important is called the *Element editor*. Following list will explain what are the extensions used for.

Element editor is an extension (shown in Fig. 3) which is used for editing properties of HMI

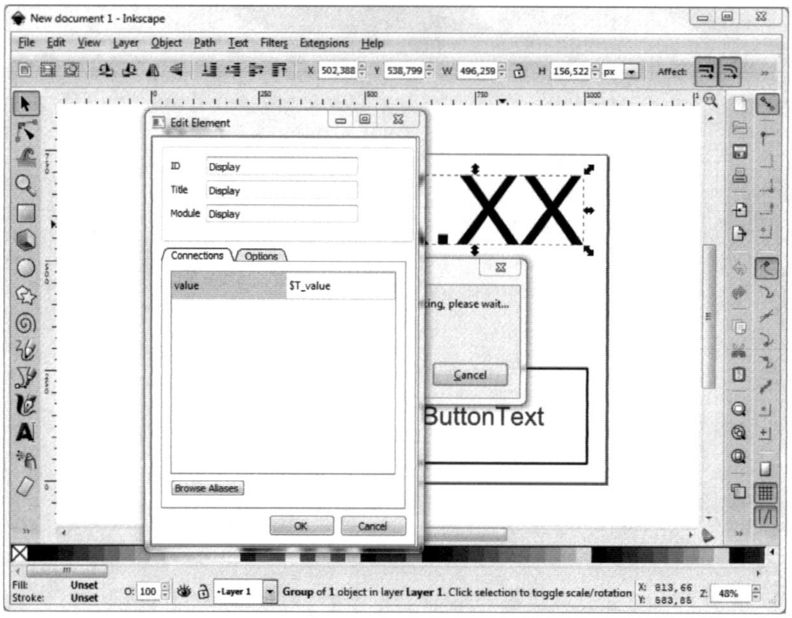

Figure 3: Inkscape with element editor extension

modules. It changes the JSON configuration inside the `<description>` element inside the SVG (HMI module). One can edit *Title, Module, Options* and *Connections* using this tool.

- *Title* is shown in the web browser after hovering the cursor above element and is also used in Connections as shortening, during the initialization is the wildcard $T replaced by the title.
- *Module* identifies uniquely the type of component and assigns the proper JavaScript functionality.
- *Connections* tab contains names of signals that trigger the element behavior.
- *Options* tab edits the additional options of the component.

HMIConfig editor is similar to the Element editor. The options for the whole visualization project are edited here. HMIConfig editor is shown after calling the Element editor with no element selected. It contains three main tabs *Connections*, *Options*, *HTML export*.

- *Connections* are the signals used in visualization assigned to the signals from the simulation/control core. If there is a proper scheme already executed in the core, the signals might be browsed online.
- *Options* tab contains an IP address of the target device and other options for the whole project.
- *HTML export* tab contain configuration of the web page exporter.

Library Browser extension just opens the windows explorer with the selection of HMI modules.

HTML Builder extension builds the web site into the target folder, in case the valid location is set in the HMI Configuration menu. This extension unzips the webpage template and inserts the content of the current SVG project. It also loads all the javascript controllers for the used modules and links them to the page. Such a exported project can be run from webserver or locally from generated file.

Group animation is the last extension (shown in Fig. 4) which is different from all the previous ones. It creates a HMI module from any group by defining its behavior (translation, rotation and changes of color, size and opacity). Such a module can be then edited by Element editor.

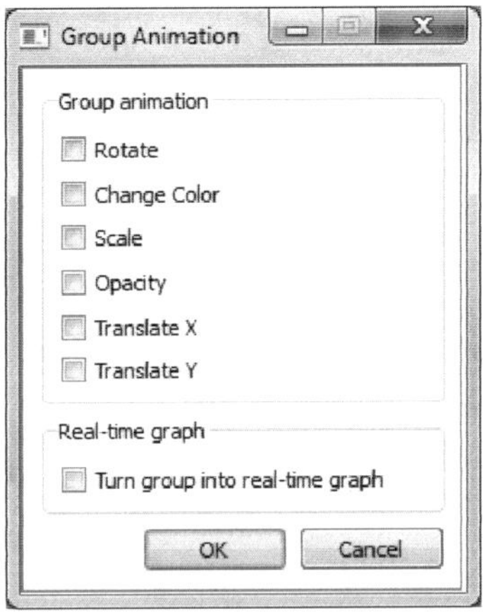

Figure 4: Group animation component

3D HMI module represents one of the modules. In this case the Inkscape editor user just create the placeholder for the Three.js 3D canvas. Then the Element editor is used to adjust connections for selected nodes and configuration for animation assigned to those nodes.

2.1.3. Communication The 3D HMI uses the client server architecture (depicted in Fig. 5). Client (Web Browser) is connected through Socket TCP/IP connection to the special server which translates data from and to the simulation / control core. This type of direct webpage communication is called WebSocket [10].

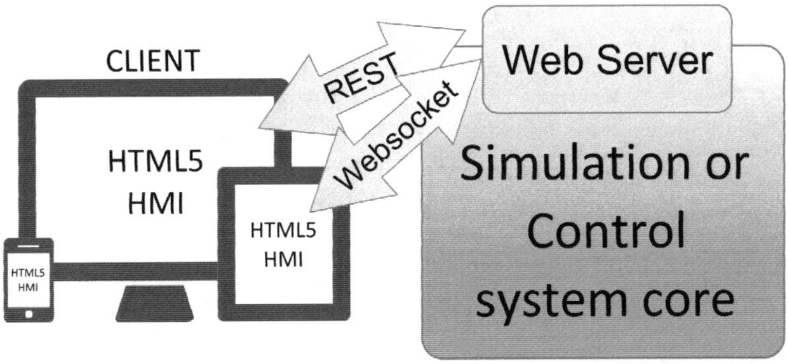

Figure 5: 3D HMI communication schema

The HMI also partially supports communication over REST web services, but this direction is planned for future version of the framework. The communication uses data messages serialized into JSON format. Current version of the simulation core and following control algorithm is done using REX Control system, so the WebSocket communication protocol is designed for the system as well. On the other hand the 3D HMI is build independently so with slight adjustments of the communication component it can be connected to different simulation and control cores.

3. Case-study: Prototype of underwater vehicle

This visualization technology was used during the development of underwater vehicle for Non Destructive Testing inspection, for a more detailed description see [11]. The vehicle was designed as a neutrally buoyant rigid body and works in two modes - the remote operated vehicle (ROV) and the autonomous underwater vehicle (AUV). An important part of this vehicle is a platform with phased array probe for weld testing. The whole vehicle is designed to be symmetrical with symmetrical arrangement of thrusters - four thrusters in the horizontal vehicle plane and four thrusters in the vertical plane. The prototype construction and thrusters configuration are depicted in Fig. 6. The mathematical model was created in Matlab/Simulink and implemented in real time control system REX. The model is used for hardware in the loop simulations. This use case shows how the mathematical model of the vehicle is extended with 3D HMI such that MIL and HIL simulations seem closer to the reality.

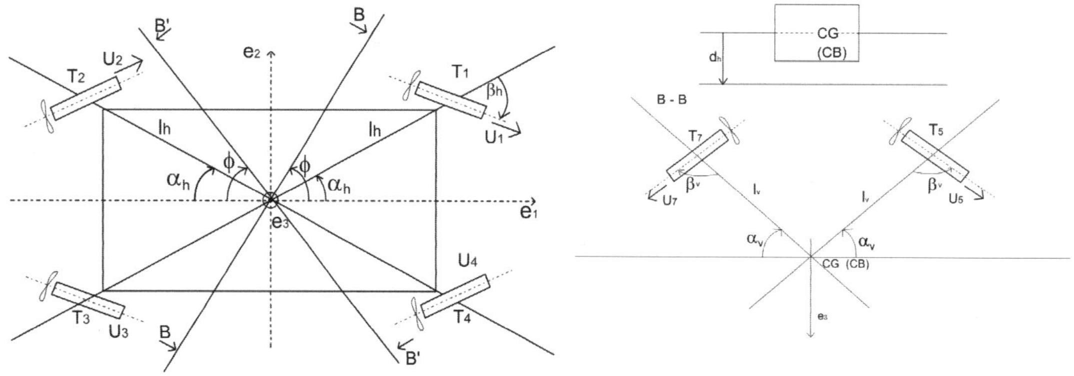

Figure 6: Symmetrical thrusters configuration.

3.1. Real-time simulation core

The real-time simulation is realized by REX Control system [12]. REX implements a software PLC supporting Function Block Diagram (FBD), Sequential Function Chart (SFC) and C-like programming language. The REX implementations of FBD and SFC are with a few exceptions compatible with IEC 61131-3 and the C-like language substitutes Structured Text language from IEC 61131-3. Since the REX Control system blocks are compatible with MATLAB/Simulink, the first drafts of the simulator and control can run directly inside MATLAB environment using REX Lib library [13].

3.2. Mathematical model

To simulate the vehicle motion in the real time and with sufficient accuracy, some useful but nonrestrictive simplifications were introduced in the construction design and motion description. The shape of the vehicle is considered almost cubic, with coincident center of gravity (CG) and center of buoyancy (CB), so the hydrodynamic effects will be very similar in all principal axes.

A low velocity of the vehicle is assumed, therefore the viscous effects can be modeled as a drag in the form of damping and the added mass as well as inertia parameters are modeled as external forces.

The dynamics of the submerged vehicle can be described by the Kirchhoff's motion equations [14]. By following considered simplifications, the translation and rotational motion can be decoupled. The dynamic equations of motion (detail described in [11]) lead to the system of nonlinear ordinary differential equations with the state variables - position and velocity of the center of gravity in the inertial frame and the rotation angles roll, pitch and yaw of the vehicle and their derivatives. The system is solved by numerical Runge-Kutta method of 4th order, which seemed to be sufficiently accurate and fast. The dynamic equations together with the numeric method are implemented in the real time control system REX [12] as a block ROUV (Remotely Operated Underwater Vehicle).

3.3. Control system

The controlled variables are the velocities in the body fixed frame - in surge and sway directions, the derivative of yaw, the depth and the angles roll, pitch and yaw. For movement in camera direction we can use the velocity in heave direction. The thrusters were collected to the virtual inputs according to their effect on the vehicle motion.

The stabilization control law is provided by six PID controllers whose parameters were determined by the pole placement method for the transfer function obtained from the linearized model. The poles were selected according to the maximal real stability radius criterion for the system to be sufficiently robust and assigned in the way to ensure similar properties for variables with similar sense. The control scheme of Remote Operational Underwater Vehicle is shown in Fig. 7, the visualization of the controlled model is in Fig. 9.

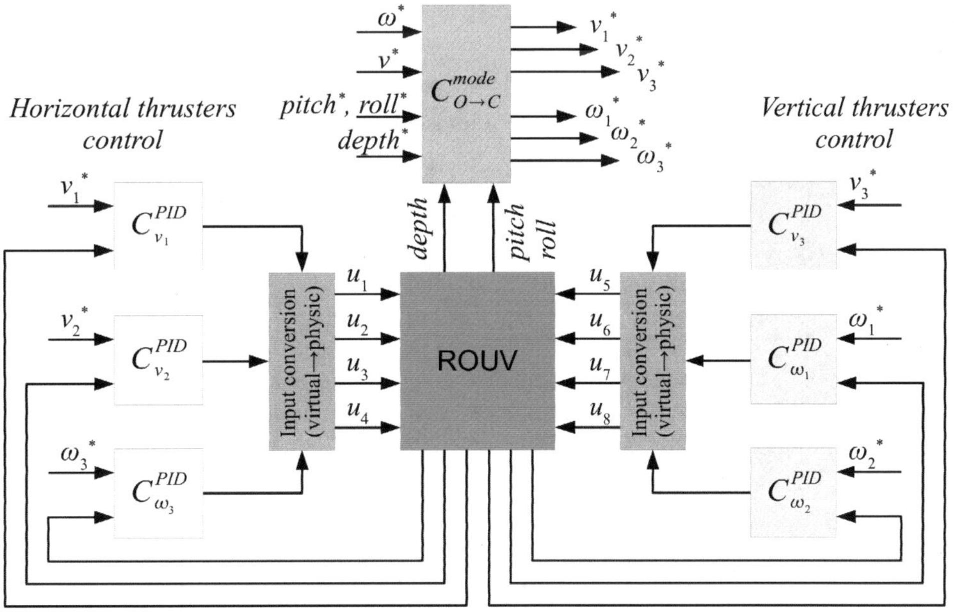

Figure 7: Control scheme of ROUV.

Journal of Physics: Conference Series **659** (2015) 012053 doi:10.1088/1742-6596/659/1/012053

3.4. 3D Human–machine interface

The 3D HMI for the simulator and as well as real device was created using the described framework. At the beginning there was CAD drawing with simple hull and proper drawing of motors. One can see (in Fig. 8) that the drawing is already imported to the Vivaty Studio (a tool for hierarchy adjustment). After the X3D sceen was created (after simplification and proper hierarchy adjustment), the layout of the webpage was designed in Inkscape editor.

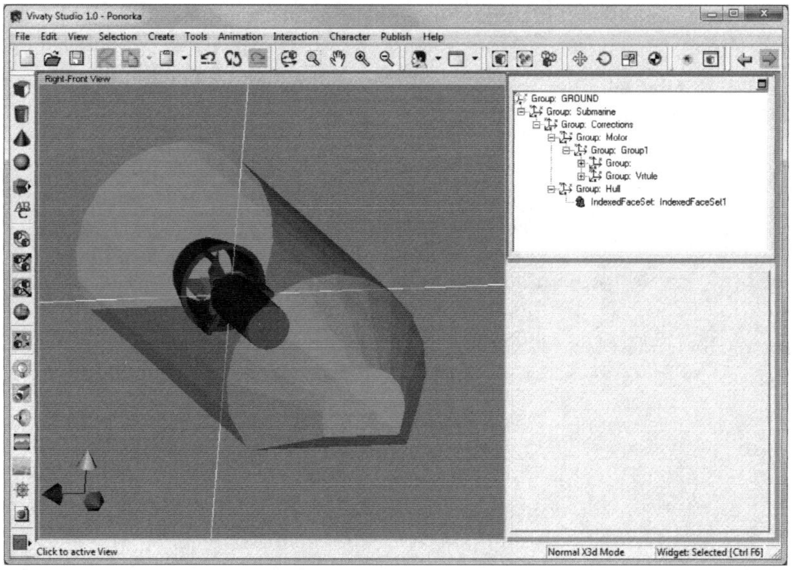

Figure 8: The CAD drawing imported into the Vivaty Studio editor

The simulation model allows user to change parameters of the UAV. One of them is positioning of the thrusters. One can see that in the CAD drawing there is only one thruster placed in the center of the hull. The framework allows user to create additional scripts (in JavaScript language) which help during initialization phase of the 3D HMI. The script for the UAV reads the parameters of the position and rotation of the thrusters (motors) and adjusts the 3D model accordingly as is shown in Fig. 9. The UAV is controlled via remote controller, thus 3D HMI contains only displays and shows the current state of the vehicle. User has the options to adjust; the camera tab, the controls tab and at last the logs tab.

- The camera tab contains a set of cameras which follows the UAV. One can choose between following mode, bird view or static mode.
- The controls tab has only one control button which is available for simulation only. It resets the model to the default state.
- The logs tab stores all the messages which are generated from the HMI (errors, warnings, info, etc.)

The initialization script also adds some helper components for the 3D HMI. First of all the path of the UAV is drawn, also the 3D HMI displays actual power generated by each propeller (motor). The UAV in motion is shown in Fig. 9.

3.5. REST-API interface

Except the WebSocket interface, the simulation and control core of REX Control System exposes REST API, thus the simulation can be also controlled via Web Services. One is able to substitute

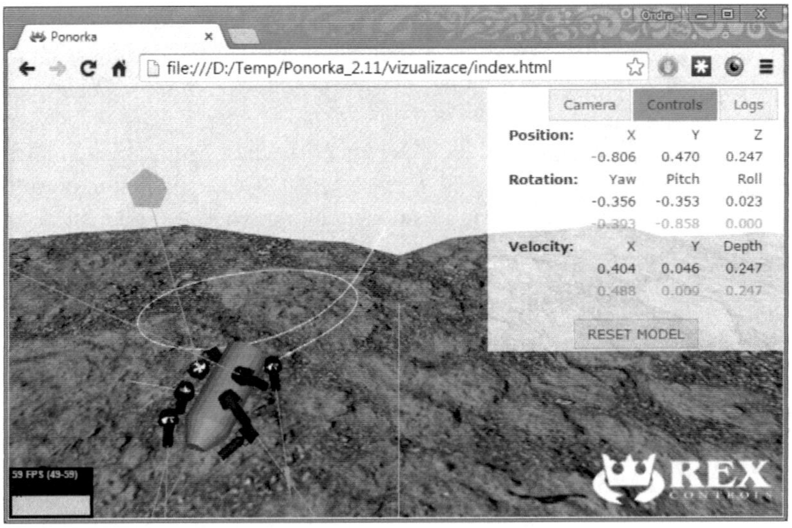

Figure 9: 3D HMI of UAV in motion

the remote control inputs and set the requested setpoints for the controllers. Each input has unique URL where one can use HTTP GET or PUT / POST request. GET is used for reading current value and PUT/POST for changing the value. For example the setpoint for the speed in X axis can be change calling PUT to the resource at `http://UAV-IP/api/uuv/speed-x`. The main advantage is that the simulation and later the real device can be easily controlled by any REST client.

4. Real prototype

The mathematical model with 3D HMI served as a base for the real UAV development. The detailed description of the UAV parameters and final version of the HMI is out of the scope of this paper. It will be shown in future paper. In Fig. 10 is depicted the real photo of the prototype.

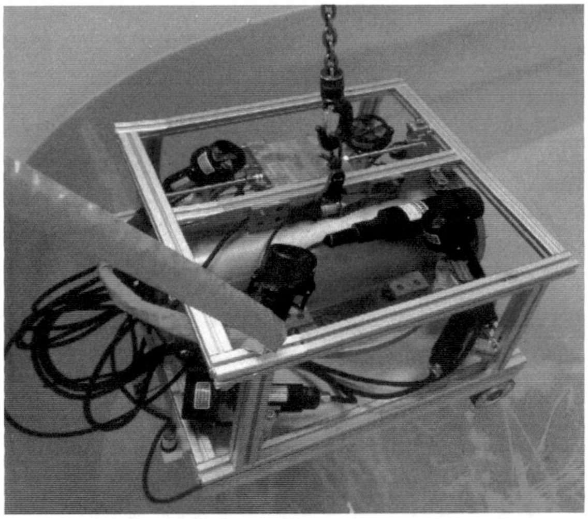

Figure 10: Real prototype of the UAV

5. Conclusion

The presented framework provides supporting software tools for 3D visualization of complex mechatronic systems. It offers a possibility of direct employment of CAD geometric model. Modern web browsers are used for the final 3D rendering to achieve platform independent solution. The human-machine interface can serve both for simulations and final prototype control which leads to shorter development time. Moreover, utilization of different software tools in the individual stages of prototype development is avoided. The introduced web-based architecture can also serve for simple creation of virtual laboratories with attractive visualization features.

The future work will be focused on further tool chain improvements. Since Vivaty Studio editor is no longer supported and Inkscape can be used only for configuration of the scene, the key node is the Blender editor. Custom extensions can be added thanks to its plugin capability with Python language support. Such an extension could directly connect to the simulation/control core via the WebSocket or REST, support simple model hierarchy adjustments, tagging groups for animation, etc. Another issue is the possibility of direct import of the STEP files without using any additional interchange formats.

Acknowledgments

This work was supported by **CIDAM** project no.: TE02000103 and by the project **LO1506** of the Czech Ministry of Education, Youth and Sports.

References

[1] Fedák V, Ďurovský F and Üveges R 2014 Analysis of Robotic System Motion in SimMechanics and MATLAB GUI Environment *MATLAB Applications for the Practical Engineer* (InTech) chap 20 ISBN 978-953-51-1719-3

[2] Martin-Villalba C, Urquia A and Dormido S 2009 Visualization and interactive simulation of modelica models for control education *Chinese Control and Decision Conference (CCDC), 2009.* pp 3076–3081

[3] Höger C, Mehlhase A, Nytsch-geusen C and Isakovic K 2012 Modelica3D - Platform Independent Simulation Visualization *Proceedings of the 9th International MODELICA Conference* (Linköping University Electronic Press; Linköpings universitet) pp 485–494

[4] Severa O and Cech M 2012 Rex - rapid development tool for automation and robotics *IEEE/ASME International Conference on Mechatronics and Embedded Systems and Applications (MESA), 2012* pp 184 –189

[5] Severa O, Jager A and Stetina M 2013 Virtual model of manipulator for advanced vessel integrity inspection build on latest W3C standards *International Conference on Process Control (PC), 2013* pp 540–545

[6] Severa O, Čech M and Balda P 2011 New tools for 3D HMI development in Java *Proceedings of the 2011 12th International Carpathian Control Conference, ICCC'2011* pp 342–346

[7] Khronos Group 2015 *WebGL - OpenGL ES 2.0 for the Web* URL `http://www.khronos.org/webgl/`

[8] The World Wide Web Consortium (W3C) 2011 *Scalable Vector Graphics (SVG) 1.1 (Second Edition)* URL `http://www.w3.org/TR/SVG11/`

[9] jQuery Foundation 2015 *jQuery* URL `http://jquery.com/`

[10] 2015 *WebSocket* URL `http://www.websocket.org`

[11] Blaha L, Schlegel M and Konigsmarkova J 2014 Design and control of underwater vehicle for NDT inspections *2014 IEEE/OES Autonomous Underwater Vehicles (AUV)* (IEEE) pp 1–6 ISBN 978-1-4799-4344-9

[12] Balda P, Schlegel M and Štětina M 2005 Advanced control algorithms + Simulink compatibility + Real-time OS = REX *IFAC Proceedings Volumes (IFAC-PapersOnline)* vol 16 pp 121–126

[13] 2015 *REX system function blocks Reference manual* URL `http://www.rexcontrols.com`

[14] Fossen T I 1994 *Guidance and control of ocean vehicles* ISBN 0471941131

Decentralized diagnosis in a spacecraft attitude determination and control system

C G Pérez[1,2,4], L Travé-Massuyès[1,3], E Chanthery[1,4] and J Sotomayor[2]

[1] CNRS, LAAS, 7 avenue du colonel Roche, F-31400 Toulouse, France
[2] Pontificia Universidad Católica del Perú, Department of Engineering, Lima 32, Perú
[3] Université de Toulouse, LAAS, F-31400 Toulouse, France
[4] Université de Toulouse, INSA, F-31400 Toulouse, France

E-mail: cgperez@laas.fr, louise@laas.fr, elodie.chanthery@laas.fr. jsotom@pucp.edu.pe

Abstract. In model-based diagnosis (MBD), structural models can provide useful information for fault diagnosis and fault-tolerant control design. In particular, they are known for supporting the design of analytical redundancy relations (ARRs) which are widely used to generate residuals for diagnosis. On the other hand, systems are increasingly complex whereby it is necessary to develop decentralized architectures to perform the diagnosis task. Decentralized diagnosis is of interest for on-board systems as a way to reduce computational costs or for large geographically distributed systems that require to minimizing data transfer. Decentralized solutions allow proper separation of industrial knowledge, provided that inputs and outputs are clearly defined. This paper builds on the results of [1] and proposes an optimized approach for decentralized fault-focused residual generation. It also introduce the concept of Fault-Driven Minimal Structurally-Overdetermined set (FMSO) ensuring minimal redundancy. The method decreases communication cost involved in decentralization with respect to the algorithm proposed in [1] while still maintaining the same isolation properties as the centralized approach as well as the isolation on request capability.

1. Introduction

With increasing complexity of industrial processes, the requirement for reliability, availability and security is growing significantly. Fault detection and isolation (FDI) are becoming a major issue in industry.

The structural approach constitutes a general framework to provide information when the system becomes complex. The main aim of the structural approach application is to identify the subsets of equations which include redundancy.

The system structure analysis, originally developed for the decomposition of large systems of equations for their hierarchical resolution, was adopted by the Fault Detection and Isolation (FDI) community [2, 3]. Structural concepts are used for analysis of system monitor ability using the concept of complete matching on a graph.

Decentralized diagnosis has received considerable attention to deal with distributed systems or with systems that may be too large to be diagnosed by one centralized site. In the same way, the decentralized solution allows proper separation of industrial knowledge, provided that inputs and outputs are clearly defined.

Decomposition has been recognized as an important leverage to manage architectural complexity of the systems to be diagnosed. Most of the approaches, however, focus on hierarchical decomposition [4], while decentralization has been explored less frequently. The majority of decentralized diagnosis methods concern discrete event systems [5, 6, 7]. In [6], the purpose of the method is to provide efficient online diagnosis to detect and isolate faults in large discrete event systems. [5] uses a decentralized approach to deal with the size of the model and to get a tractable representation of diagnosis. Along the same idea, [7] proposes a hierarchical framework that exploits different local decisions.

These ideas are taken into account for our approach of decentralization using the idea of *isolation on request*. If local diagnosis is not sufficient, a hierarchical architecture is built that allows for less ambiguous hierarchical diagnosis tests. Moreover, our proposal to use fault-driven residual generation targets decreasing computational effort.

Decentralized diagnosis methods have been applied only recently on continuous or hybrid systems. [4] presents a decentralized architecture for systems modeled in the qualitative framework. The architecture proposed is similar to that which is presented in this paper: the model of each subsystem is private and the supervisor needs no a-priori knowledge about the subsystems and their interactions. In addition, a multi-layered hierarchy of diagnosers is possible.

This paper presents the design, in a decentralized architecture, of a fault-driven residual generation scheme bridging the algorithm for computing a Fault-Driven Minimal Structurally-Overdetermined (FMSO) set of subsystems. This paper is organized as follows: Section 2 introduces the basic concepts of analytical redundancy and presents the structural approach. Section 3 presents an approach for residual focus generation. Section 4 proposes an algorithm to generate the Decentralized Diagnoser based on FMSO. Section 5 illustrates the method proposed on the Attitude Determination and Control System (ADCS) of a LEO satellite. Finally, Section 6 discusses some conclusions and brings in current work.

2. Structural approach

In spite of their simplicity, structural models can provide much useful information for fault diagnosis and fault-tolerant control design. Structural analysis is able to identify those components of the system which are or are not able to be monitored, to provide design approaches for analytic redundancy based residuals, to suggest alarm filtering strategies, as well as to identify those components whose failure can be tolerated through reconfiguration [8].

The system structural model is an abstraction of its behavior in a sense that the structure of the constraints only considers links between variables and parameters and does not consider links between the constraints themselves. The links are represented by a bipartite graph, which is independent of the nature of the constraint (e.g. quantitative, qualitative, equations, rules, etc.), variables and parameters values [9]. It is important to summarize some structural concepts.

Definition 1 (Model): The behavior model of a system is denoted by $M(z,x)$ or M for short, to be any set of equations relating the known variables z, i.e. measured variables, and the unknown variables x. The equations $e_i(z,x) \subseteq M(z,x)$, $i = 1, ..., n$ are assumed to be differential or algebraic equations in z and x, and the model is considered a differential-algebraic equation system. By example shown in Table 1, we illustrate the concepts introduced in this section. This is composed of five equations, e_1 to e_5, relating the unknown variables $x = \{x_1, x_2, x_3, x_4, x_5\}$ and the known variables $z = \{u, y, z\}$.

Definition 2 (ARR for M(z,x)): The structure of a system is an abstract representation of which variables are involved in different equations and compose the model of the system. The structural abstraction allows us to derive redundancies disregarding the actual analytical expressions of equations making up the system model.

Let $M(z,x)$ be a model. Then, a relation $r(z, \dot{z}, \ddot{z}, ...) = 0$ is an Analytical Redundancy Relation (ARR) for $M(z,x)$ if for each z consistent with $M(z,x)$ the equation is fulfilled.

Table 1. Expressions and biadjacency matrix of an illustrative example.

Equation	Expression	Unknown					Known		
		x_1	x_2	x_3	x_4	x_5	u	y	z
e_1	$e^{x_1} = au$	X					X		
e_2	$x_1^2 = bx_2$	X	X						
e_3	$y = x_2$		X					X	
e_4	$z = x_2 + x_3^2$		X	X					X
e_5	$x_3 = x_4 + x_5$			X	X	X			

Obtaining ARRs for a model $M(z, x)$ involves the elimination of unknown variables, which can be inferred from the bipartite graph. The bipartite graph actually represents which unknown variables are involved in the equations models of the system.

3. Fault-driven Residual Generation

In graph theory, a classical result from bi-partite graph states that any finite-dimensional graph can be decomposed into three subgraphs with specific properties by using the Dulmage-Mendelson (DM) canonical decomposition [10] as can be seen in Figure 1. The structurally over-determined part, represented by M^+, has more equations than variables; the structurally just-determined part, represented by M^0, has many equations than variables and the structurally under-determined part represented by M^- has more variables than equations.

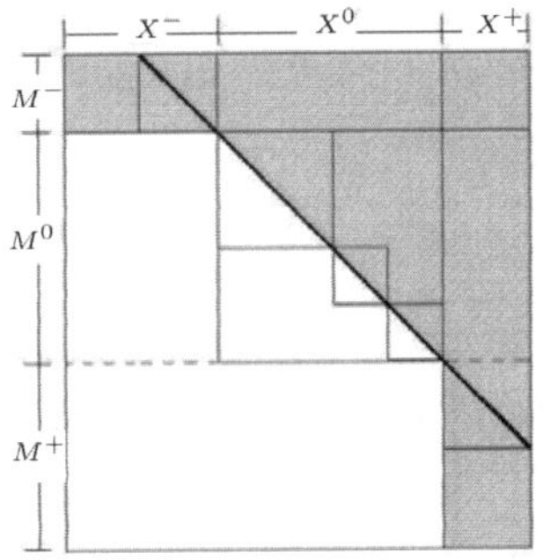

Figure 1. Dulmage-Mendelson decomposition of M.

The biadjacency matrix in Figure 1, shows a DM canonical decomposition of a bipartite graph $G(M \cup X, \mathcal{A})$. The light blue-shaded areas contain ones and zeros, while the white areas only contains zeros. The thick line represents a maximal matching in the graph, where rows and columns are rearranged. For the example, by rearranging rows and columns, the DM decomposition is:

The bold **X** denotes a complete matching. The subsets for this example are: $M^- = \{e_5\}$, $M^0 = \{e_4\}$, $M^+ = \{e_1, e_2, e_3\}$, $X^- = \{x_4, x_5\}$, $X^0 = \{x_3\}$ and $X^+ = \{x_1, x_2\}$.

The structural redundancy is used to define the ideas of structurally overdetermined (SO) set, proper structurally overdetermined (PSO) set and minimal structurally overdetermined (MSO) set. These ideas are therefore also compared to analytical redundancy [11].

Table 2. DM decomposition of illustrative example.

Equation	Unknown					Known		
	x_4	x_5	x_3	x_1	x_2	u	y	z
e_5	X	**X**	X					
e_4			**X**		X			X
e_1				**X**		X		
e_2				X	**X**			
e_3					X		X	

Definition 3 (Structurally overdetermined, SO): A set M of equations is SO if M has more equations that unknowns.

Definition 4 (Proper structurally overdetermined, PSO): A set of equations M is a PSO set if $M = M^+$. A PSO set is generically a testable subsystem but may contain smaller PSO subsets that are also testable subsystems [12]. The minimal PSO sets, called the MSO sets, are of special interest since these have the attractive properties of giving good isolation.

Definition 5 (Minimal structurally overdetermined, MSO): An SO set is a MSO set if no proper subset is also an SO set. Note that an MSO set is also a PSO set.

Structural methods have been used to perform isolation capability analysis and to find testable submodels (MSO sets). However, the number of MSO sets grows exponentially with the degree of redundancy making the task of computing MSO sets intractable for systems with high degree of redundancy. [13] introduces the concept of Test Equation Supports (TES), which is a set of equations expressing redundancy specific to a set of considered faults. The influence of faults is taken into account in the TES.

Let $F(M)$ denote the set of faults that influence any of the equations in M. Then, since a PSO set exactly characterizes a set of equations that can be used to form a test, a formal definition is then given by:

Definition 6 (Test Support, TS): Given a model M and a set of faults F, a subset of faults $\zeta \subseteq F$ is a test support if there exists a PSO set $M' \subseteq M$ such that $F(M) = \zeta$. Of special interest are minimal test supports which are naturally defined as:

Definition 7 (Minimal Test Support, MTS): Given a model, a test support is a minimal test support (MTS) if no proper subset is a test support.

Definition 8 (Test Equation Support, TES): An equation set M is a Test Equation Support (TES) if: $F(M) \neq 0$, M is PSO set and for any $M' \not\supseteq M$ where M' is a PSO set it holds that $F(M') \not\supseteq F(M)$.

Also here it is interesting to consider minimal such sets of equations.

Definition 9 (Minimal Test Equation Support, MTES): A TES M is a minimal TES (MTES) if there exists no subset of M that is a TES.

Here it is clear that there is a one to one correspondence between a TES M and TS ζ given by the relation $\zeta = F(M)$

An MSO set or MTES signifies the theoretical presence of a structural redundancy which can be used as a consistency check for a part of the system. It is important to notice that, whereas an MSO gives rise to one single residual generator, an MTES gathers all the equations leading to all the possible residual generators for the associated set of faults. Besides the fact that an MTES has a non empty TS, it represents a compact form for the whole set of MSOs sharing the same TS.

Definition 10 (Fault Driven Minimal structurally overdetermined, FMSO): A FMSO is a minimal redundancy structurally overdetermined set of M focused on a minimal set of fault of interest. $minimal|(F(M)|$

The elements of a set of MTES can become FMSOs if they have a minimal redundancy. MTES does not guarantee the minimal redundancy sets so these sets will not always be the optimal sets to develop diagnosability.

4. Decentralized Diagnoser

This section introduces the ideas required to formalize decentralized ARR generation in a structural framework. In the following, the global level refers to no decentralization and, without loss of generality, we consider two hierarchical levels, the so-called local level and hierarchical level.

4.1. Fault-driven Decentralized Ideas

A decomposition of the system M, with associated bipartite graph $G(M \cup X \cup Z, A)$, into several subsystems M_i is defined as a partition of its equations.

Let $M = \{M_1, M_2, ..., M_n\}$ with $M_i \subseteq M$: $M_i \neq \varnothing$, $\bigcup_{i=1}^{n} M_i = M$, $M_i \cap M_j = \varnothing$ if $i \neq j$.

The decomposition of the global system leads to n subsystems denoted $M_i(x_i^{local}, z_i)$, with associated subgraphs $G(M_i \cup X_i^{local} \cup Z_i, A_i)$, $i = 1, ..., n$, X_i^{local} is defined below.

Definition 11 (Local Variables): The set of local variables of the ith subsystem, denoted as X_i^{local}, is defined as the subset of vertices of X_i that are adjacent only to vertices in M_i and not to vertices in any other subsystem M_j, $j \neq i$,

$$X_i^{local} = \{u \in X_i : \nexists j (j \neq i), v \in M_j, (u, v) \in A\}$$

Definition 12 (Shared Variables): The set of shared variables, denoted as X^{shared} are defined as the subset of vertices of X that can not be considered local variables for any subsystem:

$$X^{shared} = X \setminus \left(\bigcup_{i=1}^{n} X_i^{local} \right) \tag{1}$$

Consequently: $\forall i \in \{1, ..., .n\}, X_i^{local} \cap X^{shared} = \varnothing$.

The extensions of the ideas of FMSO and MSO at the local and the hierarchical level are first given below.

Definition 13 (Local FMSO): The set of local FMSO , denoted as $Local\ FMSO_i$, is defined as the subset of FMSO of the ith subsystem that are adjacent only to local variables in M_i and not to shared variables in any other subsystem M_j, $j \neq i$,

$$Local\ FMSO_i = \{FMSO \in FMSO_i : \nexists j (j \neq i), x^{shared} \in M_j\}$$

Definition 14 (Shared FMSO): The set of *Shared FMSO* of the ith subsystem, is defined as the subset of FMSO that can not be considered *Local FMSO*:

$$Shared\ FMSO_i = FMSO_i \setminus \left(\bigcup_{i=1}^{n} Local\ FMSO_i \right) \tag{2}$$

Consequently: $\forall i \in \{1, ..., .n\}, Local\ FMSO_i \cap Shared\ FMSO_i = \varnothing$.

Definition 15 (Stored MSO): The set of Stored MSO, denoted as $Stored\ MSO_i$, is defined as the subset of MSO of the ith subsystem whose vertices X^m are adjacent to vertices X^n in *Shared FMSO* of the jth subsystem.

Definition 16 (Hierarchical FMSO): A set of *hierarchical FMSO* is a set of top level which includes the relations in M_i of the ith subsystem which are adjacent to the relations in *Shared* $FMSO_i$ and *Stored* $FMSO_i$ sets for all corresponding subordinates subsystems.

Figure 2 shows the scheme of the decentralized diagnoser proposed in this paper, for subsystems at level i of the diagnoser hierarchy. We consider two stages: the first one is offline for calculating analytical residual generators, and the second one is online and communicates directly with the system through a residual generator bank. In the following, we present the algorithm for calculating analytical residual generators in the first stage.

4.2. Algorithm

The diagnoser design is done *offline* and consists of the 6 steps below. These steps are performed for each subsystem $M_{i,j}$ $j = 1...n_i$ at each level $i = 1...n_j$, with a nested loop. Here i is the level in the hierarchy, and j the enumeration of subsystems at the level. Figure 2 illustrates the proposed algorithm for level i.

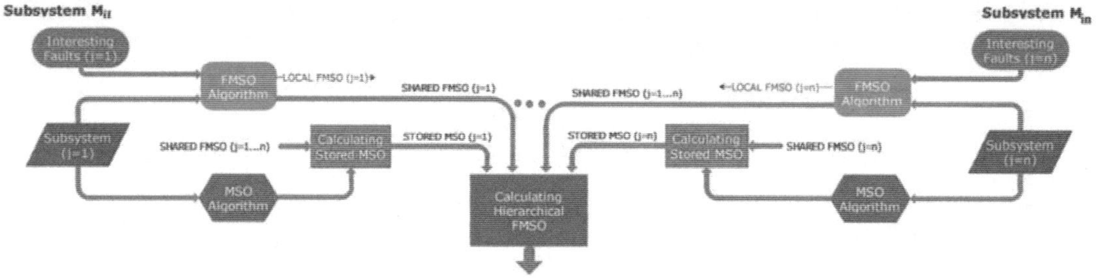

Figure 2. Architecture of diagnoser *offline*

1) At the local level (level i), calculate the FMSO of the structural model for each subsystem $M_{i,j}$, considering shared variables as observed. Branch the results in two subsets.
   Outputs:
   - *Local* $FMSO$: for the subsystem $M_{i,j}$ at level i.
   - *Shared* $FMSO$: for the subsystem $M_{i,j}$ at level i.
2) Still at local level i, compute the FMSO sets of the structural model for each subsystem $M_{i,j}$ and store the set of equations (included in FMSO set) that share variables with *Shared FMSO* of the other subsystem $M_{i,j}$.
   Output:
   - *Stored MSO* set for the subsystem $M_{i,j}$ at level i to be sent at level i+1.
3) Send the *Shared FMSO* and *Stored MSO* sets to the hierarchical level.
4) Store the set of relations included in the *Shared FMSO* and the *Stored MSO* set for each subsystem $M_{i,j}$.
   Output:
   - $M_{i,j}$ *included Rel* : relations included in the *Shared FMSO* and the *Stored MSO* set for the each subsystem $M_{i,j}$ as input at level i.
5) At the hierarchical level: compute the FMSO with the $M_{i,j}$ *included Rel* of the all subordinate local diagnosers of level i-1.
   Output:

- *Hierarchical FMSO*: FMSO of the relations included in the *Shared FMSO* and the *Stored MSO* for all subsystem subordinate local diagnosers of level *i*-1.

6) (optional) As a check, join the *local FMSO* and the *hierarchical FMSO* to find the FMSO computed for the global system in the centralized way.

Output:

- *Global FMSO*: *local FMSO* plus *hierarchical FMSO.*

In step one, the aim is find the *Shared FMSO*. *Local FMSO* are only involved in local redundancies and do not need to be analyzed in the upper hierarchical level. After the offline structural analysis, the diagnoser is implemented as a residual generator bank. With the system inputs/outputs, hierarchical and local residual generators are used online to detect and isolate faults at each level. Notice that, if a *Shared FMSO* or *Stored MSO* set is not sent at hierarchical level, it is possible that all MTES are not obtained.

The offline / online FDI process is illustrated in Figure 3. Considering the system inputs and outputs, the hierarchical and local residual generators are used online to detect and isolate faults each level.

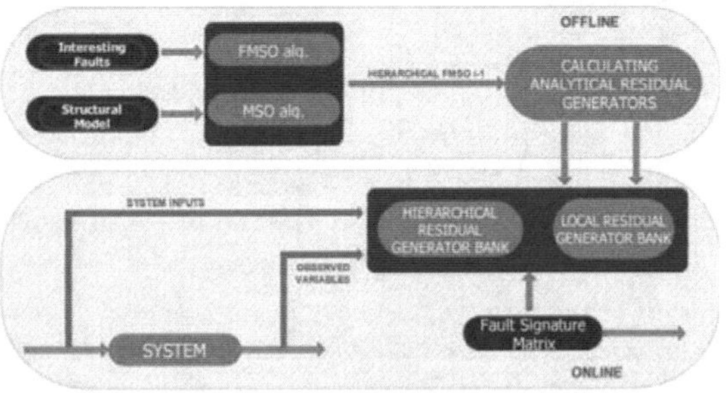

Figure 3. Scheme of a decentralized diagnoser for a subsystem at level *i*

4.3. The equivalence of centralized and decentralized diagnosis

It is desirable that properties such as detectability and isolation of faults are not altered by decentralization. This can be ensured if the set of FMSO derived with global and decentralized architectures are identical [1].

Proposition 1: The set of FMSO computed in a centralized way and the set of MTES computed in a decentralized way from the *Stored MSO* is the same.

Proof: Given a system decomposed into n subsystems denoted as $\{M_1, M_2, ..., M_n\}$. Let $\{$*Local FMSO*$_i$, *Shared FMSO*$_i$, *Stored MSO*$_i\}$, $i = 1..n$ it is not necessary to send all MSO because the necessary redundant information is in relationships that contain shared variables; i.e., the FMSO sets that share information with all *Shared FMSO* of the other subsystems. In the same way, the set of MTES computed in a centralized way and the set of MTES computed in a decentralized way ends the same, resulting in both case [1].

5. Application to the Attitude Determination and Control System of a LEO Satellite

The decentralized diagnoser has been tested on the Attitude Determination and Control System (ADCS) of a low earth-orbit satellite. The composition of the ADCS of a typical satellite is represented in figure 4:

The attitude determination subsystem (ADS) is composed of sensors which sense the rate and angular position of the satellite. An attitude estimate is achieved using a sensor fusion (rate

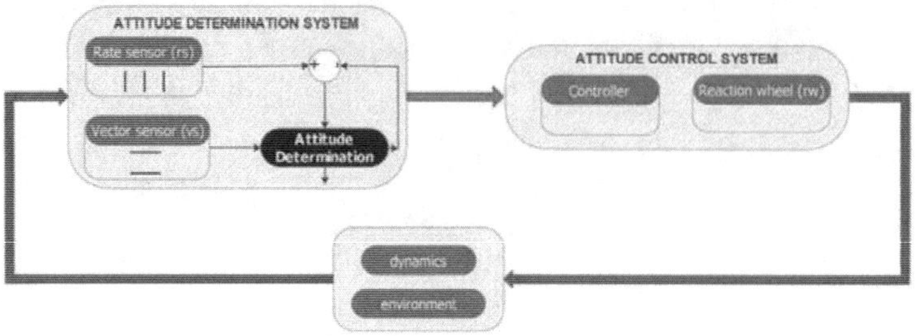

Figure 4. Attitude Determination and Control System of a typical satellite

and vector sensors) [14], which is provided as input to the attitude control subsystem (ACS). The ACS is composed of the control signal calculation and the actuators which provide the stabilizing and/or control torque to the satellite. The satellite under study is assumed to be a three-axis stabilized satellite in orbit around the earth. Here, reaction wheels and magnetorquers are considered as actuators [15].

5.1. Dynamics of the satellite

To analyze the motion of the satellite, two sets of coordinate systems are defined: a Sun-centered inertial frame with its origin at the center of the Sun and its third axis (z_i) normal to the ecliptic plane of the rotation of the Earth around the Sun (x_i, y_i) and a body-fixed frame which has its origin at the center of mass and its axes aligned with the principal axes of the satellite inertia.

The satellite is modeled as a rigid body having the moments of inertia matrix along the principal axes of rotation, I=Diag$_{3x3}\{I_x, I_y, I_z\}$. Assuming that x_b, y_b and z_b are the principal axes of inertia, the rotational motion of the satellite can be described in the body frame as follow [14]:

$$\mathbf{I}\dot{w} = \mathbf{T} - w \times (\mathbf{I}w) \tag{3}$$

where the angular velocity vector w has components w_x, w_y and w_z, each along the body axes x_b, y_b and z_b of the satellite, $\dot{w}$ is the angular acceleration, and T is the total torque acting along the body axes. The system of differential equations describing the vehicle attitude is

$$\begin{bmatrix} \dot{\psi} \\ \dot{\theta} \\ \dot{\phi} \end{bmatrix} = \frac{1}{c_\theta} \begin{bmatrix} 0 & s_\phi & c_\phi \\ 0 & c_\phi c_\theta & -s_\phi c_\theta \\ c_\theta & s_\phi s_\theta & c_\phi s_\theta \end{bmatrix} \begin{bmatrix} w_x \\ w_y \\ w_z \end{bmatrix} \tag{4}$$

where ψ, θ and ϕ denote yaw, pitch and roll angles, respectively. The state vector of the satellite is $X = [\psi, \theta, \phi, \dot{\psi}, \dot{\theta}, \dot{\phi}]$. They are the angles by which the body frame is rotated relative to a reference frame.

5.2. Attitude Determination and Control System modelling

The rate sensors of the satellite are three gyroscopes and the vector sensors are sun and star sensors. It is assumed that sensing axes of the rate gyros are aligned with each of the body axes of the satellite. The angular rate measurements from the gyros are used to solve the set of differential equations described by 4. w_x,w_y and w_z represent outputs of the three orthogonal rate gyros with their sensing axes aligned with the roll, pitch and yaw axes,respectively.

The attitude measurement from vector sensors is bounded and used to aid gyros to eliminate attitude drift error. The sensitive axes of the rate gyros are aligned with each of the body axes of the satellite. The modelling of the AD system is described in [14]. The vector and rate sensor outputs are used to estimate the state vector both independently and merged together. These preliminary estimates are then fused together to arrive at the estimate which is feedback to the ACS. This redundancy can be used to check consistency [15].

The ACS is equipped with a three reaction wheels for 3-axis control. Another external torque source is necessary to unload the wheel's angular momentum. For this satellite, magnetic torques are selected instead of thrusters that consume a large amount of fuel [16].

5.3. Fault scenarios
Faults are introduced in the system model equations. We consider faults occurring in the rate and vector sensors of the ADS and in the reaction wheels of the ACS. Each of these faults has three components corresponding to the three axes. The faults are summarized in Table 3.

Table 3. Fault scenarios of the ADCS.

Component	Subsystem	Fault
Vector sensors (vs)	ADS	$f_{vs}(f_{vs_x}, f_{vs_y}, f_{vs_z})$
Rate sensors (rs)	ADS	$f_{rs}(f_{rs_x}, f_{rs_y}, f_{rs_z})$
Reaction wheel (rw)	ACS	$f_{rw}(f_{rw_x}, f_{rw_y}, f_{rw_z})$

5.4. Structural Model of the ADCS
The structure of the ADCS is abstracted as a set of constraints on a set of variables. Related information of such modelling can be founded in [16, 14]. Most constraints are composed of three behavioral relations corresponding to three axes. From the set of variables of the system, the sensed quantities form the set of observed variables with all the rest assumed to be unobserved. The general procedure for the diagnoser design starts with assuming a small set of observed quantities, and can be optionally expanded to fulfill diagnosis and isolation capability specifications.

The bi-adjacency matrix of the ADCS is shown in figure 5. The unobserved faults and observed variables are separeted along the X-axis. The constraints that describe the behavior of the system components are described on the Y-axis. The structural models of the ADCS, ADS and ACS are represented as $(C_{ADCS}, X_{ADCS}, Z_{ADCS})$,$(C_{ADS}, X_{ADS}, Z_{ADS})$ and $(C_{ADC}, X_{ADC}, Z_{ADC})$ respectively. The structural model of the ADCS is composed of 42 constraints with 42 unobserved variables , 15 observed variables and 9 faults (modeled as variables in the bi-adjacency matrix).

The structural model of the satellite ADCS, is considered to demonstrate the proposed decentralized architecture while still maintaining the same isolation capability power as the centralized approach and the advantageous isolation on request capability.

5.5. Diagnoser Design
For reference, first we use the algorithm to determine MTS, MSO and FMSO sets to the ADCS considered globally. Constraints e_i from 1 to n are denoted $e_1...e_n$.

ADCS global diagnoser
Max fault isolability: $[frw_x], [frw_y], [frw_z], [frs_x], [frs_y], [frs_z], [fvs_x], [fvs_y], [fvs_z]$
FMSO sets: $[e_4, e_3, e_{10}, e_{13}], [e_5, e_8, e_{11}, e_{14}], [e_6, e_9, e_{12}, e_{15}], [e_7...e_{21}, e_{25}], [e_7...e_{21}, e_{26}], [e_7...e_{21}, e_{27}], [e_7...e_{21}, e_{22}, e_{28}], [e_7...e_{21}, e_{23}, e_{29}], [e_7...e_{21}, e_{24}, e_{30}]$

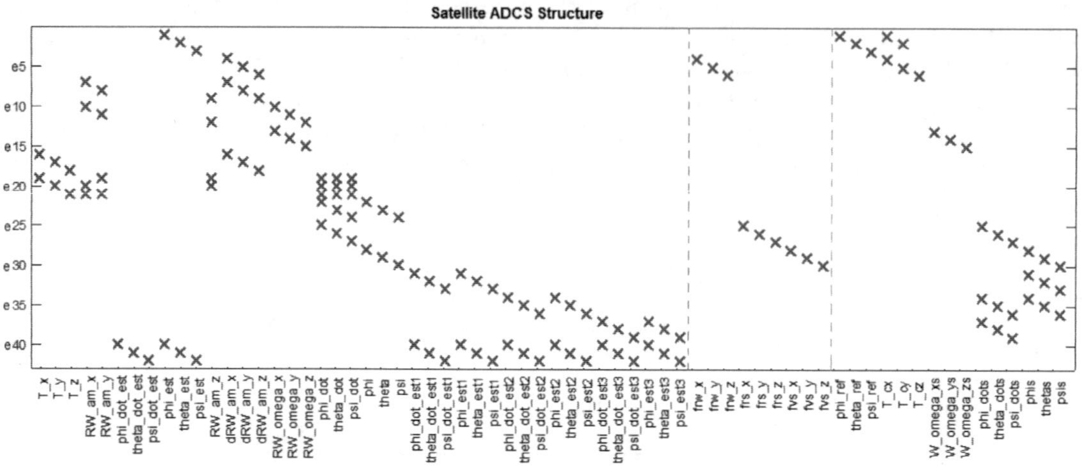

Figure 5. ADCS structure of a LEO satellite.

Num. MSO sets: 2448

ACS local diagnoser (X^{SHARED} unobserved)
Max fault isolability: $[frw_x], [frw_y], [frw_z]$
Local FMSO sets: $[e_4, e_7, e_{10}, e_{13}], [e_5, e_8, e_{11}, e_{14}], [e_6, e_9, e_{12}, e_{15}]$

ADS local diagnoser (X^{SHARED} unobserved)
Max fault isolability: $[frs_x, fvs_x], [frs_y, fvs_y], [frs_z, fvs_z]$
Local FMSO sets: $[e_{22}, e_{25}, e_{28}], [e_{23}, e_{26}, e_{29}], [e_{24}, e_{27}, e_{30}]$

According to these results, it can be seen that all faults can be isolated with a centralized global diagnoser for the ADCS. However, for the local diagnoser ADS, faults on the rate and vector sensors can not be isolated locally. It also shows that between 2448 MSO sets and 9 FMSO sets there is a large computational advantage. Because MTES does not guarantee minimum redundancy, it has not been considered. It is more efficient to consider FMSO introduced in the previous section.

The proposed decentralized architecture will now be applied to the ADCS by designing the local and hierarchical diagnosers. It will be demonstrated that the isolation capability of such a decentralized diagnoser is equivalent to the global diagnoser above.

For each local diagnoser, the shared variables X^{SHARED} are assumed to be observed. It can be assumed that for each subsystem, shared variables are calculated in other local systems. *Shared FMSO* can be derived using the algorithm with this assumption.

ADS local diagnoser (X^{SHARED} observed)
Max fault isolability: $[frs_x], [frs_y], [frs_z], [fvs_x], [fvs_y], [fvs_z]$
Shared FMSO: $[e_{19}, e_{20}, e_{21}, e_{22}, e_{28}], [e_{19}, e_{20}, e_{21}, e_{23}, e_{29}], [e_{19}, e_{20}, e_{21}, e_{24}, e_{30}], [e_{19}, e_{20}, e_{21}, e_{25}], [e_{19}, e_{20}, e_{21}, e_{26}], [e_{19}, e_{20}, e_{21}, e_{27}]$

ACS local diagnoser (X^{SHARED} observed)
Max fault isolability: $[frw_x], [frw_y], [frw_z]$
Shared FMSO: $[e_4, e_7], [e_4, e_{16}], [e_5, e_8], [e_5, e_{17}], [e_6, e_9], [e_6, e_{18}]$

In the case of ACS local diagnoser, the results demonstrate the efficiency of FMSO sets to ensure minimal redundancy ($sr = 1$) compared with MTES sets: using the algorithm, as it discussed in [1], the structural redundancy for this MTES set is equal to 10. Now, we can calculate the Hierarchical diagnoser with relations included in *Shared FMSO* and *Stored MSO* sets for every subsystem, this calculation its necessary to disambiguate faults.

ADCS Hierarchical diagnoser
Rel in Shared FMSO and Store MSO (ADCS) $= [e_1, e_2, e_3...e_{18}], [e_{19}, e_{20}, e_{21}...e_{30}]$
Fault isolability: $[frw_x], [frw_y], [frw_z], [frs_x], [frs_y], [frs_z], [fvs_x], [fvs_y], [fvs_z]$
Hierarchical FMSO sets: $[e_4, e_3, e_{10}, e_{13}], [e_5, e_8, e_{11}, e_{14}], [e_6, e_9, e_{12}, e_{15}], [e_7...e_{21}, e_{25}], [e_7...e_{21}, e_{26}], [e_7...e_{21}, e_{27}], [e_7...e_{21}, e_{22}, e_{28}], [e_7...e_{21}, e_{23}, e_{29}], [e_7...e_{21}, e_{24}, e_{30}]$
Num. MSO sets: 2448

If we calculate the Hierarchical MTES sets, we can find the same sets because they have the minimal redundancy ($sr = 1$). However, they do not always have the same results because the MTES may have no minimum structural redundancy, which does not make them fully effective.

5.6. Simulation of the decentralized ADCS diagnoser

Figure 6 and 7 show the performance of the proposed decentralized diagnoser. In 6, there is a fault injection between 100 and 105 seconds for reaction wheel 3 during reference tracking. Simultaneously, as shown in Figure 7, there is a fault injection in the rate sensor (gyro 2) between 80 and 85 seconds during reference tracking.

Faults with a 7.5 % range around their mean values were injected into the reaction wheels and rate sensors in the simulation. A much smaller set of threshold parameters is required compared to the case of the conventional diagnoser, as the residual generators are structurally decoupled from the design phase. It is noted that decentralized diagnoser correctly identifies both faults.

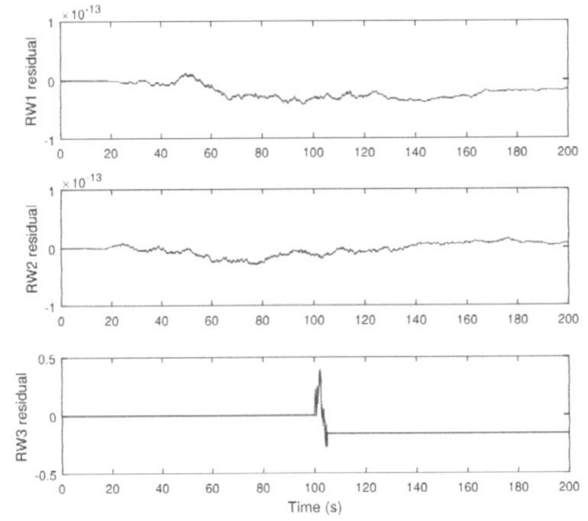

Figure 6. Reaction wheel residuals with fault injection in reaction wheel 3.

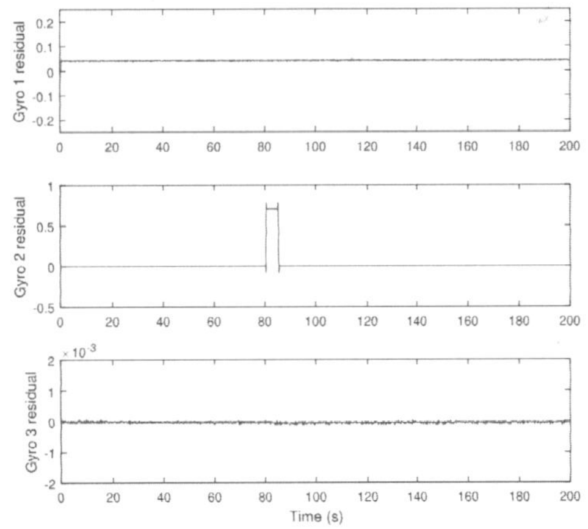

Figure 7. Rate sensor residuals with fault injection in rate gyro 2.

6. Conclusion

This paper proposes an approach for the decentralization of model-based diagnosis through structural redundancy analysis. Decentralized diagnosis is justified by many applications such as on-board systems that need to reduce the calculations required in each time cycle or large systems that require minimal communication between elements. The first contribution of the paper is the design of a fault-driven residual generation method and corresponding algorithm in a decentralized case. The second contribution is the introduction of the Fault-Driven MSO concept which can be directly used to construct an ARR or residual generator, as compared to MTES that by definition may involve several subsets leading to RRAs. Consequently, FMSO sets represent a more practical solution for generating RRAs focused on faults. The algorithm has been implemented and tested on the Attitude Determination and Control System of a LEO Satellite. This implementation illustrates the advantages of decentralized diagnosis architecture, which offers lower complexity and isolation on request capabilities with the same isolation capability power as centralized architecture.

The fault-driven residual generation based on FMSO drastically reduces the number of residual generators to be computed. However, as in the case of the ACS local diagnoser studied, all the generated FMSOs may not be necessary to achieve required isolability. Thus, it is necessary to use optimization criteria to select the better sets to develop RRAs. Current work focuses on that issue.

References

[1] Chanthery E 2015 *Fault isolation on request based on decentralized residual generation* (Laboratory for Analysis and Architecture of Systems Internal report) *Preprint* Nro.15059

[2] Staroswiecki M 1999 et al. *Proc. of the 38th IEEE Conference on Decision and Control* vol 4 (Phoenix) p 3581 - 3586

[3] Travé-Massuyès L et al. 2006 *Diagnosability Analysis Based on Component-Supported Analytical Redundancy Relations* (*IEEE Trans. Sys. Man and Cyb. Part A*) vol 36 p 1146 - 1160.

[4] Console L, Picardi C and Theseider D 2007 *A Framework for Decentralized Qualitative Model-Based Diagnosis* (20th int. joint conf. on artificial intelligence) p 286 - 291

[5] Cordier M-O and Grastien A 2007 *Exploiting independence in a decentralised and incremental approach of diagnosis* (20th int. joint conf. on artificial intelligence) p 292 - 297

[6] Pencolé Y and Cordier M 2005 *A formal framework for the decentralized diagnosis of large scale discrete event systems and its application to telecommunications networks* (Artificial Intelligence) Vol 164(1-2) p 121 - 170

[7] Wang Y, Yoo T-S and Lafortune S 2007 *Diagnosis of discrete event systems using Decentralized architectures* (*Discrete Event Dynamic Systems*) vol 17, no. 2 p 233-263

[8] Blanke M et al. 2006 Diagnosis and Fault-tolerant Control (Berlin: Springer)

[9] Said A, Djamel B and Boudjema I 2013 *Structural analysis for fault detection and isolation using the matching rank algorithm for residual generation: application on an industrial water heating system*(*J. of Control Engineering and Applied informatics*) vol 15,no. 2 p 20 - 29

[10] Dulmage A L and Meldenson N S 1963 *Two algorithms for bipartite graphs (J. of the Society for Industrial and Applied Mathematics*) vol 11 p 183 -194

[11] Krysander M 2006 *Design and Analysis of Diagnosis Systems Using Structural Methods* Ph.D. thesis Linkoping University

[12] Krysander M, Aslund J and Nyberg M 2008 *An efficient algorithm for finding minimal overconstrained subsystems for model-based diagnosis* (IEEE Trans. on Systems, Man and Cybernetics, Part A) vol 38 p 197 - 206

[13] Krysander M, Aslund J and Nyberg M 2010 *A structural algorithm for finding testable sub-models and multiple fault isolability analysis* (Proc. of the 21st Int. Workshop on Principles of Diagnosis) Dx 10

[14] Pirmoradi F, Sassani F and De Silva C *Fault detection and diagnosis in a spacecraft attitude determination system* (Acta Astronautica) vol 65 p 710 - 729

[15] Won C-H 1999 *Comparative study of various control methods for attitude control of a LEO satellite* (Aerospace Science and Technology) vol 3 p 323-333

[16] Zuliana I, Renuganth V *A study of reaction wheel configurations for a 3-axis satellite attitude control* (Advances in Space Research) vol 45 no 6 p 750 - 759

Big Data Analysis of Manufacturing Processes

Stefan Windmann[1], Alexander Maier[1], Oliver Niggemann[1], Christian Frey[2], Ansgar Bernardi[3], Ying Gu[3], Holger Pfrommer[4], Thilo Steckel[5], Michael Krüger[6], Robert Kraus[7]

[1]Fraunhofer Application Center Industrial Automation (IOSB-INA), Lemgo
[2]Fraunhofer-Institut für Optronik, Systemtechnik und Bildauswertung (IOSB), Karlsruhe
[3]DFKI GmbH, Multimedia Analysis and Data Mining, Kaiserslautern
[4]Hilscher Gesellschaft für Systemautomation mbH, Hattersheim
[5]CLAAS E-Systems KGaA mbH & Co KG, Gütersloh
[6]Karl Tönsmeier Entsorgungswirtschaft GmbH & Co. KG, Porta Westfalica
[7]Bayer Technology Services GmbH, Leverkusen

E-mail: `stefan.windmann@iosb-ina.fraunhofer.de`

Abstract. The high complexity of manufacturing processes and the continuously growing amount of data lead to excessive demands on the users with respect to process monitoring, data analysis and fault detection. For these reasons, problems and faults are often detected too late, maintenance intervals are chosen too short and optimization potential for higher output and increased energy efficiency is not sufficiently used. A possibility to cope with these challenges is the development of self-learning assistance systems, which identify relevant relationships by observation of complex manufacturing processes so that failures, anomalies and need for optimization are automatically detected. The assistance system developed in the present work accomplishes data acquisition, process monitoring and anomaly detection in industrial and agricultural processes. The assistance system is evaluated in three application cases: Large distillation columns, agricultural harvesting processes and large-scale sorting plants. In this paper, the developed infrastructures for data acquisition in these application cases are described as well as the developed algorithms and initial evaluation results.

1. Introduction

Due to increasing demands, manufacturing processes are getting more and more complex [1]. For this reason, users often fail to get an overview of the current system status since data is spread over different computer subsystems or hierarchies. The increasing complexity of manufacturing processes and the permanently growing amount of data lead to an overload of the user with respect to process monitoring, data analysis and fault detection. Therefore, problems and failures are often detected too late, maintenance intervals are not chosen correctly and optimization potential with respect to throughput and energy efficiency is not sufficiently exploited.

The following problems with respect to data acquisition, process monitoring and anomaly detection have to be solved:

- Data acquisition: Data in industrial and agricultural applications is stored in a distributed way, without sufficient time synchronization and without any definition of semantics.
- Process monitoring and anomaly detection: The amount of data and real-time requirements render a manual analysis, which is e.g. based on displayed signals, impossible, even if the

data is already integrated, time-synchronized and augmented with semantics. For these problems, algorithmic solutions exist. However, such approaches are not frequently used in industrial practice because they are not yet suitable for the required processing of big amounts of data.

In the present work, an intelligent assistance system is developed to cope with these challenges (see Fig. 1). The proposed assistance system allows to analyze the available information and to

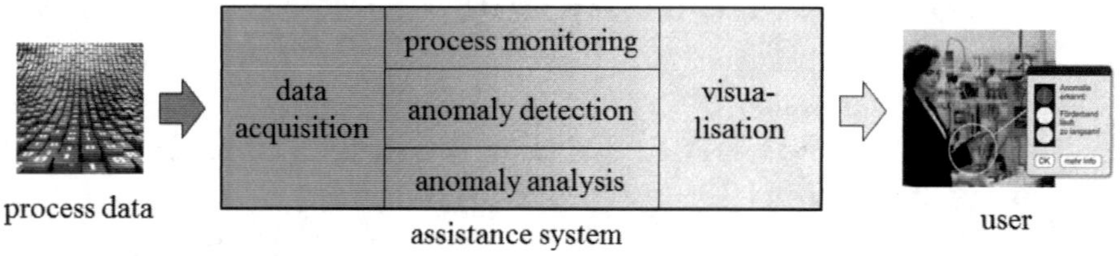

Figure 1. Data analysis of manufacturing processes

detect anomalies. Integration of the assistance system into different manufacturing systems is achieved by suitable data acquisition approaches and flexible methods for process monitoring, which can be adapted with model-learning approaches to changing process behavior. One contribution of the present work is the development of data acquisition approaches for complex distributed manufacturing processes and the integration of big data from several heterogeneous data sources. Furthermore, approaches to process monitoring and anomaly detection in complex manufacturing systems have been developed.

The developed approaches have been evaluated in three application cases, which cover a wide range of agricultural and industrial processes:

- Production processes in distillation columns.
- Anomaly detection in agricultural harvesting processes
- Optimal process control of large-scale sorting plants

The remaining part of this paper is structured as follows: The application cases, which have been investigated in this work, are outlined in section 2. State of the art with respect to data acquisition and anomaly detection in complex industrial and agricultural processes is outlined in section 3. Data acquisition in the described application cases is considered in section 4. The investigated methods for process monitoring and anomaly detection are outlined and evaluated in section 5. Finally, a conclusion and an outlook are given in section 6.

2. Application Cases

2.1. Anomaly detection in complex processes of chemical industry

Large scale continuous production processes in chemical industry typically manufacture, produce, or process materials in continuous operation with infrequent maintenance shutdowns. The continuous diagnosis of the functionality, the early detection of potential failures and the monitoring of the underlying physical process itself is essential for the cost-effective operation of such production processes. Chemical production processes are typically dominated by a complex system behavior with a high system order, e.g. industrial column processes in chemical industry can have up to 120 state variables. In this work, a large scale distillation process is considered (see Fig. 2 a)). This process is typical for chemical industry and is characterized by a high complexity and a large number of measured process states (temperatures, pump speed, pressures, etc.). Efficient plant operation requires an early detection of failures such as sensor failures or blockages of valves, pumps, etc. in order to prevent damages of the plant or production losses.

Figure 2. Application cases: a) Distillation tower b) Agricultural harvester c) Sorting plant

2.2. Process control for mobile agricultural harvesters

The productivity of agricultural harvesters (see Fig. 2 b)) depends on extremely heterogeneous and dynamic conditions such as conditions of field and harvested crop, adjustment of machine parameters and manual skills of the machine operator. For these reasons, many agricultural processes are unstable and low-deterministic. Comparison of several harvesters in the same region and with similar constraints shows performance deviations of more than 20% (tons of crop in a given time unit) between the particular machines. Modern agricultural harvesters provide extensive data streams with a multitude of sensor information with high acquisition rates. Automatic anomaly detection and analysis of performance deviations by evaluation of continuous data streams allow for economic operation of the harvesters in this application case.

2.3. Monitoring of large-scaled sorting plants

The company Tönsmeier operates a lightweight package sorting plant in Porta Westfalica (see Fig. 2 c)) where secondary raw material is extracted from garbage collections. Optimal process control of sorting plants with respect to the quality of the extracted raw materials and the energy consumption of the plant as well as the reduction of downtime of the plant require immediate detection and repair of anomalies. Typical anomalies or causes for anomalies in this application case are sensor failures, decreased pressure in separation units, material jams or too slow operation of spiral conveyors in the sorting plant.

3. State of the art

3.1. Data acquisition

The large amount of data in industrial processes is acquired at several layers of the automation pyramid. At field and control level, IEC 61131-3 function blocks are usually applied for communication (e.g. TCP/IP or SQL) [2]. In this layer, data loggers can be installed in networks to extract sensor signals and signals of actuators (see e.g. [3]). Agricultural harvesters are equipped in most cases with CAN 2.0 networks and GSM-based loggers, which send the acquired data web servers (see [4]). In distributed automation systems, Supervisory Control and Data Acquisition (SCADA) systems employ a multitude of solutions for data acquisition, which are in many cases based on web services (see [5] for an overview). Furthermore, real-time middle-ware such as CORBAe [6] and OSA+ [7] exists. An overview on agent-based control and computer-based manufacturing technology (HMS) is available in [8]. Widespread protocols at the layer of Manufacturing Execution Systems (MES) are Object Linking and Embedding for Process Control (OPC DA, OPC HDA etc.) [9] and the OPC Unified Architecture (OPC UA) [10]. The acquisition of energy data requires either a real-time bus system or technologies such as

ProfiEnergy [11]. Data is described with information models such as CAEX [12] or CIM [13]. The combination of the described methods and protocols leads in industrial practice to heterogeneous networks with industry-sector-specific structures. So far, no unified data acquisition approach exists, which allows for universal application of the considered data mining methods.

3.2. Anomaly detection

Generally speaking, two classes of algorithmic approaches exist for the detection of anomalous situations: model-based approaches and data-driven approaches.

Model-based approaches employ a model to simulate the normal behavior of the manufacturing process. If the actual measurements vary significantly from the simulation results, the behavior is classified as anomalous (see e.g. [14], [15]). Self-learning of process models from data leads to considerable simplification of model creation and configuration. Industrial and agricultural systems are in general hybrid systems, which are composed of both discrete and continuous system parts. Several algorithms for the self-learning of automata for discrete system parts exist, e.g. RTI+ [16], BUTLA [17] and OTALA [18]. Besides, a lot of research is conducted in the related field of model-learning for continuous systems and application to anomaly detection. Clustering-based methods create groups of strongly related objects and find objects which do not strongly belong to any cluster [19]. Self-organising feature maps (SOM) are capable of generating a topology-preserving mapping of a high-dimensional feature space into an output space of a lower dimensionality [20], which can be utilized as an anomaly detector. By applying the so-called UMatrix representation to a self-organizing map, it is possible to perform a classification or a clustering of the feature space [21] where the clusters correspond to typical process phases. Neural networks and regression-based methods have been used to approximate the functional dependency between continuous process variables [22]. Sensor signals are predicted according to this functional dependency and significant deviations of predicted signal values from the observations are reported as anomalies. Statistical approaches to anomaly detection are predominantly based on building a probability distribution model and considering how likely objects are under that model [23]. In most of these approaches, state variables are employed for modeling the temporal transitions of hidden process variables, which are related to the measurements with a measurement model. Common approaches are Kalman filters (e.g. [24], [25], [26]) and particle filters [27].

Data-driven anomaly detection methods have been receiving considerable attention in recent years as they do not require an analytical model, depending only on the measurable process data [28]. Examples are principal component analysis (PCA), partial least squares (PLS) regression, subspace-aided approach (SAP), etc. [29]. Furthermore, a variety of unsupervised statistical and proximity based methods for the direct detection of anomalies in raw data has been developed in recent years, e.g. k-NN anomaly detectors [30], LOF [31], LOCI and aLOCI [32]. By means of these methods, additional anomalies that are currently not obvious from domain knowledge or that cannot be discovered using model-based approaches, are detected. Fast algorithms have been developed for anomaly detection in large scale data sets, such as HBOS, LDCOF and FastLOF (see [33] for an overview).

4. Data Acquisition

Data analysis in the application cases is based on centralized and synchronized data acquisition. Big amounts of data from several heterogeneous data sources such as energy data, process data, MES data or ERP data have to be analyzed in a central infrastructure. Therefore, data acquisition in distributed manufacturing has to cope with the following challenges:

- Heterogenous manufacturing processes: In the manufacturing processes, products of different vendors, proprietary network protocols and different MES are used.

- Time synchronization: Data, which is used for process analysis must be related to the same time instance. Therefore time synchronization of data, which is acquired at different positions in the manufacturing processes, is required.
- Data integration and standard interfaces: Data shall be integrated in a system-wide infrastructure with standard interfaces. Furthermore, the interfaces should contain meta data about the acquired data.

It is outlined in the following sections, how these challenges are met in the particular application cases.

4.1. Distillation columns
In chemical distillation columns, data is gathered on different levels and for different purposes. For safety and control of the product quality, the process data is directly used on the control level for automation. Additionally, the data is used in management execution systems (MES) and enterprise resource planning (ERP) systems for the control of the general production process. Different software tools are used for the different purposes with a set of relevant and well established interfaces, namely OPC-DA, OPC-HDA, OPC-UA and SQL. Due to the sensitive nature of the production plants, the access on the process data is secured and strictly limited. Depending on the level of IT security, any form of OPC can be blocked and SQL remains as only generally allowed interface. The process data, which is available in the considered distillation columns, consists of temperatures, pressures, flow rates and control values. The measuring points are not equidistant with respect to the time and can differ, depending on the measured value.

4.2. Agricultural harvesters
The infrastructure for the agricultural application case is depicted in Fig. 3. Core of the environment are harvesters equipped with embedded data units for acquisition, pre-processing, time-synchronization and transmission of process data (the embedded system in Fig. 3) and a powerful back-end, which is composed of a Hadoop cluster and a data mining tool) for finding patterns in data being delivered by machines and other parts of the infrastructure. The embedded system is configured in a way that data for subsequent analysis is defined as well

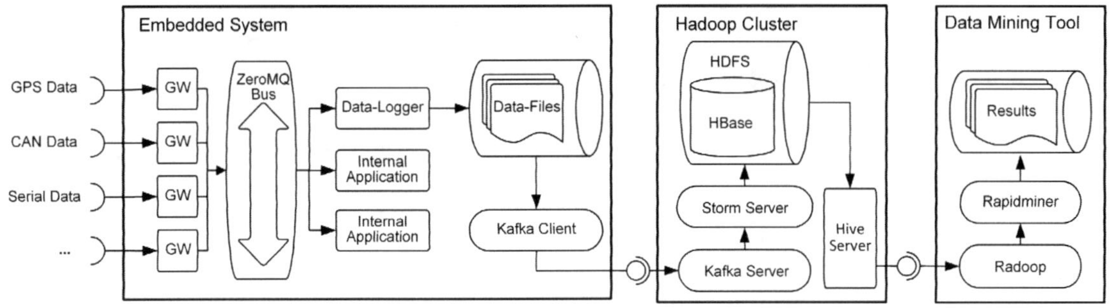

Figure 3. Infrastructure for machine analytics

as specified rules for acquisition (e.g. "record position if the driving direction deviates more than five degrees"). For 2015 harvest season, 10 combine harvesters are equipped with such technology, which provides approximately 3000 different attributes. Data is captured from each available interface (Global Positioning System (GPS) data, CAN bus, serial data, etc.) and sent via a Kafka client to the back-end. In addition, a subset of combine harvesters had been equipped with cameras to provide imagery information of the machine's surrounding, since no

Journal of Physics: Conference Series **659** (2015) 012055 doi:10.1088/1742-6596/659/1/012055

sensors or information sources are available to characterize the crop respectively the harvesting conditions in detail. On the back-end side, the Hortonworks distribution is installed to provide a Hadoop based cluster. Important components are the Kafka server for secure and reliable reception of machine data, the Storm server for fast processing of data and deposition, HDFS as distributed file system, and REST and Hive for data access by analytic tools like R and RapidMiner. The physical infrastructure consists of six servers (five data nodes and a name node). Environment data like field boundaries and digital terrain models were integrated into the cluster to serve as filters in analytic processes.

4.3. Sorting plant

Data acquisition in the sorting plant is accomplished according to Figure 4. Data is collected from three sources:

- TiTech Analysis Kits
- Syperion Flow Meters
- Programmable logic controllers (PLCs).

The TiTech Analysis Kits provide measurements with respect to the material and the composition of the volume flow. The shape and extension of the volume flow is measured by the Syperion Flow Meters. The TiTech Analysis Kits and the Syperion Flow Meters are connected to a central control unit (CCU) and an industry PC (IPC), respectively. Further data is collected from the PLCs of the sorting plant. APIs have been developed in the present work to connnect TiTech analysis kits and Syperion flow meters to the central infrastructure, respectively. The PLC provides an OPC DA interface and is connected via an OPC DA/UA wrapper in the MES API. Due to different data sources, time synchronization has been applied to obtain suitable data sets.

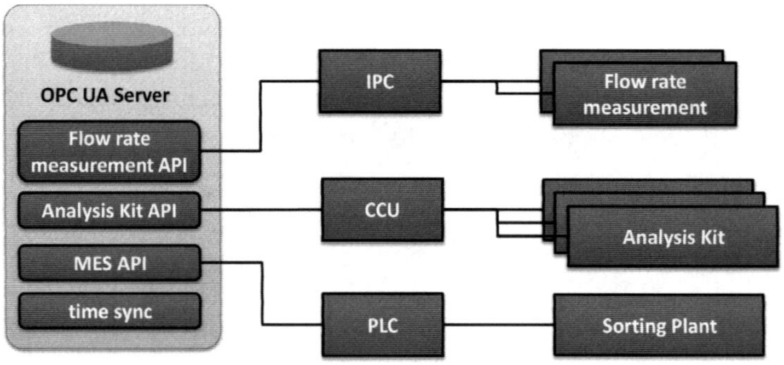

Figure 4. Data acquisition in the sorting plant.

5. Algorithms

In the present work, several model-based and data-driven anomaly detection algorithms have been investigated. Distance based methods compute the distances of potential anomalies to the nearest neighbors in the set of fault-free data. Regression models and self-organizing maps are used to predict observations and to compare the predicted values with the actual observations. Furthermore, PCA-based methods have been investigated.

5.1. Distance based approaches

Initial evaluations have been conducted with distance based methods, which are based on the k-Nearest-Neighbors (k-NN) algorithm (see e.g. [34]). In these approaches, each instance $x = (x_1, \ldots, x_n)$ in a set X of n data points is assessed with respect to the subset $N_k(x) \subset X$ of the k nearest neighbors of x. In doing so, a chosen distance function d, e.g. the Euclidean distance, is used. A score based on the average distance of nearest neighbors is the k-NN Global Anomaly Score (GAS) [30]:

$$GAS_{kNN}(x) = \frac{\sum\limits_{o \in N_k(x)} d(x,o)}{|N_k(x)|}. \tag{1}$$

Instances $x \in X$ with the highest GAS are considered as potential anomalies. The Local Outlier Factor (LOF) [35] of the instance is defined as the ratio of the average inverse GAS of its neighbors and the inverse GAS of x:

$$LOF_k(x) = \frac{\frac{1}{|N_k(x)|} \sum\limits_{o \in N_k(x)} [GAS_{kNN}(o)]^{-1}}{[GAS_{kNN}(x)]^{-1}}. \tag{2}$$

If the densities of an instance are comparable to those of its neighbors the LOF score is approximately equal to 1. An instance with a low density yields a high LOF score and can be considered an outlier. A minimal LOF value of 1.2 is chosen to label an instance as an outlier.

Fig. 5 a) shows the result of the k-NN Global Anomaly Score algorithm applied to the sensor data of three combine harvesters. The x-, y- and z-axis represent motor rotation (x), driving speed (y) and throughput (z) respectively. The colorbar displays the outlier score. The blue instances are considered normal whereas the red and yellow points indicate outliers. The outlier classification highly depends on the user-chosen parameters k and p.

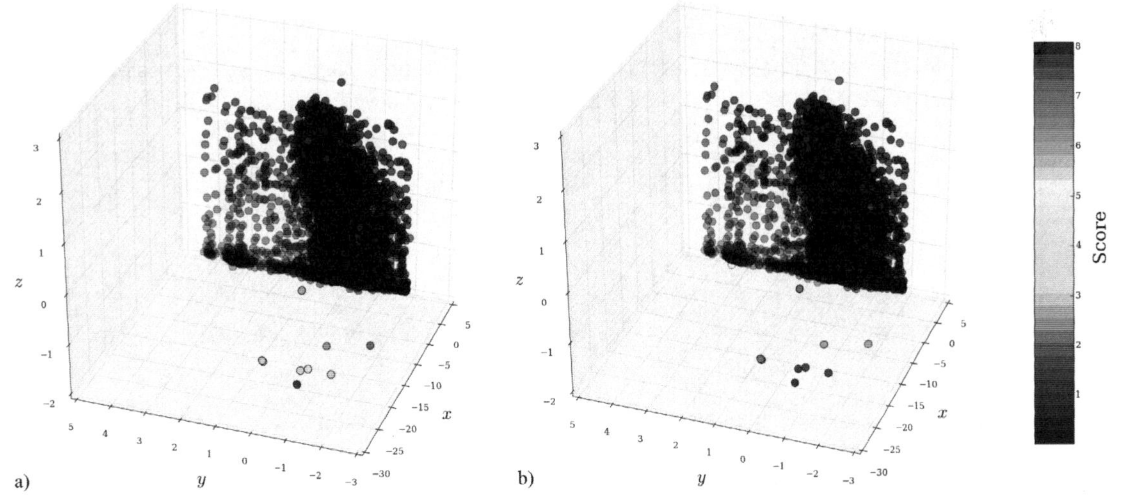

Figure 5. Outlier classification based on $k = 10$: a) GAS score b) LOF score. Outliers are indicated by red and orange circles.

Applying the LOF score on the same dataset as above, we obtain the results shown in Fig. 5 b). In this setup, LOF detects one anomaly more than GAS and allows for more reliable discrimination between anomalies and fault-free data points.

5.2. Regression models

In a regression analysis, the relationship between a specific dependent sensor signal and a set of independent sensor signals has been exploited. Application of linear regression for outlier detection is based on strongly correlated sensor signals. In this case, different attributes, which are represented by the sensor data, are usually generated by the same underlying process in closely related ways [36].

Table 1 shows the correlation between the throughput y of an agricultural harvester and some highly dependent sensor signals $x_1, \ldots, x_m$ in terms of the magnitude of the correlation coefficient.

Table 1. Absolute correlation coefficient $|\rho_{y,x}|$ between throughput y of a harvester and the correlated sensor signals x (for signals with $|\rho_{y,x}| > 0$)

Sensor signal x	$\rho_{y,x}$	Sensor signal x	$\rho_{y,x}$
peripheral reel speed	0.74	motor speed	0.47
fan speed	0.56	shredder	0.44
working position	0.55	water content in grain	0.42
return speed	0.55	average water content	0.36
threshing drum speed	0.53	crop type	0.27
motor load	0.52	lower sieve position	0.25
shaker speed	0.52	upper sieve position	0.24
revolving drum on/off	0.50		

The magnitude of the correlation coefficient has a value of 1 for completely correlated signals and a value of 0 for uncorrelated signals. Based on correlation analysis, a linear model

$$y(x_1, ..., x_m) = a_0 + \sum_{i=1}^{m} a_i x_i \quad (3)$$

has been established for m highly correlated signals with respect to y. Outliers are selected as those instances that exhibit a large deviation from the model.

Fig. 6 displays the outliers (red triangles) found from the successive linear regression method [37], which is based on this model. The x axis is general position system (GPS) latitude, the

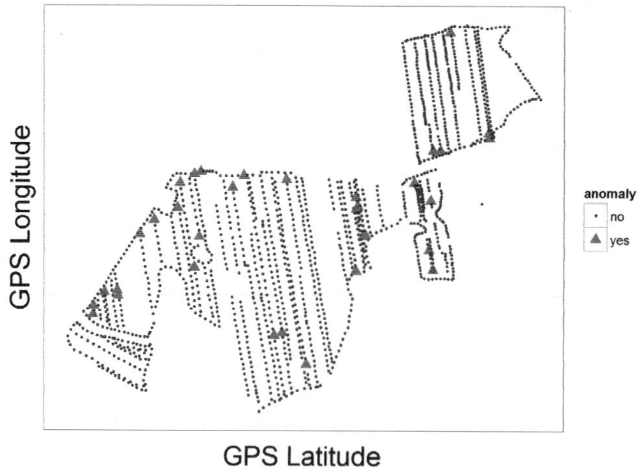

Figure 6. Outliers found from the linear successive regression method

y axis is the GPS longitude. It can be observed that the outliers often lie near corners, field boundaries or crowded regions. In the present work, the linear regression method has further been extended to quadratic regression models. Furthermore, system states with individual regression models assigned to each state have been identified. The system states have been obtained by the OTALA algorithm [18]. For each system state, a particular quadratic regression model has been determined by least squares estimation. Both model learning algorithms - OTALA and the learning of QRMs - have been parallelized by application of MapReduce technology. Initial results have been published in [38]. The MapReduce version of OTALA allows to distribute the workload on $|E|$ edges, by processing the edges E of the created automaton in parallel. For the MapReduce version of QRM, distribution of workload on $|S|$ nodes is possible by processing the states S of the automaton in parallel. Furthermore, online algorithms have been proposed, which efficiently handle novel observations to update the models, which have been created from large historical data sets. In the next step, more complex nonlinear regression models with dependent variables, such as neural networks, have to be evaluated in order to improve the prediction accuracy and to fully exploit the large amounts of data.

5.3. Self-organizing maps

In this approach, a data-driven model of the system in the form of a self-organizing map is generated. With the aid of the UMatrix transformation [21], the various process phases or operating states of the system (e.g. emptying containers or filling containers) were identified using the Watershed transformation. An an example, the experimental demo plant, illustrated in Fig. 7 a) is used, which consists essentially of two containers between which liquid is pumped around in cycles at varying pumping powers and valve positions. The obtained Watershed

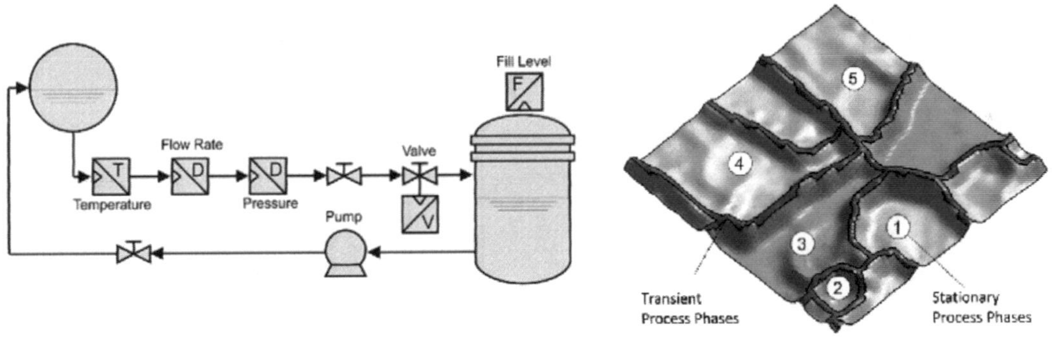

Figure 7. a) Demo plant b) self-organizing map

transformation for this process is depicted in Fig. 7 b). Based on the trained self-organizing map an online-diagnosis of the process behavior is performed. Interventions in the process behavior of the system are clearly revealed in the curve of the quantization error (Q_{error}) of the map by applying a threshold evaluation. The concept is currently evaluated to very large-scale datasets from industrial distillation columns - first results show, that the proposed concept can successfully be applied to real-world applications.

5.4. PCA-based anomaly detection

The PCA-based anomaly detection method that is used in this project [39] works on a data base. Each point in this data base represents the complete sensor data at a given time instance. This interpretation leads to a high-dimensional problem, which can be reduced by dimensionality reduction. Dimensionality reduction leads to a compact representation of system behavior. In the next step, patterns are extracted, which represent the normal behavior of the system. Fig. 8 shows the data of PCA transformed data of the sorting plant on the left hand side. The area of normal operations is depicted with green points, while known errors are shown as red points. The interference is small so that the green cluster can be used as model of the normal plant operation. Deviations of the learnt model of normal operation can be related to failures. The distance

$$d(y_{pca}(k); x_{pca}(k)) = \|y_{pca}(k) - x_{pca}(k)\|^2 \tag{4}$$

of PCA-transformed measurements $y_{pca}(k)$ with respect to PCA-reduced data points x_{pca} of the normal behavior has been assessed by application of Marr-Wavelets

$$\Psi(y_{pca}(k), x_{pca}(k)) = \frac{2}{\sqrt{3\sigma}\pi^{\frac{1}{4}}} \left(1 - \frac{2}{\sigma^2}\right) e^{-\frac{d(y_{pca}(k); x_{pca}(k))}{2\sigma^2}}. \tag{5}$$

The standard deviation σ of the Marr-Wavelets is obtained from training data. For $\Psi(y_{pca}(k); x_{pca}(k)) > 0$, a normal process state is assumed at time instance k. $\Psi(y_{pca}(k); x_{pca}(k)) \leq 0$ indicates potential faults, which are displayed to the user.

The PCA-based anomaly detection method has been used to detect and predict a process anomaly in the sorting plant. Based on a test dataset, an initiating process anomaly could be detected approximately 1-2 minutes beforehand.

Figure 8 shows the visualized PCA on the left hand side and the raw data on the right hand side. Further it can be seen that the data points pointing at a process anomaly (yellow/red) are clearly separated from the data points representing the normal behavior (green). The status of the plant (for each sorting module) is visualized using a traffic light.

As a benefit, the error cause can be determined at an early stage by analyzing which signals from the raw data lead to a certain warning.

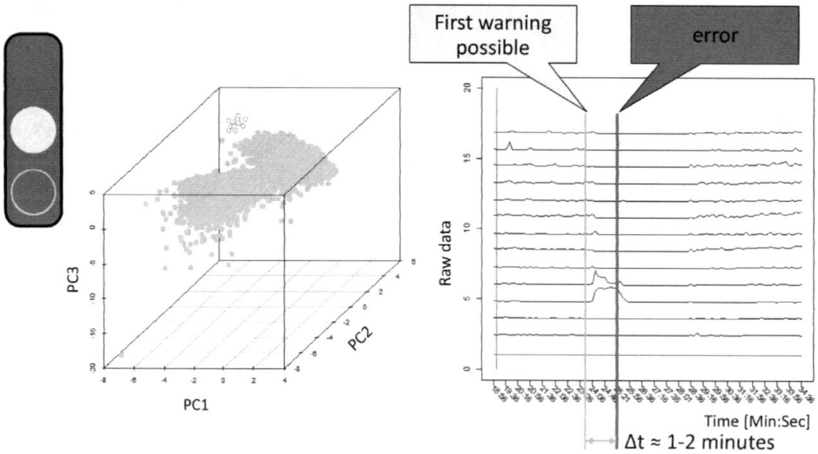

Figure 8. Using the PCA in the sorting plant, the error could be predicted approximately 1-2 minutes beforehand.

6. Conclusion and outlook

In the present work, three industrial and agricultural application cases for big data analysis have been investigated. In these application cases, heterogeneous networks and protocols required the development of convenient data connectors between the central infrastructures and the manufacturing systems. Data analysis in the central infrastructure was accomplished independently of the particular application cases. For this purpose, several application-independent anomaly detection algorithms have been investigated in the present work. The distance-based anomaly detection method LOF outperformed the GAS score in the investigated agricultural application case. Furthermore, regression-based anomaly detection methods were observed to find anomalies at reasonable positions in this application case. MapReduce algorithms have been developed for efficient learning of hybrid system models, which are based on quadratic regression models. A further step on the research agenda is the development of non-linear regression methods, which are based on neural nets and support vector machines. Self-organizing maps have been shown in the present work to be suitable for anomaly detection in a demo plant of the chemical industry and have to be extended to large-scale plants. A PCA-based method showed promising results for anomaly detection in a large-scale production plant, where process anomalies could be detected 1-2 minutes beforehand.

So far, initial evaluations of the developed algorithms have been conducted in particular application cases, respectively. However, the proposed algorithms are designed to work in each of the three application cases. Systematic evaluation of the proposed algorithms is a further step on the research agenda.

Acknowledgments

The present work was funded by the German Federal Ministry of Education and Research (BMBF) within the cooperative project AGATA (Analyse großer Datenmengen in Verarbeitungsprozessen, engl.: analysis of large amount of data in manufacturing processes) and managed by the Project Management Agency DLR. The author is responsible for the contents of this publication.

References

[1] MANUFUTURE-EU, "Factories of the Future PPP Strategic Multi-annual Roadmap," 2010.
[2] *IEC 61131-3: Speicherprogrammierbare Steuerungen, Teil 3: Programmiersprachen*, International Electrotechnical Commission Std.
[3] F. Pethig, , B. Kroll, O. Niggemann, A. Maier, T. Tack, and M. Maag, "A generic synchronized data acquisition solution for distributed automation systems," in *17th IEEE International Conference on Emerging Technologies and Factory Automaion*, 2012.
[4] *ISO 11783 Tractors and machinery for agriculture and forestry - Serial control and communications data networks*, International Organization for Standardization Std.
[5] W. Emmerich, M. Aoyama, and J. Sventek, "The Impact of Research on the Development of Middleware Technology," *ACM Transactions on Software Engineering and Methodology*, vol. 17, no. 4, pp. 19:1–19:48, 2008.
[6] Object Management Group (OMG), "Common Object Request Broker Architecture (CORBA) Specification Version 3.1," http://www.omg.org/spec/CORBA/, 2008.
[7] F. Picioroaga, A. Bechina, U. Brinkschulte, and E. Schneider, "OSA+ Real-Time Middleware, Results and Perspectives," in *7th IEEE International Symposium on Object-Oriented Real-Time Distributed Computing*, 2004.
[8] P. Leitao, "Agent-based distributed manufacturing control: A state-of-the-art survey," *Engineering Applications of Artificial Intelligence*, vol. 22, pp. 979–991, 2009.
[9] *OPC Data Access Custom Interface Standard Version 3.00*, OPC Foundation Std.
[10] M. Damm, S. Leitner, and W. Mahnke, *OPC Unified Architecture.* Springer-Verlag Berlin Heidelberg, 2009.
[11] P. . P. International, "The profienergy profile," Tech. Rep., 2010.
[12] M. Schleipen and M. Okon, "The caex tool suite - user assistance for the use of standardized plant engineering data exchange," in *IEEE Conference on Emerging Technologies and Factory Automation (ETFA)*, 2010.
[13] S. Rohjans, M. Uslar, and H. Appelrath, "Opc ua and cim: Semantics for the smart grid," in *IEEE Transmission and Distribution Conference and Exposition*, 2010.
[14] L. Christiansen, A. Fay, B. Opgenoorth, and J. Neidig, "Improved diagnosis by combining structural and process knowledge," in *Emerging Technologies Factory Automation (ETFA), 2011 IEEE 16th Conference on*, 2011, pp. 1–8.
[15] S. Faltinski, H. Flatt, F. Pethig, B. Kroll, A. Vodenčarević, A. Maier, and O. Niggemann, "Detecting

Journal of Physics: Conference Series **659** (2015) 012055 doi:10.1088/1742-6596/659/1/012055

anomalous energy consumptions in distributed manufacturing systems," in *Industrial Informatics (INDIN), 2012 9th IEEE International Conference on*, 2012, pp. 358 – 363.

[16] S. Verwer, "Efficient identification of timed automata: Theory and practice," Ph.D. dissertation, Delft University of Technology, 2010.

[17] O. Niggemann, B. Stein, A. Vodenčarević, A. Maier, and H. Kleine Buning, "Learning behavior models for hybrid timed systems," in *Twenty-Sixth Conference on Artificial Intelligence (AAAI-12)*, Jul 2012.

[18] A. Maier, "Online passive learning of timed automata for cyber-physical production systems," in *The 12th IEEE International Conference on Industrial Informatics (INDIN 2014)*. Porto Alegre, Brazil, Jul 2014.

[19] S. Haykin, *Neural Networks and Learning Machines (3rd Edition)*, 3rd ed. Upper Saddle River, New Jersey: Prentice-Hall, 2008.

[20] T. Kohonen, *Self Organization and Assoziative Memory*. Springer, 1990.

[21] C. Frey, "Monitoring of complex Industrial processes based on self-organizing maps and watershed transformations," in *IEEE International Conference on Industrial Technology (ICIT)*, 2012.

[22] A. Vodenčarević, H. Kleine Buning, O. Niggemann, and A. Maier, "Identifying behavior models for process plants," in *Emerging Technologies & Factory Automation (ETFA), 2011 IEEE 16th Conference on*, 2011, pp. 1–8.

[23] A. Shui, W. Chen, P. Zhang, S. Hu, and X. Huang, "Review of fault diagnosis in control systems," in *Control and Decision Conference, 2009. CCDC '09. Chinese*, june 2009, pp. 5324 –5329.

[24] B. C. Williams and M. M. Henry, "Model-based estimation of probabilistic hybrid automata," Tech. Rep., 2002.

[25] F. Zhao, X. D. Koutsoukos, H. W. Haussecker, J. Reich, and P. Cheung, "Monitoring and fault diagnosis of hybrid systems," *IEEE Transactions on Systems, Man, and Cybernetics, Part B*, vol. 35, no. 6, pp. 1225–1240, 2005.

[26] S. Windmann, S. Jiao, O. Niggemann, and H. Borcherding, "A Stochastic Method for the Detection of Anomalous Energy Consumption in Hybrid Industrial Systems," in *INDIN*, 2013.

[27] M. Wang and R. Dearden, "Detecting and Learning Unknown Fault States in Hybrid Diagnosis," in *Proceedings of the 20th International Workshop on Principles of Diagnosis, DX09*, Stockholm, Sweden, 2009, pp. 19–26.

[28] G. Wang, S. Yin, and O. Kaynak, "An LWPR-Based Data-Driven Fault Detection Approach for Nonlinear Process Monitoring," *IEEE Transactions on Industrial Informatics*, vol. 10, no. 4, pp. 2016–2023, 2014.

[29] S. Yin, S. Ding, A. Haghani, H. Hao, and P. Zhang, "A Comparison Study of Basic Data-driven Fault Diagnosis and Process Monitoring Methods on the Benchmark Tennessee Eastman Process," *J. Process Control*, vol. 22, no. 9, pp. 1567–1581, 2012.

[30] F. Angiulli and C. Pizzuti, *Lecture Notes in Computer Science: Principles of Data Mining and Knowledge Discovery*. Springer, 2002, ch. Fast Outlier Detection in High Dimensional Spaces, pp. 43–78.

[31] M. Breuning, H.-P. Krigel, R. Ng, and J. Sander, "LOF: Identifying Density-Based Local Outliers," in *Proceedings of the 2000 ACM SIGMOD International Conference on Management of Data*, 2000.

[32] S. Papadimitriou, H. Kitagawa, P. Gibbons, and C. Faloutsos, "LOCI: Fast Outlier Detection Using the Local Correlation Integral," in *International Conference on Data Engineering*, 2003.

[33] M. Amer and M. Goldstein, "Nearest-Neighbor and Clustering based Anomaly Detection Algorithms for RapidMiner," in *3rd RapidMinder Community Meeting and Conference*, 2012.

[34] R. Duda, P. Hart, and D. Stork, *Pattern Classification (2nd Edition)*. Wiley-Interscience, 2000.

[35] C. Phua, D. Alahakoon, and V. Lee, "Minority Report in Fraud Detection: Classification of Skewed Data," *ACM SIGKDD Explorations Newsletter*, vol. 6, no. 1, pp. 50–59, 2004.

[36] K. Chakrabarti, E. Keogh, M. Pazzani, and S. Mehrotra, "Local adaptive dimensionality reduction for indexing large time series databases," *ACM Transactions on Database Systems*, vol. 27, no. 2, pp. 188–228, 2002.

[37] C. Aggarwal, *Outlier Analysis*. Springer, 2013.

[38] S. Windmann and O. Niggemann, "MapReduce Algorithms for Efficient Generation of CPS Models from Large Historical Data Sets," in *IEEE International Conference on Emerging Technologies and Factory Automation (ETFA 2015)*, Luxembourg, 2015.

[39] J. Eickmeyer, P. Li, O. Givehchi, F. Pethig, and O. Niggemann, "Data driven modeling for system-level condition monitoring on wind power plants," in *26th International Workshop on Principles of Diagnosis (DX 2015)*, 2015.

Control and Diagnosis in Integrated Product Development – Observations during the Development of an AGV

R Stetter[1], A Simundsson[2]

[1] Hochschule Ravensburg-Weingarten, Weingarten, Germany
[2] University of Manitoba, Winnipeg, Manitoba, Canada

E-mail: stetter@hs-weingarten.de

Abstract. This paper is concerned with the integration of control and diagnosis functionalities into the development of complete systems which include mechanical, electrical and electronic subsystems. For the development of such systems the strategies, methods and tools of integrated product development have attracted significant attention during the last decades. Today, it is generally observed that product development processes of complex systems can only be successful if the activities in the different domains are well connected and synchronised and if an ongoing communication is present – an ongoing communication spanning the technical domains and also including functions such as production planning, marketing/distribution, quality assurance, service and project planning. Obviously, numerous approaches to tackle this challenge are present in scientific literature and in industrial practice, as well. Today, the functionality and safety of most products is to a large degree dependent on control and diagnosis functionalities. Still, there is comparatively little research concentrating on the integration of the development of these functionalities into the overall product development processes. The main source of insight of the presented research is the product development process of an Automated Guided Vehicle (AGV) which is intended to be used on rough terrain. The paper starts with a background describing Integrated Product Development. The second section deals with the product development of the sample product. The third part summarizes some insights and formulates first hypotheses concerning control and diagnosis in Integrated Product Development.

1. Integrated Product Development

In recent years, powerful methods and algorithms for the control, monitoring and diagnosis of industrial systems and processes have been developed and the potential resulting from the application could be proven. However, as a consequence of the multitude and variety of these methods and tools it is difficult for product developers of future systems and processes to choose the right combination of methods and tools for their specific application. The challenging tasks are to achieve a technical and economical optimum, including a sensible use of available resources and considerations of robustness, life-cycle and sustainability issues. This paper is based on the hypothesis that the well-known methodology integrated product development (IPD) could assist product developers and control or diagnosis engineers in their endeavor.

First the terms control and diagnosis will be explained before the core of IPD will be elucidated. The term **control** names activities intended to manage, command, direct or regulate the behavior of devices or systems. Control has been the core of extensive research for many decades. In recent years the techniques of predictive control have found rising attention (compare e. g. Camacho and Bordons [1] or Wang and Boyd [2]). Predictive control usually relies on dynamic models of the process, most often linear empirical models obtained by system identification. In the area of pump systems predictive control can pursue three different objectives: smoothing changes of system states, better coordination of multiple pumps and evaluating decision alternatives.

Diagnosis is usually understood as the process of estimating the object's condition. Diagnosis is carried out by the estimation of important parameters, by the detection and isolation of faults, by the identification of faults and by the determination what should be done in the case that faults occur. Over the last three decades, the growing demand for safety, reliability, and maintainability in technical systems has drawn significant research in the field of diagnosis. Such efforts have led to the development of many techniques; see for example the most recent survey works (Blanke et al.[3], Isermann [4], Witczak [5], Korbicz et al.[6]). For fault compensation, fault tolerant control methods are proposed which can generally be classified into two types, i.e. Passive Fault Tolerant Control Scheme (PFTCS) and Active Fault Tolerant Control Scheme (AFTCS) (Blanke et al. [3]).

According to Ehrlenspiel [7] integrated product development is a "holistic approach to overcome the problems that arise in product development due to the division of manpower" (Figure 1).

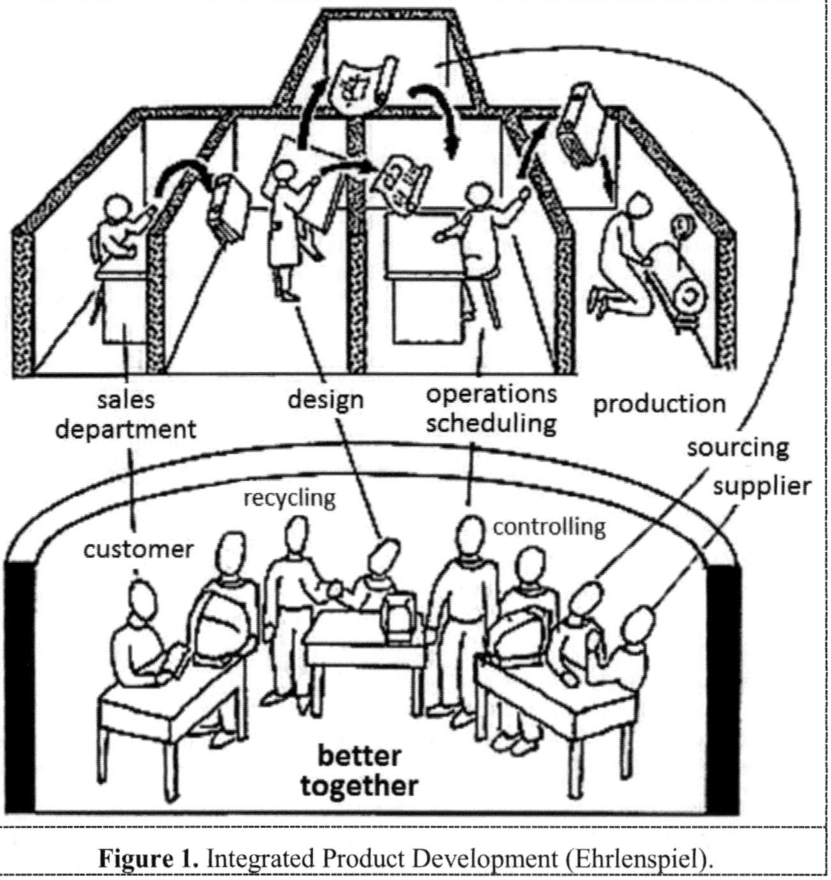

Figure 1. Integrated Product Development (Ehrlenspiel).

He states that integrated product development is an approach that includes different methods of problem solving, organizational methods of optimizing interpersonal processes and technical methods for the direct improvements of products. Traditional "over the wall" approaches are sequential by nature as illustrated in Figure 1.

The major disadvantage of this approach is that products are developed with a limited exchange of information and ideas, and people late in the sequence do not have any input to earlier decision stages. As a result, poor decisions are made, which resulted in longer "Time to Market" and higher development costs [8]. Modern parallel approaches have developed from simultaneous engineering over concurrent engineering to integrated product development. Andreasen and Hein [9] describe the main activities in the different streams market, product and (production) process), see Figure 2.

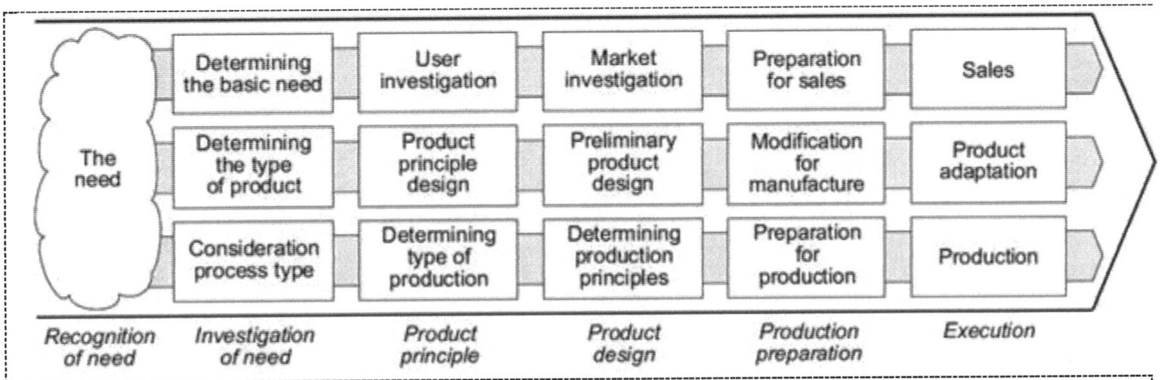

Figure 2. Integrated Product Development (Andreasen and Hein).

Integrated product development emphasizes the earliest possible involvement of all "downstream" departments into the design decisions. Now, while "upstream" work processes have to invest more time into planning, "downstream" work processes save time and costs. Several studies have demonstrated that this approach can result in a shorter "Time to Market". Furthermore an enhanced product quality can be achieved.

Integrated product development is a holistic approach aiming at the creation of complex innovative systems. Several aspects are important part of this approach; Figure 3 gives an overview.

It is important to note that the implementation of the complex approach integrated product development (IPD) cannot just be the simple use of some methods and tools. A strategic implementation of IPD is necessary; similar to concepts such as Total Quality Management (TQM). IPD requires a stringent process orientation; the implementation requires constant top management level support. Two aspects of IPD enable breakthrough success: an open multi-domain communication in connection with a shared multi-domain project responsibility. The backbone of IPD is the application of state of the art processes, methods and tools on all levels of a product development organization. One core element of IPD is the conscious and systematic use of several abstraction levels of product models, which allows the product development engineers to analyse and understand their product from several viewpoints. This also fosters a deep understanding of the essential requirements and functions of the product under development. IPD greatly emphasizes the fact that first solution ideas are in most cases not the optimum solutions. Therefore, the creation of alternative solutions on all abstraction levels is an integral element of IPD. One of the challenges in development processes of complex systems is to ensure an overview of all activities, decisions and intermediate results. IPD proposes a planning on the macro level based on phases with pre-determined results. This approach allows realistic planning and ensures the traceability or the product development progress, but also prevents inefficient planning. Cornerstones of this approach are multi-domain synchronization points, which link the different domains involved in the product development (such as mechanical engineering, control engineering, electrical engineering as well as computer soft- and hardware). IPD also emphasizes the connection of the product models of the different domains and incorporates approaches towards multi-domain product models. Such models also allow multi-domain simulation, which is a main prerequisite for domain-spanning optimum solutions. Many product examples which are now outdated such as tape recorders show that the main causes for failure lie in the product concept. IPD therefore promotes a special emphasis on abstract levels (requirements, functions, physical effects) of the product models. Last but not least, the consideration of profound changes over time also aims at preventing a concentration on solutions or technologies which can be outdated. Very often wrong hypotheses such as the size of certain market segments (e.g. computer) have led product management to wrong strategic decisions.

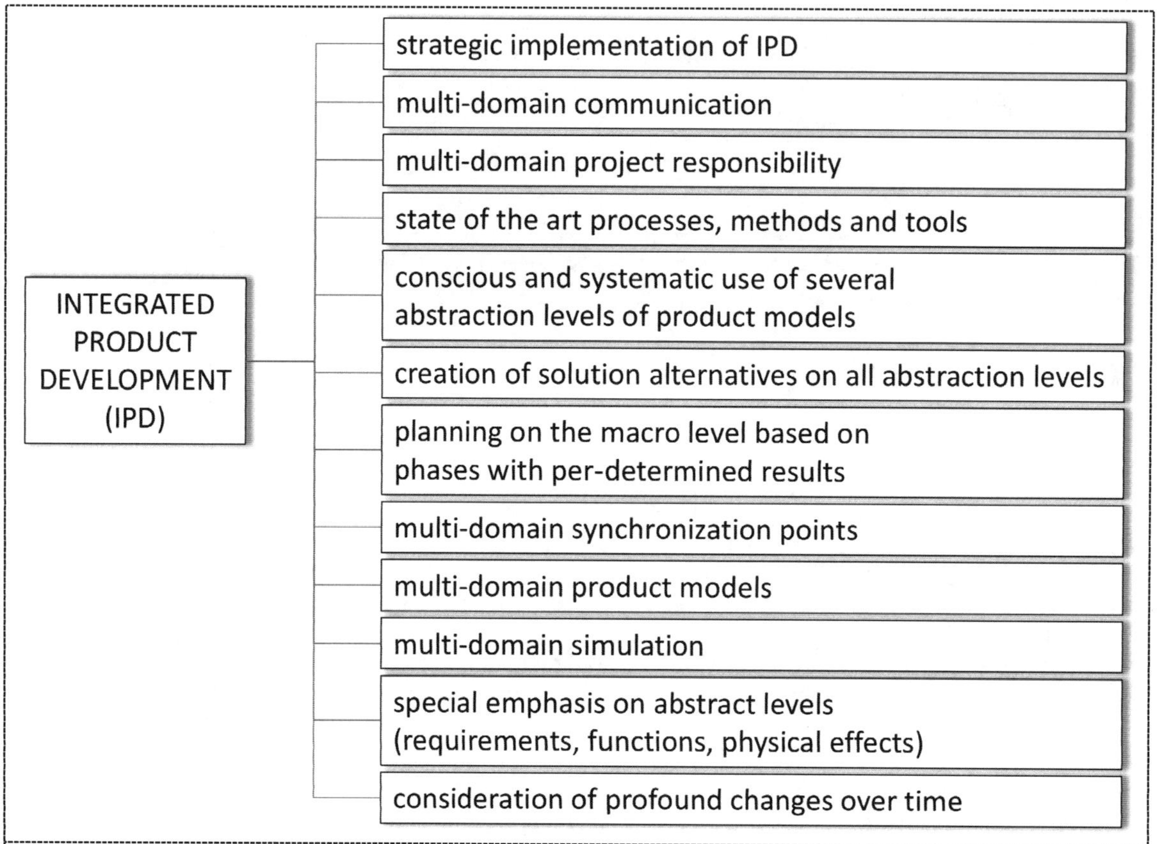

Figure 3. Important aspects of Integrated Product Development.

All the listed aspects of IPD enable more conscious and more successful product development. However, an emphasis towards control and diagnosis was not yet a central aspect of IPD (consequently it is not yet shown in Figure 3 as an important aspect – it may belong there). In order to improve the integration of these functionalities a long-term development project at the Hochschule Ravensburg-Weingarten – the development of an AGV – was observed.

2. Product Development of an Automated Guided Vehicle AGV

This section describes the product development of an AGV for outdoor used. This development process was observed and leads to some hypotheses presented in Section 3. The AGV makes use of a unique steering system based on torque differences and is suited for rough terrain because of four independent legs with wheels (Figure 4).

Journal of Physics: Conference Series **659** (2015) 012056 doi:10.1088/1742-6596/659/1/012056

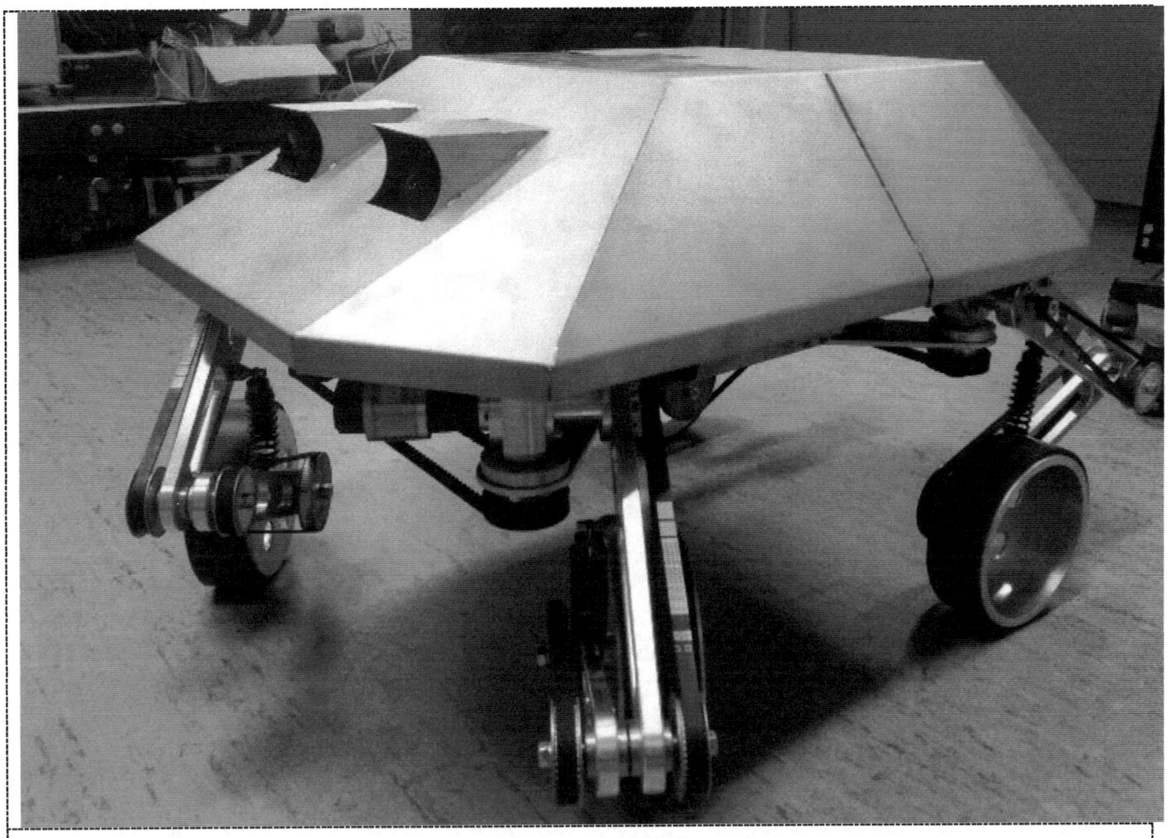

Figure 4. Automated Guided Vehicle.

The main objective of the project was to develop a Semi-Autonomous Outdoor Vehicle designed to explore and work in open areas [10]. The steering is based on torque differences.

2.1. Torque steering system

Autonomous vehicles such as mobile robots and the respective steering systems have been successfully developed and built for some years (Anderson&Jones [11], Ashmore&Barnes [12], Dillard [13]). The distinctive quality of the steering system used is the dynamic behavior. This innovative steering system that is already registered as a patent ([14], [15]) is based on the concept to use the torque of drive motors (more exactly the torque differences between wheels) to steer four independent axles of a vehicle. The principal steering system is shown in Figure 5.

In this example a vehicle consists of four drive motors which are fastened on arms that may freely rotate. These arms have no drive or brake, only an angle encoder is attached at the end of each axle. These angle encoders measure the angle of the motor and the wheel with regard to the vehicle platform. The distinct characteristic of the innovative drive system is the absence of dedicated steering motors. By means of angle encoders applied at the four steering axles and highly dynamic control algorithms it is possible to steer such vehicles only by means of the four drive motors (compare Figure 4).

Each of the wheels on the short axle can be directed into the desired position by means of the torque applied on the wheel. This could take place sequentially for each individual wheel but also simultaneously, if the control allows different torque on all wheels. This characteristic allows simpler and simultaneously more robust vehicle concepts. It is also a main advantage of this concept that the resulting vehicle is able to drive directly in any direction without time and space consuming turning maneuvers.

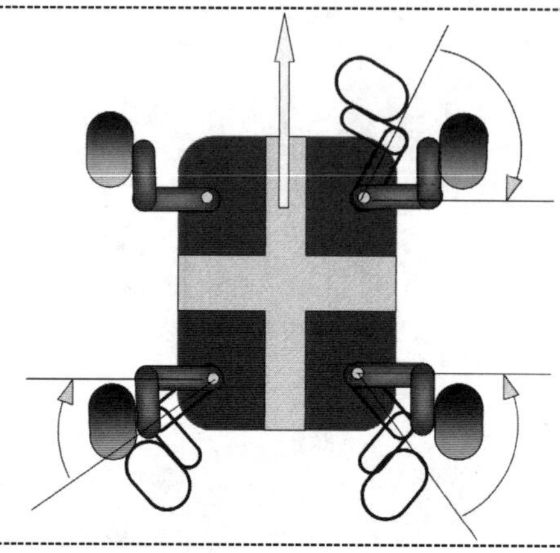

Figure 5. Principal steering system.

Furthermore, a vehicle based on the dynamic drive system is able to turn around its own center. This characteristic is very important if cameras or other equipment are mounted on such vehicles which can only be used in a certain orientation. The innovative steering system shares these advantages with Omni drive systems (Ashmore&Barnes [12]), but has reduced friction as well as easier controllability and offers the possibility to determine an exact position and orientation from an analysis of the angles of the steering axes and the angles of the drive wheels (odometry).

2.2. Realization of the Suspension System

The semi-autonomous vehicle is controlled by a small PC-ALIX that it is connected to 4 EPOS motor controllers, one per wheel. For the communication between the controllers it uses the CANopen protocol. It has also two encoders connected which read the angle of the directions from the rear and the front wheels. The front wheels are moved in parallel, due to the fact that they are connected by a pulley. The rear wheels are connected in the same way as the front wheels.

The vehicle moves on 4 wheels. The wheels are located in each corner, as it can be seen in Figure 6.

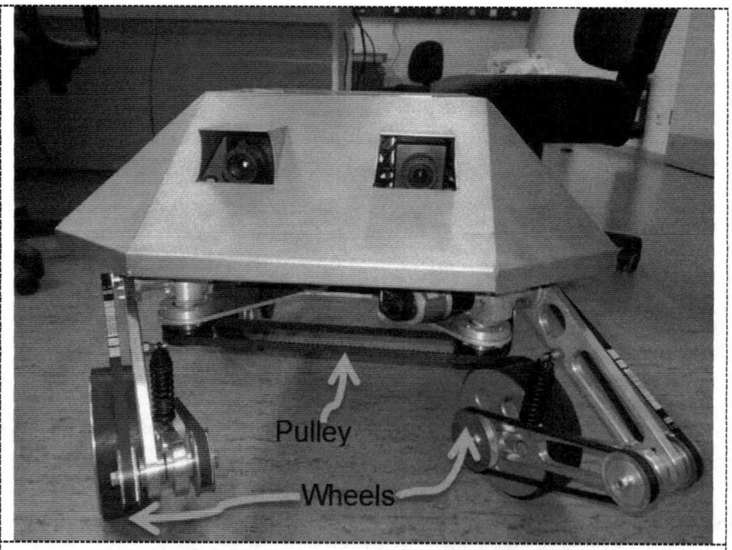

Figure 6. Collocation of the wheels and the pulley.

The diameter of the wheel has been chosen bigger to have a trim and steady movement in different terrains. For the better traction the surface of the wheel is made out of rubber. In Figure 7 it can be seen that the wheels are joined to the vehicle by a traction system made of two arms and one suspension.

Figure 7. Traction system of the vehicle.

This system permits the vehicle to absorb the different heights of the terrain and the impacts against the ground. There are two belt driven pulleys to transmit the movement from the motor to the wheel. This transmission is a reduction system, because the torque required by the vehicle is more than the torque produced by the motor. The front wheels are joined in pairs by a belt pulley system as well as the rear wheels in the same manner. The pair of wheels has the same direction at every moment. When the wheels are aligned in a straight line, they can turn almost 90 degrees in each direction. The belt pulley system can be seen in Figure 7.

The vehicle has two brakes: one on the front wheel pair and the other one on the rear wheel pair which are intended to hold the steering angle when the direction is not changed. This leads to improved dynamic behavior.

2.3. Control System

So-called EPOS controllers are used for the direct control of the drive motors; there is one EPOS per motor. An EPOS is a small-sized, fully digital motion controller [18]. The EPOS 24/5 can control brushed DC motors with digital encoders as well as brushless EC motors with digital Hall sensors and encoders. An EPOS is specially designed to be a slave in a CANopen network and it can also be operated through a RS-232 communication port. The vehicle has two other additional encoders, to know the angle between the pair of wheels and the vehicle; they are installed on the vertical axis of one wheel from each pair of wheels- one at the front and one at the rear. The master of the vehicle is a small PC-ALIX by Acer Company. It has five USB ports, HDMI, built-in WiFi and a VGA output. It has a hard disk, and a compact flash memory of 4 GB. The vehicle has two batteries of 12V each. They are connected in series giving 24V. They are used to give the power supply to the EPOS and the Encoders. As the motors are connected to the EPOS, the motors rely on the two batteries indirectly. The lithium ion battery pack is used to supply the PC-ALIX. The vehicle has two switches to control the electrical circuit and has a fuse box to provide safety to the electrical installation. An electrical block diagram can be seen in Figure 8.

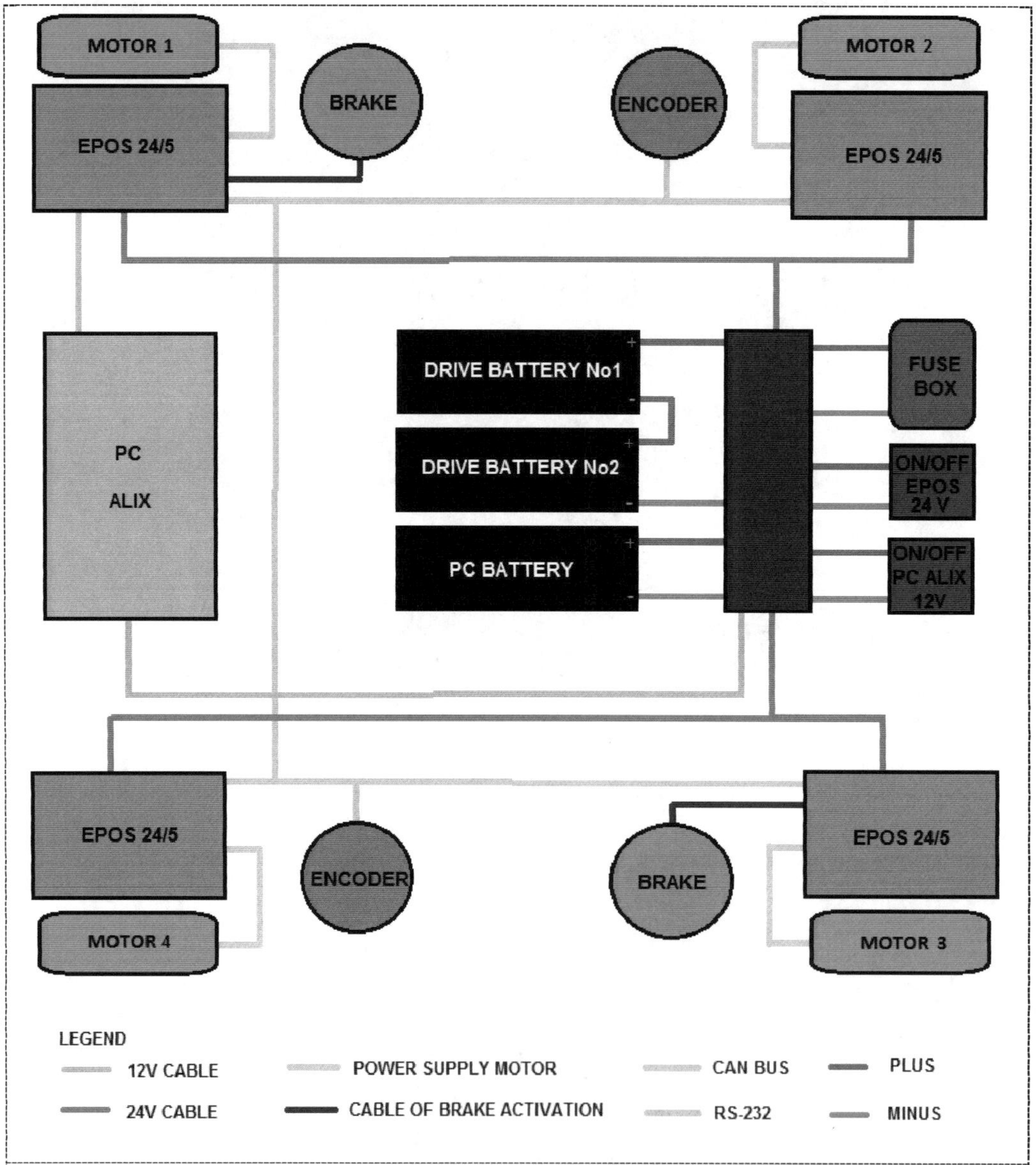

Figure 8. Electric block diagram.

Journal of Physics: Conference Series **659** (2015) 012056 doi:10.1088/1742-6596/659/1/012056

2.4. Control and Diagnosis of the vehicle

The control of the vehicle happens on different levels. A central computer receives a movement wish either directly from the user (manual driving) or improved by a superordinate intelligence (semi-autonomous driving mode) or only from a superordinate intelligence (autonomous driving mode). Based on a mathematical model of the AGV this movement wish is than translated into a speed command to the drive motors as well as desired angles for the wheels (one angle for both front wheels and one angle for both back wheels).

One critical aspect of the control of the AGV was the control of the drive motors based on a proportional-integral-derivative (PID) controller, which is a control loop feedback mechanism which adjusts processes and control outputs through error monitoring and minimization [16]. Tuning a control loop is defined as the adjustment of the PID gains to values that provide the quickest response with a satisfactory amount of oscillation for the system, therefore providing stability to the system. However, tuning is a rather difficult process and various methods can be used. It should also be noted that processes that have a certain degree of non-linearity may benefit from gain scheduling; that is, if the system contains parameters that work well at full-load conditions but have a subpar performance at start-up or no-load, this can be corrected by using different parameters in different operating regions (i.e. by gain scheduling).

One method of tuning is manual tuning. This involves setting the values of the integral (K_i) and derivative (K_d) controllers to 0, and increasing the proportional value (K_p) of the controller until the output of the loop oscillates. K_p should be then set to approximately half that value. Next, increase K_i until any offset is corrected in an acceptable time frame for the process. Finally, increase K_d (if necessary) until the loop reaches the desired output value in an acceptable time frame. Note that most systems accept a slight amount of overshoot in order to reach the desired set-point more quickly, but only in cases in which some overshoot is not an issue. If the system cannot handle any degree of overshoot, an over-damped system is required, which will have a K_p of significantly less than half the K_p that would be used to cause a slight overshoot in the system. The controller used in this project also has a position regulation with a feed forward feature, through which the effectiveness of the PID controller can be improved even further. The variables used in the feed forward mechanism used are the "Velocity Feedforward Factor" (K_w) and the "Acceleration Feedforward Factor" (K_a). K_a is calculated based on the total inertia of the drive system ($J_{tot}= J_{motor} + J_{load}$) and the motor torque constant (k_m). The motor torque constant can be found in the motor catalogue. For the motor used in this project, it is 16,9 mNm/A. The load inertia must be measured in order to calculate the total inertia, and the motor inertia can be taken as the rotor inertia, also given in the motor catalogue.

The Velocity Feedforward Factor is calculated based on a linear load current increases if the angular velocity is enhance by a $\Delta\omega$. This can easily be determined by measuring the current at two different speeds, or assuming the current is zero when the angular velocity is zero. The constants are calculated based on formulae 1, 2, and 3:

$$K_a = \frac{J_{tot}}{k_m} 1000, \left[\frac{A}{rad/s^2} \cdot 10^{-7} \right] \quad (1)$$

$$J_{tot} = J_{mot} + J_{load}, \left[gcm^2 \right] \quad (2)$$

$$K_w = \frac{30}{\pi} \frac{\Delta I}{\Delta\omega}, \left[\frac{\mu A}{rad/s} \right] \quad (3)$$

Once calculated, these values are written to the objects "Acceleration Feed-forward Factor", and the "Velocity Feedforward Factor" in the object dictionary as per instructions given in the Epos Application Note. Once these values are written to objects, the system is now ready for tuning with the tuning wizard available in Epos Studio. This approach led to a satisfactory control behavior.

Journal of Physics: Conference Series **659** (2015) 012056 doi:10.1088/1742-6596/659/1/012056

3. Observations in the project

The development of the AGV took more than one year and several different teams which members from mechanical engineering, electrical engineering and computer science were involved. In this section some observation of this process are reported.

The development processes which could be observed were much more **evolutionary** than revolutionary. This distinction was suggested by Stetter et al. [17] for mechatronic development processes. In their research they were observing evolutionary processes in different types of industrial development processes. Evolutionary processes are characterized by gradual change and development. On the contrary, revolutionary design can be understood as a sudden or complete change in something. The observed project showed several iterations and many jumps from abstract to concrete levels of product development, e. g. between task clarification and detail design. The success of the project indicates that the central advantages of an evolutionary process – flexibility – did outweigh the risk of an early focus on maybe suboptimal solution.

From later research Stetter et al. [18] also report the importance of **fast, early prototyping**. Fast, early prototyping requires conscious jumps at early stages of a product development process to concrete prototypes in the sense of physical mock-ups, tests but also software prototypes. In system engineering preliminary studies can describe similar activities. Throughout the project, the question: "How should I develop a control system, if I don't know the mechanical arrangement?" could be heard frequently from the persons responsible for control and diagnosis functionalities. The collaboration was usually much easier if some physical system was available, even if it only was a rather basic prototype.

A large difference between the **documentation needs** could be observed. Mechanical engineers needed to provide accurate technical drawings in order to have the parts to be made in the university workshop or by outside companies. Therefore, they were forced to document and rethink their solution before they could touch or test it . Also, because of the production expenses these students tried rather hard to avoid mistakes – maybe they also concentrated on "safe" solutions. On the contrary, the students in charge of computer programs initially documented nothing (later on they added some commentaries in the code) and could immediately test and improve the resulting code. Probably as a consequence, they were more open to try new and unexplored ideas.

In all disciplines some forms of **trouble shooting** could be observed. In the mechanical engineering this was mainly during the assembly process, which was not documented up front. In electrical engineering the cable routing exhibited several instances of trouble shooting. As mentioned before, a large share of the work of the computer scientist was aimed in this direction.

One important prerequisite for successful interdisciplinary work is **trust** – trust that the other disciplines will stick to the negotiated interfaces and will quickly advise as to any necessary changes. The importance of trust was also reported by Pulm&Stetter ([19], [20]), which they understand to be a willingness to rely on another party in the confidence that this reliance will provide a positive outcome. They conclude that the main function of trust in product development is achieving a shared understanding and therefore an enhanced effectiveness and efficiency. The development process of the AGV showed usually a certain time span until specialists of different disciplines were able to trust each other. This trust was enhanced, when mechanical engineers were able to show physical models and computer scientists could demonstrate successful programs.

Stetter&Pulm [21] also report that the **synchronization of the different disciplines** is consequently a major challenge for project management of mechatronic products. The consequences of this challenge could also be frequently observed during the development process of the AGV. Very often the needs for information as well as the intermediate results that were achieved were unclear for students from other scientific disciplines and common project planning was difficult and time consuming. In general Gantt chart planning was used and was in general successful. However, it was frequently unclear which level of detail is appropriate for this kind of planning. Also the controlling of mile-stones was rather difficult as the deliverables at mile-stones were usually not defined or only very roughly defined.

The main problem concerning the **abstract structure** of a mechatronic system is that the mechatronic system has to be structured in different dimensions [21]. The structures have to cover the abstract functional structure, the structure of modules, the mechanical structure (geometry), the electronic structure ("systems") in the meaning of a network, as well as the software structure (software functions). Ideally, these should be congruent, but in the project there were various reasons why they differed: functions were spread over various components; each structure had to be regarded independently in order to get an optimized system (and together in order to get an optimized product); functions were strongly interconnected and defined from different viewpoints or on different levels. While these problems could in theory be solved by a clear functional structuring, other aspects hindered the success of this approach: the complexity of the mechatronic system did not allow a complete modeling of the system; the involved disciplines had a different thinking; many problems could only be solved by bypass solutions regarding the structure. In the project the abstract structure was never really documented and evolved by trial and error.

4. Summary and Outlook

One of the main trends in the evolution of technical systems is a development towards higher automation, i.e. towards a lower amount of required human interaction with the system [22], [23]. This trend requires extremely powerful control algorithms. The control of an autonomous or semi-autonomous system has to be able to compensate for external influences such as disturbances and noise. Furthermore the control of such systems has to be able to detect and identify certain faults and has to be able to compensate for their consequences or to transform the system into a safe state. Therefore, extremely powerful diagnosis capabilities are also required for autonomous or semi-autonomous systems. Conventional product development processes do not emphasize this prominent role of control and diagnosis. Today these capabilities are frequently added to systems in rather late stages of the product development processes, thus leading to sub-optimum results. Consequently, efforts toward a stronger integration of the capabilities control and diagnosis into the product development processes are mandatory.

The main intention of this paper is to describe a product development project including control and diagnosis functionalities – the development of an novel automated guided vehicle with unique properties and unique technical solutions – as well as to report observations during this process and derive first hypotheses about a more efficient and effective integration of the development of control and diagnosis functionalities into product development processes. As these observations and hypotheses are in this stage only based on one product development process they are preliminary. The main intentions of this paper were to underline the paramount importance of a stronger integration of control and diagnosis into product development processes and to present central aspects and challenges in the integration endeavor. Further research is planned in order to test the hypotheses as well as both to widen the scope and to deepen the investigation by means of analyzing product development processes.

References

[1] Camacho, E., Bordons, C.: Model Predictive Control. Berlin: Springer, 2004.
[2] Wang, Y., Boyd, S.: Fast Model Predictive Control using Online Optimization. In: Proceedings of the 17th World Congress. The International Federation of Automatic Control. Seoul, Korea, July 6-11, 2008.
[3] Blanke, M., Kinnaert, M., Lunze, J., and Staroswiecki, M.: Diagnosis and Fault–Tolerant Control. Berlin: Springer, 2006.
[4] Isermann, R.: Fault–Diagnosis Systems: An Introduction from Fault Detection to Fault Tolerance. Berlin: Springer 2005.
[5] Witczak, M.: Modelling and Estimation Strategies for Fault Diagnosis of Non–Linear Systems: From Analytical to Soft Computing Approaches. Lecture Notes in Control and Information Sciences. Berlin: Springer 2007.

[6] Korbicz, J., J.M. Kościelny, Z. Kowalczuk and W. Cholewa: Fault Diagnosis: Models, artificial intelligence methods, applications. Springer: Berlin, 2004.
[7] Ehrlenspiel, K.: Integrierte Produktentwicklung - Methoden für Prozessorganisation, Produkterstellung und Organisation. München: Hanser, 1995.
[8] Schätz, C.: A Methodology for Production Development - The Body of Knowledge Approach. Trondheim: Norwegian University of Science and Technology, 2006.
[9] Andreasen, M. M.; Hein, L.: Integrated Product Development. New York: Springer 1987.
[10] Mandhyani, T.; Kubendra Prasad, P. K.: Development of a Semi-Autonomous outdoor vehicle. Project Report, Hochschule Ravensburg-Weingarten, 2012.
[11] Anderson, K. Jones, C.: "Autonomous Robotic Vehicle Design", Proceedings of the Southwest Regional ASEE Conference, ASEE 2005.
[12] Ashmore, M., Barnes, N.: "Omni-drive robot motion on curved paths: The fastest path between two points is not a straight line", Proceedings of the Australian Joint Conference on Artificial Intelligence, Dec., 2002, pp. 225-236.
[13] Dillard, W.: "Revisting the Autonomous Robot. Finding the Engineer within the Student", Proceedings of the 2004 AEE Conf. , Salt Lake City, 2004.
[14] Patent „Fahrzeug" Aktenzeichen DE 10 2005 057 635 A1, Deutsches Patent- und Markenamt, München.
[15] Patentanmeldung „Fahrzeug" Aktenzeichen EP 1 954 516 A1 und WO 2007/062858 A1, Europäisches Patentamt, München und World Intellectual Property Organization, Genf, Schweiz.
[16] Simundsson, A.: Summary of Project Work. Project Report, Hochschule Ravensburg-Weingarten, 2014.
[17] Stetter, R., Möhringer, S., Pulm, U.: A Comparison of Evolutionary and Revolutionary Approaches in Mechatronic Design. In: Proceedings of the 18th International Conference on Engineering Design (ICED 11) Vol. 1, 2011, pp. 221-232.
[18] Stetter, R., Möhringer, S., Günther, J., Pulm, U.: Investigation and Support of Evolutionary Design. Accepted for the 20th International Conference on Engineering Design (ICED 15), 2015.
[19] Pulm, U.; Stetter, R.: Pressure and Trust in Competitive Engineering. In: Norell Bergendahl, M.; Grimheden, M.; Leifer, L.; Skogstad, P.; Lindemann, U. (Eds.): Proceedings of the 17th International Conference on Engineering Design (ICED'09), Vol. 9., 2009, pp 69 – 80.
[20] Pulm, U.; Stetter, R.: Influences of pressure on cooperative product development. In: International Journal of CoCreation in Design and the Arts, 1745-3755, Volume 5, Issue 3, 2009, pp 193 – 210.
[21] Stetter, R.; Pulm, U.: Problems and Chances in Industrial Mechatronic Product Development. In: Norell Bergendahl, M.; Grimheden, M.; Leifer, L.; Skogstad, P.; Lindemann, U. (Eds.): Proceedings of the 17th International Conference on Engineering Design (ICED'09), Vol. 5., 2009, pp 97 – 108.
[22] Herb, R.; Herb, T.: Kohnhauser, V.: TRIZ. miVerlag, 2000.
[23] Diers, S.: Systematik in der Anwendung der Evolutionsprinzipien (TRIZ). Triz-online: 2006.

Identification of State and Measurement Noise Covariance Matrices using Nonlinear Estimation Framework

Oliver Kost, Ondřej Straka, and Jindřich Duník

European Centre of Excellence - NTIS & Department of Cybernetics
Faculty of Applied Sciences, University of West Bohemia
Univerzitní 8, 306 14 Pilsen, Czech Republic

E-mail: kost@ntis.zcu.cz, straka30@kky.zcu.cz, dunikj@kky.zcu.cz

Abstract. The paper deals with identification of the noise covariance matrices affecting the linear system described by the state-space model. In particular, the stress is laid on the autocovariance least-squares method which belongs into to the class of the correlation methods. The autocovariance least-squares method is revised for a general linear stochastic dynamic system and is implemented within the publicly available MATLAB toolbox Nonlinear Estimation Framework. The toolbox then offers except of a large set of state estimation algorithms for prediction, filtering, and smoothing, the integrated easy-to-use method for the identification of the noise covariance matrices. The implemented method is tested by a thorough Monte-Carlo simulation for various user-defined options of the implemented method.

Keywords: State estimation; Identification; Noise covariance matrices; State-space models; Software toolbox; Implementation

1. Introduction

State estimate plays a key role in many applications including various signal processors, target tracking, satellite navigation, fault detection, and adaptive and optimal control problems. It constitutes an essential part of any decision-making process. Estimation of discrete-time stochastic dynamic systems is very important field of study which, for the past seventy years, has been considerably developed. In the beginning the development was focused on linear dynamic systems and it has been further developed until today with the stress on the state estimation of nonlinear stochastic dynamic systems.

Up to now many state estimation, both local and global[1], algorithms have been proposed. The algorithms differ in both the assumptions tied with the system description and in the expected performance and computational complexity of algorithms affected by a specification of one or more user-defined estimator parameters. Whereas the system description is given, selection of the estimation algorithm and description of all its parameters a key designer task. Often, the process of selection of a suitable estimator for a given task has to be supported by a simulation study for a given set-up. To facilitate the simulation study several toolboxes and software packages have been introduced within MATLAB, such as KalmTool [1], EKF/UKF Toolbox [2], and Nonlinear Estimation Framework (NEF) [3]. The first two packages provide user with implemented algorithms of certain local estimators in stand-alone files. They are advantageous for an experienced user who want only estimation algorithms. Contrary to this, the NEF is fully object-oriented package providing the user with a collection of mutually

[1] The local estimation algorithms provide estimates in the form of conditional mean and covariance matrix. The global estimation algorithms provide estimates in the form of conditional probability density functions.

linked functions forming an easy-to-use a user interface. The NEF offers both the local and global estimation algorithms.

The state estimators, however, require knowledge of the whole system; not only the deterministic part of the model is required, but also the stochastic part. The deterministic part of the model is mostly based on various laws of physics, mathematics, kinematics, etc., whereas the description of the noise statistics might questionable or even unknown in many cases. Incorrect description of the noise statistics may cause significant worsening of estimation or control quality or even divergence of the underlying algorithm output.

Therefore, along with the estimation algorithms development, the methods for identification of the covariance matrices of the noises have been designed. Methods for identification of the noises covariance matrices can be divided into several categories; correlation methods [4, 5, 6, 7, 8], Bayesian estimation methods [9], maximum likelihood estimation methods [10], covariance matching methods [11], methods based on the minimax approach [12], subspace methods [13], prediction error methods [14], the Kalman filter working as a parameter estimator [15] or methods tied with variational Bayesian approximation [16]. Besides the noise covariance matrix estimation methods, alternative approach directly estimating the gain of a linear estimator have been developed as well [4, 17, 18]. A characterisation of the methods with their assumptions, properties and limitations can be found in e.g., [19, 12, 6, 20, 8]. But contrary the state estimation methods, there is no publicly available toolbox offering the algorithms for the noise covariance matrices identification.

The goal of the paper is to describe extension of the NEF toolbox with the correlation method for the noise covariance matrices identification. Through this linking the user can get the estimate of the state of the discrete-time linear stochastic time-invariant system only from knowledge of the deterministic part of model and control and measurement sequences.

The rest of the paper is organized as follows. Section 2 provides system description. In Section 3 description of the state estimation methods and state estimation using the toolbox NEF is introduced. Section 4 is devoted to the noise covariance matrices estimation and implementation of the method in the toolbox. Numerical illustrations and conclusions are given in Sections 5 and 6, respectively.

2. System description

Let a discrete-time linear dynamic stochastic system be considered

$$\mathbf{x}_{k+1} = \mathbf{F}\mathbf{x}_k + \mathbf{N}\mathbf{u}_k + \mathbf{M}\mathbf{w}_k, k = 0, 1, 2, \ldots T, \quad (1)$$

$$\mathbf{z}_k = \mathbf{H}\mathbf{x}_k + \mathbf{O}\mathbf{v}_k, k = 0, 1, 2, \ldots T, \quad (2)$$

where the vectors $\mathbf{x}_k \in \mathbb{R}^n$, $\mathbf{z}_k \in \mathbb{R}^p$ and $\mathbf{u}_k \in \mathbb{R}^m$ represent the system state, measurement and the input of the system at the time instant k, respectively. The matrix $\mathbf{F} \in \mathbb{R}^{n\times n}$ is system matrix, $\mathbf{H} \in \mathbb{R}^{n\times n}$ is the measurement matrix, $\mathbf{N} \in \mathbb{R}^{n\times m}$ is the input matrix, $\mathbf{M} \in \mathbb{R}^{n\times n}$ is the state noise distribution matrix, and $\mathbf{O} \in \mathbb{R}^{p\times p}$ is the measurement noise distribution matrix. The initial state of the system $\mathbf{x}_0$ has the mean $\bar{\mathbf{x}}_0$ and the covariance matrix $\mathbf{P}_0$. The vectors $\mathbf{w}_k \in \mathbb{R}^n$ and $\mathbf{v}_k \in \mathbb{R}^p$ are the noises in the state and measurement equations, which are supposed to be zero-mean and white with the covariance matrices $\mathbf{Q} \in \mathbb{R}^{n\times n}$ and $\mathbf{R} \in \mathbb{R}^{p\times p}$, respectively. State and measurement equations are assumed to be mutually independent and independent of the initial state.

3. State estimation using NEF toolbox

In this section, state estimation and the toolbox NEF is introduced.

3.1. State Estimation

A general solution to the recursive state estimation problem is given by the Bayesian recursive relations (BRRs). The BRRs provide probability density functions (PDFs) of the state conditioned by the measurements representing a full description of the state. The BRRs are given by

$$p(\mathbf{x}_k|\mathbf{z}_k;\mathbf{u}_{k-1}) = \frac{p(\mathbf{x}_k|\mathbf{z}_{k-1};\mathbf{u}_{k-1})p(\mathbf{z}_k|\mathbf{x}_k)}{\int p(\mathbf{x}_k|\mathbf{z}_{k-1};\mathbf{u}_{k-1})p(\mathbf{z}_k|\mathbf{x}_k)d\mathbf{x}_k}, \quad (3)$$

$$p(\mathbf{x}_k|\mathbf{z}_{k-1};\mathbf{u}_{k-1}) = \int p(\mathbf{x}_{k-1}|\mathbf{z}_{k-1};\mathbf{u}_{k-2})p(\mathbf{x}_k|\mathbf{x}_{k-1};\mathbf{u}_{k-1})d\mathbf{x}_{k-1}, \quad (4)$$

where $p(\mathbf{x}_k|\mathbf{z}_k;\mathbf{u}_{k-1})$ is the filtering PDF, $p(\mathbf{x}_k|\mathbf{z}_{k-1};\mathbf{u}_{k-1})$ is the one-step predictive PDF, $p(\mathbf{x}_k|\mathbf{x}_{k-1};\mathbf{u}_{k-1})$ is the transition PDF stemming from (1), and $p(\mathbf{z}_k|\mathbf{x}_k)$ is the measurement PDF stemming from (2).

The closed-form solution to the BRRs is available only for a few special cases among which the linear Gaussian system plays an important role and the solution then leads to the Kalman filter (KF). In other cases, an approximate solution has to be used. The approximate methods can be divided with respect to the validity of the resulting estimates into two groups; the local methods and the global methods. The local methods are based on such approximations allowing the BRRs solution analogous to the KF design even for nonlinear systems. The approximations involve the Gaussian approximation of the conditional PDFs and system description approximations. The local methods are represented by the extended Kalman filter (EKF), unscented Kalman filter (UKF), divided difference filters (DDFs), or the stochastic integration filter (SIF). Contrary to the local methods, the global methods are often based on a numerical solution to the BRRs resulting in an approximate (but in general non-Gaussian) conditional PDFs within the almost whole state space. Within this group, the particle filter (PF), Gaussian sum filter (GSF), or the point-mass method can be mentioned.

3.2. Nonlinear estimation framework toolbox

The NEF[2] is a software toolbox for MATLAB, devoted to the state estimation of generally nonlinear discrete-time stochastic systems [3]. The NEF contains various local and global algorithms for filtering, prediction, and smoothing. Toolbox is based on mutually linked classes, which are divided into the following categories:

- System description,
- Random variable description,
- Function definition,
- Estimator definition,
- Performance evaluation.

The *System description* allows specification of the system model in a probabilistic or a structural fashion. The *Random variable description* (RV) enables description of the characteristics of RVs related to the model or the estimate. The following PDFs are supported: uniform, beta, gamma, Gaussian, Gaussian sum, and empirical. The *Function definition* enables description of equations in the state and measurement equations or a PDFs of RVs in the case of the probabilistic system description. The *Estimator definition* contains various local and global estimation algorithms as the KF, EKF, UKF, DDFs, PF, GSF, the ensemble Kalman filter, etc. The category *Performance evaluation* enables calculation of various performance metrics related to the estimate quality (mean square error (MSE), non-credibility index [21], etc.). The NEF thus represents a powerful and complex tool for the state estimation of the nonlinear systems. In this paper, the linear systems are of interests, therefore the NEF is illustrated via the KF design and application.

In the first step, the system is modelled. The state and measurement equations are created by the class `nefLinFunction` designed for description of the linear equations (1) and (2) as:

[2] The toolbox is publicly available at ***http://nft.kky.zcu.cz/nef***.

```
f = nefLinFunction(F,N,M);
h = nefLinFunction(H,[],O);
```

Then, the state and measurement noises are modeled using the class `nefGaussianRV` designed for description of the noises **w**, **v** and the initial state $\mathbf{x}_0$ and are created by the commands:

```
w = nefGaussianRV(zeros(n,1),Q);
v = nefGaussianRV(zeros(n,1),R);
x0 = nefGaussianRV(x0mean,P0);
```

where the term `zeros(n,1)` means a zero vector of dimension $n \times 1$, `x0mean` and `P0` are the mean and covariance matrix of the initial state, respectively. The complete model of the system is created by the class `nefEqSystem` as:

```
myModel = nefEqSystem(f,h,w,v,x0);
```

In the second step, the estimator is created. The state of the linear system is estimated by the KF. The KF for the model is created by the class `nefKalman` as:

```
myKalman = nefKalman(myModel);
```

Having the model and estimator defined, the state estimation for a sequence of measurement data **z** and input data **u** is realized by a command:

```
[xEst,xPred] = myKalman.estimate(z,u);
```

The variable `xEst` is a required filtering estimate of the state and `xPred` is the predictive estimate of the state. The estimates are in the form of the Gaussian PDFs. The mean and covariance matrices of predictive estimate can be calculate by:

```
xPredMean = evalMean(xPred);
xPredVar = evalVariance(xPred);
```

The NEF can be also used for data simulation using a command:

```
[z,xTrue] = system.simulate(T,u);
```

where `T` is the simulation horizon and `xTrue` is the true state of the system. Further details of the NEF can be found in [3].

4. Extension of NEF toolbox with noise covariance matrices identification block

The estimation techniques implemented in the NEF toolbox require the complete description of the system i.e., the known model of the deterministic part (i.e., the matrices **F**, **N**, **M**, **H**, **O**) and of the stochastic part (i.e., the matrices **Q**, **R**). However, the knowledge of the stochastic part might be the limiting assumption in the estimator design. Therefore, from seventies various techniques for estimation of the properties of the noises affecting the system have been proposed [9, 10, 12, 16] prevailingly for the linear systems. Among them, the correlation methods have attracted significant attention [4, 5, 6, 7, 8].

4.1. Noise covariance matrices identification by correlation method

In contrast to other methods, the correlation methods has several quite considerable positive including characteristics the unbiased estimate and small computing time even for high-dimensional systems. The methods are based on an analysis the innovation sequence of a "non-optimal" linear estimator and were originally derived in [4] and [5] in the form of the three-step procedure for estimation of the noise covariance matrices. The three-step procedure was reformulated in one-step method providing the unbiased estimate of the noise covariance matrices in [6, 7, 22, 8] assuming a discrete-time linear time-invariant system. The method is generally called the autocovariance least-squares (ALS) method[3].

[3] Note that the ALS method has been proposed also for nonlinear systems. Then, the ALS method takes an advantage of a linearised predictor following the structure of the EKF [23], [24].

4.2. Autocovariance least-square method

The ALS method, proposed for the identification of the noise covariance matrices $\mathbf{Q}$ and $\mathbf{R}$ of linear systems, is based on an analysis of the second-order statistics of an innovation sequence produced by a linear (one-step) predictor [6, 22] of the form

$$\hat{\mathbf{x}}_{k|k} = \hat{\mathbf{x}}_{k|k-1} + \mathbf{K}\mathbf{e}_k, \tag{5}$$

$$\hat{\mathbf{x}}_{k+1|k} = \mathbf{F}\hat{\mathbf{x}}_{k|k} + \mathbf{N}\mathbf{u}_k, \tag{6}$$

where $\hat{\mathbf{x}}_{k+1|k}$ is the one-step state prediction, $\hat{\mathbf{x}}_{k|k}$ is the filtering state estimate, $\mathbf{K}$ is a user-defined predictor gain, and $\mathbf{e}_k$ is the measurement prediction error (innovation) given by

$$\mathbf{e}_k = \mathbf{z}_k - \mathbf{H}\hat{\mathbf{x}}_{k|k-1}, \forall k. \tag{7}$$

The estimation error of the predictor

$$\boldsymbol{\epsilon}_k = \mathbf{x}_k - \hat{\mathbf{x}}_{k|k-1} \tag{8}$$

evolves according to

$$\boldsymbol{\epsilon}_{k+1} = (\mathbf{F} - \mathbf{FKH})\boldsymbol{\epsilon}_k + [\mathbf{M}, -\mathbf{FKO}]\left[\begin{smallmatrix}\mathbf{w}_k\\ \mathbf{v}_k\end{smallmatrix}\right] = \bar{\mathbf{F}}\boldsymbol{\epsilon}_k + \mathbf{G}\bar{\mathbf{w}}_k. \tag{9}$$

The innovation $\mathbf{e}_k$ is related to the estimation error $\boldsymbol{\epsilon}_k$ by

$$\mathbf{e}_k = \mathbf{H}(\mathbf{x}_k - \hat{\mathbf{x}}_{k|k-1}) + \mathbf{O}\mathbf{v}_k = \mathbf{H}\boldsymbol{\epsilon}_k + \mathbf{O}\mathbf{v}_k. \tag{10}$$

The innovation sequence is, after reaching the steady-state, a zero-mean stochastic process with the autocovariance matrices defined as

$$\mathbf{C}_{e,0} = E[\mathbf{e}_k\mathbf{e}_k^T] = \mathbf{H}\mathbf{P}_\epsilon\mathbf{H}^T + \mathbf{O}\mathbf{R}\mathbf{O}^T, \tag{11}$$

$$\mathbf{C}_{e,j} = E[\mathbf{e}_{k+j}\mathbf{e}_k^T] = \mathbf{H}\bar{\mathbf{F}}^j\mathbf{P}_\epsilon\mathbf{H}^T - \mathbf{H}\bar{\mathbf{F}}^{j-1}\mathbf{FKORO}^T, \tag{12}$$

where $j = 1, 2, \ldots, L-1$ and L is the user-defined parameter defining the maximum lag (and thus, the number of equations used in the ALS). Assuming that the $\bar{\mathbf{F}}$ is stable, the matrix $\mathbf{P}_\epsilon$ is the steady-state covariance matrix of the state prediction (estimation) error (9) given by

$$(\mathbf{P}_\epsilon)_s = (\mathbf{I}_{n_x^2} - \bar{\mathbf{F}} \otimes \bar{\mathbf{F}}^T)^{-1}(\mathbf{G} \otimes \mathbf{G}^T)\boldsymbol{\Sigma}_s, \tag{13}$$

where the symbol $\otimes$ stands for the Kronecker product and the notation $\mathbf{A}_s$ means the column-wise stacking of the matrix $\mathbf{A}$ into a vector [25]. Note that the covariance matrix $\mathbf{P}_\epsilon$ is a solution to the Lyapunov equation

$$\mathbf{P}_\epsilon = \bar{\mathbf{F}}\mathbf{P}_\epsilon\bar{\mathbf{F}}^T + \mathbf{G}\boldsymbol{\Sigma}\mathbf{G}^T \tag{14}$$

with $\boldsymbol{\Sigma} = \left[\begin{smallmatrix}\mathbf{Q} & \mathbf{0}\\ \mathbf{0} & \mathbf{R}\end{smallmatrix}\right]$ stemming from (9).

The relations (11) and (12) can be rewritten in a form suitable for the least-squares method [6, 7, 22, 8]

$$\mathcal{A}\boldsymbol{\theta} = \mathbf{b}, \tag{15}$$

where the design matrix $\mathcal{A}$ is defined as

$$\mathcal{A} = [\mathbf{D}(\mathbf{M} \otimes \mathbf{M}), \mathbf{D}(\mathbf{FKO} \otimes \mathbf{FKO}) + (\mathbf{O} \otimes \boldsymbol{\Gamma})], \tag{16}$$

$$\mathbf{D} = (\mathbf{H} \otimes \mathcal{O})(\mathbf{I}_{n^2} - \bar{\mathbf{F}} \otimes \bar{\mathbf{F}})^{-1}, \tag{17}$$

with the observability matrix $\mathcal{O}$ and matrix $\mathbf{\Gamma}$ given as

$$\mathcal{O} = [\mathbf{H}^T, (\mathbf{H}\bar{\mathbf{F}})^T, \ldots, (\mathbf{H}\bar{\mathbf{F}}^{L-1})^T]^T, \quad (18)$$

$$\mathbf{\Gamma} = [\mathbf{O}, -(\mathbf{HFKO})^T, \ldots, -(\mathbf{H}\bar{\mathbf{F}}^{L-2}\mathbf{FKO})^T]^T. \quad (19)$$

The vectors of parameters and dependent variables are of the form

$$\boldsymbol{\theta} = [\mathbf{Q}_s^T, \mathbf{R}_s^T]^T, \mathbf{b} = (\mathcal{C}_e(L))_s \quad (20)$$

with

$$\mathcal{C}_e(L) = [\mathbf{C}_{e,0}^T, \mathbf{C}_{e,1}^T, \ldots, \mathbf{C}_{e,L-1}^T]^T. \quad (21)$$

The unbiased estimate of the autocovariance matrices [6], and subsequently of the vector $\mathbf{b}$, computed on the basis of the innovation sequence (7), is computed as

$$\hat{\mathbf{C}}_{e,j} = \frac{1}{T-j-\kappa} \sum_{k=\kappa}^{T-j} \mathbf{e}_{k+j}\mathbf{e}_k^T, j = 0, 1, \ldots, L-1, \quad (22)$$

$$\hat{\mathcal{C}}_e(L) = [\hat{\mathbf{C}}_{e,0}^T, \hat{\mathbf{C}}_{e,1}^T, \ldots, \hat{\mathbf{C}}_{e,L-1}^T]^T, \quad (23)$$

$$\hat{\mathbf{b}} = \left(\hat{\mathcal{C}}_e(L)\right)_s, \quad (24)$$

where κ defines the time instant in which the effect of the initial conditions of the predictor (9) on the actual state estimate is negligible.

Then, the unbiased (ALS) estimate of the vector of parameters (composed by the elements of the unknown noise covariance matrices $\mathbf{Q}$ and $\mathbf{R}$) in the least-squares sense is given by

$$\hat{\boldsymbol{\theta}} = (\mathcal{A}^T\mathcal{A})^{-1}\mathcal{A}^T\hat{\mathbf{b}} = \mathcal{A}^\dagger\hat{\mathbf{b}}. \quad (25)$$

Note that $\mathbf{K}$ is, in the ALS method, considered as a user-defined parameter. Optimal gain can not be computed as the matrices $\mathbf{Q}$ and $\mathbf{R}$ are unknown and are being estimated. It means the estimator (5) and (6) is not optimal in the MSE sense nor the innovation sequence is a white process. In principle, any predictor gain $\mathbf{K}$ can be selected if it results in the stable state prediction error transition matrix $\bar{\mathbf{F}}$ appearing in (9). A reasonable choice of $\mathbf{K}$, fulfilling the stability condition, is given by

$$\mathbf{K} = (\bar{\mathbf{P}}\mathbf{H}^T)(\mathbf{H}\bar{\mathbf{P}}\mathbf{H}^T + \mathbf{R}_A)^{-1}. \quad (26)$$

where $\bar{\mathbf{P}}$ is the solution to the Riccati equation

$$\bar{\mathbf{P}} = \mathbf{F}\bar{\mathbf{P}}\mathbf{F}^T - \mathbf{F}\bar{\mathbf{P}}\mathbf{H}^T(\mathbf{H}\bar{\mathbf{P}}\mathbf{H}^T + \mathbf{R}_A)^{-1}\mathbf{H}\bar{\mathbf{P}}\mathbf{F}^T + \mathbf{Q}_A. \quad (27)$$

and $\mathbf{Q}_A$ and $\mathbf{R}_A$ are arbitrary user-defined positive semi-definite matrices [26].

It can be seen that state estimation (non-optimal in the MSE sense) by a linear estimator, providing innovation sequence, is the inherent part of the ALS method. The NEF toolbox is the ideal package for such a task.

4.3. Implementation of ALS method to NEF

The basic idea is utilize methods and objects of the NEF and extend them with the ALS method identifying the noise covariance matrices from the knowledge of the deterministic part of the system description, control sequence, and measurements of the system. By identification of the noises covariance matrices, the user gets complete description of the system, which can subsequently be used in the NEF for the optimal state estimation of a linear system.

Specifically, the NEF was extended with category *System identification* having two classes. The classes have been designed to be consistent with and utilizing other components of the NEF toolbox.

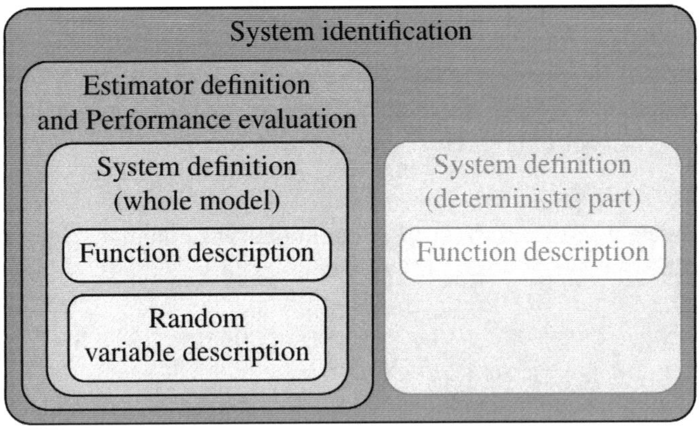

Figure 1. Structure of the NEF extended with the Identification class (including ALS method).

4.3.1. Identification class - The first class is the parent class `nefIdentificaton`. The class contains the method `gainCalculation`, which can calculate the current estimator gain according to (26). The gain is computed by the following command:

```
gain = gainCalculation(f,h,wApriori,vApriori);
```

where `f` and `h` are the objects of the class `nefLinFunction` as defined in Section 3.2 and `wApriori` and `vApriori` are instances of the class `nefGaussianRV` defined as:

```
wApriori = nefGaussianRV(zeros(n,1),QA);
vApriori = nefGaussianRV(zeros(p,1),RA);
```

with `QA` and `RA` being the user-defined matrices defining the predictor gain (26). The class design supports fast extension with other supporting methods for the indetification.

4.3.2. ALS class - The second class is `nefALS` class which contains the implemented method ALS detailed in Section 4.2. Class constructor is `nefALS` which can create object:

```
myALS = nefALS(f,h,x0,varargin);
```

where `x0` is the object of the class `nefGaussianRV` as defined in Section 3.2 and `myALS` is the new object of the class `nefALS`. The `varargin` is the MATLAB notation for input parameters consisting of two parts. The first part is the name and the second part is the specified value. The `varargin` of `nefALS` is currently the following:

- name `'numberCovEq'` with value:
  - an integer number in $\langle 2, \infty)$
    - the number coincides with the variable L as defined in (12), (20).
    - default value is $\lceil \frac{n+p}{p} \rceil$ as discussed in Appendix.
- name `'omittedInitPeriod'` with value:
  - an integer number in $\langle 0, 100 \rangle$
    - defines percentage period of the time T in which the initial conditions of the predictor (9) are significant; after that period, the predictor is assumed to be in the steady-state and the innovation sequence independent of the initial condition.
    - default value is 50 resulting in $\kappa = \lfloor \frac{T}{2} \rfloor$ in (22).

The main method of the class is `identifyQR`, which identifies the noise covariance matrices. The noise covariance matrices are identified by the following command:

```
[QEst,REst] = myALS.identifyQR(z,u,varargin);
```

where `QEst` is the identified state noise covariance matrix **Q**, `REst` is the identified measurement noise covariance matrix **R**, `z` is the vector of measurements, and `u` is the vector of control input. The possible input parameters (`varargin`) of the method `identifyQR` are these:

- name `'ALSaprioriSetting'` with values:
  - `'fast'` - sets gain **K** to zero (fast computation because of simplifications in (5-16)),
  - `'automatic'` - selects optimal **K** to minimize **Q**, **R** estimates variance (exact identification),
  - `aprioriQR` - computes **K** for the user-defined matrices $\mathbf{Q}_A$ and $\mathbf{R}_A$ (`aprioriQR` is a structure containing the user-define matrices),
- name `'knownQ'` with values:
  - `elemensQ` defining the known elements of the noise covariance matrix **Q** (if any),
- name `'knownR'` with values:
  - `elemensR` defining the known elements of the noise covariance matrix **R** (if any).

Detailed example of possible parameters is in following section.

The class `nefALS` contains a set of other auxiliary methods. The method `calcInnovationCov` calculates the innovation sequence covariance matrices (11), (12) using a command:

```
innovCov = calcInnovationCov(myALS,xPredMean,z);
```

where `innovCov` is a set of (cross-)covariance matrices of the innovation and `xPredMean` is a vector of predictive mean values as defined in Section 3.2.

The method `calculateLSDesignMatrix` calculates the matrix $\mathcal{A}$ defined in (16). The method is called by the following command:

```
LSDM = calculateLSDesignMatrix(myALS,K);
```

where `K` is the predictor gain **K** defined in (26).

The method `reductionLSDesignMatrix` reduces duplicate columns in the matrix $\mathcal{A}$ be cause of symetry of estimate covariance matrices. By the reduction the calculation is faster. The method is called by:

```
reducedLSDM = reductionLSDesignMatrix(myALS,LSDM);
```

The last important method of the class `nefALS` is the method `minimalizationCrit` which finds a gain **K** resulting in alow variance of the **Q**, **R** estimate by the following command:

```
[wALS,vALS,KALS] = minimalizationCrit(myALS);
```

where `KALS` is the gain used by the predictor (9), `wALS` and `vALS` are the `nefGaussian` object noises which were used for calculation of `KALS`. This method is employed if the input parameter `'ALSaprioriSetting'` of the method `identifyQR` is set to `'automatic'`.

All the classes and methods monitor the correctness of the input parameters including their dimensions and allowed settings. Moreover, the class `nefALS` also assesses the identifiability of the considered system with respect to the actual set-up. If any input condition is violated, the user is notified and a possible solution is proposed. For completeness, the overall illustration of the updated NEF toolbox with the identification block is given in Figure 1.

Use of the NEF toolbox for the noise covariance matrices identification by the implemented ALS method with different settings is illustrated in the next section.

5. Use case of ALS method in NEF and numerical illustration

Through this section, the linear model (1) and (2) with the following matrices:

$$n = 1, m = 1, p = 2, \tag{28}$$

$$F = 0.9, N = 0.5, M = 2, Q = 1.5, \tag{29}$$

$$\mathbf{H} = \begin{bmatrix} 1 \\ 0.5 \end{bmatrix}, \mathbf{O} = \begin{bmatrix} 2 & 0.1 \\ 0.1 & 1 \end{bmatrix}, \mathbf{R} = \begin{bmatrix} 2 & 0.5 \\ 0.5 & 3 \end{bmatrix}, \tag{30}$$

$$\bar{\mathbf{x}}_0 = 0, P_0 = 1, \tag{31}$$

is used. The matrices Q, $\mathbf{R}$ are assumed to be unknown (except of the system output simulation) and thus estimated.

5.1. Model of deterministic part of system

In the first step, the deterministic part of the system is modelled for the purposes of the ALS method by a sequence of commands:

```
f = nefLinFunction(0.9,0.5,2);
H = [1;0.5];
O = [2 0.1;0.1 1];
h = nefLinFunction(H,[],O);
x0 = nefGaussianRV(0,1);
```

5.2. System simulation

The second (optional) step is related to the measured data generation. To simulate the system, the true characteristics of the state and measurement noises has to be defined by

```
w = nefGaussianRV(0,1.5);
v = nefGaussianRV([0;0],[2 0.5;0.5 3]);
```

Then, the object representing the complete model (28)–(31) is created by:

```
myModel = nefEqSystem(f,h,w,v,x0);
```

and the object is simulated by the NEF toolbox by commands:

```
T = 2e3;
u = [sin(0:pi/(T-1):pi)];
z = myModel.simulate(T,u);
```

where `T` is the number of simulation steps, `z` is the outputted measurement sequence, and `u` is the control sequence.

It should be mentioned that the simulation part can be skipped if the measurement and control sequences are available. Also note that the noises need not be only of the Gaussian PDFs (`nefGaussianRV`), but other RVs supported by the NEF e.g., the Gaussian sum PDF (`nefGaussianSumRV`) or the uniform PDF (`nefUniformRV`) can be selected, as well. The ALS method is not, in principle, based on the Gaussian assumption.

5.3. Noise covariance matrices identification

The third and last step is the identification of the noise covariance matrices Q and $\mathbf{R}$. Noise covariance matrices identification step starts with the creation of the `nefALS` class object:

```
myALS = nefALS(f,h,x0,'numberCovEq',3);
```

Journal of Physics: Conference Series **659** (2015) 012057 doi:10.1088/1742-6596/659/1/012057

Having the simulated data `z`, the matrices Q and $\mathbf{R}$ can be identified using the method `identifyQR`. Four identification experiments[4] each with different setting of the ALS method, are considered:

Example 1 – estimate of *all* elements of Q and $\mathbf{R}$ with `'fast'` setting of the `nefALS`

```
[wEst,vEst] = myALS.identify(z,u,'ALSAprioriSetting','fast');
```

Example 2 – estimate of *all* elements of Q and $\mathbf{R}$ with `'automatic'` setting leading to the estimates with low variance

```
[wEst,vEst] = myALS.identify(z,u,'ALSAprioriSetting','automatic');
```

Example 3 – estimate of *all* elements of Q and $\mathbf{R}$ with predictor gain $\mathbf{K}$ given by the user defined matrices Q_A and $\mathbf{R}_A$

```
aprioriMatrix.Q=10;
aprioriMatrix.R=[6 0,0 6];
[wEst,vEst] = myALS.identify(z,u,'ALSAprioriSetting',aprioriMatrix);
```

Example 4 – estimation of a *subset* of elements of Q and $\mathbf{R}$ (just one element of $\mathbf{R}$) with `'fast'` setting of the `nefALS`

```
knownR =[2 0.5;0.5 NaN];
[wEst,vEst]=ALS.identify(z,u,'knownR',knownR,...
                         'ALSAprioriSetting','fast');
```

All examples were simulated using $M = 10^4$ Monte-Carlo (MC) simulations.

5.4. Identification results

The results of the identification experiments for all examples are summarized in Table 1. The results are in the form of the sample mean value and standard deviation (STD) of the estimates of the noise covariance matrices elements over all MC simulations. The table also gives the average computational time over one MC simulations.

Noise covarinace matrices		Example 1	Example 2	Example 3	Example 4
	True	Estimated [mean±standard deviation]			
Q	1.5	1.5037 ± 0.2446	1.5184 ± 0.1671	1.5213 ± 0.1865	1.5012 ± 0.2313
$R_{1,1}$	2	1.9982 ± 0.2202	1.9944 ± 0.1559	1.9918 ± 0.1669	–
$R_{1,2}$	0.5	0.4980 ± 0.2168	0.4937 ± 0.1447	0.4940 ± 0.1505	–
$R_{2,2}$	3	2.9964 ± 0.3254	2.9936 ± 0.1924	2.9926 ± 0.19687	2.9992 ± 0.3795
Identification time per simulation [sec:msec]		0:857	1:714	1:502	0:857

Table 1. ALS estimation statistics summary for all examples.

The results indicate that the ALS method provides unbiased estimates for an arbitrary stable gain of the linear predictor. The ALS method with the choice `'automatic'` selects such a gain which results in the estimates with the low covariance matrix (represented by the low STD). On the other hand, the choice `'fast'` simplifies the ALS method relations resulting in approximately 50 per cent reduction of

[4] The example are also publicly available at ***http://nft.kky.zcu.cz/nef***.

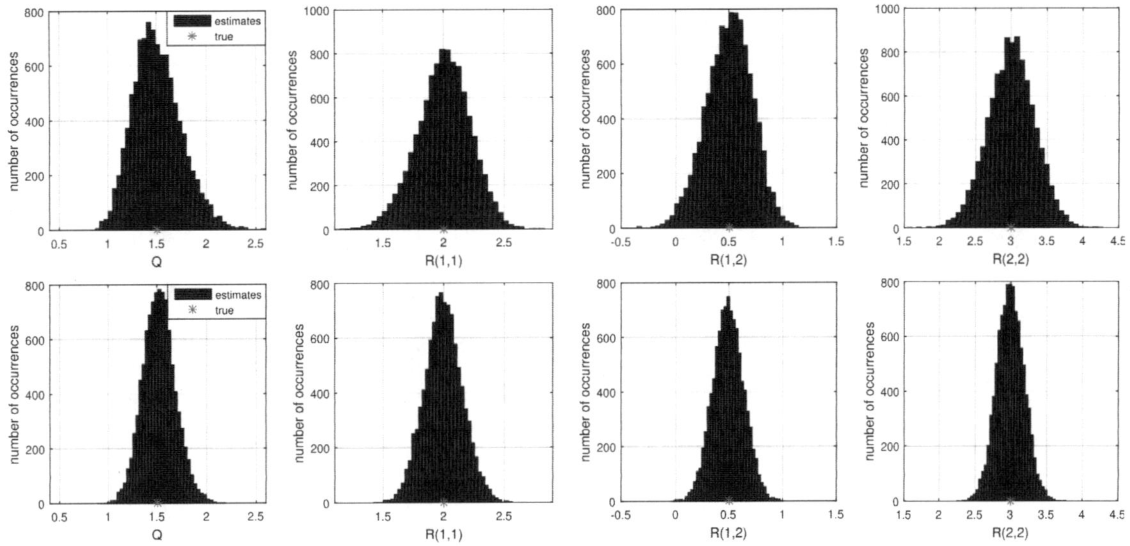

Figure 2. ALS estimates by NEF for Example 1 (upper row) and Example 2 (lower row)

the computational time. The possibility of the gain specification by the user (according to (26)) offers the trade-off between the computational complexity and effectiveness.

For completeness, the histograms of the identification experiments for Examples 1 and 2 are shown in Figure 2. The histograms illustrate the lower variance of the noise covariance matrices elements estimates of the ALS method in the set-up of Example 2 (lower row).

6. Conclusion

The paper was devoted to the identification of the noise covariance matrices of a linear stochastic dynamic system described by the state-space model. The identification method, the autocovariance least-squares method, was revised for a general state-space model and implemented within the Nonlinear Estimation Framework. Implementation of the method and the user-defined settings were discussed in detail and illustrated in examples. Usage of the toolbox and the performance of the implemented identification method were illustrated in numerical Monte-Carlo study. The introduced Nonlinear Estimation Framework is the only publicly available MATLAB toolbox offering a set of state estimation algorithms combined with the method for the system noise covariance matrices identification. The toolbox is available at ***http://nft.kky.zcu.cz/nef***.

Appendix: Number of ALS identifiable parameters

According to [22], it can be shown that the maximum number of the ALS identifiable parameters is

$$N_{maxIdentParam} = np + p^2, \text{ if } n > p. \tag{32}$$

However, with respect to the structure of $\mathbf{b}$ in (20) and symmetry of $\mathbf{C}e, 0$ (11), the ALS estimate is based on

$$N_{independentEqs} = \frac{p(p+1)}{2} + p^2(L-1) \tag{33}$$

independent equations. Then, comparing (32) with (33), the minimum number of required equations is

$$L_{min} = \lceil \frac{p+n}{p} \rceil. \tag{34}$$

Acknowledgement
This work was supported by the Czech Science Foundation, project no. GA 15-12068S.

References

[1] M. Norgaard, N. K. Poulsen, and O. Ravn, "Kalmtool for use with Matlab," in *Proceedings of the 13th IFAC Symposium on System Identification*, pp. 1490–1495, 2003. http://server.oersted.dtu.dk/personal/or/kalmtool3/.

[2] S. Sarkka and J. Hartikainen, "EKF/UKF Toolbox," 2007. http://www.lce.hut.fi/research/mm/ekfukf/.

[3] O. Straka, M. Flidr, J. Dunik, and M. Simandl, "A software framework and tool for nonlinear state estimation," in *Proceedings of the 15th IFAC Symposium on System Identification*, (Saint-Malo, France), pp. 510–515, 2009.

[4] R. K. Mehra, "Fault-on the identification of variances and adaptive filtring," *IEEE Transactions on Automatic Control*, vol. 15, no. 2, pp. 175–184, 1970.

[5] P. R. Belanger, "Estimation of noise covariance matrices for a linear-varying stochastic process," *Automatica*, vol. 10, no. 3, pp. 267–275, 1974.

[6] B. J. Odelson, M. R. Rajamani, and J. B. Rawlings, "A new autocovariance least-squares method for estimating noise covariances," *Automatica*, vol. 42, no. 2, pp. 303–308, 2006.

[7] B. M. Akesson, J. B. Jorgensen, N. K. Poulsen, and J. S. B., "A generalized autocovariance least-squares method for Kalman filter tuning.," *Journal of Process Control*, vol. 18, no. 7–8, pp. 769–779, 2008.

[8] M. Simandl and J. Dunik, "Estimation of noise covariance matrices for periodic systems," *International Jornal of Adaptive Control and Signal Processing*, vol. 25, pp. 928–942, 2009.

[9] D. G. Lainiotis, "Optimal adaptive estimation: Structure and parameters adaptation," *IEEE Transactions on Automatic Control*, vol. 16, no. 2, pp. 160–170, 1971.

[10] R. L. Kashyap, "Maximum likelihood identification of stochastic linear systems," *IEEE Transactions on Automatic Control*, vol. 15, no. 1, pp. 25–34, 1970.

[11] K. A. Myers and B. D. Tapley, "Adaptive sequential estimation wtih unknown noise statistics," *IEEE Transactions on Automatic Control*, vol. 21, no. 8, pp. 520–523, 1976.

[12] S. Verdu and H. V. Poor, "Minimax linear observers and regulators for stochastic systems with uncertain second-order statistics," *IEEE Transactions on Automatic Control*, vol. 26, no. 6, pp. 499–511, 1984.

[13] P. van Overshee and B. de Moor, *Subspace Identification for Linear Systems: Theory-Implementation-Aplications*. Kluwer Academic Publishers, 1996.

[14] L. Ljung, *System Identification: Theory for the User*. UpperSaddle River, New Jersey, USA: Prentice-Hall, 1999.

[15] D. M. Wiberg, T. D. Powel, and D. Ljungquist, "An on-line parameter estimator for quick convergence and time-varying linear system," *IEEE Transactions on Automatic Control*, vol. 45, no. 10, pp. 1854–1863, 2000.

[16] S. Sarkka and A. Nummenmaa, "Recursive noise adaptive Kalman filtering by variational bayesian approximations," *IEEE Transactions on Automatic Control*, vol. 54, no. 3, pp. 596–600, 2009.

[17] B. Carew and P. R. Belanger, "Identification of optimum steady-state gain for system with unknown noise covarinaces," *IEEE Transactions on Automatic Control*, vol. 18, no. 6, pp. 582–587, 1973.

[18] R. Bos, X. Bombois, and P. Van den Hof, "Desinging a Kalman filter when no noise covariance information is available," in *Proceedings of 16th IFAC world Congress*, (Prague, Czech Republic), 2005.

[19] R. K. Mehra, "Approaches to adaptive filtering," *IEEE Transactions on Automatic Control*, vol. 17, no. 10, pp. 693–698, 1972.

[20] J. Dunik, M. Simandl, and O. Straka, "Methods for estimating state and measurement noise covariance matices: Aspects and comparison," in *Proceedings of 15th IFAC Symposiun on System Identification*, vol. 25, (Saint-Malo, France), 2009.

[21] X. R. Li and Z. Zhao, "Evaluation of estimation algorithms part I: Incomprehensive measures of performance," *IEEE Transactions on Aerospace and Electronic Systems*, vol. 42, no. 4, pp. 1340–1358, 2006.

[22] M. R. Ramajani and J. B. Rawlings, "Estimation of the disturbance structure from data using semedefinite programing and optimal weighthing," *Automatica*, vol. 45, no. 1, pp. 142–148, 2009.

[23] F. V. Lima, M. R. Rajamani, T. A. Soderstrom, and J. B. Rawlings, "Covariance and state estimation of weakly observable systems: Application to polymerization processes," *IEEE Transactions on Control Systems Technology*, vol. 21, no. 4, pp. 1249–1257, 2013.

[24] J. Duník, O. Straka, and M. Šimandl, "Estimation of noise covariance matrices for linear systems with nonlinear measurements," in *Proceedings of the 17th IFAC Symposium on System Identification*, (Beijing, China), 10 2015.

[25] J. W. Brewer, "Kronecker products and matrix calculus in system theory," *IEEE Transactions on Circuits and Systems*, vol. 25, no. 9, pp. 772–781, 1978.

[26] D. Simon, *Optimal State Estimation: Kalman, H Infinity, and Nonlinear Approaches*. Wiley-Interscience, 2006.